Juran's Quality Handbook

About the Editors

Joseph M. Juran was the international thought leader in the quality management field for more than 70 years and continues to be considered the father of modern day quality management. He named the universal concept of the "vital few and the useful many" as the Pareto Principle, which we know today as the 80-20 rule. As an original member of the Board of Overseers, Dr. Juran helped to create the U.S. Malcolm Baldrige National Quality Award. He was the founder of Juran Institute, Inc. and author of more than 20 books.

Joseph A. De Feo is a leading quality management practitioner and successor to Dr. Juran as president of Juran, Inc. He is recognized worldwide for his expertise in enabling organizations to achieve organizational excellence. Mr. De Feo's varied areas of expertise include coaching executives to understand key factors in driving organization-wide change programs and deploying breakthrough management principles like Lean, Six Sigma, strategic planning, and business process improvement.

Juran's Quality Handbook

The Complete Guide to Performance Excellence

Joseph M. Juran

Joseph A. De Feo

Sixth Edition

New York Chicago San Francisco
Lisbon London Madrid Mexico City
Milan New Delhi San Juan
Seoul Singapore Sydney Toronto

The McGraw-Hill Companies

Cataloging-in-Publication Data is on file with the Library of Congress.

Juran's Quality Handbook, Sixth Edition

1 2 3 4 5 6 7 8 9 0 DOC/DOC 1 9 8 7 6 5 4 3 2 1 0

ISBN 978-0-07-162973-7
MHID 0-07-162973-4

The pages within this book were printed on acid-free paper.

Sponsoring Editor
Steven S. Chapman

Acquisitions Coordinator
Alexis Richard

Editorial Supervisor
David E. Fogarty

Project Manager
Deepti Narwat Agarwal, Glyph International

Copy Editor
Patti Scott

Proofreader
Surendra Nath, Glyph International

Indexer
Kevin Broccoli

Production Supervisor
Richard C. Ruzycka

Composition
Glyph International

Art Director, Cover
Jeff Weeks

Contents

Section III Applications: Most Important Methods in Your Industry

Section IV Key Functions: Your Role in Performance Excellence

Contributors

Marcos E. J. Bertin *Director, Bertin Quality Consulting, Argentina* (CHAPTER 31, ROLE OF THE BOARD OF DIRECTORS: EFFECTIVE AND EFFICIENT GOVERNANCE)

Marcos Bertin Schmidt *Director, Bertin Quality Consulting, Argentina* (CHAPTER 31, ROLE OF THE BOARD OF DIRECTORS: EFFECTIVE AND EFFICIENT GOVERNANCE)

R. Kevin Caldwell *Executive Vice President, Juran, Southbury, CT* (CHAPTER 11, LEAN TECHNIQUES: IMPROVING PROCESS EFFICIENCY)

Richard C. H. Chua, Ph.D. *Former Executive Vice President, Juran, Southbury, CT* (CHAPTER 27, THE QUALITY OFFICE: LEADING THE WAY FORWARD)

Joseph A. De Feo *President and CEO, Juran, Southbury, CT* (CHAPTER 4, QUALITY PLANNING: DESIGNING INNOVATIVE PRODUCTS AND SERVICES; CHAPTER 5, QUALITY IMPROVEMENT: CREATING BREAKTHROUGHS IN PERFORMANCE; CHAPTER 7, STRATEGIC PLANNING AND DEPLOYMENT: MOVING FROM GOOD TO GREAT; CHAPTER 8, BUSINESS PROCESS MANAGEMENT: CREATING AN ADAPTABLE ORGANIZATION; CHAPTER 9, THE JURAN TRANSFORMATION MODEL AND ROAD MAP; CHAPTER 12, SIX SIGMA: IMPROVING PROCESS EFFECTIVENESS; CHAPTER 14, CONTINUOUS INNOVATION USING DESIGN FOR SIX SIGMA; CHAPTER 16, USING INTERNATIONAL STANDARDS TO ENSURE ORGANIZATION COMPLIANCE; CHAPTER 27, THE QUALITY OFFICE: LEADING THE WAY FORWARD)

Joseph R. De Feo, Jr. *Associate, Juran, Southbury, CT* (CHAPTER 10, A LOOK AHEAD TO ECO-QUALITY FOR ENVIRONMENTAL SUSTAINABILITY)

Steven M. Doerman *Client Relationship Executive and Senior Consultant, Juran, Southbury, CT* (CHAPTER 11, LEAN TECHNIQUES: IMPROVING PROCESS EFFICIENCY; CHAPTER 20, PRODUCT-BASED ORGANIZATIONS: DELIVERING QUALITY WHILE BEING LEAN AND GREEN)

John F. Early *Executive Vice President, Juran, Southbury, CT* (CHAPTER 12, SIX SIGMA: IMPROVING PROCESS EFFICIENCY; CHAPTER 14, CONTINUOUS INNOVATION USING DESIGN FOR SIX SIGMA; CHAPTER 19, ACCURATE AND RELIABLE MEASUREMENT SYSTEMS AND ADVANCED TOOLS)

Mary Beth Edmond *Executive Vice President and Senior Nurse Executive, Juran, Southbury, CT* (CHAPTER 23, HEALTH CARE-BASED ORGANIZATIONS: IMPROVING QUALITY OF CARE AND PERFORMANCE; CHAPTER 26, EMPOWERING THE WORKFORCE TO TACKLE THE "USEFUL MANY" PROCESSES)

Alexander Eksir, D.M. *Vice President, Mission Assurance Six Sigma EMBB and Quality, Raytheon Integrated Defense Systems, Andover, MA* (CHAPTER 25, DEFENSE-BASED ORGANIZATIONS: ASSURING NO DOUBT ABOUT PERFORMANCE)

Jonathan Flanders *Vice President and Patient Safety Executive, Juran, Southbury, CT* (CHAPTER 23, HEALTH CARE-BASED ORGANIZATIONS: IMPROVING QUALITY OF CARE AND PERFORMANCE)

Bruce J. Hayes *President, Executive Advisor Group, Rockland, MA and Cofounder and Board Director, NeuraMetrics, Inc., Jacksonville, FL* (CHAPTER 29, SOFTWARE AND SYSTEMS DEVELOPMENT: FROM WATERFALL TO AGILE)

Alexander Janssen *Executive Vice President, Juran BV, The Netherlands* (CHAPTER 15, BENCHMARKING: DEFINING BEST PRACTICES FOR MARKET LEADERSHIP)

Joseph M. Juran (CHAPTER 1, ATTAINING SUPERIOR RESULTS THROUGH QUALITY; CHAPTER 2, QUALITY'S IMPACT ON SOCIETY AND THE NATIONAL CULTURE; CHAPTER 3, THE UNIVERSAL METHODS TO MANAGE FOR QUALITY; CHAPTER 5, QUALITY IMPROVEMENT: CREATING BREAKTHROUGHS IN PERFORMANCE; CHAPTER 6, QUALITY CONTROL: ASSURING REPEATABLE AND COMPLIANT PROCESSES)

Dennis J. Monroe *Vice President, Juran, Southbury, CT* (CHAPTER 13, ROOT CAUSE ANALYSIS TO MAINTAIN PERFORMANCE; CHAPTER 26, EMPOWERING THE WORKFORCE TO TACKLE THE "USEFUL MANY" PROCESSES; CHAPTER 30, SUPPLY CHAIN: BETTER, FASTER, FRIENDLIER SUPPLIERS)

Michael J. Moscynski *Associate, Juran, Southbury, CT* (CHAPTER 21, SERVICE-BASED ORGANIZATIONS: CUSTOMER SERVICE AT ITS BEST)

James Er Ralston *Vice President, Juran, Southbury, CT* (CHAPTER 17, USING NATIONAL AWARDS FOR EXCELLENCE TO DRIVE AND MONITOR PERFORMANCE; CHAPTER 23, HEALTH CARE-BASED ORGANIZATIONS: IMPROVING QUALITY OF CARE AND PERFORMANCE)

Brian A. Stockhoff, Ph.D. *Senior Consultant, Juran, Southbury, CT* (CHAPTER 10, A LOOK AHEAD: ECO-QUALITY FOR ENVIRONMENTAL SUSTAINABILITY; CHAPTER 18, CORE TOOLS TO DESIGN, CONTROL, AND IMPROVE PERFORMANCE; CHAPTER 19, ACCURATE AND RELIABLE MEASUREMENT SYSTEMS AND ADVANCED TOOLS; CHAPTER 24, CONTINUOUS PROCESS-BASED ORGANIZATIONS: QUALITY IS A CONTINUOUS OPERATION; CHAPTER 28, RESEARCH AND DEVELOPMENT: MORE INNOVATION, SCARCE RESOURCES)

Janice Doucet Thompson *Principal with JD Thompson and Associates, Sacramento, CA* (CHAPTER 9, THE JURAN TRANSFORMATION MODEL AND ROADMAP)

Angel D. Tonchev *Senior Consultant, Juran BV, The Netherlands* (CHAPTER 22, SELF-SERVICE BASED ORGANIZATIONS: ASSURING QUALITY IN A NANOSECOND)

Christo D. Tonchev *Senior Consultant, Juran BV, The Netherlands* (CHAPTER 22, SELF-SERVICE BASED ORGANIZATIONS: ASSURING QUALITY IN A NANOSECOND)

Brad Wood, Ph.D. *Director, Juran BV, The Netherlands* (CHAPTER 15, BENCHMARKING: DEFINING BEST PRACTICES FOR MARKET LEADERSHIP)

Introduction to the Sixth Edition

Preface by Joseph A. De Feo, Editor-in-Chief

Dr. Joseph M. Juran and *Juran's Quality Handbook* were unknown to me until my career led me to meet this amazing person and resource. He was my teacher, mentor, boss, and later my friend. Although I was less than half his age at the time we met, he helped me and my colleagues implement his universal methods to improve our business performance. From the Pareto Principle he named, to the universal breakthrough methods, to quality by design to develop new products, we soon learned that these methods were indeed universal. These methods improved our performance. They work in any country, industry, or organization, and even in families. These methods can and will improve financial, organizational, and cultural performance by simply implementing them across your organization.

Dr. Juran once stated, "To someone that has not seen the miracles, one will not believe that they occurred." I learned the methods he espoused and I saw the results, the miracles. As a customer of the Juran Institute in 1986, our organization applied his methods and tweaked them to fit our needs. In the end we witnessed the miracles. The company I worked for improved quality and performance dramatically. We stayed in business and competed head to head with the Japanese. I have seen many organizations improve their performance in the same way. It is time to share one more edition of the handbook with another generation.

Unfortunately, Dr. Juran left us before this sixth edition of the handbook was completed. He passed away in 2008 at the enduring age of 103! He left behind his wonderful wife, Sadie (who passed away nine months later in 2008 also at 103), a remarkable family, an organization bearing his name to promote the methods to improve performance, and many publications for future generations to learn from. He trained thousands of "practitioners" in his methods. CEOs, CFOs, COOs, quality officers, and employees from many industries, in many countries. They became disciples and applied his methods. Many are now advocates of performance excellence in their own right. At Dr. Juran's 100th birthday over 200 people from around the globe came to thank him for what he did for them.

To Dr. Juran, I thank you, and to all the disciples, congratulations. To those who are just beginning their careers as new practitioners, the best has yet to be learned.

We have chapters that were authored by Dr. Juran for this sixth edition of *Juran's Quality Handbook*. The chapters that Dr. Juran completed have only his name as the author. He asked that others "fill in the blanks." This is noted by my name appearing on Chapter 5,

Quality Improvement: Creating Breakthroughs in Performance. For chapters he did not complete, we had some of the finest practitioners—researchers—update those chapters. To ensure this edition of the handbook is truly viewed as an important body of knowledge, we added new chapters, such as Lean and Six Sigma, since the methods for managing for quality have evolved and changed since the last edition was published in 1999. In these chapters we included Dr. Juran's input related to that subject. These were from his papers and other publications. We want to appeal to new practitioners and new leaders so they can have the same benefit and chance to learn methods and witness their own miracles. We also hope that all our readers will become acquainted with a new generation of authors who someday may be gurus as well.

I must admit this was not an easy undertaking. Following in the footsteps of giants in our field, such as Dr. Juran, Dr. Blanton Godfrey, Dr. Armand Feigenbaum, Dr. W. Edwards Deming, Dr. Noriaki Kano and many more, was overwhelming at times. The good news is that Dr. Juran provided me a critique of the fifth edition. His critique provided direction, what to add, what to remove, what to research prior to his passing. Although he was not here when it was time to send this to the publishers, all readers of the handbook can rest assured that Dr. Juran is present in this edition.

I hope anyone who reads this will grant me a little leeway when providing his or her own critique. At times I wondered, did I do this book justice? In the end, our authors came through and gave me an opportunity to shine. I feel I have done my best to honor all the gurus who have gone before me and all those who have yet to come.

I would like to thank Linda Ellrodt, my assistant, who fought with me many times and in the end always won. Also I thank Brian Stockhoff, Tina Pietraszkiewicz, Jackie Allard, Jeremy Hopfer, Matt Pachniuk, Geeno Carlone and so many others for helping me pull all the materials together into one big pile of information. For the authors: Without you this book would still be in production. In addition, I thank the staff of the Juran Institute and Steve Chapman and his team at McGraw-Hill for caring so much about this text and the next generation who will read it.

To the late Dr. Bill Barnard, Dr. Frank Gryna, and Mr. Bob Hoogstoel, my mentors while growing up at the Juran Institute: Your guidance when I was a rookie made this all possible. Who would have ever thought a wood shop teacher from a humble beginning could be the co-editor someday with Dr. Juran? I believe each of you did. You encouraged me to be pragmatic, to learn how to research, to make my customers successful, and to practice what we preach. To you I offer my thanks.

To my three sons Christopher, Mark, and Joe—all who have learned that becoming a lifelong learner, as Dr. Juran was, is the best way to provide value to society. Thank you for putting up with me when so many evenings and weekends I was stuck typing on a computer.

To my wife, Monica, without you I would not be who I am today—a father, husband, and now an author of a book with a great legacy. A big thank you.

Joseph A. De Feo

Highlights of the Introduction

1. The founding concept of the handbook was to create a compendium of knowledge in the field of managing for quality. The emphasis has always been on universals—"the principles that are valid no matter what the product, the process or function" (Dr. Joseph M. Juran).

2. These universal methods deliver superior results and performance excellence in any company, organization, industry, country, or process. They can be used to improve financial, organizational, and cultural performance by simply learning and applying them.
3. This handbook is a reference text for all who are involved with creating, producing, and delivering high-quality goods, services, and processes to attain superior business results.
4. There are several tables of contents. At the beginning of the book is the list of *section and chapter headings*, each of which describes, in the broadest terms, the contents of that section or chapter.
5. The handbook is a condensation of each author's knowledge; i.e., what she or he wrote is derived from materials that are one or two orders of magnitude more voluminous than the published work. In some cases it is worthwhile to contact the author for further elaboration. Most authors have no objection to being contacted, and some of these contacts lead not only to more information but also to visits and enduring collaboration.
6. In many cases a practitioner is faced with adapting to a specific situation the knowledge derived from a totally different technology, i.e., a different industry, product, or process. Making this transition requires that he or she identify the commonality, i.e., the common principle to which both the specific situation and the derived knowledge correspond.

Dr. Juran on the Creation of *Juran's Quality Handbook*

The idea of a handbook on quality control originated late in 1944 and was part of my decision to become a freelance consultant after World War II. I had in mind a whole series of books: the *Quality Control Handbook*, which was to be a comprehensive reference book; and separate manuals on quality control for executives, engineers, foremen, and inspectors.

I prepared an extensive outline for the handbook plus a brief description for each of the other books. My publisher was McGraw-Hill, who offered me a contract (a mere one-page document!) in December 1945. Publication was in 1951. It became a flagship of the many books I have written.

The concept of the handbook was to create a compendium of knowledge in the field of managing for quality. The emphasis was on universals—"the principles that are valid no matter what the product, the process or function." I was only dimly aware that I would be contributing to the evolution of a new science—managing for quality. I ended up with fifteen chapters. I wrote six of those; other authors wrote the remaining nine, which I often edited. From the outset, the handbook became the "bible" of managing for quality and has increasingly served as an international reference book for professionals and managers in the field. With the publication of the fifth edition the name was changed to the *Juran's Quality Handbook*; the joint editors-in-chief were Dr. A. Blanton Godfrey and I. For the sixth edition, Joseph A. De Feo will be editor-in-chief. As a fine consultant and practitioner, he has earned the recognition that comes with the publication of such a book as this.

Joseph M. Juran

Remembering Dr. Joseph M. Juran

From Dr. Armand V. Feigenbaum
President and CEO, General Systems Company

The issuance of the *Juran's Quality Handbook*, sixth edition, is a continuation of Dr. Juran's economic, social and educational service to the future of quality in America as well as the world community. I knew Dr. Joseph Juran for a very long time and I want to express the deep sense of loss I feel because of his passing. Many, many people and organizations are much the better because of his work, his writings and his guidance and this importance of his influence—already very great—will continue to grow throughout the world.

I first came to meet and to know Joseph Juran before quality was recognized as a field and before quality organizations and meaningful quality literature and guidance had come into being as an area of explicit importance and attention. Already a man of high professional standing and of major business and governmental experience, the very fact of Joseph Juran placing his personal emphasis upon quality brought enormous attention and meaning to the subject of quality which previously had been thought of as a technical factor in inspection. The subsequent more than half century of the growth and evolution of quality into the importance of its global recognition and high effectiveness throughout the world owes a very great deal to the contribution of Joseph Juran. As he is no longer with us, we can, however, take some comfort that his guidance and influence will continue through the availability of the content of his writings and of the spirit of his personal commitment that continues to come through them.

From Dr. Noriaki Kano
Professor Emeritus, Tokyo University of Science

The most widely-utilized tools among those in the quality world are probably the "Pareto Chart" and the "Cause and Effect Diagram" (also known as the "Fishbone Diagram" or the "Ishikawa Diagram"). Among these two, the number of people who know the Pareto Chart was initially proposed by Dr. Juran have drastically decreased in recent years while it is well known that "Cause and Effect Diagram" was proposed by Dr. Ishikawa. The Pareto is named after the Italian Economist, Vilfredo Pareto (1848–1923) who is known for his study of income distribution. Dr. Juran called it a principle of "Vital few and Useful many" as the Pareto Principle because the cumulative curve of frequency distribution by quality defects does have a shape similar to that of income distribution Pareto revealed.

In response to Dr. Juran's article (1975) on Pareto, a motion to rename the principle to the "Juran Principle" was proposed. Dr. Juran (1975) responded that: *"I hope I may be pardoned for suggesting that such a change in name, if it comes to pass, await my journey to the Great Beyond. I hope I may also be pardoned for hoping that this journey will be delayed for some decades to come."* and then the discussion was left over. I tried to make this happen in the event of Dr. Juran's 100th year celebration, but I had to give it up without being able to receive consent from Dr. Juran. I am thinking to make a proposal to change Juran Principle and Juran Chart from current Pareto Principle and Pareto Chart again in the near future according to Dr. Juran's wish of over a quarter century ago. I am certain it is the contribution toward the Quality Innovation of Japanese products that Dr. Juran was most proud of in his life and we thank him for that.

From Ken Takatori
Acting Manager, International Relations
Union of Japanese Scientists and Engineers (JUSE)

Thank you for your email in providing JUSE an opportunity to give a tribute at the Juran 6th edition handbook issuance. First of all, the *Juran's Quality Handbook* was translated into Japanese, and taught the Japanese top management that the Quality Management can be applied as the significant "management tool." We consider Dr. Juran as a foster parent who actually raised and developed the Japanese Quality Management, whereas Dr. Deming as a birth parent.

From Dr. Lennart Sandholm
Sandholm Associates

Dr. Juran is the person, outside my family, who has had the greatest influence on my life, both professionally and personally. The first time I met him was 1965. He was my mentor for almost 40 years. According to his memories he visited Sweden 31 times. It was a delight for me to organize his courses and seminars in Sweden from 1966 to 1991. These events gave me a tremendous insight in Dr Juran's thinking and philosophy. The meetings didn't just mean being together professionally, but also personally. My wife and I had the pleasure to have Dr. Juran several times as guest in our home. A personal and warm friendship emerged. He showed a keen interest in the children. When I met him at conferences, he used to ask about the children and always recalled the names of them. I will for ever keep Dr. Juran in grateful memory.

From Madam Tang Xiaofeng
Former Dean of Juran Institutes of Shanghai
Head of Shanghai Association for Quality
President of Shanghai Academy of Quality Management

On behalf of the Shanghai Association for Quality, Shanghai Academy of Quality Management and myself, the passing of Dr. Juran still a shock and deep regret to us. Dr. Juran is a well-known great father for modern quality management worldwide. I always remember the time when I met with him in 1998 Spring, his happiness when I sent him the Chinese version of Architect of Quality in 2006. His gentlemanly speaking, mentor style behavior and his great knowledge, always make him unforgettable to me. Since 1960s, Dr. Juran and his Quality Trilogy made great contribution to the worldwide quality management, and will be remembered all the time by people. Especially the impact of his contribution to China and Shanghai quality management will last forever. Both my organization and myself are benefit from his past encouragement and support, we will commemorate him for long. In his autobiography, he used to wrote "When I'm dying, don't cry to me because my life is so splendid". With his life journey, we observed the colorfulness of the development of quality management as well as his excellent lifestyle. We sincerely wish our quality business, Mr. Joe De Feo and his Juran Institutes grow continuously under inspiration of Dr. Juran's spirit.

From Hesam Aref Kashfi
Academician, International Academy for Quality

Dr. Joseph M. Juran who deservedly won the epithet, "Father of Quality Management," founded this new science on his Trilogy and specially on Quality Planning and sincerely practiced it for the welfare of people all over the world and left us a thoughtful message: "My job of contributing to the welfare of my fellow man is the great unfinished business." He solidly originated the theory and functionally evolved this new science. If we compare Quality to a tree, which he prophesized to flourish in the 21st century, his legendary Quality Handbook is the roots, and as he joined the selected eternals due to his admirable integrity, honesty and conviction, his legacy, as his product, thanks to its authenticity and comprehensiveness, will be increasingly appreciated as the unique pioneer classic on Quality.

From Tom Pyzdek
President, Pyzdek Institute

I've known Dr. Juran since that day in 1967 when I, a quality technician at a can factory, picked up that thick green book and looked up control charts. It was the *Quality Control Handbook*, second edition and it introduced me to the wonderful, magical world of quality. The book divulged the secrets of statistics and quality management and inspired me, an 18 year old just entering college, to consider making quality a career. As the years went by, I continued reading the work and investing in new editions of the handbook. When I made my first presentation at an ASQC conference I joined Dr. Juran on an elevator trip to the opening presentation, where he delivered the keynote address. I was surprised by how small he was, physically. On the dais, however, he was a giant as he told those of us assembled there that we had a big responsibility to help America recover her position as the world's quality leader. In 1988 he and I rode the bus together to the White House for the presentation by President Reagan of the first Malcolm Baldrige National Quality Award. We chatted briefly. That evening I was waiting for a restaurant table outside the door of the room where the Baldrige Board of Overseers was about to meet. When he asked me to join the group I was delighted. I was thrilled to be able to chat with so many of the leaders in the quality movement. Of course, he sat at the head of the table, where he belonged. The last time we met was in 2004, to celebrate his 100th birthday. It was only May 6 and his birthday wasn't until December 24 and he quipped that his people weren't sure that he would make it to December. He stood tall and his voice and handshake were strong and we in the audience had no doubt that he would blow by 100 like it was no big deal. It has been over 40 years since I first picked up his book. It's still on my bookshelf. Like me, it's physically showing its age a bit these days. But like Dr. Juran, its creator, the ideas are fresh and full of meaning and value. Today, somewhere, some young person is reaching for one of works, about to be inspired by him. Inspired to consider a career of helping people and organizations achieve excellence. Or maybe just inspired to try harder, to do things better. Dr. Juran is not gone. He will never be gone.

From Paul Borawski CAE,
Executive Director and Chief Strategic Officer, American Society for Quality

I met Dr. Juran for the first time in 1986 and it didn't take but a minute or two to know I was in the presence of an extraordinary leader and a management prophet. I had many subsequent opportunities to learn from Dr. Juran. He taught us all about the science of management. The language of executives. The process of improvement. He embodied the example of disciplined devotion to excellence. I doubt we will again experience business wisdom as profound as Dr. Juran's or knowledge as practical as his teachings.

Uses of the Handbook

We set out to make this sixth edition handbook a reference for all who are involved with leading, creating, producing, and delivering high-quality products (goods and services), and processes to attain superior results. Experience with the first five editions has shown that "those who are involved" include the following:

1. *All levels in the organizational hierarchy,* from the board members, chief executives, and operational managers to the workforce. It is a mistake to assume that the sole purpose of the book is to serve the needs of just the quality managers and engineers. The purpose of the book is to serve the entire organization including the workforce at all levels of the organization.
2. The various *functions* engaged in producing products (goods, services, or information), such as research and development, market research, finance, operations, marketing and sales, human resources, supplier relations, customer service, administration, and support activities.
3. The various *specialists* associated with the processes carrying out the strategic and tactical tasks to ensure all products, and business processes are properly designed, controlled, and continually improved to meet customer and societal needs.
4. The various *industries* that make up the global economy: manufacturing, construction, services of all kinds—high-tech, transportation, communication, oil and gas, energy, utilities, financial, healthcare hospitality, government, and so on.

The handbook is also an aid to certain gatekeepers and stakeholders who, although not directly involved in leading, producing, or marketing products, nevertheless have "a need to know" about the qualities produced and the associated positive and negative side effects. These stakeholders include the executive leadership, quality officers, and engineers who are given the responsibility to improve managing for quality from day to day; customers looking to better understand how to improve their suppliers; the supply chain; the users; the public; the owners; the media; and even government regulators.

We have conducted many focus groups and learned that our readers, the practitioners, make a wide variety of uses of *Juran's Quality Handbook.* Experience has shown that usage is dominated by the following principal motives:

1. To use as a "one-stop shopping" reference guide for the methods, tools, and road maps to create a sustainable business operation driven by high-quality products.
2. To utilize as a guide to educate their specialists, such as quality management and assurance departments, systems engineering, organizational and operational effectiveness departments, finance, and the like.
3. To find special tools or methods and examples of their use on topics such as leadership's role in leading quality, incorporating the voice of the customer in the design of goods and services, reliability engineering, design of experiments, or statistical tools.
4. To study the narrative material as an aid to solving their own organizational and business problems.
5. To review subject matter for specific self-training or to secure material for teaching or training others.

Using the handbook appears to be more frequent during times of change such as when developing new business initiatives, working on new processes and projects, organizing departments and functions, or just trying out new ideas.

Organization of the Handbook

Irrespective of intended use, the information provided in this handbook must be found easily. The problem for the user becomes one of (1) knowing where to find it and (2) adapting the information to his or her specific needs. Although there is a great deal of know-how in this book, it is important to understand how it is organized. There are four sections with a total of 31 chapters. Each chapter is a topic on its own and collectively makes up a robust performance excellence system. Here is a brief look at the layout of the book.

1. The Introduction answers the question, Why a handbook on quality? The Introduction includes the preface, dedications, and acknowledgments and how to use the handbook. There is a short but important section called "Remembering, Dr. Joseph M. Juran" and his contribution to society. This part of the handbook has therefore been designed to help the reader find and apply the specific content that relates to the problem at hand. To know where to locate it requires understanding how the handbook is structured. The handbook consists of the Introduction, Sections 1 to 4, and the Appendix I and Appendix II. The sections are outlined as follows.
2. **Section I: Key Concepts: What Leaders Need to Know about Quality** (Chapters 1 through 10). This section deals with basic concepts through which quality is managed (planning, control, and improvement) and why it is crucial to the success of most organizations. It deals with the evolution of managing for quality in the past to managing for organizational performance excellence in the future.
3. **Section II: Methods and Tools: What to Use to Attain Performance Excellence** (Chapters 11 through 19). This section focuses on how to use the management of quality methods, basic tools, and advanced statistical tools to obtain results. The primary topics include:
   - Six Sigma
   - Lean techniques
   - Root Cause Analysis
   - Innovation and Design for Six Sigma
   - Benchmarking
   - International standards
   - National awards for excellence
   - Core and advanced tools to design, control, and improve
4. **Section III: Applications: Most Important Methods in Your Industry** (Chapters 20 through 25). This section includes industries that we felt have been most effective in using the methods for managing for quality and where performance excellence has worked:
   - Manufacturing
   - Service and self-service
   - Healthcare systems
   - Processing-based organizations
   - Software development
   - Defense

5. **Section IV: Key Functions: Your Role in Performance Excellence**(Chapters 26 through 31). This section illustrates the roles that key functions must play to enable the organization to attain superior results.
6. **The Appendixes**. This part of the text provides additional information and tools to aid a leader, manager, or practitioner.

How to Find It

There are four main roads for locating information in the handbook:

Tables of contents

Index

Cross-references

Internet

Table of Contents

There are two tables of contents. At the beginning of the book is the list of *sections and the chapter titles within each section,* which describes, in the broadest terms, the contents of that chapter.

Next, there is the *chapter contents* list that appears on the first page of each chapter. Each entry in the chapter contents is a *major heading* within that chapter.

In a good many cases, it will suffice merely to follow the hierarchy of chapter contents to find the information sought. In many other cases it will not. For such cases, an alternative approach is to use the index.

Use of the Index

A great deal of effort has gone into preparing the index so that the reader can locate all the material bearing on a subject. For example, the topic "Pareto principle" is found in several sections. The index entry for "Pareto principle" assembles *all* uses of the term "Pareto principle" and shows the page numbers on which they may be found. The fact that information about a single topic is found in more than one chapter (and even in many chapters) gives rise to criticism of the organization of the handbook, i.e., Why can't all the information on one topic be brought together in one place? The answer is that we require multiple and interconnected uses of knowledge, and hence these multiple appearances cannot be avoided. In fact, what must be done to minimize duplication is to make one and only one exhaustive explanation at some logical place and then to use cross-referencing elsewhere. In a sense, all the information on one topic *is* brought together—in the index.

Some key words and phrases may be explained in several places in the handbook. However, there is always one passage that constitutes the major explanation or definition. In the index, the word "defined" is used to identify this major definition, e.g., "Evolutionary operation, defined."

The index also serves to assemble all case examples or applications under one heading for easy reference. For example, Chapter 5 deals with the general approach to quality improvement and includes examples of the application of this approach. However, additional examples are found in other chapters to illustrate and support their specific topics. The index enables the reader to find these additional examples readily, since the page numbers are given.

Cross-References

The handbook makes extensive use of cross-references in the text to (1) guide the reader to further information on a subject and (2) avoid duplicate explanations of the same subject matter. The reader should regard these cross-references, wherever they occur, as extensions of the text. Cross-referencing is to either (1) specific major headings in various chapters or (2) specific figure numbers or table numbers. Study of the referenced material will provide further illumination.

A Note on Abbreviations

Abbreviations of names or organizations are usually used only after the full name has previously been spelled out, e.g., American Society for Quality (ASQ). In any case, all such abbreviations are listed and defined in the index.

The text of the handbook emphasizes the "main road" of quality management know-how, i.e., the comparatively limited number of usual situations that nevertheless occupy the bulk of the time and attention of practitioners. Beyond the main road are numerous "side roads," i.e., less usual situations that are quite diverse and require special solutions. (The term "side road" is not used in any derogatory sense. The practitioner who faces an unusual problem must nevertheless find a solution for it.) As to these side roads, the handbook text, while not complete, nevertheless points the reader to available solutions. This is done in several ways.

1. *Citations.* The handbook cites numerous papers, books, and other bibliographic references. In most cases these citations also indicate the nature of the special contribution made by the work cited to help the reader decide whether to go to the original source for elaboration.
2. *Special bibliographies.* Some chapters provide supplemental lists of bibliographical material for further reference under References. The editors have attempted to restrict the contents of these lists to items that (1) bear directly on the subject matter discussed in the text or (2) are of uncommon interest to the practitioner.
3. *Literature search.* Papers, books, and other references cited in the handbook contain further references which can be found for further study. Use can be made of available abstracting and indexing services. Various other specialized abstracting services are available on such subjects as reliability, statistical methods, research and development, and so on.
4. *The Internet.* It is now possible to find almost any book or article in print and many that are out of print in just a few minutes by using a "Web search." Using search engines, one can find thousands of articles on numerous topics or by selected authors. Many special sites focus on performance excellence and quality management in the broadest sense. A simple e-mail contact with a website author may bring forth even more unpublished works or research in progress. Sites developed by university departments doing research in quality are especially useful for searching for specific examples and new methods and tools.
5. *Author contact.* The written book or paper is usually a condensation of the author's knowledge, i.e., what he or she wrote is derived from materials that are one or two orders of magnitude more voluminous than the published work. In some cases it is worthwhile to contact the author for further elaboration. Most authors have no objection to being contacted, and some of these contacts lead not only to more information but also to visits and enduring collaboration.

6. *Societies and other sources.* Resourceful people are able to find still other sources of information relating to the problem at hand. They contact the editors of journals to discover which organizations have faced similar problems, so that they may contact these organizations. They contact suppliers and customers to learn if competitors have found solutions. They attend meetings, such as courses, seminars, and conferences of professional societies, at which there is discussion of the problem. There is hardly a problem faced by any practitioner that has not already been actively studied by others.

Adapting the Handbook to Your Special Needs

In many cases a practitioner is faced with adapting to a specific situation the knowledge derived from a totally different technology, i.e., a different industry, product, or process. Making this transition requires that he or she identify the commonality, i.e., the common principle to which both the specific situation and the derived knowledge correspond.

Often the commonality is managerial in nature and is comparatively easy to grasp. For example, the concept of self-control is a universal management concept and is applicable to any person in any organization.

Commonality of a statistical nature is even easier to grasp, since so much information is reduced to formulas that are indifferent to the nature of the technology involved. Even in technological matters, it is possible to identify commonalities despite great outward differences. For example, concepts such as process capability apply not only to manufacturing processes, but to healthcare services and administrative and support processes as well. In like manner, the approaches used to make improvements by discovering the causes of failures have been classified into specific categories that exhibit a great deal of commonality despite wide differences in technology.

In all these situations, the challenge to practitioners is to establish a linkage between their own situations and those from which the expertise was derived—the discovery of a commonality.

SECTION I

Key Concepts: What Leaders Need to Know about Quality

CHAPTER 1

Attaining Superior Results through Quality

Joseph M. Juran

About This Chapter

This chapter defines the role and importance of the *quality* of products, services, and processes in an organization. An organization that is superior to its competition in quality is considered the *market quality leader* or, as used recently, in "a state of performance excellence." The goods and services that an organization produces must meet its customers' needs. If they do, then customers purchase these goods and services, and the organization receives sales revenue as a result. All organizations can achieve superior results through the application of the universal methods to manage quality, which design, maintain, and continually improve the quality of goods and services.

High Points of This Chapter

1. Organizations that engage in a relentless pursuit of delivering high-quality products and services outperform those that do not. Customers' satisfaction depends on having the right quality of goods and services to meet their needs.
2. High-quality goods and services impact an organization in two ways. First, quality can improve financial results by delivering products, services, and processes that are superior to those of the competition. Second, the relentless pursuit of high quality transforms the culture which leads to sustainability.
3. High-quality goods and services can increase the sales revenue. Revenue can be derived from sales, budget appropriations, tuition, government agency grants, and so on.
4. Quality superiority can be translated into higher market share, but it requires a systematic approach that is moving at a pace greater than that of the competition.
5. Quality has come to mean "fitness for purpose." This means that no matter what you produce—a good or a service—it must be "fit" for its purpose. To be fit for purpose, every good or service must have the right features to satisfy customer needs with little or no failures (Six Sigma levels and greater).
6. Customers state their needs as they see them, in their language. Suppliers (producers) need to understand the real needs behind the stated needs and translate those needs into the suppliers' language.
7. Organizations that were successful at creating superiority in quality made use of numerous strategies. Our analysis shows that despite differences in these strategies among the organizations, there was a long list of common practices.
8. This handbook provides both the details of those common strategies and insight into how your organization can benefit from them.

Quality, Performance Excellence, and Superior Results

An organization that creates high-quality goods and services can be affected by this in two ways. First, quality can affect its financial results because products and services that are superior to the competitors' products are more salable, thereby increasing sales and lowering costs and thus leading to greater profitability. Second, the pursuit of high quality transforms a culture. This happens after repeated success in eliminating poor quality, process waste, and customer dissatisfaction. The transformational changes required of an organization do not happen haphazardly. They are a result of an organization's relentless pursuit to be the best in quality and implementing a systematic method to get there. This destination has had multiple names over the decades. Organizations that attain superior results, by designing and continuously improving the quality of their goods and services, are often called *world class, best practices, vanguard companies*, and most recently *performance excellence*. We define this as an organization that has *attained a state of performance excellence because its products and services exceed customers' expectations; they are regarded by their peers and have superior, sustainable results.*

This pursuit of performance excellence through quality creates high stakeholder and employee satisfaction, which enables the organization to sustain the pursuit for the long term. These organizations have reached a state of *performance excellence.*

How to Think about Quality

One of the first tasks is to provide a definition of the word "quality." We must first agree on a meaning of the word so that an organization will know how to manage "it." If one can define it, then one can manage it; if one can manage it, one can deliver it to the satisfaction of customers and stakeholders. If one does not agree on a common meaning of quality for an organization, then one will not be able to manage it efficiently.

We have seen so many meetings where leaders argue when asked the question, "Does high quality cost more, or does high quality cost less?" Seemingly they disagree. One-half agree that it costs more, and the other half feel it costs less. The fact is that some people literally do not know what the others are talking about. The culprit is the word "quality." It is spelled the same way and pronounced the same way, but has multiple meanings. To manage for results, one must agree on the definition of the word "quality" from the perspective of customers—those people who buy the goods, services, and even reputation of your organization.

At one financial services company the leaders would not support a proposal to reduce wasteful business processes because the staff had labeled them "quality improvement." Some of the leaders felt improving quality would cost more money. In their view, higher quality meant higher cost. Others felt it would cost less. The subordinates were forced to rename the proposal "productivity improvement" to secure approval and avoid confusion. Such confusion can be reduced if each organization makes clear the distinction between the multiple meanings of the word "quality." However, some confusion is inevitable as long as we use a single word to convey very different ideas.

There have been efforts to clarify matters by adding supplemental words. To date, none of these efforts has gained broad acceptance. There also have been efforts to coin a short phrase that would clearly and simultaneously define both the major meanings of the word "quality." A popular definition was first presented in the third handbook. "Quality" was defined as meaning "fitness for use." Dr. Deming used "conformance to requirements." Robert Galvin, Chairman Emeritus of Motorola, used "Six Sigma" to distinguish the high level of quality as it related to defects. Others stated that quality means world-class excellence or best-in-class and now performance excellence.

For this sixth edition we have found that many of these definitions fall short for service organizations. To find a term that is universal, and can be applied to any situation, we modified our previous definition found in prior Juran handbooks. For many decades we have said that *quality means fitness for use*. The use is defined by the customers that purchase, use, or are affected by the good or service. If an organization understands the needs of its many customers, it should be able to design goods and services that are fit for use. However, as more and more service industries use the methods of managing for quality, the prior definition of quality is not applicable enough.

We have settled on a new definition. Quality means *fitness for purpose*. So no matter what you produce—a good or a service—it must be fit for its purpose. To be fit for purpose, every good and service must have the right features to satisfy customer needs and must be delivered with few failures. It must be effective to meet the customer requirements and efficient for superior business performance.

It is unlikely that any short phrase can provide the depth of meaning needed by leaders and managers who are faced with choosing a course of action to improve quality. The best you can do it to understand the distinctions set out in Figure 1.1 and define quality based on these distinctions.

Figure 1.1 presents two of the many meanings of the word "quality." These two are of critical importance to managing for quality.

Features That Meet Customer Needs	Freedom from Failures
Higher quality enables organizations to	Higher quality enables organizations to
♣ Increase customer satisfaction	♣ Reduce error rates
♣ Make products salable	♣ Reduce rework, waste
♣ Meet competition	♣ Reduce field failures, warranty charges
♣ Increase market share	♣ Reduce customer dissatisfaction
♣ Provide sales income	♣ Reduce inspection, test
♣ Secure premium prices	♣ Shorten time to put new products on the market
♣ Reduce risk	♣ Increase yields, capacity
	♣ Improve delivery performance
Major effect is on revenue.	Major effect is on costs.
Higher quality costs more.	Higher quality costs less.

Figure 1.1 The Meaning of Quality.

Quality Impacts Revenue and Costs

First, quality has a big effect on *costs*: In this case, "quality" has come to mean freedom from troubles traceable to office errors, factory defects, field failures, and so on. "Higher quality" means fewer errors, fewer defects, and fewer field failures. When customers perceive a service or good as low-quality, they usually refer to the failures, the defects, the poor response times, etc. To increase this type of quality, an organization must master the universal of quality improvement. This is often called *breakthrough* or *Six Sigma*. It is a systematic method to reduce the number of such deficiencies or the "costs of poor quality" to create a greater level of quality and fewer costs related to it.

Second, quality has an effect on *revenue*: In this case, "higher quality" means delivery of those features of the good or service that respond better to customer needs. Such features make the product or service salable. Since the customers value the higher quality, they buy it and you get revenue from it. It is well documented that being the quality leader can also generate premium prices and greater revenue.

The authors will periodically define words that often have multiple meanings. We have provided a Glossary of Key Terms at the end of the book. This will make it easy for the practitioner to have a common ground while using the handbook to drive performance.

These important terms will be used throughout the book:

"Organizations" include any enterprise, company, operating institution, an industrial organization, a government agency, a school, a hospital, and so on.

"Revenue" means gross receipts, whether from sales, budget appropriations, tuition, government agency grants, and so on.

"Costs" refer to the total amount of dollars spent by an organization to meet customer needs. With respect to quality, costs include the expenditure to design and ensure delivery of high-quality goods and services plus the costs or losses as the result of poor quality.

The effects on costs and on revenue interact with each other. Not only do goods or services with deficiencies add to suppliers' and customers' costs, but also they discourage repeat sales. Customers who are affected by field failures are, of course, less willing to buy again from the guilty supplier. In addition, such customers do not keep this information to themselves—they publicize it so that it affects the decisions of other potential buyers, with negative effects on the sales revenue of the supplier.

The effect of poor quality on organizational finances has been studied broadly. In contrast, study of the effect of quality on revenue has lagged. This imbalance is even more surprising, since most upper managers give higher priority to increasing revenues than to reducing costs. This same imbalance presents an opportunity for improving organization economics through better understanding of the effect of quality on revenue. (See Chapter 12, Six Sigma: Improving Process Effectiveness.)

Quality, Earnings, and the Stock Market

At the most senior levels of management and among board members, there is keen interest in financial metrics such as net income and share price. It is clear that different levels of quality can greatly affect these metrics, but so do other variables. Variables such as market choices, pricing, and financial policy can influence these metrics. Separating out the market benefits of managing for quality has just become feasible.

During the early 1990s, some of the financial press published articles questioning the merits of the Malcolm Baldrige National Quality Award, Six Sigma, and other similar initiatives to improve performance. These articles were challenged with an analysis of the stock price performance of organizations known to practice these methods. The Baldrige winners were compared to that of the S&P 500 as a whole. The results were striking. The Malcolm Baldrige National Quality Award winners outperformed the S&P 500. The Baldrige winners had advanced 89 percent, as compared to only 33 percent for the broad Standard & Poor's Index of 500 stocks ("Betting to Win on the Baldie Winners" 1993, p. 8.) This set of winners became known as the "Baldie Fund."

The impact of the quality universals is also clear for organizations that are not measured by the performance of their asset values. Michael Levinson, City Manager of 2007 Award Recipient for the City of Coral Springs stated it this way: "People ask, 'Why Baldrige?' My answer is very simple: Triple A bond rating on Wall Street from all three ratings agencies, bringing capital projects in on time and within budget, a 96 percent business satisfaction rating, a 94 percent resident satisfaction rating, an overall quality rating of 95 percent, and an employee satisfaction rating of 97 percent . . . that's why we're involved with Baldrige."

Market Quality Leadership and Business Strategy

Building Market Quality Leadership

Market leadership is often the result of entering a new market first and gaining superiority that marketers call a *franchise*. Once gained, this franchise can be maintained through continuing product or service improvement and effective promotion. However, another organization may decide to redefine that market by improving the performance of the good or service—improving its quality—and gaining superiority over the market leader. Then it becomes the "quality leader" in the eyes of the customers. Organizations that have attained this leadership have usually done so on the basis of two principal strategies:

- Let nature take its course. In this approach, organizations apply their best efforts, hoping that in time these efforts will be recognized as the leader creates a failure or gives up its position.
- Help nature out by adopting a positive strategy—establish leadership as a formal business goal and then set out to reach that goal. That goal, once attained, can lead to superior results and sustain that position for long periods.

Those who decided to take action to make superior quality a formal goal soon found that they also had to answer the question, "Leadership in what?" Leadership in quality can exist in any of the multiple aspects of fitness for purpose, but the focus of the organization will differ depending on which aspects are chosen. If quality leadership is to consist of

1. Superior quality of design
2. Superior quality of conformance
3. Availability
4. Guarantees
5. Speed of field repairs

Then the organization must focus on

1. Product development of its goods and services
2. Strong quality control and systematic quality improvement
3. Operational controls
4. Reliability and maintainability programs
5. Creation of a field service capability that is rapid and free of defects

Once attained, quality leadership endures until there is clear cumulative evidence that some competitor has overtaken the leadership. Lacking such evidence, the leadership can endure for decades and even centuries. However, superior quality can also be lost through some catastrophic change.

A brewery reportedly changed its formulation in an effort to reduce costs. Within several years, its share of market declined sharply. The original formula was then restored but market share did not recover. (See "The Perils of Cutting Quality," 1982.)

In some cases, the quality reputation is built not around a specific organization but around an association of organizations. In that event, this association adopts and publicizes some mark or symbol. The quality reputation becomes identified with this mark, and the association goes to great lengths to protect its quality reputation.

The medieval guilds imposed strict specifications and quality controls on their members. Many medieval cities imposed "export controls" on selected finished goods in order to protect the quality reputation of the city (Juran 1995).

The growth of competition in quality has stimulated the expansion of strategic business planning to include planning for quality and quality leadership. (For elaboration, see Chapter 7, Strategic Planning and Deployment: Moving from Good to Great.)

One approach to superior quality is through product development in collaboration with the leading user of the goods or services—a user who is influential in the market and hence is likely to be followed. For example, in the medical field, an individual is "internationally renowned; a chairman of several scientific societies; is invited to congresses as speaker or chairman; writes numerous scientific papers" (Ollson 1986). Determining the identity of the leading user requires some analysis. (In some respects, the situation is similar to the sales problem of discovering who within the client organization is the most influential in the decision to buy.) Ollson lists 10 leader types, each playing a different role.

Quality and Share of Market

Growth in market share is often among the highest goals of upper managers. Greater market share means higher sales volume. In turn, higher sales volume accelerates return on investment disproportionally due to the workings of the break-even chart.

In Figure 1.2, to the right of the break-even line, an increase of 20 percent in sales creates an increase of 50 percent in profit, since the fixed costs do not increase. (Actually, constant costs do vary with volume, but not at all in proportion.) The risks involved in increasing market share are modest, since the technology, product or service, facilities, market, and so on are already in existence and of proved effectiveness.

Effect of Quality Superiority

Quality superiority can often be translated into higher share of market, but it may require special effort to do so. The superior quality must be clearly based on the customer needs and the benefits the customer is seeking. If the quality superiority is defined only in terms of the

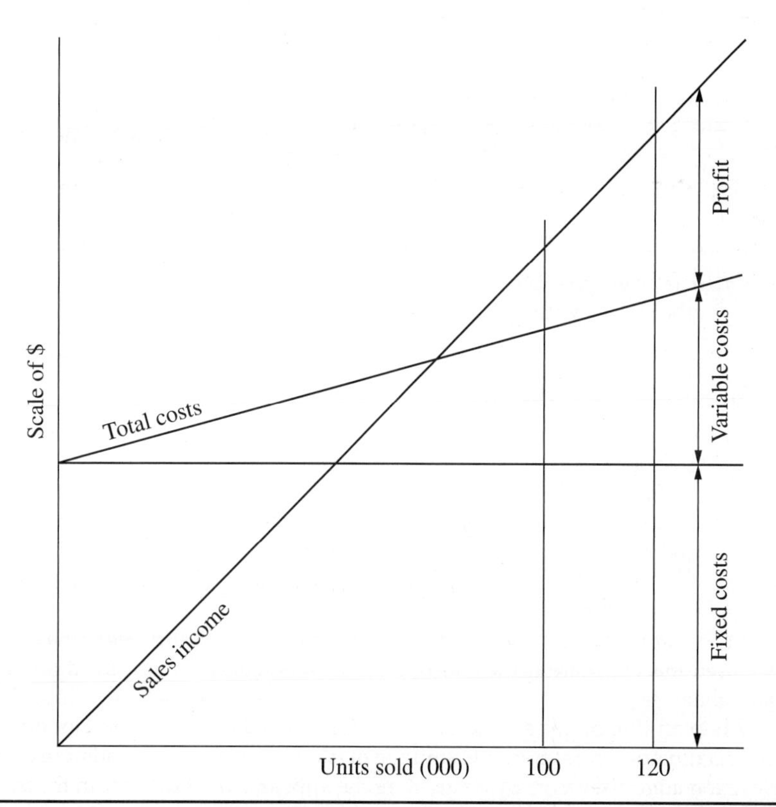

Figure 1.2 Break-even chart. (*Juran, J. M., Juran's Quality Handbook, 5th ed., McGraw-Hill, New York, 1999, p. 7.13.*)

company's internal standards, the customer may not perceive the value. Patients may be willing pay the extra cost of a trip to the Mayo Clinic in the United States rather than visit a local practice because they perceive the superior clinical outcomes available at the Mayo Clinic.

Superiority Obvious to the Buyer

In such cases, the obvious superiority can be translated into higher share of market. This concept is fully understood by marketers, and from time to time they have urged product or service developers to come up with product or service features, which can then be promoted to secure higher share of market. Examples of such cases are legion.

Superiority Translatable into Users' Economics

Some products or services are outwardly alike but have dissimilar performances. An obvious example is the difference in the electric power consumption of appliances with otherwise identical features. In this and similar examples, it is feasible to translate the technological difference into the language of money. Such translation makes it easier for amateurs in technology to understand the significance of the quality superiority.

The superior reliability of a power tool can be translated into the language of money to secure a price premium. The superior reliability could also have been used to secure higher share of market. The superior quality of a truck tire can be translated into cost per unit of distance traveled.

The initiative to translate may also be taken by the buyer. Some users of grinding wheels keep records on wheel life. This is then translated into money—grinding wheel costs per 1000 pieces processed. Such a unit of measure makes it unnecessary for the buyer to become expert in the technology of abrasives.

Collectively, cases such as the above can be generalized as follows:

- There is in fact a quality difference among competing product or services.
- This difference is technological so that its significance is not understood by many users.
- It is often possible to translate the difference into the language of money or into other forms within the users' systems of values.

Superiority Minor but Demonstrable

In some cases, quality superiority can secure added share of market even though the competitive "inferior" product is nevertheless fit for purpose.

A manufacturer of antifriction bearings refined its processes to such an extent that its product or services were clearly more precise than those of the competitors. However, the competitors' product or services were fit for purpose, so no price differential was feasible. Nevertheless, the fact of greater precision impressed the client's engineers and secured increased share of market.

In consumer goods or services, even a seemingly small difference may be translated into increased market share if the consumers are adequately sensitized to the differentials and value them.

An executive of a manufacturer of candy-coated chocolates seized on the fact that his product did not create chocolate smudge marks on consumers' hands. He dramatized this in television advertisements by contrasting the appearance of the children's hands after eating his and the competitors' (uncoated) chocolate. His share of market rose dramatically.

Superiority Accepted on Faith

Consumers can be persuaded to accept, on faith, assertions of good or service superiority, which they themselves are unable to verify. An example was an ingenious market research on electric razors. The sponsoring organization (Schick) employed an independent laboratory to conduct the tests. During the research, panelists shaved themselves twice, using two electric razors, one after the other. On one day, the Schick razor was used first and a competing razor immediately after. On the next day, the sequence was reversed. In all tests, the contents of the second razor were weighed precisely. The data clearly showed that when the Schick was the second razor, its contents weighed more than those of the competitors. The implication was that Schick razors gave a cleaner shave. Within a few months, the Schick share of market rose as follows:

- September, 8.3 percent
- December, 16.4 percent

In this case, the consumers had no way to verify the accuracy of the asserted superiority. They had the choice of accepting it on faith or not at all. Many accepted it on faith.

No Quality Superiority

If there is no demonstrable quality superiority, then share of market is determined by marketing skills. These take such forms as persuasive value propositions, attractive packaging, and so on. Price reductions in various forms can provide increases in share of market, but this is usually temporary. Competitors move promptly to take similar action. Such price reduction can have permanent effect if the underlying cost of production has also been reduced as the result of process improvements that give the company a competitive cost edge over its competitors.

Consumer Preference and Share of Market

Consumers rely heavily on their own senses to aid them in judging quality. This fact has stimulated research to design means for measuring quality by using human senses as measuring instruments. This research has led to development of objective methods for measuring consumer preference and other forms of consumer response. A large body of literature is now available, setting out the types of sensory tests and the methods for conducting them.

At first, these methods were applied to making process control and product or service acceptance decisions. But the applications were soon extended into areas such as consumer preference testing, new-product or new-service development, advertising, and marketing.

For some products or services, it is easy to secure a measure of consumer preference through "forced-choice" testing. For example, a table is set up in a department store and passersby are invited to taste two cups of coffee, A and B, and to express their preference. Pairs of swatches of carpet may be shown to panels of potential buyers with the request that they indicate their preferences. For comparatively simple consumer goods or services, such tests can secure good data on consumer preference. More complex products such as insurance or financial instruments may require more sophisticated analysis such as conjoint analysis or discrete choice methods.

The value of consumer preference data is greatly multiplied through correlation with data on share of market. Figure 1.3 shows such a correlation for 41 different packaged consumer food products. This was an uncommonly useful analysis and deserves careful study.

Each dot on Figure 1.3 represents a food product sold on supermarket shelves. Each product has competitors for the available shelf space. The competing products sell for identical

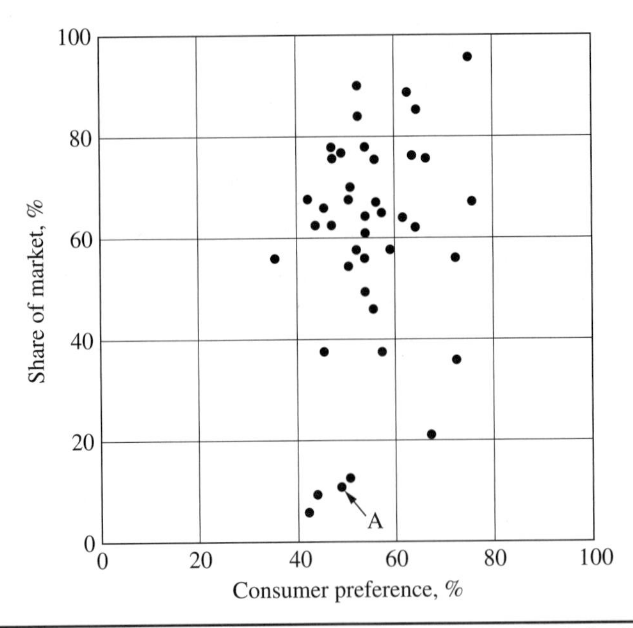

FIGURE 1.3 Consumer preference versus share of market. (*Juran, J. M., Juran's Quality Handbook, 5th ed., McGraw-Hill, New York, 1999, p. 7.17.*)

prices and are packaged in identically sized boxes containing identical amounts of product or service. What may influence the consumer?

- The contents of the package, as judged by senses and usage, which may cause the consumer to prefer product A over product B
- The marketing features such as attractiveness of the package, appeal of prior advertising, and reputation of the manufacturer

In Figure 1.3 the horizontal scale shows consumer preference over the leading competitor as determined by statistically sound preference testing. The vertical scale shows the share of market versus the leading competitor, considering the two as constituting 100 percent.

In Figure 1.3 no product showed a consumer preference below 25 percent or above 75 percent. The 75/25 preference levels mean that the product is so superior (or inferior) that three users out of four can detect the difference. Since all other factors are essentially equal, this result implies that a product that is preferred by more than 75 percent of consumers eventually takes over the entire market, and its competition disappears.

In contrast to the vacant areas on the horizontal scale of consumer preference, the vertical scale of share of market has data along the entire spectrum. One product (marked A in Figure 1.3) lies squarely on the 50 percent consumer preference line, which probably means (under forced-choice testing) that the users are guessing as to whether they prefer that product or service or that of its competitor. Yet product or service A has only 10 percent share of market and its competitor has 90 percent. In addition, this inequality in share of market has persisted for years. The reason is that the 90 percent organization was the first to bring that product to market. As a result, it acquired a "prior franchise" and has retained its position through good promotion.

The conclusion is that when competing products or services are quite similar in consumer preference, any effect of such small quality differentials is obscured by the effect of the marketing skills. In consequence, it is logical to conclude that when quality preferences are evident to the user, such quality differences are decisive in share of market, all other things being equal. When quality differences are slight, the decisive factor in share of market is the marketing skills.

As a corollary, it appears that organizations are well advised to undertake quality improvements, which will result in either (1) bringing them from a clearly weak to an acceptable preference or (2) bringing them from an acceptable preference to a clearly dominant preference. However, organizations are not well advised to undertake quality improvements that will merely make minor improvements that are not largely perceived and valued by their customers, since marketing skill is usually the dominant factor in determining the share of market when the differences in quality are small.

It is easy for technologists to conclude that what they regard as important in the good or service is also of prime concern to the user. In the carpet industry, the engineers devote much effort to improving wear qualities and other technological aspects of fitness for purpose. However, after a market research study was conducted, it was determined that consumers' reasons for selecting carpets were primarily sensory, and not durability:

- Color 56 percent
- Pattern 20 percent
- Other sensory qualities 6 percent
- Nonsensory qualities 18 percent

For more complex consumer goods or services it is feasible, in theory, to study the relation of quality to market share by securing quantitative data on (1) actual changes in buying patterns of consumers and (2) actions of suppliers that may have created these changes. In practice, such information is difficult to acquire. It is also difficult to conclude, in any one instance, why the purchase was of model A rather than B. What does emerge is "demographic" patterns, i.e., age of buyers, size of family, and so on, that favor model A rather than B. For goods or services sold through merchants, broad consumer dissatisfaction with quality can translate into *merchant preference,* with extensive damage to share of market.

A maker of household appliances was competitive with respect to product or service features, price, and promptness of delivery. However, it was not competitive with respect to field failure, and this became a major source of complaints from consumers to the merchants. Within several years the maker (B) lost all its leadership in share of market, as shown in the table below. This table stimulated the upper managers of organization B to take action to improve its product reliability.

Model Price	Leaders in Market Share during:			
	Base Year	Base Year Plus 1	Base Year Plus 2	Base Year Plus 3
High	A	C	C	C
Medium	B	B	C	C
Low	C	C	C	C
Special	B	B	B	C

Industrial Products and Share of Market

Industrial goods or services are sold more on technological performance than on sensory qualities. However, the principle of customer preference applies, as does the need to relate quality differences to customer preference and to share of market.

Quality and Competitive Bidding

Many industrial products are bought through competitive bidding. Most government agencies are required by law to secure competitive bids before awarding large contracts. Industrial organizations require their purchasing managers to do the same. The invitations to bid usually include the parameter of quality, which may be specified in detail or though performance specifications.

To prospective suppliers the ratio of awards received to bids made is of great significance. The volume of sales and profit depends importantly on this ratio. In addition, the cost of preparing bids is substantial. Finally, the ratio affects the morale of the people involved. (Members of a winning team fight with their competitors; members of a losing team fight with one another.) It is feasible to analyze the record of prior bids in order to improve the percentage of successful bids. Figure 1.4 shows such an analysis involving 20 unsuccessful bids.

To create Figure 1.4, a multifunctional team analyzed 20 unsuccessful bids. It identified the main and contributing reasons for failure to win the contract. The team's conclusions show that the installation price was the most influential factor—it was a contributing cause in 10 of the 14 cases that included bids for installation. This finding resulted in a revision of the process for estimating the installation price and an improvement in the bidding/success ratio.

Carryover of Failure-Prone Features

Market leadership can be lost by perpetuating failure-prone features of predecessor models. The guilty features are well known, since the resulting field failures keep the field service force busy restoring service. Nevertheless, there has been much carryover of failure-prone features into new models. At the least, such carryover perpetuates a sales detriment and a cost burden. At its worst, it is a cancer that can destroy seemingly healthy product or service lines.

A notorious example was the original xerographic copier. In that case the "top 10" list of field failure modes remained essentially identical, model after model. A similar phenomenon existed for years in the automobile industry.

The reasons behind this carryover have much in common with the chronic internal wastes, which abound in so many organizations:

1. The alarm signals are disconnected. When wastes continue, year after year, the accountants incorporate them into the budgets. That disconnects the alarm signals—no alarms ring as long as actual waste does not exceed budgeted waste.
2. There is no clear responsibility to get rid of the wastes. There are other reasons as well. The technologists have the capability to eliminate much of the carryover. However, those technologists are usually under intense pressure from the marketers to develop new product or service and process features in order to increase sales. In addition, they share distaste for spending their time cleaning up old problems. In their culture, the greatest prestige comes from developing the new.

The surprising result can be that each department is carrying out its assigned responsibilities, and yet the product or service line is dying. Seemingly nothing short of upper management intervention—setting goals for getting rid of the carryover—can break up the impasse.

	Bid Not Accepted due to				
Contract Proposal	Quality of Design	Product Price	Installation Price	Reciprocal Buying	Other
A1		X	X		X
A2			XX		
A3	XX	X			
A4	XX		X		
A5	XX				
A6	XX				
A7		XX			
A8		XX			
A9			XX		
A10			XX		
B1	X		X		
B2				XX	
B3				XX	
B4				XX	
B5		X	X		
B6		X	XX		
B7	XX				
B8		X	X		
B9				X	
B10	X	X	X		
Totals	7	8	10 (out of 14)	4	1

X = contributing reasons; XX = main reason
Only 14 bids were made for installation.

Figure 1.4 Analysis of unsuccessful bids. (*Juran, J. M., Juran's Quality Handbook, 5th ed., McGraw-Hill, New York, 1999.*)

Macroeconomic Influences on Results

The ability of an organization to secure revenue is strongly influenced by the economic climate and by the cultural habits that the various economies have evolved. These overriding influences affect product or service quality as well as other elements of commerce.

National Affluence and Organization

The form of a nation's economy and its degree of affluence strongly influence the approach to its problems.

Subsistence Economies

In subsistence economies the numerous impoverished users have little choice but to devote their revenue to basic human needs. Their protection against poor quality is derived more from their collective political power than from their collective economic power. Much of the world's population remains in a state of subsistence economy.

Shortages and Surpluses

In all economies, a shortage of goods (a "sellers' market") results in a relaxing of quality standards. The demand for goods exceeds the supply, so users must take what they can get (and bid up the price to boot). In contrast, a buyers' market results in a tightening of quality standards.

Life with the Risk of Failure

As societies industrialize, they revise their lifestyle in order to secure the benefits of technology. Collectively, these benefits have greatly improved the quality of life, but they have also created a new dependence. In the industrial societies, great masses of human beings place their safety, health, and even their daily well-being behind numerous "quality dikes." For instance, many pharmaceuticals often enable a person to receive quick short-term health benefits, but in the long term the illness may get worse.

For elaboration, see Chapter 2, Quality's Impact on Society and the National Culture.

Voluntary Obsolescence

As customers acquire affluence, economic organizations increasingly bring out new goods or services (and new models of old ones) that they urge prospective users to buy. Many of the users who buy these new models do so while possessing older models that are still in working order. This practice is regarded by some reformers as a reprehensible economic waste.

In their efforts to put an end to this asserted waste, the reformers have attacked the organizations who bring out these new models and who promote their sale. Using the term "planned obsolescence," the reformers imply (and state outright) that the large organizations, by their clever new models and their powerful sales promotions, break down the resistance of the users. Under this theory, the responsibility for the waste lies with the organizations that create the new models.

In the experience and judgment of the author, this theory of planned obsolescence is mostly nonsense. The simple fact, obvious to both producers and consumers, is that the consumer makes the decision (of whether to discard the old product or service and buy the new). Periodically, this fact is dramatized by some massive marketing failure.

- The early models of home refrigerators lacked many features of modern models: freezer compartments, ice cube makers, shelves in the door, and so on. As these features were added to new models, homeowners who had bought the original models became increasingly unhappy until they bought a new model despite the fact that the old model was still running. Note that the decision to buy the new model was made by the customer, not by the manufacturer.
- The latter half of the 1970s saw the introduction of recorded entertainment into the home of the consumer with the creation of the video cassette recorder (VCR). For two decades this invention was a staple in millions of people's domestic lives across the globe. The introduction of the digital video disc (DVD) player, a machine whose ultimate utility was the same as a VCR, replaced it within years, not decades. Offered improved quality and additional features but with the same basic function as a VCR, consumers chose the DVD even when they already had an operating appliance with equivocal functionality in their homes. New forms of downloadable video through the Internet are beginning to revolutionize this market again.

Involuntary Obsolescence

A very different category of obsolescence consists of cases in which long-life products contain failure-prone components that will not last for the life of the product or service. The life of these components is determined by the manufacturer's design. As a result, even though the user decides to have the failed component replaced (to keep the product or service active), the manufacturer has made the real decision because the design determined the life of the component.

This situation is at its worst when the original manufacturer has designed the product or service in such a way that the supplies, spare parts, and so on are nonstandard, so that the sole source is the original manufacturer. In such a situation, the user is locked into a single source of supply. Collectively, such cases have lent themselves to a good deal of abuse and have contributed to the consumerism movement.

Contrast in Views: Customers' and Producers'

Industrial organizations derive their revenue from the sale of their goods or services. These sales are made to "customers," but customers vary in their functions. Customers may be wholesalers, processors, ultimate users, and so on, with resulting variations in customer needs. Response to customer needs in order to sell more goods or services requires a clear understanding of just what those needs are and how the organization can meet them.

Human needs are complex and extend beyond technology into social, artistic, status, and other seemingly intangible areas. Suppliers are nevertheless obliged to understand these intangibles in order to be able to provide products or services that respond to such needs.

The Spectrum of Affluence

In all economies the affluence of the population varies across a wide spectrum. Suppliers respond to this spectrum through variations in product or service features. These variations are often called *grades*.

For example, all hotels provide overnight sleeping accommodations. Beyond this basic service, hotels vary remarkably in their offerings, and the grades (deluxe suites, four-star, and so on) reflect this variation. In like manner, any model of automobile provides the basic service of point-to-point transportation. However, there are multiple grades of automobiles. There are luxury brands such as Porsche, BMW, Mercedes, Cadillac, and Lexus; and there are more affordable ones such as GM, Hyundai, Ford, and Toyota. The higher grades supply services beyond pure transportation. They may provide more features that result in higher levels of safety, comfort, appearance, and status.

Fitness for Purpose and Conformance to Specification

Customers and suppliers sometimes differ in their definition of quality. Such differences are an invitation to trouble. To most customers, quality means those features of the product or service that respond to customer needs. In addition, quality includes freedom from failures, plus good customer service if failures do occur. One comprehensive definition for the above is "fitness for purpose."

In contrast, for years many suppliers had defined quality as conformance to specification at the time of final test. This definition fails to consider numerous factors that influence quality as defined by customers: packaging, storage, transport, installation, reliability, maintainability, customer service, and so on.

Figure 1.5 tabulates some of the differences in viewpoint as applied to long-life goods.

Aspects	Principal Views	
	Of Customers	Of Producers
What is purchased?	A product needed by the customer	Goods made by the producer
Definition of quality	Fitness for purpose during the life of the product or service	Conformance to specification on final test
Cost	Cost of use, including ♣ Purchase price ♣ Operating costs ♣ Maintenance ♣ Downtime ♣ Depreciation ♣ Loss on resale	Cost of producers
Responsibility for keeping in service	Over the entire useful life	During the warranty period
Spare parts	A necessary evil	A profitable business

FIGURE 1.5 Contrasting views: customers' and producers'.

The ongoing revolution in quality has consisted in part of revising the suppliers' definition of quality to conform more nearly to the customers' definition.

Cost of Use

For consumable goods or many services, the purchase price paid by the customer is quite close to the cost of using (consuming) the good or service. However, for long-lived product or services, the cost of use can diverge considerably from the purchase price because of added factors such as operating costs, maintenance costs, downtime, depreciation, license fees, new releases, and transaction service charges.

The centuries-old emphasis on purchase price has tended to obscure the subsequent costs of use. One result has been suboptimization; i.e., suppliers optimize their costs rather than the combined costs of suppliers and customers.

The concept of life-cycle costing offers a solution to this problem, and progress is being made in adopting this concept.

Degrees of User Knowledge

In a competitive market, customers have multiple sources of supply. In making a choice, quality is an obvious consideration. However, customers vary greatly in their ability to evaluate quality, especially prior to purchase. Figure 1.6 summarizes the extent of customer knowledge and strength in the marketplace as related to quality matters.

The broad conclusions that can be drawn from Figure 1.6 are as follows:

- Original equipment manufacturers (OEMs) can protect themselves through their technological and/or economic power as much as through contract provisions. Merchants and repair shops must rely mainly on contract provisions supplemented by some economic power.

Aspects of the Problem	Original Equipment Manufacturers (OEMs)	Dealers and Repair Shops	Consumers
Makeup of the market	A few, very large customers	Some large customers plus many smaller ones	Very many, very small customers
Economic strength of any one customer	Very large, cannot be ignored	Modest or low	Negligible
Technological strength of customer	Very high; has engineers and laboratories	Low or nil	Nil (requires technical assistance)
Political strength of customer	Modest or low	Low or nil	Variable, but can be very great collectively
Fitness for purpose is judged mainly by:	Qualification testing	Absence of consumer complaints	Successful usage
Quality specifications dominated by:	Customers	Manufacturer	Manufacturer
Use of incoming inspection	Extensive test for conformance to specification	Low or nil for dealers; in-use tests by repair shops	In-use test
Collection and analysis of failure data	Good to fair	Poor to nil	Poor to nil

FIGURE 1.6 Customer influences on quality. (*Juran, J. M., Juran's Quality Handbook, 5th ed., McGraw-Hill, New York, 1999, p. 7.5.*)

- Small users have very limited knowledge and protection. The situation of the small user requires some elaboration.

With some exceptions, small users do not fully understand the technological nature of the product or service. The user does have sensory recognition of some aspects of fitness for use: the bread smells fresh-baked, the radio set has clear reception, the shoes are good-looking. Beyond such sensory judgments, and especially concerning the long-life performance of the product or service, the small user must rely mainly on prior personal experience with the supplier or merchant. Lacking such prior experience, the small user must choose from the propaganda of competing suppliers plus other available inputs (neighbors, merchants, independent laboratories, and so on).

To the extent that the user does understand fitness for use, the effect on the supplier's revenue is somewhat as follows:

As seen by the user, the good or service or service is

- Not fit for purpose
- Fit for purpose but noticeably inferior to competitive products or services
- Fit for purpose and competitive
- Noticeably superior to competitive products or services

The resulting revenue to the supplier is

- None, or in immediate jeopardy
- Lower due to loss of market share or need to lower prices
- At market prices
- High due to premium prices or greater share of market

In the foregoing table, the terms "fitness for purpose," "inferior," "competitive," and "superior" all relate to the situation as seen by the user. (The foregoing table is valid as applied to both large customers and small users.)

As Seen by the User, the Product or Service Is	The Resulting Income to the Supplier Is
Not fit for purpose	None, or in immediate jeopardy
Fit for purpose, but noticeably inferior to competitive products	Low due to loss of market share or need to lower prices
Fit for purpose and competitive	At market prices
Noticeably superior to competitive products	High due to premium prices or greater share of market

Stated Needs and Real Needs

Customers state their needs as they see them, and in their language. Suppliers are faced with understanding the real needs behind the stated needs and translating those needs into suppliers' language.

It is quite common for customers to state their needs in the form of goods, when their real needs are for the services provided by those goods. For example:

Stated Needs of Customer	Real Needs of Customer
Food	A pleasant taste and nourishment
An automobile	Transportation, safety, comfort
A flat screen TV	Entertainment, news, movies in the home
Toothpaste	Clean teeth, sweet breath, etc.
7/24 Banking	Ability to deposit or get money anytime as needed

Preoccupation with selling goods can divert attention from the real needs of customers.

In the classic, widely read paper "Marketing Myopia," Levitt (1960) stressed service orientation as distinguished from product orientation. In his view, the railroads missed an opportunity for expansion due to their focus on railroading rather than on transportation. In like manner, the motion picture industry missed an opportunity to participate in the growing television industry due to its focus on movies rather than on entertainment (Levitt 1960).

Understanding the real needs of customers requires answers to questions such as these: Why are you buying this product or service? What service do you expect from it?

Psychological Needs

For many products or services, customer needs extend beyond the technological features of the good or service; customer needs also include matters of a psychological nature. Such needs apply to both goods and services. "A person in need of a haircut has the option of going to (1) a 'shop' inhabited by 'barbers' or (2) a 'salon' inhabited by 'hair stylists.' Either way, she/he is cut by a skilled artisan. Either way, her/his resulting outward appearance is essentially the same. What differs is her/his remaining assets and her/his sense of well-being." (Juran, 1984)

What applies to services also applies to physical goods. There are factories in which chocolate-coated candies are conveyed by a belt to the packaging department. At the end of the belt are two teams of packers. One team packs the chocolates into modest cardboard boxes destined for budget-priced merchant shops. The other team packs the chocolates into satin-lined wooden boxes destined to be sold in deluxe shops. The resulting price for a like amount of chocolate can differ by several fold. The respective purchasers encounter other differences as well: the shop decor, level of courtesy, promptness of service, sense of importance, and so on. However, the goods are identical. Any given chocolate on that conveyer belt has not the faintest idea of whether it will end up in a budget shop or in a deluxe shop.

Technologists may wonder why consumers are willing to pay such price premiums when the goods are identical. However, for many consumers, the psychological needs are perceived as real needs, and the consumers act on their perceptions. Most suppliers design their marketing strategies to respond to customers' perceived needs.

"User-Friendly" Needs

The "amateur" status of many users has given rise to the term "user-friendly" to describe a condition that enables amateurs to use technological and other complex product or services with confidence. Consider the following example.

The language of published information should be simple, unambiguous, and readily understood. Notorious offenders have included legal documents, owners' operating manuals, forms to be filled out, and so on. Widely used forms (such as federal tax returns) should be field-tested on a sample of the very people who will later be faced with filling out the forms.

Goods or services should be broadly compatible. Much of this has been done through standardization committees or through natural monopolies. An example of the lack of such compatibility during the 1980s was the personal computer—many personal computers were able to "talk" to computers made by the same manufacturer but not to computers made by other manufacturers.

The Need to Be Kept Informed

Customers sometimes find themselves in a state of uncertainty: Their train is late, and they don't know when to expect it; there is a power outage, and they don't know when power will be restored. In many such cases, the supplier organization has not established the policies and processes needed to keep customers informed. In actuality, customers, even if kept informed, usually have no choice but to wait it out. Nevertheless, being kept informed reduces the anxiety—it provides a degree of assurance that human beings are aware of the problem and that it is in the process of being solved.

The New York City subway system rules require conductors to explain all delays lasting 2 minutes or more. One survey reported that this rule was followed only about 40 percent of the time. A City Hall report concluded that "shortage of information is a significant source of public antagonism toward the Transit Authority" (Levine 1987).

In contrast, some airlines go to pains to keep their customers informed of the reasons for a delay and of the progress being made in providing a remedy.

A different category of cases involves organizations secretly taking actions adverse to quality but without informing the customer. The most frequent are those in which goods or services not conforming to specification are shipped to unwary customers. In the great majority of such cases, the products or services are fit for use despite the nonconformance. In other cases, the matter may be debatable. In still other cases, the act of shipment is at least unethical and at worst illegal.

The partnership between Firestone Tires and Ford Explorer SUVs in the late 1990s created one of the most defective and ultimately deadly relationships in modern automotive history. Firestone tires were continually failing under the frame of the Ford Explorer, often causing the SUV to flip and roll. More than 250 people lost their lives because of this defect while 3000 other incidents were reported due to this imperfection. What made this situation sordid was the fact that neither Ford nor Firestone took responsibility for this obvious problem. Ford Explorers with Firestone tires were still being sold to the general public even when, in the early stages, unusually high rates of crashes were taking place. Instead of recalling the models in question, the manufacturer let this problem persist for years until the incident rate and death count became so high that the problem could not be ignored further.

Once discovered, any secretive actions tend to arouse suspicions, even if the product or service is fit for customer use. The customers wonder, "What else has been done secretly without our being informed?"

The usual reason for not informing the customer is a failure to raise the question, What shall we tell the customers? It would help if every nonconformance document included a blank space titled "What is to be communicated to the customers?" The decision may be to communicate nothing, but at least the question has been asked.

Cultural Needs

The needs of customers, especially internal customers, include cultural needs—preservation of status, continuity of habit patterns, and still other elements of what is broadly called the *cultural pattern*. Some of the inability to discover customer needs is traceable to failure to understand the nature and even the existence of the cultural pattern.

Cultural needs are seldom stated openly—mostly they are stated in disguised form. A proposed change that may reduce the status of some employee will be resisted by that employee. The stated reasons for the resistance will be on plausible grounds, such as the effect on costs. The real reason will not emerge. No one will say, "I am against this because it will reduce my status." Discovery of the real needs behind the stated needs is an important step toward a meeting of the minds. (For elaboration on the nature of cultural patterns and the "rules of the road," see Chapter 5, Quality Improvement: Creating Breakthroughs in Performance.)

Needs Traceable to Unintended Use

Many quality failures arise because the customer uses the product or service in a manner different from that intended by the supplier. This practice takes many forms:

- Untrained patient care workers are assigned to processes requiring trained workers.
- Equipment is overloaded or is allowed to run without adherence to maintenance schedules.
- The product or service is used in ways never intended by the supplier. For instance, a screwdriver is used as a hammer. It was not designed to hit things! It can break and cause harm to the user.

All this influences the relationship between quality and revenue. The critical question is whether the product or service or service development should be based on intended use or actual use. The latter often requires adding a factor of safety during the development. For example:

- Fuses and circuit breakers are designed into electrical circuits for protection against overloads.
- Spell-check software is designed to detect grammar and spelling errors.
- Public utility invoicing may include a check of customers' prior usage to guard against errors in reading the meters.

Such factors of safety may add to the cost. Yet they may well result in an optimal overall cost by helping to avoid the higher cost arising from actual use or misuse.

Needs Related to Dissatisfaction

When products or services fail to meet the needs of their customers, a new set of customer needs arises—how to restore service and get compensated for the associated losses and inconvenience. These new needs are communicated through customer complaints, which then are acted on by special departments such as customer service or call centers. Inadequate organizational response to consumer complaints and to the terms of warranties has contributed importantly to the rise of the "consumerism" movement. (See Chapter 2, Quality's Impact on Society and the National Culture.)

Studies of how to respond to customer complaints have identified the key features of a response system that meets customer needs.

Complaints also affect product or service salability. This has been researched in studies commissioned by the U.S. Office of Consumer Affairs. The findings may be summarized as follows:

- Of customers who were dissatisfied with products or services, nearly 70 percent did not complain. The proportions varied with the type of product or service involved. The reasons for not complaining included these: the effort to complain was not worth it; the customers believed that complaining would do no good; customers lacked knowledge of how to complain.
- Over 40 percent of the complaining customers were unhappy with the responsive action taken by the suppliers. Here again the percentage varied according to the type of product or service involved.

Future salability is strongly influenced by the action taken on complaints. Figure 1.7 shows broadly the nature of consumer behavior following dissatisfaction with a purchase. This strong

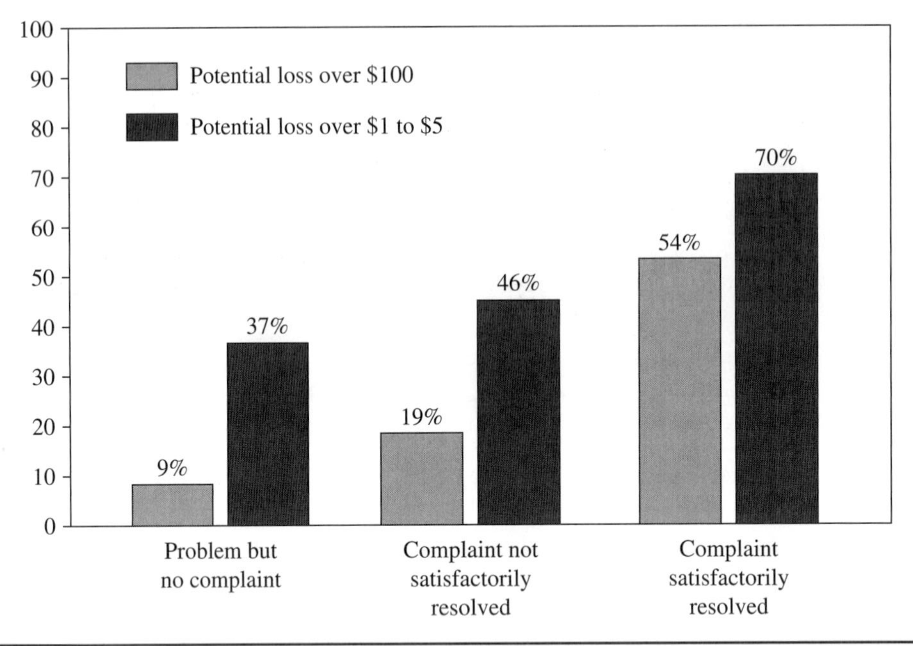

Figure 1.7 Behaviors of consumers after they experience product dissatisfaction. (*Planning for Quality, 2d ed., Juran Institute, 1990, pp. 4–12.*)

influence extends to brand loyalty. Figure 1.8 shows the extent of this influence as applied to large-ticket durable goods, financial services, and automobile services, respectively.

That same research concluded that an organized approach to complaint handling provides a high return on investment. The elements of such an organized approach may include

- A response center staffed to provide 7/24 access by consumers
- A toll-free telephone number
- A computerized database
- Special training for the personnel who answer the telephones
- Active solicitation of complaints to minimize loss of customers in the future

Discovering Hidden Customer Needs

The most simplistic assumption is that customers are completely knowledgeable as to their needs and that market research can be used to extract this information from them. In practice, customer knowledge can be quite incomplete. In some cases the customer may be the last person to find out. It is unlikely that any customer ever expressed the need for a Walkman (a miniature, portable audiotape player) before such devices came on the market. However, once they became available, many customers discovered that they needed one.

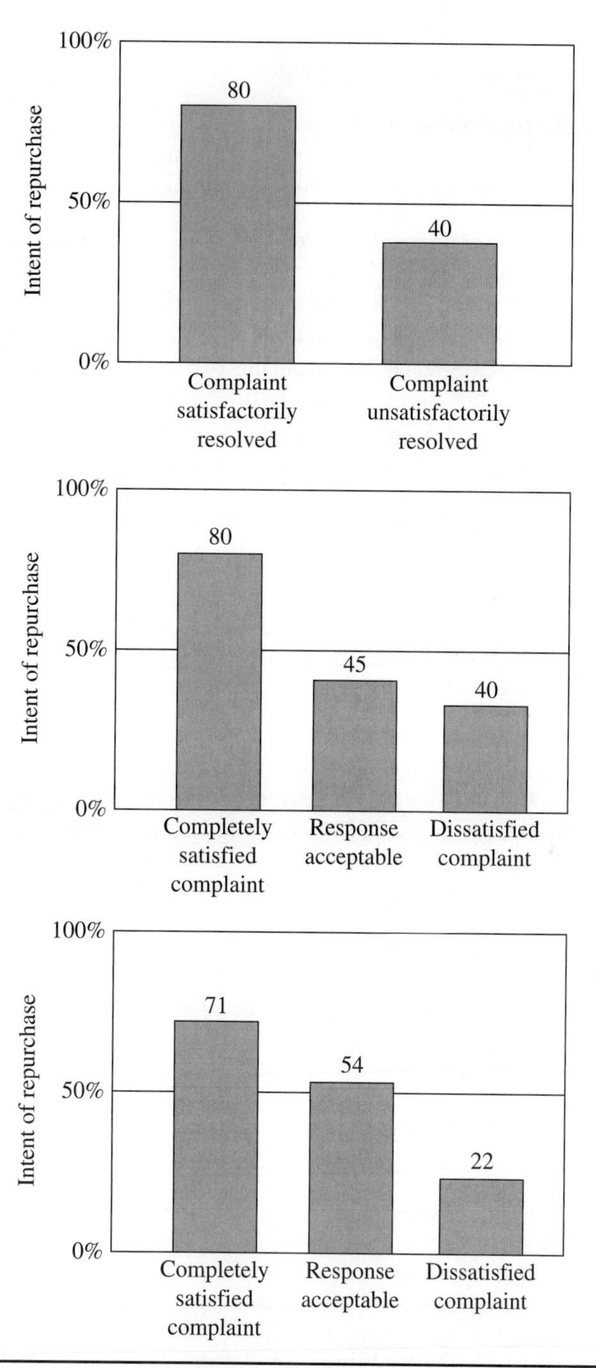

Figure 1.8 Consumer loyalty versus complaint resolution: large ticket durable goods, financial services, automotive services. (*Planning for Quality, 2d ed., Juran Institute, 1990, pp. 4–14.*)

These gaps in customer knowledge are filled in mainly by the forces of the competitive market and by the actions of entrepreneurs.

Inferior Available Product or Services

When available products are perceived as inadequate, a vacuum waiting to be filled emerges. Human ingenuity then finds ways to fill that vacuum:

1. The number of licensed New York taxicabs has remained frozen for years while the population has increased. The resulting shortage of cabs has been filled by unlicensed cabs ("gypsy cabs"), limousines, buses, and even bicycles.
2. Government instructions for filling out tax forms have been confusing to many taxpayers. One result has been the publication of some best-selling books and software on how to prepare tax returns.
3. The service provided by tradesmen has been widely regarded as expensive and untimely. One result has been the growth of a large do-it-yourself industry.

Reduction of Time for Service

Some cultures exhibit an urge to "get it over with." In such cultures, those who can serve customers in the shortest time are rewarded by a higher share of market. A spectacular example of this urge is the growth of the fast-food industry. This same need for fast food and quick service is an essential element in the urge of a producer to employ "just-in-time" manufacturing techniques.

Changes in Customer Habits

Customer habits can be notoriously fickle. Obvious examples are fashions in clothing or concerns over health. Consumerism is now driving lifestyles. Many people have reduced the consumption of beef and increased that of poultry and fish. Such shifts are not limited to consumers. Industrial organizations often launch "drives," most of which briefly take center stage and then fade away. The associated "buzzwords" similarly come and go.

Perfectionism

The human being exhibits an instinctive drive for precision, beauty, and perfection. When unrestrained by economics, this drive has created the art treasures of the ages. In the arts and in aesthetics, this timeless human instinct still prevails.

In the industrial society, there are many situations in which this urge for perfection coincides with human needs. In food and drug preparation, certain organisms must be completely eliminated, or they will multiply and create health hazards. Nuclear reactors, underground mines, aircraft, and other structures susceptible to catastrophic destruction of life require a determined pursuit of perfection to minimize dangers to human safety. So does the mass production of hazardous materials.

However, there are numerous other situations in which the pursuit of perfection is antagonistic to society, since it consumes materials and energy without adding to fitness for purpose, either technologically or aesthetically. This wasteful activity is termed *perfectionism* because it adds cost without adding value.

Quality: Brief History of Quality and the Management of It

A young recruit who joins an organization soon learns that it has in place numerous processes (systems) to manage its affairs, including managing for quality. The recruit might assume that humans have always used those processes to manage for quality and will continue to so in the future. Such assumptions would be grossly in error. The processes used to manage for quality have undergone extensive change over the millennia, and there is no end in sight.

Primitive Societies—The Family

Quality is a timeless concept. The origins of ways to manage for quality are hidden in the mists of the ancient past. Yet we can be sure that humans have always faced problems of quality. Primitive food gatherers had to learn which fruits were edible and which were poisonous. Primitive hunters had to learn which trees supplied the best wood for making bows or arrows. The resulting know-how was then passed down from generation to generation.

The nuclear human organizational unit was the family. Isolated families were forced to create self-sufficiency—to meet their own needs for food, clothing, and shelter. There was division of work among family members. Production was for self-use, so the design, production, and use of a product were all carried out by the same persons. Although the technology was primitive, the coordination was superb. The same human beings received all inputs and took all remedial action. The limiting factor for achieving quality was the primitive state of the technology.

The Village—Division of Labor

Villages were created to serve other essential human requirements such as mutual defense and social needs. The village stimulated additional division of labor and development of specialized skills. There emerged farmers, hunters, fishers, and artisans of all sorts—weavers, potters, shoemakers. By going through the same work cycle over and over, the artisans became intimately familiar with the materials used, the tools, the steps in the process, and the finished product. The cycle included selling the product to users and receiving their feedback on product performance. The experience derived from this intimate familiarity then enabled human ingenuity to take the first steps toward the evolution of technology.

The Village Marketplace—Caveat Emptor

As villages grew, the village marketplace appeared where artisans and buyers met on scheduled market days. In this setting, producers and users met face to face with the goods between them. The goods typically were natural products or were made from natural materials. The producers and purchasers had long familiarity with the products, and the quality of the products could be judged to a high degree by the unaided human senses.

Under such a state of affairs, the village magistrates tended to avoid being drawn into quality disputes between seller and buyer. This forced buyers to be vigilant so as to protect themselves against poor quality. In effect, the seller was responsible for supplying the goods, but the buyer became responsible for supplying the quality "assurance." This arrangement was known as *caveat emptor*, which is Latin for "let the buyer beware." Thus buyers learned to beware through product inspection and test. They looked closely at the cloth, smelled the fish, thumped the melon, and tasted a grape. Any failure to beware was at their own peril. In the village marketplace, caveat emptor was quite a sensible doctrine. It is widely applied to this day in villages all over the world.

A further force in the village marketplace was the fact of common residence. Producer and buyer both lived in the same village. Each was subject to scrutiny and character evaluation by the villagers. Each also was subject to village discipline. For the artisan, the stakes were high. His or her status and livelihood (and those of her or his family) were closely tied to the person's reputation as a competent and honest artisan. In this way, the concept of craftsmanship became a quiet yet powerful stimulus to maintain a high level of quality.

Effects of the Growth of Commerce

In due course villages expanded into towns and cities, and improved transport opened the way to trade among regions.

A famous example of organized multiregional trade was the Hanseatic League which was centered among the cities of northern Europe from about the 1200s to the 1600s. Its influence extended into Scandinavia and Russia as well as to the Mediterranean and Black Sea (von der Porten 1994).

Under trade among regions, producer and user could no longer meet face to face in the marketplace. Products were now made by chains of suppliers and processors. Marketing was now done by chains of marketers. The buyers' direct point of contact was now with some merchant rather than with the producer. All this reduced the quality protections inherent in the village marketplace to a point requiring invention of new forms of quality assurance. One such invention was the quality warranty.

Quality Warranties

Early quality warranties were no doubt in oral form. Such warranties were inherently difficult to enforce. Memories differed as to what was said and meant. The duration of the warranty might extend beyond the life of the parties. Thus the written warranty was invented.

An early example was on a clay tablet found amid the ruins of Nippur in ancient Babylon. It involved a gold ring set with an emerald. The seller guaranteed that for 20 years the emerald would not fall out of the gold ring. If it did fall out of the gold ring before the end of 20 years, the seller agreed to pay to the buyer an indemnity of 10 mana of silver. The date is the equivalent of 429 B.C. (Bursk et al. 1962, vol. I, p. 71).

Quality warranties are now widely used in all forms of trade and commerce. They stimulate producers to give priority to quality and stimulate sellers to seek out reliable sources of supply. So great is their importance that recent legislation has imposed standards to ensure that the wording of warranties does not mislead the buyers.

Quality Specifications

Sellers need to be able to communicate to buyers the nature of what they have to sell. Buyers need to be able to communicate to sellers the nature of what they want to buy. In the village marketplace, oral communication could take place directly between producer and buyer. With the growth of commerce, communication expanded to include chains of producers and chains of merchants who often were widely separated. New forms of communications were needed, and a major invention was the written quality specification. Now quality information could be communicated directly between designer and producer or between seller and buyer no matter how great the distance between them or how complex the nature of the product.

Like warranties, written specifications have an ancient origin. Examples have been found in Egyptian papyrus scrolls over 3500 years old (Durant 1954). Early specifications focused on defining products and the processes for producing them. In due course the concept was extended to defining the materials from which the products were made. Then as conflicts arose because sellers and buyers used different methods of test, it became necessary to establish inspection and test specifications as well.

Measurement

The emergence of inspection and test specifications led to the evolution of measuring instruments. Instruments for measuring length, volume, and time evolved thousands of years ago. Instruments have continued to proliferate, with ever-increasing precision. In recent centuries, the precision of the measurement of time has increased by more than 10 orders of magnitude (Juran 1995, Chapter 10).

Artisans and Guilds

The artisan's possession of the skills of a trade was a source of income and status as well as self-respect and respect from the community. However, as villages grew into towns and cities, the numbers of artisans grew as well. The resulting competition was perceived by artisans as threatening the benefits they derived from their trade.

To perpetuate their benefits, the artisans within a trade organized trade unions—guilds. Each guild then petitioned the city authorities to confer on guild members a monopoly on practicing their trade.

Guilds flourished for centuries during the Middle Ages until the Industrial Revolution reduced their influence. They used their monopolistic powers chiefly to provide a livelihood and security for their members. The guilds also provided extensive social services to their members. (For elaboration, see Bursk et al. 1962, vol. III, pp. 1656–1678.)

The Guild Hierarchy

Each guild maintained a hierarchy of (usually) three categories of workers: the apprentice, the journeyman, and the master. Considerable formality surrounded the entry into each category.

At the bottom was the apprentice or novice, whose entry was through an indenture—a formal contract that bound the apprentice to serve a master for a specified number of years. In turn, the master became responsible for teaching the trade to the apprentice.

To qualify for promotion, the apprentice was obliged to serve out the full term of the indenture. In addition, he or she was required to pass an examination by a committee of masters. Beyond the oral part of the examination, the apprentice was required to produce a perfect piece of work—a masterpiece—that was then inspected by the examination committee. Success in the examination led to a ceremonial admission to the status of journeyman.

The journeyman's right to practice the trade was limited. He or she could become an employee of a master, usually by the day. The journeyman also could travel to other towns, seeking employment in his or her trade. Only after admission to the rank of master could the journeyman set up shop on his or her own.

Admission to the rank of master required first that there be an opening. Guilds imposed limits on the numbers of masters in their areas. On the death or retirement of an active master, the guild would decide whether to fill that opening. If so, a journeyman would be selected and admitted, again through a formal ceremony.

Guilds and Quality Planning

Guilds were active in managing for quality, including quality planning. They established specifications for input materials, manufacturing processes, and finished products as well as for methods of inspection and test.

Guilds and Quality Control

Guild involvement in quality control was extensive. They maintained inspections and audits to ensure that artisans followed the quality specifications. They established means of "traceability" to identify the producer. In addition, some applied their "mark" to finished products as added assurance to consumers that quality met guild standards. Control by the guilds also extended to sales. The sale of poor-quality goods was forbidden, and offenders suffered a range of punishments—all the way from fines to expulsion from membership. The guilds also established prices and terms of sale and enforced them.

Guilds and Quality Improvement

An overriding guild policy was solidarity—to maintain equality of opportunity among members. To this end, internal competition among members was limited to "honest" competition. Quality improvement through product or process innovation was not considered to be "honest" competition. This limitation on quality improvement did indeed help to maintain equality among members, but it also made the guild increasingly vulnerable to competition from other cities that did evolve superior products and processes.

Guilds and External Forces

The guilds were able to control internal competition, but external competition was something else. Some external competition came in the form of jurisdictional disputes with other guilds, which consumed endless hours of negotiation. More ominous was competition from other cities, which could be in quality as well as in price and value.

The policy of solidarity stifled quality improvement and thereby became a handicap to remaining competitive. Thus the guilds urged the authorities to restrict imports of foreign goods. They also imposed strict rules to prevent their trade secrets from falling into the hands of foreign competitors. (The Venetian glass industry threatened capital punishment to those who betrayed such secrets.)

Inspection and Inspectors

The concepts of inspection and inspectors are of ancient origin. Wall paintings and reliefs in Egyptian tombs show the inspections used during stone construction projects. The measuring instruments included the square, level, and plumb bob for alignment control. Surface flatness of stones was checked by "boning rods" and by threads stretched across the faces of the stone blocks.

As shops grew in size, the function of inspection gave rise to the full-time job of inspector. In due course, inspectors multiplied in numbers to become the basis for inspection departments, which in turn gave birth to modern quality departments (Singer et al. 1954, vol. I, p. 481).

Government Involvement in Managing for Quality

Governments have long involved themselves in managing for quality. Their purposes have included protecting the safety and health of citizens, defending and improving the economics of the state, and protecting consumers against fraud. Each of these purposes includes some aspect of managing for quality.

Safety and Health of the Citizens

Early forms of protection of safety and health were after-the-fact measures. The *Code of Hammurabi* (c. 2000 B.C.) prescribed the death penalty for any builder of a house that later collapsed and killed the owner. In medieval times, the same fate awaited the baker who inadvertently had mixed rat poison with the flour.

Economics of the State

With the growth of trade between cities, the quality reputation of a city could be an asset or a liability. Many cities took steps to protect their reputation by imposing quality controls on exported goods. They appointed inspectors to inspect finished products and affix a seal to certify the quality. This concept was widely applied to high-volume goods such as textiles.

Continued growth of commerce then created competition among nations, including competition in quality. Guilds tended to stifle quality improvement, but governments favored improving the quality of domestic goods in order to reduce imports and increase exports. For example, in the late sixteenth century, James VI of Scotland imported craftsmen from the Low Countries to set up a textile factory and to teach their trade secrets to Scottish workers (Bursk et al. 1962, vol. IV, pp. 2283–2285).

Consumer Protection

Many states recognized that concerning some domestic trade practices, the rule of caveat emptor did not apply. One such practice related to measurement. The states designed official standard tools for measuring length, weight, volume, and so on. Use of these tools was then mandated, and inspectors were appointed to ensure compliance. (See, e.g., Juran 1995, Chapter 1.) The twentieth century witnessed a considerable expansion in consumer protection legislation. (For elaboration, see Juran 1995, Chapter 17.)

The Mark or Seal of Quality

A mark or seal has been applied to products over the centuries to serve multiple purposes. Marks have been used to

- Identify the producer, whether artisan, factory, town, merchant, packager, or still others. Such identification may serve to fix responsibility, protect the innocent against unwarranted blame, enable buyers to choose among multiple makers, advertise the name of the maker, and so on.
- Provide traceability. In mass production, use of lot numbers helps to maintain uniformity of product in subsequent processing, designate expiration dates, make selective product recalls, and so on.
- Provide product information, such as type and quantities of ingredients used, date when made, expiration dates, model number, ratings (such as voltage, current), and so on.
- Provide quality assurance. This was the major purpose served by the marks of the guilds and towns. It was their way of telling buyers, "This product has been independently inspected and has good quality."

An aura of romance surrounds the use of seals. The seals of some medieval cities are masterpieces of artistic design. Some seals have become world-renowned. An example is the British "hallmark" that is applied to products made of precious metals.

The Industrial Revolution

The Industrial Revolution began in Europe during the middle of the eighteenth century. Its origin was the simultaneous development of power-driven machinery and sources of mechanical power. It gave birth to factories that soon outperformed the artisans and small shops and made them largely obsolete.

The Factory System—Destruction of Crafts

The goals of the factories were to raise productivity and reduce costs. Under the craft system, productivity had been low due to primitive technology, whereas costs had been high due to the high wages of skilled artisans. To reach their goals, the factories reengineered the manufacturing processes. Under the craft system, an artisan performed every one of the numerous tasks needed to produce the final product—pins, shoes, barrels, and so on. Under the factory system, the tasks within a craft were divided up among several or many factory workers. Special tools were designed to simplify each task down to a short time cycle.

A worker then could, in a few hours, carry out enough cycles of his or her task to reach high productivity.

Adam Smith, in his book *The Wealth of Nations*, was one of the first to publish an explanation of the striking difference between manufacture under the craft system and under the factory system. He noted that pin making had been a distinct craft, consisting of 18 separate tasks. When these tasks were divided among 10 factory workers, production rose to a per-worker equivalent of 4800 pins a day, a level that was orders of magnitude higher than would be achieved if each worker were to produce pins by performing all 18 tasks (Smith 1776). For other types of processes, such as spinning or weaving, power-driven machinery could outproduce hand artisans while employing semiskilled or unskilled workers to reduce labor costs.

The broad economic result of the factory system was mass production at low costs. This made the resulting products more affordable and contributed to economic growth in industrialized countries as well as to the associated rise of a large "middle class."

Quality Control under the Factory System

The factory system required associated changes in the system of quality control. When craft tasks were divided among many workers, those workers were no longer their own customers, over and over. The responsibility of workers was no longer to provide satisfaction to the buyer (also customer, user). Few factory workers had contact with buyers. Instead, the responsibility became one of "make it like the sample" (or specification).

Mass production also brought new technological problems. Products involving assemblies of bits and pieces demanded interchangeability of those bits and pieces. Then with the growth of technology and of ever-wider commercial territories, the need for standardization emerged as well. All this required greater precision throughout—machinery, tools, measurement. (Under the craft system, the artisan fitted and adjusted the pieces as needed.)

In theory, such quality problems could be avoided during the original planning of the manufacturing processes. Here the limitation rested with the planners—the "master mechanics" and shop supervisors. They had extensive, practical experience, but their ways were empirical, being rooted in craft practices handed down through the generations. They had little understanding of the nature of process variation and the resulting product variation. They were unschooled in how to collect and analyze data to ensure that their processes had "process capability" to enable the production workers to meet the specifications. Use of such new concepts had to await the coming of the twentieth century.

Given the limitations of quality planning, what emerged was an expansion of inspection by departmental supervisors supplemented by full-time inspectors. Where inspectors were used, they were made responsible to the respective departmental production supervisors. The concept of a special department to coordinate quality activities broadly also had to await the coming of the twentieth century.

Quality Improvement

The Industrial Revolution provided a climate favorable for continuous quality improvement through product and process development. For example, progressive improvements in the design of steam engines increased their thermal efficiency from 0.5 percent in 1718 to 23.0 percent in 1906 (Singer et al. 1958, vol. IV). Inventors and entrepreneurs emerged to lead many countries into the new world of technology and industrialization. In due course, some organizations created internal sources of inventors—research laboratories to carry out product and process development. Some created market research departments to carry out the functions of entrepreneurship.

In contrast, the concept of continuous quality improvement to reduce chronic waste made little headway. One likely reason is that most industrial managers give higher priority to increasing income than to reducing chronic waste. The guilds' policy of solidarity, which stifled quality improvement, also may have been a factor. In any event, the concept of quality improvement to reduce chronic waste did not find full application until the Japanese quality revolution of the twentieth century.

The Taylor System of Scientific Management

A further blow to the craft system came from F. W. Taylor's system of "scientific management." This originated in the late nineteenth century when Taylor, an American manager, wanted to increase production and productivity by improving manufacturing planning. His solution was to separate planning from execution. He brought in engineers to do the planning, leaving the shop supervisors and the workforce with the narrow responsibility of carrying out the plans.

Taylor's system was stunningly successful in raising productivity. It was widely adopted in the United States but not so widely adopted elsewhere. It had negative side effects in human relations, which most U.S. managers chose to ignore. It also had negative effects on quality. The U.S. managers responded by taking the inspectors out of the production departments and placing them in newly created inspection departments. In due course, these departments took on added functions to become the broad-based quality departments of today. (For elaboration, see Juran 1995, Chapter 17.)

The Rise of Quality Assurance

The anatomy of "quality assurance" is very similar to that of quality control. Each evaluates actual quality. Each compares actual quality with the quality goal. Each stimulates corrective action as needed. What differs is the prime purpose to be served.

Under quality control, the prime purpose is to serve those who are directly responsible for conducting operations—to help them regulate current operations. Under quality assurance, the prime purpose is to serve those who are not directly responsible for conducting operations but who have a need to know—to be informed as to the state of affairs and hopefully to be assured that all is well.

In this sense, quality assurance has a similarity to insurance. Each involves spending a small sum to secure protection against a large loss. In the case of quality assurance, the protection consists of an early warning that may avoid the large loss. In the case of insurance, the protection consists of compensation after the loss.

Quality Assurance in the Village Marketplace

In the village marketplace, the buyers provided much of the quality assurance through their vigilance—through inspection and test before buying the product. Added quality assurance came from the craft system—producers were trained as apprentices and were then required to pass an examination before they could practice their trade.

Quality Assurance through Audits

The growth of commerce introduced chains of suppliers and merchants that separated consumers from the producers. This required new forms of quality assurance, one being quality warranties. The guilds created a form of quality assurance by establishing product and process standards and then auditing to ensure compliance by the artisans. In addition, some political authorities established independent product inspections to protect their quality reputations as exporters.

Audit of Suppliers' Quality Control Systems

The Industrial Revolution stimulated the rise of large industrial organizations. These bought equipment, materials, and products on a large scale. Their early forms of quality assurance were mainly through inspection and test. Then, during the twentieth century, there emerged a new concept under which customers defined and mandated quality control systems. These systems were to be instituted and followed by suppliers as a condition for becoming and remaining suppliers. This concept was then enforced by audits, both before and during the life of the supply contracts.

At first, this concept created severe problems for suppliers. One problem was the lack of standardization. Each buying company had its own idea of what was a proper quality control system, so each supplier was faced with designing its system to satisfy multiple customers. Another problem was that of multiple audits. Each supplier was subject to being audited by each customer. There was no provision for pooling the results of audits into some common data bank, and customers generally were unwilling to accept the findings of audits conducted by personnel other than their own. The resulting multiple audits were especially burdensome to small suppliers.

In recent decades, steps have been taken toward standardization by professional societies, by national standardization bodies, and most recently by the International Organization for Standardization (ISO). ISO's 9000 series of standards for quality control systems are now widely accepted among European organizations. There is no legal requirement for compliance, but as a marketing matter, organizations are reluctant to be in a position in which their competitors are certified as complying with ISO 9000 standards but they themselves are not.

There remains the problem of multiple audits. In theory, it is feasible for one audit to provide information that would be acceptable to all buyers. This is already the case in quality audits conducted by Underwriters Laboratories and in financial audits conducted by leading financial auditing firms. Single quality audits may become feasible in the future under the emerging process for certification to the ISO 9000 series, but considerably greater maturity would be required.

Extension to Military Procurement

Governments have always been large buyers, especially for defense purposes. Their early systems of quality assurance consisted of inspection and test. During the twentieth century, there was a notable shift to mandating quality control systems and then using audits to ensure conformance to the mandated systems. The North Atlantic Treaty Organization (NATO) evolved an international standard—the Allied Quality Assurance Publications (AQAP)—that includes provisions to minimize multiple audits. (For elaboration, see Juran 1977.)

Resistance to Mandated Quality Control Systems

At the outset, suppliers resisted the mandated quality control systems imposed by their customers. None of this could stop the movement toward quality assurance. The economic power of the buyers was decisive. Then as suppliers gained experience with the new approach, they realized that many of its provisions were simply good business practice. Thus the concept of mandated quality control systems seems destined to become a permanent feature of managing for quality.

Shift of Responsibility

Note that the concept of mandating quality control systems involves a major change in responsibility for quality assurance. In the village marketplace, the producer supplies the product, but the buyer has much of the responsibility for supplying the quality assurance. Under mandated quality control systems, the producer becomes responsible for supplying both the product and the quality assurance. The producer supplies the quality assurance by

- Adopting the mandated system for controlling quality
- Submitting the data that prove that the system is being followed

The buyers' audits then consist of seeing to it that the mandated system is in place and that the system is indeed being followed.

The Twentieth Century and Quality

The twentieth century witnessed the emergence of some massive new forces that required responsive action. These forces included an explosive growth in science and technology, threats to human safety and health and to the environment, the rise of the consumerism movement, and intensified international competition in quality.

An Explosive Growth in Science and Technology

This growth made possible an outpouring of numerous benefits to human societies: longer life spans, superior communication and transport, reduced household drudgery, new forms of education and entertainment, and so on. Huge new industries emerged to translate the new technology into these benefits. Nations that accepted industrialization found it possible to improve their economies and the well-being of their citizenry.

The new technologies required complex designs and precise execution. The empirical methods of earlier centuries were unable to provide appropriate product and process designs, so process yields were low and field failures high. Organizations tried to deal with low yields by adding inspections to separate the good from the bad. They tried to deal with field failures through warranties and customer service. These solutions were costly, and they did not reduce customer dissatisfaction. The need was to prevent defects and field failures from happening in the first place.

Threats to Human Safety and Health and to the Environment

With benefits from technology came uninvited guests. To accept the benefits required changes in lifestyle, which, in turn, made quality of life dependent on continuity of service. However, many products were failure-prone, resulting in many service interruptions. Most of these were minor, but some were serious and even frightening—threats to human safety and health as well as to the environment.

Thus the critical need became quality. Continuity of the benefits of technology depended on the quality of the goods and services that provided those benefits. The frequency and severity of the interruptions also depended on quality—on the continuing performance and good behavior of the products of technology. This dependence came to be known as "life behind the quality dikes." (For elaboration, see Chapter 10, A Look Ahead: Eco-Quality for Environmental Sustainability.)

Expansion of Government Regulation of Quality

Government regulation of quality is of ancient origin. At the outset, it focused mainly on human safety and was conducted "after the fact"—laws provided for punishing those whose poor quality caused death or injury. Over the centuries, there emerged a trend to regulation "before the fact"—to become preventive in nature.

This trend was intensified during the twentieth century. In the field of human health, laws were enacted to ensure the quality of food, pharmaceuticals, and medical devices. Licensing of practitioners was expanded. Other laws were enacted relating to product safety, highway safety, occupational safety, consumer protection, and so on.

Growth of government regulation was a response to twentieth-century forces as well as a force in its own right. The rise of technology placed complex and dangerous products in the hands of amateurs—the public.

Government regulation then demanded product designs that avoided these dangers. To the organizations, this intervention then became a force to be reckoned with. (For elaboration, see Juran 1995, Chapter 17.)

The Rise of the Consumerism Movement

Consumers welcomed the features offered by the new products but not the associated new quality problems. The new products were unfamiliar—most consumers lacked expertise in technology. Their senses were unable to judge which of the competing products to buy, and the claims of competing organizations often were contradictory.

When products failed in service, consumers were frustrated by vague warranties and poor service. "The system" seemed unable to provide recourse when things failed. Individual consumers were unable to fight the system, but collectively they were numerous and hence potentially powerful, both economically and politically. During the twentieth century, a "consumerism" movement emerged to make this potential a reality and to help consumers deal more effectively with these problems. This same movement also was successful in stimulating new government legislation for consumer protection. (For elaboration, see Juran 1995, Chapter 17.)

Intensified International Competition in Quality

Cities and countries have competed for centuries. The oldest form of such competition was probably in military weaponry. This competition then intensified during the twentieth century under the pressures of two world wars. It led to the development of new and terrible weapons of mass destruction.

A further stimulus to competition came from the rise of multinational organizations. Large organizations had found that foreign trade barriers were obstacles to export of their products. To get around these barriers, many set up foreign subsidiaries that then became their bases for competing in foreign markets, including competition in quality.

The most spectacular twentieth-century demonstration of the power of competition in quality came from the Japanese. Following World War II, Japanese organizations discovered that the West was unwilling to buy their products—Japan had acquired a reputation for making and exporting shoddy goods. The inability to sell became an alarm signal and a stimulus for launching the Japanese quality revolution during the 1950s. Within a few decades, that revolution propelled Japan into a position of world leadership in quality. This quality leadership in turn enabled Japan to become an economic superpower. It was a phenomenon without precedent in industrial history.

The Twenty-First Century and Quality

The cumulative effect of these massive forces has been to "move quality to center stage." Such a massive move logically should have stimulated a corresponding response—a revolution in managing for quality. However, it was difficult for organizations to recognize the need for such a revolution—they lacked the necessary alarm signals. Technological measures of quality did exist on the shop floors, but managerial measures of quality did not exist in the boardrooms. Thus, except for Japan, the needed quality revolution did not start until very late in the twentieth century. To make this revolution effective throughout the world, economies will require many decades—the entire twenty-first century. Thus, while the twentieth century has been the "century of productivity," the twenty-first century will be known as the "century of quality."

The failure of the West to respond promptly to the need for a revolution in quality led to a widespread crisis. The 1980s then witnessed quality initiatives being taken by large numbers of organizations. Most of these initiatives fell far short of their goals. However, a few were stunningly successful and produced the lessons learned and role models that will serve as guides for the West in the decades ahead.

Today all countries can attain superiority in quality. The methods, tools, and know-how exist. A country that is an emerging country today may provide higher quality than one that has been producing it for centuries. Today and into the foreseeable future all organizations in all industries must continue to strive for perfection. They need to be in a state of performance excellence. That is why this sixth edition handbook is subtitled *The Complete Guide to Performance Excellence.*

Lessons Learned

Organizations that were successful in their quality initiatives made use of numerous strategies. Analysis shows that despite differences among the organizations, there was much commonality. These common strategies included the following:

1. *Customers and quality have top priority.* Thus customer satisfaction was the chief operating goal embedded in the vision and strategic plans. This was written into corporate policies and scorecards.
2. *Create a performance excellence system.* All organizations that attained superior results did so with a change program or a systematic model for change. This model enables organizational breakthroughs to occur.
3. *Do strategic planning for quality.* The business plan was opened up to include quality goals and balanced scorecards, year after year.
4. *Benchmark best practices.* This approach was adopted to set goals based on superior results already achieved by others.
5. *Engage in continuous innovation and process improvement.* The business plan was opened up to include goals for improvement. It was recognized that quality is a moving target; therefore there is no end to improving processes.
6. *Offer training in managing for quality, the methods and tools.* Training was extended beyond the quality department to all functions and levels, including upper managers.
7. *Create an organization-wide assurance focus.* This focus is on improving and ensuring that all goods, services, processes, and functions in an organization are of high quality.

8. *Project by project, create multifunctional teams.* Multifunctional teams, adopted to give priority to organization results rather than to functional goals, and later extended to include suppliers and customers, are key to creating breakthroughs in current performance. They focus on the "vital few" opportunities for improvement.
9. *Empower employees.* This includes training and empowering the workforce to participate in planning and improvement of the "useful many" opportunities. Motivation was supplied through extending the use of recognition and rewards for responding to the changes demanded by the quality revolution. Measurements were developed to enable upper managers to follow progress toward providing customer satisfaction, meeting competition, improving quality, and so on. Upper managers took charge of managing for quality by recognizing that certain responsibilities were not delegable—they were to be carried out by the upper managers, personally.
10. *Build an adaptable and sustainable organization.* Quality is defined by the customers. Customers are driven by societal problems. Quality now includes safety, no harm to the environment, low cost, ease of use, etc. To succeed, all organizations must focus on attaining sustainable organizations.

Acknowledgments

This chapter has drawn extensively from the following:

De Feo, J. A., and Bernard, W. W. (2004). *Juran Institute's Six Sigma Breakthrough and Beyond.* McGraw-Hill, New York.

Gryna, F. M., Chua, R. C. H., and De Feo, J. A. (2007). *Quality Planning and Analysis*, 5th ed. McGraw-Hill, New York.

Juran, J. M. (ed.) (1995). *A History of Managing for Quality.* Sponsored by Juran Foundation, Inc. Quality Press, Milwaukee, WI.

Juran, J. M., and Godfrey, A. B. (1999). *Juran's Quality Handbook*, 5th ed. McGraw-Hill, New York.

Juran, J. M. (2004). *Architect of Quality*. McGraw-Hill, New York.

The author is grateful to the copyright holders for permission to quote from these works.

References

"Betting to Win on the Baldie Winners." (1993). *Business Week*, October 18.

Bursk, E. C., Clark, D. T., and Hidy, R. W. (1962). *The World of Business*, vol. I, p. 71, vol. III, pp. 1656–1678, vol. IV, pp. 2283–2285. Simon and Schuster, New York.

Durant, W. (1954). *The Story of Civilization, Part 1: Our Oriental Heritage*, pp. 182–183. Simon and Schuster, New York.

Firestone Tire Recall, Legal Information Center http://www.firestone-tire-recall.com/pages/overview.html

Juran, J. M. (1964, 1995). *Managerial Breakthrough*. McGraw-Hill, New York.

Juran, J. M. (1970). "Consumerism and Product or Service Quality." *Quality Progress*, July, pp. 18–27. American Society for Quality.

Paraphrased from Juran Quality Minute: Hair Nets.

Paraphrased from the Juran Quality Minutes (2004) Planning Quality and Meeting Customer's Needs, Discussion Guide, p 19. Southbury, CT. Juran, J. M. (1977). "Quality and Its Assurance—An Overview." Second NATO Symposium on Quality and Its Assurance, London.

Juran, J. M. (ed.) (1995). *A History of Managing for Quality*. Quality Press, Milwaukee, WI.
Levine, R. (1987). "Breaking Routine: Voice of the Subway." *The New York Times*, January 15.
Levitt, T. (1960). "Marketing Myopia." *Harvard Business Review*, July-August, pp. 26–28ff.
Levinson, M. S. (2008). *City Manager, City of Coral Springs*, FL. Prepared remarks Malcolm Baldrige National Quality Award Ceremony, April 23, 2008 http://www.nist.gov/speeches/levinson_042308.html
Ollson, J. R. (1986). "The Market-Leader Method; User-Oriented Development." *Proceedings 30th EOQC Annual Conference*, pp. 59–68. European Organization for Quality Control.
"The Perils of Cutting Quality." (1982). *The New York Times*, August 22.
Singer, C., Holmyard, E. J., and Hall, A. R. (eds.) (1954). *A History of Technology*, vol. I, Fig. 313, p. 481. Oxford University Press, New York.
Smith, A. (1776). *The Wealth of Nations*. Random House, New York. Originally published in 1937.
von der Porten (1994). "The Hanseatic League, Europe's First Common Market." *National Geographic*, October, pp. 56–79.

CHAPTER 2

Quality's Impact on Society and the National Culture

Joseph M. Juran

About This Chapter

The subsequent growth of commerce and of science and technology greatly expanded the extent and variety of nonnatural goods and services. As a result, human beings in many modern industrial societies live longer and safer lives. They are largely shielded from the perils that their ancestors faced. However, all those nonnatural goods and services have created a new dependence and therefore new risks. Years ago I coined the phrase "life behind the quality dikes" to designate these risks (Juran 1969). This chapter is about the effects of poor quality on our society and environment.

High Points of This Chapter

1. Human society has depended on quality since the dawn of history. In industrial societies, we place our safety, health, and even daily well-being behind numerous protective "dikes" of quality control. However, all these nonnatural goods and services have created a new dependence and therefore new risks on our society. These risks must be mitigated, or we will have many losses, both financial and cultural.

2. The goal of high quality is common to all countries. This common goal must compete with other national goals amid the massive forces—political, economic, and social—that determine the national priorities.
3. *Consumerism* is the name for the movement to help consumers solve their problems through collective action. No one knows whether the rate of consumer grievances has grown over the centuries. However, we know that the volume of grievances has grown to enormous numbers due to the growth in volume of goods and services.
4. Consumers exhibit a wide range of product knowledge, including the lowest. In consequence, actual use of the product can differ significantly from intended use. Many designed products and services are designed not for actual usage but rather for intended usage. Many lawsuits occur because people are hurt as a result of using a product in a way that was not "fit for purpose."
5. Governing bodies have established and enforced standards of quality. Some of these have been political: national, regional, local. Others have been nonpolitical: guilds, trade associations, standardization organizations, and so on. These governing bodies have attained a status that enables them to carry out programs of regulation and can become a hidden customer to the salability of our products and services. All organizations must plan for these regulations or face penalties.
6. During the twentieth century these lawsuits have, in the United States, grown remarkably in numbers. This growth in number of lawsuits has been accompanied by an equally remarkable growth in the sizes of individual claims and damages. All products and services must be created to minimize potential suits brought because of poor quality.

Life Behind the Quality Dikes

Human society has depended on quality since the dawn of history. In primitive societies this dependence was on the quality of natural goods and "services." Human life can exist only within rather narrow limits of climatic temperature, air quality, food quality, and so on. For most primitive societies, life even within these narrow limits is marginal, and human beings in most primitive societies live precariously. Hours of work are often long and exhausting. Life spans are shortened by malnutrition, disease, natural disasters, and so on. To reduce such risks, primitive societies created nonnatural aids to their mental and physical capabilities, aids such as

- Division of labor
- Communal forms of society, such as villages
- Artificial shelter, e.g., houses
- Processing of natural materials to produce nonnatural goods such as pottery, textiles, tools, and weapons
- Lessons learned from the experience of the past, such as when to plant crops and which berries are poisonous, handed down from generation to generation

The subsequent growth of commerce and of science and technology greatly expanded the extent and variety of nonnatural goods and services. As a result, human beings in many modern industrial societies live longer and safer lives. They are largely shielded from the perils that their ancestors faced. However, all those nonnatural goods and services have created a new dependence and therefore new risks. Years ago I coined the phrase "life behind the quality dikes" to designate these new risks (Juran 1969).

In industrial societies, great masses of human beings place their safety, health, and even daily well-being behind numerous protective "dikes" of quality control. For example, the daily safety and health of the citizenry now depends absolutely on the quality of manufactured products: drugs, food, aircraft, automobiles, elevators, tunnels, bridges, and so on. In addition, the very continuity of our lifestyle is built around the continuity of numerous vital services: power, transport, communication, water, waste removal, and many others. A major power failure paralyzes the lives of millions.

There are numerous minor breaks in the quality dikes—occasional failures of goods and services. These are annoying as well as costly. Far more serious are the terrifying major breaks such as occurred at Chernobyl and Bhopal.

Not only individuals but also nations and their economies live dangerously behind the dikes of quality control. National productivity relies on the quality of product and process design. National defense relies on the quality of complex weaponry. The growth of the national economy is keyed to the reliability of its systems for energy, communication, transport, and so on.

So while technology confers wonderful benefits on society, it also makes society dependent on the continuing performance and good behavior of technological goods and services. This is life behind the quality dikes—a form of securing benefits but living dangerously. Like the Dutch who have reclaimed much land from the sea, we secure benefits from technology. However, we need good dikes—good quality—to protect us against the numerous service interruptions and occasional disasters. These same risks have also led to legislation that at the outset was bitterly opposed by industrial organizations. Since then it has become clear that the public is serious about its concerns. What is encouraging is that users (whether individuals or nations) are willing to pay for good dikes.

The ability to cope with breaks in the quality dikes varies remarkably among users. Large organizations (industrial organizations, governments) employ technologists or otherwise use their economic and political strengths to plan, control, and improve quality. In contrast, individuals (consumers, the citizenry) find themselves pitted against forces that seem to them as mysterious and overpowering as natural forces appeared to their primitive ancestors.

Any one individual has only a very limited capacity to deal with these forces. However, there are very many individuals. Collectively their economic and political powers are formidable. These powers have emerged as a movement generally called *consumerism*. This movement, though loosely organized, has become influential in providing individual members of society with protection and recourse when faced with breaks in the quality dikes.

The Growth of Consumerism

Consumerism is a popular name for the movement to help consumers solve their problems through collective action. No one knows whether the rate of consumer grievances has grown over the centuries. However we know that the volume of grievances has grown to enormous numbers due to the growth in volume of goods and services. By the middle of the twentieth century, consumer frustrations had reached levels that stimulated attacks on industrial organizations for their alleged responsibility for consumers' problems. Then when most organizations failed to take appropriate action, the resulting vacuum attracted numerous contenders for leadership of a consumerism movement: government agencies, politicians, social reformers, consumer advocates, consumer associations, standardization organizations, independent test laboratories, and still others. A risk arose that a bargaining agent would emerge to intervene between industrial organizations and their customers.

Note: The text for this topic includes some extracts and paraphrases of material from *Juran's Quality Handbook* (1995, Chapter 17).

Consumer Perceptions

Starting in the 1970s, researchers began to identify the dominant consumer problems as well as the perceptions of the groups in interest: consumers, consumer organizations, government, business, insurance organizations, and so on. Table 2.1 lists the major quality-oriented consumer problems as derived from one such study.

Consumer expectations sometimes rise faster than the market rate of improvement. In addition, consumer perceptions can differ from the realities. For example, many consumers believed that quality of product was getting worse, that "products don't last as long as they used to." Yet the author's studies of specific product lines have almost always found that quality has kept improving.

Consumers are generally more negative than positive on the attitudes of business toward problems of consumers. Consumers strongly favor competition as a means of ensuring higher quality, safer products, and better prices. They also feel that most advertising is misleading, and that much is seriously misleading.

During the late 1970s consumer perceptions of quality in specific industries varied widely. The favorable perceptions included banks, department stores, small shopkeepers, telephone organizations, supermarkets and food stores, and airlines. At the other end of the spectrum, consumers had poor perceptions of car manufacturers, the advertising industry, the oil industry, local garages, auto mechanics, and used car dealers (Sentry 1976, p. 13).

While consumer perceptions are sometimes in error, the perceptions are important in their own right. People act on their perceptions, so it is important to understand what their perceptions are.

Consumers generally felt that there was much they could do to help themselves relative to quality. They felt that the necessary product information was available but the

Poor reliability of many products
Failure to live up to advertising claims
Poor quality and slow responsiveness of after-sales service and repairs
Misleading packaging and labeling
Futility of making complaints—nothing substantial will be done
Costly guarantees and warranties
Failure of organizations to handle complaints properly and quickly
Too many dangerous products, especially toys for children
The absence of reliable information about service quality, particularly in health care
Too many self-serve diagnostics
Not knowing what to do when something goes wrong with a purchased product
Too many similar, but different models to choose from

TABLE 2.1 Major Consumer Quality-Oriented Problems

information was not being used by consumers. They had similar views with respect to product safety. They generally felt that most products were safe if used properly and that many product safety problems arose because of failure to read the instructions properly (Sentry 1976).

Remedial Proposals

There are a number of proposals for remedies, amid much difference of opinion. The differences arise in part because of the impact on costs and prices (see below). In addition there are differences due to a contest for power. The various consumer organizations and government departments all feel that they should play larger roles, and that certain traditional powers of business should be restricted.

Ideally, the remedies should eliminate the causes of consumer problems at the source. The consumerism movement has been skeptical that such prevention will take place at the initiative of the industrial organizations. Hence the main proposals have related to establishing ways to enable consumers to judge beforehand whether they are about to buy trouble.

Access to Information before Purchase

Consumers could make better buying decisions if they had access to information on competitive product test data, field performance, and so on. Many industrial organizations possess such information but will not disclose it—they regard it as proprietary. They do disclose selected portions, but mainly to aid in sale of the product. The risk of bias is obvious. Consumers' needs for information extend also to after-sale service, response to complaints, and so on. Here again, the organizations regard such information as proprietary. The lack of information from industrial organizations has created a vacuum that has attracted alternative sources of product information to help consumers judge which products to buy and which to avoid. One such source is test laboratories that are independent of the organizations making and selling the products.

Under this concept a competent laboratory makes an expert, independent evaluation of product quality so that consumers can obtain the unbiased information needed to make sound purchasing decisions. Adequate consumer test services require professionals and skilled technicians, well-equipped test laboratories, acquisition of products for test, and dissemination of the resulting information. Financing of all these needs is so severe a problem that the method of financing determines the organization form and the policies of the test service.

Product Testing: Consumer-Financed

In this form the test laboratory derives its income by publishing its test results, usually in a monthly journal plus an annual summary. Consumers are urged to subscribe to the journal on the grounds that they will save money by acquiring the information needed to make better purchasing decisions. Advertisements of these test laboratories raise questions such as "Would you pay $100 for an appliance when independent tests show that a $75 appliance is just as good?"

In their operation, these consumer-financed test laboratories buy and test competitive products, evaluate their performance and failures, compare these evaluations with the product prices, and rate the products according to some scale of relative value. The ratings, test result summaries, descriptions of tests conducted, and so on are published in the laboratory's journals. The industrial organizations play no role in the testing and evaluation. In addition, the organizations are not permitted to quote the ratings, test results, or other material published in the journals.

Thus the service offered to consumers consists of

1. The laboratory's test results, which are mainly objective and unbiased.
2. Judgments of values that are subjective and carry a risk of bias. That is, the stress of the advertising (showing the consumer that some lower-priced products are as good as higher-priced products) creates a bias against higher-priced products. More importantly, the judgments are not necessarily typical of consumers' judgments.

Despite the obvious problems of financing a test service from numerous consumer subscriptions, there are many such services in existence in affluent and even developing countries. In the United States, the most widely known source of such tests is Consumers Union. The test results are published in the journal *Consumer Reports*.

Product Testing: Government-Financed

Governments have long been involved in matters of product quality, to protect the safety and health originally of the citizenry and later of the environment. For elaboration, see later, under Government Regulation of Quality.

The most recent extension has been in the area of consumer economics. Some of this has been stimulated by the consumerism movement. A by-product has been the availability of some product test results and other quality-related information. This information is made available to the public, whether in published form or on request.

Government-Subsidized Tests

In some countries, the government subsidizes test laboratories to test consumer products and to publish the results as an aid to consumers. The rationale is that there is a public need for this information and hence the costs should be borne by the public generally.

Mandated Government Certification

Under this concept, products are required by law to be independently approved for adequacy before they may be sold to the public. This concept is applied in many countries to consumer products for which human safety is critical (e.g., pharmaceuticals, foods). For other products, there has been a sharp division in practice. Generally, the market-based economies have rejected mandated government certification for (noncritical) consumer products, and they have relied on the forces of the competitive marketplace to achieve quality. In contrast, the planned economies, as exemplified by the former Soviet Union, went heavily into the setting of standards for consumer products and the use of government laboratories to enforce compliance to these standards (which had the force of law).

Product Testing: Company-Financed

In this form, industrial organizations buy test services from independent test laboratories in order to secure the mark (certificate, seal, label) of the laboratory for their products. In some product categories it is unlawful to market the products without the mark of a qualified testing service. In other cases it is lawful, but the mark is needed for economic reasons—the insurance organizations will demand extraordinarily high premiums or will not provide insurance at all.

An example of a sought-after mark is that of Underwriters Laboratories, Inc. (UL). Originally created by the National Board of Fire Underwriters to aid in fire prevention, UL (now independent) is involved in the general field of fire protection, burglary protection, hazardous chemicals, and still other matters of safety. Its activities include

- Developing and publishing standards for materials, products, and systems
- Testing manufacturers' products for compliance with these standards (or with other recognized standards)
- Awarding the UL mark to products that comply, known as "listing" the products

Numerous other laboratories are similarly involved in safety matters, e.g., steam boilers and marine safety. Some of these laboratories have attained a status in their specialty that confers a virtual monopoly on performing the tests.

Another purpose of securing a mark from an independent test service is to help market the product. Organizations vary in their views of the value of such "voluntary" marks. Strong organizations tend to feel that their own brand or mark carries greater prestige than that of the test laboratory, and that the latter has value only for weak organizations. The test services that offer this category of marks vary widely in their purpose and in their objectivity.

In some countries the voluntary mark is offered by the national standardization bodies, such as Japan Standards Association or the French AFNOR. The mark is awarded to products that meet their respective product standards. Organizations that wish to use the Japan Industrial Standards (JIS) mark or the Normale Français (NF) mark must submit their products for test and must pay for the tests. If the products qualify, the organizations are granted the right to use the marks.

Data Banks on Business Practices

Many consumer grievances are traceable to company business practices, such as evasiveness in meeting the provisions of the guarantee. The Pareto principle applies—a comparative few organizations are named in the bulk of the grievances. In this way, a data bank on company business practices can help to identify the vital few "bad guys" and aid in reducing their influence.

Consumer Education

Beyond product tests and data banks on business practices, still other forms of before-purchase information are available to consumers. Some government departments publish information describing the merits (or lack of merits) of products and product features in general. However, the most often used source of product information is advice received from relatives and friends who have experience to share. Consumers regard such advice as reliable.

The Standards Organizations

There are many of these. For example, in the United States, those of importance to consumers include

- Leading manufacturers and merchants, whose standards exert wide influence on their suppliers and competitors

- Industry bodies such as the American Gas Association (AGA) or the Association of Home Appliance Manufacturers (AHAM)
- Professional organizations such as the American Society for Testing and Materials (ASTM)
- Independent agencies such as Underwriters Laboratories (UL)
- The American National Standards Institute (ANSI), which is a recognized clearinghouse for committees engaged in setting national standards and is the official publisher of the approved standards
- The National Institute of Standards and Technology (NIST), formerly the National Bureau of Standards (NBS), the government agency that establishes and maintains standards for metrology

Standards for Consumer Products

Awarding a mark presupposes the existence of some standard against which the product can be tested on an objective basis. Provision of such standards for consumer products has not received the priorities given to standards for metrology, basic materials, and other technological and industrial needs. However, the consumerism movement has very likely stimulated the pace of developing these standards. Industry associations especially have been stimulated to undertake more of this type of activity.

A serious limitation on creating standards for consumer products is the pace of product obsolescence versus the time required to set standards. Usually, it takes years to evolve a standard due to the need for securing a consensus among the numerous parties in interest. For subject matter such as metrology or basic materials, the standards, once approved, can have a very long life. However, for consumer products the life is limited by the rate of obsolescence, and for many products the life of the standard is so short as to raise serious questions about the economics of doing it at all.

In some cases the obsolescence is traceable to the zeal of the marketers. For example, one measure of the quality of mechanical watches has been the number of jewels. Then some manufacturers began to include nonfunctional jewels to provide a basis for claiming higher quality. It became necessary to redefine the word "jewel."

A further problem in standards for consumer products is that the traditional emphasis of the standardization bodies has been on *time zero*—the condition of the product when tested prior to use. However, many products, especially the most costly, are intended to give service for years. Many consumer problems are traceable to field failures during service, yet most consumer product standards do not adequately address the "abilities"—reliability, maintainability, and so on.

Objectivity of Test Services

Unless the testing service is objective, consumers may be misled by the very organization on which they thought they could rely. The criteria for objectivity include the following:

- *Financial independence.* The income of the test service should have no influence on the test results. This independence is at its best when the income is derived from sources other than the company whose products are under test. Failing this, the payments by the company should be solely for the testing service and in no way contingent on the test results. One example of failure to meet this criterion is any test service that carries on the dual activities of (1) offering a mark based on product test and approval and (2) publishing a journal of general circulation in which organizations that receive the mark are required to place advertisements. In such cases the risk of conflict of interest is very high, so consumers should be cautious about giving credence to such marks.

- *Organizational independence.* The personnel of the test service should not be subordinate to the organizations whose products are undergoing test.
- *Technological capability.* This obvious need includes a qualified professional staff, adequate test equipment, and competent management. Whether the managers should be the sole judges of such capabilities is open to question.

So important is the question of objectivity that in cases of government controls on quality it is usual to write into the statute the need for defining criteria for what constitutes a qualified test laboratory. The administrator of the act then becomes responsible for certifying laboratories against these criteria.

The Resulting Information

Consumer test services offer consumers a wide range of information. The principal forms include these:

- Comparative data on competitive products for (1) price and (2) fitness for use, plus judgments of comparative values. In this form, the information is also a recommendation for action.
- Data on product conformance to the standards. In this form, consumers are on their own to discover competitive prices and to make a judgment on comparative fitness for use. For many consumers, it is a burden to provide this added information.
- Evidence of product conformance to the standard (through the mark). Here the consumer is largely asked to equate the standard with fitness for use and to use other means to discover competitive differences and competitive prices.

Information on conformance to standard is quite useful to industrial buyers, but less so to consumers. For consumers the optimal information consists of comparative data on fitness for use plus comparative data on cost of usage.

Traditional test services do not provide adequate information as to certain important quality problems faced by consumers: products arrive in defective condition; products fail during use; response to consumer complaints is poor.

- *Products defective on arrival.* Test services typically conduct their tests on a small sample of one or a few units of product. These nevertheless enable the test service to judge whether the product design can provide fitness for use. However, the sample is too small to provide information on how often units will be defective on arrival.
- *Products fail during use.* Traditionally, test services have evaluated consumer products at time zero—prior to use. For long-life products this is no longer good enough—there is a need for information on field failure rates. Some test services now do conduct a degree of life testing, but the number of units tested is too small to predict field failure rates. There are some efforts to secure such information through questionnaires sent to consumers. An alternative source is to secure information from the repair shops.
- *Poor response to customer complaints.* Here the situation is at its worst. The test laboratory and its instruments are irrelevant, since the needed information relates to the competence, promptness, and integrity of the service organizations.

Remedies after Purchase

Consumers who encounter product quality problems during the warranty period have a choice of approaches. They may be able to resolve the problem unaided; i.e., they study the product information and then apply their skills and ingenuity. More usually they must turn to one of the organizations directly in interest: the merchant who sold them the product or the manufacturer who made the product. If none of these provides satisfaction, consumers have still other alternatives for assistance (see below).

Warranties

Quality warranties are a major after-purchase aid to consumers. However, many consumers feel that warranties are not understandable. In addition, most feel that warranties are written mainly to protect manufacturers rather than consumers. Nevertheless, consumers are increasingly making the warranty an input to their buying decisions. This means also that warranties are increasingly important as marketing tools (Sentry 1976, pp. 14, 15).

Consumer Affairs...The Ombudsman

"Ombudsman" is a Swedish word used to designate an official whose job is to receive citizens' complaints and to help them secure action from the government bureaucracy. The ombudsman is familiar with government organization channels and is able to find the government official who has the authority or the duty to act. The ombudsman has no authority to compel action, but has the power to publicize failures to act.

The concept of the ombudsman has been applied to problems in product quality. Some organizations have created an in-house ombudsman and have publicized the name and telephone number. Consumers can phone (free of charge) to air grievances and to secure information. In the United States a more usual title is Manager (Director), Consumer Affairs (Relations). Such a manager usually carries added responsibilities for stimulating changes to improve relations with consumers on a broad basis. In one company these efforts resulted in programs to effectuate a consumer "bill of rights," which includes rights to safety, to be informed, to choose, to be heard, and to redress.

Another form is the industry ombudsman. An example is the Major Appliance Consumer Action Panel (a group of independent consumer experts) created by the Association of Home Appliance Manufacturers to receive complaints from consumers who have not been able to secure satisfaction locally.

Still another form is the Joint Industry-Consumer Complaint Board. Examples are the government-funded boards that mediate and adjudicate consumer disputes in some Scandinavian industries. The boards have no power to enforce their awards other than through publicity given to unsatisfied awards. Yet they have met with wide acceptance by and cooperation from the business people.

The concept of the ombudsman is fundamentally sound. It is widely supported by consumers and regulators as well as by a strong minority of business managers (Sentry 1976, p. 77). Some newspapers provide an ombudsman service as part of their department of Letters to the Editor.

Mediation

Under the mediation concept, a third party—the mediator—helps the contestants to work out a settlement. The mediator lacks the power of enforcement—there is no binding agreement to abide by the opinion of the mediator. Nevertheless mediation stimulates settlements. Best (1981) reports that the New York City Department of Human Affairs achieved a 60 percent settlement rate during 1977 and 1978.

The mediation process helps to open up the channels of communication and thereby to clear up misunderstandings. In addition, an experienced mediator exerts a moderating influence which encourages a search for a solution.

Arbitration

Under arbitration the parties agree to be bound by the decision of a third party. Arbitration is an attractive form of resolving differences because it avoids the high costs and long delays inherent in most lawsuits. In the great majority of consumer claims, the cost of a lawsuit is far greater than the amount of the claim. Nevertheless there are obstacles to use of the arbitration process. Both parties must agree to binding arbitration. There is a need to establish local, low-cost arbitration centers and to secure the services of volunteer arbitrators at nominal fees or no fees. These obstacles have limited the growth of use of arbitration for consumer complaints.

Consumer Organizations

There are many forms of consumer organizations. Some are focused on specific products or services such as automotive safety or truth in lending. Others are adjuncts of broader organizations such as labor unions or farm cooperatives. Still others are organized to deal broadly with consumer problems. In addition, there are broad consumer federations, national and international, that try to improve the collective strength of all local and specialized consumer groups.

Government Agencies

These exist at national, state, and local levels of government. All invite consumers to bring unresolved complaints to them as well as to report instances of business malpractice. These complaints aid the agency in identifying widespread problems, which, in turn, become the basis for

1. Conducting investigations in depth
2. Proposing new legislation
3. Issuing new administrative regulations

The agencies also try to help complaining consumers, either in an ombudsman role or by threat of legal action. However, in practice, broad government agencies are unable to become involved in specific consumer grievances due to the sheer numbers. See The Enforcement Process later, under Government Regulation of Quality.

No Remedy

Under the prevailing free-enterprise, competitive market system, many valid consumer complaints result in no satisfaction to the consumer. Nevertheless the system includes some built-in stabilizers. Organizations that fail to provide such satisfaction also fail to attract repeat business. In due course they mend their ways or lose out to organizations that have a better record of providing satisfaction. In the experience of the author, every other system is worse.

Perceptions of the Consumer Movement

There is wide agreement, including among business managers, that "the consumer movement has kept industry and business on their toes." There is also wide agreement that the consumer movement's demands have "resulted in higher prices." Despite this, most of the public feels that the "changes are generally worth the extra cost." Consumers feel strongly that the consumer advocates should consider the costs of their proposals. However, a significant minority of the consumers believe that the advocates do not consider the costs involved (Sentry 1976, pp. 39, 40, 42, 47).

Quality and the National Culture

The goal of high quality is common to all countries. This common goal must compete with other national goals amid the massive forces—political, economic, and social—that determine the national priorities. This section examines these forces and their effect on the problems of attaining quality.

The growth of international trade and of multinational organizations has required that attention be directed to understanding the impact of national culture on managing for quality. To aid in this understanding, the subject is organized under the following general subdivisions:

In all types of national economy, there are natural resources and limitations that influence the priority of goals. However, an even greater force is that of human leadership and determination. Historically, these human forces have been more significant than natural resources in determining whether goals are attained.

The words "capitalistic," "socialistic," and "developing" are simple labels for some very complex concepts. The broad definition of "capitalism" is private ownership of the means of production and distribution, as contrasted with state ownership under socialism. Yet all self-styled capitalistic countries include a degree of state ownership, e.g., in matters of health, education, transport, and communication. Similarly, the self-styled socialistic countries contain, in varying degrees, some private ownership of organizations for production of goods and services. In like manner, countries that are "developing" in the industrial sense may be developed in terms of other aspects of national maturity, e.g., political or social. The reader is urged to keep in mind that the words "capitalistic," "socialistic," and "developing" are used in a relative sense and cannot be considered as absolutes.

The subject matter of this chapter is of obvious interest and importance to those engaged (or contemplating engagement) in operations of an international nature. Such operations are becoming ever more extensive as trade barriers are progressively removed. However, removal of government barriers has little effect on cultural barriers. These remain a continuing problem until the cultural patterns (and the reasons behind them) are understood, appreciated, and taken into account.

In the economic sense, the capitalistic developed countries are the "vital few." The developing countries are the most numerous, occupy most of the land surface, and include most of the human population. However, it is the capitalistic developed countries that produce the bulk of the world's goods and services. This great importance (in the economic sense) suggests that those who engage in international trade should acquire a working knowledge of the cultures that prevail in other countries.

Quality in Global Economies

All capitalistic economies exhibit some basic similarities that influence the importance of quality in relation to other goals in the economy.

Competition in Quality

Capitalistic societies permit and even encourage competition among organizations, including competition in quality. This competition in quality takes multiple forms.

Creation of New Organizations

A frequent reason for the birth of new organizations is poor quality of goods or services. For example, a neighborhood has outgrown the capacity of the local food shop or restaurant, so the clients must wait in long queues before they can receive service. In such cases, entrepreneurs will sense a market opportunity and will create a new organization that attracts clients by offering superior service.

The ease of creating new organizations is a far greater force in quality improvement than is generally realized. All economies, whether capitalistic or socialistic, suffer poor quality during shortages of goods. Creation of new organizations is one means of alleviating shortages, and thereby of eliminating an invariable cause of poor quality.

Product Improvement

A common form of competition in quality is through improving products so that they have greater appeal to the users and can therefore be sold successfully in the face of competition from existing products. These product improvements come mainly from internal product development carried on by existing organizations. In addition, some product improvements are designed by independents that either launch new organizations or sell their ideas to existing organizations.

New Products

These may be "products" or even new systems approaches, e.g., designs that minimize user maintenance. The industrial giants of today include many members founded on new systems concepts. As with product improvements, the new products may originate through development from within or through acquisition from the outside.

Competition in quality results in duplication of products and facilities. Such duplication is regarded as wasteful by some economists. However, the general effect has been to stimulate producers to outdo one another, with resulting benefit to users.

Direct Access to Marketplace Feedback

In the capitalistic economies, the income of the organization is determined by its ability to sell its products, whether directly to users or through an intermediate merchant chain. If poor quality results in excessive returns, claims, or inability to sell the product, the manufacturers are provided with the warning signals that are a prerequisite to remedial action.

This severe and direct impact of poor quality on the manufacturers' income has the useful by-product of forcing manufacturers to keep improving their market research and early warning signals, so as to be able to respond promptly in case of trouble.

Direct access to the marketplace is not merely a matter of receiving complaints and other information about bad quality, important though that is. Even more important is the access to the marketplace before products are launched and sales programs are prepared. In the capitalistic economies, the autonomous organizations all make their own forecasts on how much they expect to sell. Their ability to thrive depends on how well they are able to realize their forecasts. The potential benefits and detriments force the organizations to pay attention to the needs of the marketplace, since it provides their income.

Protection of Society

The autonomy of capitalist organizations may permit them to misrepresent their products, sell unsafe products, damage the environment, fail to live up to their warranties, and so on, until these misdeeds become significant enough to generate extensive preventive legislation.

Cultural Differences

There are many of these, including the following:

- *Language*. Many countries harbor multiple languages and numerous dialects. These are a serious barrier to communication.
- *Customs and traditions*. These and related elements of the culture provide the precedents and premises that are guides to decisions and actions.
- *Ownership of the organizations*. The pattern of ownership determines the strategy of short-term versus long-term results, as well as the motivations of owners versus nonowners.
- *The methods used for managing operations*. These are determined by numerous factors such as reliance on system versus people, extent of professional training for managers, extent of separation of planning from execution, and careers within a single company versus mobile careers.
- *Suspicion*. In some countries, there is a prior history of hostilities resulting from ancient wars, religious differences, membership in different clans, and so on. The resulting mutual suspicions are then passed down from generation to generation.

It is clearly important to learn about the nature of a culture before negotiating with members of that culture. Increasingly, organizations have provided special training to employees before sending them abroad. Similarly, when organizations establish foreign subsidiaries, they usually train local nationals to qualify for the senior posts.

Government Regulation of Quality

From time immemorial, "governments" have established and enforced standards of quality. Some of these governments have been political: national, regional, local. Others have been nonpolitical: guilds, trade associations, standardization organizations, and so on. Whether through delegation of political power or through long custom, these governing bodies have attained a status that enables them to carry out programs of regulation as discussed next.

Standardization

With the evolution of technology came the need for standardizing certain concepts and practices.

- *Metrology*. One early application of standardization was to the units of measure for time, mass, and other fundamental constants. So basic are these standards that they are now international in scope.
- *Interchangeability*. This level of standardization has brought order out of chaos in such day-to-day matters as household voltages and interchangeability of myriad bits and pieces of an industrial society. Compliance is an economic necessity.
- *Technological definition*. A further application of standardization has been to define numerous materials, processes, products, tests, and so on. These standards are developed by committees drawn from the various interested segments of society. While compliance is usually voluntary, the economic imperatives result in a high degree of acceptance and use of these standards.

The foregoing areas of regulation are all related to standardization, and have encountered minimal resistance to compliance. Other areas do encounter resistance, in varying degrees.

Safety and Health of the Citizenry

A major segment of political government regulation has been to protect the safety and health of its citizens. At the outset the focus was on punishment "after the fact"—the laws provided punishment for those whose poor quality had caused death or injury. Over the centuries there emerged a trend to regulation "before the fact"—to become preventive in nature.

For example, in the United States there are laws that prescribe and enforce safety standards for building construction, oceangoing ships, mines, aircraft, bridges, and many other structures. Other laws aim at hazards having their origins in fire, foods, pharmaceuticals, dangerous chemicals, and so on. Still other laws relate to the qualifications needed to perform certain activities essential to public safety and health, such as licensing of physicians, professional engineers, and airline pilots. Most recently these laws have proliferated extensively into areas such as consumer product safety, highway safety, environmental protection, and occupational safety and health.

Safety and Economic Health of the State

Governments have always given high priority to national defense: the recruitment and training of the armed forces and the quality of the weaponry. With the growth of commerce, laws were enacted to protect the economic health of the state. An example is laws to regulate the quality of exported goods in order to protect the quality reputation of the state. Another example is laws to protect the integrity of the coinage. (Only governments have the right to debase the currency.) In those cases where the government is a purchaser (defense weapons, public utility facilities), government regulation includes the normal rights of a purchaser to ensure quality.

Economics of the Citizenry

Government regulation relative to the economics of the citizenry is highly controversial in market economies. Some of the resistance is based on ideological grounds—the competitive marketplace is asserted to be a far better regulator than a government bureau. Other resistance is based on the known deficiencies of the administration of government regulation (see below). Some of the growth of this category of regulation has been stimulated by the consumerism movement.

The Volume of Legislation

Collectively, the volume of quality-related legislation has grown to formidable proportions. A desk reference book (Kolb and Ross 1980) includes lists (in fine print) of appendixes as follows:

- 21 pages of exposure limits for toxic substances
- 93 pages of hazardous materials and the associated criteria for transportation
- 24 pages of American National Standards for safety and health
- 36 pages of federal record-retention requirements
- 38 pages of standards-setting organizations

In the United States, much of this legislation is within the scope of the Federal Trade Commission, which exercises a degree of oversight relative to "unfair or deceptive practices in commerce." That scope has led to specific legislation or administrative action relative to product warranties, packaging and labeling, truth in lending, and so on.

In a sense these actions all relate to representations made to consumers by industrial organizations. In its oversight the Federal Trade Commission stresses two major requirements:

1. The advertising, labeling, and other product information must be clear and unequivocal as to what is meant by the seller's representation.
2. The product must comply with the representation.

These forms of government regulation are a sharp break from the centuries-old rule of *caveat emptor* (let the buyer beware). That rule was (and is) quite sensible as applied to conditions in the village marketplaces of developing countries. However it is not appropriate for the conditions prevailing in industrialized, developed countries. For elaboration, see Juran (1970).

The Plan of Regulation

Once it has been decided to regulate quality in some new area, the approach follows a well-beaten path. The sequence of events listed below, while described in the language of regulation by political government, applies to nonpolitical government as well.

The Statute

The enabling act defines the purpose of the regulation and especially the subject matter to be regulated. It establishes the "rules of the game" and creates an agency to administer the act.

The Administrator

The post of administrator is created and given powers to establish standards and to see that the standards are enforced. To this end he or she is armed with the means for making awards and applying sanctions on matters of great importance to the regulated industries.

The Standards

The administrator has the power to set standards and may exercise this power by adopting existing industry standards. These standards are not limited to products; they may deal with materials, processes, tests, descriptive literature, advertising, qualifications of personnel, and so on.

Test Laboratories

The administrator is given power to establish criteria for judging the qualifications of "independent" test laboratories. Once these criteria are established, he or she also may have the power to issue certificates of qualification to laboratories meeting the criteria. In some cases administrators have the power to establish their own test laboratories.

Test and Evaluation

Here there is great variation. In some regulated areas, agency approval is a prerequisite to going to market, e.g., new drug applications or plans for the operation and maintenance of a new fleet of airplanes. Some agencies put great stress on surveillance, i.e., review of the organizations' control plans and adherence to those plans. Other agencies emphasize final product sampling and test.

The Seal or Mark

Regulated products are frequently required to display a seal or mark to attest to the fact of compliance with the regulations. Where the regulating agency does the actual testing, it affixes this mark; e.g., government meat inspectors physically stamp the carcasses.

More usually, the agency does not test and stamp the product. Instead, it determines, by test, that the product design is adequate. It also determines, by surveillance, that the organizations' systems of control are adequate. Any company whose system is adequate is then authorized to affix the seal or mark. The statutes always provide penalties for unauthorized use of the mark.

Sanctions

The regulatory agency has wide powers of enforcement, such as the right to

- Investigate product failures and user complaints
- Inspect organizations' processes and systems of control
- Test products in all stages of distribution
- Recall products already sold to users
- Revoke organizations' right to sell or to apply the mark
- Inform users of deficiencies
- Issue cease-and-desist orders

Effectiveness of Regulation

Regulators face the difficult problem of balance—protecting consumer interests while avoiding creation of burdens that in the end are damaging to consumer interests. In part the difficulty is inherent because of the conflicting interests of the parties. However, much of the difficulty is traceable to unwise agency policies and practices in carrying out the regulatory process. These relate mainly to the conceptual approach, setting standards, the enforcement process, and cost of regulation.

The Conceptual Approach

An example is seen in the policies employed by the National Highway Traffic Safety Administration (NHTSA) for administering two laws enacted in 1966:

1. The National Traffic and Motor Vehicle Safety Act, directed primarily at the vehicle
2. The Highway Safety Act, directed primarily at the motorist and the driving environment

Even prior to 1966, the automobile makers, road builders, and so on had improved technology to an extent that provided the motorist with the means of avoiding the "first crash," i.e., accidents due to collisions, running off the road, etc. The availability of seat belts then provided the motorist with greatly improved means of protection against the "second crash." This crash takes place when the sudden deceleration of a collision hurls the occupants against the steering wheel, windshield, and so on.

At the time NHTSA was created, the U.S. traffic fatality rate was the lowest among all industrial countries. It was also known, from overwhelming arrays of data, that the motorist was the limiting factor in traffic safety:

- Alcohol was involved in about one-half of all fatal accidents.
- Young drivers (under age 24) constituted 22 percent of the driver population but were involved in 39 percent of the accidents.

- Excessive speed and other forms of "improper" driving were reported as factors in about 75 percent of the accidents. (During the oil crisis of 1974 the mandated reduction of highway speeds resulted in a 15 percent reduction in traffic fatalities, without any change in vehicles.)
- Most motorists did not buy safety belts when they were optional, and most did not wear them when they were provided as standard equipment.

In the face of this overwhelming evidence, NHTSA paid little attention to the main problem—improving the performance of the motorists. Instead, NHTSA concentrated on setting numerous standards for vehicle design. These standards did provide some gains in safety with respect to the second crash. However, the gains were minor, while the added costs ran to billions of dollars—to be paid for by consumers in the form of higher prices for vehicles.

The policy is seen to have been one of dealing strictly with a highly visible political target—the automobile makers—while avoiding any confrontation with a large body of voters. It was safe politically, but it did little for safety. For elaboration, see Juran (1977).

Setting Standards

A major regulatory question is whether to establish design standards or performance standards.

- Design standards consist of precise definitions, but they have serious disadvantages. Their nature and numbers are such that they often lack flexibility, are difficult to understand, become very numerous, and become prohibitive to keep up to date.
- Performance standards are generally free from the above disadvantages. However, they place on the employer the burden of determining how to meet the performance standard, i.e., the burden of creating or acquiring a design. Performance standards also demand level-of-compliance officers who have the education, experience, and training needed to make the subjective judgments of whether the standard has been met.

These alternatives were examined by a presidential task force assigned to review the safety regulations of the Occupational Safety and Health Administration (OSHA). The task force recommended a "performance/hazard" concept. Under that concept, the standard would "codify into a requirement the fact that a safe workplace can be achieved only by ensuring that employees are not exposed to the hazards associated with the use of machines. Under this standard, the employer would be free to determine the most appropriate manner in which to guard against any hazard which is presented, but his compliance with the requirement is objectively measurable by determining whether or not an employee is exposed to the hazard."

The Enforcement Process

A major deficiency in the regulatory process is failure to concentrate on the vital few problems. Regulatory agencies receive a barrage of grievances: consumer complaints, reports of injuries, accusations directed at specific products, and so on. Collectively the numbers are overwhelming. There is no possibility of dealing thoroughly with each and every case. Agencies that try to do so become hopelessly bogged down. The resulting paralysis then becomes a target for critics, with associated threats to the tenure of the administrator, and even to the continued existence of the agency.

In the United States the Occupational Safety and Health Administration faced just such a threat in the mid-1970s. In response it undertook to establish a classification for its cases based on the seriousness of the threats to safety and health. It also recalled about 1000 safety regulations that were under attack for adding much to industry costs and little to worker safety.

With experience, the agencies tend to adopt the Pareto principle of the vital few and the useful many. This enables them to concentrate their resources and to produce tangible results.

Choice of the vital few is often based on quantitative data such as frequency of injuries or frequency of consumer complaints. However, subjective judgment plays an important role, and this enables influential special pleaders to secure high priority for cases that do not qualify as being among the vital few.

How to deal with the "useful many" needs for assistance has been a perplexing problem for all agencies. The most practical solution seems to have been to make clear that the agency is in no position to resolve such problems. Instead the agency provides consumers with information and educational material of a self-help nature: where to apply for assistance; how to apply for assistance; what are the rights of the consumer; what to do and not to do.

The failure of regulators to deal forthrightly with such consumer problems has no doubt contributed to the mediocre status given to regulators by the public, in response to the question, Which (of four options) would you like to be primarily responsible for the job of seeing that consumers get a fair deal?

A Rule for Choosing the Vital Few

In 1972 the author proposed the following as a quantitative basis for separating the vital few from the rest, on matters of safety:

> Any hour of human life should be as safe as any other hour.

To effectuate such a policy, it is first necessary to quantify safety nationally, on some common basis such as injuries per million worker-hours of exposure. In general, the data for such quantification are already available, although some conversions are needed to arrive at a common unit of measure.

For example, statistics on safety at school are computed on the basis of injuries per 100,000 student-days, motor vehicle statistics are on a per 100 million miles of travel basis, and so on.

The resulting national average will contain a relatively few situations that are well above the average and a great many that are below. Those above the average would automatically be nominated to membership in the vital few. Those below the average would not be so nominated; the burden of proof would be on any special pleader to show why something below the national average should take priority ahead of the obvious vital few. For elaboration, see Juran (1972).

The Costs and Values of Regulation

The costs of regulation consist largely of two major components:

1. *The costs of running the regulatory agencies.* These are known with precision. In the United States they have risen to many billions of dollars per year. These costs are paid for by consumers in the form of taxes that are then used to fund the regulatory agencies.
2. *The costs of complying with the regulations.* These costs are not known with precision, but they are reliably estimated to be many times the costs of running the regulatory agencies. These costs are in the first instance paid for by the industrial organizations, and ultimately by consumers in the form of higher prices.

The value of all this regulation is difficult to estimate. (There is no agreement on the value of a human life.) Safety, health, and a clean environment are widely believed to be enormously valuable. Providing consumers with honest information and prompt redress is likewise regarded as enormously valuable. However, such general agreements provide no guidelines for what to do in specific instances. Ideally, each instance should be examined as to its cost-value relationship. Yet the statutes have not required the regulators to do so. The regulators have generally avoided facing up to the idea of quantifying the cost-value relationship.

Until 1994 the support for studying the cost-value relationships came mainly from the industrial organizations. For example, a study of mandated vehicle safety systems found that

> "... states which employ mandatory periodic inspection programs do not have lower accident rates than those states without such requirements."
>
> "... only a relatively small portion of highway accidents—some 2 to 6 percent—are conclusively attributable to mechanical defects."
>
> "...human factors (such as excess speeds) are far more important causes of highway accidents than vehicle condition." Crain 1980.

The indifference of regulators to costs inevitably creates some regulations and rigid enforcements so absurd that in due course they become the means for securing a change in policy. The organizations call such absurdities to the attention of the media, which relish publicizing them. (The media have little interest in scholarly studies.) The resulting publicity then puts the regulators on the defensive while stimulating the legislators to hold hearings. During such hearings (and depending on the political climate) the way is open to securing a better cost-value balance.

The political climate is an important variable in securing attention to cost-versus-value considerations. During the 1960s and 1970s the political climate in the United States was generally favorable to regulatory legislation. Then during the 1980s the climate changed, and with it a trend toward requiring cost justifications. This trend then accelerated in late 1994, when the elections enabled the opponents of regulation to gain majority status in the national legislatures.

Product Safety and Product Liability

Growth of the Problem

Until the early twentieth century, lawsuits based on injuries from use of products (goods and services) were rarely filed. When filed, they were often unsuccessful. Even if they were successful, the damages awarded were modest in size.

During the twentieth century these lawsuits have, in the United States, grown remarkably in numbers. By the mid-1960s they were estimated to have reached over 60,000 annually and by the 1970s to over 100,000 per year. (Most are settled out of court.) This growth in number of lawsuits has been accompanied by an equally remarkable growth in the size of individual claims and damages. From figures measured in thousands of dollars, individual damages have grown to a point where awards in excess of $100,000 are frequent. Damages in excess of $10,000,000 are no longer a rarity.

In some fields the costs of product liability have forced organizations to abandon specific product lines or go out of business altogether.

Twenty years ago, 20 organizations manufactured football helmets in the United States. Since that time, 18 of these organizations have discontinued making this product because of high product liability costs (Grant 1994).

Several factors have combined to bring about this growth in number of lawsuits and in size of awards. The chief factors include the "population explosion" of products. The industrial society has placed large numbers of technological products into the hands of amateurs. Some of these products are inherently dangerous.

- Other products are misused. The injury rate (injuries per million hours of usage) has probably been declining, but the total number of injuries has been rising, resulting in a rise in total number of lawsuits.
- There is erosion of organization defenses. As these lawsuits came to trial, the courts proceeded to erode the former legal defenses available to organizations.

Formerly, a plaintiff's right to sue a manufacturer rested on one of two main grounds:

A *contract* for sale of the product, with an actual or implied warranty of freedom from hazards. Given the contract relationship, the plaintiff had to establish "privity," i.e., that he or she was a party to the contract. The courts in effect have abolished the need for privity by taking the position that the implied warranty follows the product around, irrespective of who is the user.

Negligence by the company. Formerly the burden of proof was on the plaintiff to show that the company was negligent. The courts have tended to adopt the principle of "strict liability" on the ground that the costs of injuries resulting from defective products should be borne "by the manufacturers that put such products on the market rather than by the injured persons who are powerless to protect themselves." In effect, if an injury results from use of a product that is unreasonably dangerous, the manufacturer can be held liable even in the absence of negligence. (Sometimes the injured persons are not powerless. Some contribute to their injuries. However, juries are notoriously sympathetic to injured plaintiffs).

Defensive Actions

The best defense against lawsuits is to eliminate the causes of injuries at their source. All company functions and levels can contribute to making products safer and to improving company defenses in the event of lawsuits. The respective contributions include the following:

- *Top management.* Formulate a policy on product safety; organize product safety committees and formal action programs; demand product dating and product traceability; establish periodic audits of the entire program; and support industry programs that go beyond the capacity of the unaided company. To this list should be added a scoreboard—a measure of the injury rate of the company's products relative to an appropriate benchmark. A useful unit of measure is the number of injuries per million hours of usage, since most major data banks on injuries are already expressed in this form or are convertible to this form.
- *Product design.* Adopt product safety as a design parameter; adopt a fail-safe philosophy of design; organize formal design reviews; follow the established codes; secure listings from the established laboratories; publish the ratings; and utilize modern design technique.
- *Manufacture.* Establish sound quality controls, include means for error-proofing matters of product safety; and train supervisors and workers in use of the product as part of the motivation plan; stimulate suggestion on product safety; and set up the documentation needed to provide traceability and historical evidence.
- *Marketing.* Provide product labeling for warnings, dangers, antidotes; train the field force in the contract provisions; supply safety information to distributors and

dealers; set up exhibits on safety procedures; conduct tests after installation, and train users in safety; publish a list of dos and don'ts relative to safety; and establish a customer relations climate that minimizes animosity and claims. Contracts should avoid unrealistic commitments and unrealistic warranties. Judicious disclaimers should be included to discourage unjustified claims.

- *Advertising*. Require technological and legal review of copy, and propagandize product safety through education and warnings. Avoid "puffing"—it can backfire in liability suits, e.g, if a product is advertised as "absolutely safe." During advertising review, one of the questions should be, How would this phrase sound in a courtroom?
- *Customer service.* Observe use of the product to discover the hazards inherent during use (and misuse); feed the information back to all concerned; and provide training and warnings to users.
- *Documentation.* The growth of safety legislation and of product liability has enormously increased the need for documentation. A great deal of this documentation is mandated by legislation, along with retention periods.

Consumers exhibit a wide range of product knowledge, including the lowest. In consequence, actual use of the product can differ significantly from intended use. For example, some stepladders include a light platform that is intended to hold tools or materials (e.g., paint) but is not intended to carry the weight of the user. Nevertheless, some users do stand on these platforms with resulting injury to themselves.

Most modern policy is to design products to stand up under actual usage rather than intended usage.

Defense Against Lawsuits

The growth of product liability lawsuits has led to reexamination of how best to defend against lawsuits once they are filed. Experience has shown the need for special preparation for such defense, including

- Reconstruction of the events that led up to the injury
- Study of relevant documents—specifications, manuals, procedures, correspondence, reports
- Analysis of internal performance records for the pertinent products and associated processes
- Analysis of field performance information
- Physical examinations of pertinent facilities
- Analysis of the failed hardware

All this should be done promptly, by qualified experts, and with early notification to the insurance company.

Whether and how to go to trial involves a great deal of special knowledge and experience.

Defense Through Insurance

Insurance is widely used as a defense against product liability. But the costs have escalated sharply, again because of the growth in number of lawsuits and size of awards. In some fields insurance has become a major factor in the cost of operations. (Soaring insurance rates have forced some surgeons to take early retirement.)

Prognosis

As of the mid-1990s there remained some formidable unsolved problems in product liability. To many observers the U.S. legal system contained some serious deficiencies:

- Lay juries lack the technological literacy needed to determine liability on technological matters.
- In most other developed countries, judges make such decisions.
- Lay juries are too easily swayed emotionally when determining the size of awards.
- In the United States, "punitive damages" may be awarded along with compensatory damages and damages for "pain and suffering." Punitive damages contribute greatly to inflated awards.
- In the United States, lawyers are permitted to work on a contingency fee basis, a concept that stimulates lawsuits. This arrangement is illegal in most countries.
- The adversary system of conducting trials places the emphasis on winning rather than on rendering justice.
- Only a minority of the award money goes to the injured parties. The majority goes to lawyers and to pay administrative expenses.

By the mid-1990s some elements of this legal system were under active review in the national Congress. However, the system that has endured these deficiencies is deeply rooted in the U.S. culture, so it is speculative whether it will undergo dramatic change. A major obstacle has been the lawyers. They have strong financial interests in the system, and they are very influential in the legislative process—many legislators are lawyers.

In most developed countries the legal system for dealing with product liability is generally free from the above asserted deficiencies. Those same countries are also largely free from the extensive damage that product liability is doing to the U.S. economy.

Personal Liability

An overwhelming majority of product liability lawsuits have been aimed at the industrial organizations; they and their insurers have the greatest capacity to pay. As a corollary, such civil lawsuits are rarely aimed at individuals, e.g., design managers or quality managers. These individuals have little cause for concern with respect to civil liability. They are not immune from lawsuits, but they are essentially immune from payment of damages.

Criminal liability is something else. Now the offense (if any) is against the state, and the state is the plaintiff. Until the 1960s, prosecution for criminal liability in product injury cases was directed almost exclusively at the corporations rather than the managers. During the 1960s and the 1970s the public prosecutors became more aggressive with respect to the persons involved. The specific targets were usually the heads of the organizations but sometimes included selected subordinate managers such as for product development or for quality.

A contributing factor has been an earlier provision of the Food, Drug and Cosmetic Act making it a crime to ship out adulterated or misbranded drugs. This provision was interpreted by the U.S. Supreme Court to be applicable to the head of a company despite the fact that she or he had not participated in the events and even had no knowledge of the goings-on.

For the great majority of industrial managers the threat of criminal liability is remote. Before there can be such liability, the manager must be found guilty of (1) having knowingly carried out illegal actions or (2) having been grossly negligent. These things must be proved to a jury beyond a reasonable doubt. It is a difficult proof. (Many guilty criminals escape conviction because of this difficulty.)

Environmental Protection

A special category of government regulation is environmental protection (EP). On the face of it, EP is a twentieth-century phenomenon. However, there is a school of thought suggesting that EP originated in a conservation movement to preserve lands that were being exploited by European colonists during the seventeenth and eighteenth centuries.

The Industrial Revolution of the mid-eighteenth century opened the way to mass production and consumption of manufactured goods at rates that grew exponentially. To support this growth required a corresponding growth in production of energy and materials. The resulting goods conferred many benefits on the societies that accepted industrialization. However, there were unwelcome by-products, and these also grew at exponential rates.

Generating the needed energy produced emissions that polluted the air and water. Nuclear power created the problem of nuclear waste disposal as well as the risk of radiation leaks. Mining for raw materials damaged the land, as did disposition of toxic wastes. Ominous threats were posed by ozone depletion and the risk of global warming. Disposition of worn-out and obsolete products grew to problems of massive proportions. All this was in addition to problems posed by the numerous inconveniences and occasional disasters caused by product failures during service. (See above under Life behind the Quality Dikes.)

Industrial organizations were generally aware that they were creating these problems, but their priorities were elsewhere. Public awareness lagged, but by the mid-twentieth century the evidence had become overwhelming. Responding to public pressures, governments enacted much legislation to avoid worsening the problem, and they provided funds to undo some of the damage.

The new legislation was at first strongly resisted by industrial organizations because of the added costs it imposed. Then as it became clear that EP was here to stay, the ingenuity of industry began to find ways to deal with the problem at the source—to use technology to avoid further damage to the environment. A striking example is Japan's achievement in energy conservation. During the 1973–1990 period, despite continuing growth in industrial production, there was no increase in energy consumption (Watanabe 1993).

Public and media preoccupation with specific instances of environmental damage has tended to stimulate allocation of funds to undo such damage. However, the long-range trend seems to be toward prevention at the source.

Recognition of the importance of EP is now evident in many ways, e.g.,

- Many countries have created new ministries to deal with the problem of EP.
- Many industrial organizations have created high-level posts for the same purpose.
- Numerous conferences are being held, including at the international level, with participation from government, industry, and academia (Strong 1993).
- An extensive and growing body of literature has emerged. Some of this is quite specific.
- Organizations have also evolved specific processes for addressing the problems of EP. These generally consist of

  Establishment of policies and goals with respect to EP

  Establishment of specific action plans to be carried out by the various company functions

  Audits to ensure that the action plans are carried out

In addition, the ingenuity of organizations has begun to find ways to reduce the costs of providing solutions. Table 2.2 lists some of the identified problems and the associated

Air pollution	Smog, tropospheric ozone, indoor air quality, volatile organic compound
Climate change	Global warming, global dimming, fossil fuels, sea level rise, greenhouse gas, ocean acidification
Conservation	Species extinction, pollinator decline, coral bleaching, Holocene extinction event, invasive species, poaching, endangered species
Consumerism	Consumer capitalism, planned obsolescence, overconsumption
Dams	Environmental impacts of dams
EMF	Electromagnetic radiation and health
Energy	Energy conservation, renewable energy, efficient energy use Renewable energy commercialization
Fishing	Blast fishing; bottom trawling; cyanide fishing; ghost nets; illegal, unreported, and unregulated fishing
Genetic engineering	Genetic pollution, genetically modified food controversies
Intensive farming	Overgrazing, irrigation, monoculture, environmental effects of meat production
Land degradation	Land pollution, desertification
Land use	Urban sprawl, habitat fragmentation, habitat destruction
Logging	Clear-cutting, deforestation, illegal logging
Mining	Acid mine drainage, mountaintop removal mining, slurry impoundments
Nanotechnology	Nanotoxicology, nanopollution
Nuclear issues	Nuclear fallout, nuclear meltdown, nuclear power, radioactive waste
Ozone depletion	CFC
Particulate matter	Sulfur oxide
Pollution	Light pollution, noise pollution, visual pollution
Resource depletion	Exploitation of natural resources, overfishing, shark finning, whaling
Soil	Soil conservation, soil erosion, soil contamination, soil salination
Thermal pollution	Urban runoff, water crisis, marine debris, ocean acidification, ship pollution, wastewater
Toxins	Chlorofluorocarbons, DDT, endocrine disrupters, dioxin, heavy metals, herbicides, pesticides, toxic waste, PCB, bioaccumulation, biomagnification
Waste	E-waste, litter, waste disposal incidents, marine debris, landfill, leachate, recycling, incineration
Water pollution	Acid rain, eutrophication, marine pollution, ocean dumping, oil spills

(*Source:* Wikipedia, 2009)

TABLE 2.2 Environmental Problems and Opportunities for U.S. Industry

opportunities for solution (Juran Institute, 2009). For more on environmental problems and quality see Chapter 10, A Look Ahead: Eco-quality for Environmental Sustainability.

Multinational Collaboration

Collaboration across cultures is a many-faceted problem. For example, a system may be designed in country A, but the subsystem designs may come from other countries. In like manner, organizations from multiple countries may supply components and carry out manufacture, marketing, installation, maintenance, and so on.

Numerous methodologies have been evolved to help coordinate such multinational activities. Those widely used include the following:

Standardization

This is accomplished through organizations such as the International Organization for Standardization (ISO) and the International Electrotechnical Commission (IEC). A special application is the Allied Quality Assurance Publication (AQAP) standards widely used by the North Atlantic Treaty Organization (NATO) countries for multinational contracting.

Contract Management

In many cases, the prime contractor provides a coordinating service for the subcontractors (who may include a consortium). [McClure (1979) relative to the F16 aircraft; see also McClure (1976).]

Technology Transfer

This is carried out in numerous well-known ways: international professional societies and their committees; conferences; exchange visits; training courses; and seminars. In large, multinational organizations, such activities are carried out within the organizations as well.

References

Best, A. (1981). *When Consumers Complain*. Columbia University Press, New York.

Crain, W. M. (1980). *Vehicle Safety Inspection Systems. How Effective?* American Enterprise Institute for Public Policy Research, Washington, DC.

Juran, J. M. (1970). "Consumerism and Product Quality." *Quality Progress*, July 1970, pp. 18–27.

Juran, J. M. (1972). "Product Safety." *Quality Progress*, July 1972, pp. 30–32.

Juran, J. M. (1977). "Auto Safety, a Decade Later." *Quality*, October 1977, pp. 26–32; November, pp. 54–60; December, pp. 18–21. Originally presented at the 1976 Conference of the European Organization for Quality.

Juran, J. M. (1995). *A History of Managing for Quality*. Quality Press, Milwaukee, WI.

McClure, J. Y. (1976) "*Quality-A Common International Goal*," ASQC Technical Conference Transactions, Milwaukee, pp. 459-466.

McClure, J. Y. (1979) "*Procurement Quality Control Within the International Environment*," ASQC Conference Transactions, Milwaukee, pp. 643–649.

O'Keefe, D. F., Jr., and Shapiro, M. H. (1975). "Personal Criminal Liability under the Federal Food, Drug and Cosmetic Act—The Dotterweich Doctrine." *Food-Drug-Cosmetic Law Journal*, January.

O'Keefe, D. F., Jr., and Isley, C. W. (1976). "Dotterweich Revisited—Criminal Liability under the Federal Food, Drug and Cosmetic Act." *Food-Drug-Cosmetic Law Journal*, February.
Sentry (1976). *Consumerism at the Crossroads*. Sentry Insurance Co., Stevens Points, WI. Results of a national opinion research survey on the subject.
Strong, M. F. (1993). "The Road from Rio." *The Bridge*, Summer, pp. 3–7.
Watanabe, C. (1993). "Energy and Environmental Technologies in Sustainable Development: A View from Japan." *The Bridge*, Summer, pp. 8–15.

CHAPTER 3

The Universal Methods to Manage for Quality

Joseph M. Juran

About This Chapter

This chapter deals with the fundamental concepts that define the subject of managing for quality. It defines key terms and makes critical distinctions between similar but different contemporary programs to improve performance. It identifies the key processes or what I called the *universals* through which quality is managed and integrated into the strategic fabric of an organization. It demonstrates that while managing for quality is a timeless concept, it has undergone frequent revolution in response to the endless procession of changes and crises faced by human societies.

High Points of This Chapter

1. "Managing for quality" is a set of universal methods that an enterprise, a business, an agency, a university, a hospital, or any organization can use to attain superior results by ensuring that all goods, services, and processes meet stakeholder needs.
2. "Quality" as stated in Chapter 1, Attaining Superior Results Through Quality has two meanings that must be clearly understood and communicated. The first relates to how well the features of the services or products you produce meet customer needs and thereby provide them satisfaction. "Quality" also means freedom from failure.

3. The Juran Trilogy embodies the universal principles needed to create high-quality goods, services, and processes.
4. The universal principles followed to manage for quality are applicable to any type of organization, including a company, an institution, an industrial organization, a government agency, a school, and a hospital.
5. Implementing processes to create innovative products and services by discovering the voice of the customer will enable every organization to understand the customers' needs better and then create or design products that meet these needs.
6. Implement processes to ensure that products conform to the design criteria when they are produced. We must control quality and predict how it will perform in the marketplace.
7. Implement a systematic approach to improving quality or creating breakthroughs to eliminate those failures that are chronic in our processes or products.

The Concept of Universals

During my studies of algebra and geometry, I stumbled across two broad ideas that I would put to extensive use in later years. One was the concept of "universals"; the other was the distinction between theory and fact.

My study of algebra exposed me for the first time to the use of symbols to create generalized models. I knew that 3 children plus 4 children added up to 7 children, and 3 beans plus 4 beans added up to 7 beans. Now by using a symbol such as x, I could generalize the problem of adding 3 + 4 and state it as a universal:

$$3x + 4x = 7x$$

This universal said that 3 + 4 always equals 7 no matter what x stands for—children, beans, or anything else. To me the concept of universals was a blinding flash of illumination. I soon found out that universals abounded, but they had to be discovered. They had various names—rules, formulas, laws, models, algorithms, patterns. Once discovered, they could be applied to solve many problems.

By 1954 in my text *Managerial Breakthrough*, I outlined the beginnings of the many universals that led to superior result. The first was the universal of control—the process for preventing adverse change. The second was the universal sequence for breakthrough improvement. The latter went on to become known as Six Sigma today. By 1986, I discovered that there was another universal. This was the planning for quality, at the strategic level and product and service design levels. I also came to realize that those three managerial processes (planning, control, and improvement) were interrelated, so I developed the Juran Trilogy diagram, to depict this interrelationship. The Juran Trilogy embodies the core processes by which we manage for quality. As a corollary, these same core processes constitute an important sector of science in managing for quality. To my knowledge there is growing awareness in our economy that mastery of those universal processes" is critical to attaining leadership in quality and superior results.

What Does Managing for Quality Mean?

For many decades, the phrase used to define quality was simply "fitness for use". It has been generally accepted that if an organization produced goods that were 'fit for use" as viewed by the customer, then those goods were considered of high quality. Throughout most of the twentieth century this definition made sense because it was easy to grasp. Simply put, if customers purchased a good and it worked, they were pleased with the quality of it. To the producers of that product it was easy to produce as long as the producer had a clear understanding of the customer requirements.

Managing for quality therefore meant to "ensure product conformance to requirements." The majority of tasks largely fell on the operations and quality departments. These functions were responsible to produce, inspect, detect, and ensure the product met requirements.

Two developments have led us to modify this time-honored definition. The first was the realization that the quality of a physical good, its fitness for use, was broader than just its conformance to specifications. Quality was also determined by the design, packaging, order fulfillment, delivery, field service, and all the service that surrounded the physical good. The operations and quality departments could not manage quality alone.

The second development was a shift in the economy from production dominated by goods to production heavily concentrated in services and information. As stated in Chapter 1, Attaining Superior Results Through Quality and to reflect these changes, the authors of this handbook (sixth edition) have chosen to use the phrase "fit for purpose" instead of "fit for use" to define the quality of a product. We will use the term "product" to refer to goods, services, and information. Regardless of whether a product is a good, a service, or information, it must be "fit for purpose" by the customers of that product. The customer is not just the end user but all those whom the product impacts, including the buyer, the user, the supplier, the regulatory agencies, and almost anyone who is affected by the product from concept to disposal. With such an expanding set of customers and their needs, the methods and tools to manage for quality must grow as well.

For the twenty-first century "managing for quality" can be defined as "a set of universal methods that any organization, whether a business, an agency, a university, or a hospital, can use to attain superior results by designing, continuously improving, and ensuring that all products, services, and processes meet customer and stakeholder needs."

Management of quality is not the only set of universal methods to manage an organization. It is one set of managerial methods that successful organizations have used and others should use if they want to assure their products: goods, services, and information meet customer requirements. This evolution will continue as more industries adopt the methods and tools used to manage the quality of goods and services. Emerging organizations and countries will create new means to adopt management methods to their unique needs. Today, a full range of industries, including hospitals, insurance organizations, medical laboratories, and financial service organizations, are managing for business superior performance excellence.

The accelerated adoption of techniques to manage for quality began in the late 1970s when U.S. businesses were badly affected by many Japanese competitors. Japanese manufactured goods were generally viewed by the purchasers of those goods as having higher quality. This led to the definition of "Japanese or Toyota quality." These terms have become synonymous with higher quality that is required to meet the needs of the customers. As consumers or customers had a better choice, it forced some U.S. organizations into bankruptcy and others to compete at a new level of performance. Eventually many American and then later European organizations regained lost markets with higher quality.

One of the first to accomplish that was Motorola. Motorola was affected by Japanese organizations such as NEC, Sony, and others. The road traveled and the improved quality resulted in Motorola becoming the first winner of the U.S. Malcolm Baldrige National Quality Award. Motorola itself evolved the universal quality improvement model and created the Six Sigma model for quality improvement. Since then American quality improved, and the quality revolution continued into a global revolution. From 1986 to today this model of quality improvement has become the most valued model for many industries around the globe. Today organizations such as Samsung, Quest Diagnostics, Oracle, and Telefonica have become more competitive and are among quality leaders in their industries.

Each of these organizations and others have all contributed to the methods to manage quality. They all used the basic tools of Six Sigma and quality management to expand into business processes and all parts of the supply chain. Now quality is not the quality department's responsibility. It is the responsibility of the entire hierarchy. Managing quality has become the way to manage an entire organization. It has become the driving force of many strategies. To be the best in my industry, to have the highest quality, and to provide the highest level of customer delight are all business strategies. If achieved, these strategies will enable these organizations to attain financial success, cultural change, and satisfied customers.

In this sixth edition of the *Juran's Quality Handbook*, we aim to provide a concise, simpler, and hopefully clear set of methods and tools to manage for "quality." This will include not only the quality of goods or services but also the quality of process and function, which lead to overall organizational quality.

As the needs of customers and society have changed, the means for meeting their needs also changed. The methods of managing quality in 1980 may not work for your organization today. What works today may not work tomorrow. Even the universals that continue to deliver superior results may one day need to be modified. This handbook will present the best of what works and the lessons learned for those that did not. One lesson learned was that many organizations that were once quality leaders failed to sustain their successful performance over time. Why did this happen? Did they fail to sustain results because of weak leadership? Was it external forces? Was it a poor execution of their strategies? These questions have haunted many professionals who have had to defend their "quality programs." We will try to provide answers to these questions and more in this version of the handbook.

In Figure 3.1 which addresses the meaning of quality, we have presented two of the many meanings of the word "quality" as they relate to goods and services. These two are of critical importance to managing for quality:

1. Quality as it relates to how well the features of a service or good meet customer needs and thereby provide them with satisfaction. In this meaning of the word, higher quality usually costs more.
2. Quality as it relates to freedom from failures. In this sense, the meaning of the word is oriented to costs, and "higher quality usually costs less."

Features Which Meet Customer Needs	Freedom from Failures
Higher quality enables organizations to	***Higher quality enables organizations to***
Increase customer satisfaction	Reduce error rates
Meet societal needs	Reduce rework, waste
Make products and services salable	Reduce failures, warranty charges
Exceed competition	Reduce customer dissatisfaction
Increase market share	Reduce inspection, test, and audits
Provide salesrevenue	Shorten time to develop new products
Secure premium prices	Increase yields, capacity
	Improve delivery performance
The major effect is on revenue	***The major effect is on costs***
Usually higher quality costs more.	Usually higher quality costs less.

FIGURE 3.1 The meaning of quality. (*Juran Institute, Inc., 2009.*)

By adopting these simple definitions of quality as it relates to goods and services one can create a systematic approach to manage quality by

- Creating processes to design goods and services to meet needs of its stakeholders (external and internal). Every organization must understand what the customers' needs are and then create or design services and goods that meet those needs.
- Creating processes to control quality. Once designed, these services and goods are produced, at which time we must ensure compliance to the design criteria.
- Creating a systematic approach for improving continuously or creating breakthroughs. Services, goods, and the processes that produce them suffer from chronic failures that must be discovered and remedied.
- Creating a functions to ensure you continue to do the three things listed above.

By designing quality, controlling it during operations, and then continuously improving on it, any organization can be on its way to becoming a "quality organization." The global quality leaders as described above are relentless in their pursuit of ensuring that all their goods and services meet or exceed their customer requirements—but not at all costs. Attaining quality that satisfies customers but not the business stakeholders is not a good business to be in. To be truly a quality organization, the products and services must be produced at costs that are affordable to the producer and its stakeholders. The quality-cost-revenue relationship, however, must be properly understood in making these judgments. Increased feature quality must generate enough revenue to cover the added costs of additional features. But higher quality from lower failures will usually reduce cost and thereby improve financial performance. For organizations that do not generate revenue, feature quality must not cost more than your budget allows, but quality improvement against failures will almost always improve financial health.

By using these two definitions of quality, and by understanding the impact of good or poor quality on an organization's performance, one can create long-term plans to maintain high quality of goods, services, processes, and financial performance. Managing over the long term also requires that the organization set up systems to ensure that the changing needs of its customers are well understood to avoid the failure to sustain performance that plagues even the most successful organizations.

Organizational Effectiveness Programs

"Organizational Effectiveness," "Lean Six Sigma," "Toyota Production System," and Total Quality Management (TQM) are "brand" names for methods, and some may find them synonymous with the universals to manage for quality. As Juran's universals of managing for quality become embedded and used in many new industries, a new brand may be formed. Most of the time these new brands are useful because they help advance the needs to improve performance. Just as the early guilds led to quality standards, society and changing customer needs also require the universals to be adapted. One common problem with the methods to manage quality was found in the service sector. Service organizations always felt that the word "quality" meant product quality. Many services do not see their products as goods. They are services. Therefore they substitute the words "service quality" with "service excellence." Over time this phrase catches on and we have a new brand. Most of the time this new brand builds positively on the previous brand. Other times the alterations to the methods result in less positive outcomes and shunning of the brand. This happened to TQM. Total Quality Management was the brand in the 1990s. It was replaced with Six

Sigma. Why? The methods of managing for quality were evolving as many organizations were trying to regain competitiveness. The problem with TQM was that it was not measurable or as business-focused as needed. Over time it lost its luster. However, there were many organizations that improved their performance immensely, and they continue with TQM today. Others move on to the new brand. At the time of this writing Lean Six Sigma and Performance Excellence are in vogue. They too will change over time. In the end it does not matter what you call your processes to manage for quality as long as you do what is needed to attain superior results. The universals live on. No matter the industry, country, or century, meeting and exceeding the needs of your customers will drive your results.

Our Glossary of Key Terms

In the world of managing for quality, there is still a notable lack of standardization of the meanings of key words. However, any organization can do much to minimize internal confusion by standardizing the definitions of key words and phrases. The basic tool for this purpose is a glossary. The glossary then becomes a reference source for communication of all sorts: reports, manuals, training texts, and so on.

The definitions of "quality" include certain key words that themselves require definition. A few are important before we continue:

1. Organizations. In this handbook we use this word to mean any entity, business, company, institution, agency, business unit, hospital, bank, Internet provider, casino, etc., that provides an output—a product, service, or information—to a customer, whether for profit or not for profit.
2. Universal management methods and tools. A universal management method, tool, or process means it can be used in any industry, any function, any organization in any culture. It is truly universal. To most employees of an organization the word "manage" means to assign resources, set goals, establish controls, and review results with respect to products, processes, and people. To organizations that are "world-class" it means a full sequence of activities that produce the intended results to meet customer and societal needs. Managerial processes not limited to just finance, human resources, technology, and operations. They also include managerial processes to understand customer needs; to design new products and services to meet those needs; to have systems and controls in place to ensure the needs are met over time; to have systems and initiatives in place to continually improve all of them: to ensure society's needs are not negatively impacted.
3. Product: goods, services, or information. These are the outputs of any process that meets the needs of your customers. To economists, products include both goods and services and can also include information. However, under popular usage, "product" sometimes means goods only. The authors will generally use "product" to refer to both goods and services.
   a. A product can also be a physical good such as a toy, a computer, or a document containing information such as a proposal, an architectural drawing, or a web site on the World Wide Web.
   b. A product can also be a service, which is work that is performed for someone else. A carpenter builds a home for a homeowner, the user; an automotive technician repairs cars for their owners; a nurse cares for patients; and a web browser provides fast information to meet the needs of its users.

4. Feature. A feature is a property or characteristic possessed by a good or service that responds to customer needs. A feature of an automobile can be the fidelity with which its stereo system meets the listening needs of its driver. This may have little or no impact on how the automobile drives and performs, but it does meet the needs of the customer in other ways. A feature can be the emergency-room fast track for critical patients in need of immediate medical attention. Features are what a company, an organization, a system, or an agency must include in the design of a good or service to meet customer needs. Features must be created through understanding exactly how to meet the most important needs.
5. Cost of poor quality (COPQ). These are the costs that would disappear in the organization if all failures were removed from a product, service, or process. They are measured as a percentage of sales or total costs.
6. Customer. A customer is anyone external to your company, organization, system, or agency who is affected by the use of the product or service. A customer receives value for the product of the organization. A customer may be the ultimate user of the product or an intermediate customer external to your organization but not the user, such as a parent who purchases a game for a child or a surgeon who implants a device in the patient. A good or service can have many customers. A common practice is to differentiate between "external" customers and "internal" customers. External customers are defined as above, and internal customers are users within the organization. The term "customer" alone will generally refer to the external customer. Customers are sometimes called *stakeholders*. This term is typically intended to encompass external customers and internal customers, shareholders, management, and employees. Since this wide range of roles will have divergent and even conflicting needs for the organization, we will generally discuss each group separately rather than place them all in one category.
7. Processor. Processors are employees, departments, functions, business units, and agencies that produce or carry out a process within the organization. To achieve superior results, an organization must clearly focus on the external customers but ensure that all processors are able to complete their work as designed, on time, every time.
8. Customer satisfaction. Customer satisfaction is the positive state of customers when their needs have been met by the good or service they purchase or use. Satisfaction is mainly driven by the features of the good or service produced.
9. Customer dissatisfaction. This is the negative state of a customer when a good or service has a failure that results in an unmet need and, consequently, customer annoyance, a complaint, a claim, or returned goods.
10. Failure. Failure is any fault, defect, failure, or error that impairs a service or product from meeting the customer needs. These can be stated as too many defects or failures (in goods or services). Failures take such forms as waiting too long for phone responses from a call center, errors on invoices, warranty claims, power outages, failures to meet delivery dates, and inoperable goods.
11. Customer loyalty. This is the delighted state of a customer when the features of the good and service meet her or his needs and are delivered free from failure. Loyalty is also relative to the offerings of the competition. A loyal customer continues to purchase or use your organization's goods and services. Loyalty is a strategic

financial measure of customer satisfaction. Creating loyal customers is the goal of a superior performer.

12. Customer disloyalty. This is the very negative state of customers who no longer want your products or services. They find better-performing products and services and then become disloyal to the producer to whom they had once been loyal.
13. Superior performers, world class, or best in class. These various labels are used in the marketplace for organizations with products that are generally accepted as having the highest quality. These organizations become the de facto comparative benchmark for others to attain. Examples include Toyota Motor Company, Samsung Electronics, the Mayo Clinic, and Google, to name a few.

Management of Quality: The Financial and Cultural Benefits

Features Effect on Revenue

Revenue can include several types of transactions: (1) money collected from selling a good or service, (2) taxes collected by a government, or (3) donations received by a charity. Whatever the source, the amount of the revenue relates in varying degrees to the ability of the good or service features produced to be valued by the recipient—the customer. In many markets, goods and services with superior features are able to attract more revenue through some combination of higher market share and premium pricing. Services and products that are not competitive with features often are sold at lower prices.

Failures Effect on Income

The customer who encounters a deficiency may take action that creates additional cost for the producer, such as file a complaint, return the product, make a claim, or file a lawsuit. The customer also may elect instead (or in addition) to stop buying from the guilty producer as well as to publicize the deficiency and its source. Such actions by multiple customers can do serious damage to a producer's revenue.

Failures Effect on Cost

Deficient quality creates excess costs associated with poor quality. "Cost of poor quality" (COPQ) is a term that encompasses all the costs that would disappear if there were no failures—no errors, no rework, no field failures, and so on. Juran Institute's research on the cost of poor quality demonstrates that for organizations that are not managing quality aggressively, the level of COPQ is shockingly high.

Calculating the costs of poor quality can be highly valuable for an organization. COPQ shows enterprise leaders just how much poor quality has inflated their costs and consequently reduced their profits. Detailed COPQ calculations provide a road map for rooting out those costs by systematically removing the poor quality that created them.

In the early 1980s, it was common for many business leaders to make a statement that their COPQ was about 20 to 25 percent of sales revenue. This astonishing number was backed up by many independent organizations calculating their own costs. For this handbook we conducted additional research to determine a more precise and current estimate for COPQ in the economy.

The task was not as easy as it may sound. Many sources were in disagreement as to what costs should be included in the total. In addition, the actual form of the statistic was presented in myriad ways: percentage of sales, percentage of operating expenses, percentage of value added, an absolute dollar value, a dollar value per employee, and even a number of deaths in the health care industry. While many sources provided hard primary data, others would cite vague "experts" or "studies." When specific sources were cited, they sometimes referred to one another in circular fashion.

Based on findings and extrapolations from published literature, as well as a report conducted by the Midwest Business Group on Health and Juran Institute, and the reasoned judgment of knowledgeable health care practitioners, it was estimated that 30 percent of all direct costs of health care are the result of poor-quality care, consisting primarily of overuse, misuse, and waste in the system. The impact of underuse on costs is not clear. With national health expenditures of roughly $1.4 trillion in 2001, the 30 percent figure translates into $420 billion spent each year as a direct result of poor quality. In addition, the indirect costs of poor quality (e.g., reduced productivity due to absenteeism) add an estimated 25 to 50 percent, or $105 to $210 billion, to the national bill. Private purchasers absorb about one-third of these costs. In fact, we estimate that poor-quality health care costs the typical employer between $1900 and $2250 per covered employee each year. Even if these figures are off by 50 percent, poor-quality health care exacts a several hundred-billion-dollar toll on our nation each year (Midwest Business Group on Health et al. 2003).

Our best synthesis suggests that by the year 2003 the COPQ was in the range of 15 to 20 percent for manufacturing organizations, with many achieving even lower levels as the result of systematic programs to reduce it. For service organizations COPQ as a percentage of sales was still a staggering 30–35 percent of sales. These numbers included the costs of redoing what had already been done, the excess costs to control poor processes, and the costs to correctly satisfy customers. Failures that occur prior to sale obviously add to a producer's costs. Failures that occur after sale add to customer's costs as well as to producer's costs. In addition, post sale failures reduce producers' future sales because customers may be less apt to purchase a poor quality service.

How to Manage for Quality: A Financial Analogy

To manage quality, it is good to begin by establishing a *vision* for the organization, along with policies, goals, and plans to attain that vision. This means that quality goals and policies must be built into the organization's strategic plan. (These matters are treated elsewhere in this handbook, especially in Chapter 7, Strategic Planning and Deployment: Moving from Good to Great.) Conversion of these goals into results (making quality happen) is then achieved through established managerial processes—sequences of activities that produce the intended results. Managing for quality makes extensive use of three such managerial processes:

- Designing or planning for quality
- Compliance, controlling or assuring quality
- Improving or creating breakthroughs in quality

These three processes are interrelated and are known as the *Juran Trilogy*. They parallel the processes long used to manage for finance. These financial processes consist of the following:

Financial planning. This process prepares the annual financial and operational budgets. It defines the deeds to be done in the year ahead. It translates those deeds into money—revenue, costs, and profits. It determines the financial benefits doing all those deeds. The final result establishes the financial goals for the organization and its various divisions and units.

Financial control. This process consists of evaluating actual financial performance, comparing this with the financial goals and taking action on the difference—the accountant's "variance." There are numerous subprocesses for financial control: cost control, expense control, risk management, inventory control, and so on.

Financial improvement. This process aims to improve financial results. It takes many forms: cost reduction projects, new facilities, and new-product development to increase sales, mergers, and acquisitions, joint ventures, and so on.

These processes are universal—they provide the basis for financial management, no matter what type of organization it is.

The financial analogy can help leaders realize that they can manage for quality by using the same processes of planning, control, and improvement. Since the concept of the trilogy is identical to that used in managing for finance, leaders are not required to change their conceptual approach.

Much of their previous training and experience in managing for finance is applicable to managing for quality.

While the conceptual approach does not change, the procedural steps differ. Figure 3.2 shows that each of these three managerial processes has its own unique sequence of

Quality Planning	Quality Control	Quality Improvement
Establish goals	Determine the control subjects	Prove the need with a business case
Identify who are the customers	Measure actual performance	Establish a project infrastructure
Determine the needs of the customers	Compare actual performance to the targets and goals	Identify the improvement projects
Develop features which respond to customers' needs	Take action on the difference	Establish project teams
Develop processes able to produce the products	Continue to measure and maintain performance	Provide the teams with resources, training, and motivation to: Diagnose the causes Stimulate remedies
Establish process controls transfer the plans to the operating forces		Establish controls to hold the gains

Figure 3.2 Managing for quality.

activities. Each of the three processes is also a universal—it follows an unvarying sequence of steps. Each sequence is applicable in its respective area, no matter what the industry, function, culture, etc.

Implementing the Juran Trilogy

The Juran Trilogy Diagram

The three processes of the Juran Trilogy are interrelated. Figure 3.3 shows this interrelationship.

The Juran Trilogy diagram is a graph with time on the horizontal axis and cost of poor quality on the vertical axis. The initial activity is quality planning. The market research function determines who the customers are and what their needs are. The planners or product realization team then develops product features and process designs to respond to those needs. Finally, the planners turn the plans they created over to operations: "Run the process, produce the features, deliver the product to meet the customers' needs."

Chronic and Sporadic

As operations proceed, soon it is evident that the processes that were designed to deliver the good or service are unable to produce 100 percent quality. Why? Because there are hidden failures or periodic failures that require rework and redoing. Figure 3.3 shows an example where more than 20 percent of the work processes must be redone owing to failures. This waste is considered chronic—it goes on and on until the organization decides to find its root causes. Why do we have this chronic waste? Because it was planned that way. The planners could not account for all unforeseen obstacles in the design process.

Under conventional responsibility patterns, the operating forces are unable to get rid of this planned chronic waste. What they can do is to carry out control—to prevent things from

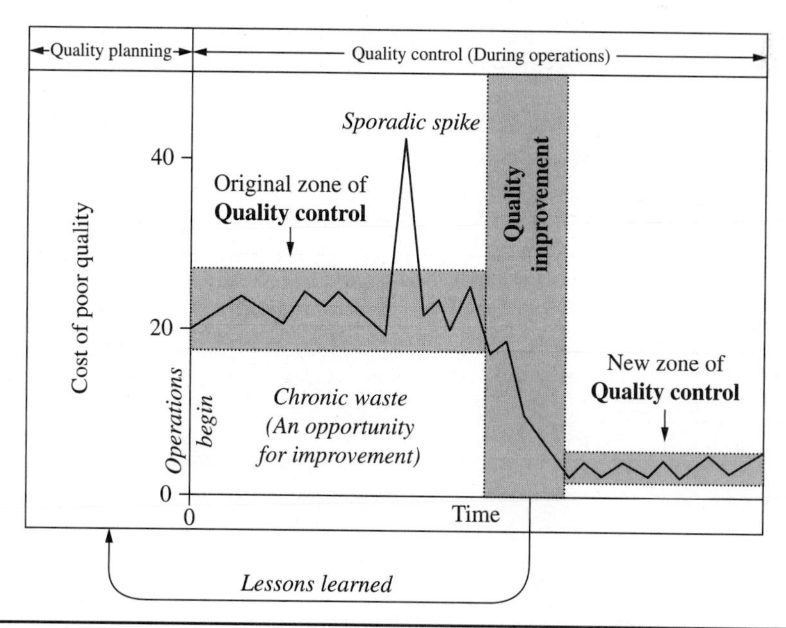

FIGURE 3.3 Juran Trilogy®.

getting worse, as shown in Figure 3.3. It shows a sudden sporadic spike that has raised the failure level to more than 40 percent. This spike resulted from some unplanned event such as a power failure, process breakdown, or human error. As a part of the control process, the operating forces converge on the scene and take action to restore the status quo. This is often called "corrective action," "troubleshooting," "fire-fighting," and so on. The end result is to restore the error level back to the planned chronic level of about 20 percent.

The chart also shows that in due course the chronic waste was driven down to a level far below the original level. This gain came from the third process in the trilogy—improvement. In effect, it was seen that the chronic waste was an opportunity for improvement, and steps were taken to make that improvement.

The Trilogy Diagram and Failures

The trilogy diagram (Figure 3.3) relates to product and process failures. The vertical scale therefore exhibits units of measure such as cost of poor quality, error rate, percent defective, service call rate, waste, and so on. On this same scale, perfection is at zero, and what goes up is bad. The results of reducing failures are reduction in the cost of poor quality, meeting more delivery promises, reduction of the waste, decrease in customer dissatisfaction, and so on.

Allocation of Time within the Trilogy

An interesting question for managers is, How do we design our functions and allocate their time relative to the processes of the trilogy?" Figure 3.4 is a model designed to show this interrelationship in a Japanese company (Itoh 1978).

In Figure 3.4 the horizontal scale represents the percentage allocation of any person's time and runs from 0 to 100 percent. The vertical scale represents levels in the organizational hierarchy. The diagram shows that the upper managers spend the great majority of their time on planning and improvement. They spend a substantial amount of time on strategic planning. The time they spend on control is small and is focused on major control subjects.

At progressively lower levels of the hierarchy, the time spent on strategic planning declines, whereas the time spent on control and maintenance grows rapidly. At the lowest levels, the time is dominated by control and maintenance, but some time is still spent on planning and improvement.

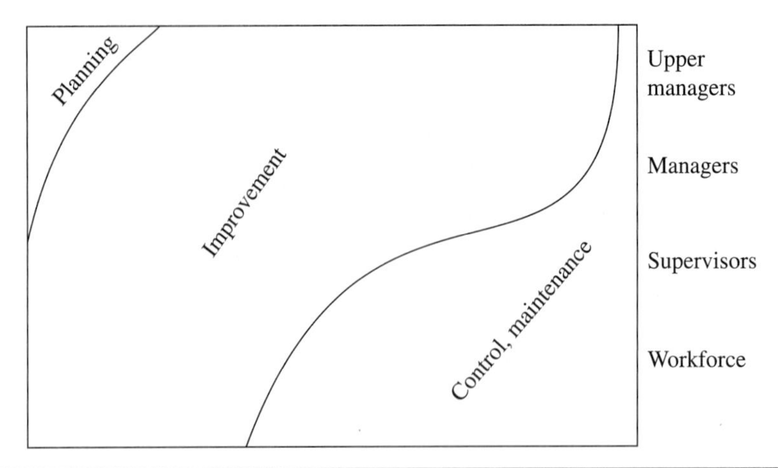

FIGURE 3.4 Itoh model. (*Adapted from Management for Quality, 4th ed., Juran Institute, Inc., 1987, p. 18.*)

Figure 3.2 shows these unvarying sequences in abbreviated form. Extensive detail is provided in other chapters of this handbook: Chapter 4, Quality Planning: Designing Innovative Products and Services; Chapter 5, Quality Improvement: Creating Breakthroughs in Performance; and Chapter 6, Quality Control: Assuring Repeatable and Compliant Processes.

References

Itoh, Y. (1978). "Upbringing of Component Suppliers Surrounding Toyota." International Conference on Quality Control, Tokyo.

Juran, J. M. (2004). "Architect of Quality," McGraw-Hill, New York, NY.

Midwest Business Group on Health, Juran Institute, Inc., and The Severyn Group, Inc. (2003). "Reducing the Costs of Poor-Quality Health Care through Responsible Purchasing Leadership."

CHAPTER 4

Quality Planning: Designing Innovative Products and Services

Joseph A. De Feo

About This Chapter

New product development processes are one of the most important business processes in an organization. It is the lifeblood of future sales, performance, and competitiveness. Traditional methods to develop new products in manufacturing usually arise within the product development functions. In service organizations new service development is done ad hoc with multiple functions contributing to the new service designs. "Quality Planning," as the term is used here, is a systematic process for developing new products (both goods and services) and processes that ensure customer needs are met. There are many methods to design innovative products. Design for Six Sigma, Design for Lean, Design for World-Class Quality, and Concurrent Engineering, Agile Design for Software are common. This chapter will focus on the methods and tools that are common to each and sometimes excluded from development functions. Quality by design methods and tools will enable an organization to develop breakthrough products and services that drive revenue.

High Points of This Chapter

1. Designing for quality and innovation is one of the three universal processes of the Juran Trilogy. It is required to achieve breakthroughs new products, services, and processes.
2. An effective design process requires a robust method and structure to create new products (goods, services, information) and ensure that these together with key operational processes—including process controls—are developed prior to the introduction of the products to the marketplace.
3. The Juran Quality by Design Model consists of following simple steps, primarily leading to a much better understanding of the customers that will benefit from the new product. It is not a statistical design method as Design for Six Sigma is considered. It is often used to design new services and processes. The steps are as follows:
   - Establish the design targets and goals.
   - Define the market and customers that will be targeted.
   - Discover the market, customers, and societal needs.
   - Develop the features of the new design that will meet the needs.
   - Develop or redevelop the processes to produce the features.
   - Develop process controls to be able to transfer the new designs to operations.
4. The Design for Six Sigma model, often called DMADV, consists of a statistical approach to design applicable to manufactured goods. It follows similar steps and incorporates some of the tools in Juran's model:
   - Define the project and the targets.
   - Measure what is critical to customers and quality (CTQs) to establish the required features.
   - Analyze the information and create a high-level design incorporating the CTQs.
   - Design by creating detailed designs, evaluate them, and optimize them before transferring them to operations.
   - Verify the design requirements and execute the final product.

Tackling the First Process of the Trilogy: Designing Innovative Products

An organization's ability to satisfy its customers depends on the robustness of the design processes because the goods you sell and the services you offer originate there.

The design process is the first of the three elements of the Juran Trilogy. It is one of the three basic functions by which management ensures the survival of the organization. The design process enables innovation to happen by designing products (goods, services, or information) together with the processes—including controls—to produce the final outputs. When design is complete, the other two elements—control and improvement—kick in to continuously improve upon the design as customer needs and technology change.

This handbook addresses two versions of the design process. In this chapter we will discuss the first version, Juran's universal quality by design model. It has been in place since 1986 and provides a structure that can be incorporated into an organization's new product development function, or it can be used independently to be carried out project by project as needed.

The second version, Design for Six Sigma (DFSS), which is referred to by the steps in the process DMADV (define, measure, analyze, design, and verify), is the most recent adaption to Juran's model. It builds upon the Six Sigma Improvement or DMAIC (define, measure, analyze, improve, and control) methodology to improve performance. DMADV was first introduced by GE. It uses elements of the Juran model and incorporates many of the statistical tools common to improvement. DFSS will be covered in detail in Chapter 14, Continuous Innovation Using Design for Six Sigma.

The Juran model is especially useful for designing products and redesigning processes simply and economically. The authors have witnessed the design of superb products, processes, and services using this model.

Examples include a prize-winning safety program for a multiple-plant manufacturer; an information system that enables both sales and manufacturing to track the procession of an order throughout the entire order fulfillment process so customers can be informed—on a daily basis—of the exact status of their order; and a redesigned accounts receivable system much faster and more efficient than its predecessor.

The DFSS model is the classic model enhanced by the addition of computers and statistical software packages, which permit the utilization of numerous design tools not easily used without a computer. The Six Sigma model is suitable for designing even complex products and for achieving extraordinary levels of quality. Although it is time consuming and expensive in the short term, when executed properly, it produces a healthy return on investment.

The Juran Quality by Design Model

Modern, structured quality design is the methodology used to plan both features that respond to customers' needs and the process to be used to make those features. "Quality by design" refers to the product or service development processes in organizations. Note the dual responsibility of those who plan: to provide the features to meet customer needs and to provide the process to meet operational needs. In times past, the idea that product design stopped at understanding the features that a product should have was the blissful domain of marketers, salespeople, and research and development people. But this new dual responsibility requires that the excitement generated by understanding the features and customer needs be tempered in the fire of operational understanding.

That is, can the processes make the required features without generating waste? To answer this question requires understanding both the current processes' capabilities and customer specifications. If the current processes cannot meet the requirement, modern design must include finding alternative processes that are capable.

The Juran Trilogy points out that the word "quality" incorporates two meanings: first, the presence of features creates customer satisfaction; second, freedom from failures about those features is also needed. In short, failures in features create dissatisfactions.

1. Removing failures is the purpose of quality improvement.
2. Creating features is the purpose of quality by design.

Kano, Juran, and others have long ago agreed that the absence of failures, that is, no customer dissatisfaction, may not lead us to the belief that satisfaction is thus in hand. We can readily conclude that dissatisfaction goes down as failures are removed. We cannot conclude that satisfaction is therefore going up, because the removal of irritants does not lead to satisfaction—it leads to less dissatisfaction.

It is only the presence of features that creates satisfaction. Satisfaction and dissatisfaction are not co-opposite terms. It is amazing how many organizations fail to grasp this point. Let's take, e.g., the typical "bingo card" seen in many hotels. These are replete with "closed-ended" questions. For example, they ask, "How well do you like this on a scale of 1 to 5?"

They do not ask, "How well do you like this?" This is the exact opposite of the question "How well don't you like it?" Therefore, any so-called satisfaction rating that does not allow for open-ended questioning such as "What should we do that we are not already doing?" or "Is there someone who provides a service we do not offer?" will always fall into a one-sided dimension of quality understanding. What, then, does a composite score of 3.5 for one branch in a chain of hotels really mean compared to another branch scoring 4.0? It means little. Their so-called satisfaction indices are really dissatisfaction indices.

So we arrive at the basic fundamental of what quality really is. As stated in Chapter 1 Attaining Superior Results Through Quality, the authors adopted a definition that Juran had postulated long before: "quality" means fitness for use, and we now have extended it to "fitness for purpose." Let's explore this concept.

First, the definition of "fitness for use" takes into account both dimensions of quality—the presence of features and the absence of failures. The sticky points are these: Who gets to decide what "fitness" means? Who decides what "purpose" means? The user decides what "use" means, and the user decides what "fitness" means. Any other answer is bound to lead to argument and misunderstanding. Providers rarely win here. Users, especially society at large, generally always win. For example, take yourself as a consumer. Did you ever use a screwdriver as a pry bar to open a paint can? Of course you did. Did you ever use it to punch holes into a jar lid so your child could watch bugs? Of course you did. Did you ever use it as a chisel to remove some wood, or metal, that was in the way of a job you were doing around the house? Of course you did. Now wait just a moment . . . a screwdriver's intended use is to drive screws!

So the word "use" has two components, *intended* use and *actual* use. When the user utilizes it in the intended way, both the provider and the user are satisfied. Conformance to specification and fitness for purpose match. But what about when the user uses it in the nonintended way, as in the screwdriver example? What, then, regarding specifications and fitness?

To delve even deeper, how does the user actually use the product? What need is it meeting for the user? Here we find another juncture: the user can create artful new uses for a product. For example:

> *"2000" Uses for WD-40.* WD-40 was formulated years ago to meet the needs of the U.S. space program. Not many know the origins of the brand name. "WD" refers to water displacement, and 40 is simply the 40th recipe the company came up with. But as the product moved into the consumer market, all kinds of new uses were uncovered by the users. People claimed it was excellent for removing scuff marks from flooring. They claimed it could easily remove price stickers from lamps, inspection stickers from windshields, and bubble gum from children's hair. The company delighted in all this. But the company didn't release all those clever new uses for public consumption. People also claimed that if they sprayed bait or lures with it, they caught more fish. Those with arthritis swore that a quick spray on a stiff elbow gave them relief. Let's not go too far. What about use where the product obviously cannot work? In Latin there is a word for this: ab-use (abuse), where the prefix "ab" simply means "not."

Some examples will help: back to the screwdriver. You could argue that using the screwdriver as a pry bar, chisel, or punch is abuse of its original designed purpose. But clearly many manufacturers have provided a product that can withstand this abuse, and so use then falls back into the "intended" column (whether this came as a result of lawsuits or from some other source). Further, a look at commercial aircraft "black boxes" (which are orange, by the way), show that they clearly survive in circumstances where the aircraft do not survive. Understanding of use in all its forms is what modern design seeks to achieve.

Last, modern design and planning, as we see over and over, seeks to create features in response to understanding customer needs. We are referring to customer-driven features. The sum of all features is the new product, service, or process.

A different type of product planning in which features meeting no stated need are put out for users to explore is beyond the scope of this chapter. 3M's Post-it Notes and the Internet are examples where we collectively did not voice needs, but which we cannot imagine life without them, once we embraced their features.

The Quality by Design Problem

The quality by design model and its associated methods, tools, and techniques have been developed because in the history of modern society, organizations rather universally have demonstrated a consistent failure to produce the goods and services that unerringly delight their customers. As a customer, everyone has been dismayed time and again when flights are delayed, radioactive contamination spreads, medical treatment is not consistent with best practices, a child's toy fails to function, a new piece of software is not as fast or user-friendly as anticipated, government responds with glacial speed (if at all), or a home washing machine with the latest high-tech gadget delivers at higher cost clothes that are no cleaner than before. These frequent, large quality gaps are really the compound result of a number of four smaller gaps, illustrated in Figure 4.1.

The first component of the quality gap is the *understanding gap*, i.e., lack of understanding of what the customer needs are. Sometimes this gap is wider because the producer simply fails to consider who the customers are and what they need. More often the gap is there because the supplying organization has erroneous confidence in its ability to understand exactly what the customer really needs. The final perception gap in Figure 4.1 also arises from a failure to understand customer needs. Customers do not experience a new suit of clothes or the continuity in service from a local utility simply based on the technical merits of the product. Customers react to how they perceive the good or service provides them with a benefit.

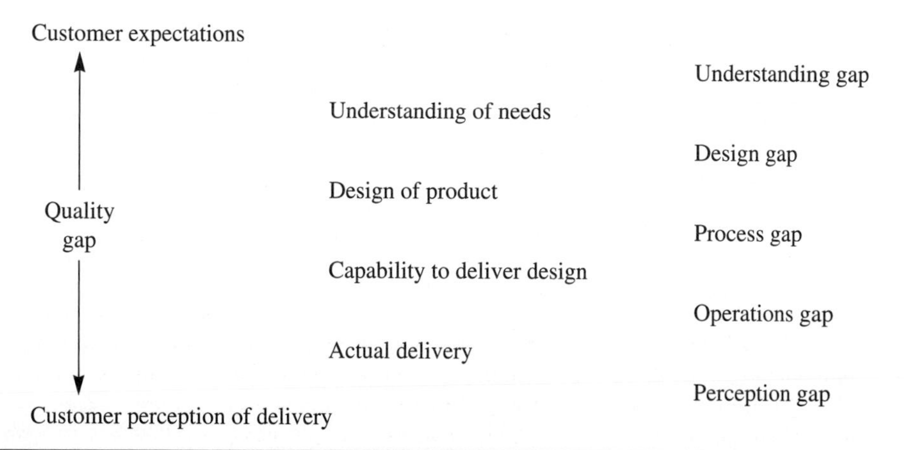

FIGURE 4.1 The quality gap. (*Inspired by A. Parasuraman, Valarie A. Zeithami, and Leonard L. Berry, "A Conceptual Model for Service Quality and Its Implications for Further Research,"* Journal of Marketing, *Fall 1985, pp. 41–50.*)

Quality Planning

1. Establish the project and design goals.
2. Identify the customers.
3. Discover the customer needs.
4. Develop the product or service features.
5. Develop the process features.
6. Develop the controls and transfer to operations.

Figure 4.2 Quality by design steps. (*Copyright 1994, Quality by Design, Juran Institute, Inc.*)

The second constituent of the quality gap is a *design gap*. Even if there were perfect knowledge about customer needs and perceptions, many organizations would fail to create designs for their goods and services that are fully consistent with that understanding. Some of this failure arises from the fact that the people who understand customers and the disciplines they use for understanding customer needs are often systematically isolated from those who actually create the designs. In addition, designers—whether they design sophisticated equipment or delicate human services—often lack the simple tools that would enable them to combine their technical expertise with an understanding of the customer needs to create a truly superior product.

The third gap is the *process gap*. Many splendid designs fail because the process by which the physical product is created or the service is delivered is not capable of conforming to the design consistently time after time. This lack of process capability is one of the most persistent and bedeviling failures in the total quality gap.

The fourth gap is the *operations gap*. The means by which the process is operated and controlled may create additional failures in the delivery of the final good or service.

Quality by design provides the process, methods, tools, and techniques for closing each of these component gaps and thereby ensuring that any final gap is at a minimum. Figure 4.2 summarizes at a high level the basic steps of quality by design. The remainder of this section will provide the details and examples for each of these steps.

Juran Quality by Design Model

We look at each of these as we step through the sequence at a high level.

Step 1: Establish the Project and Design Goals

All design should take place project by project. There is no such thing as design in general; there is only design in specific. In strategic planning, we set out the vision, mission, strategies, objectives, and so on. Each is a specific thing. In product planning, we start with a project, i.e., something to plan. We might design a new training room, a new car, a wedding, a customer toll-free hotline, or a new Internet process for bidding on travel booking (such as Priceline.com, Expedia.com). Note that each is a specific thing, and each can be clearly differentiated from anything else. A training room is not a cafeteria, a new car is not a Howitzer, a hotline is not long-distance service, and the travel booking process is not a bookstore online. This is a significant point. Without being able to differentiate what we are designing from anything else, everything collapses into vagueness. So a project is our starting point.

Step 2: Identify the Customers

Going back to the 1980s Total Quality Management (TQM) days, we learned that those who receive the product are customers in some way. If we were designing a training room, the trainees would be an important customer segment. So, too, would the custodians, because they have to clean the room, set it up in different ways, and so on. Customers of the new car include the purchasers, the insurance organizations, the dealers, the carriers, etc. Customers of the hotline include our clients, our service agents, etc. We can include as customers for the travel process the travelers, airlines, and the Web server entity. From all this emerges the basic understanding: A customer is a cast of characters, and each has unique needs that must be met.

Step 3: Discover the Customers' Needs

Wants, needs, perceptions, desires, and other emotions are all involved in our discovery of customer needs. We need to learn how to separate things and prioritize them. But at this point, we need to emphasize that not all high-priority customers (such as the car buyer) are the only ones with high-priority needs. We also stress that just because some customer entity is lower in priority doesn't mean at all that it automatically has lesser-priority needs. We need to understand the "voice of the customer" and the "voice of the market."

Take, e.g., the automobile carriers; we simply cannot overlook their needs for the car to be only so high and only so wide. If we ignored their needs, they could stop the product from reaching the cash-paying ultimate customer, our buyer. So, too, could regulators (the various states, the National Highway Transportation Safety Board, the Environmental Protection Agency, etc., impose "needs" that if unmet, could stop the process from going forward at all). So from all this, we reach another point: Customers have to be prioritized in an agreed upon way.

Step 4: Develop the Product or Service Features

The word "feature," as used in product planning, means what the product does, its characteristics, or its functionality. In structured product planning, we adopt a different definition: A feature is the thing that the customer employs to get her or his needs met. For example, in our training room, the trainees need to take notes as they learn. A feature might then be a flip chart, a white board, or a desk. Our custodians might need to move things around quite a bit, so features might include portability, size, weight, and modularity.

As our list of features grows, we soon realize that we cannot possibly have all features at the same priority level. So we need a way to put things in order, once again, and in an agreed upon way. We finalize by optimizing and agreeing on the list of features and the goals for them as well. Note what optimization means: Not all features survive product planning.

Step 5: Develop the Process Features

Because we know that the process is the thing that creates the features, we need to examine current and alternative processes to see which ones will be used to create the features. We need to be sure that the product feature goals can be accomplished via the processes we choose. In other words:

> Process capability must reconcile with product requirements. That statement is very important. No process knows its product goals; product goals come from humans. Ideal product goals would naturally reflect the various customers. But the key issue is this: Variation comes from processes; goals come from humans.

In the example of the training room, process goals might be to reset the room in 20 minutes, keep a supply of flip charts in a closet, certify the trainees to a standard, and so on. As before, we need to list all the possible routes to making the product, select the ones we will use based on some rationale, establish goals for the processes, and reach an optimum.

Step 6: Develop Process Controls and Transfer to Operations

Develop Process Controls

Control is basic to all human activity, from how the body regulates itself as to temperature and metabolism, to financial controls in how we run our organizations or homes. Control consists of three fundamentals:

> In product planning, we need to ensure that the processes work as designed within their capabilities. In the training room, e.g., controls might take the form of a checklist for resetting the room and a minimum inventory of flip charts. Control makes use of the concept of the feedback loop.

Here's an example you might keep in mind:

Did you ever check the oil in your car? The dipstick is a form of control point. Note that we begin with a control subject (volume of oil), a unit of measure (quarts or liters), a sensor (you and the dipstick), and a goal (keep the oil somewhere between "full" and "add"—inside those hash marks). Then we move on to sample the process (clean the dipstick, put it back in, remove it, and observe the oil level). Next we adjust when adjustment is called for (oil levels below add require us to add oil until we bring the oil up to somewhere between add and full, the agreed goal). If the oil is already within the hash marks, the control activity is to replace the dipstick, shut the hood, and drive on until another checkpoint is reached (perhaps next month). Note that the control activity must reflect the agreed upon goal for control. In the engine oil example, the control point was "inside the hash marks," so the control action is to bring the oil to somewhere "inside the hash marks." Many people miss this point; e.g., they add oil until the stick reads "full." This is overcontrol. Control actions must reflect control goals.

Transfer to Operations

Transfer to operations winds up the whole design process. As used here, "operations" means those who run the process, not "manufacturing." To continue the examples used earlier, operations for the training room is the activity of the trainers, the custodians, and the purchasing department. For the new car, operations includes manufacturing, transport, dealer relations, and the legal department. For the hotline, operations means the customer service agents who answer the phone. In the travel bidding process, operations include those who shop the bid or reject it and those who maintain the software that interfaces the prospect with the carriers. From the lessons of the era of productivity, the Industrial Revolution, and into the twentieth century, we have learned that the involvement of the operators is key to any well-running process.

With the development of the Ford Taurus came solid understanding of the value of a "platform" team. Designers, engineers, workers, purchasing agents, salespeople, and managers all sat under one roof to develop the car. The concept of platform teams is well ingrained in many car organizations today. The Chrysler Technical Center in Auburn Hills, Michigan, is a later example of such broad collaboration. Thus, successful transfer to operations must include the operators in the design process as early as possible.

The remainder of this section will provide details, practical guidance, and examples for each of these steps.

Juran Quality by Design Model Substeps

Step 1: Establish the Goals and the Project Team

A quality by design project is the organized work needed to prepare an organization to deliver a new or revised product, service, or process. The following steps or activities are associated with establishing a quality by design project:

1. Identify which projects are required to fulfill the organization's sales or revenue generation strategy.
2. Prepare a goal statement for each project.
3. Establish a team to carry out the project.

Identification of Projects

Deciding which projects to undertake is usually the outgrowth of the strategic and business design of an organization. (See Chapter 7, Strategic Planning and Deployment: Moving from Good to Great, for a discussion of how specific projects are deployed from an organization's vision, strategies, and goals.) Typically, design for quality projects create new or updated products that are needed to reach specific strategic goals, to meet new or changing customer needs, to fulfill legal or customer mandates, or to take advantage of a new or emerging technology.

Upper management must take the leadership in identifying and supporting the critical quality by design projects. Acting as a design council, council, or similar body, management needs to fulfill the following key roles:

1. *Setting design goals.* Marketing, sales, and similar management functions identify market opportunities and client needs currently not being met. By setting these goals, management is beginning the process to create new products, services, or processes to meet these unmet needs.
2. *Nominating and selecting projects.* The management or council selects the appropriate design projects critical to meeting strategic business and customer goals.
3. *Selecting teams.* Once a project has been identified, a team is appointed to see the project through the remaining steps of the design for quality process. A team may be defined by a project manager in the product development function.
4. *Supporting project team.* New technologies and processes are generally required to meet the new design goals. It is up to management to see that each design team is well prepared, trained, and equipped to carry out its goals. The support may include the following:

   a. Provide education and training in design tools.

   b. Provide a trained project leader to help the team work effectively and learn the design for quality process.

c. Regularly review team progress.

d. Approve revision of the project goals.

e. Identify or help with any issues that may hinder the team.

f. Provide resource expertise in data analysis.

g. Furnish resources for unusually demanding data collection such as market studies.

h. Communicate project results.

5. *Monitoring progress.* The council is responsible for keeping the quality by design process on track, evaluating progress, and making midcourse corrections to improve the effectiveness of the entire process. Once the council has reviewed the sources for potential projects, it will select one or more for immediate attention. Next, it must prepare a goal statement for the project.

Prepare Goal Statement

Once the council has identified the need for a project, it should prepare a goal statement that incorporates the specific goal(s) of the project. The goal statement is the written charter for the team that describes the intent and purpose of the project. The team goal describes

- The scope of the project, i.e., the product and markets to be addressed
- The goals of the project, i.e., the results to be achieved (sales targets)

Writing goal statements requires a firm understanding of the driving force behind the project. The goal helps to answer the following questions:

- Why does the organization want to do the project?
- What will the project accomplish once it is implemented?

A goal statement also fosters a consensus among those who either will be affected by the project or will contribute the time and resources necessary to plan and implement the project goal.

Examples include the following:

- The team goal is to deliver to market a new low-energy, fluorocarbon-free refrigerator that is 25 percent less expensive to produce than similar models.
- The team will create accurate control and minimum cost for the inventory of all stores.

While these goal statements describe what will be done, they are still incomplete. They lack the clarity and specificity required of a complete quality by design goal statement that incorporates the goal(s) of a project. Well-written and effective goal statements define the scope of the project by including one or more of the following.

Inherent Performance How the final product will perform on one or more dimensions, e.g., 24-hour response time, affects the scope of the project.

Comparative Performance How the final product will perform vis-a-vis the competition, e.g., the fastest response time in the metropolitan area, is relevant.

Customer Reaction How will customers rate the product compared with others available? For example, one organization is rated as having a better on-time delivery service than its closest rival.

Voice of Market Who are or will be the customers or target audience for this product, and what share of the market or market niche will it capture, e.g., to become the "preferred" source by all business travelers within the continental United States?

Performance Failures How will the product perform with respect to product failure, e.g., failure rate of less than 200 for every 1 million hours of use.

Avoidance of Unnecessary Constraints It is important to avoid overspecifying the product for the team; e.g., if the product is intended for airline carry-on, specifying the precise dimensions in the goal maybe too restrictive. There may be several ways to meet the carry-on market.

Basis for Establishing Quality Goals In addition to the scope of the project, a goal statement must include the goal(s) of the project. An important consideration in establishing quality goals is the choice of the basis for which the goal(s) are set.

Technology as a Basis In many organizations, it has been the tradition to establish the quality goals on a technological basis. Most of the goals are published in specifications and procedures that define the quality targets for the supervisory and nonsupervisory levels.

The Market as a Basis Quality goals that affect product salability should be based primarily on meeting or exceeding market quality. Because the market and the competition undoubtedly will be changing while the design for quality project is underway, goals should be set so as to meet or beat the competition estimated to be prevailing when the project is completed. Some internal suppliers are internal monopolies. Common examples include payroll preparation, facilities maintenance, cafeteria service, and internal transportation. However, most internal monopolies have potential competitors. There are outside suppliers who offer to sell the same service. Thus the performance of the internal supplier can be compared with the proposals offered by an outside supplier.

Benchmarking as a Basis "Benchmarking" is a recent label for the concept of setting goals based on knowing what has been achieved by others. (See Chapter 15, Benchmarking: Defining Best Practices for Market Leadership.) A common goal is the requirement that the reliability of a new product be at least equal to that of the product it replaces and at least equal to that of the most reliable competing product. Implicit in the use of benchmarking is the concept that the resulting goals are attainable because they have already been attained by others.

History as a Basis A fourth and widely used basis for setting quality goals has been historical performance; i.e., goals are based on past performance. Sometimes this is tightened up to stimulate improvement. For some products and processes, the historical basis is an aid to needed stability. In other cases, notably those involving chronically high costs of poor quality, the historical basis helps to perpetuate a chronically wasteful performance. During the goal-setting process, the management team should be on the alert for such misuse of the historical basis.

Goals as a Moving Target It is widely recognized that quality goals must keep shifting to respond to the changes that keep coming over the horizon: new technology, new competition, threats, and opportunities. While organizations that have adopted quality management methods practice this concept, they may not do as well at providing the means to evaluate the impact of those changes and revise the goals accordingly.

Project Goals Specific goals of the project, i.e., what the project team is to accomplish, are part of an effective goal statement. In getting the job done, the team must mentally start at the finish. The more focused it is on what the end result will look like, the easier it will be to achieve a successful conclusion.

Measurement of the Goal In addition to stating what will be done and by when, a project goal must show how the team will measure whether it has achieved its stated goals. It is important to spend some time defining how success is measured. Listed below are the four things that can be measured:

1. Quality
2. Quantity
3. Cost
4. Time, speed, agility

An effective quality by design project goal must have five characteristics for it to be smart and provide a team with enough information to guide the design process. The goal must be

1. Specific.
2. Measurable.
3. Agreed to by those affected.
4. Realistic—it can be a stretch, but it must be plausible.
5. Time-specific—when it will be done.

An example of a poorly written goal that is not smart might look something like this: "To design a new life insurance plan for the poor."

Contrast this with the following example: "To design and deliver a whole life plan in less than 90 days that enables poor families to ensure a level of insurance for under $500 per year (at time of introduction). The design also should allow the organization to sell the plans with an average return of between 4 and 6 percent."

The second example is smart—much more detailed, measurable, and time-specific than the first. The target or end result is clearly stated and provides enough direction for the team to plan the features and processes to achieve the goal.

New Product Policies Organizations need to have very clear policy guidance with respect to quality and product development. Most of these should relate to all new products, but specific policies may relate to individual products, product lines, or groups. Four of the most critical policies are as follows.

1. *Failures in new and carryover designs.* Many organizations have established the clear policy that no new product or component of a product will have a higher rate of failures than the old product or component that it is replacing. In addition, they

often require that any carryover design have a certain level of performance; otherwise, it must be replaced with a more reliable design. The minimum carryover reliability may be set by one or more of the following criteria: (1) competitor or benchmark reliability, (2) customer requirements, or (3) a stretch goal beyond benchmark or customer requirements.

2. *Intended versus unintended use.* Should stepladders be designed so that the user can stand on the top step without damage, even though the step is clearly labeled "Do Not Step Here?" Should a hospital design its emergency room to handle volumes of routine, nonemergency patients who show up at its doors? These are policy questions that need to be settled before the project begins. The answers can have a significant impact on the final product, and the answers need to be developed with reference to the organization's strategy and the environment within which its products are used.
3. *Requirement of formal quality by design process.* A structured, formal process is required to ensure that the product planners identify their customers and design products and processes that will meet those customer needs with minimum failures. Structured formality is sometimes eschewed as a barrier to creativity. Nothing could be more misguided. Formal quality by design identifies the points at which creativity is demanded and then encourages, supports, and enables that creativity. Formal design also ensures that the creativity is focused on the customers and that creative designs ultimately are delivered to the customer free of the destructive influences of failures.
4. *Custody of designs and change control.* Specific provision must be made to ensure that approved designs are documented and accessible. Any changes to designs must be validated, receive appropriate approvals, be documented, and be unerringly incorporated into the product or process. Specific individuals must have the assigned authority, responsibility, and resources to maintain the final designs and administer change control.

Establish Team

The cross-functional approach to complete a quality by design project is effective for several reasons:

- Team involvement promotes sharing of ideas, experiences, and a sense of commitment to being a part of and helping "our" organization achieve its goal.
- The diversity of team members brings a more complete working knowledge of the product and processes to be planned. Design of a product requires a thorough understanding of how things get done in many parts of the organization.
- Representation from various departments or functions promotes the acceptance and implementation of the new plan throughout the organization. Products or processes designed with the active participation of the affected areas tend to be technically superior and accepted more readily by those who must implement them.

Guidelines for Team Selection When selecting a team, the council identifies those parts of the organization that have a stake in the outcome. There are several places to look:

- Those who will be most affected by the result of the project
- Departments or functions responsible for various steps in the process

- Those with special knowledge, information, or skill in the design of the project
- Areas that can be helpful in implementing the plan

Step 2: Identify the Customers

This step may seem unnecessary; of course, the planners and designers know who their customers are: the driver of the automobile, the depositor in the bank account, the patient who takes the medication. But these are not the only customers—not even necessarily the most important customers. Customers comprise an entire cast of characters that needs to be understood fully.

Generally, there are two primary groups of customers: the external customers—those outside the producing organization—and the internal customers—those inside the producing organization.

Types of External Customers

The term "customer" is often used loosely; it can refer to an entire organization, a unit of a larger organization, or a person. There are many types of customers, some obvious, others hidden. Below is a listing of the major categories to help guide complete customer identification.

The Purchaser This is someone who buys the product for himself or herself or for someone else, e.g., anyone who purchases food for his or her family. The end user/ultimate customer is someone who finally benefits from the product, e.g., the patient who goes to a health care facility for diagnostic testing.

Merchants These are people who purchase products for resale, wholesalers, distributors, travel agents and brokers, and anyone who handles the product, such as a supermarket employee who places the product on the shelf.

Processors Processors are organizations and people who use the product or output as an input for producing their own product, e.g., a refinery that receives crude oil and processes it into different products for a variety of customers.

Suppliers Those who provide input to the process are suppliers, e.g., the manufacturer of the spark plugs for an automobile or the law firm that provides advice on the organization's environmental law matters. Suppliers are also customers. They have information needs with respect to product specification, feedback on failures, predictability of orders, and so on.

Potential Customers Those not currently using the product but capable of becoming customers are potential customers; e.g., a business traveler renting a car may purchase a similar automobile when the time comes to buy one for personal use.

Hidden Customers Hidden customers comprise an assortment of different customers who are easily overlooked because they may not come to mind readily. They can exert great influence over the product design: regulators, critics, opinion leaders, testing services, payers, the media, the public at large, those directly or potentially threatened by the product, corporate policymakers, labor unions, and professional associations.

Internal Customers

Everyone inside an organization plays three roles: supplier, processor, and customer. Each individual receives something from someone, does something with it, and passes it to a third individual. Effectiveness in meeting the needs of these internal customers can have a major impact on serving the external customers. Identifying the internal customers will require some analysis because many of these relationships tend to be informal, resulting in a hazy perception of who the customers are and how they will be affected. For example, if an organization decides to introduce just-in-time manufacturing to one of its plants, this will have significant effects on purchasing, shipping, sales, operations, and so on.

Most organizations try to set up a mechanism that will allow seemingly competing functions to negotiate and resolve differences based on the higher goal of satisfying customer needs. This might include conducting weekly meetings of department heads or publishing procedure manuals. However, these mechanisms often do not work because the needs of internal customers are not fully understood, and communication among the functions breaks down. This is why a major goal in the design for quality process is to identify who the internal customers are, discover their needs, and plan how those needs will be satisfied. This is also another reason to have a multifunctional team involved in the planning; these are people who are likely to recognize the vested interests of internal customers.

Identifying Customers

In addition to the general guidance just laid out, it is most often helpful to draw a relatively high-level flow diagram of the processes related to the product being planned. Careful analysis of this flow diagram often will provide new insight, identifying customers that might have been missed and refining understanding of how the customers interact with the process. Figure 4.3 is an example of such a diagram. A review of this diagram reveals that the role of "customer" is really two different roles—placing the order and using the product. These may or may not be played by the same individuals, but they are two distinct roles, and each needs to be understood in terms of its needs.

Step 3: Discover Customer Needs

The third step of quality by design is to discover the needs of both external customers and internal processors for the product. Some of the key activities required for effective discovery of customer needs include the following:

Plan to discover customers' needs.

Collect a list of customers' needs in their language.

Analyze and prioritize customers' needs.

Translate their needs into "our" language.

Establish units of measurement and sensors.

Our own experience tells us that the needs of human beings are both varied and complex. This can be particularly challenging to a design team because the actions of customers are not always consistent with what they say they want. The challenge for quality by design is to identify the most important needs from the full array of those needs expressed or assumed by the customer. Only then can the product delight the customers.

When a product is being designed, there are actually two related but distinct aspects of what is being developed: the technology elements of what the product's features will

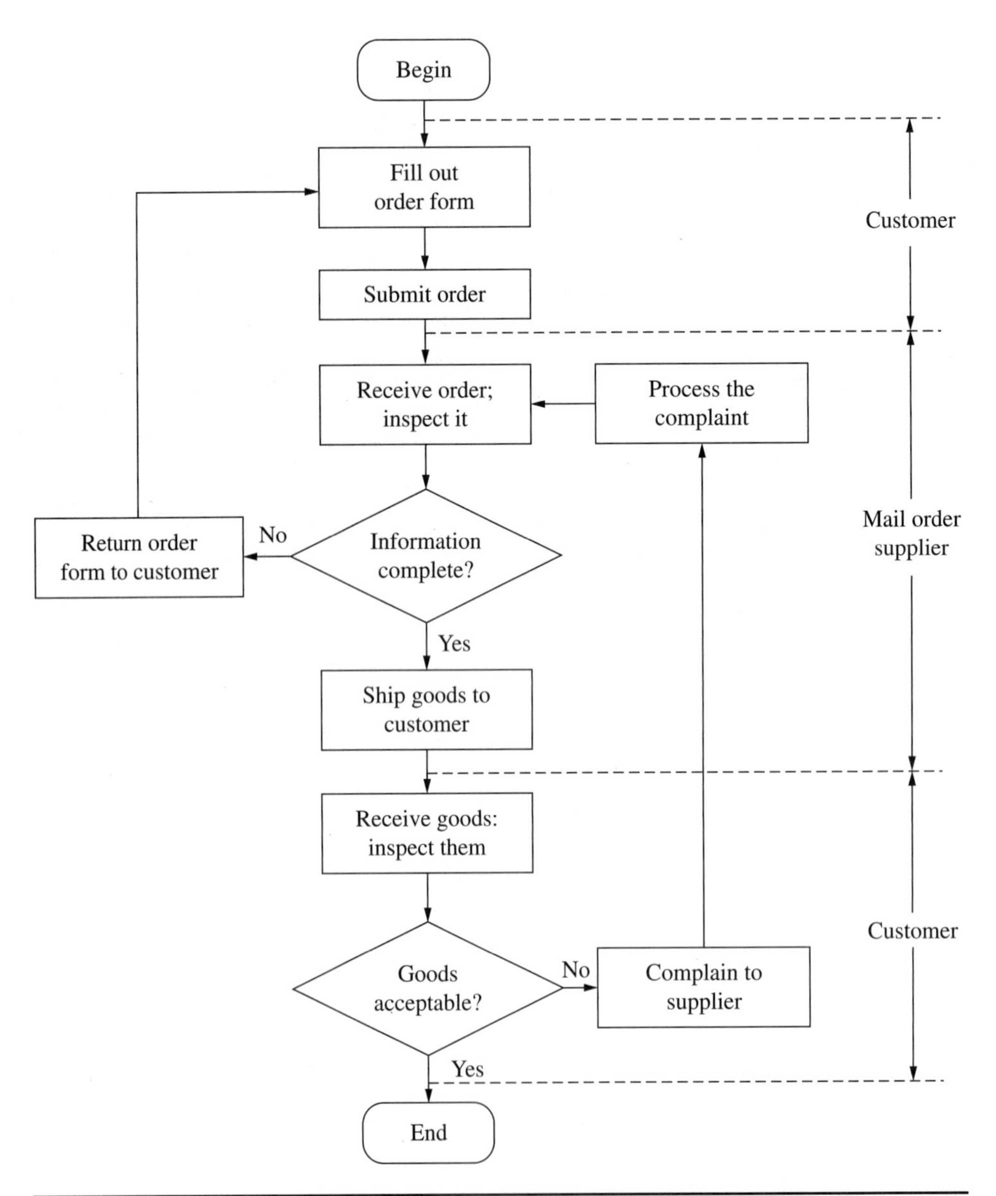

Figure 4.3 Flow diagram and customers. (*From J. M. Juran, Quality Control Handbook, 5th ed., McGraw-Hill, New York, 1999, p. 3.12.*)

actually do or how it will function and the human elements of the benefits customers will receive from using the product. The two must be considered together.

Discovering customer needs is a complex task. Experience shows that customers usually do not state, in simple terms, exactly what they want; often they do not even mention some of their most basic needs. Accuracy of bank statements, competence of a physician, reliability of a computer, and grammatical correctness of a publication may be assumed and never stated without probing.

One of the ways customers express their needs is in terms of problems they experience and their expectation that a product will solve their problems. For example, a customer may state, "I cannot always answer my telephone personally, but I do not want callers to be either inconvenienced or disgusted with nonresponsive answering systems." Or the customer may state, "My mother's personal dignity and love of people are very important to me. I want to find an extended care facility that treats her as a person, not a patient." Even when the need is not expressed in such terms, the art and science of discovering needs are to understand exactly the benefit that the customer expects.

When a product's features meet a customer's need, it gives the customer a feeling of satisfaction. If the product fails to deliver the promised feature defect-free, the customer feels dissatisfaction. Even if a product functions the way it has been designed, a competing product, by virtue of superior service or performance, may provide customers with greater satisfaction.

Stated Needs and Real Needs

Customers commonly state their needs as seen from their viewpoint and in their language. Customers may state their needs in terms of the goods or services they wish to buy. However, their real needs are the benefits they believe they will receive.

To illustrate:

Customer wishes to buy:	**Benefit customer needs might include:**
Fresh pasta	Nourishment and taste
Newest personal computer	Write reports quickly and easily Find information on the Web Help children learn math
Health insurance	Security against financial disaster Access to high-quality health care Choice in health care providers
Airline ticket	Transportation, comfort, safety, and convenience

Failure to grasp the difference between stated needs and real needs can undermine the salability of the product in design. Understanding the real needs does not mean that the planners can dismiss the customers' statements and substitute their own superior technical understanding as being the customers' real needs. Understanding the real needs means asking and answering such questions as these:

- Why is the customer buying this product?
- What service does she or he expect from it?
- How will the customer benefit from it?
- How does the customer use it?
- What has created customer complaints in the past?
- Why have customers selected competitors' products over ours?

Perceived Needs

Customers understandably state their needs based on their perceptions. These may differ entirely from the supplier's perceptions of what constitutes product quality. Planners can mislead themselves by considering whether the customers' perceptions are wrong or right rather than focusing on how these perceptions influence customers' buying habits. Although such differences between customers and suppliers are potential troublemakers, they also can be an opportunity. Superior understanding of customer perceptions can lead to competitive advantage.

Cultural Needs

The needs of customers, especially internal customers, go beyond products and processes. They include primary needs for job security, self-respect, respect of others, continuity of habit patterns, and still other elements of what we broadly call the "cultural values"; these are seldom stated openly. Any proposed change becomes a threat to these important values and hence will be resisted until the nature of the threat is understood.

Needs Traceable to Unintended Use

Many quality failures arise because a customer uses the product in a manner different from that intended by the supplier. This practice takes many forms. Patients visit emergency rooms for nonemergency care. Untrained workers are assigned to processes requiring trained workers. Equipment does not receive specified preventive maintenance.

Factors such as safety may add to the cost, yet they may well result in a reduced overall cost by helping to avoid the higher cost arising from misuse of the product. What is essential is to learn the following:

- What will be the actual use (and misuse)?
- What are the associated costs?
- What are the consequences of adhering only to intended use?

Human Safety

Technology places dangerous products into the hands of amateurs who do not always possess the requisite skills to handle them without accidents. It also creates dangerous by-products that threaten human health, safety, and the environment. The extent of all this is so great that much of the effort of product and process design must be directed at reducing these risks to an acceptable level. Numerous laws, criminal and civil, mandate such efforts.

User-Friendly

The amateur status of many users has given rise to the term "user-friendly" to describe the product feature that enables amateurs to make ready use of technological products. For example, the language of published information should be *simple, unambiguous,* and *readily understood.* (Notorious offenders have included legal documents, owners' operating manuals, administrative forms, etc. Widely used forms such as government tax returns should be field-tested on a sample of the very people who will later be faced with filling out such forms.) The language of published information should also be *broadly compatible.* (For example, new releases of software should be "upward-compatible with earlier releases.")

Promptness of Service

Services should be prompt. In our culture, a major element of competition is promptness of service. Interlocking schedules (as in mail delivery or airline travel) are another source of a growing demand for promptness. Still another example is the growing use of just-in-time manufacturing, which requires dependable deliveries of materials to minimize inventories. All such examples demonstrate the need to include the element of promptness in design to meet customer needs.

Customer Needs Related to Failures

In the event of product failure, a new set of customer needs emerges—how to get service restored and how to get compensated for the associated losses and inconvenience. Clearly, the ideal solution to all this is to plan quality so that there will be no failures. At this point, we will look at what customers need when failures do occur.

Warranties

The laws governing sales imply that there are certain warranties given by the supplier. However, in our complex society, it has become necessary to provide specific, written contracts to define just what is covered by the warranty and for how long a time. In addition, it should be clear who has what responsibilities.

Effect of Complaint Handling on Sales

While complaints deal primarily with product dissatisfaction, there is a side effect on salability. Research in this area has pointed out the following: Of the customers who were dissatisfied with products, nearly 70 percent did not complain. The proportions of these who did complain varied according to the type of product involved. The reasons for not complaining were principally (1) the belief that the effort to complain was not worth it, (2) the belief that complaining would do no good, and (3) lack of knowledge about how to complain. More than 40 percent of the complaining customers were unhappy with the responsive action taken by the suppliers. Again, percentages varied according to the type of product.

Future salability is strongly influenced by the action taken on complaints. This strong influence also extends to brand loyalty. Even customers of popular brands of large-ticket items, such as durable goods, financial services, and automobile services, will reduce their intent to buy when they perceive that their complaints are not addressed.

This same research concluded that an organized approach to complaint handling provides a high return on investment. The elements of such an organized approach may include

- A response center staffed to provide 24-hour access by consumers and/or a toll-free telephone number
- Special training for the employees who answer the telephones
- Active solicitation of complaints to minimize loss of customers in the future

Keeping Customers Informed

Customers are quite sensitive to being victimized by secret actions of a supplier, as the phrase "Let the buyer beware!" implies. When such secrets are later discovered and publicized, the damage to the supplier's quality image can be considerable. In a great many cases, the products are fit for use despite some nonconformances. In other cases, the matter may be debatable. In still other cases, the act of shipment is at least unethical and at worst illegal.

Customers also have a need to be kept informed in many cases involving product failures. There are many situations in which an interruption in service will force customers to wait for an indefinite period until service is restored. Obvious examples are power outages and delays in public transportation. In all such cases, the customers become restive. They are unable to solve the problem—they must leave that to the supplier. Yet they want to be kept informed as to the nature of the problem and especially as to the likely time of solution. Many suppliers are derelict in keeping customers informed and thereby suffer a decline in their quality image. In contrast, some airlines go to great pains to keep their customers informed of the reasons for a delay and of the progress being made in providing a remedy.

Plan to Collect Customers' Needs

Customer needs keep changing. There is no such thing as a final list of customer needs. Although it can be frustrating, design teams must realize that even while they are in the middle of the design process, forces such as technology, competition, social change, and so on can create new customer needs or may change the priority given to existing needs. It becomes extremely important to check with customers frequently and monitor the marketplace. Some of the most common ways to collect customer needs include

1. Customer surveys, focus groups, and market research programs and studies
2. Routine communications, such as sales and service calls and reports, management reviews, house publications
3. Tracking customer complaints, incident reports, letters, and telephone contacts
4. Simulated-use experiments and design processes that involve the customer
5. Employees with special knowledge of the customer: sales, service, clerical, secretarial, and supervisory who come into contact with customers
6. Customer meetings
7. User conferences for the end user
8. Information on competitors' products
9. Personal visits to customer locations; observe and discuss
10. Government or independent laboratory data
11. Changes in federal, state, and local regulations that will identify current need or new opportunity
12. Competitive analysis and field intelligence comparing products with those of competitors
13. Personal experience dealing with the customer and the product (However, it is important to be cautious about giving personal experience too much weight without direct verification by customers. The analysts must remember that looking at customer needs and requirements from a personal viewpoint can be a trap.)

Often customers do not express their needs in terms of the benefits they wish to receive from purchasing and using the product.

Collect List of Customers' Needs in Their Language

For a list of customers' needs to have significant meaning in the design of a new product, they must be stated in terms of benefits sought. Another way of saying this is to capture needs in the customer's voice. By focusing on the benefits sought by the customer rather than on the means of delivering the benefit, designers will gain a better understanding of what the customer needs and how the customer will be using the product. Stating needs in terms of the benefits sought also can reveal opportunities for improved quality that often cannot be seen when concentrating on the features alone.

Analyze and Prioritize Customer Needs

The information actually collected from customers is often too broad, too vague, and too voluminous to be used directly in designing a product. Both specificity and priority are needed to ensure that the design really meets the needs and that time is spent on designing for those needs that are truly the most important. The following activities help provide this precision and focus:

- Organizing, consolidating, and prioritizing the list of needs for both internal and external customers
- Determining the importance of each need for both internal and external customers
- Breaking down each need into precise terms so that a specific design response can be identified
- Translating these needs into the supplying organization's language
- Establishing specific measurements and measurement methods for each need

One of the best design tools to analyze and organize customers' needs is the design for quality spreadsheet.

Quality by Design Spreadsheets

Designing new products can generate large amounts of information that is both useful and necessary, but without a systematic way to approach the organization and analysis of this information, the design team may be overwhelmed by the volume and miss the message it contains.

Although planners have developed various approaches for organizing all this information, the most convenient and basic design tool is the quality by design spreadsheet. The spreadsheet is a highly versatile tool that can be adapted to a number of situations. The quality by design process makes use of several kinds of spreadsheets, such as

- Customer needs spreadsheet
- Needs analysis spreadsheet
- Product or service design spreadsheet
- Process design spreadsheet
- Process control spreadsheet

Besides recording information, these tools are particularly useful in analyzing relationships among the data that have been collected and in facilitating the stepwise conversion of customer needs into features and then features into process characteristics and plans. This conversion is illustrated in Figure 4.4. Analysis of customers and their needs provides the basis for designing the product. The summary of that design feeds the process design, which feeds the control spreadsheet.

For most design projects, simple matrix spreadsheets will suffice. For other projects, more complex quality functional deployment spreadsheets are helpful in computing design tradeoffs. All these spreadsheets are designed to allow the team to record and compare the relationships among many variables at the same time. We will illustrate some of these spreadsheets at the appropriate point in the design process. Figure 4.5 illustrates the generic layout of any one of these spreadsheets. In general, the row headings are the "whats" of the analysis—the customers to be satisfied, the needs to be met, and so on. The columns are the "hows"—the needs that, when met, will satisfy the customer, the features that will meet the needs, and so on. The bottom row of the spreadsheet generally contains specific measurable goals for the how at the top. The body of the spreadsheet expresses with symbols or numerics the impact of the how on the what, e.g., none, moderate, strong, very strong. Other columns can be added to give specific measures of the importance of the respective rows, benchmarks, and so on.

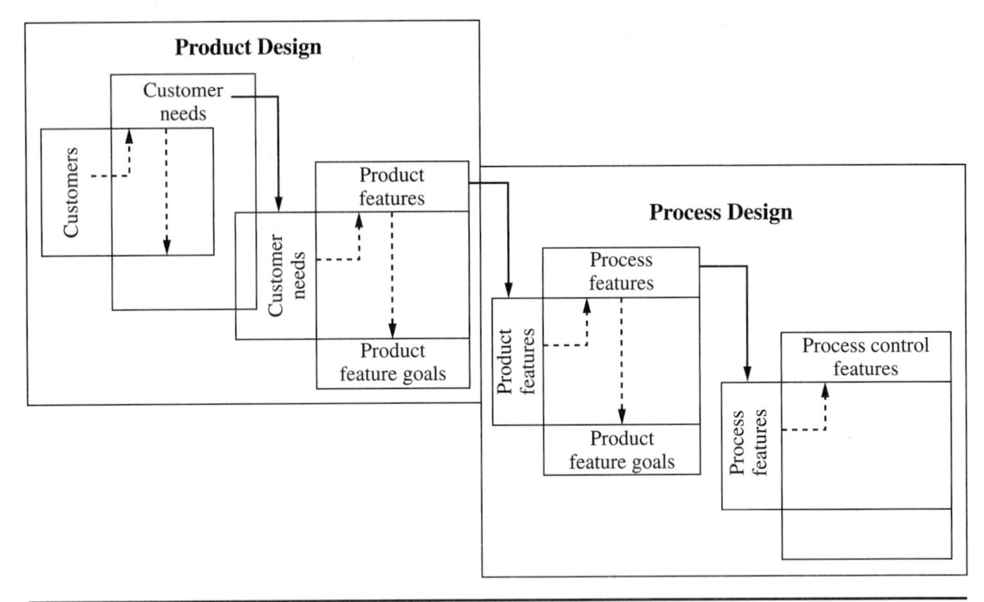

Figure 4.4 Sequence of activities. (*Juran Institute, Inc. Used by permission.*)

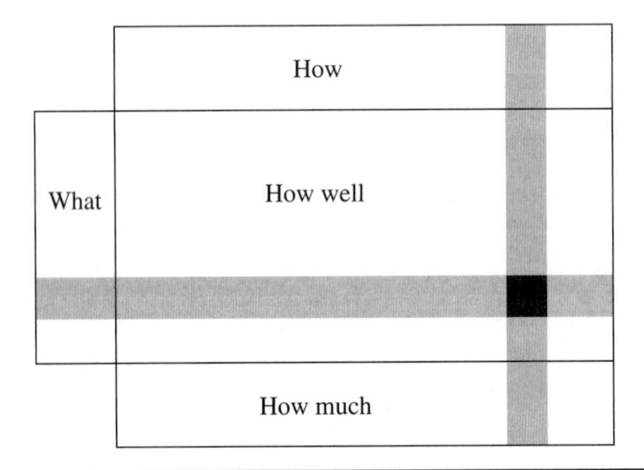

Figure 4.5 Planning spreadsheet. (*Juran Institute, Inc. Copyright 1994.*)

Customer Needs Spreadsheet

Figure 4.6 provides a simple example of a customer needs spreadsheet. The left column lists, in priority order, all the external and internal customers. The column headings are the various needs that have been discovered. By either checking or entering a designation for importance, it is possible to create a simple but comprehensive picture of the importance of meeting each need. All product development must operate within a budget. Prioritizing the customers and their needs ensures that the budget is focused on what is most important.

Precise Customer Needs

Once the needs that must be met have been prioritized, they must be described in sufficiently precise terms to design a product based on them. A customer needs spreadsheet

Customer Needs Spreadsheet

Customers	Customer Needs							
	Attractive	Informative/ well-written articles	Catchy cover lines	Stable circulation	It sells	Enough time	Material complete	No last minute changes
Readers	●	●	●					
Advertisers	●	○	●	●	●			
Printers						●	●	●
Typesetters						●	●	●
Color separators						●	●	●
Newsstand	●	○	●	○	●			

Legend ● Very strong ○ Strong △ Week

FIGURE 4.6 Customer needs spreadsheet. (*Juran Institute, Inc. Copyright 1994.*)

helps assemble this analysis. At this point, customer needs are probably a mixture of relatively broad expectations such as "ease of use" and more specific requests such as "access on Saturday." Figure 4.7 illustrates how broad needs (called *primary*) are broken into succeeding levels of specificity (secondary, tertiary, etc.). Note that primary and secondary do not mean more and less important; they mean, respectively, less specific and more specific. Each need

Primary Need	Secondary Need	Tertiary Need
Convenience	Hours of operation	Open between 5:00 and 9:00 p.m. Saturday hours
	Transportation access	Within three blocks of bus stop Ample parking
	Short wait times	Urgent appointment within 24 hours Routine appointment within 14 days Waiting time at appointment less than 15 minutes
	Complementary services available	Pharmacy on site Lab on site

FIGURE 4.7 Needs analysis spreadsheet for medical office. (*Juran Institute, Inc.*)

must be broken down to the level at which it can (1) be measured and (2) serve as an unambiguous guide for product design. In some cases two levels of detail may suffice; in others four or five may be required. Figure 4.7 illustrates how this might be done for the primary need "convenience" associated with a group medical practice.

Translate Their Needs into "Our" Language

The precise customer needs that have been identified may be stated in any of several languages, including

- The customer's language
- The supplier's ("our") language
- A common language

An old aphorism claims that the British and Americans are separated by a common language. The appearance of a common language or dialect can be an invitation to trouble because both parties believe that they understand each other and expect to be understood. Failure to communicate because of the unrecognized differences can build additional misunderstanding that only compounds the difficulty. It is imperative, therefore, for planners to take extraordinary steps to ensure that they properly understand customer needs by systematically translating them. The need to translate applies to both internal and external customers. Various organization functions employ local dialects that are often not understood by other functions.

Vague terminology constitutes one special case for translation that can arise even (and often especially) between customers and suppliers who believe they are speaking the same dialect. Identical words have multiple meanings. Descriptive words do not describe with technological precision.

Aids to Translation

Numerous aids are available to clear up vagueness and create a bridge across languages and dialects. The most usual listed are the following: A glossary is a list of terms and their definitions. It is a published agreement on the precise meanings of key terms. The publication may be embellished by other forms of communication, such as sketches, photographs, and videotapes.

Samples can take many forms, such as physical goods (e.g., textile swatches, color chips, audio cassettes) or services (e.g., video recordings to demonstrate "samples" of good service—courtesy, thoughtfulness, etc.). They serve as specifications for features. They make use of human senses beyond those associated with word images.

A special organization to translate communications with external customers may be required because of the high volume of translation. A common example is the order-editing department, which receives orders from clients. Some elements of these orders are in client language. Order editing translates these elements into supplier language, e.g., product code numbers, supplier acronyms, and so on.

Standardization is used by many mature industries for the mutual benefit of customers and suppliers. This standardization extends to language, products, processes, and so on. All organizations make use of short design actions for their products, such as code numbers, acronyms, words, phrases, and so on. Such standardized nomenclature makes it easy to communicate with internal customers.

Measurement is the most effective remedy for vagueness and multiple dialects—"Say it in numbers." This is the first, but not the last, point in the design process where measurement is critical. Design for quality also requires measurement of features, process features, process capability, control subjects, and so on.

Establish Units of Measurement and Sensors

Sound quality by design requires precise communication between customers and suppliers. Some of the essential information can be conveyed adequately by words. However, an increasingly complex and specialized society demands higher precision for communicating quality-related information. The higher precision is best attained when we say it in numbers.

Quantification Requires a System of Measurement Such a system consists of a unit of measurement, which is a defined amount of some quality feature and permits evaluation of that feature in numbers, e.g., hours of time to provide service, kilowatts of electric power, or concentration of a medication.

A sensor, which is a method or instrument of measurement, carries out the evaluation and states the findings in numbers in terms of the unit of measure, e.g., a clock for telling time, a thermometer for measuring temperature, or an X-ray to measure bone density.

By measuring customer needs, one has established an objective criterion for whether the needs are met. In addition, only with measurement can one answer questions such as these: Is our quality getting better or worse? Are we competitive with others? Which one of our operations provides the best quality? How can we bring all operations up to the level of the best?

Units of Measure for Features

The first task in measurement is to identify the appropriate unit of measurement for each customer need. For features, we know of no simple, convenient, generic formula that is the source of many units of measure. The number and variety of features are simply enormous. In practice, each product feature requires its own unique unit of measure. A good starting point is to ask the customers what their units of measure are for evaluating product quality. If the supplier's units of measure are different, the stage is set for customer dissatisfaction, and the team will need to come up with a unit of measure acceptable to both parties. Even if the customers have not developed an explicit unit of measure, ask them how they would know whether their needs were met. Their response may carry with it an implicit unit of measure.

Application to Goods

Units of measure for quality features of goods make extensive use of "hard" technological units. Some of these are well known to the public: time in minutes, temperature in degrees, or electric current in amperes. Many others are known only to the specialists. There are also "soft" areas of quality for goods. Food technologists need units of measure for flavor, tenderness, and still other properties of food. Household appliances must be "handsome" in appearance. Packaging must be "attractive." To develop units of measure for such features involves much effort and ingenuity.

Application to Services

Evaluation of service quality includes some technological units of measure. A widespread example is promptness, which is measured in days, hours, and so on. Environmental pollutants (e.g., noise, radiation, etc.) generated by service organizations are likewise measured using technological units of measure.

Service quality also involves features such as courtesy of service personnel, decor of surroundings, and readability of reports. Since these features are judged by human beings, the units of measure (and the associated sensors) must be shown to correlate with a jury of customer opinion.

The Ideal Unit of Measure

The criteria for an ideal unit of measure are summarized below. An ideal unit of measure

- Is understandable
- Provides an agreed upon basis for decision-making
- Is conducive to uniform interpretation
- Is economical to apply
- Is compatible with existing designs of sensors, if other criteria also can be met

Measuring Abstractions

Some quality features seem to stand apart from the world of physical things. Quality of service often includes courtesy as a significant quality feature. Even in the case of physical goods, we have quality features, such as beauty, taste, aroma, feel, or sound. The challenge is to establish units of measure for such abstractions.

The approach to dealing with abstractions is to break them up into identifiable pieces. Once again, the customer may be the best source to start identifying these components. For example, hotel room appearance is certainly a quality feature, but it also seems like an abstraction. However, we can divide the feature into observable parts and identify those specifics that collectively constitute "appearance," e.g., the absence of spots or bare patches on the carpet, clean lavatory, linens free from discoloration and folded to specified sizes, windows free of streaks, bedspreads free of wrinkles and hanging to within specific distances from the floor, and so on. Once units of measure have been established for each piece or component, they should be summarized into an index, e.g., number of soiled or damaged carpets to total number of hotel rooms, number of rooms with missing linens to total number of rooms, or number of customer complaints.

Establish the Sensor

To say it in numbers, not only do we need a unit of measure, but also we need to evaluate quality in terms of that unit of measure. A key element in making the evaluation is the sensor.

A *sensor* is a specialized detecting device or measurement tool. It is designed to recognize the presence and intensity of certain phenomena and to convert this sense knowledge into information. In turn, the resulting information becomes an input to decision-making because it enables us to evaluate actual performance.

Technological instruments are obviously sensors. So are the senses of human beings. Trends in some data series are used as sensors. Shewhart control charts are sensors.

Precision and Accuracy of Sensors

The *precision* of a sensor is a measure of the ability of the sensor to reproduce its results over and over on repeated tests. For most technological sensors, this reproducibility is high and is also easy to quantify.

At the other end of the spectrum are the cases in which we use human beings as sensors: inspectors, auditors, supervisors, and appraisers. Human sensors are notoriously less precise than technological sensors. Such being the case, planners are well advised to understand the limitations inherent in human sensing before making decisions based on the resulting data.

The *accuracy* of a sensor is the degree to which the sensor tells the truth—the extent to which its evaluation of some phenomenon agrees with the "true" value as judged by an established standard. The difference between the observed evaluation and the true value is the *error*, which can be positive or negative.

For technological sensors, it is usually easy to adjust for accuracy by recalibrating. A simple example is a clock or watch. The owner can listen to the time signals provided over the radio. In contrast, the precision of a sensor is not easy to adjust. The upper limit of precision is usually inherent in the basic design of the sensor. To improve precision beyond its upper limit requires a redesign. The sensor may be operating at a level of precision below that of its capability owing to misuse, inadequate maintenance, and so on. For this reason, when choosing the appropriate sensor for each need, planners will want to consider building in appropriate maintenance schedules along with checklists of actions to be taken during the check.

Translating and Measuring Customer Needs

The customer need for performance illustrates how high-level needs break down into myriad detailed needs. Performance included all the following detailed, precise needs:

Product Design Spreadsheet All the information on the translation and measurement of a customer need must be recorded and organized. Experience recommends placing these data so that they will be close at hand during product design. The example in Figure 4.8 shows a few needs all prepared for use in product design. The needs, their translation, and their measurement are all placed to the left of the spreadsheet. The remainder of the spreadsheet will be discussed in the next section.

Step 4: Develop the Product or Service Features

Once the customers and their needs are fully understood, we are ready to design the organization. Most organizations have some process for designing and bringing new products to market. In this step of the quality by design process, we will focus on the role of quality in product development and how that role combines with the technical aspects of development and design appropriate for a particular industry. Within product development, product design is a creative process based largely on technological or functional expertise.

The designers of products traditionally have been engineers, systems analysts, operating managers, and many other professionals. In the quality arena, designers can include any whose experience, position, and expertise can contribute to the design process. The outputs of product design are detailed designs, drawings, models, procedures, specifications, and so on.

The overall quality objectives for this step are two:

1. Determine which features and goals will provide the optimal benefit for the customer.
2. Identify what is needed so that the designs can be delivered without failures.

In the case of designing services, the scope of this activity is sometimes puzzling. For example, in delivering health care, where does the product of diagnosing and treating end and the processes of laboratory testing, chart reviews, and so on begin? One useful way to think about the distinction is that the product is the "face to the customer." It is what the customer sees and experiences. The patient sees and experiences the physician interaction, waiting time, clarity of information, and so on. The effectiveness and efficiency of moving blood samples to and around the laboratory have an effect on these features but are really features of the process that delivers the ultimate product to the customer.

Those who are designing physical products also can benefit from thinking about the scope of product design. Given that the customer's needs are the benefits that the customer wants from the product, the design of a piece of consumer electronics includes not only the

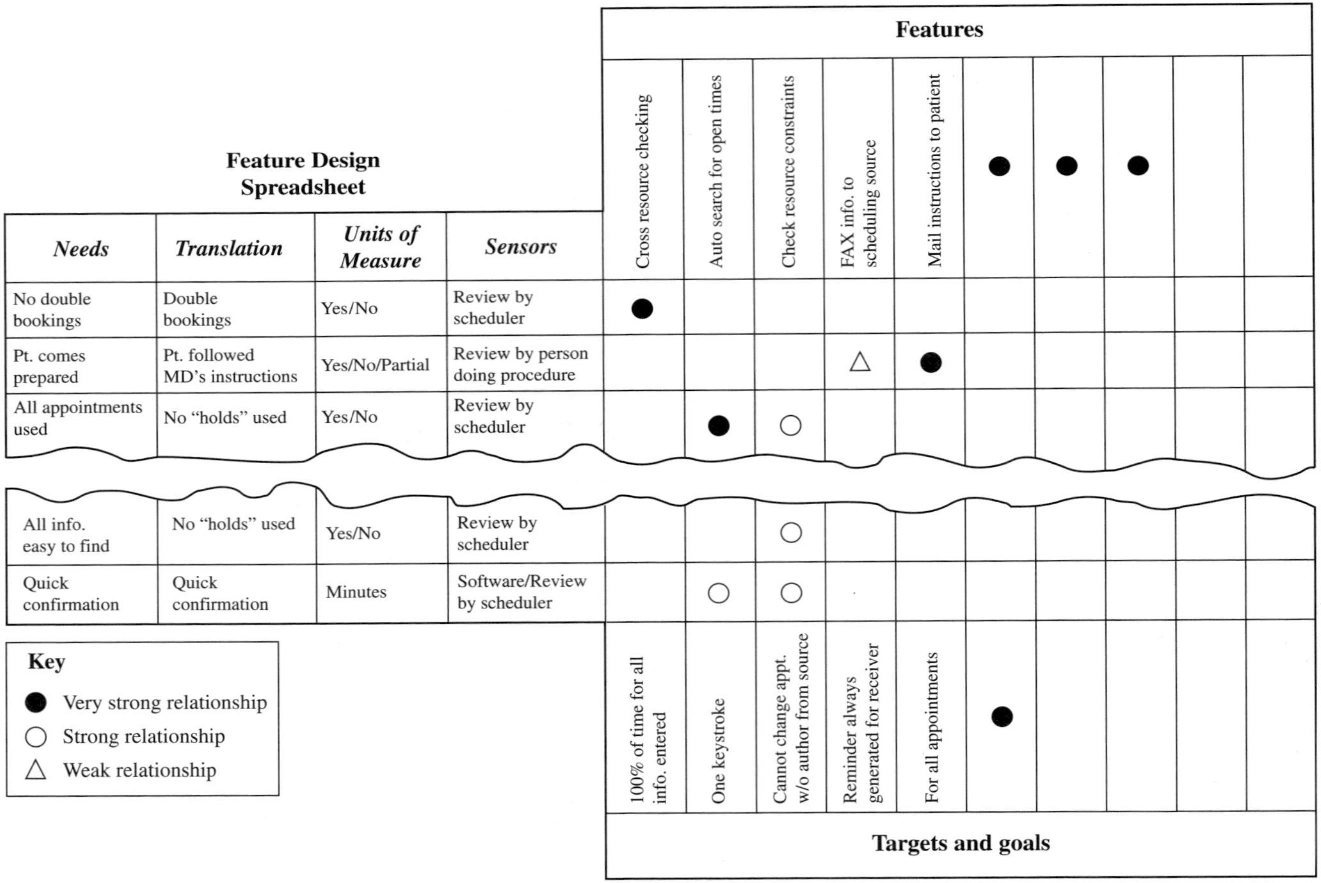

Feature Design Spreadsheet

Needs	*Translation*	*Units of Measure*	*Sensors*	Cross resource checking	Auto search for open times	Check resource constraints	FAX info. to scheduling source	Mail instructions to patient	●	●	●		
No double bookings	Double bookings	Yes/No	Review by scheduler	●									
Pt. comes prepared	Pt. followed MD's instructions	Yes/No/Partial	Review by person doing procedure				△	●					
All appointments used	No "holds" used	Yes/No	Review by scheduler		●	○							
All info. easy to find	No "holds" used	Yes/No	Review by scheduler			○							
Quick confirmation	Quick confirmation	Minutes	Software/Review by scheduler		○	○							
Targets and goals				100% of time for all info. entered	One keystroke	Cannot change appt. w/o author from source	Reminder always generated for receiver	For all appointments	●				

Key

- ● Very strong relationship
- ○ Strong relationship
- △ Weak relationship

Figure 4.8 Product design spreadsheet. (*Juran Institute, Inc. Copyright 1999.*)

contents of the box itself but also the instructions for installation and use and the help line for assistance. There are six major activities in this step:

1. Group together related customer needs.
2. Determine methods for identifying features.
3. Select high-level features and goals.
4. Develop detailed features and goals.
5. Optimize features and goals.
6. Set and publish final product design.

Group Together Related Customer Needs

Most quality by design projects will be confronted with a large number of customer needs. Based on the data developed in the preceding steps, the team can prioritize and group together those needs that relate to similar functionality. This activity does not require much time, but it can save a lot of time later. Prioritization ensures that the scarce resources of product development are spent most effectively on those items that are most important to the customer. Grouping related needs together allows the design team to "divide and conquer," with subteams working on different parts of the design. Such subsystem or component approaches to design, of course, have been common for years. What may be different here is that the initial focus is on the components of the customers' needs, not the components of the product. The component design for the product will come during the later activities in this step.

Determine Methods for Identifying Features

There are many complementary approaches for identifying the best product design for meeting customers' needs. Most design projects do not use all of them. Before starting to design, however, a team should develop a systematic plan for the methods it will use in its own design. Here are some of the options.

Benchmarking This approach identifies the best in class and the methods behind it that make it best. See Chapter 15, Benchmarking: Defining Best Practices for Market Leadership, for details.

Basic Research One aspect of research might be a new innovation for the product that does not currently exist in the market or with competitors. Another aspect of basic research looks at exploring the feasibility of the product and features. While both these aspects are important, be careful that fascination with the technological abilities of the product does not overwhelm the primary concern of its benefits to the customer.

Market Experiments Introducing and testing ideas for features in the market allows one to analyze and evaluate concepts. The focus group is one technique that can be used to measure customer reactions and determine whether the features actually will meet customer needs. Some organizations also try out their ideas, on an informal basis, with customers at trade shows and association meetings. Still others conduct limited test marketing with a prototype product.

Creativity Developing features allows one to dream about a whole range of possibilities without being hampered by any restrictions or preconceived notions. Design for quality is a proven, structured, data-based approach to meeting customers' needs. But this does not mean it is rigid and uncreative. At this point in the process, the participants in design must be encouraged and given the tools they need to be creative so as to develop alternatives for design. After they have selected a number of promising alternatives, they will use hard analysis and data to design the final product.

Design teams can take advantage of how individuals view the world: from their own perspective. Every employee potentially sees other ways of doing things. The team can encourage people to suggest new ideas and take risks. Team members should avoid getting "stuck" or taking too much time to debate one particular idea or issue. They can put it aside and come back to it later with a fresh viewpoint. They can apply new methods of thinking about customers' needs or problems, such as the following:

- *Change in key words or phrases.* For example, call a "need" or "problem" an "opportunity." Instead of saying, "Deliver on time," say, "Deliver exactly when needed."
- *Random association.* For example, take a common word such as "apple" or "circus" and describe your business, product, or problem as the word. For example, "Our product is like a circus because . . . "
- *Central idea.* Shift your thinking away from one central idea to a different one. For example, shift the focus from the product to the customer by saying, "What harm might a child suffer, and how can we avoid it?" rather than "How can we make the toy safer?"
- *Putting yourself in the other person's shoes.* Examine the question from the viewpoint of the other person, your competitor, your customer—and build their case before you build your own.
- *Dreaming.* Imagine that you had a magic wand that you could wave to remove all obstacles to achieving your objectives. What would it look like? What would you do first? How would it change your approach?
- *The spaghetti principle.* When you have difficulty considering a new concept or how to respond to a particular need, allow your team to be comfortable enough to throw out a new idea, as if you were throwing spaghetti against the wall, and see what sticks. Often even "wild" ideas can lead to workable solutions.

The initial design decisions are kept as simple as possible at this point. For example, the idea of placing the control panel for the radio on the steering wheel would be considered a high-level product feature. Its exact location, choice of controls, and how they function can be analyzed later in greater detail. It may become the subject of more detailed features as the design project progresses.

Standards, Regulations, and Policies This is also the time to be certain that all relevant standards, regulations, and policies have been identified and addressed. While some of these requirements are guidelines for how a particular product or product feature can perform, others mandate how they must perform. These may come from inside the organization, and others may come from specific federal, state, or local governments; regulatory agencies; or industry associations. All features and product feature goals must be analyzed against these requirements prior to the final selection of features to be included in the design.

It is important to note that if there is a conflict when evaluating features against any standards, policies, or regulations, it is not always a reason to give up. Sometimes one can work to gain acceptance for a change when it will do a better job of meeting customer needs. This is especially true when it comes to internal policies. However, an advocate for change must be prepared to back up the arguments with appropriate data.

Criteria for Design As part of the preparation for high-level design, the design team must agree on the explicit criteria to be used in evaluating alternative designs and design features. All designs must fulfill the following general criteria:

- Meet the customers' needs
- Meet the suppliers' and producers' needs
- Meet (or beat) the competition
- Optimize the combined costs of the customers and suppliers

In addition to the preceding four general criteria, the team members should agree explicitly on the criteria that they will use to make a selection. (If the choices are relatively complex, the team should consider using the formal discipline of a selection matrix.) One source for these criteria will be the team's goal statement and goals. Some other types of criteria that the team may develop could include

- The impact of the feature on the needs
- The relative importance of the needs being served
- The relative importance of the customers whose needs are affected
- The feasibility and risks of the proposed feature
- The impact on product cost
- The relationship to competitive features uncovered in benchmarking
- The requirements of standards, policies, regulations, mandates, and so on

As part of the decision on how to proceed with design, teams also must consider a number of other important issues regarding what type of product feature will be the best response to customers' needs. When selecting features, they need to consider whether to

- Develop an entirely new functionality
- Replace selected old features with new ones
- Improve or modify existing features
- Eliminate the unnecessary

Select High-Level Features and Goals

This phase of quality by design will stimulate the team to consider a whole array of potential features and how each would respond to the needs of the customer. This activity should be performed without being constrained by prior assumptions or notions as to what worked or did not work in the past. A response that previously failed to address a customer need or solve a customer problem might be ready to be considered again because of changes in technology or the market.

The team begins by executing its plan for identifying the possible features. It should then apply its explicit selection criteria to identify the most promising features.

The product design spreadsheet in Figure 4.8 is a good guide for this effort. Use the right side of the spreadsheet to determine and document the following:

- Which features contribute to meeting which customer needs
- That each priority customer need is addressed by at least one product feature
- That the total impact of the features associated with a customer need is likely to be sufficient for meeting that need

- That every product feature contributes to meeting at least one significant customer need
- That every product feature is necessary for meeting at least one significant customer need (i.e., removing that feature would leave a significant need unmet)

Team Sets Goals for Each Feature In quality terms, a goal is an aimed-at quality target (such as aimed-at values and specification limits). As discussed earlier, this differs from quality standards in that the standard is a mandated model to be followed that typically comes from an external source. While these standards serve as "requirements" that usually dictate uniformity or how the product is to function, product feature goals are often voluntary or negotiated. Therefore, the quality by design process must provide the means for meeting both quality standards and quality goals.

Criteria for Setting Product Feature Goals As with all goals, product feature goals must meet certain criteria. While the criteria for establishing product feature goals differ slightly from the criteria for project goals verified in step 1, there are many similarities. Product feature goals should encompass all the important cases and be

- Measurable
- Optimal
- Legitimate
- Understandable
- Applicable
- Attainable

Measuring Features Goals Establishing the measurement for a product feature goal requires the following tasks:

- Determine the unit of measure: meters, seconds, days, percentages, and so on.
- Determine how to measure the goal (i.e., determine what the sensor is).
- Set the value for the goal.

The work done in measuring customer needs should be applied now. The two sets of measurements may be related in one of the following ways:

- Measurement for the need and for the product feature goal may use the same units and sensors. For example, if the customer need relates to timeliness measured in hours, one or more features normally also will be measured in hours, with their combined effects meeting the customer need.
- Measurement for the product feature may be derived in a technical manner from the need measurement. For example, a customer need for transporting specified sizes and weights of loads may be translated into specific engineering measurements of the transport system.
- Measurement for the product feature may be derived from a customer behavioral relationship with the product feature measure. For example, automobile manufacturers have developed the specific parameters for the dimensions and structure of an automobile seat that translate into the customer rating it "comfortable."

Since we can now measure both the customer need and the related product feature goals, it is possible for the quality by design team to ensure that the product design will go a long way toward meeting the customers' needs, even before building any prototypes or conducting any test marketing.

For large or complex projects, the work of developing features is often divided among a number of different individuals and work groups. After all these groups have completed their work, the overall quality by design team will need to integrate the results. Integration includes

- Combining features when the same features have been identified for more than one cluster
- Identifying and resolving conflicting or competing features and goals for different clusters
- Validating that the combined design meets the criteria established by the team

Develop Detailed Features and Goals

For large and highly complex products, it will usually be necessary to divide the product into a number of components and even subcomponents for detailed design. Each component will typically have its own design team that will complete the detailed design described below. To ensure that the overall design remains integrated, consistent, and effective in meeting customer needs, these large, decentralized projects require

- A steering or core team that provides overall direction and integration
- Explicit charters with quantified goals for each component
- Regular integrated design reviews for all components
- Explicit integration of designs before completion of the product design phase

Once the initial detailed features and goals have been developed, then the technical designers will prepare a preliminary design, with detailed specifications. This is a necessary step before a team can optimize models of features using a number of quality by design tools and ultimately set and publish the final features and goals.

It is not uncommon for quality by design teams to select features at so high a level that the features are not specific enough to respond to precise customer needs. Just as in the identification of customers' primary needs, high-level features need to be broken down further into terms that are clearly defined and can be measured.

Optimize Features and Goals

Once the preliminary design is complete, it must be optimized. That is, the design must be adjusted so that it meets the needs of both customer and supplier while minimizing their combined costs and meeting or beating the competition.

Finding the optimum can be a complicated matter unless it is approached in an organized fashion and follows quality disciplines. For example, there are many designs in which numerous variables converge to produce a final result. Some of these designs are of a business nature, such as design of an information system involving optimal use of facilities, personnel, energy, capital, and so on. Other such designs are technological, involving optimization of the performance of hardware. Either way, finding the optimum is made easier through the use of certain quality disciplines.

Finding the optimum involves balancing the needs, whether they are multiorganizational needs or within-organization needs. Ideally, the search for the optimum should be done through the participation of suppliers and customers alike. There are several techniques that help achieve this optimum.

Design Review Under this concept, those who will be affected by the product are given the opportunity to review the design during various formative stages. This allows them to use their experience and expertise to make such contributions as

- Early warning of upcoming problems
- Data to aid in finding the optimum
- Challenge to theories and assumptions

Design reviews can take place at different stages of development of the new product. They can be used to review conclusions about customer needs and hence the product specifications (characteristics of product output). Design reviews also can take place at the time of selecting the optimal product design. Typical characteristics of design reviews include the following:

- Participation is mandatory.
- Reviews are conducted by specialists, external to the design team.
- Ultimate decisions for changes remain with the design team.
- Reviews are formal, scheduled, and prepared for with agendas.
- Reviews will be based on clear criteria and predetermined parameters.
- Reviews can be held at various stages of the project.

Ground rules for good design reviews include

- Adequate advance design review of agenda and documents
- Clearly defined meeting structure and roles
- Recognition of interdepartmental conflicts in advance
- Emphasis on constructive—not critical—inputs
- Avoidance of competitive design during review
- Realistic timing and schedules for the reviews
- Sufficient skills and resources provided for the review
- Discussion focus on untried/unproved design ideas
- Participation directed by management

Multifunctional Design Teams Design teams should include all those who have a vested interest in the outcome of the design of the product along with individuals skilled in product design. Under this concept, the team, rather than just the product designers, bears responsibility for the final design.

Structured Negotiation Customers and suppliers are tugged by powerful local forces to an extent that can easily lead to a result other than the optimum. To ensure that these negotiating sessions proceed in as productive a fashion as possible, it is recommended that ground rules be established before the meetings. Here are some examples:

- The team should be guided by a spirit of cooperation, not competition, toward the achievement of a common goal.

- Differences of opinion can be healthy and can lead to a more efficient and effective solution.
- Everyone should have a chance to contribute, and every idea should be considered.
- Everyone's opinions should be heard and respected without interruptions.
- Avoid getting personal; weigh the pros and cons of each idea, looking at its advantages before its disadvantages.
- Challenge conjecture; look at the facts.
- Whenever the discussion bogs down, go back and define areas of agreement before discussing areas of disagreement.
- If no consensus can be reached on a particular issue, it should be tabled and returned to later on in the discussion.

Create New Options Often teams approach a product design with a history of how things were done in the past. Optimization allows a team to take a fresh look at the product and create new options. Some of the most common and useful quality tools for optimizing the design include the following:

Competitive analysis provides feature-by-feature comparison with competitors' products. (See below for an example.)

Salability analysis evaluates which features stimulate customers to be willing to buy the product and the price they are willing to pay. (See below for an example.)

Value analysis calculates not only the incremental cost of specific features of the product but also the cost of meeting specific customer needs and compares the costs of alternative designs. (See below for an example.)

Criticality analysis identifies the "vital few" features that are vulnerable in the design so that they can receive priority for attention and resources.

Failure mode and effect analysis (FMEA) calculates the combined impact of the probability of a particular failure, the effects of that failure, and the probability that the failure can be detected and corrected, thereby establishing a priority ranking for designing in failure prevention countermeasures.

Fault-tree analysis aids in the design of preventive countermeasures by tracing all possible combinations of causes that could lead to a particular failure.

Design for manufacture and assembly evaluates the complexity and potential for problems during manufacture to make assembly as simple and error-free as possible.

Design for maintainability evaluates particular designs for the ease and cost of maintaining them during their useful life.

Competitive Analysis Figure 4.9 is an example of how a competitive analysis might be displayed. The data for a competitive analysis may require a combination of different approaches such as laboratory analysis of the competitors' products, field testing of those products, or in-depth interviews and on-site inspections where willing customers are using a competitor's product.

Note that by reviewing this analysis, the design team can identify those areas in which the design is vulnerable to the competition as well as those in which the team has developed an advantage. Based on this analysis, the team will then need to make optimization choices about whether to upgrade the product. The team may need to apply a value analysis to make some of these choices.

Product Feature & Goal	Check if Product Feature Is Present			Feature Performance vs. Goal (*)			Identify if Significant Risk or Opportunity
	Product A	Product B	Ours	Product A	Product B	Ours	
Retrieve messges from all touch tone phones easily	Yes	Yes	Yes	4	5	4	—
Change message from any remote location	Yes	No	Yes	3	N/A	5	O
2 lines built in	No	No	Yes	N/A	N/A	4	O
Below Add Features in Competitors' Product Not Included in Ours	**Check if Product Feature Is Present**			**Feature Performance vs. Goal (*)**			**Identify if Significant Risk or Opportunity**
	Product A	**Product B**	**Ours**	**Product A**	**Product B**	**Ours**	
No cassette used to record message	Yes	Yes		4	N/A		R
Telephone and answering machine in one unit	Yes	Yes		3	4		R

Legend (*)
1 = Poor
2 = Fair
3 = Satisfactory
4 = Good
5 = Excellent

Figure 4.9 Competitive analysis. (*Juran Institute, Inc. Copyright 1994.*)

Salability Analysis An example of salability analysis is shown in Figure 4.10. This analysis is similar to a competitive analysis, except that the reference point is the response of customers to the proposed design rather than a comparison with the features of the competitors' designs. Note, however, that elements of competitive and salability analyses can be combined, with the salability analysis incorporating customer evaluation of both the proposed new design and existing competitive designs.

Complex products, such as automobiles, with multiple optional features and optional configurations offer a unique opportunity to evaluate salability. Observed installation rates of options on both the existing car line and competitors' cars provide intelligence on both the level of market demand for the feature and the additional price that some segments of the market will pay for the feature, although the other segments of the market may place little or no value on it.

Value Analysis Value analysis has been quite common in architectural design and the development of custom-engineered products, but it also can be applied successfully to other environments as well, as illustrated in Figure 4.11. By comparing the costs for meeting different customer needs, the design team can make a number of significant optimization decisions. If the cost for meeting low-priority needs is high, the team must explore alternative ways to meet those needs and even consider not addressing them at all if the product is highly price-sensitive. If very important needs have not consumed much of the expense, the team will want to make certain that it has met those needs fully and completely. While low expense for

Name of Product: Car Repair Service — Tune-Up	How Do Customers Rate Product? Poor Fair Satisfactory Good Excellent	Basis for Rating Prior Use vs. Opinion	How Do Customers See Differences between Our Products and Competing Products? Positively (+) Negatively (–) No difference (O)	Would Customers Buy if Price Were Not Important? Yes No	Would Customer Buy if Price Were Important?		Of All Products Listed, Prioritize Which Would Customers Buy and Its Basis? Price Features	Identify if Significant Risk or Opportunity
					Price	Yes No		
Ours —	E	U		Y	$175	Y	2-F	
Competitor A —	G	O	+	N	$145	Y	3-P	O
Competitor B —	E	U	O	Y	$175	Y	1-F	R

Product Feature: Pick-Up and Delivery of Car to Be Repaired Product Feature Goal: Same Day Service	How Would Customers Rate Featurs? Poor Fair Satisfactory Good Excellent	Basis for Rating Prior Use vs. Opinion	How Do Customers See Diff. between Our Features against Competing Features? Positively (+) Negatively (–) No Difference (O)	Does the Addition of the Feature Make the Product: More Salable (+) Less Salable (–) No Difference (O)	Identify if Significant Risk or Opportunity
Ours — Offered	G	U		O	
Competitor A — Not offered	S	O	+	—	O
Competitor B — Offered. Also provides loaner car to customer	E	U	—	+	R

Figure 4.10 Salability analysis. (*Juran Institute, Inc. Copyright 1994. Used by permission.*)

Customer Need (listed in priority order)	Product Feature and Goals						Cost of Meeting Need
	Walk in appointments handled by nurse, 5 days a week	Board certified obstetrician, 2 days a week	Social worker, 5 days a week	Nutritional counselor, 5 days a week	On-site billing clerk takes medicaid insurance from all eligible patients	On-site laboratory–most results under 1 hour	
Convenient to use	60,000	30,000	10,000	10,000	20,000	40,000	170,000
Confidence in staff		70,000	10,000	15,000			95,000
Reasonable cost						25,000	25,000
Sensitivity			15,000	5,000			20,000
Informed choices			5,000	15,000			20,000
Cost for feature	60,000	100,000	40,000	45,000	20,000	65,000	330,000

FIGURE 4.11 Value analysis spreadsheet. (*Juran Institute, Inc. Copyright 1994. Used by permission.*)

meeting a high-priority need is not necessarily inappropriate, it does present the designers with the challenge of making certain that lower-priority needs are not being met by using resources that could be better directed toward the higher-priority needs. It is not uncommon for products to be overloaded with "bells and whistles" at the expense of the fundamental functionality and performance.

Set and Publish Final Product Design

After the design has been optimized and tested, it is time to select the features and goals to be included in the final design. This is also the stage where the results of product development are officially transmitted to other functions through various forms of documentation. These include the specifications for the features and product feature goals as well as the spreadsheets and other supporting documents. All this is supplemented by instructions, both oral and written. To complete this activity, the team must first determine the process for authorizing and publishing features and product feature goals. Along with the features and goals, the team should include any procedures, specifications, flow diagrams, and other spreadsheets that relate to the final product design. The team should pass along results of experiments, field testing, prototypes, and so on that are appropriate. If an organization has an existing process for authorizing product goals, it should be reexamined in light of recent experience. Ask these questions: Does the authorization process guarantee input from key customers—both internal and external? Does it provide for optimization of the design? If an organization has no existing goal authorization process, now is a good time to initiate one.

Step 5: Develop the Process Features

Once the product is designed and developed, it is necessary to determine the means by which the product will be created and delivered on a continuing basis. These means are, collectively, the *process*. *Process development* is the set of activities for defining the specific means to be used by operating personnel for meeting product quality goals. Some related concepts include

- Subprocesses: Large processes may be decomposed into these smaller units for both the development and operation of the process.
- Activities: These are steps in a process or subprocess.
- Tasks: These comprise detailed step-by-step description for execution of an activity.

For a process to be effective, it must be goal-oriented, with specific measurable outcomes; systematic, with the sequence of activities and tasks fully and clearly defined and all inputs and outputs fully specified; and capable, i.e., able to meet product quality goals under operating conditions; and legitimate, with clear authority and accountability for its operation.

The 11 major activities involved in developing a process are as follows:

- Review product goals.
- Identify operating conditions.
- Collect known information on alternate processes.
- Select general process design.
- Identify process features and goals.
- Identify detailed process features and goals.
- Design for critical factors and human error.
- Optimize process features and goals.
- Establish process capability.
- Set and publish final process features and goals.
- Set and publish final process design.

Review Product Goals

Ideally, this review will be relatively simple. Product quality goals should have been validated with the prior participation of those who would be affected. In many organizations, however, product design and process design often are executed by different teams. There is no real joint participation on either group's part to contribute to the results that both teams are expected to produce. This lack of participation usually reduces the number of alternative designs that could have been readily adopted in earlier stages but become more difficult and more expensive to incorporate later. In addition, those who set the product goals have a vested interest in their own decisions and exhibit cultural resistance to proposals by the process design team to make changes to the product design. If the product and process design efforts are being performed by different groups, then review and confirmation of the product quality goals are absolutely critical.

Review of product quality goals ensures that they are understood by those most affected by the process design. The review helps achieve the optimum. Process designers are able to present product designers with some realities relative to the costs of meeting the quality goals. The review process should provide a legitimate, unobstructed path for challenging costly goals.

Identify Operating Conditions

Seeking to understand operating conditions requires investigation of a number of dimensions.

User's Understanding of the Process By "users," we mean either those who contribute to the processes in order to meet product goals or those who employ the process to meet their own needs. Users consist, in part, of internal customers (organization units or persons) responsible for running the processes to meet the quality goals. Operators or other workers are users. Process planners need to know how these people will understand the work to be done. The process must be designed either to accommodate this level of understanding or to improve the level of understanding.

How the Process Will Be Used Designers always know the intended use of the process they develop. However, they may not necessarily know how the process is actually used (and misused) by the end user. Designers can draw on their own experiences but usually must supplement these with direct observation and interviews with those affected.

The Environments of Use Planners are well aware that their designs must take account of environments that can influence process performance. Planners of physical processes usually do take account of such environmental factors as temperature, vibration, noise level, and so on. Planners who depend heavily on human responses, particularly those in the service areas, should address the impact of the environment on human performance in their process designs. For example, a team designing the process for handling customer inquiries should consider how environmental stress can influence the performance of the customer service representatives. This stress can result from large numbers of customer complaints, abusive customers, lack of current product information, and so on.

Collect Known Information on Alternative Processes Once the goals and environment are clear, the design team needs reliable information on alternative processes available for meeting those goals in the anticipated environment.

Process Anatomy At the highest level, there are some basic process anatomies that have specific characteristics that planners should be aware of. A *process anatomy* is a coherent structure that binds or holds the process together. This structure supports the creation of the goods or the delivery of the service. The selection of a particular anatomy also will have a profound influence on how the product is created and the ability of the organization to respond to customers' needs. Figure 4.12 illustrates these.

The Assembly Tree The *assembly tree* is a familiar process that incorporates the outputs of several subprocesses. Many of these are performed concurrently and are required for final assembly or to achieve an end result at or near the end of the process. This kind of process anatomy is widely used by the great mechanical and electronic industries that build automotive vehicles, household appliances, electronic apparatus, and so on. It is also used to define many processes in a hospital, such as in the case of performing surgery in the operating room. The branches or leaves of the tree represent numerous suppliers or in-house departments making parts and components. The elements are assembled by still other departments.

In the office, certain processes of data collection and summary also exhibit features of the assembly tree. Preparation of major accounting reports (e.g., balance sheet, profit statement) requires assembly of many bits of data into progressively broader summaries that finally converge into the consolidated reports. The assembly tree design has been used at

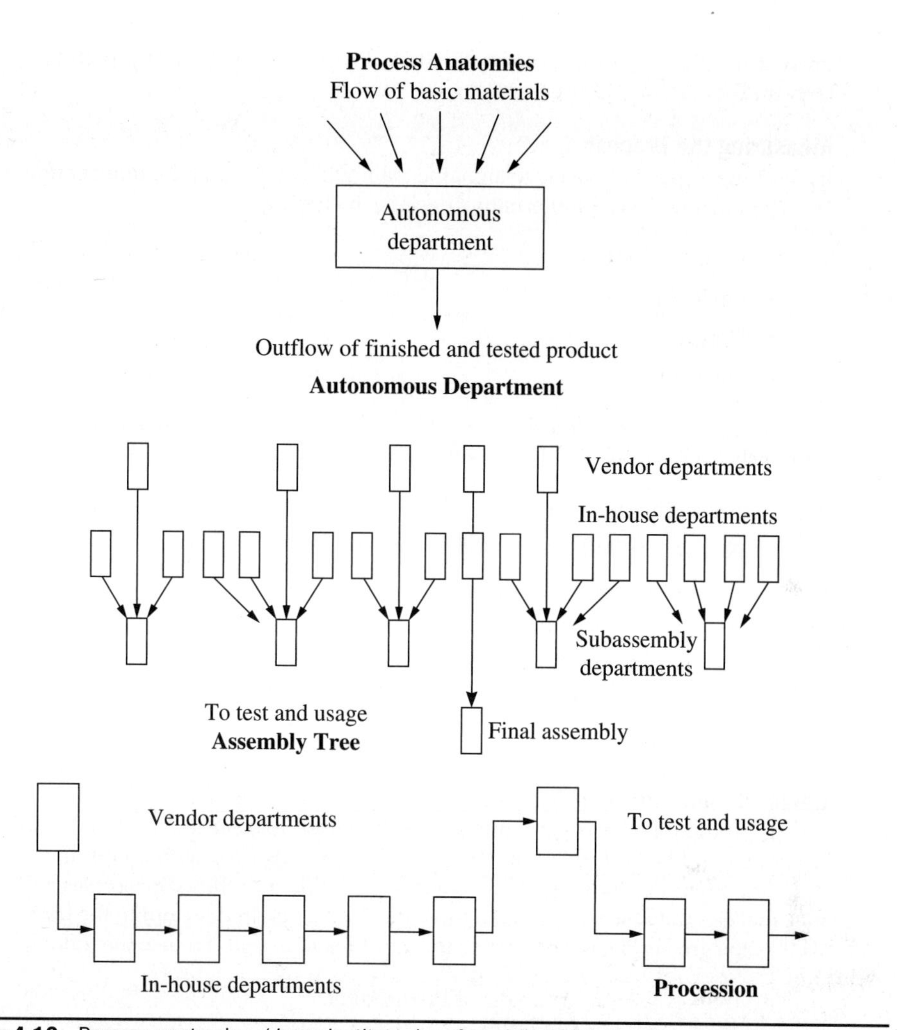

FIGURE 4.12 Process anatomies. (*Juran Institute, Inc. Copyright 1994. Used by permission.*)

both the multifunctional and departmental levels. In large operations, it is virtually mandatory to use staff specialists who contribute different outputs at various multifunctional levels. An example of this is the budget process. While it is not mandatory to use staff specialists for large departmental processes, this is often the case. This can be illustrated by the design department, where various design engineers contribute drawings of a project that contribute to the overall design.

Business Process Quality Management to Hold the Gains

Increasingly, many planners are applying a fourth, less traditional form of management known as *business process quality management* to their major processes. This new, alternative management form has come about in response to an increased realization that many of today's business goals and objectives are becoming even more heavily dependent on large, complex, cross-functional business processes. Process quality management emphasizes that there are several critical processes that are crucial to an organization if it is to maintain and grow its business. (See Chapter 8, Business Process Management: Creating an Adaptable Organization, for a full discussion.)

Measuring the Process

In selecting a specific process design, the team will need to acquire information on the effectiveness and efficiency of alternative designs, including

- Deficiency rates
- Cycle time
- Unit cost
- Output rate

To acquire the needed data, the planners must typically use a number of different approaches, including

- Analyzing the existing process
- Analyzing similar or related processes
- Testing alternative processes
- Analyzing new technology
- Acquiring information from customers
- Simulating and estimating
- Benchmarking

Select General Process Design

Just as product design began with a high-level description expanded to the details, process design should begin by describing the overall process flow with a high-level process flow diagram. From this diagram it will be possible to identify the subprocesses and major activities that can then be designed at a more detailed level. In developing the high-level flow as well as the greater detail later, the team should ensure that it meets the following criteria:

- Deliver the quality goals for the product.
- Incorporate the countermeasures for criticality analysis, FMEA, and fault-tree analysis.
- Meet the project goals.
- Account for actual use, not only intended use.
- Be efficient in consumption of resources.
- Demand no investments that are greater than planned.

While some process designs will largely repeat existing designs and other process designs will represent "green field" or "blank sheet" redesigns, most effective process redesigns are a combination of the tried and true existing processes with some significant quantum changes in some parts of the process.

The preceding criteria should be the guidelines for whether a particular part of the process should be incorporated as it is, improved, or replaced with a fundamentally different approach.

This is the point in process design to think as creatively as possible, using some of the same techniques discussed under product development. Consider the impact of radically different anatomies. Would the customer be served better with dedicated, multispecialty units or with highly specialized expert functionality accessed as needed? What approach is mostly likely to reduce failures? How can cycle time by cut dramatically? Is there a new technology that would allow us to do it differently? Can we develop such a technology?

Once the high-level flow is completed, each activity and decision within the flow diagram needs to be fully documented with a specification of the following for each:

- Inputs
- Outputs
- Goals for outputs
- Cycle time
- Cost
- General description of the conversion of inputs to outputs

Clear specification of these factors makes it possible to divide up the work of detailed design later and still be confident that the final design will be consistent and coordinated.

Once the initial new process flow is completed, it should be reviewed for opportunities to improve it, such as these:

- Eliminate sources of error that lead to rework loops.
- Eliminate or reduce redundant subprocesses, activities, or tasks.
- Decrease the number of handoffs.
- Reduce cycle time.
- Replace tasks, activities, or processes that have outputs with defects.
- Correct sequencing issues in the process to reduce the amount of activity or rework.

Testing Selected Processes

One of the key factors for a successful design is the incorporation of the lessons learned from testing the product, the features, and the overall process and subprocesses to ensure that they meet quality goals. Testing should be conducted throughout the entire quality by design process to allow for changes, modifications, and improvements to the plan before it is transferred to operations. Testing is performed at various points to analyze and evaluate alternate designs of the overall process and subprocesses.

Comparisons or Benchmarks

Other units inside and outside the organization may already be using a process similar to the one designed. The process can be validated by comparing it with existing similar processes.

Test Limitations

All tests have some limitations. The following are common limitations that should be understood and addressed.

Differences in Operating Conditions Dry runs and modular testing obviously differ from operating conditions. Even pilot tests and benchmarks will differ in some details from the actual, full implementation. Some common differences between conditions for testing and conditions for full-scale use include

- People operating the process
- Customers of the process

- Extreme values and unusual conditions
- Interactions with other processes and other parts of the organization

Differences in Size Especially with critical failures, such as breakdown of equipment, loss of key personnel, or any other potential failure, as in the case of complications in a surgical procedure, a test might not be large enough to allow these rare failures to occur with any high degree of certainty.

Other Effects Sometimes designing a new process or redesigning an existing process may create or exacerbate problems in other processes. For example, improved turnaround time in approving home loans may create a backlog for the closing department. Such interactions among processes might not occur in an isolated test.

Identify Process Features and Goals

A *process feature* is any property, attribute, and so on that is needed to create the goods or deliver the service and achieve the product feature goals that will satisfy a customer need. A *process goal* is the numeric target for one of the features.

Whereas features answer the question "What characteristics of the product do we need to meet customers needs?" process features answer the question "What mechanisms do we need to create or deliver those characteristics (and meet quality goals) over and over without failures?" Collectively, process features define a process. The flow diagram is the source of many of, but not all, these features and goals.

As the process design progresses from the macro level down into details, a long list of specific process features emerges. Each of these is aimed directly at producing one or more features. For example:

- Creating an invoice requires a process feature that can perform arithmetic calculations so that accurate information can be added.
- Manufacturing a gear wheel requires a process feature that can bore precise holes into the center of the gear blank.
- Selling a credit card through telemarketing requires a process feature that accurately collects customer information.

Most process features fall into one of the following categories:

- Procedures—a series of steps followed in a regular, definite order
- Methods—an orderly arrangement of a series of tasks, activities, or procedures
- Equipment and supplies—"physical" devices and other hard goods that will be needed to perform the process
- Materials—tangible elements, data, facts, figures, or information (these, along with equipment and supplies, also may make up inputs required as well as what is to be done to them)
- People—numbers of individuals, skills they will require, goals, and tasks they will perform
- Training—skills and knowledge required to complete the process
- Other resources—additional resources that may be needed
- Support processes—secretarial support, occasionally other support, such as outsources of printing services, copying services, temporary help, and so on

		Process Features			
Product Feature	Product Feature Goal	Spray delivery capacity	Crew Size	Certified materials	Scheduling forecast on PC to determine to/from and work needed
Time to perform job	Less than one hour 100 percent of time	○	●		●
Guaranteed appointment time	99 percent of jobs within 15 min. of appointment				●
All materials environmentally safe	All naturally occuring/no synthetics			●	
	Process Feature Goals	10 gallons per minute	One person per 10,000 sq. ft. of yd.	100% approved by State Dept. of Agriculture	Forecast time always within 10 percent of actual

Legend ● Very strong ○ Strong △ Weak

Figure 4.13 Process design spreadsheet. (*Juran Institute, Inc. Copyright 1994. Used by permission.*)

Just as in the case of product design, process design is easier to manage and optimize if the process features and goals are organized into a spreadsheet indicating how the process delivers the features and goals. Figure 4.13 illustrates such a spreadsheet.

The spreadsheet serves not only as a convenient summary of the key attributes of the process, it also facilitates answering two key questions that are necessary for effective and efficient process design. First, will every product feature and goal be attained by the process? Second, is each process feature absolutely necessary for at least one product feature; i.e., are there any unnecessary or redundant process features? Also, verify that one of the other process features cannot be used to create the same effect on the product.

Often high-level process designs will identify features and goals that are required from organization wide macro processes. Examples might include cycle times from the purchasing process, specific data from financial systems, and new skills training. Because the new process will depend on these macro processes for support, now is the time to verify that they are capable of meeting the goals. If they are not, the macro processes will need to be improved as part of the process design, or they will need to be replaced with an alternative delivery method.

Identify Detailed Process Features and Goals

In most cases, it will be most efficient and effective for individual subteams to carry out the detailed designs of subprocesses and major activities. These detailed designs will have the process features and goals as their objectives and criteria. Each subprocess team will develop the design to the level at which standard operating procedures can be developed, software coded, equipment produced or purchased, and materials acquired.

Design for Critical Factors and Human Error

One key element of process design is determining the effect that critical factors will have on the design. "Critical factors" are those aspects which present serious danger to human life, health, and the environment or risk the loss of very large sums of money. Some examples of such factors involve massive scales of operations: airport traffic control systems, huge construction projects, systems of patient care in hospital, and even the process for managing the stock market. Design for such factors should obviously include ample margins of safety as to structural integrity, fail-safe provisions, redundancy systems, multiple alarms, and so on. Criticality analysis and failure-mode and effect analysis (see Chapter 19, Accurate and Reliable Measurement Systems and Advanced Tools) are helpful tools in identifying those factors which require special attention at this point.

Workers vary in their capabilities to perform specific tasks and activities. Some workers perform well, whereas others do not perform nearly as well. What is consistent about all workers is that they are a part of the human family, and human beings are fallible. Collectively, the extent of human errors is large enough to require that the process design provides for means to reduce and control human error. Begin by analyzing the data on human errors, and then apply the Pareto principle. The vital few error types individually become candidates for special process design. The human errors that can be addressed by process design fall into these major classes:

- Technique errors arising from individuals lacking specific, needed skills
- Errors aggravated by lack of feedback
- Errors arising from the fact that humans cannot remain indefinitely in a state of complete, ready attention

Principles of Mistake Proofing

Research has indicated that there are a number of different classifications of error proofing methods, and these are spelled out below. *Elimination* consists of changing the technology to eliminate operations that are error-prone. For example, in some materials handling operations, the worker should insert a protective pad between the lifting wire and the product so that the wire will not damage the product. Elimination could consist of using nylon bands to do the lifting.

Optimize Process Features and Goals

After the planners have designed for critical factors and made modifications to the plan for ways of reducing human error, the next activity is to optimize first the subprocesses and then the overall process design. In step 4, develop product, the concept of optimization was introduced. The same activities performed for optimizing features and product feature goals also apply to process planning. Optimization applies to both the design of the overall process and the design of individual subprocesses.

Establish Process Capability

Before a process begins operation, it must be demonstrated to be capable of meeting its quality goals. The concepts and methods for establishing process capability are discussed in detail in Chapter 20, Product-Based Organizations: Delivering Quality While Being Lean and Green, under Process Capability. Any design project must measure the capability of its process with respect to the key quality goals. Failure to achieve process capability should be followed by systematic diagnosis of the root causes of the failure and improvement of the process to eliminate those root causes before the process becomes operational.

Reduction in Cycle Time

Process capability relates to the effectiveness of the process in meeting customer needs. One special class of needs may relate to subprocess cycle time—the total time elapsed from the beginning of a process to the end. Reducing cycle time has almost become an obsession for many organizations. Pressures from customers, increasing costs, and competitive forces are driving organizations to discover faster ways of performing their processes. Often these targeted processes include launching new products, providing service to customers, recruiting new employees, responding to customer complaints, and so on. For existing processes, designers follow the well-known quality improvement process to reduce cycle time. Diagnosis identifies causes for excessive time consumption. Specific remedies are then developed to alleviate these causes.

Set and Publish Final Process Features and Goals

After the design team has established the flow of the process, identified initial process features and goals, designed for critical processes and human error, optimized process features and goals, and established process capabilities, it is ready to define all the detailed process features and goals to be included in the final design. This is also the stage where the results of process development are officially transmitted to other functions through various forms of documentation. These include the specifications for the features and product feature goals as well as the spreadsheets and other supporting documents. All this is supplemented by instructions, both oral and written.

Filling out the process design spreadsheet is an ongoing process throughout process development. The spreadsheet should have been continually updated to reflect design revisions from such activities as reviewing alternative options, designing for critical factors and human error, optimizing, testing process capability, and so on. After making the last revision to the process design spreadsheet, it should be checked once more to verify the following:

- Each product feature has one or more process features with strong or very strong relation. This will ensure the effective delivery of the product feature without significant defects. Each product feature goal will be met if each process goal is met.
- Each process feature is important to the delivery of one or more features. Process features with no strong relationship to other features are unnecessary and should be discarded.

The completed process design spreadsheet and detailed flow diagrams are the common information needed by managers, supervisors, and workers throughout the process. In addition, the design team must ensure that the following are specified for each task within the process:

- Who is responsible for doing it
- How the task is to be competed
- Its inputs
- Its outputs
- Problems that can arise during operations and how to deal with them
- Specification of equipment and materials to be used
- Information required by the task
- Information generated by the task
- Training, standard operating procedures, job aids that are needed

Step 6: Develop Process Controls and Transfer to Operations

In this step, planners develop controls for the processes, arrange to transfer the entire product plan to operational forces, and validate the implementation of the transfer. There are seven major activities in this step.

1. Identify controls needed.
2. Design feedback loop.
3. Optimize self-control and self-inspection.
4. Establish audit.
5. Demonstrate process capability and controllability.
6. Plan for transfer to operations.
7. Implement plan and validate transfer.

Once design is complete, these plans are placed in the hands of the operating departments. It then becomes the responsibility of the operational personnel to manufacture the goods or deliver the service and to ensure that quality goals are met precisely and accurately. They do this through a planned system of quality control. Control is largely directed toward continuously meeting goals and preventing adverse changes from affecting the quality of the product. Another way of saying this is that no matter what takes place during production (change or loss of personnel, equipment or electrical failure, changes in suppliers, etc.), workers will be able to adjust or adapt the process to these changes or variations to ensure that quality goals can be achieved.

Identify Controls Needed

Process control consists of three basic activities:

- Evaluate the actual performance of the process.
- Compare actual performance with the goals.
- Take action on the difference.

Detailed discussions of these activities in the context of the feedback loop are contained in Chapter 6, Quality Control: Assuring Repeatable and Compliant Processes.

Process Features

Much control consists of evaluating those process features that most directly affect the features, e.g., the state of the toner cartridge in the printer, the temperature of the furnace for smelting iron, or the validity of the formulas used in the researcher's report. Some features become candidates for control subjects as a means of avoiding or reducing failures. These control subjects typically are chosen from previously identified critical factors or from conducting FMEA, fault-tree analysis (FTA), and criticality analysis. Process controls are associated with the decision: Should the process run or stop?

- Setting the standards for control, i.e., the levels at which the process is out of control and the tools, such as control charts, that will be used to make the determination
- Deciding what action is needed when those standards are not met, e.g., troubleshooting
- Designating who will take those actions

	Process Controls						
Product feature	Control subject	Sensor	Goal	Measurement frequency	Sample size	Criterion	Responsibility
Process feature 1							
Process feature 2							
– – –							
Wave solder	Solder temperature	Thermo- couple	505°F	Continuous	n/a	≥510°F decrease heat 500°F increase heat	Operator
	Conveyor speed	ft/min meter	4.5 ft/min	1/hour	n/a	≥5 ft/min reduce speed ≤4 ft/min increase speed	Operator
	Alloy purity	Lab chem analysis	1.5% max total contaminants	1/month	15 grams	≥1.5% drain bath, replace solder	Process engineer

Figure 4.14 Control spreadsheet. (*From J. M. Juran,* Quality Control Handbook, *5th ed., McGraw-Hill, New York, 1999, p. 3.48.*)

A detailed process flow diagram should be used to identify and document the points at which control measurements and actions will be taken. Then each control point should be documented on a control spreadsheet similar to Figure 4.14.

Training

Workers should be trained to make the product conformance decisions and should also be tested to ensure that they make good decisions. Specifications must be unequivocally clear.

The quality audit and audit of control systems are treated elsewhere in detail; see, e.g., Chapter 20, Product-Based Organizations: Delivering Quality While Being Lean and Green, under Audit of Operations Quality. While the audit of a control system is a function independent of the design team, the design team does have the responsibility for ensuring that adequate documentation is available to make an effective audit possible and that there are provisions of resources and time for conducting the audit on an ongoing basis.

Demonstrate Process Capability and Controllability

While process capability must be addressed during the design of the process, it is during implementation that initial findings of process capability and controllability must be verified.

Plan for Transfer to Operations

In many organizations, receipt of the process by operations is structured and formalized. An information package is prepared consisting of certain standardized essentials: goals to be met, facilities to be used, procedures to be followed, instructions, cautions, and so on. There are also supplements unique to the project. In addition, provision is made for briefing and training the operating forces in such areas as maintenance, dealing with crisis, and so on.

The package is accompanied by a formal document of transfer of responsibility. In some organizations, this transfer takes place in a near-ceremonial atmosphere.

The Structured Approach Has Value It tends to evolve checklists and countdowns that help ensure that the transfer is orderly and complete. If the organization already has a structure for transfer, project information may be adapted to conform to established practice. If the organization has a loose structure or none at all, the following material will aid in design the transfer of the project.

Regardless of whether the organization has a structure, the team should not let go of the responsibility of the project until it has been validated that the transfer has taken place and everyone affected has all the information, processes, and procedures needed to produce the final product.

Transfer of Know-How During process design, the planners acquire a great deal of know-how about the process. The operating personnel could benefit from this know-how if it were transferred. There are various ways of making this transfer, and most effective transfers make use of several complementary channels of communication, including

- Process specifications
- Briefings
- On-the-job training
- Formal training courses
- Prior participation

Audit Plan for the Transfer As part of the plan for formal transfer, a separate audit plan should be developed as a vehicle for validating the transfer of the plan. This kind of audit is different from the control audits described previously. The purpose of this audit is to evaluate how successful the transfer was. For the audit to have real meaning, specific goals should be established during the design phase of the transfer. Generally, these goals relate to the quality goals established during the development of the product, features, and process features. The team may decide to add other goals inherent to the transfer or to modify newly planned quality goals during the first series of operations. For example, during the first trial runs for producing the product, total cycle time may exceed expected goals by 15 percent. This modification takes into account that workers may need time to adjust to the plan. As they become more skilled, gain experience with the process, and get more comfortable with their new set of responsibilities, cycle time will move closer to targeted quality goals. The audit plan for the transfer should include the following:

- Goals to meet
- How meeting the goals will be measured
- The time phasing for goals, measurement, and analysis
- Who will audit
- What reports will be generated
- Who will have responsibility for corrective action for failure to meet specific goals

Implement Plan and Validate Transfer

The final activity of the quality by design process is to implement the plan and validate that the transfer has occurred. A great deal of time and effort has gone into creating the product plan, and validating that it all works is well worth the effort.

Frequently Used Design Tools

- *Affinity diagrams.* This diagram clusters together items of similar type, is a prelude to a cause-effect diagram used in quality improvement, and is used in quality design to group together similar needs or features.
- *Benchmarking.* This technique involves openly sharing and investigating the best practices of organizations, largely for business and internal processes (not for competitive or proprietary manufacturing). In today's world, this has improved from "industrial tourism" to research, largely through participation in online databases.
- *Brainstorming.* This popular technique obtains group ideas as to cause (for improvement) or as to features (for planning).
- *Carryover analyses.* Usually a matrix depicts the degree of carryover of design elements, with particular regard to failure proneness.
- *Competitive analyses.* Usually a matrix depicts a feature-by-feature comparison to the competition, with particular regard to "best-in-class" targets.
- *Control chart.* This is a widely used depiction of process change over time. The most popular is the Shewhart control chart for averages.
- *Criticality analyses.* Usually a matrix depicts the degree of failure of a feature or component against the ranking of customer needs, along with responsibilities detailed for correction.
- *Data collection: focus group.* This popular technique places customers in a setting led by a trained facilitator to probe for the understanding of needs.
- *Data collection: market research.* Any of a variety of techniques aim at answering the three fundamental questions: What is important to the users? What is the order of the items of importance? How well do we do in meeting them in that order, as compared to the competition?
- *Data collection: surveys.* This passive technique elicits answers to preset questions about satisfaction or needs. Usually it is "closed-ended," with meager space for comments or answers to open-ended questions. Poor return rates are a hallmark of this technique, along with the suspicion that those with dissatisfactions respond at higher rates.
- *Failure mode and effect analyses.* Otherwise called FMEA, the matrix presents the probability of failure, significance of the failure, and ease of detection, resulting in a *risk priority number* (RPN). Higher RPNs are attacked first. This is used in both improvement and design settings, although the chief use is as a design tool.
- *Fault-tree analyses.* A graphical presentation of the modes of failure shows events that must occur together ("and") or separately ("or") in order to have the failure occur. Usually this is shown vertically, with the "ANDed" and "ORed" events cascading as branches on a tree.
- *Flow diagram.* This extremely popular depiction of a process uses standard symbols for activities and flow directions. It originated in software design during the 1950s and evolved into the process mapping widely used today.
- *Glossary.* The glossary is the chief weapon used to remove the ambiguity of words and terms between customers and providers. This is a working dictionary of in-context usage, e.g., the meaning of "comfortable" as it applies to an office chair.

- *Design network.* A tree diagram depicts the events that occur either in parallel or sequentially in the design of something. Usually the network is shown with the total time needed to complete the event, along with earliest start and subsequent stop dates. It is used to manage a particularly complex design effort. Like techniques include the program evaluation and review technique (PERT) and critical path method (CPM). Today's spreadsheetlike project management software usually combines the key features of each.
- *Process analysis technique.* This process flowchart technique also shows the time necessary to do each task, the dependencies the task requires (such as access to the computer network), and the time "wasted" in between tasks. Usually it is interview-driven and requires a skilled process expert.
- *Process capability.* This term is given to any number of tools, usually statistical, that thereby reveal the ability of a process to repeat itself and the ability of the process to meet its requirements.
- *Salability analyses.* This is another matrix tool used to depict the price willing to be borne, or the cost needed to deliver, a given feature of a product.
- *Scatter diagram.* This is a graphical technique of plotting one variable against another, to determine corelationship. It is a prelude to regression analyses to determine prediction equations.
- *Selection matrix.* This matrix tool shows the choices to be made ranked according to agreed upon criteria. It is used in both improvement and design settings.
- *Customer needs spreadsheet.* This spreadsheet tool depicts the relationship between customer communities and the statements of need. Needs strongly relating to a wide customer base subsequently rise in priority when features are considered. Advanced forms of this spreadsheet and others appear as the "house of quality," or quality function deployment (QFD); see the section in this chapter about Design for Six Sigma.
- *Needs analysis spreadsheet.* This spreadsheet tool is used to "decompose" primary statements of need into other levels. Thus, "economical" for a new car purchaser might break down further to purchase price, operating costs, insurance costs, fuel economy, and resale value. Decomposing needs has the principal benefit of single-point response and measurement if taken to the most elemental level.
- *Product design spreadsheet.* This is a continuation of the customer needs spreadsheet, further developing the features and feature goals that map to the customer needs. The features with the strongest relationship to needs are elevated in priority when considering the processes used to make them.
- *Tree diagram.* Any of a variety of diagrams depict events that are completed in parallel or simultaneously as branches of a tree. This technique is less refined than the design network, but useful to understand the activities from a "big picture" perspective.
- *Value analysis.* This is a matrix depiction of customer needs and costs required to support or deliver a given feature to meet that need. It is a close cousin to salability analysis.

Design for Six Sigma

Product and service design is the creation of a detailed description for a physical good or service, together with the processes to actually produce that good or service. In quality theory terms, product design means establishing quality goals and putting in place the means to reach those goals on a sustained basis. In Six Sigma terms, product design [Design for Six

Sigma (DFSS)] means contemporaneously creating a design for a product and includes the process to produce it in such a way that defects in the product and the process are not only extremely rare, but also predictable. What is more, defects are rare and predictable, even at the point when full-scale production begins. To achieve this level of excellence and its attendant low costs and short cycle times, as well as soaring levels of customer satisfaction, requires some enhancements to traditional design methods. For example, each DFSS design project starts with an identification of customers and a detailed analysis and understanding of their needs. Even "redesign" starts at the beginning because all successful designs are based on customer needs, and in this world of rapid change, customer needs—and even customers—have a way of rapidly changing. Another example is the widespread intensive use of statistical methods in DFSS. The power of the information gained from statistical analyses provides the means to achieve Six Sigma levels of quality, which are measured in parts per million. DFSS is carried out in a series of phases known as DMADV.

DMADV stands for: define, measure, analyze, design, and verify. The discussion that follows does not cover all the details of procedures and tools used in DMADV; that would require many hundreds of pages, and they can be found elsewhere in published form. We will, however, attempt to acquaint the reader with what any manager needs to know about the purpose, the issues, the questions, and the sequence of steps associated with the respective phases of DMADV.

A "new" codification of the process for developing quality products is known as Design for Six Sigma. It combines the concept of quality design with the popular goal of Six Sigma quality. The DFSS process directs the designers of the product to create their designs so that manufacturing can produce them at Six Sigma quality levels. In the case of services, it means developing the service process so that it can be delivered at Six Sigma quality levels.

DFSS is targeted at design activities that result in a new product, a new design of an existing product, or the modification of an existing design. It consists of five phases in the following sequence: define, measure, analyze, design, verify. Figure 4.15 expands on the activities of each phase. (See Chapter 14, Continuous Innovation Using Design for Six Sigma, for more details on DFSS.)

Define	Measure	Analyze	Design	Verify
• Initiate the project • Scope the project • Plan and manage the project	• Discover and prioritize customer needs • Develop and prioritize CTQs • Measure baseline performance	• Develop design alternative • Develop high-level design • Evaluate high-level design	• Optimize detail level design parameters • Evaluate detail level design • Plan detail design verification tests • Verify detail and design of product • Optimize process performance	• Execute pilot/ analyze results • Implement production process • Transition to owners

FIGURE 4.15 Major activities in DFSS.

References

Designs for World Class Quality. (1995). Juran Institute, Wilton, CT.

Juran, J. M. (1992). *Quality by Design*. Free Press, New York.

Parasuraman, A., Zeithami, V. A., and Berry, L. L. (1985). "A Conceptual Model for Service Quality and Its Implications for Further Research." *Journal of Marketing*, Fall, pp. 41–50.

Veraldi, L. C. (1985). "The Team Taurus Story." MIT Conference paper, Chicago, August. 22. Center for Advanced Engineering Study, MIT, Cambridge, MA.

CHAPTER 5

Quality Improvement: Creating Breakthroughs in Performance

Joseph M. Juran and Joseph A. De Feo

About This Chapter

The purpose of this chapter is to explain the nature of breakthrough and its relation to attaining superior results. This chapter deals with the universal and fundamental concepts that define the methods to create "breakthroughs in current performance." The Six Sigma Model for Performance Improvement, popularized by Motorola and GE, is the most widely used method for attaining breakthrough. Although this is covered in detail in Chapter 12, Six

Sigma: Improving Process Effectiveness, we will focus on setting a foundation, presenting key terms, and making critical distinctions between similar, but different, contemporary methods to improve performance. This chapter will focus on leadership's role in creating a strategy that enables the organization to continue to improve year after year.

High Points of This Chapter

1. A breakthrough in current performance aims to eliminate failures such as excessive number of defects, excessive delays, excessively long time cycles, and the high costs of poor quality due to poorly performing processes.
2. The Juran Universal Sequence for Breakthrough, identified in the 1950s, consists of six steps to achieve superior results. The steps are
   a. Nominate and identify problems. (Management does this.)
   b. Establish a project and team. (Management does this.)
   c. Diagnose the cause(s). (The project team does this.)
   d. Remedy the cause(s). (The project team plus the work group where the cause[s] originate do this.)
   e. Hold the gains. (The project team and affected operating forces do this.)
   f. Replicate results and nominate new projects. (Management does this.)
3. All improvement happens project by project. To achieve breakthrough requires leaders to define goals and projects that are resourced to ensure completion and results.
4. It is upper management's responsibility to mandate breakthrough. Specifically, upper management must
   a. Establish multifunctional councils or steering teams to prioritize projects.
   b. Nominate and select breakthrough projects.
   c. Create project charters that include problem and goal statements.
   d. Provide resources, especially people and time, to carry out the project.
   e. Assign teams, team leaders, facilitators, "Black Belts" to projects.
   f. Review progress, remove barriers, and manage cultural resistance.
   g. Provide recognition and rewards.
5. Project selection requires expertise and practice on the part of management, so "doable" projects are identified so that the team clearly understands both the problem and the goal.
6. To attain a breakthrough in current performance requires two "journeys": the diagnostic journey and the remedial journey. These journeys represent the application of the fact-based method to solve the performance problems.
7. The diagnostic journey proceeds as follows:
   a. From problem to symptoms of the problem
   b. From symptoms to theories of causes of the symptoms
   c. From theories to testing of the theories
   d. From tests to establishing root cause(s) of the symptoms

8. The remedial journey proceeds as follows:
   a. From root cause(s) to design of remedies of the cause(s)
   b. From design of remedies to testing and proving the remedies under operating conditions
   c. From workable remedies to dealing with predictable resistance to change
   d. From dealing with resistance to establishing new controls on the remedies to hold the gains
9. There have been numerous efforts to create simpler and less intensive improvement methods. Most of them failed to deliver the results. The Six Sigma DMAIC Improvement Model has gained wide acceptance and is the most widely used. This will be covered in more detail in Chapter 12, Six Sigma: Improving Process Effectiveness. It follows these basic steps:
   a. Select the problem and launch a project. (Management does this.)
   b. Define the problem. (Champions and Management do this.)
   c. Measure the magnitude of the symptoms. (The project team does this.)
   d. Analyze information to discover the root cause(s). (The project team does this.)
   e. Improve by providing a remedy for the cause(s). (Project teams do this.)
   f. Control to hold the gains. (Project team and departments do this.)
10. All projects and teams will encounter obstacles when making changes. Objections will be raised by various sources. There may be delaying tactics or rejection by a manager, the work force, or the union. We refer to this as resistance to change. All managers must understand how to overcome this resistance.

The Universal Sequence for Breakthrough

Improvement happens every day, in every organization—even among the poor performers. That is how businesses survive—in the short term. Improvement is an activity in which every organization carries out tasks to make incremental improvements, day after day. Improvement is different from breakthrough improvement. Breakthrough requires special methods and support to attain significant changes and results. It also differs from planning and control. Breakthrough requires taking a "step back" to discover what may be preventing the current level of performance from meeting the needs of its customers. This chapter focuses on attaining breakthrough improvement and how leaders can create a system to increase the rate of improvement. By attaining just a few (the Pareto principle) vital breakthroughs year after year, the organization can outperform its competitors and meet stakeholder needs.

As used here, "breakthrough" means "the organized creation of beneficial change and the attainment of unprecedented levels of performance." Synonyms are "quality improvement" or "Six Sigma improvement." Unprecedented change may require attaining a Six Sigma level (3.4 ppm) or 10-fold levels of improvement over current levels of process performance. Breakthrough results in significant cost reduction, customer satisfaction enhancement, and superior results that will satisfy stakeholders.

"The concept of a universal sequence evolved from my experience first in Western Electric Organization (1924–1941) and later during my years as an independent consultant, starting in 1945. Following a few preliminary published papers, a universal sequence was published in book form (Juran 1964). This sequence then continued to evolve based on experience gained from applications by operating managers.

The creation of the Juran Institute in 1979 led to the publication of the videocassette series *Juran on Breakthrough* (Juran 1981). This series was widely received and was influential in launching breakthrough initiatives in many organizations. These organizations then developed internal training programs and spelled out their own versions of a universal sequence. All of these have much in common with the original sequence published in 1964. In some cases, the organizations have come up with welcome revisions or additions."

Breakthrough means change: a dynamic, decisive movement to new, higher levels of performance. In a truly static society, breakthrough is taboo, forbidden. There have been many such societies, and some have endured for centuries. During those centuries, their members either suffered or enjoyed complete predictability. They knew precisely what their station in life was—the same as that lived out by their forebears—but this predictability was, in due course, paid for by a later generation. The price paid was the extinction of the static society through conquest or another takeover by some form of society that was on the move. The threat of extinction may well have been known to the leaders of some of these static societies. Some gambled that the threat would not become a reality until they were gone. It was well stated in Madame de Pompadour's famous letter to Louis XV of France: "After us, the deluge."

History is vital to today's leaders. The threat to the static society stems from basic human drives: the drive for more of everything—knowledge, goods, power, and wealth. The resulting competition is what makes breakthrough important (Juran 1964).

There is an unvarying sequence of events by which we break out of the old levels of performance and into the new. The details of this sequence are important. The starting point is the attitude that a breakthrough is both desirable and feasible. In human organizations, there is no change unless there is first an advocate of change. If someone does not want change, there is a long, hard road before change is finally achieved. The first step on that road is someone's belief that a change—a breakthrough—is desirable and feasible. That change is desirable is mainly an act of faith or belief. Feasibility requires some digging. This leads to the second step.

The second step is to see whether a breakthrough is likely to happen if we mobilize for it—a feasibility study or demonstration project. This study will help separate the problem into major parts, the vital few from the useful many. I call this the Pareto analysis. These vital few problems then become the subject of a drive for new knowledge. But the creation of new knowledge does not just happen—we must organize for it. This leads to the next step.

Organization for breakthrough in knowledge is next. It requires that we appoint or create two systems: one that directs or guides the breakthrough, and one that does the fact-gathering and analysis. We call them the steering arm and the diagnostic arm, respectively. For breakthrough in knowledge, both of these arms are necessary. Neither one alone is sufficient. When both are in place, diagnosis begins. Facts are collected and examined, and new knowledge gained. At this stage, a breakthrough in knowledge has been achieved.

However, a breakthrough in knowledge does not automatically create a breakthrough in performance. Experience has shown that the technical changes needed usually affect the status, habits, beliefs, etc., of the people involved. Anthropologists have given the name "cultural pattern" to this collection of human beliefs, practices, etc.

Breakthroughs in the cultural pattern are in this way an added essential step. Before new levels of performance can be reached, we must discover the effect of the proposed changes on the cultural pattern and find ways to deal with the resistances generated. This turns out at times to be a difficult and important problem.

Finally, a breakthrough in performance can be achieved. This is the result we had set out to attain. To sustain it, we must rely on controls to maintain the status quo until another breakthrough comes along.

Two Kinds of Breakthrough

Breakthrough can be aimed at both sides of quality.

1. *Having higher-quality product and service features* provides customer satisfaction and revenue for the producing organization. These product features drive revenue.
2. *Achieving freedom from failures will reduce* customer dissatisfaction and nonvalue-added waste. To the producing organization, reducing the product failures, which reduce costs, is a target for breakthrough.

Breakthrough is applicable to any industry, problem, or process. To better understand why so many organizations create extensive quality improvement programs such as Lean Six Sigma we must contrast planning versus improvement. In the previous chapter, we discussed the quality planning process to design features.

Breakthrough to reduce excess failures and deficiencies may consist of such actions as

- Increase the yield of production processes
- Reduce error rates of administrative reports
- Reduce field failures
- Reduce claim denials
- Reduce the time it takes to perform critical patient clinical procedures

The result in both cases is performance improvement, which can lead to performance excellence. However, the rate of improvement required to attain market leadership needs to move at a revolutionary rate, and this often eludes most organizations. The methods and tools used to secure superior results are fundamentally different from day-to-day improvement methods, and for subtle reasons.

Creating breakthrough to increase revenue starts by setting strategic goals, such as new product development goals to provide best-in-class features, or reducing cycle times to beat the competition. Meeting such new goals requires a systematic "quality planning" process (Juran 1999). Multiple levels of quality planning are needed. An organization needs to plan new products or to design for quality. Other forms of quality planning are to design for manufacturing, Design for Six Sigma, and even Design for Green and Lean.

Quality planning differs from most product and service development methods in that it is carried out through a universal series of steps focusing on understand the "voice of the customers" (internal and external) and incorporating it into the design of the product. The best design methods always begin with an identification of who we are designing for. In other words, who are the "customers?" This is often followed by determining the needs of those customers, then developing the product or service features required to meet those needs, and so on. Collectively, this series of steps is the "quality planning or quality by design roadmap." Creating breakthroughs in design is covered in Chapter 4, Quality Planning: Designing Innovative Products and Services.

Many organizations maintain an organized approach for evolving new products and services, year after year. Under this organized approach

- Product development projects are a part of the business plan.
- A new product development function maintains business surveillance over these projects.
- Full-time product and process development departments are equipped with personnel, laboratories, and other resources to carry out the technological work.

- There is clear responsibility for carrying out the essential technological work.
- A structured procedure is used to process the new developments through the functional departments.
- The continuing existence of this structure favors new product development on a year-to-year basis.

This special organizational structure, while necessary, is not sufficient to ensure good results. In some organizations, the cycle time for getting new products to market is lengthy, the new models compete poorly in the market, or new chronic wastes are created. Such weaknesses usually are traceable to weaknesses in the planning process.

In the case of too many nonvalue-added tasks or too high a cost associated with chronic waste, the product or service is already in production, the goals are already in place. The processes for meeting those goals and the means to maintain them are being carried out by the workforce. However, the resulting products (goods and services) do not always meet the goals. Consequently, the approach to reducing these nonvalue-added tasks or chronic waste is different from the design or planning methods. Instead, to attain breakthroughs in current levels of performance, we must first have management commit to a program of quality improvement such as Six Sigma. This program can provide the means to identify the problems and then discover their causes. The organization must make the time to carry out a diagnosis of the current process. Once the causes are uncovered, remedies can be applied to remove the causes. It is this approach—to attain breakthroughs—that is the subject of this chapter.

Continuing to attain breakthrough is needed to meet the changing needs of customers, which are a moving target. Competitive prices are also a moving target. However, breakthroughs in improvement usually lag behind breakthroughs in design. They have progressed at very different rates. The chief reason is that many upper managers give a higher priority to increasing revenue from other means than on focusing resources on attaining breakthroughs by achieving unprecedented levels of performance in this way. This difference in priority is usually reflected in the respective organizational structures. An example is seen in the approach to new product development.

Historically, the efforts to meet the competition and improve performance proceeded along two lines based on two very different philosophies:

- Political leaders focused on traditional political solutions—import quotas, tariffs, legislation on "fair trade," and so on.
- Business leaders increasingly became convinced that the necessary response to competition was to become more competitive. This approach required applying the lessons learned from the role models across the entire national economy. Such a massive scaling up has extended well into the twenty-first century.

The experience of recent decades has led to an emerging consensus that managing for quality (planning, control, and improvement) is one of the most cost-effective means to deal with the threats and opportunities, and provide a means of actions that need to be taken. As it relates to breakthrough, the high points of this consensus include the following:

- Global competition has intensified and has become a permanent unpleasant fact. A needed response is to create a high rate of breakthrough, year after year.
- Customers are increasingly demanding improved products from their suppliers. These demands are then transmitted through the entire supplier chain. The demands may go beyond product breakthrough and extend to improving the system of managing for quality.

- The chronic wastes can be huge in organizations that do not have a strategic program aimed at reducing them. In many organizations during the early 1980s, about a third of all work consisted of redoing what was done previously, due to deficiencies. By the end of the 1990s, this number improved to only 20 to 25 percent (estimated by the authors). The emerging consensus is that such waste should not continue, since it reduces competitiveness and profitability.
- Breakthroughs must be directed at all areas that influence an organization's performance: all business, transactional, and manufacturing processes.
- Breakthroughs should not be left solely to voluntary initiatives; they should be built into the strategic plan and DNA of a system. They must be mandated.
- Attainment of market leadership requires that the upper managers personally take charge of managing for quality. In organizations that did attain market leadership, the upper managers personally guided the initiative. The authors are not aware of any exceptions.

Unstructured Reduction of Chronic Waste

In most organizations, the urge to reduce costs has been much lower than the urge to increase sales. As a result:

- The business plan has not included goals for the reduction of chronic waste.
- Responsibility for such breakthroughs has been vague. It has been left to volunteers to initiate action.
- The needed resources have not been provided, since such breakthroughs have not been a part of the business plan.

The lack of priority by upper managers is traceable in large part to two factors that influence the thinking processes of many upper managers:

- Not only do many upper managers give top priority to increasing sales, but some of them even regard cost reduction as a form of lower-priority work that is not worthy of the time of upper managers. This is especially the case in high-tech industries.
- Upper managers have not been aware of the size of the chronic waste, nor of the associated potential for high return on investment. The "instrument panel or scorecards" available to upper managers have stressed performance measures such as sales, profit, cash flow, and so on, but not the size of chronic waste and the associated opportunities. The managers have contributed to this unawareness by presenting their reports in the language of specialists rather than in the language of management—the language of money.

Breakthrough Models and Methods

Breakthrough addresses the question, How do I reduce or eliminate things that are wrong with my products, services, or processes and the associated customer dissatisfaction? Breakthrough models must address problems that create customer dissatisfaction, products and services of poor quality, and failures to meet the specific needs of specific customers, internal and external.

Based on my research, attaining breakthroughs in current performance by reducing customer-related problems has one of the greatest returns on investment and usually comes down to correcting just a few types of things that go wrong, including

- Excessive number of defects
- Excessive numbers of delays or excessively long time cycles
- Excessive costs of the resulting rework, scrap, late deliveries, dealing with dissatisfied customers, replacement of returned goods, loss of customers and clients, loss of goodwill, etc.
- High costs and ultimately high prices, due to the waste

Effective breakthrough models require that

- Leaders mandate it, project by project year after year
- Projects be assigned to teams that must discover root causes of the problems to sustain the gains
- Teams devise remedial changes to the "guilty" processes to remove or deal with the cause(s)
- Teams work with functions to install new controls to prevent the return of the causes
- Teams look for ways to replicate the remedies to increase the effect of the breakthrough
- All teams must follow a systematic fact-based method, which requires making two journeys:
  - *The diagnostic journey.* From symptoms (evidence a problem exists) to theories about what may cause the symptom(s); from theories to testing of the theories; from tests to establishing root cause(s). Once the causes are found, a second journey takes place.
  - *The remedial journey.* From root causes to remedial changes in the process to remove or deal with the cause(s); from remedies to testing and proving the remedies under operating conditions; from workable remedies to dealing with resistance to change; from dealing with resistance to establishing new controls to hold the gains.
- Regardless of what your organization calls or brands its improvement model, breakthrough results only occur after the completion of both journeys.

It has been more than 50 years since I first published articles on the universal sequence for breakthrough. Over that stretch of time, I have witnessed many models and many organizations trying to simplify, reengineer, and rename this simple method called breakthrough. Some have worked; some have not.

The most recent success is Six Sigma or Six Sigma DMAIC. Six Sigma has become the most effective "brand" of improvement since the Motorola Corporation first began using the quality improvement method I espoused in the late 1970s. Six Sigma methods and tools employ many of these universal principles. They have been combined with the rigor of statistical and technological tools to collect and analyze data.

GE's former chairman, Jack Welch, defined Six Sigma in this way: "Six Sigma is a quality program that, when all is said and done, improves your customers' experiences, lowers your costs and builds better leaders" (Welch 2005).

We will discuss Six Sigma in detail and fill in the blanks on its steps: Define, Measure, Analyze, Improve, and Control in Chapter 12, Six Sigma: Improving Process Effectiveness.

A Breakthrough Improvement Case

The following is the outline of the anatomy of a breakthrough improvement project. Because this book is written as a guide, we will limit our detailed discussions to some of the more important activities that are carried out by management. Each of the following topics that are outlined contains a large body of technical knowledge, tools, and techniques.

Identify a project (Management does this):

- Nominate projects.
- Evaluate projects.
- Select a project.
- Determine if this is a design project, an improvement project, or another type, such as a lean project.

Establish the project (the champions do this):

- Prepare a problem statement and a goal statement.
- Select a facilitator, or in a Six Sigma Program, a Black Belt or expert (see Chapter 12, Six Sigma: Improving Process Effectiveness).
- Select and launch a team.

Diagnose the cause (the project team and Black Belts do this):

- Analyze symptoms.
- Confirm and quantify or modify the goal.
- Formulate theories of causes.
- Test likely theories of causes.
- Identify root cause(s).

Remedy the cause (the project team and the workgroup where the cause[s] originate do this, perhaps with assistance from many others who are affected by, or who contribute to, the remedy):

- Evaluate alternative remedial changes.
- Design the solution, remedy, and changes needed to eliminate the root causes.
- Design new controls to hold the gains.
- Design for the culture (prevent or overcome resistance to the remedial changes).
- Prove the effectiveness of the remedy under operating conditions.
- Implement the remedial changes.

Hold the gains (the project team and the affected operating forces do this):

- Design and implement effective controls.
- Mistake-proof the process, as necessary.
- Audit the controls.

Replicate results and nominate new projects (Management does this):

- Replicate the results (clone, perhaps with modifications, the remedy).
- Nominate new projects based on lessons learned from the last project.
- Organize leaders into "performance excellence" or "quality councils."
- Select problems or new goals that need to be improved, and establish projects for them.
- Create project charters: problem and goal statements.
- Provide resources: training, staff, expertise, coaching, and especially time to complete the improvement.
- Assign teams and projects to teams to stimulate remedies and controls.
- Review progress and provide recognition and rewards.

The Mysterious Damage to Linoleum in Manufactured Housing

Here is a brief case of a straightforward, relatively simple (yet valuable) project that illustrates the breakthrough improvement methodology.

Nearly half of the residential single-family dwelling units built in the United States are manufactured on moving production lines. The modular units are transported to remote locations, joined together there, and set upon prepared foundations on the home purchaser's lot. It is hard to tell the difference between an assembled manufactured house and a stick-built house once they are finished and landscaped.

A large manufacturer of modular housing units was dissatisfied with the level of very expensive rework some of its factories in various locations around the country were experiencing. Customer dissatisfaction was rising; profits were eroding. Quality councils consisting of the general manager and all direct reports were formed at each factory. They received training in quality improvement, identified the most expensive rework, formed and trained teams in quality improvement, and set them to reducing the amount of rework. This is the story of one such improvement project. We begin by identifying the problem.

Identify the Problem: One Factory's Quality Council Listed and Prioritized Its Rework Problems Using the Pareto Analysis

The Pareto distribution (arranged in descending order of cumulative percent) of their most costly rework types during the past six months looked as follows (learn more about the Pareto analysis in Chapter 18, Core Tools to Design, Control, and Improve Performance):

- Replacing damaged linoleum: 51 percent
- Repairing cut electrical wires in walls: 15 percent
- Replacing missing fixtures at the site: 14 percent
- Repairing leaks in water pipes: 12 percent
- Repairing cracks in drywall: 8 percent

Based on the Pareto analysis, the quality council selected public enemy number one: replacing damaged linoleum. This problem is expensive to repair. Often, walls had to be removed and new linoleum laid, followed by replacing the wall. The next step was to establish a legitimate project with responsibility to resolve the problem.

Establish the Project

- A problem statement was formulated: *"The excess number of occurrences of replacing damaged linoleum accounts for 51 percent of all rework."*
- The goal statement was provided to the team as to the direction they should take: *"Reduce the number of occurrences of replacing damaged linoleum."*

Note that both the problem and goal are described and the variable and unit of measure in the problem statement and goal statement are identical. This is important because the problem statement tells the team what problem it is trying to solve. The rest of the project focuses on whatever the council selects as the problem; if they do not match, the team may carry out the goal and not solve the problem. The council chartered a project team consisting of representatives of the workstations where the linoleum was installed and where the damaged linoleum was observed. The council appointed a worker in one of those workstations to be the project team leader. The project team leader received training not only in quality improvement, but also in leading a project team. A trained facilitator coached the team in the breakthrough improvement methodology. The team began its diagnostic journey: the journey from symptom to cause.

Diagnose the Cause

The team's first task was to analyze the symptoms. (Symptoms are outward evidence of the problem.) The primary symptom was, of course, the number of occurrences of replacing damaged linoleum. Secondary symptoms were the cost of replacement, the various types of damage, the location where the damage showed up, downtime due to replacement, overtime to do the replacement, and the like. The symptoms were analyzed by *defining* them, *quantifying* them, and *visualizing* them. What follows is an analysis of the primary symptom.

Various types of damage were identified and defined as gouges, scrapes, cuts, gaps, and smears. A flow diagram was constructed showing all operations in all workstations that related to linoleum or replacement of linoleum. The flow diagram also identified the workstations where damage showed up. Several Pareto analyses were performed. The first Pareto was "by type of damage." It showed

- Gouges (dents): 45 percent
- Scrapes: 30 percent
- Cuts: 21 percent
- Gaps: 4 percent
- Smears: 2 percent

Accordingly, the team now focused temporarily on the top priority: gouges.

A second Pareto analysis of gouges by location in the house was performed. It showed which areas of the home had the most occurrences of damage:

- Kitchen: 38 percent
- Interior hall: 21 percent
- Bathroom 1: 18 percent
- Bathroom 2: 14 percent
- Laundry: 9 percent

Now the team refocused their attention on gouges in kitchens to the temporary exclusion of all the other symptoms. The Pareto principle states that for any given effect (an output of a process or a symptom in this case), there are a number of contributors. These contributors make unequal contributions. By far, a relatively few contributors make the greatest contribution. These are called the "vital few." Some contributors occur less often and are called the "useful many." Following the Pareto analysis, the team concentrated on the vital few contributors to the problem to get the greatest return for the least effort.

A third Pareto analysis of gouges in kitchens by work shift showed no difference in occurrences between shifts, indicating that "shift" is not a contributor to gouges in kitchens. Based on the experience of the team, they next generated a list of theories (or hypotheses) about what causes gouges in kitchens. They generated a long list of theories. The most compelling ones were

- Dropping heavy, sharp objects (tools)
- Dragging objects across the floor
- Grit on boots of employees
- Careless employees
- No protection for the new linoleum

In the manufactured housing industry, it is known that the first three theories, if they in fact occur, cause gouges in linoleum. Those theories did not need testing. What about "lack of protection"? The only way to test that theory is first to correlate gouges in kitchens with the presence or absence of protection. The team arranged that all reports of linoleum damage or replacement would include an indication if protection was "present."

When this was done, it was discovered that virtually all cases of gouge damage to linoleum occurred when floor protection was missing. Furthermore, the team discovered that there was no formal control plan for protection to be installed. Consequently, no quality inspections or quality assurance checks revealed that no controls were being exercised and that none even existed! Floor protection was a haphazard phenomenon at best.

Remedy the Cause

Workers, purchasing personnel, and engineers went to work to select and procure material that was strong and economical to lay on freshly installed linoleum. All agreed that the operator would be responsible for laying it immediately after each job, and that supervisors would check to see that it happened. Incidents of gouge damage—and other types of damage—to linoleum went down dramatically. (It seemed several damage types had common causes, one of which was no protection.) For a few weeks, damage to linoleum almost entirely disappeared. Celebrations were held. The plant manager began to look forward to granting bigger bonuses—and getting one himself!

At the weekly meeting of the factory management team a few weeks later, the quality manager reported the mysterious reappearance of gouge damage. This news was greeted with incredulity and disappointment. "We thought we had gouge damage licked!" Indeed they had, except for a couple of "small details."

Hold the Gains

When the team investigated, it discovered that (1) no formal control plan for providing protection had been devised and published; (2) there had been a turnover of workers in the various workstations who had not been trained in the procedure; and (3) the new workers had not been trained because there was no published plan; what's more, there also was no formal training program (with controls to ensure that training actually happened). Consequently, no training could or did take place. It became apparent that the "factory" operated more like a construction site under a roof, with standards upheld by the skill and pride of artisans. A factory, by contrast, is characterized by more formal procedures and controls. All this was a valuable lesson learned for all concerned, and led to a number of additional new improvement and planning projects, new attitudes toward the work, and a maturing of the plant as it evolved from construction site to factory. Controls and training were formalized.

Initiatives of the Past

In response to a crisis or economic downturns, many organizations, especially in the United States, undertake "improvement" initiatives to improve their performance. For various reasons, many of these initiatives fall far short of their goals. Some of the methods selected were doomed from the beginning; however, a few organizations made stunning breakthroughs, improved their performance, and became the role models, the market leaders in best practices.

The methods used by these role models have been analyzed and provide us with some lessons learned—the actions that are needed to attain breakthrough and market leadership and what methods and tools must be used to enable those results to happen.

Breakthrough Lessons Learned

My analysis of the actions taken by the successful organizations shows that most of them carried out many or all of the following tasks or strategies:

1. They enlarged the business plan at all levels to include annual goals for breakthrough and customer satisfaction.
2. They implemented a systematic process for making breakthroughs and set up special infrastructure or organizational machinery to carry out that process.
3. They adopted the big Q concept—they applied the breakthrough methods to all business processes, not just the manufacturing processes.
4. They trained all levels of personnel, including upper management, in the methods and tools to carry out their respective goals.
5. They enabled the workforce to participate in making breakthroughs in their daily work practices.
6. They established measures and scorecards to evaluate progress against the breakthrough goals.
7. The managers, including the upper managers, reviewed progress against the breakthrough goals.
8. They expanded the use of recognition and revised the reward system to recognize the changes in job responsibilities and using the new methods and tools.
9. They renewed their programs every few years to include changes to their programs as their performance improved.
10. They created a "rate of improvement" that exceeded the competition's.

The Rate of Breakthrough Is Most Important

The tenth lesson learned is an important one. Just having a system of breakthrough may not be enough. This lesson learned demonstrated that the annual rate of breakthrough

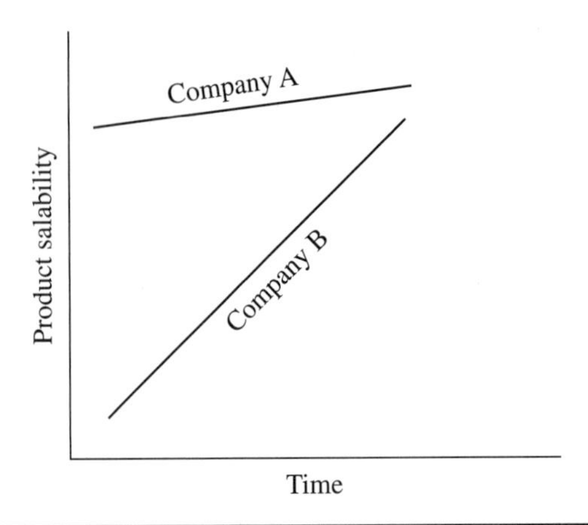

Figure 5.1 Two contrasting rates of improvement. (*Juran Institute, Inc., 2009.*)

determines which organizations will emerge as the market leaders. Figure 5.1 shows the effect of differing rates of breakthrough.

In this figure, the vertical scale represents product salability, so what goes up is good. The upper line shows the performance of Organization A, which at the outset was the industry leader. Organization A kept getting better, year after year. In addition, Organization A was profitable. Organization A seemed to face a bright future.

The lower line shows that Organization B, a competitor, was at the outset not the leader. However, Organization B is improving at a rate much faster than that of Organization A. Organization A is now threatened with loss of its leadership when Organization B surpasses them. The lesson is clear:

The most decisive factor in the competition for market leadership is the rate of breakthrough an organization maintains.

–Joseph M. Juran

The sloping lines of Figure 5.1 help to explain why Japanese goods attained market leadership through quality in so many products. The major reason was that the Japanese organizations' rate of breakthrough was for decades revolutionary when compared with the evolutionary rate of the West. Eventually, they had to surpass the evolutionary rate of the Western organizations. The result was an economic disaster for many U.S. organizations in the early 1980s. Today, U.S. automobile manufacturers have made great strides in quality while Toyota has had recalls. Figure 5.2 shows my estimate of the rates of breakthrough in the automobile industry from 1950–1990.

There are also lessons to be learned from the numerous initiatives to improve competitiveness during the 1980s, some of which failed to produce bottom-line results. The introduction of quality circles, employee involvement teams, TQM, reengineering, and National Quality Awards all were methods used to respond to the Japanese quality revolution. Some were not sustainable and failed. Each of them may have helped the organization that used them at that point in time. An important lesson does stand out. The initiatives showed us that attaining a revolutionary rate of breakthrough is not simple at all. It takes a strategic focus to sustain market leadership. Only the National Quality Awards continue today in most parts of the world. Organizations that made statements like, Quality is dead or TQM

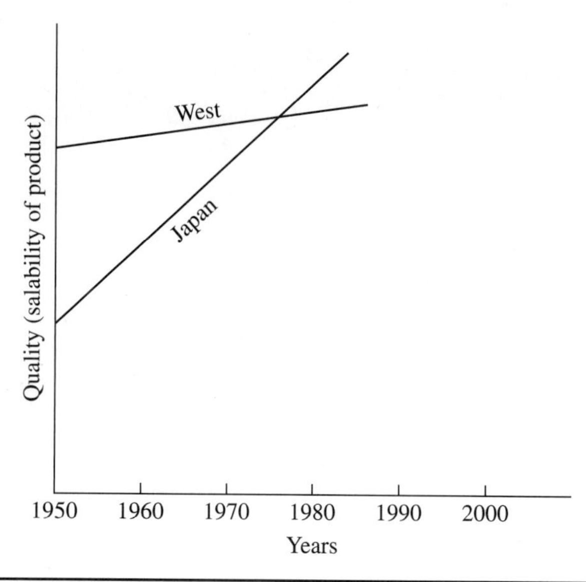

FIGURE 5.2 Estimate of rates of quality improvement in the automobile industry. (*Juran Institute, Inc., 1994. Used by permission.*)

did not work, blamed the methodology for their failures. This was only partially true. In some cases, the wrong method was selected and in others, their own management did not deal with the numerous obstacles and cultural resistance that prohibited these methods from working in the first place. These obstacles and the means to manage them will be discussed throughout this chapter.

The Breakthrough Fundamentals

Creating breakthroughs rests on just a few fundamental concepts. For most organizations and managers, annual breakthrough is not only a new responsibility; it is also a radical change in the style of management—a change in the organization's culture. Therefore, it is important to grasp the basic concepts before getting into the breakthrough process itself.

Breakthrough Distinguished from Design and Control

Breakthrough improvement differs from design (planning) and control. The trilogy diagram (Figure 5.3) shows this difference. In this figure, the chronic waste level (the cost of poor quality) was originally about 23 percent of the amount produced. This chronic waste was built into the process—"It was planned that way." Later, a breakthrough improvement project reduced this waste to about 5 percent. Under my definition, this reduction in chronic waste is a breakthrough—it attained an unprecedented level of performance.

Figure 5.3 also shows a "sporadic spike"—a sudden increase in waste to about 40 percent. Such spikes are unplanned—they arise from various unexpected sources. The personnel promptly got rid of that spike and restored the previous chronic level of about 23 percent. This action did not meet the definition of a breakthrough. It did not attain an unprecedented level of performance. It removed the spike and returned performance to the planned level. This is referred to as root-cause analysis, taking corrective action, or "firefighting."

Creating breakthroughs in current performance differs from creating breakthroughs in design. Current performance has well-known customer needs and targets. New product or

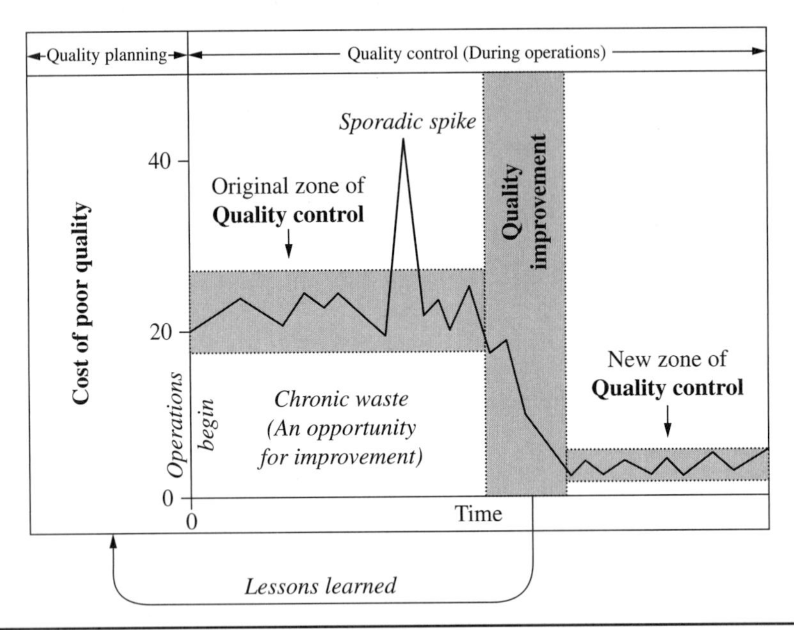

FIGURE 5.3 The Juran Trilogy®. (*Juran Institute, Inc., 1986.*)

service development is trying to create something new that meets a new need of the customer. It is new, innovative, and requires proper planning.

All three—design, control, and improvement—result in better performance, and all use teams to get there. It is only the steps that must be carried out that are different. This is analogous to a carpenter, electrician, and a plumber. Each tradesperson works on building a home or solving a problem (leaky pipes, rotted wood, failed circuit breaker). They have common methods and use similar tools, but at different times and for different purposes.

All Breakthrough Takes Place Project by Project

There is no such thing as breakthrough in a general way. All breakthrough takes place project by project and in no other way.

As used here, "breakthrough" means "the solving of a chronic problem by scheduling (launching a project) to find a solution." Since the word breakthrough has multiple meanings, the organization should create a glossary and educate all employees on what it means. The definition is helped by presenting a few examples that were carried out successfully in your organization.

Breakthrough Is Applicable Universally

The huge numbers of projects carried out during the 1980s and 1990s demonstrated that breakthrough is applicable to all:

- Service industries as well as manufacturing industries
- Business processes as well as manufacturing processes
- Support activities as well as operations
- Software- and information-based industries

During this period, breakthrough was applied to virtually all industries, including government, education, and health. In addition, breakthrough has been applied successfully to the entire spectrum of organization functions: finance, product development, marketing, legal, and so on.

In one organization, the legal vice president doubted that breakthrough could be applied to legal work. Yet within two years, she reduced by more than 50 percent the cycle time of filing for a patent. (For elaboration and many more case examples, see the Juran Institute e-Lifeline at www.juran.com.)

Breakthrough Expands to Many Parameters

Published reports of breakthroughs show that the effects have extended to all parameters:

- *Productivity*. The output per person-hour.
- *Cycle time*. The time required to carry out processes, especially those that involve many steps performed sequentially in various departments. Chapter 8, Business Process Management: Creating an Adaptable Organization, elaborates on breakthrough as applied to such processes.
- *Human safety*. Many projects improve human safety through mistake-proofing, fail-safe designs, and so on.
- *The environment*. Similarly, many projects have been directed at protecting the environment by reducing toxic goals and so on.

Some projects provide benefits across multiple parameters. A classic example was the color television set. The Japanese Matsushita Organization had purchased an American color television factory (Quasar). Matsushita then made various breakthroughs, including

- Product redesign to reduce field failures
- Process redesign to reduce internal defect rates
- Joint action with suppliers to improve purchased components

The results of these and other changes are set out in the before and after data:

	1974	1977
Fall-off rate, i.e., defects (on assembled set) requiring repair	150 per 100 sets	4 per 100 sets
Number of repair and inspection personnel	120	15
Failure rate during the warranty period	70%	10%
Cost of service calls	$22 million	$4 million

The manufacturer benefited in multiple ways: lower costs, higher productivity, more reliable deliveries, and greater salability. The ultimate users also benefited—the field failure rate was reduced by more than 80 percent.

The Backlog of Breakthrough Projects Is Never-Ending

The existence of a huge backlog of problems to solve is evident from the numbers of breakthroughs actually made by organizations that carried out successful initiatives during the 1980s and 1990s. Some reported making breakthroughs by the thousands, year after year. In very large organizations, the numbers are higher still, by orders of magnitude.

The backlog of breakthrough projects exists in part because of internal and external factors. Internally, the planning of new products and processes has long been deficient. In effect, the planning process has been a dual hatchery. It hatched out new plans. It also hatched out new chronic wastes, and these accumulated year after year. Each such chronic waste then became a potential breakthrough project.

A further reason for a huge backlog is the nature of human ingenuity—it seems to have no limit. Toyota Motor Corporation has reported that its 80,000 employees offered four million suggestions for breakthrough during a single year—an average of 50 suggestions per person per year (Sakai 1994).

Externally, the constantly changing needs of customers and our society will always challenge the status quo. Targets today are not good enough for tomorrow. This creates a never-ending backlog of projects.

Breakthrough Does Not Come Free

Breakthrough and the resulting reduction of chronic waste do not come free—they require an expenditure of effort in several forms. It is necessary to create an infrastructure to mobilize the organization's resources toward annual breakthrough. This involves setting specific goals to be reached, choosing projects to be tackled, assigning responsibilities, following progress, and so on.

There is also a need to conduct extensive training in the nature of the breakthrough improvement methods and tools, how to serve on breakthrough teams, how to use the tools, and so on.

In addition to all this preparatory effort, each breakthrough improvement project requires added effort to conduct diagnoses to discover the causes of the chronic waste and provide remedies to eliminate the causes. This is the time it takes for all the people involved in the team to solve the problem.

The preceding adds up to a significant front-end outlay, but the results can be stunning. They *have* been stunning in the successful organizations, the role models. Detailed accounts of such results have been widely published, notably in the proceedings of the annual conferences held by the U.S. National Institute for Standards and Technology (NIST), which administers the Malcolm Baldrige National Award.

Reduction in Chronic Waste Is Not Capital-Intensive

Reduction in chronic waste seldom requires capital expenditures. Diagnosis to discover the causes usually consists of the time of the breakthrough project teams. Remedies to remove the causes usually involve fine-tuning the process. In most cases, a process that is already producing more than 80 percent good work can be raised to the high 90s without capital investment. Such avoidance of capital investment is a major reason why reduction of chronic waste has a high return on investment (ROI).

In contrast, projects to create breakthroughs in product design and development to increase sales can involve costly outlays to discover customer needs, design products and processes, build facilities, and so on. Such outlays are largely classified as capital expenditures and thereby lower the ROI estimates. There is also a time lag between investing in design and receiving revenue from the sale of the new designs.

The Return on Investment for Breakthrough Improvement Is High

This is evident from results publicly reported by national award winners in Japan (Deming Prize), the United States (Baldrige Award), Europe, and elsewhere. More and more organizations have been publishing reports describing their breakthroughs, including the gains made.

It has been noted that the actual return on investment from breakthrough projects has not been well researched. My own research conducted by examining papers published by organizations found that the average breakthrough project yielded about $100,000 of cost reduction (Juran 1985). The organizations were large—sales in the range of over $1 billion per year.

I have also estimated that for projects at the $100,000 level, the investment in diagnosis and remedy combined runs to about $15,000 or 15 percent. The resulting ROI is among the highest available to managers. It has caused some managers to quip, "The best business to be in is breakthrough." Today, breakthrough projects return many more dollars, but the cost of attaining breakthrough has not changed from the 15 percent investment level.

I am astounded by some of the recent organizations that have become world quality leaders using the project-by-project approach of Six Sigma. One of them is Samsung Electronics.

Samsung Electronics Co. (SEC) of Seoul, Korea, has perfected its fundamental improvement approach using Six Sigma as a tool for innovation, efficiency, and quality. SEC was founded in 1969 and sold its first product, a television receiver, in 1971. Since that time, the company has used tools and techniques such as total quality control, total process management, product data management, enterprise resource management, supply chain management, and customer relationship management. Six Sigma was added to upgrade these existing innovations and improve SEC's competitive position in world markets. The financial benefits made possible by Six Sigma, including cost savings and increased profits from sales and new product development, are expected to approach $1.5 billion.

SEC completed 3,290 Six Sigma improvement projects in the first two years; 1,512 of these were Black Belt-level projects. By the third year, 4,720 projects are expected to be completed, 1,640 of them by Black Belts.

SEC's Six Sigma projects have also contributed to an average of 50 percent reduction in defects. There is no thought of improvement in quality and productivity without Six Sigma. These impressive numbers have certainly played a major role in Samsung's recent growth. Some indications of this include the following:

- By 2001, SEC had earned a net income of $2.2 billion on total revenues of $24.4 billion. Market capitalization stood at $43.6 billion.
- According to SEC's 2001 annual report, SEC now is one of the top 10 electronic and electrical equipment manufacturing companies in the world, with the best operating profit ratios and superior fiscal soundness.
- The report also says the debt-to-equity ratio is lower than that of any top ranking company, and the shareholders' equity-to-net-assets ratio surpasses the average.
- SEC says its technological strengths, Six Sigma quality initiatives, and product marketability helped increase its share of the memory chip market in 2001 to 29 percent, monitors to 21 percent, and microwave ovens to 25 percent of those sold worldwide.

Despite a downturn in the world economy and a reduction in exports to the United States, credit for SEC's current operating profit margin of 8.5 percent is due mostly to quality improvements and Six Sigma deployment.

SEC's quality and innovative strategy helped it reach the number-one position in the *BusinessWeek* 2002 information technology guide. The guide noted SEC's computer monitors, memory chips, telephone handsets, and other digital products, focusing on four Standard & Poor's criteria: shareholder return, return on equity, revenue growth, and total revenues.

The *BusinessWeek* ranking was also due to SEC's employees' belief that quality is the single most important reason for the company's higher sales, lower costs, satisfied customers, and profitable growth. Only a few years ago, SEC's products were virtually unknown by Americans or were known as the cheaper, lower-quality substitute for Japanese brands. This perception is changing. The U.S. market now represents 37 percent of SEC's total sales.

The Major Gains Come from the Vital Few Projects

The bulk of the measurable gains comes from a minority of the breakthrough projects—the "vital few." These are multifunctional in nature, so they need multifunctional teams to carry them out. In contrast, the majority of the projects are in the "useful many" category and are carried out by local departmental teams. Such projects typically produce results that are orders of magnitude smaller than those of the vital few.

While the useful many projects contribute only a minor part of the measurable gains, they provide an opportunity for the lower levels of the hierarchy, including the workforce, to participate in breakthrough. We will discuss the useful many projects in Chapter 11, Lean Techniques: Improving Process Efficiency. In the minds of many managers, the resulting gain in work life is as important as the tangible gains in operating performance.

Breakthrough—Some Inhibitors

While the role-model organizations achieved stunning results through breakthrough, most organizations did not. Some of these failures were due to honest ignorance of how to mobilize for breakthrough, but there are also some inherent inhibitors to establishing breakthrough on a year-to-year basis. It is useful to understand the nature of some of the principal inhibitors before setting out.

Disillusioned by the Failures

The lack of results mentioned earlier has led some influential journals to conclude that breakthrough initiatives are inherently doomed to failure. Such conclusions ignore the stunning results achieved by the role-model organizations. (Their results prove that these are achievable.) In addition, the role models have explained how they got those results, thereby providing lessons learned for other organizations to follow. Nevertheless, the conclusions of the media have made some upper managers wary about going into breakthrough.

Higher Quality Costs More

Some managers hold a mindset that "higher costs more." This mindset may be based on the outmoded belief that the way to improve is to increase inspection so that fewer defects escape to the customer. It also may be based on the confusion caused by the two meanings of the word.

Higher in the sense of improved product features (through product development) usually requires capital investment. In this sense, it does cost more. However, higher in the sense of lower chronic waste usually costs less—a lot less. Those who are responsible for preparing proposals for management's approval should be careful to define the key words—which kind are they talking about?

The Illusion of Delegation

Managers are busy people, yet they are constantly bombarded with new demands on their time. They try to keep their workload in balance through delegation. The principle that "a good manager is a good delegator" has wide application, but it has been overdone as applied to breakthrough. The lessons learned from the role-model organizations show that going into annual breakthrough adds minimally about 10 percent to the workload of the entire management team, including the upper managers.

Most upper managers have tried to avoid this added workload through sweeping delegation. Some established vague goals and then exhorted everyone to do better—"Do it right the first time." In the role-model organizations, it was different. In every such organization, the upper managers took charge of the initiative and personally carried out certain nondelegable roles.

Employee Apprehensions

Going into breakthrough involves profound changes in a organization's way of life—far more than is evident on the surface. It adds new roles to the job descriptions and more work to the job holders. It requires accepting the concept of teams for tackling projects—a concept that is alien to many organizations and that invades the jurisdictions of the functional departments. It requires training on how to do all this. Collectively, the megachange disturbs the peace and breeds many unwanted side effects.

To the employees, the most frightening effect of this profound set of changes is the threat to jobs and/or status. Reduction of chronic waste reduces the need for redoing prior work and hence, the jobs of people engaged in such redoing. Elimination of such jobs then becomes a threat to the status and/or jobs of the associated supervision. It should come as no surprise if the efforts to reduce waste are resisted by the workforce, the union, the supervision, and others.

Nevertheless, breakthrough is essential to remaining competitive. Failure to go forward puts all jobs at risk. Therefore, the organization should go into breakthrough while realizing that employee apprehension is a logical reaction of worried people to worrisome proposals. A communication link must be opened to explain the why, understand the worries, and search for optimal solutions. In the absence of forthright communication, the informal channels take over, breeding suspicions and rumors.

Additional apprehension has its origin in cultural patterns. (The preceding apprehensions do not apply to breakthrough of product features to increase sales. These are welcomed as having the potential to provide new opportunities and greater job security.)

Securing Upper Management Approval and Participation

The lessons learned during the 1980s and 1990s included a major finding: Personal participation by upper managers is indispensable to getting a high rate of annual breakthrough. This finding suggests that advocates for initiatives should take positive steps to convince the upper managers of

- The merits of planning for annual breakthrough
- The need for active upper management to provide resources
- The precise nature of the needed upper management participation

Proof of the Need

Upper managers respond best when they are shown a major threat or opportunity. An example of a major threat is seen in the case of Organization G, a maker of household appliances.

Model Number	2000	2001	2002	2003
1	G	G	R	R
2	R	R	R	R
3	G	G	G	R
4	T	R	R	R

Table 5.1 Suppliers to a Major Customer

Organization G and its competitors, R and T, were all suppliers to a major customer involving four models of appliances. (See Table 5.1.) This table shows that in 2000, Organization G was a supplier for two of the four models. Organization G was competitive in price, on-time delivery, and product features, but it was definitely inferior in the customer's perception of the chief problem being field failures. By 2002, lack of response had cost Organization G the business on model number 1. By 2003, Organization G also had lost the business on model number 3.

Awareness also can be created by showing upper managers other opportunities, such as cost reduction through cutting chronic waste.

The Size of the Chronic Waste

A widespread major opportunity for upper managers is to reduce the cost of poor quality or the costs associated with poorly performing processes. In most cases, this cost is greater than the organization's annual profit, often much greater. Quantifying this cost can go far toward proving the need for a radical change in the approach to breakthrough. An example is shown in Table 5.2. This table shows the estimated cost of poor quality for an organization in a process industry using the traditional accounting classifications. The table brings out several matters of importance to upper managers:

The order of magnitude. The total of the costs is estimated at $9.2 million per year. For this organization, this sum represented a major opportunity. (When such costs have never before been brought together, the total is usually much larger than anyone would have expected.)

The areas of concentration. The table is dominated by the costs of internal failures—they are 79.4 percent of the total. Clearly, any major cost reduction must come from the internal failures.

Category	Amount, $	Percent of Total
Internal failures	7,279,000	79.4
External failures	283,000	3.1
Appraisal	1,430,000	15.6
Prevention	170,000	1.9
	9,162,000	100.0

Table 5.2 Analysis of Cost of Poor Quality

COPQ versus Cost Reduction

Company X wanted to reduce operating costs by 10 percent. It began with a mission to have each executive identify where costs could be cut in business units. The executives created a list of 60 items, including things like eliminating quality audits, changing suppliers, adding new computer systems, reducing staff in customer services, and cutting back R&D.

The executives removed functions that provide quality and services to meet customer needs. They bought inferior parts and replaced computer systems at great expense. They disrupted their organization, particularly where the customers were most affected, and reduced the potential for new services in the future.

After accomplishing this, most of the executives were rewarded for their achievements. The result? Their cost reduction goal was met, but they had dissatisfied employees, upset customers, and an organization that still had a significant amount of expense caused by poor performance.

The financial benefit to the bottom line of an organization's balance sheet by improving the cost of quality is not always fully appreciated or understood. This misunderstanding stems from old misconceptions that improving quality is expensive.

However, this misconception is partially true. For example, if an organization provides a service to clients for a given price and a competitor provides the same basic service with enhanced features for the same price, it will cost your organization more to add those features that the competitor already provides.

If your organization does not add those features, it may lose revenue because customers will go to a competitor. If you counteract by reducing the price, you may still lose revenue. In other words, the quality of your competitor's service is better.

For your organization to remain competitive, it will have to invest in developing new features. This positively affects revenue. To improve quality, features have to be designed in—or in today's terminology, a new design must be provided at high Sigma levels.

Because of this historical misconception, organizations do not always support the notion that improving quality will affect costs and not add to them. They overlook the enormous costs associated with poor performance of products, services, and processes—costs associated with not meeting customer requirements, not providing products or services on time, or reworking them to meet the customer needs. These are the costs of poor quality (COPQ) or the cost of poorly performing processes (COP^3).

If quantified, these costs will get immediate attention at all management levels. Why? When added together, costs of poor quality make up as much as 15 to 30 percent of all costs. Quality in this complete sense, unlike the quality that affects only income, affects costs. If we improve the performance of products, services, and processes by reducing deficiencies, we will reduce these costs. To improve the quality of deficiencies that exist throughout an organization, we must apply breakthrough improvements.

A Six Sigma program focused on reducing the costs of poor quality due to low Sigma levels of performance and on designing in new features (increasing the Sigma levels) will enable management to reap increased customer satisfaction and bottom-line results. Too many organizations reduce costs by eliminating essential product or service features that provide satisfaction to customers, while ignoring poor performance that costs the bottom line and shareholders millions of dollars.

A Better Approach

Company Y approached its situation differently than did Company X, as described at the beginning of this section. The executives identified all costs that would disappear if everything worked better at higher Sigma levels. Their list included costs associated with credits or allowances given to customers because of late delivery, inaccuracy or errors in billings, scrap and rework, and accounts payable mistakes caused by discount errors and other mistakes.

When this company documented its costs of poor quality, the management team was astounded by the millions of dollars lost due to poor quality of performance within the organization.

This total cost of poor quality then became the target. The result? Elimination of waste and a return to the bottom line from planned cost reductions and more satisfied customers. Why? Because the company eliminated the reasons these costs existed in the first place. There were process and product deficiencies that caused customer dissatisfaction. Once these deficiencies were removed, the quality was higher and the costs were lower.

While responding to customer demands for improved quality in everything an organization does is becoming essential, organizations should not overlook the financial impact of poor performance. In fact, these costs should be the driver of the project selection process for Six Sigma.

In other words, the cost of poor quality provides proof of why changes must be made. The need to improve an organization's financial condition correlates directly with the process of making and measuring quality improvements. Regardless of the objective you start with, enhancing features as well as reducing costs of poor quality will affect the continuing financial success of an operation.

While there is a limit to the amount quality can be improved when cost effectiveness and savings are measured against the costs of achieving them, it's not likely this will occur until you approach Five or Six Sigma levels. A business must pursue the next level of quality based on what is of critical importance to its customers. If customers demand something, chances are it must be done to keep the business. If they do not, there's time to plan.

Driving Bottom-Line Performance

If you accept the reality that customers and the marketplace define quality, then your organization must have the right product or service features and lower your deficiencies to create loyal customers.

With a competitive price and market share strongly supported by fast cycle time, low warranty costs, and low scrap and rework costs, revenue will be higher and total cost lower. The substantial bonus that falls to the profit column comes, in effect, from a combination of enhancing features and reducing the costs of poor quality.

Before getting into specific ways to identify, measure, and account for the impact of costs of poor quality on financial results, look at what to do first if you are trying to understand how the costs of quality can drive a financial target.

For example, if your organization sets a cost reduction target to save $50 million, there is a simple methodology to determine how many improvement projects it will take to reach that goal. The organization can then manage the improvement initiative more effectively if it puts some thought behind how much activity it can afford. The answer will help determine how many experts or Black Belts are needed to manage the improvements and how much training will be required.

The methodology includes the following six steps:

1. Identify your cost reduction goal of $50 million over the next two years—$25 million per year.
2. Using an average return of $250,000 for each improvement, calculate how many projects are needed to meet the goal for each year. For this example, we would need an incredible 200 projects—100 per year.

3. Calculate how many projects per year can be completed and how many experts will be required to lead the team. If each project can be completed in four months, that means one Black Belt on two projects per four months. Hence, one Black Belt can complete six projects in one year. We will then need about 17 black belts.
4. Estimate how many employees will be involved on a part-time basis to work with the Black Belts to meet their targets. Assume four per Black Belt per four months. We would need about 200 employees involved at some level each year, possibly as little as 10 percent of their time.
5. Identify the specific costs related to poor performance, and select projects from this list that are already causing your organization to incur at least $250,000 per deficiency. If you haven't created this list, use a small team to identify the costs and create a Pareto analysis prior to launching any projects.
6. Use this method and debate each variable among the executive team to ensure the right amount of improvement can be supported. All organizations make improvements, but world-class organizations improve at a faster rate than their competition.

Where to Find Costs of Poor Performance

To put targets of opportunity into perspective, look at the traditional costs of poor quality and, even more critically, the hidden costs of poor quality, as shown in Figures 5.2 and 5.3. The hidden costs must be quantified to get a complete picture of losses due to poor performance. These costs of poor quality could disappear entirely if every activity were performed without deficiency every time.

Three major categories of costs of poor quality exist in organizations. You can focus your efforts better if you put them into the following three categories:

- Appraisal and inspection costs
- Internal failure costs
- External failure costs

Appraisal and Inspection Costs

Appraisal and inspection costs are costs associated with inspection—checking or assuring that deficiencies are discovered before customers are affected.

Examples include

- Testing products or checking documents before providing them to customers
- Reviewing documents and correcting errors before mailing
- Inspecting equipment or supplies
- Proofreading reports or correspondence
- Auditing customer bills prior to sending invoices
- Retooling due to poor design

Discovering deficiencies at this stage avoids serious failure costs later and helps develop more effective and efficient inspection methods. There will always be some costs in this category because some level of auditing will be needed to ensure consistent performance. The point is to avoid excessive costs.

Internal Failure Costs

Failure costs within an organization are attributed to the repair, replacement, or discarding of defective work the customer does not see.

Examples include

- Replacing metal stampings that do not meet specifications during production
- Repainting scratched surfaces
- Making up for unplanned computer downtime
- Replacing components damaged when being moved from one station to another
- Rewriting parts of a proposal
- Working overtime to make up for slippage
- Correcting database errors
- Stocking extra parts to replace defective components
- Scrapping products that do not meet specifications
- Spending excess accounts-payable time to correct supplier invoice errors
- Engineering change notices to correct errors in specifications or drawings

These costs may affect customer service indirectly.

External Failure Costs

External failure affects customers directly; these usually are the most expensive failures to correct. External failure costs may result from

- Satisfying warranty claims
- Investigating complaints
- Offsetting customer dissatisfaction with a recovery strategy
- Collecting bad debts
- Correcting billing errors
- Processing complaints
- Expediting late shipments by purchasing more expensive means of transportation
- Replacing or repairing damaged or lost goods
- Housing stranded passengers from cancelled flights
- Paying interest or losing discounts for late payments to vendors
- Providing on-site assistance to customers when field problems occur
- Providing credits and allowances to clients for lack of performance or late deliveries

Efforts to correct external failures usually focus on regaining customer confidence or lost sales. Both are debatable costs that may or may not be fully calculated.

Interpreting the Costs of Poor Quality

The costs of poor quality at this stage are determined by educated estimates used to guide organizational decisions. They should not be part of a monthly financial analysis, although understanding these costs may affect the way financial and cost accounting data are compiled and interpreted.

The precision required to identify the costs of poor quality varies depending on how data are used. When used to help select an improvement project, data need not be as precise as those used in developing new budgets for a process after it has been approved.

When you are evaluating projects, data on poor quality help identify, charter, and support projects with the greatest potential for reducing costs. Black Belts and teams may select some projects because of the impact on customers or internal culture, but data must show where costs are highest so that focus can be concentrated on the vital few.

The amount of cost reduction provided by a remedy is another indicator of project effectiveness. When planning for a remedy, a task force should develop supportable estimates of costs that will be eliminated by the remedy and use those estimates to develop a budget for the revised process.

There are four major steps in measuring the costs of poor quality:

1. Identify activities resulting from poor quality.
2. Decide how to estimate costs.
3. Collect data and estimate costs.
4. Analyze results and decide on the next steps.

Identify Activities Resulting from Poor Quality

Activities are categorized as resulting from poor quality only if they exist solely because of deficiencies assessed when doing appraisals, inspections, and internal or external cost estimates.

A project team usually begins by measuring the obvious costs of a problem's primary symptom, such as discarded supplies, customer complaints, or erroneous shipments. After a flow diagram of the process in question has been created and further analysis has been conducted, additional activities are usually identified as those required, for example, to dispose of and replace returned items.

Efforts to identify remedial activities are generally more global since the focus is on costs of poor quality throughout an organization. This effort is best undertaken by one or a small number of analysts working with a team of midlevel and senior managers experienced in key areas.

The task force usually launches its efforts by identifying major organizational processes and their customers. For each process, the task force brainstorms major activities associated with poor quality and expands the list through carefully constructed interviews with individuals representing different levels within the most critical functions. At this point, the objective is to prepare a list of activities related to poor quality, not estimate costs.

Project teams and task forces find it easier to explain what they are looking for if they have a full list of typical examples associated with poor quality. The examples described earlier fall into major categories of poor quality costs. Using key words such as rework, waste, fix, return, scrap, complaint, repair, expedite, adjust, refund, penalty, waiting, and excess usually stimulates a healthy response, too.

Decide How to Estimate Costs

When a specific activity related to poor quality is identified, two strategies help estimate its costs: total resources and unit costs. These strategies can be used individually or together.

An example of the total resource approach is how an operational unit calculated the human resource time to process customer complaints and the dollar value of that time. This approach requires two pieces of data: total resources consumed in a category and the percentage of those resources consumed for activities associated with poor quality.

An example of the unit cost approach is when a project team calculates the annual cost of correcting erroneous shipments. To find that cost, the team should estimate the cost of correcting an average erroneous shipment and how many errors occurred in one year, and then multiply the average cost by the annual number of errors.

Data for calculating the total resources used in a category might come from a variety of sources, such as accounting, time reporting, other information systems, informed judgment, special time reporting, special data collections, and unit costs. These sources are described in the section "Calculating Resources Used."

Collect Data and Estimate Costs

Procedures for collecting data on costs of poor quality are generally the same as those for any good data collection:

- Formulate questions to be answered.
- Know how data will be used and analyzed.
- Determine where data will be collected.
- Decide who will collect it.
- Understand data collectors' needs.
- Design a simple data collection form.
- Prepare clear instructions.
- Test forms and procedures.
- Train data collectors.
- Audit results.

To estimate the costs of poor quality, it is sometimes necessary to collect personal opinions and judgments about relative magnitudes of time spent or costs. Even though precise numeric data is not required for such estimates, it is important to plan carefully. The manner in which opinions are solicited affects responses.

Sampling works when the same activity is performed often in different parts of an organization. All field sales offices, for example, perform similar functions. If a company has 10 field sales offices, estimates from one or two would provide a reasonable value for calculating overall costs of poor quality.

Analyze Results and Decide on the Next Steps

Collecting data on costs of poor quality helps make decisions such as

- Selecting the most important quality improvement projects
- Identifying the most costly aspects of a specific problem
- Identifying specific costs to be eliminated

The Results

Of note is the fact that every organization that has adopted Six Sigma and integrated the discipline throughout its operations has produced impressive savings that were reflected on the bottom line. More customers were satisfied and became loyal, and revenues, earnings, and operating margins improved significantly.

For example, Honeywell's cost savings have exceeded $2 billion since implementing Six Sigma in 1994. At General Electric, the Six Sigma initiative began in 1996 and produced more than $2 billion in benefits in 1999. Black & Decker's Six Sigma productivity savings rose to about $75 million in 2000, more than double the prior year's level, bringing the total saved since 1997 to over $110 million.

A more revealing insight into the cost of poor quality as a function of Six Sigma performance levels is the following:

- When +/− 3 Sigma of the process that produces a part is within specification, there will be 66,807 defects per million parts produced. If each defect cost $1,000 to correct, then the total COPQ would be $66,807,000.
- When an organization improves the process to within +/− 4 Sigma, there will be only 6210 defects per million at a COPQ of $6,210,000.
- At +/− 5 Sigma, the cost of defects declines to $233,000 per million, a savings of $66,574,000 more than the savings at a process capability of +/− 3 Sigma.
- At the near perfection level of +/− 6 Sigma, defects are almost eliminated at $3400 per million parts produced.

After all data are collected and tabulated, and decisions are made, no study of the cost of poor quality should end without a continuing action plan to eliminate a major portion of the costs that have been identified. There is no need to use a complex accounting method for measuring costs because it would be expensive and waste valuable effort. Simple methods are sufficient.

The most important step in developing useful COPQ data is simply to identify activities and other factors that affect costs. Any consistent and unbiased method for estimating costs will yield adequate information that will identify key targets for quality improvement. More refined estimates may be needed for specific projects when diagnosing the cause of a specific problem or identifying specific savings.

Calculating Resources Used

Data for calculating the total resources used in an expense category come from a variety of sources, as the following sections explain.

Accounting Categories

Financial and cost accounting systems often contain specific categories that can be allocated partly or totally to costs of poor quality. Typical examples include scrap accounts, warranty costs, professional liability, discarded inventory, and total department operating costs.

Time Reporting

Many organizations routinely ask employees to report how much time they spend on specific activities. This makes it possible to assign some or all of the time in a category to a specific cost of poor quality.

Other Information Systems

Other information systems include cost accounting, activity-based cost accounting, materials management, sales, or similar reports.

Data for calculating the percentage of resources used for cost of poor quality activities can be obtained through a variety of techniques, including

- *Informed judgment.* Supervisors and experienced employees can make adequate judgments about what proportion of a department's time is spent on an activity. This is especially true if the unit performs very few distinct functions or if the effort consumes a very large or small portion of total time.
- *Special time reporting.* This method has been used to calculate costs for processing computer complaints. A special short-term collection of time distribution data may be appropriate if a department performs many different functions, activity is neither unusually small nor large, or there is uncertainty or significant disagreement among informed individuals as to the percentage of time or money allocated to a specific activity. A significant disagreement would typically be one of more than 10 percent of the total amount allocated.
- *Special data collections.* Besides collecting data on how much employee time is spent on an activity, an organization might also collect data on the amount of time a computer network is inoperative, the volume of items consumed or discarded, or the amount of time special equipment or other resources are not used.

In all these examples, the general calculation to determine costs of poor quality is:

Cost of poor quality = (cost of total resources in a category) × (percentage of resources in category used for activities related to poor quality)

Unit Cost

An example of this strategy occurs when a project team calculates the annual cost of correcting erroneous shipments. To find out the cost, the team should estimate the cost of correcting an average erroneous shipment, estimate how many such errors occurred in one year, and then multiply the average cost by the annual number of errors.

Focusing on unit cost requires two pieces of data: the number of times a particular deficiency occurs and the average cost for correcting and recovering from that deficiency when it does occur.

This average cost, in turn, is computed from a list of resources used to make corrections, on the amount used of each resource, and on the cost of each resource unit.

Unit cost is often the most appropriate strategy when deficiencies occur rarely and may be costly, when deficiencies are complex and require the participation of many departments to correct, or when deficiencies occur frequently and correcting them is so routine that those involved may not realize their pervasiveness.

Data on frequency of a deficiency may come from any of the following:

- Quality assurance
- Warranty data
- Customer surveys
- Field service reports
- Customer complaints

- Management engineering studies
- Internal audit reports
- Operational logs
- Special surveys

Estimating the cost of a single occurrence usually requires some analysis. A flowchart showing various rework loops associated with a deficiency can often help identify all-important resources used.

When searching for resources, consider hours worked by occupation and level, contracted services, materials and supplies, capital equipment and facilities, and cost of money for borrowed or uncollected funds.

To find out how much of each resource is used, check the following sources:

- Time reporting systems
- Cost accounting systems
- Various administrative logs
- Management engineering studies
- Informed judgment
- Special data collections

When a team has identified the amount of each resource used, it is ready to calculate the cost for each and add up costs for all resources. The finance or engineering functions typically will have standard methods for calculating the unit costs a team might require.

Here are hints to remember when calculating unit costs:

- Include benefits as well as wages and salaries.
- Include allocated capital costs for major equipment and facilities. While this is a minor consideration for many activities that can be safely ignored, it is vital for some activities.

Do not be misled by the argument that capital costs are fixed and would exist even if deficiencies did not occur. This is a typical example of the cost of poor quality being hidden by standard practices. If computers were used more efficiently, it would be possible to process more jobs without buying additional equipment. Idle capital or misused capital resources are a cost of poor quality just as surely as discarded paper from a faulty print job.

Be sure to include penalties or misused discounts for late payments and premium prices paid for rush orders or shipments.

Other Methods

Still other methods can be developed for special projects. For example, with regard to lost supplies, the organization should calculate the cost that would have been consumed if there had been no defects and the cost of supplies had actually been consumed. The difference between the two is the cost of poor quality. This type of approach might also be applied in comparing actual outcomes with the best outcomes others have achieved.

Special circumstances may lead a team to develop still other approaches that are appropriate to the specific problem. For example, a greater investment in prevention would be cost-effective.

The Potential Return on Investment

A major responsibility of upper managers is to make the best use of the organization's assets. A key measure of judging what is best is return on investment (ROI). In general terms, ROI is the ratio of (1) the estimated gain to (2) the estimated resources needed. Computing ROI for projects to reduce chronic waste requires assembling estimates such as

- The costs of chronic waste associated with the projects
- The potential cost reductions if the projects are successful
- The costs of the needed diagnosis and remedy

Many proposals to go into breakthrough have failed to gain management support because no one has quantified the ROI. Such a goal is a handicap to the upper managers—they are unable to compare (1) the potential ROI from breakthrough with (2) the potential ROI from other opportunities for investment.

Managers and others who prepare such proposals are well advised to prepare the information on ROI in collaboration with those who have expertise in the intricacies of ROI. Computation of ROI can be complicated because two kinds of money are involved—capital and expenses. Each is money, but in some countries (including the United States) they are taxed differently. Capital expenditures are made from after-tax money, whereas expenses are paid out of pretax money.

This difference in taxation is reflected in the rules of accounting. Expenses are written off promptly, thereby reducing the stated earnings and hence, the income taxes on earnings. Capital expenditures are written off gradually—usually over a period of years. This increases the stated earnings and hence, the income taxes on those earnings. This means it is advantageous for proposals to go into breakthrough because breakthrough is seldom capital-intensive. (Some upper managers tend to use the word *investment* as applying only to capital investment.)

Getting Cost Figures

Organization accounting systems typically quantify only a minority of the costs of poor quality. The majority are scattered throughout the various overheads. As a result, specialists have looked for ways to supply what is missing. Their main efforts toward solution have been as follows:

- *Make estimates*. This is the "quick and dirty" approach. It is usually done by sampling, and involves only a modest amount of effort. It can, in a few days or weeks, provide (a) an evaluation of the approximate cost of chronic waste and (b) indicate where this is concentrated.
- *Expand the accounting system*. This is much more elaborate. It requires a lot of work from various departments, especially accounting, and it runs into a lot of calendar time, often two or three years.

In my experience, estimates involve much less work, can be prepared in far less time, and yet are adequate for managerial decision making.

—J.M. Juran

Note that the demand for "accuracy" of the cost figures depends on the use to which the figures will be put. Balancing the books demands a high degree of accuracy. Making managerial decisions sometimes can tolerate a margin of error. For example, a potential breakthrough project has been estimated to incur about $300,000 in annual cost of poor quality. This figure is challenged. The contesting estimates range from $240,000 to $360,000—quite a wide range. Then someone makes an incisive observation: "It doesn't matter which estimate is correct. Even at the lowest figure, this is a good opportunity for breakthrough, so let's tackle it." In other words, the managerial decision to tackle the project is identical despite a wide range of estimate.

Languages in the Hierarchy

A subtle aspect of securing upper management approval is the choice of language. Industrial organizations make use of two standard languages—the language of money and the language of things. (There are also local dialects, each peculiar to a specific function.) However, as seen in Figure 5.4, use of the standard languages is not uniform.

Figure 5.4 shows the use of standard languages in different levels of a typical hierarchy. At the apex, the principal language of the top management team is the language of money. At the base, the principal language of the first-line supervisors and the workforce is the language of things. In between, the middle managers and the specialists need to understand both principal languages—*the middle managers should be bilingual.*

It is quite common for chronic waste to be measured in the language of things: percent errors, process yields, hours of rework, and so on. Converting these measures into the language of money enables upper managers to relate them to the financial measures that have long dominated the management "instrument panel."

Years ago, I was invited to visit a major British manufacturer to study its approach to managing for quality and to provide a critique. I found that the organization's cost of poor quality was huge, that it was feasible to cut this in two in five years, and that the resulting return on investment would be much greater than that of making and selling the organization's products. When I explained this to the managing director, he was most impressed—it

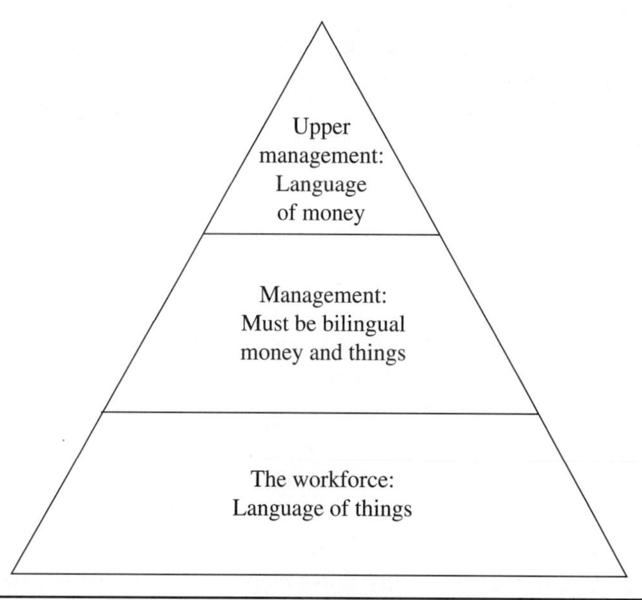

FIGURE 5.4 Common languages in the hierarchy. (*Juran Institute, Inc., 1994.*)

was the first time that the problem of chronic waste had been explained to him in the language of return on investment. He promptly convened his directors (vice presidents) to discuss what to do about this opportunity.

Presentations to Upper Managers

Presentations to upper managers should focus on the goals of the upper managers, not on the goals of the advocates. Upper managers are faced with meeting the needs of various stakeholders: customers, owners, employees, suppliers, the public (e.g., safety, health, environment), and so on. It helps if the proposals identify specific problems of stakeholders and estimate the benefits to be gained.

Upper managers receive numerous proposals for allocating the organization's resources: invade foreign markets, develop new products, buy new equipment to increase productivity, make acquisitions, enter joint ventures, and so on. These proposals compete with each other for priority, and a major test is ROI. It helps if the proposal to go into breakthrough includes estimates of ROI.

An explanation of proposals is sometimes helped by converting the supporting data into units of measure that are already familiar to upper managers. For example:

- Last year's cost of poor quality was five times last year's profit of $1.5 million.
- Cutting the cost of poor quality in half would increase earnings by 13 cents per share of stock.
- Thirteen percent of last year's sales orders were canceled due to poor quality.
- Thirty-two percent of engineering time was spent in finding and correcting design weaknesses.
- Twenty-five percent of manufacturing capacity is devoted to correcting problems.
- Seventy percent of the inventory carried is traceable to poor quality.
- Twenty-five percent of all manufacturing hours were spent in finding and correcting defects.
- Last year's cost of poor quality was the equivalent of our operation making 100 percent defective work during the entire year.

Experience in making presentations to upper management has provided some useful dos and don'ts:

- *Do* summarize the total of the estimated costs of poor quality. The total will be big enough to command upper management's attention.
- *Do* show where these costs are concentrated. A common grouping is in the form of Table 5.2. Typically (as in that case), most of the costs are associated with failures, internal and external. Table 5.2 also shows the fallacy of trying to start by reducing inspection and test. The failure costs should be reduced first. After the defect levels come down, inspection costs can be reduced as well.
- *Do* describe the principal projects that are at the heart of the proposal.
- *Do* estimate the potential gains, as well as the return on investment. If the organization has never before undertaken an organized approach to reducing related costs, then a reasonable goal is to cut these costs in two within a space of five years.
- *Do* have the figures reviewed in advance by those people in finance (and elsewhere) to whom upper management looks for checking the validity of financial figures.

- *Don't* inflate the present costs by including debatable or borderline items. The risk is that the decisive review meetings will get bogged down in debating the validity of the figures without ever discussing the merits of the proposals.
- *Don't* imply that the total costs will be reduced to zero. Any such implication will likewise divert attention from the merits of the proposals.
- *Don't* force the first few projects on managers who are not really sold on them or on unions who are strongly opposed. Instead, start in areas that show a climate of receptivity. The results obtained in these areas will determine whether the overall initiative will expand or die out.

The needs for breakthrough go beyond satisfying customers or making cost reductions. New forces keep coming over the horizon. Recent examples have included growth in product liability, the consumerism movement, foreign competition, legislation, and environmental concerns of all sorts. Breakthrough has provided much of the response to such forces.

Similarly, the means of convincing upper managers of the need for breakthrough go beyond reports from advocates. Conviction also may be supplied by visits to successful organizations, hearing papers presented at conferences, reading reports published by successful organizations, and listening to the experts, both internal and external. However, none of these is as persuasive as results achieved within one's own organization.

A final element of presentations to upper managers is to explain their personal responsibilities in launching and perpetuating breakthrough.

Mobilizing for Breakthrough

Until the 1980s, breakthrough in the West was not mandated—it was not a part of the business plan or a part of the job descriptions. Some breakthrough did take place, but on a voluntary basis. Here and there, a manager or a nonmanager, for whatever reason, elected to tackle some breakthrough project. He or she might persuade others to join an informal team. The result might be favorable, or it might not. This voluntary, informal approach yielded few breakthroughs. The emphasis remained on inspection, control, and firefighting.

The Need for Formality

The crisis that followed the Japanese revolution called for new strategies, one of which was a much higher rate of breakthrough. It then became evident that an informal approach would not produce thousands (or more) breakthroughs year after year. This led to experiments with structured approaches that in due course helped some organizations become role models.

Some upper managers protested the need for formality: "Why don't we just do it?" The answer depends on how many breakthroughs are needed. For just a few projects each year, informality is adequate; there is no need to mobilize. However, making breakthroughs by the hundreds or the thousands requires a formal structure.

As it has turns out, mobilizing for breakthrough requires two levels of activity, as shown in Figure 5.5. The figure shows the two levels of activity. One of these mobilizes the organization's resources to deal with the breakthrough projects collectively. This becomes the responsibility of management. The other activity is needed to carry out the projects individually. This becomes the responsibility of the breakthrough teams.

Activities by Management	Activities by Project Teams
Establish infrastructure: quality councils	Verify problem
Select problems; determine goals and targets	Analyze symptoms
	Theorize as to causes
Create project charters and assign teams	Test theories
Launch teams and review progress	Discover causes
Provide recognition and rewards	Stimulate remedies and controls

FIGURE 5.5 Mobilizing for breakthrough.

The Executive "Quality Council"

The first step in mobilizing for breakthrough is to establish the organization's council (or similar name). The basic responsibility of this council is to launch, coordinate, and "institutionalize" annual breakthrough. Such councils have been established in many organizations. Their experiences provide useful guidelines.

Membership and Responsibilities

Council membership is typically drawn from the ranks of senior managers. Often, the senior management committee is also the council. Experience has shown that councils are most effective when upper managers are personally the leaders and members of the senior councils.

In large organizations, it is common to establish councils at the divisional level as well as at the corporate level. In addition, some individual facilities may be so large as to warrant establishing a local council. When multiple councils are established, they are usually linked together—members of high-level councils serve as chairpersons of lower-level councils. Figure 5.6 is an example of such linkage.

Experience has shown that organizing councils solely in the lower levels of management is ineffective. Such organization limits breakthrough projects to the "useful many" while neglecting the "vital few" projects—those that can produce the greatest results. In addition, councils solely at lower levels send a message to all: "Breakthrough is not high on upper management's agenda."

It is important for each council to define and publish its responsibilities so that (1) the members agree on their goal and (2) the rest of the organization can become informed relative to upcoming events.

Many councils have published their statements of responsibility. Major common elements have included the following:

- Formulate the policies, such as focus on the customer has top priority, breakthrough must go on year after year, participation should be universal, or the reward system should reflect performance on breakthrough.
- Estimate the major dimensions, such as the status of the company's quality compared with its competitors, the extent of chronic waste, the adequacy of major business processes, or the results achieved by prior breakthroughs.
- Establish processes for selecting projects, such as soliciting and screening nominations, choosing projects to be tackled, preparing goal statements, or creating a favorable climate for breakthrough.

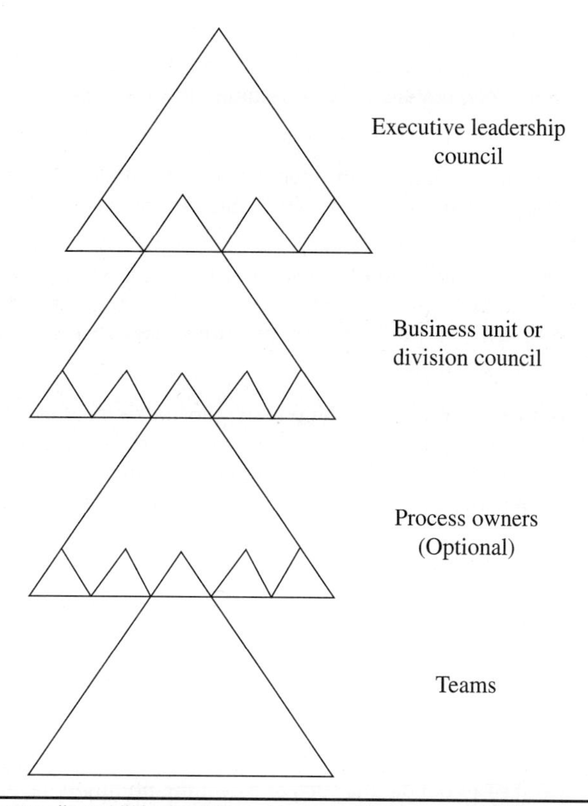

Figure 5.6 Quality councils are linked together. (*Juran Institute, Inc., 1994.*)

- Establish processes for carrying out the projects, such as selecting team leaders and members or defining the role of project teams.
- Provide support for the project teams, such as training time for working on projects, diagnostic support, facilitator support, or access to facilities for tests and tryouts.
- Establish measures of progress, such as effect on customer satisfaction, effect on financial performance, or extent of participation by teams.
- Review progress, assist teams in the event of obstacles, and ensure that remedies are implemented.
- Provide for public recognition of teams.
- Revise the reward system to reflect the changes demanded by introducing annual breakthrough.

Councils should anticipate the troublesome questions and, to the extent feasible, provide answers at the time of announcing the intention to go into annual breakthrough. Some senior managers have gone to the extent of creating a videotape to enable a wide audience to hear the identical message from a source of undoubted authority.

Leaders Must Face Up to the Apprehensions about Elimination of Jobs

Employees not only want dialogue on such an important issue, they also want assurance relative to their apprehensions, notably the risk of job loss due to improvements. Most upper managers have been reluctant to face up to these apprehensions. Such reluctance is understandable. It is risky to provide assurances when the future is uncertain.

Nevertheless, some managers have estimated in some depth the two pertinent rates of change:

- The rate of creation of job openings due to attrition: retirements, offers of early retirement, resignation, and so on. This rate can be estimated with a fair degree of accuracy.
- The rate of elimination of jobs due to reduction of chronic waste. This estimate is more speculative—it is difficult to predict how soon the breakthrough rate will get up to speed. In practice, organizations have been overly optimistic in their estimates.

Analysis of these estimates can help managers judge what assurances they can provide, if any. It also can shed light on the choice of alternatives for action: retrain for jobs that have opened up, reassign to areas that have job openings, offer early retirement, assist in finding jobs in other organizations, and/or provide assistance in the event of termination.

Assistance from the Quality and/or Performance Excellence Functions

Many councils secure the assistance of the performance excellence and quality departments. These are specialists that are skilled in the methods and tools to attain high quality. They are there to

- Provide inputs needed by the council for planning to introduce breakthrough
- Draft proposals and procedures
- Carry out essential details such as screening nominations for projects
- Develop training materials
- Develop new scorecards
- Prepare reports on progress

It is also usual for the quality directors to serve as secretaries of the council.

Breakthrough Goals in the Business Plan

Organizations that have become the market leaders—the role models—all adopted the practice of enlarging their business plan to include quality goals. In effect, they translated the threats and opportunities faced by their organizations into goals, such as

- Increase on-time deliveries from 83 to 100 percent over the next two years.
- Reduce the cost of poor quality by 50 percent over the next five years.

Such goals are clear—each is quantified, and each has a timetable. Convincing upper managers to establish such goals is a big step, but it is only the first step.

Deployment of Goals

Goals are merely a wish list until they are deployed—until they are broken down into specific projects to be carried out and assigned to specific individuals or teams who are then provided with the resources needed to take action. Figure 5.7 shows the anatomy of the

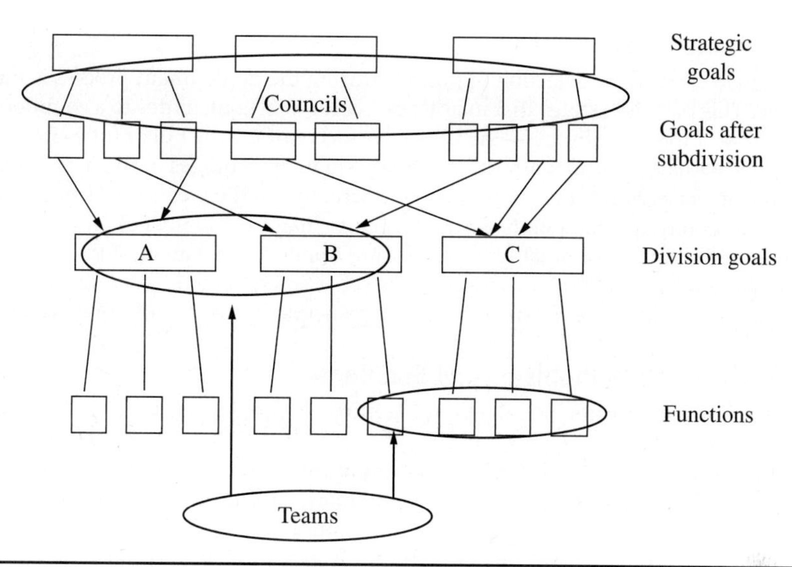

FIGURE 5.7 Anatomy of the deployment process. (*Juran Institute, Inc.*)

deployment process. In the figure, the broad (strategic) goals are established by the council and become a part of the organization's business plan. These goals are then divided and allocated to lower levels to be translated into action. In large organizations, there may be further subdivision before the action levels are reached. The final action level may consist of individuals or teams.

In response, the action levels select breakthrough projects that collectively will meet the goals. These projects are then proposed to the upper levels along with estimates of the resources needed. The proposals and estimates are discussed and revised until final decisions are reached. The end result is an agreement on which projects to tackle, what resources to provide, and who will be responsible for carrying out the projects.

This approach of starting at the top with strategic goals may seem like a purely top-down activity. However, the deployment process aims to provide open discussion in both directions before final decisions are made, and such is the way it usually works out.

The concept of strategic goals involves the vital few matters, but it is not limited to the corporate level. Goals also may be included in the business plans of divisions, profit centers, field offices, and still other facilities. The deployment process is applicable to all of these. (For added discussion of the deployment process, see Chapter 7, Strategic Planning and Deployment: Moving from Good to Great.)

The Project Concept

As used here, a project is a chronic problem scheduled for solution. The project is the focus of actions for breakthrough. All breakthrough takes place project by project and in no other way.

Some projects are derived from the goals that are in the organization's business plan. These are relatively few in number, but each is quite important. Collectively, these are among the vital few projects (see "Use of the Pareto Principle"). However, most projects are derived not from the organization's business plan but from the nomination-selection process, as discussed later.

Use of the Pareto Principle

A valuable aid to the selection of projects during the deployment process is the Pareto principle. This principle states that in any population that contributes to a common effect, a relative few of the contributors—the vital few—account for the bulk of the effect. The principle applies widely in human affairs. Relatively small percentages of the individuals write most of the books, commit most of the crimes, own most of the wealth, and so on.

Presentation of data in the form of a Pareto diagram greatly enhances communication of the information, most notably in convincing upper management of the source of a problem and gaining support for a proposed course of action to remedy the problem. (For an account of how Dr. Juran came to name the Pareto principle, see Appendix in this Handbook.)

The Useful Many Problems and Solutions

Under the Pareto principle, the vital few projects provide the bulk of the breakthrough, so they receive top priority. Beyond the vital few are the useful many problems. Collectively, they contribute only a minority of the breakthrough, but they provide most of the opportunity for employee participation. The useful many projects are made through the application of workplace improvement teams, quality circles, the lean 5S tools, or self-directed work teams. See Chapter 26, Empowering the Workforce to Tackle the "Useful Many" Processes.

The Nomination and Selection Process

Most projects are chosen through the nomination and selection process, involving several steps:

- Project nomination
- Project screening and selection
- Preparation and publication of project goal statements

Sources of Nominations

Nominations for projects can come from all levels of the organization. At the higher levels, the nominations tend to be extensive in size (the vital few) and multifunctional in their scope. At lower levels, the nominations are smaller in size (the useful many) and tend to be limited in scope to the boundaries of a single department.

Nominations come from many sources. These include

- *Formal data systems*, such as field reports on product performance, customer complaints, claims, returns, and so on; accounting reports on warranty charges and on internal costs of poor quality; and service call reports. (Some of these data systems provide for analyzing the data to identify problem areas.)
- *Special studies*, such as customer surveys, employee surveys, audits, assessments, benchmarking against competitors, and so on.
- *Reactions from customers* who have run into product dissatisfactions are often vocal and insistent. In contrast, customers who judge product features to be not competitive may simply (and quietly) become ex-customers.
- *Field intelligence* derived from visits to customers, suppliers, and others; actions taken by competitors; and stories published in the media (as reported by sales, customer service, technical service, and others).

- *The impact on society*, such as new legislation, extension of government regulation, and growth of product liability lawsuits.
- *The managerial hierarchy*, such as the council, managers, supervisors, professional specialists, and project teams.
- *The workforce* through informal ideas presented to supervisors, formal suggestions, ideas from circles, and so on.
- *Proposals* relating to business processes.

Effect of the Organizationwide or Big Q Concept

Beginning in the 1980s and continuing for the near future, the scope of nominations for projects broadened considerably under the big Q concept. The breadth of the big Q concept is evident from the wide variety of projects that have already been tackled:

- Improve the precision of the sales forecast.
- Reduce the cycle time for developing new products.
- Increase the success rate in bidding for business.
- Reduce the time required to fill customers' orders.
- Reduce the number of sales cancellations.
- Reduce the errors in invoices.
- Reduce the number of delinquent accounts.
- Reduce the time required to recruit new employees.
- Improve the on-time arrival rate (for transportation services).
- Reduce the time required to file for patents.

The Nomination Process

Nominations must come from human beings. Data systems are impersonal—they make no nominations. Various means are used to stimulate nominations for breakthrough projects:

- *Call for nominations.* Letters or bulletin boards are used to invite all personnel to submit nominations, either through the chain of command or to a designated recipient, such as the secretary of the council.
- *Make the rounds.* In this approach, specialists (such as engineers) are assigned to visit the various departments, talk with the key people, and secure their views and nominations.
- *The council members themselves.* They become a focal point for extensive data analyses and proposals.
- *Brainstorming meetings.* These are organized for the specific purpose of making nominations.

Whatever the method used, it will produce the most nominations if it urges use of the big Q concept—the entire spectrum of activities, products, and processes.

Nominations from the Employees at All Levels

The workforce is potentially a source of numerous nominations. Workers have extensive residence in the workplace. They are exposed to many local cycles of activity. Through this exposure, they are well poised to identify the existence of problems and to theorize about their causes. As to the details of goings-on in the workplace, no one is better informed than the workforce. "That machine hasn't seen a maintenance man for the last six months." In addition, many workers are well poised to identify opportunities and to propose new ways.

Workforce nominations consist mainly of local useful many projects along with proposals of a human relations nature. For such nominations, workers can supply useful theories of causes as well as practical proposals for remedies. For projects of a multifunctional nature, most workers are handicapped by their limited knowledge of the overall process and of the interactions among the steps that collectively make up the overall process.

In some organizations, the solicitation of nominations from the workforce has implied that such nominations would receive top priority. The effect was that the workforce was deciding which projects the managers should tackle first. It should have been made clear that workers' nominations must compete for priority with nominations from other sources.

Joint Projects with Suppliers and Customers

All organizations buy goods and services from suppliers; over half the content of the finished product may come from suppliers. In earlier decades, it was common for customers to contend. The supplier should solve his problems. Now there is growing awareness that these problems require a partnership approach based on

- Establishing mutual trust
- Defining customer needs as well as specifications
- Exchanging essential data
- Direct communication at the technical level as well as the commercial level

Project Screening

Calls for nominations can produce large numbers of responses—numbers that are beyond the digestive capacity of the organization. In such cases, an essential further step is screening to identify those nominations that promise the most benefits for the effort expended.

To start with a long list of nominations and end up with a list of agreed-upon projects requires an organized approach—an infrastructure and a methodology. The screening process is time-consuming, so the council usually delegates it to a secretariat, often the department. The secretariat screens the nominations—it judges the extent to which the nominations meet the criteria set out below. These judgments result in some preliminary decision-making. Some nominations are rejected. Others are deferred. The remainder is analyzed in greater depth to estimate potential benefits, resources needed, and so on.

The councils and/or the secretariats have found it useful to establish criteria to be used during the screening process. Experience has shown that there is a need for two sets of criteria:

- Criteria for choosing the first projects to be tackled by any of the project teams
- Criteria for choosing projects thereafter

Criteria for Projects

During the beginning stages of project-by-project breakthrough, everyone is in a learning state. Projects are assigned to project teams, who are in training. Completing a project is a part of that training. Experience with such teams has evolved a broad set of criteria:

- The project should deal with a *chronic problem*—one that has been awaiting a solution for a long time.
- The project should be *feasible*. There should be a good likelihood of completing it within a few months. Feedback from organizations suggests that the most frequent reason for failure of the first project has been failure to meet the criterion of feasibility.
- The project should be *significant*. The end result should be sufficiently useful to merit attention and recognition.
- The results should be *measurable*, whether in money or in other significant terms.
- The first projects should be winners.

Additional criteria to select projects are aimed at what will do the organization the most good:

- *Return on investment*. This factor has great weight and is decisive, all other things being equal. Projects that do not lend themselves to computing return on investment must rely for their priority on managerial judgment.
- *The amount of potential breakthrough*. One large project will take priority over several small ones.
- *Urgency*. There may be a need to respond promptly to pressures associated with product safety, employee morale, and customer service.
- *Ease of technological solution*. Projects for which the technology is well developed will take precedence over projects that require research to discover the needed technology.
- *Health of the product line*. Projects involving thriving product lines will take precedence over projects involving obsolescent product lines.
- *Probable resistance to change*. Projects that will meet a favorable reception take precedence over projects that may meet strong resistance, such as from the labor union or from a manager set in his or her ways.

Most organizations use a systematic approach to evaluate nominations relative to these criteria. This yields a composite evaluation that then becomes an indication of the relative priorities of the nominations. (For more detail and an example of a project selection matrix, see Chapter 12, Six Sigma: Improving Process Effectiveness.)

Project Selection

The result of the screening process is a list of recommended projects in their order of priority. Each recommendation is supported by the available information on compatibility with the criteria and potential benefits, resources required, and so on. This list is commonly limited to matters in which the council has a direct interest.

The council reviews the recommendations and makes the final determination on which projects are to be tackled. These projects then become an official part of the organization's business. Other recommended projects are outside the scope of the direct interest of the council. Such projects are recommended to appropriate subcouncils, managers, and so on. None of the preceding prevents projects from being undertaken at local levels by supervisors or by the workforce.

Vital Few and Useful Many

Some organizations completed many projects. Then, when questions were raised—"What have we gotten for all this effort?"—they were dismayed to learn that there was no noticeable effect on the bottom line. Investigation then showed that the reason could be traced to the process used for project selection. The projects actually selected had consisted of

- *Firefighting projects*. These are special projects for getting rid of sporadic "spikes." Such projects did not attack the chronic waste and hence, could not improve financial performance. (See Chapter 13, Root Cause Analysis to Maintain Performance)
- *Useful many projects*. By definition, these have only a minor effect on financial performance but have great effect on human relations.
- *Projects for improving human relations*. These can be quite effective in their field, but the financial results are usually not measurable.

To achieve a significant effect on the bottom line requires selecting the "vital few" projects as well as the "useful many." It is feasible to work on both, since different people are assigned to each.

There is a school of thought emerging that contends that the key to market leadership is "tiny breakthroughs in a thousand places"—in other words, the useful many Another school urges focus on the vital few. In my experience, neither of these schools has the complete answer. Both are needed—at the right time.

The vital few projects are the major contributors to leadership and to the bottom line. The useful many projects are the major contributors to employee participation and to the quality of work life. Each is necessary; neither is sufficient.

The vital few and useful many projects can be carried out simultaneously. Successful organizations have done just that by recognizing that while there are these two types of projects, they require the time of different categories of organization personnel.

The interrelation of these two types of projects is shown in Figure 5.8. In this figure, the horizontal scale is time. The vertical scale is chronic waste. What goes up is bad. The useful

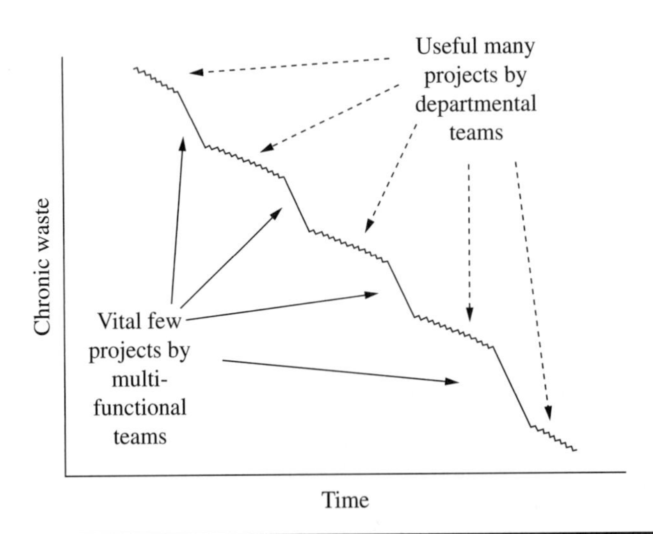

FIGURE 5.8 Interrelation of projects. (*Juran Institute, Inc., 1994.*)

many breakthroughs collectively create a gradually sloping line. The vital few breakthroughs, though less frequent, contribute the bulk of the total breakthrough.

Cost Figures for Projects

To meet the preceding criteria (especially that of return on investment) requires information on various costs:

- The cost of chronic waste associated with a given nomination
- The potential cost reduction if the project is successful
- The cost of the needed diagnosis and remedy

Costs versus Percent Deficiencies

It is risky to judge priorities based solely on the percentage of deficiencies (errors, defects, and so on). On the face of it, when this percentage is low, the priority of the nomination also should be low. In some cases this is true, but in others it can be seriously misleading.

Elephant-Sized and Bite-Sized Projects

There is only one way to eat an elephant: bite by bite. Some projects are "elephant-sized"; that is, they cover so broad an area of activity that they must be subdivided into multiple "bite-sized" projects. In such cases, one project team can be assigned to "cut up the elephant." Other teams are then assigned to tackle the resulting bite-sized projects. This approach shortens the time to complete the project, since the teams work concurrently. In contrast, use of a single team stretches the time out to several years. Frustration sets in, team membership changes due to attrition, the project drags, and morale declines.

A most useful tool for cutting up the elephant is the Pareto analysis. For an application, see the paper mill example earlier, under "Use of the Pareto Principle." For elephant-sized projects, separate goal statements are prepared for the broad coordinating team and for each team assigned to a bite-sized project.

Replication and Cloning

Some organizations consist of multiple autonomous units that exhibit much commonality. A widespread example is the chains of retail stores, repair shops, hospitals, and so on. In such organizations, a breakthrough project that is carried out successfully in one operating unit logically becomes a nomination for application to other units. This is called cloning the project.

It is quite common for the other units to resist applying the breakthrough to their operation. Some of this resistance is cultural in nature (not invented here, and so on). Other resistance may be due to real differences in operating conditions. For example, telephone exchanges perform similar functions for their customers. However, some serve mainly industrial customers, whereas others serve mainly residential customers.

Upper managers are wary of ordering autonomous units to clone breakthroughs that originated elsewhere. Yet cloning has advantages. Where feasible, it provides additional breakthroughs without the need to duplicate the prior work of diagnosis and design of remedy.

What has emerged is a process as follows:

- Project teams are asked to include in their final report their suggestions as to sites that may be opportunities for cloning.
- Copies of such final reports go to those sites.
- The decision of whether to clone is made by the sites.

However, the sites are required to make a response as to their disposition of the matter. This response is typically in one of three forms:

- We have adopted the breakthrough.
- We will adopt the breakthrough, but we must first adapt it to our conditions.
- We are not able to adopt the breakthrough for the following reasons.

In effect, this process requires the units to adopt the breakthrough or give reasons for not doing so. The units cannot just quietly ignore the recommendation.

A more subtle but familiar form of cloning is done through projects that have repetitive application over a wide variety of subject matter.

A project team develops computer software to find errors in spelling. Another team evolves an improved procedure for processing customer orders through the organization. A third team works up a procedure for conducting design reviews. What is common about such projects is that the result permits repetitive application of the same process to a wide variety of subject matter: many different misspelled words, many different customer orders, and many different designs.

Project Charters: Problem and Goal Statements for Projects

Each project selected should be accompanied by a written problem and goal statement that sets out the intended focus and the intended result of the project. Upon approval, this statement defines the actions required of the team assigned to carry out the project.

The Purpose of the Project Charter

The problem and goal statement serves a number of essential purposes:

- It defines the problem and the intended result and so helps the team know when it has completed the project.
- It establishes clear responsibility—the goal becomes an addition to each team member's job description.
- It provides legitimacy—the project becomes official organization business. The team members are authorized to spend the time needed to carry out the goal.
- It confers rights—the team has the right to hold meetings, ask people to attend and assist the team, and request data and other services germane to the project.

Perfection as a Goal

There is universal agreement that perfection is the ideal goal—complete freedom from errors, defects, failures, and so on. The reality is that the absence of perfection is due to many kinds of such deficiencies and that each requires its own breakthrough project. If a organization tries to eliminate all of them, the Pareto principle applies

- The vital few kinds of deficiencies cause most of the trouble but also readily justify the resources needed to root them out. Hence, they receive high priority during the screening process and become projects to be tackled.
- The remaining many types of deficiencies cause only a small minority of the trouble. As one comes closer and closer to perfection, each remaining kind of deficiency

becomes rarer and rarer and hence, receives lower and lower priority during the screening process.

All organizations tackle those rare types of failure that threaten human life or that risk significant economic loss. In addition, organizations that make breakthroughs by the thousands year after year tackle even the mild, rare kinds of deficiency. To do so, they enlist the creativity of the workforce.

Some critics contend that publication of any goal other than perfection is proof of a misguided policy—a willingness to tolerate defects. Such contentions arise out of a lack of experience with the realities. It is easy to set goals that demand perfection now. Such goals, however, require organizations to tackle failure types so rare that they do not survive the screening process.

Nevertheless, there has been progress. During the twentieth century, there was a remarkable revision in the unit of measure for deficiencies. In the first half of the century, the usual measure was in percent defective, or defects per hundred units. By the 1990s, many industries had adopted a measure of defects per million units and use Sigma metrics and calculations. The leading organizations now do make thousands of breakthroughs year after year. They keep coming closer to perfection, but it is a never-ending process.

While many nominated projects cannot be justified solely on their return on investment, they may provide the means for employee participation in the breakthrough process, which has value in its own right.

The Project Team

For each selected project, a team is assigned. This team then becomes responsible for completing the project. Why a team? The most important projects are the vital few, and they are almost invariably multifunctional in nature. The symptoms typically show up in one department, but there is no agreement on where the causes lie, what the causes are, or what the remedies should be. Experience has shown that the most effective organizational mechanisms for dealing with such multifunctional problems are multifunctional teams.

Some managers prefer to assign problems to individuals rather than to teams. ("A camel is a horse designed by a committee.") The concept of individual responsibility is in fact quite appropriate if applied to control. ("The best form of control is self-control.") However, breakthrough, certainly for multifunctional problems, inherently requires teams. For such problems, assignment to individuals runs severe risks of departmental biases in the diagnosis and remedy.

A process engineer was assigned to reduce the number of defects coming from a wave soldering process. His diagnosis concluded that a new process was needed. Management rejected this conclusion on the grounds of excess investment. A multifunctional team was then appointed to restudy the problem. The team found a way to solve the problem by refining the existing process (Betker 1983).

Individual biases also show up as cultural resistance to proposed remedies. However, such resistance is minimal if the remedial department has been represented on the project team.

Appointment of Teams/Sponsors

Project teams are not attached to the chain of command on the organization chart. This can be a handicap in the event that teams encounter an impasse. For this reason, some organizations assign council members or other upper managers to be sponsors (or "champions") of specific projects. These sponsors follow team progress (or lack thereof). If the team does run into an impasse, the sponsor may be able to help the team get access to the proper person in the hierarchy.

Teams are appointed by sponsors of the projects, process owners, local managers, or others. In some organizations, workforce members are authorized to form teams (circles and so on) to work on breakthrough projects. Whatever the origin, the team is empowered to make the breakthrough as defined in the goal statement.

Most teams are organized for a specific project and are disbanded on completion of the project. Such teams are called ad hoc, meaning "for this purpose." During their next project, the members will be scattered among several different teams. There are also "standing" teams that have continuity—the members remain together as a team and tackle project after project.

Team Responsibilities

A project team has responsibilities that are coextensive with the goal statement. The basic responsibilities are to carry out the assigned goal and to follow the universal breakthrough process. In addition, the responsibilities include

- Proposing revisions to the goal statement
- Developing measurements as needed
- Communicating progress and results to all who have a need to know

Membership

The team is selected by the sponsor after consulting with the managers who are affected. The selection process includes consideration of (1) which departments should be represented on the team, (2) what level in the hierarchy team members should come from, and (3) which individuals in that level.

The departments to be represented should include:

- *The ailing department*. The symptoms show up in this department, and it endures the effects.
- *Suspect departments*. They are suspected of harboring the causes. (They do not necessarily agree that they are suspect.)
- *Remedial departments*. They will likely provide the remedies. This is speculative, since in many cases, the causes and remedies come as surprises.
- *Diagnostic departments*. They are needed in projects that require extensive data collection and analysis.
- *On-call departments and subject matter experts (SMEs)*. They are invited in as needed to provide special knowledge or other services required by the team.

This list includes the usual sources of members. However, there is need for flexibility.

Choice of level in the hierarchy depends on the subject matter of the project. Some projects relate strongly to the technological and procedural aspects of the products and processes. Such projects require team membership from the lower levels of the hierarchy. Other projects relate to broad business and managerial matters. For such projects, the team members should have appropriate business and managerial experience.

Finally comes the selection of individuals. This is negotiated with the respective supervisors, giving due consideration to workloads, competing priorities, and so on. The focus is on the individual's ability to contribute to the team project. The individuals need

- *Time* to attend the team meetings and to carry out assignments outside the meetings—"the homework."
- A *knowledge base* that enables the individual to contribute theories, insights, and ideas, as well as job information based on his or her hands-on experience.
- *Training* in the breakthrough process and the associated tools. During the first projects, this training can and should be done concurrently with carrying out the projects.

Most teams consist of six to eight members. Larger numbers tend to make the team unwieldy as well as costly. (A convoy travels only as fast as the slowest ship.)

Should team members all come from the same level in the hierarchy? Behind this question is the fear that the biases of high-ranking members will dominate the meeting. Some of this no doubt takes place, especially during the first few meetings. However, it declines as the group dynamics take over and as members learn to distinguish between theory and fact.

Once the team is selected, the members' names are published, along with their project goal. The act of publication officially assigns responsibility to the individuals as well as to the team. In effect, serving on the project team becomes a part of the individuals' job descriptions. This same publication also gives the team the legitimacy and rights discussed earlier.

Membership from the Workforce

During the early years of using breakthrough teams, organizations tended to maintain a strict separation of team membership. Teams for multifunctional projects consisted exclusively of members from the managerial hierarchy plus professional specialists. Teams for local departmental projects (such as quality circles and employee involvement teams) consisted exclusively of members from the workforce. Figure 5.9 compares the usual features of these teams with those of multifunctional teams.

Experience then showed that as to the details of operating conditions, no one is better informed than the workforce. Through residence in the workplace, workers can observe local changes and recall the chronology of events. This has led to a growing practice of securing such information by interviewing the workers. The workers become can be "on call" or full-time team members. In a hospital, a doctor can be considered in the same way. Removing a worker tied directly to the production of a product must minimize their time away from their work or patients.

One result of all this experience has been a growing interest in broadening worker participation generally. This has led to experimenting with project teams that make no distinction as to rank in the hierarchy. These teams may become the rule rather than the exception. (For further discussion on the trends in workforce participation, see Chapter 26, Empowering the Workforce to Tackle the "Useful Many" Processes.

Upper Managers on Teams

Some projects, by their very nature, require that the team include members from the ranks of upper management. Here are some examples of breakthrough projects actually tackled by teams that included upper managers:

- Shorten the time to put new products on the market.
- Improve the accuracy of the sales forecast.
- Reduce the carryover of prior failure-prone features into new product models.

Feature	Department Teams or Quality Circles	Breakthrough Teams
Primary purpose	To improve departmental processes and human relations	To improve performance by creating breakthroughs across multiple departments
Secondary purpose	To improve quality	To improve teamwork and participation
Scope of project	Within a single department	Across multiple departments
Size of project	One of the useful many	One of the vital few
Membership	From a single department	From multiple departments
Basis of membership	Voluntary or mandatory	Mandatory
Hierarchical status of members	The manager, staff in any department	Management, subject matter experts, and the workforce
Continuity	Team remains intact, project after project	Team is ad hoc, disbands after project is completed

Figure 5.9 Contrast departmental teams and quality circles to multifunctional teams. (*From Making Quality Happen, Juran Institute, Inc., 1998.*)

- Establish a teamwork relationship with suppliers.
- Develop the new measures needed for strategic planning.
- Revise the system of recognition and rewards for breakthrough.

There are some persuasive reasons urging all upper managers to personally serve on some project teams. Personal participation on project teams is an act of leadership by example. This is the highest form of leadership. Personal participation on project teams also enables upper managers to understand what they are asking their subordinates to do, what kind of training is needed, how many hours per week are demanded, how many months it takes to complete the project, and what kinds of resources are needed. Lack of upper management understanding of such realities has contributed to the failure of some well-intentioned efforts to establish annual breakthrough.

Model of the Infrastructure

There are several ways to show in graphic form the infrastructure for breakthrough—the elements of the organization, how they relate to each other, and the flow of events. Figure 5.10 shows the elements of the infrastructure in pyramid form. The pyramid depicts a hierarchy consisting of top management, the autonomous operating units, and the major staff functions. At the top of the pyramid are the corporate council and the subsidiary councils, if any. Below these levels are the multifunctional breakthrough teams. (There may be a committee structure between the councils and the teams.)

At the intradepartmental level are teams from the workforce—circles or other forms. This infrastructure permits employees in all levels of the organization to participate in breakthrough projects, the useful many as well as the vital few.

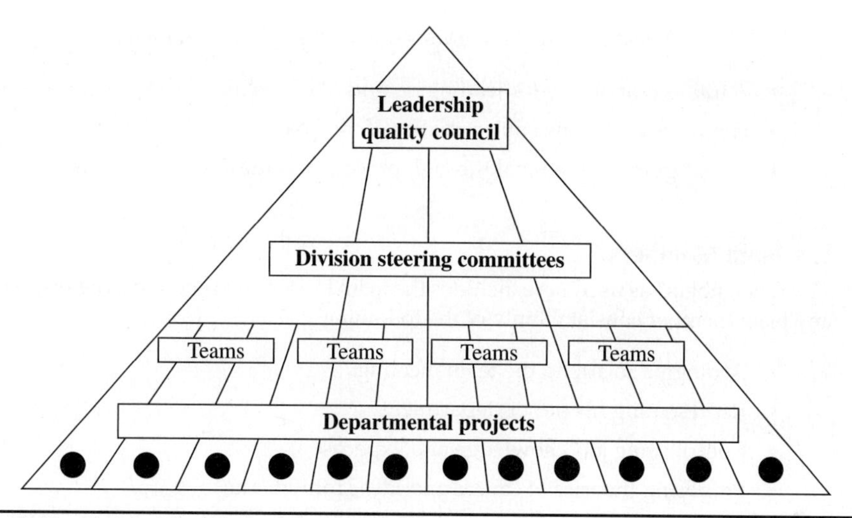

Figure 5.10 Model of the infrastructure for breakthrough quality improvement. (*Juran Institute, Inc.*)

Team Organization

Breakthrough teams do not appear on the organization chart. Each "floats"—it has no personal boss. Instead, the team is supervised *impersonally* by its goal statement and by the breakthrough roadmap.

The team does have its own internal organizational structure. This structure invariably includes a team *leader* (chairperson and so on) and a team *secretary*. In addition, there is usually a *facilitator*.

The Team Leader

The leader is usually appointed by the sponsor—the council or other supervising group. Alternatively, the team may be authorized to elect its leader.

The leader has several responsibilities. As a team member, the leader *shares* in the responsibility for completing the team's goal. In addition, the leader has administrative duties. These are *unshared* and include

- Ensuring that meetings start and finish on time
- Helping the members attend the team meetings
- Ensuring that the agendas, minutes, reports, and so on are prepared and published
- Maintaining contact with the sponsoring body

Finally, the leader has the responsibility of *oversight*. This is met not through the power of command—the leader is not the boss of the team—it is met through the power of leadership. The responsibilities include

- Orchestrating the team activities
- Stimulating all members to contribute
- Helping to resolve conflicts among members
- Assigning the homework to be done between meetings

To meet such responsibilities requires multiple skills, which include

- A trained capability for leading people
- Familiarity with the subject matter of the goal
- A firm grasp of the breakthrough process and the associated tools

The Team Members

"Team members" as used here includes the team leader and secretary. The responsibilities of any team member consist mainly of the following:

- Arranging to attend the team meetings
- Representing his or her department
- Contributing job knowledge and expertise
- Proposing theories of causes and ideas for remedies
- Constructively challenging the theories and ideas of other team members
- Volunteering for or accepting assignments for homework

Finding the Time to Work on Projects

Work on project teams is time consuming. Assigning someone to a project team adds about 10 percent to that person's workload. This added time is needed to attend team meetings, perform the assigned homework, and so on. Finding the time to do all this is a problem to be solved, since this added work is thrust on people who are already fully occupied.

No upper manager known to me has been willing to solve the problem by hiring new people to make up for the time demanded by the breakthrough projects. Instead, it has been left to each team member to solve the problem in his or her own way. In turn, the team members have adopted such strategies as

- Delegating more activities to subordinates
- Slowing down the work on lower-priority activities
- Improving time management on the traditional responsibilities
- Looking for ongoing activities that can be terminated. (In several organizations, there has been a specific drive to clear out unneeded work to provide time for breakthrough projects.)

As projects begin to demonstrate high returns on investment, the climate changes. Upper managers become more receptive to providing resources. In addition, the successful projects begin to reduce workloads that previously were inflated by the presence of chronic wastes.

Facilitators and Black Belts

Most organizations make use of internal consultants, usually called "facilitators" or "Black Belts," to assist teams. A facilitator like a Black Belt does not have to be a member of the team and may not have any responsibility for carrying out the team goal. (The literal meaning of the word facilitate is "to make things easy.") The prime role of the facilitator is to help the team to carry out its goal. The usual roles of facilitators consist of a selection from the following:

Explain the organization's intentions. The facilitator usually has attended briefing sessions that explain what the organization is trying to accomplish. Much of this briefing is of interest to the project teams.

Assist in team building. The facilitator helps the team members to learn to contribute to the team effort: propose theories, challenge the theories of others, and/or propose lines of investigation. Where the team concept is new to an organization, this role may require working directly with individuals to stimulate those who are unsure about how to contribute and to restrain the overenthusiastic ones. The facilitator also may evaluate the progress in team building and provide feedback to the team.

Assist in training. Most facilitators have undergone training in team building and in the breakthrough process. They usually have served as facilitators for other teams. Such experiences qualify them to help train project teams in several areas: team building, the breakthrough roadmap, and/or use of the tools.

Relate experiences from other projects. Facilitators have multiple sources of such experiences:

- Project teams previously served on
- Meetings with other facilitators to share experiences in facilitating project teams
- Final published reports of project teams
- Projects reported in the literature

Assist in redirecting the project. The facilitator maintains a detached view that helps him or her sense when the team is getting bogged down. As the team gets into the project, it may find itself getting deeper and deeper into a swamp. The project goal may turn out to be too broad, vaguely defined, or not doable. The facilitator usually can sense such situations earlier than the team and can help guide it to a redirection of the project.

Assist the team leader. Facilitators provide such assistance in various ways:

- Assist in planning the team meetings. This may be done with the team leader before each meeting.
- Stimulate attendance. Most nonattendance is due to conflicting demands made on a team member's time. The remedy often must come from the member's boss.
- Improve human relations. Some teams include members who have not been on good terms with each other or who develop friction as the project moves along. As an "outsider," the facilitator can help to direct the energies of such members into constructive channels. Such action usually takes place outside the team meetings. (Sometimes the leader is part of the problem. In such cases, the facilitator may be in the best position to help.)
- Assist on matters outside the team's sphere of activity. Projects sometimes require decisions or actions from sources that are outside the easy reach of the team. Facilitators may be helpful due to their wider range of contacts.

Support the team members. Such support is provided in multiple ways:

- Keep the team focused on the goal by raising questions when the focus drifts.
- Challenge opinionated assertions by questions such as, Are there facts to support that theory?
- Provide feedback to the team based on perceptions from seeing the team in action.

Report progress to the councils. In this role, the facilitator is a part of the process of reporting on progress of the projects collectively. Each project team issues minutes of its meetings. In due course, each also issues its final report, often including an oral presentation to the council.

However, reports on the projects collectively require an added process. The facilitators are often a part of this added reporting network.

The Qualifications of Facilitators and Black Belts

Facilitators undergo special training to qualify them for these roles. The training includes skills in team building, resolving conflicts, communication, and management of quality change; knowledge relative to the breakthrough processes, for example, the breakthrough roadmap and the tools and techniques; and knowledge of the relationship of breakthrough to the organization's policies and goals. In addition, facilitators acquire maturity through having served on project teams and providing facilitation to teams. This topic is covered in more detail in Chapter 12, Six Sigma: Improving Process Effectiveness.

This prerequisite training and experience are essential assets to the facilitator. Without them, he or she has great difficulty winning the respect and confidence of the project's team.

Most organizations are aware that to go into a high rate of breakthrough requires extensive facilitation. In turn, this requires a buildup of trained facilitators. However, facilitation is needed mainly during the startup phase. Then, as team leaders and members acquire training and experience, there is less need for facilitator support. The buildup job becomes a maintenance job.

This phased rise and decline has caused most organizations to avoid creating full-time facilitators or a facilitator career concept. Facilitation is done on a part-time basis. Facilitators spend most of their time on their regular job.

In many larger organizations, Black Belts are full-time specialists. Following intensive training in the breakthrough process, these persons devote all their time to the breakthrough activity. Their responsibilities go beyond facilitating project teams and may include

- Assisting in project nomination and screening
- Conducting training courses in the methods and tools
- Coordinating the activities of the project team with those of other activities in the organization, including conducting difficult analyses
- Assisting in the preparation of summarized reports for upper managers

A team has no personal boss. Instead, the team is supervised impersonally. Its responsibilities are defined in

- *The project charter.* This goal statement is unique to each team.
- *The steps or universal sequence for breakthrough.* This is identical for all teams. It defines the actions to be taken by the team to accomplish its goal.

The project team has the principal responsibility for the steps that now follow—taking the two "journeys." The diagnostic and remedial journeys are as follows:

- The diagnostic journey from symptom to cause. It includes analyzing the symptoms, theorizing as to the causes, testing the theories, and establishing the causes.
- The remedial journey from cause to remedy. It includes developing the remedies, testing and proving the remedies under operating conditions, dealing with resistance to change, and establishing controls to hold the gains.

Diagnosis is based on the factual approach and requires a firm grasp of the meanings of key words. It is helpful to define some of these key words at the outset.

Leaders Must Learn Key Breakthrough Terminology

A "defect" is any state of unfitness for use or nonconformance to specification. Examples are illegible invoices, scrap, and low mean time between failures. Other names include "error," "discrepancy," and "nonconformance."

A "symptom" is the outward evidence that something is wrong or that there is a defect. A defect may have multiple symptoms. The same word may serve as a description of both defect and symptom.

A "theory" or "hypothesis" are unproved assertions as to reasons for the existence of defects and symptoms. Usually, multiple theories are advanced to explain the presence of defects.

A "cause" is a proved reason for the existence of a defect. Often there are multiple causes, in which case they follow the Pareto principle—the vital few causes will dominate all the rest.

A "dominant cause" is a major contributor to the existence of defects and one that must be remedied before there can be an adequate breakthrough.

"Diagnosis" is the process of studying symptoms, theorizing as to causes, testing theories, and discovering causes.

A "remedy" is a change that can eliminate or neutralize a cause of defects.

Diagnosis Should Precede Remedy

It may seem obvious that diagnosis should precede remedy, yet biases or outdated beliefs can get in the way.

For example, during the twentieth century, many upper managers held deep-seated beliefs that most defects were due to workforce errors. The facts seldom bore this out, but the belief persisted. As a result, during the 1980s, many of these managers tried to solve their problems by exhorting the workforce to make no defects. (In fact, defects are generally over 80 percent management-controllable and under 20 percent worker-controllable.)

Untrained teams often try to apply remedies before the causes are known. ("Ready, fire, aim.") For example:

- An insistent team member "knows" the cause and pressures the team to apply a remedy for that cause.
- The team is briefed on the technology by an acknowledged expert. The expert has a firm opinion about the cause of the symptom, and the team does not question the expert's opinion.
- As team members acquire experience, they also acquire confidence in their diagnostic skills. This confidence then enables them to challenge unproved assertions.
- Where deep-seated beliefs are widespread, special research may be needed.

In a classic study, Greenridge (1953) examined 850 failures of electronic products supplied by various organizations. The data showed that 43 percent of the failures were traceable to product design, 30 percent to field operation conditions, 20 percent to manufacture, and the rest to miscellaneous causes.

Institutionalizing Breakthrough

Numerous organizations have initiated breakthrough, but few have succeeded in institutionalizing it so that it goes on year after year. Yet many of these organizations have a long history of annually conducting product development, cost reduction, productivity breakthrough, and so on. The methods they used to achieve such annual breakthrough are well known and can be applied to breakthrough. They are

- Enlarge the annual business plan to include goals for breakthrough.
- Make breakthrough a part of everyone's job description. In most organizations, the activity of breakthrough has been regarded as incidental to the regular job of meeting the goals for cost, delivery, and so on. The need is to make breakthrough a part of the regular job.
- Establish upper management audits that include review of progress on breakthrough.
- Revise the merit rating and reward system to include a new parameter—performance on breakthrough—and give it proper weight.
- Create well-publicized occasions to provide recognition for performance on breakthrough.

Review Progress

Scheduled, periodic reviews of progress by upper managers are an essential part of maintaining annual breakthroughs. Activities that do not receive such review cannot compete for priority with activities that do receive such review. Subordinates understandably give top priority to matters that are reviewed regularly by their superiors.

There is also a need for regular review of the breakthrough process. This is done through audits that may extend to all aspects of managing for quality. (Refer to Chapter 16, Using International Standards to Ensure Organization Compliance.)

Much of the database for progress review comes from the reports issued by the project teams. However, it takes added work to analyze these reports and to prepare the summaries needed by upper managers. Usually, this added work is done by the secretary of the council with the aid of the facilitators, the team leaders, and other sources such as finance.

As organizations gain experience, they design standardized reporting formats to make it easy to summarize reports by groups of projects, by product lines, by business units, by divisions, and for the corporation. One such format, used by a large European organization, determines for each project:

- The original estimated amount of chronic waste
- The original estimated reduction in cost if the project were to be successful
- The actual cost reduction achieved
- The capital investment
- The net cost reduction
- The summaries are reviewed at various levels. The corporate summary is reviewed quarterly at the chairperson's staff meeting (personal communication to the author).

Evaluation of Performance

One of the objectives of progress review is evaluation of performance. This evaluation extends to individuals as well as to projects. Evaluation of individual performance on breakthrough projects runs into the complication that the results are achieved by teams. The problem then becomes one of evaluating individual contribution to team efforts. This new problem has as yet no scientific solution. Thus, each supervisor is left to judge subordinates' contributions based on inputs from all available sources.

At higher levels of an organization, the evaluations extend to judging the performance of supervisors and managers. Such evaluations necessarily must consider results achieved on multiple projects. This has led to an evolution of measurement (metrics) to evaluate managers' performance on projects collectively. These metrics include

- Numbers of breakthrough projects: initiated, in progress, completed, and aborted
- Value of completed projects in terms of breakthrough in product performance, reduction in costs, and return on investment
- Percentage of subordinates active on project teams
- Superiors then judge their subordinates based on these and other inputs.

Training for Breakthrough

Throughout this chapter, there have been numerous observations on the needs for training employees. These needs are extensive because all employees must understand the methods and tools employed to attain breakthrough. Project-by-project breakthrough may be new to the organization, turnover may be high, or employees may be assigned new responsibilities. To carry out these new responsibilities requires extensive training.

So far in this decade, many organizations made significant investments in training their workforces in the methods and tools to attain performance excellence. According to *iSix Sigma* and the American Society for Quality (ASQ), more than 100,000 people were trained as Black Belts. Another 500,000 may have been trained as Green Belts. A new certification process has been added at the ASQ and at many firms like the Juran Institute to ensure that these exerts are qualified and competent to drive results. A Black Belt training program may consist of up to six weeks of training plus time to be certified.

This trend has been reversed from the 1990s. Training budgets were cut to reduce costs. Today, training is an investment in the future. This will benefit us as organizations move into the future.

Acknowledgments

This chapter of the handbook has drawn extensively from various training materials published by the Juran Institute, Inc., and Juran, J. M., ed. (1955), A History of Managing for Quality, sponsored by the Juran Foundation, Inc., ASQ Press, Milwaukee. I am also grateful to the Juran Institute, Inc.

References

Betker, H. A. (1983). "Breakthrough Program: Reducing Solder Defects on Printed Circuit Board Assembly." *The Juran Report*, No. 2, November, pp. 53–58.

Greenridge, R. M. C. (1953). "The Case of Reliability vs. Defective Components et al." *Electronic Applications Reliability Review*, No. 1, p. 12.

Juran, J. M. (1964). *Managerial Breakthrough*. McGraw-Hill, New York. Revised edition, 1995.

Juran, J. M. (1975). "The Non-Pareto Principle; Mea Culpa." Progress. May, pp. 8–9.
Juran, J. M. (1981). "Juran on Breakthrough," a series of 16 videocassettes on the subject. Juran Institute, Inc., Wilton, CT.
Juran, J. M. (1985). "A Prescription for the West—Four Years Later." European Organization for Quality, 29th Annual Conference. Reprinted in *The Juran Report*, No. 5, Summer 1985.
Juran, J. M. (1993), "Made in USA, a Renaissance in Quality." *Harvard Business Review*, July-August, pp. 42–50.
Juran, J. M., and Godfrey, A. B. (1999). "Juran's Quality Handbook, Fifth Edition," McGraw-Hill, NY.
Juran, J. M. "Juran on Quality Leadership," A video package, Juran Institute, Inc., Wilton, CT
Welch, J. (2005). Winning. Harper Collins, New York, NY

CHAPTER 6

Quality Control: Assuring Repeatable and Compliant Processes

Joseph M. Juran

About This Chapter

This chapter describes the compliance process or simply the "control process." "Control" is a universal managerial process to ensure that all key operational processes are stable—to prevent adverse change and to "ensure that the planned performance targets are met." Control includes product control, service control, process control, and even facilities control. To maintain stability, the control process evaluates actual performance, compares actual performance to targets, and takes action on any differences.

High Points of This Chapter

1. The quality control process is a universal managerial process for conducting operations so as to provide stability—to prevent adverse change and to "maintain the status quo." Quality control takes place by use of the feedback loop.

2. Each feature of the product or process becomes a control subject—a center around which the feedback loop is built. As much as possible, human control should be done by the workforce—the office clerical force, factory workers, salespersons, etc.
3. The flow diagram is widely used during the planning of quality controls. The weakest link in facilities control has been adherence to schedule.
4. To ensure strict adherence to schedule requires an independent audit.
5. Knowing which process variable is dominant helps planners during allocation of resources and priorities.
6. The design for process control should provide the tools needed to help the operating forces distinguish between real change and false alarms. It is most desirable to provide umpires with tools that can help to distinguish between special causes and common causes. An elegant tool for this purpose is the Shewhart control chart (or just control chart). The criteria for self-control are applicable to processes in all functions and all levels, from general manager to nonsupervisory worker.
7. Responsibility for results should be keyed to controllability. Ideally the decision of whether the process conforms to process quality goals should be made by the workforce.
8. To make use of self-inspection requires meeting several essential criteria: Quality is number one; mutual confidence, self-control, training, and certification are the others. Personnel who are assigned to make product conformance decisions should be provided with clear definitions of responsibility as well as guidelines for decision-making.
9. The proper sequence in managing is first to establish goals and then to plan how to meet those goals, including the choice of the appropriate tools. The planning for quality control should provide an information network that can serve all decision-makers.

Compliance and Control Defined

Compliance or quality control is the third universal process in the Juran Trilogy. The others are quality planning in and quality improvement, which are discussed in Chapters 4 and 5, respectively. The Juran Trilogy diagram (Figure 6.1) shows the interrelation of these processes.

Figure 6.1 is used in several other chapters in this handbook to describe the relationships between planning, improvement, and control—the fundamental managerial processes in quality management. What is important for this chapter is to concentrate on the two "zones of control."

In Figure 6.1, we can easily see that although the process is in control in the middle of the chart, we are running the process at an unacceptable level of performance and "waste." What is necessary here is not more control, but improvement—actions to change the level of performance.

After the improvements have been made, a new level of performance has been achieved. Now it is important to establish new controls at this level to prevent the performance level from deteriorating to the previous level or even worse. This is indicated by the second zone of control.

The term "control of quality" emerged early in the twentieth century (Radford 1917, 1922). The concept was to broaden the approach to achieving quality, from the then-prevailing

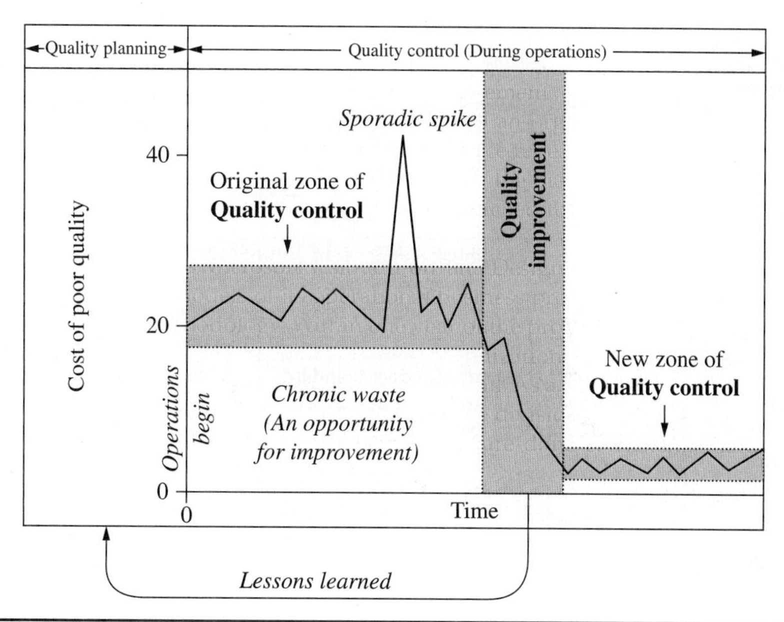

Figure 6.1 Juran Trilogy diagram. (*Juran Institute, Inc., 1986.*)

after-the-fact inspection (detection control) to what we now call "prevention (proactive control)." For a few decades, the word "control" had a broad meaning, which included the concept of quality planning. Then came events that narrowed the meaning of "quality control." The "statistical quality control" movement gave the impression that quality control consisted of using statistical methods. The "reliability" movement claimed that quality control applied only to quality at the time of test but not during service life.

In the United States, the term "quality control" now often has the meaning defined previously. It is a piece of a "performance excellence, operational excellence, business excellence, or total quality program," which are now used interchangeably to comprise the all-embracing term to describe the methods, tools, and techniques to manage the quality of an organization.

In Japan, the term "quality control" retains a broad meaning. Their "total quality control" is equivalent to our term "business excellence." In 1997, the Union of Japanese Scientists and Engineers (JUSE) adopted the term Total Quality Management (TQM) to replace Total Quality Control (TQC) to more closely align themselves with the more common terminology used in the rest of the world.

Figure 6.2 shows the input-output features of this step.

In Figure 6.2, the input is operating process features, or key control characteristics, developed to produce the product features, or key product characteristics, required to meet customer needs. The output consists of a system of product and process controls, which can provide stability to the operating process.

A key product characteristic is a product characteristic for which reasonably anticipated variation *could significantly affect* a product's safety, compliance to government regulations, performance, or fit.

Key product characteristics (KPCs) are *outputs from a process that are measurable* on, within, or about the product itself. They are the outputs perceived by the customer.

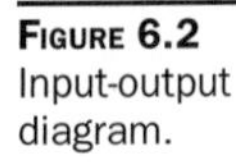
FIGURE 6.2 Input-output diagram.

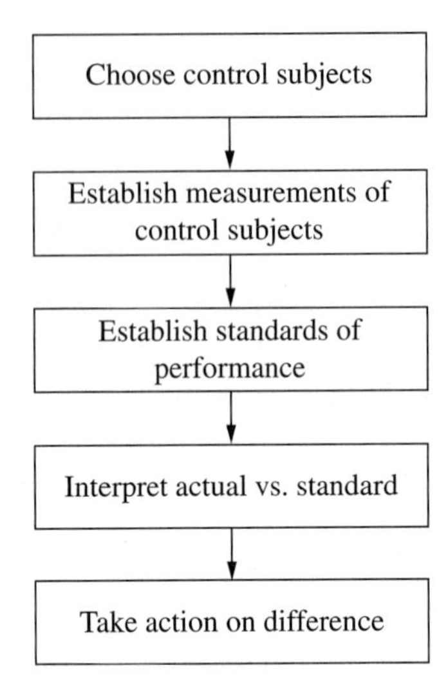

Examples of KPCs include

- KPCs "On:" the product: width, thickness, coating adherence, surface cleanliness, etc.
- KPCs "Within:" the product: hardness, density, tensile strength, mass, etc.
- KPCs "About:" the product: performance, weight, etc.

In general, key control characteristics (KCCs) are *inputs that affect the outputs* (KPCs). They are unseen by the customer and are measurable only when they occur.

A KCC is

- A process parameter for which variation must be controlled around some target value to ensure that variation in a KPC is maintained around its target values during manufacturing and assembly
- A process parameter for which reduction in variation will reduce the variation of a KPC
- Directly traceable to a KPC
- Particularly significant in ensuring a KPC achieves target value
- Not specified on a product drawing or product documentation

The Relation to Quality Assurance

Quality control and quality assurance have much in common. Each evaluates performance. Each compares performance to goals. Each acts on the difference. However, they also differ from each other. Quality control has as its primary purpose maintaining control. Performance is evaluated during operations, and performance is compared to targets during operations. In the process, metrics are utilized to monitor adherence to standards. The resulting information is received and used by the employees.

The main purpose of quality assurance is to verify that control is being maintained. Performance is evaluated after operations, and the resulting information is provided to both the employees and others who have a need to know. Results metrics are utilized to determine

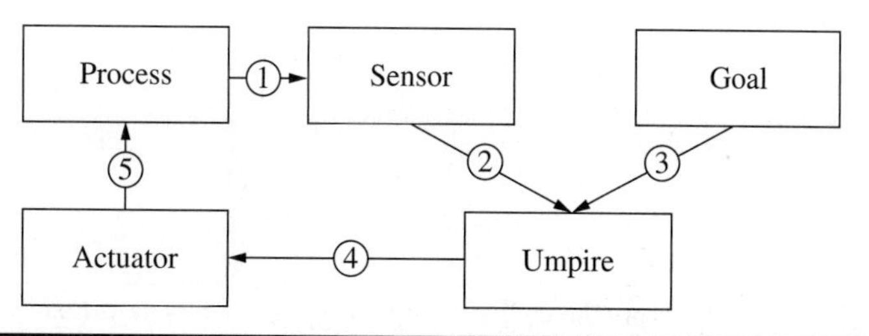

Figure 6.3 Feedback loop.

conformance to customer needs and expectations. Others may include leadership, plant, functional; corporate staffs; regulatory bodies; customers; and the general public.

The Feedback Loop

Quality control takes place by use of the feedback loop. A generic form of the feedback loop is shown in Figure 6.3.

The progression of steps in Figure 6.3 is as follows:

1. A sensor is "plugged in" to evaluate the actual quality of the control subject—the product or process feature in question. The performance of a process may be determined directly by evaluation of the process feature, or indirectly by evaluation of the product feature—the product "tells" on the process.
2. The sensor reports the performance to an umpire.
3. The umpire also receives information on the quality goal or standard.
4. The umpire compares actual performance to standard. If the difference is too great, the umpire energizes an actuator.
5. The actuator stimulates the process (whether human or technological) to change the performance so as to bring quality into line with the quality goal.
6. The process responds by restoring conformance.

Note that in Figure 6.3 the elements of the feedback loop are functions. These functions are universal for all applications, but responsibility for carrying out these functions can vary widely. Much control is carried out through automated feedback loops. No human beings are involved. Common examples are the thermostat used to control temperature and the cruise control used in automobiles to control speed.

Another form of control is self-control carried out by employees. An example of such self-control is the village artisan who performs every one of the steps of the feedback loop. The artisan chooses the control subjects based on understanding the needs of customers, sets the quality targets to meet the needs, senses the actual quality performance, judges conformance, and becomes the actuator in the event of nonconformance.

This concept of self-control is illustrated in Figure 6.4. The essential elements here are the need for the employee or work team to know what they are expected to do, to know how they are actually doing, and to have the means to regulate performance. This implies that they have a capable process and have the tools, skills, and knowledge necessary to make the adjustments and the authority to do so.

A further common form of feedback loop involves office clerks or factory workers whose work is reviewed by umpires in the form of inspectors. This design of a feedback loop is largely the result of the Taylor Management System adopted in the early twentieth century.

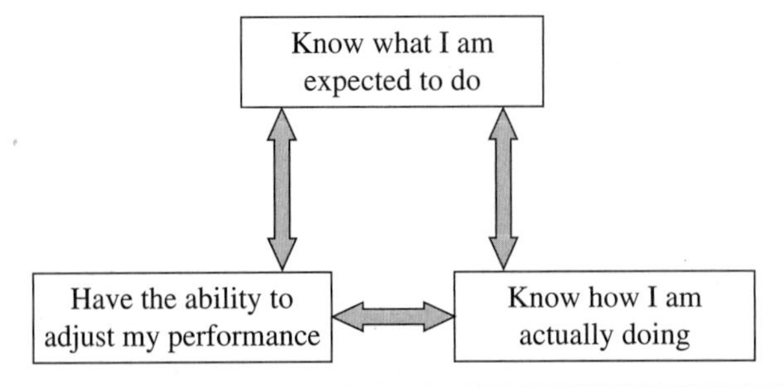

Figure 6.4 Concept of self-control. (*The Juran Institute, Inc.*)

It focused on the separation of planning for quality from the execution or operations. The Taylor Management System emerged a century ago and contributed greatly to increasing productivity. However, the effect on quality was largely negative. The negative impact resulted in large costs associated with poor quality, products and services that have higher levels of failure, and customer dissatisfaction.

The Elements of the Feedback Loop

The feedback loop is a universal. It is fundamental to maintaining control of every process. It applies to all types of operations, whether in service industries or manufacturing industries, whether for profit or not. The feedback loop applies to all levels in the hierarchy, from the chief executive officer to the members of the workforce. However, there is wide variation in the nature of the elements of the feedback loop.

In Figure 6.5 a simple flowchart is shown describing the control process with the simple universal feedback loop imbedded.

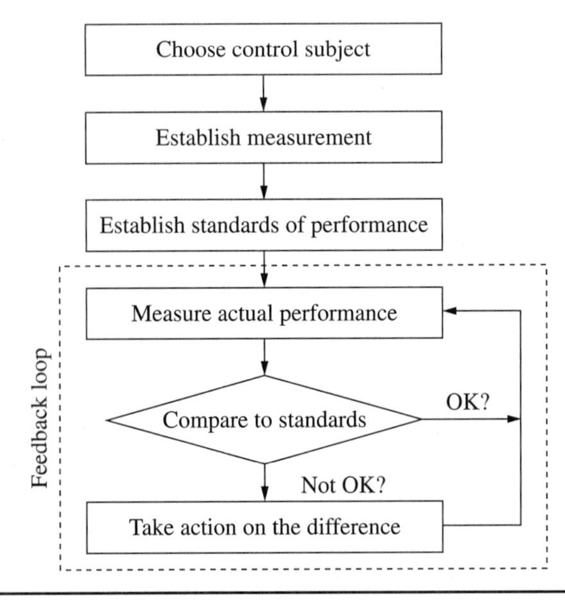

Figure 6.5 Simple flowchart describing the control process.

The Control Subjects

Each feature of the product (goods and services) or process becomes a control subject (the specific attribute or variable to be controlled)—a center around which the feedback loop is built. The critical first step is to choose the control subject. To choose control subjects, you should identify the major work processes and products, define the objectives of the work processes; succinctly define the work processes; identify the customers of the process, and then select the control subjects (KPCs and/or KCCs). Control subjects are derived from multiple sources, which include

- Stated customer needs for product features
- Translated "voice of the customer" needs into product features
- Defined process features that create the product or service features
- Industry and government standards and regulations (i.e., Sarbanes Oxley, ISO 9000, etc.)
- Need to protect human safety and the environment (i.e., OSHA, ISO 14000)
- Need to avoid side effects such as irritations to stakeholders, employees, or to a neighboring community
- Failure mode and effects analyses
- Control plans
- Results of design of experiments

At the staff level, control subjects consist mainly of product and process features defined in technical specifications and procedures manuals. At managerial levels, the control subjects are broader and increasingly business-oriented. Emphasis shifts to customer needs and to competition in the marketplace. This shift in emphasis then demands broader control subjects, which, in turn, have an influence on the remaining steps of the feedback loop.

Establish Measurement

After choosing the control subjects, the next step is to establish the means of measuring the actual performance of the process or the quality level of the goods or services being created. Measurement is one of the most difficult tasks of management and is discussed in almost every chapter of this handbook. In establishing the measurement, we need to clearly specify the means of measuring (the sensor), the accuracy and precision of the measurement tool, the unit of measure, the frequency of measuring, the means by which data will be recorded, the format for reporting the data, the analysis to be made on the data to convert it to usable information, and who will make the measurement. In establishing the unit of measure, one should select a unit of measure that is understandable, provides an agreed-upon basis for decision-making, is customer focused, and can be applied broadly.

Establish Standards of Performance: Product Goals and Process Goals

For each control subject it is necessary to establish a standard of performance—a target or goal (also metrics, objectives, etc.). A standard of performance is an aimed-at target toward which work is expended. Table 6.1 gives some examples of control subjects and the associated goals.

The prime goal for products and services is to meet customer needs. Industrial customers often specify their needs with some degree of precision. Such specified needs then become goals for the producing company. In contrast, consumers tend to state their needs in vague terms. Such statements must then be translated into the language of the producer in order to become product goals.

Control Subject	Goal
Vehicle mileage	Minimum of 25 mi/gal highway driving
Overnight delivery	99.5% delivered prior to 10:30 A.M. next morning
Reliability	Fewer than three failures in 25 years of service
Temperature	Minimum 505°F; maximum 515°F
Purchase-order error rate	No more than 3 errors/1000 purchase orders
Competitive performance	Equal or better than top three competitors on six factors
Customer satisfaction	90% or better rate, service outstanding or excellent
Customer retention	95% retention of key customers from year to year
Customer loyalty	100% of market share of over 80% of customers

TABLE 6.1 Control Subjects and Associated Quality Goals

Other goals for products that are also important are those for reliability and durability. Whether the products and services meet these goals can have a critical impact on customer satisfaction, loyalty, and overall costs. The failures of products under warranty can seriously affect the profitability of a company through both direct and indirect costs (loss of repeat sales, word of mouth, etc.).

The processes that produce products have two sets of goals:

- To produce products and services that meet customer needs. Ideally, each and every unit produced should meet customer needs (meet specifications).
- To operate in a stable and predictable manner. In the dialect of the quality specialist, each process should be "in a state of control." We will later elaborate on this, in the section "Process Conformance."

Quality targets may also be established for functions, departments, or people. Performance against such goals then becomes an input to the company's scorecard, dashboard, and reward system. Ideally such goals should be

- *Legitimate*. They should have undoubted official status.
- *Measurable*. They can be communicated with precision.
- *Attainable*. As evidenced by the fact that they have already been attained by others
- *Equitable*. Attainability should be reasonably alike for individuals with comparable responsibilities.

Quality goals may be set from a combination of the following bases:

- Goals for product and service features and process features are largely based on technological analysis.
- Goals for functions, departments, and people should be based on the need of the business and external benchmarking rather than historical performance.

In the later 2000s quality goals used at the highest levels of an organization have become commonplace. Establishing long-term goals such as reducing the costs of poor quality or becoming best in class have become a normal part of strategic business plans. The emerging practice is to establish goals on "metrics that matter," such as meeting customers' needs, exceeding the competition, maintaining a high pace of improvement, improving the effectiveness of business processes, and setting stretch goals to avoid failure-prone products and processes.

Measure Actual Performance

A critical step in controlling quality characteristics is to measure the actual performance of a process as precisely as possible. To do this requires measuring with a "sensor." A sensor is a device or a person that makes the actual measurement.

The Sensor

A "sensor" is a specialized detecting device. It is designed to recognize the presence and intensity of certain phenomena and to convert the resulting data into "information." This information then becomes the basis of decision-making. At the lower levels of an organization, the information is often on a real-time basis and is used for daily control. At higher levels, the information is summarized in various ways to provide broader measures, detect trends, and identify the vital few problems.

The wide variety of control subjects requires a wide variety of sensors. A major category is the numerous technological instruments used to measure product features and process features. Familiar examples are thermometers, clocks, and weight scales. Another major category of sensors is the data systems and associated reports, which supply summarized information to the managerial hierarchy. Yet another category involves the use of human beings as sensors. Questionnaires, surveys, focus groups, and interviews are also forms of sensors.

Sensing for control is done on an organization level. Information is needed to manage for the short- and long-term. This has led to the use of computers to aid in the sensing and in converting the resulting data into information.

Most sensors provide their evaluations in terms of a unit of measure—a defined amount of some feature—which permits evaluation of that feature in numbers or pictures. Familiar examples of units of measure are degrees of temperature, hours, inches, and tons. A considerable amount of sensing is done by human beings. Such sensing is subject to numerous sources of error. The use of pictures as a standard to comparison can help reduce human errors. Also of vital importance to alleviate human errors is the application of detailed instructions.

Compare to Standards

The act of comparing to standards is often seen as the role of an umpire. The umpire may be a person or a technological device. Either way, the umpire may be called on to carry out any or all of the following activities:

- Compare the actual process performance to the targets.
- Interpret the observed difference (if any); determine if there is conformance to the target.
- Decide on the action to be taken.
- Stimulate corrective action.
- Record the results.

These activities require elaboration and will be examined more closely in an upcoming section.

Take Action on the Difference

In any well-functioning control system we need a means of taking action on any difference between desired standards of performance and actual performance. For this we need an actuator. This device (human or technological or both) is the means for stimulating action to restore conformance. At the operations or employee level, it may be a keypad for giving orders to a centralized computer database, a change in a new procedure, a new specification document, or a new setting of a dial to adjust a machine to the right measure. At the management level, it may be a memorandum to subordinates, a new company policy, or a team to change a process.

The Key Process

In the preceding discussion we have assumed a process. This may also be human or technological or both. It is the means for producing the product and service features, each of which requires control subjects to ensure conformance to specifications. All work is done by a process. A process consists of inputs, labor, technology, procedures, energy, materials, and outputs.

Taking Corrective Action

There are many ways of taking corrective action to troubleshoot a process and return to the "status quo." A popular example of a root cause and corrective action method is the so-called PDCA or PDSA Cycle (first popularized by Walter Shewhart and then by Dr. Deming as the Deming Wheel) as shown in Figure 6.6. Deming (1986) referred to this as the Shewhart cycle, which is the name many still use when describing this version of the feedback loop.

In this example, the feedback loop is divided into four steps labeled Plan, Do, Check, and Act (PDCA) or Plan, Do, Study, Act (PDSA). This model is used by many health care and service industries. These steps correspond roughly to the following:

- "Plan" includes choosing control subjects and setting goals.
- "Do" includes running and monitoring the process.
- "Check" or "Study" includes sensing and umpiring.
- "Act" includes stimulating the actuator and taking corrective action.

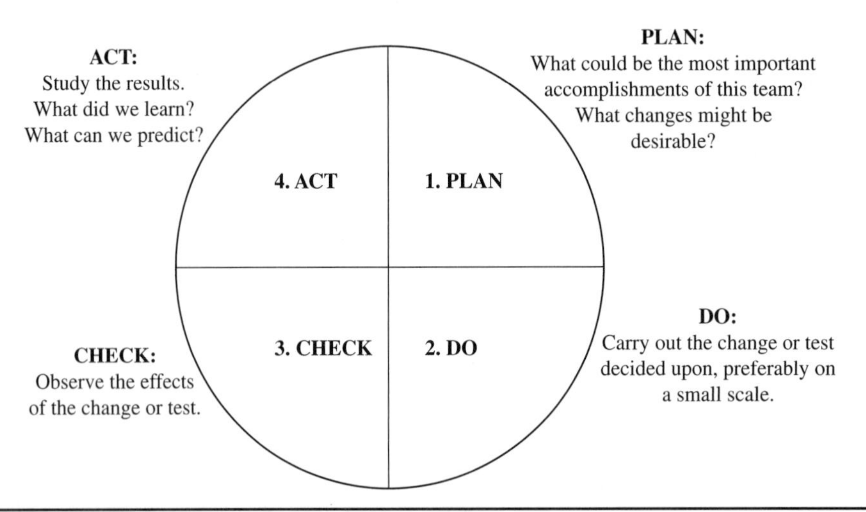

FIGURE 6.6 The PDCA Cycle. (*Shewart and Deming, 1986.*)

An early version of the PDCA cycle was included in W. Edwards Deming's first lectures in Japan (Deming 1950). Since then, additional versions have been used, like PDSA, PDCA, RCCA, and so on.

Some of these versions have attempted to label the PDCA cycle in ways that make it serve as a universal series of steps for both control and improvement. The authors feel that this confuses matters, since two very different processes are involved. Our experience is that all organizations should define two separate methods. One is to take corrective action on a "sporadic change" in performance.

RCCA, PDSA, and PDCA differ from improvement methods like Six Sigma in that the scope of the problem lends itself to a simpler, less complex analysis to find the root cause of a "sporadic problem." RCCA analytical and communication tools contribute to the reduction of day-to-day problems that plague processes. Tools utilized for analysis and diagnosis of sporadic spikes typically take the form of graphical tools with less emphasis on statistical applications. Often many organizations that have been trained in RCCA and the like do not have the right tools and methods to solve chronic problems. It is best to use the Six Sigma D-M-A-I-C improvement methods. This is described in Chapter 12, Six Sigma: Improving Process Effectiveness.

The Pyramid of Control

Control subjects run to large numbers, but the number of "things" to be controlled is far larger. These things include the published catalogs and price lists sent out, multiplied by the number of items in each; the sales made, multiplied by the number of items in each sale; the units of product produced, multiplied by the associated numbers of quality features; and so on for the numbers of items associated with employee relations, supplier relations, cost control, inventory control, product and process developments, etc.

A study in one small company employing about 350 people found that there were more than a billion things to be controlled (Juran 1964, pp. 181–182).

There is no possibility for upper leaders to control huge numbers of control subjects. Instead, they divide up the work of control using a plan of delegation similar to that shown in Figure 6.7.

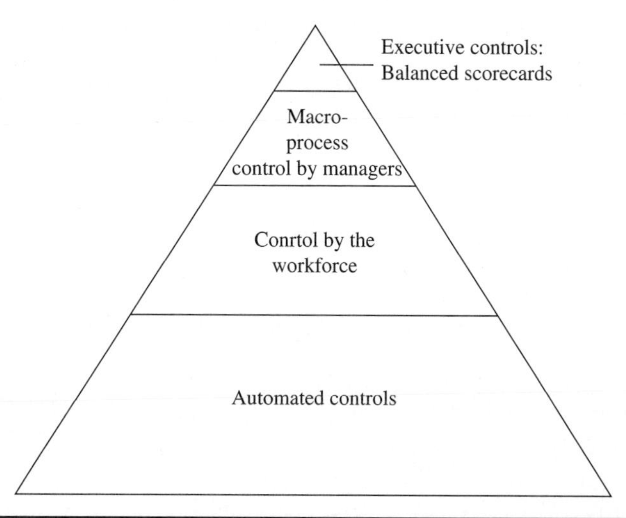

Figure 6.7 The pyramid of control. (*Making Quality Happen, Juran Institute, Inc., Senior Executive Workshop, p. F-5.*)

This division of work establishes three areas of responsibility for control: control by nonhuman means, control by the workforce, and control by the managerial hierarchy.

Control by Technology (Nonhuman Means)

At the base of the pyramid are the automated feedback loops and error-proofed processes, which operate with no human intervention other than maintenance of facilities (which, however, is critical). These nonhuman methods provide control over a great majority of things. The control subjects are exclusively technological, and control takes place on a real-time basis.

The remaining controls in the pyramid require human intervention. By a wide margin, the most amazing achievement in quality control takes place during a biological process that is millions of years old—the growth of the fertilized egg into an animal organism. In human beings the genetic instructions that program this growth consist of a sequence of about three billion "letters." This sequence—the human genome—is contained in two strands of DNA (the double helix), which "unzip" and replicate about a million billion times during the growth process from fertilized egg to birth of the human being.

Given such huge numbers, the opportunities for error are enormous. (Some errors are harmless, but others are damaging and even lethal.) Yet the actual error rate is of the order of about one in ten billion. This incredibly low error rate is achieved through a feedback loop involving three processes (Radman and Wagner 1988):

- A high-fidelity selection process for attaching the right "letters," using chemical lock-and-key combinations
- A proofreading process for reading the most recent letter, and removing it if incorrect
- A corrective action process to rectify the errors that are detected

Control by the Employees (Workforce)

Delegating such decisions to the workforce yields important benefits in human relations and in conduct of operations. These benefits include shortening the feedback loop; providing the workforce with a greater sense of ownership of the operating processes, often referred to as "empowerment"; and liberating supervisors and leaders to devote more of their time to planning and improvement.

It is feasible to delegate many quality control decisions to the workforce. Many organizations already do. However, to delegate process control decisions requires meeting the criteria of "self-control" or "self-management." (See later in this chapter in the sections "Self-Control and Controllability.")

Control by the Managerial Hierarchy

The peak of the pyramid of control consists of the "vital few" control subjects. These are delegated to the various levels in the managerial hierarchy, including the upper leaders.

Leaders should avoid getting too deep into making decisions on quality control. Instead, they should

- Make the vital few decisions.
- Provide criteria to distinguish the vital few decisions from the rest. For an example of providing such criteria see Table 6.3.
- Delegate the rest under a decision-making process that provides the essential tools and training.

	At Workforce Levels	At Managerial Levels
Control goals	Product and process features in specifications and procedures	Business-oriented, product salability, competitiveness
Sensors	Technological	Data systems
Decisions to be made	Conformance or not?	Meet customer needs or not?

(*Making Quality Happen, Juran Institute, Inc., Senior Executive Workshop, p. F-4, Southbury, CT.*)

Table 6.2 Contrast of Quality Control and Two Levels–Workforce and Upper Management

The distinction between vital few matters and others originates with the control subjects. Table 6.2 shows how control subjects at two levels—workforce and upper management—affect the elements of the feedback loop.

Planning for Control

Planning for control is the activity that provides the system—the concepts, methodology, and tools—through which company personnel can keep the operating processes stable and thereby produce the product features required to meet customer needs. The input-output features of this system (also plan, process) were depicted in Figure 6.2.

Critical to Quality (CTQs): Customers and Their Needs

The principal customers of control systems are the company personnel engaged in control—those who carry out the steps that enable the feedback loop. Such personnel require (1) an understanding of what is critical to quality (CTQ), customers' needs, and (2) a definition of their own role in meeting those needs. However, most of them lack direct contact with customers. Planning for control helps to bridge that gap by supplying a translation of what customers' needs are, along with defining responsibility for meeting those needs. In this way, planning for quality control includes providing operating personnel with information on customer needs (whether direct or translated) and defining the related control responsibilities of the operating personnel. Planning for quality control can run into extensive detail.

Who Plans for Control? Planning for control has in the past been assigned to

- Product development staff
- Quality engineers and specialists
- Multifunctional design teams
- Departmental leaders and supervisors
- The workforce

Planning for control of critical processes has traditionally been the responsibility of those who plan the operating process. For noncritical processes, the responsibility was usually assigned to quality specialists from the quality department. Their draft plans were then submitted to the operating heads for approval.

Recent trends have been to increase the use of the team concept. The team membership includes the operating forces and may also include suppliers and customers of the operating process. The recent trend has also been to increase participation by the workforce.

Compliance and Control Concepts

The methodologies of compliance and control are built around various concepts, such as the feedback loop, process capability, self-control, etc. Some of these concepts are of ancient origin; others have evolved in this and the last centuries. During the discussion of planning for control, we will elaborate on some of the more widely used concepts.

The Process Map or Flow Diagram

The usual first step in planning for control is to map out the flow of the operating process, as discussed in Chapter 4, Quality Planning: Designing Innovative Products and Services. Figure 6.8 is an example of a flow diagram.

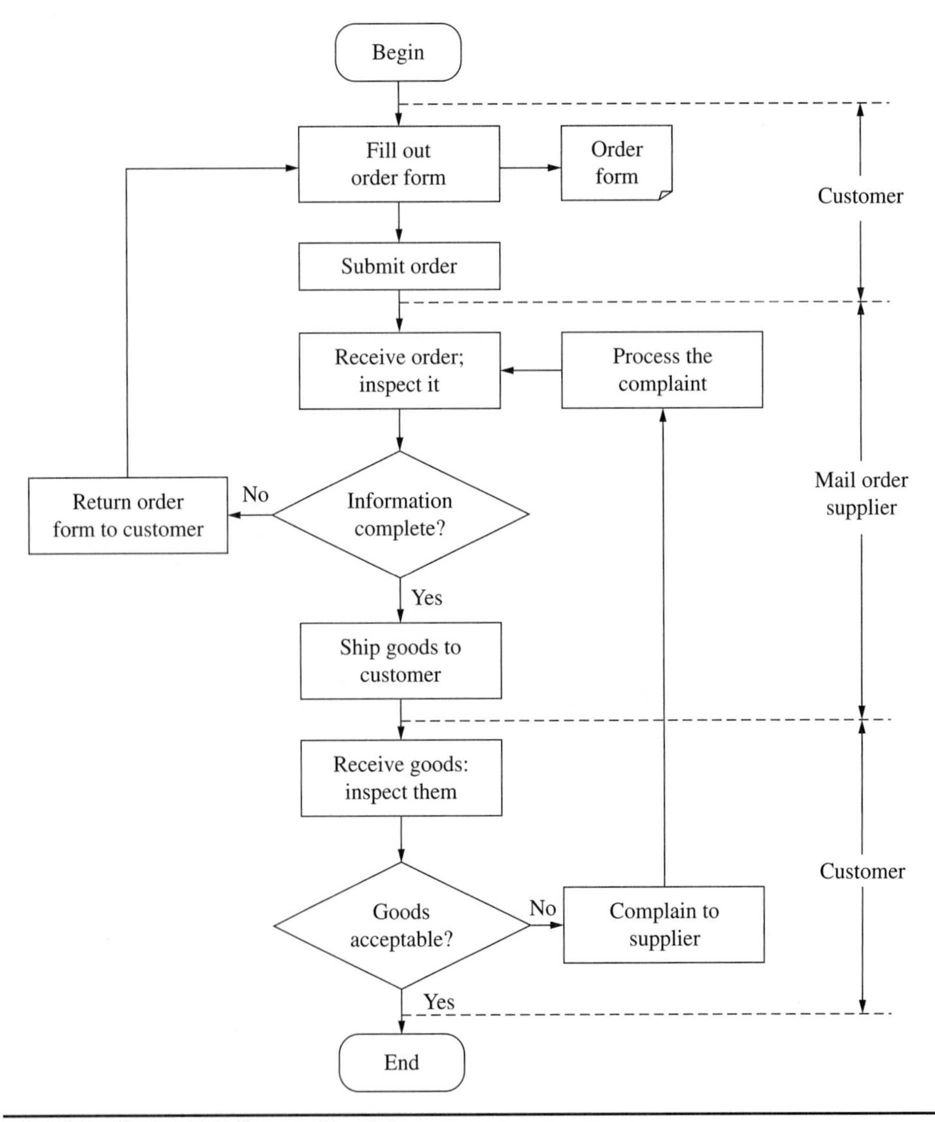

Figure 6.8 The flow diagram.

The flow diagram is widely used during the planning of quality controls. It helps the planning team to

- Understand the overall operating process. Each team member is quite knowledgeable about his or her segment of the process, but less so about other segments and about the interrelationships.
- Identify the control subjects around which the feedback loops are to be built. The nature of these control subjects was discussed previously in the section "The Control Subjects."

The Control Spreadsheet

The work of the planners is usually summarized on a control spreadsheet. This spreadsheet is a major planning tool. An example can be seen in Figure 6.9.

In this spreadsheet the horizontal rows are the various control subjects. The vertical columns consist of elements of the feedback loop, plus other features needed by the operating forces to exercise control so as to meet the quality goals.

Some of the contents of the vertical columns are unique to specific control subjects. However, certain vertical columns apply widely to many control subjects. These include unit of measure, type of sensor, quality goal, frequency of measurement, sample size, criteria for decision-making, and responsibility for decision-making.

Who Does What?

The feedback loop involves multiple tasks, each of which requires a clear assignment of responsibility. At any control station there may be multiple people available to perform those tasks. For example, at the workforce level, a control station may include setup specialists, operators, maintenance personnel, inspectors, etc. In such cases it is necessary to agree on who should make which decisions and who should take which actions. An aid to reaching such agreement is a special spreadsheet similar to Figure 6.9.

In this spreadsheet, the essential decisions and actions are listed in the leftmost column. The remaining columns are headed up by the names of the job categories associated with the

	Process Control Features							
Control Subjects	**Unit of Measure**	**Type of Sensor**	**Goal**	**Frequency of Measurement**	**Sample Size**	**Criteria for Decision-Making**	**Responsibility for Decision-Making**	...
Wave solder conditions solder temperature	Degree F (°F)	Thermo-couple	505°F	Continuous	N/A	510°F reduce heat 500°F increase heat	Operator	...
Conveyor speed	Feet per minute (ft/min)	ft/min	4.5 ft/min	1 per hour	N/A	5 ft/min reduce speed 4 ft/min increase speed	Operator	...
Alloy purity	% Total contaminates	Lab chemical analysis	1.5% max	1 per month	15 grams	At 1.5%, drain bath, replace solder	Process engineer	...

Figure 6.9 The control spreadsheet.

control station. Then, through discussion among the cognizant personnel, agreement is reached on who is to do what.

This spreadsheet is a proven way to find answers to the long-standing, but vague, question, Who is responsible for quality? This question has never been answered because it is inherently unanswerable. However, if the question is restated in terms of decisions and actions, the way is open to agree on the answers. This clears up the vagueness.

Test and Control Stations

A "control station" is an area in which quality control takes place. In the lower levels of an organization, a control station is usually confined to a limited physical area. Alternatively, the control station can take such forms as a patrol beat or a "control tower." At higher levels, control stations may be widely dispersed geographically, as in the scope of a manager's responsibility.

A review of numerous control stations shows that they are usually designed to provide evaluations and/or early warnings in the following ways:

- At changes of jurisdiction, where responsibility is transferred from one organization to another
- Before embarking on some significant irreversible activity, such as signing a contract
- After creation of a critical quality feature
- At the site of dominant process variables
- At areas ("windows") that allow economical evaluation to be made

Stages of Control

The flow diagram not only discloses the progression of events in the operating process, it also suggests which stages should become the centers of control activity. Several of these stages apply to the majority of operating processes.

Setup (Startup) Control

The end result of this form of control is the decision of whether or not to "push the start button." Typically this control involves:

- A *countdown* listing the preparatory steps needed to get the process ready to produce. Such countdowns sometime come from suppliers. Airlines provide checklists to help travelers plan their trips; electric power organizations provide checklists to help householders prepare the house for winter weather.
- *Evaluation* of process and/or product features to determine whether, if started, the process will meet the goals.
- *Criteria* to be met by the evaluations
- *Verification* that the criteria have been met
- *Assignment* of responsibility. This assignment varies, depending largely on the criticality of the quality goals. The greater the criticality, the greater is the tendency to assign the verification to specialists, supervisors, and "independent" verifiers rather than to nonsupervisory workers.

Running control

This form of control takes place periodically during the operation of the process. The purpose is to make the "run or stop" decision—whether the process should continue to produce a product or whether it should stop.

Running control consists of closing the feedback loop, over and over again. The process and/or product performance is evaluated and compared with goals. If the product and/or process conforms to goals, and if the process has not undergone some significant adverse change, the decision is to "continue to run." If there is nonconformance, or if there has been a significant change, corrective action is in order.

The term "significant" has meanings beyond those in the dictionary. One of these meanings relates to whether an indicated change is a real change or is a false alarm due to chance variation. The design for process control should provide the tools needed to help the operating forces distinguish between real changes and false alarms. Statistical process control (SPC) methodology is aimed at providing such tools. (See Chapter 18, Core Tools to Design, Control, and Improve Performance.)

Product Control

This form of control takes place after some amount of product has been produced. The purpose of the control is to decide whether the product conforms to the product quality goals. Assignment of responsibility for this decision differs from company to company. However, in all cases, those who are to make the decision must be provided with the facilities and training that will enable them to understand the product quality goals, evaluate the actual product quality, and decide whether there is conformance.

Since all this involves making a factual decision, it can, in theory, be delegated to anyone, including members of the workforce. In practice, this delegation is not made to those whose assigned priorities might bias their judgment. In such cases, the delegation is usually to those whose responsibilities are free from such bias, for example, "independent" inspectors. Statistical quality control (SQC) is a methodology frequently employed to yield freedom from biases.

Facilities Control

Most operating processes employ physical facilities: equipment, instruments, and tools. Increasingly, the trend has been to use automated processes, computers, robots, etc. This same trend makes product quality more and more dependent on maintenance of the facilities. Leading organizations have moved to elements of total productive maintenance (TPM). The extent of application varies by company, but TPM and reliability-centered maintenance (RCM) support sound facilities control. The elements of design for facilities control are well known:

- Establish a schedule for conducting facilities maintenance.
- Establish a checklist—a list of tasks to be performed during a maintenance action.
- Train the maintenance forces to perform the tasks.
- Assign clear responsibility for adherence to schedule.
- Enhance management of critical spares.
- Standardize preventive maintenance tasks and frequency on equipment.
- Optimize efficiency related to maintenance staffing and organization.
- Increase mechanical interface with equipment from operator.

The weakest link in facilities control has been adherence to schedule. To ensure strict adherence to schedule requires an independent audit.

In cases involving introduction of a new technology, a further weak link is training the maintenance forces (White 1988).

During the 1980s, the automakers began to introduce computers and other electronics into their vehicles. It soon emerged that many repair shop technicians lacked the technological education base needed to diagnose and remedy the associated field failures. To make matters worse, the automakers did not give high priority to standardizing the computers. As a result, a massive training backlog developed.

Concept of Dominance

Control subjects are so numerous that planners are well advised to identify the vital few so that they will receive appropriate priority. One tool for identifying the vital few is the concept of dominance.

Operating processes are influenced by many variables, but often one variable is more important than all the rest combined. Such a variable is said to be the "dominant variable." Knowledge of which process variable is dominant helps planners during allocation of resources and priorities. The more usual dominant variables include

- *Setup-dominant.* Some processes exhibit high stability and reproducibility of results over many cycles of operation. A common example is the printing process. The design for control should provide the operating forces with the means for precise setup and validation before operations proceed.
- *Time-dominant.* Here, the process is known to change progressively with time, for example, depletion of consumable supplies, heating up, and wear of tools. The design for control should provide means for periodic evaluation of the effect of progressive change and for convenient readjustment.
- *Component-dominant.* Here, the main variable is the quality of the input materials and components. An example is the assembly of electronic or mechanical equipment. The design for control should be directed at supplier relations, including joint planning with suppliers to upgrade the quality of the inputs.
- *People-dominant.* In these processes, quality depends mainly on the skill and knack possessed by the workers. The skilled trades are well-known examples. The design for control should emphasize aptitude testing of workers, training and certification, quality rating of workers, and error-proofing to reduce employee errors.
- *Information-dominant.* Here, the processes are of a "job-shop" nature so that there is frequent change in what product is to be produced. As a result, the job information changes frequently. The design for control should concentrate on providing an information system that can deliver accurate, up-to-date information on just how this job differs from its predecessors.

Process Capability

One of the most important concepts in the quality planning process is "process capability." The prime application of this concept is during planning of the operating processes.

This same concept also has applications in quality control. To explain this, a brief review is in order. All operating processes have an inherent uniformity for producing products. This uniformity can often be quantified, even during the planning stages. The process planners can use the resulting information for making decisions on adequacy of processes, choice of alternative processes, need for revision of processes, and so forth, with respect to the inherent uniformity and its relationship to process goals.

Applied to planning for quality control, the state of process capability becomes a major factor in decisions on frequency of measuring process performance, scheduling maintenance of facilities, etc. The greater the stability and uniformity of the process, the less the need for frequent measurement and maintenance.

Those who plan for quality control should have a thorough understanding of the concept of process capability and its application to both areas of planning—planning the operating processes as well as planning the controls.

Process Conformance

Does the process conform to its quality goals? The umpire answers this question by interpreting the observed differences between process performance and process goals. When current performance does differ from the quality goals, the question arises, What is the cause of this difference?

Special and Common Causes of Variation

Observed differences usually originate in one of two ways: (1) the observed change is caused by the behavior of a major variable in the process (or by the entry of a new major variable) or (2) the observed change is caused by the interplay of multiple minor variables in the process.

Shewhart called (1) and (2) "assignable" and "nonassignable" causes of variation, respectively (Shewhart 1931). Deming later coined the terms "special" and "common" causes of variation (Deming 1986). In what follows we will use Deming's terminology.

"Special" causes are typically sporadic, and often have their origin in single variables. For such cases, it is comparatively easy to conduct a diagnosis and provide remedies. "Common" causes are typically chronic and usually have their origin in the interplay among multiple minor variables. As a result, it is difficult to diagnose them and to provide remedies. This contrast makes clear the importance of distinguishing special causes from common causes when interpreting differences. The need for making such distinctions is widespread. Special causes are the subject of quality control; common causes are the subject of quality improvement.

The Shewhart Control Chart

It is most desirable to provide umpires with tools that can help to distinguish between special causes and common causes. An elegant tool for this purpose is the Shewhart control chart (or just control chart) shown in Figure 6.10.

In Figure 6.10, the horizontal scale is time and the vertical scale is quality performance. The plotted points show quality performance as time progresses.

The chart also exhibits three horizontal lines. The middle line is the average of past performance and is, therefore, the expected level of performance. The other two lines are statistical "limit lines." They are intended to separate special causes from common causes, based on some chosen level of probability, such as 1 chance in 100.

Points within Control Limits

Point A on the chart differs from the historical average. However, since point A is within the limit lines, this difference could be due to common causes (at a probability of more than 1 in 100).

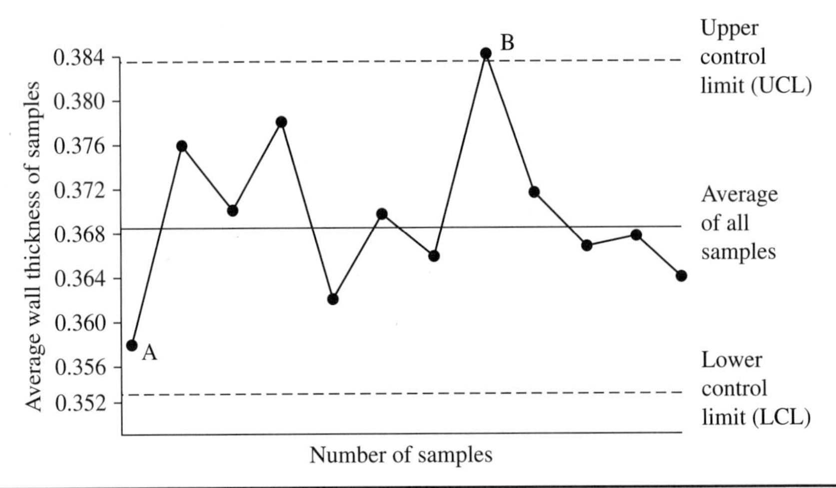

Figure 6.10 Shewhart control chart. (*"Quality Control," Leadership for the Quality Century, Juran Institute, Inc.*)

Hence, we assume that there is no special cause. In the absence of special causes, the prevailing assumptions include

- Only common causes are present.
- The process is in a state of "statistical control."
- The process is doing the best it can.
- The variations must be endured.
- No action need be taken—taking action may make matters worse (a phenomenon known as "hunting" or "tampering").

The preceding assumptions are being challenged by a broad movement to improve process uniformity. Some processes exhibit no points outside of control chart limits, yet the interplay of minor variables produces some defects.

In one example, a process in statistical control was nevertheless improved by an order of magnitude. The improvement was by a multifunctional improvement team, which identified and addressed some of the minor variables. This example is a challenge to the traditional assumption that variations due to common causes must be endured (Pyzdek 1990).

In other cases the challenge is more subtle. There are again no points outside the control limits, but in addition, no defects are being produced. Nevertheless, the customers demand greater and greater uniformity. Examples are found in business processes (precision of estimating), as well as in manufacturing (batch-to-batch uniformity of chemicals, uniformity of components going into random assembly). Such customer demands are on the increase, and they force suppliers to undertake projects to improve the uniformity of even the minor variables in the process. There are many types of control charts.

Points Outside of Control Limits

Point B also differs from the historical average, but is outside of the limit lines. Now the probability is against this being the result of common causes, less than 1 chance in 100. Hence, we assume that point B is the result of special causes. Traditionally such "out-of-control" points become nominations for corrective action.

Ideally, all such nominations should stimulate prompt corrective action to restore the status quo. In practice many out-of-control changes do not result in corrective action. The usual reason is that the changes involving special causes are too numerous—the available personnel cannot deal with all of them. Hence, priorities are established based on economic significance or on other criteria of importance. Corrective action is taken for the high-priority cases; the rest must wait their turn. Some changes at low levels of priority may wait a long time for corrective action.

A further reason for failure to take corrective action is a lingering confusion between statistical control limits and quality tolerances. It is easy to be carried away by the elegance and sensitivity of the control chart. This happened on a large scale during the 1940s and 1950s. Here are two examples from my personal experience:

- A large automotive components factory placed a control chart at every machine.
- A viscose yarn factory created a "war room" of more than 400 control charts.

In virtually all such cases the charts were maintained by the quality departments but ignored by the operating personnel. Experience with such excesses has led leaders and planners to be wary of employing control charts just because they are sensitive detectors of change. Instead, the charts should be justified based on value added. Such justifications include

- Customer needs are directly involved.
- There is risk to human safety or the environment.
- Substantial economics are at stake.
- The added precision is needed for control.

Statistical Control Limits and Tolerances

For most of human history, targets and goals consisted of product features or process features, usually defined in words. Words such as "the color is red" and "the length is long enough" are targets, but are open to too much interpretation. The growth of technology stimulated the growth of measurement, plus a trend to define targets and goals in precise numbers. In addition, there emerged the concept of limits, or "tolerances," around the targets and goals. For example:

- Ninety-five percent of the shipments shall meet the scheduled delivery date.
- The length of the bar shall be within 1 mm of the specified number.
- The length of time to respond to customers is 10 minutes, plus or minus two minutes.

Such targets had official status. They were set by product or process designers, and published as official specifications. The designers were the official quality legislators—they enacted the laws. Operating personnel were responsible for obeying the quality laws—meeting the specified goals and tolerances.

Statistical control limits in the form of control charts were virtually unknown until the 1940s. At that time, these charts lacked official status. They were prepared and published by quality specialists from the quality department. To the operating forces, control charts were a mysterious, alien concept. In addition, the charts threatened to create added work in the form of unnecessary corrective action. The operating personnel reasoned as follows: It has always been our responsibility to take corrective action whenever the product becomes nonconforming. These charts are so sensitive that they detect process changes that do not result in nonconforming products. We are then asked to take corrective action even when the products meet the quality goals and tolerances.

So there emerged a confusion of responsibility. The quality specialists were convinced that the control charts provided useful early-warning signals that should not be ignored. Yet the quality departments failed to recognize that the operating forces were now faced with a confusion of responsibility. The latter felt that so long as the products met the quality goals there was no need for corrective action. The upper leaders of those days were of no help—they did not involve themselves in such matters. Since the control charts lacked official status, the operating forces solved their problem by ignoring the charts. This contributed to the collapse in the 1950s of the movement known as "statistical quality control."

The 1980s created a new wave of interest in applying the tools of statistics to the control of quality. Many operating personnel underwent training in "statistical process control." This training helped to reduce the confusion, but some confusion remains. To get rid of the confusion, leaders should

- Clarify the responsibility for corrective action on points outside the control limits. Is this action mandated or is it discretionary?
- Establish guidelines on actions to be taken when points are outside the statistical control limits but the product still meets the quality tolerances.

The need for guidelines for decision-making is evident from Figure 6.11. The guidelines for quadrants A and C are obvious. If both process and product conform to their respective goals, the process may continue to run. If neither process nor product conform to their respective goals, the process should be stopped and remedial action should be taken. The guidelines for quadrants B and D are often vague, and this vagueness has been the source of a good deal of confusion. If the choice of action is delegated to the workforce, the leaders should establish clear guidelines.

		Product	
		Conforms	Does not conform
Process	Does not conform	B vague	C clear
	Conforms	A clear	D vague

FIGURE 6.11 Areas of decision-making. (*"Making Quality Happen," Juran Institute, Inc. Used by permission.*)

Numerous efforts have been made to design control chart limits in ways that help operating personnel detect whether product quality is threatening to exceed the product quality limits.

Self-Control and Controllability

Workers are in a state of self-control when they have been provided with all the essentials for doing good work. These essentials include

- Means of knowing what the goals are.
- Means of knowing what their actual performance is.
- Means for changing their performance in the event that performance does not conform to goals. To meet this criterion requires an operating process that (1) is inherently capable of meeting the goals and (2) is provided with features that make it possible for the operating forces to adjust the process as needed to bring it into conformance with the goals.

These criteria for self-control are applicable to processes in all functions and at all levels, from general manager to nonsupervisory worker.

It is all too easy for leaders to conclude that the above criteria have been met. In practice, however, there are many details to be worked out before the criteria can be met. The nature of these details is evident from checklists, which have been prepared for specific processes in order to ensure that the criteria for self-control are met. Examples of these checklists include those designed for product designers, production workers, and administrative and support personnel. Examples of such checklists can be found by referring to the subject index of this handbook.

If all the criteria for self-control have been met at the worker level, any resulting product nonconformances are said to be worker-controllable. If any of the criteria for self-control have not been met, then management's planning has been incomplete—the planning has not fully provided the means for carrying out the activities within the feedback loop. The nonconforming products resulting from such deficient planning are then said to be management-controllable. In such cases it is risky for leaders to hold the workers responsible for quality.

Responsibility for results should, of course, be keyed to controllability. However, in the past, many leaders were not aware of the extent of controllability as it prevailed at the worker level. Studies conducted by Juran during the 1930s and 1940s showed that at the worker level, the proportion of management-controllable to worker-controllable nonconformances was of the order of 80 to 20. These findings were confirmed by other studies during the 1950s and 1960s. That ratio of 80 to 20 helps to explain the failure of so many efforts to solve the organizations' quality problems solely by motivating the workforce.

Effect on the Process Conformance Decision

Ideally, the decision of whether the process conforms to process quality goals should be made by the workforce. There is no shorter feedback loop. For many processes, this is the actual arrangement. In other cases, the process conformance decision is assigned to nonoperating personnel—independent checkers or inspectors. The reasons include

- The worker is not in a state of self-control.
- The process is critical to human safety or to the environment.
- Quality does not have top priority.
- There is a lack of mutual trust between the leaders and the workforce.

Product Conformance: Fitness for Purpose

There are two levels of product features, and they serve different purposes. One of these levels serves such purposes as

- Meeting customer needs
- Protecting human safety
- Protecting the environment

Product features are said to possess "fitness for use" if they are able to serve the above purposes.

The second level of product features serves purposes such as

- Providing working criteria to those who lack knowledge of fitness for use
- Creating an atmosphere of law and order
- Protecting innocents from unwarranted blame

Such product features are typically contained in internal specifications, procedures, standards, etc. Product features that are able to serve the second list of purposes are said to possess conformance to specifications, etc. We will use the shorter label "conformance."

The presence of two levels of product features results in two levels of decision-making: Is the product in conformance? Is the product fit for use? Figure 6.12 shows the interrelation of these decisions to the flow diagram.

The Product Conformance Decision

Under prevailing policies, products that conform to specification are sent on to the next destination or customer. The assumption is that products that conform to specification are also fit for use. This assumption is valid in the great majority of cases.

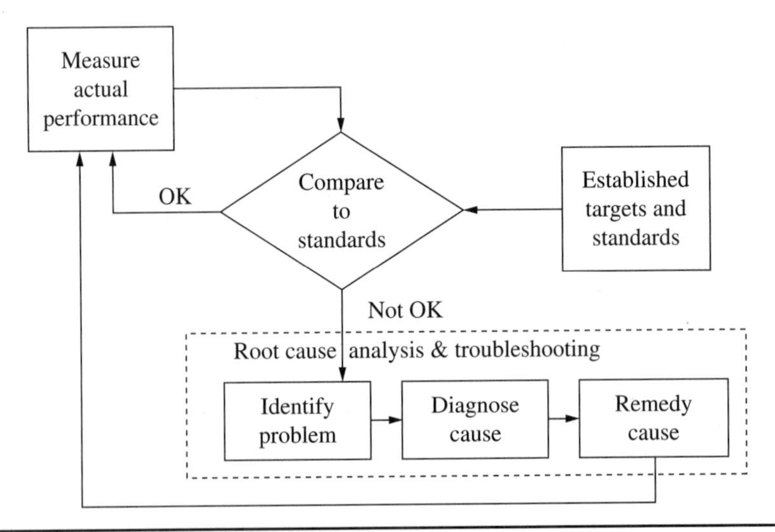

FIGURE 6.12 Interrelation of QC and RCA.

The combination of large numbers of product features, when multiplied by large volumes of product, creates huge numbers of product conformance decisions to be made. Ideally, these decisions should be delegated to the lowest levels of organization—to the automated devices and the operating workforce. Delegation of this decision to the workforce creates what is called "self-inspection."

Self-Inspection

We define "self-inspection" as a state in which decisions on the product are delegated to the workforce. The delegated decisions consist mainly of: Does product quality conform to the quality goals? What disposition is to be made of the product?

Note that self-inspection is very different from self-control, which involves decisions on the *process*.

The merits of self-inspection are considerable:

The feedback loop is short; the feedback often goes directly to the actuator—the energizer for corrective action.

Self-inspection enlarges the job of the workforce—it confers a greater sense of job ownership. Self-inspection removes the police atmosphere created by use of inspectors, checkers, etc.

However, to make use of self-inspection requires meeting several essential criteria:

- *Quality is number one.* Quality must undoubtedly be the top priority.
- *Mutual confidence.* The leaders must have enough trust in the workforce to be willing to make the delegation, and the workforce must have enough confidence in the leaders to be willing to accept the responsibility.
- *Self-control.* The conditions for self-control should be in place so that the workforce has all the means necessary to do good work.
- *Training.* The workers should be trained to make the product conformance decisions.
- *Certification.* The recent trend is to include a certification procedure. Workers who are candidates for self-inspection undergo examinations to ensure that they are qualified to make good decisions. The successful candidates are certified and may be subject to audit of decisions thereafter.

In many organizations, these criteria are not fully met, especially the criterion of priority. If some parameter other than quality has top priority, there is a real risk that evaluation of product conformance will be biased. This problem happens frequently when personal performance goals are in conflict with overall quality goals. For example, a chemical company found that it was rewarding sales personnel on revenue targets without regard to product availability or even profitability. The salespeople were making all their goals, but the company was struggling.

The Fitness for Purpose Decision

The great majority of products do conform to specifications. For the nonconforming products there arises a new question: Is the nonconforming product nevertheless fit for use?

A complete basis for making this decision requires answers to questions such as

- Who are the user(s)?
- How will this product be used?

- Are there risks to structural integrity, human safety, or the environment?
- What is the urgency for delivery?
- How do the alternatives affect the producer's and the user's economics?

To answer such questions can involve considerable effort. Organizations have tried to minimize the effort through procedural guidelines. The methods in use include:

- *Treat all nonconforming products as unfit for use.* This approach is widely used for products that can pose risks to human safety or the environment—products such as pharmaceuticals or nuclear energy.
- *Create a mechanism for decision-making.* An example is the material review board so widely used in the defense industry. This device is practical for matters of importance, but is rather elaborate for the more numerous cases in which little is at stake.
- *Create a system of multiple delegation.* Under such a system, the "vital few" decisions are reserved for a formal decision-making body such as a material review board. The rest are delegated to other people.

Table 6.3 is an example of a table of delegation used by a specific company. (Personal communication to one of the authors.)

Disposition of Unfit Product

Unfit product is disposed of in various ways: scrap, sort, rework, return to supplier, sell at a discount, etc. The internal costs can be estimated to arrive at an economic optimum. However, the effects go beyond money: schedules are disrupted, people are blamed, etc. To

	Amount of Product or Money at Stake Is:	
Effect of Nonconformance Is on	**Small**	**Large**
Internal economics only	Department head directly involved, quality engineer	Plant managers involved, quality manager
Economic relations with supplier	Supplier, purchasing agent, quality engineer	Supplier, manager
Economic relations with client	Client, salesperson, quality engineer	Client (for marketing, manufacturing, technical, quality)
Field performance of the product	Product designer, salesperson, quality engineer	Client (managers for technical, manufacturing, marketing, quality)
Risk of damage to society or of nonconformance to government regulations	Product design manager, compliance officer, lawyer, quality managers	General manager and team of upper managers

*For those industries whose quality mission is really one of conformance to specification (for example, atomic energy, space), the real decision-maker on fitness for use is the client or the government regulator.

Table 6.3 Multiple Delegations of Decision-Making on Fitness for Purpose*

minimize the resulting human abrasion, some organizations have established rules of conduct, such as

- Choose the alternative that minimizes the total loss to all parties involved. Now there is less to argue about, and it becomes easier to agree on how to share the loss.
- Avoid looking for blame. Instead, treat the loss as an opportunity for quality improvement.
- Use "charge backs" sparingly. Charging the vital few losses to the departments responsible has merit from an accounting viewpoint. However, when applied to the numerous minor losses, this is often uneconomic as well as detrimental to efforts to improve quality.

Failure to use products that meet customer needs is a waste. Sending out products that do not meet customer needs is worse. Personnel who are assigned to make product conformance decisions should be provided with clear definitions of responsibility as well as guidelines for decision-making. Leaders should, as part of their audit, ensure that the processes for making product conformance decisions are appropriate to company needs.

Corrective Action

The final step in closing the feedback loop is to actuate a change that restores conformance with quality goals. This step is popularly known as "troubleshooting" or "firefighting."

Note that the term "corrective action" has been applied loosely to two very different situations, as shown in Figure 6.1. The feedback loop is well designed to eliminate sporadic nonconformance, like that "spike" in Figure 6.1; the feedback loop is not well designed to deal with the area of chronic waste shown in the figure. Instead, the need is to employ the quality improvement process discussed in Chapter 5, Quality Improvement: Creating Breakthroughs in Performance.

We will use the term "corrective action" in the sense of troubleshooting—eliminating sporadic nonconformance.

Corrective action requires the journeys of diagnosis and remedy. These journeys are simpler than for quality improvement. Sporadic problems are the result of adverse change, so the diagnostic journey aims to discover what has changed. The remedial journey aims to remove the adverse change and restore conformance.

Diagnosing Sporadic Change

During the diagnostic journey, the focus is on what has changed. Sometimes the causes are not obvious, so the main obstacle to corrective action is diagnosis. The diagnosis makes use of methods and tools such as

- Forensic autopsies to determine with precision the symptoms exhibited by the product and process
- Comparison of products made before and after the trouble began to see what has changed; also comparison of good and bad products made since the trouble began
- Comparison of process data before and after the problem began to see what process conditions have changed
- Reconstruction of the chronology, which consists of logging on a time scale (of hours, days, etc.) (1) the events that took place in the process before and after the sporadic change—that is, rotation of shifts, new employees on the job, maintenance

actions, etc., and (2) the time related to product information—that is, date codes, cycle time for processing, waiting time, move dates, etc.

Analysis of the resulting data usually sheds a good deal of light on the validity of the various theories of causes. Certain theories are denied. Other theories survive to be tested further.

Operating personnel who lack the training needed to conduct such diagnoses may be forced to shut down the process and request assistance from specialists, the maintenance department, etc. They may also run the process "as is" in order to meet schedules and thereby risk failure to meet the quality goals.

Corrective Action—Remedy

Once the cause(s) of the sporadic change is known, the worst is over. Most remedies consist of going back to what was done before. This is a return to the familiar, not a journey into the unknown (as is the case with chronic problems). The local personnel are usually able to take the necessary action to restore the status quo.

Process designs should provide means to adjust the process as required to attain conformance with quality goals. Such adjustments are needed at startup and during running of the process. This aspect of design for process control ideally should meet the following criteria:

- There should be a known relationship between the process variables and the product results.
- Means should be provided for ready adjustment of the process settings for the key process variables.
- A predictable relationship should exist between the amount of change in the process settings and the amount of effect on the product features.

If such criteria are not met, the operating personnel will, in due course, be forced to cut corners in order to carry out remedial action. The resulting frustrations become a disincentive to putting high priority on quality.

The Role of Statistical Methods in Control

An essential activity within the feedback loop is the collection and analysis of data. This activity falls within the scientific discipline known as "statistics." The methods and tools used are often called "statistical methods." These methods have long been used to aid in data collection and analysis in many fields: biology, government, economics, finance, management, etc. Chapter 18, Core Tools to Design, Control, and Improve Performance, and Chapter 19, Accurate and Reliable Measurement Systems and Advanced Tools, contain a thorough discussion of the basic and advanced statistical methods.

Statistical Process Control (SPC)

The term has multiple meanings, but in most organizations, it is considered to include basic data collection; analysis through such tools as frequency distributions, Pareto principle, Ishikawa (fish bone) diagram, Shewhart control chart, etc.; and application of the concept of process capability.

Advanced tools, such as design of experiments and analysis of variance, are a part of statistical methods but are not normally considered part of statistical process control.

The Merits

These statistical methods and tools have contributed in an important way to quality control and to the other processes of the Juran Trilogy—quality improvement and quality planning. For some types of quality problems, the statistical tools are more than useful—the problems cannot be solved at all without using the appropriate statistical tools.

The SPC movement has succeeded in training a great many supervisors and workers in basic statistical tools. The resulting increase in statistical literacy has made it possible for them to improve their grasp of the behavior of processes and products. In addition, many have learned that decisions based on data collection and analysis yield superior results.

The Risks

There is a danger in taking a tool-oriented approach to quality instead of a problem-oriented or results-oriented approach. During the 1950s, this preoccupation became so extensive that the entire statistical quality control movement collapsed; the word "statistical" had to be eliminated from the names of the departments.

The proper sequence in managing is first to establish goals and then plan how to meet those goals, including choosing the appropriate tools. Similarly, when dealing with problems—threats or opportunities—experienced leaders start by first identifying the problems. They then try to solve those problems by various means, including choosing the proper tools.

During the 1980s, numerous organizations did, in fact, try a tool-oriented approach by training large numbers of their personnel in the use of statistical tools. However, there was no significant effect on the bottom line. The reason was that no infrastructure had been created to identify which projects to tackle, to assign clear responsibility for tackling those projects, to provide needed resources, to review progress, etc.

Leaders should ensure that training in statistical tools does not become an end in itself. One form of such assurance is through measures of progress. These measures should be designed to evaluate the effect on operations, such as improvement in customer satisfaction or product performance, reduction in cost of poor quality, etc. Measures such as numbers of courses held or numbers of people trained do not evaluate the effect on operations and hence, should be regarded as subsidiary in nature.

Information for Decision-Making

Quality control requires extensive decision-making. These decisions cover a wide variety of subject matter and take place at all levels of the hierarchy. The planning for quality control should provide an information network that can serve all decision-makers. At some levels of the hierarchy, a major need is for real-time information to permit prompt detection and correction of nonconformance to goals. At other levels, the emphasis is on summaries that enable leaders to exercise control over the vital few control subjects. In addition, the network should provide information as needed to detect major trends, identify threats and opportunities, and evaluate the performance of organization units and personnel.

In some organizations, the quality information system is designed to go beyond control of product features and process features; the system is also used to control the quality performance of organizations and individuals, such as departments and department heads. For example, many organizations prepare and regularly publish scoreboards showing summarized quality performance data for various market areas, product lines, operating functions, etc. These performance data are often used as indicators of the quality performance of the personnel in charge.

To provide information that can serve all those purposes requires planning that is directed specifically at the information system. Such planning is best done by a multifunctional team whose mission is focused on the quality information system. That team properly includes the customers as well as the suppliers of information. The management audit of the quality control system should include assurance that the quality information system meets the needs of the various customers.

The Quality Control System and Policy Manual

A great deal of quality planning is done through "procedures," which are really repetitive-use plans. Such procedures are thought out, written out, and approved formally. Once published, they become the authorized ways of conducting the company's affairs. It is quite common for the procedures relating to managing for quality to be published collectively in a "quality manual" (or similar title). A significant part of the manual relates to quality control.

Quality manuals add to the usefulness of procedures in several ways:

- *Legitimacy*. The manuals are approved at the highest levels of organization.
- *Easy to find*. The procedures are assembled into a well-known reference source rather than being scattered among many memoranda, oral agreements, reports, minutes, etc.
- *Stable*. The procedures survive despite lapses in memory and employee turnover.

Study of company quality manuals shows that most of them contain a core content, which is quite similar from company to company. Relative to quality control, this core content includes procedures for

- Applying the feedback loop to process and product control
- Ensuring that operating processes are capable of meeting the quality goals
- Maintaining facilities and calibration of measuring instruments
- Relating to suppliers on quality matters
- Collecting and analyzing the data required for the quality information system
- Training the personnel to carry out the provisions of the manual
- Auditing to ensure adherence to procedures

The need for repetitive-use quality control systems has led to an evolution of standards at industry, national, and international levels. For elaboration, see Chapter 16, Using International Standards to Ensure Organization Compliance. For an example of developing standard operating procedures, including the use of videocassettes, see Murphy and McNealey (1990). Workforce participation during the preparation of procedures helps to ensure that the procedures will be followed.

Format of Quality Manuals

Here, again, there is much commonality. The general chapters of the manual include

1. An official statement by the general manager. It includes the signatures that confer legitimacy.
2. The purpose of the manual and how to use it.

3. The pertinent company (or divisional, etc.) quality policies.
4. The organizational charts and tables of responsibility relative to the quality function.
5. Provision for audit of performance against the mandates of the manual.

Additional chapters of the manual deal with applications to functional departments, technological products and processes, business processes, etc. For elaboration, see Juran (1988, pp. 6.40–6.47).

Leaders are able to influence the adequacy of the quality control manual in several ways:

- Participate in defining the criteria to be met by the manual
- Approve the final draft of the manual to make it official
- Periodically audit the up-to-date-ness of the manual as well as conformance to the manual

Provision for Audits

Experience has shown that control systems are subject to "slippage" of all sorts. Personnel turnover may result in loss of essential knowledge. Entry of unanticipated changes may result in obsolescence. Shortcuts and misuse may gradually undermine the system until it is no longer effective.

The major tool for guarding against deterioration of a control system has been the audit. Under the audit concept, a periodic, independent review is established to provide answers to the following questions: Is the control system still adequate for the job? Is the system being followed?

The answers are obviously useful to the operating leaders. However, that is not the only purpose of the audit. A further purpose is to provide those answers to people who, though not directly involved in operations, nevertheless have a need to know. If quality is to have top priority, those who have a need to know include the upper leaders.

It follows that one of the responsibilities of leaders is to mandate establishment of a periodic audit of the quality control system.

Tasks for Leaders

1. Leaders should avoid getting too deeply involved in making decisions on quality control. They should make the vital few decisions, provide criteria to distinguish the vital few from the rest, and delegate the rest under a decision-making process.
2. To eliminate the confusion relative to control limits and product quality tolerance, leaders should clarify the responsibility for corrective action on points outside the control limits and establish guidelines on action to be taken when points are outside the statistical control limits but the product still meets the quality tolerances.
3. Leaders should, as part of their audit, ensure that the processes for making product conformance decisions are appropriate to company needs. They should also ensure that training in statistical tools does not become an end in itself. The management audit of the quality control system should include assurance that the quality information system meets the needs of the various customers.

4. Leaders are able to influence the adequacy of the quality control manual in several ways: participate in defining the criteria to be met, approve the final draft to make it official, and periodically audit the up-to-dateness of the manual as well as the state of conformance.

References

Deming, W. E. (1950). "Elementary Principles of the Statistical Control of Quality." Nippon Kagaku Gijutsu Renmei (Japanese Union of Scientists and Engineers), Tokyo.

Deming, W. E. (1986). "Out of the Crisis." MIT Center for Advanced Engineering Study. Cambridge, MA.

Juran, J. M. (1964). *Managerial Breakthrough*. McGraw-Hill, New York.

Murphy, R. W., and McNealey, J. E. (1990). "A Technique for Developing Standard Operating Procedures to Provide Consistent Quality." 1990 Juran IMPRO Conference Proceedings, pp. 3D1–3D6.

Pyzdek, T. (1990). "There's No Such Thing as a Common Cause." ASQC Quality Congress Transactions, pp. 102–108.

Radford, G. S. (1917). "The Control of Quality." *Industrial Management*, vol. 54, p. 100.

Radford, G. S. (1922). *The Control of Quality in Manufacturing*. Ronald Press Company, New York.

Radman, M., and Wagner, R. (1988). "The High Fidelity of DNA Duplication." *Scientific American*, August, pp. 40–46.

Shewhart, W. A. (1931). *Economic Control of Quality of Manufactured Product*. Van Nostrand, New York, 1931. Reprinted by ASQC, Milwaukee, 1980.

White, J. B. (1988). "Auto Mechanics Struggle to Cope with Technology in Today's Cars." *The Wall Street Journal*, July 26, p. 37.

CHAPTER 7

Strategic Planning and Deployment: Moving from Good to Great

Joseph A. De Feo

About This Chapter

This chapter describes the process by which an organization must create a "vision" and aligned strategic plan to be the market "quality" leader. The strategic planning and deployment process explains how an organization can integrate and align the methods to attain performance excellence. It addresses such important issues as how to align strategic goals with the organization's vision and mission, how to deploy those goals throughout the organization, and how to derive the benefits of strategic planning.

High Points of This Chapter

1. Strategic planning (SP) is the systematic approach to defining long-term business goals and planning the means to achieve them. An organization transformation based on the management of quality should be integrated with the strategic plans of the organization.

2. The strategic plan enables organizations to deploy all goals, including quality improvement goals, to the organization. It provides the basis for senior management to make sound strategic choices and prioritize the organization's focus and other change activities.
3. Activities not aligned with the organization's strategic goals should be changed or eliminated.
4. In this chapter we will define the strategic planning process and its deployment tasks and describe the systematic approach to deploying "quality goals."
5. This chapter will also explain the specific roles of leaders when implementing and ensuring the success of the strategic plan and its deployment.

Strategic Planning and Quality: The Benefits

Strategic planning (SP) is the systematic approach to defining long-term business goals and planning the means to achieve them. Once an organization has established its long-term goals, effective strategic planning enables it, year by year, to create an annual business plan, which includes the necessary annual goals, resources, and actions needed to move toward those goals.

Many organizations have created a vision to be the best performers by creating and producing high-quality products and services for their customers. By doing so, they have outperformed those that did not. This performance is not just related to the quality of their goods and services, but to the business itself: more sales, fewer costs, and better culture through employee satisfaction and ultimately better market success for its stakeholders.

It is necessary to incorporate these goals into the strategic planning process and into the annual business plans. This will ensure that the new focus becomes part of the plan and does not compete with the well-established priorities for resources. Otherwise, the best-intended desired changes will fail.

Many leaders understand the meaning of strategic planning as it relates to the creation of the strategic plan and the financial goals and targets to be achieved. Often, they do not include the deployment of strategic "quality" goals, subgoals, and annual goals or the assignment of the resources and actions to achieve them. We will try to highlight this difference and use the term "strategic planning and deployment" throughout this chapter. Many organizations have overcome failures of change programs and have achieved long-lasting results through strategic deployment.

Six Sigma, Lean Six Sigma, and in prior years TQM all became pervasive change processes and were natural candidates for inclusion in the strategic plan of many organizations. The integration of these "quality and customer-driven" methods with strategic planning is important for their success.

Organizations have chosen different terms for this process. Some have used the Japanese term "hoshin kanri." Others have partially translated the term and called it "hoshin planning." Still others have used a rough translation of the term and called it "policy deployment." In an earlier version of the United States Malcolm Baldrige National Quality Award, this process was called "strategic quality planning." Later this award criterion was renamed "strategic planning."

Whether the upper managers should align quality with the plan is a decision unique to each organization. What is decisive is the importance of integrating major change initiatives or quality programs into the strategic plan. The potential benefits of strategic planning and deployment are clear:

- The goals become clear—the planning process forces clarification of any vagueness.
- The planning process then makes the goals achievable.
- The monitoring process helps to ensure that the goals are reached.
- Chronic wastes are "scheduled" to be reduced through the improvement process.
- Creation of new focus on the customers and quality is attained as progress is made.

What Is Strategic Planning and Deployment?

It is a systematic approach to integrating customer-focused, systemwide quality and business excellence methods into the strategic plan of the organization. Strategic planning is the systematic process by which an organization defines its long-term goals with respect to quality and customers, and integrates them—on an equal basis—with financial, human resources, marketing, and research and development goals into one cohesive business plan. The plan is then deployed throughout the entire organization.

As a component of an effective business management system, strategic planning enables an organization to plan and execute strategic organizational breakthroughs. Over the long term, the intended effect of such breakthroughs is to achieve competitive advantage or to attain a status of "quality leadership."

Strategic planning has evolved during the past decades to become an integral part of many organizational change processes, like Six Sigma or Operational Excellence (OpEx). It is now is part of the foundation that supports the broader system of managing the business of the organization. A simple strategic planning and deployment model is shown in Figure 7.1. This is what will be used throughout this chapter.

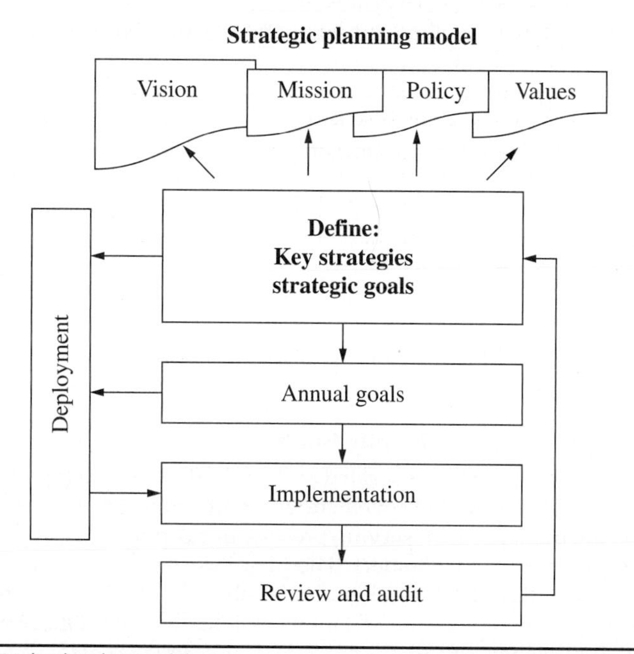

FIGURE. 7.1 Strategic planning model.

	1988–1996 investments	Value on 12/1/97	Percent change
All recipients	$7,496.54	$33,185.69	342
Standard & poor's 500	$7,496.54	$18,613.28	148

Data: National institute of standards and technology

FIGURE 7.2 Malcolm Baldrige National Quality Award winner performance. (*BusinessWeek, March 16, 1998, p. 60.*)

Strategic planning and deployment also is a key element of the U.S. Malcolm Baldrige National Quality Award and the European Foundation for Quality Management (EFQM) Award, as well as other international and state awards. The criteria for these awards stress that customer-driven quality and operational performance excellence are key strategic business issues, which need to be an integral part of overall business planning. A critical assessment of the Malcolm Baldrige National Quality Award winners demonstrates that those organizations that won the award outperformed those that did not (Figure 7.2).

From 1995–2002, quality demonstrated just how profitable it can be. The "Baldrige Index" outperformed the S&P 500 stock index for eight straight years, in certain years beating the S&P by wide margins of 4:1 or 5:1. The index was discontinued in 2004 when Baldrige began to recognize and award small businesses and educational entities along with their normal categories for National Quality Awards. The additions of smaller organizations skewed the "Baldrige Index," yet the results from the original study, when the playing fields were level, speaks volumes; quality pays off.

Godfrey (1997) has observed that to be effective, strategic deployment should be used as a tool, a means to an end, not as the goal itself. It should be an endeavor that involves people throughout the organization. It must capture existing activities, not just add to already overflowing plates. It must help senior managers face difficult decisions, set priorities, and not just start new initiatives but eliminate many current activities that add no value.

Strategic Planning Today

The approach used to establish organization-wide financial goals has evolved into a more robust strategic plan. To be effective in the global marketplace, large organizations must create a strategic plan that includes the elements discussed in the following sections.

Quality and Customer Loyalty Goals

These major goals are incorporated and supported by a hierarchy of goals at lower levels: subgoals, projects, etc. Improvement goals are goals aimed at creating a breakthrough in performance of a product, serving process, or people by focusing on the needs of customers, suppliers, and shareholders. The plan incorporates the "voice of the customer" and aligns them to the plan. This alignment enables the goals to be legitimate and balances the financial goals (which are important to shareholders) with those of importance to the customers. It also eliminates the concern that there are two plans, one for finance and one for quality.

A systematic, structured methodology for establishing annual goals and providing resources must include the following:

- *A provision of rewards.* Performance against improvement goals is given substantial weight in the system of merit rating and recognition. A change in the structure that includes rewarding the right behaviors is required.
- *Required and universal participation.* The goals, reports, reviews, etc., are designed to gain participation from within the organization's hierarchy. This participation involves every employee at every level, providing support for the change initiative and helping achieve the desired results.
- *A common language.* Key terms, such as quality, benchmarking, and strategic quality deployment, acquire standard meanings so that communication becomes more and more precise.
- *Training.* It is common for all employees to undergo training in various concepts, processes, methods, tools, etc. Organizations that have so trained their workforce, in all functions, at all levels, and at the right time, are well poised to outperform organizations in which such training has been confined to the quality department or managers.

Why Strategic Deployment? The Benefits

The first question that often arises in the beginning stages of strategic planning in an organization is, Why do strategic planning in the first place? To answer this question requires a look at the benefits that other organizations have realized from strategic planning. They report that it

- Focuses the organization's resources on the activities that are essential to increasing customer satisfaction, lowering costs, and increasing shareholder value (see Figure 7.2)
- Creates a planning and implementation system that is responsive, flexible, and disciplined
- Encourages interdepartmental cooperation
- Provides a method to execute breakthroughs year after year
- Empowers leaders, managers, and employees by providing them with the resources to carry out the planned initiatives
- Eliminates unnecessary and wasteful initiatives that are not in the plan
- Eliminates the existence of many potentially conflicting plans—the finance plan, the marketing plan, the technology plan, and the quality plan
- Focuses resources to ensure financial plans are achievable

Why Strategic Deployment? The Risks

Different organizations have tried to implement total quality management systems as well as other change management systems. Some organizations have achieved stunning results; others have been disappointed by their results, often achieving little in the way of bottom-line savings or increased customer satisfaction. Some of these efforts have been classified as failures. One of the primary causes of these disappointments has been the inability to incorporate these "quality programs" into the business plans of the organization.

Other reasons for failure are that

- Strategic planning was assigned to planning departments, not to the upper managers themselves. These planners lacked training in concepts and methods, and were not among the decision-makers in the organization. This led to a strategic plan that did not include improvement goals aimed at customer satisfaction, process improvement, etc.

- Individual departments had been pursuing their own departmental goals, failing to integrate them with the overall organizational goals.
- New products or services continued to be designed with failures from prior designs that were carried over into new models, year after year. The new designs were not evaluated or improved and hence, were not customer-driven.
- Projects suffered delays and waste due to inadequate participation and ended before positive business results were achieved.
- Improvement goals were assumed to apply only to manufactured goods and manufacturing processes. Customers became irritated not only by receipt of defective goods; they were also irritated by receiving incorrect invoices and late deliveries. The business processes that produce invoices and deliveries were not subject to modern quality planning and improvement because there were no such goals in the annual plan to do so.

The deficiencies of the past strategic planning processes had their origin in the lack of a systematic, structured approach to integrate programs into one plan. As more organizations became familiar with strategic quality deployment, many adopted its techniques, which treat managing for change on the same organizationwide basis as managing for finance.

Launching Strategic Planning and Deployment

Creating a strategic plan that is quality- and customer-focused requires that leaders become coaches and teachers, personally involved, consistent, eliminate the atmosphere of blame, and make their decisions on the best available data.

Juran (1988) has stated, "You need participation by the people that are going to be impacted, not just in the execution of the plan but in the planning itself. You have to be able to go slow, no surprises, use test sites in order to get an understanding of what are some things that are damaging and correct them."

The Strategic Deployment Process

The strategic deployment process requires that the organization incorporate customer focus into the organization's vision, mission, values, policies, strategies, and long- and short-term goals and projects. Projects are the day-to-day, month-to month activities that link quality improvement activities, reengineering efforts, and quality planning teams to the organization's business objectives.

The elements needed to establish strategic deployment are generally alike for all organizations. However, each organization's uniqueness will determine the sequence and pace of application and the extent to which additional elements must be provided.

There exists an abundance of jargon used to communicate the strategic deployment process. Depending on the organization, one may use different terms to describe similar concepts. For example, what one organization calls a vision, another organization may call a mission (see Figure 7.3).

The following definitions of elements of strategic planning are in widespread use and are used in this chapter:

Selected definitions	
Mission	What business we are in
Vision	Desired future state of organization
Values	Principles to be observed to meet vision or principle to be served by meeting vision
Policy	How we will operate and our commitment to customers and society

FIGURE 7.3 Organizational vision and mission. (*Juran Institute, Inc.*)

Vision. A desired future state of the organization or enterprise. Imagination and inspiration are important components of a vision. Typically, a vision can be viewed as the ultimate goal of the organization, one that may take five or even ten years to achieve.

Mission. This is the purpose of or the reason for the organization's existence and usually states, for example, what we do and whom we serve.

- The presence of Jet Blue at JFK International is unmatched. Measured by number of passengers booked, Jet Blue carries almost the equivalent of every other airline conducting business at JFK. With their entrenchment in the United States' largest travel market, Jet Blue ensures itself profitability even in difficult markets. "Our mission is to bring humanity back to air travel."

Strategies. The means to achieve the vision. Strategies are few and define the key success factors, such as price, value, technology, market share, and culture, that the organization must pursue. Strategies are sometimes referred to as "key objectives" or "long-term goals."

Annual goals. What the organization must achieve over a one- to three-year period; the aim or end to which work effort is directed. Goals are referred to as "long term" (two to three years) and "short term" (one to two years). Achievement of goals signals the successful execution of the strategy.

- Jet Blue aims to preserve the core Jet Blue experience of unique, low-cost, high-quality flights while adding optional product offerings for all customers.

Ethics and values. What the organization stands for and believes in.

- For the fourth year in a row, Jet Blue was ranked number one in customer service for low-cost carriers by J.D. Power & Associates. It is this exceptional customer service that continues to drive Jet Blue and set it apart. Partnerships with Sirius XM, and Direct TV, and improved leg room all make the flight experience for every customer a more enjoyable experience.

Policies. A guide to managerial action. An organization may have policies in a number of areas: quality, environment, safety, human resources, etc. These policies guide day-to-day decision-making.

Initiatives and projects. These should be multifunctional teams launched to address a deployed goal, and whose successful completion ensures that the strategic goals are achieved. An initiative or project implies assignment of selected individuals to a team, which is given the responsibility, tools, and authority to achieve the specific goal or goals.

- After six years of planning and three years of construction, Jet Blue's Terminal 5 opened at JFK. Terminal 5 offers Jet Blue customers their own parking lot and road for improved access to the airliner. It comprises twenty-six gates, affords the highest in modern amenities and concession offerings, and due to its proximity to the runway allows Jet Blue to be more efficient in their processes. Terminal 5 only advances the company's stake in the New York travel market.

Deployment plan. To turn a vision into action, the vision must be broken apart and translated into successively smaller and more specific parts—key strategies, strategic goals, etc.—to the level of projects and even departmental actions. The detailed plan for decomposition and distribution throughout the organization is called the "deployment plan." It includes the assignment of roles and responsibilities, and the identification of resources needed to implement and achieve the project goals (Figure 7.4).

Scorecards and key performance indicators. Measurements that are visible throughout the organization for evaluating the degree to which the strategic plan is being achieved.

- By the end of 2008, Jet Blue was the seventh largest passenger carrier in the United States and conducted 600 flights daily.

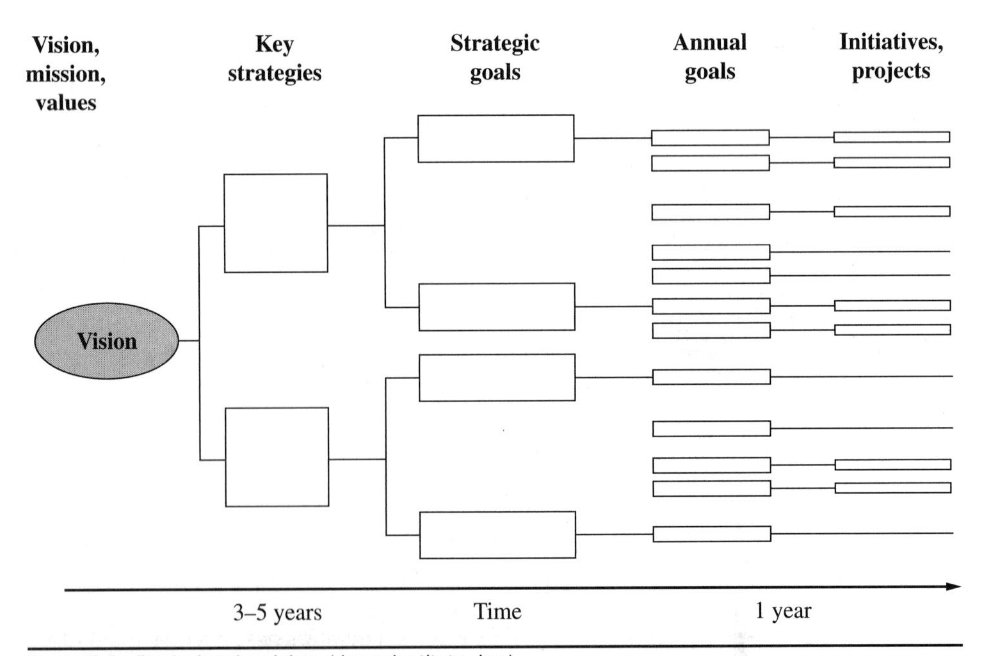

FIGURE 7.4 Deploying the vision. (*Juran Institute, Inc.*)

Developing the Elements of Strategic Planning and Deployment

Establish a Vision

Strategic deployment begins with a vision that is customer-focused. In the organizations we know that are successfully making the transition to a more collaborative organization, the key to success is developing and living by a common strategic vision. When you agree on an overall direction, you can be flexible about the means to achieve it (Tregoe and Tobia 1990).

"Really powerful visions are simply told. The Ten Commandments, the Declaration of Independence, a Winston Churchill World War II speech—all present messages that are so simple and direct you can almost touch them. Our corporate strategies should be equally compelling."

A vision should define the benefits a customer, an employee, a shareholder, or society at large can expect from the organization:

Here are a few examples:

- Samsung, the world's largest manufacturer of high-quality digital products is guided by a singular vision: "to lead the digital convergence movement."

  Samsung believes that through technology innovation today, we will find the solutions we need to address the challenges of tomorrow. From technology comes opportunity—for businesses to grow, for citizens in emerging markets to prosper by tapping into the digital economy, and for people to invent new possibilities. It's our aim to develop innovative technologies and efficient processes that create new markets, enrich people's lives, and continue to make Samsung a trusted market leader.
- Sentara Health (based in the mid-Atlantic states): We have commitment to grow as one of the nation's leading health care organizations by creating innovative systems of care that help people achieve and maintain their best possible state of health.
- Kaiser Permanente (a large U.S.-based health care system): "We are committed to providing our members with quality, cost-effective health care. Our physicians and managers work together to improve care, service, and the overall performance of our organization."

Each of the preceding visions offers a very different view of the direction and character of the organization. Each conveys a general image to customers and employees of where the organization is headed. For the organization, the vision provides, often for the first time in its history, a clear picture of where it is headed and why it is going there.

Good vision statements should also be compelling and shared throughout the organization. It is often a good idea to make the vision a stretch for the organization but possible of being achieved within three to five years, and to state a measurable achievement (e.g., being the best). In creating the vision, organizations should take into account its customers, the markets in which it wants to compete, the environment within which the organization operates, and the current state of the organization's culture.

Vision statements, by themselves, are little more than words. Publication of such a statement does not inform the members of an organization what they should do differently from what they have done in the past. The strategic deployment process and the strategic plan become the basis for making the vision a reality. The words of the vision are just a reminder of what the organization is pursuing. The vision must be carried out through deeds and actions.

Some common pitfalls when forming a vision are

- Focusing the vision exclusively on shareholders as customers
- Thinking that once a strategic plan is written it will be carried out with no further work
- Failing to explain the vision as a benefit to customers, employees, suppliers, and other stakeholders
- Creating a vision that is either too easy or too difficult to achieve
- Failing to consider the effects that the rapid changes taking place in the global economy will have three to five years into the future
- Failing to involve key employees at all levels in creating the vision
- Failing to benchmark competitors or to consider all possible sources of information on future needs, internal capabilities, and external trends

Agree on Your Mission

Most organizations have a mission statement. A mission statement is designed to address the question, What business(es) are we in? A mission is often confused with a vision and even published as one. A mission statement should clarify the organization's purpose or reason for existence. It helps clarify who your organization is.

The following are some examples:

- **Samsung:** Everything we do at Samsung is guided by our mission: to be the best "digital- e-company."
- **Amazon.com:** Our vision is to be earth's most customer-centric company, to build a place where people can come find and discover anything they might want to buy online.
- **Dell:** To be the most successful computer company in the world and delivering the best customer experience in the markets they share.
- **eBay** pioneers communities built on commerce, sustained by trust, and inspired by opportunity. eBay brings together millions of people every day on a local, national, and international basis through an array of websites that focus on commerce, payments, and communications
- **Facebook** is a social utility that helps people communicate more efficiently with their friends, family members, and coworkers. The company develops technologies that facilitate the sharing of information through the social graph, the digital mapping of people's real-world social connections. Anyone can sign up for Facebook and interact with the people they know in a trusted environment.
- Google's mission is to organize the world's information and make it universally accessible and useful.
- The **Ritz-Carlton Hotel** is a place where the genuine care and comfort of our guests is our highest mission.
- **Sentara Health:** We will focus, plan, and act on our commitments to our community mission, to our customers, and to the highest quality standards of health care to achieve our vision for the future.

In the Sentara example, the references to leadership and the future may lead the reader to confuse this mission statement (what business we are in) with a vision statement (what we aim to become). Only the organization itself can decide whether these words belong in its mission statement. It is in debating such points that an organization comes to a consensus on its vision and mission.

Together, a vision and a mission provide a common agreed-upon direction for the entire organization. This direction can be used as a basis for daily decision-making.

Develop Long-Term Strategies or Goals

The first step in converting the vision into an achievable plan is to break the vision into a small number of key strategies (usually four or five). Key strategies represent the most fundamental choices that the organization will make about how it will go about reaching its vision. Each must contribute significantly to the overall vision. For example:

- Xerox initiated their Leadership Through Quality program as part of a broader corporate focus on quality. More than 100,000 employees were trained over a three-year period in a six-step process. Empowered employees started a number of initiatives, many involving environmental and quality improvements, yielding millions of dollars in added profits each year. Xerox management credits the success of new environmental initiatives primarily to employees using quality management practices. Cross-function teams are formed to focus *on a variety of issues."*

Responsibility for executing these key strategies is distributed (or deployed) to key executives within the organization, the first step in a succession of subdivisions and deployments by which the vision is converted into action.

In order to determine what the key strategies should be, one may need to assess five areas of the organization and obtain the necessary data on

- Customer loyalty and customer satisfaction
- Costs related to poor quality or products, services, and processes
- Organization culture and employee satisfaction
- Internal business processes (including suppliers)
- Competitive benchmarking

Each of these areas, when assessed, can form the basis for a balanced business scorecard (see "The Scorecard" later in this chapter). Data must be analyzed to discover specific strengths, weaknesses, opportunities, and threats as they relate to customers, quality, and costs. Once complete, the key strategies can be created or modified to reflect measurable and observable long-term goals.

Develop Annual Goals

An organization sets specific, measurable strategic goals that must be achieved for the broad strategy to be a success. These quantitative goals will guide the organization's efforts toward achieving each strategy. As used here, a goal is an aimed-at target. A goal must be specific. It must be quantifiable (measurable) and is to be met within a specific period. At first, an organization may not know how specific the goal should be. Over time, the measurement systems will improve and the goal setting will become more specific and more measurable.

Despite the uniqueness of specific industries and organizations, certain goals are widely applicable. There are seven areas that are minimally required to ensure that the proper goals are established. They are:

Product performance. Goals in this area relate to product features that determine response to customer needs, for example, promptness of service, fuel consumption, mean time between failures, and courteousness. These product features directly influence product salability and affect revenues.

Competitive performance. This has always been a goal in market-based economies, but seldom a part of the business plan. The trend to make competitive performance a long-term business goal is recent but irreversible. It differs from other goals in that it sets the target relative to the competition, which, in a global economy, is a rapidly moving target. For example: All of our products will be considered the "best in class" within one year of introduction as compared to products of the top five competitors.

Business improvement. Goals in this area may be aimed at improving product deficiencies or process failures, or reducing the cost of poor quality waste in the system. Improvement goals are deployed through a formal structure of quality improvement projects with assignment of associated responsibilities. Collectively, these projects focus on reducing deficiencies in the organization, thereby leading to improved performance.

Cost of poor quality. Goals related to quality improvement usually include a goal of reducing the costs due to poor quality or waste in the processes. These costs are not known with precision, though they are estimated to be very high. Nevertheless, it is feasible, through estimates, to bring this goal into the business plan and to deploy it successfully to lower levels. A typical goal is to reduce the cost of poor quality by 50 percent each year for three years.

Performance of business processes. Goals in this area have only recently entered the strategic business plan. These goals relate to the performance of major processes that are multifunctional in nature, for example, new product development, supply-chain management, and information technology, and subprocesses, such as accounts receivable and purchasing. For such macroprocesses, a special problem is to decide who should have the responsibility for meeting the goal? We discuss this later under "Deployment to Whom?"

Customer satisfaction. Setting specific goals for customer satisfaction helps keep the organization focused on the customer. Clearly, deployment of these goals requires a good deal of sound data on the current level of satisfaction/dissatisfaction and what factors will contribute to increasing satisfaction and removing dissatisfaction. If the customers' most important needs are known, the organization's strategies can be altered to meet those needs most effectively.

Customer loyalty and retention. Beyond direct measurement of customer satisfaction, it is even more useful to understand the concept of customer loyalty. Customer loyalty is a measure of customer purchasing behavior between customer and supplier. A customer whose needs for a product offered by supplier A and who buys solely from that supplier is said to display a loyalty with respect to A of 100 percent. A study of loyalty opens the organization to a better understanding of product salability from the customer's viewpoint and provides the incentive to determine how to better satisfy customer needs. The organization can benchmark to discover the competition's performance, and then set goals to exceed that performance (see Figure 7.5).

The goals selected for the annual business plan are chosen from a list of nominations made by all levels of the hierarchy. Only a few of these nominations will survive the screening process and end up as part of the organizationwide business plan. Other nominations may instead enter the business plans at lower levels in the organization. Many nominations will be deferred because they fail to attract the necessary priority and, therefore, will get no organization resources.

Upper managers should become an important source of nominations for strategic goals, since they receive important inputs from sources such as membership on the executive

Product performance (customer focus). This relates to performance features that determine response to customer needs, such as promptness of service, fuel consumption, MTBF, and courtesy. (Product includes goods and services.)

Competitive performance. Meeting or exceeding competitive performance has always been a goal. What is new is putting it into the business plan.

Performance improvement. This is a new goal. It is mandated by the fact that the rate of quality improvement decides who will be the quality leader of the future.

Reducing the cost of poor quality. The goal here relates to being competitive as to costs. The measures of cost of poor quality must be based on estimates.

Performance of business processes. This relates to the performance of major multifunctional processes such as billing, purchasing, and launching new products.

FIGURE 7.5 Quality goals in the business plan. (*Juran Institute, Inc.*)

council, contacts with customers, periodic reviews of business performance, contacts with upper managers in other organizations, shareholders, and employee complaints.

Goals that affect product salability and revenue generation should be based primarily on meeting or exceeding marketplace quality. Some of these goals relate to projects that have a long lead time, for example, a new product development involving a cycle time of several years, computerizing a major business process, or a large construction project that will not be commissioned for several years. In such cases, the goal should be set so as to meet the competition estimated to be prevailing when these projects are completed, thereby "leap-frogging" the competition.

In industries that are natural monopolies (e.g., certain utilities), the organizations often are able to make comparisons through use of industry databanks. In some organizations there is internal competition as well—the performances of regional branches are compared with each other.

Some internal departments may also be internal monopolies. However, most internal monopolies have potential competitors—outside suppliers who offer the same services. The performance of the internal supplier can be compared with the proposals offered by an outside supplier.

A third and widely used basis for setting goals has been historical performance. For some products and processes, the historical basis is an aid to needed stability. For other cases, notably those involving high chronic costs of poor quality, the historical basis has done a lot of damage by helping to perpetuate a chronically wasteful performance. During the goal-setting process, upper managers should be on the alert for such misuse of the historical data. Goals for chronically high cost of poor quality should be based on planned breakthroughs using the breakthrough improvement process described in Chapter 5, Quality Improvement: Creating Breakthroughs in Performance.

Articulate Ethics and Values

Corporate values reflect an organization's culture.

Simply said: Culture is a set of habits and beliefs that a group of people have in common—for example, facing similar questions and problems because they operate in a similar business.

This is the internal element of culture; the other one is external-oriented: In what environment does this group of people operate and how does this affect them? How do they interact with the environment? (Bool, 2008)

Social responsibility is obviously a value that is focused on this second element: the interaction of the group (corporation) with its environment.

Some organizations create value statements to further define themselves. Values are what an organization stands for and believes in. A list of values must be supported with actions and deeds from management, lest its publication create cynicism in the organization. Training and communication of values for all employees becomes a prerequisite to participation in the planning process. Organization-published values are policies that must be changed to support the values of the organization.

Samsung's value statements are an example of this:

- We will devote our human resources and technology to create superior products and services, thereby contributing to a better global society.
- Our management philosophy represents our strong determination to contribute directly to the prosperity of people all over the world. The talent, creativity, and dedication of our people are key factors to our efforts, and the strides we've made in technology offer endless possibilities to achieving higher standards of living everywhere.
- At Samsung we believe that the success of our contributions to society and to the mutual prosperity of people across national boundaries truly depends on how we manage our company.
- Our goal is to create the future with our customers.

Communicate Organization Policies

"Policy" as used here is a guide to managerial action. Published policy statements are the result of a good deal of deliberation by management, followed by approval at the highest level. The senior executive team or quality council plays a prominent role in this process.

Policy declarations are a necessity during a period of major change, and organizations have acted accordingly. Since the 1980s we have seen an unprecedented surge of activity in publishing "quality policies." While the details vary, the published policies have much in common from organization to organization. For instance, most published quality policies declare the intention to meet the needs of customers. The wording often includes identification of specific needs to be met, for example, "The organization's products should provide customer satisfaction."

Most published policies include language relative to competitiveness in quality, for example, "Our organization's products shall equal or exceed the competition."

A third frequent area of published quality policy relates to quality improvement, declaring, for example, the intention to conduct improvement annually.

Some quality policy statements include specific reference to internal customers or indicate that the improvement effort should extend to all phases of the business. For example:

"Helix Energy Solutions Group, Inc., (Helix) is fully committed to being the leading provider of select life-of-field solutions. A primary goal of Helix is to achieve the highest standards of quality in all business units' practices and operations without compromise. Our objective is to continually improve our organization performance, while offering our customers a safe, cost-effective, and professional service."

Enforcement of policies is a new problem due to the relative newness of documented quality policies. In some organizations, provision is made for independent review of adherence to policies. ISO 9000, the international standard for quality assurance, requires a quality policy as a declaration of intent to meet the needs of customers. An audit process is mandated to ensure that the policy is carried out.

Leadership

A fundamental step in the establishment of any strategic plan is the participation of upper management acting as an executive "council." Membership typically consists of the key executives. Top-level management must come together as a team to determine and agree upon the strategic direction of the organization. The council is formed to oversee and coordinate all strategic activities aimed at achieving the strategic plan. The council is responsible for executing the strategic business plan and monitoring the key performance indicators. At the highest level of the organization, an executive council should meet monthly or quarterly.

The executives are responsible for ensuring that all business units have a similar council at the subordinate levels of the organization. In such cases, the councils are interlocked, that is, members of upper-level councils serve as chairpersons for lower-level councils (see Figure 7.6).

If a council is not in place, the organization should create one. In a global organization, processes are too complex to be managed functionally. A council ensures a multifunctional

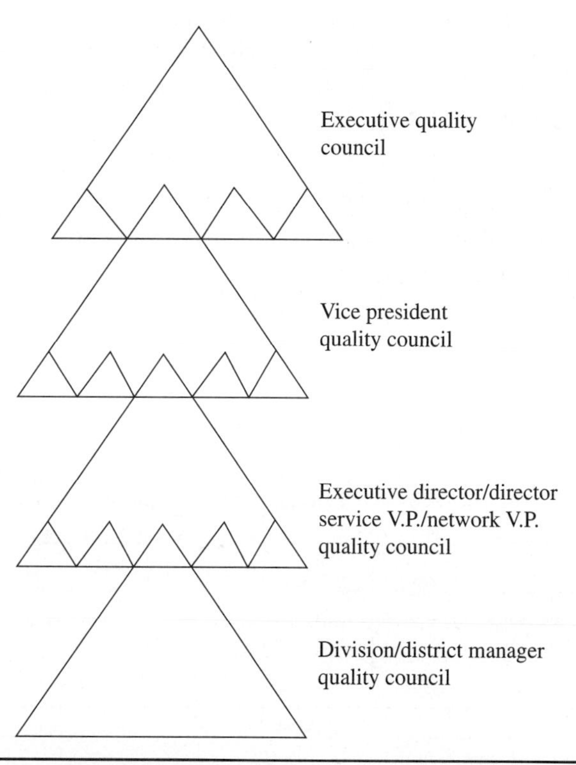

Figure 7.6 How quality councils are linked together. (*Juran Institute, Inc., 1994.*)

team working together to maximize process efficiency and effectiveness. Although this may sound easy, in practice it is not. The senior management team members may not want to give up the monopolies they have enjoyed in the past. For instance, the manager of sales and marketing is accustomed to defining customer needs, the manager of engineering is accustomed to sole responsibility for creating products, and the manager of manufacturing has enjoyed free rein in producing products. In the short run, these managers may not easily give up their monopolies to become team players.

Deploy Goals

The deployment of long- and short-term goals is the conversion of goals into operational plans and projects. "Deployment" as used here means subdividing the goals and allocating the subgoals to lower levels. This conversion requires careful attention to such details as the actions needed to meet these goals, who is to take these actions, the resources needed, and the planned timetables and milestones. Successful deployment requires establishment of an infrastructure for managing the plan. Goals are deployed to multifunctional teams, functions, and individuals (see Figure 7.7).

Subdividing the Goals

Once the strategic goals have been agreed upon, they must be subdivided and communicated to lower levels. The deployment process also includes dividing up broad goals into manageable pieces (short-term goals or projects). For example:

- An airline's goal of attaining 99 percent on-time arrivals may require specific short-term (eight to twelve months) initiatives to deal with such matters as:
    - The policy of delaying departures in order to accommodate delayed connecting flights
    - The decision-making of gate agents at departure gates
    - The availability of equipment to clean the plane
    - The need for revisions in departmental procedures to clean the plane
    - The state of employee behavior and awareness

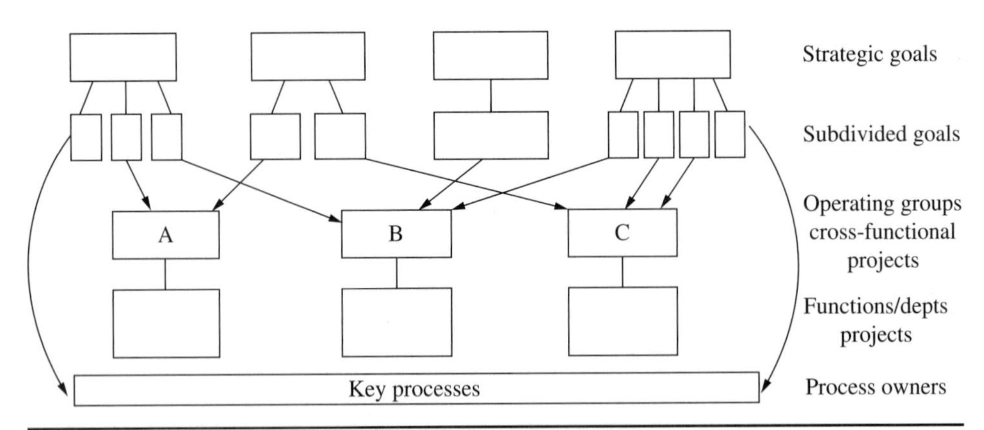

Figure 7.7 Deployment of strategic goals. (*Juran Institute, Inc.*)

- A hospital's goal of improving the health status of the communities they serve may require initiatives that
  - Reduce incidence of preventable disease and illness
  - Improve patient access to care
  - Improve the management of chronic disease conditions
- Develop new services and programs in response to community needs

Such deployment accomplishes some essential purposes:

- The subdivision continues until it identifies specific deeds to be done.
- The allocation continues until it assigns specific responsibility for doing the specific deeds.

Those who are assigned responsibility respond by determining the resources needed and communicating this to higher levels. Many times, the council must define specific projects, complete with team charters and team members, to ensure goals are met (see Figure 7.8). (For more on the improvement process, see Chapter 5, Quality Improvement: Creating Breakthroughs in Performance.)

Deployment to Whom?

The deployment process starts by identifying the needs of the organization and the upper managers. Those needs determine what deeds are required. The deployment process leads to an optimum set of goals through consideration of the resources required. The specific projects to be carried out address the subdivided goals. For example, in the early 1980s, the goal of having the newly designed Ford Taurus/Sable become "Best in Class" was divided

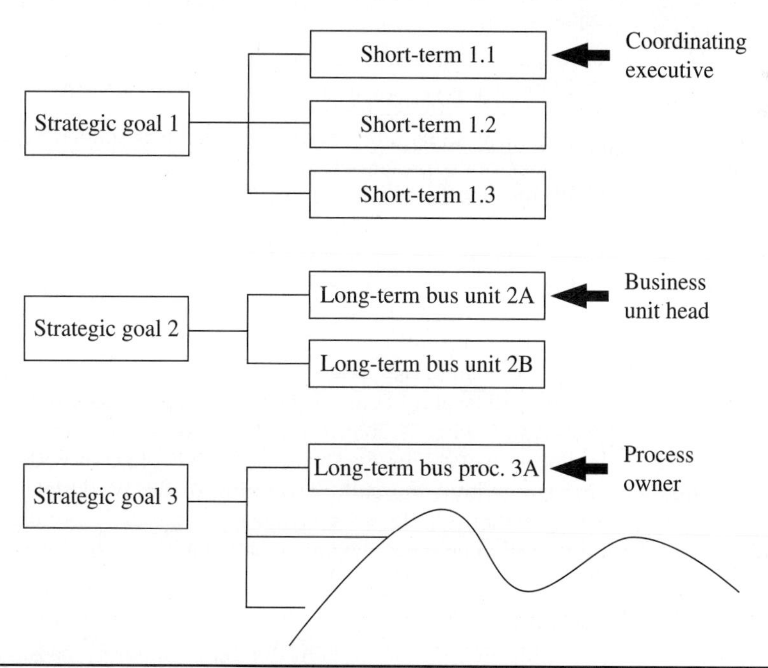

Figure 7.8 Subgoals. (*Juran Institute, Inc.*)

into more than 400 specific subgoals, each related to a specific product feature. The total planning effort was enormous and required more than 1500 project teams.

To some degree, deployment can follow hierarchical lines, such as corporate to division and division to function. However, this simple arrangement fails when goals relate to cross-functional business processes and problems that affect customers.

Major activities of organizations are carried out by the use of interconnecting networks of business processes. Each business process is a multifunctional system consisting of a series of sequential operations. Since it is multifunctional, the process has no single "owner"; hence, there is no obvious answer to the question, Deployment to whom? Deployment is thus made to multifunctional teams. At the conclusion of the team project, an owner is identified. The owner (who may be more than one person) then monitors and maintains this business process. (See Chapter 8, Business Process Management: Creating an Adaptable Organization.)

Communicating the Plan: "Catch Ball"

Once the goals have been established, the goals are communicated to the appropriate organization units. In effect, the executive leadership asks their top management, What do you need to support this goal? The managers at this level discuss the goal and ask their subordinates a similar question, and so on. The responses are summarized and passed back up to the executives. This process may be repeated several times until there is general satisfaction with the final plan.

This two-way communication process is called "catch ball," a term coined by the Japanese. Catch ball includes the following:

- Clear communication of what top management proposes as the key focus areas of the strategic plan for the coming business year
- Identification and nomination by managers at various lower levels of other areas for organization attention
- Decisions as to what departments and functions should do about the areas that have been identified in the plan

This two-way communication requires that the recipients be trained in how to respond. The most useful training is prior experience in quality improvement. Feedback from organizations using catch ball suggests that it outperforms the process of unilateral goal-setting by upper managers.

For example, Boeing Aerospace Systems has been very successful in introducing its strategic quality plan, with its mission, vision, key strategies, and strategic goals. To review and refine mission statements, strategies, and the overall vision of the organization, Boeing conducts yearly assessments drawing from customer satisfaction assessments, human resource assessments, supplier assessments, risk assessments, and financial assessments. By essentially taking feedback from all facets of their business (customers, workforce, suppliers, community, and shareholders), Boeing is able to enact improved implementation plans and better manage their allocation of resources. The identification of needs within the infrastructure, addressing problems within the culture/training of the workforce, and modifying institutionalized processes for the better are all results of continually deploying assessments and consistent communication between the management and its workforce.

A Useful Tool for Deployment

The tree diagram is a graphic tool that aids in the deployment process (see Figure 7.8). It displays the hierarchical relationship of the goals, long-term goals, short-term goals, and projects, and indicates where each is assigned in the organization. A tree diagram is useful in visualizing the relationship between goals and objectives or teams and goals. It also provides a visual way to determine if all goals are supported.

Measure Progress with KPI

There are several reasons why measurement of performance is necessary and why there should be an organized approach to it.

- Performance measures indicate the degree of accomplishment of objectives and, therefore, quantify progress toward the attainment of goals.
- Performance measures are needed to monitor the continuous improvement process, which is central to the changes required to become competitive.
- Measures of individual, team, and business unit performance are required for periodic performance reviews by management.

Once goals have been set and broken down into subgoals, key measures (performance indicators) need to be established. A measurement system that clearly monitors performance against plans has the following properties:

- Indicators that link strongly to strategic goals and to the vision and mission of the organization
- Indicators that include customer concerns; that is, the measures focus on the needs and requirements of internal and external customers
- A small number of key measures of key processes that can be easily obtained on a timely basis for executive decision-making
- The identification of chronic waste or cost of poor quality

For example, Poudre Valley Health Systems (PVHS) established measures of their processes early in the implementation of their business plan and were able to monitor and quantify the following:

- Improve and maintain employee satisfaction to the top 10 percent of vacancy rate in all U.S. organizations.
- Strengthen overall service area market share by establishing market strategies specific to service area needs. By breaking down service areas to primary/local and total/national market shares, PVHS aims to control 65 percent of their primary market share and 31.8 percent of total market share by 2012.
- Support facility development by opening a cancer center.
- Enhance physician relations by initiating a physician engagement survey tool and reaching a goal of 80 percent satisfaction.
- Strengthen the company's financial position by achieving a financial flexibility unit of 11 and meeting a five-year plan.

The best measures of the implementation of the strategic planning process are simple, quantitative, and graphical. A basic spreadsheet that describes the key measures and how they will be implemented is shown in Figure 7.9. It is simply a method to monitor the measures.

Annual quality goals	Specific measurements	Frequency	Format	Data source	Name

FIGURE 7.9 Measurement of quality goals. (*Juran Institute, Inc.*)

As goals are set and deployed, the means to achieve them at each level must be analyzed to ensure that they satisfy the objective that they support. Then the proposed resource expenditure must be compared with the proposed result and the benefit/cost ratio assessed. Examples of such measures are

- Financial results
    - Gains
    - Investment
    - Return on investment
- People development
    - Trained
    - Active on project teams
- Number of projects
    - Undertaken
    - In process
    - Completed
    - Aborted
- New product or service development
    - Number or percentage of successful product launches
    - Return on investment of new product development effort
    - Cost of developing a product versus the cost of the product it replaces
    - Percent of revenue attributable to new products
    - Percent of market share gain attributable to products launched during the last two years

- Percent of on-time product launches
- Cost of poor quality associated with new product development
- Number of engineering changes in the first twelve months of introduction
- Supply-chain management
    - Manufacturing lead times—fill rates
    - Inventory turnover
    - Percent on-time delivery
    - First-pass yield
    - Cost of poor quality

The following is an example of measures that one bank used to monitor teller quality:

- Speed
    - Number of customers in the queue
    - Amount of time in the queue (timeliness)
    - Time per transaction
    - Turnaround time for no-wait or mail transactions
- Accuracy
    - Teller differences in adding up the money at the end of the day
    - Amount charged off/amount handled

Once the measurement system is in place, it must be reviewed periodically to ensure that goals are being met.

Reviewing Progress

A formal, efficient review process will increase the probability of reaching the goals. When planning actions, an organization should look at the gaps between measurement of the current state and the target it is seeking. The review process looks at gaps between what has been achieved and the target (see Figure 7.10).

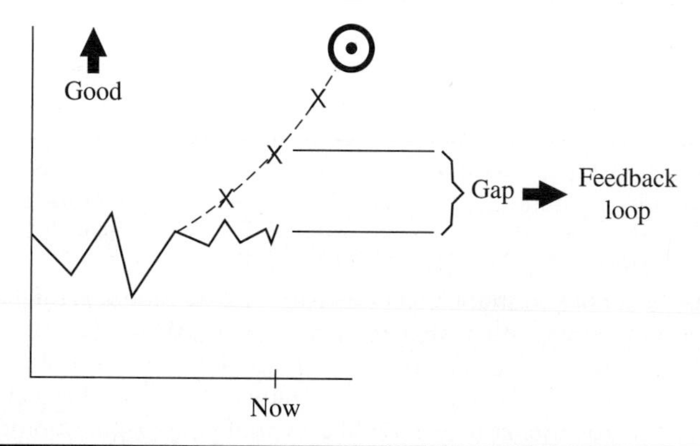

Figure 7.10 Review. (*Juran Institute, Inc.*)

Projects	Project leaders	Baseline measurements	Targets	Initial plan	Review points				Review leader
					Resources	Analysis	Plan	Results	

FIGURE 7.11 Progress review plan. (*Juran Institute, Inc.*)

Frequent measurements of strategic deployment progress displayed in graphic form help identify the gaps in need of attention. Success in closing those gaps depends on a formal feedback loop with clear responsibility and authority for acting on those differences. In addition to the review of results, progress reviews are needed for projects under way to identify potential problems before it is too late to take effective action. Every project should have specific, planned review points, much like those in Figure 7.11.

Organizations today include key performance indicators as discussed in the following sections.

Product and Service Performance

There may be several product or service features. For the great majority of product features, there exist performance metrics and technological sensors to provide objective product evaluation.

Competitive Quality

These metrics relate to those qualities that influence product salability—for example, promptness of service, responsiveness, courtesy of pre-sale and after-sale service, and order fulfillment accuracy. For automobiles, qualities include top speed, acceleration, braking distance, and safety. For some product features, the needed data must be acquired from customers through negotiation, persuasion, or purchase. For other product features, it is feasible to secure the data through laboratory tests. In still other cases, it is necessary to conduct market research.

Trends must now be studied so that goals for new products can be set to correspond to the state of competition anticipated at the time of launch.

Some organizations operate as natural monopolies, for example, regional public utilities. In such cases, the industry association gathers and publishes performance data. In the case of internal monopolies (e.g., payroll preparation, transportation), it is sometimes feasible to secure competitive information from organizations that offer similar services for sale.

Performance on Improvement

This evaluation is important to organizations that go into quality improvement on a project-by-project basis. Due to lack of commonality among the projects, collective evaluation is limited to the summary of such features as:

- *Number of projects*. Undertaken, in-process, completed, aborted.
- *Financial results*. Amounts gained, amounts invested, returns on investment.
- *Persons involved as project team members*. Note that a key measure is the proportion of the organization's management team that is actually involved in improvement projects. Ideally, this proportion should be over 90 percent. In the great majority of organizations, the actual proportion has been less than 10 percent.

Cost of Poor Quality

As stated in prior chapters, the "cost of poor quality" are those costs that would disappear if our products and processes were perfect and generated no waste. Those costs are huge. Our research indicates that 15-25% of all work performed consisted of redoing prior work because products and processes were not perfect.

The costs are not known with precision. In most organizations, the accounting system provides only a minority of the information needed to quantify this cost of poor quality. It takes a great deal of time and effort to extend the accounting system so as to provide full coverage. Most organizations have concluded that such effort is not cost-effective.

The gap can be filled somewhat by estimates that provide upper managers with approximate information as to the total cost of poor quality and the major areas of concentration. These areas of concentration then become the target for quality improvement projects. Thereafter, the completed projects provide fairly precise figures on quality costs before and after the improvements.

Product and Process Failures

Even though the accounting system does not provide for evaluating the cost of poor quality, much evaluation is available through measures of product and process deficiencies, either in natural units of measure or in money equivalents—for example, cost of poor quality per dollar of sales, dollar of cost of sales, hour of work, or unit shipped. Most measures lend themselves to summation at progressively higher levels. This feature enables goals in identical units of measure to be set at multiple levels: corporate, division, and department.

Performance of Business Processes

Despite the wide prevalence and importance of business processes, they have been only recently controlled as to performance. A contributing factor is their multifunctional nature. There is no obvious owner and hence, no clear, sole responsibility for their performance. Responsibility is clear only for the subordinate microprocesses. The system of upper management controls must include control of the macroprocesses. That requires establishing goals in terms of cycle times, deficiencies, etc., and the means for evaluating performances against those goals.

The Scorecard

To enable upper managers to "know the score" relative to achieving strategic quality deployment, it is necessary to design a report package, or scorecard. In effect, the strategic plan dictates the choice of subjects and identifies the measures needed on the upper management scorecard.

The scorecard should consist of several conventional components:

- Key performance indicators (at the highest levels of the organization)
- Quantitative reports on performance, based on data
- Narrative reports on such matters as threats, opportunities, and pertinent events
- Audits conducted (see "Business Audits" later in this chapter)

These conventional components are supplemented as required to deal with the fact that each organization is different. The end result should be a report package that assists upper managers to meet the quality goals in much the same way as the financial report package assists the upper managers to meet the financial goals.

The council has the ultimate responsibility for designing such a scorecard. In large organizations, design of such a report package requires inputs from the corporate offices and divisional offices alike. At the division level, the inputs should be from multifunctional sources.

The report package should be specially designed to be read at a glance and to permit easy concentration on those exceptional matters that call for attention and action. Reports in tabular form should present the three essentials: goals, actual performances, and variances. Reports in graphic form should, at the least, show the trends of performances against goals. The choice of format should be made only after learning the preferences of the customers, that is, the upper managers.

Managerial reports are usually published monthly or quarterly. The schedule is established to coincide with the meetings schedule of the council or other key reviewing body. The editor of the scorecard is usually the director of quality (quality manager, etc.), who is usually also the secretary of the council.

Scorecards have become an increasing staple in corporations across the globe, so much so that they have moved beyond their initial purpose. Scorecards have now been created not just to document an organization's bottom line but to judge how green an organization actually is. A "Climate Counts" Organization Scorecard rates organizations across different industry sectors in their practices to reduce global warming and create greener business practices. Organizations that are making concerted efforts to alleviate these causes receive higher scores. Like regular scorecards, the information is available to the public, and the opportunity to further a positive public image is at hand. Items include

- Leading indicators (e.g., quality of purchased components)
- Concurrent indicators (e.g., product test results, process conditions, and service to customers)
- Lagging indicators (e.g., data feedback from customers and returns)
- Data on cost of poor quality

The scorecard should be reviewed formally on a regular schedule. Formality adds legitimacy and status to the reports. Scheduling the reviews adds visibility. The fact that upper managers personally participate in the reviews indicates to the rest of the organization that the reviews are of great importance.

Many organizations have combined their measurements from financial, customer, operational, and human resource areas into "instrument panels" or "balanced business scorecards."

Business Audits

An essential tool for upper managers is the audit. By "audit," we mean an independent review of performance. "Independent" signifies that the auditors have no direct responsibility for the adequacy of the performance being audited.

The purpose of the audit is to provide independent, unbiased information to the operating managers and others who have a need to know. For certain aspects of performance, those who have a need to know include the upper managers.

To ensure quality, upper management must confirm that

- The systems are in place and operating properly
- The desired results are being achieved

Growing to encompass a broad range of fields, quality audits are now utilized in a plethora of industries, including science. The Royal College of Pathologists implements quality audits on a number of their research reports. The quality audit ensures that individuals and teams are meeting the procedures and standards expected of them and that their work is in line with the mission of the study.

These audits may be based on externally developed criteria, on specific internal objectives, or on some combination of both. Three well-known external sets of criteria to audit organization performance are those of the United States' Malcolm Baldrige National Award for Excellence, the European Foundation Quality Management Award (EFQM), and Japan's Deming Prize. All provide similar criteria for assessing business excellence throughout the entire organization.

Traditionally, quality audits have been used to provide assurance that products conform to specifications and that operations conform to procedures. At upper-management levels, the subject matter of quality audits expands to provide answers to such questions as

- Are our policies and goals appropriate to our organization's mission?
- Does our quality provide product satisfaction to our clients?
- Is our quality competitive with the moving target of the marketplace?
- Are we making progress in reducing the cost of poor quality?
- Is the collaboration among our functional departments adequate to ensure optimizing organization performance?
- Are we meeting our responsibilities to society?

Questions such as these are not answered by conventional technological audits. Moreover, the auditors who conduct technological audits seldom have the managerial experience and training needed to conduct business-oriented quality audits. As a consequence, organizations that wish to carry out quality audits oriented to business matters usually do so by using upper managers or outside consultants as auditors.

Juran (1998) has stated:

> One of the things the upper managers should do is maintain an audit of how the processes of managing for achieving the plan is being carried out. Now, when you go into an audit, you have three things to do. One is to identify what are the questions to which we need answers. That's nondelegable; the upper managers have to participate in identifying these questions. Then you have to put together the information that's needed to give the answers to those questions. That can be delegated, and that's most of the work, collecting and analyzing the data. And there's the decisions of what to do in light of those answers. That's nondelegable. That's something the upper managers must participate in.

Audits conducted by executives at the highest levels of the organization where the president personally participates are usually called "The President's Audit" (Kondo 1988). Such audits can have major impacts throughout the organization. The subject matter is so fundamental in nature that the audits reach into every major function. The personal participation of the upper managers simplifies the problem of communicating to the upper levels and increases the likelihood that action will be forthcoming. The very fact that the upper managers participate in person sends a message to the entire organization relative to the priority placed on quality and to the kind of leadership being provided by the upper managers—leading, not cheerleading (Shimoyamada 1987).

Lessons Learned

There are some important lessons learned about the risks in implementing strategic deployment:

- Pursuing too many objectives, long term and short term, at the same time will dilute the results and blur the focus of the organization.
- Excessive planning and paperwork will drive out the needed activities and demotivate managers.
- Trying to plan strategically without adequate data about customers, competitors, and internal employees can create an unachievable plan or a plan with targets so easy to achieve that the financial improvements are not significant enough.
- If leaders delegate too much of the responsibility, there will be a real and perceived lack of direction.
- For an organization to elevate quality and customer focus to top priority creates the impression that it is reducing the importance of finance, which formerly occupied that priority. This perceived downgrading is particularly disruptive to those who have been associated with the former top-priority financial goals.

When embarking on strategic planning, the biggest disruption is created by imposing a structured approach on those who prefer not to have it. Resistance to the structured approach will be evident at the outset. The single most important prerequisite for embarking on a long-term, effective, organizationwide improvement effort is the creation of an environment conducive to the many changes that are necessary for success. Organizations have aggressively sought to eliminate these barriers that have taken years or decades to establish. The process of change takes time, however, and change will occur only as an evolutionary process.

References

Bool, H. Social Responsibility & The Corporate Values Statement, http://ezinearticles.com/?id=1871566, 2008

Godfrey, A. B. (1997). "A Short History of Managing Quality in Health Care." In Chip Caldwell, ed., *The Handbook for Managing Change in Health Care*. ASQ Quality Press, Milwaukee, WI.

Godfrey, A. B. (1998). "Hidden Costs to Society." *Quality Digest*, vol. 18, no. 6.

Jet Blue Airways, Terminal 5, JFK International Airport, http://phx.corporate-ir.net/External.File?item=UGFyZW50SUQ9MzMzODAzfENoaWxkSUQ9MzE2NTMwfFR5cGU9MQ==&t=1, 2008.

Juran, J. M. (1988). *Juran on Planning for Quality*. Free Press, New York.

Kondo, Y. (1988). "Quality in Japan." In J. M. Juran, ed., *Juran's Quality Control Handbook,* 4th ed. McGraw-Hill, New York. (Kondo provides a detailed discussion of quality audits by Japanese top managers, including The President's Audit. See Chapter 35F, "Quality in Japan," under "Internal QC Audit by Top Management.")

Shimoyamada, K. (1987). "The President's Audit: QC Audits at Komatsu." *Quality Progress,* January, pp. 44–49. (Special Audit Issue).

Treqoe, B., and Tobia, P. (1990). "Strategy and the New American Organization." *Industry Week.* August 6.

CHAPTER 8

Business Process Management: Creating an Adaptable Organization

Joseph A. De Feo

About This Chapter

Success in achieving superior results depends heavily on managing such large, complex, multifunctional business processes, as product development, the revenue cycle, invoicing, patient care, purchasing, materials procurement, supply chain, and distribution, among others. In the absence of management's attention over time, many processes may become too slow, obsolete, overextended, redundant, excessively costly, ill defined, and not adaptable to the demands of a constantly changing environment. For processes that have suffered this neglect (and this includes many processes for reasons that will be discussed later in this chapter) quality of output falls far short of the quality required for competitive performance. This chapter focuses on helping an organization maintain its sustainability and adaptability by ensuring proper day-to-day ownership of important business processes. Business process ownership happens after an organization masters all processes of the Juran Trilogy.

High Points of This Chapter

1. Creating a sustainable and adaptable organization requires that all key business processes are managed on a day-to-day basis. This occurs when an organization has fully deployed its vision and mission.
2. Strategic goals tied to the organization vision, which are shared by executive leadership and deployed throughout the organization in the form of key business objectives are commonplace.
3. Multifunctional, full-time business process teams, owned and supported by the management system (education, communication, performance management, recognition and reward, compensation, and new career path structures) are identified.
4. The charter of each team is to continuously and dramatically improve the effectiveness and efficiency of each major business process to which it is assigned on an ongoing, daily managed basis.
5. Process owners are empowered and held accountable to act in support of these key business objectives.
6. Skills in performance excellence methods and project management to enable many of its schedules, costs, and work plans being coordinated and implemented throughout the organization is required.
7. Executive management promotion of the importance, impact, progress, and success of the Business Process Management (BPM) effort throughout the organization and to external stakeholders is a prerequisite.
8. Organization leaders who have adopted BPM as a management tool know that process management is a continuous managerial focus, not a single event or a quick fix. They also know that a constant focus on business processes is essential to the long-term success of their organization.

Why Business Process Management?

The dynamic environment in which business is conducted today is characterized by what has been referred to as "the six Cs": *change, complexity, customer demands, competitive pressure, cost impacts, and constraints.* All have a great impact on an organization's ability to meet its stated business goals and objectives. Organizations have responded to these factors by developing new products and services. They also have carried out numerous "breakthrough projects" and are at a maturity level conducive to process ownership.

> A business process is the logical organization of people, materials, energy, equipment, and information into work activities designed to produce a required end result (product or service) (Pall 1986).

There are three principal dimensions for measuring process performance: effectiveness, efficiency, and adaptability:

1. The process is *effective* if the output meets customer needs. It is *efficient* when it is effective at the least cost.
2. The process is *adaptable* when it remains effective and efficient in the face of the many changes that occur over time.

On the surface, the need to maintain high-quality processes would seem obvious. To understand why good process quality is the exception and not the rule requires us to look closely at how processes are designed and what happens to them over time.

BPM has become a critical component of information technology (IT) programs. Without having good business process management system, an IT system can fail. All disciplined IT implementations must include well-developed BPM processes. With technology, BPM allows organizations to abstract business processes from the technology infrastructure and go far beyond automating business processes or solving business problems. BPM enables business to respond to changing consumer, market, and regulatory demands faster than competitors, creating a competitive advantage. In the IT world, BPM is often called a BPM life cycle.

For reasons of history, the business organization model has evolved into a hierarchy of functionally specialized departments. Management direction, goals, and measurements are deployed from the top downward through this vertical hierarchy. However, the processes that yield the products of work—in particular, products that customers buy (and that justify the existence of the organization)—flow horizontally across the organization through functional departments (Figure 8.1). Traditionally, each functional piece of a process is the responsibility of a department, whose manager is held accountable for the performance of that piece. However, no one is accountable for the entire process. Many problems arise from the conflict between the demands of the departments and the demands of the overall major processes.

In a competition with functional goals, functional resources, and functional careers, cross-functional processes are starved for attention. As a result, the processes as operated are often neither effective nor efficient, and they are certainly not adaptable.

A second source of poor process performance is the natural deterioration to which all processes are subject in the course of their evolution. For example, at one railroad, the company telephone directory revealed that there were more employees with the title "rework clerk" than with the title "clerk." Each of the rework clerks had been put in place to guard against the recurrence of some serious problem that arose. Over time, the imbalance in titles was the outward evidence of processes that had established rework as the organization's norm.

The rapidity of technological evolution, in combination with rising customer expectations, has created global competitive pressures on costs and quality. These pressures

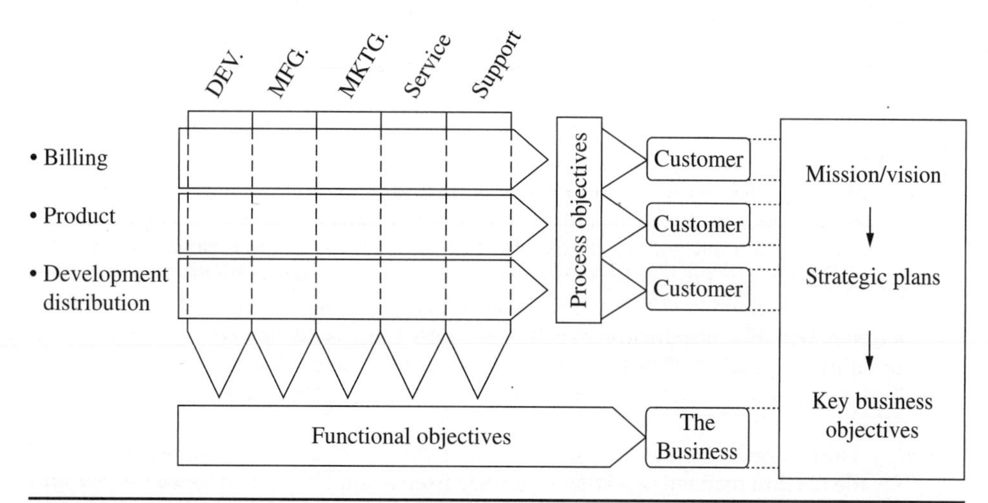

Figure 8.1 Horizontal flow through functional departments.

have stimulated an exploration of cross-functional processes—to identify and understand them and to improve their performance. There is now much evidence that, within the total product cycle, a major problem of poor process performance lies with BPM technologies. Functional objectives frequently conflict with customer needs, served as they must be by cross-functional processes. Furthermore, the processes generate a variety of waste (e.g., missed deadlines, factory scrap). It is not difficult to identify such products—generating invoices, preparing insurance policies, or paying a claim—that take over 20 days to accomplish in less than 20 minutes of actual work. Processes are also not easily changed in response to the continuously changing environment. To better serve customer needs, there is a need to restore these processes to effectiveness, efficiency, and adaptability.

The Origins of BPM

The IBM Corporation was among the first American companies to see the benefits of identifying and managing business processes. The spirit of IBM's first efforts in managing these processes in the early 1980s was expressed as follows: "Focus for improvement must be on the job process" (Kane 1986). BPM has long been practiced in the manufacturing arena. In product manufacturing, the plant manager "owns" a large part of the manufacturing process. This manager has complete responsibility for operating his or her part of the manufacturing process and is accountable for the results. As an owner, the manager is expected to control, improve, and optimize the manufacturing process to meet customer needs and business needs (e.g., cost, cycle time, waste elimination, and value creation). In pursuit of these targets, managers of the manufacturing process have developed some indispensable concepts and tools, including defining process requirements, documenting a step-by-step process, establishing process measurements, removing process defects, and ensuring process optimization. In fact, much of the science of industrial engineering is concerned with these tasks. Recognizing the value of these tools in manufacturing and their applicability to business processes, the IBM senior management committee directed that BPM methodology be applied to all major business processes (such as product development, business planning, distribution, billing, and market planning)—and not just to the manufacturing process.

Around the same time, other North American companies, including AT&T, Ford Motor Company, Motorola, Corning, and Hewlett-Packard, also began applying BPM concepts to their business processes. In all these companies, emphasis was placed on cross-functional and cross-organizational processes. Applying BPM methodology resulted in breaking down functional barriers within the processes. In each case, a new, permanent managerial structure was established for the targeted process.

By mid-1985, many organizations and industries were managing selected major business processes with the same attention commonly devoted to functions, departments, and other organizational entities. Early efforts bore such names as Business Process Management, Continuous Process Improvement, and Business Process Quality Improvement.

Michael Hammer (1990) raised the visibility of business processes to a level that created a frenzy with the introduction of BPR—Business Process Reengineering—in the early 1990s. In subsequent years, BPR has often been associated with drastic change and downsizing initiatives, rather than improving processes and resulted in many failed reengineering efforts.

The emergence of BPM in the new millennium, post-BPR, has resulted in renewed focus for the need to manage workflows and has been a solid, yet silent, business revolution. To

Organization Behaviors	Functional Organization	Process-Centric Organization
Managers manage:	Resources work	Customers and results
Teams produce:	Independently	Collaboratively
Organization dynamics and self-reorganization:	Rigid to adapt, frequent reorganization	Flexible to new demands
Resources focus:	Meeting job requirements	Best results, customers
Knowledge dissemination:	Islands of information	Integrated across the organization
Culture:	Closed	Open

TABLE 8.1 Functional vs. Process Organization

understand why an entire organization would evolve from process improvement to business process management and ownership, we must understand the primary characteristics of the business process and the benefits brought about by BPM. The traditional "functional organization" is a remnant of the Industrial Revolution in which the guiding principle for organizing by function is the distribution of work by labor specialization.

Functional organizations may not disappear completely but rather be transformed into the context for managing processes that bring value to customers. Technological superiority, innovation, or longevity are no longer what makes or breaks organizations—it is how well organizations are organized to respond to and serve their customers. As an organization learns how to manage across functions via project by project improvement, it moves toward BPM. BPM is a process to sustain the changes made from all those improvement projects.

The only way to achieve such sustainable customer satisfaction and results is to become an adaptable organization. To be adaptable means your organization can respond quickly to the changing needs of customers, technology, and innovation by competitors.

Table 8.1 highlights important cultural differences between a functional organization and a process-centric one.

BPR should be mentioned as part of this family of methodologies. Like the methodologies mentioned previously in this chapter, BPR accomplishes a shift of managerial orientation from function to process. According to the consultants who first described BPR and gave it its name, BPR departs from the other methodologies in emphasizing radical change of processes rather than on incremental change. Furthermore, BPR frequently seeks to change more than one process at a time. Because of the economic climate of the early 1990s, and the outstanding payback that some writers attribute to BPR, its popularity grew rapidly for a time.

However, there is evidence, including the testimony of Michael Hammer, one of the most widely published researchers on BPR, that in many early applications, the lure of rapid improvement caused some managers (and their consultants), who ignored human limitations, to impose too much change in too short a time, with a devastating effect on long-term organization performance. Furthermore, in many early applications, users became so fascinated by the promise of radical change that they changed everything, overlooking elements of the existing process design that worked perfectly well and would have been better carried over as part of the new design. Such a carryover would have saved time, reduced demand on the designers, and produced a better result.

BPM Defined

The methodology described here is one that has been introduced with increasing success by a number of prominent corporations, including the ones already mentioned. Although it may vary in name and details from organization to organization, methodology possesses a core of common features that distinguishes it from other approaches to managing quality. That core of features includes a conscious orientation toward customers and their needs, a specific focus on managing a few key cross-functional processes that most affect satisfaction of customer needs, a pattern of clear ownership—accountability for each key process, a cross-functional team responsible for operating the process, and application at the process level of quality-management processes—quality control, quality improvement, and quality planning. In this chapter, the methodology is referred to as process quality management, or BPM.

The BPM Methodology

BPM is initiated when executive management selects key processes, identifies owners and teams, and provides them with process goal statements. After the owners and the team are trained in performance excellence methods and tools, they work through the three phases of BPM methodology: planning, transfer, and operational management.

The *planning phase,* in which the process design (or redesign) takes place, and is the most time consuming of the three phases, involves five steps:

1. Defining the present process.
2. Determining customer needs and process flow.
3. Establishing process measurements.
4. Conducting analyses of measurement and other data.
5. Designing the new process. The output is the new process plan.

The *transfer phase* is the second phase, in which the plans developed in the first phase are handed off from the process team to the operating forces and put into operation.

The operational management phase is the third phase of BPM. Here, the working owner and team first monitor new process performance, focusing on process effectiveness and efficiency measurements. They apply quality control techniques, as appropriate, to maintain performance. They use quality improvement techniques to rid the process of chronic deficiencies. Finally, they conduct a periodic executive management review and assessment to ensure that the process continues to meet customer and business needs and remains competitive.

Note: BPM is not a one-time event; it is itself a continuous process carried out in real-time.

A Case Study: Unisys is a worldwide information technology services and solutions organization with a client base spread over 100 countries. The organization offers a rich portfolio of business solutions led by its expertise in systems integration, outsourcing, infrastructure services, server technology, and consulting. Unisys Global Infrastructure Services (GIS) provides value-added services needed by organizations to design, integrate, and manage their distributed IT infrastructures including desktop environments, servers, networks, and mobile/wireless systems. One of the key divisions in GIS is Infrastructure Managed Services (IMS), which drives services-based solutions that enable Unisys clients' infrastructures to be managed and continuously improved for business value and cost management.

Defining a vision: the IMS division began to develop an organization-unique, process-based methodology named Unify. The goal was to introduce repeatability and consistency in the way clients are serviced around the world. Developing the methodology required IMS to map and document approximately 600 processes. IMS searched for a BPM solution with this in mind.

Using such technologies as Microsoft Visio add-on, and Designer (process simulation software) IMS creates complete models of its operational business processes. IMS began the task of implementing its BPM methodology. Within one year of the project start date, they capture and manage their key business processes, including organizations, resources and roles, and related information, and view them at various levels of detail. As a fully dynamic solution that is specifically designed for business users, BPM helped the division to identify optimal ways to increase organization efficiency, ensure that activities are done consistently, and reduce training requirements.

Another approach to BPM is based on five categories of BPM activities: design, modeling, execution, monitoring, and optimization.

Design

Process design encompasses both the identification of existing processes and the design of future processes. Areas of focus include representation of the process flow, the actors within it, alerts and notifications, escalations, standard operating procedures, service level agreements, and task handover mechanisms. Good design reduces the number of problems over the lifetime of the process. Whether or not existing processes are considered, the aim of this step is to ensure that a correct and efficient theoretical design is prepared. The proposed improvement could be in human-to-human, human-to-system, and system-to-system workflows and might target regulatory, market, or competitive challenges that businesses face.

Modeling

Modeling takes the theoretical design and introduces combinations of variables (e.g., changes in rent or materials costs that determine how the process might operate under different circumstances). It also involves running a "what-if analysis" on the processes: *"What if I have 75 percent of resources to do the same task?" "What if I want to do the same job for 80% of the current cost?"*

Execution

One way to automate processes is to develop or purchase an application that executes the required steps of the process; however, in practice, these applications rarely execute all the steps of the process accurately or completely. Another approach is to use a combination of software and human intervention; however this approach is more complex, making the documentation process difficult.

As a response to these problems, software has been developed that enables the full business process (as developed in the process design activity) to be defined in a computer language that can be directly executed by the computer. The system will either use services in connected applications to perform business operations (e.g., calculating a repayment plan for a loan) or will ask for human input when a step is too complex to automate. Compared to either of the previous approaches, directly executing a process definition can be more straightforward and is, therefore, easier to improve. However, automating a process definition requires a flexible and comprehensive infrastructure, which typically rules out implementing these systems in a legacy IT environment.

Business rules have been used by systems to provide definitions for governing behavior, and a business rule engine can be used to drive process execution and resolution.

Monitoring

Monitoring encompasses the tracking of individual processes so that information on their state can be easily seen and statistics on the performance of one or more processes can be provided. An example of the tracking is being able to determine the state of a customer order (e.g., order arrival, awaiting delivery, invoice paid) so that operational problems can be identified and corrected.

In addition, this information can be used to work with customers and suppliers to improve their connected processes. Examples of the statistics are the generation of measures on how quickly a customer order is processed or how many orders were processed in the last month. These measures tend to fit into three categories: cycle time, defect rate, and productivity.

The degree of monitoring depends on what information the business wants to evaluate and analyze and how business wants it to be monitored—in real time, near real time, or ad hoc. Here, business activity monitoring extends and expands the monitoring tools in generally provided by BPMS.

Process mining is a collection of methods and tools related to process monitoring. process mining aims to analyze event logs extracted through process monitoring and to compare them with an a priori process model. Process mining allows process analysts to detect discrepancies between the actual process execution and the a priori model as well as to analyze bottlenecks.

Optimization

Process optimization includes retrieving process performance information from modeling or monitoring phase, identifying potential or actual bottlenecks and the potential opportunities for cost savings or other improvements, and applying those enhancements to the design of the process. This creates greater business value overall.

Meir H. Levi from the Interfacing Technologies Corporation, Montreal, Canada, states that "the awareness of business processes is the most important management paradigm today. The idea of the 'process organization' is gaining strong momentum; the process 'option' is now becoming a mandatory requirement. The integration of the Process Framework into the management structure introduces clear focus on consistent and collaborative ways to achieve results that directly impact the bottom line; hence, delighted customers and stakeholders."

Deploying BPM

Selecting Key Process(es)

Organizations operate dozens of major cross-functional business processes. From these, a few key processes are selected as the BPM focus. An organization's strategic plan provides guidance in selecting key processes. (See Chapter 7, Strategic Planning and Deployment: Moving from Good to Great.)

There are several approaches to selecting key business processes:

- The Critical Success Factor approach holds that, for any organization, relatively few (no more than eight) factors can be identified as "necessary and sufficient" for attaining its mission and vision. Once identified, these factors are used to select the key business processes and rank them by priority (Hardaker and Ward 1987).
- The Balanced Business Scorecard (Kaplan and Norton 1992) measures business performance in four dimensions: financial performance, performance in the eyes of the customer, internal process performance, and performance in organization learning and innovation. Performance measures are created and performance targets are set for each dimension. Using these measures to track performance provides a "balanced" assessment of business performance. Processes that create imbalances in the scorecard are identified as processes that need attention most—the key processes.
- Another approach is to invite upper management to identify a few (four to six) organization-specific critical selection criteria to use in evaluating the processes. Examples of such criteria are the effect on business success, the effect on customer satisfaction, the significance of problems associated with the process, the amount of resources currently committed to the process, the potential for improvement, the affordability of adopting BPM, and the effect of process on the schedule. Using these criteria and a simple scoring system (such as "low, medium, or high"), managers evaluate the many processes from the long list of the organization's major business processes (10 to 25 of them) and, by comparing the evaluations, identify the key processes. (The long list may be prepared in advance in a process identification study conducted separately, often by the chief quality officer, and often with the support of a consultant.

Whatever approach is used to identify key processes, the process map can be used to display the results. The "process map" is a graphic tool for describing an organization in terms of its business processes and their relationships to the organization's principal stakeholders. The traditional organization chart answers the question: "Who reports to whom?" The process map answers the question: "How does the organization's work get done?"

Organizing: Assigning Ownership, Selecting the Team, and BPM Infrastructure

Because certain major cross-functional business processes, the *key processes*, are critical to business success, the quality council sees to it that those processes are organized in a special way. After selecting key processes, the quality council appoints a process owner, who is responsible for making the process effective, efficient, and adaptable, and is accountable for its performance (Riley 1989, Riley et al. 1994).

For large complex processes, especially in large organizations, a two-tier ownership arrangement is used most often. An appointed executive owner operates as a sponsor, champion, and supporter at the upper management level and is accountable for process results. At the operating level, a working owner, usually a first- or second-level manager, leads the process-management team responsible for day-to-day operation. Owner assignments—executive owner and working owner—are ongoing. The major advantages of this structure are that there is, at the same time, "hands-on" involvement and support of upper management and adequate management of the process details.

The process-management team is a peer-level group that includes a manager or supervisor from each major function within the process. Each member is an expert in a segment of the process. Ideally, BPM teams have no more than eight members, and the individuals

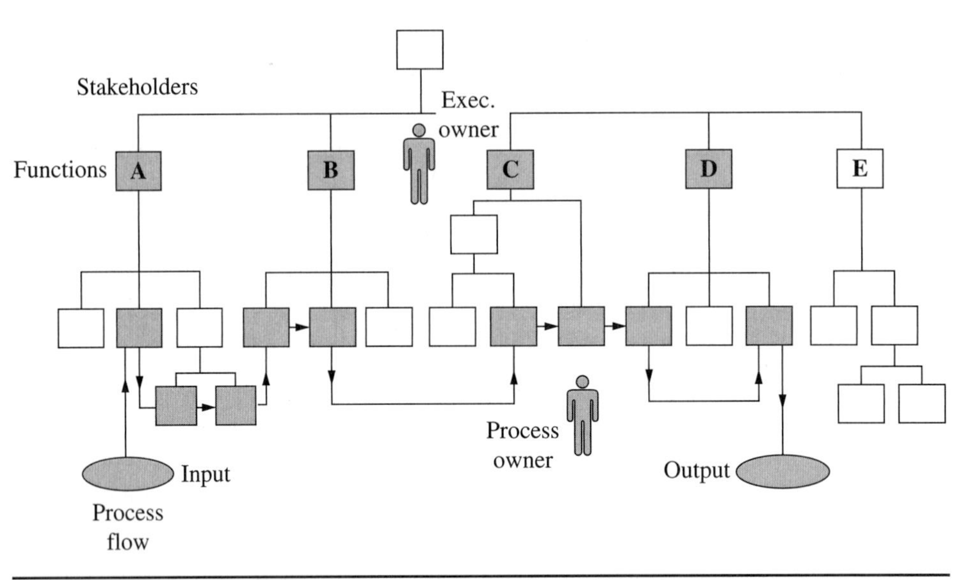

FIGURE 8.2 Diagram of a multifunctional organization and one of its major processes.

chosen should be proven leaders. The team is responsible for managing and continuously improving the process. The team shares with the owner the responsibilities for effectiveness and efficiency. Most commonly, team assignments are ongoing.

From time to time, a process owner creates an ad hoc team to address some special issue (human resources, information technology, activity-based costing, etc.). The mission of such a project-oriented team is limited, and the team disbands when the mission is complete. The ad hoc team is different from the process-management team.

Figure 8.2 is a simplified diagram of a multifunctional organization and one of its major processes. The shaded portions include the executive owner, the working owner, the BPM team, and the stakeholders—functional heads at the executive level who have work activities of the business process operating within their function. Customarily, stakeholders are members of the quality council, along with the executive owner. Taken together, this shaded portion is referred to as the BPM Infrastructure.

Establishing the Team's Mission and Goals

The preliminary process mission and improvement goals for the process are communicated to the owners (executive and working levels) and the team by the quality council. To do their jobs most effectively, the owners and the team must make the mission and goals their own. They do this by defining the process, the first step of the planning phase.

The Planning Phase: Planning the New Process

The first phase of BPM is planning, which consists of five steps: (1) defining the process, (2) discovering customer needs and flowcharting the process, (3) establishing measurements of the process, (4) analyzing process measurements and other data, and (5) designing (or redesigning) the process. The output of the planning phase is the new process plan.

Defining the Current Process

The owner(s) and the team collaborate to define the process precisely. In accomplishing this, the starting point and principal reference is the process documentation developed by the quality council during the selection of key processes and identification of owners and teams. This documentation includes preliminary statements of mission and goals.

Effective mission and goal statements explicitly declare

- The purpose and scope of the process
- "Stretch" targets for customer needs and business needs

(The purpose of the stretch target is to motivate aggressive process improvement activity.) For example, a mission statement for the Special-Contract Management Process is to provide competitive special pricing and supportive terms and conditions for large information systems procurements that meet customer needs for value, contractual support, and timeliness at affordable cost.

The goals for the same process are to

- Deliver approved price and contract support document within 30 days of date of customer's letter of intent.
- Achieve a yield of special-contract proposals (percent of proposals closed as sales) of not less than 50 percent.

The team must reach a consensus on the suitability of these statements, propose modifications for the quality council's approval, if necessary, and also document the scope, objectives, and content. Based on available data and collective team experience, the team will document process flow, process strengths and weaknesses, performance history, measures, costs, complaints, environment, and resources. This will probably involve narrative documentation and will certainly require the use of flow diagrams.

Bounding the business process starts with inventorying the major subprocesses—six to eight of them is typical—that the business process comprises. The inventory must include the "starts with" subprocess (the first subprocess executed), the "ends-with" subprocess (the last executed), and the major sub processes in between. If they have significant effect on the quality of the process output, activities upstream of the process are included within the process boundary. To provide focus and avoid ambiguity, it is also helpful to list subprocesses that are explicitly excluded from the business process. The accumulating information on process components is represented in diagram form, which evolves from a collection of subprocesses to a flow diagram as the steps of the planning phase are completed.

Figure 8.3 shows a high-level diagram of the special-contract process that resulted from process analysis but before the process was redesigned. At the end of the process definition step, such a diagram is not yet a flow diagram, as there is no indication of the sequence in which the subprocesses occur. Establishing those relationships as they presently exist is the work of Step 2.

Discovering Customer Needs and Flowcharting the Process

For the process to work well, the team must identify all customers, determine their needs, and prioritize the information. Priorities enable the team to focus its attention and spend its energies where they will be most effective.

Determining customer needs and expectations requires ongoing, disciplined activity. Process owners must ensure that this activity is incorporated in the day-to-day conduct of the business process as the customer requirements subprocess and assign accountability for

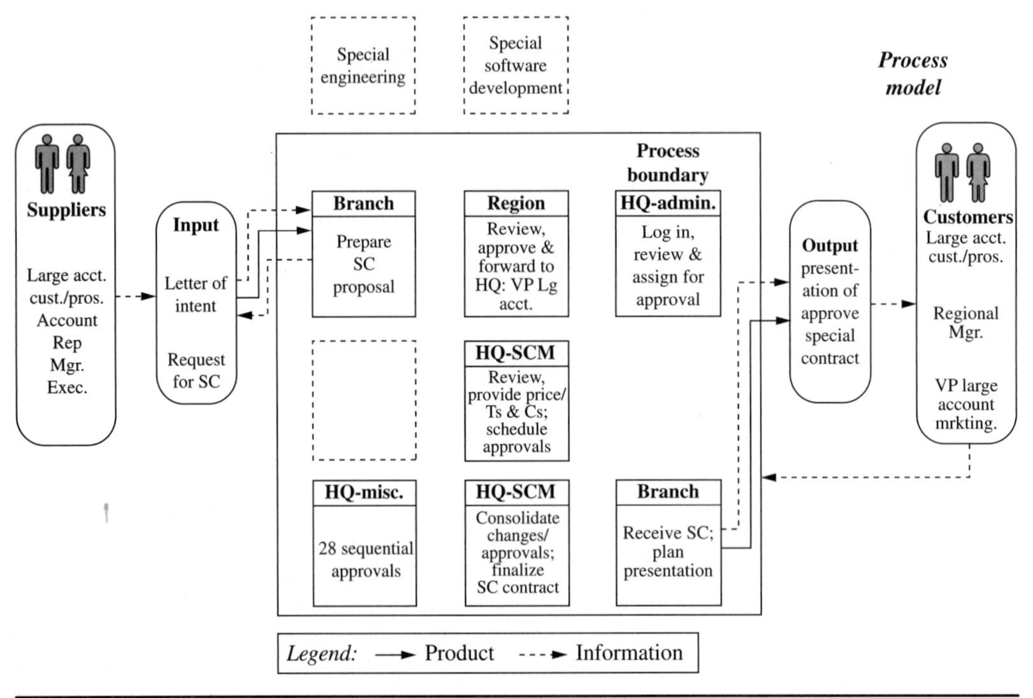

FIGURE 8.3 High-level diagram of the special-contract process.

its performance. The output of this vital activity is a continually updated customer requirement statement.

On the process flowchart, it is usual to indicate key suppliers and customers and their roles in the process as providers or receivers of materials, product, information, and the like. Although the diagram can serve a number of specialized purposes, the most important here is to create a common, high-level understanding among the owner and the team members of how the process works—how the subprocesses relate to each other and to the customers and suppliers and how information and product move around and through the process. In creating the process flowchart, the team will also verify the list of customers and may, as understanding of the process deepens, add to the list of customers.

The process flowchart is the team's primary tool for analyzing the process to determine whether it can satisfy customer needs. By walking through the chart together, step by step, sharing questions and collective experience, the team determines whether the process is correctly represented, making adjustments to the diagram as necessary to reflect the process as it presently operates.

When the step is complete, the team has a starting point for analyzing and improving the process. In Figure 8.4, the product flow is shown by solid lines and the information flow by dotted lines.

Establishing Process Measurements

What gets measured gets done. Establishing, collecting, and using the correct measures is critical in managing business process quality. "Process capability," "process performance," and other process measures have no practical significance if the process they

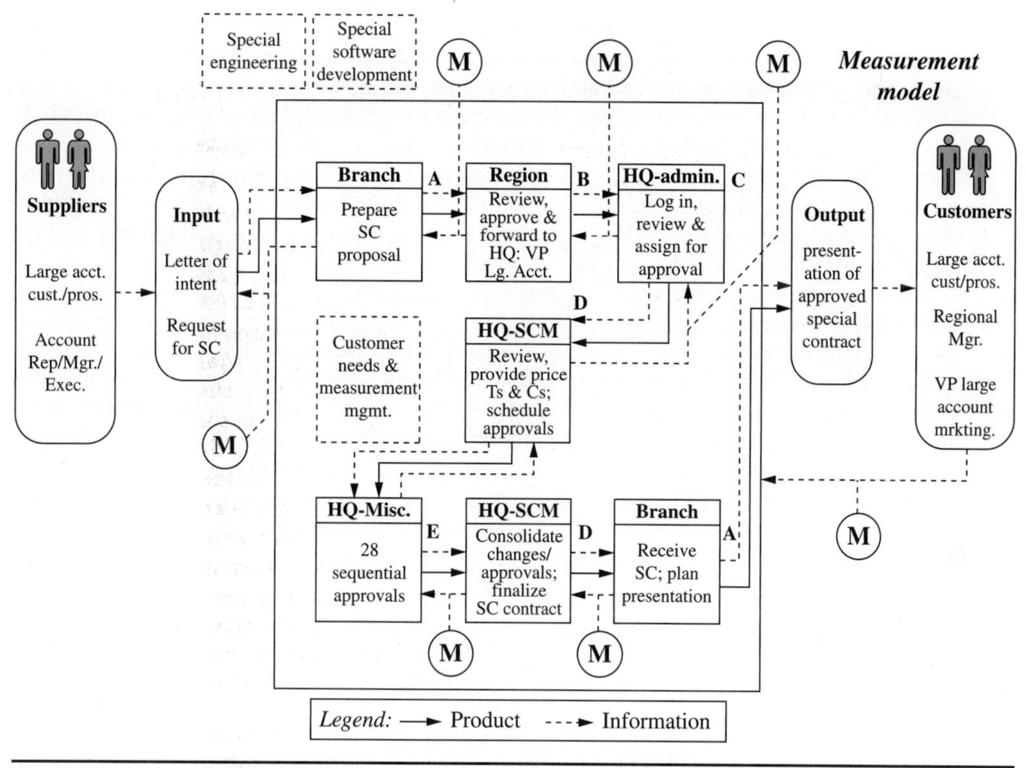

FIGURE 8.4 Flowchart of the special-contract process including process control points.

purport to describe is not managed. To be managed, the process must fulfill certain minimum conditions:

1. It has an owner.
2. It is defined.
3. Its management infrastructure is in place.
4. Its requirements are established.
5. Its measurements and control points are established.
6. It demonstrates stable, predictable, and repeatable performance.

A process that fulfills these minimum conditions is said to be *manageable*. Manageability is the precondition for all further work in BPM.

Of these criteria, (1) through (4) have already been addressed in this chapter. Criteria (5) and (6) are addressed as follows.

Process Measurements

In deciding what aspects of the process to measure, we look for guidance to the process mission and to our list of customer needs. Process measures based on customer needs provide a way of measuring process effectiveness. For example, if the customer requires delivery of an order within 24 hours of order placement, we incorporate into our order-fulfillment process a measure such as "time elapsed between receipt of order and delivery of order," and a system for collecting, processing, summarizing, and reporting information from the data generated. The statistic reported to the executive owner will be one such as "percent of orders delivered within 24 hours," a statistic that summarizes on-time performance. The team will also need data on which to base analysis and correction of problems and continuous improvement of the process. For this purpose, the team needs data from which they can compute such descriptive statistics as distribution of delivery times by product type, among others. The uses to which the data will be put must be thought through carefully at the time of process design to minimize redesign of the measures and measurement systems.

Process measures based on cost, cycle time, labor productivity, and process yield measure process efficiency. Suppose that a goal for our order-fulfillment process is to reduce order-picking errors to one error per thousand order lines. Managing that goal requires identifying order-picking errors in relation to the number of order lines picked. For inadvertent order-picking errors—that is, when they happen, the picker is unaware of them—measuring them requires a separate inspection to identify errors. In a random audit on a sample of picked orders, an inspector identifies errors and records them. As with delivery-time measurement, the team must think through all the uses it will make of these measurements. To report an estimated error rate, the data needed are the number of errors and the number of order lines inspected. To improve process performance in this category, the data must help the team identify error sources and determine their root cause. For that to occur, each error must be associated with time of day, shift, product type, and size of package so that the data can be stratified to test various theories of root cause.

Although process adaptability is not a measurement category, it is an important consideration for process owners and teams. Adaptability is discussed later in the chapter.

Process measurements must be linked to business performance. If certain key processes must run exceptionally well to ensure organization success, it follows that collective success of the key processes is good for the organization's performance. Process owners must take care to select process measures that are strongly correlated with traditional business indicators, such as revenue, profit, return on investment, earnings per share, productivity per employee, and so on. In high-level business plan reviews, managers are motivated and rewarded for maintaining this linkage between process and organization performance measures because of the two values that BPM supports: organization success is good, and BPM is the way we will achieve organization success.

Figure 8.5 shows some typical process measurements and the traditional business indicators with which they are linked. To illustrate, "percent of sales quota achieved" is a traditional business indicator relating to the business objective of improving revenue. The special-contract management process has a major impact on the indicator, as more than 30 percent of U.S. revenue comes from that process. Therefore, the contract close rate (ratio of the value of firm contracts to the total value of proposals submitted) of the special-contract management process is linked to percent of sales quota and other traditional revenue measures, and is, therefore, a measure of great importance to management. Measurement points appear on the process flow diagram.

Control Points

Process measurement is also a part of the control mechanisms established to maintain planned performance in the new process. To control the process requires that each of a few selected process variables be the control subjects of a feedback control loop. Typically, there will be five to six control points at the macro-process level for variables associated with: external output, external input, key intermediate products, and other high-leverage process points. Control points in the special-contract management process are represented graphically

The traditional business view		The process view	
Business objective	**Business indicator**	**Key process**	**Process measure**
	Percent of sales quota achieved	Contract management	Contract close rate
	Percent of revenue plan achieved	Product development	Development cycle time
Higher revenue	Value of orders cancelled after shipment	Account management	Backlog management and system assurance timeliness
	Receivable days outstanding		Billing quality index
Reduce costs	S, G & A	Manufacturing	Manufacturing cycle time
	Inventory turns		

Figure 8.5 Typical process measurements and the traditional business indicators.

in Figure 8.4. Feedback loop design and other issues surrounding process control are covered in detail in Chapter 6, Quality Control: Assuring Repeatable and Compliant Processes.

Process Variability, Stability, and Capability

As in all processes, business processes exhibit variability. The tools of statistical process control such as Shewhart charts (see Chapter 6) help the team to minimize process variation and assess process stability.

Evaluation of process capability is an important step in process quality improvement. Process capability is a measure of variation in a process operating under stable conditions. The phrase "Under stable conditions" means that all variations in the process are attributable to random causes. The usual criterion for stability is that the process, as plotted and interpreted on a Shewhart control chart, is "in control."

Statistical process control, process capability, and associated tools are useful components of the process team's tool kit. They are covered in detail in Chapter 18, Core Tools to Design, Control, and Improve Performance.

The output of the measurement step is a measurement plan, a list of process measurements to be made and the details of making each one, including who will make it, how it will be made, and on what schedule.

Analyzing the Process

Process analysis is performed for the following purposes:

- Assess the current process for its effectiveness and efficiency.
- Identify the underlying causes of any performance inadequacy.
- Identify opportunities for improvement.
- Make the improvements.

Business process	Sub-process	Activity	Task
Procurement	Supplier selection	Supplier survey	Documentation of outside supplier
Development engineering	Hardware design	Engineering change	Convening the change board
Office administration	Providing administrative support services	Managing calendars	Making a change to existing calendar

FIGURE 8.6 Three levels of decomposition.

First, referring to the process flowchart, the team breaks the process into its component activities using a procedure called "process decomposition," which consists of progressively breaking apart the process, level by level, starting at the macro level. As decomposition proceeds, the process is described in ever-finer detail.

As the strengths and weaknesses of the process are understood at one level, the BPM team's interim theories and conclusions will help decide where to go next with the analysis. The team will discover that certain subprocesses have more influence on the performance of the overall business process than others (an example of the Pareto principle). These more significant subprocesses become the target for the next level of analysis.

Decomposition is complete when the process parts are small enough to judge as to their effectiveness and efficiency. Figure 8.6 shows examples from three levels of decomposition (subprocess, activity, and task) of three typical business processes (procurement, development engineering, and office administration).

Measurement data are collected according to the measurement plan to determine process effectiveness and efficiency. The data are analyzed for effectiveness (conformance to customer needs) and long-term capability to meet current and future customer requirements.

The goal for process efficiency is that all key business processes operate at minimum total process cost and cycle time while still meeting customer requirements.

Process *effectiveness* and *efficiency* are analyzed concurrently. Maximizing effectiveness and efficiency together means that the process produces high quality at low cost; in other words, it can provide the most *value* to the customer.

"Business process adaptability" is the ability of a process to readily accommodate changes both in the requirements and the environment while maintaining its effectiveness and efficiency over time. To analyze the business process, the flow diagram is examined in four steps and modified as necessary.

The Process Analysis Summary Report is the culmination and key output of this process analysis step. It includes the findings from the analysis, that is, the reasons for inadequate process performance and potential solutions that have been proposed and recorded by owner and team as analysis progressed. The completion of this report is an opportune time for an executive owner/stakeholder review.

The owner/stakeholder reviews can be highly motivational to owners, teams, stakeholders. Of particular interest is the presentation of potential solutions for improved process operation. These have been collected throughout the planning phase and stored in an idea bin. The design suggestions are now documented and organized for executive review as part of the process analysis summary report presentation.

In reviewing potential solutions, the executive owner and the quality council provide the selection criteria for acceptable process design alternatives. Knowing upper management's criteria for proposed solutions helps to focus the process-management team's design efforts and makes a favorable reception for the reengineered new process plan more likely.

Designing (or Redesigning) the Process

In Process Design, the team defines the specific operational means for meeting stated product goals. The result is a newly developed Process Plan. Design changes fall into five broad categories: workflow, technology, people and organization, physical infrastructure, and policy and regulations.

In the design step, the owner and the team must decide whether to create a new process design or to redesign the existing process. Creating a new design might mean radical change; redesign generally means incremental change with some carryover of existing design features.

The team will generate many design alternatives, with input from both internal and external sources. One approach to generating these design alternatives from internal sources is to train task level performers to apply creative thinking to the redesign of their process.

Ideas generated in these sessions are documented and added to the idea bin. Benchmarking can provide a rich source of ideas from external sources, including ideas for radical change. Benchmarking is discussed in detail in Chapter 15.

In designing for process effectiveness, the variable of most interest is usually process cycle time. In service-oriented competition, lowest process cycle time is often the decisive feature. Furthermore, cycle–time reduction usually translates to efficiency gains as well. For many processes, the most promising source of cycle–time reduction is introducing new technology, especially information technology.

Designing for speed creates surprising competitive benefits: growth of market share and reduction of inventory requirements. Hewlett-Packard, Brunswick Corp., GE's Electrical Distribution and Control Division, AT&T, and Benetton are among the companies who have reported stunning achievements in cycle-time reduction for both product development and manufacturing (Dumaine 1989). In each of the companies, the gains resulted from efforts based on a focus on major processes. Other common features of these efforts included the following:

- Stretching objectives proposed by top management
- Absolutely adhering to the schedule, once agreed to
- Applying state-of-the art information technology
- Reducing management levels in favor of empowered employees and self-directed work teams
- Putting speed in the culture

In designing for speed, successful redesigns frequently originate from a few relatively simple guidelines: eliminate handoffs in the process, eliminate problems caused upstream of activity, remove delays or errors during handoffs between functional areas, and combine steps that span businesses or functions. A few illustrations are provided as follows:

- *Eliminate handoffs in the process.* A "handoff" is a transfer of material or information from one person to another, especially across departmental boundaries. In any process involving more than a single person, handoffs are inevitable. It must be recognized, however, that the handoff is time consuming and full of peril for process integrity—the missed instruction, the confused part identification, the obsolete specification, the miscommunicated customer request. In the special-contract management process, discussed previously in the chapter, the use of concurrent review boards eliminated the 28 sequential executive approvals and associated handoffs.

- *Eliminate problems caused upstream of activity.* Errors in order entry at a U.S. computer organization were caused when sales representatives configured systems incorrectly. As a result, the cost of the sales-and-order process was 30 percent higher than that of competitors, and the error rates for some products were as high as 100 percent. The cross-functional redesign fixed both the configurations problem and sales-force skills so that on-time delivery improved at significant cost savings (Hall et al. 1993).
- *Remove delays or errors during handoffs between functional areas.* The processing of a new policy at a U.K. insurance organization involved 10 handoffs and took at least 40 days to complete. The organization implemented a case-manager approach by which only one handoff occurred, and the policy was processed in less than 7 days (Hall et al. 1993).
- *Combine steps that span businesses or functions.* At a U.S. electronics equipment manufacturer, as many as seven job titles in three different functions were involved in the nine steps required to design, produce, install, and maintain hardware. The organization eliminated all but two job titles, leaving one job in sales and one job in manufacturing (Hall et al. 1993).

Process design testing is performed to determine whether the process design alternative will work under operating conditions. Design testing may include trials, pilots, dry runs, simulations, etc. The results are used to predict new process performance and cost/benefit feasibility.

Successful process design requires employee participation and involvement. To overlook such participation creates a lost opportunity and a barrier to significant improvement. The creativity of the first-line work force in generating new designs can be significant.

Creating the New Process Plan

After we have redefined a key process, we must document the new process and carefully explain the new steps. The new process plan now includes the new process design and its control plan for maintaining the new level of process performance. The new process plan for the special-contract management process, shown as a high-level process schematic, is shown in Figure 8.7.

The Transfer Phase: Transferring the New Process Plan to Operations

There are three steps in the transfer phase: (1) planning for implementation problems, (2) planning for implementation action, and (3) deploying the new process plan.

Planning for Implementation Problems

A major BPM effort may involve huge expenditures and precipitate fundamental change in an organization, affecting thousands of jobs. All of this poses major management challenges. All of the many changes must be planned, scheduled, and completed so that the new process may be deployed to operational management. Figure 8.8 identifies specific categories of problems to be addressed and the key elements that are included.

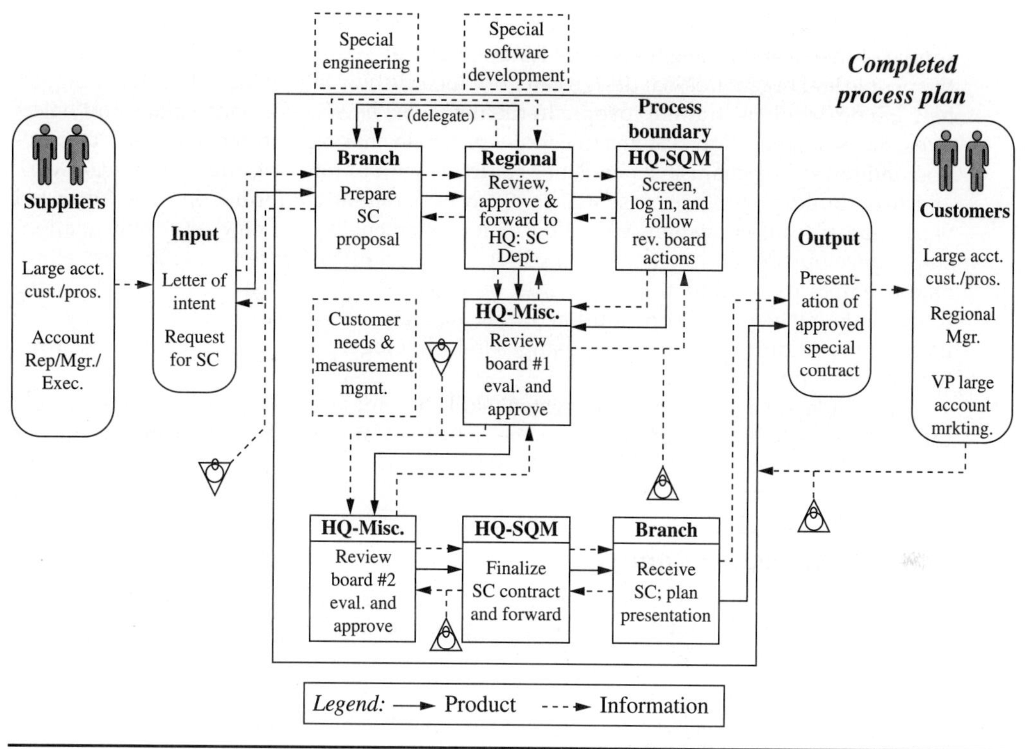

FIGURE 8.7 High-level process schematic.

Category	Key elements
Workflow	Process anatomy (macro/micro; cross-functional; intra-functional; inter-departmental; intra-departmental)
Technology	Information technology; automation
People and organization	Jobs; job description; training and development, performance management; compensation (incentive-based or not); recognition/reward; union involvement; teams; self-directed work teams; reporting relationships; de-layering.
Infrastructure (physical)	Location; space; layout; equipment; tools; furnishings
Policy/regulations	Government; community; industry; company; standards; culture
New-process design issues	Environmental, quality, costs, sourcing

FIGURE 8.8 Specific categories of problems to be addressed.

Of the five categories listed in Figure 8.8, "people and organization" is usually the source of the most challenging change issues in any BPM effort. Implementation issues in the people and organizational design category include new jobs, which are usually bigger; new job descriptions; training people in the new jobs; new performance plans and objectives; new compensation systems (incentive pay, gain sharing, and the like); new recognition and reward mechanisms; new labor contracts with unions; introduction of teamwork and team-building concepts essential to a process orientation; formation of self-directed work teams; team education; reduction of management layers; new reporting relationships; development and management of severance plans for those whose jobs are eliminated; temporary continuation of benefits; outplacement programs; and new career paths based on knowledge and contribution, rather than on promotion within a hierarchy. The list goes on. Additionally, there are changes in technology, policy, and physical infrastructure to deal with.

The importance of change management skills becomes clear. Deploying a new process can be a threat to those affected. The owner and the team must be skilled in overcoming resistance to change.

Creating Readiness for Change

Change happens when four conditions are combined. First, the current state must be seen as unsatisfactory, even painful; it must constitute a tension for change. Second, there must be a satisfactory alternative, a vision of how things can be better. Third, some practical steps must be available to reach the satisfactory state, including instruction in how to take the steps, and support during the journey. Fourth, to maintain the change, the organization and individuals must acquire skills and reach a state of self-efficacy.

These four conditions reinforce the intent to change. Progress toward that change must be monitored continuously to make the change permanent. In the operational management phase, operational controls, continuous improvement activity, and ongoing review and assessment all contribute to ensuring that the new process plan will continue to perform as planned.

Planning for Implementation Action

The output of this step is a complex work plan, to be carried out by the owner and the BPM team. They will benefit from skills in the techniques of Project Management.

Deploying the New Process Plan

Before actually implementing the new process, the team tests the process plan. They test selected components of the process and may carry out computer simulations to predict the performance of the new process and determine its feasibility. Also, tests help the team refine the "roll out" of the process and decide whether to conduct parallel operation (old process and new process running concurrently). The team must decide how to deploy the new process. There are several options

- Horizontal deployment, function by function.
- Vertical deployment, top down, all functions at once.
- Modularized deployment, activity by activity, until all are deployed.
- Priority deployment, subprocesses and activities in priority sequence, those having the highest potential for improvement going first.

- Trial deployment, a small-scale pilot of the entire process, then expansion for complete implementation. This technique was used in the first redesign of the special-contract management process, that is, a regional trial preceded national expansion. The insurance organization USAA conducts all pilot tests of new process designs in their Great Lakes region. In addition to "working the bugs out of the new design before going national," USAA uses this approach as a "career-broadening experience for promising managers," and to "roll out the new design to the rest of the organization with much less resistance" (Garvin 1995).

Full deployment of the new process includes developing and deploying an updated control plan. Figure 8.9 lists the contents of a new process plan.

Operational Management Phase: Managing the New Process

The Operational Management Phase begins when the process is put into operation. The major activities in operational management are (1) process quality control, (2) process quality improvement, and (3) periodic process review and assessment.

Business Process Metrics and Control

"Process control" is an ongoing managerial process, in which the actual performance of the operating process is evaluated by measurements taken at the control points, comparing the measurements to the quality targets, and taking action on the difference. The goal of process control is to maintain performance of the business process at its planned level. (See Chapter 6, Quality Control: Assuring Repeatable and Compliant Processes.)

Process plan
Process purpose or mission
Process goals and targets
Process management infrastructure (process owner, team, stakeholder)
Process contract
Process description and model
Customer requirements (customer list, customer needs, requirements statement
Process flow
Measurement plan
Process analysis summary report
Control plan
Implementation action plan
Resource plan
Schedules and timeline

FIGURE 8.9 Contents of a new process plan.

Business Process Improvement

By monitoring process performance with respect to customer requirements, the process owner can identify gaps between what the process is delivering and what is required for full customer satisfaction. These gaps are targets for process quality improvement efforts. They are signaled by defects, complaints, high costs of poor quality, and other deficiencies. (See Chapter 6, Quality Control: Assuring Repeatable and Compliant Processes.)

Periodic Process Review and Assessment

The owner conducts reviews and assessments of current process performance to ensure that the process is performing according to plan. The review should include a review and an assessment of the process design itself to protect against changes in the design assumptions and anticipated future changes such as changes in customer needs, new technology, or competitive process designs. It is worthwhile for the process owner to establish a schedule for reviewing the needs of customers and evaluating and benchmarking the present process.

As customer needs change, process measures must be refined to reflect these changes. This continuous refinement is the subject of a measurement management subprocess, which is established by the owners and the team and complements the customer's needs subprocess. The two processes go hand in hand.

The business process management category in the Malcolm Baldrige National Quality Award criteria provides a basis for management review and assessment of process performance.

Other external award criteria from worldwide sources, as well as many national and international standards, serve as inspiration and guidance for owners and teams contemplating process reviews.

The criteria of the Malcolm Baldrige National Quality Award have come to be regarded as the de facto definition of performance excellence. Business process management is an important concept within the performance excellence framework.

Organizations have learned not to limit managerial attention to the financial dimension. They have gained experience in defining, identifying, and managing the quality dimension. They are accustomed to thinking strategically—setting a vision, mission, and goals, all in alignment. And they will have experience reviewing progress against those goals.

The quality improvement process, which began in Japan in the 1950s and was widely deployed in the United States in the early 1980s, was an important step beyond functional management. Organizations found that quality improvement required two new pieces of organization machinery—the quality council and the cross-functional project team. The Quality Council usually consists of the senior management team; to its traditional responsibility for management of finance the responsibility for the management of quality is added. The project team recognizes that, in a functional organization, responsibility for reducing chronic deficiencies has to be assigned to a cross-functional team.

BPM is a natural extension of many of the lessons learned in early quality improvement activities. It requires a conceptual change—from reliance on functional specialization to an understanding of the advantages of focusing on major business processes. It also requires an additional piece of organization machinery: an infrastructure for each of the major processes.

The Future of BPM Combined with Technology

Business process management (BPM) is being combined with service-oriented architectures (SOAs) technologies and performance excellence tools such as Lean and Six Sigma to accelerate improvements and results. At the same time, these tools are increasing organizational flexibility and technology-enabled responsiveness. Many successful organizations have found that the linkages are clear.

According to IBM (*Reference: "Aligning Business Process Management, Service-Oriented Architecture, and Lean Six Sigma for Real Business Results)*, early adopters who have worked their way past cultural and organizational barriers are seeing impressive performance and financial results, seen as follows:

- Improved responsiveness to market challenges and changes through aligned and significantly more flexible business and technical architectures
- Improved ability to innovate and achieve strategic differentiation by driving change into the market and tuning processes to meet the specific needs of key market segments
- Reduced process costs through automation and an improved ability to monitor, detect, and respond to problems by using real-time data, automated alerts, and planned escalation
- Significantly lower technical implementation costs through shared process models and higher levels of component reuse
- Lower analysis costs and reduced risk through process simulation capabilities and an improved ability to gain feedback and buy-in prior to coding

The rewards can be great, especially for those who take action now.

References

Dumaine, B. (1989). "How Managers Can Succeed through Speed." Fortune, February. 13, pp. 54–60.

Garvin, D. A. (1995). "Leveraging Processes for Strategic Advantage." *Harvard Business Review*, September/October, vol. 73, no. 5, pp. 77–90.

Hall, G., Rosenthal, J., and Wade, J. (1993). "How to Make Reengineering Really Work." *Harvard Business Review*, November/December, vol. 71, no. 6, pp. 199–130.

Kane, E. J. (1986). "IBM's Focus on the Business Process." *Quality Progress*, April, p. 26.

Kaplan, R. S., and Norton, D. P. (1992). "The Balanced Scorecard—Measures that Drive Performance." *Harvard Business Review*, January/February, vol. 7, no. 1, pp. 71–79, reprint #92105.

Levi, M. H. "Transformation to Process Organization." Interfacing Technologies Corporation, http://www.interfacing.com/uploads/File/The%20Business%20Process-Meir%20Levi.pdf

Pall, G. A. (1987). *Quality Business Process Management*. Prentice-Hall, Inc., Englewood Cliffs, NJ.

Riley, J. F., Jr. (1989). *Executive Quality Focus: Discussion Leader's Guide*. Science Research Associates, Inc., Chicago.

Riley, J. F., Jr., Pall, G. A., and Harshbarger, R. W. (1994). *Reengineering Processes for Competitive Advantage: Business Process Quality Management* (BBPM), 2nd ed., Juran Institute, Inc., Wilton, CT.

CHAPTER 9

The Juran Transformation Model and Roadmap

Joseph A. De Feo and Janice Doucet Thompson

About This Chapter

Creating the state of performance excellence in organizations will enable our global society to avoid technological failures from harming the environment and, ultimately, its people. Transforming an organization from one culture to another is not an easy task. However, it can happen when an organization creates systematically significant, sustainable, and beneficial change. We provided the universal principles in Chapters 1 through 8. This chapter outlines the means to pull the universal into one roadmap to create a culture of performance excellence. The systematic approach that we call the Juran Transformation Model can enable any organization to transform itself by knowing what to expect. Transformation usually requires six organizational breakthroughs before a state of performance excellence can be attained.

High Points of This Chapter

1. A breakthrough is defined as the purposeful creation of significant, sustainable beneficial change. It is often associated with process improvement targets. Transformation requires that the organization attain breakthroughs. The required breakthroughs are leadership and management, organization structure, current performance, culture, and adaptability and sustainability.

2. Organizational change is important for three reasons (each of which can destroy an organization): (1) Costs of poorly performing processes are too high, (2) dealing with continual societal changes, and (3) without change, organizations die.
3. All organizations must be thought of as open systems. Open systems depend on successful transactions with the organization's external environment and proper coordination of the organization's various specialized internal functions.
4. In attempting to create these breakthroughs, problems that appear in one work area often have their origin upstream in the process. Therefore, people in a given work location suffering from a performance problem cannot necessarily solve it by themselves.
5. Performance excellence can only be attained with the active participation, not only of individuals who created the problems, but also individuals affected by the problem and those who create remedial changes to the problem (usually those who are the source of the problem and perhaps others).
6. Organizational change attempted in isolation from the whole organization and without systems thinking can easily create more problems than existed previously.
7. The approach can provide the model and roadmap for transformational change.

Transforming a Culture

Changing a culture is difficult and usually unsuccessful unless a comprehensive approach exists to achieve and sustain it. The Juran Transformation Model and Roadmap describes five separate and unique types of breakthroughs that must occur in an organization before sustainability is attained. Without these breakthroughs, an organization attains superior results, but the results may not be sustainable for long periods of time. If performance excellence is the state in which an organization attains superior results through the application of the universal quality management methods, then an organization must ensure that these methods are used successfully. The journey from where your organization is to where it wants to go may require a transformational change. This change will result in the ability of the organization to sustain its performance, attain world class status, and market leadership.

The five breakthroughs are listed as follows:

1. Leadership and management
2. Organization and structure
3. Current performance
4. Culture
5. Adaptability and sustainability

The Juran Transformation Model

The Juran Transformation Model (Figure 9.1) is based on over 60 years of experience and research from Dr. Juran and the Juran Institute. The five breakthroughs, when complete, help produce a state of performance excellence. Each breakthrough addresses a specific

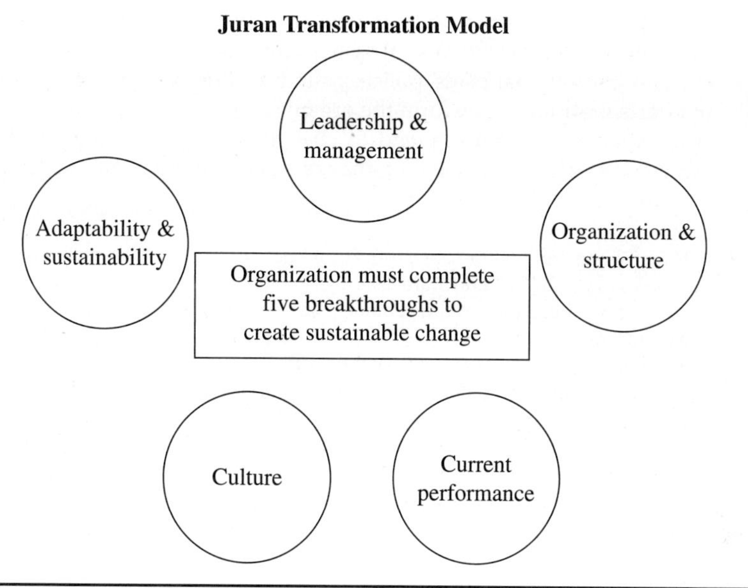

FIGURE 9.1 Juran Transformation Model. (*Juran Institute, Inc. 2009.*)

organizational subsystem that must change. Each is essential for supporting organizational life; none by itself is sufficient. In effect, the breakthroughs all empower the operational subsystem whose mission is to achieve technological proficiency in producing the goods, services and information for which customers will pay for or use. There is some overlap and duplication of activities and tasks among the different breakthrough types. This is to be expected because each subsystem is interrelated with all the others, and each is affected by activities in the others. The authors acknowledge that some issues in each type of breakthrough may have already been addressed by the reader's organization—so much the better. If this is the case, if you did not start your organization's performance excellence journey from the beginning, pick up the journey from where your organization presently finds itself. Closing the gaps will likely be part of your organization's next strategic business planning cycle. To close the gaps, design strategic and operational goals and projects to reach those goals and deploy them to all functions and levels.

Breakthrough and Transformational Change

Breakthroughs can occur in an organization at any time, usually as the result of a specific initiative, such as a specific improvement project (e.g., a Six Sigma, improvement project; a design of a new service, or the invention of a new technology). These changes can produce sudden explosive bursts of beneficial change for your organization and society. But they may not be enough to cause the culture to change or sustain itself to the changes that occurred. This is because it may not have happened for the right reason. It was not purposeful. It came about through chance. Change by "chance" is not predictable or sustainable. What an organization needs is predictable change.

Today's organizations operate in a state of perpetual, unpredictable change that requires the people in them to produce continuous adaptive improvements as pressure mounts for new improvements to be made from the outside. These improvements may take months or even years to accomplish because it is the cumulative effect of many coordinated and interrelated organizational plans, policies, and breakthrough projects. Taken together, these diligent efforts gradually transform the organization.

Organizations that do not intend to change usually will when a crisis—or a fear of impending crisis—triggers a need for change within an organization. Consider the following scenario:

> Two of the largest competitors have introduced new products that are better than ours. Consequently, sales of products X and Y are heading steadily down, and taking our market share along. Our new product introduction time is much slower than the competition, making the situation even worse. The new plant can't seem to do anything right. Some equipment is often down, and, even when in operation, produces too many costly defective items.
>
> Too many of our invoices are returned because of errors, with the resulting postponement of revenue and a growing number of unsatisfied customers, not to mention the hassle and costs of rework. Accounts receivable have been much too high and are gradually increasing. We are becoming afraid that the future may offer additional threats we need to ward off or, more importantly, plan for so they can be prevented altogether. Leadership must take action, or the organization is going to experience a loss of market share, customer base, and revenue.

Breakthroughs Are Essential to Organizational Vitality

There are four important reasons why an organization cannot survive very long without the medicinal renewing effects of continual breakthrough:

1. *The costs of* poor quality (COPQ) continue to increase if they are not tackled. They are too high. One reason is that organizations are plagued by a continuous onslaught of crises precipitated by mysterious sources of chronic high costs of poorly performing processes. As we stated in Chapters 1 and 5, the total chronic levels of COPQ have been reported to be as high as 20 percent or more of the costs of goods sold. This number varies by type of industry and organization. It is not unusual for these costs at times to exceed profit or be a major contributor to losses. In any case, the average overall level is appalling (because it is substantial and *avoidable*), and the toll it takes on the organization can be devastating. COPQ is a major driver of many cost-cutting initiatives, not only because it can be so destructive if left unaddressed but also because savings realized by reducing COPQ directly affect the bottom line. Furthermore, the savings continue, year after year, as long as remedial improvements are irreversible, or controls are placed on reversible improvements.
2. It makes good business sense that mysterious and chronic causes of waste must be discovered, removed, and prevented from returning. Breakthrough improvement becomes the preferred initial method of attack because of its ability to uncover and remove specific root causes and to hold the gains—it is designed to do just that. One could describe breakthrough improvement methodology as applying the scientific method to solving performance problems. Breakthrough improvement methodology closely resembles the medical model of diagnosis and treatment.
3. *Chronic and continuous change.* Another reason why breakthroughs are required for organizational survival is the state of chronic accelerating change found in today's business environment. Unrelenting change has become so powerful and so pervasive that no constituent part of an organization finds itself immune from its effects for

long. Because any or all components of an organization can be threatened by changes in the environment, if an organization wishes to survive, it is most likely to be forced into creating basic changes that are powerful enough to bring about accommodation with new conditions. Performance breakthroughs, consisting as it does of several specific types of breakthrough in various organization functions, is a powerful approach that is capable of determining countermeasures sufficiently effective to prevail against the inexorable forces of change. An organization may have to re-invent itself. It may even be driven to reexamine, and perhaps modify, its core products, business, service or even its customers.

4. *Without continuous improvement, organizations die.* Another reason why breakthroughs are essential for organizational survival is found in knowledge derived from scientific research into the behavior of organizations. Leaders can learn valuable lessons about how organizations function and how to manage them by examining open systems theory. Among the more important lessons taught by this theory is the notion of *negative entropy*. Negative entropy refers to characteristics that human organizations share with biological systems such as the living cell, or the living organism (which is a collection of cells). *Entropy* is the tendency of all living things—and all organizations—to head toward their own extinction. Negative entropy consists of countermeasures that living systems and social systems take to stave off their own extinction. Organisms replace aging cells, heal wounds, and fight disease. Organizations build up reserves of energy (backlogs and supplies) and constantly replace expended energy by acquiring more energy (sales and raw materials) from their environment. Eventually, living organisms lose the race. So do organizations if they do not continually adapt, heal "wounds" (make performance breakthrough improvements), and build up reserves of cash and goodwill. The Juran Transformation Model is a means by which organizations can stave off their own extinction.

Systems Thinking and Transformational Change

Organizations are like living organisms. They consist of a number of subsystems, each of which performs a vital specialized function that makes specific, unique, and essential contributions to the life of the whole. A given individual subsystem is devoted to its own specific function such as design, production, management, maintenance, sales, procurement, and adaptability. One cannot carry the biological analogy very far because living organisms separate subsystems with physical boundaries and structures (e.g., cell walls, the nervous system, the digestive system, the circulatory system, etc.). Boundaries and structure of subsystems in human organizations, on the other hand, are not physical; they are repetitive events, activities, and transactions. The repetitive patterns of activities are, in effect, the work tasks, procedures, and processes carried out by organizational functions. Open systems theorists call these patterns of activities roles. A role consists of one or more recurrent activities out of a total pattern of activities which, in combination, produce the organizational output.

Roles are maintained and carried out in a repetitive, relatively stable manner by means of mutually understood sets of expectations and feedback loops, shown in Figure 9.2, The Triple Role Open Systems theory and Juran's model focuses particularly on the technical methods, human relationships, organization structures, and interdependence of functional roles associated with these activities and transactions. Detailed knowledge of the repetitive transactions between the organization and its environment, and also within the organization itself, is essential in accomplishing breakthroughs because these transactions determine the effectiveness and efficiency of performance.

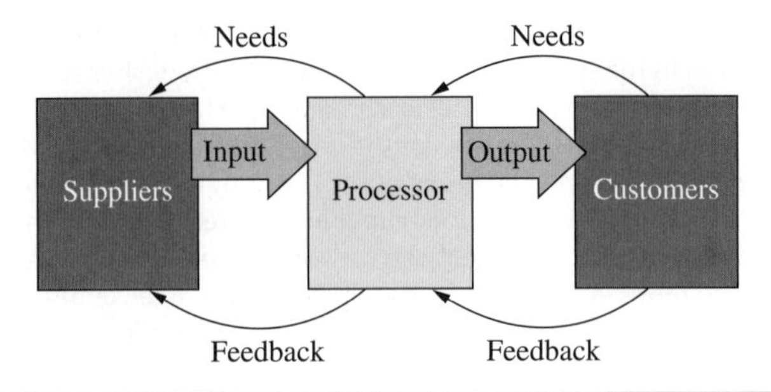

FIGURE 9.2 The triple role. (*Juran Institute, Inc. 2009, p. 8.*)

Figure 9.2 shows a model that applies equally to an organization as a whole, to individual subsystems and organizational functions (e.g., departments and workstations within the organization, and to individual organizational members performing tasks in any function or level. All these entities perform three more or less simultaneous roles, acting as supplier, processor, and customer. Acting as a processor, charged with the duty of transforming imported energy, organizations receive raw materials—goods, information, and/or services—from their suppliers, who may be located inside or outside the organization. The processor's job consists of transforming the received things into a new product of some kind—goods, information, or service. In turn, the processor supplies the product to his or her customers who may be located within or outside the organization.

Each of these roles requires more than merely the exchange of things. Each role is linked by mutually understood expectations (i.e., specifications, work orders, and procedures) and feedback as to how well the expectations are being met (i.e., complaints, quality reports, praise, and rewards). Note that in the diagram, the processor must communicate (shown by arrows) to the supplier a detailed description of his or her needs and requirements. In addition, the processor provides the supplier with feedback on the extent to which the expectations are being met. This feedback is part of the control loop and helps to ensure consistent adequate performance by the supplier. The customer bears the same responsibilities to his or her processors who, in effect, are also suppliers (not of the raw materials but of the product).

When defects, delays, errors, or excessive costs occur, causes can be found somewhere in the activities performed by suppliers, processors, and customers, in the set of transactions between them, or perhaps in gaps in the communication of needs and feedback. Breakthrough efforts must uncover the precise root causes by deep probing and exploration. If the causes are really elusive, discovering them may require placing the offending repetitive process under a microscope of unprecedented power and precision, as is done in Six Sigma. Performance excellence initiatives require that all functions and levels be involved, at least to some extent, because each function's performance is interrelated and dependent to some degree on all other functions. Moreover, a change in the behavior of any one function will have some effect on all the others, even though it may not be apparent at the time. This interrelatedness of all functions has practical day-to-day implications for a leader at any level, that is, the imperative of using "systems thinking" when making decisions, particularly decisions to make changes.

Because an organization is an open system, its life depends on (1) successful transactions with the organization's external environment and (2) proper coordination of the organization's various specialized internal functions and their outputs.

The proper coordination and performance of the various internal functions is dependent on the management processes of planning, controlling, and improving and on human factors such as leadership, organizational structure, and culture. To manage in an open system (such as an organization), management at all levels must think and act in systems terms. Managers must consider the impact of any proposed change not only upon the whole organization but also the impact on the interrelationships of all the parts. Failure to do so, even when changing seemingly little things, can make some pretty big messes. Leaders need to reason as follows: "If there is to be a change in x, what is required (inputs) from all functions to create this change, and how will x affect each of the other functions, and the total organization as well (ultimate output/results)?" Organizations will not change until the people in them change, regardless of the breakthrough approach.

There are three important lessons learned from the experience of the authors:

1. *All organizations need a systematic approach to ensure that change happens.* The problems that appear in one function or step in a process often have their origin upstream from that function or step in the process. People in a given work area cannot necessarily solve the problem in their own work area by themselves—they need to involve others in the problem-solving process. Without systematic involvement of the other functions, suboptimization will occur. Suboptimization results in excess costs and internal customer dissatisfaction—the exact opposite of what is intended.
2. *Change can only be created with active participation of all employees from the top on down and over time.* This includes not only individuals who are the source of a problem but also those affected by the problem and those who will initiate changes to remedy the problem (usually those who are the source of the problem, and perhaps others).
3. *Functional change alone is not sufficient to transform an organization.* Breakthroughs attempted in isolation or within a structure from the whole organization and without systems thinking can easily create more problems than existed at the start of the breakthrough attempt.

Attempts to bring about substantial organizational change such as performance excellence requires not only changing the behavior of individuals (as might be attempted by training) but also of redefining their roles in the social system. This requires, among other things, changing the expectations that customers have for their processors and changing expectations that processors have for their suppliers. In other words, performance breakthroughs require a capability of organizational design to produce consistent, coordinated behavior to support specific organizational goals. Modifications will likely also be made to other elements that define roles such as job descriptions, job fit, work procedures, control plans, other elements of the quality system, and training. To achieve a breakthrough, it is not sufficient simply to train a few Black Belts in the martial arts as experts and complete a few projects. Although this will probably result in some improvement, it is unlikely to produce long-term culture change and sustainability. The authors believe that too many organizations are settling for simple improvements when they should be striving for breakthroughs.

As we have seen, attaining a performance excellence state consists of achieving and sustaining beneficial changes. It is noteworthy that having a bright idea for a change does not, by itself, make change actually happen. People must understand why the change is needed and see the impact it will have on them before they can change what they do and perhaps how they do it. Beneficial change is often resisted, sometimes by the very persons who could benefit most from it, especially if they have been successful in doing things the usual way. Leading change can be a perplexing and challenging undertaking. Accordingly, individuals trying to implement change should acquire know-how in how to do it.

Breakthroughs in Leadership and Management

Breakthroughs in leadership occur when managers answer two basic questions:

1. How does management set performance goals for the organization and motivate the people in the organization to reach them and be held accountable?
2. How do managers best use the power of the workforce and other resources in the organization and how should they best manage them?

Issues with leadership are found at *all* levels, not just at the top of an organization. A breakthrough in leadership and management results in an organization characterized by unity of purpose and shared values as well as a system that enables engagement of the workforce.

Each work group knows what its goals are and, specifically, what performance is expected from the team and the individuals. Each individual knows specifically what he or she is to contribute to the overall organizational mission and how his or her performance will be measured. Few erratic or counterproductive behaviors occur. Should such behaviors occur, or should conflict arise, guidelines to behavior and decision-making are in place to enable relatively quick and smooth resolution of the problem. There are two major elements to leadership: (1) leaders must decide and clearly communicate where they want their employees to go; and (2) leaders must entice them to follow the path by providing an understanding of why this is a better way. In this handbook, the words "leader" and "manager" do not necessarily refer to different persons. Indeed, most leaders are managers, and managers should be leaders. The distinctions are matters of intent and activities, not players. Leadership can and should be exercised by managers; leaders also need to manage. If leadership consists of influencing others in a positive manner that attracts others, it follows that those at the top of the managerial pyramid (CEOs and C-Suite) can be the most effective leaders because they possess more formal authority than anyone else in an organization. In fact, top managers are usually the most influential leaders. If dramatics change, such as introducing Lean Six Sigma into an organization, the most effective approach by far is for the CEO to lead the charge. Launching Lean Six Sigma is helped immensely if other leaders, such as union presidents also lead the charge. The same can be said if senior and middle managers, first-line supervisors, and leaders of non-management work crews "follow the leader" and support a performance excellence program by word and actions. Leadership is not dictatorship because dictators make people afraid of behaving in "incorrect" ways, and perhaps they occasionally provide public treats (e.g., free gasoline, for example, as has happened in Turkmenistan); freeing prisoners; or staging public spectacles that, together with propaganda, are designed to make people follow the leader. Dictators do not really get people to want to behave "correctly" (what the dictator says is correct); the people merely become afraid not to.

The Roles of Leaders to Attain a Breakthrough in Leadership

Strategic Planning and Deployment: Moving from Good to Great

The first step in strategic planning is to determine the organization's mission. (What business are we in? What services do we provide?) Next, a vision for the desired future state of the organization is formulated and published (e.g., "We will become the supplier of choice, worldwide, of product X or service Y."). After proclaiming the basic reason for the organization's existence, and the overall general goal the organization seeks to achieve in the future, senior management generates a few key strategies the organization is to implement to fulfill the mission and realize the vision (e.g., ensure a reliable source of high-quality raw materials, ensure a stable well-qualified workforce at all levels for the foreseeable future, and/or reduce our overall costs of poor quality by 50 percent of last year's annual cost by the end of the year). Now the process becomes more precise. For each key strategy, a small number of quantified strategic goals (targets) are listed, a few that can be accomplished by the resources and people available. These quantified strategic goals are further divided into goals for this year, goals for the next two years, and so on. Finally, for each quantified strategic goal, a practical number of operational goals are established that describe exactly who is to do exactly what to reach each specific strategic goal. Normally, operational goals are specific projects to be accomplished (such as Six Sigma projects), specific performance targets to be reached by each function or work group, for example.

Strategic deployment is the process of converting goals into specific precise actions, each action designed to realize a specific goal. Deployment occurs in two phases: one phase is *during* the strategic planning process; the other phase is *after* the strategic plan is completed. During the strategic planning process, after the management team determines its key strategies, the team circulates these strategies to others in the organization: department heads, functional heads, process owners, and the like. They, in turn, may circulate the strategies further out to supervisors, team leaders, and so on. These individuals, in turn, may circulate the strategies to everyone they supervise. Each party is asked to contribute ideas and suggestions concerning what activities could be undertaken to carry out the strategies, what the specific quantified strategic goals should be, and what resources would be required. These responses are conveyed to the senior management team, who use the responses to promulgate more specific strategic and operational goals. This exchange of proposed activities to reach goals may take place several times. Some individuals call these iterations and reiterations "catch ball." With each cycle, various goals are refined, becoming more specific, more practical, and quantified. Finally, a set of precise strategic and operational goals emerges, each with owners. In addition, metrics are devised by which to measure performance toward goals and to provide managers at all levels with a scorecard of progress. Most significantly, these goals have been established with the participation of leaders who will be responsible for carrying them out and be accountable for the results.

Emerging from this event is an organization united in its commitment to reaching the same goals. All functions and levels have been included. This is highly significant because leadership is not considered to be something exercised by one person at the top of an organization. It is ideally performed at any level and in any function, by anyone who influences others. With a well-deployed strategic plan, specific acts of leadership (attempts to influence others) should be relatively consistent from leader to leader, from function to function, and from time to time. Decisions made at different levels or in different functions should not conflict with one another very often. That, at least, is the ideal.

Providing Employee Empowerment and Self-Control

When managers do everything they can to provide the means for everyone to be empowered or attain a state of self-control, this also will greatly enhance their credibility and the level of trust followers will feel toward them. This will happen because when an individual is in self-control, that individual has at his or her disposal all of the elements necessary to be successful on his or her job. When a leader does this, followers will feel gratitude and respect toward that leader and will be inclined to follow that leader because "My leader comes through for me. My leader doesn't just talk; my leader delivers!"

A brief review of these elements follows because they can be so instrumental in demonstrating leadership. A person is in a state of self-control if that person

- *Knows exactly what is expected:* the standard of performance for the process; who does what and who decides what and to know how he or she is doing compared to the standards
- Receives timely feedback to have the ability to *regulate the process*
- Has a capable process that includes the necessary tools, equipment, materials, maintenance, time, and the authority to *adjust the process when it is nonconforming*

A person in a state of self-control has, at his or her disposal, all the means necessary to perform work tasks successfully. Management must provide the means because only management controls the required resources needed to put someone in self-control. Persons who have long been suffering from lack of self-control and its associated inability, through no fault of their own, to perform as well as they would like, are especially grateful to a leader/manager who relieves them from the suffering by making self-control possible. These persons come to respect and trust such a manager, and they tend to become an enthusiastic follower of that manager, mindful of the good things—including enhanced self-confidence and self-esteem—that have flowed from that manager.

Performing Periodic Audits

Conducting periodic audits performed by leaders and managers is a superb method of demonstrating commitment to and support for an effort to change. Leaders and managers, especially senior executive managers, enhance their credibility and power to lead by personally walking around the organization and talking to the people about what they do, and how they do it. A management audit has both formal and informal aspects. The formal aspect consists of asking each person being audited to answer certain specific written questions and to produce data and other evidence of performance that conforms to the formal controls. The informal aspect is simply talking with the people who are being audited about what is on their mind and sharing with them what is on the manager's mind. The management audit is roughly the equivalent of senior generals visiting the troops in the field. It is a chance for managers to demonstrate their interest in how things are going: what is going well and/or what needs corrective action. It is a splendid opportunity to listen to what people have to say and to show respect for them. If managers follow up on the suggestions and complaints they hear, that is yet another way to demonstrate that they care enough about "the troops" to provide them with needed support and assistance. It grants to anyone in the organization a direct line of communication with the top, something that makes many people feel important, and motivates them to keep performing at their best. Importantly, the managers' ability to lead is reinforced.

Conferring Public Rewards and Recognition

Leaders can assist their followers to take desired new norms and patterns of behavior upon themselves as their own, if doing so is rewarding to the followers consistently and over time. The effect of rewards and recognition can be magnified when

- Rewards and recognition are awarded in public, with fanfare and ceremony.
- Leaders are in the presence of individuals whose behavior the leader is seeking to influence.
- The award is accompanied by an explanation of its connection to a specific desired new behavior toward which the leader is attempting to extract from the followers. For example, after launching a Six Sigma initiative, your organization decides to

have an all-organization special assembly to recognize the seven original Six Sigma project teams. Each team makes a presentation of its just-completed project, complete with slides, handouts, and exhibits.

Carrying Out the Nondelegable Managerial Practices Stated Previously in the Handbook

- Creating and serving on an executive council to lead and coordinate the performance breakthrough activities
- Forming policy to allow time to participate in breakthrough teams
- Establishing organizational infrastructure
- Providing resources (especially time)
- Reviewing the progress of performance toward goals, including the progress of projects
- Removing obstacles, dissolving resistance, and providing support and other corrective action if progress is too slow

Creating and serving on an executive council to lead and coordinate breakthrough activities. An executive council probably exists already in your organization.

Provide Resources to Continually Innovate and Improve

Breakthroughs are produced by teams, project by project. These teams are assigned goals to attain by using project charters. Each project is formally chartered, in writing, by the Executive Council. The Executive Council also provides the project teams with the people and other resources the teams need to carry out their missions.

The manager's role is one of managing the organization so that high standards are met, proper behavior is rewarded—or individuals are held accountable—facilities and processes are maintained, and employees are motivated and supported. Performance toward goals is measured and tracked for all functions and levels (i.e., overall organization, function, division, department, work group, and individual). Performance metrics are regularly summarized, reported, and reviewed to compare actual performance with goals. Management routinely initiates corrective action to address poor performance or excessively slow progress toward goals. Actions may include establishing performance breakthrough improvement projects, providing additional training or support, clearing away resistance, providing needed resources, and performing disciplinary action. Leaders and management must do the following:

- Create and maintain systems and procedures that ensure the best, most efficient, and effective performance of an organization in all functions and levels.
- Reward (and hold people accountable, if necessary) appropriate behavior.
- Consistently uphold and demonstrate high standards.
- Focus on stability.

Breakthroughs in Organizational Structure

Creating a breakthrough in organizational structure does the following:

- Designs and puts into place the organization's operational systems (i.e., quality system, orientation of new employees, training, communication processes, and supply chains)

- Designs and puts into practice a formal structure that integrates each function with all the others and sets forth relative authority levels and reporting lines (e.g., organization charts and the means to manage across it)
- Aligns and coordinates the respective interdependent individual functions into a smooth functioning, integrated organization

Creating a breakthrough in organization structure is a response to the basic question: "How do I set up organizational structures and processes to reap the most effective and efficient performance toward our goals?"

Trends in this area are clear. More and more work is performed by project teams. Job tasks may be described by *team project* descriptions rather than, or in addition to, *individual* job descriptions. Performance evaluation is often related to the accomplishments of one's team instead of or in addition to, one's individual accomplishments.

Management structure consists of cross-functional *processes* that are managed by process owners, as well as vertical *functions* that are managed by functional managers. Where both vertical and horizontal responsibility exists, potential conflicts are resolved by matrix mechanisms that require negotiated agreements by the function manager and the cross-functional (horizontal) process owner.

Unity and consistency in the operation of *both* cross-functional processes and vertical functions is essential to creating performance breakthroughs and is essential to continued organizational survival. All members of leadership teams at all levels simply must be in basic agreement as to goals, methods, priorities, and styles. This is especially vital when attempting performance breakthrough improvement projects because the causes of so many performance problems are cross functional, and the remedies to these problems must be designed and carried out cross functionally. Consequently, one sees in a Lean or Six Sigma implementation, for example, quality or executive councils, steering committees, champions (who periodically meet as a group), cross-functional project teams, project team leaders, Black Belts, and Master Black Belts. These roles all involve dealing with change and teamwork issues. There is also a steady trend toward fewer authority or administrative levels and shorter reporting lines.

> The rate of change in the business world is not going to slow down anytime soon. If anything, competition in most industries will probably speed up over the next few decades. Enterprises everywhere will be presented with even more terrible hazards and wonderful opportunities, driven by the globalization of the economy along with related technological and social trends (John P. Kotter 1996).

There are three accepted basic types of organization for managing any function work and one newer, emerging approach. The most traditional and accepted organization types are functional, process, and matrix. They are important design baselines because these organizational structures have been tested and studied extensively and their advantages and disadvantages are well known. The newer, emerging organizational designs are network organizations.

Function-Based Organization

In a function-based organization, departments are established based on specialized expertise. Responsibility and accountability for process and results are usually distributed piecemeal among departments. Many firms are organized around functional departments that have a well-defined management hierarchy. This applies both to the major functions (e.g., human resources, finance, operations, marketing, and product development) and also to sections within a functional department. Organizing by function has certain advantages—clear responsibilities and efficiency of activities within a function. A function-based organization typically develops and nurtures talent and fosters expertise and excellence within the functions.

Therefore, a function-based organization offers several long-terms benefits. However, this organizational form also creates "walls" between the departments. These walls—sometimes visible, sometimes invisible—often cause serious communications barriers. However, function-based organizations can result in a slow, bureaucratic decision-making apparatus as well as the creation of functional business plans and objectives that may be inconsistent with overall strategic business unit plans and objectives. The outcome can be efficient operations *within* each department but with less-than-optimal results delivered to external (and internal) customers.

Business Process–Managed Organizations

Many organizations are beginning to experiment with an alternative to the function-based organization in response to today's "make it happen fast" world. Businesses are constantly redrawing their lines, work groups, departments, and divisions, even entire companies, trying to increase productivity, reduce cycle-time, enhance revenue, or increase customer satisfaction. Increasingly, organizations are being rotated 90 degrees into processed-based organizations.

In a process organization, reporting responsibilities are associated with a process, and accountability is assigned to a process owner. In a process-based organization, each process is provided with the functionally specialized resources necessary.

This eliminates barriers associated with the traditional function-based organization, making it easier to create cross-functional teams to manage the process on an ongoing basis.

Process-based organizations are usually accountable to the business unit or units that receive the benefits of the process under consideration. Therefore, process-based organizations are usually associated with responsiveness, efficiency, and customer focus.

However, over time, pure process-based organizations run the risk of diluting and diminishing the skill level within the various functions. Furthermore, a lack of process standardization can evolve, which can result in inefficiencies and organizational redundancies. Additionally, such organizations frequently require a matrix-reporting structure, which can result in confusion if the various business units have conflicting objectives. The matrix structure is a hybrid combination of functional and divisional archetypes.

Merging Functional Excellence with Process Management

What is required, however, is an organization that identifies and captures the benefits of supply chain optimization in a responsive, customer-focused manner while promoting and nurturing the expertise required to manage and continuously improve the processes on an ongoing basis.

This organization will likely be a hybrid of functional and process-based organizations, with the business unit accountable for objectives, priorities, and results, and the functional department accountable for process management and improvement and resource development.

According to the late Dr. Frank Gryna, the Center for Quality at the University of Tampa, Florida, the organization of the future will be influenced by the interaction of two systems that are present in all organizations: the technical system (equipment, procedures) and the social system (people, roles)—thus the name "sociotechnical systems" (STSs).

Much of the research on sociotechnical systems has concentrated on designing new ways of organizing work, particularly at the workforce level. For example, supervisors are emerging as "coaches"; they teach and empower rather than assign and direct. Operators are becoming "technicians"; they perform a multiskilled job with broad decision-making, rather than a narrow job with limited decision-making. Team concepts play an important role in these new approaches. Some organizations now report that, within a given year, 40 percent of their people participate on a team; some organizations have a goal of 80 percent. Permanent teams (e.g., process team, self-managing team) are responsible for all output parameters, including quality; ad hoc teams (e.g., a quality project team) are typically responsible for improving quality. The literature on organizational forms in operations and other functions is extensive and increases continuously. For a discussion of research conducted on teams, see Katzenbach and Smith (1993). Mann (1994) explains how managers in process-oriented operations need to develop skills as coaches, developers, and "boundary managers." The attributes associated with division managers, functional managers, process managers, and customer service network managers are summarized in Table 9.1. There is emerging evidence that divisional and functional organizations may not have the flexibility to adapt to a rapidly changing marketplace or to technological changes.

Design a system that promotes employee empowerment and involvement. Traditional management was based on Frederick Taylor's teachings of specialization. At the turn of the twentieth century, Taylor recommended that the best way to manage manufacturing organizations was to standardize the activity of general workers into simple, repetitive tasks and then closely supervise them (Taylor 1947). Workers were "doers"; managers were "planners." In the first half of the twentieth century, this specialized system resulted in large productivity increases and a very productive economy. As the century wore on, workers became more educated, and machinery and instruments more numerous and complicated. Many organizations realized the need for more interaction among employees. The training and experience of the workforce was not being used. Experience in team systems, where employees worked together, began in the latter half of the twentieth century, although team systems did not seriously catch on until the mid-1970s as pressure mounted on many organizations to improve performance. Self-directed teams began to emerge in the mid-1980s. For maximum effectiveness, the work design should require a high level of employee involvement.

Attributes of Roles	Division Manager	Function Manager	Process Manager	Network Leader
Strategic orientation	Entrepreneurial	Professional	Cross-functional	Dynamic
Focus objectives	Customer adaptability	Internal efficiency	Customer effectiveness	Variable adaptability, speed
Operational responsibility	Cross-functional	Narrow, parochial	Broad, pan-organizational	Flexible
Authority	Less than responsibility	Equal to responsibility	Equal to responsibility	Ad hoc, based on leadership
Interdependence	May be high	Usually high	High	Very high
Personal style	Initiator	Reactor	Active	Proactive
Ambiguity of task	Moderate	Low	Variable	Can be high

(*Sources*: The first two columns are adapted from the work of Financial Executive Research Foundation, Morristown, NJ. The last two columns represent the work of Edward Fuchs.)

Table 9.1 Attributes of Various Roles

Empowerment and Commitment

Workers who have been working under a directive command management system where the boss gives orders and the worker carries them out cannot be expected to adapt instantly to a highly participative, high-performance work system. There are too many new skills to learn and too many old habits to overcome. According to reports from numerous organizations that have used high-performance work systems, such systems must evolve. This evolution is carefully managed, step by step, to prepare team members for the many new skills and behaviors required of them.

The first stage of involvement is the consultative environment, in which the manager consults the people involved, asks their opinions, discusses their opinions, then takes unilateral action. A more advanced state of involvement to appoint a special team or project team to work on a specific problem, such as improving the cleaning cycle on a reactor. This involvement often produces in team members' pride, commitment, and sense of ownership.

An example of special quality teams is the "blitz team" from St. Joseph's Hospital in Paterson, NJ. Teams had been working for about a year as a part of the total quality management (TQM) effort there. Teams were all making substantial progress, but senior management was impatient because the TQM was moving too slowly. Recognizing the need for the organization to produce quick results in the fast-paced marketplace, the team developed the blitz team method (from the German word for lightning). The blitz team approach accelerated the standard team problem-solving approach by adding the services of a dedicated facilitator. The facilitator reduced elapsed time in three areas: problem-solving focus, data processing, and group dynamics.

Because the facilitator was very experienced in the problem-solving process, the team asked the facilitator to use that experience to provide more guidance and direction than is normally the style on such teams. The result was that the team was more focused on results and took fewer detours than usual. In the interest of speed, the facilitator took responsibility for the processing of data between meetings, thus reducing the time that elapsed between team meetings. Furthermore, the facilitator managed the team dynamics more skillfully than might be expected of an amateur in training within the organization. The team went from first meeting to documented root causes in one week. Some remedies were designed and implemented within the next few weeks.

The team achieved the hospital's project objectives by reducing throughput delays for emergency room (ER) patients. ER patients are treated more quickly, and worker frustrations have been reduced (Niedz 1995). Special teams can focus sharply on specific problems. The team's success depends on assigning team people who are capable of implementing solutions quickly.

Project Teams

Employees need time to organize work that should be accomplished by the team. Time is necessary to organize and make sure the team members know what they are doing, why they are doing it, how to organize the work, and who will be involved. However, schedules are so short; there is never time to organize the work team.

Many teams start working, believing they know what to do, and taking direct action. They do not have time to get support from others, or to determine the right goals for the team, or to create and implement a plan that allows them to achieve the proposed goals or to plan how to work together. Nevertheless, what these teams all have in common is that no one understands in the same way what they are doing, why, how, and with whom. Linking the effort of the team to five critical success factors can solve this problem.

Leadership Style

Empowered team members share leadership responsibilities, sometimes willingly and sometimes reluctantly. Decision-making is more collaborative, with consensus as the objective. Teams work toward win–win agreements. Teamwork is encouraged. Emphasis is more on problem solution and prevention, rather than on blame. During a visit to Procter & Gamble's plant in Foley, FL, the host employee commented that in the past he would not have believed he would ever be capable of conducting this tour. His new leadership roles had given him confidence to relate to customers and other outsiders.

Citizenship

Honesty, fairness, trust, and respect for others are more readily evident. In mature teams, members are concerned about each other's growth in the job (i.e., members reaching their full potential). Members share their experiences more willingly and coach each other, as their goal is focused on the team success, rather than on their personal success. Members recognize and encourage each other's (and the team's) successes more readily.

Reasons for High Commitment

As previously stated, empowered team members have the authority, capability, desire, and understanding of the organization's goals. In many organizations, they believe that this makes members feel and behave as if they were owners and makes them more willing to accept greater responsibility. Empowered team members also have greater knowledge, which further enhances their motivation and willingness to accept responsibility.

Means of Achieving High Performance

It has been observed that as employees accept more responsibility and have more motivations, and greater knowledge, they freely participate more toward the interests of the business. They begin to truly act like owners, displaying greater discretionary effort and initiative. Empowered team members have the authority, the capability, and the desire and understand the organization's direction. Consequently, members feel and behave as if they were owners and are willing to accept greater responsibility. They also have greater knowledge, which further enhances their motivation and willingness to accept responsibility.

Enough progress has been made with various empowered organizations that we can now observe some key features of successful efforts. These have come from experiences of various consultants, visits by the authors to other companies, and published books and articles. These key features can help us learn how to design new organizations or redesign old ones to be more effective. The emphasis is on key features, rather than a prescription of how each organization is to operate in detail. This list is not exhaustive, but it is a helpful checklist, useful for a variety of organizations.

Focus on External Customers

The focus is on the external customers, their needs, and the products or services that satisfy those needs.

- The organization has the structure and job designs in place to reduce variation in process and product.
- There are few organizational layers.
- There is a focus on the business and customers.
- Boundaries are set to reduce variances at the source.
- Networks are strong.

- Communications are free flowing and unobstructed.
- Employees understand who the critical customers are, what their needs are, and how to meet customer needs with their own actions. Thus, all actions are based on satisfying the customer. The employees (e.g., operator, technicians, and plant manager) understand that they work for the customer rather than for the plant manager.
- Supplier and customer input are used to manage the business.

In empowered organizations, managers create an environment to make people great, rather than control them. Successful managers are said to "champion" employees and make them feel good about their jobs, their organization, and themselves. When he was head of the Nissan plant in Smyrna, TN, Marvin Runyon stressed that "management's job is to provide an environment in which people can do their work" (Bernstein 1988).

Organization and Knowledge Management

Broken down into its simplest form, the learning process consists of observation–assessment–design–implementation, which can vary along two main dimensions:

- *Conceptual learning*. The process of acquiring a better understanding of cause and effect relationship, leading to "know-why."
- *Operational learning*. The process of obtaining validation of action outcome links, leading to "know-how."

Professor M. Lapré, Assistant Professor of Operations Management at Owen Graduate School of Management at Vanderbilt University, Nashville, TN, and L. Van Wassenhove, the Henry Ford Chaired Professor of Manufacturing at INSEAD (Institut Européen d'Administration des Affaires), a multicampus international graduate business school and research institution, show that it is possible to accelerate factories' learning curves through focused quality and productivity improvement efforts.

Breakthroughs in Current Performance

Breakthroughs in current performance (or improvement) do the following:

- Significantly improve current levels of results that an organization is currently attaining. This happens when a systematic project-by-project improvement system of discovering root causes of current chronic problems and implements solutions to eliminate them.
- Devise changes to the "guilty" processes and reduce the costs of poorly performing processes.
- Install new systems and controls to prevent the return of these root causes.

A system to attain breakthroughs in current performance addresses the question "How do we reduce or eliminate things that are wrong with our products or processes, and the associated customer dissatisfaction and high costs (waste) that consumes the bottom line?" A breakthrough improvement program addresses *quality* problems—failures to meet specific important needs of specific customers, internal and external. (Other types of problems are addressed by other types of breakthroughs.) Lean, Six Sigma, Lean Six Sigma, Root Cause Corrective Action, and other programs need to be part of a systematic approach to improve current performance. These methods address a few specific types of things that always go wrong:

- Excessive number of defects
- Undue number of delays
- Unnecessary long cycle times
- Unwarranted costs of the resulting rework, scrap, late deliveries, dissatisfied customers, replacement of returned goods, loss of customers, and loss of goodwill

Lean and Six Sigma teams are all methods to improve performance. They are all project based and require multifunctional teams to improve current levels of performance. Each requires a systematic approach to complete the projects.

A systematic approach to improving performance of processes is to

- Define the problem (performed by the champions and executive council)
- Measure (performed by the project team)
- Analyze (performed by the project team)
- Improve (performed by the project team, often with help of others)
- Control (performed by the project team and the operating forces is)

Breakthroughs in current levels of performance problems are attained using these methods. The Lean and Six Sigma method will place your ailing processes under a microscope of unprecedented precision and clarity and make it possible to understand and control the relationships between input variables and desired output variables.

Your organization does have a choice as to what "system" to bring to bear on your problems: a "conventional" weapon system (quality improvement) or a "nuclear" system (Six Sigma). The conventional system is perfectly effective with many problems and much cheaper than the more elaborate and demanding nuclear system. The return on investment is considerable from both approaches, but especially so from Six Sigma if your customers are demanding maximum quality levels.

Breakthroughs in current performance solve problems such as excessive number of defects, excessive delays, excessively long time cycles, and excessive costs.

Breakthroughs in Culture

The result of completing many improvements creates a habit of improvement in the organization. Each improvement starts to create a quality culture because collectively it does the following:

- Creates a set of new behavior standards and social norms that best supports organizational goals and climate.
- Instills in all functions and levels the values and beliefs that guide organizational behavior and decision-making.
- Determines organizational cultural patterns such as style (e.g., informal versus formal, flexible versus rigid, congenial versus hostile, entrepreneurial/risk-taking versus passive/risk adverse, rewarding positive feedback versus punishing negative feedback), extent of internal versus external collaboration, and high energy/morale versus low energy/morale. Performance breakthrough in culture is a response to the basic question: "How do I create a social climate that encourages organization members to align together eagerly toward the organization's performance goals?"

As employees continue to see their leadership "sticking to it" culture change happens. An organization is not yet at a sustainable level yet or transformational change. There are still issues that must be addressed, including

- Reviewing the organization's vision, mission and values
- Orienting new employees and training practices
- Rewarding and recognizing policies and practices
- Human resource policies and administration
- Quality and customer satisfaction policies
- Fanatic commitment to customers and their satisfaction
- Commitment to continuous improvement
- Standards and conduct codes, including ethics
- No "sacred cows" regarding people, practices, and core business content
- Community benefit and public relations

An organization's culture exerts an extraordinarily powerful impact on organizational performance. The culture determines what is right or wrong, what is legitimate or illegitimate, and what is acceptable or unacceptable. Consequently, a breakthrough in the culture is profoundly influential in achieving a performance breakthrough. It is also probably the most difficult and time-consuming breakthrough to make happen. It is also so widely misunderstood that attempts to pull it off often fail.

A breakthrough in culture (1) creates a set of behavior standards, and a social climate that supports organizational goals, (2) instills in all functions and levels the values and beliefs that guide organizational behavior and decision-making, and (3) determines organizational cultural patterns such as *style* (informal versus formal, flexible versus rigid, authoritarian top–down versus participative collaboration, management driven versus leadership driven, and the like), the organization's *caste system* (the relative status of each function), and the *reward structure* (who is rewarded for doing what).

Culture Defined

Your organization is a society. A society is "an enduring and cooperating social group whose members have developed organized patterns of relationships through interaction with each other" . . . a group of people engaged in a common purpose," according to Webster's. A society consists of habits and beliefs ingrained over long periods of time. Your workplace is a society, and, as such, it is held together by the shared *beliefs* and *values* that are deeply embedded in the personalities of the society's members. (A workplace whose workforce is segmented into individuals or groups who embody conflicting beliefs and values does not hold together. Various social explosions will eventually occur, including resistances, revolts, mutinies, strikes, resignations, transfers, firings, divestitures, and bankruptcies.)

Society members are rewarded for conforming to their society's beliefs and values—its norms—and they are punished for departing from them. Not only do norms encompass values and beliefs, they also include enduring systems of relationships, status, customs, rituals, and practices.

Societal norms are so strong and deeply embedded that they lead to customary patterns of social behavior sometimes called "cultural patterns." In the workplace, one can identify performance-determining cultural patterns such as participative versus authoritarian management styles, casual versus formal dress, conversational styles ("Mr./Ms." and "Sir/ Madam" versus first names), and a high trust level that makes it safe to say what you really think versus low trust level/suspiciousness that restricts honest or complete communication and breeds game playing, deceit, and confusion.

What Does Culture Have to Do with Managing an Organization?

To achieve a performance breakthrough, it is desirable—if not necessary—that the organization's norms and cultural patterns support the organization's performance goals. Without this support, performance goals may well be diluted, resisted, indifferently pursued, or simply ignored. For these reasons, the characteristics of your organization's culture are a vital matter that your management needs to understand and be prepared to influence. As we shall see, this is easier said than done; but it *can* be done.

A timely example of the influence of culture on an organization's performance is provided by J. M. Juran. Here are excerpts from his description of a management challenge currently facing managers as it has for many years: getting acceptance on the shop floor for statistical control charts, typically a key element in the Control Phase of Six Sigma. (Control charts detect the pattern of variation exhibited by a repetitive process. They can provide a great deal of information about the performance of a process—information unobtainable from any other source. Control charts are widely used in manufacturing and in all kinds of repetitive transactional processes such as those found in hospitals and offices. Among other things, control charts inform the employee if and when to adjust the process, a feature that largely replaces the traditional practice of the employee making this decision. On top of that, control charts are based on the laws of probability and statistics, topics that are widely misunderstood or regarded as impenetrable mysteries.)

> There has been great difficulty in getting production operators and supervisors to accept control charts as a shop tool. I believe this to be a statement of fact, based on extensive firsthand observation of the shockingly high mortality rate of control charts when actually introduced on the shop floor. This difficulty is not merely a current phenomenon. We encountered it back in the late 1920s in the pioneering effort to use control charts on the production floor of the Hawthorne Works of the Western Electric Organization. Neither is it merely an American phenomenon, since I have witnessed the same difficulty in Western Europe and in Japan as well. . . . It is my belief that the failure of the control chart to secure wide acceptance on the factory floor is due mainly to lack of adaptation into the culture of the factory, rather than to technical weaknesses in the control chart. . . . There are a number of problems created by the control chart, as viewed by the shop supervisor:
>
> The control chart lacks "legitimacy" (i.e., it is issued by a department not recognized as having industrial legislative powers).
>
> The control chart conflicts with the specification, leaving the operator to resolve the conflict.
>
> The control chart is in conflict with other forms of data collection and presentation, leaving the operator to resolve the conflict.
>
> The control chart calls for a pattern of operator action that differs from past practice, but without solving the new problems created as a result of disturbing this past practice.

Legitimacy of the Metrics and the Control Chart

The human passion for "law and order" does not stop at the organization's gate. Within the plant, there is the same human need for a predictable life, free from unpleasant surprises. Applied to the workforce, this concept of law and order resolves into various principles:

- There must be one and only one personal supervisor (boss) to whom an employee is responsible.
- There is no limit to the number of impersonal bosses (manuals, drawings, routines), but each boss must be legitimate; that is, it must have clear official status.

- When there is a conflict between the orders of the personal boss and an impersonal boss, the former prevails.
- When there is a conflict between something "legitimate" and something not established as legitimate, the former prevails.

Dr. Juran stated, "There can be no quarrel with these principles, since they are vital to law and order on the factory floor. . . . ". . . Introduction of control charts to the factory floor results in a series of changes in the cultural pattern of the shop:

- A new source of industrial law is opened up, without clear evidence of its legitimacy.
- This new industrial law conflicts with long-standing laws for which there has been no clear repeal through recognized channels of law.
- New sources of factual information are introduced without clear disposition of old sources.
- New duties are created without clear knowledge of their effect on employees who are to perform those duties.

Conclusions

The introduction of modern techniques has an impact on the factory in two aspects:

1. The technical aspect, involving changes in processes, instrument records, and other technical features of the operation
2. The social aspect, involving changes in humans, status, habits, relationships, scale of values, language, and other features of the cultural pattern of the shop

The main resistance to change is due to the disturbance of the cultural pattern of the shop.

—J.M. Juran

How Are Norms Acquired?

New members of a society—a baby born into a family or a new employee hired into the workplace—are carefully taught who is who and what is what. In short, these new members are taught the norms and the cultural patterns of that particular society. In time, they discover that complying with the norms and cultural patterns can be satisfying and rewarding. Resistance or violation of the norms and cultural patterns can be very dissatisfying because it brings on disapproval, condemnation, and possibly punishment. If an individual receives a relatively consistent pattern of rewards and punishments over time, the beliefs and the behaviors being rewarded gradually become a part of that individual's personal set of norms, values, and beliefs. Behaviors that are consistently disapproved or punished will gradually be discarded and not repeated. The individual will have become socialized.

How Are Norms Changed?

Note that socialization can take several years to take hold. This is an important prerequisite for successfully changing an organization's culture that must be understood and anticipated by agents of change, such as senior management. The old patterns must be extinguished and replaced by new ones. This takes time and consistent, persistent effort. These are the realities. Consider what the anthropologist Margaret Mead has to say about learning new behaviors and beliefs:

> An effective way to encourage the learning of new behaviors and attitudes is by consistent prompt attachment of some form of satisfaction to them. This may take the form of consistent praise, approval, privilege, improved social status, strengthened integration with one's group, or material reward. It is particularly important when the desired change is such that the advantages are slow to materialize—for example, it takes months or even years to appreciate a change in nutrition, or to register the effect of a new way of planting seedlings in the increased yield of an orchard. Here the gap between the new behavior and results, which will not reinforce the behavior until they are fully appreciated, has to be filled in other ways.

She continues:

> The learning of new behaviors and attitudes can be achieved by the learner's living through a long series of situations in which the new behavior is made highly satisfying—without exception if possible—and the old not satisfying.
>
> New information psychologically available to an individual, but contrary to his customary behavior, beliefs, and attitudes, may not even be perceived. Even if he is actually forced to recognize its existence, it may be rationalized away, or almost immediately forgotten."
>
> . . . as an individual's behavior, beliefs, and attitudes are shared with members of his cultural group, it may be necessary to effect a change in the goals or systems of behavior of the whole group before any given individual's behavior will change in some particular respect. This is particularly likely to be so if the need of the individual for group acceptance is very great—either because of his own psychological make-up or because of his position in society.

Implications for achieving breakthroughs in culture are as follows:

To be most effective, the entire management team at all levels must share, exhibit, and reinforce desired new cultural norms and patterns of behavior—and the norms must be consistent, uninterrupted, and persistent.

Do not expect cultural norms or behavior to change simply because you publish the organization's stated values in official printed material or describe them in speeches or exhortations. Actual cultural norms and patterns may bear no resemblance at all to the values described to the public or proclaimed in exhortations. The same is true of the actual flow of influence compared to the flow shown on the organization chart. (New employees rapidly learn who is really who and what is really what, in contrast to and in spite of the official publicity.)

A forceful leader-manager can, by virtue of his or her personality and commitment, influence the behavior of individual followers in the *short term* with rewards, recognition, and selective exclusion from rewards. The authors know of organizations who, in introducing a Six Sigma or similar effort, have presented messages to their employees along the following lines:

> The organization cannot tell you what to believe, and we are not asking you to believe in our new Six Sigma initiative, although we hope you do. We can, however, expect you to behave in certain ways with respect to it. Therefore, let it be known that you are expected to support it, or at least get out of its way, and not resist. Henceforth, rewards and promotions will go to those who energetically support and participate in the Six Sigma activities. Those who do not support it and participate in it will not be eligible for raises or promotions. They will be left behind, and perhaps even replaced with others who do support it.

This is fairly strong language. Such companies often achieve some results in the short term. However, should a forceful leader depart without causing the new initiative to become embedded in the organization's cultural norms and patterns (to the extent that individual members have taken on these new values and practices as their own), it is not unusual for the new thrust to die out for lack of consistent and persistent reinforcement.

Resistance to Change

Curiously, even with such reinforcement, change—even beneficial change—will often be resisted. The would-be agent of change needs to understand the nature of this resistance and how to prevent or overcome it.

The example of the control chart case drew the conclusion that the main resistance to change is due to the disturbance of the cultural pattern of the shop when a change is proposed or attempted. People who are successful—and therefore comfortable—functioning in the current social or technical system do not want to have their comfortable existence disrupted, especially by an "illegitimate" change.

When a technical or social change is introduced into a group, group members immediately worry that their secure status and comfort level under the new system may be very different (worse) than under the current system. Threatened with the frightening possibility of losing the ability to perform well or losing status, the natural impulse is to resist the change. Group members have too much at stake in the current system. The new system will require them not only to let go of the current system willingly but also to embrace the uncertain, unpredictable new way of performing. This is a tall order. It is remarkable how profoundly even a tiny departure from cultural norms will upset society members.

What Does Resistance to Change Look Like?

Some resistance is intense, dramatic, and even violent. Dr. Juran reminds us of some examples: When fourteenth-century European astronomers postulated a sun-centered universe, this idea flew in the face of the prevailing cultural beliefs in an Earth-centered universe. This belief had been passed down for many generations by their ancestors, religious leaders, grandparents, and parents. (Furthermore, on clear days, one could see with one's own eyes the sun moving around the Earth.) Reaction to the new "preposterous" unacceptable idea was swift and violent. If the sun-centered believers are correct, then the Earth-centered believers are incorrect—an unacceptable, illegitimate, wrong-headed notion. To believe in the new idea required rejecting and tossing out the old. But the old was deeply embedded in the culture. So the "blasphemous" astronomers were burned at the stake.

Another example from Dr. Juran: When railroads converted from steam-powered to diesel-powered locomotives in the 1940s, railroad workers in the United States objected. It is unsafe, even immoral, they protested, to trust an entire trainload of people or valuable goods to the lone operator required to drive a diesel. Locomotives had "always" been operated by two people, an engineer who drove, and a fireman who stoked the fire. If one *were* incapacitated, the other could take over. But what if the diesel engineer had a heart attack and died? So intense were the resulting strikes that an agreement was finally hammered out to keep the fireman on the job in the diesels! Of course the railroad workers were really protesting the likely loss of their status and jobs.

Norms Helpful in Achieving a Cultural Transformation

Transforming a culture requires a highly supportive workforce. Certain cultural norms appear to be instrumental in providing the support needed. If these norms are not now part of your culture, some breakthroughs in culture may be required to implant them. Some of the more enabling norms are as follows:

A belief that the quality of a product or process is at least of equal importance, and probably of greater importance than the mere quantity produced. This belief results in decisions favoring quality: defective items do not get passed on down the line or out the door; chronic errors and delays are corrected.

A fanatical commitment to meeting customer needs. Everyone knows who his or her customers are (those who receive the results of their work), and how well he or she is doing at meeting those needs (They *ask*.). Organization members, if necessary, drop everything and go out of their way to assist customers in need.

A fanatical commitment to stretch goals and continuous improvement. There is always an economic opportunity for improving products or processes. Organizations who practice continuous improvement keep up with, or become better than, competitors.

Organizations that do not practice continuous improvement fall behind and become irrelevant or worse—go out of business. Six Sigma product design and process improvement is capable, if executed properly, of producing superb economical designs and nearly defect-free processes to produce them, resulting in very satisfied customers and sharply reduced costs. The sales and the savings that follow show up directly on the organization's bottom line.

A customer-oriented code of conduct and code of ethics. This code is published, taught in new employee orientations, and taken into consideration in performance ratings and in distributing rewards. Everyone is expected at all times to behave and make decisions in accordance with the code. The code is enforced, if needed, by managers at all levels. The code applies to everyone, even board members—perhaps especially to them considering their power to influence everyone else.

A belief that continuous adaptive change is not only good but necessary. To remain alive, organizations must develop a system for discovering social, governmental, international, or technological trends that could impact the organization. In addition, organizations will need to create and to maintain structures and processes that enable a quick, effective response to these newly discovered trends.

Given the difficulty of predicting trends in the fast-moving contemporary world, it becomes vital for organizations to have such processes and structures in place and operating. If you fail to learn and appropriately adapt to what you learn, your organization can be left behind very suddenly and unexpectedly and end up in the scrap heap. The many rusting, abandoned factories the world over testify to the consequences of not keeping up and consequently being left behind.

Policies and Cultural Norms

Policies are guides for managerial action and decision-making. Organization manuals typically begin with a statement of the organization's quality policy. This statement rates the relative worth that organization members should place on producing high-quality products, as distinguished from the mere quantity of products produced. ("High-quality products" are goods, services, or information that meets important customer needs at the lowest optimum cost with few, if any, defects, delays, or errors.) High-quality products produce customer satisfaction, sales revenue, repeat demand or sales, and low costs of poor quality (unnecessary waste). Here, in that one sentence, are reasons for attempting quality improvement. Including a value statement in your organization's quality manual reinforces some of the instrumental cultural norms and patterns essential for achieving a "quality culture" and, ultimately, performance breakthroughs.

Keep in mind that if the value statement, designed to be a guide for decision-making, is ignored and not enforced, it becomes worthless, except perhaps as a means of deceiving customers and employees in the short term. You can be sure, however, that customers and employees will soon catch on to the truth and dismiss the quality policy, waving it away as a sham that diminishes the whole organization and degrades management credibility.

Human Resources and Cultural Patterns

Human resources plays a significant role in reinforcing cultural norms. It does so by several means that include

- *Recruiting*. Advertisements contain descriptions of desirable traits (e.g., dependable, energetic, self-starter, creative, analytic), as well as characterizations of the organization (e.g., service oriented, customer oriented, committed to being a world leader in quality, progressive, world class, and equal opportunity). Organizational values are often featured in these messages.
- *Orientation and training*. It is customary when providing new employees with an introduction to an organization to review with them expected modes of dress, behavior, attitudes, and traditional styles of working together.
- *Publishing employee handbooks*. The handbooks distributed to new employees, and to everyone annually, are replete with descriptions of organizational history, traditional policies and practices, and expectations for organization members. All of these topics express directly or indirectly detailed elements of the official culture.
- *Reward and recognition practices*. In our rapidly changing world, management teams find themselves agonizing over what kind of employee behavior should be rewarded. Whatever the behavior is, and when it is rewarded, the reward reinforces the cultural norms embodied in that behavior, and it should induce more of the same behavior from the ones who are rewarded, as well as attract others to do the same.
- *Career path and promotion practices*. If you track the record of those promoted in an organization, you are likely to find either (1) behavior that conforms to the traditional cultural norms in their background or (2) behavior that resembles desired new cultural norms required for a given organizational change, such as launching a Six Sigma effort. In the former case, management wants to preserve the current culture; in the latter case, management wants to create breakthroughs in culture and bring about a new culture that is at least somewhat altered. In both cases, the issue of the relationship of the person being promoted to the organizational culture is a significant factor in granting the promotion.

Breakthroughs in Adaptability and Sustainability

Creating a breakthrough in adaptability and sustainability requires

- Creating structures and processes that uncover and predict changes or trends in the environment that are potentially promising or threatening to the organization
- Creating processes that evaluate information from the environment and refer it to the appropriate organizational person or function
- Participation in creating an organizational structure that facilitates rapid adaptive action to exploit the promising trends or avoid the threatening disasters
- A response to the question "How do I prepare my organization to respond quickly and effectively to unexpected change?"

The survival of an organization, like all open systems, depends on its ability to detect and react to threats and opportunities that present themselves from within and from outside. To detect potential threats and opportunities, an organization must not only gather data and information about what is happening but also discover the (often) elusive meaning and significance the data hold for the organization. Finally, an organization must take appropriate action to minimize the threats and exploit the opportunities gleaned from the data and information.

To do all this will require appropriate organizational structures, some of which may already exist (an intelligence function, using an adaptive cycle, an Information Quality Council) and a data quality system. The Information Quality Council acts, among other things, as a "voice of the market." Dates are defined as "facts" (such as name, address, and age) or "measurements of some physical reality, expressed in numbers and units of measure that enable our organization to make effective decisions by." These measurements are the raw material of information, which is defined as "answers to questions" or the "meaning revealed by the data, when analyzed." The typical contemporary organization appears to the authors to be awash in data but bereft of useful information. Even when an organization possesses multiple databases, much doubt exists regarding the quality of the data and, therefore, the organization's ability to tell the truth about the question it is supposed to answer.

Managers dispute the reliability of reports, especially if the messages contained in the data are unfavorable. Department heads question the accuracy of financial statements and sales figures, especially when they bring bad tidings.

Often, multiple databases will convey incongruent or contradictory answers to the same question. This is because each individual database has been designed to answer questions couched in a unique dialect or based on the unique definitions of terms used by one particular department or function, but not all functions. Data often are stored (hoarded?) in isolated unpublicized pockets, out of sight of the very people in other functions who could benefit from them if they knew they existed. Anyone who relies on data for making strategic or operational decisions is rendered almost helpless if the data are not available or are untrustworthy. How can a physician decide on a treatment if X-rays and test results are not available? How can the sales team plan promotions when it does not know how its products are selling compared to the competition? What if these same sales people knew that the very database that could answer their particular questions already exists but is used for the exclusive benefit of another part of the organization? It is clear that making breakthroughs in adaptability is difficult if one cannot get necessary data and information or if one cannot trust the truthfulness of the information one does get. Some organizations for which up-to-date and trustworthy data are absolutely critical go to great lengths to get useful information. However, in spite of their considerable efforts, many organizations nevertheless remain plagued by chronic data quality problems.

The Route to Adaptability: The Adaptive Cycle and Its Prerequisites

Creating a breakthrough in adaptability creates structures and processes that do the following:

- Detect changes or trends in the internal or external environment that are potentially threatening or promising to the organization.
- Interpret and evaluate the information.
- Refer the distilled information to empowered functions or persons within the organization who take action to ward off threats and exploit opportunities. This is a continuous perpetual cycle.
- Take action to ward off threats and exploit opportunities. This is a continuous, perpetual cycle.

The cycle might more precisely be conceptualized as a *spiral*, as it goes round and round, never stopping (see Figure 9.3). Several prerequisite actions are needed to set the cycle in

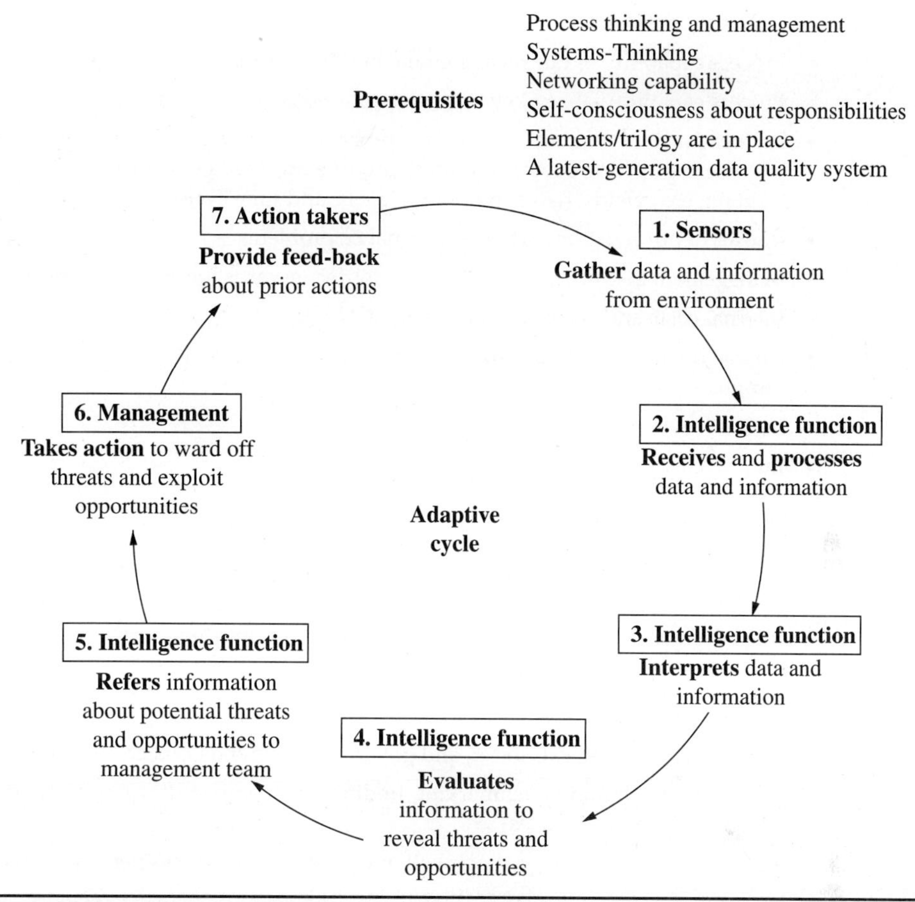

Figure 9.3 Adaptive cycle—to detect and to react to organizational threats and opportunities. (*De Feo and Barnard 2004, p. 291.*)

motion and create breakthroughs in adaptability. Although each prerequisite is essential, and all are sufficient, perhaps the most crucial is the Information Quality Council and the data quality system. Everything else flows from timely trustworthy data—data that purport to describe truthfully the aspects of reality that is vital to your organization.

Prerequisites for the Adaptive Cycle: Breakthroughs

- Leadership and management
- Organization structure
- Current performance
- Culture

A Journey around the Adaptive Cycle

An intelligence function gathers data and information from the internal and external environment. At minimum, we need to know some of the following basic things.

From the Internal Environment

- Process capability of our measurement and data systems
- Process capability of our key repetitive processes
- Performance of our key repetitive processes (human resources, sales, design, engineering, procurement, logistics, production, storage, transportation, finance, training, etc.; yields, defect types and levels, and time cycles)
- Causes of our most important performance problems
- Management instrument panel information: score cards (performance toward goals)
- Internal costs and costs of poor quality (COPQ)
- Characteristics of our organizational culture (how much does it support or subvert our goals)
- Employee needs
- Employee loyalty

From the External Environment

- Customer needs, now and in the future (what our customers or clients and potential customers or clients want from us or our products)
- Ideal designs of our products (goods, services, and information)
- Customer satisfaction levels
- Customer loyalty levels
- Scientific, technological, social, and governmental trends that can affect us
- Market research and benchmarking findings (us compared to our competition; us compared to best practices)
- Field intelligence findings (how well our products or services perform in use)

You may add to this list other information of vital interest to your particular organization. This list may seem long. It may seem expensive to get all this information. (It can be.) You may be tempted to wave it away as excessive or unnecessary. Nevertheless, if your organization is to survive, there appears to be no alternative but to gather this kind of information, and on a regular, periodic basis. Fortunately, as part of routine control and tracking procedures already in place, your organization probably gathers much of this data and information. Gathering the rest of the information is relatively easy to justify, given the consequences of being unaware of, or deaf or blind to, vital information.

Information about internal affairs is gathered from routine production and quality reports, sales figures, accounts receivable and payable reports, monthly financial reports, shipment figures, inventories, and other standard control and tracking practices. In addition, specially designed surveys—written and interviews—can be used to gain insights into such matters as the state of employee attitudes and needs. A number of these survey instruments are available off the shelf in the marketplace. Formal studies to determine the capability of your measurement systems and your repetitive processes are routinely conducted if you are using Six Sigma in your organization. Even if you do not use Six Sigma, such studies are an integral part of any contemporary quality system. Score cards are very widely utilized in organizations that carry out annual strategic planning and deployment. The scores provide management with a dashboard, or instrument panel, which indicates warnings of trouble in specific organizational areas. Final reports of operational projects from quality improvement teams, Six Sigma project teams, and other projects undertaken as part of executing the annual strategic business plan, are excellent sources of "lessons learned" and ideas for future projects. The tools and techniques for conducting COPQ studies on a continuing basis are widely available. The results of COPQ studies become powerful drivers of new breakthrough projects because they identify specific areas in need of improvement. In sum, materials and tools for gathering information about your organization's internal functioning are widely available and easy to use.

Gathering information about conditions in the external environment is somewhat more complex. Some approaches require considerable know-how and great care. Determining customer needs is an example of an activity that sounds simple but actually requires some know-how to accomplish properly. First, it is proactive. Potential and actual customers are personally approached and asked to describe their needs in terms of benefits they want from a product, services, or information. Many interviewees will describe their needs in terms of a problem to be solved or a product feature. Responses like these must be translated to describe the benefits the interviewee wants, not the problem to be solved or the product feature they would like. Tools and techniques for determining ideal designs of current and future products or services are also available. They require considerable training to acquire the skills, but the payoffs are enormous. The list of such approaches includes Quality Planning, Design for Six Sigma (DFSS), TRIZ, a technique developed in Russia for projecting future customer needs and product features. Surveys are typically used to get a feel for customer satisfaction. A "feel" may be as close as you can get to knowledge of customer feelings and perceptions. These glimpses can be useful if they reveal distinct patterns of perceptions whereby large proportions of a sample population respond very favorably or very unfavorably to a given issue. Even so, survey results can hardly be considered "data," although they have their uses if suitable cautions are kept in mind. The limitations of survey research methodology cloud the clarity of results from surveys. (What really is the precise difference between a rating of "2" and a rating of "3"? A respondent could answer the same question different ways at 8:00 A.M. and at 3:00 P.M, for example.) (A satisfaction score increase from one month to another could be meaningless if the group of individuals polled in the second month is not the exact same group that was polled the first month. Even if they were the same individuals, the first objection raised above would still apply to confound the results.)

A more useful approach for gauging customer "satisfaction," or more precisely, their detailed responses to the products or services they get from you, is the customer loyalty study, which is conducted in person with trained interviewers every six months or so on the same people. The results of this study go way beyond the results from a survey. Results are quantified and visualized. Customers and former customers are asked carefully crafted standard questions about your organization's products and performance. Interviewers probe the responses with follow-up questions and clarifying questions. From the responses, a number of revealing pieces of information are obtained and published graphically. Not only do you learn the features of your products or services that cause the respondents happiness and unhappiness but also such things as how much improvement of defect X (late deliveries, for example) it would take for former customers to resume doing business with you. Another example:

You can graphically depict the amount of sales (volume and revenue) that would result from given amounts of specific types of improvements. You can also learn what specific "bad" things you'd better improve, and the financial consequences of doing so or not doing so. Results from customer loyalty studies are powerful drivers of strategic and tactical planning, and breakthrough improvement activity.

Discovering scientific, technological, social, and governmental trends that could affect your organization simply requires plowing through numerous trade publications, journals, news media, websites, and the like, and networking as much as possible. Regular searches can be subcontracted so you receive, say, published weekly summaries of information concerning very specific types of issues of vital concern to you. Although there are numerous choices of sources of information concerning trends, there appears to be little choice of whether to acquire such information. The trick is to sort out the useful from the useless information.

A basic product of any intelligence function is to discover how the sales and performance of our organizations' products, services, and sales compare with our competitors and potential competitors. Market research and field intelligence techniques are standard features in most commercial businesses, and books on those topics proliferate.

Many organizations undertake benchmarking studies to gather information on world-class best practices. They study the inner workings of repetitive processes such as design, warehousing, operating oil wells, and mail order sales—almost anything. The processes studied are not necessarily those of your competitors; they need only be the very best (efficient, effective, and most economical). Benchmarking studies are classic intelligence detective work, and are often conducted on a subcontract basis with organizations that specialize in benchmarking. The results are typically published and shared with all participants. When you have discovered best practices, you can compare your performance with them and describe gaps between theirs and yours, thus identifying breakthrough opportunities.

Completing the adaptive cycle, will enable the organization to attain a breakthrough in adaptability and lead to sustainability. Skipping a breakthrough may not indicate a problem in the short term, only in the long term. Consider the economic crisis that hit the global economy in 2008. There were many global organizations that we considered leaders in their markets—when business was good. During the crisis, so many top performers of the past went out of business, were merged with others, or went into bankruptcy only to emerge a different organization. Why did so many organizations have trouble? Our theory was that although these organizations were good at responding to their customer needs they were not watching societies needs. This led to a lack of information that, if it was available, would have provided enough time to "batten down the hatches," to ride the crisis out. To avoid this from happening, creating a high performing, adaptable organization may lead to better performance when things are not so good.

Sustainability

The second part of this breakthrough is sustainability. Sustainability has two important meanings. The first is to sustain the benefits of the transformational changes that took place. The second is to assure the organization is sustainable from an environmental point of view. At the time of publication, we felt we would only focus on long-term results. As more organizations take on the environmental issues that will plague us in the future, sustainability will focus on both. Chapter 10 elaborates on ecoquality.

Sustainability is the return to evaluate performance annually based on the findings of the Information Council. With this information, leaders can adjust the organization to ensure that it stays ahead of its customers and can sustain itself for the long term.

A Transformation Roadmap

The Juran Transformation Roadmap

There are five phases in the Juran Transformation Roadmap, each one corresponding to the breakthroughs that are described in this chapter. Each phase is independent, but the beginning and end of each phase are not clearly delineated. Each organization reacts differently to changes.

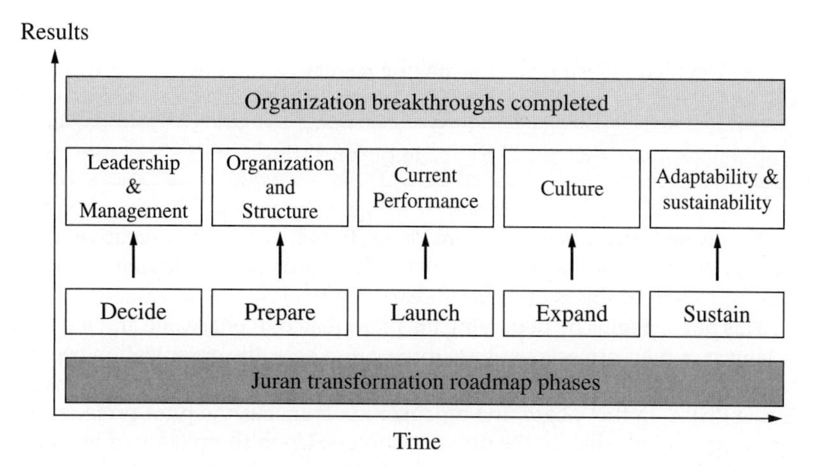

FIGURE 9.4 Juran roadmap and breakthroughs.

This means that one business unit in an organization may remain in one phase longer than another unit. These phases once again are a managerial guide to change, not a prescription.

The five phases of the Transformation Model and Roadmap are shown in the Figure 9.4. The road starts at the Decide Phase. This phase begins when someone on the executive team decides that something must be done or else the organization will not meet shareholder expectations or will not meet its plan and ends with a clear plan for change.

In the Decide Phase, the organization will need to create new information or better information than it may have had about itself. This information can come from a number of reviews or assessments. Our experience shows that the more *new* information an organization has, the better its planning for change. Some of the important areas that should be reviewed are as follows:

- Conduct a Customer Loyalty Assessment to determine what they like or dislike about your products and services.
- Identify the areas of strength and uncover possible problems in the organization's performance.
- Understand employee attitudes toward the proposed changes.
- Understand the key business processes and how the changes will affect them.
- Conduct a cost analysis of poorly performing processes to determine the financial impact of these costs on the bottom line.
- Conduct a world-class quality review of all business units to understand the level of improvement needed in each unit.

A comprehensive review of the organization prior to launch is essential for success. We show a typical review that we recommend to all organizations embarking upon a Six Sigma transformational initiative. From these assessments and reviews, the executive team now has qualitative and quantitative information to define the implementation plan for its organization.

The deployment plan must include the following items:

- Infrastructure that is needed to steer the changes
- Methodology and tools that will be used throughout the implementation

- Goals and objectives of the effort
- Detailed milestones for achieving results

The conclusion of this phase results in the breakthrough in leadership and management

The second phase is the Prepare Phase. In this phase, the executive team begins to prepare for the changes that will take place. It focuses on developing a pilot effort to try the change in a few business units before carrying it out in the organization as a whole.

This phase begins by deploying the plan created in phase one and it ends after a successful launch of pilot projects in phase three. From here, the organization begins to identify the improvement projects that must be carried out to meet the desired goals established in the Decide Phase. In this phase, the organization launches the pilot projects, reviews the projects' progress, and enables the projects' success. Upon completion of the pilot projects, executives evaluate what has worked and what has not. Then executives either abandon their efforts or change the plan and expand it throughout the organization.

The following actions can be taken for your organization:

- Identify the areas of strength and uncover possible problems in the organization's performance from phase one.
- Identify value streams and key business processes that need improvement.
- Select multifunctional pilot or demonstration projects and create project charters.
- Create a training plan and set of learning events to train the teams.
- Communicate the steps taken in this phase to the workforce.

The conclusion of this phase results in a breakthrough in organizational structure.

The third phase is the Launch Phase. In this phase, the executive team begins demonstration projects in a few business units before carrying them out in the total organization. Each project will require a project charter, a team and an effective launch, reviewing the progress and maintaining the gains before results are attained. The length of this phase depends on the number of projects and results expected. For most organizations, this phase completion takes less than one year. As each project is completed, and results are attained, leaders can then evaluate the lessons learned and expand by launching more projects.

The conclusion of this phase results in a breakthrough in current performance

Expansion can take months or years, depending on the size of the organization. An organization of 500 employees will require less time to deploy a plan across the organization than an organization of 50,000. The Expand Phase may take three to five years. Note that positive financial results will occur long before cultural changes take place. Staying in the Expand Phase is not a bad thing. An organization must continue to implement its plan, business unit by business unit, until the organization has had enough time to implement the desired changes. The Final Phase is the Sustain Phase when the organization has a fully integrated operation. All improvement and Six Sigma goals are aligned with the strategy of the organization. Key business processes are defined and well managed, and process owners are assigned to manage them. Employee performance reviews and compensation are in line with the changes required. Those who comply with the change are rewarded. The executives and business unit heads conduct regular reviews and audits of the change process. This may result in a discussion or even a change in the strategy of the organization.

The organization may have learned more about its capabilities and more about its customers that may lead to a change in strategy.

The conclusion of this phase results in a breakthrough in culture.

The Sustain Phase also lasts as long as the organization is meeting its strategic and financial goals. Deviations from expected results, possibly due to macroeconomic events outside the organization, require a review of the scorecard to determine what has changed. When this is determined, the organization makes the changes, continues, and sustains itself at the current level.

The conclusion of this phase results in a breakthrough in adaptability and sustainability.

Lessons Learned in Deploying the Transformation Road Map

As you begin your journey down this road, note the many lessons learned from organizations that have led a change process and failed initially. These failures can be avoided by suitable planning, listed as follows:

- All organizations and their units are at different levels of maturity regarding performance.
- Champions and internal experts (such as Six Sigma Black Belts) become drivers who propel their organization to superior performers or best in class.
- Extensive training in tools and techniques for all employees ensures that learning has taken place and that they can use the tools to improve performance.
- Systematic application and deployment through proven methodologies such as Six Sigma Improvement (DMAIC) and Design (DFSS) are necessary to create a common language and create results in current performance.
- Focusing improvements on the customer first will enable cost reduction, and delighted customers will enable breakthrough bottom-line results.
- Significant increase in customer satisfaction happens only when you improve the processes and services that impact them.
- No organization has ever successfully implemented a plan without the leadership and commitment of the executive team—they are the ones who control the resources and provide the communication that will change the culture.

With this road map and the lessons learned, all organizations should be able to achieve sustainable results well into the future. If more organizations get on board with positive, customer-focused change initiatives, we will be able to create a global society that reduces our dependence on the quality dikes we have built over the years.

As your organization continues to renew itself annually through the strategic planning process, this cycle of improvement should continue. Barring any leadership changes or crisis, your organization should be on its way to attaining superior and sustainable results.

References

Bernstein, P. (1988). The Trust Culture, SAM Advanced Management Journal, pp. 4–8.

De Feo, J. A., and Barnard, W. W. (2004). *Juran Institute's Six Sigma Breakthrough and Beyond: Quality Performance Breakthrough Methods*. McGraw-Hill, New York.

Juran Institute, Inc. (2009). *Quality 101: Basic Concepts and Methods for Attaining and Sustaining High Levels of Performance and Quality*, version 4. (*Source*: Juran Institute, Inc., Southbury, CT.)

Katzenbach, J. R., and Smith, D. K. (1993). Wisdom of Teams: Creating the High Performance Organization, Harvard Business School Press, Boston, MA.

Kotter, J. P. (1996). *Leading Change*. Harvard Business School Press, Cambridge, MA.

Mann, D. W. (1994). "Reengineering the Manager's Role," ASQC Quality Congress Transactions 1994, American Society for Quality, Milwaukee, WI, pp. 155–159.

Neidz, B. A. (1995). "The Blitz Team" IMPRO95 Conference Proceedings, Juran Institute, Inc. Southbury, CT.

Taylor, F. W. (1947). The Principles of Scientific Management, Harper and Row, New York.

CHAPTER 10

A Look Ahead: Eco-Quality for Environmental Sustainability

Joseph R. De Feo, Jr. and Brian A. Stockhoff

About This Chapter

As we move forward into the twenty-first century, managing for quality is breaking new ground. Product developers need to design products and services that meet the newest concern from its customers—the need for sustainable and ecologically friendly products. As a society and as businesses, we need to not merely maintain the status quo but to break through self-imposed constraints and fundamentally shift to a new landscape—a new zone of quality—and to design processes and products for ecological quality from the start. This chapter focuses on what we believe is the next addition to the management of quality. We call it "eco-quality."

High Points of This Chapter

1. *Understanding Climate Change.* Increasing atmospheric carbon dioxide (CO_2) levels have been linked to climate change and a variety of environmental problems. Such phenomena as melting ice caps, freshwater shortages, and species extinctions are

implicated as examples of breaks in the "quality dikes" created by humans in the course of technological advancement.

2. *Societal responsibility.* Once-separate societies have begun to band together ideologically on environmental issues, taking visible, concerted action to shore up the quality dikes. In particular, legislation, agreements, treaties, and accords are being put into place to incentivize and set limits as to what is appropriate and inappropriate activity.
3. *Corporate responsibility.* Industries are beginning to invest in programs and initiatives to address—and reduce—the environmental impact associated with all life cycle stages of products and processes. Organizations have a responsibility to shareholders and the community to prepare for changes, including understanding their carbon profile and having a plan in place to reduce it. Five life cycle stages are identified as part of a cradle-to-grave assessment to arrive at a comprehensive carbon profile.
4. The concept of eco-quality and the methods to attain it are new and being tested. We focus on four tools being tested here.

Quality and Sustainability: An Introduction

Organizations large and small that have gained success in the past and want to thrive in the future are being challenged to find—and capitalize upon—opportunities to meet their own strategic goals while also meeting societal needs. More and more organizations are being encouraged to look at the entire landscape unfolding before them, from a perspective of a balanced array of outcomes characterized by the new "triple bottom line" of people, planet, and profits (Savitz and Weber 2006). As we go forward into the twenty-first century, organizations cannot focus only on profits and their bottom line; they also must take into consideration people and our planet. Quality Management has always taken people into consideration; now as we go forward a third dimension has been added that encompasses environmental sustainability and stewardship.

Quality and environmental sustainability are becoming increasingly interdependent. Organizations of all sizes are looking at ways to increase efficiencies and productivity without compromising the integrity of the environment. As a result, as we see it, there is a paradigm shift evolving within the quality management arena as quality and environmental sustainability are merging toward a partnership. This partnership makes perfect sense; the performance excellence we strive for in a business environment extends to the larger, natural environmental that provides the context in which businesses operate.

This partnership accelerated with the creation of the ISO (International Organization for Standardization) 14000 Environmental Management System, a companion system to ISO 9000 (Chapter 16). With increased environmental awareness, organizations are looking for innovative ways to reach their strategic goals while keeping within societal, environmental constraints. The worldwide issues of global warming and sustainability are on the minds of many millions of people throughout the world, and the widespread adoption of ISO 14000 reflects this. According to the most recent report at the time of printing this handbook, there were over 14,000 sites worldwide certified to ISO14000. Of these, the majority were in the following countries:

- Japan (2600)
- Germany (1600)

- UK (1200)
- Sweden (650)
- Taiwan (500)
- United States (590)
- Netherlands (475)
- Korea (460)
- Switzerland (400)
- France (360)

This partnership between quality management and environmental sustainability will bring positive change for both business and the environment.

A number of years ago Dr. Joseph Juran coined the phrase "life behind the quality dikes." (Juran 1969). These quality dikes, as explained by Dr. Juran, are a way of securing benefits while living dangerously. These benefits are the results of technological advances, and we are kept safe from harmful byproducts of technology by these quality dikes. Dr. Juran went on to say that there are minor breaks in these quality dikes—occasional failures of goods and services. As he stated, these failures are annoying as well as costly. More significant failures also can be cited, such as the Chernobyl and Bhopal disasters. These are extreme examples, but they pale in comparison with the potential impending tsunami of the effects of global warming.

If this approaching tsunami plays out as many believe it will, it will spring more than just a few leaks in the dikes, posing instead a much more significant threat to our environment. This situation has slowly been gaining momentum for several years with much discussion regarding the effects of greenhouse gases (GHG), in particular CO_2, on temperature and climate. Carbon dioxide, which is generated from various sources both man-made and natural, has physical and correlative properties that make it a prime suspect in global temperature fluctuations. Unlike the events at Chernobyl and Bhopal, which originated in point sources with effects confined to a general geographic area, global warming is, by definition, more far reaching—it is worldwide. Although technology-induced catastrophes such as Chernobyl and Bhopal are plainly different in many regards from the global warming issue, they have something very much in common in that technology and people have the capacity to create change, for better and for worse.

Global Warming

Global warming is one of those topics that tend to divide people into two separate schools of thought. On one hand, there are those who believe that global temperature fluctuations reflect normal, common-cause variation, or perhaps represent part of a natural environmental or physical cycle the earth is now experiencing, and has experienced in our distant past. The other school of thought contends that humans are causing the earth's temperature to increase as a result of our technologies, specifically through the increase GHG. These gases are emitted primarily through natural sources and human (technological) activities, and they contribute to the "greenhouse effect"; GHGs include water vapor, methane, nitrous oxide, hydrofluorocarbons, perfluorocarbons, sulfur hexafluoride, and carbon dioxide. Significantly, the U.S. Environmental Protection Agency issued in early 2009 a proposed finding that GHGs contribute to air pollution that may endanger public health or welfare.

Of these major contributors, CO_2 is judged to have the most far reaching and consequential effects on our environment. Atmospheric CO_2 levels are at record highs, and links have been suggested with rising sea levels, water shortages around the world, depletion of fisheries, extinction of species, and numerous other phenomena. For example, polar ice caps have been melting, with the Arctic sea ice minimum dropping 7.5 percent per decade between 1979 and 2006 (NASA 2009). Large areas of the world's regions are expected to suffer a substantial decrease in fresh water by midcentury (U.N. Environment Programme 2009). Atlantic cod stocks have collapsed from a likely combination of overfishing, natural and human environmental impacts, including temperature shifts (Cascorbi and Stevens 2004) and recovery has been minimal. Estimates of species extinction over the next several decades have been placed at over 30 percent, potentially threatening over a million species (Thomas et al. 2004).

As developing economies step onto the global stage, they are expected to contribute to CO_2 emissions at alarming rates. China, considered to be the "world's factory," has about 1.35 billion people and is only 30 percent mobilized and working in factories. China surpassed the United States in carbon dioxide emissions in 2006 (Aufhammer and Carson 2008), and the country's share of global emissions is projected to rise from 18 percent in 2005 to 33 percent in 2030 (Garnaut 2008, Table 3.2). To understand in more detail from where CO_2 is originating, see the pie charts depicting worldwide CO_2 emissions by region and sector (Figure 10.1a, b), and by sector within the United States (Figure 10.2).

Whether the earth is simply experiencing another environmental cycle or humans directly are contributing to an increase in the earth's temperature perhaps is a moot point. If there is something society can be doing to mitigate potentially damaging effects to the environment, then we as citizens of this planet have a responsibility to our children and future generations to take action to preserve and to sustain our environment.

Although formal steps are being taken, many believe that international cooperation is not keeping pace with the world's ever-growing interdependence and threats to the environment.

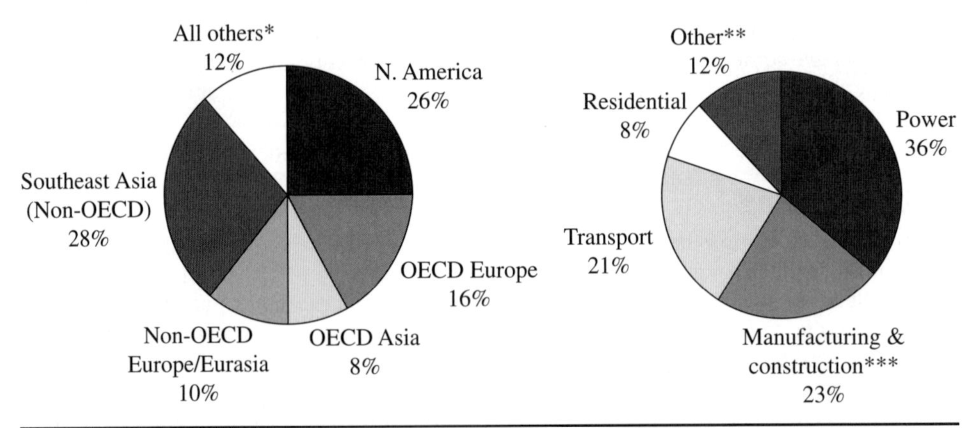

Figure 10.1 Worldwide CO_2 emissions by (a) region and (b) sector. *All others include Africa, the Middle East, and Central and South America. OECD: Organization for Economic Co-Operation and Development. **Other includes commercial and public services. ***Manufacturing and Construction includes other energy industries (e.g., oil refineries, coal mining, oil and gas extraction, and other energy-producing industries. [*Intergovernmental panel on climate change (IPCC, 2005).*]

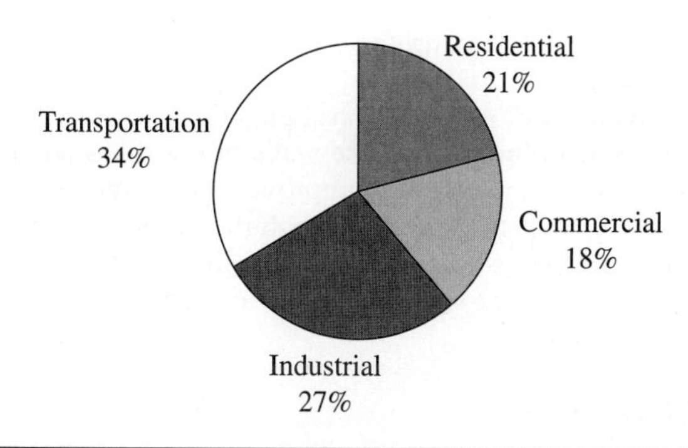

FIGURE 10.2 Energy-related carbon dioxide emissions by end-use sector, 2007. (*Energy Information Association, 2008.*)

We have not inherited the Earth from our fathers. We are borrowing it from our children.

—Native American saying

Societal Responsibility

Social interactions among people in their everyday lives have expanded through technological advances and the capacity to impact others across distances. Historically, "society," in the sense of the general public, was defined functionally by the physical constraints of distance. An extended family was the only "society" of concern to an individual in prehistoric days, with subsequent expansion to tribes, villages, cities, cultures, and civilizations. To be sure, society in the sense of humanity always existed, but it had little practical impact to an individual that neither intermingled with nor directly depended on others that lived more than the next valley away. Societies were small, and so were perceived responsibilities.

Two factors—technology and the sheer number of people—have worked together to facilitate the mingling and the interdependence of people. Technologies as simple as the wheel and as complex as the Internet allow people and ideas to move vast distances with relative ease. Populations once limited to local impact now are sufficiently large to affect many others; fishing in international waters is one example; shipping heavy metal-laden electronic components to other continents for recycling is another.

Once-separate societies have begun to band together ideologically on environmental issues. Although no universal consensus exists on many aspects of the environmental impact of various human activities, governments, grass-roots organizations, and other assemblages are more vocally conversing and taking visible, concerted action to shore up the quality dikes Dr. Juran so astutely saw decades ago. These efforts appear certain to result in widespread sea changes in legislation controlling aspects of quality we have long taken for granted. Legislation is necessary, perhaps, to achieve emerging societal goals, but is control alone sufficient?

Consider what legislation, agreements, treaties, and accords are intended to do. Although they may incentivize, they act as constraints by setting out limits as to appropriate and inappropriate activity. This is the essence of control—to maintain performance within certain boundaries. Through continuous, incremental innovation and the elimination of sporadic, special causes of poor performance against environmental standards, the dikes gradually will be strengthened.

What else can be done? A good control plan eventually makes itself unnecessary. Once all the leaks have been plugged and the walls of the dikes strengthened, the dikes themselves become the limitation to quality improvement. In the same fashion, we need to think beyond mere control, to instead break through the self-imposed constraints and fundamentally shift to a new landscape—a new zone of quality—and to design processes and products for ecological quality from the start. This is eco-quality.

Corporate Responsibility

What is the role of corporations in this context of ecoquality? The global marketplace increasingly is focused on the environment, and customer needs now include social responsibility. As a part of this, organizations are beginning to invest in programs and initiatives to address—and reduce—the environmental impact associated with all life cycle stages of their products and processes. For example, U.S. Fortune 500 corporations and their global counterparts are beginning to recognize the importance of understanding and improving the environmental impact of internal technologies and business practices. With the belief that corporate sustainability (including environmental dimensions) creates long-term shareholder value, Dow Jones established the Sustainability Indexes in 1999, providing the first tracking of financial performance of leading sustainability-driven organizations worldwide.

Another international program is the Carbon Disclosure Project (CDP). The CDP is a nonprofit organization with the mission to provide information to investors and stakeholders regarding the opportunities and risks to commercial operations presented by climate change. The CDP is a special project of the Rockefeller Philanthropy Advisors, an organization formed to help donors create thoughtful, effective philanthropy throughout the world having U.S. IRS 501(c)(3) charitable status, with the sole purpose of providing a coordinating secretariat for the participating investors. The CDP seeks to create long-lasting relationships between shareholders and corporations regarding the implications for shareholder value and commercial operations presented by climate change. The primary goal of the CDP is to facilitate a dialogue, supported by quality information, from which a rational response to climate change will emerge.

CDP Risks and Opportunities

A major objective of the CDP is to identify strategic risks and opportunities and their implications for businesses. The following is an example of risk-related questions from a questionnaire sent from the CDP to their clients (Carbon Disclosure Project 2008).

- Regulatory—What is your company's exposure to regulatory risks related to climate change?
- Physical—What is your company's exposure to physical risks from climate change?
- General—What is your company's exposure to risks in general as a result of climate change?
- Management—Has your company taken or planned action to manage the general and regulatory risks and/or adapt to the physical risks identified?
- Financial and business implications—How do you assess the current and/or future financial effects of the risks identified and how those risks might affect your business?

These questions reflect a concerted effort by industry to capture the risks and associated opportunities that present themselves via climate change, or global warming, as we have come to know it. As governments move toward stricter regulations of CO_2 levels, they will be demanding evidence that organizations have sustainable practices in place. Additionally, carbon quotas, caps, and similar legislation are in various stages of planning and implementation worldwide. Organizations have shareholder and community responsibility to prepare for upcoming changes, including understanding their carbon profile and having a plan in place to reduce it if at all possible.

Many global organizations have been proactive for a number of years promoting the design of new products for sustainability; two are highlighted below.

Spain's Telefónica Environmental Footprint

Corporate responsibility is exemplified in the Spanish company, Telefónica. In 2008, Telefónica implemented an Operational Control Standard across the entire Telefónica Group, with the objective of introducing best practices of environmental management to its fixed and mobile telephony operations, thereby minimizing Telefónica's environmental footprint.

As a leader in global telecom, Telefónica strives continuously to minimize the impact of its activities on the environment. It has an Environmental Management System and an Operational Control Standard that reflects best practice in the environmental management field. They developed the Standard by identifying the environmental concerns surrounding the activities and processes involved in rolling out our networks and then framing good environmental management practices specifically for each network. They felt that impact control would help minimize the company's footprint as and when the Standard is put into practice in the countries where they do business. The environmental concerns identified in Telefónica's operations are

- Energy use
- Waste (particularly electrical and electronic equipment and batteries)
- Radio wave emissions
- Environmental and visual impact
- Noise

Another globally recognized organization is Volvo. The Volvo Environment Prize is awarded for "Outstanding innovations or scientific discoveries which in broad terms fall within the environmental field." The Volvo Environment Prize is awarded by an independent foundation, which was instituted in 1988. Laureates represent all fields of environmental and sustainability studies and initiatives.

Product and Process Life Cycle Analysis

Earlier we introduced the concept of product or process life cycle stages. Here, we will develop this concept more fully. The life cycle is important because it is necessary to analyze all aspects of products and supporting processes from the cradle to the grave to arrive at a comprehensive carbon profile. This can only be achieved through a collaborative effort between commercial operations and supply and distribution chains. After a baseline carbon profile is established, the objective is to institute best practices that support the environmental sustainability of products and processes that support the end products. Although this is straightforward in theory, currently there is no single centralized or standardized set of data for the life cycle activities and processes to be included in quantifying product or process CO_2 emissions.

However, we can start by identifying five major phases of the life cycle of a product with its supporting processes that could be evaluated and analyzed. We can determine the probable range of CO_2 emissions generated by these various stages throughout the life cycle of the product and its supporting processes. The five basic life cycle stages of a typical commercial operation are as follows:

1. Product/process design (identification of supply-chain members)
2. Manufacturing process
3. Production operations
4. Supply-chain system
5. Final disposal (end of life)

Understandably, the initial focus of an organization in establishing its carbon profile is primarily focused on manufacturing and production and supply-chain activities that are under the most direct control. With maturity, however, design and end-of-life contributions should be considered more fully and addressed. Ultimately, the creation and development of an environmental sustainability process will benefit an organization's customers, shareholders, and society at large.

In support of meeting the responsibility corporations have to shareholders and society in terms of environmental sustainability, we recommend the following activities:

- Develop a data set for the life-cycle activities and processes to be included in estimating CO_2 emissions.
- Encourage active collaboration to foster industry partnerships to further expand the environmental sustainability of products and deliver continuous improvement.
- Continue to evaluate and analyze CO_2 emissions and other environmental metrics of commercial operations to ensure that we are not just shifting the environmental burden.
- Educate manufacturers and consumers from all industries regarding the relative environmental impacts of their products and supporting processes.

Origins of Quality and the Environment

As we move further into the twenty-first century, the quality movement is breaking new ground. As stated previously, we need to think beyond merely controlling hazards (building and maintaining quality dikes), to instead break through constraints and fundamentally improve the environmental dimension of existing processes and products. Additionally, we need to acknowledge the new customer need of social responsibility, and design future processes and products for ecological quality from the start. Eco-Quality embodies these concepts.

Quality and its relationship to the environment is not new. In 1969, Dr. Juran stated his concerns about the effect of poor quality, and the technology that developed it, on the environment. Technology in this context was the means by which organizations met customer needs and how people interact with the physical, chemical, and biological world. As society makes technological advances, there must be means of controlling them so that the advances provide benefit rather than harm. In his analogy of quality dikes (Chapter 2, Quality's Impact on Society and the National Culture), Dr. Juran viewed quality as protecting mankind from the surging water of technological advances. On occasion there are leaks in the dike, and we must repair them before they go further out of control.

In stating his concerns about the environment and how it relates to quality management, Dr. Juran was well ahead of his time, nearly 30 years ago. Dr. Juran's concern for natural resources and the sustainability of the environment was based on a sincere desire to bring a better quality of life to not just his children and grandchildren but all of humanity. Dr. Juran intuitively made the connection between quality and environmental sustainability but did not give it a name. In recognition of his contribution, the Juran Institute refers to this as Eco-Quality.

Eco-Quality Defined

Eco-Quality is not a replacement for designing a product and service that must be "fit for purpose." It is an extension on what fit for purpose will mean in the future. We believe that customers, of their own volition and through pressure from society and lobbyists, will create a new landscape for *Quality and Performance Excellence*, a new zone of quality that incorporates the dimension of *Environmental Sustainability* in partnership with the *Management of Quality*. We now have the knowledge and experience to combine quality design, control and improvement tools with best practices for environmental sustainability. Eco-Quality is intended to enable clients across industries to respond to demands from customers, regulatory agencies and shareholders for accountability in producing products and services fit for ecological use, focusing on understanding carbon profiles and reducing them to appropriate levels.

Eco-Quality and Performance Excellence

An effective performance excellence program in the future will include fulfillment of customer Eco-Quality needs. This is in alignment with the Juran Trilogy, encompassing the distinct processes of quality design, quality control, and quality improvement. Starting with a complete needs assessment of a client's products and supporting processes, a best fit methodology per the Trilogy is determined. The program core is quality improvement of processes, accomplished via a detailed accounting of carbon emissions and sources. The outcome is a baseline carbon footprint, with corresponding recommendations to improve process efficiency, eliminate waste associated with CO_2, and control emissions over the long term through continuous improvement methods. The triple bottom line of people, planet, and profits goes from red to green by listening to the mounting voice of the customer, reducing negative impact to the environment, and providing a return on investment through improved efficiencies and cost reduction.

Methods and Tools for Eco-Quality

A number of methods and tools are being used to move toward eco-friendly, eco-quality products.

ISO 14000 Environmental Management System

The ISO 14000 standard requires that organizations establish an *environmental management* system (Chapter 16, Using International Standards to Ensure Organization Compliance). It is applicable to any business, regardless of size, location, or industry. The purpose of the standard is to reduce a business' environmental footprint and to decrease pollution and waste that a business produces. The most recent version of ISO 14001 was released in 2004 by the *International Organization for Standardization* (ISO).

The ISO 14000 environmental management standards exist to help organizations minimize how their operations negatively affect the environment. In structure it is similar to *ISO 9000 quality management* and both can be implemented side by side. In order for an organization to be awarded an ISO 14001 certificate, they must be externally audited by an audit body that has been accredited by an accreditation body.

An effective Environmental Management System meeting the requirements of ISO 14001:2004 is a management tool enabling an organization of any size or type to do the following:

- Identify and control the environmental impact of its activities, products, or services.
- Improve its environmental performance continually.
- Implement a systematic approach to setting environmental objectives and targets, to achieving these, and to demonstrating that they have been accomplished.

Life Cycle Assessments

This is a "cradle-to-grave" analysis of the environmental impacts of a product or service caused or necessitated by its existence, from birth to death. Not limited to greenhouse gases (see carbon footprint, below), it encompasses many forms of damage such as ozone depletion, desertification, and resource depletion. The objective of a life cycle analysis is to encourage informed and appropriate choices by providing fair comparison of products and services in terms of negative environmental impact.

The ISO 14000 environmental management standards define four phases of a life cycle assessment:

1. Goal and scope—description of the objectives, functional unit, system boundaries, method of assessment, and impact categories included in the assessment
2. Life cycle inventory—detailed listing of inputs and outputs (e.g., materials, energy, water, chemicals, emissions, radiation) in terms of elementary flow to and from processes and the environment; relies heavily on software for data collection and modeling
3. Life cycle impact assessment—characterization of potential impacts, normalization to a common unit of measure, and weighting of impact categories
4. Interpretation—sensitivity and overall analysis and conclusions regarding major contributing factors; assessment relative to the goal and scope

Life cycle assessments can be used as a comparative tool, for example, to compare plastic versus glass versus aluminum beverage containers for environmental impact, with the results used for marketing purposes, or new product design. A recent study reports life cycle assessments being used predominantly to support business strategy (18 percent) and R&D (18 percent) as inputs to product or process design (15 percent), for educational purposes (13 percent), and for labeling or product declarations (11 percent) (Cooper and Fava 2006).

Carbon Footprinting

A carbon footprint (or profile) is the combined total of all greenhouse gas emissions caused directly and indirectly by an individual, event, organization, or product (The Carbon Trust 2009). This is frequently reported as being "CO_2 equivalent" with carbon dioxide used as a convenient, common currency; a carbon footprint therefore need not be strictly confined to CO_2 alone. This is an expansive definition and includes many sources over which an individual or organization has varying degrees of control. From a practical perspective, it is useful to classify the CO_2 equivalents according to the degree of control. Common categories are

- Emissions from activities, products, and services under direct control
- Emissions from activities, products, and services under indirect control
- Emissions from electricity usage

Understanding an organization's carbon footprint is important for two reasons, already alluded to. First, customers, suppliers, shareholders, government agencies, and other third parties increasingly request this information from businesses. For example, organizations engaged in carbon neutrality "cap and trade" or those developing green marketing messages will need comprehensive, accurate, and verifiable reporting of GHG emissions, especially as this may become part of the public record. Second, from the adage "you cannot manage what you do not measure," measuring a carbon footprint is a necessary step toward reducing and controlling it, ultimately achieving gains in the triple bottom line.

Energy Audits

An energy audit is an inspection and analysis of the energy flow through a building, process, or system, one that is carried out to improve energy efficiency and reduce overall consumption. Although energy audits are not new (efficiency has long been an issue in corporate accounting offices), the "pollution" factor is gaining in prominence as an impetus. Because a large proportion of energy typically comes from carbon-based fossil fuels, carbon dioxide is a natural byproduct of energy use, and energy use therefore is a major contributor to a carbon footprint.

An energy audit consists of the following types of information:

- Building information—type of building (e.g., office, school), prior modifications, current conservation measures, occupancy profile
- Building characteristics—gross floor space, ceiling height, exterior wall area, number and placement of doors, insulation type and thickness, glass area, heating and cooling methods
- Electricity usage—metering method, demand patterns (including peak, average, and minimum), energy cost, service cost
- Nonelectricity energy usage—other sources such as natural gas, liquefied petroleum, kerosene, coal, wood, steam
- HVAC system—heating, ventilation, and air conditioning units, sensors and controls; air flow and pressure
- Hot water—energy source, temperature at origin and point of use, distance from heater to point of use, insulation, recirculation
- Lighting—area, lighting type (incandescent, fluorescent, mercury vapor, high-pressure sodium, metal halide), wattage, output, operating hours, controls

Based on the audit results, opportunities are identified to eliminate energy waste, and reduce CO_2 emissions and operating costs. Many governments now sponsor programs to encourage "green building" and provide information to assist in energy audits, e.g., as part of the EPA's Energy Star program.

The End Game

Just as no single factor is implicated in climate change, no single player is driving the ball on social change; it is a collective effort. As organizations forge ahead and put together plans to meet the needs their customers, it is easy to dismiss the once-solitary voices calling for change. This would be a mistake. Compelling expectations originate from multiple sources and perspectives:

- Customers—sensitive to the environmental impact of products and services they purchase

- Shareholders—demanding accountability, transparency, and favorable return on investment
- Legislators—pursuing legal incentives and constraints
- Scientific community—seeking evidence-based action
- Suppliers and distributors—looking forward and back to manage their "cradle-to-grave" chain

Ignoring these factors will not make them go away; instead, a real possibility exists that organizations failing to heed these influences will sooner go away. We are all faced with this environmental challenge in one way or another.

The future belongs to those who are planning for it today.

—African proverb

References

Aufhammer, M., and Carson, R. T. (2008). "Forecasting the Path of China's CO_2 Emissions Using Province-Level Information." *Journal of Environmental Economics and Management*, vol. 55, no. 3, pp. 229–247.

Carbon Disclosure Project (2008). CDP6 Questionnaire, 1 February. Carbon Disclosure Project, London.

Carbon Trust (2009). "What Is a Carbon Footprint?" The Carbon Trust, London. Retrieved November 19, 2009 from http://www.carbontrust.co.uk/solutions/CarbonFootprinting/what_is_a_carbon_footprint.htm

Cascorbi, A., and Stevens, M. M. (2004). "Seafood Watch Seafood Report: Atlantic Cod." Final Report, July 29. Monterey Bay Aquarium, Monterey, CA.

Cooper, J. S., and Fava, J. (2006), "Life Cycle Assessment Practitioner Survey: Summary of Results." *Journal of Industrial Ecology*, vol. 10, no. 4, pp. 12–14.

Energy Information Administration (2008). "Emissions of Greenhouse Gases Report: Carbon Dioxide Emissions." Report # DOE/EIA-0573(2007), Energy Information Administration, U.S. Department of Energy, Washington, D.C.

Garnaut, R. (2008). "The Garnaut Climate Change Review (2008)." Commonwealth of Australia, Canberra Australian Capital Territory.

Intergovernmental Panel on Climate Change (IPCC), 2005.

Juran, J. M. (1969). "Mobilizing for the 1970s." Quality Progress, August, pp. 8–17.

NASA (2009). Earth Observatory: Arctic Sea Ice. Retrieved November 17, 2009 from http://earthobservatory.nasa.gov/Features/SeaIce/page3.php

Savitz, A. W., and Weber, K. (2006). The Triple Bottom Line. Jossey-Bass, San Francisco.

Thomas, C. D., Cameron, A., Green, R. E., Bakkenes, M., Beaumont, L. J., Collingham, Y. C., Erasmus, et al. (2004). "Extinction Risk from Climate Change." *Nature*, vol. 427, pp. 145–148.

U.N. Environment Programme (2009). UNEP Climate Change Presentation: Science. Retrieved 11/18/09 from http://www.unep.org/climatechange/Science/

Wirtenberg, J., Lipsky, D., and Russell, W. G. (2009). *The Sustainable Enterprise Field Book: When It All Comes Together*. Amacom Books, New York.

SECTION II

Methods and Tools: What to Use to Attain Performance Excellence

CHAPTER 11

Lean Techniques: Improving Process Efficiency

Steven M. Doerman and R. Kevin Caldwell

About This Chapter

Lean is the process of optimizing organizational systems by eliminating, or at least reducing, the "waste" within them. Anything that does not provide value to the customer or the organization can be considered waste. This chapter introduces the Lean Methods and Tools and their relationship to managing quality and superior results. Lean methods and tools can provide significant improvements in organizational efficiency. In the past decade, Lean has experienced a rebirth in manufacturing-based industries as well as service and health care-based organizations.

High Points of This Chapter

1. Lean is based on creating a "pull system" to produce faster rather than the traditional "push" systems used by most organizations. One of the main goals of Lean is to always pull from the customer demand, not push to the customer.
2. Value Stream Mapping is an important Lean tool. It maps and documents all the tasks (material and information flow) and the metrics associated with them

(cycle time, costs) within a process, including inherent waste. This provides the guidance to select the right problems and solve them as process improvement projects.

3. There is a standardized approach and set of tools, such as rapid improvement events, or *kaizens* (Japanese word for "improvement") to attack embedded wastes and increase the velocity of a process. Improving velocity exposes the problems—waste—and eliminates them, thereby making the processes faster, better, and cheaper.
4. 6S (sort, set in order, shine, standardize, sustain, and safety) is a Lean method to achieve a highly effective workplace that is clean and well organized. The benefits of an efficient workplace include prevention of defects; prevention of accidents, and elimination of time wasted searching for tools, documentation, and other ingredients to produce goods or services.
5. The integration of Lean and Six Sigma has become known as Lean Six Sigma. Lean focuses on efficiency and Six Sigma focuses on how effectiveness can lead to faster results than either method applied independent of the other.

A Truly Lean Introduction

Lean is the process of optimizing systems to reduce costs and improve efficiency by eliminating product and process waste. The emphasis is on eliminating non-value-added activities such as producing late services, defective products, excess inventory charges and excess finished goods inventory, excess internal and external transportation of products, excessive inspection, and idle time of equipment or workers due to poor balance of work steps in a sequential process. The goal of Lean has long been a goal of industrial engineering—to improve the efficiency of all processes.

As Shuker states in his article "The Leap to Lean," creating a lean organization encompasses the delivery of goods and services using less of everything: less waste, less human effort, less manufacturing space, less investment in tools, less inventory, and less engineering time to develop a new product, and less motion, for example. Lean manufacturing was a process management philosophy derived mostly from the War Manpower Commission, a World War II U.S. agency, which led to the Toyota Production System (TPS) and from other sources. The War Manpower Commission is renowned for its focus on reducing the original Toyota seven deadly wastes: overproduction, wait time, transportation, processing methods, inventory, motion, and defects (sometimes called the eight deadly wastes) in order to improve overall customer satisfaction. The eighth deadly waste was the waste of people's unused creativity. Lean is often linked with Six Sigma because of that methodology's emphasis on reduction of process variation (or its converse smoothness) and Toyota's combined usage (with TPS). Although Lean concepts began in manufacturing operations, it has been successfully applied in many industries as diverse as hospital patient care, internal auditing, and insurance customer service. Lean principles can be applied in most processes because mosty all contain waste that a customer is not willing to pay for, nor is the business willing to accept higher costs because of them. For additional information the TPS please reference Spear and Bowen's article in the Harvard Business Review entitled Decoding the DNA of the Toyota Production System.

For many, Lean is the set of TPS "tools" that assist in the identification and steady elimination of waste (*muda* in Japanese terminology), the improvement of quality in production time, and costs. This and other Japanese terms used by Toyota are strongly represented in the Lean vernacular. To solve the problem of waste, Lean has several tools at its disposal, including continuous process improvement (*kaizen*) 6S, and mistake proofing (*poka-yoke*). In this way, Lean can be seen as taking a very similar approach to other improvement methodologies.

The second, and complementary approach to Lean, which is also promoted by the TPS, is the focus upon improving the "flow" or smoothness of work (thereby steadily eliminating *mura*, unevenness) through the system and not upon waste reduction per se. Techniques to improve flow include "production leveling," "pull production" (by means of *kanban*, signboard or billboard), and the *Heijunka* box (achieving smoother production flow).

Lean implementation and the TPS are therefore focused on getting the right things to the right place, at the right time, and in the right quantity to achieve perfect work flow while minimizing waste and being flexible and able to change. More importantly, all of these concepts have to be understood, appreciated, and embraced by the actual employees who build the products and therefore own the processes that deliver the value. The cultural and managerial aspects of a Lean organization are just as, and possibly more, important than the actual tools or methodologies of production itself.

Lean in Nonmanufacturing-Based Industries

Lean methods and tools have made their way into most industries. A method that was used in manufacturing to reduce waste is now used to improve cycle time, flow, and velocity, improve workplace department performance and, yes, reduce waste in hospitals, insurance companies, financial services, and more. Here is one example from a hospital (Volland 2005):

Adapted from *A Case Study: Now That's Lean*

Jennifer Volland
Reprinted with Permission from Medical Imaging Magazine.

In the hopes of improving workflow and patient throughput, the Nebraska Medical Center (Omaha) began implementation of Lean Six Sigma in December 2002. As a 735-bed nonprofit hospital, the center is the largest teaching hospital in Nebraska with both academic and private practice physicians. One of the first Six Sigma projects for the organization was in the Interventional Radiology (IR) department, where such invasive procedures are performed.

A project team—which included the lead nurse scheduler, lead technologist, and department manager—was assembled to address patient throughput problems. Physician involvement was initiated early with ongoing input and information sharing for process improvements.

The project team defined physicians who referred patients into the IR department as their primary customer. They quickly realized that current volumes supported by the department did not fully meet the needs of referring physicians. Patients were lost to other healthcare systems that could accommodate the additional patients within the community, resulting in loss of revenue and market share.

The project team measured the cycle time of each step to determine where to best focus improvement efforts. Reducing holding room (HR) time quickly became evident as an area of opportunity. A patient's HR time averaged 151 minutes with a standard deviation of 242.4 minutes (February 4-19, 2003). Upon further examination, however, many more problems were identified. First, patient flow coordination from the HR into one of three procedure rooms was problematic because of different equipment in the rooms. Often, the nurse scheduler was pulled to function as the department appointment scheduler as well as the person coordinating patient flow. The duality of tasks created problems for timeliness in appointment scheduling with the referring clinics and flow of patients through the HR.

Changes made during the Lean Six Sigma implementation had a significant impact on the amount of time patients spent in the HR. The amount of time a patient spent in the HR, after the improvements, averaged 32.7 minutes with a standard deviation of 37.71 minutes (March 17-24, 2003). Follow-up monitoring during the control phase showed sustained improvements, with the HR time leveraging 31.02 minutes and a standard deviation of 24.86 minutes (October 29-December 16, 2003).

Lean techniques applied within the IR department resulted in improved processes and an ability to better meet customer expectations. As a result of the project, referring clinics were successfully able to feel the impact of changes for improved interventional radiologists within the department. Not only were the changes significant, but, post-project, the department as been able to successfully sustain the gains made in the HR.

Reducing Waste Alone Is Not Lean

It is not enough to just believe "if I eliminate the nonvalued waste we will be Lean." This is only one aspect of a Lean organization. Although the elimination of waste may seem like a simple and clear subject, it is noticeable that waste is often very conservatively identified. This then hugely reduces the potential of an organization. Although the elimination of waste is the goal of Lean, the TPS defines three types of waste: *muri* or overburden, *mura* or unevenness, and *muda* or non-value-added work.

Muri is all the unreasonable work that management imposes on workers and machines because of poor organization, such as carrying heavy weights, moving things around, dangerous tasks, and even working significantly faster than usual. *Muri* is pushing a person or a machine beyond its natural limits.

Mura focuses on implementing and eliminating fluctuation at the scheduling or operations level, such as quality and volume.

Muda is discovered after the process is in place and is dealt with reactively rather than proactively with *muri* and *mura*. It is seen through variation in output (which as mentioned earlier) can blend well with Six Sigma applications. It is the role of management to examine the *muda*, or waste, in the processes and eliminate the deeper causes by considering the connections to the *muri* and *mura* of the system. The *muda* (waste) and *muri* (overburden) must be fed back to the *mura* planning stage for the next project.

More often than not, most organizations improperly only focus on *muda* or non-value-added waste and fail to understand this approach is reactive and will only partially position the organization for success (if at all). One must ensure that all three waste types are addressed.

Muri can be avoided through standard work disciplines. To achieve this, a standard condition or output must be defined. Then every process and function must be reduced to its simplest elements for examination and later recombination. This is done by taking simple work elements and combining them, one by one, into standard work sequences.

Mura is avoided by using Just-in-Time (JIT) systems that are based on little or no inventory by supplying the production process with the right part, at the right time, in the right amount, and first-in, first out component flow. JIT systems create a "pull system" in which each subprocess withdraws its needs from the preceding subprocesses, and ultimately from an outside supplier. When a preceding process does not receive a request or withdrawal, it does not make more parts.

To properly manage outcomes in a Lean organization, you must ensure that all three types of waste are managed and controlled. Demand and capacity must be balanced to that demand must be fully understood. Current state conditions must be understood in order to move to future state pull production and the elimination of non-value added activities creating waste. Standard work must be institutionalized, which alleviates overburdening associates as they perform activities. These activities will create the model for cultural transformation from a batch-and-queue operation to an operation with synchronous flow, team-based activities, and a true focus on the customer mindset.

Lean Manufacturing Case Study

AGC Flat Glass North America, a wholly owned subsidiary of the world's second-largest glass producer, Asahi Glass Company, operates 45 facilities throughout North America, and all were experiencing pressure to provide the lowest total cost product with rapid order fulfillment in a highly competitive market. In September 2006, AGC launched an initiative to drive operational excellence and improve profitability. This initiative was coined JPI (Jikko Process Improvement) by AGC and is based on the principles of the TPS and Lean enterprise.

One of the first facilities to implement the JPI process was AGC Hebron, a fabrication facility located near Columbus, Ohio. Hebron serves the Ohio market and neighboring states. Hebron receives glass from one of AGC's primary glass facilities and transforms these raw materials into a number of end products, including single-pane products, sealed insulated units for window manufacturing, and tempered (heat-treated) glass for safety applications. The Hebron fabrication processes include cutting, tempering, and insulating unit assembly. An initial assessment of the facility was performed, and the results indicated that manufacturing lead times were exceeding seven days with wide swings up to weeks in some cases. Excess inventory made it nearly impossible to quickly find a specific job or determine what to fabricate next. There was also a concern for employee safety, specifically increased risk of injury attributed to the large cut-glass inventory. Wide swings in product demand placed on manufacturing also served to complicate the business. Some days, the plant capacity was underutilized while other days customer demand exceeded capacity by twofold.

A cross-functional team was formed to drive the improvement efforts. Team members included sales, production control, purchasing, production employees, corporate JPI members, and a transformation coach. In the first days, the team was introduced to the concepts of the TPS and Lean manufacturing.

One of the first things the team quickly developed was a "Current State Map," a valuable tool to understand the actual situation on the production floor and in-order fulfillment activities. Once completed, the current state map clearly told the present story and set a firm direction for future improvement.

The first step to improve the efficiency of the workplace focused on implementing the 6S (sort, set-in-place, sweep/shine, standardize, self-discipline and safety) process. After the initial training, the team began to attack waste; sorting unnecessary items from needed items, implementing visual control for tools and materials, cleaning everything, and putting in place a robust auditing system to sustain the gains. From there, the team focused on their "Current State Map." Points of delay and inventory builds were addressed and, in most cases, eliminated. Equipment was relocated to aid product flow, which reduced movement and product queues. To further consolidate inventory, over half the material-handling racks used to store glass were removed. The reduction in inventory in a matter of days translated to improved lead times to the customer. At this point, the Hebron team adopted the motto, "There is no tomorrow." A key to a Lean enterprise and the TPS inherent in this philosophy is the idea that customer delivery requirements will be met and that all products can and will be produced in a single day to customer demand and pull. This expectation was well within the plant capabilities for cycle times. The team also studied demand patterns compared to the demonstrated capacity. Once this relationship was understood and lead times were reduced, the plant could successfully be level loaded, thus further solidifying delivery reliability to levels above 99 percent on time. This percentage was well above historic levels. The improved product flow quickly identified quality issues that were previously hidden by excess work in process. In the weeks that followed, a number of other enhancements were included such as improved equipment maintenance to assure reliability, mistake proofing methods, *kanbans* for supply replenishment, and a focus on faster changeovers. During the time the physical changes were occurring, another important transformation took place—the culture slowly changed. The plant began running differently. Employees knew what the customer needed by the hour and produced accordingly. Orders moved seamlessly through the operations without heroic efforts, making work life easier and, more importantly, safer.

Within weeks, the customers began to see and feel the changes. The new Hebron customer complaints turned to customer compliments. Overall demand steadily increased as past customers lost due to service issues began to return and new customers began to come to Hebron for their glass needs. The financial results followed as Hebron experienced a turn around in profitability. Commenting on profitability, Jerry Hackler, Hebron's Operations Manager remarked, "The effect of the bottom line came quickly. Even in the early months the facility generated more operating income on fewer sales, a clear indication of the cost improvement impact."

History of Lean

The history of manufacturing and the introduction of Lean are summarized in Figure 11.1. The Lean mission is to have the following throughout the entire supply chain to win the marketplace:

- Shortest possible lead time
- Optimum level of strategic inventory
- Highest practical customer service levels
- Highest possible quality (low defect rate)
- Lowest possible waste (low cost of poor quality)

This is accomplished by synchronizing the flow of work (both internal and external to the organization) to the "drumbeat" of the customer's requirements. All kinds of waste are driven out (time, material, labor, space, and motion). The overall intent is to reduce variation and drive out waste by letting customers pull value through the entire value stream (or supply chain).

In their book Lean Thinking, Womack and Jones state that the key principles of Lean are to

- Specify value in the eyes of the customer; the voice of the customer
- Identify the value stream for each product
- Make value flow without interruptions
- Reduce defects in products and deficiencies in processes
- Let customers pull value
- Pursue perfection—Six Sigma levels
- Drive out variation (short and long term)

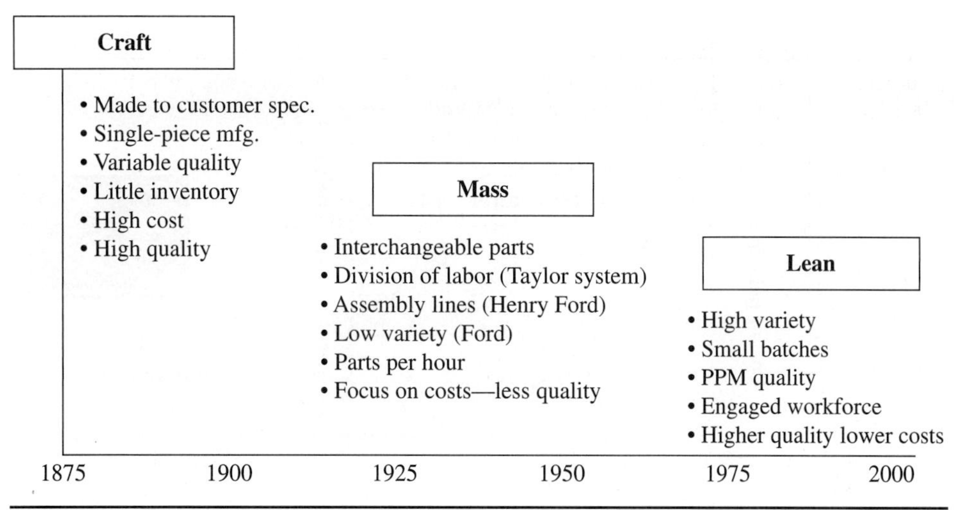

FIGURE 11.1 History of manufacturing.

The Relationship of Lean to Managing for Quality

One key component of being a Lean organization is the need to create "value" as seen from the eyes of the customers. The operational definition of value is the benefit the customer gains from using the product or service. Value is created by the customer. Providing value to the customer is why the producer exists. Lean starts with defining value in terms of products/services and benefits provided to the customer at the right time at an appropriate price. Anything that does not provide value to the customer can be considered waste (see Figure 11.2).

If we review the Juran Trilogy® in Figure 11.3, we can see that Lean supports the definition of quality in that all products and services must be "fit for purpose." Customers define quality as both the features and freedom from failures. Therefore, because Lean is about creating value by eliminating nonvalue, it is important to include in Lean the management of quality. Lean is used in quality control because it enables work to be standardized, leading to better compliance. Lean is used in improvement to decrease the costs of nonperforming processes in the form of waste reduction. Most recently, Lean methods are being used in quality planning to design for Lean. Designing for Lean is similar to designing for quality. An organization now must design a product or service so that it can flow easily with little disruption from customer need to customer use.

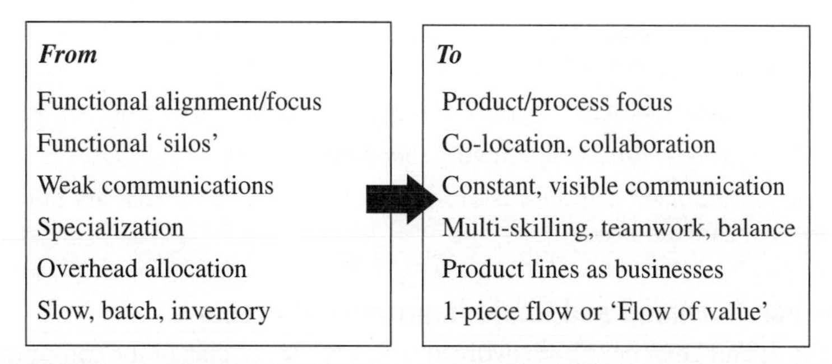

FIGURE 11.2 Lean characteristics.

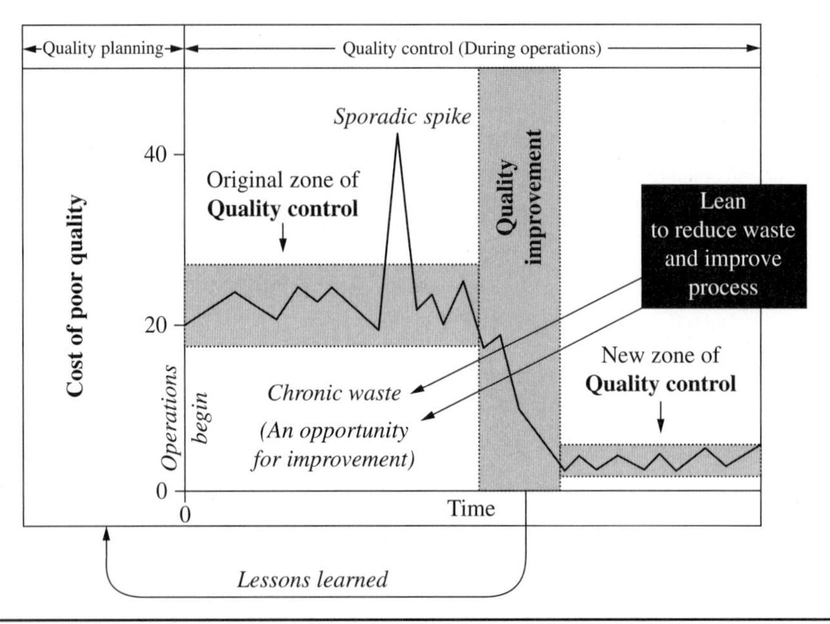

FIGURE 11.3 Lean and the Juran Trilogy®. (*Juran Institute, Inc., Southbury, CT.*)

The Eight Wastes

Taiischi Ohno (1988) identified seven types of waste that exist in most processes and organizational systems. These identifiable wastes lead to the cost of poor quality if they are not dealt with and removed. Lean practitioners and experts must focus on reducing or eliminating these wastes, part of a *kaizen* or Rapid Improvement Event.

The following includes Ohno's seven types of waste, which were focused on production in addition to the eighth waste (which seems to have no origin) directed at all processes.

1. *Overproduction*—making or doing more than is required or earlier than needed
2. *Waiting*—for information, materials, people, and maintenance
3. *Transport*—moving people or goods around or between sites
4. *Poor process design*—too many/too few steps, nonstandardization, and inspection rather than prevention
5. *Inventory*—raw materials, work in progress, finished goods, papers, and electronic files
6. *Motion*—inefficient layouts at workstations, in offices, poor ergonomics
7. *Defects*—errors, scrap, rework, nonconformance
8. *Underutilized personnel resources and creativity*—ideas that are not listened to, skills that are not used

The Lean Roadmap and Rapid Improvement Events

Six Sigma and Lean have both evolved over decades as part of the continuing revolution of quality, excellence, and breakthrough performance. Motorola created the term "Six Sigma" as it worked to raise the standard for improvement to new heights. Lean grew out of the experiences of the TPS.

Now Lean and Six Sigma have evolved to reflect today's core business challenges: the challenge to execute and to maximize value, as well as respond to "nanosecond customer" needs. Joe De Feo of Juran, refers to the speed at which today's demanding customers expect results. Lean and Six Sigma are now used for sustainable competitive advantage across all industries and cultures.

Every organization wants to be Lean and have

- The shortest possible process lead times for providing products and services
- The optimum level of strategic inventory and human resources
- The highest practical customer service level
- The highest possible quality (low defect rate)
- The lowest possible waste (low COPQ, cost of poor quality) . . . throughout the entire value chain

Although there have been numerous techniques and tools utilized in Lean implementation, most Lean practitioners did not have a Lean model until the collaboration with Six Sigma DMAIC (Define, Measure, Analyze, Improve, Control). The Juran Lean Roadmap in Figure 11.4

Define value

1. Define stakeholder value and critical to quality (CTQ).
2. Map high-level process.
3. Assess for 6S.

Measure value

1. Measure customer demand.
2. Plan for data collection.
3. Create a value stream attribute map.
4. Determine pace, Takt Time and manpower.
5. Identify replenishment and capacity constraints.
6. Implement 6S (S1–S3).

Analyze process—flow

1. Analyze the value stream attribute map.
2. Analyze the process load and capacity.
3. Perform value added/non-value added analysis.
4. Apply Lean problem-solving.

Improve process—pull

1. Conduct rapid improvement events (RIE).
2. Design the process changes and flow.
3. Feed, balance, and load the process.
4. Standardize work tasks.
5. Implement new process.

Maintain control

1. Stabilize and refine value stream.
2. Complete process and visual controls.
3. Identify mistake-proofing opportunities.
4. Implement 6S (S4–S6).
5. Monitor results and close out project.

FIGURE 11.4 Lean Six Sigma roadmap and substeps.

is an example of a model designed to carryout "Lean projects or events." It provides the five DMAIC steps as in Six Sigma and includes the lean tasks. This set of steps provides a Lean or "Lean Six Sigma" practitioner with a reminder to focus both on efficiency and effectiveness.

Figure 11.5 provides a tool grid to demonstrate tools that can be used at every step in the method. Each of the tools in this grid can be found in this chapter as well as in Chapter 18, Core Tools to Design, Control, and Improve Performance and Chapter 19, Accurate and Reliable Measurement Systems and Advanced Tools.

Rapid Improvement Events or *Kaizens*

Rapid Improvement Events (RIE) or *kaizens* are typically one-week focused efforts that are facilitated and conducted by Lean Experts or Black Belts to enable Lean teams to analyze the value streams and quickly develop/implement solutions in a short time-frame. These events have application in offices, service organizations, health care arenas, and manufacturing operations and consistently yield tremendous, real-time improvement. *Kaizen* is the Japanese word for incremental improvement. It has become associated with the use of small teams carrying out improvements on a regular basis. It is often used as a name for all encompassing continuous improvement methods. We have chosen to use it as it is defined: a small improvement that is made on a regular basis. RIE or *kaizen* teams are multifunctional so that all aspects of the process and problems associated with them are considered and soutions developed will be understood and accepted by all. Rapid improvement teams are fast because Lean is easier than, say, Six Sigma. Rapid improvement teams are fast because they tackle focused projects bit by bit. They also tackle problems where the data are typically readily available.

This technique is a good tool to involve all levels of the workforce. It can help build an empowered and engaged workforce. RIEs can be used to identify and solve departmental problems as well.

What Do RIE and *Kaizen* Teams Do

A Lean Expert or a Black Belt works with management to select the area to focus the improvement on. They then carryout the following preparations for the events.

1. One to three weeks prior to conducting the event the expert assembles the team, facilitates development of a charter and gathers as much data as possible surrounding the area to be improved. The type of data depends on the area selected but typically includes a manufacturing area of focus:
    a. Process flow diagrams for each product or product family (if available)
    b. Yields by operation
    c. Setup time by operation
    d. Changeover time by operation
    e. Average WIP (work in progress) inventory levels between operations
    f. Average materials inventory
    g. Average finished goods inventory
    h. Cycle times by operation
    i. Average daily customer demand by end item
    j. Monthly customer demand by end item

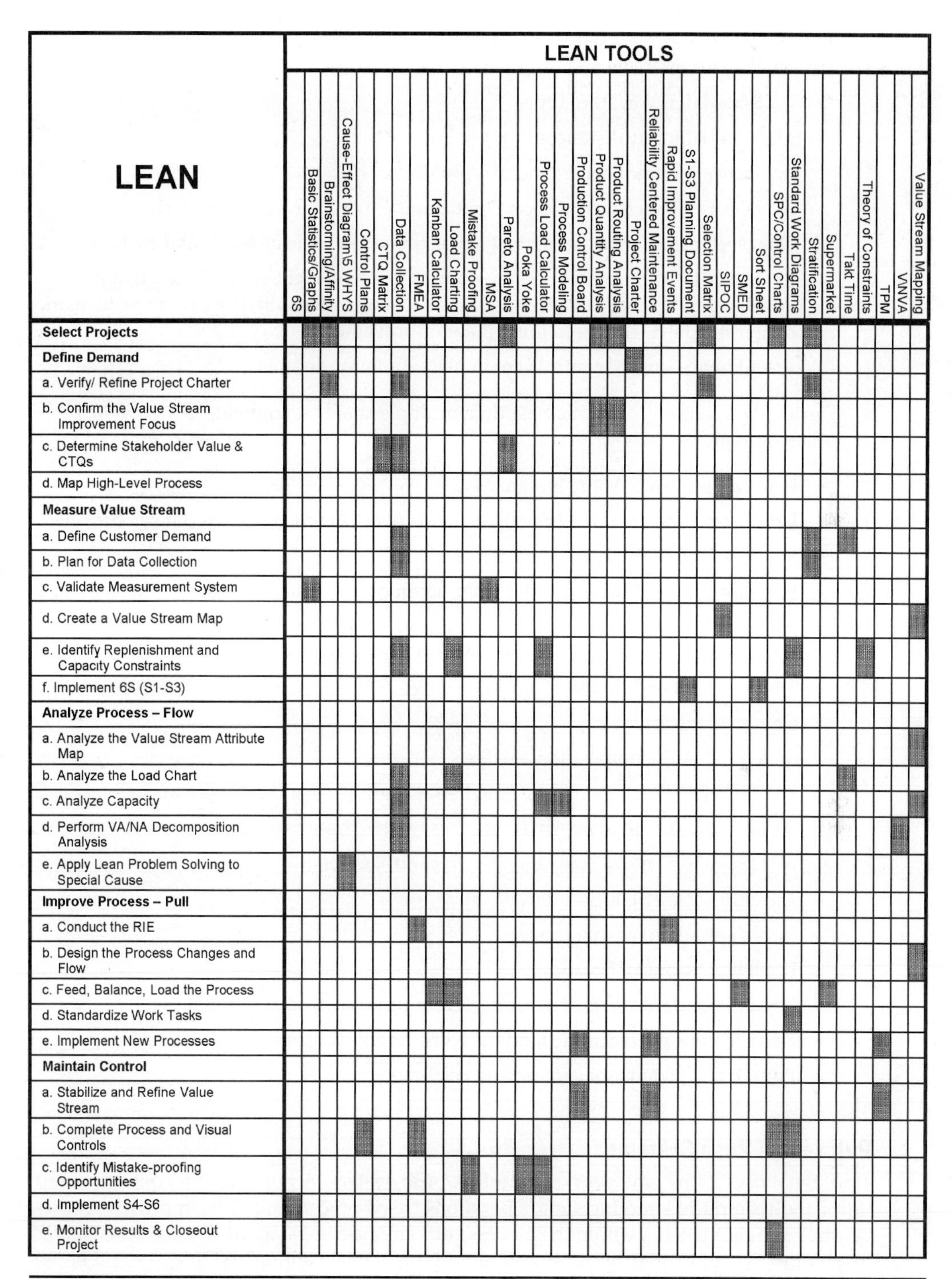

LEAN TOOLS

LEAN	6S	Basic Statistics/Graphs	Brainstorming/Affinity	Cause-Effect Diagram/5 WHYS	Control Plans	CTQ Matrix	Data Collection	FMEA	Kanban Calculator	Load Charting	Mistake Proofing	MSA	Pareto Analysis	Poka Yoke	Process Load Calculator	Process Modeling	Production Control Board	Product Quantity Analysis	Product Routing Analysis	Project Charter	Reliability Centered Maintenance	Rapid Improvement Events	S1-S3 Planning Document	Selection Matrix	SIPOC	SMED	Sort Sheet	SPC/Control Charts	Standard Work Diagrams	Stratification	Supermarket	Takt Time	Theory of Constraints	TPM	VNVA	Value Stream Mapping
Select Projects		X	X										X					X	X					X				X		X						
Define Demand																				X																
a. Verify/ Refine Project Charter			X				X																	X						X						
b. Confirm the Value Stream Improvement Focus																		X	X																	
c. Determine Stakeholder Value & CTQs						X	X						X																							
d. Map High-Level Process																									X											
Measure Value Stream																																				
a. Define Customer Demand							X																							X		X				
b. Plan for Data Collection							X																							X						
c. Validate Measurement System		X										X																								
d. Create a Value Stream Map																									X											X
e. Identify Replenishment and Capacity Constraints							X			X					X														X				X			
f. Implement 6S (S1-S3)																							X				X									
Analyze Process – Flow																																				
a. Analyze the Value Stream Attribute Map																																				X
b. Analyze the Load Chart							X			X																						X				
c. Analyze Capacity							X								X	X																				X
d. Perform VA/NA Decomposition Analysis							X																												X	
e. Apply Lean Problem Solving to Special Cause				X																																
Improve Process – Pull																																				
a. Conduct the RIE								X														X														
b. Design the Process Changes and Flow																																				X
c. Feed, Balance, Load the Process									X	X																X					X					
d. Standardize Work Tasks																													X							
e. Implement New Processes																	X				X													X		
Maintain Control																																				
a. Stabilize and Refine Value Stream																	X				X													X		
b. Complete Process and Visual Controls					X			X																				X	X							
c. Identify Mistake-proofing Opportunities											X			X	X																					
d. Implement S4-S6	X																																			
e. Monitor Results & Closeout Project																												X								

Figure 11.5 Lean methods and tools.

k. List of suppliers including items supplied, amounts, annual dollar value, and delivery frequency
l. Material move/store times
m. Material move distances
n. Inspection frequencies and sample sizes
o. DPMO (Defects per Million Opportunities) or Sigma levels of each process

2. One week prior to the event, the team is trained in basic methods and tools of Lean.
3. Event week—the team begins by validating the current state Value Stream Maps and develop "Future State" maps, define customer demand, pace, balance the work, define standard work, and implement improvements.
4. After Event—ensure controls are in place; monitor progress.

During the event the teams may conduct multiple small assignments. Some of the more important ones are

- Begin current state Value Stream Map
- Understand the data that is available and collect as much needed data as possible.
- Ensure the availability of equipment
- Implement S1, S2, and S3 of 6S (sort, set in order, shine, standardize, sustain, and safety)
- Validate value stream maps—understand the "before" values
- Study current conditions
- Complete the following:
  - VA/NVA Decomposition Analysis
  - Current State Load Charts, Spaghetti Diagrams, Standard Worksheets
  - Review Current State Analyses
  - Design the Future State and design control sheets
- Develop Future State Standard Work
- Implement changes (big moves)
- Implement Control Boards
- Review standard work, standard work-in-process, needed fixtures, etc
- Finalize flow, procedures, standard work, and Production Control Board
- Present results to management and celebrate

Pull versus Push Systems

Traditional operations have worked within a push system. A push system computes start times and then pushes products into operations based on demand. This approach ignores constraints or bottlenecks within the process and can cause unbalanced flow and excess WIP inventories. A pull system, by contrast, only produces when authorized to do so and based on the process status.

Pull systems produce faster than push systems, and, by nature, pull production controls and enhances flow. The goal should always be to pull to customer demand.

Lean Value Stream Management

Lean focuses on finding value streams. These value streams consist of all activities required to bring a product from conception to commercialization. They can include all key business processes such as design, order taking, scheduling, production, sales, marketing, and delivery. Understanding the value stream allows one to see value-added steps, non-value-added but needed steps, and non-value-added steps. Value-added activities *transform* or shape material or information into something that meets customer requirements. Non-value-added activities take time or resources, but they do not add value to the customer's requirement (but they may meet the organization's requirements). The value stream improvement journey typically starts with training the team on key concepts of Lean and mapping the *current state* using value stream maps that document materials and information flow as well as any pertinent information on the process (such as cycle times, downtime, capacity, wait times, yield, and inventory levels).The goal is to identify all the necessary components to bring a product to commercialization, as well as all waste inherent in the process. Improvements are identified from here. The desired future state is then documented as a *future state value stream map*, and the improvements are implemented to drive toward the desired future state goal.

Value streams can be mapped for a single product or service but, more often, a process supports more than one single-ended item. When products share the same design and fabrication processes, they are called a *product family*. In practice, value stream maps are frequently developed around a product family. It is not uncommon for maps to commingle with other product families as they progress through the process.

As mentioned above, a value stream comprises all the tasks currently required to move the product family though its process. There are three typically mapped cycles: Concept to launch (the design cycle), raw materials to customer (the build cycle), and delivery to recycling (the sustain cycle). The build cycle is the most commonly mapped.

An example of a value stream map for a paint line showing both the current state and future state are shown in Figures 11.6 and 11.7. There are a number of excellent sources for the techniques of mapping the value stream such as Learning to See (Rother and Shook 2003), Value Stream Management (Tapping, Luyster, and Shuker 2002), and Creating Mixed Model Value Streams (Duggan 2002). To be most effective, mapping should include all process steps involved, including suppliers and customers. Specific attributes, including information flow, for each step should be well documented and verified. These data should be as realistic as possible and show variation within the attributes if it exists. These data will be the starting point developing the future state map, which incorporates improvements and waste reduction.

Impact of Demand

The impact of demand on an operation cannot be understated. A key component to satisfying the customer is understanding their demands of the product. This is one of the single most important elements within the value stream. It is important to understand the pattern of demand as well, whether growing or declining, seasonal, or stationary. The producer must react quickly and effectively to changing demand to assure delivery reliability and cost effective operations. Demand variability can be mix driven, quantity driven, or as often the case, both. Demand variability can adversely affect delivery reliability, product quality, inventory costs, and total cost, among others, all with negative consequence to the customer. Demand is also utilized to determine Takt Time (from the German *Taktzeit*), the rate at which customers buy a single unit. Takt Time is discussed later in the chapter. Changing demand causes changes in Takt Time, which causes changes to required resources. If this flux is not understood and managed correctly, many of the adverse effects mentioned will quickly become a reality. It is recommended that if demand varies significantly, multiple Value

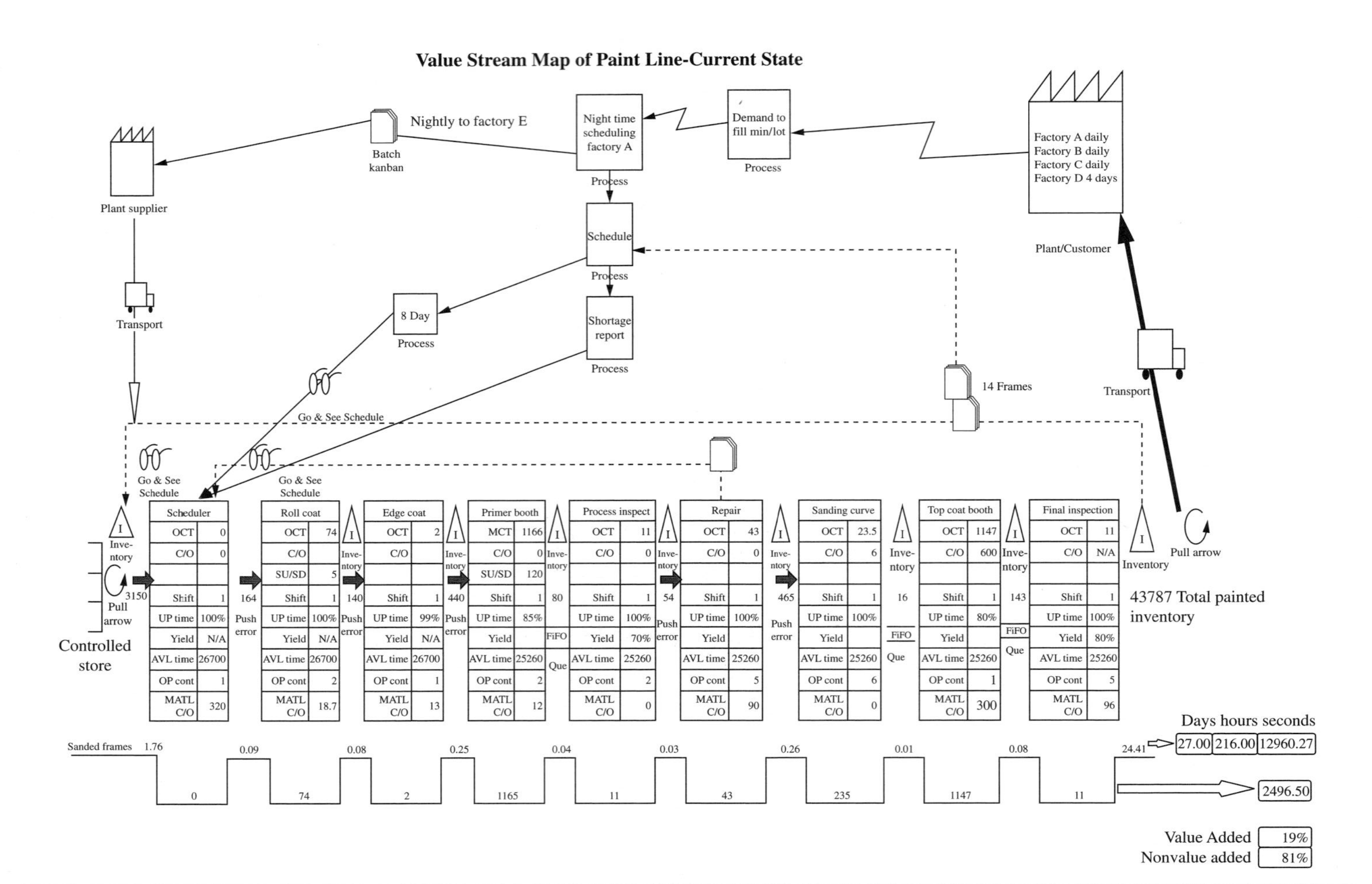

FIGURE 11.6 Value stream map for paint line—current state.

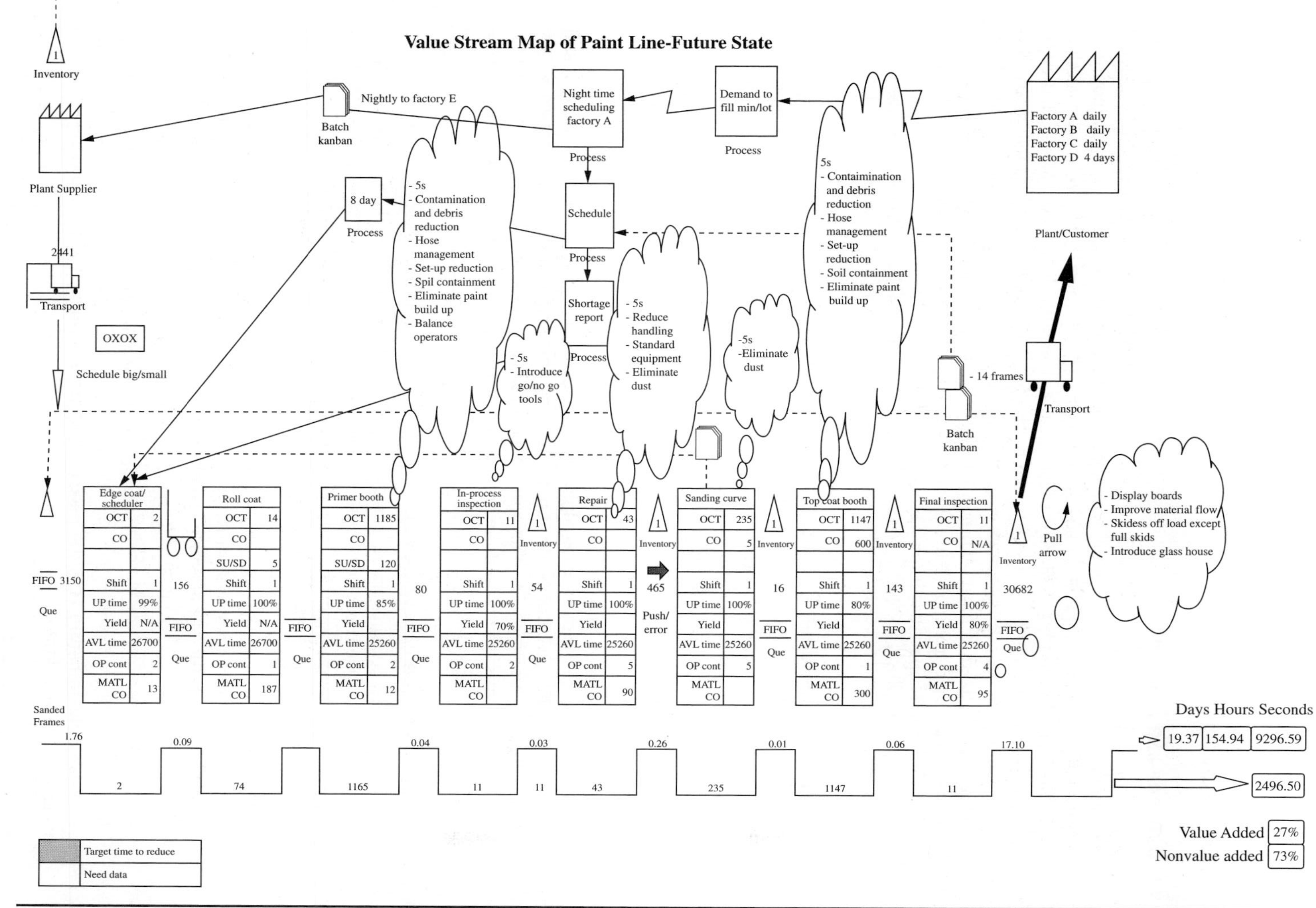

Edge coat/scheduler		Roll coat		Primer booth		In-process inspection		Repair		Sanding curve		Top coat booth		Final inspection	
OCT	2	OCT	14	OCT	1185	OCT	11	OCT	43	OCT	235	OCT	1147	OCT	11
CO		CO		CO		CO		CO		CO	5	CO	600	CO	N/A
		SU/SD	5	SU/SD	120										
Shift	1	Shift	1	Shift	1	Shift	1	Shift	1	Shift	1	Shift	1	Shift	1
UP time	99%	UP time	100%	UP time	85%	UP time	100%	UP time	100%	UP time	100%	UP time	80%	UP time	100%
Yield	N/A	Yield	N/A	Yield		Yield	70%	Yield		Yield		Yield		Yield	80%
AVL time	26700	AVL time	26700	AVL time	25260	AVL time	25260	AVL time	25260	AVL time	25260	AVL time	25260	AVL time	25260
OP cont	2	OP cont	1	OP cont	2	OP cont	2	OP cont	5	OP cont	5	OP cont	1	OP cont	4
MATL CO	13	MATL CO	187	MATL CO	12	MATL CO		MATL CO	90	MATL CO		MATL CO	300	MATL CO	95

FIGURE 11.7 Value stream map for paint line—future state.

Streams be developed; each with the specific Takt Time and specific resources to match customer expectations.

Capacity and Demand

Capacity and demand must balance to ensure proper flow. With too little capacity, you have unhappy customers; with too much capacity, you have waste. Capacity is the amount of output that a system is capable of sustaining over a given time. It is loosely calculated as Available Time divided by the longest Cycle Time. Theoretical capacity (also called engineered capacity or maximum capacity) can be thought of as output at the ideal state. This may be nameplate output information of a machine. It operates under perfect conditions, which are not realized in most facilities. On the other hand, demonstrated capacity can be calculated based on current, real-life situations. The difference between theoretical capacity and demonstrated capacity is improvement opportunity.

Demand should not be confused with capacity. Demand is the customer's requirements and is independent of the producer's abilities.

Value/Non-Value-Added Decomposition Analysis

The main goal of Lean is to identify and eliminate waste. This can be accomplished once we have a solid understanding of the process as it currently is. This is the first step to improvement; determine what is of value to the customer and what is not. As mentioned above, anything that does not provide value to the customer can be considered waste. If constructed carefully, the current state map will provide a wealth of opportunity for improvement. The basic premise of value/non-value-added decomposition analysis simply is to ask the question, is the customer willing to pay for this? This should be performed for each process step. If not, what can be done to reduce the waste or completely eliminate the waste all together? In some cases, due to the current capability of the process, a non-value-added activity is still required, at least for the time being. An example of a non-value-added, but necessary, task would include inspections or other quality checks. This activity will remain in place to ensure customer satisfaction until the process can be made robust enough not to require the non-value-added activity.

Flow and Takt Time

The concept of flow requires the rearrangement of mental thoughts regarding "typical" production processes. One must not think of just "functions" and "departments." We need to redefine how functions, departments, and organizations work to make a positive contribution to the value stream. Flow production requires that we produce at the customer's purchase rate and if necessary, make every product every day to meet customer's orders, i.e., to meet the pace or "drumbeat." The pace or drumbeat is determined by Takt Time. Takt Time comes from the German word for meter, as in music, which establishes the pace, or beat, of the music. It is the time that reflects the rate at which customers buy one unit.

$$\text{Takt Time} = \frac{\text{Available time (in a day)}}{\text{Average daily demand}}$$

For example, in Figure 11.8, the pace or Takt Time is calculated for the demand shown during a 10-day period.

Takt Time Calculation Example

To be practical, Takt Time may need to be modified, depending on the variability of the process. When modifying Takt Time beyond the simple equation, another name should be used,

Determine pace

Over 10 days	Demand
1	30
2	40
3	50
4	60
5	10
6	30
7	40
8	20
9	60
10	40
10	**380**

Per day:

$$\frac{\text{Time available in period (840 min.)}}{\text{Average demand (38)}} = \textbf{22.1 minutes}$$

Based on 2 shifts of 7 hours

FIGURE 11.8 Takt Time calculation example.

such as Cell Takt, Machine Takt, or Practical Takt. Although modifiers may be planned, they are still waste, or planned waste. Manpower staffing requirements can then be determined as follows:

$$\text{Minimum staffing required} = \frac{\text{Total labor time in process}}{\text{Takt Time}}$$

6S—A Plan for Neat and Clean Workplaces

Many workplace departments are dirty and disorganized. The benefits of an efficient and effective workplace include the means to prevent defects; accidents; and the elimination of time wasted searching for tools, documentation, and other important items to complete a work process. By focusing on the removal of the dirtiness and organizing the workplace departments, they will perform work safer, faster, and cheaper.

A simple tool called 6S now provides us with a framework to create a neat and clean workplace. Its steps are as follows:

- *Sort.* Remove all items from the workplace that are not needed for current operations.
- *Set in order.* Arrange workplace items so that they are easy to find, to use, and to put away.
- *Shine.* Sweep, wipe, and keep the workplace clean.
- *Standardize.* Make "shine" become a habit.
- *Sustain.* Create the conditions (e.g., time, resources, rewards) to maintain a commitment to the 6S approach.
- *Safety.*

Decades ago, industries producing critical items (e.g., health care, aerospace) learned that clean and neat workplaces are essential in achieving extremely low levels of defects. The quality levels demanded by the Six Sigma approach now provide the same impetus.

Perhaps the significance of the 6S approach is its simplicity. The benefits are obvious: The tools are the simplest work-simplification tools and are easy to understand and apply. Simple tools sometimes get dramatic results, and that is what has happened with 6S. For elaboration of the five steps (excluding safety), see The Productivity Press Development Team (1996); Figure 11.9.

6S should be implemented throughout the improvement process and sustained into the future, adjusting as needed. 6S provides a solid foundation for most all Lean tools and techniques.

A note on safety: Once the first 5Ss are firmly in place, a remarkable thing happens, the workplace becomes safer. Very often, no additional effect is required to achieve this benefit. With the work area sustaining organization and cleanliness, a 50 percent reduction in work related safety incidences can occur. Combining the 5S with a formal safety program can deliver amazing results, and it is called 6S for "success."

Inventory Analysis

Inventory is the amount of stock of any item or resource in an organization. In manufacturing inventory normally includes raw materials, finished goods, component parts, supplies, and Work in Progress (WIP). The purpose of inventory is to manage variation (demand, delivery, and the process itself), ease production scheduling, reduce setups, and balance the

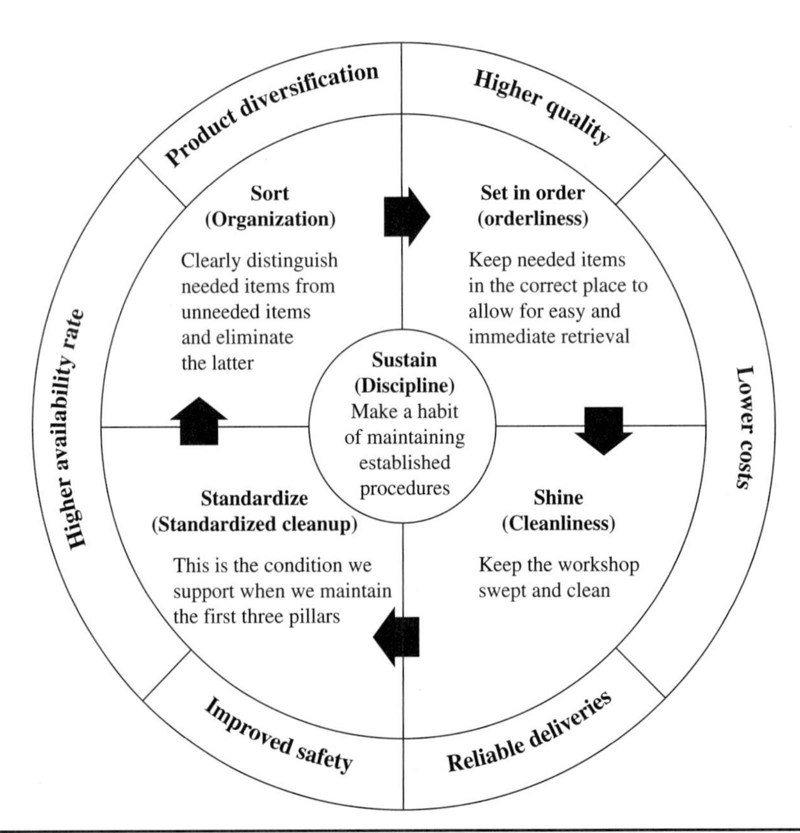

FIGURE 11.9 The 5S concept. (*The Productivity Press Development Team (1996). Reprinted with permission of Productivity Press.*)

quantity of the economic order. Although a certain volume of inventory can have strategic value, inventory is most often viewed as waste. Waste is the cash tied up in the materials and labor, and waste in storage and movement. Inventory is also open to damage, theft, and obsolescence. The aim of Lean is to reduce, if not eliminate, inventory.

There is a place for inventory besides in the hands of the customer. Inherent variation occurs in every process daily. Strategic inventories can compensate for process efficiencies and buffer customer demand fluctuations. Inventory is strategically placed and is set with calculated minimum-maximum stocking levels to ensure optimum flow through the process. When calculating stocking levels, one should consider customer demand (and variation), quantity consumed during replenishment, cycle time intervals for replacement, and impact of flow disruptions.

A regularly overlooked source of waste related to inventory is inventory inaccuracies. The differences between actual counts and recorded counts (commonly known as "book to actual") can be costly to both the producer and the customer. Measuring this difference can be the first step in improving accuracy. Another approach to improvement is cycle counting. Cycle counting is a physical inventory-taking task in which inventory is counted frequently rather than once or twice a year. Benefits of a more perpetual approach include more accurate inventory records, less overproduction, and less stockouts and can be prioritized based on value.

Inventory in all its forms should be eliminated, or at least minimized. When developing the process improvements, the Lean practitioner should review each point of inventory and ensure continuous flow, and, if necessary, set a countermeasure inventory against variation. The educational society American Production & Inventory Control Society (APICS, now the Educational Society for Resource Management) provides an excellent source of information supporting resource management. Using the Lean Inventory Analysis Tool can reduce the inventory by matching it to the level of demand that occurs in your supply chain.

Little's Law

In our quest to achieve a Lean environment, we are fortunate to have a very simple, yet powerful, relationship known as Little's law. Simply stated, Little's law is a straightforward mathematical relationship among WIP, lead time, and the process' throughput. Little's, law:

$$\text{WIP} = \text{TP} \times \text{LT}$$

where WIP = work-in-process, TP = throughput, and LT = lead time.

Rewritten:

$$\text{LT} = \frac{\text{WIP}}{\text{TP}}$$

This relationship shows that by reducing WIP, we can directly improve time to the customer through reduce lead time. It also states that if WIP inventories are allowed to vary, so will lead times. In other words, if WIP is held constant. so will lead times (see Figure 11.10).

Managing and Eliminating Constraints

A constraint is anything that limits a system from achieving higher performance or throughput. Constraints can come in many forms, including:

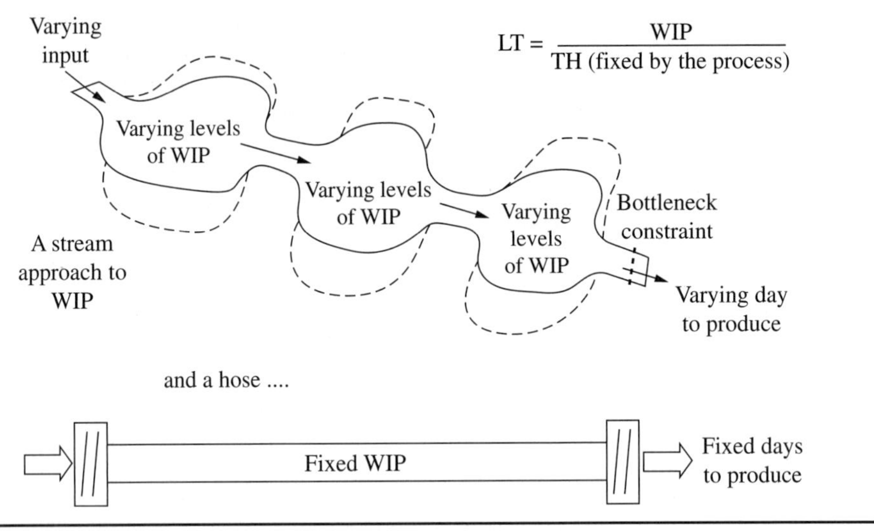

Figure 11.10 Little's law.

- *Equipment*. capacity, speed, capability
- *Labor*. supply, skills
- *Information*. speed, accuracy
- *Suppliers*. reliability, quality

This is an important concept when evaluating the current state value stream. When evaluating the value stream, special attention should be paid to the constraint. An improvement in any other area is, by definition, a waste; improvement should occur at the constraint. Once this resource is no longer a constraint, another resource will be the rate-limiting step. Focus should then move to the new constraint. The goal for a manufacturing organization is to drive the constraint to sales.

Goldratt's Theory of Constraints (Goldratt 1992) offered a five-step process for addressing constraints, involving the following:

- Identifying the constraint
- Deciding how to exploit the constraint
- Subordinating all else to the above decision
- Elevating the performance of the constraint
- Moving to the next constraint and go back to step 1

As we can see, this is an ongoing process to drive continuous improvement.

Improving the Process and Implementing Pull Systems

Once Takt Time has been calculated, each constraint (such as long setup times) should be identified and managed (or eliminated) to enable smaller batch sizes. Ideally, this leads to single-piece flow. If this reduction can be achieved, it will eliminate overproduction and excess inventories. Pull production scheduling techniques are used so that customer demand pulls demand through the value stream (from supplier to production to the customer). In pull production, materials are staged at the point of consumption. As they are consumed, a signal is sent back to previous steps in the production process to pull forward sufficient materials to replenish only what has been consumed.

The steps for improvement teams (or *kaizen* teams) to Lean out an operation are as follows:

- Determining the pace (Takt Time and manpower)
- Establishing sequence and replenishment (product family turnover and setup/changeover required)
- Designing the line or process (proximity, sequence, interdependence)
- Feeding the line or process (strategic inventory, standard WIP)
- Balancing the line or process (load, standard work)
- Stabilizing and refining (6S, continuous improvement)

Competitive pressures to reduce lead time are now a driving force to analyze processes for improvement. A flow diagram or preferably a Value Stream Map can reveal a wealth of sources for improvement such as:

- The number of functions and how they interact
- The extent to which the same macroprocess is used for the vital few customers and the useful many
- The existence of rework
- The extent and location of bottlenecks, such as numerous needs for signatures
- The location and amount of inventory

Numerous ways have been found to shorten the cycle time for processes. These include:

- Providing a simplified process for the useful many applications
- Reducing the number of steps and handoffs
- Eliminating wasteful "loops"
- Reducing changeover time
- Managing the constraint or bottleneck resource
- Reducing inventory

Physical Design and Proximity

As the Lean practitioner continues to evaluate the value stream for opportunities, it is not uncommon to find movement to be a waste. This is due to sequential operations not being in close physical proximity, as is often the case with departmentalized facilities. Simply moving processes closer together can improve flow and reduce waste of all types. When we expand this idea and group all the interdependent assets into a "cell," the benefits can be become even more significant. The cellular design will minimize space; a 50 percent reduction is common.

Cells should also be designed so that the steps are interdependent and run to the same Takt Time or pace. This approach will reduce inventory, reduce cycle times, and provide immediate quality feedback.

Another approach to aligning resources is the idea of group technology. Group technology is the process of examining all items produced by an organization to identify those with sufficient similarity that common design or manufacturing plans can be used. This would reduce the number of new designs or new manufacturing plans. In addition to the savings in resources, group technology can improve both the quality of design and the quality of conformance by using proven designs and manufacturing plans. In many companies, only 20 percent of the parts initially thought to require a new design actually need it; of the remaining new parts, 40 percent could be built from an existing design, and the other 40 percent could be created by modifying an existing design. Relocating production machines can also benefit from the group technology concept. Machines are grouped according to the parts they make and can be sorted into cells of machines, each cell producing one or several part families.

Balancing the Process

When designing improvements into a future state, smooth and sequenced flow is critical. The design should be balanced from step to step. Make process steps interdependent, and run to the same Takt Time with minimum inventory and the smallest lot sizes possible. In addition to reduced lead time as calculated by Little's law, this approach provides immediate quality feedback. As operations approach a continuous flow and single-piece processing, wastes will be quickly eliminated. Allocation of resources (people and equipment) to accomplish a series of tasks is minimized toward the idle point. Often, by combining work, the process can reduce the required resources by balancing new combined cycle times as close as possible to one another.

Kanbans: Signal to Produce

As mentioned earlier, nonstrategic inventory, excessive transporting, waiting, and overproduction are all forms of waste. An effective way to control these wastes is to use a signaling system to authorize production and motion within the value stream. This is sometimes, but not always, a card. The signaling device, whatever its type, is called a *kanban*. The device is used to conrol strategic inventory levels, standard WIP and is the trigger for a pull process. Some producers use marked-up floors to identify where the materials should be stored and in what quantity. When the space is empty, the supplying operation is approved to replenish the inventory. Containers can also be used as signaling tools; for example, when a container is empty, this triggers production of the upstream operation. Hopp and Spearman (2000) provide a detailed explanation of the design and applications of *kanban* systems.

Setup Reduction or SMED

In some processes, the waste associated with changeover from one product (process) type to the next scheduled can be sizeable. This was the case at Toyota, which promoted the work of Shigeo Shingo (1989) to reduce the changeover time for stamping presses from four hours to three minutes. The methods for reducing changeover were called "single minute exchange of die" (SMED). SMED is a set of techniques used to perform equipment setup and changeover operations in fewer than 10 minutes, or dramatically reduced from current levels. These principles can be applied to all types of changeovers.

The benefits of SMED include decreased inventory, improved capacity and throughput, and improved on-time delivery to the customer. The longer the setup time, the more likely the operation is to store inventory. Like equipment maintenance breakdowns, changeovers cost productivity which can not be recouped. Faster changeovers also improve flexibility to produce wider ranges of products at reduced costs (scrap, labor, and skills).

The primary steps to faster changeovers include:

- Moving as much of the work of change over from *internal* activity (which requires production to stop) to *external* activity (which can be completed without stopping production).
- Streamline the internal activity with the same principles as production: minimizing motion and travel, adjacency, and balancing. Then streamline external activity.
- Eliminate the need for adjustments and trial runs.
- Streamline external activity.

Although originally developed for changing capital equipment configurations for different product runs, the same principles have been applied to improving lead times for service and knowledge work—for example, staging the data for insurance underwriters so that they can began a new case immediately rather than having to retrieve the needed data, minimizing the time for a customer service representative to open a new case by prepopulating key fields in the case documentation, or organizing all audit data in a standard format to facilitate switching from one study to another.

Reliability and Maximizing Equipment Performance

Reliability is the ability to supply a product or service on or before it is promised. Within operations, this normally directly ties to a resource being able to consistently produce the quantity and quality demanded by the customer. To ensure quantity, the asset must be available when called upon. Maintenance excellence is the mindset to maximize resources through the highest levels of equipment consistency and dependability. Maintenance excellence is based on a sound philosophy of guiding performance, combined with a strong tactical approach for implementation. The overall philosophy is called total productive maintenance (TPM) and the tactical approach, reliability-centered maintenance (RCM).

Maintaining equipment is generally recognized as being essential, but pressures for production can result in delaying scheduled maintenance. Sometimes, the delay is indefinite, the equipment breaks down, and maintenance becomes reactive instead of preventive.

The planning should determine how often maintenance is necessary, what form it should take, and how processes should be audited to ensure that maintenance schedules are followed. Prioritizing maintenance activities is discussed as follows in RCM.

In the event of objections to the proposed plan for maintenance on the grounds of high cost, data on the cost of poor quality from the process can help to justify the maintenance plan.

Total Productive Maintenance (TPM)

Equipment maintenance used to be carried out by the operator. After work was organized and more specialized, maintenance was turned over to specialists. This was typically a small group of highly trained individuals who could fix nearly any problem with the equipment. It has become imperative to return as much of the routine maintenance responsibilities to operators. TPM looks into the value stream for improvements. TPM identifies the sources of losses and drives toward the elimination of all of them and focuses on zero losses (including quality losses) for productivity.

The operator forms the core of TPM and is the process expert. They are in the best position to help drive improvement in accidents, defects, and breakdowns. TPM is a philosophy based on total employee involvement, which is called autonomous maintenance. Operators are trained to stop abnormalities and other sources of accelerated deterioration. Operators will also perform daily checks for cleanliness, carry out routine lubrication, and tighten fasteners. Training is the key and should be incorporated with 6S mentioned earlier.

Reliability-Centered Maintenance (RCM)

TPM sets the overall philosophy and standards for maintenance. To complement this, a planning method is needed, a way to prioritize resources and actions. This is called reliability-centered maintenance (RCM). The goal of RCM is to ensure process reliability through data collection, analysis, and detailed planning. Like TPM, if properly deployed, RCM will drive down inventories, shorten lead times, provide more stable operations, and improve job satisfaction.

Prioritization is the foundation of RCM. The basic premise is to allocate resources as effectively as possible to eliminate unplanned downtime, reduce deteriorating quality, or ensure planned output. Assets are prioritized into one of three categories: reactive, preventive, and predictive. The reactive maintenance approach is to run to failure. These assets could include noncritical components, redundant equipment, small simple items, and assets with low failure rates. Examples would include electric solenoids, relay coils, lamps, and all breakdowns. The priority for this class is low; allow for running to failure. The next step is preventive maintenance. This set of assets has a known failure pattern and is often a time-based relationship. Consumables also fall into this group. Motor brushes, bearings and gears, filters, and most normal planned maintenance actions are some examples. Here, a planned schedule can be generated based on the number of cycles or a time interval, performing maintenance activities (hopefully) before failure. The final class is predictive maintenance. This category is the highest priority in terms of planning and assigning of resources. These resources are the most critical to the operations and the ones required to provide customer satisfaction. This group also includes assets with random failure patterns, assets not normally subject to wear, and replacement components with long lead times for replenishment. The group is analyzed based on condition. Methods such as vibration analysis, lubrication analysis, temperature, current signature, and high-speed videos can determine machine conditions. If successfully implemented, RCM can deliver significant business benefits. Experience with Juran's principles has shown that reactive maintenance costs are two to three times higher than preventative; and preventative is two to three times higher than predictive.

Measuring improvement in reliability should include several dimensions. The most encompassing is overall equipment effectiveness (OEE). This measures the cumulative effect of all losses due to equipment condition–machine availability, machine efficiency, and machine quality performance. Figure 11.11 shows a calculation OEE. Other measures for maximizing equipment performance include those found in Figure 11.12(a) and (b).

OEE calculation:

$$\text{Machine availability (MA)} = \frac{\text{Actual running time}}{\text{Planned running time}}$$

$$\text{Machine efficiency (ME)} = \frac{\text{Cycle time X units produced}}{\text{Uptime}}$$

$$\text{Machine quality performance (MQ)} = \frac{\text{Number of good units}}{\text{Total units produced}}$$

$$\textbf{OEE = MA X ME X MQ}$$

FIGURE 11.11 Calculation of overall equipment effectiveness.

Maintainability – Mean time to repair (MTTR)

$$\text{MTTR} = \frac{\text{Sum of downtime for repair}}{\text{number of repairs}}$$

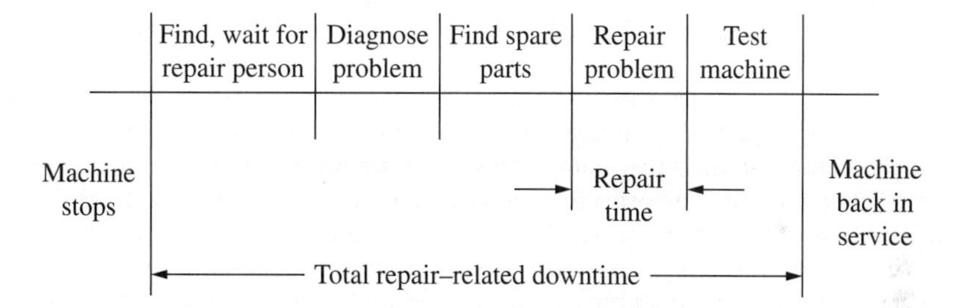

Reliability – Simple measures

$$\text{Machine availability (MA)} = \frac{\text{Actual running time}}{\text{planned running time}}$$

$$\text{Mean time between failure (MTBF)} = \frac{\text{Total running time}}{\text{Number of failures}}$$

FIGURE 11.12 (a) Maintainability: mean time to repair (MTTR), (b) Simple measures of reliability.

Mistake Proofing the Process

An important element of prevention is designing the process to be error free through "mistake proofing" (the Japanese call it *poka-yoke*).

A widely used form of mistake proofing is the design (or redesign of the machines and tools, the "hardware") to make human error improbable, or even impossible. For example,

Principle	Objective	Example
Elimination	Eliminating the possibility of error	Redesigning the process or product so that the task is no longer necessary
Replacement	Substituting a more reliable process for the work	Using robotics (e.g., in welding or painting)
Facilitation	Making the work easier to perform	Color coding parts
Detection	Detecting the error before further processing	Developing computer software that notifies the worker when a wrong type of keyboard entry is made (e.g., alpha versus numeric)
Mitigation	Minimizing the effect of the error	Using fuses for overload circuits

Table 11.1 Summary of Mistake-Proofing Principles

components and tools may be designed with lugs and notches to achieve a lock-and-key effect, which makes it impossible to misassemble them. Tools may be designed to sense the presence and correctness of prior operations automatically or to stop the process on sensing depletion of the material supply. For example, in the textile industry, a break in a thread releases a spring-loaded device that stops the machine. Protective systems (e.g., fire detection) can be designed to be "fail safe" and to sound alarms as well as all-clear signals.

In a classic study, Nakajo and Kume (1985) discuss five fundamental principles of mistake proofing developed from an analysis of about 1000 examples collected mainly from assembly lines: elimination, replacement, facilitation, detection, and mitigation (see Table 11.1).

Mistake proofing is both a proactive and reactive tool. As Figure 11.13 shows, the upper portion of the chart (prevent defects) highlights a proactive effort, whereas the lower part of the chart (mitigate errors) assumes a reactive effort because a problem already exists. It is better to use mistake proofing in a proactive mode. Stop defects from ever occurring by mistake-proofing products and processes at the design stage. However, the next best alternative is to prevent defects from passing along to the next operation, reactive mode.

Mistake proofing can, of course, result in defect-free work. The advantage can also include eliminating many inspection operations and requiring an immediate response when problems do arise. For more information on mistake proofing reference Mistake Proofing for Operators from The Productivity Press Development Team (1997).

Summary

Competitive pressures compounded with increased customer expectations with respect to quality, service, and price has prompted many businesses to seek creative solutions. These businesses are experiencing pressure to provide the lowest total cost of a product or a service with rapid order fulfillment in highly competitive markets. Lean implementation provides the tool kit and the methodology for organizations to focus on getting the right things, to the right place, at the right time, in the right quantity to achieve perfect work flow while minimizing waste and being flexible and being able to change. Value proposition from Lean implementation includes increases in customer satisfaction, cost reduction, and increase in shareholder value. Lean implementation increases operating profit and decreases inventory (a large draw on cash) and capital expenditures. In short, it is the right thing to do.

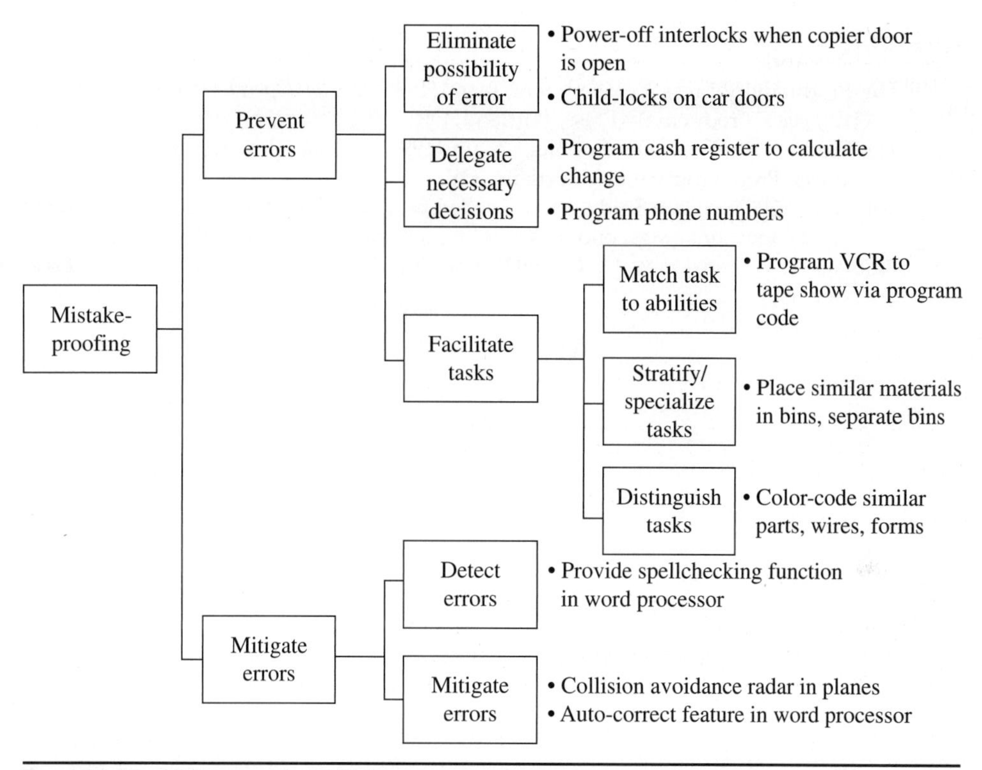

FIGURE 11.13 Mistake-proofing guidelines.

References

Duggan, K. (2002). *Creating Mixed Model Value Streams*. Productivity Press, New York.

Goldratt, E. M. (1992). *The Goal*, 2nd ed. North River Press, Great Barrington, MA.

Hopp, W., and Spearman, M. (2000). *Factory Physics*, 2nd ed. Irwin McGraw-Hill, New York.

Nakajo, T., and Kume, H. (1985). "A Case History: Development of a Foolproofing Interface Documentation System." *IEEE Transactions on Software Engineering*, vol. 19, no. 8, pp. 765–773.

Nicholas, J. (1998). *Competitive Manufacturing Management*. Irwin McGraw-Hill, New York.

Ohno, T. (1988). *Toyota Production System: Beyond Large-Scale Production*. Productivity Press, Portland, OR.

Rother, M., and Shook, J. (2003). *Learning to See*. The Lean Enterprise Institute, Cambridge, MA.

Rubrisch, R., and Watson, M. (2000). *Implementing World Class Manufacturing*. WCM Associates, Fort Wayne, IN.

Shingo, S. (1989). *A Study of the Toyota Production System from an Industrial Engineering Viewpoint*. Productivity Press, Portland, OR.

Shuker, T. J. (2000). "The Leap to Lean," *Annual Quality Congress Proceeding*, ASQ, Milwaukee, pp. 105–112.

Spear, S., and Bowen, H. K. (1999). "Decoding the DNA of the Toyota Production System," *Harvard Business Review*, September-October, pp. 96–106.

Tapping, D., Luyster, T., and Shuker, T. (2002). *Value Stream Management*. Productivity Press, New York.

The Productivity Press Development Team (1996). *5S for Operators—5 Pillars of the Visual Workplace*. Productivity Press, Portland, OR.

The Productivity Press Development Team (1997). *Mistake Proofing for Operators: The ZQC System*. Productivity Press, Portland, OR.

Volland, J. (2005). "Case Study: Now That's Lean." *Medical Imaging Magazine*, January 2005 http://www.imagingeconomics.com/issues/articles/mi_2005-01_07.asp.

Womack, J. P., and Jones, D. T. (2003). *Lean Thinking* (revised and updated). Free Press, New York.

CHAPTER 12

Six Sigma: Improving Process Effectiveness

Joseph A. De Feo and John F. Early

About This Chapter

Organizations have a number of methods available to deal with process and performance problems. We have defined a number of them in this handbook. This chapter is about the use of Six Sigma and Lean Six Sigma. They have become some of the most effective and widely used improvement programs in the history of quality. Six Sigma is a multifunctional, organizationwide method to improve process effectiveness and customer satisfaction. From its inception at the Motorola Company in the early 1980s through the release of this handbook (2010), it has been the choice of most organizations when needing to improve performance. Lean Six Sigma is a combination of Lean methods (which improve efficiency) and Six Sigma (which improves effectiveness). This combination has also led to a unified effort to improve organization performance.

High Points of This Chapter

1. Six Sigma and Lean Six Sigma have developed into one of the most widely recognized and effective methods for creating breakthrough improvement. Both have evolved from the basis of Juran's Universal on Quality Improvement.
2. Six Sigma methods focus on identifying and meeting the needs of customers first and the business second. In this way, revenues increase and costs decrease, improving results.

3. Many large organizations like Samsung Electronics, General Electric, and Honeywell have experienced great success employing Six Sigma and Lean Six Sigma methods since its inception at Motorola in the 1980s. Today, organizations like Naples Community Hospital Florida, The Mayo Clinic, Bank of America, Telefónica in Spain, and hundreds of others have adopted Lean Six Sigma as their improvement method of choice.
4. Six Sigma and Lean Six Sigma methods help both traditional manufacturers of goods as well as producers of services and information to improve their bottom line and increase customer satisfaction.
5. The two primary Six Sigma methods are DMAIC to improve processes and products (the focus of this chapter) and DMADV (Design for Six Sigma) to help ensure that products and processes function well from the voice of the customer through the delivery of goods. (This will be the focus of Chapter 14, Continuous Innovation Using Design for Six Sigma.)
6. The five steps to carryout a Six Sigma DMAIC project are discussed in detail.
7. A successful Six Sigma deployment depends on a clear understanding of roles, responsibilities, structures, and training requirements of the employee.

Six Sigma: A New Global Standard for Improvement

Six Sigma and Lean Six Sigma (which adds Lean tools to the basic methodology) are quality improvement methods with value-added enhancements of computers and an increasing array of statistical and other software packages. See more on breakthrough in Chapter 5, Quality Improvement: Creating Breakthroughs in Performance. For simplicity, we will refer to the full range of quality improvement methods and tools simply as Six Sigma for this chapter. (See Figure 12.1, Six Sigma and the Juran Trilogy.®)

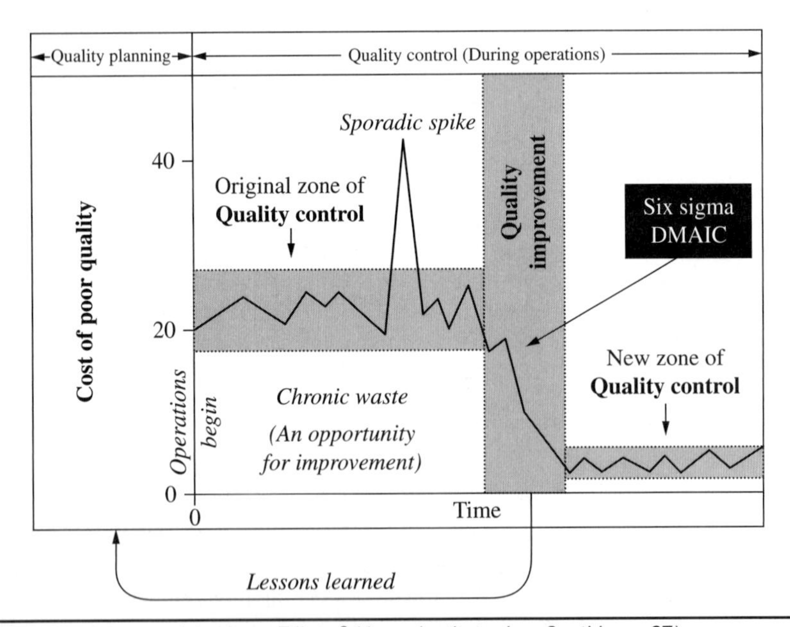

FIGURE 12.1 Six Sigma and the Juran Trilogy.® (*Juran Institute, Inc. Southbury, CT.*)

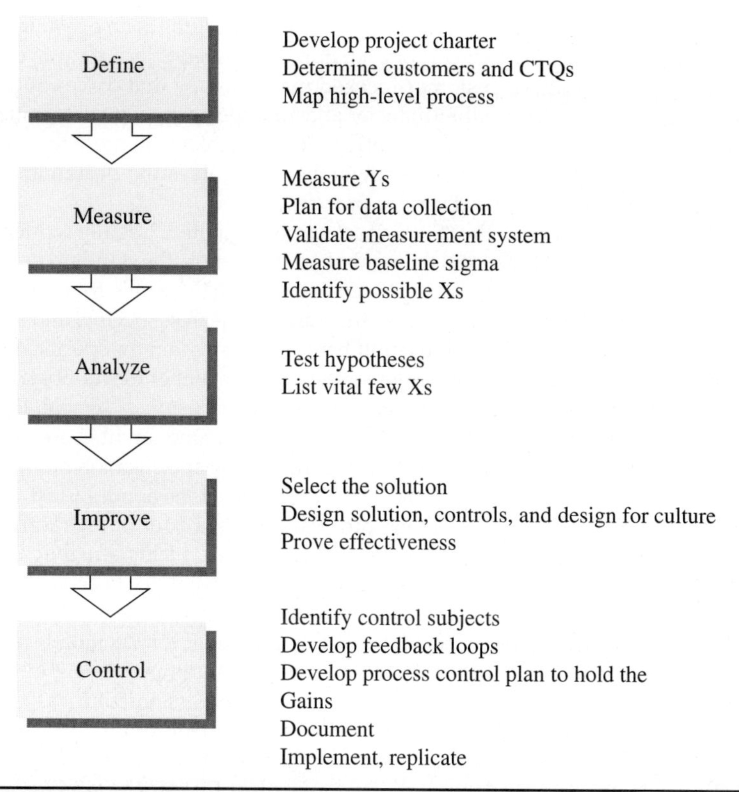

FIGURE 12.2 Six Sigma phases and steps.

If solutions to your problems are elusive, or if you must attain quality levels measured in parts per million or approaching perfection, Six Sigma will place your ailing process under a microscope to find solutions. Figure 12.2 presents the Six Sigma or DMAIC steps and tools most often used with it. The DMAIC steps are

1. **Define** the problem as clearly as one can in words.
2. **Measure** the current level of performance and voice of the customers.
3. **Analyze** collected data to determine the cause(s) of the problem.
4. **Improve** by selecting the right solutions to solve the problem.
5. **Control** to hold the gains.

With these fundamental steps, Six Sigma is enabling many organizations around the world to succeed in achieving performance breakthroughs where they had failed before. The smart companies recognize this as not simply a "fix" to one-time problems, but truly a new way of doing business. Business challenges do not go away in a free marketplace; rather, they continually change in degree and form. Organizations worldwide are under continuing pressure to control costs, maintain high levels of safety and quality, and meet growing customer expectations. This breakthrough improvement process of Six Sigma has been adopted by many companies, including Samsung Electronics, General Electric, Honeywell, and other organizations, as the most effective method for achieving these and other goals.

More than just a formal program or discipline, Six Sigma is an operating philosophy that can be shared beneficially by everyone: customers, shareholders, employees, and suppliers. Fundamentally, it is also a customer-focused methodology that drives out waste, raises levels of quality, and improves the financial and time performance of organizations to breakthrough levels. Six Sigma's target for perfection is to achieve no more than 3.4 defects, errors, or mistakes per million opportunities, whether it involves the design and production of a product or a customer-oriented service process.

It is from this target that the "Six Sigma" name originated. Usually written as a small sigma in the Greek alphabet, sigma (σ) is the symbol used to denote the standard deviation or measure of variation in a process. A process with less variation will be able to fit more standard deviations, or "sigmas," between the process center and the nearest specification limit than a process that is highly variable. The greater the number of sigmas within the specifications, the fewer the defects. The smaller the variation, the lower the cost. The higher the number of sigmas, the more consistent the process of delivering a good, product, or customer service. Figure 12.3 demonstrates a Six Sigma level of performance. This means that one can fit in six standard deviations, or six sigmas, between the process center and the nearest specification limit.

Most organizations operate at the Three Sigma level, or about 66,000 defects per million opportunities (DPMO) for most of their processes and at a Four or Five Sigma level in some of the mission-critical processes. Comparisons of Sigma levels, yields, and the corresponding defect rates are shown in Table 12.1. It would be foolish, however, to try to achieve Six Sigma levels of performance for every process in the organization. This is because not all processes are equally important. For example, the process for requesting time off for vacation is not as critical as the order fulfillment process. What really counts is significant improvement in the mission-critical areas—that is, critical as defined by the customer. These mission-critical aspects of the product, service, or process are called "critical-to-quality" requirements or CTQs for short.

Six Sigma Is Customer Focused—Organization Examples of Success

Why does Six Sigma work as well as it does? In large part, it is because of a strong emphasis on the customer. While the saying "the customer is always right" is not literally true, customers hold the key that can unlock unrealized potential in your business. Basically, the DMAIC process translates a customer's needs into actionable, operational terms and defines the critical

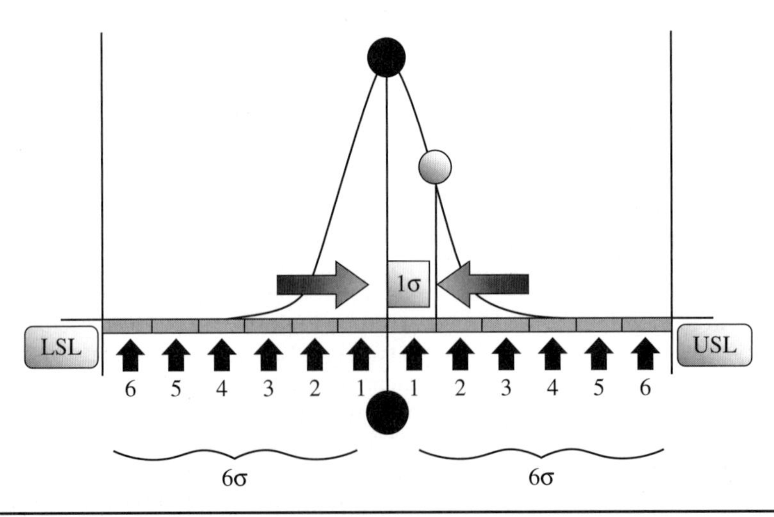

FIGURE 12.3 Six Sigma level of performance.

Process Sigma (Short Term)	Long-Term Yield	Defects Per Million
6	99.99966%	3.4
5.5	99.9968%	32
5	99.9767%	230
4.5	99.8650%	1,340
4	99.3790%	6,200
3.5	97.725%	22,700
3	93.319%	66,800
2.5	84.13%	158,000
2	69.15%	308,000
1.5	50%	499,000
1	31%	691,000
0.5	16%	841,000

TABLE 12.1 Sigma Level, Yield, and Defect Level

processes and tasks that must be done well to meet the customer needs. Although the details vary, depending on the analysis and improvement interventions that follow, Six Sigma consistently will drive the performance of products, services, and processes to breakthrough levels, that is, to new and sustained levels of performance. Breakthroughs are achieved not by massive teams or flashy initiatives, but by using a steady and concerted project-by-project approach. In this manner, the Six Sigma approach will help organizations:

- Improve cycle times, quality, and cost
- Improve effectiveness and efficiency of processes, including e-commerce
- Design products and services that will sell well
- Reduce chronic waste, or the cost of poor quality (COPQ)
- Grow profits by improving revenue and reducing costs

In short, Six Sigma is financially rewarding. Our experience indicates returns on investment (ROI) are achievable ranging from 10:1 to more than 100:1.

Samsung Electronics

When the decision was made by Samsung Electronics Company, Ltd., Vice Chairman and CEO Jong-Yong Yun to position the company for the future, the catalyst was Six Sigma. Samsung Electronics began its journey with training as the first essential step to prepare for implementing the methodology. Starting initially in manufacturing operations and R&D in 2000, the company expanded to transactional business processes and the entire supply chain, ultimately obtaining significant savings and financial benefits in all sixteen of its business units in South Korea and internationally. Ten years later, the methodology's philosophy and methods are being integrated still more deeply throughout the company by developing the internal specialists needed to teach, implement, maintain, and grow this competence in the future. No single person, nor any operation in Samsung Electronics, is exempted from the process, and the company is not looking back.

General Electric

Mr. Jack Welch, General Electric's retired CEO, was one of the first high visibility executives who became a Six Sigma leader and advocate. As an international business role model, he was vocal in expressing his views as to what leaders must do to achieve superior results. GE became an early adopter of Six Sigma, and through its demonstrated success and bottom-line results, enabled Mr. Welch to vault Six Sigma from the mail room to the boardroom. In his book *Winning*, he said "Six Sigma, originally focused on reducing waste and elevating the quality in our products and processes, has delivered billions of dollars to GE's bottom line in savings. Six Sigma has grown from an internally focused activity to an outside focus—also improving the productivity and efficiency of our customers' operations. Increasing the intimacy between GE and its customer base is making everyone more productive and helps all of us grow through tough economic environments." "Today," Mr. Welch explained, "Six Sigma has evolved to an even larger role in GE. Its rigorous process discipline and relentless customer focus has made it the perfect training ground and vehicle for the future leadership of GE. Our best and brightest employees are moving into Six Sigma assignments. I'm confident that when the board picks a successor to Jeffrey Immelt twenty years from now, the man or woman chosen will be someone with Six Sigma in his or her blood. Six Sigma has become the language of leadership in our company in GE. Its rigorous process discipline and relentless customer focus has made it the perfect training ground and vehicle for the future leadership of GE."

Six Sigma Works for Production, Service, and Transactional Processes

The Six Sigma movement gained interest in health care, financial services, legal services, engineering, consulting, and almost all organizations. In addition to achieving major improvement in manufacturing goods, managing inventory, delivering products, and managing repetitive processes, the Six Sigma methods have migrated to transactional processes. Processes that avoided continuous improvement because, as many stated, "the tools did not apply to us" have joined the Six Sigma bandwagon. Processes like completing an invoice, writing a contract, and boarding passengers on an airline, banking, hospitals, insurance, government, and other service organizations have tried Six Sigma. Most succeeded in

- Optimizing equipment usage
- Experiencing fewer rejects or errors
- Cutting response times to customer inquiries
- Reducing inspection, maintenance, inventory, and supply chain costs
- Creating more satisfied customers, external as well as internal

When implemented strategically, Six Sigma also

- Helps turn over working capital faster
- Reduces capital spending
- Makes existing capacity available and new capacity unnecessary
- Fosters an environment that motivates employees
- Improves morale, teamwork, and career potential

Telefónica

One of the biggest names in business in Spain and in the Spanish- and Portuguese-speaking world has a long tradition of quality management practices and achievements. So when the company embarked on a pilot Six Sigma program towards the end of 2000, the scale and ambition of the effort reflected the company's experience of business improvement initiatives. Between March and July 2001, some twenty-one first-phase projects were completed. Efficiency savings from these projects to date amount to more than 22 million euros; customer satisfaction levels are at all-time highs. A further eighty projects—phase two of the experimental stage—will be up and running by the end of 2001. Telefónica has already committed itself to 300 Six Sigma projects next year and estimates that it will have conducted 3000 projects within the next three years (European Quality, 2002).

The Six Sigma Model for Improvement has been widely used to address repetitive production-like processes and ones that address repetitive transactional processes.

We need to clearly establish the difference between production (aka, manufacturing) and service or transactional processes. All processes are transformations that result in the change of state of one or more things that can be physical objects or services. *Production processes* directly transform raw materials or semifinished goods into a final physical product (aka, goods). The output of production processes is a transformed physical product; these processes are deterministic, workflow-oriented, highly procedural, and, therefore, highly repeatable. Because of this, production processes are well suited for representation by the traditional, workflow-based triple role of input-process-output (I-P-O) or supplier-input-process-output-customer (SIPOC) models.

A process to produce goods is a series of work activities performed by people and other resource-consuming assets in order to transform given input(s) into output(s).

A service process or transactional process (sometimes also called people or paper processes) directly transforms one state or condition of one or more things (objects, abstractions such as information, data, symbolic representations, etc.) into another. One execution of a transactional process results in a transaction, the outcome of which, in turn, may be a change of state in a number of things (physical objects such as inventories, data and information, people, etc.). Examples of transactional processes include:

- Value-added service processes related to production (transporting, installing, storing, repairing, maintaining, etc.)
- Support or back-office processes in manufacturing and service organizations (selling, purchasing, subcontracting, warehousing, billing, human resources, etc.)
- Value-added processes in service industries (banking, insurance, transportation, health care, hospitality, education, etc.)
- Value-added processes in the public sector (including the military) and the not-for-profit sector (legislative and administrative processes, planning, command and control, fundraising, etc.)

The output of transactional processes is a change of state or condition, defined by the transaction. These processes are information (communication)-driven in that successive executions of a transactional process depend on the informational inputs (requests, offers, etc.) received at the outset of each execution. Accordingly, successive executions may be different with different results. Therefore, these processes are not always repeatable, but are self-regulating and highly adaptable. A transactional process is a logical set of customer-supplier tasks that drive work activities performed by people.

Transactional process characteristics that differentiate them from production processes may include:

- Scarcity of measurement data; available measurements are primarily discrete (attribute)
- Measurement system is partially or entirely I/T-defined (e.g., reporting)
- The definition of quality includes information quality

- Dominant variables: people and information
- High-cost labor
- Disproportionately large financial leverage

DMADV (Design) vs. DMAIC (Improvement)

As stated in the "High Points of This Chapter" section, there is another Six Sigma methodology for *designing and developing* a new product, service, or process with no defects. For elaboration on Design for Six Sigma, see Chapter 14, Continuous Innovation Using Design for Six Sigma.

Design for Six Sigma, DFSS for short, follows the D-M-A-D-V steps. DMADV is different from DMAIC as follows:

1. *Define.* Provides the goals and direction to design a new product or service with development of a team charter.
2. *Measure.* Translates customer needs into a CTQ. A CTQ is what is critical to quality in the eyes of the external customer. DMADV may deal with many CTQs in one design project. Six Sigma DMAIC typically focuses on only one CTQ that is creating customer dissatisfaction or related to the problem at hand.
3. *Analyze.* Understand the information collected from the voice of the customers and define the design features that collectively will be developed into a concept and then into one or more high-level designs. DMAIC focuses on identifying the root causes of the customer dissatisfaction and the problem at hand.
4. *Design.* In this step, the final product or service design is developed. A detailed design with associated design elements is completed and the critical-to-process variables are identified, from which the process for creating and delivering the good or service is developed.
5. *Verify.* The new design plans are implemented and the organization prepares for full-scale rollout and puts control mechanisms in place. In DMAIC, we control the process to hold the gains. In DMADV, we verify that the project goals are met, that the customer receives the value expected, and assure that control is effective to deliver on the CTQs and product design.

Key Roles to Deploying Six Sigma Successfully

Deploying a Six Sigma program requires building a suitable infrastructure, as described in Chapter 5, Quality Improvement: Creating Breakthroughs in Performance, and Chapter 9, The Juran Transformation Model and Roadmap. A number of key roles are important, as shown in Figure 12.4. Each role is essential, yet, by itself, insufficient to produce the improvement an organization expects from Six Sigma. Each role requires knowledge of the methods and tools. In addition, the Six Sigma community led by the American Society for Quality has established a standard curriculum and certification process for the roles of Green, Black, and Master Black Belts. Certification is granted upon completing subject matter training, carrying out a number of significant projects, and passing written and oral reviews.

The key roles to drive Six Sigma are

- Leadership
- Champion

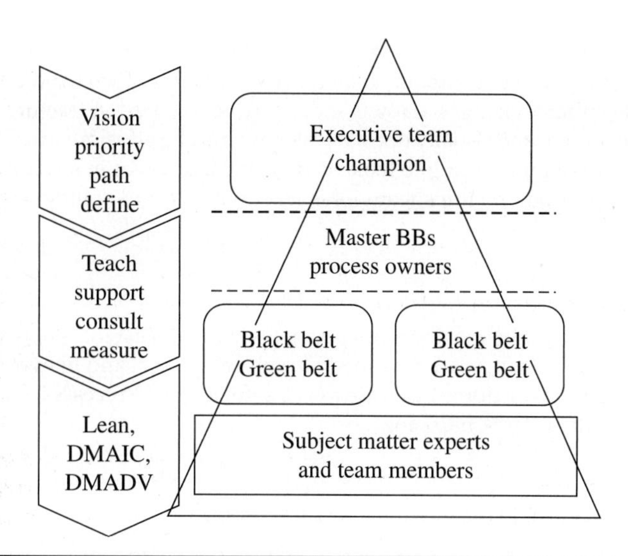

FIGURE 12.4 Key roles for Six Sigma. (*Juran Institute, Inc.*)

- Master Black Belt
- Black Belt
- Green Belt
- Project Team and Subject Matter Experts (SMEs)
- Process Owner

Leadership's Role

The roles of *all* the members of the organization leadership team to create annual breakthrough has been presented in Chapter 5, Quality Improvement: Creating Breakthroughs in Performance. Here is a summary of their roles when acting as a steering team:

- *Setting improvement goals.* Identify the best opportunities to improve performance and set strategic and annual goals for the organization. Establish accountability for meeting goals.
- *Establish infrastructure to enable Six Sigma Projects to happen.* Establish or revise management systems for selecting and assigning projects, organizational reporting of project progress, accountability of the various roles, performance appraisal, reward, and recognition.
- *Appoint Champions.* They can sponsor projects and ask the right questions at each phase of DMAIC of the Six Sigma project.
- *Support projects and monitor progress.* Enable project teams to carry out their project goals. Provide the necessary training, resources, facilities, budgets, time, and most importantly, management support. Monitor progress of projects and keep them on track.
- *Provide organizational support to deal with resistance to change that occurs when implementing breakthroughs.*
- *Become educated and receive training in the methods of Six Sigma to be able to support and evaluate the work of all the other roles.*

All members of the executive team and managers at all levels should be committed to the Six Sigma effort, agree to support it, and act with unified focus and consistency to facilitate the gradual cultural changes that will inevitably be required. A fractured executive and management team can, and usually does, wreak havoc and confusion on a Six Sigma effort, drains the energy out of those trying to make it succeed, and leaves in its wake disillusionment and meager results. The executive team loses its credibility and ability to lead (assuming it ever had any credibility).

Role of Champions

Champions are usually members of management (or at least folks with organizational clout). The ideal Champion is one who wants to sponsor a project and likes change. See Chapter 26, Empowering the Workforce to Tackle the "Useful Many" Processes, for more on Champions as change leaders. The Champion

- Identifies improvement projects that meet strategic goals
- Is responsible for creating a project charter
- Identifies and selects competent Belts and team members
- Mentors and advises on prioritizing, planning, and launching Six Sigma projects
- Removes organizational obstacles that may impede the work of the Belts or project teams
- Provides approval and support to implement improvements designed by the project teams
- Provides recognition and rewards to the Black Belts and teams upon successful completion of their projects
- Communicates with executive management and peers as to the progress and results associated with the Six Sigma efforts
- Removes barriers the teams encounter
- Understands and upholds the Six Sigma methodology

In general, Champions manage, support, defend, protect, fight for, maintain, uphold, and function as an advocate for Six Sigma. Usually, a strong Champion can be found behind every successful project. Weaker Champions are usually associated with weaker results.

After helping the steering team select projects, the Champions mentor and support the overall process. Once criteria are established and business unit managers and Champions are identified, projects are selected for their potential in breakthrough improvement. This means evaluating opportunities for strategic relevance, operational efficiency, product and service quality related to customer satisfaction or dissatisfaction, and bottom-line savings.

Six Sigma project teams are supported by the Champions and leadership of each business unit. As influential members of management, they are expected to promote the application, acceptance, and evolution of the process within their business units in the following ways:

- Project selection
- Leadership reviews
- Project support
- Resource allocation
- Career development

Role of Master Black Belts

A Master Black Belt receives training and coaching beyond that of a Black Belt. Master Black Belts are qualified to train Black Belts. The role of Master Black Belt includes:

- Acting as internal Six Sigma consultant, trainer, and expert on Six Sigma
- Managing and facilitating multiple projects—and their Black Belts
- Supporting and advising Champions and executive management
- Providing technical support and mentoring as needed

Everyone else in the organization—those who are not Champions, Master Black Belts, or Black Belts—become either a Green Belt or a team member (some organizations call them Yellow Belts). Suffice it to say that the different colored belts vary according to the amount of skill they will need, the formal training received, and the active roles each takes in participating in Six Sigma activities. In an ideal situation, all organization members receive training at some minimal level and are awarded the appropriate belt. Everyone feels included, and everyone understands what Six Sigma is all about, and just as important, what it is not about. No one is left to wonder what Six Sigma is all about or to resent or resist it. This unifies the organization behind the Six Sigma effort and significantly reduces pockets of resistance.

Role of Black Belts

Black Belts are on-site implementation experts with the ability to develop, coach, and lead cross-functional process improvement teams. They mentor and advise management on Six Sigma issues. Black Belts have an in-depth understanding of Six Sigma philosophy, theory, strategy, tactics, and Six Sigma tools. Each project is targeted to save at least $250,000 ROI per project. Black Belts are expected to guide three to six projects per year, which increases further the ROI of Six Sigma.

The training required to be certified as a Black Belt is rigorous and demanding. An illustrative list of topics would include:

Critical team leadership and facilitation skills	Correlation and regression
Six Sigma methodology	Hypothesis testing using attribute and variables data
Core improvement tools	ANOVA: Analysis of variance
Use of an appropriate statistical software package	DOE: Design of experiments
Measurement system analysis	EVOP: Evolutionary operations
Determining process capability	Lean enterprise principles and tools
Process mapping	Mistake-proofing
Quality function deployment	SPC: Statistical process control
FMEA: Failure mode, effect, and criticality analysis	Process control plans
Basic statistical methods	Transfer to operations

Armed with this training—usually delivered in four weeklong sessions with four- to five-week intervening intervals—the Black Belt is full-time and devoted to carrying out a real Six Sigma project. When Black Belt training has been completed, employees are able to

- Develop, coach, and lead cross-functional teams
- Mentor and advise management on prioritizing, planning, and launching projects

- Disseminate tools and methods to team members
- Achieve results that match the company's business strategies with a positive benefit to financial performance

Role of Green Belts

Employees who become members of each project team often enter the process by becoming Green Belts. A Green Belt requires four to eight days of training in the overall Six Sigma improvement methods and tools. They become key team members on a Black Belt-level project or can be leaders of smaller scoped projects.

Each week in the classroom is followed by four to five weeks of practical application on the same projects back in their business units. If properly selected, these initial projects will produce significant bottom-line savings and, typically, return more than the entire training investment. Each project is targeted to save at least $100,000 to $250,000 ROI per project.

The total number of employees trained in Six Sigma throughout the world must be in the hundreds of thousands by now. More and more companies, like Samsung and GE, are planning for these employees to move up the ranks to top management levels. In the final analysis, success in achieving results with this process depends on whether top management, particularly CEOs, accept responsibility for their nondelegable roles.

Mr. Bob Galvin at Motorola, Mr. Larry Bossidy at AlliedSignal—now Honeywell—and Mr. Jack Welch at GE were role models for making Six Sigma and opportunities for Black Belt employees a vital part of the culture during their tenure as CEOs. Top management can overcome the powerful forces in any organization that may resist unity of direction. The answer is to find a universal improvement process like Six Sigma that fits all functions in an organization. Six Sigma is an extremely healthy and productive cultural change that takes time to complete. It is not free. It requires resources and training, but customer satisfaction, quality products and services, and a highly competitive organization produce a significant return on investment, satisfaction all employees have from being on a winning team, and pride in being part of such an organization. (For more on leadership's role in promoting change, see Chapter 26, Empowering the Workforce to Tackle the "Useful Many" Processes.)

Roles of Project Team Members and Subject Matter Experts

The members of the Six Sigma team can come from throughout the organization and are often subject matter experts from the various functional departments that are involved in the operation or maintenance of the process under study. Team members are expected to attend all team meetings, contribute to the work process, and complete assignments given to them by the project leader between meetings. Often, the subject matter experts (SMEs) are of greatest value assisting the team

- When identifying key aspects of the problem and evaluating the appropriate goal for the project (Define phase)
- During the process flow diagramming activity by contributing their expertise (Measure phase)
- Collecting data about the parts of the process that they are most familiar with (Measure and Analyze phases)
- Identifying possible causes of the problem (Measure phase)
- Identifying possible failure modes and ranking their severity, occurrence, and detection during completion of the PFMEA (Measure phase)
- Developing possible solutions to the proven causes (Improve phase)
- Identifying control subjects for ongoing measurements of the product and process (Control phase)

Process Owners

Process owners are usually at the high supervisory or managerial level of the organization and are directly responsible for the successful creation of the product (goods, services, or information). They are typically not core team members, but may be called upon to assist the team with specific tasks as needed. Some of the most important needs for support from process owners occur during the Improve and Control phases when the team is

- Defining possible solutions to the proven causes of the problem
- Planning for dealing with cultural resistance
- Conducting pilot evaluations of possible solutions
- Implementing the selected improvements
- Designing the control plan and applying it to the everyday maintenance of the process performance
- Disbanding the team after project completion and turning full responsibility back to the operating forces

The Juran Transformation Roadmap

Any deployment of quality improvement methods, whether it is called Lean Six Sigma, Performance Excellence, Lean or by some other nomenclature, requires a methodical approach to be successful. A phased approach that starts small and then expands has been shown to be the most effective. See Figure 12.5 for a description of such an approach using the Juran Transformation Roadmap.

Decide

During the Decide phase of deployment, upper management is becoming familiar with the Six Sigma methodology (or whatever methodology is being considered) and evaluating how well the approach fits with their organization's strategies and goals, particularly those related to performance excellence. A decision must be made whether Six Sigma, or some other approach, best fits the organization's needs.

The upper managers must then decide what roadmap to follow. The one recommended here is an option, but variations on this could work as well. The important thing is that the managers have a roadmap to follow so that the deployment will be done in a methodical way.

A decision must be made at this point whom to select as a training partner. It would be quite rare for an organization to have qualified internal resources to train, consult with, and mentor the resources being developed during the deployment, so it is almost always necessary to contract outside resources. It's important to select a partner that fits well with the organization's culture, business style, and desires for implementation flexibility. The first training delivered by the partner selected should be an executive briefing, which is attended by the entire upper management team. For more on the importance of leading change from the highest levels of the organization, see Chapter 26, Empowering the Workforce to Tackle the "Useful Many" Processes.

Finally and very importantly, management must decide what they will stop doing. Resources in any organization are finite, and laying a Six Sigma deployment over lots of other projects that are underway is a recipe for failure. The current initiatives should be evaluated and prioritized, and only the vital few continued. For the organization to successfully weave Six Sigma into its culture, it must become the method of choice for creating breakthrough improvement. Launching Six Sigma projects that mirror others already underway will lead to confusion over ownership of the problem and, likely, failure to effectively solve it.

Decide	Prepare	Launch	Expand	Sustain
Learn about Lean and Six Sigma. Assess organization status-identify strategies, goals, and projects. Determine if Lean Six Sigma is to become part of the strategy. Decide on a "roadmap" to follow. Select external training/consulting partner. Attend an executive briefing. Discuss other initiatives and their impact on resources.	Organize the steering team to manage the process. Nominate and appoint champions. Nominate and select pilot projects. Train the Champions and Black Belts Decide on the use of Green Belts and train them. Develop first wave plan. Select projects for first wave.	Support and mentor black and Green Belts. Support pilot project teams. Allow time to work on projects. Develop a project nomination process. Develop on-going cost of poor quality metric. Integrate participation reward and recognition. Establish assessments/meas-urements for on-going project selection.	Support infrastructure and review progress. Support expansion of all types and number of teams to other business units. Mandate improvement to all levels: other Belts. Begin product development and design teams. Create key macro-business process teams. Identify benchmarking opportunities.	Integrate process measures and move toward process owners. Fully Integrate Lean Six Sigma goals into next year's business plan. Deploy Lean Six Sigma to all business units. Enable employee participation with training and resources. Act on audits of business systems to drive new projects. Continue to assess culture and act on gaps. Sustain breakthrough performance.

FIGURE 12.5 The Juran Transformation roadmap. (*Juran Institute, Inc.*)

Consider the following case in point: A Juran client that has undertaken a Lean Six Sigma (LSS) deployment failed to do this "stop doing" exercise as fully as they should have. As a result, a couple of the first-wave projects devolved into turf battles over whose solution to or analysis of the problem was best. Since the other "non–Six Sigma" projects were initiated by the process owners before the LSS deployment, the Six Sigma teams ultimately had to abandon their projects for lack of implementation support and be assigned another project to complete their training requirements. These events led to tarnishing the reputation of the LSS deployment and reduced its chances for ultimate success. One key factor leading to this outcome was lack of involvement and buy-in at the highest levels of the organization.

Prepare

During this second phase of the deployment, upper management begins to define the support infrastructure. The steering team or teams are established. A team should be designated at the corporate level to oversee the wide deployment, and ultimately at divisional and even unit levels to oversee the deployment regionally and locally. For a graphic depiction of this arrangement, see Figure 12.6.

In addition to the steering teams being developed, this is the time to select and train the initial group of project Champions. These people then can participate with the steering team in nominating the initial projects to be undertaken (see the discussion on nominating projects in the "Select the Problem" section later in the chapter).

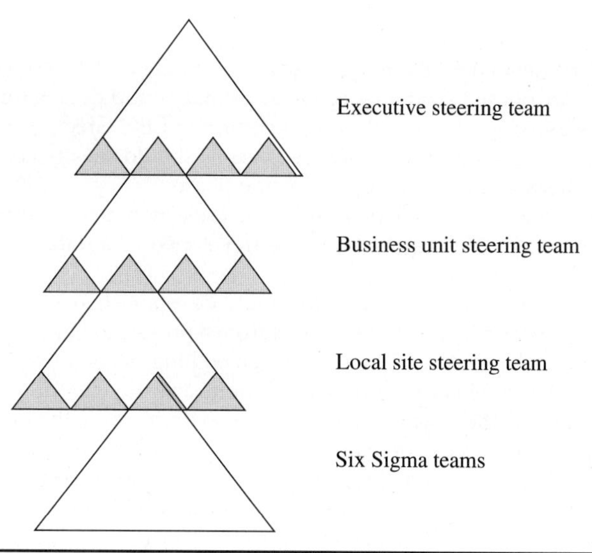

FIGURE 12.6 Linking steering teams together. (*Juran Institute, Inc.*)

The next step in this phase is very important: training the initial wave of Belts. There are a number of ways to approach this, including:

- Only a Green Belt workshop at this phase and then Black Belt training during the Expand phase
- A combined Green Belt and Black Belt workshop with the whole class attending ten days of training over a four-week period and the Black Belt participants attending the full twenty days over that period. See Figure 12.7 for a sample schedule for this method.
- Two workshops: a ten-day Green Belt and a twenty-day Black Belt. This method is preferable if there are sufficient candidates to fill both workshops.

In any case, there should be four to five weeks between training weeks for the students to complete project work. By using this approach, within a month or two of the completion of training, the first projects should be complete.

	Week 1 define and team skills	Week 2 measure	Week 3 analyze	Week 4 improve/ control
Day 1	GB/BB	GB/BB	GB/BB	GB/BB
Day 2	GB/BB	GB/BB	GB/BB	GB/BB
Day 3	BB Only	GB/BB	GB/BB	BB Only
Day 4	BB Only	BB Only	BB Only	BB Only
Day 5	BB Only	BB Only	BB Only	BB Only

FIGURE 12.7 Combined GB/BB training schedule. (*Juran Institute, Inc.*)

Launch

Due to the practicum period between weeks of training noted above, the Launch phase overlaps the Prepare phase in some aspects. The primary effort during this phase is the execution of the first wave of projects, including mentoring of the Green Belts (GBs) and Black Belts (BBs) by the designated coaches. Those coaches should ideally come from the designated training partner identified during the Decide phase.

Also during this phase, the ongoing project selection method should be institutionalized and the ongoing COPQ metric should be developed as a data source to help with future project selections.

A reward and recognition program should be established at this time as well. Monetary rewards, sometimes a share of the savings from a project, are often but not universally used. Intangible rewards, such as desirable career pathing, recognition, and pride in a job well done, can be effective instead of or in addition to any monetary rewards.

This is the time the organization should also decide how the improvement projects will be tracked and measured. There are a number of commercially available products, such as Power Steering, Minitab's Quality Companion, and i-nexus, which are widely used to track six sigma deployments. Internally developed solutions based on Microsoft Access or SharePoint are also widely used. Effective applications include a project "hopper" or "pipeline" for nominating future projects to be considered by the steering teams for execution.

Expand

This phase includes what the name implies: expansion of the methodology to other divisions, additional (often deeper) levels of the organization, and additional project methodologies (e.g., value stream improvement, DFSS). Expansion to different types of processes that may not have been considered during the initial wave of projects will also begin during this phase. This would include key macro business processes, for example, order processing, strategic planning, and price setting (see Figure 12.8).

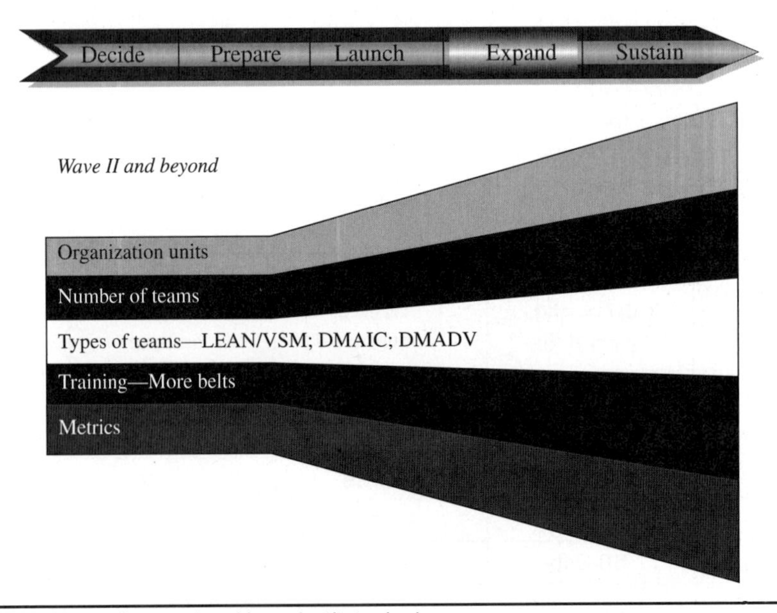

FIGURE 12.8 Areas of expansion. (*Juran Institute, Inc.*)

Benchmarking other divisions, companies, and industries for best practices in Six Sigma implementation is also a key activity of this deployment phase. This is aimed at bringing your deployment to a world-class level.

During this phase, companies will often decide to begin developing their own internal resources to conduct the workshops and coach projects going forward. The development of Master Black Belts and Lean Masters during this phase is usually facilitated by the same training partner the organization has employed thus far.

Sustain

The Sustain phase is intended to solidify the results and methods implemented during the prior phases of deployment. This is the time when the deployment becomes ingrained in the corporate culture. The six sigma goals are integral to the organization's yearly strategy deployment and are widely deployed to all business units.

All employees should now have access to training in at least basic six sigma principles and tools and be empowered to improve quality in the workplace every day. Six Sigma should be a way of corporate life.

A key sustaining feature is the conducting of audits on a regular basis. Management should review the audit results and take action on any gaps identified. Often, these gaps will point to additional projects that should be undertaken. The audits may also point to gaps in adoption of the process by the corporate culture. The best way to address these remaining gaps is to up the level of involvement and extend it to all levels (if it hasn't already been) from the top of the organization to the bottom. Ultimately, there may be some who just refuse to get on board. In these cases, the ultimate decision may be that they should find employment elsewhere.

The Six Sigma DMAIC Steps

Experience with applying the five DMAIC steps shows that the team's DMAIC journey needs to be preceded by management selecting the project. The DMAIC journey with its five steps and some of the critical activities in each step are shown in Figure 12.2. Here is a brief explanation of each step. For detail on the tools used in each step, refer to Chapter 18, Core Tools to Design, Control, and Improve Performance, and Chapter 19, Accurate and Reliable Measurement Systems and Advanced Tools.

Select the Opportunity

In the Select phase, potential projects are identified. Nominations can come from various sources, including customers, reports, and employees. To avoid suboptimization, management has to evaluate and select the projects. While evaluation criteria for project selection are many, the most frequent basis should be the costs of poor quality (COPQ) at the organization or division level. Other criteria include impact on customer loyalty, employee effectiveness, and conformance with regulatory or other requirements. The project problem and goal statements are prepared and included in a team charter, which is confirmed by management. Management selects the most appropriate personnel for the project, assures that they are properly trained, and assigns the necessary priority. Project progress is monitored to ensure success.

Select: Deliverables

- List of potential projects.
- ROI and contribution to strategic business objective(s) for each potential project.

- List of potential projects.
- Evaluation of projects.
- Selected the project.
- Project problem, goal statements, and a team charter for each project.
- Formal project team(s) headed by Black Belt

Select: Questions to be Answered

1. What customer-related issues confront us?
2. What mysterious, costly quality problems do we have that should be solved?
3. What are the likely benefits to be reaped by solving each of these problems?
4. Which of problems deserves to be tackled first, second, etc.?
5. What formal problem statement and goal statement should we assign to each project team?
6. Who should be the project team members and leader (Black Belt) for each project?

Define Phase

The Define phase completes the project definition begun with the charter developed during selection. The team confirms the problem, goal, and scope of the project. The completed definition includes the following:

- Identify key customers related to the project.
- Determine customer needs with respect to the project in the voice of the customer (VOC).
- Translate the VOC into CTQ requirement statements.
- Define a high-level process flow to define the project limits.

Define: Deliverables

- Confirmed project charter
- Voice of the customer
- CTQ statements
- A high-level flow, usually in the form of a supplier-input-process-output-customer (SIPOC) diagram

Define: Questions to be Answered

1. Exactly what is the problem, in measurable terms?
2. What is the team's measurable goal?
3. What are the limits of the project? What is in and what is out of scope?
4. What resources are available—team members, time, finances—to accomplish the project?
5. Who are the customers related to this project?
6. What are their needs and how do we measure them in practical terms?

Measure Phase

The project team begins process characterization by measuring baseline performance (and problems) and documenting the process as follows:

- Understand and map the process in detail.
- Measure baseline performance.
- Map and measure the process creating the problem.
- Plan for data collection
- Measure key product characteristics (outputs; Ys) and process parameters (inputs; Xs).
- Measure key customer requirements (CTQs).
- Measure potential failure modes.
- Measure the capability of the measurement system.
- Measure the short-term capability of the process.

Map the Process

Focusing on the vital one (or few) outputs (Ys) identified by the Pareto analysis, graphically depict the process that creates it (them) by mapping the process with a flow diagram in order to understand the process anatomy.

Determine Baseline Performance

Measure the actual performance (outputs; Ys), such as costs of poor quality, number of defects, and cycle times of the process(es), which creates the problem to discover—by Pareto analysis—which vital outputs (Ys) make the greatest contribution to the problem.

Identify Key Input and Output Variables

To understand the process in more detail, analyze the flow diagram to identify KPIVs (Xs) and KPOVs (Ys) associated with each process step, and indicate which of these process steps add value and which don't. Focus on the most significant variables. In order to narrow further the focus of the project team, and to prioritize which specific KPIVs (Xs) and KPOVs (Ys) the team will examine, create a functional deployment matrix (FDM) utilizing the list of KPIVs and KPOVs generated in the process analysis. The FDM will identify the KPIVs and KPOVs that have the greatest impact on key customer requirements. It will also translate customer requirements into product design specifications (desired Ys) and, in turn, translate design specifications into appropriate part, process, and production requirements (desired Xs, CTQs).

Measure Potential Failure Modes

Referring to the analyzed process flow diagram (PFD) and the FDM, for each process step, perform a failure mode and effect analysis (FMEA) by listing potential process defects (Ys) that could occur, their effects (Ys; KPOVs) and their potential causes (KPIVs, Xs). (An additional source of ideas of possible KPIVs is the cause–effect diagram, which displays brainstormed possible causes for a given effect.) In addition, rate the severity of each effect, the likelihood of its occurrence, and the likelihood of its being detected should it occur. Upon completing the analysis, you will be able to identify those potential process failures that have the most risk associated with them. These results are used to further focus the project on those variables most in need of improvement.

Plan Data Collection for Short-Term Capability Study

- In preparation for determining the capability of the measurement system (which measures the KPIVs and KPOVs upon which the project team has focused) and the short-term capability of the process, create a sampling and data collection plan.
- In preparation for determining the short-term capability of the process, determine the capability of the measurement system (which will be used to measure KPIVs and KPOVs) to provide consistently accurate and precise data upon which the project team can depend to "tell the truth" about the process.
- If the measurement system is found to be not capable, take corrective action to make it so.
- If the measurement system is found to be capable, proceed with the next step—determining if the process is in statistical control with respect to given variables (Ys).

Measure the Short-Term Capability of the Process

- In preparation for measuring short-term capability of the process to meet given specifications (Ys), ascertain whether the process is in statistical control with respect to the given output (Y) of interest. A good way to measure process stability is to use a control chart to plot the process data and discover any indications of instability.
- If the process is not in statistical control—that is, if control charts detect special causes of variation in the process—take action to remove the special causes of variation before proceeding with the process baseline performance measurement.
- If the process is in statistical control—that is, the control charts do not detect special causes of variation in the process—perform a short-term capability study to provide baseline data of the ability of the process to consistently produce a given output (Y).

Confirm or Modify the Goal

In light of discoveries made during measurement of the current process performance, determine if the problem and goal statements are still appropriate for this project.

- Evaluate the project's problem statement and goal statement.
  - Does the problem statement and the goal statement meet the criteria of an effective problem and goal statement with clearly defined boundaries?
  - Are the same variables and units of measure found in the problem statement also found in the goal statement?
  - Can the project be handled by a single team?
  - Does it avoid unnecessary constraints but still specify clearly any necessary global constraints, such as organizational strategy?
  - Are there any points that need clarification or modification?
  - Are the team members representative of departments, divisions, or work units affected by the project? The detailed process flow diagram, particularly if it is constructed in "swim lane" fashion, can help with this.
- Verify that the problem truly exists. If the problem has not been measured, the team must do so at this point.

- Validate project goal(s). Verify that the basis for the project goal(s) is (are) from one or more of the following:
  - Technology
  - Market
  - Benchmarking
  - History
- Modify the problem statement and goal statement if either does not meet the criteria above.
- Obtain confirmation from the leadership team, Champion, Black Belt, or quality council on any necessary changes to the project goal or to team membership.
- Create a glossary (list of operational definitions) for your project that will serve as a "dictionary" for important terms relating to your project. Select a team member to act as glossary chief with the responsibility of maintaining the project glossary.

List Theories of Root Cause Based on the Process Flows and Measures

The team needs to develop a comprehensive and creative list of theories of root cause. A root cause is a factor that affects the outcome, would eliminate or reduce the problem if it were removed or mitigated. Tools typically used include Cause-Effect (fish bone or Ishikawa) diagrams, FMEA, and Fault Tree Analysis.

Measure: Deliverables

- Baseline performance metrics describing outputs (Ys)
- Process flow diagram; key process input variables (KPIVs); key process output variables (KPOVs); cause–effect diagram; function deployment matrix (FDM)—optional; potential failure mode and effect analysis (FMEA) (to get clues to possible causes [Xs] of the defective outputs [Ys])
- Data collection plan, including sampling plan
- Gage reproducibility and repeatability or attribute measurement system analysis (to measure the capability of the measurement system itself)
- Capability measurement in terms of defect rates, capability indexes, and/or Sigma levels.
- Confirmed or modified project goal.
- Prioritized list of theories of cause based on Cause-Effect analysis, FMEA, or similar tools

Measure: Questions to be Answered

1. How well is the current process performing with respect to the specific Ys (outputs) identified to Pareto analyses?
2. What data do we need to obtain in order to assess the capability of (a) the measurement system(s) and (b) the production process(es)?
3. What is the capability of the measurement system(s)?
4. Is the process in statistical control?
5. What is the capability of the process(es)?

6. Does the project goal need to be modified?
7. What are all the possible root causes for the problem?

Analyze Phase

In the Analyze phase, the project team analyzes past and current performance data. Key information questions formulated in the previous phase are answered through this analysis. Hypotheses on possible cause–effect relationships are developed and tested. Appropriate statistical tools and techniques are used: histograms, box plots, other exploratory graphical analysis, correlation and regression, hypothesis testing, contingency tables, analysis of variance (ANOVA), and other graphical and statistical tests may be used. In this way, the team confirms the determinants of process performance (i.e., the key or "vital few" inputs that affect response variable[s] of interest are identified). It is possible that the team may not have to carryout designed experiments (DOEs) in the next (Improve) phase if the exact cause–effect relationships can be established by analyzing past and current performance data.

Procedure to analyze response variables (outputs, Ys) and input variables (Xs):

- Perform graphical analysis using tools such as histograms, box plots, and Pareto analysis.
- Visually narrow the list of important categorically discrete input variables (Xs).
- Learn the effects of categorically discrete inputs (Xs) on variable outputs (Ys) and display the effects graphically.
- Perform correlation and regression to
    - Narrow the list of important continuous input variables (Xs) specifically to learn the "strength of association" between a specific variable input (Xs) and a specific variable output (Ys).
- Calculate confidence intervals to
    - Learn the range of values that, with a given probability, include the true value of our estimated population's parameter, which has been calculated from a sample (e.g., the population's center and/or spread).
    - Analyze relationships between specific Ys and Xs, to prove cause–effect relationships.
    - Confirm the vital few determinants (Xs) of process performance (Ys).
- Perform hypothesis testing using continuous variables data to
    - Answer the question, Is our population actual standard deviation the same as or different from its target mean? Perform 1 variance test.
    - Answer the question, Is our population actual mean the same as or different from its target mean? Perform 1-sample t-tests.
    - Answer the questions, Is our population mean the same or different after a given treatment as it was before the treatment? or Is the average response at level 1 of the X factor the same or different as it is at level 2 of that factor? Perform 2 sample t-tests, or if there is a natural pairing of the response variable, paired t-tests.
    - Answer the question, Are several (>2) means the same or different? Perform analysis of variance.

Note: The above tests are referred to as parametric tests because they assume normally distributed response data and, in the case of ANOVA, equality of variances across all levels of the factor. For a discussion of nonparametric (also referred to as "distribution tree") tests to use when assumptions of normality and or equality of variances are violated, see Chapter 19, Accurate and Reliable Measurement Systems and Advanced Tools.

Perform hypothesis testing using attribute data to

- Answer the question, Is the proportion of some factor (e.g., defectives) in our sample the same or different from the target proportion?" Perform a Minitab test and calculation of confidence interval for one proportion.
- Answer the question, Is proportion1 the same or different from proportion2? Perform the binomial proportions test and calculation of confidence interval for two proportions.
- Answer the question, Is a given output (Y) independent of or dependent on a particular input (X)?" (This involves testing the theory that a given X is an important causal factor that should be included in our list of vital few Xs.) Perform a chi-squared test of independence (also called a contingency table).

Analyze: Deliverables

- Histograms, box plots, scatter diagrams, Pareto analysis, correlation and regression analyses (to analyze response variables [Ys])
- Results of hypothesis testing (to analyze input variables [Xs])
- List of vital few process inputs (Xs) that are proven root causes of the observed problem.

Analyze: Questions to be Answered

1. What patterns, if any, are demonstrated by current process outputs (Ys) of interest to the project team?
   - Analyze response variables (outputs; Ys).
   - Analyze input variables (Xs).
   - Analyze relationships between specific Ys and Xs, identifying cause–effect relationships.
2. What are the key determinants of process performance (vital few Xs)?
3. What process inputs (Xs) seem to determine each of the outputs (Ys)?
4. What are the vital few Xs on which the project team should focus?

Improve Phase

In the Improve phase, the project team seeks to quantify the cause–effect relationship (mathematical relationship between input variables and the response variable of interest) so that process performance can be predicted, improved, and optimized. The team may utilize DOEs if applicable to the particular project. Screening experiments (fractional factorial designs) are used to identify the critical or "vital few" causes or determinants. A mathematical model of process performance is then established using 2k factorial experiments.

If necessary, full factorial experiments are carried out. The operational range of input or process parameter settings is then determined. The team can further fine-tune or optimize process performance by using such techniques as response surface methods (RSM) and evolutionary operation (EVOP). (See Chapter 19, Accurate and Reliable Measurement Systems and Advanced Tools, for a full discussion of DOE.)

Procedures to define, design, and implement improvements include

1. Plan designed experiments.
2. Conduct screening experiments to identify the critical, vital few process determinants (Xs).
3. Conduct designed experiments to establish a mathematic model of process performance.
4. Optimize process performance.
5. Evaluate improvements.
6. Design the improvement.

Plan Designed Experiments

- Learn about DOEs in preparation for planning and carrying out experiments to improve the "problem" process.
- In preparation for designing factorial experiments, learn about randomized block design.
- Design in detail the experiments required by the project.

Conduct Fractional Factorial Screening Experiments

- Perform fractional factorial screening experiments to reduce even further the list of input variables to the vital few that strongly contribute to the outputs of interest. (A relatively large number of factors [Xs] are examined at only two levels in a relatively small number of runs.)

Conduct Further Experiments, If Necessary, to Develop Mathematical Model and Optimize Performance

- Perform 2k factorial experiments. Multiple factors (Xs, identified by screening experiments) are examined at only two levels to obtain information economically with relatively few experimental runs. Precise mathematical relationships between Xs and Ys are discovered by constructing equations that predict the effect on output Y of a given causal factor X. In addition, not only are the critical factors (X) identified, but also the level at which each factor performs the best and any significant interactions among the factors.
- If necessary, perform full factorial experiments. More information than is provided by 2k factorial experiments may be required. A full factorial experiment produces the same type of information as a 2k factorial does, but does so by examining multiple factors (Xs) at multiple levels.
- If necessary, and in addition, utilize RSM and/or EVOPs techniques to further assist in determining optimal process parameters.
- Using results of experiments, derive mathematical models of the process and establish optimal settings for process parameters (Xs) to achieve desired (Ys).

Evaluate Alternatives and Choose Optimal Improvements

- Identify a broad range of possible improvements.
- Agree on criteria against which to evaluate the improvements and on the relative weight each criterion will have. The following criteria are commonly used:
  - Total cost
  - Impact on the problem
  - Benefit–cost relationship
  - Cultural impact or resistance to change
  - Implementation time
  - Risk
  - Health, safety, and the environment
- Evaluate the improvements using agreed-upon criteria.
- Agree on the most suitable improvements.

Design the Improvements

- Evaluate the improvements against the project goal.
- Verify that it will meet project goals.
- Identify the following customers of the improvements:
  - Those who will create part of the improvements
  - Those who will operate the revised process
  - Those served by the improvements
- Determine customer needs with respect to the improvements.
- Determine the following required resources: people, money, time, materials.
- Specify the procedures and other changes required.
- Assess human resource requirements, especially training.
- Verify that the design of the improvement meets customer needs.
- Plan to deal with any cultural resistance to change.

Improve: Deliverables

- Plan for designed experiments
- Reduced list of vital few inputs (Xs)
- Mathematical prediction model(s)
- Established process parameter settings
- Designed improvements
- Implementation plan
- Plans to deal with cultural resistance

Improve: Questions to be Answered

1. What specific experiments should be conducted to arrive ultimately at the discovery of what the optional process parameter settings should be?

2. What are the vital few inputs (Xs, narrowed down still further by experimentation) that have the greatest impact on the outputs (Ys) of interest?
3. What is the mathematical model that describes and predicts relationships between specific Xs and Ys?
4. What are the ideal (optimal) process parameter settings for the process to produce output(s) at Six Sigma levels?
5. Have improvements been considered and selected that will address each of the vital few Xs proven during the Analyze phase?
6. Has expected cultural resistance to change been evaluated and plans made to overcome it?
7. Has a pilot plan been developed and executed and the solutions appropriately adjusted based on the results?
8. Have all solutions been fully implemented along with required training, procedural changes, and revisions to tools and processes?

Control Phase

The project team designs and documents the necessary controls to ensure that gains from the improvement effort can be sustained once the changes are implemented. Sound quality principles and techniques are used, including the concepts of self-control and dominance, the feedback loop, mistake-proofing, and statistical process control. Process documentations are updated (e.g., the failure mode and effects analysis), and process control plans are developed. Standard operating procedures (SOP) and work instructions are revised accordingly. The measurement system is validated, and the improved process capability is established. Implementation is monitored, and process performance is audited over a period to ensure that the gains are held. The project team reports the goal accomplished to management, and upon approval, turns the process totally over to the operating forces and disbands.

Quality Control is discussed in great detail in Chapter 6, Quality Control: Assuring Repeatable and Compliant Processes, so only the elements that are unique to DMAIC and highlights of the Control step are discussed here. The activities required to complete the Control step include:

1. Design controls and document the improved process.
2. Design for culture
3. Validate the measurement system.
4. Establish the process capability.
5. Implement and monitor.

Design Controls and Document Improved Process

- Update FMEA to ensure that no necessary controls have been overlooked.
- Mistake-proof the improvement(s), if possible:
  - Identify the kind(s) of tactic(s) that can be incorporated into the improvements to make it mistake-proof. Some options include:
    - Designing systems to reduce the likelihood of error
    - Using technology rather than human sensing
    - Using active rather than passive checking

        - Keeping feedback loops as short as possible
        - Designing and incorporating the specific steps to mistake-proof as part of the improvements
- Design process quality controls to ensure that your improved levels of inputs (Xs) and outputs (Ys) are achieved continuously. Place all persons who will have roles in your improved process into a state of self-control to ensure that they have all the means necessary to be continuously successful.
- Provide the means to measure the results of the new process:
    - Control subjects:
        - Output measures (Ys)
        - Input measures and process variables (Xs)
        - Establish the control standard for each control subject.
        - Base each control standard on the actual performance of the new process
- Determine how actual performance will be compared to the standard.
    - Statistical Process Control
- Design actions to regulate performance if it does not meet the standard. Use a control spreadsheet to develop an action plan for each control subject.
- Establish self-control for individuals:
    - They know exactly what is expected (product standards and process standards).
    - They know their actual performance (timely feedback).
    - They are able to regulate the process because they have
        - A capable process
        - The necessary materials, tools, skills, and knowledge
        - The authority to adjust the process

Design for Culture to Minimize or Overcome Resistance

- Identify likely sources of resistance (barriers) and supports (aids). Resistance typically arises because of
    - Fear of the unknown
    - Unwillingness to change customary routines
    - The need to acquire new skills
    - Unwillingness to adopt a remedy "not invented here"
    - Failure to recognize that a problem exists
    - Failure of previous solutions
    - Expense
- Rate the barriers and aids according to their perceived strengths.
- Identify the countermeasures needed to overcome the barriers. Consider:
    - Providing participation
    - Providing enough time
    - Keeping proposals free of excess baggage
    - Treating employees with dignity

    - Reversing positions to better understand the impact on the culture
    - Dealing with resistance seriously and directly
- Install statistical process control (SPC) where necessary to ensure that your process remains stable and predictable, and runs in the most economic manner.
- Consider introducing 5s standards to make the workplace function smoothly with maximum value-added activity and minimum nonvalue-added activity.

Validate Measurement System

Utilize commercially available software such as Minitab to evaluate measurement system capability (as in the Measure phase) to ensure that the measurements utilized to evaluate control subjects can be depended on to tell the truth.

Establish Process Capability

- Prove the effectiveness of the new, improved process to ensure that the new controls work and to discover if your original problem has improved, and ensure that no new problems have inadvertently been created by your improvement(s).
- Decide how the improvements will be tested:
    - Agree on the type of test(s).
    - Decide when, how long, and who will conduct the test(s).
    - Prepare a test plan for each improvement.
    - Identify limitations of the test(s):
        - Develop an approach to deal with limitations.
        - Conduct the test.
        - Measure results.
        - Adjust the improvements if results are not satisfactory.
        - Retest, measure, and adjust until satisfied that the improved process will work under operating conditions.
- Utilizing control charts, ensure that the new process is in statistical control with respect to each individual control subject. If not, improve the process further until it is.
- When, and only when, the process is in statistical control, utilize Minitab Capability Analysis—as in the Measure phase—to determine process capability for each individual control subject.

Implement the Controls and Monitor

- Transfer to the operating forces all the updated control plans, etc., and train the people involved in the process in the new procedure.
- Develop a plan for transferring the control plan to the operating forces. The plan for transferring should indicate:
    - How, when, and where the improvements will be implemented
    - Why the changes are necessary and what they will achieve
    - The detailed steps to be followed in the implementation
- Involve those affected by the change in the planning and implementation.

- Coordinate changes with the leadership team, Black Belt, Champion, executive council, and the affected managers.
- Ensure preparations are completed before implementation, including:
  - Written procedures
  - Training
  - Equipment, materials, and supplies
  - Staffing changes
  - Changes in assignments and responsibilities
  - Monitoring the results.
- Periodically audit the process, and also the new controls, to ensure that the gains are being held.
- Integrate controls with a balanced scorecard.
- Develop systems for reporting results. When developing systems for reporting results, determine:
  - What measures will be reported
  - How frequently
  - To whom (should be a level of management prepared to monitor progress and respond if gains are not held)
- Document the controls. When documenting the controls, indicate:
  - The control standard
  - Measurements of the process
  - Feedback loop responsibilities (who does what if controls are defective)
- After a suitable period, transfer the audit function to the operating forces and disband the team (with appropriate celebrations and recognition).

Control: Deliverables

- Updated FMEA, process control plans, and standard operating procedures
- Validated capable measurement system(s)
- Production process in statistical control and able to get as close to Six Sigma levels as is optimally achievable, at a minimum accomplishing the project goal
- Updated project documentation, final project reports, and periodic audits to monitor success and hold the gains

Control: Questions to be Answered

- What should be the plan to assure the process remains in statistical control and produces defects only at or near Six Sigma levels?
- Is our measurement system capable of providing accurate and precise data with which to manage the process?
- Is our new process capable of meeting the established process performance goal?

- How do we ensure that all people who have a role in the process are in a state of self-control (have all the means to be successful on the job)?
- What standard procedures should be in place, and followed, to hold the gains?

Training and Certification of Belts

The introduction of Six Sigma in the past decade led to an insurgence in the certification of Belts. This was largely due to a lesson learned from the Total Quality Management era. During TQM, many so-called experts were trained in the "methods of TQM." Unfortunately, few were trained in the tools to collect and analyze data. As a result, numerous organizations did not benefit from the TQM program.

Motorola introduced a core curriculum that all Six Sigma practitioners needed to learn. That evolved into a certification program that went beyond the borders of Motorola. As a result, there are many "certifiers" that will provide a certification as a Master Black Belt, Black Belt, Green Belt, and so on. Most certifications state that the person certified is an "expert" in the skills of Six Sigma or Lean or both. Certification did lead to improved performance, but also to some weak experts due to no oversight of the certifiers, many of which were consulting companies or universities not well versed in the methods or tools of Six Sigma and Lean.

The American Society for Quality (ASQ) for many years offered certification for quality technicians, quality auditors, quality engineers, and quality managers. As the Six Sigma movement grew, the ASQ and its affiliates around the world began to certify Black Belts. Although not perfect, the ASQ is in a better position to monitor the certifications than self-serving firms. Certification must be based on legitimacy to be effective. Having too many firms certifying Belts will only lead to a weaker certification process.

ASQ's Certified Quality Engineer (CQE) program is for people who want to understand the principles of product and service quality evaluation and control. (ASQ, 2009) For a detailed list of the CQE body of knowledge, the reader is referred to the certification requirements for Certified Quality Engineer at www.asq.org.

ASQ also offers a certification for quality officers at the quality management level, called Certified Manager of Quality/Organizational Excellence. ASQ views the Certified Manager of Quality/Organizational Excellence as "a professional who leads and champions process-improvement initiatives—everywhere from small businesses to multinational corporations—that can have regional or global focus in a variety of service and industrial settings. A Certified Manager of Quality/Organizational Excellence facilitates and leads team efforts to establish and monitor customer/supplier relations, supports strategic planning and deployment initiatives, and helps develop measurement systems to determine organizational improvement. The Certified Manager of Quality/Organizational Excellence should be able to motivate and evaluate staff, manage projects and human resources, analyze financial situations, determine and evaluate risk, and employ knowledge management tools and techniques in resolving organizational challenges" (ASQ, 2009).

Note: No matter what organization you use to certify your experts, here are some lessons learned about certification:

- One project is not enough to make someone an expert.
- Passing a written test that is not proctored is no guarantee the person who is supposed to be taking the test is actually taking it.
- If you get someone in your organization to sign off on the success of the Belt project, you need independent evidence that the person is knowledgeable about the methods of Six Sigma.
- Select a reputable certifying body.

References

Alonso, F., Rafael de Ansorena, and Gregorio del Rey. (2002). "How Telefonica Makes Its Management Connections." *European Quality*, v. 8, Number 6.

Chua, R., and Jong-Yong Yun (2002). "Samsung Uses Six Sigma to Change Its Image." *Six Sigma Forum Magazine*, November.

De Feo, J., and Barnard, W. (2004). "Six Sigma Breakthrough and Beyond," New York, McGraw-Hill.

Juran Institute, Inc. (2008). Six Sigma Black Belt, version 2.1, Southbury, CT.

CHAPTER 13

Root Cause Analysis to Maintain Performance

Dennis J. Monroe

About This Chapter

In this chapter, we will discuss three approaches to finding the root causes of sporadic problems. Root Cause Analysis from Juran has been used for the past 60 years as the basic four-step method that we call RCCA. Plan, Do, Check, Act, also known as PDCA or the Shewhart Cycle (sometimes called the Deming Cycle), and Plan, Do, Study, Act have been adopted by many service organizations—for example, healthcare. We will try to define a recent phenomenon called "Just Do Its." This is the latest method that attempts to avoid long, drawn-out methods of diagnosing problems and instead "just do it." It is the highest risk with one of the smallest gains used in today's best organizations.

High Points of This Chapter

1. Juran's RCCA approach is an effective method for identifying and addressing the root cause of sporadic problems in products and processes. It follows four basic steps: identify a problem, diagnose the cause, remedy the cause, and hold the gains.

2. Root cause analysis is one of the most critical skills all employees need to understand. The heart of this is the "cause and effect" relationships that occur in all organizations.
3. PDSA and PDCA are similar and useful methods for root cause analysis, planning, and executing tests of potentially beneficial changes. They are used in services and health care organizations. The four steps are Plan, Do, Check, Act and Plan, Do, Study, Act.
4. Just Do Its can be a useful approach when the need to solve a problem is urgent, the penalties for risks of failed "solutions" are low, and the rewards of effective (experience-based) solutions are high.

Introduction

The heart of all organizational problem-solving is understanding enough of the causes of the problem and finding an acceptable solution to it. Finding the "root cause or causes" of an organizational problem is the single most important determinant of success or failure of any problem-solving method. This is true of all types of problems: chronic problems that have been resistant to solution, day-to-day sporadic problems that occur infrequently but have a tendency to recur, problems that involve identifying and eliminating waste—all require effective root cause analysis and identification to reduce the risks and resistance associated with changing a process.

Why Do We Need to Know About Root Causes of Problems

Though effective root cause analysis is at the heart of all problem-solving methods, such as Six Sigma, Lean, RCCA, and Plan, Do, Study/Check Act, its purpose is to solve problems that occur due to special or assignable causes. A special cause is one that occurs during daily operations. It is sporadic because it happens periodically, but it can cause havoc on the organization if it persists. These problems differ from the larger, chronic problems that quality improvement programs such as Six Sigma focus on. For more on root cause analysis used in solving the other types of problems noted earlier, see Chapter 12, Six Sigma: Improving Process Effectiveness. A comparison of the different methods to discover root causes and their uses is shown in Table 13.1.

When to Apply Root Cause Analysis

The Juran Trilogy® has been discussed in detail in Chapter 3, The Universal Methods to Manage for Quality, but the question remains: At what stage of the trilogy does basic RCCA apply? To answer this question, refer to Figure 13.1.

As seen in the figure, the application of the basic RCCA process is a control activity: restoration of the process performance to a previously acceptable level. Although control also has elements of the broad sense of continuous improvement, for the purposes of the present discussion, we will apply the definitions of the trilogy strictly; that is, control is restoring process performance to a previously acceptable level. Improvement, on the other hand, is defined as changing the very nature of the process, creating breakthrough and moving to a new and better level of performance with reduced waste and cost of poor quality. The former primarily deals with special causes, the latter with common causes of variation.

Method	Purpose	Risk	Benefits	Level of Difficulty
Six Sigma DMAIC	Solve large, chronic, multifunctional problems	Low	High ROI (25:1)	High: Large scope problems require difficult diagnosis and expert skills
Juran's Breakthrough Model	Solve large, chronic, multifunctional problems	Low	High ROI (25:1)	High: Large scope problems require difficult diagnosis and expert skills
RCCA	Solve sporadic day-to-day problems	Low	Moderate ROI (5:1)	Low: Sporadic problems require finding out what changed; skills easy to gain by all staff
PDCA	Solve sporadic day-to-day problems	Low	Moderate ROI (5:1)	Low: Sporadic problems require finding out what changed; skills easy to gain by all staff
Lean Problem-Solving	Solve sporadic day-to-day problems	Med	Moderate ROI (1:1)	Low: Purpose is to identify waste and its causes, which tend to be well understood
PDSA	Solve sporadic day-to-day problems	Med	Moderate ROI (1:1)	Easy: Many services do not use tools to analyze data; rather, they move from symptom to solution
Just Do It	Make daily decisions based on what is already known	High	Moderate ROI (0:0)	Easy: Since this is not recommended, it is easy to do; there are no methods other than instinct

Table 13.1 Purpose of Root Cause Corrective Action vs. Other Methods

Juran's RCCA

Juran's four-step process for RCCA is an outgrowth of the work of Dr. J. M. Juran in which he described the universal process for quality improvement, shown in Figure 13.2. This universal, as Dr. Juran said in Chapter 3, The Universal Methods to Manage for Quality, and Chapter 5, Quality Improvement: Creating Breakthroughs in Performance, "is now referred to as "Six Sigma." The third universal is that of "control—the process for preventing adverse change." To ensure that all processes are in a state of control requires three basic elements:

- The means to know the actual performance of the process
- The ability to compare the actual performance to the targets or quality goals
- The means to act on the difference to maintain control

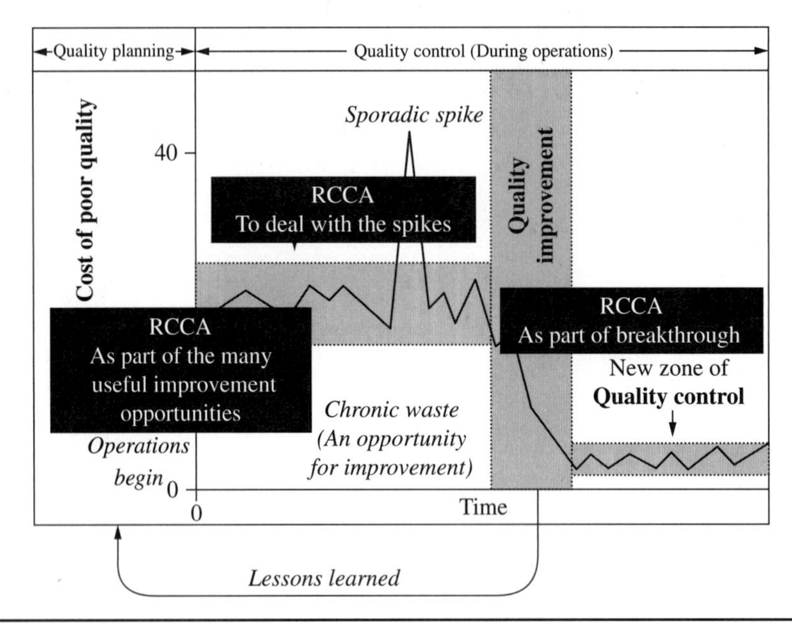

FIGURE 13.1 RCCA and the Juran Trilogy®.

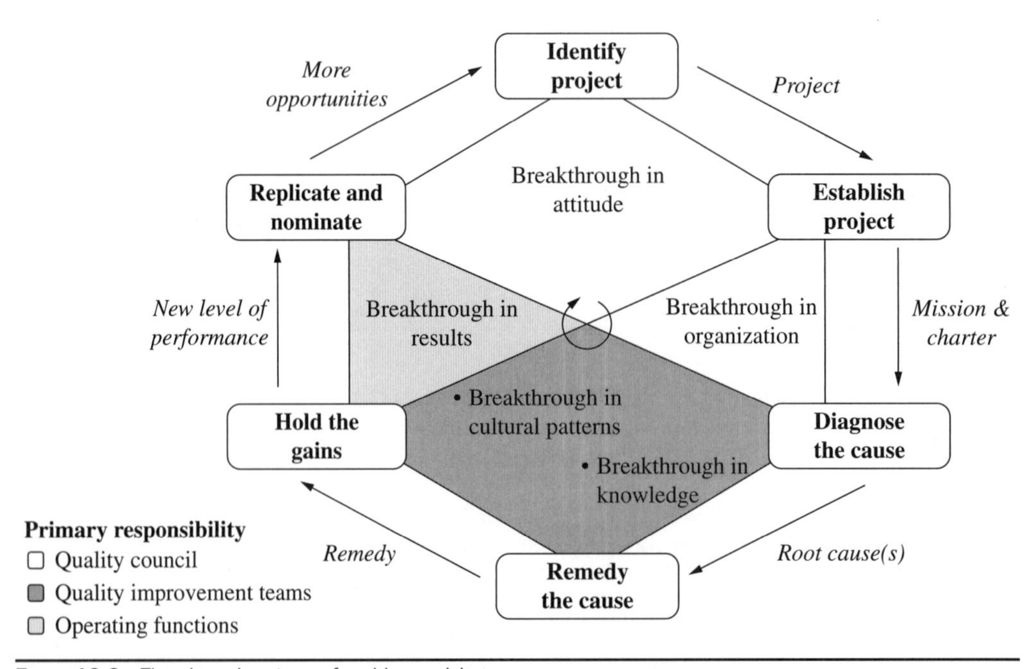

FIGURE 13.2 The six major steps of problem solving.

The third step requires a means and a method to determine what correct action should be taken. There have been many versions to act on the difference for centuries. Walter Shewhart coined the term PDCA (Plan, Do, Check, Act) as a means to set up the control functions. Practitioners of PDCA still need to know how to perform the action.While there are many tools to aid in root cause analysis, a simple method is needed to solve daily, sporadic, small-scope problems.

The Juran RCCA method described here is a simplification of the universal process for improvement described by Dr. Juran, and consists of four steps:

1. Identify a problem.
2. Diagnose the cause.
3. Remedy the cause.
4. Hold the gains.

"Quality control can be defined as the maintenance or restoration of the operating status quo as measured by [meeting] the acceptable level of defects and provision of customer needs" (Monroe, 2009). The mechanism of controlling quality is depicted in Figure 13.3. The troubleshooting portion of the control feedback loop is where RCCA is needed. When a measurement of a control subject is outside the established standard of acceptability, some means for identifying the cause is needed. Once the root cause or causes are identified, a remedy must be put in place that will eliminate them. After the cause is eliminated, the control feedback loop continues to monitor the process so the cause and problem do not recur.

The four-step RCCA approach described above has several substeps that must be undertaken to effectively diagnose and remedy the cause:

1. Identify a problem.
    - Is the problem sporadic or chronic (if the latter, apply breakthrough methods)?
    - Establish responsibility to solve it, if it is not already established in a control plan.
    - Prepare a problem statement.

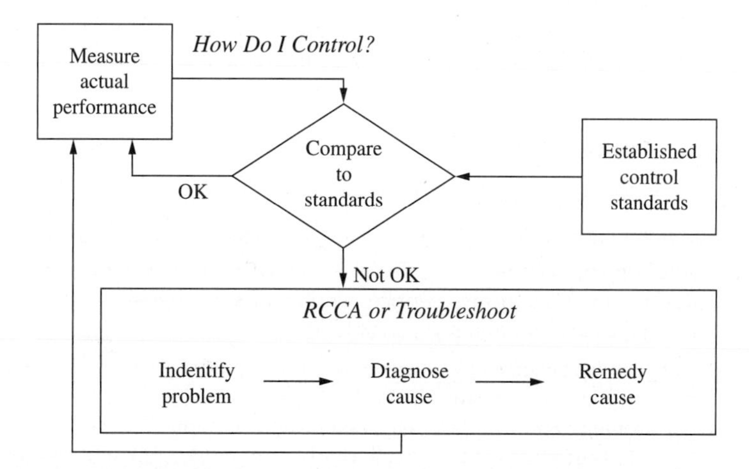

FIGURE 13.3 The control feedback loop.

2. Diagnose the cause.
   - Analyze symptoms.
   - Formulate theories.
   - Test theories.
   - Identify root cause(s).
3. Remedy the cause.
   - Design and implement the remedy.
4. Hold the gains.
   - Adjust controls.

Each of the steps and substeps of Juran's RCCA approach is discussed in more detail in the following sections.

The Medical Analogy

The Juran RCCA approach is analogous to the approach a physician takes in treating an ill patient. First, the doctor will want to understand what is wrong: What's the problem? Without a clear understanding of the problem, it will be impossible to solve.

Next, the doctor will want to know more about the outward evidence that the problem exists: the symptoms. He might take the patient's temperature, ask what kind of discomfort the patient is experiencing, look into the patient's throat and ears, and so on.

Based on the observed symptoms, the doctor will formulate tentative diagnoses—theories about what could be causing the patient's illness. At this point the doctor is still unsure what the true cause of the illness is, so he will order tests to determine which of his tentative diagnoses is true. Perhaps blood will be drawn from the patient for analysis; perhaps the patient will be given an MRI exam or other diagnostic tests.

Once the data about the possible causes of the illness have been gathered, the doctor is ready to settle on a final diagnosis based on the facts. Now, hopefully, the true root cause of the patient's illness is known and the doctor can apply an appropriate remedy. Perhaps the patient will be given medication, prescribed physical therapy, recommended to make certain lifestyle changes—whatever the appropriate remedy is to alleviate the proven cause of the illness.

Finally, the doctor might say, "Come back and see me in two weeks." This is the activity of holding the gains intended to ensure that the patient is continuing with the prescribed regimen and the remedy is effective: The patient is getting better.

Elements of Effective RCCA

These are the necessary elements for effective root cause corrective action:

- *A Problem.* A problem is outward evidence that something is wrong and warrants a solution, for example, a visible performance deficiency in the output of an important design, manufacturing, service, or business processes. Time and resources are needed to analyze and solve problems.
- *Data and Information.* We cannot solve the problem until we have the hard facts that prove what the root cause is. Without data, we are merely guessing at the causes of the problem, and our efforts to solve it will be hampered by our lack of knowledge. More importantly, we will create doubt and greater risk will be introduced into our system.

- *Tools*. When a problem arises, there are many questions that need answers. Those answers will come from data found within our processes. At times, we are often faced with a great deal of data, but little information or facts. We can use tools to help us organize and understand the data. They are invaluable aids to effective root cause analysis.
- *Structure*. A logical and structured approach is needed to guide the RCCA process. This structure becomes the "guide or boss," not the people trying to solve the problem. At a minimum, this structure needs to use and involve multiple functions to discover root causes. This structure will allow us to "torture the data until it confesses." Data contains information. We need a means to extract it.

An almost unlimited number of tools are available to the problem-solving team, but those most often used for basic RCCA are:

- *Affinity Method and Brainstorming*. Brainstorming is a quality tool intended to stimulate creativity. It is useful because it helps the team consider a full range of theories about possible causes. The affinity process helps organize those theories.
- *Cause and Effect Diagrams*. An effective way to organize and display the various theories of potential causes of a problem.
- *Data Collection* methods are used to gather information about a quality problem. A typical data collection form is arranged for easy use and includes clear instructions.
- *Failure Mode and Effects Analysis (FMEA)* is a structured methodology for identifying potential failure modes in a process or design and assessing the risk associated with the failures. It helps identify the most likely possible causes and helps design a more robust remedy.
- *Graphs and Charts* are pictorial representations of quantitative data. They can summarize large amounts of information in a small area and communicate complex situations concisely and clearly.
- *Histograms* are a graphic summary of variation in a set of data. The pictorial nature of the histogram enables us to see patterns that are difficult to see in a simple table of numbers.
- *Box Plots*, like histograms, provide a graphic summary of the pattern of variation in a set of data. The box plot is especially useful when working with small sets of data or when comparing many different distributions.
- *Juran's Pareto Analysis* is a ranked comparison of factors related to a quality problem. It helps identify and focus on the vital few factors.
- *Mistake Proofing* is a proactive approach to reducing defects by eliminating the opportunity to create a defect by designing and implementing creative devices and procedures.
- *Process Control Plans* summarize the plan of action for a process out of control. Its purpose is to document the actions necessary to bring the process back into control and assist the process owners in holding the gains achieved by the problem solving.
- *Scatter Diagram* is a graphic presentation of the relationship between two variables. In root cause corrective action, scatter diagrams are usually used to explore cause-effect relationships in the diagnostic journey.

- *Stratification* is the separation of data into categories. Its most frequent use is during the diagnostic journey to identify which categories contribute the most to the problem being solved.

Phase 1: Identify the Problem

It has been said that a problem well defined is half solved. The clear identification and definition of problems to be addressed in the RCCA project is an early key to success. In practice, a well-constructed control plan with effective feedback loops will identify problems to be addressed by RCCA nearly in real time. For further discussion of control activities, see the section "Hold the Gains."

Nominate Projects

RCCA projects are identified by monitoring the process and detecting an out-of-the-norm condition. Not all problems are serious or complicated enough to warrant full-blown RCCA activities. RCCA project nominations can come from:

- Analysis of data collected from the operating process, control charts, automated data collection, and employee observations
- Information received from the internal or external customers, expressions of dissatisfaction, warranty returns, quality notices, and corrective action requests
- Information received from the operating forces, problem observations, "fire fighting" reports, and internal corrective action requests

The most often used tools during this step are the brainstorming and affinity process, data collection, and flow diagrams.

Select the Problem

Select the problem to be addressed. Once the data and information about potential problems to address have been gathered, tools must be applied to select the most important problems to address. Data collection and Juran's Pareto analysis are most often used to identify the vital few problems to address.

Once the problem for action has been selected, the nature of the problem must be stated clearly and concisely. A good problem statement should have the following characteristics, summarized by the acronym MOMS:

- *Measureable.* The problem must be stated in terms that can be measured, either by using an existing measurement system or creating a new one. Although the problem may not have been measured to date, the problem-solving team must be able to conceptualize how it could be measured in quantifiable terms.
- *Observable.* The problem must been seen and evidenced by its symptoms. Symptoms are the outward evidence that the problem exists.
- *Manageable.* The problem statement must be narrow enough in scope that the team can solve it with a reasonable application of resources over a reasonable period of time. "Boil the ocean" projects should be avoided.
- *Specific.* The problem statement should focus on specific products, services, or information; specific parts of the organization; or specific aspects of a larger problem.

In addition to the MOMS guidelines, problem statements should never include implications of a cause, blame for the problem occurring, or suggested solutions.

Is the Problem Sporadic or Chronic in Nature

This is a fork in the road. If the problem involves a process that has gone out of control—for example, a fire has erupted and is burning—apply the RCCA process and tools discussed in this chapter to restore process control. If the problem is one that has been around for a while (chronic) and plaguing the operation with higher-than-tolerable COPQ, consider using the more sophisticated breakthrough improvement methods described in Chapter 11, Lean Techniques: Improving Process Efficiency and Chapter 12, Six Sigma: Improving Process Effectiveness.

Tools most often used at this step are data collection and Juran's Pareto analysis.

Establish Responsibility

Select the team that will be charged with solving the problem. Utilize a team, rather than an individual, because the outcome will likely be better. The old adage that the whole is greater than the sum of the parts applies. Make the team cross-functional because the additional expertise realized from people of different functions within the organization will allow the problem to be solved faster and better.

Tools most often used at this step are process flow diagrams (particularly matrix or swim lane diagrams) and Juran's Pareto analysis.

Prepare a Goal Statement

Typically, the goal statement for an RCCA project is simple: Eliminate the root cause or causes of the problem and restore control. In some cases, complete elimination may not be possible or practical; then the goal should be to reduce the impact of the causes so that the undesirable effects are minimized. In this case, the goal may be stated in terms of a percentage improvement, reduction in defect levels, etc.

Phase 2: Diagnose the Cause

Analyze Symptoms

Analysis of symptoms is an important step in finding the root cause of the problem because this activity enables us to understand the current situation. How often is the problem occurring? How severe is it? What types of failures contribute most to the problem being analyzed? At what point in the process is the failure most often observed? These types of questions about the current situation must be answered to help us better understand where the root cause(s) may lie.

Think of it this way: If you were asked what route to take if driving to Cleveland, what would your response be? Typically, people will respond from their own frame of reference: "Well, go to route 224 east, and then take I-77 north . . ." This misses an important point, however; the question didn't specify a starting point. Depending where one starts their journey to Cleveland, the route to get there will be entirely different. The point is that unless you know where you are starting from (the symptoms of the problem), it's difficult to map a route to where you want to be (achievement of the goal). This is why it's so important to do a thorough analysis of symptoms as a first step in diagnosing the cause. This analysis will be of great help when the team gets to the point of brainstorming possible causes and will result in a more thorough list of possible causes than would otherwise be achieved.

Tools that are often used at this step are data collection, process flow diagrams, Juran's Pareto analysis, and stratification.

Confirm the Goal

Often, things discovered during the analysis of symptoms about the detailed nature, severity, or vital symptoms of the problem will lead the team to modify the goal toward either a more aggressive outcome or less, depending on the amount of opportunity that is recognized during the analysis of symptoms. This should not be done lightly or on a whim, but only in the face of data and information that clearly indicate that the original goal needs to be modified.

It's not uncommon for the goal set during the Identify phase to be based on a best guess because insufficient information was available at the time to establish a more firm goal. If a thorough analysis of symptoms has been made, the appropriate goal should now be clear. Tools most commonly used to assist in confirmation of the goal are data collection, flow diagrams, graphs and charts, and Juran's Pareto analysis.

Formulate Theories

A theory is simply an unproven statement of the cause of a certain condition. A student receiving a poor grade on an exam may tell his or her parents that the cause is that the teacher included material on the exam that was not discussed in class. But the parents may consider this only a theory. The parents may consider a number of other theories as well, such as the student did not read required chapters that explained the material or the student did not attend class every day. In the same way, when determining the cause of a quality problem, there must be speculation about its many possible causes. Jumping to conclusions before considering many theories and proving which one is correct could mean wasting time and resources on an inappropriate solution.

The formulation of theories follows a thought process moving from creative to empirical, divergent to convergent. Beginning with brainstorming, the team and any subject matter experts will attempt to identify as many causes as possible. Next, the team will organize these brainstormed theories into logical groups, probably using the affinity process. Finally, the group will begin to hone in on the most likely root causes by using cause and effect diagramming, FMEA, and possibly other prioritization tools.

These most likely theories of causes are the input for the next step of the diagnostic journey.

Test Theories to Identify Root Cause(s)

"Before beginning to test theories, the team should be very clear on exactly which theories are being tested. A copy of the cause-effect diagram is an excellent guide for the team at this point. Diagram the theories that will be tested with a particular set of data. If the data demonstrate that a theory is not important, that theory can be crossed off as a possible cause. The cause-effect diagram also helps identify related theories that can be tested together. When the theories to be tested are stated clearly and precisely as they are understood, it is time to plan for collecting data to test them (Juran Institute, Inc., 2008)."

Theories are tested by assessing data that have been collected to answer questions regarding the truth or falsity of a given theory. The theory is assumed to be false unless the data indicate it to be otherwise. Once the data have been collected, appropriate analysis tools must be applied to convert the data into information. Information then becomes the answer to the question. This process is sometimes referred to as "torturing the data until it confesses."

The project team should recognize that rarely does the answer to one question constitute the end of the exploration. Testing of theories is typically an iterative process. The answer to one question leads to another question, and another, and another. Each time an answer is discovered, the team should ask again, why? Why does the analysis look the way it does? Why is the upper level (not root) cause we have proven occurring? When the "why" questions reach a level that has no more answers or goes beyond a level of cause that can be controlled, the team has arrived at the (operationally defined) root cause.

As an example, take the case of a problem the National Park Service experienced several years ago concerning the Jefferson Memorial. The stone in the monument was crumbling due to frequent washing to remove bird droppings. The initial (mistaken) approach the Park Service took was to reduce by half the number of times the stone was cleaned. This saved some money and reduced the magnitude of the stone erosion, but it's easy to see how the "solution" led to other problems. People visiting the monument were dissatisfied with the unclean conditions.

So the Park Service undertook a more thorough analysis to find the root cause of the problem. They first asked, "Why are there so many bird droppings?" Of course, they considered several theories to answer the question. Perhaps the birds were attracted to food dropped by visitors. Perhaps they were attracted to the good roosting places in the structure. Perhaps there was an abundant natural food supply. Could they immediately determine which of these theories was true? Of course not. It was necessary that they visit the place where the problem was taking place (what the Japanese call the "gemba"), collect data about the possible causes, and identify the true cause of the proliferation of birds in the monument. It turned out that the third theory was true; hundreds of fat spiders were providing an ample food source for the birds. But was the investigation complete? No, it was not because the investigators had not yet reached the root cause of the problem.

The next question to be answered was, "Why are there so many spiders?" A number of theories could have been forwarded about this question too.

- The crevices in the monument provide a good place to spin webs.
- There are insects there that provide food for the spiders.
- The spiders are attracted to and hide in the shadows inside the monument.

Further data-gathering proved that the second theory was true. Inside the Jefferson Memorial were thousands of tiny midges (a small flying insect that spiders eat). The investigators were nearing the root cause of the problem, but were not there yet.

"Why are there so many midges?" they asked. Possible answers included:

- Midges were attracted to a food supply inside the monument.
- The Jefferson Memorial, like many others in Washington, D.C., is near a body of water (the Potomac River), and the midges lay their eggs in the water.
- The midges are attracted to the lights that illuminate the memorial at night.

The second theory actually did explain why so many midges were in the vicinity of the monument, but not why they were on and inside it. Investigation revealed that the midges came out at sunset each evening in a "mating frenzy" at just the time that the lights were turned on. They were attracted to the illumination of the monument and took up residence where the spiders could feast on them. Now the investigators had found the true root cause of the problem: illuminating the monument each night at dusk. They delayed the lighting by one hour (the remedy), the midge population was dramatically reduced, and the food chain was broken. Now the Park Service could substantially reduce the washings and, therefore, the crumbling of the stone (the original problem). This application of the remedy to the true root cause resulted in many multiples of savings compared to the original solution of just reducing the washings. The solution was also one that could be replicated to other D.C. monuments to reap additional savings (The Juran Quality Minute: Jefferson Memorial).

One may ask, "How will I know when to stop asking 'Why?'" In other words, when have the investigators drilled down deeply enough to conclude they are at the level of the root cause?

There are two questions that will help you decide whether you have found the root cause:

1. Do the data suggest any other possible causes? After each data collection and analysis, it is usually possible to discard some theories and place more confidence in others. Theorizing is not a one-time activity, however. Each data display—the Pareto diagram, histogram, scatter diagram, or other chart—should always be examined by asking whether it suggests additional theories. If you have competing plausible theories that are consistent with the new data and cannot be discarded based on other data, then you have not arrived yet at the root cause.
2. Is the proposed root cause controllable in some way? Some causes are beyond our ability to control, like the weather. The effects of the weather can be controlled by turning up the heat or running a humidifier, but the weather cannot be controlled directly. So no useful purpose is served by testing theories about why the weather is cold.

Tools most often used during the steps of formulating and testing theories are data collection, flow diagrams, graphs and charts, histograms, Juran's Pareto analysis, scatter diagrams, and stratification.

These steps of formulating and testing theories complete the diagnosis of the problem's root cause. Some may ask, why should I go to all that trouble just to find the root cause of the problem? Why is it important? Denise Robitaille, an ASQ fellow and leading expert in root cause analysis provides useful answers in an article entitled "Four Things You Should Get From Root Cause Analysis." Emphasis on effective root cause analysis has gotten increased attention in several sectors. Registrars, for example, are requiring more substantial evidence of root cause analysis as part of responses to their requests for corrective action. All of this is good news. Except, my personal experience is that although people understand that they're required to do root cause analysis, they don't comprehend three issues:

- What root cause analysis is
- How to conduct effective root cause analysis
- What the results of root cause analysis should yield

Let's start by reviewing what root cause analysis is. It's an in-depth investigation into the cause of an identified problem. It asks why something happened. It should also investigate how something could have gone wrong, which will help to identify contributing factors and interim breakdowns.

There are two important things to remember at the outset. Root cause analysis is focused on cause, and the ultimate intent is to use the information to develop a corrective action plan. This perception is relevant to the next two issues people need to know.

People don't know how to do root cause analysis. They still treat it like it's a haphazard activity. Organizations fail to train individuals in good investigative techniques. They perpetuate a culture of blame: "Let's find out who screwed up." And they simply don't treat root cause analysis like a controlled process.

Apart from the five whys there are many other tools that can be used. There are flowcharts, brainstorming, fish bone diagrams, Pareto charts, and design of experiment—just to name a few. Several tools should be used in concert to achieve the most productive results. For example, use brainstorming or the five whys to conjecture what could have gone wrong, then organize the results in a fish bone diagram that will direct you to the areas where you'll find the evidence you need to objectively conclude what the root cause of the problem really is. Organizations have to stop assigning people to do root cause analysis without giving them the necessary training and tools.

Finally, individuals need to understand what the expected outcome of this process is. It's great to say that we're going to conduct root cause analysis. Do people have any idea what they're supposed to do when they figure out the cause?

You should be able to get four things from root cause analysis:

- Uncover the root cause or causes of the problem. That's the primary output of this process.
- Identify weaknesses or other contributing factors, which, in and of themselves, are not necessarily nonconformances. They may be the outcome of shortsighted decisions to curtail activities so that efficiency or cost savings is perceived. You may have, for example, decided to wait until the first point of use to test components. The time savings experienced at the receiving process could result in costly delays and scheduling snafus that dwarf any savings that had been anticipated. It wasn't a bad idea at the time, but it may have contributed to late deliveries.
- Better understand the process surrounding the problem, as well as supporting processes. If you don't, you haven't done a thorough root cause analysis. Without that heightened comprehension of the process, you can't understand interrelations, interdependencies, or other factors that are reliant on the outcome of seemingly unrelated processes. This takes us to the final outcome.
- Create an architecture into which you can build your corrective action plan. Corrective action isn't just one activity. It needs to be a plan, reflective of all aspects of the problem. If you've done a good root cause analysis, you'll have identified not only the root cause, but the many different factors that need to be addressed to ensure that the problem doesn't recur, that you don't inadvertently create a new problem, and that your organization experiences some benefit from the action taken.

Your root cause analysis will let you see what processes may need to be modified, what documents and forms will have to be revised, who will require training, and a myriad of other considerations that go into a typical project plan.

Without root cause analysis, effective corrective action is impossible. Without corrective action, root cause analysis is a waste of time.

Phase 3: Remedy the Cause

Now that the project team has discovered the root cause(s) of the problem, the task is to restore control to the process. This is done by applying appropriate remedies that will directly affect the cause and eliminate it, or at least drastically reduce its undesirable effects.

Evaluate Alternative Solutions

Like the formulate theories step, this step moves from creative to empirical, divergent to convergent thinking. Beginning with brainstorming, the team, subject matter experts, and

Solution Selection Matrix Rank possible solutions 1–10. 10 = fully meets criteria						Updated: 10/12/09		
		Possible solutions						
Criteria	**Weight**	**A**	**B**	**C**	**D**	**E**	**F**
Low cost	3	9	8	10	7	9	7
High effectiveness	2	8	10	9	10	9	9
Low risk	2	8	8	7	9	9	7
Low resistance	1	9	8	5	8	9	10
Minimal process disruption	2	6	7	7	6	8	7
	Total score	80	82	81	79	88	77

FIGURE 13.4 Solution selection matrix.

process owners will attempt to identify as many alternatives for solutions as possible. Creativity is essential at this point, as often, solutions must be quite novel to fully address the root cause. Next, the team will evaluate these brainstormed potential solutions to determine which solution or combination of solutions will best address and eliminate the cause(s).

The team may construct flow diagrams of possible solution implementations to visualize which will act most effectively. They may also use a criteria-based selection matrix to assist their decision-making process and help them arrive at the best solutions (see Figure 13.4). The solution selection matrix can help the team optimize the ultimate solution by combining the best potential solutions from the matrix.

Tools most often used in this step are brainstorming, data collection, selection matrices, and flow diagrams.

Design and Implement the Remedy

Once the team selects a remedy, it designs the remedy by performing four tasks:

1. Ensure that the remedy achieves the project goals. Review project goals to verify that the remedy will achieve the desired results and that all involved are in agreement on this point. This is a final check before moving ahead.
2. Determine the required resources. Make every effort to determine, as accurately as possible, what resources are required to implement the proposed remedy. These resources include:
    - People
    - Money
    - Time
    - Materials
3. Specify the procedures and other changes required. Before implementing the remedy, describe explicitly what procedures will be required to adopt the proposed remedy. Any changes that need to be made to existing organizational policies, procedures, systems, work patterns, reporting relationships, and other critical operations must also be described. Any surprises down the line may sabotage the remedy.

4. Assess human resource requirements. The success of any remedy depends on the people who will implement the required changes. Often, it will be necessary to train or retrain staff. Explore fully all training requirements, as well as the training resources needed.

Once these tasks have been performed, a flow diagram can be created to help specify the new procedures clearly.

As the team is designing the remedy, they should take into account the need to mistake-proof the remedy. They should consider and develop a variety of techniques to avoid, prevent, or reduce inadvertent errors that may occur even with the improved process.

The final action of this step is to implement the remedy. Depending on the complexity of the problem being addressed and the solutions to be implemented, a formal implementation plan may be needed. At a minimum, procedures, process standards, or work instructions will need to be modified to institutionalize the change.

Phase 4: Hold the Gains

This phase is the most important one in the RCCA process for ensuring that the problem does not recur or, if it does, that the recurrence is recognized and remedied quickly. If a recurrence is recognized, it should be an indication to the project team that their job is not finished—they've missed a root cause during the course of their problem solving or designed and implemented an ineffective remedy.

If the problem solving has been done methodically, as described here, and a broad range of possible causes and remedies were considered during the formulate theories and evaluate alternatives steps, the remedy should be robust and the cause and problem should not recur. The controls put in place to hold the gains will indicate whether this is so.

Redesign Controls

The primary activity in designing controls is the development of a control plan. Hopefully, an effective control plan for the process in question is already in place and will only require modification to add control subjects related to the problem's solution.

The first step in building an effective control plan is selecting appropriate control subjects. Control subjects are those features of the product or process that will be measured to determine whether the process is remaining in control. Each control subject's performance is monitored using the feedback loop described in Figure 13.3. A control plan matrix is used to keep track of the function of the feedback loop and to plan for action if the process or product does not meet standards. An important purpose of the process control matrix is to alert the process operator when the process is out of control and what to do to get it back under control.

In this matrix (see Figure 13.5), the horizontal rows describe the control elements for each subject. The vertical column headings indicate each element of the control activity:

- *Control subject.* Those features of the product or process that will be measured to determine whether the process is remaining in control
- *Subject goal or standard.* The acceptable limits of performance for the product or process. Often, these are control limits on an SPC chart and are the primary basis for determining if the process is stable or out of control.
- *Unit of measure.* How will the measurement be stated? Inches? Millimeters? Percent defective?

Process control plan for:						Date:		Revision level:		Approved by:
Control subject	Subject goal (standard)	Unit of measure	Sensor	Frequency of measurement	Sample size	Where measurement recorded	Measured by whom	Criteria for taking action	What actions to take	Who decides

FIGURE 13.5 Control plan matrix.

- *Sensor*. What device, person, or combination of the two will be used to obtain the measurement?
- *Frequency of measurement*. How often will the control subject be measured (e.g., hourly, daily, weekly, etc.)?
- *Sample size*. How many measurements will be taken at the stated frequency?
- Where are measurement recorded (logbook, chart, database, etc.)?
- *Measured by whom*. Who is responsible for applying the sensor to the control subject and obtaining and recording the measurements?
- *Criteria for taking action*. This generally includes whatever process performance is outside the subject goal or standard. This variation is usually due to special causes and would prompt the troubleshooting part of the feedback loop.
- *What actions to take*. Knowing the cause of the out-of-control condition helps the assigned person take the appropriate action to bring the process back into conformance with the subject goal.
- *Who decides*. Who will make the call on the action to be taken?
- *Who acts*. Specific action(s) to be taken by the actor on the control subject to bring the process back into conformance with the subject goal.
- *Where action recorded*. Identifies where the actions taken to resolve the issue will be recorded. This recording is useful for analysis of similar problems in the future.

Implement Controls

Once a suitable control plan has been designed, implementation is a matter of training process owners and operators in its use. If SPC is a part of the plan, specific training on the proper use, interpretation of, and appropriate response to control charts must be included. The process owners also become the owners of the control plan, so their involvement in its implementation is essential.

Audit Controls

For a short time after the controls are in place, the project team, in conjunction with the process owners and operators, should monitor their effectiveness. This will provide the opportunity to recognize any ineffective elements of the plan and modify accordingly.

By following the above four-phase approach to RCCA, project teams should consistently identify the root cause(s) and apply appropriate remedies in a relatively short time. During the time that it takes to identify and alleviate the causes, an interim action may be needed to ensure that defective products, services, or information do not reach the customer. These actions are sometimes referred to as containment. They should be designed to be effective and temporary until the root cause of the problem can be determined and alleviated.

Plan-Do-Study-Act (PDSA)

PDSA is another problem-solving approach many use to find and address root causes of problems. The method was originally proposed by Dr. Walter Shewhart (as PDCA, Plan-Do-Check-Act) in his book *Economic Control of Quality of Manufactured Product* (1931) and later espoused by W. Edwards Deming. Deming referred to the method as the Shewhart Cycle, but many, particularly after Deming achieved fame, refer to it as the Deming Cycle.

The method differs from the root cause analysis method described previously in that it is primarily a guide for identifying root causes through experimentation. This implies that the analysis of symptoms and theorizing of causes are done before the cycle actually starts, and then iterative experiments are performed to drill down to the root causes of the problem being addressed.

The PDSA method is particularly popular in health care organizations, probably due to its promotion by the Institute for Healthcare Improvement (IHI) as a method for finding causes and stimulating improvement.

The work done prior to the actual PDSA cycle starts by "setting aims," which is analogous to the establishment of the goal in the Juran RCCA process. The piece of stating the problem to be solved, however, seems to be absent, so one might wonder how the activity of the team becomes focused. Then the team gathers knowledge about the process they are attempting to improve upon so they can come up with good ideas for changes to the process.

". . . [T]he more complete the appropriate knowledge, the better the improvements will be when the knowledge is applied to making changes. Any approach to improvement, therefore, must be based on building and applying knowledge. This view leads to a set of fundamental questions, the answers to which form the basis of improvement:

- What are we trying to accomplish?
- How will we know that a change is an improvement?
- What changes can we make that will result in improvement?" (Langley, et al. 1996)

In contrast to Juran's RCCA, the PDSA approach seeks to identify changes that might improve the process or outcomes of it, then implements those changes to see if they are effective in producing an improvement. The PDSA cycle is the method applied to this trial of changes. In a manner of thinking, PDSA seeks to confirm or refute ideas of problem causes by trial and error of solutions.

"These questions [above] provide a framework for a "trial and learning' approach. The word 'trial' suggests that a change is going to be tested. The term 'learning' implies that criteria have been identified that will be used to study and learn from the trial" (Langley, et al. 1996).

The PDSA approach follows these phases and steps:

1. Plan:
    - Define the change to be tested.
    - Design the experiment to test the change.
2. Do:
    - Carry out the experimental plan.
    - Collect data about the effectiveness of the change.
3. Study:
    - Analyze the data from the experiment.
    - Summarize what was learned.

4. Act:
   - Determine what permanent changes are to be implemented.
   - Determine what additional changes need to be tested.

Clearly, this approach has some advantages:

- It can yield results quickly if the experimenters are good at selecting solutions that will yield true improvement.
- It follows an experimental approach, which can yield a great deal of useful knowledge.
- It is widely accepted, particularly within health care and other organizations that typically rely on experimentation to determine beneficial changes (e.g., development of medications).

One might also note some disadvantages:

- Results can be slow to come if the experimenters are not good at selecting solutions that will yield true improvement.
- Changes that do not succeed may not yield a lot of useful information.
- Experimentation, unless it is done in a laboratory setting, can be disruptive to the process and can be resource-intensive.
- Experimentation can be costly in many cases.

Based on these pros and cons, the project team should choose the methodology that best fits their work style and organization's needs.

Just Do Its (JDIs)

As the name implies, Just Do Its (JDIs) do not really include an analysis of the root cause of the problem because that root cause is usually readily apparent in what is sometimes referred to as a "blinding flash of the obvious." So analyzing the root cause in this case is done entirely by observation.

A number of years ago, consultants transitioned from the old way of teaching using overhead projectors to the new computerized method: constructing the materials to be taught in a presentation graphics program and projecting them using a liquid crystal display (LCD) projector. As the transition from the old way to the new way progressed, fewer and fewer meeting rooms had overhead projectors available, and more and more had LCD projectors. A problem arose for some training providers: If the trainer arrived at the training room prepared to show slides on an overhead projector and none was available, the training had to be either postponed or done in a less-than-desirable fashion, reading from and referring to printed materials only. What was the obvious cause of this problem? The consulting organization had not provided the consultant with the proper tools (either a laptop or some digital media that could be used on the training room PC) to do the job in the new environment. The JDI in this case, of course, was to provide the trainer with the needed tools.

Another situation where the JDI approach may be appropriate is when the need for a solution is urgent and delaying can have serious repercussions.

Such an example of an urgent need for a solution occurred in London in 1854. There had been a terrible outbreak of cholera, which ultimately claimed more than 500 lives in a period of ten days. Dr. John Snow came to the rescue. After analyzing the pattern of occurrence of the deaths using a concentration diagram, Snow recognized that most of the deaths were grouped around the Broad Street pump. Even though he did not recognize the root cause was bacteria in the water, Snow went directly to a solution and had the handle removed from the pump. Within days the cholera outbreak was over. (The Juran Quality Minute: London Cholera Epidemic)

To implement JDIs without a thorough analysis and discovery of the root cause of the problem, three factors must be present:

- The need for change must be urgent. Don't use the JDI approach just because it is quick and easy.
- The change must carry a low cost of failure. What if you're wrong? The price to pay for making the change must be low, preferably zero. Dr. Snow had little if anything to lose by removing the pump handle. The worst that would happen is people would have to travel farther to get their water.
- The change must have a significant potential reward. The decision here is, "Well, what if I'm right? Things will be a lot better if the change is effective."

JDI's used at the appropriate times and in the right situations can be a beneficial and effective method of attaining some quick wins.

Summary

Effective root cause analysis is a key to the success of the control activities of producers of goods, services, and information. If it is absent, problems will frequently recur and deficiencies will continue to reach the customer. A few different methods were discussed in this chapter, and multiple variations are available. The important thing is not which specific method is used, but that the producer has some effective method for identifying these assignable causes of day-to-day problems.

References

The list and descriptions of tools in this section are edited from the pertinent sections of Juran Institute, Inc. (2008) "Root Cause, Corrective Action, Version 1.2," Southbury, CT.

Juran Institute, Inc. (1994). *Quality Improvement Tools, Second Edition*, Wilton, CT., Ch. 1, p. 5 and p. 13.

Monroe, D. (2009). "Process Variables Controlled," *Quality Magazine*, November.

Juran Institute, Inc. (2007). *Six Sigma Manufacturing Black Belt*, Volume 2, Section 10a, p. 14.

Paraphrased from Juran Institute, Inc. "The Quality Minutes: The Jefferson Memorial."

Robitaille, D. (2009). "Four Things You Should Get from Root Cause Analysis," November 10, *Quality Digest The Quality Insider* (used by permission).

Langley, G. G., Nolan, K. M., Norman, C. L., Provost, L. P., and Nolan, T. W. (1996). *The Improvement Guide: A Practical Approach to Enhancing Organizational Performance*, Jossey- Bass, New York, pp. 3–4.

Paraphrased from Juran Institute, Inc. "The Quality Minutes: London Cholera Epidemic."

CHAPTER 14

Continuous Innovation Using Design for Six Sigma

Joseph A. De Feo and John F. Early

About This Chapter

Creating new services and goods that meet customer needs will lead to an increase in sales and revenue for an organization. These new services and products, if designed effectively, will have the features that meet their unique needs. There are three critical aspects of the design that will make a difference in how well a new good or service is received by the customers and the marketplace. The first is the overall innovativeness of the design. Does it create a sense of surprise and delight? The second is the effectiveness of the product features of the new good or service in delivering the benefits that the customer is seeking. The third is the ability of the organization to deliver all the innovation and features without any deficiency. Organizations that have effective methods to design for both innovation and cost will ultimately create products or services that are salable. This chapter provides a review of contemporary design and development methods with a focus on continuous innovation to drive top line performance of an organization. This is not about quality in the product development function. It is about methods that organizations must use to be innovative when developing new products, services, and key processes.

High Points of This Chapter

1. Innovation is key to the survival of all organizations. Innovation, like continuous improvement, is the result of a systematic approach, not a haphazard one.

2. Continuous innovation (CI) is different from product development. Continuous innovation must happen in all areas of an organization, from creating products, services, or processes used to meet internal and external customer needs to designing new facilities' or office environments.
3. There have been many improvements in the methods used to design and develop products and services in the past decade. Design for Manufacturing, Design for Assembly, Design for Lean, Design for Environment, and Six Sigma all have become models to meet critical to quality customer needs—and lead to innovative products.
4. Continuous innovation using the steps of Design for Six Sigma or DMADV, as it is often referred to, is similar to the Juran Quality By Design model (see Chapter 4, Quality Planning: Designing Innovative Products and Services) and has become the basis for what we call "continuous innovation of goods, services, and processes."
5. Creating the habit of innovation requires that management create an infrastructure similar to that of continuous improvement. Set goals, select projects, and educate teams to create innovative goods and services—project by project.
6. Continuous innovation using Design for Six Sigma consists of carrying out five steps:
   a. *Define* the goals and objectives for the new good, service, or process.
   b. *Measure* and discover hidden customers needs.
   c. *Analyze* the customer needs and determine the innovative features that will meet those needs.
   d. *Design* by combining the features, thereby creating new products, services, or processes that incorporate the features.
   e. *Verify* that the new innovation meets the customers' and organization's needs.

Continuous Innovation and the Juran Trilogy®

We have previously explored the Juran Trilogy® as it relates to quality planning. Designing for customer needs always leads to higher-quality products and services as well as innovative outcomes because an effective design process uncovers hidden customer needs. This discovery and the subsequent solving of the problems that kept customer needs hidden lead to innovation (see Figure 14.1). This chapter addresses the use of the DMADV steps and tools for creating continuous innovation (CI). Adapting the most effective models such as the Quality by Design used by the FDA and Design for Six Sigma (DFSS) model used by many such as GE, Samsung, and Microsoft, organizations can create the *habit of innovation* which is similar to creating the *habit of improvement*. Deploying a CI program will ensure organization adaptability and sustainability in meeting societal and business needs.

CI using the Design for Six Sigma model and tools, which arose out of GE Medical's adaptation of the Juran quality planning model described in Chapter 4, Quality Planning: Designing Innovative Products and Services, is a powerful engine available for those who want to truly plan quality into their products, typically goods rather than services or processes.

Juran referred to the quality planning design steps (see Figure 4.2) as a framework for planning (designing) new products and services (or revisions). These steps apply to both the manufacturing and service sectors and to products for both external and internal customers.

Planning an effective solution for an improvement project (see Chapter 3, The Universal Methods to Manage for Quality) may require one or more steps of this quality planning process. Early and Colletti (1999) and Juran (1988) provide extensive discussions of the steps. These quality planning steps must be incorporated with the technological tools for the product being developed. Designing an automobile requires automotive engineering

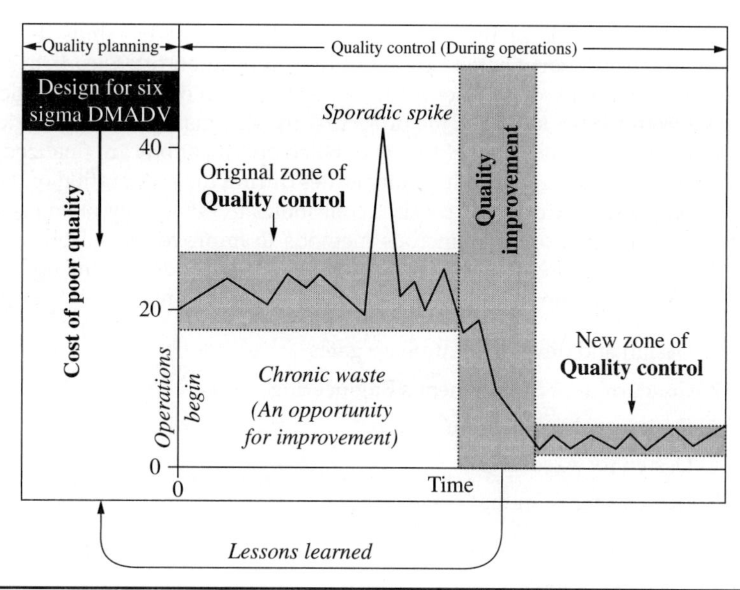

FIGURE 14.1 Design for Six Sigma and the Juran Trilogy.® (*Juran Institute, Inc. Southbury, CT.*)

disciplines; designing a path for treating diabetes requires medical disciplines. But both need the tools of quality planning to ensure that customer needs are met.

The road map is presented in greater detail in Figure 4.4. It is useful, however, to present an overview now to explain briefly the steps (Early and Colletti 1999).

New designs or innovations happen when one discovers hidden customer needs. Some examples include the following:

Abrasive cloth:	Lower internal cost of polishing parts due to better durability of cloth
Automobile:	Less effort in closing door; better "sound" when door closes
Dishwasher:	Greater durability because heavier parts make up the appliance
Electronics:	Simplicity all-in-one device, e.g., iPhone, iPod,
Software:	Understandable owner's manual
Fibers:	Lower number of breaks in processing fibers
Tire valve:	Higher productivity when tire manufacturer uses valve in a vulcanizing operation
Photographic film:	Fewer process adjustments when processing film due to lower variability
Commodity product:	Delivery of orders within 24 hours rather than the 48-hour standard requirement
Home mortgage application:	Decision in shorter time than that of competition

Traditionally, the main activities to capitalize on these insights were executed *sequentially*. For example, the planning department studied customer desires and then presented the results to design; design performed its tasks and handed the results to engineering; engineering created the detailed specifications; and the results were then given to manufacturing. Unfortunately, the sequential approach results in a minimum of communication between the departments as the planning proceeds—each department hands its output "over the wall" to the next department. This lack of communication often leads to problems for the next internal customer department. To prevent this from occurring, activities are organized as a team from the beginning of the project. Thus, e.g., manufacturing works *simultaneously* with design and engineering before the detailed specifications are finalized. This approach allows the team to address producibility issues during the preparation of the specifications.

Creating new products and services contributes to the vitality of an organization. Many organizations have adopted numerous methods to improve the salability of their designs. From the1980s to the present there were a number of newly adopted methods based on Juran's Quality by Design to improve product salability. Many continue today to pay dividends:

1. Design and development phase gates
2. Concurrent or Simultaneous Engineering
3. Design for Manufacture
4. Design for Assembly
5. Design for Six Sigma

In this past decade a number of new methods have popped up. Most recently there have been promising methods such as

1. Design for Environment
2. Lean Design
3. Sustainable Design

Today, Design for Six Sigma is a systematic methodology to provide the means to attain new services and innovative designs. The steps for designing new products and services that lead to innovation are as follows:

- Discover the customers and their needs.
- Gather and research information, and observe the behaviors of these customers.
- Generate and then design solutions to meet their needs.
- Design the solution and validate that the needs are met.
- Transfer the design to operations.

Along the way, these steps force people to "think outside the box." They force people to gain new information in a structured and organized way, arriving sometimes at revolutionary means to create new services.

DFSS Works for Goods, Services, and Transactional Processes

The Design for Six Sigma model has been used within new product introduction (NPI) processes for a wide variety of physical goods including electronics, chemicals, sophisticated industrial equipment, transportation equipment, and a plethora of consumer goods. It has also been used successfully to develop high-quality new services in insurance, health care, banking, and public service.

In the design phase of DFSS, a multifunctional team develops both the detailed product design down to the full engineering drawings and the process design for delivering the product, including all equipment, work instructions, work cell organization, etc. The difference between product design and process design is fairly clear when physical goods are produced. It is sometimes less clear for services where the two are intertwined.

Making and acting on the distinction between the design of the service and the design of the process that delivers that service has proved to be very helpful. The *service design* is the flow of activity as experienced by the customer. The service *process design* is the flow of activity required to make the customer experience possible.

For example, the service for paying a customer's insurance claim will have features related to timeliness, ease of use, responsiveness, and transparency. These are what the customer sees, feels, hears, and touches. To deliver that seamless flow of activity to the customer, the production process will include features related to data processing, information access, payment procedures and policies, and interpersonal skills of individuals interacting with the customer during the process. The behind-the-scenes production process is largely invisible to the customer. In fact, when these invisible production processes become visible to the customer, it is usually because they have broken down and failed to deliver the seamless service as designed.

Experience shows that it is a useful division of the work to first design the customer-experienced service and then design the process that makes it possible. Teams that try to design both the service and the process as a single step usually subordinate the customer experience to the exigencies of operations.

An Example of Designing for Services

In an example from the service sector, the quality planning process was applied to replanning the process of acquiring corporate and commercial credit customers for a major affiliate of a large banking corporation. Here is a summary of the steps in the quality planning process.

1. *Establish the project.* A goal of $43 million of sales revenue from credit customers was set for the year.
2. *Identify the customers.* This step identified 10 internal customer departments and 14 external customer organizations.
3. *Discover customers' needs.* Internal customers had 27 needs; external customers had 34 needs.
4. *Develop the product.* The product had nine product features to meet customers' needs.
5. *Develop the process.* To produce the product features, 13 processes were developed.
6. *Develop process controls and transfer to operations.* Checks and controls were defined for the processes, and the plans were placed in operation.

The revised process achieved the goal on revenue. Also, the cost of acquiring the customers was only one-quarter of the average of other affiliates in the bank. Quality planning generates a large amount of information that must be organized and analyzed systematically. The alignment and linkages of this information are essential for effective quality planning for a product. A useful tool is the quality planning spreadsheet or matrix (basically, a table). Figure 14.2 shows five spreadsheets corresponding to steps in the quality planning process. Note how the spreadsheets interact and build on one another; they cover both quality planning for the product and quality planning for the process that creates the product. The approach is often called *quality function deployment* (QFD). Thus QFD is a technique for documenting the logic of translating customer needs into product and process characteristics. The use of spreadsheets in the quality planning process unfolds later in this chapter.

These six quality planning steps apply to a new or modified products (goods or services) or process in any industry. In the service sector the "product" could be a credit card approval, a mortgage approval, a response system for call centers, or hospital care. Also the product may be a service provided to internal customers. Endres (2000) describes the application of

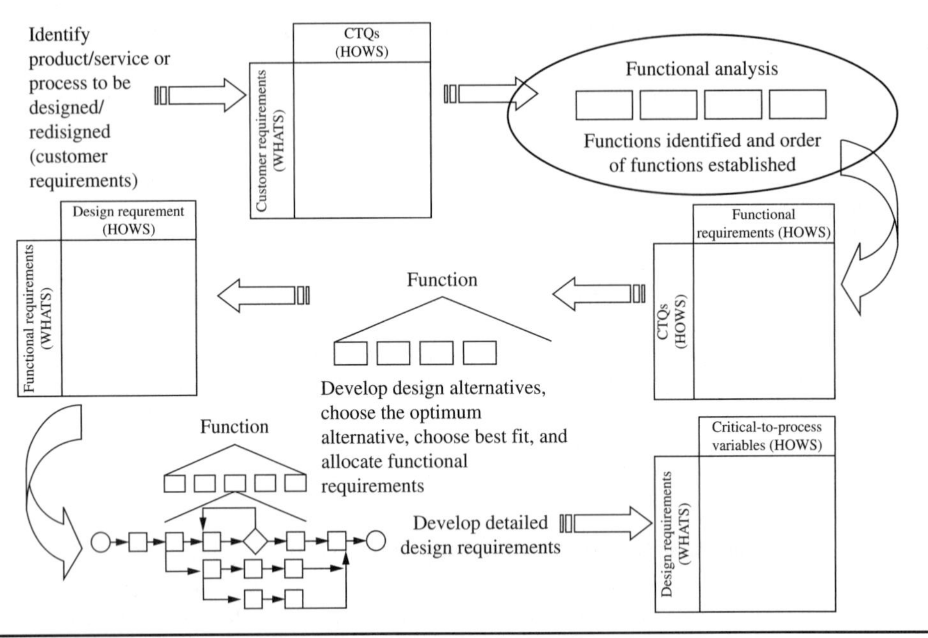

Figure 14.2 How to design matrices. (*Juran Institute, Inc. Used by permission.*)

the six quality planning steps at the Aid Association for Lutherans insurance company and the Stanford University Hospital.

CI Requires Understanding Customer Needs and Solving Their Problems

Designing innovative and superior quality services and products requires gaining a clear understanding of the customers' needs and translating those needs into services aimed at meeting them. This information is the driver of most innovation. Yet most do not recognize it as such.

Innovation has everything to do with creating something new. In competitive business situations, success often comes to the best innovators. Many organizations have design and development functions that create annual plans to develop new models and new services. Sometimes these functions design the good or service internally to the organization and then look for customers to sell it to. Other innovation comes from solving societal problems. And still other organizations look for customer problems to solve; as a result they create something new, something innovative. It is the latter that we have found to be the most economical and therefore provides the greatest return on its investment.

To create continuous innovation, an organization must design to meet customers' unmet (often hidden) needs. To do this one must

1. Capture the voice of the customers—the potential new customers or existing ones.
2. Discover hidden customers and needs. It is hidden customers or hidden needs that must be found.
3. Design solutions to meet those needs. This usually means solving a challenge or contradiction.
4. Use a systematic approach to ensure innovation happens—continuously.

5. Have tools to capture the information and use it to ensure that the good or service is produced efficiently.
6. Use multifunctional staff to carry out the systematic process to ensure the good or service can be produced as planned.

One can learn about innovation, which means "making something new," by studying innovations and innovative methods from the past.

Polaroid Camera

The conventional photographic process involves exposing light-sensitive material, which in turn must be developed, fixed, and printed and the print developed and fixed, a procedure that can take hours (or days if the processing facility is far from the place where the photograph was taken). In 1947, a remarkable new system of developing and taking pictures was introduced by U.S. physicist Edwin Herbert Land (1909–1991). Land had left Harvard after his freshman year to conduct his own research on the polarization of light. Two years later, he invented a sheet polarization filter that could be used on camera lenses to eliminate reflection and glare. In 1937, Land founded the Polaroid Corporation to manufacture and market his filters, lamps, window shades, and sunglasses. In February 1947, he introduced Polaroid instant film for use in his own Polaroid Land Camera. The Land Camera (U.S. Patent 2,543,181) was first offered for sale on November 26, 1948. Polaroid film processes chemicals in a flat, hermetically sealed compartment attached to the photosensitive paper. A pair of pressure rollers spreads the chemicals uniformly across the paper when exposed, and the completed print is ready a minute later. In 1963, Polaroid introduced Polacolor, a full-color film that could be processed in less than a minute.

Life Savers Candy

In 1912, when candy maker Clarence Crane first marketed Crane's Peppermint Life Savers, life preservers were just beginning to be used on ships—the round kind with a hole in the center for tossing to a passenger who had fallen overboard. But that is not the whole story. Crane had been basically a chocolate maker. Chocolates were hard to sell in summer, however, so he decided to try to make a mint that would boost his summertime sales. At that time most of the mints available came from Europe, and they were square. Crane was buying bottles of flavoring in a drugstore one day when he noticed the druggist using a pill-making machine. It was operated by hand and made round, flat pills. Crane had his idea. The pill-making machines worked fine for his mints, and he was even able to add the life preserver touch by punching a tiny hole in the middle. In 1913, Crane sold the rights to his Life Savers candy to Edward Noble for only $2900. Noble then sold Life Savers in many flavors, including the original peppermint. Clarence Crane may have regretted that decision to sell, for Life Savers earned the new manufacturer many millions of dollars.

iPod

The iPod originated with a business idea dreamed up by Tony Fadell, an independent inventor. Fadell's idea was to take an MP3 player, build a Napster music sales service to complement it, and build a company around it. It resulted in Apple creating the iPod.

Segway

This new means of transportation meant reimagining virtually every piece of conventional wisdom about the last century of transportation, from how it moves, to the fuel it uses, to how you control it. The result is electric transportation that doesn't look, feel, or move like anything that has come before. And of all the conventional wisdom we've left in pieces behind us, none has been shattered more fully than the belief that we must choose between "more" and "less." In 2001, Dean Kamen announced the arrival of the first self-balancing, zero emissions personal transportation vehicle: the Segway® Personal Transporter (PT).

Founded on the vision to develop highly efficient, zero emissions transportation solutions using dynamic stabilization technology, Segway focused its research and development on creating devices that took up a minimal amount of space, were extremely maneuverable, and could operate on pedestrian sidewalks and pathways. Today, Segway continues to develop safe, unique transportation solutions that address urban congestion and pollution.

Two Types of Innovation

There are two basic types of innovation. The first, type I, does happen, but rarely. Type I is something completely new. And new things under the sun do not occur as often as we think they do. The first automobile and internal combustion engine were certainly new innovations, but even they built on the wheel, cart, and other existing technologies.

Things such as nuclear power, radio, phones, electricity in the home, and manned flight are certainly good examples of something that was pretty close to new under the sun. All the great, really new innovations can often be traced back to a genius, a lucky accident, or both.

We know the names of many of the geniuses—Fermi, Wright, Edison, Benz, and Ford. However, this is not an endless list, and while lucky accidents are good, they are too chancy. Type II innovation presents a better way.

Type II innovation is much more common than type I. This second type of innovation can be reduced to three general approaches:

- Making something that already exists larger
- Making something that already exists smaller
- Combining one thing that exists with something else that exists

The simplicity of type II is profound. It can create dramatic breakthroughs and change the way we live. Most of what we see and consider to be great innovations were derived from the three methods of type II innovations listed.

For example, the mobile phone or PDA in your pocket was once a fair-sized wooden box on the wall. The phone has been made smaller from the original wall model hardwired to the outside world. The phone has also been "combined" with a radio, calculator, computer, TV, and music player. The flat-screen television evolved from a device that was once considered a piece of furniture and that took up more room than an easy chair. Over time, the TV's depth and height have been "made smaller," and its width has been "made larger." Add the appropriate technology, and you have your flat-screen display.

An example is Web-based learning. Web-based learning came about when transparencies were replaced by electronic slides such as PowerPoint. This led to improved quality of presentation graphics, then added animation, placed on the Internet, with voice-over IP, and video, thus delivering Web-based learning.

The "bigger/smaller/combination" approach sounds simple when you look backward. But the trick is doing it in the present, as an innovation for the future. However, it is still much easier than becoming a genius.

The good news is you can get better at type II innovation. As good as we are today, we can also get better with practice.

The next time you are in a serious brainstorming meeting and need an innovation for a new product, service, marketing strategy, or similar task, put up three new header columns, and attack them one at a time.

A header is the place where you will hang your ideas. The three headers are, of course, "make it bigger," "make it smaller," and "combine it with." The "it" is whatever product or service or whatever you are working on. Have fun with it. Remember not to critique or scrub the ideas until after the generation of ideas is done. Most people are surprisingly good at type II innovation. Morph some of the wild ideas into something that is doable. The great innovator Henry Ford said, "If you think you can or can't, you're right."

Innovators are not born that way. If you have your heart set on being the next Thomas Edison, you are probably going a bit too far. But whatever your innovation quotient is now, you can make it better with practice and by using a methodology that causes innovation to happen.

For instance, how many times do we hear "Think outside the box"? That's all well and good, but what box? Few of us recognize that the box is in fact ourselves. Learning to temporarily let go, be foolish for a moment, and be comfortable with ambiguity is necessary for innovation.

Getting beyond our "boxed" selves is a skill that can be learned and improved with technique, practice, and courage. For example, imagining oneself as someone else and seeing everything through his or her eyes can be a great technique.

Arriving at this level of letting go will require a systematic methodology. Many methods have been used in developing simpler and better products. These design processes incorporate early involvement teams.

The teams are composed of a broad spectrum of employees, customers, and suppliers who work together through a systematic process of looking and thinking outside the box to solve problems. The results are significant, and new products can be discovered.

This concept of push innovations (e.g., toys and foods) is a short-term exercise that continues to flood the market with new products. Some are good and last a long time; many are short-lived. If you are trying to innovate to solve a customer or societal problem, the outcome of a purposeful design process often leads to products that benefit society for many years. Drug development is a good example. Aspirin has been around for more than 100 years. New drugs that reduce cholesterol will also be here for decades.

Why do some products last so long and others do not? This answer lies in the methods used to design or create the innovation. Innovation requires a systematic process and set of tools to create customer-focused, need-driven designs.

Designing world-class services and products requires gaining a clear understanding of the customers' needs and translating those needs into services aimed at meeting them. The process goes on to design and optimize the features and then develop and execute the new designs. This process is sometimes referred to as the service development process, the design process, or the DFSS process.

Random, innovative ideas, no matter how clever, will not deliver economic success unless they meet a customer need better than the current method or fulfill a previously unknown or unmet need. The talented design people we have working for our organizations give us excellent designs when we specify who wants it and what it is that they want—the "they" being the customers who make up a market segment.

The problem with most failed new products and services is not poor design. The problem is that the product or service did not have customers waiting and ready for the things that were actually produced. The question is whether there is a way to reliably get around this problem of good design. There are also innovations that are replaced or that evolve quickly. Foods based on fad diets and toys based on television shows come and go. Other innovations, such as the computer, stay for generations. Why do some innovative products and services splash onto the scene and evaporate while others last? The answer often lies in the reason for wanting to create them in the first place.

DFSS was developed to precisely fill this methodological void. DFSS is a rich concept with a well-developed core methodology. The process entails a five-phase service or product development method, and the phases are as follows:

Define

In the define phase, top management has to look critically at the business. It would help to revisit the organization's strategic plan. (If you do not have an up-to-date strategic plan, you should get one.) Management provides the design team with specific guidance on the need for the new service or product; management should not, however, design the product. It is okay to provide a high-level concept, but leave the design to the designers.

Measure

The measure phase is all about discovering and exploring customers and their needs—especially any unmet needs. This is the heart of DFSS. How do you ask a target audience for what they want in a service or product that does not exist? You cannott, at least not directly. It is best to focus on needs. Again, let the designers design the product, not the customer.

The team then transforms the customers' needs into something more technical. We will call these critical-to-quality characteristics (CTQs). In the CTQs, we transform the needs as articulated by the customer into words and phrases we can measure. The CTQs become the targets for the designers. This step makes it possible to design a product or service that will interest a target group of customers. (Recall that this was the failing of most unsuccessful products or services.)

Analyze

In the analyze phase, the designers try several concept designs with potential to meet the CTQs developed in the measure phase. The concepts are now traceable to one or more CTQs, which in turn are traceable to one or more customer needs. The team develops and matches functional requirements of the concept design to the CTQs. The analyze phase is the exciting part for most designers, but the foundation was laid during the define and measure phases.

Design

The detail design follows. In the design phase, we take the winning concept design and fill in all the details. When inevitable choices and tradeoffs must be made, we have ready-made selection criteria: the CTQs. The CTQs are like having the customer beside us at every decision point. We will develop and match the functional requirements from analysis to the design requirements of the detail design.

Verify

When the team is satisfied with the details of the design, they are ready to verify meeting the business needs given to them by management in the define phase and the customers' needs provided during the measure phase. Complete planning for procurement, production, delivery, advertising, warranty, and other items is also completed during the verify phase.

Innovation can be enhanced. Most innovation will flourish if organizations can develop their own creative talents. Type II innovation is the key—encouraging all employees to think in terms of making something bigger, smaller, or combined with something else. DFSS then helps us to identify customers, learn their needs, and deliver products or services that meet those needs. Innovation cannot be commanded. But innovation can certainly be encouraged and managed to achieve an organization's goals by assigning teams to solve customer problems, by creating new goods and services to solve them.

Evolution of Design and Innovation Methods

Quality by Design

Quality by Design was a concept first outlined by Dr. Juran in various publications, most notably *Juran on Quality by Design*, by Dr. Juran and the Juran Institute. It stated that quality must be planned into products, and that most quality crises and problems relate to the way in which quality was planned in the first place. While Quality by Design principles have been used to advance product and process quality in every industry, and particularly the automotive industry, they have most recently been adopted by the U.S. Food and Drug Administration (FDA) as a vehicle for the transformation of how drugs are discovered, developed, and commercially manufactured. The Food and Drug Administration defines Quality by Design as the level of effectiveness of the design function in determining a product's operational requirements (and their incorporation into design requirements) that can be converted into a finished product in a production process. Today Quality by Design has evolved into numerous other methods. Here are some of the most popular:

Concurrent Engineering

Concurrent Engineering was a popular new product development process in which all individuals responsible for development and production were involved at the earliest stages of product design. Some 70 to 80 percent of a product's cost is locked in at these early stages of development, when the product's configuration is determined and choices are made for the manufacturing processes and materials from which the product will be made. If a product is to end up cost-competitive, it is absolutely essential that cost be a consideration when these decisions are made.

One of the earliest forms of Design for Quality was the Design for Manufacturing and Assembly (DFMA) from University of Massachusetts Profs. Boothroyd and Dewhurst. They created a methodology and later software technology that help guide design teams through this critical stage of product development with cost information, even before prototype design models are created.

Design for Manufacture

Design for Manufacture (DFM) is a systematic approach that allows engineers to anticipate manufacturing costs early in the design process, even when only rough geometries are available on the product being developed. Given the large number of process technologies and materials available, few design engineers have detailed knowledge of all the major shape-forming processes. Consequently, engineers tend to design for manufacturing processes with which they are familiar. DFM methodology encourages individual engineers and concurrent development teams to investigate additional processes and materials and to develop designs that may be more economical to produce. With more information about viable processes and materials, users can quantify manufacturing costs for competing design alternatives and decide which design is best.

Design for Manufacture provides guidance in the selection of materials and processes and generates piece part and tooling cost estimates at any stage of product design. DFM is a critical component of the DFMA process that provides manufacturing knowledge into the cost reduction analysis of Design for Assembly.

Design for Assembly

Design for Assembly (DFA) is a methodology for evaluating part designs and the overall design of an assembly. It is a quantifiable way to identify unnecessary parts in an assembly and to determine assembly times and costs. Using DFA software, product engineers assess the cost contribution of each part and then simplify the product concept through part reduction strategies. These strategies involve incorporating as many features into one part as is economically feasible. The outcome of a DFA-based design is a more elegant product with fewer parts that is both functionally efficient and easy to assemble. The larger benefits of a DFA-based design are reduced part costs, improved quality and reliability, and shorter development cycles.

Design for Environment

Meeting the needs of an increasingly eco-conscious marketplace, DFMA allows product designers to conduct an environmental assessment during the concept stage of design, where they can evaluate the impact of material selection as well as account for the end-of-life status of their product.

The analysis prompts designers to select, from the DFMA database, the materials they prefer to use or avoid, then reveals the proportions (by weight) of those materials in the product. It also estimates and designates the proportions of product that go to different end-of-life destinations, including reuse, recycling, landfill and incineration. These measures help manufacturers meet such requirements as the European Union's Restriction of Hazardous Substances (RoHS) regulations.

Sustainable Design

Sustainable Design (also called Environmental Design, Environmentally Sustainable Design, Environmentally Conscious Design, etc.) is a method of designing physical goods that comply with the principles of economic, social, and ecological sustainability. The intention of Sustainable Design is to prevent negative environmental impact by identifying potential impacts and applying creative or best practices to prevent or mitigate them. Manifestations of sustainable designs require no nonrenewable resources, impact the environment minimally, and relate people with the natural environment.

Design for Six Sigma

The evolution of many lessons learned has led to the development of DFSS. It is focused on creating new or modified designs that are capable of significantly higher levels of performance (approaching Six Sigma). The define, measure, analyze, design, verify (DMADV) sequence is a design methodology applicable to developing new or revised products, services, and processes. Although DFSS implies to design to the lowest level of defects possible, Six Sigma, it is more than that. The steps in DFSS enable one to understand the customers and their needs. DFSS actually focuses on both sides of quality: the right features and the fewest failures.

Design for Six Sigma—DMADV Steps

Table 14.1 summarizes the main activities within each of the DMADV steps. These are discussed in more detail in this section. Experience with applying the five DMADV steps has led us to believe that it is useful to define a step to select the project before the team actually begins its DMADV journey

Select the Opportunity

The select phase in DFSS is more strategic than for quality improvement or DMAIC projects (see Chapter 12, Six Sigma: Improving Process Effectiveness). A target for a new product or capability is identified as part of the strategic and annual business planning processes. When a major opportunity is identified, leadership will determine that it is best served with a new design or redesign of something that exists. Typically this means that a new or emerging market has been targeted; it may also mean that customer needs in an existing market are shifting, or that competition has shifted, and a new approach is required.

This type of project selection is different from a DMAIC project in which specific deficiencies or wastes are targeted for an existing product or process. Rarely is an existing

Define	Measure	Analyze	Design	Verify
Agree to opportunity	Identify customers	Develop alternative designs	Develop detailed designs	Execute manufacturing/ operations verification
Agree to goals	Discover customer needs	Complete functional analysis	Integrate designs	Execute pilot and ramp-up
Agree to scope	Translate needs into CTQs	Select best-fit design	Model predictions of performance	Execute control plan
Establish project plan	Establish design scorecard	Specify functional requirements	Optimize design parameters	Finalize design scorecard
Assign resources		Specify subsystem functional requirements	Develop statistical tolerances	Transition to operational owners and validate
		Complete high-level design review	Specify process features and detailed operations	
		Validate with customer	Design complete control plan	
		Update design scorecard	Complete design verification test	
			Validate with customer	
			Complete design review	
			Update design scorecard	

TABLE 14.1 Major Activities in Phases of DFSS

product or process so broken that the initial analysis in DMAIC leads to the conclusion that a total redesign is required. A major health insurer reached that conclusion with respect to payment of claims. Instead of multiple improvement projects, it redesigned the entire claims payment service so as to raise customer satisfaction from 75 to 93 percent, improve timeliness by a factor of 10, and reduce costs by more than one-half.

The project opportunity and goal statements are prepared and included in a team charter, which is confirmed by management. Unlike the rather simple and direct goal statements for a DMAIC project, the DMADV goal statement may, in fact, be multiple statements about the market to be served by the new product and the economic returns to be achieved, such as market penetration, growth, and profitability. Management selects the most appropriate team of personnel for the project, ensures that they are properly trained, and assigns the necessary priority. Project progress is monitored to ensure success.

Select: Deliverables

- Make a list of potential projects.
- Calculate the return on investment and contribution to strategic business objective(s) for each potential project.
- Identify potential projects.
- Evaluate projects and select a project.
- Prepare project opportunity statement and a team charter.
- Select and launch team.
- Formal project team leader should be a qualified practitioner or black belt.

Select: Questions to Be Answered

1. What new market opportunities do we have?
2. What new emerging customers or customer needs can we go after?
3. What are the likely benefits to be reaped by gaining or increasing that business?
4. Which of our list of opportunities deserves to be tackled first, second, etc.?
5. What formal opportunity statement and goal statement should we assign to each project team?
6. Who should be the project team members and leader (black belt) for each project?

Define Phase

A project begins with the define phase when it is officially launched by the management team. It may be necessary for the management team or Champion to work closely with the project design team to refine the design opportunity. This refinement will lead to an accurate scope of the project and will ensure a common understanding of the objectives and deliverables. Experience has shown that projects that fail to deliver the expected results frequently get off track at the start, when the project is being defined.

A key task in the define phase is to create the initial business case that validates the selection rationale and establishes the business justification through reduced product cost, increased sales, or entirely new market opportunities. The initial business casework is conducted under the auspices of the management team, and then it is validated and updated continuously by the design team through the subsequent phases of the design project. The management team selects a black belt to lead the design project. The Champion, who is the management sponsor with vested interest in the success of the design, in conjunction with the black belt, is responsible for selecting a cross-functional team that will conduct all the activities to complete the design and carry it into production.

Define: Deliverables

- Initial business case is developed.
- Design strategy and project are established; leaders and team are selected.
- Project charter is drafted, including project opportunity statement and design objectives.
- Team is launched and a list of customers defined: market customers, nonmarket customers—users, regulators, stakeholders etc.—and internal customers.

Define: Questions to Be Answered

1. What are the design goals or objectives of the project?
2. What are the specific goals of the project team?
3. What is the business case that justifies the project?
4. What charter will the team members receive from management empowering them to carryout the project?
5. What will be the project plan?
6. How will the project be managed?
7. Who will be the customers of this project?

Measure Phase

The measure phase in the DMADV sequence is mainly concerned with identifying the key customers, determining what their critical needs are, and developing measurable critical quality (CTQ) requirements necessary for a successfully designed product. An initial assessment of our markets and customer segmentation by various factors is required to identify the key customers. This assessment is often completed by the marketing organization and is then reviewed and verified by the design team. However, it is the design team's responsibility to complete the customer needs analysis and compile the results into a prioritized tabulation of customer needs. The design team transforms the critical customer needs into measurable terms from a design perspective. These translated needs become the measurable CTQs that must be satisfied by the design solution. Competitive benchmarking and creative internal development are two additional sources to generate CTQs. These methods probe into design requirements that are not generally addressed or possibly even known by the customer. The result is a set of CTQs stated in specific technical requirements for design in the voice of the organization that become the measurable goals (specifications) for product performance and ultimate success.

The project team may use several means to set the goals for each CTQ. Some tools include competitive benchmarking, competitive analysis, value analysis, criticality analysis, and stretch objectives for current performance. The result is a combination of customers' stated requirements, and requirements that may not be generally addressed or known by the customer. The measure phase ends with the assessment of the current baseline performance against the enumerated CTQs and performance of risk assessments. To establish these baselines, typical process capability methods and tools are utilized. These include the following:

1. Establish the ability of the measurement system to collect accurate data using measurement system analysis (MSA).
2. Measure the stability of the current or surrogate process(es) using statistical process control techniques.
3. Calculate the capability and sigma level of the current or surrogate process(es).
4. To evaluate risk, the team may make use of tools such as design failure mode effects analysis (DFMEA) and process failure mode effects analysis (PFMEA).

Another tool employed by some design project teams is the set of quality function deployment (QFD) matrices. (See Figure 14.2.) Each matrix lists vertically some objectives to be fulfilled (the "what") and then horizontally the means to fulfill the objectives (the "how"). Within the body of the matrix are indicators for how well each objective is met by the respective means. For example, the first matrix displays how well each of the customer needs is addressed by the specific CTQs. As a group, the matrices are tied together, with the means (how) of one matrix becoming the objectives (what) of the next. In this way the customer needs are tied seamlessly to the CTQs, to the functional requirements, to the design requirements, finally to the process requirements, and ultimately to the control requirements. In this way nothing critical is lost and no extraneous matters are introduced.

The QFD matrix (or simpler versions) is meant to highlight the strengths and weaknesses that currently exist. In particular, the weaknesses represent gaps that the design team must shrink or overcome. The demand on the team then is to provide innovative solutions that will economically satisfy customer needs. Keeping this matrix up to date provides a running gap analysis for the team.

Discover Customer Needs

- Plan to collect customer needs from internal customers and external customers.
- Collect list of customers' needs in their language.
- Discover and prioritize customer needs in terms of the customer-perceived benefit.

Translate and Prioritize Customer Needs

- Translate needs and benefits from the voice of the customer (VOC) into voice of the producer as CTQ requirements
- Establish measurement for all prioritized CTQs, including units of measure, sensor, and validation.
- Establish targets and upper and lower specification limits for all CTQs.
- Establish target permissible defect rate (DPMO, Sigma) for each CTQ.

Establish Baseline and Design Scorecard

Once the prioritized list of CTQs is produced, the design team proceeds to determine the baseline performance of relevant existing product and production process. The current baseline performance is determined in terms of multiple components:

- Measurement systems analysis
- Product capability
- Production process capability
- Risk assessment by using tools such as product FMEA
- Competitive performance

Finally, a design scorecard is created that tracks the design evolution toward a Six Sigma product performance. This tool is used in the attempt to predict what the final product performance and defect levels will be after integration of all the design elements. The design scorecard is updated throughout the project to ensure that objectives are met.

Measure: Deliverables

In summary, the key deliverables that are required to complete the measure phase are

- Prioritized list of customer needs
- Prioritized list of CTQs
- Current baseline performance
- Design scorecard

Measure: Questions to Be Answered

1. What customer needs must the new product meet?
2. What are the critical product and process requirements that will enable the customer needs to be met?
3. How capable is our current product and production process of meeting these requirements?
4. How capable must any new product and production process be to meet these requirements?

Analyze Phase

The main purpose of the analyze phase is to select a high-level design and develop the design requirements that will be the targets for performance of the detailed design. This is sometimes referred to as system-level design versus the subsystem or component design levels.

The design team develops several high-level alternatives that represent different functional solutions to the collective CTQ requirements. A set of evaluation criteria is then developed, against which the design alternatives will be analyzed. The final configuration selected may be a combination of two or more alternatives. As more design information is developed during the course of the project, the design may be revisited and refined.

In developing the high-level design, the team establishes the system's functional architecture. The flow of signals, flow of information, and mechanical linkages indicate the relationship among the subsystems for each design alternative. Hierarchical function diagrams, functional block diagrams, function trees, and signal flow diagrams are commonly used to illustrate these interrelationships. Where possible, models are developed and simulations run to evaluate the overall system functionality.

The requirements for each subsystem are expressed in terms of their functionality and interfaces. The functionality may be expressed as the system transfer function, which would represent the desired behavior of the system or subsystem. Interfaces are described in terms of the input and output requirements and the controls (feedback, feed-forward, automatic controls). These specifications will be provided to the detail design teams in the design phase.

In the analyze phase, DMADV analysis tools enable the design team to assess the performance of each design alternative and to test the differences in performance of the competing design alternatives. The results of the these tests lead to the selection of the best-fit design, which is then the basis to move into the next phase, detailed design. These analyses are accomplished using graphical analysis and statistical tools, some of which are

- Competitive analysis
- Value analysis
- Criticality analysis
- Fault-tree analysis
- Risk analysis
- Capability analysis
- High-level design matrices from QFD
- TRIZ
- Updated design scorecard

One of the significant advances affecting this process is the availability of several statistical analysis tools. These software applications, running on desktops or laptop computers, speed up the number crunching required to perform the preceding analysis. This availability has also made it necessary for individuals who would not normally use these tools to be trained in the use and interpretation of the results.

Analyze: Deliverables

Develop a high-level product or service and process design and detail design requirements.

- Design alternatives
- Functional analysis
- Best alternative selected
- Best-fit analysis
- High-level quantitative design elements
- High-level resource requirements and operating ranges
- High-level design capability analysis and prediction
- Detail design requirements for subsystems/modules
- Key sourcing decisions
- Initial product introduction resources and plans
- Updated design scorecard
- QFD design matrices

Analyze: Questions to Be Answered

1. What design alternatives could be employed in the new product or process service?
2. Which is the "best" alternative?
3. What are the requirements for the detailed design?
4. Has customer feedback been obtained?
5. Does the high-level design pass a business and technical design review?
6. Has the design been validated with customers?

Design Phase

The design phase builds upon the high-level design requirements to deliver a detailed optimized functional design that meets operational manufacturing and service requirements. Detail designs are carried out on the subsystems and eventually integrated into the complete functional system (product). DMADV tools focus on optimizing the detail-level design parameters.

In particular, designed experiments and/or simulations serve several purposes. One purpose is to determine the best set of features (optimum configuration) to employ. Another purpose can be to obtain a mathematical prediction equation that can be used in subsequent modeling and simulations. Experiments are typically are designed at differing levels of complexity, from minimal-run screening experiments to multilevel replicated design. Screening experiments typically try to establish which factors influence the system, providing somewhat limited results for modeling. More detailed experiments, including response surface and mixture designs, are conducted to determine system performance more accurately and produce a mathematical equation suitable for prediction and modeling applications. More complex products will often require nonlinear response surface models as well as mixture and multiple-response models.

During the design phase, the design team is also concerned about the processes that must be developed to provide the service or build the product. During the measure phase, the team examines the current capability of the business to deliver the product or service at the expected quality levels (approaching Six Sigma). During the design phase, the team continually updates the design scorecard with the results of designed experiments, benchmarking results, process capability studies, and other studies to track the design performance against the established goals, continuing the gap analysis that runs throughout the project. The product design is also reevaluated against the manufacturing or operational capability. Product designs may be revised as needed to ensure reliable, capable manufacturing and operations.

Part of the design for operations includes the validation of tolerances for each parameter. Designed experiments can contribute to developing these tolerances, and statistical tolerancing can also validate them.

To conclude the design phase requires the goals of the design for performance to be verified through testing of prototype, preproduction models, or initial pilot samples or pilot runs. The design team documents the set of tests, experiments, simulations, and pilot builds required to verify the product/service performance in a design verification test (DVT) plan. Upon completion of the several iterations that occur during the DVT and pilot runs, the design is solidified and the results of testing are summarized. A design review meeting marks the conclusion of the design phase, when the results of the DVT are reviewed. The design scorecard is updated, and each area of the development plan (quality plan, procurement plan, manufacturing plan, etc.) is adjusted as necessary.

Design: Deliverables

- Optimized design parameters (elements)—nominal values that are most robust
- Prediction models
- Optimal tolerances and design settings
- Detailed functional design
- Detailed designs and design drawings
- Detailed design for operations/manufacturing
- Standard operating procedures, standard work, and work instructions
- Reliability/lifetime analysis results
- Design verification test results
- Updated design scorecard

Design: Questions to Be Answered

1. What detailed product design parameters minimize variation in product performance?
2. What tolerances both are practical and ensure performance?
3. How do we ensure optimum product reliability?

4. How do we ensure simplicity and ease of manufacture or operations?
5. What detailed process parameters consistently and predictably minimize production process variation around target values?

Verify Phase

The purpose of the verify phase in the DMADV sequence is to ensure that the new design can be manufactured or service delivered and field supported within the required quality, reliability, and cost parameters. Following DVT, a ramp-up to full-scale production is accomplished via the manufacturing verification test (MVT) or operations verification test (OVT). The objective of this series of tests is to uncover any potential production or support issues or problems. The operations process is typically exercised through one or more pilot runs. During these runs, appropriate process evaluations occur, such as capability analyses and measurement systems analyses. Process controls are verified and adjustments are made to the appropriate standard operating procedures, inspection procedures, process sheets, and other process documentation. These formal documents are handed off to downstream process owners (e.g., manufacturing, logistics, and service). They should outline the required controls and tolerance limits that should be adhered to and maintained by manufacturing and service. These documents come under the stewardship of the company's internal quality systems. One of the considerations of the design team is to ensure that the project documentation will conform to the internal requirements of the quality system.

The design team should ensure that appropriate testing in a service and field support environment is accomplished to uncover potential lifetime or serviceability issues. These tests will vary greatly, depending on the product and industry. These tests may be lengthy and possibly not conclude before production launch. The risks associated with not having completed all tests depend on the effectiveness of earlier testing and the progress of final MVT/OVT tests that are underway. A final design scorecard should be completed, and all key findings should be recorded and archived for future reference. The team should complete a final report that includes a look back at the execution of the project. Identifying and discussing the positive and not-so-positive events and issues will help the team learn from any mistakes made and provide the basis for continuing improvement of the DFSS sequence.

Verify: Deliverables

- Verify product/process performance against project targets.
- Pilot build is complete.
- Pilot tests are completed and results are analyzed.
- All operational and control documentation, procedures, controls, and training are complete.
- Scale-up decision(s) are made.
- Full-scale processes are built and implemented.
- Business results are determined/analyzed.
- Processes are transitioned to owners.
- DFSS project is closed.

Verify: Questions to Be Answered

1. Is the product or process meeting the specifications and requirements?
2. Is the production process "owned" by the business?

Examples of Continuous Innovation Process Using Design for Six Sigma

Example 1: A Design for Six Sigma (DMADV) Project*

Project Background

The current process to look up, retrieve, and interpret product engineering information such as component specification drawings and product structures has been in place since 1998. This system is complex and expensive to maintain. From the beginning, this process has had many shortcomings from the point of view of the primary users—the manufacturing plants. These shortcomings cost the company money in lost productivity and high system maintenance costs.

DMADV Process Implementation

With the long history of complaints and a limited customer base, areas of improvement were not difficult to determine. To provide focus for our team, a survey was developed and analyzed to prioritize customer groups and customer needs as well as their performance expectations for the new system. The needs became the customer CTQ items.

We worked with our customers to determine baseline capability against four criteria:

- Accuracy of information
- Fast retrieval of information
- Easy retrieval of information
- Easy-to-interpret information

From this list we constructed a quality function deployment flow-down matrix to convert the CTQs to product feature alternatives that support customer needs. The current process was mapped at high and then more detailed levels to identify areas of improvement.

A high-level design was prepared, and high-level capability was estimated. Next a more detailed design was developed, simulated, documented, and verified.

Results

- Accuracy level unchanged (Six Sigma–capable)
- A 451 percent improvement in average print access/printout time (from 1.5 to 6 sigma)
- 100 percent improvement in virtual viewing/inquiry capability
- 300 percent improvement in drawing line weight differentiation
- Final expected savings: not insignificant

Project Details and Selected Slides

Problem Statement Plant quality and customer service/technical support personnel find that our current system to find and view product component and assembly information is cumbersome to access, interpret, and maintain.

Project Definition The purpose is to provide faster access to product engineering information in a consolidated format using a single user-friendly interface.

---

*Adapted from the final report of a Six Sigma design project led by Dave Kinsel at a Juran Institute client; with acknowledgment of thanks.

Mission Statement The project team will develop a user interface and training system to provide faster single-point access for plant quality managers, engineering, customer service, and technical support to product structures and related component and assembly specifications by July 2004.

The slides shown in Figures 14.3 through 14.20 highlight the project for each phase: define, measure, analyze, design, and verify.

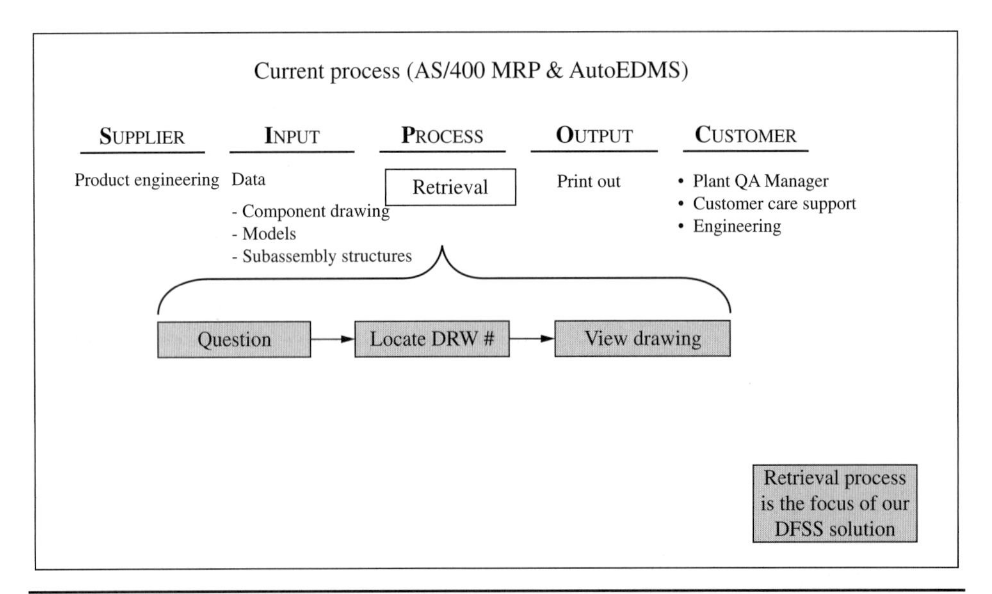

FIGURE 14.3 SIPOC (high-level process map).

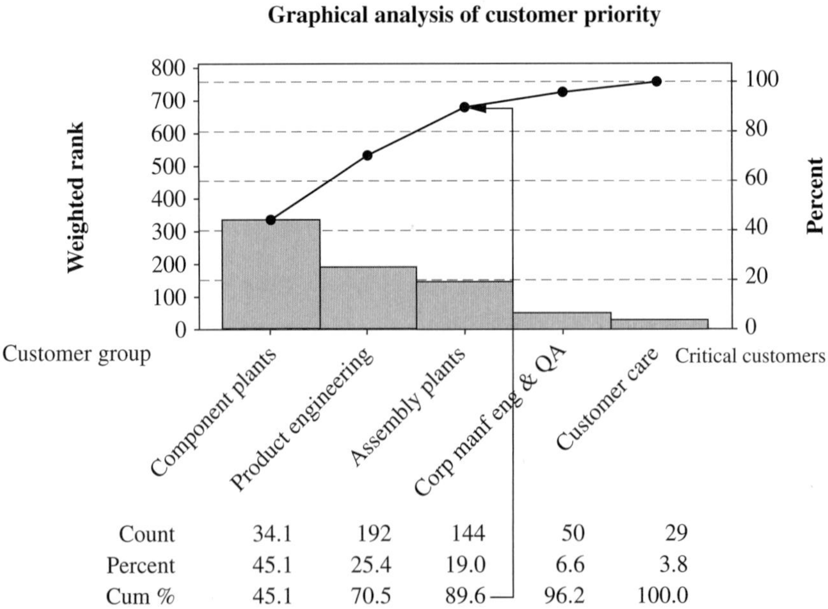

Graphical analysis of customer priority

Weighted rank: 0–800; Percent: 0–100

Customer group: Component plants, Product engineering, Assembly plants, Corp manf eng & QA, Customer care — Critical customers

Customer group	Component plants	Product engineering	Assembly plants	Corp manf eng & QA	Customer care
Count	34.1	192	144	50	29
Percent	45.1	25.4	19.0	6.6	3.8
Cum %	45.1	70.5	89.6	96.2	100.0

FIGURE 14.4 Pareto of customer prioritization.

	Customer weighting	Customer needs (based on survey results)	Accuracy of information	Speed to retrieve	Ease of retrevial	Format
		Need weighting	**592**	**343**	**275**	**260**
Prioritized customers (based on # users in group & frequency of need)			Association table customer Wt × Need Wt.			
Component plant	341		201872	116963	93775	88660
Product engineering	192		113664	65856	52800	49920
Assembly plant	144		85248	49392	39600	37440
		Totals:	*400,784*	*232,211*	*186,175*	*176,020*

Figure 14.5 Flow-down customer versus customer needs.

Count	400784	232211	186175	176020
Percent	40.3	23.3	18.7	17.7
Cum %	40.3	63.6	82.3	100.0

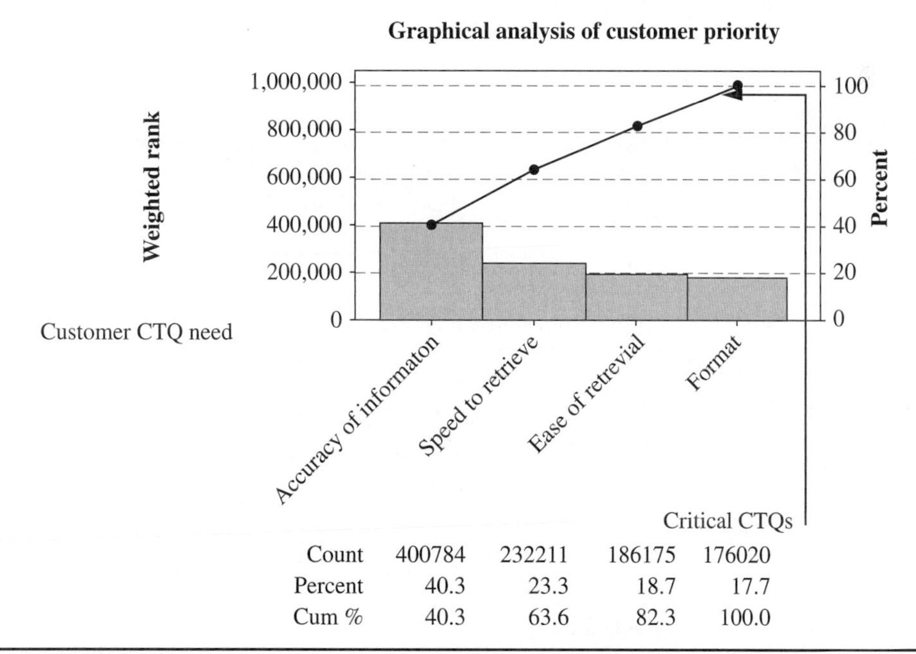

Figure 14.6 The vital few CTQs.

Need/ Expectation	Priority	Characteristic	Measure/ Sensor	Target	Upper specification limit	Allowable defect rate
Accuracy of information	400784	Drawing represents part number correctly	Match: Y or N/ visual	Y	Must match	3.4 DPMO
Fast retrieval of information	232211	Time to find a component drawing and print	Time/ Stopwatch	1.7 min	+1.6 min	10,700 DPMO
Easy retrieval of information	186175	Number of user inputs to locate a drawing	Number of inputs/visual	10	+3	3.4 DPMO
Information is easy to interpret (format)	176020	Different line weights are apparent on drawing	Number of multiple line weights/ visual	3	+1	3.4 DPMO

FIGURE 14.7 Translation of customer needs into measurable CTQs.

Baseline CTQ Capability Analysis

Drawing Access Spec (combined data from Dim and Non-Dim Lookup)

Normality Test

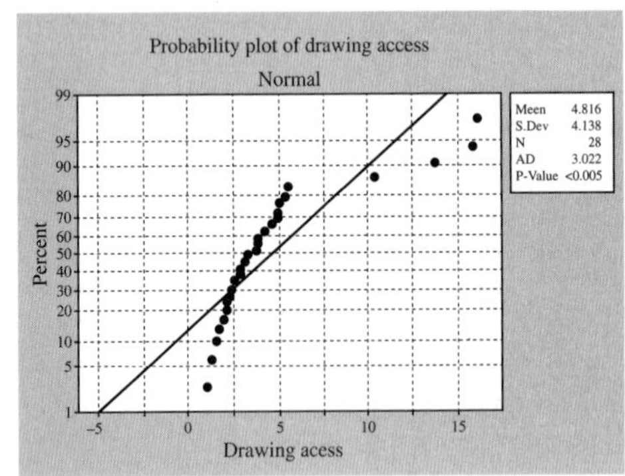

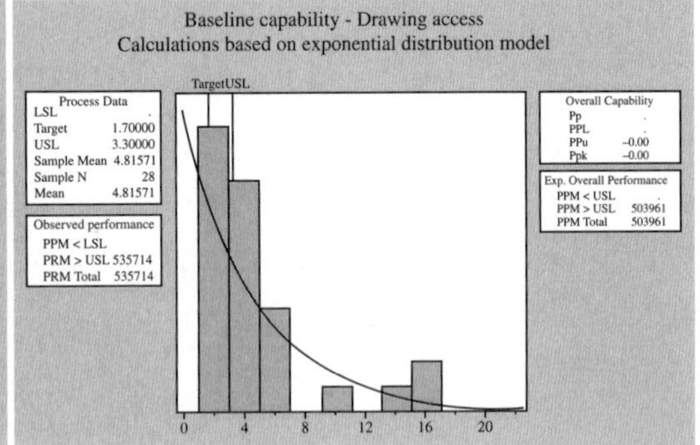

Alpha: 0.05

Ho: Data are normal

Ha: Data are not normal

P-value: <0.005, therefore reject Ho.

Data are not normal

Conclusion:

Use 1 sample Wilcoxon for data statistical analysis

Conclusion:

Baseline capability based on 500,000 DPMO is 1.4 sigma

FIGURE 14.8 Baseline CTQ.

Baseline CTQ Capability Analysis

Drawing Access Spec (combined data from Dim and Non-Dim Lookup)

Customer expectations are 1.70 minutes or faster to access and print a drawing, from CTQ survey. Is the current system running at the customers' expectations? This is the primary CTQ on our design scorecard.
See Appendix A for survey data analysis.

Ho = Sample median is equal to 1.70 minutes
Ha = Sample median is not equal to 1.70 minutes
Alpha = 0.05

Wilcoxon Signed Rank Test: Drawing Access

```
Test of median = 1.700 versus median not = 1.700

                         N
                       for   Wilcoxon             Estimated
                   N  Test  Statistic       P        Median
Drawing Access    28    28      393.5   0.000         3.750
```

P value = .000 < 0.05
Therefore reject Ho
Ha = Sample median is not equal to 1.70 minutes

Ho = Sample median is equal to 1.70 minutes
Ha = Sample median is greater than 1.70 minutes
Alpha = 0.05

Wilcoxon Signed Rank Test: Drawing Access

```
Test of median = 1.700 versus median not > 1.700

                         N
                       for   Wilcoxon             Estimated
                   N  Test  Statistic       P        Median
Drawing Access    28    28      393.5   0.000         3.750
```

P value = .000 < 0.05
Therefore reject Ho
Ha = Sample median is greater than 1.70 minutes

Conclusion: The current system is not meeting customer expectations

FIGURE 14.9 Baseline CTQ.

CTQs					
Description	Spec/Target LSL	Spec/Target USL	Current capability	High-level Capability	Feature Capability (From verification testing)
Information is accurate	0 Errors (6 Sigma)		0 Errors (6 Sigma)		
Fast retrieval	0 sec	3.3 min	4.82 min (1.4 Sigma)		
Easy retrieval (mimimal inputs)	8	12	17 (0 Sigma)		
Easy to Interpret format (number) of line weights)	2	4	1 (0 Sigma)		

FIGURE 14.10 Design scorecard.

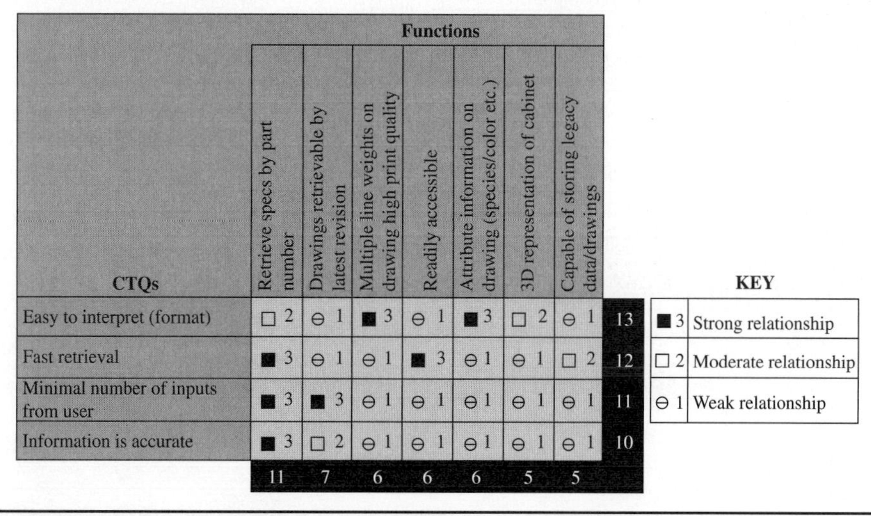

CTQs	Functions: Retrieve specs by part number	Drawings retrievable by latest revision	Multiple line weights on drawing high print quality	Readily accessible	Attribute information on drawing (species/color etc.)	3D representation of cabinet	Capable of storing legacy data/drawings	
Easy to interpret (format)	□ 2	⊖ 1	■ 3	⊖ 1	■ 3	□ 2	⊖ 1	13
Fast retrieval	■ 3	⊖ 1	⊖ 1	■ 3	⊖ 1	⊖ 1	□ 2	12
Minimal number of inputs from user	■ 3	■ 3	⊖ 1	⊖ 1	⊖ 1	⊖ 1	⊖ 1	11
Information is accurate	■ 3	□ 2	⊖ 1	⊖ 1	⊖ 1	⊖ 1	⊖ 1	10
	11	7	6	6	6	5	5	

KEY	
■ 3	Strong relationship
□ 2	Moderate relationship
⊖ 1	Weak relationship

FIGURE 14.11 QFD flow down: CTQs versus functions.

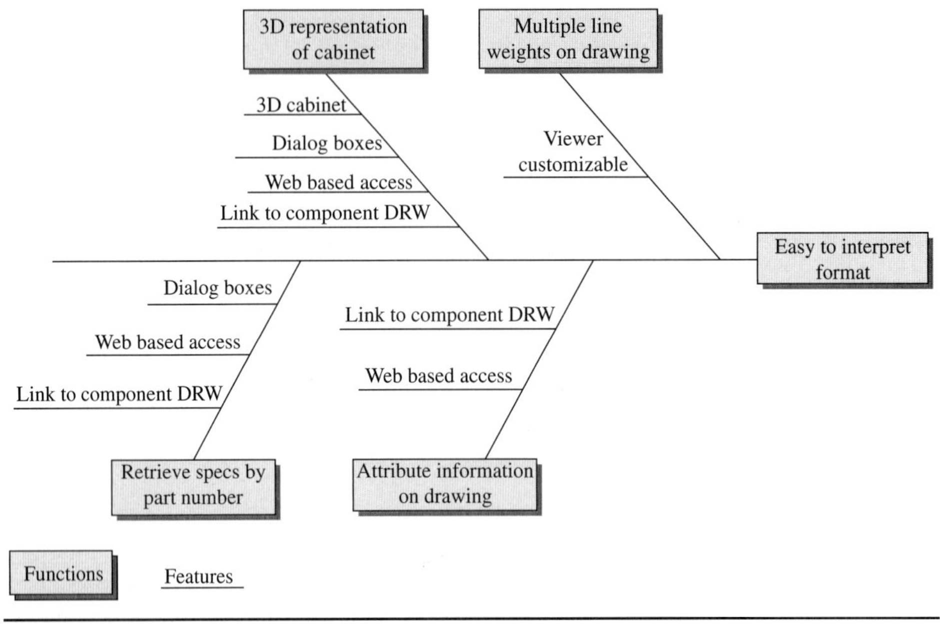

Figure 14.12 Easy to interpret function/feature diagram.

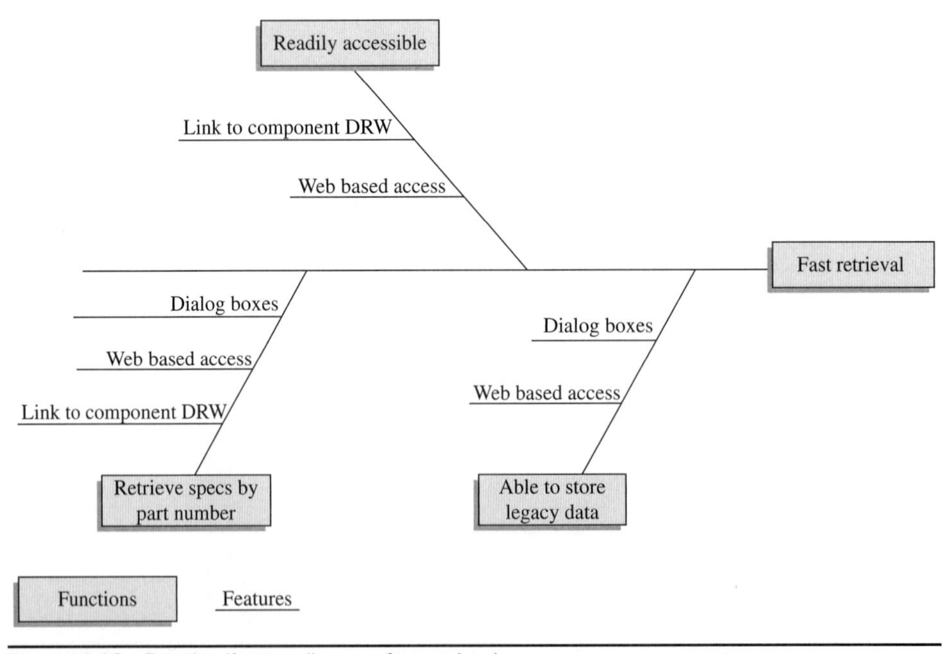

Figure 14.13 Function/feature diagram, fast retrieval.

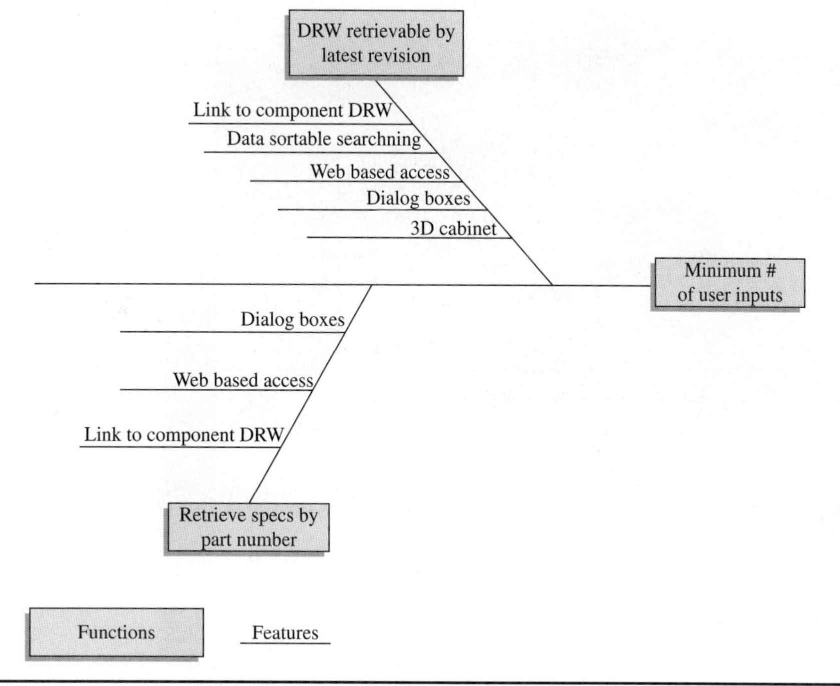

FIGURE 14.14 Function/feature diagram, minimum number of user inputs.

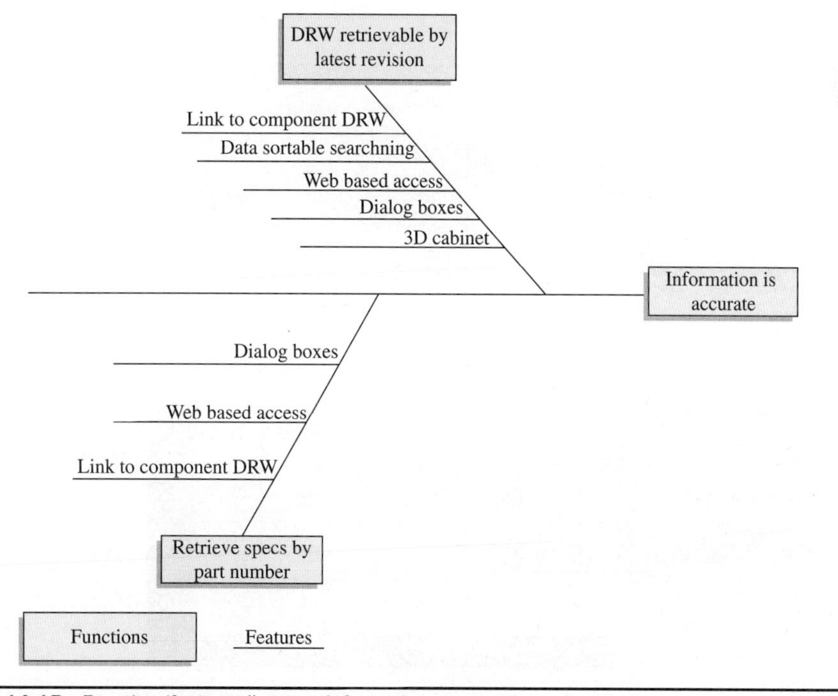

FIGURE 14.15 Function/feature diagram, information accurate.

Functions	Link to component drawings	Web-based access	Dialog boxes	Virtual cab viewing	Date sortable searching	Viewer customization	
	Features						
Drawings retrievable by latest revision	■ 3	□ 2	□ 2	□ 2	■ 3	⊖ 1	13
3D representation of cabinet	□ 2	□ 2	□ 2	■ 3	⊖ 1	⊖ 1	11
Readily accessible	■ 3	■ 3	⊖ 1	⊖ 1	⊖ 1	⊖ 1	10
Retrieve specs by part number	□ 2	□ 2	■ 3	⊖ 1	⊖ 1	⊖ 1	10
Attribute information on drawing (species/color etc.)	■ 3	□ 2	⊖ 1	⊖ 1	⊖ 1	⊖ 1	9
Capable of storing legacy data/drawings	⊖ 1	□ 2	□ 2	⊖ 1	⊖ 1	⊖ 1	8
Multiple line weights on drawing high print quality	⊖ 1	⊖ 1	⊖ 1	⊖ 1	⊖ 1	■ 3	8
	15	14	12	10	9	9	

KEY	
■ 3	Strong relationship
□ 2	Moderate relationship
⊖ 1	Weak relationship

FIGURE 14.16 QFD flowdown, functions versus features.

Features	Use case-based training materials	JSP web page	Browser plug-in	Link from list	Java applet input box	Link from thumbnails	
	Alternatives (X's)						
Web based access	■ 3	■ 3	■ 3	■ 3	■ 3	■ 3	18
Virtual cab viewing	■ 3	□ 2	■ 3	■ 3	⊖ 1	■ 3	15
Link to component drawing	■ 3	■ 3	□ 2	■ 3	■ 3	⊖ 1	15
Dialog boxes	■ 3	■ 3	⊖ 1	⊖ 1	■ 3	⊖ 1	12
Date sortable searching	■ 3	■ 3	⊖ 1	⊖ 1	⊖ 1	⊖ 1	10
Viewer customization	■ 3	⊖ 1	■ 3	⊖ 1	⊖ 1	⊖ 1	10
	18	15	13	12	12	10	

KEY	
■ 3	Strong relationship
□ 2	Moderate relationship
⊖ 1	Weak relationship

FIGURE 14.17 QFD flowdown, features versus alternatives.

Verification CTQ Capability Analysis

Drawing Access Spec (combined data from dim and non-dim lookup)

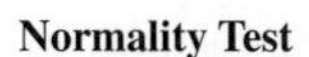

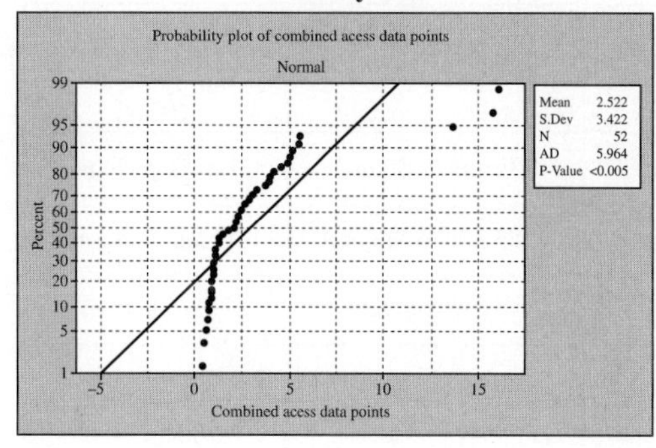

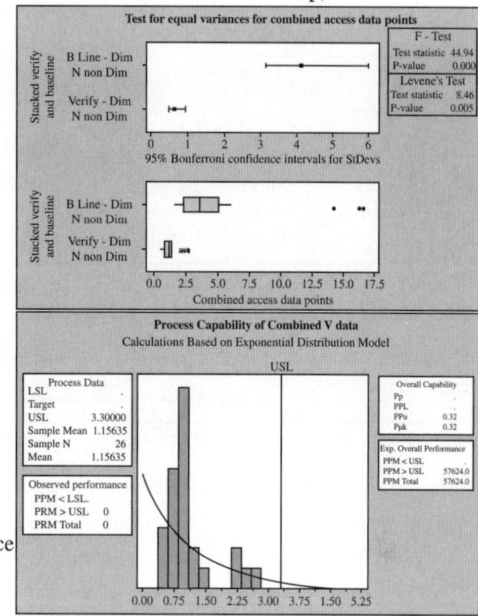

Alpha: 0.05
Ho: Data are normal
Ha: Data are not normal
P-value: <0.005, therefore reject Ho.
Data are not Normal. Perform Levenes test for equal variance
Conclusion:
Use Mann -Whitney for data statistical analysis

Conclusion:
Baseline capability based on 0 DPMO is 6 sigma

Figure 14.18 Verification CTQ capability analysis.

Verification CTQ Capability Analysis

Drawing Access Spec (combined data from dim and non-dim lookup)

Customer Expectations are 1.70 minutes or faster to access and print a drawing, from CTQ Survey. Is the new system running at the customers expectations? This is the primary CTQ on our design scorecard.

Ho = Sample median is equal to 1.70 minutes
Ha = Sample median is not equal to 1.70 minutes
Alpha = 0.05

Wilcoxon Signed Rank Test: Combined V data

```
Test of median = 1.700 versus median not = 1.700

                         N
                        for   Wilcoxon            Estimated
                    N  Test  Statistic       P       Median
Combined V data    26    26       53.5   0.002        1.033
```

P value = .002 < 0.05
Therefore reject Ho
Ha = Sample median is not equal to 1.70 minutes

Ho = Sample median is equal to 1.70 minutes
Ha = Sample median is less than 1.70 minutes
Alpha = 0.05

Wilcoxon Signed Rank Test: Combined V data

```
Test of median = 1.700 versus median < 1.700

                         N
                        for   Wilcoxon            Estimated
                    N  Test  Statistic       P       Median
Combined V data    26    26       53.5   0.001        1.033
```

P value = .001 < 0.05
Therefore reject Ho
Ha = Sample median is less than 1.70 minutes

Conclusion: DDL and the design of the user interface will meet customer expectations for drawing access time.

Figure 14.19 Verification CTQ capability analysis.

CTQs					
Description	Spec/Target		Current capability	High level capability	Feature capability (from verification testing)
	LSL	USL			
Information is accurate	0 Errors (6 Sigma)		0 Errors (6 Sigma)	0 Errors (6 Sigma)	0 Errors (6 Sigma)
Fast retrieval	0 sec	3.3 min	4.82 min (1.4 Sigma)	0.75 min (6 Sigma)	1.07 min (6 Sigma)
Easy retrieval (minimal inputs)	8	12	17 (0 Sigma)	10 (6 Sigma)	10 (6 Sigma)
Easy to interpret format (number of line weights)	2	4	1 (0 Sigma)	3 (6 Sigma)	3 (6 Sigma)

FIGURE 14.20 Design scorecard updated and verified.

This example depicts well how the DFSS process takes place and can be used by the practitioner as a guide for their own projects.

Example 2: A Design for Six Sigma (DMADV) Project

This second project is an example of DFSS applied to a new product development and how that application can result in a more successful product being brought to market because it better meets customer needs. Due to the sensitive competitive nature of such a project, the example has intentionally been made generic for presentation here.

Project Background

The project was chartered to design a new, more competitive consumer medical device. The following sections detail the project background, important business considerations, and the customer characteristics.

Development Goals

Provide an improved consumer device that optimally meets feature and benefit requirements of the product line.

Product Description

The product is a medical device for use by patients with specific conditions that lend themselves to use of self-monitoring systems.

Process(es) within Scope

- Industrial design
- Packaging configuration
- Device color, texture
- Device configuration
- Device ergonomics, ease of use
- Launch schedule

Market Strategy

The current market for this device mirrors the market for the higher-level devices it is used with. However, only 40 to 50 percent of our users of the higher-level device report using this device's current version. Among our competition this device is a much higher source of revenue, and this would imply that they can produce it at a lower cost.

Financial Strategy

Of today's similar devices 98 percent go into kits. Therefore, reducing cost is important.

Technology Strategy

No off-the-shelf original equipment manufacturer (OEM) devices provide multiple capabilities. However, the project can leverage prior development efforts to implement enhanced capabilities in this design.

Product Strategy

The strategy is to provide a device that maximizes customer acceptance across all major higher-level device platforms.

Design Project Approach

Leverage existing device development from [device name], especially internal mechanism with reduced bounce (associated with pain).

Figure 14.21 shows the process the team followed, each step completed. and the associated timing. Also, you can see the actual results of the project as measured by the marketing and financial strategies above. The project in fact exceeded the cost reduction goal of $0.13 per device with the actual reduction of $0.335 per device.

Results Achieved and Champion Approval

Black Belt: Joe Black Belt
Project: Device
CTQs: Ease of Use, Performance, Cost

Champion:________________
Approval:________________
Date:________________
Estimated Benefits: Save $0.13 per device, 6% increase in sales
Actual Benefits: Saved €0.23 ($0.335) per device, sales TBD

Define	Measure	Analyze (High-level design)	Design	Verify
Start Date: Jan 2010 *End Date: Jun 2010* ☑ Design concept ☑ High-level process map ☑ Project charter ☑ Business case ☑ Formal project approval	*Start Date: Jan 2010* *End Date: Sep 2010* ☑ CTQs ☑ Performance objective(s): Project Y(s) ☑ Possible Xs ☑ Data collection plan ☑ Measurement system analysis ☑ Baseline product performance ☑ Baseline product performance	☑ *Start Date: Sep 2010* ☑ *End Date: Jan 2011* ☑ Functional requirements ☑ QFD house of quality ☑ Design alternatives ☑ Selection criteria ☑ High-level design ☑ Design capability ☑ Transfer function	☑ *Start Date: Jan 2011* ☑ *End Date: Jul 2011* ☑ Detail design requirements ☑ Design optimization ☑ Robust desing ☑ Tolerancing ☑ Reliability study ☑ Cost analysis ☑ DVT	☑ *Start Date: Jan 2011* ☑ *End Date: Mar 2012* ☑ MVT ☑ Process pilot test ☑ Sustainable process controls ☑ Validated: ☑ Control system ☑ Monitoring plan ☑ Response plan ☑ Standardize & translate ☑ Validated business case ☑ Formal champion approval

☐ Not complete
✓ Complete
❖ Not applicable

Figure 14.21 DFSS applied to medical device design.

It may also be noted from Figure 14.21 that the total project took a little more than two years. Particularly for the application of DFSS to new product design, as in this case, it is common for DMADV projects to take a good bit longer than DMAIC projects.

Lessons Learned

1. Competitive benchmarking is the continuous process of measuring products, services, and practices against those of the toughest competitors or leading companies.
2. Quality can be the decisive factor in lost sales, and sometimes its impact can be quantified.
3. Customer complaints resolved with less than complete customer satisfaction will result in significant lost sales.
4. Planning for product quality must be based on meeting customer needs, not just meeting product specifications.
5. In-depth market research can identify suddenly arising customer needs.
6. Planning for quality must recognize a spectrum of customers with different needs.
7. For some products, we need to plan for perfection; for other products, we need to plan for value.
8. Life-cycle cost is the total cost to the user of purchasing, using, and maintaining a product over its life.
9. Quality superiority can be translated into a higher market share or a premium price.
10. Quality planning for a new or modified product follows these steps: establish the project, identify the customers, discover customers' needs, develop the product, develop process features, develop process controls, and transfer the plans to operations. The measurement process must be applied during all steps.

References

Early, J. F., and Colleti, O. J. (1999). "The Quality Planning Process," *Juran's Quality Handbook*, 5th ed., McGraw-Hill, New York.

Endres, A. (2000). *Implementing Juran's Roadmap for Quality Leadership*, John Wiley and Sons, Inc., New York.

Gryna, F. M., Chua, R. C. H., and De Feo, J. A. (2007). *Quality Planning and Analysis,* 5th ed. McGraw-Hill, New York.

Juran, J. M. (1988). *Juran on Planning for Quality*, The Free Press, New York.

Juran, J. M., and Godfrey, A. B. (eds.) (1999). *Juran's Quality Handbook*, 5th ed., McGraw-Hill, New York.

CHAPTER 15

Benchmarking: Defining Best Practices for Market Leadership

Brad Wood and Alexander Janssen

About This Chapter

This chapter defines benchmarking and how it can be used as an effective tool to drive organizational performance and aid in strategic planning. A structured and well-established step-by-step process and managerial requirements for successful benchmarking are included.

High Points of This Chapter

1. Benchmarking is a systematic and continuous process that facilitates the measurement and comparison of performance and the identification of best practices that enable superior performance.

2. The main objective of benchmarking is to identify superior performance internal or external to your organization, determine the nature of and reasons for this performance, and determine if there are gaps in your organization to this performance.
3. Organizations may benchmark for many different reasons, but the strongest driver should be to improve organizational performance.
4. Benchmarking can be classified in many different ways, but the same principles and processes must be applied for benchmarking to be successful.
5. Benchmarking can provide vital input to the performance improvement process as depicted by the close interactions with the Juran Trilogy.
6. Benchmarking can have direct input to the strategic planning process by providing factual foundations to frame the vision, goals, and plans for world-class leadership.
7. A well-structured and systematic process such as the Juran 7-Step Benchmarking Process is essential to realize successful benchmarking.
8. Critical to success are a clear scope and objectives, good definitions, thorough validation, effective normalization, clear reporting, and a willingness to share information on best practices.
9. It is essential that benchmarking always be conducted within a legal and ethical framework.
10. The resources required to conduct a benchmarking study and to act upon the findings must be provided by management if any real value is to be gained from benchmarking.

Benchmarking: What It Is and What It Is Not

Benchmarking has been in existence for a great many years. The concept of one individual observing how another performs a given task to and then applying any learning from that to adapt and improve how the task is executed is one of the fundamental ways in which human beings learn and develop. In the context of business, learning from one's competitors has also been in existence for as long as business has. However, the application of learning from best practices to the business environment in a structured, methodical, and indeed legal and ethical way is relatively new. Xerox Corp. is most commonly credited with developing the modern form of benchmarking, and it is fair to say that the majority of today's benchmarking practices are built upon the approach developed by them in the 1970s.

Although their story has previously been well told in a multitude of management texts, it is still worthy of brief comment here to set the scene. A combination of poor product quality, high overheads, and increasing competition from a growing number of Japanese organizations had left Xerox in a precarious position in the late 1970s. A visit to Japan provided the wake-up call that change was essential if they were to survive (Nadler and Kearns 1993). They put in place a series of benchmarking activities aimed at identifying the best-performing organizations in various aspects of their business and determining what it was that these organizations were doing that enabled a superior performance. Most famous is the benchmarking of logistics operations that they undertook with L.L. Bean (Camp 1989). With this, modern benchmarking was born.

Benchmarking has evolved to become an essential element of the business performance improvement tool kit and is now frequently used by many organizations in a wide range of different industries. But despite this, it remains one of the most widely misunderstood improvement tools. It means many different things to many different people, and all too frequently benchmarking projects fail to deliver on their promise of improvement or real results.

However, executed correctly, benchmarking can provide a powerful focus for organizations, driving home the facts and convincing the organization of the need to embark upon improvement strategies. Benchmarking is a tool that enables the identification and ultimately the achievement of excellence, based upon the realities of the business environment rather than on internal standards and historical trends.

Benchmarking is not what we would term "industrial tourism" in which superficial industrial visits are undertaken in the absence of any point of reference and do not assist in the improvement process. It is impossible to acquire detailed knowledge after only a quick glance or one short visit, and it is rare for such visits to result in an action plan that will lead to improvement. In the absence of prior benchmarking, it is also difficult to identify which organizations should be visited, and so there is a real risk that visits are made to organizations that are perceived as being the best or at least better, when the reality may be very different. However, there is a valuable role to be played by this type of site visit, when it is conducted following a structured benchmarking analysis and the organization being visited has been identified as a best performer.

Benchmarking also should not be considered a personal performance appraisal tool. The focus should be on the organization and the individuals within it. Failure to adopt this philosophy will only lead to resistance and will undoubtedly add roadblocks to a successful benchmarking journey.

Nor should benchmarking be a momentary glimpse, but rather it should be considered a continuous process. Organizations must change performance rapidly to remain competitive in business environments today. This fast-paced tempo is further accelerated in sectors where benchmarking is commonplace, where businesses rapidly and continuously learn from one another. A prime example comes from the oil and gas industry where organizations have to respond to ever-increasing business, technological, and regulatory demands. The majority of the key players in this industry are participating in focused benchmarking consortia on an annual basis. It is also much more than a competitive analysis. Benchmarking goes further than examining the pricing and features of competitors' products or services. It considers not only the output but also the process by which the output was obtained. Benchmarking is also much more than market research, considering the business practices in place that are enabling the satisfaction of customer needs and thus realizing superior business performance. It provides evidence-based input offering a powerful focus for management, driving home the facts and convincing the organization of the need to embark on improvement activities.

Participating in benchmarking should also not be viewed as a stand-alone activity. To succeed, it must be part of a continuous improvement strategy, it must be conducted regularly, and it should be enveloped in the continuous improvement culture of an organization. Like any other project, it has to have the full support of senior management, the resources necessary to fulfill the objectives, and a robust project plan that is adhered to.

Finally, benchmarking should not be viewed as the answer in itself. It is a means to an end. An organization will not improve performance by benchmarking alone. It must act upon the findings of the benchmarking to improve. The output of benchmarking should provide input to decision making or improvement action planning. This requires detailed consideration of the benchmarking analysis, formulation of learning points, and development of action plans in order to implement change and realize improvements.

So how can we define benchmarking? A scan of the literature will quickly reveal myriad definitions (Anand and Kodali 2008), each offering a slight variance on a common theme. Rather than repeat these here, we prefer to offer our own definition:

> *Benchmarking* is a systematic and continuous process that facilitates the measurement and comparison of performance and the identification of best practices that enable superior performance.

This definition is deliberately generic so that it can encompass all types of benchmarking. In this context, measurement and comparison may be between organizations, business units, business functions, and business processes, products, or services. The benchmarking may be internal or external, between competitors, within the same industry, or cross-industry. Regardless of the category of benchmarking this definition still applies.

Objectives of Benchmarking

The objectives of benchmarking can be summarized as follows:

1. Determine superior performance levels.
2. Quantify any performance gaps.
3. Identify best practices.
4. Evaluate reasons for superior performance.
5. Understand performance gaps in key business areas.
6. Share knowledge of working practices that enable superior performance.
7. Enable learning to build foundations for performance improvement.

When one is talking of superior performance, ultimately of course the aim should be for world-class performance. However, in reality it is often difficult to be able to ensure that the world's leading performers are participating in a given benchmarking exercise. Instead, benchmarking partners should be selected carefully to ensure the output will provide the required added value.

Once superior performance has been determined, the gap between this and the performance level of the benchmarker is quantified. The working practices enabling superior performance are identified and the enablers evaluated. This knowledge is then shared between benchmarkers to enable the learning to be taken away and implemented as part of a performance improvement program.

Thus benchmarking can be viewed as a two-phase process, where phase one is a positioning analysis aimed at identifying gaps in performance and phase two is focused upon learning from those best practices that enable superior performance.

Why Benchmark?

There are two good reasons for organizations to benchmark themselves. First, it will help them stay in business by offering opportunities to become better than other similar organizations, competitors or not. Second, it ensures that an organization is continually striving to improve its performance through learning. Benchmarking opens minds to new ideas from sources either within the same industry or from many other unrelated industries, identifying how those who have demonstrated performance leadership work.

Yet many organizations benchmark simply to be able to demonstrate to stakeholders, be they customers, shareholders, lenders, regulators, etc., that the organization is performing at an acceptable level. Of course this is a perfectly legitimate reason for benchmarking, although the real potential value of the technique is missed by narrowing the focus in this way.

Benchmarking also provides a very effective input to an organization's strategic planning processes by establishing credible goals and realistic targets based upon external references.

To really grasp the intent of benchmarking, an organization should be benchmarking not only to demonstrate good performance but also to identify ways in which it can change its practice to significantly improve its performance. Those organizations with a strong performance improvement culture will be benchmarking continuously as this provides them with objective evidence of where to focus improvement activities, how much they should be improving, and what changes to their working practices they might consider to realize improvements.

Classifying Benchmarking

There are many different ways to classify benchmarking (Table 15.1), and the literature is full of different classifications (Anand and Kodali 2008) that make it very confusing for someone new to the topic to really understand what benchmarking is and which approach is best for her or him. The fact of the matter is that there is an underlying process that can be considered generic to almost all types of benchmarking. However, to provide some clarity on the differences in classification, we have considered benchmarking in terms of what it is that is to be benchmarked, who the benchmarking is going to involve, and how the benchmarking is to be conducted:

- Subject matter and scope (what)
- Internal and external, competitive and noncompetitive benchmarking (who)
- Data and information sources (how)

Subject Matter and Scope (What)

Benchmarking is often categorized according to what it is that is being benchmarked. Typical categories include

- Functional benchmarking
- Process benchmarking
- Business unit or site (location) benchmarking
- Project benchmarking
- Generic benchmarking
- Business excellence models

Classification Criteria		
Subject Matter (What)	**Participants (Who)**	**Data Sources (How)**
Functional benchmarking Process benchmarking Business unit or site (location) Benchmarking Projects benchmarking Generic benchmarking Business excellence models	Internal benchmarking External benchmarking Competitive benchmarking Noncompetitive benchmarking (same industry and cross-industry)	Database benchmarking Survey benchmarking Self-assessment benchmarking One-to-one benchmarking Consortium benchmarking

(*Source:* Juran Institute, Inc. Copyright 1994. Used by permission.)

Table 15.1 Ways in Which Benchmarking Is Often Classified

Functional Benchmarking

Functional benchmarking describes the process whereby a specific business function forms the focus for the benchmarking. In the context of the organization, this may involve benchmarking several different business units or site locations. Typical examples of functional benchmarking include the analysis of the procurement, finance, Internet technology (IT), safety, operations, or maintenance functions. The analysis focuses upon all aspects of the function rather than on the processes involved and the specific activities conducted.

Process Benchmarking

In process benchmarking the focus of the study is upon a specific business process or a part thereof. Examples include product development, invoicing, order fulfillment, contractor management, and customer satisfaction management. Process benchmarking will often involve several functional groups and may also involve many different site locations. There is often a lot of overlap between what is termed *functional benchmarking* and *process benchmarking* (e.g., a benchmarking of the procurement process may look very similar to a benchmarking of the procurement function). Many business processes are not specific to any one industry and so can benefit from broadening participation in the analysis to organizations from a multitude of industries.

Business Unit or Site (Location) Benchmarking

Benchmarking individual business units or site locations against one another is often (but not always) seen in internal benchmarking studies within a single organization. The performance of each unit is analyzed and compared to that of other units. This analysis may incorporate all activities of each unit in their entirety or may be confined to selected functional groups or business processes. For example, Juran manages an annual benchmarking consortium comparing the performance of many of the world's oil and gas processing facilities. Each of the key business processes is included in the analysis, and participants come from a wide range of different organizations.

Project Benchmarking

This type of benchmarking focuses upon projects undertaken by organizations. Because projects vary widely in their nature, these studies are normally tailored for specific project types. For example, one may benchmark oil pipeline construction projects, software implementation projects, facility decommissioning projects, etc. Normally included are all the business processes pertaining to the project being analyzed, although the scope may often be limited to a subset of processes. For example, a construction project benchmarking may focus specifically upon contractor selection, procurement, and commissioning.

Generic Benchmarking

Generic benchmarking considers all business processes required to achieve a certain level of performance in a given area. The focus is upon the result and what is required to achieve it. For example, a hospital may undertake a generic benchmarking exercise to identify ways in which it can reduce treatment waiting times. In so doing it may benchmark across a number of different industries where customer waiting times are of paramount importance, e.g., insurance claims processing, vessels clearance procedures for major waterways (e.g., Suez Canal), calamity response times for the different emergency services (police, fire, ambulance). Inevitably there is a lot of overlap between process benchmarking and generic benchmarking although in the latter there is often less emphasis on gap analysis and greater emphasis upon a detailed consideration of working practices.

Business Excellence Models

Business excellence models have been developed to provide a framework by which organizations can holistically measure and therefore improve their performance. The purpose of their design is such that they encompass all key aspects of an organization that drive performance. Two of the most well-known models are the Baldridge Award (see Chapter 17) and the European Foundation for Quality Management (EFQM) Excellence Model. These models are similar to each other in many respects, and both identify a number of critical success criteria for realizing superior performance.

Although these models are designed to support a self-assessment process, they also lend themselves to providing an excellent framework for comparative benchmarking, although they are infrequently used for this purpose. Benchmarking in this way using such models is essentially a form of generic benchmarking whereby all elements required for excellence are considered. Furthermore, a requirement of both the Baldridge and EFQM models is that organizations be able to demonstrate benchmarking activity.

Internal and External, Competitive and Noncompetitive Benchmarking (Who)

Benchmarking studies are frequently classified by type of participant. Depending on the type of benchmarking being undertaken, it is not always possible to have any control over participant selection. But where this is possible, the selection of others to benchmark with is one of the first and often the most difficult tasks at the outset of any benchmarking study. Potential participants will be identified according to a range of criteria, the main one being the perceived performance level (where superior or world-class performance is the aim).

The four main types of benchmarking are

1. Internal benchmarking
2. External benchmarking
3. Competitive benchmarking
4. Noncompetitive benchmarking

Each of these has specific benefits and drawbacks that need to be considered when selecting the most suitable benchmarking approach. These include

- The similarity between the participants in terms of the subjects to be benchmarked
- The level of control over the benchmarking process
- The cost and time input required to conduct the benchmarking
- The degree of openness that is possible and the level of confidentiality necessary
- The potential for learning and therefore performance improvement

Internal Benchmarking

Internal benchmarking is the comparison of performance and practices of similar operations within the same organization. Depending upon the size of the organization and the nature of its business, this may or may not be feasible for the organization would need to have duplicate groups conducting the same activities. Should this be the case, internal benchmarking is often a popular first step as it allows organizations to prepare themselves for broader benchmarking activities within the safety of their own environment where they have full control over the process. This is likely also to be the least costly and time-consuming way to benchmark. But the potential for finding performance leaders is much smaller, and the opportunity for learning is usually more limited.

External Benchmarking

External benchmarking involves participants from different organizations. The opportunity for learning is normally greater than that achievable by internal benchmarking, but there is obviously a requirement to share information outside of the organization. This brings with it some potential restraints. There will almost certainly be limits on the data organizations are willing to share, especially if the other participants are competitors, and there will of course be a need for stricter confidentiality. External benchmarking is further categorized according to the nature of the participants who can be competitors (competitive benchmarking) or not (noncompetitive benchmarking). These are considered below.

Competitive Benchmarking

Competitive benchmarking is a form of external benchmarking in which the participants are all in competition with one another. By definition, the participants in a competitive benchmarking program are from the same industry, and the focus is normally upon industry-specific processes. For example, Juran has studied patient safety performance between different hospitals. This normally brings with it a high degree of sensitivity which needs to be carefully managed for a successful outcome to be realized; but when it is conducted properly, the results can be very valuable. Conversely, topics that are not directly related to the core business of the competing organizations are usually less sensitive to benchmark between competitors. But ironically these more generic topics are not normally those that an organization wishes to benchmark with its competitors as greater value is more frequently gained from benchmarking such topics cross-industry. There are also likely to be some subject areas that most organizations will not be willing to benchmark with competitors such as proprietary processes or products and innovations that provide competitive advantage.

Noncompetitive Benchmarking

Noncompetitive benchmarking is a form of external benchmarking in which the participants are not in direct competition with one another. They may be from within the same industry or cross-industry. For example, an organization operating a container port in the United States may benchmark with another in Europe. Although they are in the same industry, they are unlikely to be competitors as they are operating in different markets. Juran manages an annual global benchmarking consortium for gas pipelines. The participants are all in the same industry and primarily interested in benchmarking those processes specific to their industry. But they are not in direct competition with one another as they operate in totally different marketplaces, delineated by geographical region. This means that they are very willing to share knowledge and practices openly for mutual benefit without fear of giving anything away that may impact their competitiveness.

In cross-industry noncompetitive benchmarking, those subject areas that are not industry-specific are most commonly analyzed, and it is in this classification that most generic benchmarking studies sit. These tend to be support processes such as administration, human resources, R&D, finance, procurement, IT, and health, safety, and environmental (HSE). Cross-industry external benchmarking potentially offers the greatest opportunities for learning and performance improvement for a number of reasons. First, the pool of potential participants is much bigger. Second, participants thought to be superior performers in the subject area being benchmarked can be identified and invited to participate. Third, the willingness to share knowledge will be greatest where there is no fear of competitive sensitivity.

Data and Information Sources (How)

Benchmarking can also be classified according to the source of the data used in the comparative analysis. Such classification can be made in many ways, but the list below addresses the main categories found:

- Database benchmarking
- Survey benchmarking
- Self-assessment benchmarking
- One-to-one benchmarking
- Consortium benchmarking

Database Benchmarking

In this type of benchmarking, data from a participant are compared to an existing database containing performance data. An analysis is performed and the results are provided to the participant. Benchmarking in this way normally requires a third party to administer the database and produce the analyses. The development of the Internet in recent years has led to the growth in this type of benchmarking as it can be easily administered online. The participating organization can submit its data via an online questionnaire and receive a report of the analyses online, usually in a very short time. A quick search of the Internet will reveal the large number and wide range of online benchmarking databases available.

This type of benchmarking is also sometimes offered by consultancy organizations that have accumulated performance data pertaining to specific activities. For example, Juran has been benchmarking in the oil and gas industry since 1995 and during this time has developed a comprehensive database of performance figures relating to this industry. And because the data have been well defined and thoroughly validated during the collection process it is extremely reliable. It is therefore an excellent data source against which oil and gas organizations can be benchmarked.

Many organizations start out on their benchmarking journey by purchasing data from proprietors of such databases. Although this type of benchmarking can be very useful in providing fast feedback on performance, it can have drawbacks. The participant has no control over the content of the analysis and has to accept the metrics that are used to determine performance. Often the source of the data is not disclosed; thus it may be difficult for the benchmarker to assure himself or herself of its relevance. The metrics used may not be clearly defined and may not be validated effectively, resulting in poor data quality and flawed analyses. Care should therefore be taken when entering into this type of benchmarking, and the participant should realize the potential shortcomings. For best results, only a bona fide consultant with a sterling reputation and a good track record should be sought.

Survey Benchmarking

This term is used to describe benchmarking exercises conducted via the completion of a survey or a review process. Typically, a survey document is sent to participating organizations to be completed and returned. Sometimes the survey documents are sent to organizations without their prior agreement to participate, in the hope that they will complete the survey and return it. Of course, this approach is nearly always less successful, with a relatively poor return rate.

The survey may be organized by a third-party consultant or by one of the participating organizations, although in the latter case there will be greater restrictions on what data can be shared directly between participants to ensure compliance with antitrust legislation. Sometimes there may be a fee involved for organizations to participate, and sometimes a single organization or even a consultancy may sponsor the entire exercise, in which case the output for the other participants is often less sophisticated.

The potential drawbacks of this approach are similar to those encountered with database benchmarking in that each organization has minimal control over the benchmarking process, the metrics may not be defined adequately, and the validation of submitted data may be limited. Nonetheless, this type of approach can provide a useful albeit limited comparative analysis with limited effort required on the part of each participating organization.

The survey process can be extended to include a review element, whereby the benchmarking coordinator (normally a third-party consultant) will visit each of the participating organizations as part of the survey process. This allows the consultant to delve more deeply into specific areas to gain richer data (often qualitative data) which can better inform the learning process of the participants, particularly in the area of assessing working practices that underpin superior performance.

Self-Assessment Benchmarking

As previously discussed, self-assessment is an integral part of many performance excellence models. These self-assessments can be used for generic benchmarking between organizations across all industries. These models provide an excellent framework for comparative benchmarking whereby all elements required for excellence are considered. The analysis will often focus on not just quantitative data analysis but also a qualitative view of working practices. However, there is an inherent weakness in the process associated with the subjective nature of self-assessment. Sometimes third parties (consultants) are employed to oversee the process to introduce some level of objectivity or even to conduct the assessments.

One-to-One Benchmarking

This type of benchmarking is probably most commonly reported in the literature, but as we pointed out, benchmarking is not industrial tourism, whereby relatively superficial site visits are conducted between organizations to explore performance. Such exchanges rarely bring fruitful insights, and there is always the uncertainty of whether the organization being visited is really a superior performer.

However, if having conducted a benchmarking study, an organization is identified as delivering superior performance levels in the subject area of interest, then a one-to-one benchmarking can investigate specific areas in much greater depth and deliver rich information pertaining to the drivers of superior performance. This approach is common in consortium benchmarking where, having received the benchmarking analysis, two organizations will agree to benchmark further one to one to obtain more detailed understanding in specific working practices. A good example of this comes from a cross-industry procurement benchmarking consortium managed by Juran. Following participation in the study, two of the participants agreed to undertake a one-to-one benchmarking. One organization was particularly strong in contract tendering and contractor selection, while the other demonstrated superiority in strategic procurement. Cooperating in this way led to a greater understanding of leading working practices in each of these areas and improved performance for both parties.

Consortium Benchmarking

Without doubt, this form of benchmarking has the greatest potential to deliver improved performance for its participants. A consortium is formed between participants, usually (but not always) supported by a third-party facilitator. They agree on the participants to be invited; the subjects to be benchmarked; the methodology to be followed; the metrics (and their definitions) to be used; the validation criteria; the nature of the analysis, reporting, and deliverables; and the time scales to be adhered to. Thus the participants have a very high level of control over the entire process, and the outcomes from the process are normally reliable data, thorough analysis, and valuable results. This approach does require a great effort on behalf of each of the participants to achieve the desired outcome. It is therefore more time-consuming and often more costly to undertake, but the added value is normally far in excess of that achieved through other benchmarking approaches.

We have demonstrated that benchmarking can be classified in many different ways according to the subject matter of the analysis, the nature of the benchmarking participants, data sources, and the methodologies employed. However, differentiating in this way is largely academic, and although these different approaches have their inherent pros and cons and some are clearly more effective than others, they all should have the same ultimate objective—to provide learning on how to improve business performance.

Benchmarking and Performance Improvement

The Juran Trilogy®, described in detail in Chapter 1, Attaining Superior Results Through Quality, and Chapter 2, Quality's Impact on Society and the National Culture, provides a model for effectively managing for quality to achieve performance breakthroughs. The ways in which benchmarking can provide vital input to the performance improvement process can be demonstrated by examining its interaction with the Juran Trilogy® (Figure 15.1).

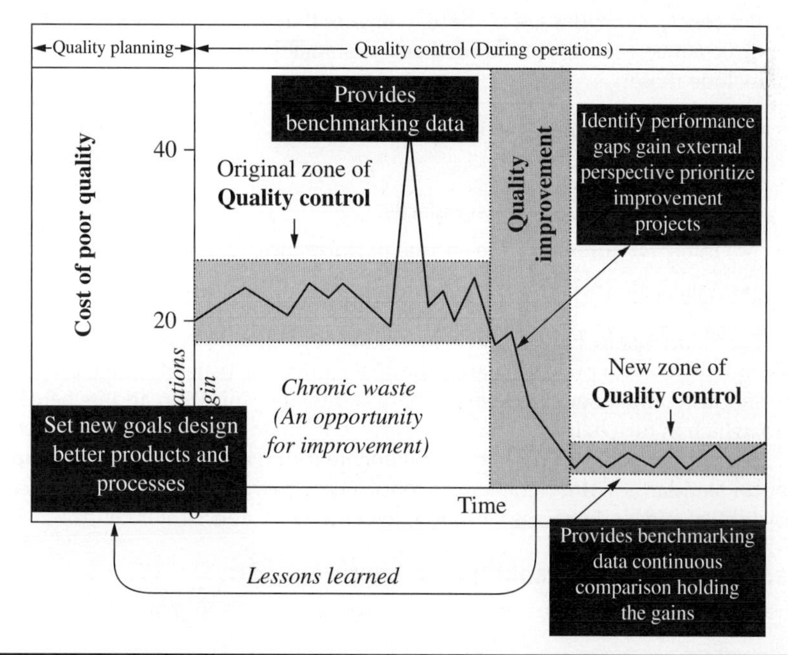

Figure 15.1 Benchmarking and the Juran Trilogy.® (*Juran Institute, Inc. Copyright 1994. Used by permission.*)

Benchmarking findings often include quantified gap analyses in key performance areas of the business unit or process being assessed. Management will often use the Pareto principle, often referred to as the *80/20 rule*, in order to focus the right resources on the right improvement opportunities. Resources will be allocated to the "vital few" areas for improvement (typically with a high return on investment, or quick payback), as opposed to the "trivial many" areas, with less improvement potential.

Benchmarking and Designing New Products

The main objective of designing for quality is to prepare organizations so that they are able to meet their performance goals. In so doing they must do the following:

1. Identify customers.
2. Determine customer needs.
3. Develop product (service) features required to meet customer needs.
4. Establish quality goals that meet customer needs.
5. Develop processes that deliver the product (service) features.
6. Prove that the organization's processes can meet the goals.

Benchmarking can input to this process by providing the vehicle whereby organizations can learn form best practices and incorporate that learning into designing new and improved business processes. It also enables performance planning goals to be established based upon the reality of what is achievable by other benchmarked organizations. Hidden customer needs may be revealed by the benchmarking and product or service features identified.

Benchmarking and Quality Control

Quality control activities are in place to ensure that an organization is in a position to meet its performance goals in a controlled and sustainable way. The key activities in quality control include these:

- Identify what needs to be controlled.
- Measure.
- Establish performance standards.
- Interpret differences (i.e., actual performance versus the standard).
- Take action where differences arise.

The relationship between benchmarking and quality control is two-way. The output from quality control provides data to be analyzed in the benchmarking process. This enables continuous comparisons to be made between the organization and its benchmarking partners, which in turn helps to ensure that any gains made through performance improvement can be maintained in the longer term. Benchmarking allows organizations to challenge the control standards with vigor. Improved process performance results in increased process capabilities, reduced process variation, fewer defects, and tighter controls.

Benchmarking and Breakthrough Improvement

The main objectives of the improvement process can be summarized:

- Demonstrate the need for improvement.
- Identify specific projects for improvement.

Process	Objective	Relationship with Benchmarking
Planning	Plan and prepare to meet performance goals.	• Learn form best practices • Set achievable goals • Design better processes
Control	Ensure performance goals are met.	• Provide data for benchmarking • Make continuous comparisons • Hold gains made
Improvement	Improve performance to significantly superior levels.	• Gain an external perspective • Identify performance gaps • Prioritize improvement projects

(*Source:* Juran Institute, Inc. Copyright 1994. Used by permission.)

Table 15.2 Benchmarking and the Juran Trilogy®

- Diagnose problems to find root causes.
- Provide remedies.
- Prove that the remedies are effective in delivering a breakthrough in performance.
- Provide new controls to hold the gains made.

Benchmarking can support this process by providing an external perspective of what levels of performance are achievable and what working practices are required to achieve these levels of performance. It enables organizations to measure the gaps in their performance compared to the superior performance of other benchmarkers. This in turn allows organizations to identify where their performance is weakest and where it is strongest, thereby offering a useful input into the prioritization of potential performance improvement projects. Table 15.2 summarizes how benchmarking relates to the Juran Trilogy®.

Benchmarking and Strategic Planning

An organization's goals all too often fall short of stakeholder expectations. A primary contributor to this failure is that goal setting is based upon past trends and current practices in many organizations. Organizations are often inward-looking, and the external perspective is frequently overlooked. Customer expectations are driven by the standards of the best providers in the industry and by their experiences with superior providers in other industries. Benchmarking can capture these external references to provide the basis for comparative analysis and learn from best practices. Thus benchmarking can have a direct input to the strategic planning process described in Chapter 7, Strategic Planning and Deployment: Moving from Good to Great. Ways in which benchmarking can help to shape an organization's strategic direction are shown in Figure 15.2.

Essential are a well-defined process, a clear understanding of the scope of what is to be benchmarked and why, and a systematic approach that is thoroughly planned. If conducted correctly, benchmarking can contribute to the effectiveness of strategic performance management by providing information on what the best-practice organizations are already achieving. Through the establishment of evidence-based best practices, benchmarking can provide the factual foundations to enable an organization to frame its vision, goals, and plans to realize world-class leadership.

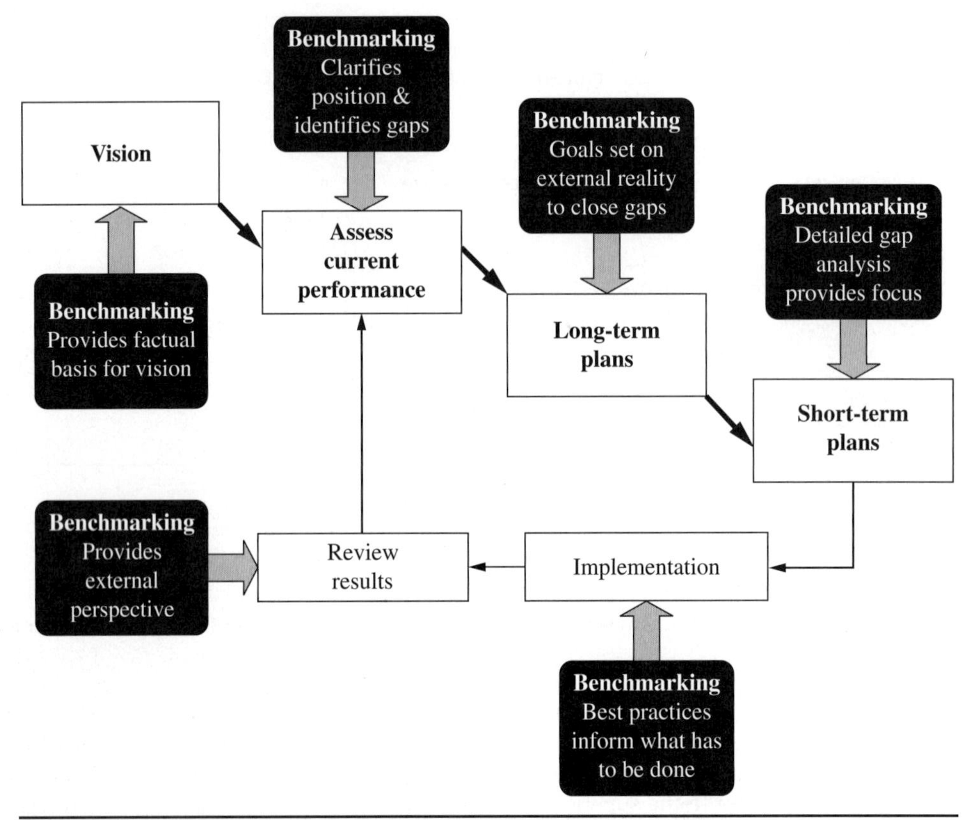

Figure 15.2 Benchmarking and strategic planning. (*Juran Institute, Inc. Copyright 1994. Used by permission.*)

Benchmarking and Vision Development

Depicted in Figure 15.2 is a typical strategic planning process for performance improvement which begins with an organization's vision for the future. The vision will always be influenced to some extent by the organization's business environment and what others have been able to achieve. Benchmarking supplies detailed analyses of that environment and provides a factual basis of what it is to be world class, thereby helping to bring the organization's vision into focus.

Assessing current performance and measuring the gap between this and the vision are critical to an organization's long-term sustainability. Many sources for measuring an organization's current performance exist, including market research, competitor analysis, and of course benchmarking. Benchmarking will clearly define an organization's current performance, clarify its position in relation to both the external business environment and the vision, and identify the performance gaps. This enables the organization to make adjustments to the strategy to close the gap between reality and the future vision.

Benchmarking and Long- and Short-Term Planning

Long-term plans or key strategies derived from the vision will comprise strategic goals addressing all aspects of the organization's performance including business process performance, product or service performance, customer satisfaction, the cost of poor quality, and

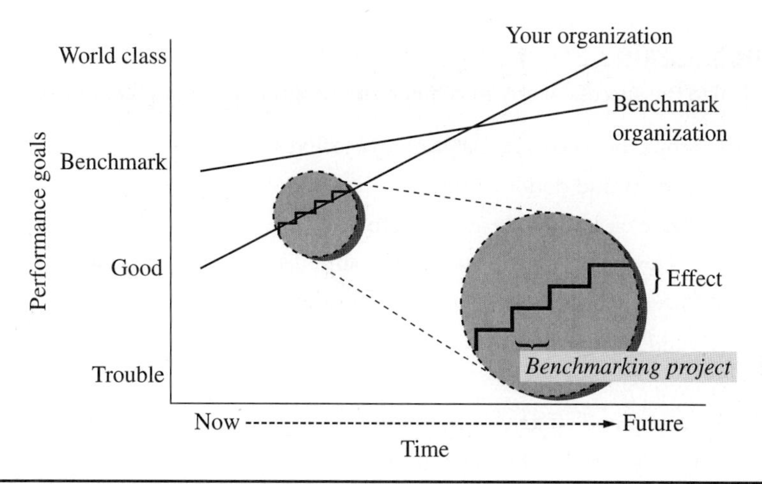

FIGURE 15.3 Benchmarking over time. (*Juran Institute, Inc. Copyright 1994. Used by permission.*)

the organization's competitive performance. By necessity, these strategic goals will be constantly evolving. Benchmarking analyses enable an organization to set these goals on the external reality and ensure it focuses on closing the gaps between actual and envisioned performance.

The findings from benchmarking enable organizations to understand exactly how much improvement is required for attainment of superior performance. Frequent and regular benchmarking supports the establishment of specific and measurable short-term plans, based upon reality rather than historical performance, resulting in step-by-step improvements in performance over time (Figure 15.3). The objective is for the organization to overtake the performance leaders, turning a performance gap into superior performance leadership.

An implementation process is required to convert the long- and short-term plans into operational plans. This requires organizations to determine exactly how their specific strategic goals are to be met, identify the actions required to enable this, determine who has the responsibility for carrying out the actions, calculate and allocate the resources required, and plan, schedule, and control the implementation. The output from benchmarking once again provides the external perspective and feeds into this process by offering information relating to the best practices that have been identified.

Organizations should review their performance on a regular basis to determine progress against the goals set and to measure the gap between the current state and the vision. Benchmarking is the perfect vehicle to support this review process by providing objective evidence of current performance, determining the gaps in performance levels being achieved by other organizations, identifying best practices, and offering the opportunity to learn from the leading performers.

Thus it is clear that benchmarking is a powerful tool that can contribute significantly to an organization's ability to effectively and strategically manage its performance. It forces organizations and their managers to consider the broader perspective, to look outside of their comfort zones, to learn from those identified as excellent performers; and it fuels the drive for change. By revealing what the best-performing organizations are already achieving and by establishing a factual base for best practice, benchmarking enables organizations to manage their performance to achieve world-class leadership.

The Benchmarking Process

Critical to the success of any benchmarking program are a number of key factors:

1. Scope the study and determine objectives.
2. Identify and define all metrics.
3. Agree on a schedule and stick to it.
4. Ensure resources are available to support the benchmarking.
5. Provide support to participants throughout the process.
6. Validate all data.
7. Normalize the data.
8. Clearly and effectively report the findings.
9. Enable sharing of best practices.

Irrespective of the type of benchmarking undertaken, it is essential that a well-structured and systematic process be followed to realize these critical success factors. There are many such benchmarking processes described in the literature (Anand and Kodali 2008), but the pioneering model from which most others have been formulated is that used by Xerox and described by Camp (1989). Camp's 10-step benchmarking process was also described in the fifth edition of this handbook (Camp and DeToro 1999). Since that time Juran has published its own 7-Step Benchmarking Process, which was developed over a period of many years and has formed the basis of a multitude of annual benchmarking consortia since 1995. Although described here in terms of external consortium benchmarking, the process is generic and equally applicable in principle to all types of benchmarking.

The Juran 7-Step Benchmarking Process depicted in Figure 15.4 is divided into two phases. Phase 1 is a positioning analysis providing the benchmarker with a comprehensive study of the relative performance of all the benchmarking participants and a thorough consideration of the performance gaps to the top-performing superior or "best in class"

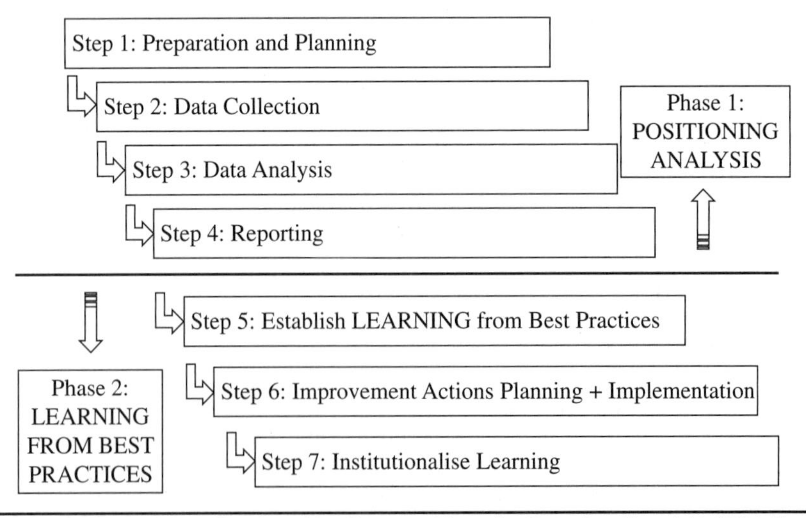

Figure 15.4 The Juran 7-Step Benchmarking Process. (*Juran Institute, Inc. Copyright 1994. Used by permission.*)

organizations. The focus of phase 2 is upon learning from the phase 1 findings, adopting and adapting best practices, and developing improvement programs to implement changes required. Each of the steps in the process is described below.

Planning and Project Setup

Step 1 is to recognize the need for benchmarking, to clearly understand what is to be benchmarked and why, to determine the benchmarking methodology that is going to deliver the analysis required, and to identify who is to be benchmarked. A benchmarking project is no different from any other project. To succeed, thorough preparation and planning at the outset is essential. Often a business case will need to be made to justify the need for the benchmarking project.

Critical at this stage of designing the benchmarking is to clearly define the scope of the benchmarking envelope, what is to be benchmarked and what is to be excluded. The metrics to be used can then be agreed upon, and these too must be clearly and unambiguously defined to ensure comparability of data collected. Finally, the most appropriate vehicle for data collection must be determined.

Once the benchmarking topic has been well defined, the participants with whom the benchmarking will be conducted must be determined. As mentioned earlier, ideally those organizations that are known to be superior performers will be identified as participants in the benchmarking. However, the participants will be dependent upon the type of benchmarking being conducted as well as the way in which the participants are selected; but of course the ultimate aim is to benchmark with the recognized performance leaders.

During this initial planning step participants will also aim to

- Identify and agree on the key performance indicators (KPIs) to be used to assess performance.
- Create a metrics model that clearly demonstrates the interrelationships among the metrics in use.
- Develop clear and unambiguous definitions for all the metrics used.
- Produce a data collection document as a vehicle for participants to collect and submit their data and conduct some initial validation of the data prior to submission.
- Agree on the project time schedule, milestones, and deadlines.

Data Collection and Normalization

Once the precise KPIs and associated definitions are identified, a method for collecting data from each participant must be developed. Commonly a data collection document is produced and issued to all participants to enable them to collect and submit their data. Data submissions are increasingly conducted online via secure Web portals. Proprietary spreadsheets are also frequently used because they are widely available (all participants are likely to have access to them); they are easy to use, have very powerful calculating capability, and can be tailored to provide automated functionality for validation and calculation. The data collection document must be designed to be easy and quick to populate by the user, to provide a suite of validation checks to maximize data quality and minimize errors.

Participant Support during the Benchmarking Process

To facilitate the data submission and validation process, it is a good idea to operate a help desk that is available during the entire project duration. This is often provided when third-party consultants are facilitating the benchmarking. The desk can provide professional advice on how to fill the data collection document and answer questions related to specific program-related matters (e.g., interpretation of definitions used). The objective is to provide a swift response to participants so as not to delay them in the data collection process. Of course, by providing clear and thorough guidance notes and a well-structured data collection document, the need for participants to seek help will be minimized. Nonetheless, a help desk can be an extremely valuable and an essential source of benchmarking support, especially for newly established benchmarking programs. When required, a list of Frequently Asked Questions (FAQs) can also be developed and forwarded to all participants.

Data Validation

Use of valid data is key to the success of any benchmarking program, where the adage "Garbage in, garbage out" has never been more appropriate. Incorrect or inaccurate data can easily result in misguided conclusions and inappropriate actions and can lead to the failure of any improvement program. Furthermore, endless rounds of clarification will lead to frustration by the participants and can delay the benchmarking process. Thus a high degree of emphasis should be placed on data validation. In Juran's benchmarking programs they adopt a two-phase approach, with the initial application of a suite of automated checks followed by a number of manual checks.

Automated checks are an integral part of the data collection document which is designed in such a way that it is easy to populate and has a number of built-in validation checks, thereby maximizing data quality and minimizing errors. The built-in automated error checks aim to prevent the input of spurious data and enable users to conduct their own first-pass manual check of the data prior to submission. Thorough initial checking by the users themselves significantly reduces the time and effort required for subsequent validation.

Once the data are submitted, the facilitator should conduct a number of manual checks according to a rigorous data validation process. Juran typically employs a three-step process:

1. Data completeness
2. Data integrity
3. Data consistency

These checks should be carried out by an experienced individual who understands not only the benchmarking process but also the nature of the data being submitted and the interrelationship between different data points. All data are first checked to ensure that they are complete. A check of the integrity of the data should then be carried out by comparing different interrelated data to ensure that the expected relationship between these data is observed. Finally a range of intelligent triangulation checks can be conducted to further ensure consistency between data provided and any available historical data sets. Any anomalies should be raised one on one with the relevant participant to ensure that corrected data are provided. Where either there are a large number of apparent errors or participants are experiencing difficulty in obtaining the data required, a data clinic may be held, attended by participants and designed to clarify any confusions relating to the data required.

Data Normalization

The single biggest problem in any benchmarking exercise is how to compare benchmarked subjects on a like-for-like basis (i.e., how to compare apples with pears). In some circumstances the benchmarkers will be similar enough to enable direct comparisons of performance between them. However, more typically the subjects being benchmarked will all be different from one another, be they organizations as a whole, business units, different sites, different functional groups, business processes, or products. No two subjects will be identical, although the extent of difference between them will vary considerably depending upon what and who is being benchmarked. Thus to be able to compare differences in performance levels requires some intervention. Some form of data normalization is usually required to enable like comparisons to be made between what may be very different subjects. Without it, direct comparisons of performance are normally impossible and may lead to misinformed conclusions. Normalization can be made on the basis of a wide range of factors including scope, scale, contractual arrangements, regulatory requirements, and geographical and political differences.

One solution is to organize benchmarkers into categories or peer groups with other benchmarkers or data sets with similar characteristics. The key is to be able to identify the factors that are driving the performance and then develop a method by which these drivers can be considered when comparing performance metrics. In its simplest form this may involve stratifying the data according to underlying criteria. For example, if a health authority wishes to compare death rates of people in different regions, it may stratify these according to gender or age. Another example comes from the chemical industry. A series of chemical organizations may decide to benchmark their performance in the field of managing for the environment, and in so doing they may wish to compare emission levels for a variety of polluting gases (e.g., oxides of nitrogen and sulfur, carbon dioxide and methane). These data could be stratified according to their harmful impact to the environment through the use of a standardized measurement such as the environmental impact unit (EIU). A further example might be those organizations comparing the efficiency of their R&D activities using a KPI that measures the percentage of their sales attributable to new products or services (e.g., products that have been on the market for less than 2 years).

But of course, even within these groups there may be differences between the benchmarking subjects. To conduct a valid comparison of performance, these differences in characteristics need to be taken into consideration in the analysis. The most effective way of doing this is through normalization of the performance data.

Normalization is essentially the process of converting metrics into a form that enables their comparison on a like-for-like basis, accounting for all (or as much as possible) the variation between the benchmarking subjects. It is that the normalizing factor used be truly a driver for the performance being benchmarked. For example, in benchmarking the operating costs of the invoicing function of an organization, perhaps a suitable normalizing factor is the number of invoices raised. For example, the costs could be compared on a per invoice raised basis. However, perhaps some invoices are more complicated to produce than others (e.g., they may contain more line items or be for a higher total value that requires more checks before the invoice is raised), so this way of normalizing may not be appropriate after all.

The most common way of doing this is by looking at performance per unit or per hour. For example, if we are measuring the cost of manufacturing a motorcar, we might compare the cost per vehicle produced; or if we are looking at the time taken to treat a hospital patient with a given ailment, we might consider the number of patients examined per hour.

In some cases a simple measurement per unit is not sufficient to accommodate the variation observed between benchmarking subjects, and a more sophisticated approach has to be developed. In such cases the use of weighting factors that represent the variation of the different benchmarking subjects is often a very effective means of normalization. Weighting factors may be developed in relation to costs, time, and efficacy. An example of a highly effective weighting factor is the Juran Complexity Factor (JCF). The JCF was developed to enable like-for-like comparisons to be made between oil and gas production facilities of very different size and design. The normalizing factor takes into consideration the equipment present in the facility and the time it takes to operate and maintain this equipment under normal conditions. The JCF is then used to normalize all cost performance between facilities in the benchmarking. This enables organizations to directly benchmark their facilities with those of other organizations even though they may be very different in design and size.

The efficacy of any normalization method should be fully tested before it is implemented. As mentioned, for a normalizing factor to be effective, it must be representative of the driving force for the performance subjects being benchmarked. Thus there must be a good relationship between the performance metric and the normalizing factor. A good way of testing this is to examine the correlation between the normalizing factor and the performance metric being normalized. There should be a strong direct relationship between the two. For example, an increase in the normalizing factor should lead to an increase in the metric being benchmarked (and vice versa), although this relationship may or may not be linear.

Analysis and Identification of Best Practices

The aim of the analysis is to determine the findings from the data collected in the benchmarking in conjunction, where appropriate, with other pertinent data and information from a number of different sources including the public domain, the participants themselves, and any previous editions of the benchmarking study. The level of analysis will be dependent upon the scope and objectives agreed upon at the commencement of the benchmarking.

It is essential that the analysis be impartial and totally objective. It must also be aligned to the benchmarking objectives; and to be of value, it must indicate the benchmarker's strengths and weaknesses, determine, and where possible quantify, the gaps to the best performers, and identify as far as possible the reasons for these gaps. It is important that the metrics be considered collectively and not in isolation as the results from one metric may help to explain those of another. The strategies and working practices of each of the participants should also be explored and used to determine how they may influence performance.

The performance data and any normalization data streams are analyzed to compare participant performance and determine performance gaps. It is also important to consider the level of statistical testing of the data to ensure that comparisons being made are statistically significant and the conclusions drawn thereafter are valid.

Quantitative analyses are typically made in relation to the top quartile (i.e., the boundary to the 25th percentile), the best in class (i.e., the single best performer), or the average (mean) of the benchmarking population. There are pros and cons to comparisons of each of these criteria. Analysis of the gap to the best in class is probably the most common and on the face of it seems the most obvious; after all the objective is to close the gap to the best performer. However, making comparisons to the data of a single benchmarker at a single point in time will always carry a risk that there is error in the data value (although the validation process should minimize this error) or that the performance level reported is not sustainable in the longer term and is therefore not realistic. In contrast, comparison to the top quartile and in particular to the average is more stable and reliable as these comprise data from more than one participant.

Reasons for apparent differences in performance should be considered during the analyses. With multinational or global benchmarking studies, it is important to consider the impact that may be attributed to differences in geographical location. For example, when one is analyzing costs, it is clear that cost levels (e.g., salaries) in the West cannot be easily compared to the East, Russia, Africa, or Latin America. In addition, fluctuations in exchange rates between currencies can have a dramatic effect. Likewise different tax regimes, regulatory requirements, political policies, and cultural differences can all significantly influence performance.

Report Development

Once the analysis is complete, it must be reported to the benchmarking participants. The content of the report and the medium used for reporting were agreed to at the outset of the benchmarking exercise and in part are determined by the type of benchmarking being undertaken.

Reports may be delivered online, electronically or in paper hard-copy format. Whatever the medium selected, the report must present the benchmarking findings in a clear, concise, and easily understood form. Optimum use of color, diagrams, pictures, and charts should be made to facilitate communication of the findings. Charts and tables should be annotated to provide guidance to the reader. The analysis should be reported in full together with recommendations for the focus of performance improvement efforts required to close gaps.

One very important point that must be addressed is the level of data anonymization that will be employed in the report. There is always a tradeoff between confidentiality and learning opportunity. The higher the level of confidentiality, the lower the potential for learning. If the identity of the superior performers cannot be revealed, then the opportunity to learn from them is minimized. However, compliance with antitrust legislation is always a primary requirement. Thus the report must always be in line with any confidentiality agreement made between the benchmarkers and must meet legal requirements. But to maximize the learning potential, the degree of openness should also be maximized.

Unfortunately, many benchmarking exercises will stop at this point. But to maximize the value gained from benchmarking, organizations must go further to try to understand the practices that enable the leaders to attain their superior performance levels. This is the purpose of phase 2 of the Juran 7-Step Benchmarking Process.

Learning from Best Practices

A benchmarking program must go well beyond the comparison of performance data. The transfer of knowledge from the best practitioners to the other benchmarkers is critical. This is essential to maximize the effectiveness of knowledge transfer, which leads to highly successful change/process improvement programs. This can be achieved in a number of ways:

- Internal forums
- One-to-one benchmarking
- Best practice forums

Internal Forums

Organizations participating in the benchmarking should fully review the benchmarking report and consider the findings in detail. Thereafter many organizations find it beneficial to organize an internal forum attended by all parties within the organization affected by the benchmarking to discuss the findings openly and determine first actions required to begin the process of gap closure and performance improvement. In cases where an organization may have numerous participants in the benchmarking (e.g., an organization may benchmark several different business units simultaneously), these internal forums can be an excellent platform for sharing knowledge among those business units. Juran's oil and gas experts have had the opportunity to attend and facilitate a series of internal knowledge exchange forums in the oil and gas industry (e.g., Pemex Mexico in 2004 and Qatar Petroleum and Saudi Aramco in 2008), where up to 500 members of staff from different departments within one organization gathered with the sole aim of sharing knowledge and best practices related to vital topics and processes.

One-to-One Benchmarking

It is common for organizations to benchmark one to one following participation in a group benchmarking study. Having identified superior performers in various areas of the benchmarking, organizations collaborate on a one-to-one basis to explore specific issues in greater detail perhaps by on-site visits or further data exchange and analyses, to maximize the learning outcomes.

Best Practice Forums

This involves the sharing of best practices between top performing organizations to the mutual benefit of all benchmarkers. Of course, when one is benchmarking with true competitors, the options for this may be limited and alternative approaches may be required to establish learning.

Once findings have been reported to all participating organizations, a best practice forum can be organized, attended by all participating benchmarking organizations. The best practitioners in each of the elements of the benchmarking model are asked to make presentations to the closed forum. Prior to the forum, all participants would be invited to submit any questions they might have for the best practitioners, which they would like to be addressed by the best practitioners in their presentations.

The objective of this forum is the identification of master class opportunities and the transfer of knowledge from the best practitioners to the other benchmarking partners. Best practitioners will present to their peers the "whys and the hows" of their best practices. The intention is that audience members will learn from these presentations which will help them subsequently to formulate their own improvement programs. Participants should leave the best practice forums in a position to develop clear action plans for the implementation of improvement programs.

Improvement Action Planning and Implementation

Once the learning points have been ascertained, each organization needs to develop and communicate an action plan for changes required to realize improvements. Here the learning from the benchmarking will feed into the organization's strategic plan and be implemented using its performance improvement processes. But how can organizations translate benchmarking findings into action plans that will lead to performance improvement?

The output of a benchmarking exercise should become input for action planning. A typical output from a benchmarking exercise will include a series of performance gaps between a participant and the best practitioners in key business processes. Often organizations react with disbelief and denial when they are confronted with the performance gaps translated into monetary terms: Comments such as "It is impossible to save this much, these numbers cannot be right!" are commonplace. It is of paramount importance, for both the credibility of the benchmarking findings and the subsequent level of managerial buy-in, that one additional internal journey be embarked upon prior to moving to action planning. The organization needs to truly understand the performance gaps identified. Therefore it must eliminate any distorting elements from the gaps presented. To render the performance gaps actionable, the organization needs to break them down into controllable and noncontrollable gaps. Noncontrollable gaps are those relating to aspects of an organization's activities that are not under the direct control of that organization at the time. For instance, these could include start-up costs, one-off expenditures, extraordinary incidentals, regulations that one has to

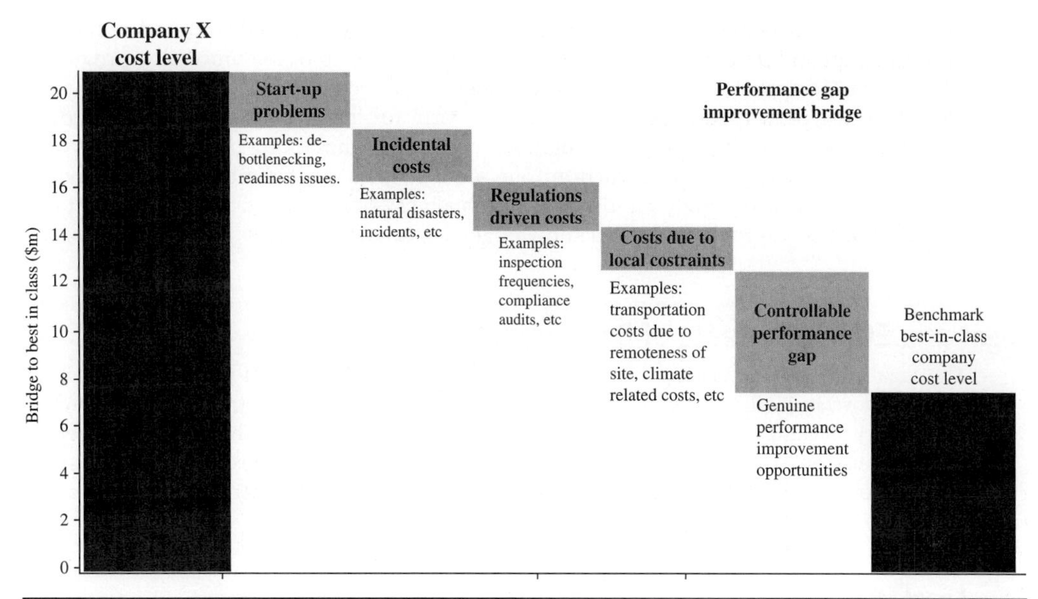

Figure 15.5 Performance bridge analysis from "gross to net performance gap." (*Juran Institute, Inc. Copyright 1994. Used by permission.*)

comply with, site-specific operational issues (e.g., climate, geography, topography), and geopolitical and safety-related costs.

An actionable performance gap should be free from noncontrollable elements (see Figure 15.5). This will allow management to do the following:

1. Assess a performance gap that they can relate to and therefore buy into.
2. Prioritize improvement areas and distinguish the "vital few" versus the "useful many" opportunities for improvement (Juran and Godfrey 1995).
3. Allocate resources to fix the problems and bridge the gaps including an accountable project manager, budget, time frame, and targets. Managers and employees will then be empowered to get things done.
4. Put controls in place by embedding the requirement to action the improvements into managerial and individual employee target setting, compensation schemes, and business planning. This will enhance the chances of a successful implementation.

Thus, benchmarking findings will have been embedded into performance improvement action plans and integrated into routine business cycles, helping to ensure that resources are focused, individuals have bought into the process, and goals are achievable.

Institutionalizing Learning

Finally, the learning gained and the improvements to performance realized must be fully embedded to ensure all gains are rolled out throughout the organization and are sustained over time. Benchmarking may take place at the corporate, operational, or functional level within an organization, and it is important that each of these levels be linked via a cascading series of goals, interlinked to ensure systematic progress toward attaining the vision.

As improvement opportunities arise, they should be embedded into and replicated throughout participant organizations. Juran supports this step of the process by providing ongoing 24/7 support in the form of a members-only secure website accessible only by benchmarking participants. This provides a platform for information sharing and knowledge management well beyond the scope of a normal benchmarking program. It provides the opportunity for peers to learn from one another on a real-time basis "as and when required." In our increasingly rapidly changing business environment, it can be of great use to have direct access to the collective knowledge of specialists facing similar challenges.

Legal and Ethical Aspects of Benchmarking

The legality of benchmarking is governed by competition (antitrust) law and intellectual property law, and all benchmarkers must be aware of the legal and ethical implications of their benchmarking activity. While the ethos of benchmarking is the sharing of knowledge and information to the mutual benefit of all participants, organizations must not lose sight of the potential value of their corporate knowledge and therefore the necessity to adequately control its use.

Legal Issues

During the benchmarking process, knowledge and information are often shared among the participants. Should this be subsequently applied by one of the benchmarking participants as part of the performance improvement process, there arises the possibility of infringing the intellectual property rights of the benchmarker who offered the information. Often organizations fail to recognize the risks of such infringements, albeit inadvertent, and the potential for conflict between benchmarking and intellectual property law (Boulter 2003).

Care has to be taken when exchanging information in a benchmarking study. This is particularly the case if the benchmarking is between competitors and if price information and business secrets are being benchmarked. Of course, the vast majority of benchmarking activities steer clear of both these topics; nonetheless they may still be considered to constitute anticompetitive behavior.

Although competition laws as depicted by antitrust laws in the United States and the Treaty of Rome (Article 81) in the European Union are well documented, their applicability to benchmarking is a gray area. They are primarily aimed at prohibiting cartel formation and price fixing. Other countries also have legal requirements pertaining to competition, and so the legal situation is further complicated where benchmarking exercises span a number of different countries. As a result, any organization considering entering into a benchmarking exercise is strongly urged to seek legal advice before doing so.

The Benchmarking Code of Conduct

The Benchmarking Code of Conduct was first developed by the International Benchmarking Clearinghouse, a service of the American Productivity and Quality Center (APQC) in 1992 (see www.apqc.org). In 1996 a European version of the code was developed (see www.efqm.org), based upon the American version, to comply with European competition law. Neither document is legally binding, but they do lay down the principles for ethical and legal benchmarking. The main principles address legality, confidentiality, and information exchange, and all benchmarking programs should ensure that participants comply with these.

Confidentiality

Essential in all benchmarking studies is the requirement for some degree of confidentiality. The strictness of the level of confidentiality will be dependent upon the sensitivity of the subjects being benchmarked, the requirement to comply with competition law, and the degree of willingness by the participants to share data and information openly. It is clear that great care must be exercised when benchmarking prices. In many cases costs are considered an indicator of prices, and therefore strict confidentiality is normally also expected when comparing costs.

The degree of confidentiality exercised in benchmarking studies can vary enormously. At one end of the scale participants are totally unaware of whom they are benchmarking with as the identities of the other participants are withheld and only the third-party facilitator is aware of who each participant is. Unfortunately a major drawback of such a strict level of confidentiality is that the learning potential is greatly reduced. If participants do not know the identity of the better performers in a benchmarking process, how can they possibly learn from the findings? The whole object of the benchmarking is lost, and the study becomes nothing more than a league table.

Therefore a more pragmatic approach is preferred whereby sensitive data (e.g., costs) can be anonymized whereas other less sensitive data can be shared more openly. And with the skillful support of a third-party facilitator the participants can still maximize the learning potential from the study.

Irrespective of the level of confidentiality and anonymity decided upon, it is essential that all parties in a benchmarking study, including the facilitator (be they a consultant or a participating organization), sign a confidentiality agreement. This agreement will be legally binding and will spell out how the data, information, and findings of the study will be shared, used, and disseminated by all parties.

Managing for Effective Benchmarking

For any benchmarking initiative to succeed, it must be managed effectively.

Each organization participating in a benchmarking exercise will normally establish a benchmarking team comprising individuals from a range of disciplines who will manage the benchmarking activities from outset to completion. The size of this team will be dependent upon the size of the organization and the scope and scale of the benchmarking. A project owner should be established who will lead the team and act as a focal point both within the organization and with the other participants and the facilitator. The owner will be responsible for briefing the team on the project findings and ensuring that all the necessary resources and support is forthcoming from senior management. The team will be responsible for delivering the benchmarking project, setting internal targets and ensuring they are met.

While senior managers are unlikely to be involved directly in conducting the benchmarking, they play a key role in ensuring it is executed successfully. Key roles of senior management are to do the following:

- Set benchmarking goals.
- Integrate benchmarking into the organization's strategic plan.
- Act as a role model.
- Establish the environment for change.
- Create the infrastructure for benchmarking.
- Monitor progress.

Set Benchmarking Goals

The intent to benchmark must be established and clearly communicated within the organization, identifying the reasons for and objectives of benchmarking. A benchmarking policy should be developed and documented.

Integrate Benchmarking into the Organization's Strategic Plan

Benchmarking must form a fundamental element of the organization's business plan. The benchmarking direction must be set and the findings communicated throughout the organization. The findings must also be incorporated into the organizational goals and driven down through the organization.

Act as a Role Model

Senior managers must openly demonstrate their commitment to the benchmarking effort even though they are unlikely to be involved day to day in the benchmarking exercise. They must remove any roadblocks that may derail the benchmarking team, and they must also commit the resources required to conduct the benchmarking and fully realize the benefits from it. This is likely to mean both a financial commitment and a time commitment, releasing individuals from the regular positions to perform their role in the benchmarking team.

Establish the Environment for Change

Managers must demonstrate a willingness to accept the findings of the benchmarking and to act upon them, creating a change environment to realize the potential improvements in performance.

Create the Infrastructure for Benchmarking

There must be a commitment from the outset to provide the resources required for successful benchmarking. Where necessary, training must be provided to those involved in the process as well as reward and recognition for benchmarking team members and those members of the organization whose efforts have led to superior performance as determined by the benchmarking. However, it is essential that any poor performance identified by the benchmarking not be punished. As we mentioned, benchmarking should not be used as a personnel appraisal tool!

Monitor Progress

There must be a commitment from the outset to provide the resources required for successful benchmarking. Where necessary, training must be provided to those involved in the process.

References

Anand, G., and Kodali, R. (2008). "Benchmarking the Benchmarking Models." *Benchmarking: An International Journal*, vol. 15, no. 3, pp. 257–291.

Boulter, L. (2003). "Legal Issues in Benchmarking." *Benchmarking: An International Journal*, vol. 10, no. 6, pp. 528–537.

Camp, R. C. (1989). *Benchmarking: The Search for Industry Best Practices That Lead to Superior Performance*. ASQC Quality Press, Milwaukee, WI.

Camp, R. C., and De Toro, I. J. (1999). "Benchmarking," in Juran, J. M., Godfrey, A. B., Hoogstoel, R. E., and Schilling, E. G. (eds.), *Juran's Quality Handbook*. 5th ed.. McGraw-Hill, New York.

Juran, J. M., and Godfrey, A. B. (1995). *Managerial Breakthrough*. New York, Barnes & Noble.

Kearns, D. T., and Nadler, D. A. (1993). "Prophets in the Dark—How Xerox Reinvented Itself and Beat Back the Japanese." Harper Business, New York.

Spendolini, M. J. (1992). *The Benchmarking Book*. Amocom, New York.

Stapenhurst, T. (2009). *The Benchmarking Book: A How to Guide to Best Practice for Managers and Practitioner*. Elsevier, Butterworth-Heinemann.

Tonchev, A., and Tonchev, C. (2005). "Single Index Measures Operational Performance of Hydrocarbon Facilities." *Oil and Gas Journal*.

Wood, B. (2009). "7 Steps to Better Benchmarking." *Business Performance Management*, Penton Media Inc. October 2009.

Case Study

This case study describes the experiences of an oil terminal that was able to realize significant improvements in performance driven by its participation in a benchmarking consortium.

In 2005 the terminal benchmarked its performance for the first time. Prior to benchmarking, the terminal believed that it would be among the highest performers. However, the findings of the study, which examined 2004 performance, indicated that its performance was fourth quartile in many areas including operations and maintenance expenditure and workers' time, and their total expenditure was nearly twice as high as the average for the benchmarking group. Even more staggering was the 9-fold gap to the best-performing terminal. The biggest performance gap lay in maintenance expenditure and workers' time which were both the highest of the study. Of particular note were very high third-party (contractors and services) costs and workers' time.

The terminal managers' first reaction was one of denial, but after comprehensively reviewing the findings and recovering from the shock of the reality of their relative performance, they established a performance improvement team led by a newly appointed performance improvement manager. An action plan was developed that focused upon the weaknesses identified in the benchmarking exercise. The first step they took was to differentiate the costs they felt they could control from those they could not. For example, there were some costs that management agreed could not be reduced, at least in the initial term, due to the remote location of the facility. These costs were therefore parked for the time being, and focus was placed on those areas that were controllable and represented the biggest performance gaps.

The team conducted an organizational design review and developed a restructuring that implemented clear accountabilities for key areas including operations, maintenance, integrity management, and projects. They focused upon their main weakness which was maintenance. A series of step change programs were initiated to realize rapid improvements in performance. Changes made were characterized by the following:

- A move from inefficient hierarchies to more efficient self-directing work teams
- A change in shift working patterns with a move from 8- to 12-hour shifts that resulted in a 30 percent increase in efficiency
- Improved control of third-party service providers and contractors resulting in a 30 percent reduction in contractor spending

- Reengineering of work planning and execution processes
- Optimized maintenance work frequencies through the adoption of a risk-based approach to maintenance

The terminal was benchmarking its performance on an annual basis and by 2007 began to see the fruits of its performance improvement initiative. The 2007 benchmarking program, which examined terminal performance in 2006, revealed some marked improvements. The total expenditure and workers' time were now third quartile with total expenditure now reduced to 25 percent higher than the group average and the gap to the lowest-cost terminal reduced to a factor of 3.5. Overall, the terminal's total storage and loading costs were reduced by some 27 percent during this period.

Greatest improvements were made in the areas of operations and maintenance. The terminal was now a first-quartile performer in terms of operations expenditure and workers' time. Between 2005 and 2007, the terminal's maintenance expenditure had shifted from fourth to third quartile and maintenance workers' time from fourth to second quartile. This shift represented a very significant 58 percent reduction in costs and a 50 percent reduction in workers' time for maintenance.

A number of conclusions can be drawn from our benchmarking experience and in particular the findings of this case study.

First and most importantly, it is possible to compare the performance of different terminals on a like-for-like basis. However, to do this, it is critical that the right normalization method be used and that the efficacy of the method be checked to ensure a healthy correlation with the subjects being measured.

Second, benchmarking is a powerful improvement tool. Furthermore, it can provide a wake-up call for complacency, as in the terminal described here; your performance may not be quite as good as you think when compared to that of your peers. However, it is also important to understand that to realize any improvement in performance, it is essential to act upon the findings of a benchmarking exercise.

Third, it is clear that if a structured and well-researched approach to performance is adopted, then significant gains are achievable.

Finally, it is important to understand the importance of benchmarking on a continuous basis for two main reasons. In so doing, a terminal will be able to trend its own performance over time and therefore to determine the effectiveness of its performance improvement activities. Also your peers will be striving to improve their performance, and so the benchmark is constantly moving. Keeping abreast of these improvements and staying in touch with the leaders will require a continuous approach.

CHAPTER 16

Using International Standards to Ensure Organization Compliance

Joseph A. De Feo

About This Chapter

International and global standards that provide requirements or give guidance on good management practice are among the standards most widely used in many organizations around the globe. Of the many standards published, a few have achieved truly global status and are now integrated with the world economy and in the organizations that use them. This chapter discusses the standards that support the management of quality:

- ISO 9000 for Quality Management Systems
- ISO 14000 for Environmental Managements Systems
- cGMPs for Pharmaceutical and Medical Device
- ISO/TS 16949: Automotive Industry
- CMMi for Software Quality
- AS9100: Aerospace

This chapter focuses on the importance of these standards as a means to ensuring the quality of products and services. We realize there are many more standards that are available for many industries. It is not our intention to single these out as the best; we have merely selected them because of their relationship to managing quality. ISO stands for International Organization for Standardization.

High Points of This Chapter

1. ISO 9000 standards have had great impact on the implementation of international trade and quality systems by organizations worldwide. The standards have been applied in a wide range of industry/economic sectors and government regulatory areas. The ISO 9000 standards deal with the management systems used by organizations to ensure quality in: design, production, delivery, and support products.
2. To maintain its registered status, the supplier organization must pass periodic surveillance audits by a registrar. Surveillance audits are often conducted semiannually. The audits may be less comprehensive than a full audit. If so, a full audit is performed every few years.
3. The ISO 14000 is a standard for an environmental management system. It is applicable to any business, regardless of size, location, or industry. The purpose of the standard is to reduce the environmental footprint of a business and to decrease the pollution and the waste a business produces.
4. cGMP refers to the Current Good Manufacturing Practice regulations enforced by the U.S. Food and Drug Administration (FDA). cGMPs provide for systems that ensure proper design, monitoring, and control of manufacturing processes and facilities.
5. CMMI stands for the Capability Maturity Model Integration and is used as a benchmark for comparison and as an aid to understanding for software development.
6. AS9100 is a widely adopted and standardized quality management system for the aerospace industry.

International Standards Overview

Standards exist principally to facilitate international trade and to avoid harming customers and society. In the prestandardization era (before 1980), there were various national and multinational standards. Standards for electrical, mechanical, and chemical process compatibility have been around for decades. Other standards such as military standards were developed for the military and other groups for the nuclear power industry, and, to a lesser extent, for commercial and industrial use. These standards have commonalities and historical linkages. However, they were often not consistent in terminology or content for widespread use in international trade. As a result, organizations were left to recreate their own standards or adapt the existing ones. This only led to even less commonality. In the 1980s as most of the organizations in the industrialized world began to improve quality and safety at record paces there became a need to fill a void. That void was a common quality management system that would be a nonbinding "contract" between the customer and the supplier. This void was filled by the ISO 176 Technical Committee in the form of the ISO 9000 set of standards. This was later followed by filling a similar void for environmental standards with ISO 14000. Many organizations globally began using these standards as a "certified" standard for performance. Although their intent was important, the standards became more of an opportunity to get a certificate of compliance that could be used to impress customers, rather than a set of requirements that ensured that customer needs are met.

Certain industry/economic sectors then began developing industrywide quality system standards, based upon the verbatim adoption of ISO 9000, together with industrywide supplemental requirements. The automotive industry (QS 9000), the pharmaceutical and medical devices industry (cGMPs), government regulatory agencies, and military procurement agencies (AS 9100 and the Mission Assurance Provisions, MAP), are adopting this approach in many places worldwide. Even software development uses the CMMi standard of software quality created in the early 1990s at Carnegie Mellon University to ensure a common approach to manage software quality. The standards play an important—but not always understood—role in managing quality.

We will include a brief discussion on the following standards and or industry practices:

- ISO 9000 for Quality Management Systems
- ISO 14000 for Environmental Managements Systems
- cGMPs for Pharmaceutical and Medical Devices
- ISO/TS 16949: Automotive Industry
- CMMI for Software Quality
- AS9100 and MAP in the U.S. defense industry

ISO 9000 Quality Management System Standard

The ISO 9000 standards have had great impact on international trade and quality systems implementation by organizations worldwide. The international standards have been adopted as national standards by over 100 countries. They have been applied in a wide range of industry/economic sectors and government regulatory areas. ISO 9000 standards deal with management systems used by organizations to ensure quality in: design, production, delivery, and support products. The standards apply to all generic product categories: hardware, software, processed materials, and services. The complete set of ISO 9000 family of standards provides quality management guidance, quality assurance requirements, and supporting technology for an organization's quality management system. The standards provide guidelines or requirements on what features are to be present in the management system of an organization but do not prescribe how the features are to be implemented. This nonprescriptive character gives the standards their wide applicability for various products and situations. Upon implementing ISO 9000, an organization can be registered as a Certified Quality Management System.

The standards in the ISO 9000 family were created and are produced and maintained by Technical Committee 176 of the International Organization for Standardization (ISO). The first meeting of ISO/TC176 was held in 1980. ISO 8402, the vocabulary standard, was first published in 1986. The initial ISO 9000 series was published in 1987, consisting of the following:

- Fundamental concepts and road map guideline standard ISO 9000
- Three alternative requirements standards for quality assurance (ISO 9001, ISO 9002, or ISO 9003)
- Quality management guideline standard ISO 9004

Since 1987, additional standards have been published. The ISO 9000 family now contains a variety of standards supplementary to the original series. In particular, revisions of

Clause Titles	
1	Scope
2	Normative reference
3	Definitions
4	Quality system requirements
4.1	Management responsibility
4.2	Quality system
4.3	Contract review
4.4	Design control
4.5	Document and data control
4.6	Purchasing
4.7	Control of customer-supplied product
4.8	Product identification and traceability
4.9	Process control
4.10	Inspection and testing
4.11	Control of inspection, measuring and test equipment
4.12	Inspection and test status
4.13	Control of nonconforming product
4.14	Corrective and preventive action
4.15	Handling, storage, packaging, preservation and delivery
4.16	Control of quality records
4.17	Internal quality audits
4.18	Training
4.19	Servicing
4.20	Statistical techniques

Table 16.1 Clauses of ISO 9001 and Their Typical Structures

the basic ISO 9000 series, ISO 9000 through ISO 9004, were published in 1994, 2000, and, most recently, in 2008 under the name ISO 9000:2008. This section is written in relation to the 2008 revisions after an initial introduction to the original standard. Table 16.1 displays the ISO 9000:2008 list of requirements.

ISO 9000 has been adopted and implemented worldwide for quality assurance purposes in both two-party contractual situations and third-party certification/registration situations. Their use grew in the 1990s and early 2000s but has since slowed. ISO 9000 will grow again as the newest update in eight years with the release of ISO 9000:2008. The infrastructure of certification and registration bodies, accreditation bodies, course providers, consultants, and auditors trained and certified for auditing to these standards followed a similar pattern. Mutual recognition arrangements between and among nations continue to develop, with the likelihood of recognizing ISO-sponsored quality system accreditation in the near future. Periodic surveillance audits that are part of the third-party certification/registration arrangements worldwide provide continuing motivation for supplier organizations to maintain their quality systems in complete conformance and to improve the systems to continually meet their objectives for quality.

The market for quality management and quality assurance standards itself grew rapidly, partly in response to trade agreements such as the European Union (EU), the General

Agreement on Tariffs and Trade (GATT), and the North American Free Trade Association (NAFTA). These agreements all depend upon standards that implement the reduction of nontariff trade barriers. The ISO 9000 family occupies a key role in implementing such agreements.

External Driving Forces

The driving forces that have resulted in widespread implementation of the ISO 9000 standards can be summed up in one phrase: the globalization of business. Expressions such as the "postindustrial economy" and "the global village" reflect profound changes in recent decades. These changes include the following:

- New technology in virtually all industry/economic sectors
- Worldwide electronic communication networks
- Widespread worldwide travel
- Dramatic increase in world population
- Depletion of natural resource reserves, arable land, fishing grounds, and fossil fuels
- More intensive use of land, water, energy, and air
- Widespread environmental problems/concerns
- Downsizing of large organizations and other organizations, flattened organizational structure and outsourcing of functions outside the core functions of the organization
- Number and complexity of language, culture, and legal and social frameworks encountered in the global economy
- Diversity a permanent key factor
- Developing countries becoming a larger proportion of the total global economy; there are new kinds of competitors and new markets

These changes have led to increased economic competition, increased customer expectations for quality, and increased demands upon organizations to meet more stringent requirements for quality of their products.

Globalization of business is a reality even for many small- and medium-size organizations. These smaller organizations, as well as their large counterparts, now find that some of their prime competitors are likely to be based in another country. Fewer and fewer businesses are able to survive by considering only competition within the local community. This affects the strategic approach and the product planning of organizations of all sizes.

Internal Response to the External Forces

Organizations everywhere are dealing with the need to change. There is now greater focus on human resources and organizational culture and on empowering and enabling people to do their jobs. ISO 9000 implementation involves establishing policy, setting objectives for quality, designing management systems, documenting procedures, and training for job skills. All of these elements are parts of clarifying what people's jobs are.

Organizations have adopted performance excellence programs that include business process management as a means of adapting to changing customer needs. This concept is emphasized in the ISO 9000 standards. Metrics are being used increasingly to characterize product quality and customer satisfaction more effectively.

Organizations are implementing better product design and work-process design procedures, and improved production strategies. Benchmarking and competitive assessment are used increasingly.

An important question that is often asked is "In this world of rapid change, how can a single family of standards, ISO 9000, apply to all industry and economic sectors, all products, and all sizes of organizations?"

ISO 9000 standards are founded on the concept that the assurance of consistent product quality is best achieved by simultaneous application of two kinds of standards:

- Product standards (technical specifications)
- Quality system (management system) standards

Product standards provide the technical specifications that apply to the characteristics of the product and, often, the characteristics of the process by which the product is produced. Product standards are specific to the particular product: both its intended functionality and its end-use situations that the product may encounter.

The management system is the domain of the ISO 9000 standards. It is by means of the distinction between product specifications and management system features that the ISO 9000 standards apply to all industry/economic sectors, all products, and all sizes of organizations.

Distinctions between Juran Trilogy® and ISO Standards

The ISO 9000 family standards contain requirements and guidelines. ISO 9001 (versus 9002, 9003, 9004, and so on) is a requirement standard for that system. It is a quality management system model to be used for quality assurance purposes for providing confidence in product and service quality. A requirements standard becomes binding upon an organization wherever the organization:

- Is explicitly called up in a contract between the organization and its customer
- Seeks and earns third-party certification and registration

ISO 9004 is an example of a guideline standard. Guideline standards are advisory documents. They are phrased in terms of the word "should," meaning that they are recommendations.

All of the ISO 9000 family standards are generic, in the sense that they apply to any product or any organization. All of the ISO 9000 family standards are nonprescriptive in the sense that they describe what management system functions shall or should be in place; but they do not prescribe how to carry out those functions.

ISO 9004 is similar to many National Awards for Excellence in that it provides a model for organizationwide quality management. The major difference is that the National Award Criteria are business focused and the ISO 9000 standards are not. Why? Because the standards do not include many of the enablers and influencers that will assure that all processes in an organization are compliant. The standards do not include the full scope of managing for quality as defined by Dr. Joseph Juran. They were not intended to be. As a result of complaints, we often hear ISO 9000 did not do what we expected it to do, whereas others said it was great for them. The ones that stated it did not work had an expectation that the standard alone, once implemented, would guarantee improved quality and better financial performance. They were not satisfied. They also did not know that the standard does not include provisions for these other tasks that must happen beyond the product and service production processes. If ISO 9004 were the registration standard, more organizations would see the benefit of registration. That is because ISO 9004 is a quality management system and ISO 9001 is only an assurance system (see Table 16.2). This is a subset of what is needed to manage for quality. The organizations that stated that ISO 9004 worked for them used the standard as a building block to a better system. They filled in the gaps where the standard was not designed to do. As a result, these organizations saw ISO Standards as an important part of their performance excellence program.

Juran defined the process of planning, control, and improvement as essential to manage for quality. Quality Assurance is important since it provides information on how our system is performing to plans. Quality Control as Juran described is different than Quality Assurance. Control is about what to control, Assurance is about proving that what you controlled was indeed controlled (see Figure 16.1).

One of the most pressing needs in the early years of ISO/TC176 work was to internationally harmonize the meanings of terms such as "quality control" and "quality assurance." These two terms, in particular, were used with diametrically different meanings among various nations, and even within nations. The term "quality management" was introduced into the ISO 9000 standards as the umbrella term for quality control and quality assurance. The term "quality management" was defined, included in ISO 8402, and adopted internationally. This, in turn, enabled agreement on harmonized definitions of the meanings of each of the terms "quality control" and "quality assurance."

According to ISO 9000:2008 the standards require the following before certification can take place:

> The organization shall establish, document, implement and maintain a quality management system and continually improve its effectiveness in accordance with the requirements of this International Standard.
>
> The organization shall:
>
> determine the processes needed for the quality management system and their application throughout the organization,
>
> determine the sequence and interaction of these processes,

The Prime Focus of	
Quality Management	**Quality Assurance**
• *Achieving* results that satisfy the requirements for quality	• *Demonstrating* that the requirements for quality have been (and can be) achieved
• Motivated by stakeholders *internal* to the organization, especially the organization's management	• Motivated by stakeholders, especially customers, *external* to the organization
• Goal is to satisfy *all stakeholders*	• Goal is to satisfy all *customers*
• Effective, efficient, and continually improving overall quality-related *performance* is the intended result	• *Confidence* in the organization's products is the intended result
• Scope covers all activities that affect the total quality-related *business results* of the organization	• Scope of demonstration covers activities that directly affect quality-related *process and product results*

TABLE 16.2 Quality Management and Quality Assurance

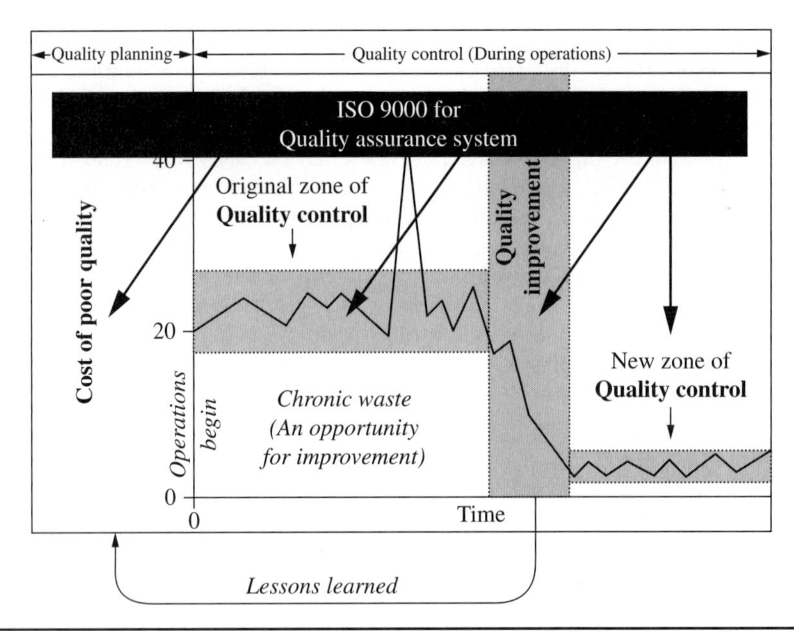

Figure 16.1 ISO 9000 and the Juran Trilogy. (*Juran Institute, Inc., Southbury CT.*)

determine criteria and methods needed to ensure that both the operation and control of these processes are effective,

ensure the availability of resources and information necessary to support the operation and monitoring of these processes,

monitor, measure where applicable, and analyze these processes, and

implement actions necessary to achieve planned results and continual improvement of these processes.

These processes shall be managed by the organization in accordance with the requirements of this International Standard. Where an organization chooses to outsource any process that affects product conformity to requirements, the organization shall ensure control over such processes. The type and extent of control to be applied to these outsourced processes shall be defined within the quality management system.

Quality System Certification/Registration

The earliest users of quality assurance requirements standards were large customer organizations such as electric power providers and military organizations. These customers often purchase complex products to specific functional design. In such situations, quality assurance requirements are called up in a two-party contract where the providing organization (i.e., the supplier) is referred to as the "first party" and the customer organization is referred to as the "second party." Such quality assurance requirements typically include provisions for the providing organization to have internal audits sponsored by its management to verify that its quality system meets the contract requirements. These are first-party audits. Such contracts typically also include provisions to have external audits sponsored by the management of the customer organization to verify that the supplier organization's quality system meets the contract requirements. These are second-party audits. Within a contractual arrangement between two such parties, it is possible to tailor the requirements, as appropriate, and to maintain an ongoing dialogue between customer and supplier.

When such assurance arrangements become a widespread practice throughout the economy, the two-party, individual-contract approach becomes burdensome. There develops a situation where each organization in the supply chain is subject to periodic management system audits by many customers and is itself subjecting many of its subsuppliers to such audits. There is a lot of redundant effort throughout the supply chain because each organization is audited multiple times for essentially the same requirements. The conduct of audits becomes a significant cost element for both the organizations performing the audit and the organizations being audited.

Certification/Registration-Level Activities

The development of quality system certification/registration is a means to reduce the redundant, non-value-adding effort of these multiple audits. A third-party organization, which is called a "certification body" in some countries, or a "registrar" in other countries (including the United States), conducts a formal audit of a supplier organization to assess conformance to the appropriate quality system standard, say, ISO 9001 or ISO 9002. When the supplier organization is judged to be in complete conformance, the third party issues a certificate to the supplying organization and registers the organization's quality system in a publicly available register. Thus, the terms "certification" and "registration" carry the same marketplace meaning because they are two successive steps signifying successful completion of the same process.

To maintain its registered status, the supplier organization must pass periodic surveillance audits by the registrar. Surveillance audits are often conducted semiannually. They may be less comprehensive than the full audit. If so, a full audit is performed every few years.

In the world today, there are hundreds of certification bodies/registrars. Most of them are private, for-profit organizations. Their services are valued by the supplier organizations they register, and by the customer organizations of the supplier organizations, because the registration service adds value in the supply chain. It is critical that the registrars do their work competently and objectively and that all registrars meet standard requirements for their business activities. They are, in fact, supplier organizations that provide a needed service product in the economy.

Accreditation-Level Activities

To ensure competence and objectivity of the registrars, systems of registrar accreditation have been set up worldwide. Accreditation bodies audit the registrars for conformity to standard international guides for the operation of certification bodies. The quality system of the registrar comes under scrutiny by the accreditation body through audits that cover the registrar's documented quality management system, the qualifications and certification of auditors used by the registrar, the record keeping, and other features of the office operations. In addition, the accreditation body witnesses selected audits done by the registrar's auditors at the facility of the client supplier organization.

Mutual International Acceptance

Various other countries have implemented these three areas of activity, too:

1. Accreditation of certification bodies/registrars
2. Certification of auditors
3. Accreditation of auditor training courses

Various bilateral mutual recognition agreements are in place between certain countries whereby, for example, the certification of an auditor in one country carries over into automatic recognition of that certification in another country. In other situations, a memorandum of understanding has been negotiated between, say, the accreditation bodies in two countries, whereby they enter into a cooperative mode of operation preliminary to entering into a formal mutual recognition agreement. Under a memorandum of understanding, the accreditation bodies may jointly conduct the audit of a registrar, and the auditors may jointly document the results of the audit. However, each of the accreditation bodies would make its own decision whether to grant or continue the accreditation, as the case may be.

In principle, there should be no need for a supplier organization to obtain more than one certification/registration. A certificate from a registrar accredited anywhere in the world should, in principle, be accepted by customer organizations anywhere else in the world. In practice, it takes time to build infrastructure comparable in any country. It takes additional time (measured in years) for that infrastructure to mature in its operation and for confidence to build in other countries. Of course, not all countries decide to set up their own infrastructure but may choose to have their supplier organizations who wish to become registered do so by employing the services of an accredited registrar from another country.

Indeed, many registrar organizations have established operations internationally and provide services in many countries. Such registrars often seek accreditation in multiple countries because their customers (supplier organizations) look for accreditation under a system with which they are familiar and have developed confidence.

At the present time, there is a multiplicity of arrangements involving single or multiple accreditations of registrars, single or multiple certifications of auditors, and single or multiple accreditations of training courses. The overall system is moving toward widespread mutual recognition, but the ultimate test of credibility is the marketplace willingness to accept a single certification and a single accreditation.

The International Organization for Standardization (ISO), in January 1995 reaffirmed its support for the Quality System Assessment Recognition (QSAR) and approved a plan of action for setting the program in motion. This effectively laid the foundation for a voluntary system aimed at encouraging worldwide acceptance of ISO 9000 certificates.

The current status where registrars and course providers may have multiple accreditations, and auditors may have multiple certifications, may seem to have more redundancy than is necessary. If we step back and compare the current situation to the alternative of widespread second-party auditing of the quality systems of supplier organizations, it must be acknowledged that the present situation is better because there is

- Much less redundancy of auditing
- Much improved consistency of auditing
- The potential for even less redundancy and further improved consistency through the use of international standards and guides as criteria and through mutual harmonization efforts driven by the marketplace

Formal International Mutual Recognition

For the United States, there is one further complication. Almost alone among the countries of the world, the U.S. standards system is a private sector activity. The American National Standards Institute (ANSI), a private sector organization, is the coordinating body for standards in the United States. Under the ANSI umbrella, many organizations produce and maintain numbers of American national standards. Most of these standards relate to product technical specifications. Among the largest U.S. producers of standards are such organizations as the

American Society of Testing and Materials (ASTM), the American Society of Mechanical Engineers (ASME), and the Institute of Electrical and Electronics Engineers (IEEE), but there are many other organizations that produce American national standards applicable to specific products or fields of activity. The ANSI system provides a consistent standards development process that is open, fair, and provides access to all parties that may be materially affected by a standard. The success of the U.S. system is attested to by the predominance of the U.S. economy internationally and the widespread adoption of U.S. standards for multinational or international use.

However, there are three levels of activities and infrastructure in relation to conformity assessment in international trade. Two levels have already been discussed: the certification/registration level and the accreditation level. The third level is recognition. At the recognition level, the national government of country A affirms to the government of country B that A's certification and accreditation infrastructure conforms to international standards and guides. In most countries of the world where the standards system is run by a government or semigovernment agency and the accreditation activities are carried out by that agency, the recognition level is virtually automatic. In the United States, various government agencies may be called upon to provide the formal recognition.

For example, in dealing with the European Union (EU) on products that fall under one of the EU directives that regulate products that have health, safety, and environmental risks, the EU insists upon dealing through designated government channels. The relevant U.S. government agency varies from one EU directive to another. In many areas, the recognition responsibility will come under the recently authorized National Voluntary Conformity Assessment System Evaluation (NVCASE) program to be run by the Department of Commerce through the National Institute of Standards and Technology. The NVCASE program had not come into operation at the time of this writing.

Conformity Assessment and International Trade

The conformity assessment approach of the EU typifies what is happening in many parts of the world. For a regulated product to be sold in any EU country, it must bear the "CE" mark. Under the EU's modular approach, to qualify to be able to use the mark, the supplier organization must produce evidence of conformity in four areas:

1. Technical documentation of product design
2. Type testing
3. Product surveillance (by samples, or by each product)
4. Surveillance of quality assurance

Depending on the directive, the EU will offer suppliers various routes (modules) to satisfy the requirements. These routes range from "Internal Control of Production," which focuses on the product surveillance aspects, to "Full Quality Assurance," which typically focuses on certification/registration to ISO 9001 and relies upon the ISO 9001 requirements for capability in product design. In most modules, the manufacturer must submit product units, and/or product design technical information, and/or quality system information to a certification body that has been designated by the government as a "notified body." In some modules, the notified body must also provide for product tests where required. Several modules involve certification to ISO 9001, ISO 9002, or ISO 9003.

Implementing this modular approach to conformity assessment for regulated products by the European Union (then called the European Community) was the largest, single, early impetus to the rapid spread of certification/registration to ISO 9001 or ISO 9002 worldwide. For example, about half of the dollar volume of U.S. trade with Europe is in regulated products. Nevertheless, global trends in technology and in requirements for quality, and the cost savings of third-party versus widespread second-party auditing, as discussed previously, are powerful additional incentives and staying power for sustained international use and growth of third-party quality system certification/registration.

Moreover, for a supplier organization it is not effective to attempt to have two quality management systems, one for regulated products and another for nonregulated products. Consequently, there are multiple incentives for large numbers of supplier organizations, engaged directly or indirectly in international trade, to operate a quality management system that conforms to ISO 9001 or ISO 9002, as appropriate.

Guiding Principles

There are many registrars; each is registering many supplier quality systems. Each supplier is dealing with many customers. It is impractical to adequately monitor the operations of such a system solely by periodic audits conducted by an accreditation body. Consequently, the guiding principle should be that primary reliance must be placed on the concept of "truth in labeling," by means of which every customer has routine, ready access to the information upon which to judge all four elements of the scope of a supplier's registered quality system.

ISO 14000 Environmental Management System

The ISO 14000 is a standard for an environmental management system. It is applicable to any business, regardless of size, location, or industry. The purpose of the standard is to reduce the environmental footprint of a business and to decrease the pollution and waste a business produces. The most recent version of ISO 14001 was released in 2004 by the International Organization for Standardization (ISO).

The ISO 14000 environmental management standards exist to help organizations minimize how their operations negatively affect the environment. In structure it is similar to ISO 9000 quality management, and both can be implemented side by side. In order for an organization to be awarded an ISO 14001 certificate, the organization must be externally audited by an audit body that has been accredited by an accreditation body.

An effective environmental management system that meets the requirements of ISO 14001:2004 is a management tool enabling an organization of any size or type to do the following:

- Identify and control the environmental impact of its activities, products, or services
- Improve its environmental performance on a continual basis
- Implement a systematic approach to setting environmental objectives and targets, to achieving these ends, and to demonstrating that they have been achieved.

Certification to Standard

Certification auditors need to be accredited by the International Registrar of Certification Auditors. The certification body has to be accredited by the Registrar Accreditation Board in the United States, or the National Accreditation Board in Ireland.

The ISO 14000 family addresses various aspects of environmental management. The very first two standards, ISO 14001:2004 and ISO 14004:2004, deal with environmental management systems (EMS). ISO 14001:2004 provides the requirements for an EMS, and ISO 14004:2004 gives general EMS guidelines.

Other standards and guidelines in the ISO 14000 family address specific environmental aspects, including labeling, performance evaluation, life cycle analysis, communication, and auditing.

The standards consist of the following elements:

- ISO 14001 environmental management systems—requirements with guidance for use.
- ISO 14004 environmental management systems—general guidelines on principles, systems and support techniques.
- ISO 14015 environmental assessment of sites and organizations.
- ISO 14020 series (14020-14025) environmental labels and declarations.
- ISO 14031 environmental performance evaluation—guidelines.
- ISO 14040 series (14040-14049), Life Cycle Assessment, LCA, discusses preproduction planning and environment goal setting.
- ISO 14050 terms and definitions.
- ISO 14062 discusses making improvements to environmental impact goals.
- ISO 14063 environmental communication—guidelines and examples.
- ISO 19011 specifies one audit protocol for both 14000 and 9000 series standards together. This replaces ISO 14011—how to tell if your intended regulatory tools worked. Using ISO 19011 is now the only recommended way to determine this.

How ISO 14000 Works

- ISO 14001:2004 does not specify levels of environmental performance. If it specified levels of environmental performance, they would have to be specific to each business activity and this would require a specific EMS standard for each business; that is not the intention.
- ISO has many other standards dealing with specific environmental issues. The intention of ISO 14001:2004 is to provide a framework for a holistic, strategic approach to the organization's environmental policy, plans, and actions.
- ISO 14001:2004 gives the generic requirements for an environmental management system. The underlying philosophy is that whatever the organization's activity, the requirements of an effective EMS are the same.

This establishes a common reference for communicating about environmental management issues between organizations and their customers, regulators, the public, and other stakeholders.

Because ISO 14001:2004 does not lay down levels of environmental performance, the standard can to be implemented by a wide variety of organizations, whatever their current level of environmental maturity. However, a commitment to compliance with applicable environmental legislation and regulations is required, along with a commitment to continual improvement—for which the EMS provides the framework.

ISO 14000 Standards

ISO 14004:2004 provides guidelines on the elements of an EMS and its implementation and discusses principal issues involved.

ISO 14001:2004 specifies the requirements for such an environmental management system. Fulfilling these requirements demands objective evidence that can be audited to demonstrate that the environmental management system is operating effectively in conforming to the standard.

What Can Be Achieved?

ISO 14001:2004 is a tool that can be used to meet internal objectives: to assure management that it is in control of the organizational processes and activities having an impact on the environment and to assure employees that they are working for an environmentally responsible organization.

ISO 14001:2004 can also be used to meet the following external objectives:

- Provide assurance on environmental issues to external stakeholders—such as customers and the community. Regulatory agencies comply with environmental regulations, support the organization's claims, and communicate about its own environmental policies, plans, and actions.
- Provide a framework for demonstrating conformity via suppliers' declarations of conformity, assessment of conformity by an external stakeholder—such as a business client—and for certification of conformity by an independent certification body.

Importance of ISO 14000 Standards to the Management of Quality

Chapter 2, Quality's Impact on Society and the National Culture, and Chapter 10, A Look Ahead: Eco-Quality for Environmental Sustainability, of this handbook outlined the importance of organizations meeting the expanding needs of its customers. As we move into the next decade, customers will require suppliers to demonstrate that they are actively concerned about the environment and that the products or services are produced free from environmental hazards. This will place the importance of this standard on a par with quality standards. As more customers demand confidence in an organization's ability to prove that the organization is worthy, pressure will be placed on the organizations to be certified in ISO 14000 standards. For more information on this standard, please refer to www.iso.org.

Industry-Specific Adoptions and Extensions of ISO 9000 Standards

In some sectors of the global economy, there are industry-specific adoptions and extensions of the ISO 9000 standards. These situations are a classic example of a problem opportunity. As problems, these adaptations and extensions strain the goal of nonproliferation. As opportunities, they have been found effective in a very few industries where there are special circumstances and where appropriate ground rules can be developed and implemented consistently. These special circumstances have been characterized by the following:

- Industries where the product impact on the health, safety, or environmental aspects is potentially severe; consequently, most nations have regulatory requirements regarding a supplier's quality management system
- Industries that have had well-established, internationally deployed industry-specific or supplier-specific quality system requirements documents prior to publication of the ISO 9000 standards

Fortunately, in the very few instances shown so far, the operational nonproliferation criteria of the ISO/IEC directives have been followed.

Medical Device Industry

Circumstance 1 relates to the medical device manufacturing industry. For example, in the United States, the Food and Drug Administration (FDA) developed and promulgated the Good Manufacturing Practice (GMP) regulations. The GMP operates under the legal imprimatur of the FDA regulations, which predate ISO 9000 standards. The FDA regularly inspects medical device manufacturers for their compliance with the GMP requirements. Many of these requirements are quality management system requirements that parallel the subsequently published ISO 9002:1987 requirements. Other GMP regulatory requirements relate more specifically to health, safety, or environmental aspects. Many other nations have similar regulatory requirements for such products.

In the United States, the FDA has created revised GMPs that parallel closely the ISO 9000 standard plus specific regulatory requirements related to health, safety, or the environment. Expanding the scope of ISO 9000 to include quality system requirements related to product design reflects the recognition of the importance of product design and the greater maturity of quality management practices in the medical device industry worldwide. Similar trends are taking place in other nations, many of which are adopting ISO 9001 verbatim for their equivalent of the GMP regulations.

Current Good Manufacturing Practices (cGMPs) for human pharmaceuticals affect every American. Consumers expect that each batch of medicines they take will meet quality standards so that they will be safe and effective. Most people, however, are not aware of cGMPs, or how FDA assures that drug manufacturing processes meet these basic objectives. Recently, FDA has announced a number of regulatory actions taken against drug manufacturers based on the lack of cGMPs.

What Are cGMPs?

cGMP refers to the Current Good Manufacturing Practice regulations enforced by the FDA. cGMPs provide for systems that ensure proper design, monitoring, and control of manufacturing processes and facilities. Adherence to the cGMP regulations ensures the identity, strength, quality, and purity of drug products by requiring that manufacturers of medications adequately control their manufacturing operations. This includes establishing strong quality management systems, obtaining appropriate quality raw materials, establishing robust operating procedures, detecting and investigating product quality deviations, and maintaining reliable testing laboratories. This formal system of controls at a pharmaceutical organization, if adequately put into practice, helps to prevent instances of contamination, mixups, deviations, failures, and errors. This ensures that drug products meet their quality standards.

cGMP requirements were established to be flexible in order to allow each manufacturer to decide individually how to best implement the necessary controls by using scientifically sound design, processing methods, and testing procedures. The flexibility in these regulations allows companies to use modern technologies and innovative approaches to achieve higher quality through continual improvement. Accordingly, the "C" in cGMP stands for "Current," requiring companies to use technologies and systems that are up-to-date in order to comply with the regulations. Systems and equipment that may have been "top-of-the-line" to prevent contamination, mix-ups, and errors 10 or 20 years ago may be less than adequate by today's standards.

It is important to note that cGMPs are minimum requirements. Many pharmaceutical manufacturers are already implementing comprehensive, modern quality systems and risk management approaches that exceed these minimum standards.

Why Are cGMPs Important to Software Development?

A consumer usually cannot detect (through smell, touch, or sight) that a drug product is safe or if it will work. Although cGMPs require testing, testing alone is not enough to ensure quality. In most instances, testing is done on a small sample of a batch (e.g., a drug manufacturer may test 100 tablets from a batch that contains 2 million tablets) so that most of the batch can be used for patients rather than be destroyed by testing. Therefore, it is important that drugs are manufactured under conditions and practices required by cGMP regulations to ensure that quality is built into the design and manufacturing process at every step. Facilities that are in good condition, equipment that is properly maintained and calibrated, employees who are qualified and fully trained, and processes that are reliable and reproducible are a few examples of how cGMP requirements help to ensure the safety and efficacy of drug products.

How Does the FDA Determine if an Organization Is Complying with cGMP Regulations?

The FDA inspects pharmaceutical manufacturing facilities worldwide using scientifically and cGMP-trained individuals whose job it is to evaluate whether the organization is following cGMP regulations. The FDA also relies upon reports of potentially defective drug products from the public and from the industry. The FDA will often use these reports to identify sites for which an inspection or investigation is needed. Most companies that are inspected are found to be fully compliant with the cGMP regulations.

cGMPs. In August 2002, the FDA announced the pharmaceutical cGMPs for the twenty-first Century Initiative. In that announcement, the FDA explained the agency's intent to integrate quality systems and risk management approaches into its existing programs to encourage industry to adopt modern and innovative manufacturing technologies. The cGMP initiative was spurred by the fact that since 1978, when the last major revision of the cGMP regulations was published, there have been many advances in manufacturing science and in our understanding of quality systems. In addition, many pharmaceutical manufacturers are already implementing comprehensive, modern quality systems and risk management approaches. This guidance is intended to help manufacturers implementing modern quality systems and risk management approaches to meet the requirements of the agency's cGMP regulations. The agency also saw a need to harmonize cGMPs with other non-U.S. pharmaceutical regulatory systems and with FDA's own medical device quality systems regulations. This guidance supports these goals. It also supports the objectives of the Critical Path Initiative, which intends to make the development of innovative medical products more efficient so that safe and effective therapies can reach patients sooner.

cGMPs for the twenty-first Century Initiative steering committee created a Quality System Guidance Development working group (QS working group) to compare current cGMP regulations, which call for specific quality management elements, to other existing quality management systems. The QS working group mapped the relationship between cGMP regulations (parts 210 and 211 and the 1978 Preamble to the CGMP regulations and various quality system models, such as the Drug Manufacturing Inspections Program (i.e., systems-based inspectional program), and the Environmental Protection Agency's Guidance for Developing Quality Systems for Environmental Programs, ISO Quality Standards, other quality publications, and experience from regulatory cases. The QS working group determined that, although the cGMP regulations do provide great flexibility, they do not incorporate explicitly all of the elements that today constitute most quality management systems.

cGMP regulations and other quality management systems differ somewhat in organization and in certain constituent elements; however, they are very similar and share some underlying principles. For example, cGMP regulations stress quality control. More recently developed quality systems stress quality management, quality assurance, and the use of risk management tools, in addition to quality control. The QS working group decided that it would be very useful to examine exactly how the CGMP regulations and the elements of a modern, comprehensive quality system fit together in today's manufacturing world. This guidance is the result of that examination.

In ISO, a new technical committee, ISO/TC210, has been formed specifically for medical device systems. TC210 has developed standards that provide supplements to ISO 9001 clauses. These supplements primarily reflect the health, safety, and environment aspects of medical devices and tend to parallel regulatory requirements in various nations.

ISO/TS 16949: Automotive Industry

In the years preceding publication of the 1987 ISO 9000 standards, various original equipment manufacturers (OEMs) in the automotive industry had developed organization-specific proprietary quality system requirements documents. These requirements were part of OEM contract arrangements for purchasing parts, materials, and subassemblies from the thousands of organizations in their supply chain. The OEMs had large staffs of second-party auditors to verify that these OEM-specific requirements were being met.

Upon publication of ISO 9001:1994, the major U.S. OEMs began implementation of an industrywide common standard—QS-9000—that incorporates ISO 9001 verbatim plus industry-specific supplementary requirements. Some of the supplementary requirements are really prescriptive approaches to some of the generic ISO 9001 requirements; others are additional quality system requirements that have been agreed on by the major OEMs; a few are OEM specific.

On December 14, 2006, all QS9000 certifications were terminated. With QS9000, the middle certification between ISO 9001 and ISO/TS 16949 were no longer valid; businesses had a choice between either ISO 9001 or TS16949. QS 9000 is considered to have been superseded by ISO/TS 16949.

ISO/TS 16949:2009, in conjunction with ISO 9001:2008, defines quality management system requirements for design and development, production and, when relevant, installation and service of automotive-related products.

ISO/TS 16949:2009 applies to sites of the organization where customer-specified parts are manufactured for production and/or service.

Supporting functions, whether on-site or remote (such as design centers, corporate headquarters and distribution centers), form part of the site audit as they support the site, but they cannot obtain stand-alone certification to ISO/TS 16949:2009. ISO/TS 16949:2009 can be applied throughout the automotive supply chain.

Computer Software

The global economy has become permeated with electronic information technology (IT). The IT industry now plays a major role in shaping and driving the global economy. As in past major technological advances, the world seems fundamentally very different, and paradoxically, fundamentally the same. Computer software development occupies a central position in this paradox.

First, note that computer software development is not so much an industry as it is a discipline.

Second, many IT practitioners emphasize that computer software issues are complicated by the multiplicity of ways that computer software quality may be critical in a supplier organization's business. For example:

- The supplier's product may be complex software whose functional design requirements are specified by the customer.
- The supplier may actually write most of its software product, or may integrate off-the-shelf packaged software from subsuppliers.
- The supplier may incorporate computer software/firmware into its product, which may be primarily hardware and/or services.
- The supplier may develop and/or purchase from subsuppliers software that will be used in the supplier's own design and/or production processes of its product.

However, it is important to acknowledge that hardware, processed materials, and services often are involved in a supplier organization's business in the same multiple ways.

What, then, are the issues in applying ISO 9001 to computer software development? There is general consensus worldwide that

- The generic quality management system activities and associated requirements in ISO 9001 are relevant to computer software, just as they are relevant in other generic product categories (hardware, other forms of software, processed materials, and services).
- There are some things that are different in applying ISO 9001 to computer software.

There is at this time no worldwide consensus as to which things, if any, are different enough to make a difference and what to do about any things that are different enough to make a difference.

ISO/TC176 developed and published ISO 9000-3:1991 as a means of dealing with this important, paradoxical issue. ISO 9000-3 contains guidelines for applying ISO 9001 to the development, supply, and maintenance of (computer) software and has been useful and widely used. ISO 9000-3 offers guidance that goes beyond the requirements of ISO 9001, and it makes some assumptions about the life cycle model for software development, supply, and maintenance. In the United Kingdom, a separate certification scheme (TickIT) for software development has been operating for several years, using the combination of ISO 9001 and ISO 9003. The scheme has received both praise and criticism from various constituencies worldwide. Those who praise the scheme claim that it:

- Addresses an important need in the economy to provide assurance for customer organizations that the requirements for quality in software they purchase (as a separate product, or incorporated in a hardware product) will be satisfied
- Includes explicit provisions beyond those for conventional certification to ISO 9001 to ensure competency of software auditors, their training, and audit program administration by the certification body.
- Provides a separate certification scheme and logo to exhibit this status publicly. Critics claim that the scheme
  - Is inflexible and attempts to prescribe a particular life cycle approach to computer software development that is out of tune with current best practices for developing many types of computer software.

- Includes unrealistically stringent auditor qualifications in the technology aspects of software development, qualifications whose technical depth is not necessary for effective auditing of management systems for software development.
- Is almost totally redundant with conventional third-party certification to ISO 9001, under which the certification body/registrar already is responsible for competency of auditors. Accreditation bodies verify the competency as part of accreditation procedures.
- Adds substantial cost beyond conventional certification to ISO 9001 and provides little added value to the supply chain.

In the United States, a proposal to adopt a TickIT-like software scheme was presented to the ANSI/RAB (Registrar Accreditation Board) accreditation program. The proposal was rejected, primarily on the basis that there was not consensus and support in the IT industry and the IT-user community.

CMMI: Software and Systems Development

Another standard that has gained popularity is the Capability Maturity Model (CMM), a service mark owned by Carnegie-Mellon University (CMU) and refers to a development model elicited from actual data. The data were collected from organizations that contracted with the U.S. Department of Defense, which funded the research, and that became the foundation from which CMU created the Software Engineering Institute (SEI). Like any model, SEI is an abstraction of an existing system. Unlike many models that are derived from academia, this model is based on observation rather than on theory.

When it is applied to an existing organization's software development processes SEI allows an effective approach toward improving them. Eventually, it became clear that this model could be applied to other processes. This gave rise to a more general concept that is applied to business processes and to developing people.

CMM was originally developed as a tool for objectively assessing the ability of the processes of government contractors to perform a contracted software project. CMM is based on the process maturity framework first described in the 1989 book, *Managing the Software Process* by Watts Humphrey. It was later published in a report in 1993 (Technical Report CMU/SEI-93-TR-024 ESC-TR-93-177 February 1993, Capability Maturity Model SM for Software, Version 1.1) and as a book authored by Xiaoqing Liu et al. in 1995.

Although CMM comes from the field of software development, it is used as a general model to aid in improving organizational business processes in diverse areas; for example, in software engineering, system engineering, project management, software maintenance, risk management, system acquisition, information technology (IT), services, business processes generally, and human capital management. The CMM has been used extensively worldwide in government, commerce, industry and software development organizations.

An organization may be assessed by an SEI-authorized lead appraiser and would then be able to claim that they have been assessed as CMM level X, where X is from 1 to 5 (maturity levels). Maturity Level 1 is Initial; Maturity Level 2 is Managed; Maturity Level 3 is Defined; Maturity Level 4 is Quantitatively Managed; and Maturity Level 5 is Optimizing (read further for more explanation of the levels). Although sometimes called CMM certification, the SEI does not use this term due to certain legal implications.

In the 1970s, the use of computers became more widespread, more flexible, and less expensive. Organizations began to adopt computerized information systems, and the demand for software development grew significantly. The processes for software development were in their infancy, with few standard or "best practice" approaches defined.

As a result, growth was accompanied by growing pains: project failure was common, the field of computer science was still in its infancy, and the ambitions for project scale and complexity exceeded market capability to deliver.

In the 1980s, several U.S. military projects involving software subcontractors ran over budget and were completed much later than planned, if they were completed at all. In an effort to determine the reason for this, the U.S. Air Force funded a study at the SEI.

The Standard CMMI Appraisal Method for Process Improvement (SCAMPI) is the official SEI method that provides benchmark-quality ratings relative to CMMI models. The CMMI model is used as a "ruler" to measure an organization's process definition, as the model is a collection of process best practices assimilated into process areas. The SCAMPI appraisal methodology is used to measure how well an organization has institutionalized the process definition into their everyday way of doing business. SCAMPI appraisals are used to identify strengths and weaknesses of current processes, reveal development/acquisition risks, and determine capability and maturity level ratings. These appraisals are mostly used either as part of a process improvement program or for rating prospective suppliers. The appraisal method consists of preparation; on-site activities; preliminary observations, findings, and ratings; final reporting; and follow-up activities.

Active development of the model by SEI began in 1986 when Watts Humphrey joined the SEI at Carnegie Mellon University in Pittsburgh, Pennsylvania, after retiring from IBM. At the request of the U.S. Air Force he began formalizing his Process Maturity Framework to aid the U.S. Department of Defense in evaluating the capability of software contractors as part of awarding contracts.

In the United States and other nations, the compatibility of the ISO 9000 standard, and the ISO 14000 standard is one part of the standardization job. Implementation requires that similar harmonization and compatibility be established in each nation in the infrastructure of accreditation bodies, certification/registration bodies, and auditor certification bodies, operating under internationally harmonized guidelines. As of this writing, the ISO 14000 infrastructure is in its infancy. However, Humphrey's approach differed because of his unique insight that organizations mature their processes in stages based on solving process problems in a specific order. Humphrey based his approach on the staged evolution of a system of software development practices within an organization, rather than measuring the maturity of each separate development process independently. CMM has thus been used by different organizations as a general and powerful tool for understanding and then improving general business process performance.

The CMM model proved useful to many organizations, but its application in software development has sometimes been problematic. Applying multiple models that are not integrated within and across an organization could be costly in terms of training, appraisals, and improvement activities. The CMMI project was formed to sort out the problem of using multiple CMMs.

For software development processes, CMM has been superseded by CMMI, though the CMM continues to be a general theoretical process capability model used in the public domain.

What Is the Capability Maturity Model?

A maturity model can be used as a benchmark for comparison and as an aid to understanding; for example, for comparative assessment of different organizations where there is something in common that can be used as a basis for comparison. In the case of CMM, the basis for comparison would be the organizations' software development processes.

CMM involves the following aspects:

Maturity levels. A five-level process maturity continuum where the uppermost (fifth) level is a notional ideal state, where processes would be systematically managed by a combination of process optimization and continuous process improvement.

Key process areas. A key process area (KPA) identifies a cluster of related activities that, when performed collectively, achieve a set of goals considered important.

Goals. The goals of a KPA summarize the states that must exist for that KPA to have been implemented in an effective and lasting way. The extent to which the goals have been accomplished indicates how much capability the organization has established at that maturity level. The goals signify the scope, boundaries, and intent of each KPA.

Common features. Common features include practices that implement and institutionalize a KPA. There are five types of common features: Commitment to Perform, Ability to Perform, Activities Performed, Measurement and Analysis, and Verifying Implementation.

Key practices. Key practices describe the elements of infrastructure and practice that contribute most effectively to the implementation and institutionalization of KPAs.

Levels of the CMM

There are five levels defined along the continuum of the CMM[9], and, according to the SEI: "Predictability, effectiveness, and control of an organization's software processes are believed to improve as the organization moves up these five levels. While not rigorous, the empirical evidence to date supports this belief." The five levels are

Level 1: Chaos or Ad hoc

It is characteristic of processes at this level that they are (typically) undocumented and in a state of dynamic change, tending to be driven in an ad hoc, uncontrolled, and reactive manner by users or events. This provides a chaotic or unstable environment for the processes.

Level 2: Repeatable

It is characteristic of processes at this level that some processes are repeatable, possibly with consistent results. Process discipline is unlikely to be rigorous, but where it exists, it may help to ensure that existing processes are maintained during times of stress.

Level 3: Defined

It is characteristic of processes at this level that there are sets of defined and documented standard processes established and subject to some degree of improvement over time. These standard processes are in place (i.e., they are the as-is processes) and are used to establish consistency of process performance across the organization.

Level 4: Managed

It is characteristic of processes at this level that, using process metrics, management can effectively control the as-is process (e.g., for software development). In particular, management can identify ways to adjust and adapt the process to particular projects without measurable losses of quality or deviations from specifications. Process capability is established from this level.

Level 5: Optimized

It is a characteristic of processes at this level that the focus is on continually improving process performance through both incremental and innovative technological changes/improvements.

At maturity level 5, processes are concerned with addressing statistical common causes of process variation and changing the process (e.g., to shift the mean of the process performance) to improve process performance. This would be done at the same time as maintaining the likelihood of achieving the established quantitative process-improvement objectives.

Within each of these maturity levels are key process areas (KPAs) which characterize that level, and there are five definitions identified for each KPA:

1. Goals
2. Commitment
3. Ability
4. Measurement
5. Verification

CMM provides a theoretical continuum along which process maturity can be developed incrementally from one level to the next. Skipping levels is not allowed.

CMM was originally intended as a tool to evaluate the ability of government contractors to perform a contracted software project. It has been used for and may be suited to that purpose, but critics pointed out that process maturity according to CMM was not necessarily mandatory for successful software development. There were/are real-life examples where the CMM was arguably irrelevant to successful software development, and these examples include many companies (also called commercial-off-the-shelf or COTS firms or software package firms). Such firms would have included, for example, Claris, Apple, Symantec, Microsoft, and IBM Lotus. Though these companies may have successfully developed their software, they would not necessarily have considered or defined or managed their processes as the CMM described as level 3 or above and so would have fitted levels 1 or 2 of the model. On the face of it, this did not impede the successful development of their software.

AS9100: Aerospace Standards

AS9100 is a widely adopted and standardized quality management system for the aerospace industry. It was released in October 1999, by the Society of Automotive Engineers and the European Association of Aerospace Industries. AS9100 replaces the earlier AS9000 and fully incorporates the entirety of the current version of ISO 9000 while adding additional requirements relating to quality and safety. Major aerospace manufacturers and suppliers worldwide require compliance and/or registration to AS9100 as a condition of doing business with them.

Prior to the adoption of an aerospace specific quality standard, various corporations typically used ISO 9000 and their own complementary quality documentation/requirements, such as the Boeing Corporation's D1-9000 or the automotive Q standard. This created a patchwork of competing requirements that were difficult to enforce and/or comply with. The major American aerospace manufacturers combined their efforts to create a single, unified quality standard, resulting in AS9000. Upon the release of AS9000, companies such as Boeing stopped using their previous quality supplements in preference to compliance to AS9000.

During the rewrite of ISO 9000 for the year 2000 release, the AS group worked closely with the ISO organization. As the year 2000 revision of ISO 9000 incorporated major organizational and philosophical changes, AS9000 underwent a rewrite as well. It was released as AS9100 to the international aerospace industry at the same time as the new version of ISO 9000. AS9100 Revision C was released in January, 2009

Standardization Is Here to Stay

Standards are here to stay. Many industries are working together with various standards bodies to periodically improve their standards and mandate as many systems as possible to ensure the safety and quality of our products. For instance, a new International Standard—ISO 31000:2009, *Risk Management—Principles and Guidelines*—was developed to help organizations manage risk effectively. ISO 31000 provides principles, framework, and a process for managing any form of risk in a transparent, systematic, and credible manner within any scope or context.

In addition, the ISO has published a standard to facilitate implementation of quality management systems, based on ISO 9001:2000, by the medical device industry. The key objectives of ISO 13485:2003 are to maximize the probability that a medical device organization will meet regulatory quality management system requirements worldwide, will provide safe and effective medical devices, and will meet customer requirements, —Ed Kimmelman, convener of the working group that developed the new standard.

ISO 13485:2003, Medical Devices—Quality Management Systems—Requirements for Regulatory Purposes, is based on quality management system requirements currently contained in medical device regulations around the world as well as those appropriate requirements contained in ISO 9001:2000. The new standard is used by organizations involved in the design, production, installation, and servicing of medical devices as well as in the design, development, and provision of related services. It can also be used by external certification bodies to assess an organization's ability to meet requirements. The new standard, which replaces ISO 13485:1996, is the work of ISO technical committee ISO/TC 210, Quality Management and Corresponding General Aspects for Medical Devices, working group WG 1, Application of Quality Systems to Medical Devices, in conjunction with members of the Global Harmonization Task Force (Study Group 3), conceived in 1992 in an effort to achieve greater uniformity between national medical device regulatory systems.

> Standardization will embrace common operational sequences, part-dimensional strategies, and guidelines for equipment use.
>
> —Barney and De Feo (2005)

Standards are becoming a way of life for global organizations. Get ready; there is more to come.

References

Barney M., and De Feo, J. A. (2008). "The Future of Manufacturing. Quality Digest, Chico CA."

Carnegie Mellon Software Engineering Institute. (2006). CMMI Executive Overview." Sponsored by U.S. Department of Defense, Carnegie Mellon University. Pittsburgh, PA.

Department of Health and Human Services (2004). "Final Report: Pharmaceutical CGMPS for the 21st Century—Risk-Based Approach." U.S. Food and Drug Administration, September.

Gasiorowski-Denis, E. (2003). Quality Management Systems for Medical Device Industry, ISO Management Systems, November-December 2003, http://www.iso.org/iso/medical_device_ims6_2003.pdf.

Humphrey, W. S. (1989). Managing for Software Process, Addison-Wesley, Reading, MA.

International Standard ISO 9001. (2008). 4th ed., November 15, 2008, www.iso.org

Liu, Xiaoqing. (1993). Technical Report CMU/SEI-93-TR-024 ESC-TR-93-177, Capability Maturity Model SM for Software, Version 1.1.

Software Engineering Institute. (2007). CMMI Version 1.2, Overview, Carnegie Mellon University, Pittsburgh, PA.

U. N. Secretariat, Geneva. Also reprinted in a number of other countries and languages, and in the *ISO 9000 Handbook*, op cit., chap. 11. *The ISO 9000 Family of International Standards* 11.27 034003-x_Ch11_Juran 3/6/00 1:38 PM, page 27 034003-x_Ch11_Juran 3/6/00 1:38 PM, page 28.

CHAPTER 17

Using National Awards for Excellence to Drive and Monitor Performance

James Er Ralston

About This Chapter

This chapter explores the history and structure of quality awards, including a more thorough discussion of some of the most influential awards like the Malcolm Baldrige National Quality in the United States, the EFQM Excellence Award, and the Deming Prize of Japan. Many other national quality awards around the world are discussed in some detail with a list of references. Although many quality awards are nationally based and sponsored, this chapter also explores the outgrowth to other quality awards including those sponsored by regions, states, and individual organizations. Finally, the chapter discusses how organizations are using these awards as a basis for assessment and continuous improvement to attain superior organizational results. We also compare these awards to other approaches and programs including ISO 9000, Lean, and Six Sigma programs.

High Points of This Chapter

1. There is evidence that organizational improvement can be advanced by adopting quality and performance excellence award models and criteria as a means for pursuing excellence.
2. Quality and performance excellence awards programs have grown rapidly over the past 30 years. Many national, regional, and local awards programs are in place to benefit their served organizations.
3. To sustain quality leadership performance, many organizations use quality and performance excellence awards in cycles of improvement by conducting periodic assessments to identify and prioritize the vital few opportunities for improvement.
4. Many organizations have found synergistic benefits when using quality and performance excellence awards programs, along with international standards and/or accreditation agencies.
5. In recent years, many quality and performance excellence award recipients have acknowledged use of Lean and Six Sigma tools and methodologies as facilitators for achieving high levels of performance.

Over the last two decades, national quality and performance excellence awards have become a major influence on attaining superior organization results. In fact, one could argue that had it not been for the recognition Motorola, Inc. received as one of the first recipients of the Malcolm Baldrige National Quality Award, it is possible that the practice of Six Sigma methodology may have never gained the level of exposure and practice that it has today.

National quality awards in particular have greatly influenced the way many organizations manage their systems for quality and performance excellence. The prestige of winning a national quality award has provided increased recognition to the award recipients' role model approaches, and created additional incentives for many organizations to apply for the award and stick to their approaches. However, the real benefit stated by many is not the award itself but the organizational improvement that can occur from adopting award models and criteria as a means for pursuing excellence.

Most quality award programs include the following components:

Application Criteria

The application process consists of questions that seek information on the approaches used and the results achieved by the organization. The Baldrige criteria, for example, contain over 100 questions which are categorized into seven sections or categories. Many organizations have found tremendous benefit in conducting self-assessments using the award criteria without even submitting the application into the awards process. For example, more than one million copies of the criteria are downloaded from the Baldrige website each year.

Scoring System

A scoring system provides a means for determining the level of quality leadership and maturity attained by an organization. Some scoring systems provide component scores to aid organizations in pinpointing significant strengths or weakness. For example, organizations might find that they have systematic approaches but that they have not improved these approaches over the years.

Examination and Judging Process

This defines how organizations are evaluated and how award recipients are selected. Many awards processes have several stages of evaluation whereby advancement in the process is an indicator of the level of quality leadership. In the Baldrige Awards process, for example, organizations that receive a site visit are sometimes referred to as "semifinalists," representing the final group of organizations from which the award recipients are selected.

History of the National Quality Awards

The Deming Prize, introduced in Japan in the 1950s, was one of the first National Quality Awards to achieve acclaim. More than 30 years later, the introduction of the Malcolm Baldrige National Quality Award in the United States and the European Quality Award in 1988 generated much attention, not just in the United States and Europe, but also worldwide. Since then, national quality awards have sprung up all around the world. Regional, state, local, and internal organization quality awards have followed, most patterned after the structure of the Baldrige Award or the European Quality Award (now known as the EFQM Excellence Award).

Since its introduction in 1988, over 1300 U.S. organizations have applied for the Baldrige Award. In the first year of the Baldrige Award, 66 organizations applied for the award. Over the past 20 years, the number of Baldrige Award applicants has not fluctuated greatly with a peak of 106 applicants in 1991 to a low of 26 applicants in the 1997. The increase in applicants since 1997 can be attributed largely to the expansion of eligibility to new sectors including education and health care in 1999 and to nonprofits (including government entities) in 2006 (see Figures 17.1 and 17.2 and Table 17.1).

The Baldrige Program also reports numbers on the number of applicants for other quality awards in the United States through state, local, and regional awards. Reporting began in 1991 with a total of 217 applicants. The number grew steadily for the next several years as more state, regional, and local programs came on board, reaching a peak of 1015 in 1999.

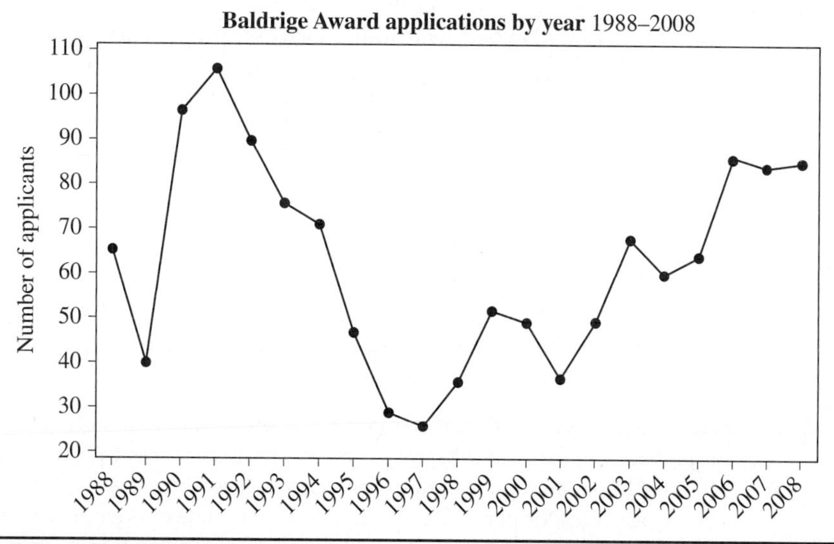

FIGURE 17.1 Total Baldrige Award applications by year. (*Baldrige National Quality Program and the Alliance for Performance Excellence.*)

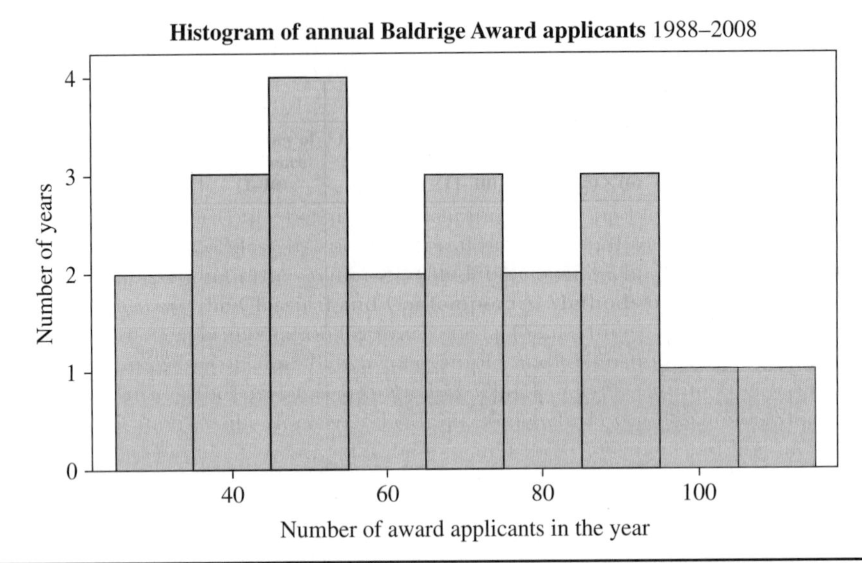

FIGURE 17.2 Histogram of annual Baldrige Award applicants. (*Baldrige National Quality Program and the Alliance for Performance Excellence.*)

Since then, the numbers have declined with just 167 applications reported in 2008. This decline can be attributed to the discontinuance of some of these programs along with an overall reduction in the number of applicants for continuing programs (see Figure 17.3).

When Harry Hertz, director of the Baldrige program, was contacted for input on this edition of the *Juran's Quality Handbook*, he responded that the key component of this chapter "would be to trace the evolution of Award Criteria, as an indication of the evolving definition of quality." The evolution he is referring to is reflected in the continued revision of the criteria for various quality awards over the past 20 years. All of the award programs have advanced their criteria far beyond their initial focus on product and service quality, and some of the awards have changed their names accordingly. For example, the European Quality Award is now the EFQM Excellence Award, reflecting a broader expansion of the awards focus beyond traditional views of product and service quality to the overall quality of performance in organizational processes and results.

The Baldrige criteria were first introduced in 1988 with a specific focus on product and service quality, primarily in the manufacturing sector. The criteria have evolved significantly since then to a comprehensive focus on overall organizational competitiveness and sustainability. Like the criteria of today, the original criteria had seven categories, although the categories have changed significantly since then (see Table 17.2).

A review of the criteria during the subsequent years highlights some of the more meaningful changes.

1991 Same categories as 1988; new item for Competitive Comparisons and Benchmarks, Employee Well-being and Morale, Business Process, and Support Service Quality

1995 More focus on strategic business drivers; increased emphasis on financial data in setting priorities for performance improvement; increased emphasis on workforce development; more emphasis on continuous learning; process focus includes support processes

Year		Manufacturing	Service	Small Business	Education	Health Care	Non Profit	Total Apps	Total Stage 2	Total Stage 3	Total Awards
1988	Applications	45	9	12				66			
	Consensus										
	Site Visits	10	2	1						13	
	Awards	2	0	1							3
1989	Applications	23	6	11				40			
	Consensus										
	Site Visits	8	2	0						10	
	Awards	2	0	0							2
1990	Applications	45	18	34				97			
	Consensus								26		
	Site Visits	6	3	3						12	
	Awards	2	1	1							4
1991	Applications	38	21	47				106			
	Consensus								28		
	Site Visits	9	5	5						19	
	Awards	2	0	1							3
1992	Applications	31	15	44				90			
	Consensus								30		
	Site Visits	7	5	5						17	
	Awards	2	2	1							5

TABLE 17.1 Summary of Annual Baldrige Award Applications by Sector and Number Information Provided by the Baldrige National Quality Program at the National Institute of Standards and Technology, United States Department of Commerce (*Continued*)

Year		Manufacturing	Service	Small Business	Education	Health Care	Non Profit	Total Apps	Total Stage 2	Total Stage 3	Total Awards
1993	Applications	32	13	31				76			
	Consensus								29		
	Site Visits	4	5	4						13	
	Awards	1	0	1							2
1994	Applications	23	18	30				71			
	Consensus								29		
	Site Visits	6	5	3						14	
	Awards	0	2	1							3
1995	Applications	18	10	19	191	461		471			
	Consensus								27		
	Site Visits	7	4	2						13	
	Awards	2	0	0							2
1996	Applications	13	6	10				29			
	Consensus	10	5	7					22		
	Site Visits	5	2	2						9	
	Awards	1	1	2							4
1997	Applications	9	7	10				26			
	Consensus	8	5	5					18		
	Site Visits	5	3	3						11	
	Awards	2	2	0							4
1998	Applications	15	5	16				36			
	Consensus	11	3	7					21		
	Site Visits	5	1	3						9	
	Awards	2	0	1							3

1999[2]	Applications	4	11	12	16[2]	9[2]		52			
	Consensus	3	3	5	5	2			18		
	Site Visits	3	2	2	2	1				10	
	Awards	1	2	1	0	0					4
2000	Applications	14	5	11	11	8		49			
	Consensus	7	3	5	5	3			23		
	Site Visits	4	1	2	1	1				9	
	Awards	2	1	1	0	0					4
2001	Applications	7	4	8	10	8		37			
	Consensus	7	2	3	7	4			23		
	Site Visits	2	2	3	4	2				13	
	Awards	1	0	1	3	0					5
2002	Applications	8	3	11	10	17		49			
	Consensus	5	2	6	4	10			27		
	Site Visits	2	2	3	0	4				11	
	Awards	1	0	1	0	1					3
2003	Applications	10	8	12	19	19		68			
	Consensus	6	8	3	7	12			36		
	Site Visits	3	3	2	2	3				13	
	Awards	1	2	1	1	2					7
2004	Applications	8	5	8	17	22		60			
	Consensus	6	4	3	6	15			34		
	Site Visits	3	2	2	2	4				13	
	Awards	1	0	1	1	1					4

Table 17.1 Summary of Annual Baldrige Award Applications by Sector and Number Information Provided by the Baldrige National Quality Program at the National Institute of Standards and Technology, United States Department of Commerce (*Continued*)

Year		Manufacturing	Service	Small Business	Education	Health Care	Non Profit	Total Apps	Total Stage 2	Total Stage 3	Total Awards
2005	Applications	1	6	8	16	33		64			
	Consensus	1	3	3	8	21			36		
	Site Visits	1	1	2	3	7				14	
	Awards	1	1	1	2	1					6
2006	Applications	3	4	8	16	45	10[3]	76[3]			
	Consensus	2	2	3	10	22	10		39[4]		
	Site Visit	0	1	3	3	6	2			134	
	Awards	0	1	1	0	1					3
2007	Applications	2	4	7	16	42	13	84			
	Consensus[5]	2	4	7	16	42	12		83		
	Site Visits	0	0	2	1	7	4			14	
	Awards	0	0	1	0	2	2				5
2008	Applications	3	5	7	11	43	16	85			
	Consensus[5]	3	5	7	11	43	16		85		
	Site Visits	1	0	2	2	7	1			13	
	Awards	1	0	0	1	1	0				3
Total	Applications							1309			
	Consensus								634		
	Site Visits									263	
	Awards										79

1. 1995 Health Care and Education Pilot Program—these numbers are not included in the total applicants for 1995.
2. 1999 was the 1st year that education and health care organizations were eligible to apply for the Baldrige Award.
3. 2006 Nonprofit Pilot Program—these numbers are not included in the total applicants for 2006.
4. Does not include the 10 consensus/2 site visit reviews from the Nonprofit Pilot.
5. Beginning in 2007, all applicants received a consensus review.

Table 17.1 Summary of Annual Baldrige Award Applications by Sector and Number Information Provided by the Baldrige National Quality Program at the National Institute of Standards and Technology, United States Department of Commerce

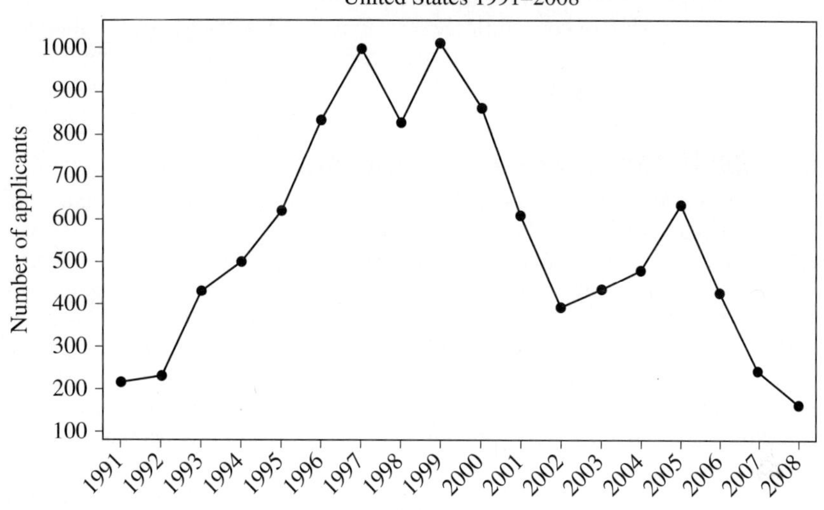

FIGURE 17.3 Total Baldrige-based applications by year for the United States. (*Baldrige National Quality Program and theAlliance for Performance Excellence.*)

1988 Baldrige Categories	2009 Baldrige Categories
Leadership	Leadership
Information and analysis	Strategic planning
Strategic quality planning	Customer focus
Human resource utilization	Measurement, analysis, and knowledge management
Quality assurance of products and services	Workforce focus
Results from quality assurance of products and services	Process management
Customer satisfaction	Results

TABLE 17.2 Comparison of 1988 and 2009 Baldrige Criteria Categories

1999 More comprehensive coverage of strategy-driven performance, addressing the needs of all stakeholders; greater focus on the systems view of performance management and the alignment of key components; increased focus on organizational and employee learning and knowledge sharing

2000 Revised the core values and added two new values: Managing for Innovation and Systems Perspective

2001 Added the Organizational Profile to provide better organizational perspective; increased use of internet and e-commerce led to revisions in Criteria

2005 Increased focus on leadership responsibilities for ethical stewardship and good governance; added focus on financial planning and stewardship of resources for support processes

2009 Increased focus on customer engagement, and the ability to identify and deliver relevant product offerings to customers now and in the future; probing of the relationship of core competencies to mission, strategy, and sustainability; probing the contribution to societal responsibilities, including environmental, social, and economic systems

The Malcolm Baldrige National Quality Award

The Malcolm Baldrige National Quality Award recognizes U.S. organizations for their achievements in quality and business performance. The award also raises awareness about the importance of quality and performance excellence as a competitive edge.

To receive an award, an organization must have a role-model organizational management system that ensures continuous improvement in the delivery of products and/or services, demonstrate efficient and effective operations, and provide a way of satisfying and responding to stakeholders. Awards are presented annually in six categories of eligibility: manufacturing, service, small business, education, health care, and nonprofit.

The award is named after Malcolm Baldrige, who was U.S. Secretary of Commerce from 1981 until his death in a rodeo accident in July 1987. Baldrige was a proponent of quality management as a key to the country's prosperity and long-term strength. In recognition of his contributions, the U.S. Congress named the award in his honor.

Creation of the Malcolm Baldrige National Quality Award

During the 1980s, there was a growing interest in the United States in promoting total quality. Many U.S. leaders felt that a national quality award, similar to the Deming Prize of Japan, would help stimulate the quality efforts of U.S. organizations.

Several individuals and organizations proposed such an award, leading to a series of hearings before the House of Representatives Subcommittee on Science, Research, and Technology. Dr. Joseph M. Juran was one of the persons who testified in the hearings regarding the potential benefits of such an award. Finally, on January 6, 1987, the Malcolm Baldrige National Quality Improvement Act of 1987 was passed. The act was signed by President Ronald Reagan on August 20, 1987 and became Public Law 100-107. This act provided for the establishment of the Malcolm Baldrige National Quality Award Program. The purpose of this program was to help improve quality and productivity by doing the following (House Resolution 812, U.S. Congress):

- Helping stimulate American organizations to improve quality and productivity for the pride of recognition while obtaining a competitive edge through increased profits
- Recognizing the achievements of those organizations that improve the quality of their goods and services and provide an example to others
- Establishing guidelines and criteria that can be used by business, industrial, governmental, and other organizations in evaluating their own quality improvement efforts
- Providing specific guidance for other American organizations that wish to learn how to manage for high quality by making available detailed information on how winning organizations were able to change their cultures and achieve eminence

The Baldrige National Quality Program is part of the National Institute of Standards and Technology (NIST), an agency of the U.S. Department of Commerce. The American Society for Quality (ASQ) assists in administering the award program under contract to NIST (Baldrige criteria).

A Board of Overseers advises the Department of Commerce on the Baldrige National Quality Program. The board is appointed by the Secretary of Commerce and consists of distinguished leaders from all sectors of the U.S. economy. Dr. Joseph M. Juran was a member of the original Board of Overseers.

> The Board of Overseers evaluates all aspects of the Program, including the adequacy of the Criteria and processes for determining Award recipients. An important part of the board's responsibility is to assess how well the Program is serving the national interest. Accordingly, the board makes recommendations to the Secretary of Commerce and to the Director of NIST regarding changes and improvements in the Program.
>
> The Board of Examiners evaluates Award applications and prepares feedback reports. The Panel of Judges, part of the Board of Examiners, makes Award recommendations to the Director of NIST. The board consists of leading experts from U.S. businesses and education, health care, and nonprofit organizations. NIST selects members through a competitive application process. For 2008, the board of examiners consisted of about 570 members. Of these, 12 (who are appointed by the Secretary of Commerce) served as Judges, and approximately 100 served as Senior Examiners. The remainder served as Examiners. All members of the board must take part in an Examiner Preparation Course.
>
> The Foundation for the Malcolm Baldrige National Quality Award was created to foster the success of the Program. The Foundation's main objective is to raise funds to permanently endow the Award Program. Prominent leaders from U.S. organizations serve as Foundation Trustees to ensure that the Foundation's objectives are accomplished. A broad cross-section of organizations throughout the United States provides financial support to the Foundation. (Baldrige criteria)

Criteria for the Malcolm Baldrige National Quality Award

The Baldrige performance excellence criteria are a framework that any organization can use to improve overall performance. For 2009–2010, there are three different criteria for the various segments served by the awards program:

2009–2010	Criteria for Performance Excellence (referred to as the Business/Nonprofit Criteria)
2009–2010	Education Criteria for Performance Excellence
2009–2010	Health Care Criteria for Performance Excellence

A copy of the Baldrige criteria can be downloaded for free from the Baldrige Program website at <http://www.quality.nist.gov/Criteria.htm>. Individual print copies of the Criteria can be obtained free of charge by contacting the Baldrige National Quality Program. Bulk orders can be purchased through the American Society for Quality (ASQ), P.O. Box 3005, Milwaukee, WI 53201-3005 or 600 North Plankinton Avenue Milwaukee, WI 53203.

Each version of the Criteria is customized for the respective sector of focus. For the purpose of simplicity, the following explanation and description focuses on the Business/Nonprofit Criteria. The seven categories that making up the award criteria are connected and integrated as depicted in Figure 17.4.

The framework consists of three main sections:

1. The Organizational Profile at the top establishes the context for the way an organization operates. The system operations are in the center and are composed of six Baldrige categories that define the operations and the results achieved.
2. Three categories (Leadership, Strategic Planning, and Customer Focus) represent the leadership triad. These categories are placed together to emphasize the importance of a leadership focus on strategy and customers. Three more categories (Workforce Focus, Process Management, and Results) make up the results triad. This signifies that an organization's workforce and key processes accomplish the work of the organization that yields the overall performance results.

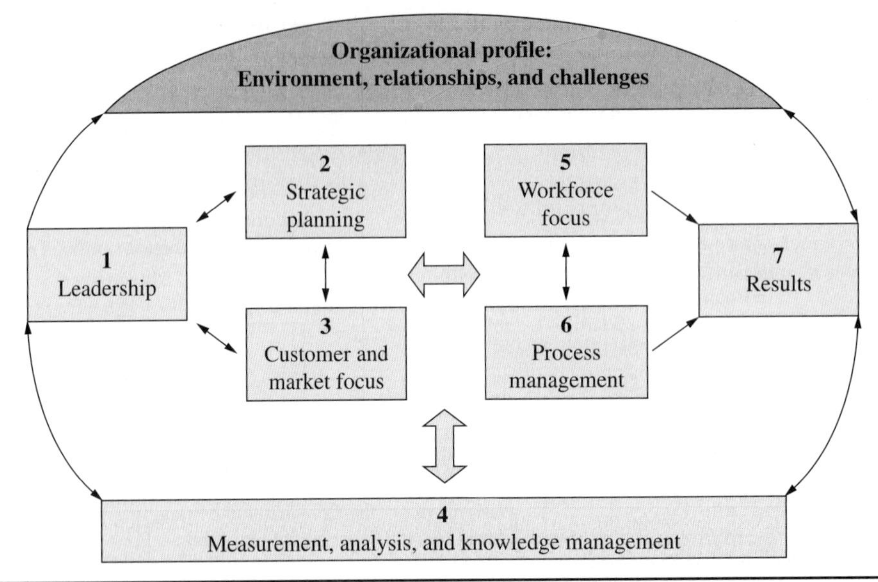

FIGURE 17.4 Framework for Baldrige Criteria for Performance Excellence. (*Baldrige National Quality Program.*)

3. The system foundation is shown at the bottom of the frame, composed of the Baldrige Category for Measurement, Analysis, and Knowledge Management, signifying the importance of this category to the effective management of the organization and to a fact-based knowledge-driven system for improving performance and competitiveness.

Categories of the Baldrige Award

- *Leadership.* Examines how senior executives guide the organization and how the organization addresses its responsibilities to the public and practices good governance and citizenship.
- *Strategic planning.* Examines how the organization sets strategic directions and how it develops key action plans.
- *Customer and market focus.* Examines how the organization engages its customers for long-term marketplace success; builds a customer-focused culture; and listens to the voice of its customers and uses this information to improve and identify opportunities for innovation.
- *Measurement, analysis, and knowledge management.* Examines the management, effective use, analysis, and improvement of data and information to support key organization processes and the organization's performance management system.
- *Workforce focus.* Examines how the organization engages, manages, and develops its workforce to develop its full potential and how the workforce is aligned with the organization's objectives.
- *Process management.* Examines aspects of how work systems are designed and how key work processes are designed, managed, and improved.
- *Results.* Examines the organization's performance and improvement in its key business areas: product and service, customer-focused, financial and marketplace,

workforce-focused, process effectiveness, and leadership. The category also examines how the organization performs relative to competitors and other organizations with similar product offerings.

Within each category of the Baldrige criteria are the requirements and questions that are used as part of an assessment or an awards application. There are 18 criteria, as seen in Figure 17.5.

There are 36 areas to address contained within the 18 criteria. The numerous requirements are expressed as individual criteria questions. The format for the criteria is shown in Figures 17.6 and 17.7.

CRITERIA FOR PERFORMANCE EXCELLENCE—ITEM LISTING

P	**Preface: Organiztional Profile**		
	P.1 Organizational description		
	P.2 Organizational situation		
Categories and Items			**Point values**
1	**Leadership**		**120**
	1.1 Senior leadership	**70**	
	1.2 Governance and societal responsibilities	**50**	
2	**Strategic Planning**		**85**
	2.1 Strategy development	**40**	
	2.2 Strategy deployment	**45**	
3	**Customer Focus**		**85**
	3.1 Customer engagement	**40**	
	3.2 Voice of the customer	**45**	
4	**Measurement, Analysis, and Knowledge Management**		**90**
	4.1 Measurement, analysis, and improvement of organizational performance	**45**	
	4.2 Management of information, knowledge, and information technology	**45**	
5	**Workforce Focus**		**85**
	5.1 Workforce engagement	**45**	
	5.2 Workforce environment	**40**	
6	**Process Management**		**85**
	6.1 Work systems	**35**	
	6.2 Work processes	**50**	
7	**Results**		**450**
	7.1 Product outcomes	**100**	
	7.2 Customer-focused outcomes	**70**	
	7.3 Financial and market outcomes	**70**	
	7.4 Workforce-focused outcomes	**70**	
	7.5 Process effectiveness outcomes	**70**	
	7.6 Leadership outcomes	**70**	
	TOTAL POINTS		**1,000**

FIGURE 17.5 Item listing from the Baldrige Criteria for Performance Excellence, including total point values per item and category. (*Baldrige National Quality Program.*)

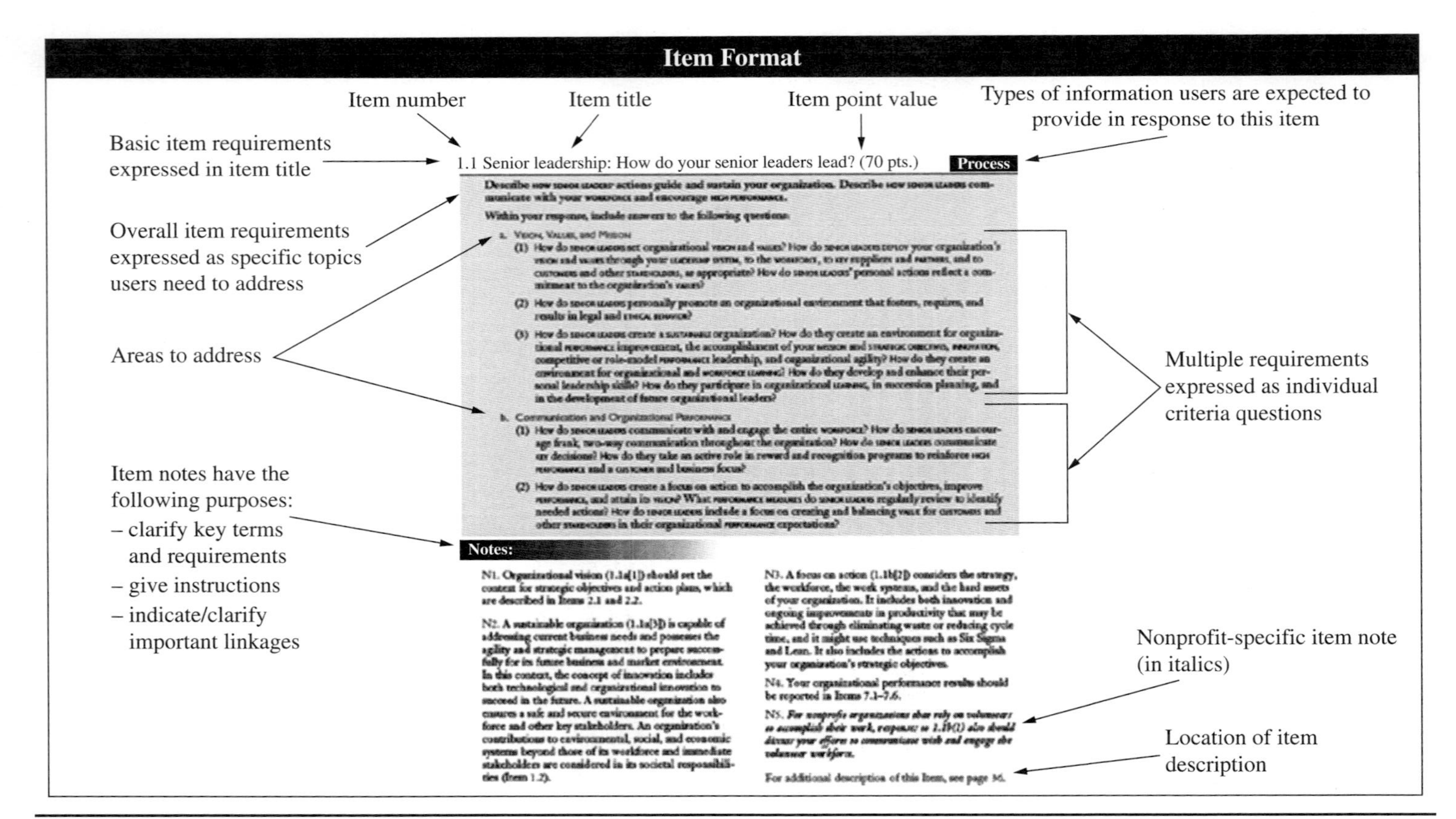

FIGURE 17.6 Example of item criteria, including areas to address and multiple requirements. (*Baldrige National Quality Program.*)

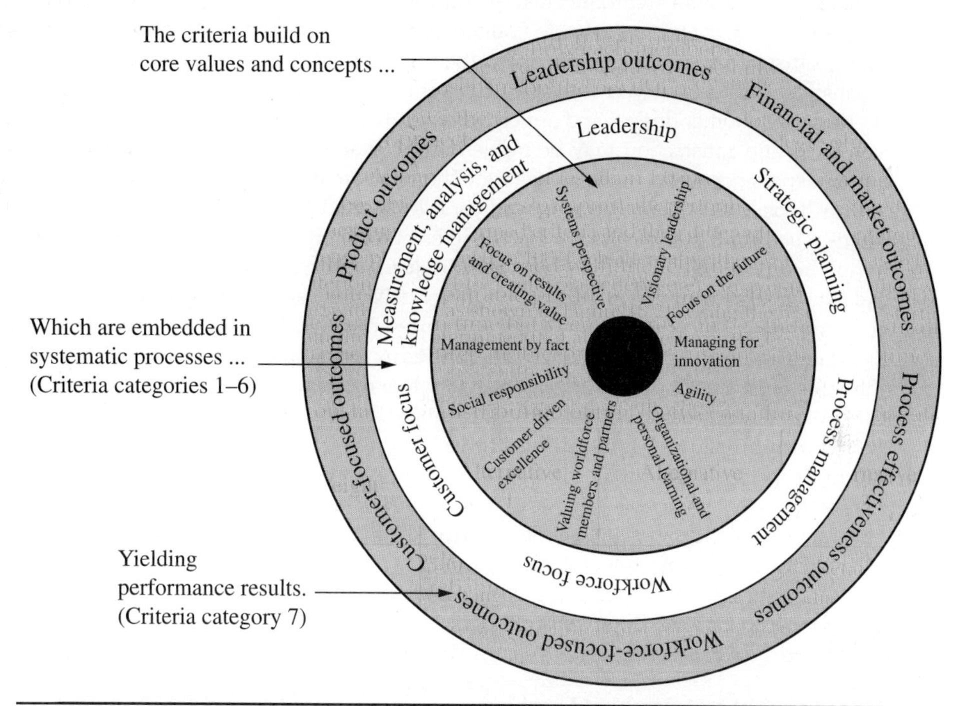

Figure 17.7 Role of core values and concepts in the Baldrige criteria. (*Baldrige National Quality Program Criteria for Performance Excellence.*)

Malcolm Baldrige National Quality Award Core Values

There are 11 core values and concepts embodied in the award criteria.

Visionary Leadership

Your organization's senior leaders should set directions and create a customer focus, clear and visible values, and high expectations. These directions, values, and expectations should balance the needs of all your stakeholders. Your leaders should ensure the creation of strategies, systems, and methods for achieving performance excellence, stimulating innovation, building knowledge and capabilities, and ensuring organizational sustainability. The defined values and strategies should help guide all of your organization's activities and decisions. Senior leaders should inspire and encourage your entire workforce to contribute, to develop and learn, to be innovative, and to embrace change. Senior leaders should be responsible to your organization's governance body for their actions and performance. The governance body should be responsible ultimately to all your stakeholders for the ethics, actions, and performance of your organization and its senior leaders. Senior leaders should serve as role models through their ethical behavior and their personal involvement in planning, communicating, coaching the workforce, developing future leaders, reviewing organizational performance, and recognizing members of your workforce. As role models, they can reinforce ethics, values, and expectations while building leadership, commitment, and initiative throughout your organization.

Customer-Driven Excellence

Performance and quality are judged by an organization's customers. Thus, your organization must take into account all product features and characteristics and all modes of customer access and support that contribute value to your customers. Such behavior leads to customer acquisition, satisfaction, preference, and loyalty; to positive referrals; and, ultimately, to business expansion. Customer-driven excellence has both current and future components: understanding today's customer desires and anticipating future customer desires and marketplace potential. Value and satisfaction may be influenced by many factors throughout your customers' overall experience with your organization. These factors include your organization's customer relationships, which help to build trust, confidence, and loyalty. Customer-driven excellence means much more than reducing defects and errors, merely meeting specifications, or reducing complaints. Nevertheless, these factors contribute to your customers' view of your organization and thus, also are important parts of customer-driven excellence. In addition, your organization's success in recovering from defects, service errors, and mistakes is crucial for retaining customers and engaging customers for the long term. A customer-driven organization addresses not only the product and service characteristics that meet basic customer requirements but also those features and characteristics that differentiate the organization from its competitors. Such differentiation may be based on innovative offerings, combinations of product and service offerings, customization of offerings, multiple access mechanisms, rapid response, or special relationships. Customer-driven excellence is thus a strategic concept. It is directed toward customer retention and loyalty, market share gain, and growth. It demands constant sensitivity to changing and emerging customer and market requirements and to the factors that drive customer engagement. It demands close attention to the voice of the customer. It demands anticipating changes in the marketplace. It demands a customer-focused culture. Therefore, customer driven excellence demands organizational agility.

Organizational and Personal Learning

Achieving the highest levels of organizational performance requires a well-executed approach to organizational and personal learning that includes sharing knowledge via systematic processes. Organizational learning includes both continuous improvement of existing approaches and significant change or innovation, leading to new goals and approaches. Learning needs to be embedded in the way your organization operates. This means that learning (1) is a regular part of daily work, (2) is practiced at personal, work unit, and organizational levels, (3) results in solving problems at their source (root cause), (4) is focused on building and sharing knowledge throughout your organization, and (5) is driven by opportunities to affect significant, meaningful change and to innovate. Sources for learning include employees' and volunteers' ideas, research and development (R&D), customers' input, best-practice sharing, and benchmarking. Organizational learning can result in (1) enhancing value to customers through new and improved products and customer services, (2) developing new business opportunities/models, (4) reducing errors, defects, waste, and related costs, (5) improving responsiveness and cycle time performance, (6) increasing productivity and effectiveness in the use of all your resources, and (7) enhancing your organization's performance in fulfilling its societal responsibilities. The success of members of your workforce depends increasingly on having opportunities for personal learning and for practicing new skills. Leaders' success depends on access to these kinds of opportunities, as well. In organizations that rely on volunteers, their personal learning is important, and their learning and skill development should be considered with employees'. Organizations invest in personal learning through education, training, and other opportunities for continuing growth and development. Such opportunities might include job rotation and increased pay for demonstrated knowledge and skills. On-the-job training offers a cost-effective way to cross-train and to better link training to your organizational needs and priorities. Education and training programs may have multiple modes, including computer- and Web-based learning and distance learning. Personal learning can result in the following:

- A more engaged, satisfied, and versatile workforce that stays with your organization
- Organizational cross-functional learning
- Building your organization's knowledge assets
- An improved environment for innovation; thus, learning is directed not only toward better products but also toward being more responsive, adaptive, innovative, and efficient—giving your organization marketplace sustainability and performance advantages and giving your workforce satisfaction and the motivation to excel

Valuing Workforce Members and Partners

An organization's success depends increasingly on an engaged workforce that benefits from meaningful work, clear organizational direction, and performance accountability and that has a safe, trusting, and cooperative environment. Additionally, the successful organization capitalizes on the diverse backgrounds, knowledge, skills, creativity, and motivation of its workforce and partners. Valuing the people in your workforce means being committed to their engagement, satisfaction, development, and well-being. Increasingly, this involves more flexible, high-performance work practices tailored to varying workplace and home life needs. Major challenges in the area of valuing members of your workforce include the following:

- Demonstrating a leader's commitment to their success.
- Providing recognition that goes beyond the regular compensation system.
- Offering development and progression within your organization.
- Sharing your organization's knowledge so your workforce can better serve your customers and contribute to achieving your strategic objectives.
- Creating an environment that encourages risk taking and innovation.
- Creating a supportive environment for a diverse workforce. Organizations need to build internal and external partnerships to better accomplish overall goals.

Internal partnerships might include labor-management cooperation. Partnerships with members of your workforce might entail developmental opportunities, cross training, or new work organizations, including high-performance work teams. Internal partnerships also might involve creating network relationships among your work units or between employees and volunteers to improve flexibility, responsiveness, and knowledge sharing. External partnerships might be made with customers, suppliers, and education or community organizations. Strategic partnerships or alliances are increasingly important kinds of external partnerships. Such partnerships might offer entry into new markets or a basis for new products or customer-support services.

Partnerships might permit the blending of your organization's core competencies or leadership capabilities with the complementary strengths and capabilities of partners to address common issues. Such partnerships may be a source of strategic advantage for your organization. Successful internal and external partnerships develop longer term objectives, thereby creating a basis for mutual investments and respect. Partners should address key requirements for success, means for regular communication, approaches to evaluating progress, and means for adapting to changing conditions. In some cases, joint education and training could offer a cost-effective method for workforce development.

Agility

Success in today's ever-changing, globally competitive environment demands agility—a capacity for rapid change and flexibility. Organizations face shorter cycles for the introduction of new/improved products, and nonprofit and government organizations are increasingly being asked to respond rapidly to new or emerging social issues. Major improvements in response times often require new work systems, simplification of work units and processes, or the ability for rapid changeover from one process to another. A cross-trained and empowered workforce is a vital asset in such a demanding environment. A major success factor in meeting competitive challenges is the design-to-introduction (product or service feature initiation) or innovation cycle time. To meet the demands of rapidly changing markets, organizations need to carry out stage-to-stage integration (such as concurrent engineering) of activities from research or concept to commercialization or implementation. All aspects of time performance now are more critical, and cycle time has become a key process measure. Other important benefits can be derived from this focus on time; time improvements often drive simultaneous improvements in work systems, organization, quality, cost, supply-chain integration, and productivity.

Focus on the Future

Creating a sustainable organization requires understanding the short- and long-term factors that affect your organization and marketplace. The pursuit of sustainable growth and sustained-performance leadership requires a strong future orientation and a willingness to make long-term commitments to key stakeholders—your customers, workforce, suppliers, partners, and stockholders; the public; and your community. Your organization's planning should anticipate many factors, such as customers' expectations, new business and partnering opportunities, workforce development and hiring needs, the increasingly global marketplace, technological developments, changes in customer and market segments, new business models, evolving regulatory requirements, changes in community and societal expectations and needs, and strategic moves by competitors. Strategic objectives and resource allocations need to accommodate these influences. A focus on the future includes developing your leaders, workforce, and suppliers; accomplishing effective succession planning; creating opportunities for innovation; and anticipating societal responsibilities and concerns.

Managing for Innovation

Innovation means making meaningful changes to improve your organization's products, services, programs, processes, operations, and business model to create new value for the organization's stakeholders. Innovation should lead your organization to new dimensions of performance. Innovation is no longer strictly the purview of research and development departments; innovation is important for all aspects of your operations and all work systems and work processes. Organizations should be led and managed so that innovation becomes part of the learning culture. Innovation should be integrated into daily work and should be supported by your performance improvement system. Systematic processes for innovation should reach across your entire organization. Innovation builds on the accumulated knowledge of your organization and its people. Therefore, the ability to rapidly disseminate and capitalize on this knowledge is critical to driving organizational innovation.

Management by Fact

Organizations depend on the measurement and analysis of performance. Such measurements should derive from business needs and strategy, and they should provide critical data and information about key processes, outputs, and results. Many types of data and information are needed for performance management. Performance measurement should include customer, product, and process performance; comparisons of operational, market, and competitive performance; supplier, workforce, partner, cost, and financial performance; and governance and compliance outcomes. Data should be segmented by markets, product lines, and workforce groups, for example, to facilitate analysis. Analysis refers to extracting larger meaning from data and information to support evaluation, decision-making, improvement, and innovation. Analysis entails using data to determine trends, projections, and cause and effect that might not otherwise be evident. Analysis supports a variety of purposes, such as planning, reviewing your overall performance, improving operations, accomplishing change management, and comparing your performance with competitors' or with "best practices" benchmarks.

A major consideration in performance improvement and change management involves the selection and use of performance measures or indicators. The measures or indicators you select should best represent the factors that lead to improved customer, operational, financial, and societal performance. A comprehensive set of measures or indicators tied to customer and organizational performance requirements provides a clear basis for aligning all processes with your organization's goals. Measures and indicators may need to support decision-making in a rapidly changing environment. Through the analysis of data from your tracking processes, your measures or indicators themselves may be evaluated and changed to better support your goals.

Societal Responsibility

An organization's leaders should stress responsibilities to the public, ethical behavior, and the need to consider societal well-being and benefit. Leaders should be role models for your organization in focusing on ethics and the protection of public health, safety, and the environment. The protection of health, safety, and the environment includes your organization's operations, as well as the life cycles of your products. Also, organizations should emphasize resource conservation and waste reduction at the source. Planning should anticipate adverse impacts from production, distribution, transportation, use, and disposal of your products.

Effective planning should prevent problems, provide a forthright response if problems occur, and make available information and support needed to maintain public awareness, safety, and confidence. For many organizations, the product design stage is critical from the point of view of public responsibility. Design decisions impact your production processes and often the content of municipal and industrial waste. Effective design strategies should anticipate growing environmental concerns and responsibilities. Organizations should not only meet all local, state, and federal laws and regulatory requirements, but they should treat these and related requirements as opportunities for improvement beyond mere compliance.

Organizations should stress ethical behavior in all stakeholder transactions and interactions. Highly ethical conduct should be a requirement of and should be monitored by the organization's governance body. "Societal well-being and benefit" refers to leadership and support—within the limits of an organization's resources—of publicly important purposes. Such purposes might include improving education and health care in your community, pursuing environmental excellence, being a role model for socially important issues, practicing resource conservation, performing community service, improving industry and business practices, and sharing nonproprietary information. Leadership as a role-model organization also entails influencing other organizations, private and public, to partner for these purposes. Managing societal responsibilities requires the organization to use appropriate measures and leaders to assume responsibility for those measures.

Focus on Results and Creating Value

An organization's performance measurements need to focus on key results. Results should be used to create and balance value for your key stakeholders—your customers, workforce, stockholders, suppliers, and partners; the public; and the community. By creating value for your key stakeholders, your organization builds loyalty, contributes to growing the economy, and contributes to society. To meet the sometimes conflicting and changing aims that balancing value implies, organizational strategy explicitly should include key stakeholder requirements. This will help ensure that plans and actions meet differing stakeholder needs and avoid adverse impacts on any stakeholders. The use of a balanced composite of leading and lagging performance measures offers an effective means to communicate short and long-term priorities, monitor actual performance, and provide a clear basis for improving results.

Systems Perspective

The Baldrige criteria provide a systems perspective for managing your organization and its key processes to achieve results—and to strive for performance excellence. The seven Baldrige criteria categories, the core values, and the scoring guidelines form the building blocks and the integrating mechanism for the system. However, successful management of overall performance requires organization specific synthesis, alignment, and integration. Synthesis means looking at your organization as a whole and building on key business attributes, including your core competencies, strategic objectives, action plans, and work systems. Alignment means using the key linkages among requirements given in the Baldrige criteria categories to ensure consistency of plans, processes, measures, and actions. Integration builds on alignment so that the individual components of your performance management system operate in a fully interconnected manner and deliver anticipated results.

A systems perspective includes your senior leaders' focus on strategic directions and on your customers. It means that your senior leaders monitor, respond to, and manage performance based on your results. A systems perspective also includes using your measures, indicators, core competencies, and organizational knowledge to build your key strategies. It means linking these strategies with your work systems and key processes and aligning your resources to improve your overall performance and your focus on customers and stakeholders. Thus, a systems perspective means managing your whole organization, as well as its components, to achieve success (see Figure 17.7).

The Baldrige Scoring System

The Baldrige scoring system is based on a 0-1000 point scale. The available points are distributed amongst the categories and items as shown in the Figure 17.5, Item Listing. There is a heavy focus on results with Category 7, Business Results, making up 450 of the total 1000 points available. Table 17.3 shows the distribution of applicant scores for the period 1988-2007. During an assessment or award application review, examiners will assign scores based upon scoring guidelines. There are two evaluation dimensions, with scoring guides for each: (1) Process and (2) Results. The scoring guidelines for Process Categories 1-6 are shown in Figure 17.8.

Process refers to the work methods your organization uses and improves to address the item requirements in Categories 1 through 6. The scoring guidelines for Process Categories 1-6 are shown in Figure 17.8. Four factors used to evaluate process are: Approach, Deployment, Learning, and Integration.

- *Approach* refers to the methods used to accomplish the process, the appropriateness of the methods to the item requirements and the organization's operating environment, the effectiveness of your use of the methods, and the degree to which the approach is repeatable and based on reliable data and information (i.e., systematic).
- *Deployment* refers to the extent to which your approach is applied in addressing item requirements relevant and important to your organization; your approach is consistently applied; your approach is used (executed) by all appropriate work units.

Percentage of Applicants in Scoring Band by Year									
Band*	**1988/1989**	**1990**	**1991**	**1992**	**1993**	**1994**			
0–125	0	0	2.8	0	2	2			
126–250	1	7.2	13.2	12	8	10			
251–400	9	18.6	35.8	30	24	28			
401–600	43	52.6	34	40	47	50			
601–750	33	19.6	14.2	18	19	10			
751–875	14	2.1	0	0	0	0			
876–1000	0	0	0	0	0	0			
Band*	**1995**	**1996**	**1997**	**1998**	**1999**	**2000**	**2001**	**2002**	**2003**
0–250	6	3	4	3	13	12	11	10	17
251–350	15	7	19	22	23	27	16	23	27
351–450	11	14	15	33	29	14	16	37	27
451–550	32	41	27	17	25	31	41	18	22
551–650	30	31	35	22	10	14	16	12	7
651–750	6	3	0	3	0	2	0	0	0
751–875	0	0	0	0	0	0	0	0	0
876–1000	0	0	0	0	0	0	0	0	0
Band*	**2004**	**2005**	**2006**	**2007**					
0–275	3	5	1	11					
276–375	22	17	13	27					
376–475	40	34	41	31					
476–575	20	28	30	25					
576–675	15	14	13	6					
676–775	0	2	1	0					
776–875	0	0	0	0					
876–1000	0	0	0	0					

*Scoring bands revised.
(*Source:* Baldrige National Quality Program.)

TABLE 17.3 Distribution of Written Scores for Award Applicants, 1988–2007

- *Learning* refers to refining your approach through cycles of evaluation and improvement; encouraging breakthrough change to your approach through innovation; sharing refinements and innovations with other relevant work units and processes in your organization.
- *Integration* refers to the extent to which your approach is aligned with your organizational needs identified in the Organizational Profile and other Process Items; your measures, information, and improvement systems are complementary

PROCESS SCORING GUIDELINES

For use with categories 1–6

Score	Process
0% or 5%	• No SYSTEMATIC APPROACH to item requirements in evident; information is ANECDOTAL. (A) • Little or no DEPLOYMENT of any SYSTEMATIC APPROACH is evident. (D) • An improvement orientation is not evident; improvement is achieved through reacting to problems. (L) • No organizational ALIGNMENT is evident; individual areas or work units operate independently. (I)
10%, 15%, 20%, or 25%	• The beginning of a SYSTEMATIC APPROACH to the BASIC REQUIREMENTS of the item is evident. (A) • The APPROACH is in the early stages of DEPLOYMENT in most areas or work units, inhibiting progress in achieving the BASIC REQUIREMENTS of the item. (D) • Early stages of a transition from reacting to problems to a general improvement orientation are evident. (L) • The APPROACH is ALIGNED with other areas or work units largely through joint problem solving. (I)
30%, 35%, 40%, or 45%	• An EFFECTIVE, SYSTEMATIC APPROACH, responsive to the BASIC REQUIREMENTS of the item, is evident. (A) • The APPROACH is DEPLOYED, although some areas or work units are in early stages of DEPLOYMENT. (D) • The beginning of a SYSTEMATIC APPROACH to evaluation and improvement of KEY PROCESSES in evident. (L) • The APPROACH is in the early stages of ALIGNMENT with your basic organizational needs identified in response to the organizational profile and other process items. (I)
50%, 55%, 60%, or 65%	• An EFFECTIVE, SYSTEMATIC APPROACH, responsive to the OVERALL REQUIREMENTS of the item, is evident. (A) • The APPROACH is well DEPLOYED, although DEPLOYMENT may vary in some areas or work units. (D) • A fact-based, SYSTEMATIC evaluation and improvement PROCESS and some organizational LEARNING, including INNOVATION, are in place for improving the efficiency and EFFECTIVENESS of KEY PROCESSES. (L) • The APPROACH is ALIGNED with your organizational needs identified in response to the organizational profile and other process items. (I)
70%, 75%, 80%, or 85%	• An EFFECTIVE, SYSTEMATIC APPROACH, responsive to the MULTIPLE REQUIREMENTS of the item, is evident. (A) • The APPROACH is well DEPLOYED, with no significant gaps. (D) • Fact-based, SYSTEMATIC evaluation and improvement and organizational LEARNING, including INNOVATION, are KEY management tools; there is clear evidence of refinement as a result of organizational-level ANALYSIS and sharing. (L) • The APPROACH is INTEGRATED with your organizational needs identified in response to the organizational profile and other process items. (I)
90%, 95%, or 100%	• An EFFECTIVE, SYSTEMATIC APPROACH, fully responsive to the MULTIPLE REQUIREMENTS of the item, is evident. (A) • The APPROACH is fully, DEPLOYED without significant weaknesses or gaps in any areas or work units. (D) • Fact-based, SYSTEMATIC evaluation and improvement and organizational LEARNING through INNOVATION are KEY organization-wide tools, refinement and INNOVATION, backed by ANALYSIS and sharing, are evident throughout the organization. (L) • The APPROACH is well INTEGRATED with your organizational needs identified in response to the organizational profile and other process items. (I)

FIGURE 17.8 Process scoring guidelines. A Process Item score of 50 percent represents an approach that meets the overall requirements of the item that is deployed consistently and to most work units, that has been through some cycles of improvement and learning, and that addresses key organizational needs. (*Baldrige National Quality Program, 2009–2010 Criteria for Performance Excellence.*)

across processes and work units; your plans, processes, results, analyses, learning, and actions are harmonized across processes and work units to support organizationwide goals.

- *Results* refers to your organization's outputs and outcomes in achieving the requirements in Items 7.1-7.6 (Category 7). The scoring guidelines for Results Category are shown in Figure 17.9. The four factors used to evaluate results are Levels, Trends, Comparisons, and Integration (LeTCI):
  - *Levels* refers to your current level of performance.
  - *Trends* refers to the rate of your performance improvements or the sustainability of good performance (i.e., the slope of trend data) and the breadth (i.e., the extent of deployment) of your performance results.

RESULTS SCORING GUIDELINES

For use with categories 1–6

Score	Process
0% or 5%	• There are no organizational PERFORMANCE RESULTS and/or poor RESULTS in areas reported. (Le) • TREND data either are not reported or show mainly adverse TRENDS. (T) • Comparative information is not reported. (C) • RESULTS are not reported for any areas of importance to the accomplishment of your organization's MISSION. No PERFORMANCE PROJECTIONS are reported. (I)
10%, 15%, 20%, or 25%	• A few organizational PERFORMANCE RESULTS are reported, and early good PERFORMANCE LEVELS are evident in a few areas. (Le) • Some TREND data are reported, with some adverse TRENDS evident. (T) • Little or no comparative information is reported. (C) • RESULTS are reported for a few areas of importance to the accomplishment of your organization's MISSION. Limited or no PERFORMANCE PROJECTIONS are reported. (I)
30%, 35%, 40%, or 45%	• Good organizational PERFORMANCE LEVELS are reported for some areas of importance to the item requirements. (Le) • Some TREND data are reported, and a majority of the TRENDS presented are beneficial. (T) • Early stages of obtaining comparative information are evident. (C) • RESULTS are reported for many areas of importance to the accomplishment of your organization's MISSION. Limited PERFORMANCE PROJECTIONS are reported. (I)
50%, 55%, 60%, or 65%	• Good organizational PERFORMANCE LEVELS are reported for most areas of importance to the item requirements. (Le) • Beneficial TRENDS are evident in areas of importance to the accomplishment of your organization's MISSION. (T) • Some current PERFORMANCE LEVELS have been evaluated against relevant comparisons and/or BENCHMARKS and show areas of good relative PERFORMANCE. (C) • Organizational PERFORMANCE RESULTS are reported for most KEY CUSTOMER, market, and PROCESS requirements. PERFORMANCE PROJECTIONS for some high-priority RESULTS are reported. (I)
70%, 75%, 80%, or 85%	• Good to excellent organizational PERFORMANCE LEVELS are reported for most areas of importance to the item requirements. (Le) • Beneficial TRENDS have been sustained over time in most areas of importance to the accomplishment of your organization's MISSION. (T) • Many to most TRENDS and current PERFORMANCE LEVELS have been evaluated against relevant comparisons and/or BENCHMARKS and show areas of leadership and very good relative PERFORMANCE. (C) • Organizational PERFORMANCE RESULTS are reported for most KEY CUSTOMER, market, PROCESS, and ACTION PLAN requirements, and they include some PROJECTIONS of your future PERFORMANCE. (I)
90%, 95%, or 100%	• Excellent organizational PERFORMANCE LEVELS are reported for most areas of importance to the item requirements. (Le) • Beneficial TRENDS have been sustained over time in all areas of importance to the accomplishment of your organization's MISSION. (T) • Evidence of industry and BENCHMARK leadership is demonstrated in many areas. (C) • Organizational PERFORMANCE RESULTS fully address KEY CUSTOMER, market, PROCESS, and ACTION PLAN requirements, and they include PROJECTIONS of your future PERFORMANCE. (I)

FIGURE 17.9 Results scoring guidelines. A Results Item score of 50 percent represents a clear indication of goodlevels of performance, beneficial trends, and appropriate comparative data for the results areas covered in the item and important to the organization's business or mission. Performance projections are present for some high-priority results. (*Baldrige National Quality Program, 2009–2010 Criteria for Performance Excellence.*)

- *Comparisons* refers to your performance relative to appropriate comparisons, such as competitors or organizations similar to yours; your performance relative to benchmarks or industry leaders.
- *Integration* refers to the extent to which: your results measures (often through segmentation) address important customer, product, market, process, and action plan performance requirements identified in your Organizational Profile and in Process Items; your results include valid indicators of future performance; your results are harmonized across processes and work units to support organization-wide goals. (*Source:* 2009–2010 Criteria for Performance Excellence, Baldrige National Quality Program.)

Baldrige Award Eligibility

There are six eligibility categories for the Baldrige Award: manufacturing, service, small business, nonprofit, healthcare, and education. Eligibility is determined by submittal of an Eligibility Certification Package. Some of the criteria used in considering eligibility include subunit and parent organization relationships, location of the applicant, and previous winners over the previous five years.

Over the lifespan of the Baldrige Awards program, there has been a dramatic shift in the distribution of applicants by sector. In the first several years, only the manufacturing, services, and small business categories existed. The largest number of applicants in the early years came from the manufacturing sector. In recent years, there have been relatively few manufacturing applicants, with the newer categories of education and health care in particular contributing the greatest number of applicants. From 2005–2008, health care applicants represented about half of the total number of applicants (see Table 17.4).

Year	Manufacturing	Service	Small Business	Education	Health Care	Non-profit	TOTAL	State, Regional, and Local Applications*
1988	45	9	12	N/A	N/A	N/A	66	N/A
1989	23	6	11	N/A	N/A	N/A	40	N/A
1990	45	18	34	N/A	N/A	N/A	97	N/A
1991	38	21	47	N/A	N/A	N/A	106	217
1992	31	15	44	N/A	N/A	N/A	90	234
1993	32	13	31	N/A	N/A	N/A	76	433
1994	23	18	30	N/A	N/A	N/A	71	499
1995	18	10	19	N/A	N/A	N/A	47	621
1996	13	6	10	N/A	N/A	N/A	29	833
1997	9	7	10	N/A	N/A	N/A	26	1,000
1998	15	5	16	N/A	N/A	N/A	36	830
1999	4	11	12	16	9	N/A	52	1015
2000	14	5	11	11	8	N/A	49	862
2001	7	4	8	10	8	N/A	37	609
2002	8	3	11	10	17	N/A	49	395
2003	10	8	12	19	19	N/A	68	437
2004	8	5	8	17	22	N/A	60	481
2005	1	6	8	16	33	N/A	64	635
2006	3	4	8	16	45	10	86	426
2007	2	4	7	16	42	13	84	243
2008	3	5	7	11	43	16	85	167
TOTAL	349	178	349	131	203	23	1318	9937

*Incomplete data, with a varying number of programs reporting each year.

For 1988–2006, results of Stage 1-Independent Review scoring. For 2007, results of Consensus scoring. There is no rescoring of applicants after site visit.

(*Source:* Baldrige National Quality Program.)

TABLE 17.4 Number of Baldrige Applications by Category, 1988–2008

The Baldrige Award Process

The steps for the Baldrige Award process are as follows:

Eligibility

Organizations submit an Eligibility Certification Package. The deadline for submittal is usually in early April of each year.

Application

Organizations submit an Award Application Package in either CD/PDF format or on paper. The application includes a written response to criteria requirements in 50 or fewer pages. In addition to narrative text, applications are typically full of charts, graphs, tables, and other forms of results. The deadline for submittal is usually in mid-to-late May of each year.

Application Review

There is an independent review of the application by at least six members of the Board of Examiners, followed by a joint review by a team of Examiners, led by a Senior Examiner.

Site Visit Review

The Panel of Judges, composed of 12 members from the Board of Examiners, reviews the scoring data for all applicants and selects those that should receive site visits. Site visit teams, usually composed of 6–8 Examiners, are then deployed to verify and clarify findings from the consensus review.

Judges' Review

The Panel of Judges reviews the applications and consensus and site visit reports to make recommendations of award recipients to the Director of NIST.

Award Recipients

Award recipients are notified by the Secretary of Commerce. Award recipients are required to share information about their successful performance strategies with other U.S. organizations. The principle mechanism for sharing information is the annual Quest for Excellence Conference.

Baldrige Award Winners

Through 2008, there have been 79 Baldrige Award recipients. The Baldrige Program website includes contact information and profiles for each award recipient. Application

2008 Cargill Corn Milling North America, Poudre Valley Health System, and Iredell-Statesville Schools

2007 PRO-TEC Coating Co., Mercy Health Systems, Sharp HealthCare, City of Coral Springs, and U.S. Army Research, Development and Engineering (ARDEC)

2006 Premier, Inc. MESA Products Inc., and North Mississippi Medical Center

2005 Sunny Fresh Foods Inc., DynMcDermott Petroleum Operations, Park Place Lexus, Jenks Public Schools, Richland College, and Bronson Methodist Hospital

2004 The Bama Organizations, Texas Nameplate Company Inc., Kenneth W. Monfort College of Business, and Robert Wood Johnson University Hospital Hamilton

2003 Medrad Inc., Boeing Aerospace Support, Caterpillar Financial Services Corp., Stoner Inc., Community Consolidated School District 15, Baptist Hospital Inc., and Saint Luke's Hospital of Kansas City

2002 Motorola Inc. Commercial, Government and Industrial Solutions Sector, Branch Smith Printing Division, and SSM Health Care
2001 Clarke American Checks Inc., Pal's Sudden Service, Chugach School District, Pearl River School District, and University of Wisconsin-Stout
2000 Dana Corp.-Spicer Driveshaft Division, KARLEE Company Inc., Operations Management International Inc., and Los Alamos National Bank
1999 STMicroelectronics Inc.-Region Americas, BI, The Ritz-Carlton Hotel Co. L.L.C., and Sunny Fresh Foods
1998 Boeing Airlift and Tanker Programs, Solar Turbines Inc., and Texas Nameplate Co. Inc.
1997 3M Dental Products Division, Solectron Corp., Merrill Lynch Credit Corp., and Xerox Business Services
1996 ADAC Laboratories, Dana Commercial Credit Corp., Custom Research Inc., and Trident Precision Manufacturing Inc.
1995 Armstrong World Industries Building Products Operation and Corning Telecommunications Products Division
1994 AT&T Consumer Communications Services, GTE Directories Corp., and Wainwright Industries Inc.
1993 Eastman Chemical Co. and Ames Rubber Corp.
1992 AT&T Network Systems Group/Transmission Systems Business Unit, Texas Instruments Inc. Defense Systems & Electronics Group, AT&T Universal Card Services, The Ritz-Carlton Hotel Co., and Granite Rock Co.
1991 Solectron Corp., Zytec Corp., and Marlow Industries
1990 Cadillac Motor Car Division, IBM Rochester, Federal Express Corp., and Wallace Co. Inc.
1989 Milliken & Co. and Xerox Corp. Business Products and Systems
1988 Motorola Inc., Commercial Nuclear Fuel Division of Westinghouse Electric Corp., and Globe Metallurgical Inc.

The Baldrige Foundation sponsors the Performance Excellence website, a point of access to information about the Baldrige Award community. The website (www.baldrigepe.org) includes links to the following sites:

- The Baldrige Foundation
- The Baldrige National Quality Program
- The Alliance for Performance Excellence (further discussed in the chapter on state and local awards)

The Juran Institute thanks the Baldrige National Quality Program at the National Institute of Standards and Technology for use of text and graphics from the Criteria for Performance Excellence (Gaithersburg, MD 2009).

The EFQM Excellence Award

The EFQM Excellence Award is given to Europe's best-performing organizations and not-for-profit organizations. This award is open to organizations and organizations that are based in the geographic region of Europe, with no regard for political country boundaries. Originally called the European Quality Award, the first award was presented in 1988 by the King of Spain at the EFQM Forum in Madrid.

The EFQM Excellence Award is one of many forms of recognition and other services provided by EFQM, a not for profit membership foundation. EFQM consists of a network of over 600 private and public organizations of varying size and sectors. EFQM is funded by member fees and through fees for services without ties to any political bodies or governments. EFQM offers four types of services: assessment, training, recognition, and sharing.

EFQM describes its Excellence model as a "nonprescriptive framework for understanding the connections between what an organization does, and the results it is capable of achieving. It is used to structure a logical and systematic review of any organization, permitting comparisons to be made with similar or very different kinds of organization. It is also used to define what capabilities and resources are necessary in order to deliver the organization's strategic objectives."

The EFQM Excellence Model was updated in 2010. The most previous version was from 2003. The 2010 version impacted changes to the Fundamental Concepts, the Model, and the RADAR elements (**R**esults, **A**pproach, **D**eployment, **A**ssessment, and **R**eview). These changes were more of an update rather than a major overhaul, whereas the fundamental structure and concepts remained the same. One of the main objectives of the update was to recognize and consider some emerging trends and topics, including

- Attention to external networks, and utilizing networks for the purposes of innovation and creativity
- Putting a more equally balanced focus on the needs of people with the needs of the organization
- Increased focus on fiscal and societal responsibilities
- Conscious use of risk management and flexibility for achieving quick and timely actions in response to changing needs

EFQM Excellence Model 2010 is depicted in Figure 17.10, integrating eight concepts, nine criteria, and the RADAR tool. The small circles on the outer perimeter of the figure represent the eight Fundamental Concepts. The large circle in the center represents the RADAR elements. The nine boxes inside the inner circle represent the model criteria.

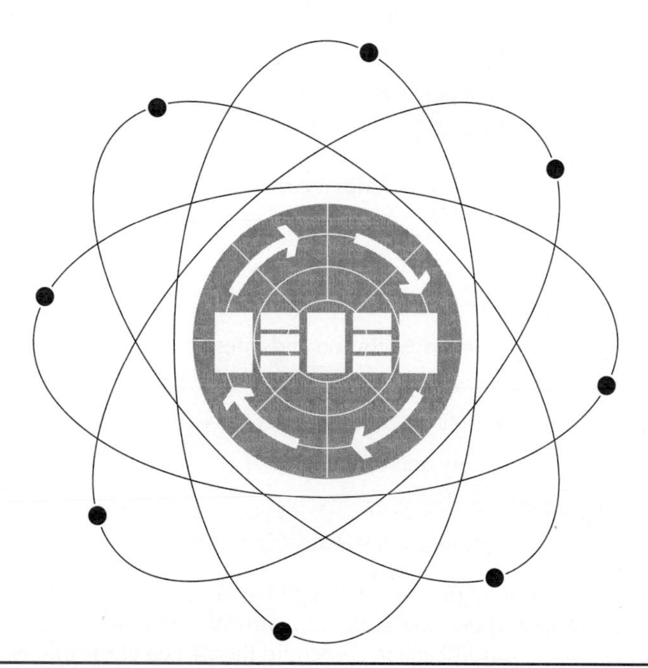

Figure 17.10 The 2010 EFQM Excellence Model. (*Copyright 2009 EFQM.*)

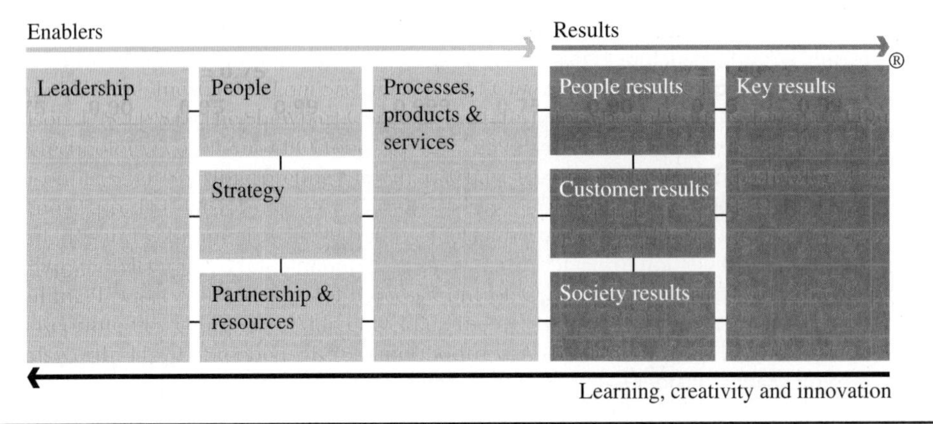

FIGURE 17.11 Criteria for the EFQM Excellence Model. (*Copyright 2009 EFQM.*)

The EFQM Excellence Model Criteria contains five enabling categories and four results categories (see Figure 17.11).

Similar to Baldrige, the criteria are weighted in a manner showing balance between the approaches and results categories. The revised weighting for the 2010 Model is as follows:

- Leadership (10 percent)
- People (10 percent)
- Strategy (10 percent)
- Partnerships and resources (10 percent)
- Processes, products, and services (10 percent)
- People results (10 percent)
- Customer results (15 percent)
- Society results (10 percent)
- Key results (15 percent)

Also like Baldrige, EFQM includes a core set of values that are described as Fundamental Concepts, listed as follows:

- Achieving balanced results
- Adding value for customers
- Leading with vision, inspiration and integrity
- Managing by processes
- Succeeding through people
- Nurturing creativity and innovation
- Building partnerships
- Taking responsibility for a sustainable future

In the 2010 version of the model, EFQM has shown a direct link between the eight Fundamental Concepts and the 32 subcriterion parts of the model. The concepts are defined in a manner that shows clear linkage to content in the criteria of the model (see Figure 17.12).

	1					2				3					4					5					6		7		8		9	
Criterion	Leadership					Strategy				People					Partnerships & resources					Processes, products and services					Customer results		People results		Society results		Key results	
Sub-criterion	A	B	C	D	E	A	B	C	D	A	B	C	D	E	A	B	C	D	E	A	B	C	D	E	A	B	A	B	A	B	A	B
Achieving balanced results		X	X			X		X	X							X	x		X						X	X	X	X	X	X	X	X
Adding value for customers			X				X														X	X	X	X	X	X		x				
Leading with vision, inspiration and integrity	X			X	X			X					X														X		X			X
Managing by process		X					x		X			X				X	x	x	X	X		x				X		x		x		X
Succeeding through people	X			X						X	X	X	x	X													X	X	x			
Nurturing creativity & innovation			X				x		X			X						X	X	X	X				x	X	x	X	x	X	x	X
Building partnerships			X			x	x								X							x			x	X	x	X	x	X	x	X
Taking responsibility for a sustainable future	X	X	X		X	x	x	X					X	X			X				x		X	x			x		X	X	x	

X (shaded) = Text from fundamental concept directly reflected in subcriterion
x = Adaptation of text from fundamental concept appears in the subcriterion

FIGURE 17.12 Integration of fundamental concepts into the EFQM model. (*Copyright 2009 EFQM.*)

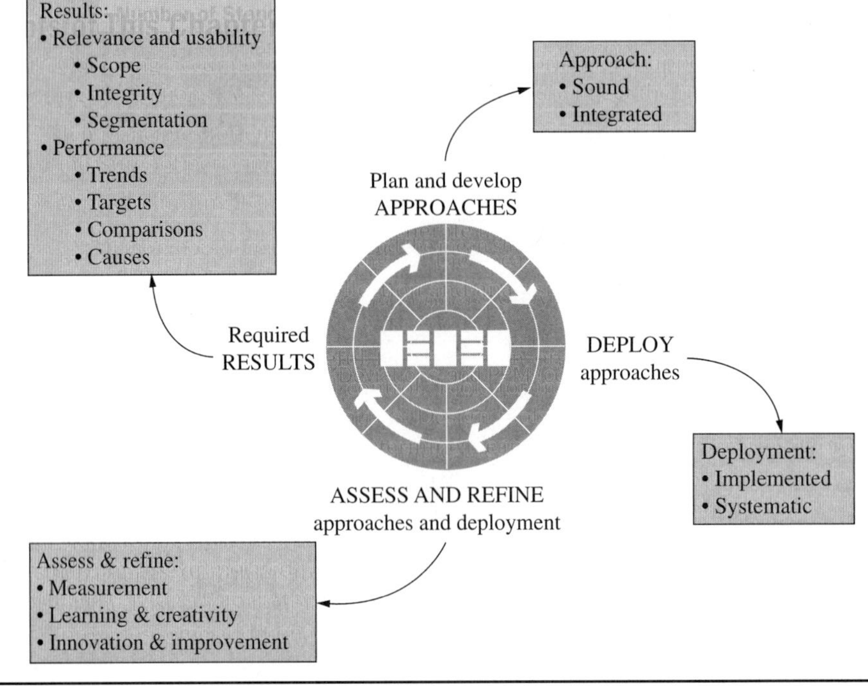

FIGURE 17.13 The EFQM RADAR. (*Copyright 2009 EFQM.*)

EFQM RADAR

The EFQM RADAR provides scoring guidelines to be used when assessing an organization using the EFQM criteria. RADAR refers to different aspects of assessing an organization relative to the model. RADAR-based scores are the basis for determining EFQM award winning levels of performance.

As displayed in Figure 17.13, the RADAR guidelines take into consideration:

- How approaches are planned and developed
- How approaches are deployed
- How approaches and deployment are assessed and refined
- How results are relevant and usable, and show favorable results relative to trends, targets, comparisons, and causes

EFQM Process

All EFQM applicants need to qualify to be accepted into the Award process, either with a qualification file for new applicants or by showing progress for reapplying candidates. Applicants are required to submit a 30- to 75-page submission document to be assessed. The document serves as a starting point for the assessor leading to a site visit conducted by a team of four to eight assessors facilitating selected interviews with applicant employees at all levels.

EFQM Recognition

EFQM provides recognition through tiered levels known as EFQM Levels of Excellence. The main purpose of the tiered levels, or schemes, is to help organizations improve their competitiveness through excellence. Exceptional organizations are recognized across a continuum of levels.

Committed to Excellence

Committed to Excellence is for organizations, or organizational units, that are at the beginning of their journey to excellence. For these organizations, the emphasis is on creating passion and commitment among internal stakeholders to generate the necessary momentum. Committed to Excellence provides the organization with a practical and simple way to build on their knowledge and experience of the EFQM Excellence Model and RADAR logic.

Recognized for Excellence is the next level and is for organizations that are well on their way to organizational excellence. These organizations, or organizational units, have experience in implementing excellence concepts and management frameworks. It recognizes the successful efforts they have made to implement excellence and good practice. Recognized for Excellence gives organizations the opportunity to identify the strengths and areas for improvement from an external point of view

At the June 2006 EFQM Learning Edge Conference in Rome, Italy, the Recognized for Excellence level was revised by introducing three-star, four-star and five-star recognition. These levels of recognition are awarded to organizations that have achieved more than 300, 400, or 500 points, respectively.

The EFQM Excellence Award is the top level of the EFQM Levels of Excellence. An independent jury selects the recipients. This decision is based on recommendations from assessor teams after on-site visits to the applicants. Two dimensions are used in making the decisions: the RADAR scoring profile and the type and strength of role model the organization was able to demonstrate. After reviewing all applicants for one of the awards, the jury first selects finalists. From the finalists, the jury then selects the best performers as prizewinners. If an applicant is really outstanding, the jury may decide to declare the organization as the award winner (see Tables 17.5 and 17.6).

Year	No. of Finalists	No. of Prizewinners	No. of Award Winners
1992		3	1
1993	2	1	1
1994	2	2	1
1995	3	1	1
1996	3	3	1
1997	4	4	2
1998	10	7	3
1999	18	5	4
2000	11	8	3
2001	14	5	2
2002	9	6	1
2003	6	9	4
2004	3	9	2
2005	5	9	2
2006	7	6	4
2007	**8**	**3**	**4**

(*Source:* Copyright 2009 EFQM.)

TABLE 17.5 Annual Number of EFQM Finalists, Prizewinners and Award Winners

2007		
Lauaxeta Ikastola Sociedad Cooperativa	Spain	Award winner
The Cedar Foundation	UK	Award winner
Villa Massa S.r.l.	Italy	Award winner
Tobermore Concrete Products Ltd.	UK	Award winner
2006		
BMW	Germany	Award winner
Grundfos	Denmark	Award winner
TNT Express GmbH-Germany	Germany	Award winner
St. Mary's	UK	Award winner
2005		
TNT Express Information and Communication Services	UK	Award winner
FirstPlus Financial Group Plc	UK	Award winner
2004		
Kocaeli Chamber of Industry	Turkey	Award winner
YELL	UK	Award winner
2003		
Bosch Sanayi ve Ticaret AS	Turkey	Award winner
Runshaw College	UK	Award winner
Maxi Coco-Mat SA	Greece	Award winner
Edinburgh International Conference Centre	UK	Award winner
2002		
Springfarm Architectural Mouldings Ltd.	UK	Award winner
2001		
St Mary's College Northern Ireland	UK	Award winner
Zahnarztpraxis	Switzerland	Award winner
2000		
Nokia Mobile Phones, Europe and Africa	Finland	Award winner
Inland Revenue, Accounts Office Cumbernauld	UK	Award winner
Burton-Apta Refractory Manufacturing Ltd.	Hungary	Award winner
1999		
Yellow Pages	UK	Award winner
Volvo Cars Gent	Belgium	Award winner
DiEU	Denmark	Award winner
Servitique Network Services	France	Award winner

Table 17.6 Annual EFQM Award Winners

1998		
TNT Express United Kingdom Ltd.	UK	Award winner
Landhotel Schindlerhof	Germany	Award winner
Beko Ticaret	Turkey	Award winner
1997		
SGS Thomson Microelectronics	Italy	Award winner
Beksa	Turkey	Award winner
1996		
BRISA	Turkey	Award winner
1995		
Texas Instruments Europe	France	Award winner
1994		
D2D (Design to Distribution) Ltd.	UK	Award winner
1993		
Milliken European Division	Belgium	Award winner
1992		
Rank Xerox	UK	Award winner

Created from information on www.EFQM.org. Refer to the website for a complete list of all finalists, prizewinners, and award winners.
(*Source:* Copyright 2009 EFQM.)

Table 17.6 (*Continued*)

The Deming Prize

The Deming Prize was established in 1951 by the Japanese Union of Scientists and Engineers (JUSE) in honor of W. Edwards Deming's contributions to statistical quality control in Japan after World War II. It was originally designed to recognize Japanese organizations for major advances in quality improvement. Today the Deming Prize is available to non-Japanese organizations and to individuals recognized as having made major contributions to the advancement of quality.

There are four categories of recognition in the Deming Prize as shown in Table 17.7.

The Deming Prize website defines Total Quality Management (TQM) as "a set of systematic activities carried out by the entire organization to effectively and efficiently achieve organization objectives so as to provide products and services with a level of quality that satisfies customers, at the appropriate time and price."

The website provides further explanation of this definition by elaborating on many key words and phrases. For example, it explains that "'systematic activities' mean organized activities to achieve the organization's mission (objectives) that are led by strong management leadership and guided by established clear mid- and long-term vision and strategies as well as appropriate quality strategies and policies." (http://www.juse.or.jp/e/deming/)

The Deming Prize committee conducts the examination and awards the Deming Prize. Committee members are TQM experts from various industries and academia. There are five subcommittees that administer and execute the awards process. The Deming Prize

The Deming Prize for Individuals	For individuals or groups
	Given to those who have made outstanding contributions to the study of total quality management (TQM) or statistical methods used for TQM, or those who have made outstanding contributions in the dissemination of TQM
The Deming Distinguished Service Award for Dissemination and Promotion (Overseas)	For individuals whose primary activities are outside Japan
	Given to individuals who have made outstanding contributions in the dissemination and promotion of TQM. Examination will be carried out every 3-5 years.
The Deming Application Prize	For organizations or divisions of organizations that manage their business autonomously
	Given to organizations or divisions of organizations that have achieved distinctive performance improvement through the application of TQM in a designated year
The Quality Control Award for Operations Business Units	For operations business units of an organization
	Given to operations business units of an organization that have achieved distinctive performance improvement through the application of quality control/management in the pursuit of TQM in a designated year

(*Source:* http://www.juse.or.jp/e/deming/)

TABLE 17.7 Deming Prize Categories

does not prescribe a specific model to follow. The Deming Prize committee evaluates whether applicant activities are likely to achieve higher performance based upon their situation and circumstances (see Table 17.8).

Other Quality Awards

Although the exact number is unknown, there are many countries around the globe that have implemented national quality awards. Most are similar in structure and with a business model and awards process similar to the Baldrige or EFQM.

GEM

In recent years, many national quality award programs have collaborated to form the Global Excellence Model Council (GEM). The GEM council is made up of the chief executives of several national quality award programs. Current GEM members include the following:

- Australian Business Excellence Awards
- EFQM Excellence Awards (Europe)
- Ibero-American Excellence Award (Brazil, Mexico, Spain, et. al.)
- CII-EXIM Bank Award for Business Excellence (India)
- Japan Quality Award
- Singapore Quality Award
- Malcolm Baldrige National Quality Award (United States)

1951	Fuji Iron & Steel Co., Ltd. Showa Denko K.K. Tanabe Seiyaku Co., Ltd. Yawata Iron & Steel Co., Ltd.
1952	Asahi Chemical Co., Ltd. Furukawa Electric Co., Ltd. Nippon Electric Co., Ltd. Shionogi & Co., Ltd. Takeda Chemical Industries, Co., Ltd. Toyo Spinning Co., Ltd. Kyushu Cloth Industry Co., Ltd.
1953	Kawasaki Steel Corp. Shin-etsu Chemical Industry Co., Ltd. Sumitomo Metal Mining Co., Ltd. Tokyo Shibaura Electric Co., Ltd.
1954	Nippon Soda Co., Ltd. Toyo Bearing Manufacturing Co., Ltd. Toyo Rayon Co., Ltd.
1955	Asahi Glass Co., Ltd. Hitachi, Ltd. Honshu Paper Manufacturing Co., Ltd.
1956	Fuji Photo Film Co., Ltd. Konishiroku Photo Industry Co., Ltd. Mitsubishi Electric Corp. Tohoku Industry, Co., Ltd.
1957	—
1958	Kanegafuchi Chemical Industry Co., Ltd. Kureha Chemical Industry Co., Ltd. Matsushita Electronics Corp. Nippon Kokan K.K. <S>Nakayo Communication Equipment Co., Ltd.
1959-1960	Asahi Special Glass Co., Ltd. Kurake Spinning Co., Ltd. Nissan Motor Co., Ltd. <S>Towa Industry Co., Ltd.
1961	Nippondenso Co., Ltd. Teijin, Ltd. <S>Nihon Radiator Co., Ltd.
1962	Sumitomo Electric Industries, Ltd.
1963	Nippon Kayaku Co., Ltd.
1964	Komatsu Manufacturing Co., Ltd
1965	Toyota Motor Co., Ltd.

TABLE 17.8 Deming Application Prize Winners, <D> Deming Application Prize for Divisions, <S> Deming Application Prize for Small Companies. (These categories were abolished in 1995.)

1966	Kanto Auto Works, Ltd. <D>Matsushita Electric Industrial Co., Ltd., Electric Components Division
1967	Shinko Wire Co., Ltd. <S>Kojima Press Industry Co., Ltd.
1968	Bridgestone Tire Co., Ltd. Yanmer Diesel Engine Co., Ltd. <S>Chugoku Kayaku Co., Ltd.
1969	<S>Shimpo Industry Co., Ltd.
1970	Toyota Auto Body Co., Ltd.
1971	Hino Motors, Ltd.
1972	Aisin Seiki Co., Ltd. <S>Saitama Chuzo Kogyo K.K.
1973	<S>Sanwa Seiki Manufacturing, Co., Ltd. <S>Saitama Kiki Manufacturing, Co., Ltd.
1974	<S>Horikiri Spring Manufacturing, Co., Ltd. <S>Kyodosokuryosha Co., Ltd.
1975	Ricoh Co., Ltd. <S>K.K. Takebe Tekkosho <S>Tokai Chemical Industries, Ltd. <S>Riken Forge Co., Ltd.
1976	Sankyo Seiki Manufacturing Co., Ltd. Pentel Co., Ltd. <S>Komatu Zoki, Ltd. <D>Ishikawajima-Harima Heavy Industries Co., Ltd., Aero-Engine & Space Operations
1977	Aisin-Warner, Ltd.
1978	Tokai Rika Co., Ltd. <S>Chuetsu Metal Works Co., Ltd.
1979	Nippon Electric Kyusyu, Ltd. Sekisui Chemical Co., Ltd. Takenaka Komuten Co., Ltd. Tohoku Ricoh Co., Ltd. <S>Hamanakodenso Co., Ltd.
1980	Kayaba Industry Co., Ltd. Komatsu Forklift Co., Ltd. Fuji Xerox Co., Ltd. The Takaoka Industrial Co., Ltd. <S>Kyowa Industrial Co., Ltd.
1981	<S>Aiphone Co., Ltd. <S>Kyosan Denki Co., Ltd. <D>Tokyo Juki Industrial Co., Ltd., Industrial Sewing Machine Division

Table 17.8 Deming Application Prize Winners, <D> Deming Application Prize for Divisions, <S> Deming Application Prize for Small Companies. (These categories were abolished in 1995.) (*Continued*)

1982	Kajima Corp. Nippon Electric Yamagata Ltd. Rhythm Watch Co., Ltd. Yokogawa Hewlett-Packard <S>Aisin Chemical Co., Ltd. <S>Shinwa Industrial Co., Ltd.
1983	Shimizu Construction Ltd. The Japan Steel Works, Ltd. <S>Aisin Keikinzoku Co., Ltd.
1984	Komatsu Zenoah Co. The Kansai Electric Power Co., Inc. Yasukawa Electric Manufacturing Co., Ltd. <S>Anjo Denki Co., Ltd. <S>Hokuriku Kogyo Co., Ltd.
1985	Nippon Carbon Co., Ltd. Nippon Zeon Co., Ltd. Toyoda Gosei Co., Ltd. Toyoda Machine Works, Ltd. <S>Comany Inc. <S>Hoyo Seiki Co., Ltd. <S>Uchino Komuten Co., Ltd. <D>Texas Instruments Japan Limited, Bipolar Department
1986	Hazama-Gumi, Ltd. Toyoda Automatic Loom Works, Ltd. <S>Nitto Construction Co., Ltd. <S>Sanyo Electric Works Ltd.
1987	Aichi Steel Works, Ltd. Aisin Chemical Co., Ltd. Daihen Corporation Co., Ltd. NEC IC Microcomputer Systems, Ltd.
1988	Aisin Keikinzoku, Co., Ltd. Asmo Co., Ltd. Fuji Tekko Co., Ltd. <D>Joban Kosan Co., Ltd., Joban Hawaiian Center
1989	Aisin Sinwa Co., Ltd. Itoki Kosakusyo Co., Ltd. Maeda Corporation NEC Tohoku, Ltd. TOTO Ltd. <O>Florida Power & Light Company (U.S.A) <S>Ahresty Corporation <S>Toyooki Kogyo Co., Ltd.

Table 17.8 (*Continued*)

1990	Aisin Hoyo Co., Ltd. Amada Wasino Co., Ltd. NEC Shizuoka, Ltd.
1991	NEC Kansai Ltd. Nachi-Fujikoshi Corp. Hokushin Industries Inc. <S>Sinei Industries Co., Ltd. <S>Niigata Toppan Printing Co., Ltd. <O>Philips Taiwan, Ltd. (Taiwan)
1992	Aisan Industry Co., Ltd. JATCO Corporation
1993	NTT Data Communications Systems Co.
1994	Maeda Seisakusho Co., Ltd. <O>AT & T Power Systems (U.S.A.) <S>AW Industries Co., Ltd. <S>NT Techno Corp. <S>Kouritsu Sangyosha Ltd., Partnership <S>Diamond Electric Mfg. Co., Ltd.
1995	Ishikawajima-Harima Heavy Industries Co., Ltd., Nuclear Power Division Mtex Matsumura Corporation Kikuchi Metal Stamping Co., Ltd. Toyoseiki Co., Ltd.
1996	Aisin-Shinei Co., Ltd. Ando Electric Co. Konica Corporation, Hino Production Division NEC Musen-Denshi Co., Ltd. Fuji Photo Optical Co., Ltd.
1997	Aisin Kiko Co., Ltd. Kojima Press Co., Ltd. Toyo Glass Co., Ltd.
1998	Aisin AW Seimitsu Co., Ltd. Ando Electric Engineering Service Co., Ltd. Itoki All Steel Co., Ltd. Okinawa Sekiyu Seisei Co., Ltd. Sanden Corporation Sundaram-Clayton Limited, Brakes Division (India) Fujimi Koken Co., Ltd.
1999	Miyama Kogyo Co., Ltd.
2000	Kanehide Aluminum Industry Co., Ltd. Sanden Butsuryu Co., Ltd. Sanwa Tech Co., Ltd. GC Corporation

Table 17.8 Deming Application Prize Winners, <D> Deming Application Prize for Divisions, <S> Deming Application Prize for Small Companies. (These categories were abolished in 1995.) (*Continued*)

2001	Sanden System Engineering Co., Ltd. Sundaram Brake Linings Ltd. (India) Thai Acrylic Fibre Co., Ltd. (Thailand) Thai Carbon Black Public Co., Ltd. (Thailand)
2002	The Siam Cement (Thung Song) Co., Ltd. (Thailand) TVS Motor Company Ltd. (India)
2003	GC Dental Products Corp. Brakes India Ltd., Foundry Division (India) Mahindra and Mahindra Ltd., Farm Equipment Sector (India) Rane Brake Linings Ltd. (India) The Siam Refractory Industry Co., Ltd. (Thailand) Sona Koyo Steering Systems Ltd. (India) Thai Paper Company Ltd. (Thailand)
2004	CCC Polyolefins Company Ltd. (Thailand) Indo Gulf Fertilisers Ltd. (India) Lucas-TVS Ltd. (India) Siam Mitsui PTA Company Ltd. (Thailand) SRF Ltd., Industrial Synthetics Business (India) Thai Ceramic Company Ltd. (Thailand)
2005	Hosei Brake Industry Co., Ltd. Krishna Maruti Limited, Seat Division (India) Rane Engine Valves Limited (India) Rane TRW Steering Systems Limited, Steering Gear Division (India)
2006	Nishizawa Electric Meters Manufacturing Co., Ltd. Sanden International (Singapore) PTE Limited(Singapore) Sanden International (U.S.A.), Inc. (U.S.A.)
2007	Asahi India Glass Limited, Auto Glass Division (India) Rane (Madras) Limited (India)
2008	Tata Steel Limited (India)

(*Source:* http://www.juse.or.jp/e/deming/)

Table 17.8 (*Continued*)

GEM's stated mission is that the council

- Maintains a leading edge position on Excellence Models
- Senses business trends and external factors that impact the Excellence Models
- Explores opportunities for new products and activities
- Coordinates and shares specific award activities

(http://www.excellencemodels.org/)

FUNDIBEQ

FUNDIBEQ (Ibero-American Foundation for Quality Management) is an international non-profit organization that promotes and develops overall quality management in Ibero-America. FUNDIBEQ promotes the Ibero-American Excellence Model for Management (IEM) and sponsors the Ibero-American Quality Award. The strategic aim of FUNDIBEQ is to facilitate opportunities that exist within Ibero-America to "take a big step towards greater and more compact competitiveness." The Ibero-American Foundation for Quality Management was founded in March 1998. Some of the services provided by FUNDIBEQ are as follows:

- Linking local quality associations and foundations and converting them into strategic allies of FUNDIBEQ
- Implementing the Ibero-American Model of Management Excellence and organizing the Ibero-American Quality Award
- Organizing Ibero-American Quality Conventions

FUNDIBEQ's partners include quality award organizations from the following countries: Argentina, Brazil, Chile, Columbia, Cuba, the Dominican Republic, Ecuador, Mexico, Paraguay, Peru, Portugal, Spain, Uruguay, and Venezuela. (http://www.fundibeq.org/English/ingles2.html#modelo)

Table 17.9 gives a listing of national quality awards.

State and Local Quality Awards

In the United States, the Alliance for Performance Excellence, sponsored by the Baldrige Foundation, was started in 2003 to form an association of state and local programs and is a nonprofit network of state and local Baldrige-based award programs. The Alliance provides potential award applicants and examiners, promote the use of the Baldrige criteria, and disseminate information regarding the award process and concepts. The stated mission/vision of the Alliance is to

- Enhance and facilitate the success of state Baldrige-based award processes
- Advance organizational excellence and U.S. competitiveness through state. The Alliance for Performance Excellence is sponsored by the Baldrige Foundation. It was started in 2003 to form an association of state and local programs.

Prior to the creation of the MBNQA, several states had U.S. Senate Productivity Awards formed in the early and mid 1980s. Programs existed in Alabama, Maryland, New Mexico, Nevada, and Virginia. There were also programs in Wyoming and Connecticut created in 1986 and 1987 initially using other criteria. All of these programs eventually converted to using the Baldrige criteria (Belter 2009).

Organizing efforts began to use the Baldrige criteria in several states in the late 1980s. By 1992, there were at least nine Baldrige-based programs in the states of Connecticut, Delaware, Florida, Maine, Massachusetts, Minnesota, New Mexico, New York, North Carolina, and Wyoming,

In the next five years, programs were formed in the states of Arkansas, Arizona, California, Georgia, Idaho, Iowa, Illinois, Kansas, Kentucky, Louisiana, Michigan, Missouri, Mississippi, Nebraska, New Hampshire, New Jersey, Nevada, Oklahoma, Oregon, Rhode Island, South Carolina, Tennessee, Texas, Utah, Vermont, Washington, and Wisconsin. Individuals formed additional programs in the states of Connecticut, Kentucky, and Maryland. A non-Baldrige-based program was established in Indiana.

Country/Region	Award Name	Year Founded	Website Address
Australia	Australian Business Excellence Awards	1998	www.saiglobal.com/improvement/business-excellence-awards
Canada	Canada Awards for Excellence	1992	http://www.nqi.ca/
China	National Quality Award	2001	http://nqa.csd.org.tw/eng/main.htm
Egypt	The National Awards for Excellence	2005	www.nationalawards-eg.com
Europe	EFQM Excellence Award	1988	www.efqm.org
Hungary	National Quality Prize	1996	Unable to locate
Ibero-America	Ibero-American Excellence Award	2000	http://www.fundibeq.org/English/ingles2.html#subir
India	CII-EXIM Bank Award for Business Excellence	1994	http://www.cii-iq.in/CII-Exim%20Bank%20Award%20for%20Excellence.htm
India	Rajiv Gandhi National Quality Award	1991	http://www.bis.org.in/other/rgnqa_geninfo.htm
Jamaica	National Quality Award		http://www.jbs.org.jm/nqa_pro.htm
Japan	Deming Prize	1951	http://www.juse.or.jp/e/deming/
Japan	Japan Quality Award	1995	www.jqac.com/website.nsf/newmainpagee?openpage
Romania	Joseph M. Juran Romanian Quality Award"	2000	Unable to locate
Singapore	Singapore Quality Award	1994	www.spring.gov.sg/sqa.aspx
Sri Lanka	Sri Lanka National Quality Award	1995	http://www.nsf.ac.lk/slsi/training/National%20Quality%20Awards.html
United States	Malcolm Baldrige National Quality Award	1987	www.quality.nist.gov

Table 17.9 Directory of National Quality Awards (Countries around the Globe)

During this same time period, local and regional award programs were formed in the cities of Memphis, Austin, Lancaster, Pittsburgh, Philadelphia, Houston, Western Texas, Cincinnati, Dayton, Colorado Springs, and other locations. The Greater Memphis Chamber of Commerce continues to operate its regional Baldrige-based award program, and there are local programs operating in Lancaster, Memphis, and Austin.

During the last 10 years, four state programs were organized in Ohio, Colorado, Pennsylvania, and Alaska. The Alliance for Performance Excellence started its efforts in 2003 to form an association of the state and local programs.

Since the mid-1990s, several state programs have failed, primarily due to a lack of program funding—Idaho, Maine, Oregon, Utah, Wyoming, New Jersey, and Nevada. The initial programs in Kentucky, New York, North Carolina, and Washington failed, but volunteer efforts established new programs. (*A Short History of State and Local Programs by Hilary Belter*)

The Alliance website lists 46 different state and local quality awards. While some states do not have an active quality award, a few states have both a state award and regional or local awards.

From 1996 until 2009, 41 of the 55 Baldrige Award recipients were also state award recipients (Baldrige Program FAQs, Baldrige National Quality Program).

Some of the local and regional awards as of the date of this publication include

- Pioneer Valley Business Excellence Award of the Affiliated Chambers of Commerce of Greater Springfield, Massachusetts
- Organizations of Noteworthy Excellence (ONE) Awards, a local award program for nonprofits in Cincinnati, Ohio
- Memphis Regional Chamber Quality Cup Award, a regional award program sponsored by the Mid-South Quality Productivity Center in Tennessee
- Performance Excellence Association of the Mid-South—PerfX, a local award program for large and small businesses and organizations in the Greater Memphis and tristate areas of Tennessee
- University of Texas Center for Performance Excellence Program, a regional award program in Texas
- Connecticut Aware for Excellence (CAFE)

An up-to-date list of state and local quality award programs can be found at www.baldrigepe.org/alliance/programs.aspx.

Quality Awards in Organizations

Many organizations have instituted internal quality awards, often using national quality award programs as a model. Business units, facilities, departments, and other units of the organization can be nominated or apply for the organization's award. This can be an effective way help to deploy quality management systems throughout the organization. For example, assessments and findings from internal quality awards programs can be part of the information-gathering phase of strategic planning. Documented approaches and results from multiple organizational applications or assessments can help to drive sharing of best practices and benchmarks.

Various approaches can be used for assessing organizations for an internal awards program. In addition to the standard approaches typically used by national quality awards programs, an internal assessment could be based upon self-assessment with examiner verification, or it could be based upon a site visit only by an examining team. Organizations may choose to use internal examiners for their awards process or may use external third-party examiners.

As example, Cargill, an international provider of agricultural, food, and risk-management products and services, has created a framework for continuous improvement called Business Excellence based on the the Malcolm Baldrige National Quality Award. Cargill created an internal awards recognition program with the highest recognition being the Chairman's Award for Business Excellence (Cargill 2008).

Using Quality Awards as a System Assessment Tool

Many organizations use the criteria from national quality awards for the benefit of getting a thorough organizational assessment, often with no intent of even applying for award recognition. There various methods that can be used to complete an assessment, based upon many factors including the size and geography of the organization, the number of facilities, and the availability of internal expertise to conduct an assessment.

Written Responses

Without completing a formal application, organizations can use a question-and-answer approach to respond to the criteria questions from an awards program. There could be multiple responses for each question if more than one input is desired for the response. Some more advanced formats for written responses could seek more probing information for each question. For example, a Baldrige-based written response questionnaire may be formatted to seek a specific response for approach, deployment, learning, and integration for each process-related question. This provides richer information for the purposes of scoring the application using the awards process scoring guidelines.

Survey

This approach can be used to gather assessment input from a large number of people. Questions are designed to gather the collective input on the performance of the organization as it relates to the awards criteria. The Baldrige National Quality Program provides a free survey for this purpose called, "Are We Making Progress." This is a 40-question survey which can be completed in about 10 minutes. A survey can be used as the primary information gathering method, or it can be used as supplemental or additional information as part of more comprehensive assessment (Baldrige National Quality Program 2008).

Application

A formal application can be prepared that simulates an actual awards process application. This requires more effort than simple written responses, but it can also be more revealing because of the thought process of the writer in attempting to use consistent language, identify linkages, and the most important strengths of the organization. For organizations that are just starting out, they may decide to answer only the higher-level questions rather than the more specific multiple requirements in awards criteria. For example, if assessing to the Category 3 section of the Baldrige criteria, Customer Focus, the self-assessment team may only respond to the item-level questions as shown:

3.1 Customer Engagement: How do you engage customers to serve their needs and build relationships?

3.2 Voice of the Customer: How do you obtain and use information from your customers?

This is a much simpler process whereas the multiple requirements from Category 3, Customer Focus, include 35 separate questions compared to only two questions at the item level.

Interview

In this approach, the assessment team would schedule interviews with key individuals to respond to criteria questions. During the interviews, the assessment team will record notes about the approaches used and results attained for further evaluation after all of the information has been collected. This approach requires less preparation time than written responses. The disadvantage is that the respondents may not provide complete information during the interview setting, depending upon their understanding of the question and their recall of all the relevant information in a single interview. There is also less learning on the part of the respondents when compared to written responses because of the level of engagement is less.

Focus Group

This is similar to the interview approach except that multiple people are involved in the sessions for different sections of the criteria review. For example, an assessment for Item 1.1, Senior Leadership, of the Baldrige criteria may involve several members of the senior leadership team in a focus group setting. As in the interview approach, the assessment team will record notes about the approaches used and results attained for further evaluation after all of the information has been collected. The advantage of this approach over individual interviews is because of the shared knowledge of the group and the ability to build on the responses of others to provide complete assessment information.

Collaborative Assessment

This is a variation on a focus group approach. Instead of gathering specific answers to the approach questions, respondents will provide their opinions about strengths and opportunities for improvement for each section of the assessment criteria. Responses are captured in real time for review by the focus group. At the end of each focus group, each individual can provide his or her rating of the importance of each opportunity for improvement. This collaborative approach to assessment has the advantage of involving many people in gathering the information and building consensus on both strengths and opportunities for improvement within the organization (Hoyt and Ralston 2001).

Cycles of Improvement

"Once and done" is almost always a wasted effort. It is not enough to reach a certain level of quality leadership. The real goal is to sustain quality leadership performance. Taking a longer term view, many organizations will use quality awards as an annual or semiannual cycle to assess and improve. These organizations recognize that achieving and sustaining quality leadership is a journey.

A typical cycle of improvement might consist of the following:

- *Assess*. This involves a comparison of actual performance to the awards criteria and scoring guidelines. This can be accomplished through an awards application or by some other form of self or third-party assessment.
- *Plan*. This involves evaluating the results of the assessment to identify and prioritize the vital few opportunities for improvement.
- *Improve*. This involves carrying out improvement projects and activities to close the vital few opportunities for improvement.
- *Repeat*. Continue the cycle for multiple iterations.

Quite often, national quality award winners have gone through multiple cycles of the awards process prior to receiving the award. These organizations have used the application process and the feedback reports to continually identify the top priorities for improvement planning and to monitor progress toward improvement. Because not all gaps can be addressed simultaneously, improvement projects are launched to address the vital few areas for improvement. As organizations address these gaps from one cycle, they can refocus during the next cycle to the next level of vital gaps. This project-by-project focus over several years allows the organization to achieve breakthrough levels of improvement in many to most key areas of importance in the organization. The awards process cycle provides a schedule for assessing, prioritizing, and improving.

Multiple cycles of improvement can be achieved using several assessment approaches.

National Quality Awards: One approach is to apply annually for a national quality award, using feedback each year to monitor progress and reprioritize improvement goals. This approach keeps the organization on a set cycle providing a fixed timetable for making important improvements.

Self-Assessments: Another approach is to begin with self-assessments for the first few iterations of improvement. One advantage to this approach is that the organization is not tied to the awards process cycle, with the possible of quicker cycles of improvement if desired. This approach is also attractive to the organizations that aspire to "award-winning" levels of performance with no intention of applying for award recognition.

State and Local Quality Awards: Many organizations have applied for state and/or local quality awards for early cycles of improvement while preparing for a national quality award application. It is often easier to get a site visit with alternative awards processes, which can provide a more thorough assessment of the current state than can be obtained through an application-only feedback report.

Applying for Quality Awards

Most national quality awards require a written application with response to the questions in the award criteria. Although there is often a limit to the length of the application (the Baldrige Award has a limit of 50 pages, for example), the applications are typically comprehensive and require significant time and skill to prepare. A well-written application will not necessarily improve the chances of receiving an award; however, a poorly written application can diminish the chances of winning if the examiners are not provided with an accurate understanding of the organization's approaches and results. Furthermore, writing a strong application is important because it provides examiners with better information from which to provide valuable feedback.

Some commonly used approaches for writing an application are as follows:

Contract Writer: Many organizations will hire a contract writer or consulting organization to write the application. An experienced application writer may have added knowledge about how to write and format the application responses. This can help the organization to tell their story better. A potential disadvantage is that senior leaders may not be appropriately engaged in the process and could miss out on the learning that occurs from attempting to fully respond to the criteria requirements.

Single Internal Application Writer: An individual from within the organization may be assigned to write the application. An advantage to this approach is consistency in writing style and familiarity of the entire application, thus facilitating key linkages and alignment within the application. A potential disadvantage is that others will not benefit from the learning and ownership that occurs from writing the application and that it is difficult for any one person to know or gather all the information required for the application.

Multiple Internal Application Writers: Many organizations will involve many people in writing the application response, including key senior leaders and executives. A common approach is to have category leads or champions that own the response for their respective categories and/or items. The category/item ownership typically includes the responsibility for driving improvement in the assigned categories. One of the advantages of this approach is that the assignment extends beyond writing the application and creates an infrastructure for continuous assessment and improvement. A potential disadvantage is that multiple writers may become too narrowly focusing on their assigned section of the criteria and not seek the alignment and integration that is required. This could carry over into inconsistent approaches and lack of continuity in the written application.

Application Software: There are some software applications on the market that facilitate the application process. These applications provide prompts for responding to multiple requirements of the criteria and structuring the responses to create and application. Some of the software applications also provide the ability to compare your responses to a typical response of a high-performing organization and/or templates to assist with responses to some of the criteria requirements.

Available software to assist with applications at the time of publication includes.

- EasyApp by Total Quality Inc.
- IAS 2008 by Stevens Group Inc.
- Performance Organizer by JIT Software Limited

The decision to apply for a quality award is usually a major decision because of the resources required to complete the application. High-performing organizations will commit additional resources to a site visit when that is a part of the application process. Many organizations will begin with self-assessments or apply for a local or state awards program for a few years prior to applying for a national quality award. It is very rare for an organization to receive a national quality award during the first one or two applications submitted. Some organizations also decide to apply every other year, using the in-between years as periods to implement improvements based upon feedback from the prior application.

Impact on Performance Excellence Systems

One of the most significant impacts from national quality awards is the use of award frameworks and models to shape models of organizational excellence. Many organizations that have never applied for a national quality award have adopted internal excellence models that are the same as, or very similar to, award models or frameworks.

According to a report by Booz Allen Hamilton, a leading consulting firm, "The Baldrige Award enjoys very broad, positive recognition among leaders in each of the Baldrige Award-eligible sectors. . . . More than 70 percent of leaders surveyed among Fortune 1000 organizations said they are likely to use the Criteria for Performance Excellence" (Booz Allen Hamilton 2003).

Evidence of national quality award influence may show up in organizational manual and procedures; it may show up on strategic planning processes; and it may be demonstrated in balanced scorecard categories. For example, some U.S.-based organizations have designed their balance scorecard to match the Baldrige Items from Category 7, Results (see Table 17.10).

A typical approach would be to select key areas of importance that are similar to national quality award categories. The next step would be to define the policies and procedures that define organizational practices and processes within each area of focus. Organization specific terms, methods, and tools will be defined as a part of the documented quality management system. Organizational assessment tools may be developed that assess approaches and results specific to the organization-defined areas of focus, policies, and procedures.

Strategic Focus	Measure of Performance
Product results	Product defect rate User satisfaction rating
Customer results	Customer loyalty rating Customer complaints
Financial and market measures	Total revenue Net profit
Workforce measures	Employee turnover Employee cross training index
Process measures	On-time shipping Product lead time
Leadership measures	Regulatory compliance rating Community support

Table 17.10 A Balanced Scorecard Example Using Category 7 Results Items for Categories of Strategic Focus

Relationship to International Standards and Accreditation Agencies

ISO 9000 is a series of five international standards first published in 1987 by the International Organization for Standardization (ISO) in Geneva, Switzerland. Organizations can use the standards to help determine what is needed to maintain an efficient quality conformance system. For example, the standards describe the need for an effective quality system, for ensuring that measuring and testing equipment is calibrated regularly, and for maintaining an adequate record-keeping system. ISO 9000 registration determines whether an organization complies with its own quality system.

The Baldrige program states that "the purpose, content, and focus of the Baldrige Award and ISO 9000 are very different. Overall, ISO 9000 registration covers less than 10 percent of the Baldrige Award criteria."

Some organizations have used their ISO system to complete much more because they have expanded the use of their standards as an organizationwide system. This has led to some organizations stating that ISO can cover almost 50 percent of the award criteria for manufacturing organizations (NIST 2003).

The Joint Commission, previously known as the Joint Commission on Accreditation of Healthcare Organizations (JCAHO) is an independent, not-for-profit organization, established more than 50 years ago. The Joint Commission, which is governed by a board that includes physicians, nurses, and consumers, sets the standards by which health care quality is measured in America and around the world. Joint Commission provides accreditation services to hospitals and other health care services. Many health care providers have found that participating in a national quality awards program such as the Baldrige Award, has enhanced their standing and ability to maintain Joint Commission accreditation.

Relationship to Continuous Improvement Programs

Quality awards are one of many programs that organizations may choose to drive improvement in organizational performance. Some of the other approaches include total quality management (TQM), Six Sigma, Lean, and others. To many, these may seem to be competing approaches. This line of thinking is driven by the fact that each requires an investment in learning and applying the approach. However, many other organizations see these approaches as being complementary.

Total Quality Management

Total quality management (TQM) is still popular in many parts of the world. It is an organizationwide system to manage quality. Practices continue to evolve and are often influenced by changes in the criteria for national quality awards. Dr. Juran once stated that the Baldrige Criteria for Performance Excellence is the embodiment of those philosophies and practices called TQM.

Six Sigma

Six Sigma is a methodology used for reducing deficiencies in a product, process, or service by determining the sources of variation and systematically reducing or eliminating the causes of variation. The origin and recognition of Six Sigma improvement methodologies has a linkage to the early days of the Malcolm Baldrige National Quality Award. The late Bill Smith, a reliability engineer at Motorola, is widely credited with originating Six Sigma and getting Motorola's senior leaders to adopt this approach to achieving higher levels of product reliability performance. Not coincidentally, Motorola won the Malcolm Baldrige National Quality Award shortly after the rollout of Six Sigma. Receiving the Baldrige Award requires the winning organization to present its concepts to the world. Thus, as Six Sigma was maturing as an approach for improvement, quality professionals at Motorola were describing their methods to their colleagues and learning how far Motorola had advanced in comparison to other organizations.

In recent years, many Baldrige Award recipients have espoused the use of Six Sigma tools and methodologies as enabling them to achieve high levels of performance. Based upon the results achieved by organizations that use Six Sigma and have been recognized by awards programs, there is plenty of evidence that the approaches are compatible and supportive of each other for building performance excellence systems and results. Honeywell International, Inc., a U.S.-based manufacturer that produces a variety of consumer products, engineering services, and aerospace systems for a wide variety of customers, blended their Baldrige-based model called Honeywell Quality Value (HQV) with Six Sigma concepts. Edward M. Romanoff, Honeywell International's communications director for Six Sigma Plus and productivity, said that "HQV provides the framework for how one should run the business in total; Six Sigma gives you the quantitative specifics of what and how to improve" (Green 2000). Jim Hinton, President and CEO, of Presbyterian Healthcare Services (PHS) shares that PHS established a strategy of national excellence in 2002 and chose the Malcolm Baldrige National Quality Award as their quality framework as "the best way to get better faster." Along the journey came the recognition that to produce higher quality at lower cost there was a need for greater standardization and a relentless focus on process improvement. *"This is when we recognized that we needed an additional tool in our toolkit and add Lean Six Sigma." The focus of Lean Six Sigma is for improved efficiency and effectiveness. Using MBNQA framework for performance excellence coupled with Lean Six Sigma has taken us to another level of building reliability.*

Lean is an approach for enhancing customer value and eliminating wastes in a process. Lean is derived from, and sometimes referred to, as the Toyota Production System. Like Six Sigma, many Baldrige Award recipients in recent years have ***used*** Lean tools and approaches to improve ***their*** performance. In fact, many organizations ***use*** both ***Six Sigma*** and Lean approaches, with some using the terminology of Lean Six Sigma to acknowledge a blended approach ***to improve*** products, services, and processes. Although many organizations use Lean as an efficiency tool set, it usually begins by getting the quality right and then focus***ing*** on efficiency. Quality and performance excellence practitioners will need to determine the best methods to manage quality in a Lean environment.

Because most awards processes focus on systematic approaches for designing, controlling, managing, and improving processes, the introduction of Lean and Six Sigma programs are essential to an organization's ability to respond to the criteria of a national quality awards program.

References

Baldrige National Quality Program. (2008). Survey: Are We Making Progress?

Baldrige Performance Excellence. The Alliance for Performance Excellence. Available from Internet: <http://www.baldrigepe.org/>; Accessed 26 March 2009.

Baldrige Performance Excellence. The Foundation for the Malcolm Baldrige National Quality Award. Available from Internet: <http://www.baldrigepe.org/>; Accessed 26 March 2009.

Belter, M. (2009). A Short History of State and Local Programs.

Booz, A. (2003). Report, Assessment of Leadership Attitudes about the Baldrige National Quality Program. Assessment of Leadership Attitudes about the Baldrige National Quality Program submitted to NIST, p. 7.

Cargill Corn Milling North America. (2008). Application Summary, 2008 Manufacturing Recipient of the Malcolm Baldrige Award.

EFQM. Brussels, Belgium. Available from Internet: http://www.efqm.org/; Accessed 09 April 2009

EFQM Excellence Model. (2010). EFQM, Brussels, Belgium 2009.

Gemoets, P. EFQM Transition Guide, EFQM, Brussels, Belgium 2009.

Green, R. (2000). "Reshaping Six Sigma at Honeywell." *Quality Digest*, December.

Global Excellence Model Council. Brussels, Belgium. Available from Internet: http://www.excellencemodels.org; Accessed 25 March 2009.

Hoyt, G., and Ralston, E. (2001). *White Paper: Building on Baldrige: American Quality for the 21st Century*. Diane Publishing Co., Darby, PA.

Ibero American Foundation for Quality Management. Madrid (España). Available from Internet: http://www.fundibeq.org/DePortada/APremio.html; Accessed 23 April 2009.

International Organization of Standardization. Geneva, Switzerland. Available from Internet: http://www.iso.org/; Accessed 23 April 2009.

Japanese Union of Scientists and Engineers, The Deming Prize. Tokyo, Japan. Available from Internet: http://www.juse.or.jp/e/deming/; Accessed 24 March 2009.

Link, A. N., and Scott, J. T. (2001). Economic Evaluation of the Baldrige National Quality Program, National Institute of Standards and Technology.

NIST. (2001). The Malcolm Baldrige National Quality Improvement Act of 1987-Public Law 100-107, National Institute of Standards and Technology, Gaithersburg, MD. Available from Internet: <http://www.baldrige.nist.gov/Improvement_Act.htm>; Accessed 23 April 2009.

NIST. (2003). Frequently Asked Questions about the Malcolm Baldrige National Quality Award, National Institute of Standards and Technology, Gaithersburg, MD. Available from Internet: <http://www.nist.gov/public_affairs/factsheet/baldfaqs.htm>; Accessed 16 April 2009.

NIST. (2009). Baldrige Criteria for Performance Excellence, National Institute of Standards and Technology, Gaithersburg, MD. Available from Internet: <http://www.baldrige.nist.gov/Criteria.htm>; Accessed 23 April 2009.

The Joint Commission. Oakbrook Terrace, IL. Available from Internet: http://www.jointcommission.org/; Accessed 23 April 2009.

The Center for Performance Excellence of The University of Texas at Austin, Austin, TX. Available from Internet: http://www.utexas.edu/cee/cpe/awards/index.php?page=overview; Accessed 23 April 2009.

CHAPTER 18

Core Tools to Design, Control, and Improve Performance

Brian A. Stockhoff

About This Chapter

There is a growing understanding among organizations in all industries that an organized approach to attaining superior performance is essential for success in the competitive marketplace. Clearly, there is a great need for training on the methods and tools that will build upon that approach. In this chapter, we provide an introduction to each tool that is used with each planning, control, and improvement method described in this handbook. Organizations do not have to use, or master, the multiple tools that exist. However, there are a small number of tools that form the basis of the "core tools," tools that are used more often and by most employees. Other, more advanced, and complex tools are important but are used less often or for specific purposes. These tools are discussed in Chapter 19, Accurate and Reliable Measurement Systems and Advanced Tools. For each of the core tools we present, what they are, why we use them, steps for tool creation and use, and an example to demonstrate its use to enable the reader to better understand the application and potential relevance to their own situation.

High Points of This Chapter

1. Obtaining accurate, reliable, and relevant information happens when asking the right questions. Ask the right question, you will get the right data. But what do you

do with it, how much do you need, what tool should you use. These are some of the questions this chapter will answer.

2. The core tools are used and integrated within the structure of design, control, and improvement methods. This is a useful starting place for managers and teams to master these tools to be well prepared for many organizational problems they are likely to face.
3. There are many tools that are useful in managing an organization and are available for process improvement, design, and control. Tools for improvement require the testing of theories and finding root causes.
4. Design tools require the collection of opinions and specifications and then determining the means to develop new services or products that are reliable.
5. Control requires the use of statistical tools to help distinguish from common and special causes of variation to reduce risk, thereby facilitating appropriate intervention.
6. Mastering the top core tools can lead to a great improvement in the ability of an organization to attain needed data and information to be successful. Other tools are less popular and are used for special cases.
7. This chapter covers the core tools from A to Z.

Introduction

Dr. Kaoru Ishikawa, distinguished quality leader from Japan, wrote the classic *Guide to Quality Control* (1972). This book is generally credited as being the first training manual of problem-solving tools specifically presented for use in quality improvement. In the first publication of this book, it was used as a training reference for factory workers who were members of quality control (QC) circles. QC circles consist of a group of employees that work together to improve the performance of their work area.

In this handbook, we expand on Ishikawa's "seven quality tools" of work to include tools that find utility more broadly across design, control, and improvement methods. There remain many other useful tools; this list is not exhaustive, nor could it or any list be so. However, the list here is a useful starting place, for managers and teams to master these tools so that they are well prepared for many organizational problems they are likely to face.

The core tools are used and integrated within the structure of design, control, and improvement methods. Each method, such as Lean or Six Sigma (both improvement methods) use tools to complete each step in the method. Figure 18.1 shows an example of how each tool is used in an application map (in matrix form), with each column corresponding to a tool, and each row corresponding to a process step, with improvement expanded in detail. At each intersection is a symbol indicating the frequency of use of that tool at that process step (frequent, infrequent, and very rarely).

The process map is a valuable guide to problem-solving teams in the following ways:

- The map reminds the team that there is a structured order to the problem-solving process and helps keep the team on track.
- At a given step, if the team is at a loss what to do next, one of the frequently used tools may suggest the next action to take.

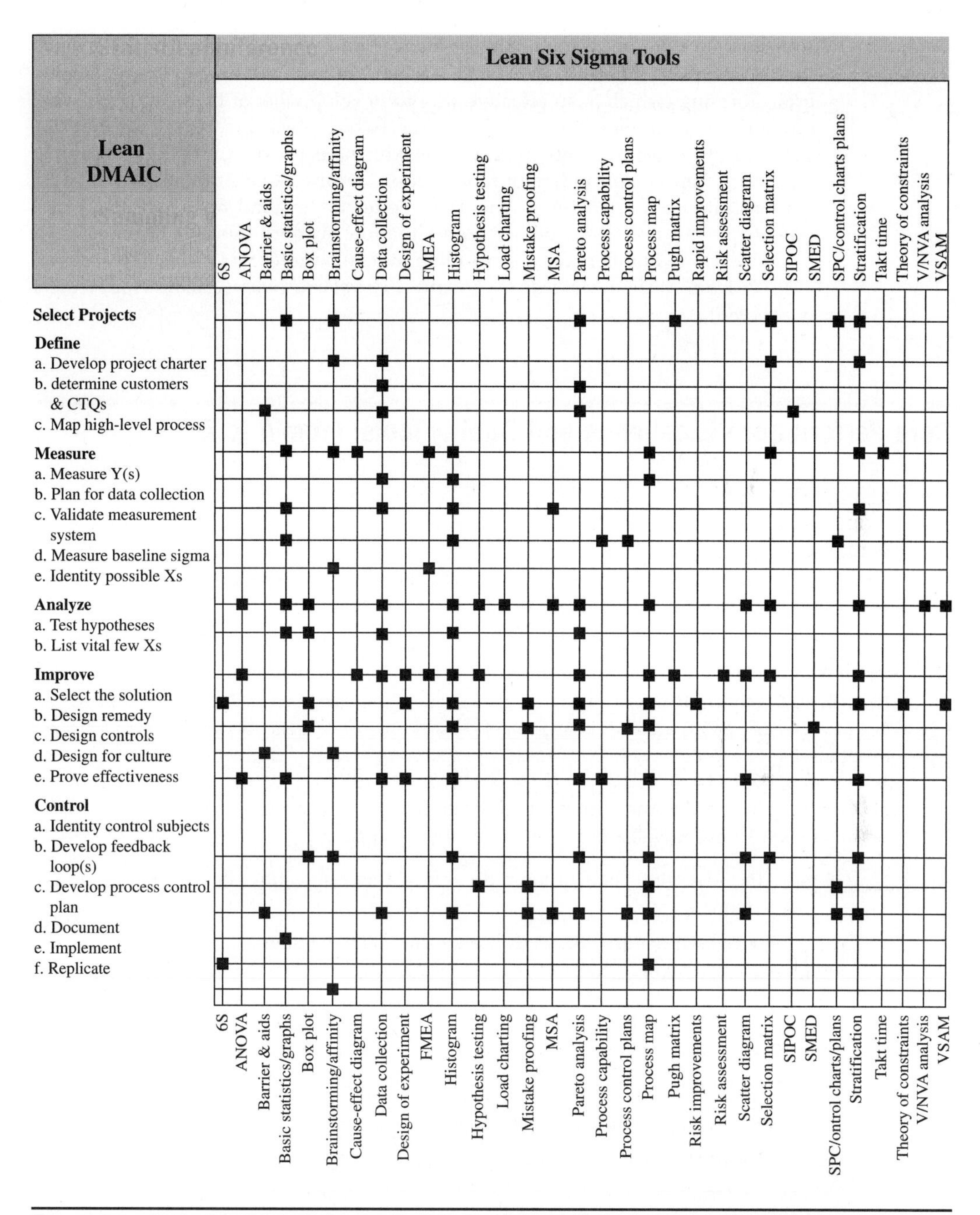

FIGURE 18.1 Application for quality improvement tools. (*Juran Institute, Inc. Used with permission.*)

- At any given step, using a tool that is indicated as one being rarely used is a signal to the team to reconsider its course of action. A convenient example is the use of brainstorming (which is an effective way to develop a list of theories, ideas, and opinions of group members) to test theories (which always requires data, not the opinions of the team members). After introducing each tool (what it is, and why we use it), steps are provided for tool creation and use. An example is provided to demonstrate its use, enabling readers to better understand the application and potential relevance to his or her own situation. A more thorough treatment is available from a number of texts, including the course notes for *Quality Improvement Tools* or from *Modern Methods for Quality Control and Improvement* (Wadsworth et al. 1986).

Core Performance Excellence and Quality Tools: From A to Z

Affinity Diagram

Purpose

The affinity process takes many items and sorts them into meaningful groups. It is used when soliciting variable information from customers or employees.

Steps to Create

1. Brainstorm ideas:
   a. Set a time limit.
   b. Record each idea on adhesive notes or 3 × 5 cards.
   c. Clarify ideas and eliminate duplicates.
2. Display the unsorted ideas on a table or stick them on a wall.
3. Sort the ideas into like groups; do this without speaking, based on individual perception.
   a. Arrange ideas into meaningful categories of "like issues."
   b. If one person does not like the placement of an idea, he or she can move it.
   c. If one idea seems to belong in more than one place, make a duplicate card.
   d. Continue sorting until a consensus is reached; aim for 5 to 10 groups.
   e. Consider breaking large groups into smaller ones.
4. Create a title or heading for each category.
5. Transfer the groups into an organized affinity diagram.
6. Discuss groupings and understand how they relate to each other; if necessary, move items to complete a consensus affinity diagram.

Example A customer focus group identified, through brainstorming, various positive attributes they would like to see in a child's toy. Each idea was written on an adhesive note and placed onto a board. Through sorting, it was determined that there were three distinct groups of attributes; these groups were given brief, descriptive names as shown in Figure 18.2.

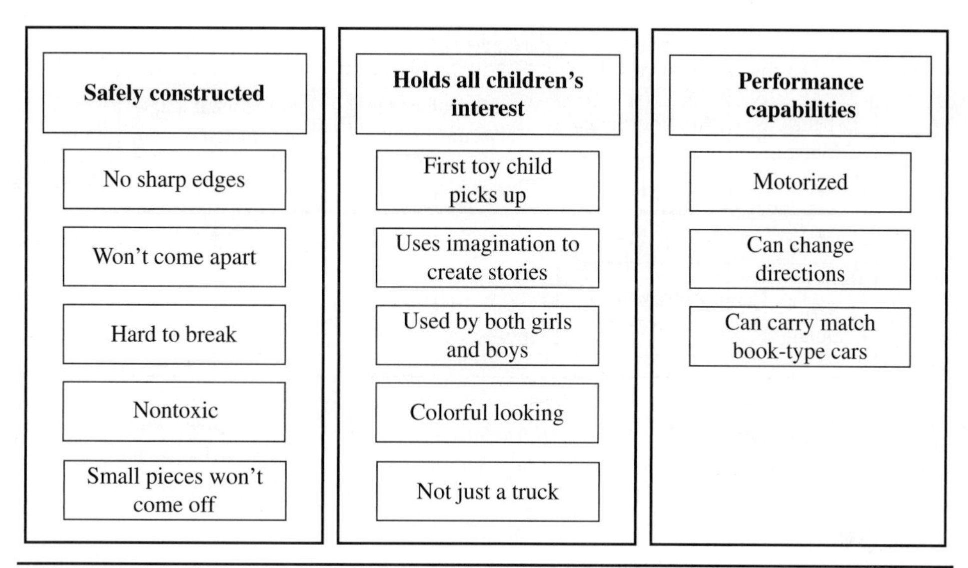

Figure 18.2 Affinity diagram. (*Lean Six Sigma Pocket Guide. Juran Institute, Inc., ©2009. Used with permission.*)

Barriers and Aids Chart

Purpose

A barriers and aids chart is a graphical means to define obstacles to improvement and listing corresponding means of alleviating these. The chart helps identify and overcome technical problems and cultural resistance that slow or prevent quality improvement.

Steps to Create

The major steps are to identify all likely sources of resistance (barriers) and support (aids), rate the barriers and known aids according to their perceived strengths, and identify countermeasures needed to overcome barriers. More specifically,

- Place a clear description of the objective (remedy) at the far right of the surface being used. Draw a heavy arrow pointing to the objective.
- Brainstorm a list of potential barriers.
- Select the vital few barriers that should be overcome and place them above the heavy arrow, labeled as barriers, with smaller arrows pointing down.
- Brainstorm a list of existing aids for overcoming the selected barriers.
- Select aids that will help overcome the barriers and place each one opposite the barrier(s) it will help overcome. Label them as aids.
- Identify any barriers that will not have adequate aids.
- Draw a horizontal line below the aids with a "countermeasures" label.
- Design the countermeasure for barriers without adequate aids, placing them opposite the barriers.
- Review the chart for missing vital-few barriers, the effectiveness of aids, and the effectiveness of countermeasures.

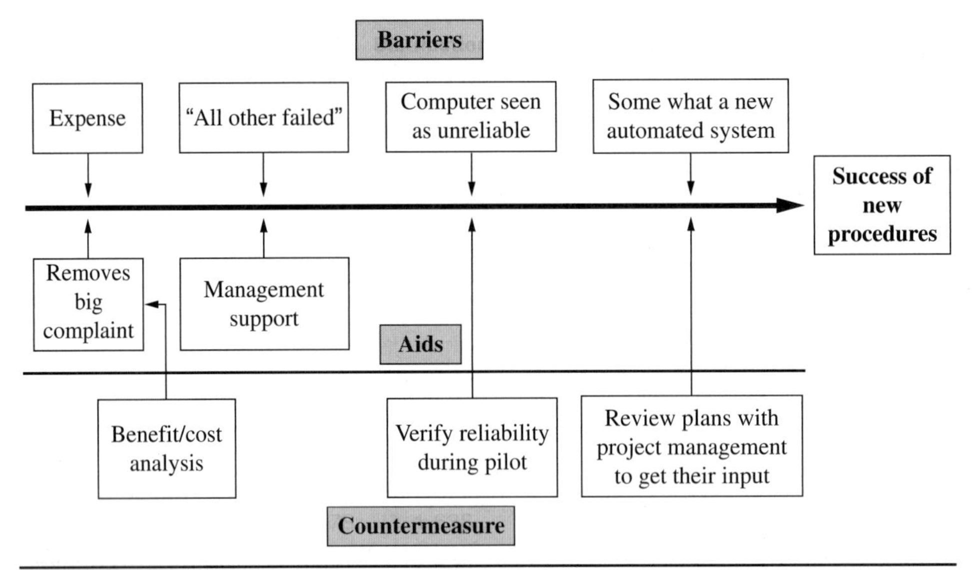

Figure 18.3 Barrier and aids chart. (*Lean Six Sigma Pocket Guide. Juran Institute, Inc., ©2009. Used with permission.*)

Example New Procedure for Engineering Studies. An improvement team developed a new procedure for completing engineering studies. When planning for cultural resistance to the remedy, the team came up with a list of barriers and aids most likely to influence its success. It also developed some countermeasures to overcome some of the barriers. The major barriers, aids, and countermeasures are shown in Figure 18.3. Note that while this chart is relatively simple, there often will be dozens of concerns to consider. The major barriers are placed at the top of the chart, the desired result at the right—much as in a cause-effect diagram.

The team brainstormed a list of existing aids that might help overcome the major barriers. Next, they reviewed the list and selected those that they believed might overcome the effects of some of the major barriers. Each aid is positioned underneath the barrier it is expected to overcome.

Finally, some barriers were not adequately overcome by existing aids, so the team developed specific countermeasures that it would implement. Those are placed below the aids (see Figure 18.3).

Basic Statistics

Purpose

Statistics are necessary to analyze and interpret data collected on a problem. Descriptive statistics help characterize problems and provide a starting point for more advanced statistical methods, such as hypothesis testing.

Measures of Central Tendency

- Mean—average value of a list of numbers
- Median—middle value in a sequential list of numbers
- Mode—value that occurs most often in a list of numbers

Measures of Dispersion Dispersion or distribution refers to the scattering of data around the central tendency. Its measures are given as follows:

- Range—difference between maximum and minimum values
- Variance—average squared deviation from each data point from the mean
- Standard deviation—square root of the variance

Types of Data Types of data direct the type of analysis to be done:

- Continuous data can be measured to an infinite level, e.g., time, temperature, thickness
- Categorical data falls into two categories:
  - Ordinal—can be arranged into some natural order (e.g., short, medium, tall)
  - Nominal—cannot be arranged into any natural order (e.g., colors, departments)

Sampling At times, it may be necessary to collect data on a sample of the population rather than use data from the entire population. The purpose of sampling is to draw conclusions about the population using the sample. This is known as statistical inference.

Key considerations for sampling are

- Sampling scheme: random, stratified
- Precision required (+/–?)
- Amount of variation in the characteristic
- Confidence level (e.g., 95 percent)
- Sample size

Qualities of a good sample include the following:

- Freedom from bias—bias is the presence or influence of any factor that causes the population or process being sampled to appear different from what it actually is.
- Representative—the data collected should accurately reflect a population or process. Representative sampling helps avoid biases specific to segments under investigation.
- Random—in a random sample, data are collected in no predetermined order, and each element has an equal chance of being selected for measurement. Random sampling helps avoid biases specific to the time and order of data collection, operator, or data collector.

Example A project team in a clinical setting was curious to know what the wait times of their patients were. The team decided to plot the times of the last 500 patients on a graph. The X-axis represented the total wait times and the Y-axis represented the total count of patients. The team concluded that most of the data centered around 60 minutes; however, some patients waited as little as 47 minutes and others waited as long as 75 minutes (see Figure 18.4).

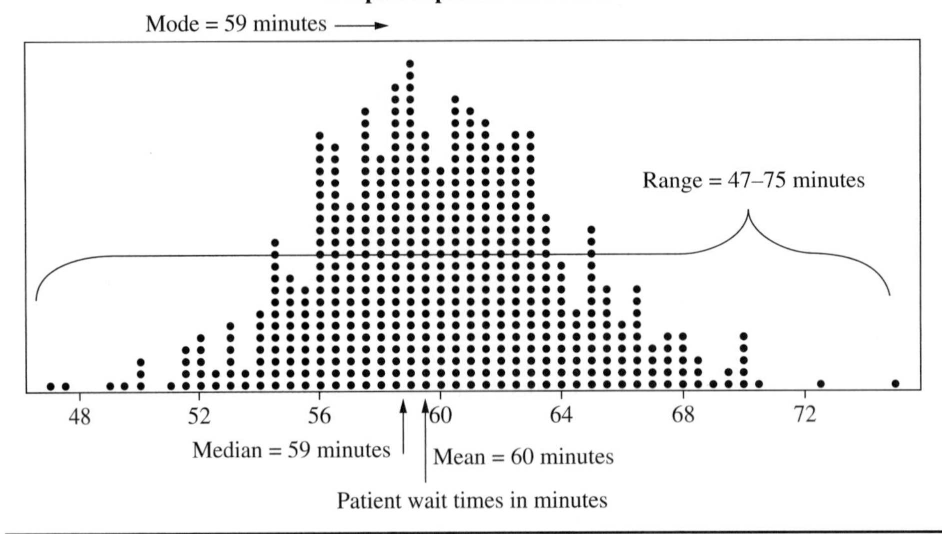

FIGURE 18.4 Central tendency and dispersion dotplot. (*Juran Institute, Inc. Used with permission.*)

Benefit/Cost Analysis

Purpose

A benefit/cost analysis characterizes the pros and cons of a solution. It is particularly useful when trying to make a business case for a quality improvement, or to decide among several alternatives.

Steps to Create

- Estimate one-time costs.
- Estimate additional annual operating costs.
- Estimate annual cost savings.
- If possible, calculate how much of the problem is likely to be eliminated because of each proposed remedy. Otherwise, rank the impact of alternatives.
- Assess the impact on customer satisfaction. Alternatives that reduce customer satisfaction should be discarded.
- Calculate net annual operating costs. A negative number means that net savings are expected.
- Calculate annual costs of one-time costs.
- Calculate total annual costs as the algebraic sum of net annual costs and the annual costs of one-time costs.
- Review data and rank the alternatives.

Example *Computerized vs. manual solution.* After identifying the root causes of a quality problem, one team decided that both computerized and manual solutions were feasible and met he customer

Remedy alternative	Manual	Computerized
One-time costs	$7,500	$134,000
Annual cost of one-time costs	$1,500	$26,800
Additional annual operating costs	0	$17,000
Annual cost savings	$1,462,200	$1,562,200
Net annual operating costs (savings)	($1,462,200)	($1,545,000)
Total annual costs (savings)	($1,460,500)	($1,518,200)
Problem impact	70%	75%
Customer satisfaction impact	Low	Low
Benefit/cost assessment rank team average	**1.7**	**1.3**

Figure 18.5 Cost/benefit analysis. (*Lean Six Sigma Pocket Guide. Juran Institute, Inc., ©2009. Used with permission.*)

criteria. To compare alternatives on a financial basis, the team applied a cost/benefit analysis, as shown in Figure 18.5. Using this information, the team ranked the two remedies. The better rank of the computerized solution (closer to 1) suggests that this would be the better of the two alternatives.

Box Plot

Purpose

This is a graphic, five-number summary of variation in a data set. The data are summarized by the smallest value, second quartile, median, third quartile, and largest value. The box plot can be used to display the variation in a small sample of data or for comparing the variation among groups.

Steps to Create

1. Collect the raw data and convert it to an ordered data set by arranging the values from the lowest to the highest.
2. Decide on the type of box plot you wish to construct.
3. Calculate the appropriate summaries.

   Depth of the median = $d(M) = (n + 1)/2$

   Depth of the first quartile = $d(\text{Q1}) = (n + 2)/4$

   Depth of the third quartile = $d(\text{Q3}) = (3n + 2)/4$

   Upper adjacent = the largest observation that is less than the third quartile – $(1.5 \times \text{IQR})$ (IQR = interquartile range)

   Lower adjacent = the smallest observation that is greater than the first quartile $- 1.5 \times \text{IQR}$
4. Draw and label horizontal axis.
5. Draw and label vertical axis.

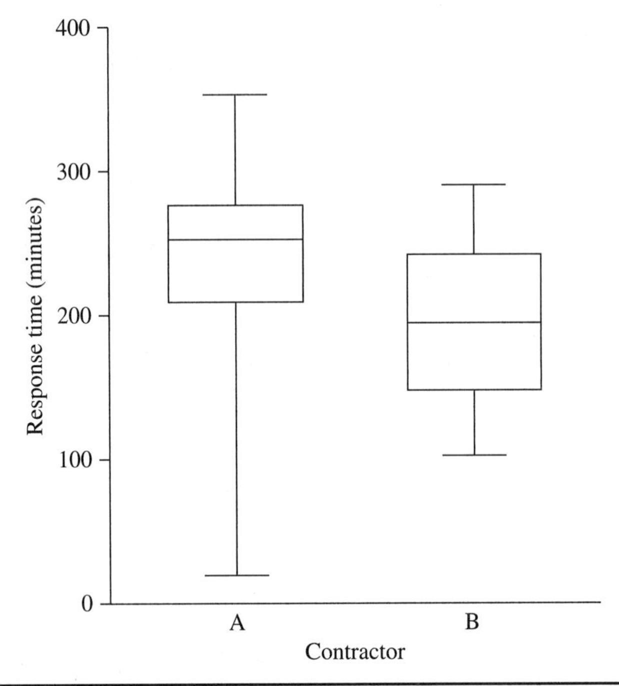

FIGURE 18.6 Basic box plot. (*Juran Institute, Inc. Used with permission.*)

Example Due to complaints that photocopiers were breaking down and failing to work, a study was conducted to see which contractor responded to the maintenance calls the quickest. The faster the contractor could respond, the faster the machines were up and running. For each contractor, the team gathered data from the last 10 calls and graphed the information to produce the box plots shown in Figure 18.6. Contractor B was not only quicker in responding to the calls, but the variation in response times were smaller, showing that contractor B was also more consistent than contractor A.

Brainstorming

Purpose

Creating a group technique for generating constructive and creative ideas from all participants. Use of this tool should provide new ideas or new applications and novel use of existing ideas. The technique is outlined in Figure 18.7.

- Good ideas are not praised or endorsed. All judgment is suspended initially in preference to generating ideas.
- Thinking must be unconventional, imaginative, or even outrageous. Self-criticism and self-judgment are suspended.
- To discourage analytical or critical thinking, team members are instructed to aim for a large number of new ideas in the shortest possible time.
- Team members should "hitchhike" on other ideas, by expanding them, modifying them, or producing new ones by association.

FIGURE 18.7 Brainstorming. (*Juran Institute, Inc. Used with permission.*)

Steps to Create

1. Phrase the statement to brainstorm.
2. Prepare for brainstorming.
   a. Communicate statement ahead of time.
   b. Provide appropriate surfaces for contributions.
3. Introduce session.
   a. Review conceptual rules.
      i. No criticism or evaluation of any kind.
      ii. Be unconventional.
      iii. Aim for quantity of ideas in a short time.
      iv. "Hitchhike" on others' ideas.
   b. Review the practical rules.
      i. Make contributions in turn.
      ii. Take only one idea per turn.
      iii. You may pass.
      iv. Do not provide explanations.
4. Warm up.
5. Brainstorm.
   a. Write issue where it will be visible to all.
   b. Have another person write all contributions where they will be visible.
   c. Stop before fatigue sets in.
6. Process ideas.

Example Prior to a meeting, members of a focus group were provided the following statement: "What are positive attributes you would like to see in a child's toy?" In the meeting, the team used adhesive notes to record and post ideas on a wall, making one contribution per turn. After all the ideas were recorded, they were processed to clarify and eliminate duplicates. (Refer to the Affinity Diagram section earlier for additional processing of ideas.)

Cause-Effect Diagram

Purpose

This tool, developed by Kaoru Ishikawa, is frequently called the Ishikawa diagram in his honor. Its purpose is to organize and display the interrelationships of various theories of the root cause of a problem. By focusing attention on the possible causes of a specific problem in a structured, systematic way, the diagram enables a problem-solving team to clarify its thinking about those potential causes and enables the team to work more productively toward discovering the true root cause or causes.

Steps to Create

1. Define clearly the effect (the Y) for which the cause must be identified.
2. Place the effect or symptom being explained at the right, enclosed in a box. Draw the central spine as a thick line pointing to it.

3. Use brainstorming or a rational step-by-step approach to identify the possible causes (the Xs).
4. Each of the major areas of potential (not less than two and normally not more than five) should be placed in a box and connected to the central spine by a line at an angle of about 70 degrees.
5. Add potential for each main area, placing them on the horizontal lines.
6. Add subsidiary causes for each cause already entered.
7. Continue adding possible causes until each branch reaches a potential root cause.
8. Check the logical validity of each causal chain. It should read "negative," i.e., a flat tire caused the car to swerve, a nail caused the tire to go flat, a person left a nail on the driveway.
9. Check for completeness.

Example A team was tasked with identifying the causes of bad photocopies (the effect). The team considered possible causes related to the 5Ms (man, machine, materials, methods, and measurement) and initially structured their fishbone diagram using these categories (they also briefly considered the 5Ps of plant, product, people, policies, and procedures). However, after further brainstorming, theteam realized that their possible causes fell into categories of handling, liquid, copying paper, environment, original, and copying machine. The team restructured their original diagram into that shown in Figure 18.8.

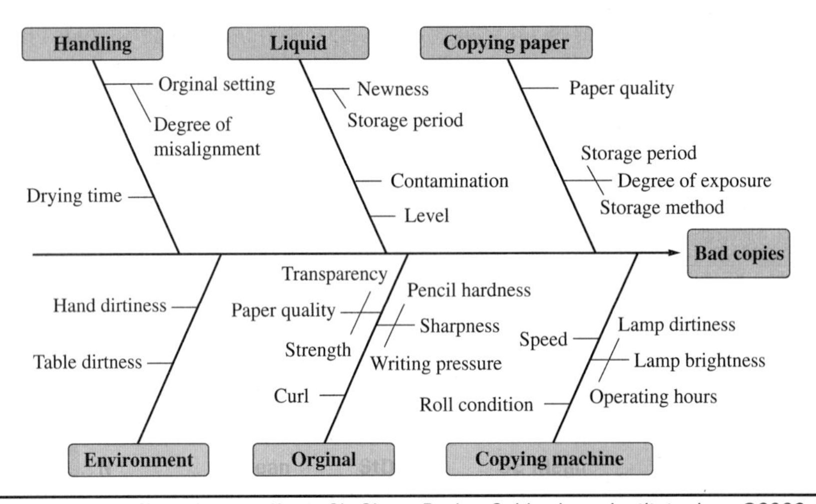

FIGURE 18.8 Cause-effect diagram. (*Lean Six Sigma Pocket Guide. Juran Institute, Inc., ©2009. Used with permission.*)

Check Sheets

Purpose

Check sheets are used to collect and analyze data. They are a type of graph or chart that is formatted to allow immediate conclusions to be drawn regarding the data, including patterns and trends.

Steps to Create

1. Title the top of the sheet with the name of the item or process being analyzed.
2. Determine if an immediate analysis can be completed or if observation of the subject is necessary to gather the data. If it is completed, record the time period.

3. Provide any additional information about the analysis that will be helpful for others to review (i.e., who is conducting the analysis, the date and time of the analysis, and where and why the analysis is being done).
4. List in a column on the right-hand side of the sheet what subcategories or items are being tallied. If more than one part is being analyzed, create separate sections with individual headers.
5. Record the information in bundles of up to five tallies per bundle, as shown in the following example.

Example Customer service for a large television manufacturer noticed higher than usual complaints about failing parts on three types of televisions. The repair department began a quick assessment of televisions that were returned for repair. The repair department used a check sheet to determine which parts needed replacing and in what quantity. The following check sheet was created for television set models 1013, 1017, and 1019 (see Figure 18.9).

COMPONENTS REPLACED BY LAB

Enter a mark for each component replaced. Mark like the following: / // /// //// HHT

Time Period: 22 Feb to 27 Feb 1988
Repair Technician: Bob

TV SET MODEL 1013

Integrated circuits	HHT
Capacitors	HHT HHT HHT HHT HHT //
Resistors	//
Transformers	////
Commands	
CRT	/

TV SET MODEL 1017

Integrated circuits	///
Capacitors	HHT HHT HHT HHT HHT //
Resistors	/
Transformers	//
Commands	HHT HHT HHT ///
CRT	/

TV SET MODEL 1019

Integrated circuits	/
Capacitors	HHT HHT HHT HHT ///
Resistors	/
Transformers	//
Commands	
CRT	/

FIGURE 18.9 Components replaced by lab. (*Juran Institute, Inc. Used with permission.*)

Control Plan

Purpose
A control plan is used to hold the gain obtained through a quality improvement project. The plan specifies control subjects to be monitored, measurement of these subjects, and actions to take based on explicit criteria.

Steps to Create
1. Identify variables that affect the remedy and the customer directly or indirectly.
2. Establish the standard that will trigger action. The best standard is a control limit from a control chart because it is achievable.
3. Establish how each control variable will be measured. Enter this information on the same line as the control variable under the column "How measured." Make similar entries for each of the following steps.
4. Establish where and when the measurements will be made and how they will be recorded, including the type of control chart used.
5. Decide who will analyze the ongoing measurement—that is, who will determine that the process is out of control.
6. Decide who will diagnose and eliminate the assignable cause for the out-of-control condition.
7. Decide what steps can be taken to bring the process back into control. Although it may not be possible to foresee all problems, identifying specific actions ahead of time will make control much more effective.
8. Review the matrix to verify that
   - All critical control variables have been identified.
   - The control plan will bring the process back into control quickly.
   - The control plan makes maximum use of self-control.

Example Improvement Team 3 completed their analysis of the process and concluded that four main subjects would be monitored to sustain the overall improvements. If the process becomes unstable and the criteria for taking action is met, management and team members know exactly who is responsible and properly trained to address the out-of-control situation. A detailed description of the action taken to bring the process back into control is later documented in the plan (see Figure 18.10).

Customer Needs Spreadsheet

Purpose
To help analyze and prioritize customer needs. Often, information that is collected from customers is too broad, too vague, and too voluminous to be used directly in designing a product. Prioritizing customers and their needs ensures that a team focuses on what is most important to a design and ensures that budgeted resources are allocated accordingly.

Steps to Create
1. Create a multicolumn spreadsheet.
2. Label the first column "Customers." List, in priority order, the vital few customers. Include groups of "useful many" customers that, collectively, can be considered as the vital few.

Process control plan for: improvement team 3				Date: 11/22/2009				Revision level: 2.1		Approved by: Champion		
Control subject	Subject goal	Unit of measure	Sensor	Frequency of measurement	Sample size	Recording of measurement/ tool used	Measured by whom	Criteria for taking action (i.e. when to take action)	What actions to take	Who decides	Who acts	Record of action taken
Spray delivery capacity	10 gallons per minute	Gallons per minute	Water meter	During start-up of every job	Each job	Gauge	Foreman	>11 gallons and <9 gallons	Reduce flow speed Increase flow speed	On-site worker assigned to "decision" role for project	Specialist II	None
Crew size	One person per 100,000 sq. t. of yard	Number of workers per 10,000 sq feet	Foreman	During start of daily run for each job and yard	Each job	Visual count	Foreman	> = 12 and < = 3 workers call office	See office manager	One-site foreman	Specialist II	None
Schedule forecast no PC to determine to/from	Forecast times always within 10% of actual	Number of workers	Foreman	Every job	Each job	Visual count	Foreman	Actual vs. estimated # of workers varies by 7	Adjust program such that variance –3	On-site foreman	Supervisor	TBD
Schedule forecast on PC to determine work need	Forecast times always within 10% of actual	Location of job	Foreman	Every job	Each job	Visual count	Foreman	Actual vs. estimated # of workers varies by 7	Adjust program such that variance –3	On-site foreman	Supervisor	TBD

FIGURE 18.10 Process control plan. (*Juran Institute, Inc. Used with permission.*)

3. Label the top row, "Customer Needs," and list all discovered needs in the columns below. Enter one need for each column.
4. Correlate the relationships between customers and needs.
   a. Create a legend to define the relationship.
   b. Base the relationship on solid evidence.
   c. More than one customer can be addressed by the same need.
   d. Enter the appropriate value where needs and customers intersect.
   e. Review the spreadsheet and add any additional customers or needs that have been left off the list.
5. Go back and summarize the data you have collected.
6. Analyze each need in terms of
   a. Strength of the relationship between needs and customers.
   b. Customer's importance
7. Determine criteria and prioritize the needs from most critical to least critical.

Example An analysis was conducted for the design and production of a new magazine. A list of customers and their needs was determined from a previous analysis and placed in a Customer Needs spreadsheet. The left column lists, in priority order, all external and internal customers. Column headings are the various needs that have been discovered. Upon completion of the analysis, the team determined that they would need to focus on two main areas of the magazine in order to ensure the most optimal results. First, they would need make the magazine attractive to their readers and create catchy cover lines. Second, they would need to ensure that the content is complete and free of errors so that production of the magazine could begin on schedule (see Figure 18.11).

Customers	Customer needs							
	Attractive	Informative and well-written articles	Catchy cover lines	Stable circulation	It sells	Enough complete	Material complete	No last minute changes
Readers	●	●	○					
Advertisers	●	○	●	●	●			
Printers						●	●	●
Typesetters						●	●	●
Color separators						●	●	●
Newsstand	●	○	●	○	●			

Legend
● Very strong relationship
○ Strong relationship
△ Weak relationship

FIGURE 18.11 VOC/customer needs spreadsheet. (*Lean Six Sigma Pocket Guide, Juran Institute, Inc., 2009. Used with permission.*)

Failure Mode and Effects Analysis

Purpose

A failure mode and effects analysis (FMEA) helps identify possible ways in which failures can occur and the effects of these failures. The tool also helps prioritize these based on risk and track subsequent actions to reduce risk. Many types of risk exist, but the most common are design (or product) and process FMEA.

Steps to Create

1. Create a nine-column spreadsheet.
2. Create an assigned-value table.
3. In column 1, list all possible modes of failure. Each item should be on a separate line.
4. In the next column, identify all possible causes of failure for each mode.
5. In column 3, determine the effect each failure will have on the customer, the overall product, other components, and the entire system.
6. Note: For steps 6, 7, and 8, use values established in the assigned-values table.
7. Evaluate the frequency of occurrence. Enter the appropriate integer in column 4.
8. Evaluate the degree of severity of the effect of each failure. Record the appropriate value in column 5.
9. Evaluate the chance of detection for each cause of failure. Place this number in column 6.
10. Calculate the risk priority factor by multiplying columns 4, 5, and 6. Enter the result in column 7.
11. Design an action/remedy for only those vital few causes with the highest risk factors. Reduce the level of failure to a rate that is acceptable.
12. Validate each action/remedy.

Example A bank established an improvement team to improve its services related to new checking accounts. The team looked at a variety of different components of this service, including the printing of new checks. Because failures in this component could directly and indirectly affect customers, an FMEA approach was taken to identify and characterize potential failures (see Figure 18.12).

Flow Diagram/Process Map

Purpose

A graphic representation of the sequence of steps needed to produce some output. The output may be a physical product, a service, information, or a combination of the three. The symbols of a flow diagram are specific to function and are explained in Figure 18.13.

Steps to Create

1. Discuss how you intend to use the flow diagram or process map.
2. Decide on the desired outcome of the session.

Product: New checking account
Component: Printing new checks

1	2	3	4	5	6	7	8	9
Mode of failure	**Cause of failure**	**Effect of failure**	**Frequency of occurrence (1–10)**	**Degree of severity (1–10)**	**Chance of detection (1–10)**	**Risk priority (1–1000) (4) × (5) × (6)**	**Design action**	**Design validation**
Checks being printed incorrectly	Incorrect information on application form	Checks have to be re-issued	4	6	8	192	Clerk reviews information with customer	Clerk initials form after review
	Data entry error	Ditto	8	6	5	240	Review step in software	Run software
	Information entered in the wrong field on application field	Ditto	5	6	2	60		

Note following assigned values

Column/value	1	2	3	4	5	6	7	8	9	10
4. Frequency (errors per 10,000 customers)	< 2	4	8	10	15	20	25	30	35	< 35
5. Severity for customer	Trivial				Cause complaint			Major time or $		Loss of customer
6. Detection	Certain				Possible					None

FIGURE 18.12 Failure mode and effects analysis. (*Lean Six Sigma Pocket Guide, Juran Institute, Inc., 2009. Used with permission.*)

3. Define the boundaries of the process. Show the first and last steps, using appropriate flow diagram symbols.
4. Document each step in sequence, starting with the first (or last) step. Lay out the flow consistently from the top to bottom or left to right.
5. When you encounter a decision or branch point, choose one branch and continue.
6. If you encounter an unfamiliar segment, make a note and continue.
7. Repeat steps 4, 5, and 6 until you reach the last (or first) step in the process.
8. Go back and make a flow diagram for the other branches from the decision symbols.
9. Review the completed chart to see if you have missed any decision points or special cases.
10. Fill in unfamiliar segments and verify accuracy.
    a. Observe process.
    b. Interview knowledgeable people.
11. Analyze the flow diagram.

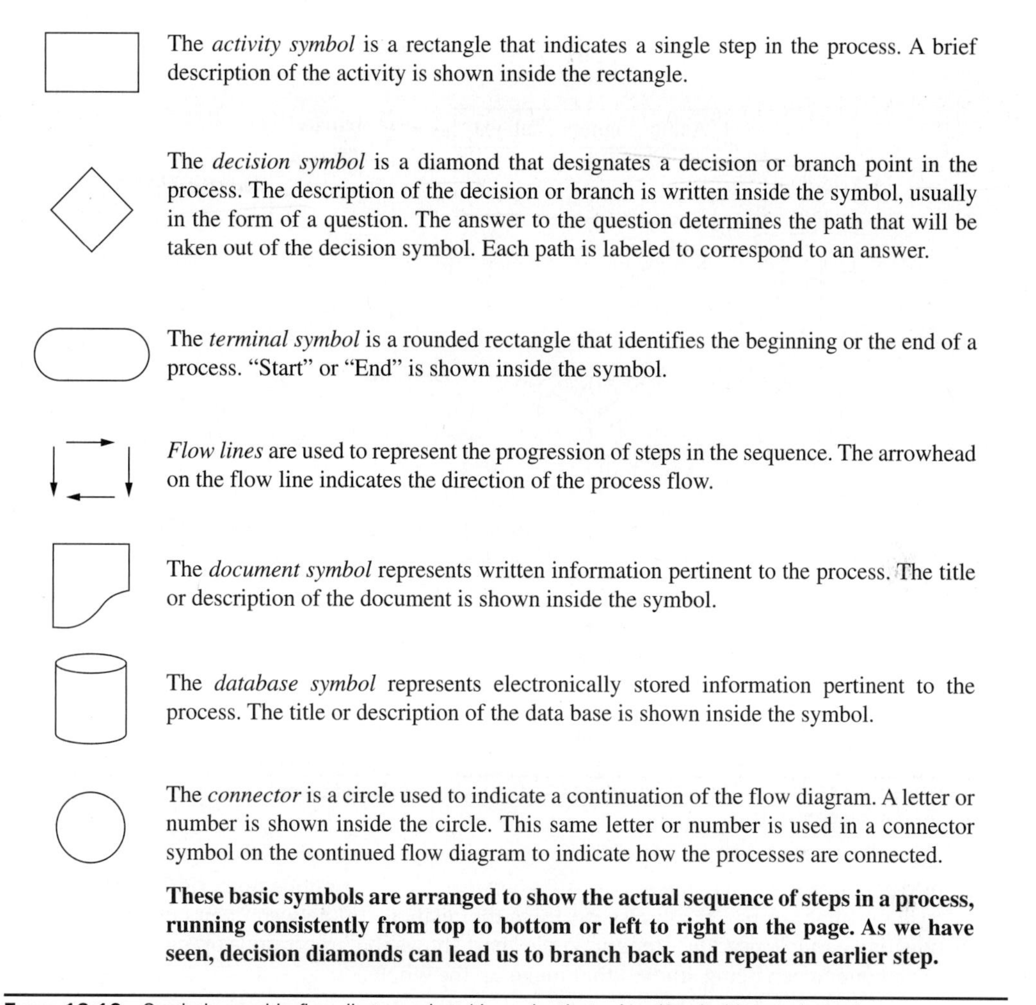

FIGURE 18.13 Symbols used in flow diagramming. (*Juran Institute, Inc. Used with permission.*)

Example A team was tasked with improving the process for distribution of technical manuals and realized they needed to understand the current process steps and boundaries better. Drawing upon the expertise of people directly involved in the process, the team developed the flow diagram shown in Figure 18.14. This baseline process map also served later to help the team identify potential problems contributing to errors and delays in distribution.

Graphs and Charts

Purpose

A broad class of tools used to summarize quantitative data in pictorial representations. Three types of graphs and charts that prove especially useful in quality improvement are line graphs, bar graphs, and pie charts. A line graph connects points that represent pairs of

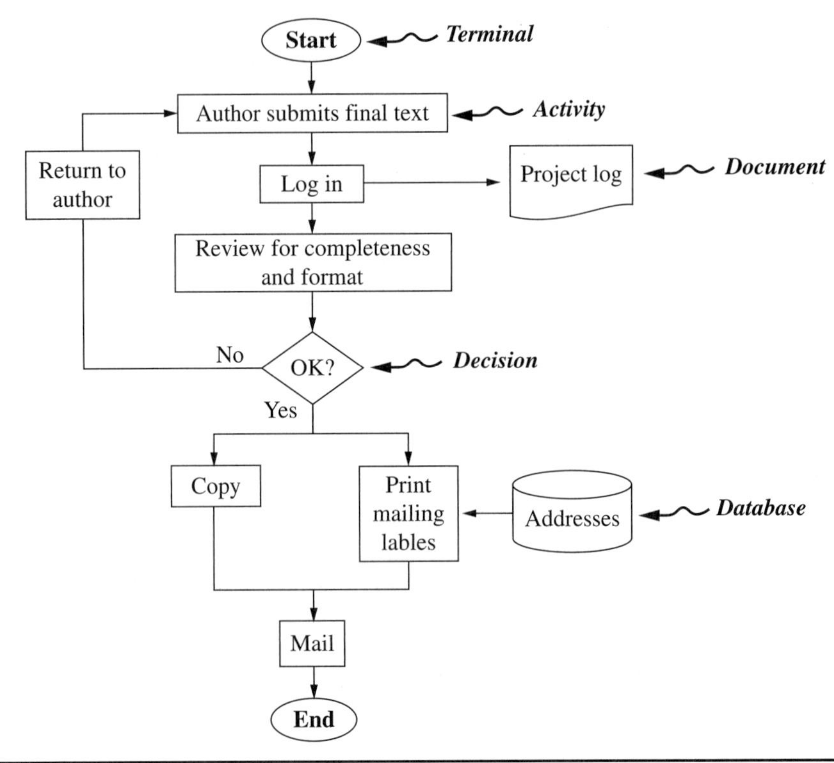

Figure 18.14 Flow diagram/process map. (*Lean Six Sigma Pocket Guide, Juran Institute, Inc., 2009. Used with permission.*)

numeric data to display the relationship between two continuous numerical variables (e.g., cost and time). A bar graph also portrays the relationship between pairs of variables, but only one variable need be numeric. A pie chart shows the proportions of the various classes of a phenomenon being studied that make up the whole.

Steps to Create

Line Graphs

1. Determine the range of the vertical axis and the size of each increment. Label the vertical axis.
2. Do the same for the horizontal axis.
3. Draw axes and, if needed, a grid.
4. Plot each data point.
5. Connect the points with a line.
6. Label and title the graph.

Bar Graphs

1. Determine the range of the vertical axis and the size of each increment. Label the vertical axis.
2. Choose a simple, grouped or stacked bar graph.

3. Determine the number of bars. Draw and label the horizontal axis.
4. Determine the order of the bars.
5. Draw the bars.
6. Give the graph a title.

Pie Charts

1. Determine the percentage for each category.
2. Convert the percentage values into degrees.
3. Draw and circle with a compass and mark the segments of the pie chart with a protractor.
4. Label the segments and title the chart.

When creating graphs and charts, keep the following in mind:

Graphic integrity. A graph must not lie. It should be constructed so that the viewer is not misled. Rather than relying solely on the graphics, look at the written data to ensure that the true information is conveyed by the graph.

Consistent scale. Numeric scales must show regular intervals. Different graphs that might be compared to each other should all be drawn to the same scale.

Ease of reading. How well a graph is understood and remembered depends on how easy it is to read. Use labels to improve clarity. Place labels close to the object being identified.

Consistency of symbols. When two or more graphs are to be compared, it is important to maintain consistency along many dimensions to minimize confusion in interpreting the graphs.

Simplicity. Do not obscure information with unnecessary decoration. Before adding text or decoration to a graph, ask, "What additional value or information am I adding?"

Example Line graph. The hours a computer operating system was not available is plotted over time for two different computing centers (see Figure 18.15*a*).

Bar graph. Customer complaints stratified by type (cosmetic, dimensional, electrical) are shown across different months (see Figure 18.15*b*).

Pie chart. The proportion of time sales staff spent in different activities is shown as different sized slices of the overall pie.

Histogram

Purpose

A histogram is a graphic summary of variation in a set of data. Four concepts related to variation in a set of data underlie the usefulness of the histogram: (1) values in a set of data almost always show variation, (2) variation displays a pattern, (3) patterns of variation are difficult to see in simple numerical tables, and (4) patterns of variation are easier to see when the data are summarized pictorially in a histogram. Analysis consists of identifying and classifying the pattern of variation displayed by the histogram (such as the shape, the location of the center, or the spread of the data from the center) and relating what is known about the characteristic pattern to the physical conditions under which the data were created to explain what might have given rise to the pattern in those conditions. Figure 18.16 illustrates some common patterns.

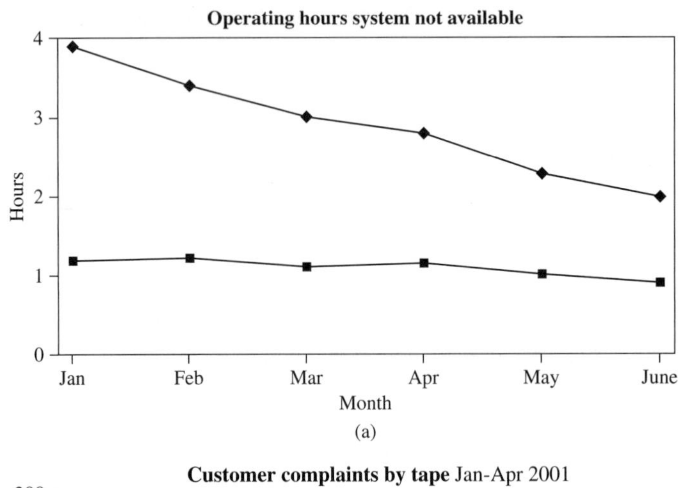

(a)

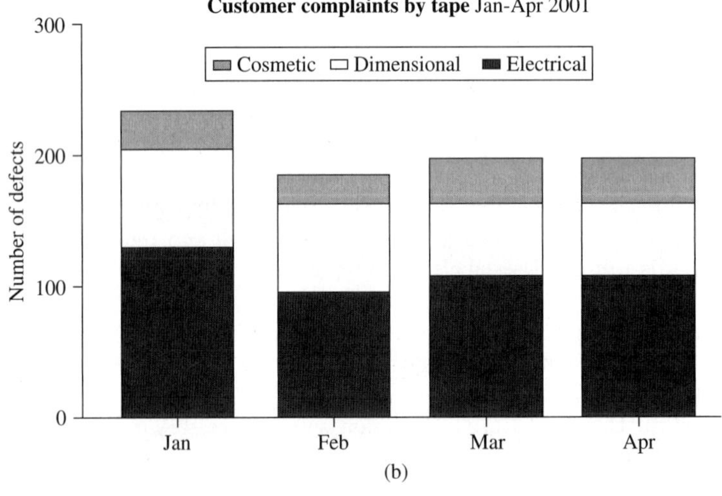

(b)

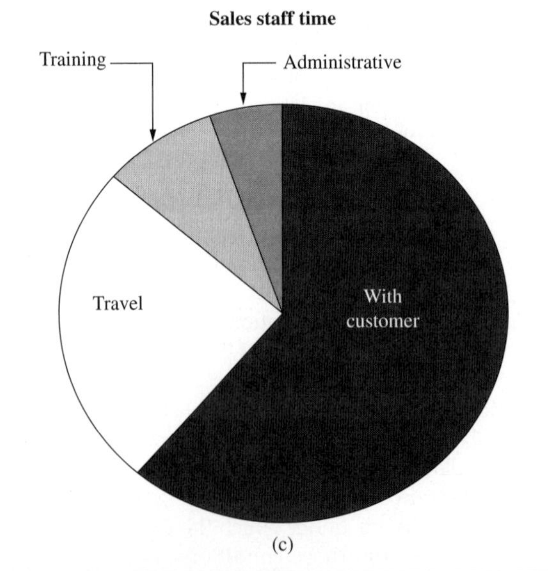

(c)

FIGURE 18.15 (a) Line graph, (b) bar graph, (c) pie chart. (*Lean Six Sigma Pocket Guide, Juran Institute, Inc., 2009. Used with permission.*)

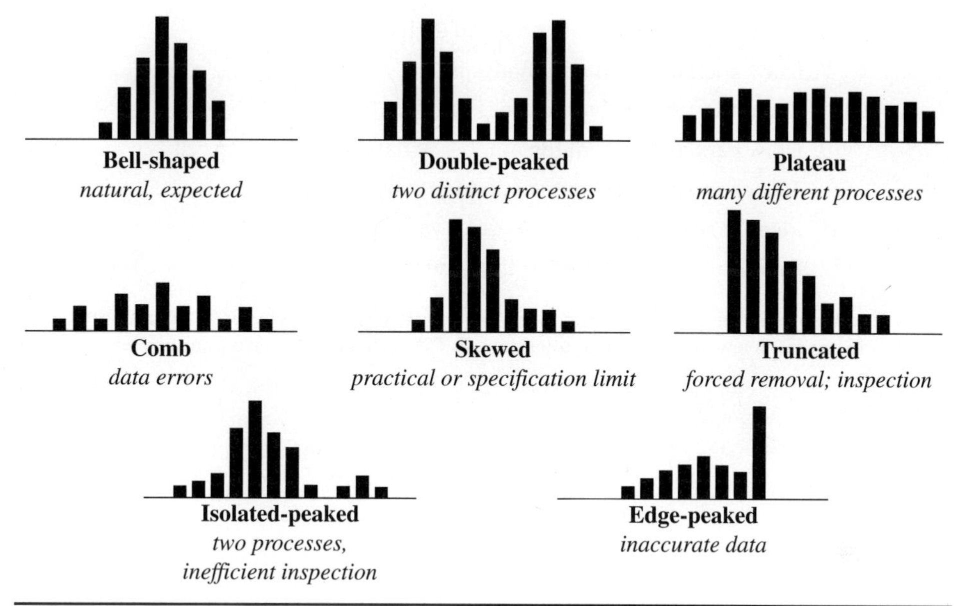

FIGURE 18.16 Histograms. (*Lean Six Sigma Pocket Guide, Juran Institute, Inc., 2009. Used with permission.*)

Steps to Create

1. Determine the high value, the low value and the range.

$$\text{Range} = (\text{high value}) - (\text{low value})$$

2. Decide on the number of cells.

Data Points	Number of Cells*
20–50	6
51–100	7
101–200	8
201–500	9
501–1000	10
Over 1000	11–20

*Less than 40 only as a result of stratification.

3. Calculate the approximate cell width.

$$\text{Approximate cell width} = (\text{range})/(\text{number of cells})$$

4. Round the cell width to a convenient number.
5. Construct the cells by listing the cell boundaries.

6. Tally the number of data points in each cell.
7. Draw and label the horizontal axis.
8. Draw and label the vertical axis.
9. Draw bars to represent the number of data points in each cell.
10. Title chart and indicate total observations.
11. Identify and classify the pattern of variation.
12. Develop an explanation for the pattern.

Example The data in Figure 18.17*a* show the days that elapsed between an interdepartmental request for an interview and the actual interview. In Figure 18.17*b*, the histogram helped a team recognize

14	15	18	19	13
12	18	22	14	18
15	17	19	15	17
18	20	10	15	15
20	14	17	20	21
15	24	15	18	14
23	13	23	21	20
18	21	18	15	15

(a)

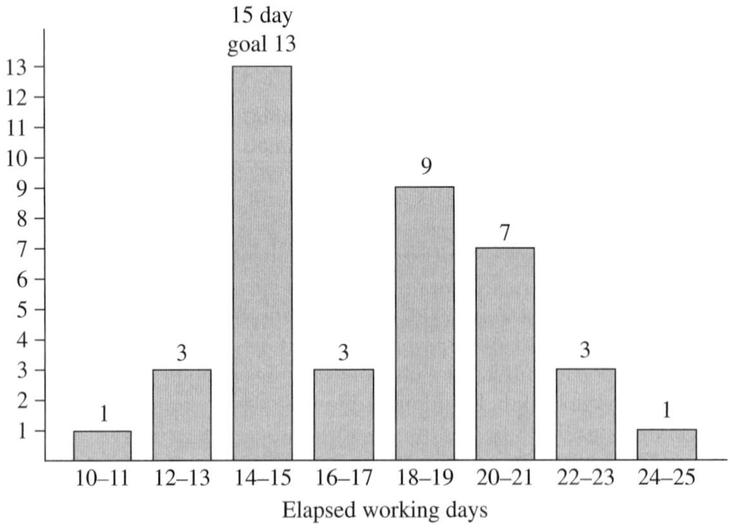

15 day goal 13

10–11	12–13	14–15	16–17	18–19	20–21	22–23	24–25
1	3	13	3	9	7	3	1

Elapsed working days

(b)

FIGURE 18.17 (a) Data showing elapsed time (in working days). (b) Histograms of elapsed working days. (*Receipt of Request to Preliminary Review and Contact with Manager, Juran Institute Inc. Used with permission.*)

the unacceptable range of time elapsed from request to interview. It also provided the team with a vivid demonstration of a human-created phenomenon—the rush at day 15 to get as many requests completed within the 15-day goal. The histogram directed the team's attention to steps needed to reduce the duration (and with it the spread) of the process.

Pareto Analysis

Purpose

A tool used to establish priorities, dividing contributing effects into the "vital few" and "useful many." A Pareto diagram includes three basic elements: (1) contributors to the total effect, ranked by the magnitude of contribution, (2) magnitude of the contribution of each expressed numerically, and (3) cumulative percent of the total effect of the ranked contributors.

Steps to Create

Pareto charts are not as commonly present in software as are similar types of graphical analysis tools. The steps for creating the chart manually help us to understand the different elements.

1. Total the data of each contributor and sum these to determine the grand total.
2. Reorder the contributors from the largest to the smallest.
3. Determine the cumulative percent of the total for each contributor.
4. Draw and label the left vertical axis from 0 to the grand total or just beyond.
5. Draw and label the horizontal axis. List contributors from largest to smallest, going from left to right.
6. Draw and label the right vertical axis from 0 to 100 percent Line up the 100 percent with the grand total on the left axis.
7. Draw bars to represent the magnitude of each contributor's effect.
8. Draw a line graph to represent the cumulative percent of the total.
9. Analyze the diagram. Look for a break point on the cumulative percent graph.
10. Title the chart; label the "vital few" and the "useful many."

Example To assist in the decision process of selecting new improvement projects, a Pareto chart was created to better understand the types of questions received from customers Based on this information, it was decided to charter a team to address queries falling into categories A, B, and C (see Figure 18.18).

Planning Matrix and Tree Diagram

Purpose

A tree diagram is a graphical method used to identify all the parts that are needed to create a final objective. The planning matrix is an extension of the tree diagram; it shows all the factors, components, and tasks required to achieve the final objectives. The diagram specifies who will complete each step and when. It may also specify for each step who will help, what the budget will be, who the team contact is for work done outside the team, and the status of the task.

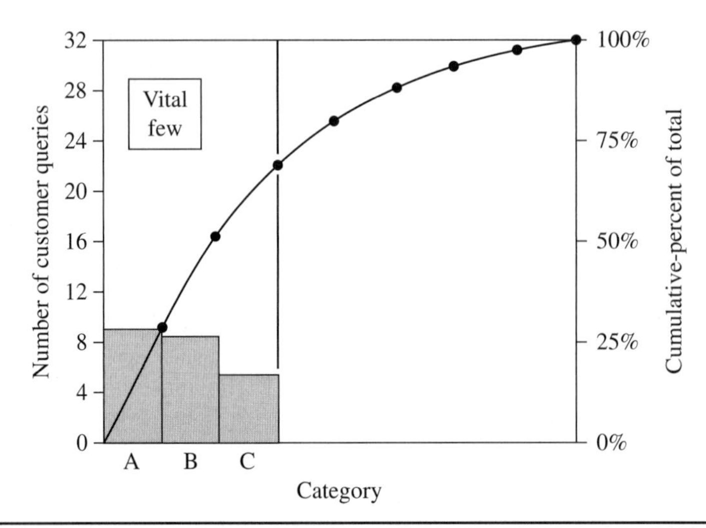

Figure 18.18 Pareto analysis. (*Lean Six Sigma Pocket Guide, Juran Institute, Inc., 2009. Used with permission.*)

Steps to Create

1. Use a tree diagram to identify all the tasks needed to complete a specific piece of work.
2. List each task on an adhesive note and post on the wall or a flip chart in a vertical column.
3. Label other columns with "who" and "when."
4. Work through the task one by one, taking the following steps.
   a. Discuss and identify the most appropriate person or group of persons to do the work.
   b. Agree on the necessary completion date.
5. Agree on how the team will monitor progress on the plan. Possibilities include the following.
   a. An agenda item at each meeting, obtain a brief report on active tasks.
   b. Have the team leader (or a designated member) check with each responsible person before the meeting and enter on the agenda only those tasks that require discussion.
6. Transfer the matrix to standard paper and include it with future team minutes and agendas.

Example An improvement team determined that a new medical record-tracking process was needed to reduce lost and missing paper records. To help plan for the creation of this new process, the team broke the process down into four major parts, and subsequently broke these parts down further into the elements necessary and sufficient to complete the major step (answering the question "What needs to be done?"). Using this tree diagram, the team proceeded to develop it further into a planning matrix by the addition of "who" and "when" information (see Figure 18.19).

New medical-record tracking procedures

		What	Who	When
New medical-record tracking process	Use new out-guides	Design new guides		
		Order and inventory		
	Use new medical record covers	Design new covers		
		Order and inventory covers		
		Install covers		
	Computer print out same-day appointment	Specify system change		
		Write new program		
		Test		
		Install		
	Medical record staff inserts same-day slips and uses them	Write procedure		
		Prepare training		
		Conduct training		
		Conduct dry run		

Figure 18.19 Planning matrix and tree diagram. (*Lean Six Sigma Pocket Guide, Juran Institute, Inc., 2009. Used with permission.*)

Pugh Matrix

Purpose

This is a useful tool for comparing several alternative concepts against preestablished criteria and allows you to

- Compare alternative solutions against project critical to quality characteristics or requirements (CTQs).
- Create strong alternative solutions from weaker ones.
- Arrive at an optimum solution that may be a hybrid or variant of other solutions.

Steps to Create

Enter the criteria. Customer CTQs must be included. Business criteria—such as time to market, complexity, and ability to patent—can also be added.

1. Weight the criteria in terms of importance.
2. Rate each alternative as better (+), worse (–), or same (s) at achieving criteria as compared to datum alternative.
3. Compare the number of positives, negatives, and same (s) between alternatives. Can you create a new alternative that leverages the best of the initial alternatives?

Example An example template for the construction of a Pugh matrix is shown in Figure 18.20.

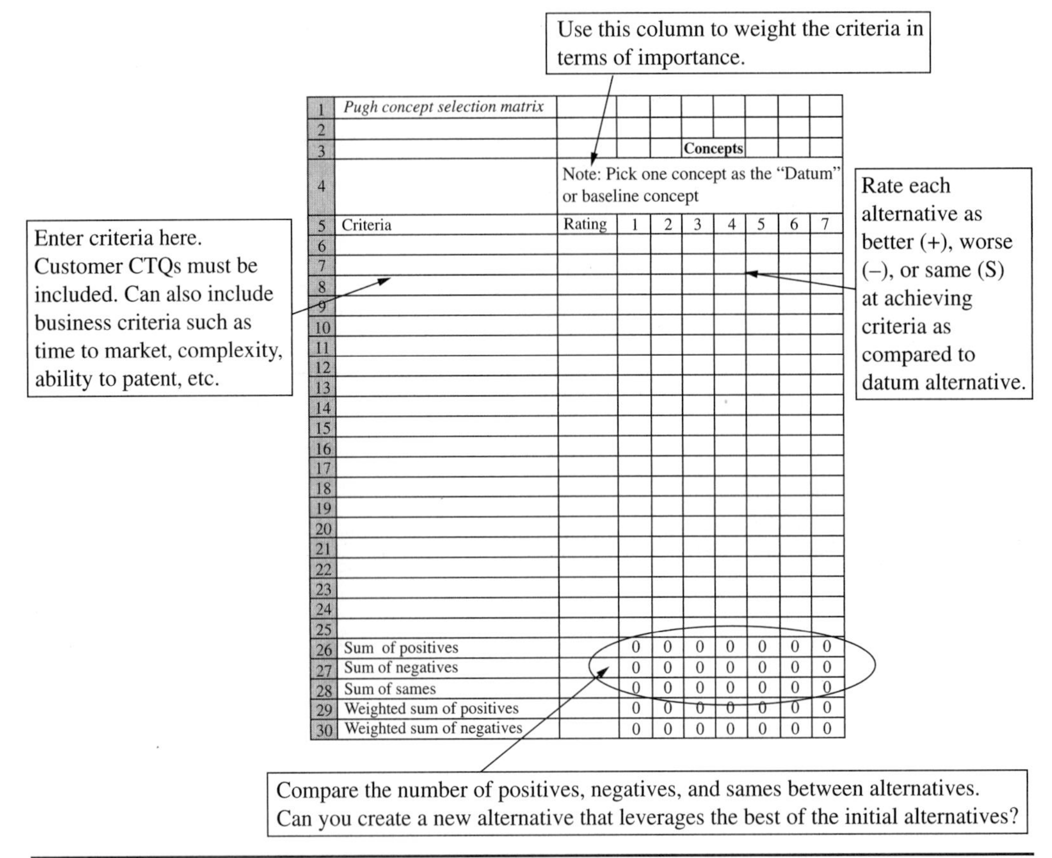

Figure 18.20 Pugh matrix. (*Lean Six Sigma Pocket Guide, Juran Institute, Inc., 2009. Used with permission.*)

Selection Matrix

Purpose

This tool assists in reducing a set of alternatives based on explicit criteria. Selection matrices often are applied when deciding among different solutions that each may have positive and negative attributes such that the "best" solution is not readily apparent. Unlike methods such as voting techniques that can introduce considerable subjectivity, selection matrices help introduce establish objectivity to the decision process.

Steps to Create

1. Agree on the criteria to be used to evaluate the alternatives.
2. Each team member allocates a total of 100 points among the criteria.
3. Calculate the average number of points allocated to each criterion.
4. Review and agree on the weights.
5. Assemble the list of alternatives to be evaluated.

6. Rate each alternative. Each team member rates each alternative according to how well it meets each criterion, using a scale of 1 (worst possible) to 5 (best possible).
7. Calculate each team member's average rating for each alternative.
8. Give each team member a table showing how each member rated each alternative.
9. Discuss the ratings and reach consensus on the next steps.

This matrix can be used to select projects and, as in this example, to select solutions.

Example A team was evaluating possible alternatives to improve the performance of saws used to cut wood for high-end cabinetry. The team brainstormed criteria and then differentiated these through a weighting method in which the weights summed to 100. Each alternative then was scored as to how well it met each individual criterion (scored 1 to 5, with 5 a better score). An average score then was calculated for each alternative, with results as shown in Figure 18.21. Using the results as a starting point for discussion, the team ultimately determined that alternative 1 was the best option.

Criterion	Weight	Alternative 1	Alternative 2	Alternative 3
Remedy Name		Overhaul and change speed	Replace equipment	Retain and change speed
Total cost	20	5	2	4
Impact on the problem	10	3	5	4
Benefit and cost relationship	30	4	3	5
Cultural impact and resistance to change	20	4	5	1
Implementation time	2	5	4	2
Uncertainty about effectiveness	6	4	5	3
Health and safety	10	4	5	5
Environment	2	3	3	3
Average rating		**4.10**	**3.74**	**3.68**

Figure 18.21 Selection matrix. (*Lean Six Sigma Pocket Guide, Juran Institute, Inc., 2009. Used with permission.*)

SIPOC

Purpose

SIPOC stands for supplier, input, process, output, and customer. It is a high-level map showing a process's primary suppliers, the inputs received from them, and the process that adds value to those inputs. That process produces an output that is intended to meet or exceed customer requirements. The SIPOC is typically used at the early stages of a project to help characterize a process and to identify appropriate team members. This model is applicable to both product and service processes.

Steps to Create

1. Define the process, name it, and define the start and stop points.
2. Identify suppliers and the critical inputs the process receives from them.
3. Identify the customers of the process (those who receive the outputs) and the outputs of the process that respond to customer needs.
4. Identify the five to eight major process steps that produce the output.
5. Validate the process map by working with the key functions that perform the major steps.

Example An improvement team was created to address the order-receiving process. To help identify the high-level steps and the scope of their project, the team created the SIPOC shown in Figure 18.22, beginning the process with receiving the order and following the process through to the time the product is scheduled for production.

Supplier	Input	Process	Output	Customer
Store location	Electronic order	Receive order	Queue-created file	Order sorter system
Order sorter system	Queue-created file	Hold order in queue	Purchase order TIF file	Order entry
Order entry	Purchase order TIF file	Enter order in system	Order ready for scheduling	Order checker
Order checker	Electronic order	Check order	Scheduling form	Scheduling
Scheduling	Scheduling form	Schedule production	Paper work	Production plant

FIGURE 18.22 SIPOC, a Six Sigma tool. (*Juran Institute, Inc. Used with permission.*)

Statistical Process Control

Purpose

The daily life of many employees involves operating a process within intended boundaries, that is, to maintain it according to specifications established through quality planning and improvement. Historically, this has relied heavily on inspection, with detection and elimination of nonconforming product after the fact. In contrast, the concept of control over a process entails predicting its performance, within certain limits. Rather than merely detecting nonconforming output ("inspecting quality into a product"), control is forward looking and seeks incremental but continuous improvement by identifying and eliminating special causes that create unpredictable variation (and potentially, but not necessarily, nonconformity to specification).

Statistical process control is the application of statistical methods to the measurement and analysis of variation in a process. A process is a collection of activities that converts inputs into outputs or results. Through use of control charts, statistical process control assists in detecting special (or assignable) causes of variation in both in-process parameters and end-of-process (product) parameters. The objective of a control chart is not to achieve a state of statistical control as an end in itself but to reduce variation.

Before proceeding with the steps to create a control chart, further discussion is warranted regarding common and special cause variation in the context of process control. A statistical control chart compares process performance data to computed "statistical control limits," drawn as limit lines on the chart. The process performance data usually consist of groups of measurements (called rational subgroups) from the regular sequence of production while preserving the order of the data. A prime objective of a control chart is detecting special (or assignable) causes of variation in a process. Knowing the meaning of "special causes" and distinguishing them from common (random or chance) causes is essential to understanding the control chart concept.

There are two kinds of process variations: (1) common (random or chance), which are inherent in the process, and (2) special (or assignable), which cause excessive variation (see Table 18.1). Ideally, only common causes are present in a process because they represent a stable and predictable process that leads to minimum variation. A process that is operating without special causes of variation is said to be in a state of statistical control. The control chart for such a process has all of the data points within the statistical control limits and exhibits no discernible patterns.

Random (Common) Causes	Assignable (Special) Causes
Description	
Consists of many individual causes	Consists of one or just a few individual causes
Any one random cause results in a minute amount of variation (but many random causes act together to yield a substantial total).	Any one assignable cause can result in a large amount of variation.
Examples are human variation in setting control dials, slight vibration in machines, NS slight variation in raw material.	Examples are operator blunder, a faulty setup, or a batch of defective raw materials.
Interpretation	
Random variation cannot be eliminated from a process economically.	Assignable variation can be detected; action to eliminate the causes is usually economically justified.
An observation within the control limits of random variation means that the process should not be adjusted.	An observation beyond control limits means that the process should be investigated and corrected.
With only random variation, the process is sufficiently stable to use sampling procedures to predict the quality of total production or do process optimization studies.	With assignable variation present, the process is not sufficiently stable to use sampling procedures for prediction.

(*Source: Quality Planning and Analysis,* Juran Institute, Inc., Copyright 2007. Used with permission.)

Table 18.1 Distinctions between Random and Assignable Causes of Variation

The control chart distinguishes between common and special causes of variation through the choice of control limits. These are calculated by using the laws of probability so that highly improbable causes of variation are presumed to be due to special causes not to random causes. When the variation exceeds the statistical control limits, it is a signal that special causes have entered the process and the process should be investigated to identify these causes of excessive variation. Random variation within the control limits means that only common (random) causes are present; the amount of variation has stabilized, and minor process adjustments (tampering) should be avoided. Note that a control chart detects the presence of a special cause but does not find the cause—that task must be handled by subsequent investigation of the process.

Steps to Create

Setting up a control chart requires taking the following steps:

1. Choosing the characteristic to be charted.
2. Giving high priority to characteristics that are currently running with a high defective rate. A Pareto analysis can establish priorities.
3. Identifying process variables and conditions that contribute to the end-product characteristics to define potential charting applications from raw materials through processing steps to final characteristics. For example, the pH, salt concentration, and temperature of a plating solution are process variables contributing to plating smoothness.
4. Verifying that the measurement process has sufficient accuracy and precision to provide data that does not obscure variation in the manufacturing or service process. The observed variation in a process reflects the variation in the manufacturing process and also the combined variation in the manufacturing and measurement processes. Anthis (1991) described how the measurement process was a roadblock to improvement by hiding important clues to the sources of variation in a manufacturing process. Dechert (2000) explained how large measurement variation can be controlled and result in effective statistical process control methods.
5. Determining the earliest point in the production process at which testing can be done to get information on assignable causes so that the chart serves as an effective early-warning device to prevent defectives.
6. Choosing the type of control chart. Table 18.2 compares three basic control charts. Schilling (1990) provides additional guidance in choosing the type of control chart to use.
7. Deciding on a central line to be used as the basis of calculating the limits. The central line may be the average of past data, or it may be a desired average (i.e., a standard value). The limits are usually set at threes, but other multiples may be chosen for different statistical risks.
8. Choosing the "rational subgroup." Each point on a control chart represents a subgroup (or sample) consisting of several units of product. For process control, rational subgroups should be chosen so that the units within a subgroup have the greatest chance of being alike and the units between subgroups have the greatest chance of being different.
9. Providing a system for collecting the data. If the control chart is to serve as a day-to-day shop tool, it must be simple and convenient to use. Measurement must be

Statistical Measure Plotted	Average $\bar{X}$ and Range R	Percentage Nonconforming (p)	Number of Nonconformities (c)
Type of data required	Variable data (measured values of a characteristic)	Attribute data (number of defective units of product)	Attribute data (number of defects per unit of product)
General field of application	Control of individual characteristics	Control of overall fraction defective of a process	Control of overall number of defects per unit
Significant advantages	Provides maximum use of information available from data Provides detailed information on process average and variation for control of individual dimensions	Data required are often already available from inspection records Easily understood by personnel Provides an overall picture of quality	Same advantages as p chart but also provides a measure of defectiveness
Significant disadvantages	Not understood unless training is provided; can cause confusion between control limits and tolerance limits. Cannot be use with go/no go type of data	Does not provide detailed information for control of individual characteristics Does not recognize different degrees of defectiveness in units of product	Does not provide detailed information for control of individual characteristics
Sample size	Usually four or five	Use given inspection results or samples of 25, 50, or 100	Any convenient unit of product such as 100 feet of wire or one television set

(*Source: Quality Planning and Analysis,* Juran Institute, Inc., Copyright 2007. Used with permission.)

TABLE 18.2 Comparison of Some Control Charts

simplified and kept error free. Indicating instruments must be designed to give prompt, reliable readings. Better yet, instruments should be designed that can record as well as indicate. Recording of data can be simplified by skillful design of data or tally sheets. Working conditions are also a factor.

10. Calculating the control limits and providing specific instructions for the interpretation of the results and the actions that various production personnel are to take (see below). Control limit formulas for the three basic types of control charts are given in Table 18.3. These formulas are based on $\pm 3\sigma$ and use a central line equal to the average of the data used in calculating the control limits. Values of the A2, D3, and D4 factors used in the formulas are given in Table 18.4. Each year, *Quality Progress* magazine publishes a directory that includes software for calculating sample parameters and control limits and for plotting the data. The general rule of thumb is to collect 20 to 30 samples (rational subgroups) before attempting to establish control limits.
11. Plotting the data and interpreting the results.

Chart for	Central Line	Lower Limit	Upper Limit
Averages $\bar{X}$	$\bar{\bar{X}}$	$\bar{\bar{X}} - A_2\bar{R}$	$\bar{\bar{X}} + A_2\bar{R}$
Ranges R	$\bar{R}$	$D_3\bar{R}$	$D_4\bar{R}$
Proportion nonconforming p	$\bar{p}$	$\bar{p} - 3\sqrt{\frac{\bar{p}(1-\bar{p})}{n}}$	$\bar{p} + 3\sqrt{\frac{\bar{p}(1-\bar{p})}{n}}$
Number of nonconformities c	$\bar{c}$	$\bar{c} - 3\sqrt{\bar{c}}$	$\bar{c} + 3\sqrt{\bar{c}}$

(*Source: Quality Planning and Analysis*, Juran Institute, Inc., Copyright 2007. Used with permission.)

TABLE 18.3 Control Chart Limits—Attaining a State of Control

Factors for $\bar{X}$ and R Control Charts;* Factors for Estimating s from R†				
Number of Observations in Sample	A_2	D_3	D_4	**Factor for s Estimate from $\bar{R}$: $d_2 = \bar{R}/s$**
2	1.880	0	3.268	1.128
3	1.023	0	2.574	1.693
4	0.729	0	2.282	2.059
5	0.577	0	2.114	2.326
6	0.483	0	2.004	2.534
7	0.419	0.076	1.924	2.704
8	0.373	0.136	1.864	2.847
9	0.337	0.184	1.816	2.970
10	0.308	0.223	1.777	3.078
11	0.285	0.256	1.744	3.173
12	0.266	0.284	1.717	3.258
13	0.249	0.308	1.692	3.336
14	0.235	0.329	1.671	3.407
15	0.223	0.348	1.652	3.472

$$\begin{cases} \text{Upper control limit for } \bar{X} = \text{UCL}_{\bar{X}} = \bar{\bar{X}} + A_2\bar{R} \\ \text{Lower control limit for } \bar{X} = \text{LCL}_{\bar{X}} = \bar{\bar{X}} - A_2\bar{R} \end{cases}$$

$$\begin{cases} \text{Upper control limit for } R = \text{UCL}_R = D_4\bar{R} \\ \text{Lower control limit for } R = \text{LCR}_R = D_3\bar{R} \end{cases}$$

TABLE 18.4 Factors for $\bar{X}$ and R Control Charts

Stage	Step	Method
Preparatory	State purpose of investigation.	Relate to quality system
	Determine state of control.	Attributes chart
	Determine critical variables.	Fishbone
	Determine candidates for control.	Pareto
	Choose appropriate type of chart.	Depends on data and purpose
	Decide how to sample.	Rational subgroups
	Choose subgroup size and frequency.	Sensitivity desired
Initiation	Ensure cooperation.	Team approach
	Train user.	Log actions
	Analyze results.	Look for patterns
Operational	Assess effectiveness.	Periodically check usage and relevance
	Keep up interest.	Change chart, involve users
	Modify chart.	Keep frequency and nature of chart current with results
Phaseout	Eliminate chart after purpose is accomplished.	Go to spot checks, periodic sample inspection, overall *p*, *c* charts

(*Source:* Schilling 1990.)

Table 18.5 Life Cycle of Control Chart Applications

The control chart is a powerful statistical concept, but its use should be kept in perspective. The ultimate purpose of an operations process is to make product that is fit for use—not to make product that simply meets statistical control limits. Once the charts have served their purpose, many should be taken down and the effort shifted to other characteristics needing improvement. Schilling (1990) traces the life cycle of control chart applications (Table 18.5). A given application might employ several types of control charts. Note that, in the phaseout stage, statistical control has been achieved, and some of the charts are replaced with spot checks.

Types of Control Charts

Traditional Shewhart control charts (named for Dr. Walter A. Shewhart; see the *Juran's Quality Handbook* (1999), Section 45 (Juran and Godfrey 1999), for a historical account of their development) are divided into two categories: variable charts (those using continuous, measurement data), and attribute charts (those using count data). Selecting the proper type of control chart is shown in Figure 18.23; the different types are described further below.

Regardless of the specific chart type or statistic (e.g., average, range, standard deviation, proportion), control limits are established such that it would be very unlikely that the values would fall outside if the process were stable; usually this is set at plus or minus three standard deviations.

Examples of Control Charts for Variables Data In these charts, the mean and either range or standard deviation are the typical statistics that are monitored. These statistics are monitored in a pair of charts. The averages chart plots the sample averages, specifically, the average of each rational subgroup (if the rational subgroup size is one, then an individual and moving range chart (X-mR) (also known as an I-mR chart) of individuals is used instead). The range chart or standard deviation chart plots the range or standard deviation of rational subgroups. The specific subtypes are as follows:

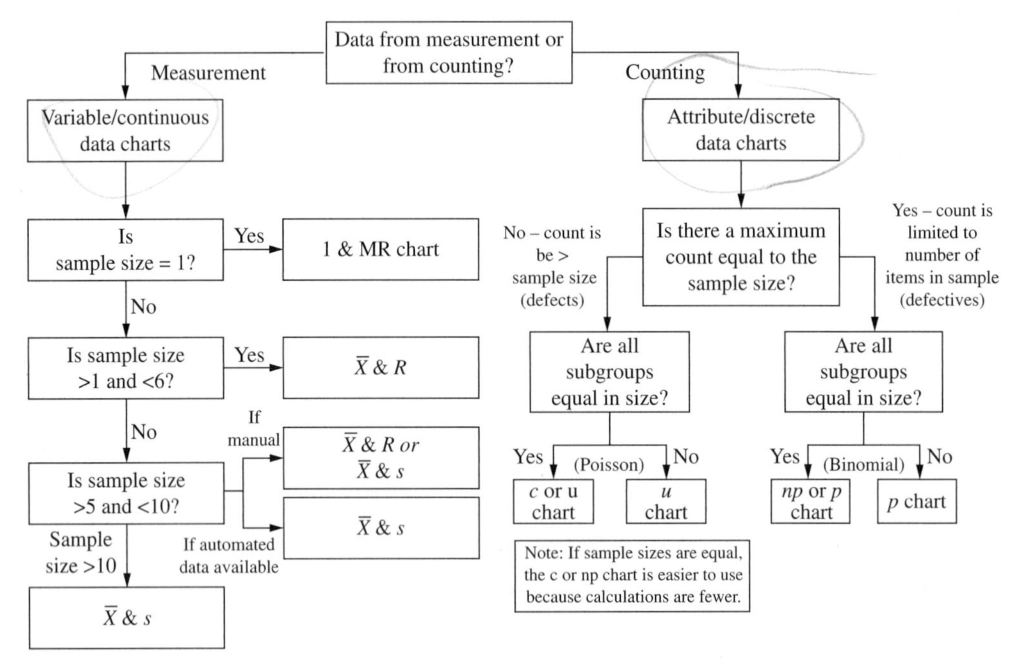

Note that "measurement" implies a continuous scale is used to assess a characteristic. "Counting" occurs when outcomes are tabulated as pass/fail, accept/reject, or some other categorization, even if the original measurement was continuous. Also, control limits depend on number of observations in each sample. When subgroup sizes vary, control limits will not be straight lines on any of the charts.

FIGURE 18.23 Flow chart of control chart selection. (*Juran Institute, Inc. Used with permission.*)

$\bar{X}$ and R chart. Also called the "average and range" chart. $\bar{X}$ refers to the average of a rational subgroup and measures the central tendency of the response variable over time. R is the range (difference between the highest and lowest values in each subgroup), and the R chart measures the gain or loss in uniformity within a subgroup which represents the variability in the response variable over time. Note that, because specification limits apply to individual values rather than averages (averages inherently vary less than the component individual values), control limits cannot be compared to specification limits which should not be placed on a control chart for averages.

$\bar{X}$ and s chart. The average and standard deviation chart is similar to the $\bar{X}$ and R chart, but the standard deviation (instead of the range) is used in the s chart. Although an s chart is statistically more efficient than the range for subgroup sizes greater than 2, a range chart is easier to compute and understand and is traditionally used for subgroup sizes smaller than about 10.

X-mR chart. Also known as an I-mR chart, this charts individual measures and a moving range. It is used when the rational subgroup size = 1 (such that there are no multiple measures from which to obtain an average).

Z-mR chart. This is similar to the X-mR chart, except that the individual values are standardized through a Z transformation. This is useful for short runs in which there are fewer than the recommended 20 to 30 needed to establish one of the preceding charts (see short-run control charts in Chapter 19, Accurate and Reliable Measurement Systems and Advanced Tools).

Individuals chart. Also called a run chart, this is an alternative to the $\bar{X}$ and I chart, and is simply a plot of individual values against time. In the simplest case, specification limits are added to the chart; in other cases, ±3σ limits of individual values are added. A chart of individual values is not as sensitive as the $\overline{X}$ chart, however.

By way of example of variable control charting, refer to the $\overline{X}$ and R charting in Figure 18.24. The upper part of the figure displays the individual observations for two machines, N-5 and N-7. For each machine, the data consist of 10 samples (with six units

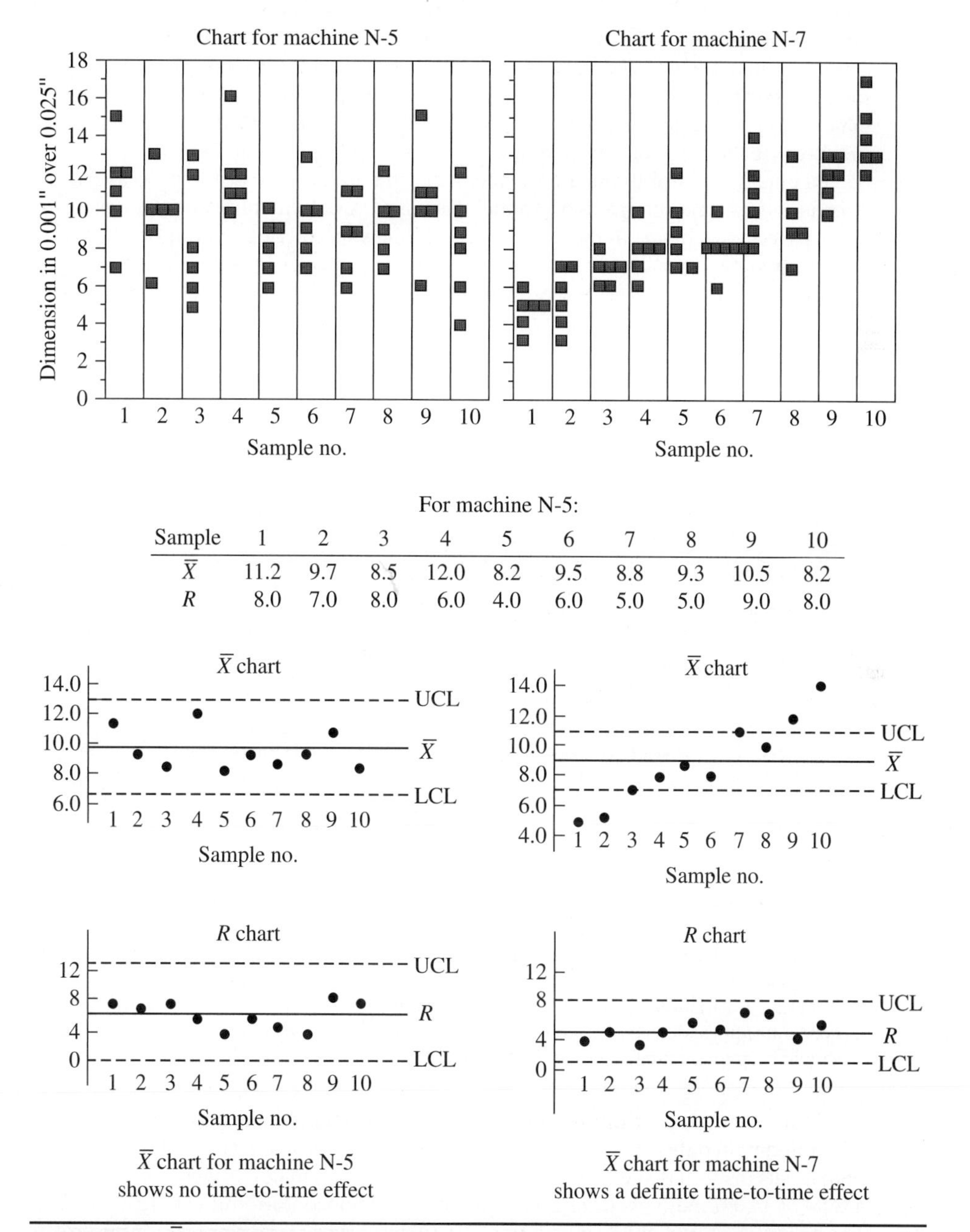

Sample	1	2	3	4	5	6	7	8	9	10
$\overline{X}$	11.2	9.7	8.5	12.0	8.2	9.5	8.8	9.3	10.5	8.2
R	8.0	7.0	8.0	6.0	4.0	6.0	5.0	5.0	9.0	8.0

FIGURE 18.24 $\overline{X}$ and R charts confirm suggested machine differences. (*Quality Planning & Analysis. Juran Institute, Inc., ©2007. Used with permission.*)

in each rational subgroup) plotted in time order of production. The lower portion shows the $\overline{X}$ and R charts for each machine. For machine N-5, all points fall within the control limits, so that (based on this rule), the process appears to be free of assignable causes of variation, and is "in control." However, machine N-7 has both within-sample variation (seen in the range chart) and between-sample variation (seen in the chart of sample averages). The $\overline{X}$ chart indicates some factor (special cause) such as tool wear is present that results in larger values of the characteristic with the passing of time (note the importance of preserving the order of measurements when collecting data).

Interpreting variables charts. Place the charts for $\overline{X}$ and R (or s) one above the other so that the average and range for any one subgroup are on the same vertical line. Observe whether either or both indicate lack of control for that subgroup. Usually, the R (or s) chart is interpreted first because the range or standard deviation is used in calculating limits for the $\overline{X}$ chart.

Rs outside control limits are evidence that the uniformity of the process has changed. Typical causes are a change in personnel, increased variability of material, or excessive wear in the process machinery. If the R or s chart exhibits a special cause variation, then the within-subgroup variation will contain both common and special cause variation, and its use in calculating control limits for the $\overline{X}$ chart will result in excessively large control limits (reducing its ability to detect out-of-control conditions). A single out-of-control R can be caused by a shift in the process that occurred while the subgroup was being taken.

$\overline{X}$s outside the control limits are evidence of a general change affecting all pieces after the first out-of-limits subgroup. The log kept during data collection, the operation of the process, and the worker's experience should be studied to discover a variable that could have caused the out-of-control subgroups. Typical causes are a change in material, personnel, machine setting, tool wear, temperature, or vibration.

Look for unusual patterns and nonrandomness. Nelson (1984, 1985) provides eight tests to detect such patterns on control charts using 3σ control limits (see Figure 18.25). Each of the zones shown is 1σ wide. (Note that test 2 in Figure 18.25 requires nine points in a row, Other authors suggest seven or eight points in a row (see Nelson 1985 for elaboration).

Ott and Schilling (1990) provide a definitive text on analysis after the initial control charts by presenting an extensive collection of cases with innovative statistical analysis clearly described.

Examples of Control Charts for Attribute Data Whereas control charts for variables data require numerical measurements (e.g., line width from a photoresist process), control charts for attribute data require only a count of observations of a characteristic (e.g., the number of nonconforming items in a sample). These also are called categorical data because units are classified into groups such as pass and fail.

p *chart.* Also called a proportions chart, this tracks the proportion or percentage of nonconforming units (percentage defective) in each sample over time.

np *chart.* This chart is used to track the number of nonconforming (defective) units in each sample over time. An np chart should only be used when the number of units sampled is constant (or nearly so).

c *chart.* Used to track the number of nonconformities (i.e., defects, rather than defective units as in the p chart).

u *chart.* A variation of the c chart, and analogous to the np chart, this chart tracks the number of nonconformities (defects) per unit in a sample of n units. As with the np chart, the number of units should be approximately constant.

As an example of attribute control charting, the fraction nonconforming (p) chart can be illustrated with data on magnets used in electrical relays. For each of 19 weeks, the number of magnets inspected and the number of nonconforming magnets were recorded. There was a total of 14,091 magnets tested. The total number nonconforming was 1030, or 7.3 percent The resulting control chart (calculating control limits based on average sample size of 741.6)

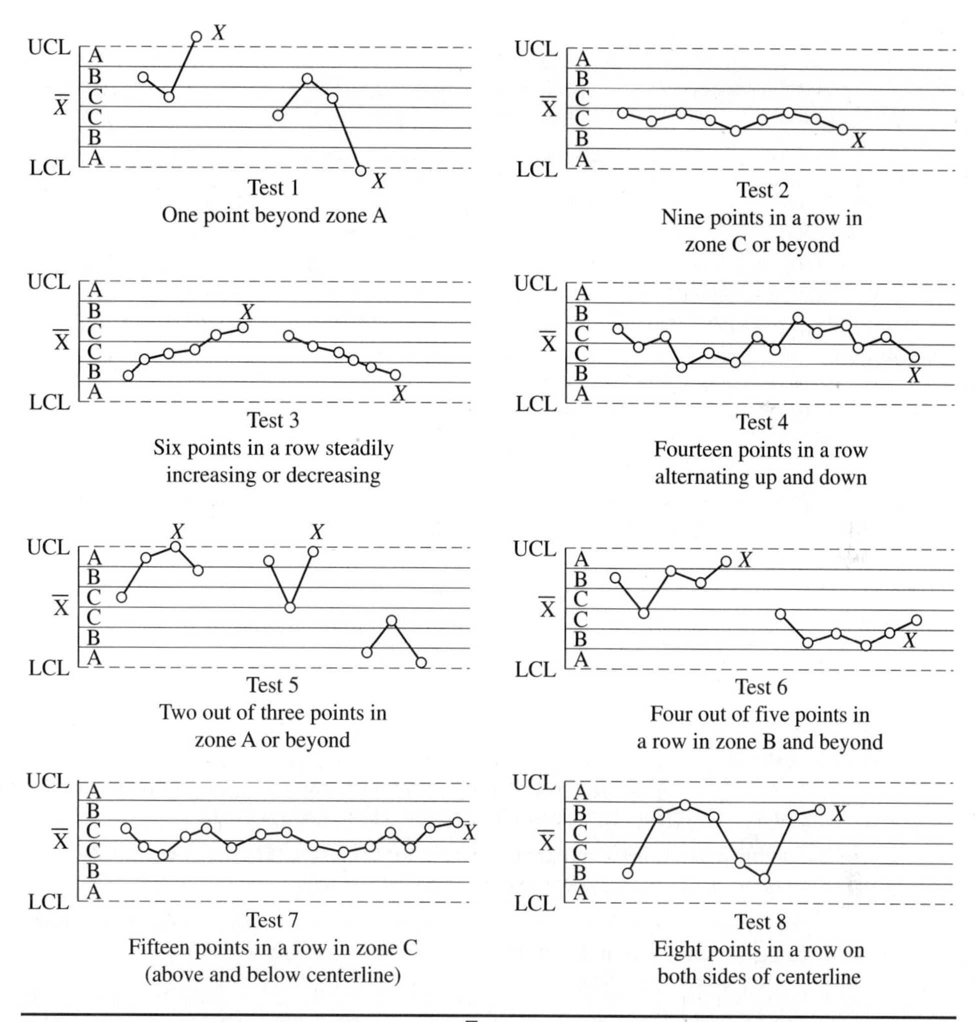

FIGURE 18.25 Tests for special causes applied $\overline{X}$ to control charts. (*Nelson 1984.*)

is shown in Figure 18.26. Note that several points fall beyond the control limits, suggesting that there are special cause(s) at work. In the case of the unusually low point for the last sample, it may be useful to identify and reinforce any special cause of the exceptionally good quality. The same rules as described above in Figure 18.25 also apply to attribute charts.

Stratification

Purpose

Stratification is the separation of data into categories. The most frequent use is during problem analysis to identify which categories contribute to the problem being solved. However, stratification can be applied when identifying projects, analyzing symptoms, testing hypotheses, and developing solutions. Stratification helps answer questions as to the frequency of defects, factors that may be contributing to a quality problem, and the degree to which results may differ across groups (strata).

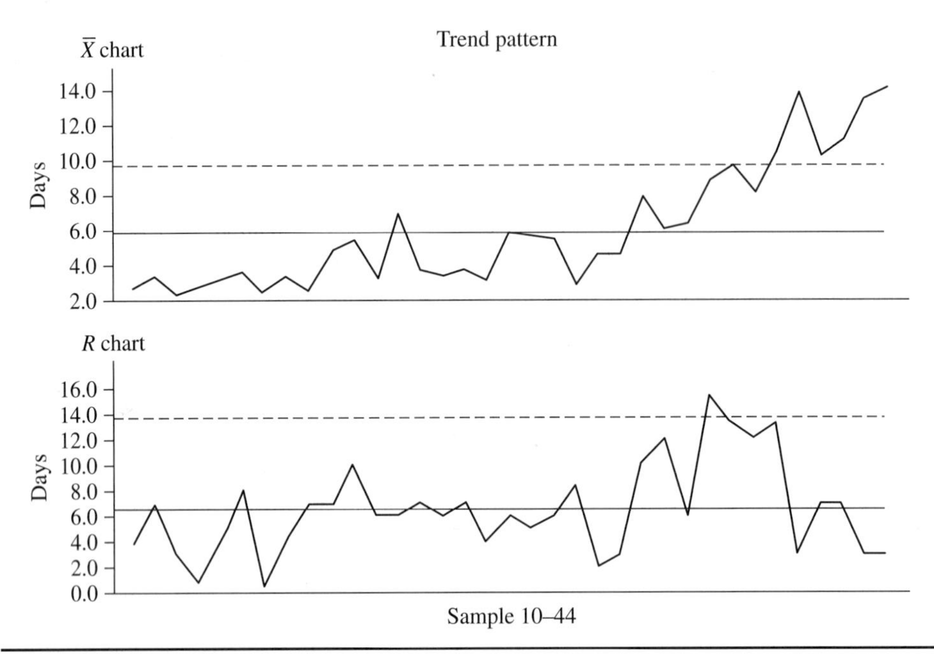

FIGURE 18.26 Average and range control charts. (*Quality Planning & Analysis. Juran Institute, Inc., ©2007. Used with permission.*)

Steps to Create

1. Select the stratification variables. If new data are to be collected, be certain that all potential stratification variables are collected as identifiers.
2. Establish categories that are to be used for each stratification variable. The categories may be either discrete values or ranges of values.
3. Sort observations into the categories of one of the stratification variables. Each category will have a list of the observations that belong to it.
4. Calculate the phenomenon being measured for each category. These calculations can be a count of the number of observations in the category, an average value for those observations or a display (like a histogram) for each category.
5. Display the results. Bar graphs are usually the most effective.
6. Prepare and display the results for other stratification variables. Repeat steps 2 through 5. Do second-stage stratification as appropriate.
7. Plan for addition confirmation.

Example A manufacturer of mechanical equipment had recently received a rash of complaints about pins (stock number 128B) coming loose from press-fit sockets. The sockets were produced internally by the manufacturer under good statistical process control. The steel pins that fit into the sockets were purchased from three different suppliers. (see Figure 18.27)

The quality improvement team looking into the complaints measured the diameter of 120 pins from inventory, 40 from each of the three suppliers. The nominal value for the pin diameter was 10 mm. The upper specification limit was 10.2 mm, and the lower limit was 9.8 mm.

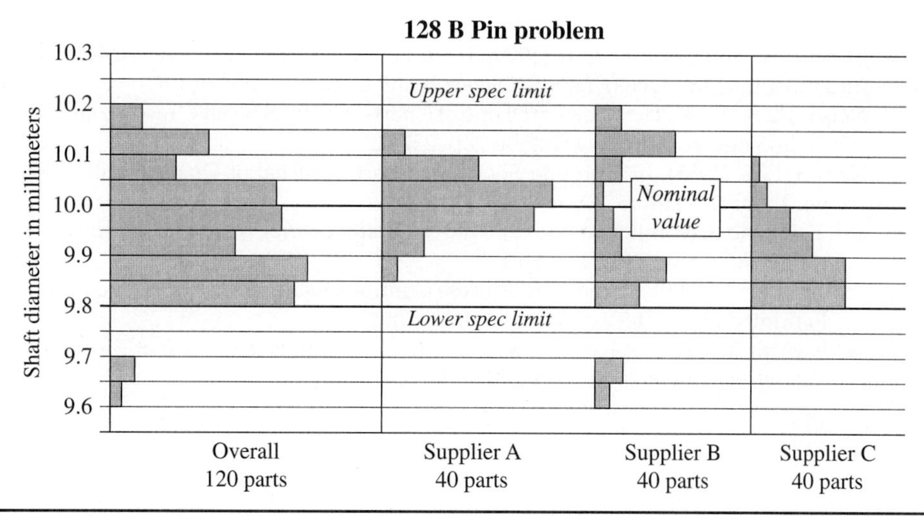

Figure 18.27 Stratification of the 128B pin problem. (*Lean Six Sigma Pocket Guide, Juran Institute, Inc., ©2009. Used with permission.*)

To get a better understanding of the data, the team produced a histogram of all 120 parts. The histogram showed that the pin diameter measurements had a broad, multipeaked distribution, with most of the data between the lower specification limit and the nominal value. Because most of the pins were smaller than nominal, there was indeed a good chance of a loose fit.

This summary histogram, however, did not tell the team much about what the cause of the problem was. So the team decided to stratify the data by supplier and to plot new histograms.

On the basis of the histograms on the previous page, the team drew the following conclusions:

- Supplier A has good controls on its process. Most of the product is close to the nominal value, and because the inherent variability in the process is smaller than the width of the specification limits, there is little chance of producing a part outside the limits.
- Supplier B appears to be running two distinct processes, neither of which has been set up to produce pins with diameters close to the nominal. The shape of the distribution for supplier B looks like the sum of two distributions similar to that of supplier A, one of which has been shifted up a bit, the other shifted down.
- Supplier C has a process that is highly variable and not set up to produce pins at the nominal value. The abruptly ended (or truncated) nature of the distribution suggests that the supplier is using inspection to screen out off-spec pins.

References

Anthis, D. L., Hart, R. F., and Stanula, R. J. (1991). The Measurement Process: Roadblock to Product Improvement, Quality Engineering, vol. 3, no. 4, pp. 461–470.

Dechert, J., Case, K. E., and Kautiainen, T. L. (2000). Statistical Process Control in the Presence of Large Measurement Variation, Quality Engineering, vol. 12, no. 3, pp. 417–423.

Ishikawa, K. (1972). *Guide to Quality Control*. Asian Productivity Organization, Tokyo.
Juran, J. M., and Godfrey, A. B. (1999). *Juran's Quality Handbook*. 5th ed., McGraw-Hill, NY.
Juran Institute, Inc. (2009). *Lean Six Sigma Pocket Guide*. Southbury, CT.
Nelson, L. S. (1984). The Shewhart Control Chart-Tests for Special Causes, Journal of Quality Technology, vol. 16, no. 4 October, pp. 237–239.
Nelson, L. S. (1985). Interpreting Shewhart Charts, Journal of Quality Technology, vol. 17, no. 2, pp. 114–116.
Ott, E. R., and Schilling, E. G. (1990). *Process Quality Control Troubleshooting and Interpretation of Data*. McGraw-Hill, New York.
Schilling, E. G. (1990). Elements of Process Control, Quality Engineering, vol. 2, no. 2, p. 132. Reprinted by courtesy of Marcel Dekker, Inc.
Wadsworth, H. M., Stephens, K. S., and Godfrey, A. B. (1986). *Modern Methods for Quality Control and Improvement*. Wiley, New York.

CHAPTER 19

Accurate and Reliable Measurement Systems and Advanced Tools

John F. Early and Brian A. Stockhoff

About This Chapter

In this chapter, we will discuss the use of tools to obtain information that will support the needed decision-making that leads to superior results. We will offer a framework for planning and gathering data that are required to answer the questions that lead to superior results. We provide guidelines for ensuring the relevance and accuracy of the data collected. Finally, we build on the core tools described in Chapter 18, Core Tools to Design, Control, and Improve Performance, and provide brief summaries of the most useful tools for collecting, analyzing, and presenting statistical data so that they clearly answer strategic and operational questions and provide a sound basis for decisions that deliver superior performance, the highest quality, loyal customers, and ultimately superior results.

High Points of This Chapter

1. Obtaining accurate, reliable, and relevant information happens when asking the right questions. Ask the right question, and you will get the right data.
2. Ten principles for effective measurement can help to develop accurate and reliable measures of performance.
3. When planning for data collection, the key issue is not how to collect data rather, the key issue is how to generate useful information. This is accomplished by a thorough understanding of the measurement system and comparative advantages among data collection and analysis choices, and beginning with the end in mind.
4. There are many tools useful to manage an organization that provides process improvement, design, and control. Tools for improvement require the testing of theories and finding root causes. Design tools require the collection of opinions and specifications, and then determine means to develop new services or products that are reliable. Control requires the use of statistical tools to help distinguish between common and special causes of variation to reduce risk, thereby facilitating appropriate intervention.
5. Chapter 18, Core Tools to Design, Control, and Improve Performance, covered some of the basic tools for planning, improving, and controlling quality. This chapter provides some additional tools for more detailed and complex analysis.

Measurement and Superior Results

In other chapters of this handbook, the reader will have seen many times the need for accurate, reliable, and relevant data in order to make the necessary decisions to achieve superior results. First and foremost, information must be directly *relevant* to the question being asked. If we wish to retain loyal customers, we require data on the actual loyalty and spending of the set of existing and potential customers. We will also need data that demonstrate the causative factors for customers' behavior. Next, the data must be *accurate*. As we will see, accuracy has two components: freedom from bias and sampling error (the uncertainty associated with using a sample of the whole) that is small enough to support the decision we must make. If we wish to improve the clinical status of the asthmatic population in a health plan, for example, we need to demonstrate that our measurement of that status is free from significant bias and that the samples we use to estimate the status are large enough so that our uncertainty arising from the samples size is small compared to the improvement we wish to achieve. Finally, the data must be *reliable*. Reliability encompasses both accuracy and relevance, but goes a step further, ensuring that the measurements will continue to be accurate and relevant on an ongoing basis within the operating and business environment so that we can continue to rely on it for decision-making. For example, if we have established that a process for manufacturing a complex electronic connector is capable of meeting customer requirements and that key variables of temperature, pressure, and speed must be controlled within proven limits, then our business success depends on those measures of temperature, pressure, and speed continuing to be both accurate and relevant in order to yield defect-free connectors.

Measurement and Analysis and the Juran Trilogy®

The Juran Trilogy® of Quality Planning, Quality Improvement, and Quality Control each rely on a foundation of accurate, reliable, and relevant information. While each is a distinct managerial process, they share a common need for information, apply many of the same tools, and often use the same data to ensure their ends. However, each has unique information requirements that form the basis for our pursuit of measurement and analysis for quality. The following sections list some of the key questions for each phase of the Juran Trilogy.

Quality Planning Measurement Questions

- What level of product market performance is required to meet strategic objectives?
- How does our product perform vis-à-vis the competition?
- How does our product perform with respect to customer expectations?
- What is the magnitude of the customer demand for the product (good or service) or process, and how much are they willing to pay?
- Who are the customers for the product (good or service) or process?
- How important is each customer?
- What are the needs/benefits for each significant customer?
- What is the relative importance of each of these needs and benefits?
- What is the impact of each product feature on each customer need?
- What are the mathematical tradeoffs among the various product features?
- What is the impact of each process feature/parameter on delivery of the product feature?
- What is the capability of the process to deliver the product?
- What are the optimal tolerances for the target of each product and process feature?
- How much of the strategic objective for the planning project was achieved?

Quality Improvement Measurement Questions

- What are the most important deficiencies driving customer disloyalty?
- What are the largest detailed categories of costs of poor quality?
- How much improvement in cost and customer loyalty is needed to meet strategic objectives?
- What are the major contributors to the identified problem?
- How much does each theory of cause contribute to the overall problem?
- Which theories are proven as root causes?
- How much improvement will the proposed remedy create?
- How much improvement did the project finally achieve?

Quality Control Measurement Questions

- What variables have the largest impact on the variability of the process?
- What is the normal random variation for the control variables?
- Is a variable exhibiting a sporadic spike in its variation due to an assignable cause?

Ten Principles of Effective Measurement

Quality measurement is central to quality control and improvement: "What gets measured, gets done." Before embarking on the details for good measurement and analysis, we need to consider the following principles that can help to develop effective measurements for quality:

1. Define the purpose and use that will be made of the measurement. An example of particular importance is the application of measurements in quality improvement. Final measurements must be supplemented with intermediate measurements for diagnosis.
2. Emphasize customer-related measurements; be sure to include both external and internal customers.
3. Focus on measurements that are useful—not just easy to collect. When quantification is too difficult, surrogate measures can at least provide a partial understanding of an output.
4. Provide for participation from all levels in both the planning and implementation of measurements. Measurements that are not used will eventually be ignored.
5. Provide for making measurements as close in time as possible to the activities they affect. This timing facilitates diagnosis and decision-making.
6. Provide not only concurrent indicators but also leading and lagging indicators. Current and historical measurements are necessary, but leading indicators help to look into the future.
7. Define in advance plans for data collection and storage, analysis, and presentation of measurements. Plans are incomplete unless the expected use of the measurements is carefully examined.
8. Seek simplicity in data recording, analysis, and presentation. Simple check sheets, coding of data, and automatic gauging are useful. Graphical presentations can be especially effective.
9. Provide for periodic evaluations of the accuracy, integrity, and usefulness of measurements. Usefulness includes relevance, comprehensiveness, level of detail, readability, and interpretability.
10. Realize that measurements alone cannot improve products and processes.

Measurements must be supplemented with the resources and training to enable people to achieve improvement.

Planning for Measurement and Data Collection

"Begin with the end in mind" is an appropriate maxim when starting any effort that requires collection of data. The "end" in this case is obtaining the information needed to effectively and efficiently plan, control, or improve. Before launching into a discourse on statistical analysis, we first will consider the need to plan for data collection. Part of this planning process is to consider the source of a set of data that we desire to analyze to solve a problem; the most common sources are historical data, newly collected operational data, and data from planned experimentation. These sources each have their advantages and drawbacks. Regardless of the source, all data need careful review before proceeding with an analysis and communication of the information gained from it.

Planning for Collection and Analysis of Data

In collecting and analyzing data, quality teams are seeking the answer to questions such as, How often does the problem occur? or What is causing the problem? In other words, they are seeking information. However, although good information always is based on data (the facts), simply collecting data does not necessarily ensure that useful information has been obtained. The key issue, then, is not, How do we collect data? Rather, the key issue is, How do we generate useful information? Although most organizations have vast stores of data about their operations, frequently, the data needed to provide truly useful information do not exist. The all-too-common practice at many organizations is to go "data diving," looking at much of or all of the data available to learn whatever they can about the process. While this practice can yield some useful information, it is inherently wasteful and can add time to the execution of a project. The process of planning the data collection with the end in mind that is described here is far more efficient and effective.

Information generation begins and ends with questions. To generate information, we need to

- Formulate precisely the question we are trying to answer. See general examples from each of the Juran Trilogy® processes above.
- Collect data relevant to that question.
- Analyze the data to determine the factual answer to the question.
- Present the data in a way that clearly communicates the answer to the question.

Learning to "ask the right questions" is the key skill in effective data collection. Accurate, precise data, collected through an elaborately designed statistical sampling plan, is useless if it is not relevant to answering a question that someone cares about.

Notice in Figure 19.1 how this planning process "works backwards" through the model. We start by defining the question. Then, rather than diving into the details of data collection,

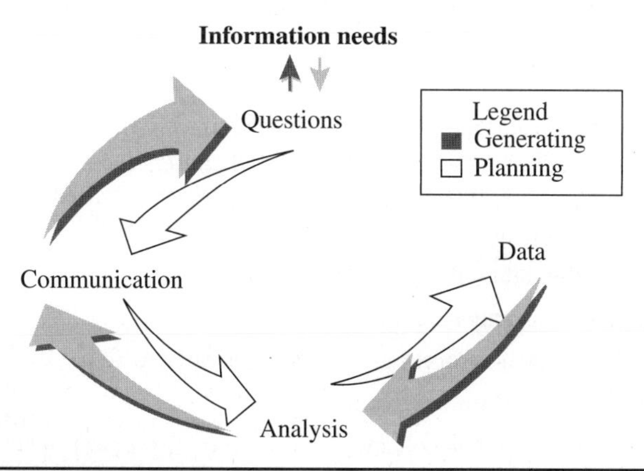

Figure 19.1 Planning for data collection.

we consider how we might communicate the answer to the question and what types of analysis we will need to perform. This helps us define our data needs and clarifies which characteristics are most important in the data. With this understanding as a foundation, we can deal more coherently with the where, who, how, and what of data collection.

To generate useful information, planning for good data collection, analysis, and communication proceeds through the following steps and associated considerations.

1. Establish data collection objectives and formulate the question in a specific statement:
   - What is your goal for collecting data?
   - What process or product will you monitor to collect the data?
   - What is the "theory" you are trying to test?
   - What is the question you are attempting to answer?
2. Decide what to measure, with consideration as to how the data will be communicated and analyzed:
   - What data do you need?
   - What type of measure is it? Time and physical measures such as length, mass, volume, and temperature are common; other measures include rankings (e.g., low-medium-high), ratios (e.g., speed), and indexes (e.g., case-mix adjusted hospital length of stay, refractive index). See the section "Types of Measures" for a discussion of scales of measurement.
   - What type of data is it? Variables data (readings on a scale of measurement) may be more expensive than attributes data (go or no-go data), but the information is much more useful.
   - What is the operational definition of each measure? An operational definition is a detailed description of a process, activity, or project term written to ensure common understanding among the members of a group.
   - How will the data be communicated and analyzed?
   - Are past data available that are applicable (however, bear in mind the hazards of historical data sets, discussed below)?
3. Decide how to measure a population or sample:
   - What measurement tool will you use? Calipers, Likert scale survey?
   - What is your sampling strategy? Simple random sampling? Stratified random sampling?
   - How much data will be collected? Calculate sample size considering the desired precision of the result, statistical risk, variability of the data, measurement error, economic factors, etc.
   - What is the measurement method?
4. Collect the data with a minimum of bias.
5. Define comprehensive data collection points:
   - Where in the process can we get appropriate data?
6. Select and train unbiased collectors:
   - Understand data collectors and their environment.
   - Who in the process can give us these data?

- How can we collect these data from these sources with minimum effort and least chance for error?

7. Design, prepare, and then test data collection methods, forms, and instructions:
   - What additional information should be captured for future analysis, reference, or traceability?
   - Conduct a measurement systems analysis (MSA) to confirm that the measures are accurate.
8. Audit the collection process and validate the results.
9. Screen the data.
10. Analyze the data.
11. Evaluate assumptions for determining the sample size and analyzing the data. Take corrective steps (including additional observations) if required.
12. Apply graphical and statistical techniques to evaluate the original problem.
13. Determine if further data and analysis are needed.
14. Consider a sensitivity analysis (e.g., by varying key sample estimates and other factors in the analysis and noting the effect on final conclusions).
15. Review the conclusions of the data analysis to determine if the original technical problem has been evaluated or if it has been changed to fit the statistical methods.
16. Present the results:
    - Write a report, including an executive summary.
    - State the conclusions in meaningful form by emphasizing results in terms of the original problem rather than the statistical indexes used in the analysis.
    - Present the results in graphic form where appropriate. Use simple statistical methods in the body of the report, and place complicated analyses in an appendix.
17. Determine if the conclusions of the specific problem apply to other problems or if the data and calculations could be a useful input to other problems.

Types of Measures

In planning for data collection, one needs to be clear about the characteristics of the data being collected and the implications of those characteristics for the questions to be answered. Two classifications of data types are useful here: the mathematical distinctions and the substantive quality questions answered.

The mathematical distinctions are typically known as measurement scales and are part of a system of measurement. The most useful scale is the *ratio scale* in which we record the actual amounts of a parameter such as weight. Ratio scales are also referred to as *continuous variables* data. An *interval scale* records ordered numbers but lacks an arithmetic origin such as zero—clock time is an example.

An *ordinal scale* records information in ranked categories—an example is customer preference for the flavor of various soft drinks. An unusual example of a measurement scale is the Wong-Baker FACES pain rating scale used widely in hospitals for children to communicate the intensity of pain felt to nurses (Wong and Baker 1998). The scale shows six faces to which a child can point, ranging from a very happy face (to indicate no hurt) to a very sad face (hurts most).

Finally, the *nominal scale* classifies objects into categories without an ordering or origin point—for example, the classification good or no-good, individual gender, color, the production shift, product, or geographic location.

Ordinal and nominal scales constitute a type of data referred to as *discrete* or *categorical* data.

The type of measurement scale determines the statistical analysis that can be applied to the data. In this regard, the ratio scale is the most powerful scale. For elaboration, see Emory and Cooper (1991).

For quality purposes, there are five general classes of quality measures:

- Defects (deficiencies, failures)
- Costs of poor quality
- Product and process features
- Customer needs
- Customer behavior

Units of measure for product deficiencies usually take the form of a fraction:

$$\frac{\text{Number of occurrences}}{\text{Opportunity for occurrence}}$$

The numerator may be in such terms as number of defects produced, number of field failures, or cost of warranty charges. The denominator may be in such terms as number of units produced, dollar volume of sales, number of units in service, or length of time in service. The deficiencies are determined by comparing the product delivered to its specification. In physical products, those specifications are in terms of physical dimensions, electrical or physical properties, or performance characteristics. In service products, the most common specification is in terms of timeliness. Other specifications usually relate to the actual performance versus the rules or specifications for the service—see the discussion of service features below.

Costs of poor quality are usually denominated in the currency of organization, but may also be expressed as fractions of sales, total costs, or gross margin.

Units of measure for product features are more difficult to create. The number and variety of these features may be large. Sometimes inventing a new unit of measure is a fascinating technical challenge. In one example, a manufacturer of a newly developed polystyrene product had to invent a unit of measure and a sensor to evaluate an important product feature. It was then possible to measure that feature of both the product and of competitors' products before releasing the product for manufacture. In another case, the process of harvesting peas in the field required a unit of measure for tenderness and the invention of a "tenderometer" gauge. A numerical scale was created, and measurements were taken in the field to determine when the peas were ready for harvesting.

Timeliness of execution is typically one important feature for a service product. Generally, the content of the service will also have certain performance features. A repair service will have features on the effectiveness and reliability of the repair.

Financial services will measure such features as the eligibility of the customer to receive a service or a specific return or interest rate. They also have specifications for calculating returns, payments, and value. These rules yield results that can be measured. The rules are extensively applied through automated decision engines, but the accuracy of these automated methods need to be validated, and there is often a human element in the setup and execution as well.

Health care has both process and outcome quality measures. The first describe the application of established standards of care for a given set of symptoms and signs. The outcomes measure the success of the treatment in restoring heath, avoiding further adverse episodes, and the safety of the patient from adverse events within the care setting, such as medication errors, falls, or procedural error.

Insurance pays claims according to the coverage of the policy and the nature of the insurable event. The insurance policy incorporates these rules for reimbursement for loss.

		Company X		Company A		Company B	
Attribute	**Relative Importance %**	**Rating**	**Weighted Rating**	**Rating**	**Weighted Rating**	**Rating**	**Weighted Rating**
Safety	28	6	168	5	140	4.5	126
Performance	20	6	120	7	140	6.5	130
Quality	20	6	120	7	140	4	80
Field service	12	4	48	8	96	5	60
Ease of use	8	4	32	6	48	5	40
Company image	8	8	64	4	32	4	32
Plant service	4	7.5	30	7.5	30	5	20
Total			582		626		488

(*Source: Quality Planning and Analysis*, Copyright 2007. Used by permission.)

TABLE 19.1 Multiattribute Study

"Claim engines" do most of the calculations, but require human specification and input. The accuracy of these payments can be expressed in monetary terms as well as in percent-defective terms.

Often a number of important product features exist. To develop an overall unit of measure, we can identify the important product features and then define the relative importance of each feature. In subsequent measurement, each feature receives a score. The overall measure is calculated as the weighted average of the scores for all features. This approach is illustrated in Table 19.1. In using such an approach for periodic or continuous measurement, some cautions should be cited (Early 1989). First, the relative importance of each feature is not precise and may change greatly over time. Second, improvement in certain features can result in an improved overall measure but can hide deterioration in one feature that has great importance.

The Sensor

The sensor is the means used to make the actual measurement. Most sensors are designed to provide information in terms of units of measure. For operational control subjects, the sensors are usually technological instruments or human beings employed as instruments (e.g., inspectors, auditors); for managerial and service subjects, the sensors are often data systems. Choosing the sensor includes defining how the measurements will be made—how, when, and who will make the measurements—and the criteria for taking action.

Clearly, sensors must be economical and easy to use. In addition, because sensors provide data that can lead to critical decisions on products and processes, sensors must be both accurate and precise, as discussed in the section "Measurement System Analysis."

Historical Data, Operational Data, and Experimental Data

Historical data are data that we already have and that may seem relevant to a question or problem at hand. Data often are saved during the production process, for example. If a satisfactory process goes out of control after some years of operation, it frequently is suggested that it would save both time and expense to analyze the historical data statistically rather than collect new data or perform a planned experiment to obtain new data that could lead to process correction. Thus, we have available data that may consist of measurements Y (such as a process yield, e.g., the strength of a material produced) and associated process variables $x1, x2, \ldots, xk$ (such as $x1$ = pressure and $x2$ = acid concentration, with $k = 2$). If such data do not exist, we might set up a data collection scheme to collect new operational data.

Historical or new operational data can both be invaluable for the following reasons:

- It is less time consuming and expensive to collect. Especially when multiple theories are at issue, even very lean eighth-fraction screening experimental designs can be prohibitive for some processes if they look at seven or more factors.
- For some types of operations that have a significant human performance component, the mere act of collecting new data, not to mention conducting experiments, can have unintended consequences on human behavior and, hence, the process—the famous Hawthorne effect.
- For out-of-control situations in previously stable processes, the information question is "what changed," which usually is a specific unique occurrence of an assignable cause that does not require significant experimentation.
- Substantial chronic random variation typically has root causes that are at least identifiable, and often quantifiable, from operational data.
- Although caveats need to be observed, operational data can be helpful in developing and testing the theories that will be used ultimately in an experiment.
- When dealing with either operational data or experimental data, the same pitfalls apply if one fails to test all the possible causes or extends results beyond the actual measured operating range.

Nevertheless, historical and operational data have potential drawbacks that include

- The x's may be highly correlated with each other in practice; hence, it may not be possible to separate the effects among them.
- The x's may cover a very small part of the possible operating range, so small that any indications of changes in Y attributable to changes in the x's may be overwhelmed by the size of the variability of the process.
- Other variables that affect the output of the process (e.g., time of day, atmospheric conditions, operator running the process, etc.) may not have been held constant and may in fact be the real causes of changes observed in the process.

In such cases, experimental data may be superior. Experiments are run at each of a number of combinations of settings that are selected in advance by statistical design criteria for each variable $x1, \ldots, xk$.

Measurement System Analysis

Control of a process, design of a new product, and elimination of chronic random variation all require accurate measurement of both the desired results and the contributing factors. A good measurement system that provides this critical information should have the following attributes:

- *Minimal bias.* Bias is the difference between the average measured value and a reference value. A reference value, in turn, is an agreed-upon standard, such as a traceable national standard. The reference standard is used to calibrate a measurement system, thereby bringing the reported measure in line with the accepted, known value. Bias sometimes is referred to as "accuracy." However, because accuracy has several meanings in the literature, "bias" is the recommended term in the present context.

- *Repeatability.* Repeatability is the variation in measurements obtained with one measurement instrument when used several times by an appraiser while measuring the identical characteristic on the same part.
- *Reproducibility.* Reproducibility is the variation in the average of the measurements made by different appraisers using the same measuring instrument when measuring the identical characteristic on the same part.
- *Stability.* Stability (or drift) is the total variation in the measurements obtained with a measurement system on the same master or parts when measuring a single characteristic over an extended period. A measurement system is stable if the same results are obtained at different points in time.
- *Linearity.* Linearity is the difference in bias values at different points along the expected operating range of a measurement instrument.
- *Precision.* Repeatability, reproducibility, and stability tend to be random, and the three together are often referred to as "precision." Refer to Figure 19.2 for a graphic depiction of the difference between bias and precision.

These five sources of measurement variation are illustrated in Figure 19.3, and are generally consistent with the definitions provided by the AIAG *Measurement Systems Analysis Reference Manual* (Automotive Industry Action Group, 2003).

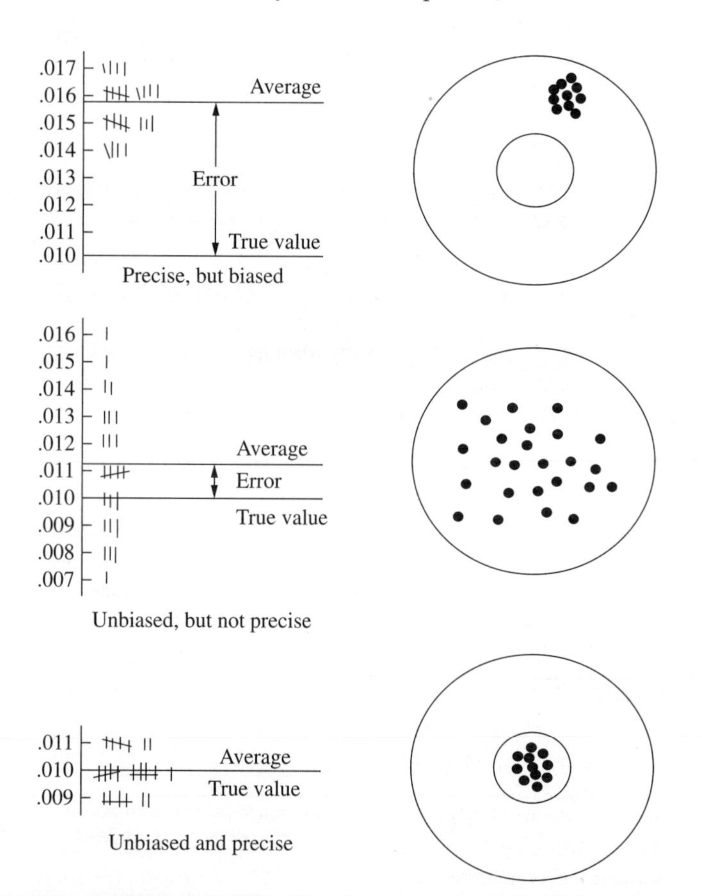

Figure 19.2 Bias and precision. (*Quality Planning and Analysis, Copyright 2007. Used by permission.*)

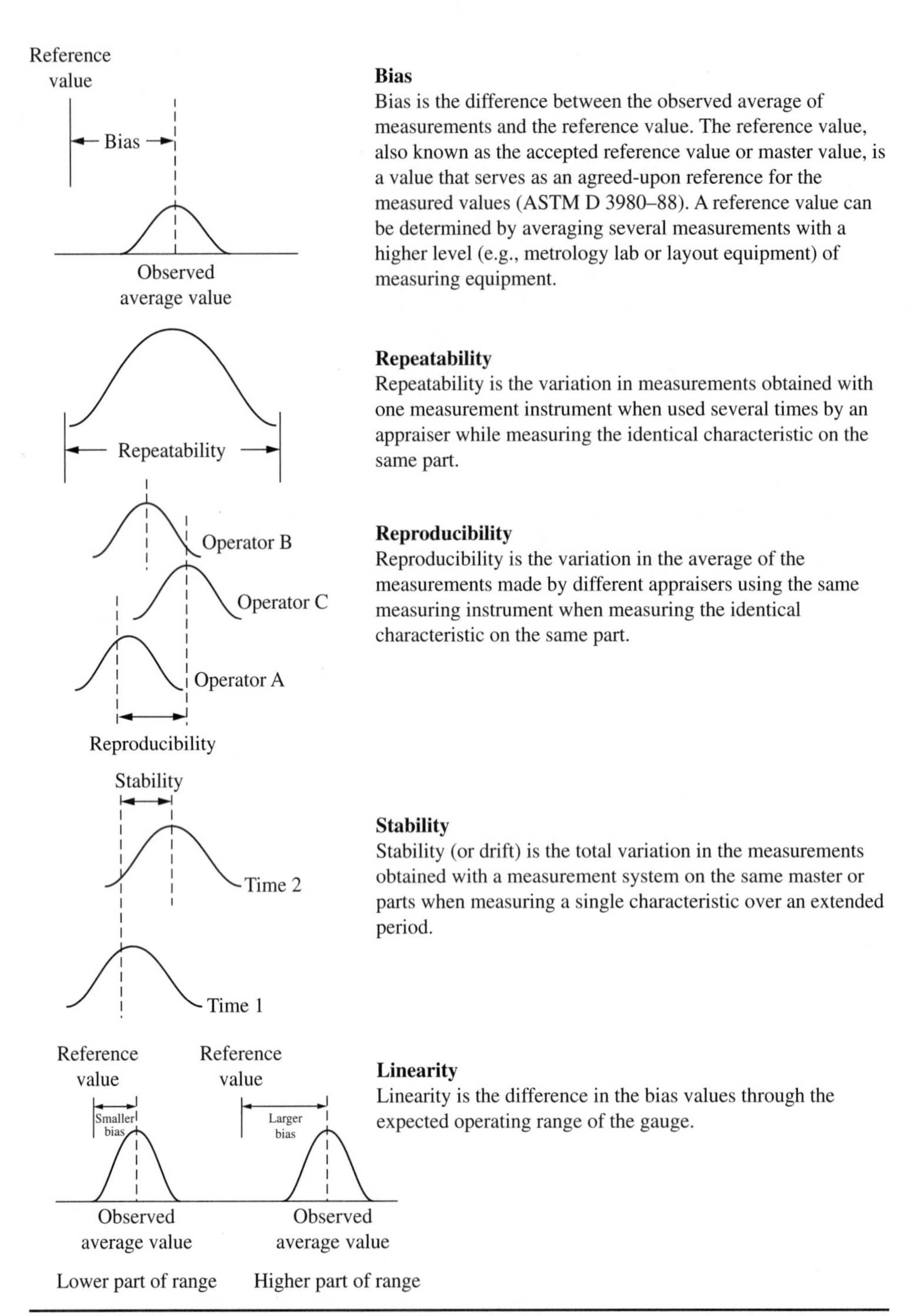

FIGURE 19.3 Five sources of measurement variation. (*Reprinted with permission from the MSA Manual DaimlerChrysler, Ford, General Motors Supplier Quality Requirements Task Force.*)

Any statement of bias and precision must be preceded by three conditions:

1. Definition of the test method. This definition includes the step-by-step procedure, equipment to be used, preparation of test specimens, test conditions, etc.
2. Definition of the system of causes of variability, such as material, analysts, apparatus, laboratories, days, etc. American Society for Testing and Materials (ASTM) recommends that modifiers of the word "precision" be used to clarify the scope of the precision measure. Examples of such modifiers are single-operator, single-analyst, single-laboratory-operator-material-day, and multilaboratory.
3. Existence of a statistically controlled measurement process. The measurement process must have stability for the statements on bias and precision to be valid. This stability can be verified by a control chart.

Effect of Measurement Error On Acceptance Decisions. Error of measurement can cause incorrect decisions on (1) individual units of product and on (2) lots submitted to sampling plans. In one example of measuring the softening point of a material, the standard deviation of the test precision is 2°, yielding two standard deviations of ±4°. The specification limits on the material are ±3°. Imagine the incorrect decisions that are made under these conditions.

Two types of errors can occur in the classification of a product: (1) a nonconforming unit can be accepted (the consumer's risk) and (2) a conforming unit can be rejected (the producer's risk). In a classic paper, Eagle (1954) showed the effect of precision on each of these errors.

The probability of accepting a nonconforming unit as a function of measurement error (called test error, σ_{TE}, by Eagle) is shown in Figure 19.4. The abscissa expresses the test error as the standard deviation divided by the plus-or-minus value of the specification range (assumed equal to two standard deviations of the product). For example, if the measurement error is one-half of the tolerance range, the probability is about 1.65 percent that a nonconforming unit will be read as conforming (due to the measurement error) and therefore will be accepted.

Figure 19.5 shows the percentage of conforming units that will be rejected as a function of the measurement error. For example, if the measurement error is one-half of the plus-or-minus tolerance range, about 14 percent of the units that are within specifications will be rejected because the measurement error will show that these conforming units are outside specification.

The test specification can be adjusted with respect to the performance specification (see Figures 19.4 and 19.5). Moving the test specification inside the performance specification reduces the probability of accepting a nonconforming product but increases the probability of rejecting a conforming product. The reverse occurs if the test specification is moved outside the performance specification. Both risks can be reduced by increasing the precision of the test (i.e., by reducing the value of σ_{TE}).

Hoag et al. (1975) studied the effect of inspector errors on type I (α) and type II (β) risks of sampling plans (see the section "Hypothesis Testing" for definitions of type I and II risks). For a single sampling plan and an 80 percent probability of the inspector detecting a defect, the real value of β is two to three times that specified, and the real value of α is about one-fourth to one-half of that specified.

Case et al. (1975) investigated the effect of inspection error on the average outgoing quality (AOQ) of an attribute sampling procedure. They concluded that the AOQ values change and significant changes can occur in the shape of the AOQ curve.

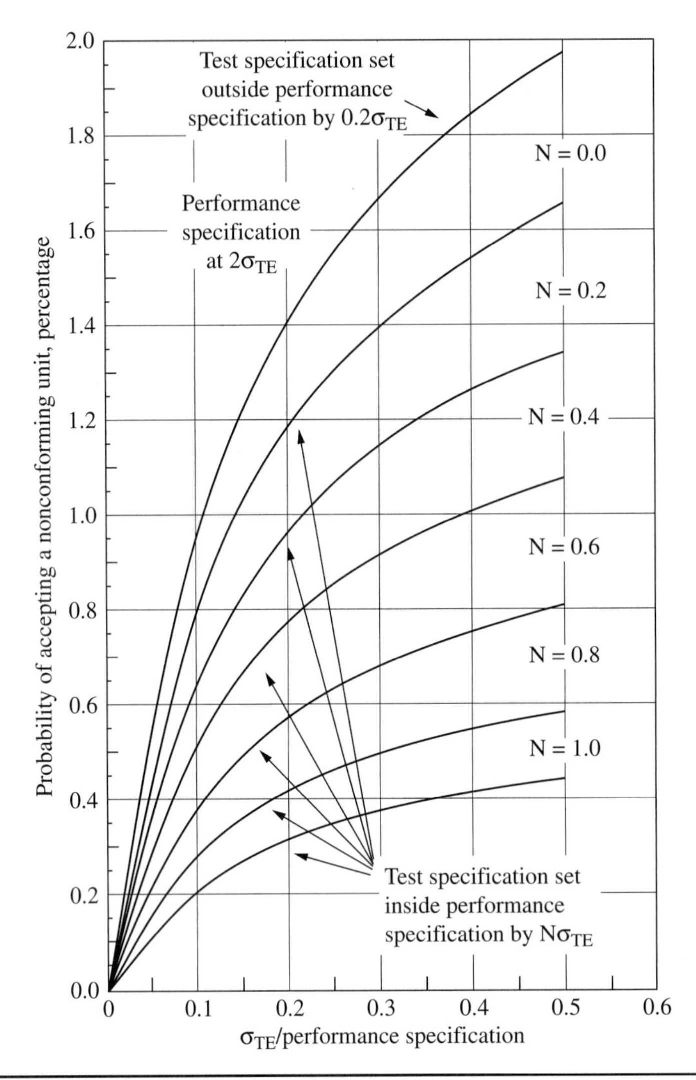

Figure 19.4 Probability of accepting a nonconforming unit.

The Automotive Industry Action Group (1995, p. 77) presents the concept of a gauge performance curve to determine the probability of accepting or rejecting a part when the gauge repeatability and reproducibility (R&R) are unknown.

All these investigations concluded that measurement error can be a serious problem.

Components of Variation. In drawing conclusions about measurement error, it is worthwhile to study the causes of variation in observed values. The relationship is

$$\sigma_{\text{observed}} = \sqrt{\sigma^2_{\text{causeA}} + \sigma^2_{\text{causeB}} + \ldots + \sigma^2_{\text{causeN}}}$$

The formula assumes that the causes act independently.

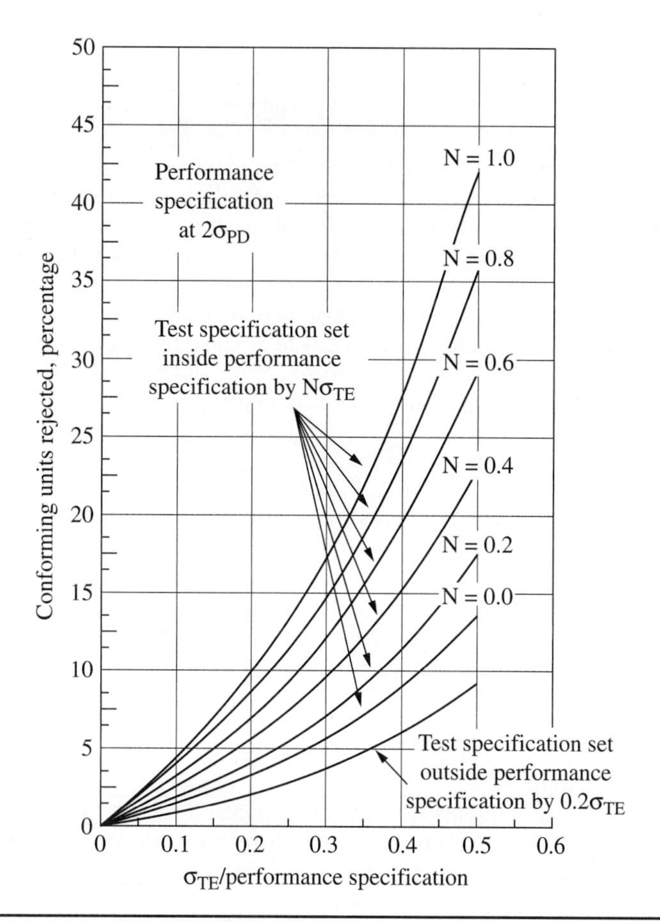

Figure 19.5 Conforming units rejected (percentage).

It is valuable to find the numerical values of the components of observed variation because the knowledge may suggest where effort should be concentrated to reduce variation in the product. A separation of the observed variation into product variation plus other causes of variation may indicate important factors other than the manufacturing process. Thus, if it is found that the measurement error is a large percentage of the total variation, this finding must be analyzed before proceeding with a quality improvement program. Finding the components (e.g., instrument, operator) of this error may help to reduce the measurement error, which in turn may completely eliminate a problem.

Observations from an instrument used to measure a series of different units of product can be viewed as a composite of (1) the variation due to the measuring method and (2) the variation in the product itself. This value can be expressed as

$$\sigma_O = \sqrt{\sigma_P^2 + \sigma_E^2}$$

where $\sigma_O = \sigma$ of the observed data
$\sigma_P = \sigma$ of the product
$\sigma_E = \sigma$ of the measuring method

Solving for σ_P yields

$$\sigma_P = \sqrt{\sigma_O^2 + \sigma_E^2}$$

The components of measurement error often focus on repeatability and reproducibility (R&R). Repeatability primarily concerns variation due to measurement gauges and equipment; reproducibility concerns variation due to human "appraisers" who use the gauges and equipment. Studies to estimate these components are often called "gauge R&R" studies.

A gauge R&R study can provide separate numerical estimates of repeatability and reproducibility. Two methods are usually used to analyze the measurement data. Each method requires a number of appraisers, a number of parts, and repeat trials of appraisers measuring different parts. For example, an R&R study might use three appraisers, ten parts, and two trials.

One method analyzes averages and ranges of the measurement study data. This method requires minimum statistical background and does not require a computer. The second method is the analysis of variance, ANOVA (see "Statistical Tools for Improvement"). Compared to the first method, ANOVA requires a higher level of statistical knowledge to interpret the results, but can evaluate the data for possible interaction between appraisers and parts. ANOVA is best done on a computer using Minitab or other software. Overall, the ANOVA method is preferred to analyzing the averages and ranges. Detailed illustrations of each method are provided in the Automotive Industry Action Group booklet *Measurement Systems Analysis* (1995). Also, see Tsai (1988) for an example using ANOVA and considering both no interaction and interaction of operators and parts. Burdick and Larsen (1997) provide methods for constructing confidence intervals on measures of variability in R&R studies.

When the total standard deviation of repeatability and reproducibility is determined from ANOVA, a judgment must then be made on the adequacy of the measurement process. A common practice is to calculate 5.15σ ($\pm 2.575\sigma$) as the total spread of the measurements that will include 99 percent of the measurements. If 5.15σ is equal to or less than 10 percent of the specification range for the quality characteristic, the measurement process is viewed as acceptable for that characteristic; if the result is greater than 10 percent, the measurement process is viewed as unacceptable. Engel and DeVries (1997) examine how the practice of comparing measurement error with the specification interval relates to making correct decisions in product testing.

Reducing and Controlling Errors of Measurement. Steps can be taken to reduce and control errors for all sources of measurement variation. The systematic errors that contribute to bias can sometimes be handled by applying a numerical correction to the measured data. If an instrument has a bias of 0.001, then, on average, it reads 0.001 too low. The data can be adjusted by adding 0.001 to each value of the data. Of course, it is preferable to adjust the instrument as part of a calibration program.

In a calibration program, the measurements made by an instrument are compared to a reference standard of known accuracy (a calibration program should include provisions for periodic audits). If the instrument is found to be out of calibration, an adjustment is made.

A calibration program can become complex for these reasons:

- The large number of measuring instruments
- The need for periodic calibration of many instruments

- The need for many reference standards
- The increased technological complexity of new instruments
- The variety of types of instruments (i.e., mechanical, electronic, chemical, etc.)

Precision in measurement can be improved through either or both of the following procedures:

- Discovering the causes of variation and remedying these causes. A useful step is to resolve the observed values into components of variation (see earlier). This process can lead to the discovery of inadequate training, perishable reagents, lack of sufficient detail in procedures, and other such problems. This fundamental approach also points to other causes for which the remedy is unknown or uneconomic (i.e., basic redesign of the test procedure).
- Using multiple measurements and statistical methodology to control the error in measurement. The use of multiple measurements is based on the following relationship:

$$\sigma_{\bar{X}} = \frac{\sigma}{\sqrt{n}}$$

As in all sampling schemes, halving the error in measurement requires quadrupling (not doubling) the number of measurements.

As the number of tests grows larger, a significant reduction in the error in measurement can be achieved only by taking a still larger number of additional tests. Thus, the cost of the additional tests versus the value of the slight improvement in measurement error becomes an issue. The alternatives of reducing the causes of variation must also be considered.

For an in-depth discussion of reducing other forms of measurement error, see Automotive Industry Action Group (2003) and Coleman et al. (2008).

A successful measurement system analysis (MSA) is critical not only for control but also for validating the measures used in quality planning and improvement, as illustrated by this Six Sigma improvement project (DMAIC) (courtesy of Steve Wittig and Chris Arquette at a Juran Institute client forum). It also illustrates the use and importance of attribute MSA studies for discrete variables.

Background. The paint line has a first run yield of 74 percent. This means that 26 percent of all frames need to be reworked at least once. Defects due to finish issues account for 15 percent, and material (wood) issues account for 11 percent. This project looks only at finish defects because this is readily within our control. Any rework is nonvalue-added and contributes to wasted paint/primer, labor, utilities, work in progress, capacity, and more hazardous waste. Our goal is to improve first-run yield to 90 percent for finish defects.

Summary of MSA Effort. The paint line is old and somewhat neglected. Our first MSA results were expectedly poor, and the appraisers were contributing to the defect rate by rejecting good frames. We improved this by continued training of the appraisers by quality control. We did two more MSAs with acceptable results. This will need to be an ongoing test/train routine. Figures 19.6 to 19.8 are attribute MSA results, and Figures 19.9 and 19.10 are results of a variable MSA.

Validate measurement system attribute data analysis – MSA 1

Sample #	Expert	Operator 1		Operator 2		Operator 3	
		Try 1	Try 2	Try 1	Try 2	Try 1	Try 2
1	Blister	Blister	Blister	Blister	Blister	Blister	Blister
2	Good	Light Ed.	Good	Good	Good	Good	Good
3	Drip	Drip	Drip	Dirt	Dirt	Drip	Drip
4	Dirt	Contam	Dirt	Dirt	Dirt	Dirt	Dirt
5	Contam	Contam	Contam	Good	Good	Over run	Over run
6	Blister	Blister	Blister	Blister	Blister	Dirt	Blister
7	Good	Good	Good	Good	Good	Good	Good
8	Dirt	Light Ed.	Contam	Good	Good	Dirt	Dirt
9	Good	Good	Drip	Dirt	Good	Drip	Drip
10	Good	Orange P.	Good	Good	Good	Good	Good
11	Dirt	Dirt	Good	Dirt	Dirt	Dirt	Dirt
12	Good	Contam	Light Ed.	Good	Good	Over run	Over run
13	Good	Light Ed.	Light Ed.	Good	Good	Light Ed.	Over run
14	Contam	Contam	Contam	Good	Good	Good	Good
15	Drip	Drip	Light Ed.	Dirt	Good	Drip	Drip
16	Light Ed.	Light Ed.	Light Ed.	Good	Good	Good	Good
17	Dirt	Contam	Contam	Good	Dirt	Dirt	Dirt
18	Dirt	Contam	Contam	Dirt	Dirt	Dirt	Dirt
19	Blister	Blister	Good	Good	Blister	Blister	Blister
20	Good	Good	Good	Good	Good	Orange P.	Orange P.

Figure 19.6 Baseline attribute MSA on appraisers' accept/reject decisions. (*Juran Institute, Inc.*)

Attribute data analysis-MSA 1 results

Within appraiser

Assessment agreement

Appraiser	# Inspected	# Matched	Percent (%)	95.0% Cl
1	20	11	55.0	(31.5, 76.9)
2	20	16	80.0	(56.3, 94.3)
3	20	18	90.0	(68.3, 98.8)

Matched: Appraiser agrees with him/herself across trials.

Each appraiser vs standard

Assessment agreement

Appraiser	# Inspected	# Matched	Percent (%)	95.0% Cl
1	20	8	40.0	(19.1, 63.9)
2	20	11	55.0	(31.5, 76.9)
3	20	12	60.0	(36.1, 80.9)

Matched: Appraisers' assessment across trials agrees with standard.

Between appraisers

Assessment agreement

# Inspected	# Matched	Percent (%)	95.0% Cl
20	2	10.0	(1.2, 31.7)

Matched: All appraisers' assessments agree with each other.

All appraisers vs standard

Assessment agreement

# Inspected	# Matched	Percent (%)	95.0% Cl
20	2	10.0	(1.2, 31.7)

Matched: All appraisers' assessments agree with standard.

Note: 38% were called bad that were good. 22% were called good that were bad. This potentially could yield an improvement in the defect rate by 16%.

Figure 19.7 Results of baseline attribute MSA. Results are not acceptable. (*Juran Institute, Inc.*)

Attribute data analysis-MSA 2 results

Within appraiser

Assessment agreement

Appraiser	# Inspected	# Matched	Percent (%)	95.0% Cl
1	20	16	80.0	(56.3, 94.3)
2	20	19	95.0	(75.1, 99.9)
3	20	20	100.0	(86.1, 100.0)

Matched: Appraiser agrees with him/herself across trials.

Each appraiser vs standard

Assessment agreement

Appraiser	# Inspected	# Matched	Percent (%)	95.0% Cl
1	20	15	75.0	(50.9, 91.3)
2	20	19	95.0	(75.1, 99.9)
3	20	18	90.0	(68.3, 98.8)

Matched: Appraisers' assessment across trials agrees with standard.

Between appraisers

Assessment agreement

# Inspected	# Matched	Percent (%)	95.0% Cl
20	13	65.0	(40.8, 84.6)

Matched: All appraisers' assessments agree with each other.

All appraisers vs standard

Assessment agreement

# Inspected	# Matched	Percent (%)	95.0% Cl
20	13	65.0	(40.8, 84.6)

Matched: All appraisers' assessments agree with standard.

Conclusion: MSA is greatly improved over first. Continue Q.C. training of inspectors to bring up agreement between all appraisers and standard.

Figure 19.8 Attribute MSA after improvement. (*Juran Institute, Inc.*)

Measurement system analysis Sheen Gage study

Gage R&R Source	**VarComp**	**%Contribution (of VarComp)**
Total Gage R&R	0.0519	0.69
Repeatability	0.0298	0.39
Reproducibility	0.0221	0.29
Operator	0.0028	0.04
Operator* measurement	0.0193	0.26
Part-to-part	7.4851	99.31
Total variation	7.5370	100.00

Source	**StdDev (SD)**	**Study Var (5.15*SD)**	**%Study Var (%SV)**	
Total Gage R&R	0.22776	1.1730	8.30	
Repeatability	0.17248	0.8883	6.28	
Reproducibility	0.14874	0.7660	5.42	
Operator	0.05294	0.2726	1.93	
Operator* measurement		0.13900	0.7159	5.06
Part-to-part	2.73590	14.0899	99.66	
Total variation	2.74536	14.1386	100.00	

Number of distinct categories = 17

Figure 19.9 Results of baseline variable data MSA on sheen gage results are acceptable. (*Juran Institute, Inc.*)

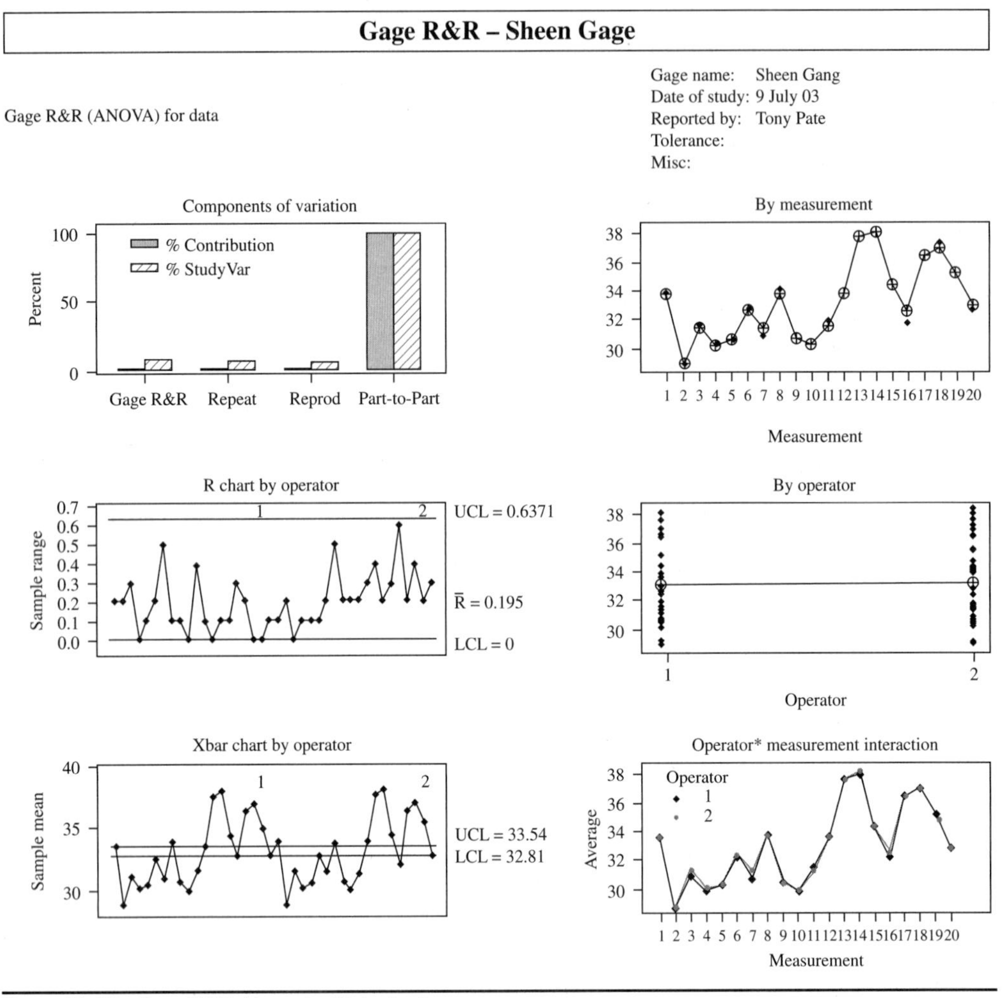

Figure 19.10 Gage R&R. (*Juran Institute, Inc.*)

Data Screening

As a practical matter, many data sets contain some instances of incorrectly transcribed values, values from points where an experiment went awry for some reason (such as equipment malfunction), and the measurement system failed, or other factors led to observational error. Procedures for finding these problems are called data screening and should be performed.

Data Screening Methods. Numerous tests are available to detect outliers, that is, "observation(s) [or (a)subset of observations] which appear to be inconsistent with the remainder of that set of data" (Barnett and Lewis 1994). One of the most common methods of data screening is to classify observations as outliers if they are outside an interval of L multiples of the standard deviation about the mean. The number L is commonly taken to be 2.5, 3, or 4. The larger L is, the less likely it is that outliers will be detected, while the smaller L is, the more good observations one will wrongly detect as potential outliers. For example, because approximately 99.73 percent of a population lies within ±3 standard deviations from the mean, application of $L = 3$ will yield (100) (0.0027) = 0.27 percent of the observations being further than 3 standard deviations from the mean even if there are no true outliers in the data set (this assumes a normal distribution for the observations). As the data set being considered becomes larger, the more possible outliers one will identify, even if there are no problems with the data. For this reason, outliers identified in this way should be deleted from the analysis only if they can be traced to specific causes (such as recording errors, experimental errors, and the like). Otherwise, there is a substantial risk of eliminating data that are, in a sense, "trying to tell you something."

Typically, one adjusts L based on the size of the data set to be screened; with $n = 1000$ points, $L = 3$ is reasonable; with $n = 100$, $L = 2.5$ can be used, and only (100) (0.0124) = 1.24 outliers will be expected to be found if the data have no problems. After bad data are deleted or replaced (this is desirable if the experiment can be rerun under comparable conditions to those specified in the experimental plan), the data should be screened again. With the "worst" points removed/corrected, less extreme cases may come to be identified as possible outliers, and should again be investigated.

Another commonly used method is to visualize the data in some way (e.g., to plot variables in box plots or scatter plots). Points visually distant from the others should be scrutinized and eliminated as outliers if (and only if) they are reasonably attributable to some specific cause unrelated to the question at hand. Regression analysis can be helpful as an additional step, using residuals (the differences between the observed and predicted values) to flag potential data points that are unusual (and may have undue leverage on the regression model). Regression is discussed later in this chapter.

Summarization of Data

A mantra for the data analyst is that the first three steps of any data analysis are to (1) plot the data, (2) plot the data, and (3) plot the data. Clearly important, many of the most practical methods of summarizing data are quite simple in concept. Depending on the goals of the data summarization, sometimes one method will provide a useful and complete summarization. More often, two or more methods will be needed to obtain the clarity of description that is desired. Several key methods are plots versus time order of data, frequency distributions and histograms, and sample characteristics such as measures of central tendency/location (mean, median, mode) and measures of dispersion (range, standard deviation, variance) displayed graphically

Plots versus Time Order of Data. Plotting the output Y against the time order in which the data were obtained (essentially a scatter plot of Y versus time) can reveal several possible phenomena:

- A few observations are far from the others. They should be investigated as to their cause and, if erroneous, corrected or discarded.
- There are trends or cycles within a time period—a day, week, etc. This may represent such phenomena as warming of a machine, operator fatigue, seasonal demand, customer timing preferences, or similar time-related trends.
- Variability decreases or increases with time; this may be due to a learning curve or raw material characteristics, as when one lot of material is used up and the next lot has lesser or greater heterogeneity. It may also reflect changes in customer behavior for services.

While the preceding trends may be apparent even in a plot of the original observations Y versus time, they are often more easily spotted in plots of the residuals of the observations after a regression analysis (see "Correlation and Regression Analysis" later in this chapter) or using a control chart.

Histograms. A frequency distribution is a tabulation of data arranged according to size. Presenting data in this form clarifies the central tendency and the dispersion along the scale of measurement, as well as the relative frequency of occurrence of the various values (i.e., the shape of the distribution of data). The shape of the distribution may suggest some theories of cause for variation in the process, and reduce the likelihood of others (see Chapter 18, Core Tools to Design, Control, and Improve Performance). Histograms usually require at least 40 data points to provide useful insight.

Box Plots. These also display frequency distribution and indications with regard to central tendency and the dispersion along the scale of measurement. They provide less rich detail than histograms, but can be used with as few as eight data points, and facilitate the comparison of many distributions. (See Chapter 18, Core Tools to Design, Control, and Improve Performance.)

Sample Characteristics. Descriptive statistics such as the mean (average), median, mode, range, variance, and standard deviation provide numerical ways of summarizing data, and should be used in conjunction with graphical displays of the type discussed previously.

Analysis

The emphasis in this section is on statistical tools used in the analysis of data for quality improvement, control, and planning. Statistics, for our purposes, is the use of a small sample of data to infer properties of a larger population or universe in which we are interested. Statistics is grounded in probability. Probability is a measure that describes the chance that an event will occur. Based on appropriately collected data, statistics and probability are used to understand explicitly the accuracy of the information we have for managing quality and assess the risks of both acting and not acting on the basis of that data.

The following are some types of problems that can benefit from statistical analysis:

- Determining the usefulness of a limited number of test results in predicting the true value of a product characteristic
- Determining the number of tests required to provide adequate data for evaluation
- Comparing test data between two alternative designs
- Predicting the amount of product that will fall within specification limits
- Predicting system performance
- Controlling process quality by early detection of process changes
- Planning experiments to discover the factors that influence a characteristic of a product or process (i.e., exploratory experimentation)
- Determining the quantitative relationship between two or more variables

The Concept of Statistical Variation

Variety is the so-called "spice of life," and this is no less true when it comes to statistics. The concept of variation is that no two items are perfectly identical. Variation is a fact of nature and a bane of industrial life. For example, even "identical" twins vary slightly in height and weight at birth. The dimensions of an integrated chip vary from chip to chip; cans of tomato soup vary slightly from can to can; the time required to assign a seat at an airline check-in counter varies from passenger to passenger. To disregard the existence of variation (or to rationalize falsely that it is small) can lead to incorrect decisions on major problems. Statistics helps to analyze data properly and draw conclusions, taking into account the existence of variation.

Statistical variation—variation due to random causes—is much greater than most people think. Often, we decide what action to take based on the most recent data point, and we forget that the data point is part of a history of data.

In order to make decisions and improve processes, statistical variation must be taken into account. Variation can be visualized through the use of histograms, box plots, and similar tools. Frequently, such tools are sufficient to draw practical conclusions because differences in central tendency are large and variation is relatively small. However, statistical tools become necessary when the picture (quite literally) is less clear.

Building on the foundation of descriptive statistics, we start with an overview of the probability distributions that underlie many statistical tools and are used to model data and allow estimation of probabilities. Terms are defined as they are encountered, including further discussion of enumerative and analytical studies. Following an introduction to statistical inference and hypothesis testing, specific methods are discussed by way of example.

Probability Distributions

Before diving in, we should make a distinction between a sample and a population. A population is the totality of the phenomenon under study. A sample is a limited number of items taken from that population. Measurements are made on the smaller subset of items, and we can calculate a sample statistic (e.g., the mean). A sample statistic is a quantity computed from a sample to estimate a population parameter. Samples for statistics must be random. Simple random samples require that every element of the population have the same equal probability of selection for the sample. More complex sampling, such as stratified sampling, requires still requires that each element have a known, but not necessarily equal, chance of selection.

A probability distribution function is a mathematical formula that relates the values of the characteristic with their probability of occurrence in the population. The collection of these probabilities is called a probability distribution. The mean (μ) of a probability distribution often is called the expected value. Some distributions and their functions are summarized in Figure 19.11. Distributions are of two types:

Continuous (for "Variable" Data). When the characteristic being measured can take on any value (subject to the fineness of the measuring process), its probability distribution is called a "continuous probability distribution." For example, the probability distribution of the resistance data in Table 19.2 is an example of a continuous probability distribution because the resistance could have any value, limited only by the fineness of the measuring instrument. Most continuous characteristics follow one of several common probability distributions: the normal distribution, the exponential distribution, or the Weibull distribution.

Discrete (for "Attribute" Data). When the characteristic being measured can take on only certain specific values (e.g., integers 0, 1, 2, 3), its probability distribution is called a "discrete probability distribution." For example, the distribution of the number of defects r in a sample of five items is a discrete probability distribution because r can be only 0, 1, 2, 3, 4, or 5 (and not 1.25 or similar intermediate values). The common discrete distributions are the Poisson and binomial.

Distribution	Form	Probability function	
Normal	μ	$y = \frac{1}{\sigma\sqrt{2\pi}} e^{-\frac{(x-\mu)^2}{2\sigma^2}}$ μ = Mean σ = Standard deviation	Applicable when there is a concentration of observations about the average and it is equally likely that observations will occur above and below the average. Variation in observations is usually the result of many small causes.
Exponential	μ	$y = \frac{1}{\mu} e^{-\frac{x}{\mu}}$	Applicable when it is likely that more observations will occur below the average than above.
Weibull	$\beta = 1/2$ $\alpha = 1$ $\beta = 1$ $\beta = 3$ X	$y = \alpha\beta(X-\gamma)^{\beta-1}e^{-\alpha(X-\gamma)^{\alpha}}$ α = Scale parameter β = Shape parameter γ = Location parameter	Applicable in describing a wide variety of patterns in variation, including departures from the normal and exponential.
Poisson*	$p = .01$ $p = .03$ $p = .05$ r	$y = \frac{(np)^r e^{-np}}{r!}$ n = Number of trials r = Number of occurrences p = Probability of occurrence	Same as binomial but particularly applicable when there are many opportunities for occurrence of an event but a low probability (less than .10) on each trial.
Binomial*	$p = .1$ $p = .3$ $p = .5$ r	$y = \frac{n!}{r!(n-r)!} p^r q^{n-r}$ n = Number of trials r = Number of occurrences p = Probability of occurrence $q = 1 - p$	Applicable in defining the probability of r occurrences in n trials of an event that has constant probability of occurrence on each independent trial.

Figure 19.11 Summary of common probability distributions. (*Quality Planning and Analysis, Copyright 2007. Used by permission.*)

3.37	3.34	3.38	3.32	3.33	3.28	3.34	3.31	3.33	3.34
3.29	3.36	3.30	3.31	3.33	3.34	3.34	3.36	3.39	3.38
3.35	3.36	3.30	3.32	3.33	3.35	3.35	3.34	3.32	3.38
3.32	3.37	3.34	3.38	3.36	3.37	3.36	3.31	3.33	3.30
3.35	3.33	3.38	3.37	3.44	3.32	3.36	3.32	3.29	3.35
3.38	3.39	3.34	3.32	3.30	3.39	3.36	3.40	3.32	3.33
3.29	3.41	3.27	3.36	3.41	3.37	3.36	3.37	3.33	3.66
3.31	3.33	3.35	3.34	3.35	3.34	3.31	3.36	3.37	3.35
3.40	3.35	3.37	3.35	3.32	3.36	3.38	3.35	3.31	3.34
3.35	3.36	3.39	3.31	3.31	3.30	3.35	3.33	3.35	3.31

(*Source: Quality Planning and Analysis,* Copyright 2007. Used by permission.)

Table 19.2 Resistance of 100 Coils, Ω

Statistical Inference

Statistical inference is the process of estimating, through sampling and application of statistical methods, certain characteristics of a population. In the world of quality, these estimates and statistical conclusions are used to draw practical conclusions, typically providing the practitioner confidence in taking subsequent action (or inaction) to improve a process.

Sampling Variation and Sampling Distributions

Suppose that a battery is to be evaluated to ensure that life requirements are met. A mean life of 30 hours is desired. Preliminary data indicate that the life follows a normal distribution and that the standard deviation is equal to 10 hours. A sample of four batteries is selected at random from the population and tested. If the mean of the four is close to 30 hours, it is concluded that the population of batteries meets the specification. Figure 19.12 plots the distribution of individual batteries from the population, assuming that the true mean of the population is exactly 30 hours.

If a sample of four is life-tested, the following lifetimes might result: 34, 28, 38, and 24, giving a mean of 31.0 hours. However, this random sample is selected from the many batteries made by the same process. Suppose that another sample of four is taken. The second sample of four is likely to be different from the first sample. Perhaps the results would be 40, 32, 18, and 29, giving a mean of 29.8 hours. If the process of drawing many samples (with four in each sample) is repeated over and over, different results would be obtained in most samples. The fact that samples drawn from the same process can yield different sample results illustrates the concept of sampling variation.

Returning to the problem of evaluating the battery, a dilemma exists. In the actual evaluation, let's assume only one sample of four can be drawn (e.g., because of time and cost limitations). Yet the experiment of drawing many samples indicates that samples vary. The question is, How reliable is the single sample of four that will be the basis of the decision? The final decision can be influenced by the "luck" of which sample is chosen. The key point is that the existence of sampling variation means that any one sample cannot always be relied upon to give an adequate decision. The statistical approach analyzes the results of the sample, taking into account the possible sampling variation that could occur.

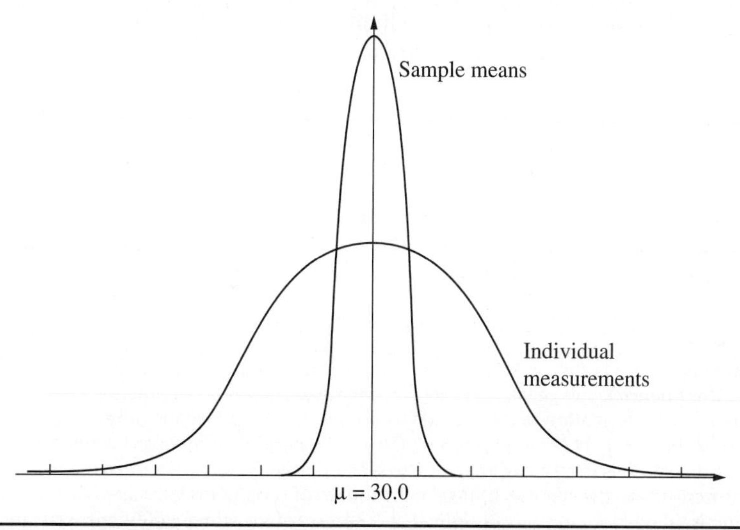

FIGURE 19.12 Distributions of individual measurements and sample means. (*Juran Institute, Inc., 1994.*)

Formulas have been developed to define the expected amount of sampling variation. In particular, the central limit theorem states that if $x_1, x_2, \ldots x_n$ are outcomes of a sample of n independent observations of a random variable x, then the mean of the samples of n will approximately follow a normal distribution, with mean μ and standard deviation $\sigma\bar{X} = \sigma\sqrt{n}$. When n is large ($n > 30$), the normal approximation is very close. For smaller samples, a modified Student-T distribution applies. The central limit theorem is very helpful to much practical statistical work. First, the variation of means is smaller than the variation of the underlying population, which makes conclusions easier. Second, because means are approximately normally distributed, we can apply the wide variety of techniques that rely on the assumption of normality.

Statistical Tools for Improvement

This concept of a sampling distribution is fundamental to the two major areas of statistical inference, estimation and tests of hypotheses, which are discussed next.

Statistical Estimation: Point Estimation and Confidence Intervals

Estimation is the process of analyzing a sample result to predict the corresponding value of the population parameter. In other words, the process is to estimate a desired population parameter by an appropriate measure calculated from the sample values. For example, the sample of four batteries previously mentioned had a mean life of 31.0 hours. If this is a representative sample from the process, what estimate can be made of the true average life of the entire population of batteries? The estimation statement has two parts:

1. The point estimate is a single value used to estimate the population parameter. For example, 31.0 hours is the point estimate of the average life of the population.
2. The confidence interval is a range of values that include (with a preassigned probability called a confidence level*) the true value of a population parameter. Confidence limits are the upper and lower boundaries of the confidence interval. Confidence limits should not be confused with other limits (e.g., control limits, statistical tolerance limits).

Table 19.3 summarizes confidence limit formulas for common parameters. The following example illustrates one of these formulas.

Problem Twenty-five specimens of brass have a mean hardness of 54.62 and an estimated standard deviation of 5.34. Determine the 95 percent confidence limits on the mean. The standard deviation of the population is unknown.

Solution Note that when the standard deviation is unknown and is estimated from the sample, the t distribution in Table 19.4 must be used. The t value for 95 percent confidence is found by entering the table at 0.975 and 25 – 1, or 24, degrees of freedom† and reading a t value of 2.064.

---

*A confidence level is the probability that an assertion about the value of a population parameter is correct. Confidence levels of 90, 95, or 99 percent are usually used in practice.

†A mathematical derivation of degrees of freedom is beyond the scope of this book, but the underlying concept can be stated. Degrees of freedom (DF) is the parameter involved when, for example, a sample standard deviation is used to estimate the true standard deviation of a universe. DF equals the number of measurements in the sample minus some number of constraints estimated from the data to compute the standard deviation. In this example, it was necessary to estimate only one constant (the population mean) to compute the standard deviation. Therefore, DF = 25 – 1 = 24.

Mean of a normal population (standard deviation known)	$\bar{X} \pm Z_{\alpha/2}\frac{\sigma}{\sqrt{n}}$ where $\bar{X}$ = sample average Z = normal distribution coefficient σ = standard deviation of population n = sample size
Mean of a normal population (standard deviation unknown)	$\bar{X} \pm t_{\alpha/2}\frac{s}{\sqrt{n}}$ where t = distribution coefficient (with $n - 1$ degrees of freedom) s = estimated σ (s is the sample standard deviation)
Standard deviation of a normal population	Upper confidence limit $= s\sqrt{\frac{n-1}{x^2_{\alpha/2}}}$ Lower confidence limit $= s\sqrt{\frac{n-1}{x^2_{1-\alpha/2}}}$ where x^2 = chi-square distribution coefficient with $n - 1$ degrees of freedom $1 - \alpha$ = confidence level
Population fraction defective	See charts: *Ninety-five percent confidence belts for population proportion* and *Binomial Distribution* at the end of this chapter, pages 670-672.
Difference between the means of two normal populations (standard deviations σ_1 and σ_2 known)	$(\bar{X}_1 - \bar{X}_2) \pm Z_{\alpha/2}\sqrt{\frac{\sigma_1^2}{n_1} + \frac{\sigma_2^2}{n_2}}$
Difference between the means of two normal populations ($\sigma_1 = \sigma_2$ but unknown)	$(\bar{X}_1 - \bar{X}_2) \pm t_{\alpha/2}\sqrt{\frac{1}{n_1} + \frac{1}{n_2}}$ $\times\sqrt{\frac{\Sigma(X - \bar{X}_1)^2 + \Sigma(X - \bar{X}_2)^2}{n_1 + n_2 - 2}}$
Mean time between failures based on an exponential population of time between failures	Upper confidence limit $= \frac{2rm}{x^2_{\alpha/2}}$ Lower confidence limit $= \frac{2rm}{x^2_{1-\alpha/2}}$ where r = number of occurrences in the sample (i.e., number of failures) m = sample mean time between failures DF = $2r$

(*Source: Quality Planning and Analysis*, Copyright 2007. Used by permission.)

Table 19.3 Summary of Confidence Limit Formulas (1 – α) (Confidence Level

Distribution of *t*

Value of *t* corresponding to certain selected probabilities (i.e., tail areas under the curve). To illustrate: the probability is .975 that a sample with 20 degrees of freedom would have $t = +2.086$ or smaller.

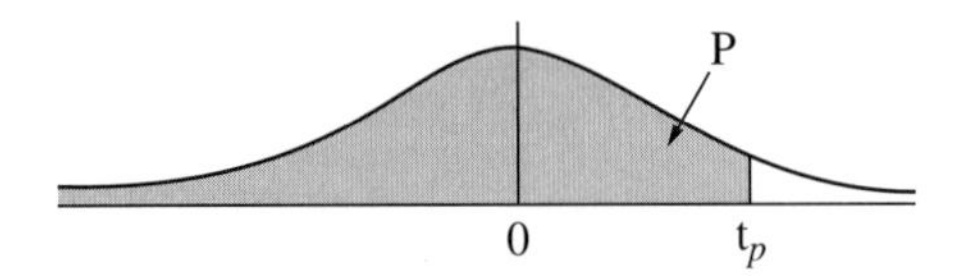

DF	$t_{.60}$	$t_{.70}$	$t_{.80}$	$t_{.90}$	$t_{.95}$	$t_{.975}$	$t_{.99}$	$t_{.995}$
1	0.325	0.727	1.376	3.078	6.314	12.706	31.821	63.657
2	0.289	0.617	1.061	1.886	2.920	4.303	6.965	9.925
3	0.277	0.584	0.978	1.638	2.353	3.182	4.541	5.841
4	0.271	0.569	0.941	1.533	2.132	2.776	3.747	4.604
5	0.267	0.559	0.920	1.476	2.015	2.571	3.365	4.032
6	0.265	0.553	0.906	1.440	1.943	2.447	3.143	3.707
7	0.263	0.549	0.896	1.415	1.895	2.365	2.998	3.499
8	0.262	0.546	0.889	1.397	1.860	2.306	2.896	3.355
9	0.261	0.543	0.883	1.383	1.833	2.262	2.821	3.250
10	0.260	0.542	0.879	1.372	1.812	2.228	2.764	3.169
11	0.260	0.540	0.876	1.363	1.796	2.201	2.718	3.106
12	0.259	0.539	0.873	1.356	1.782	2.179	2.681	3.055
13	0.259	0.538	0.870	1.350	1.771	2.160	2.650	3.012
14	0.258	0.537	0.868	1.345	1.761	2.145	2.624	2.977
15	0.258	0.536	0.866	1.341	1.753	2.131	2.602	2.947
16	0.258	0.535	0.865	1.337	1.746	2.120	2.583	2.921
17	0.257	0.534	0.863	1.333	1.740	2.110	2.567	2.898
18	0.257	0.534	0.862	1.330	1.734	2.101	2.552	2.878
19	0.257	0.533	0.861	1.328	1.729	2.093	2.539	2.861
20	0.257	0.533	0.860	1.325	1.725	2.086	2.528	2.845
21	0.257	0.532	0.859	1.323	1.721	2.080	2.518	2.831
22	0.256	0.532	0.858	1.321	1.717	2.074	2.508	2.819

TABLE 19.4 Distribution of *t*

23	0.256	0.532	0.858	1.319	1.714	2.069	2.500	2.807
24	0.256	0.531	0.857	1.318	1.711	2.064	2.492	2.797
25	0.256	0.531	0.856	1.316	1.708	2.060	2.485	2.787
26	0.256	0.531	0.856	1.315	1.706	2.056	2.479	2.779
27	0.256	0.531	0.855	1.314	1.703	2.052	2.473	2.771
28	0.256	0.530	0.855	1.313	1.701	2.048	2.467	2.763
29	0.256	0.530	0.854	1.311	1.699	2.045	2.462	2.756
30	0.256	0.530	0.854	1.310	1.697	2.042	2.457	2.750
40	0.255	0.529	0.851	1.303	1.684	2.021	2.423	2.704
60	0.254	0.527	0.848	1.296	1.671	2.000	2.390	2.660
120	0.254	0.526	0.845	1.289	1.658	1.980	2.358	2.617
∞	0.253	0.524	0.842	1.282	1.645	1.960	2.326	2.576

(*Source: Introduction to Statistical Analysis*, Copyright 1969, Used by permission.)

TABLE 19.4 (*Continued*)

$$\text{Confidence limits} = \bar{X} = \pm t\frac{s}{\sqrt{n}}$$

$$= 54.62 \pm (2.064)\frac{5.34}{\sqrt{25}}$$

$$= 52.42 \text{ and } 56.82$$

There is 95 percent confidence that the true mean hardness of the brass is between 52.42 and 56.82.

Determination of Sample Size

The only way to obtain the true value of a population parameter such as the mean is to measure (with a perfect measurement system) each and every individual within the population. This is not realistic (and is unnecessary when statistics are properly applied), so samples are taken instead. But how large a sample should be taken? The answer depends on (1) the sampling risks desired (alpha and beta risk, discussed further below and defined in Table 19.5), (2) the size of the smallest true difference that is desired to be detected, and (3) the variation in the characteristic being measured.

For example, suppose it was important to detect that the mean life of the battery cited previously was 35.0 hours (recall that the intended value is 30.0 hours). Specifically, we want to be 80 percent certain of detecting this difference (this is the "power" of the test, and has a corresponding risk of $\beta = 0.2$; this means we are willing to take a 20 percent chance of failing to detect the five-hour difference when, in fact, it exists). Further, if the true mean was

Null hypothesis (H_0): Statement of no change or no difference. This statement is assumed true until sufficient evidence is presented to reject it.
Alternative hypothesis (H_a): Statement of change or difference. This statement is considered true if H_0 is rejected.
Type I error: The error in rejecting H_0 when it is true or in saying there is a difference when there is no difference.
Alpha risk: The maximum risk or maximum probability of making a type I error. This probability is preset, based on how much risk the researcher is willing to take in committing a type I error (rejecting H_0 wrongly), and it is usually established at 5% (or .05). If the *p*-value is less than alpha, reject H_0.
Significance level: The risk of committing a type I error.
Type II error: The error in failing to reject H_0 when it is false or in saying there is no difference when there really is a difference.
Beta risk: The risk or probability of making a type II error or overlooking an effective treatment or solution to the problem.
Significant difference: The term used to describe the results of a statistical hypothesis test where a difference is too large to be reasonably attributed to chance.
p-value: The probability of obtaining different samples when there is really no difference in the population(s)—that is, the actual probability of committing a type I error. The *p*-value is the actual probability of incorrectly rejecting the null hypothesis (H_0) (i.e., the chance of rejecting the null when it is true). When the *p*-value is less than alpha, reject H_0. If the *p*-value is greater than alpha, fail to reject H_0.
Power: The ability of a statistical test to detect a real difference when there really is one, or the probability of being correct in rejecting H_0. Commonly used to determine if sample sizes are sufficient to detect a difference in treatments if one exists. Power = $(1 - \beta)$, or 1 minus the probability of making a type II error.

(*Source: Quality Planning and Analysis*, Copyright 2007. Used by permission.)

TABLE 19.5 Hypothesis Testing Definitions

30.0 hours, we want to have only a 5 percent risk of wrongly concluding it is not 30.0 hours (a risk of $\alpha = 0.05$). Then, using the following formula:

$$n = \left[\frac{(Z_{\alpha/2} + Z_{\beta})_{\sigma}}{\mu - \mu_{o}}\right]^2$$

we plug in our values to obtain

$$n = \left[\frac{(1.96 + 0.84)10}{35 - 30}\right]^2 = 31.4$$

The required sample size is 32 (Gryna et al., 2007, p. 605).

Note that sample size sometimes is constrained by cost or time limitations; in addition, rules of thumb exist to estimate sample size. However, these potentially lead to gross under- or oversampling, with wasted time and effort. The recommended approach is to use power and sample size calculators (available online and in statistical software; these readily apply formulas appropriate for different sampling situations) in order to enter data collection and hypothesis testing with full knowledge of the statistically appropriate sample size.

Hypothesis Testing

A hypothesis, as used here, is an assertion about a population. Typically, the hypothesis is stated as a pair of hypotheses as follows: the null hypothesis (H_0) and an alternative hypothesis, H_a. The null hypothesis, H_0, is a statement of no change or no difference—hence, the term "null." The alternative hypothesis is the statement of change or difference—that is, if we reject the null hypothesis, the alternative is true by default.

For example, to test the hypothesis that the mean life of a population of batteries equals 30 hours, we state:

$$H_0: \mu = 30.0 \text{ hours}$$

$$H_a: \mu \neq 30.0 \text{ hours}$$

A hypothesis test is a test of the validity of the assertion, and is carried out by analyzing a sample of data. Sample results must be carefully evaluated for two reasons. First, there are many other samples that, by chance alone, could be drawn from the population. Second, the numerical results in the sample actually selected can easily be compatible with several different hypotheses. These points are handled by recognizing the two types of sampling errors, already alluded to above.

The Two Types of Sampling Errors. In evaluating a hypothesis, two errors can be made

- Reject the null hypothesis when it is true. This is called a type I error, or the level of significance. The maximum probability of a type I error is denoted by α.
- Fail to reject the null hypothesis when it is false. This is called type II error, and the probability is denoted by β.

These errors are defined in terms of probability numbers and can be controlled to desired values. The results possible in testing a hypothesis are summarized in Table 19.6. Definitions are found in Table 19.5. For additional detail on sampling errors in the context of quality, see Gryna at al (2007).

	Suppose the H_0 Is	
Suppose Decision of Analysis Is	**True**	**False**
Fail to reject H_0	Correct decision $p = 1 - \alpha$	Wrong decision $p = \beta$
Reject H_0	Wrong decision $p = \alpha$	Correct decision $p = 1 - \beta$

(*Source: Quality Planning and Analysis*, Copyright 2007. Used by permission.)

TABLE 19.6 Type I (α) Error and Type II (β) Error

Steps to Hypothesis Testing. As emphasized earlier, it is important to plan for data collection and analysis; an investigator ideally should arrive at the point of actual hypothesis testing with elements such as sample size already defined. Hypothesis testing often is an iterative process, however, and as mentioned above in the opening discussion of data collection, further data may be needed after initial collection, for example, to bolster sample sizes to obtain the desired power so that both type I and type II errors are defined in advance.

Generally, then, the steps to test a hypothesis are as follows:

1. State the practical problem.
2. State the null hypothesis and alternative hypothesis.
3. Choose a value for α (alpha). Common values are 0.01, 0.05, and 0.10.
4. Choose the test statistic for testing the hypothesis.
5. Determine the rejection region for the test (i.e., the range of values of the test statistic that results in a decision to reject the null hypothesis).
6. Obtain a sample of observations, compute the test statistic, and compare the value to the rejection region to decide whether to reject or fail to reject the hypothesis.
7. Draw the practical conclusion.

Common Tests of Hypotheses. No single means of organizing hypothesis tests can convey all the information that may be of interest to an investigator. Table 19.7 summarizes some common tests of hypotheses in terms of the formulas. Table 19.8 categorizes tests according to the question being asked and type of data. Figure 19.13 provides similar information but in the form of a roadmap to assist in deciding what hypothesis test(s) are appropriate. Readers may find that the combination of these presentations will provide the best understanding of what is a multifaceted topic.

The hypothesis testing procedure is illustrated through the following example.

1. State the practical problem. To investigate a problem with warping wood panels, it was proposed that warping was caused by differing moisture content in the layers of the laminated product before drying. The sample data shown in Table 19.9 were taken between layers 1-2 and 2-3. Is there a significant difference in the moisture content?
2. State the null hypothesis and alternative hypothesis:

$$H_o: \mu 1\text{-}2 = \mu 2\text{-}3$$

$$H_a: \mu 1\text{-}2 \neq \mu 2\text{-}3$$

3. Choose a value for α. In this example, a type I error (α) of 0.05 will be assumed.
4. Choose the test statistic for testing the hypothesis.

Because we have two samples and desire to test for a difference in the means, a two-sample t-test is appropriate. (Note: A probability plot or test for normality will confirm the assumption of normality in the data. Also, an equal variance test concludes variances are approximately equal.)

1. Determine the rejection region for the test.

Hypothesis	Test Statistic and Distribution
H_o: $\mu = \mu_0$ (the mean of a normal population is equal to a specified value μ_0; σ is known)	$Z = \frac{\bar{X} - \mu_0}{\sigma / \sqrt{n}}$ Standard normal distribution
H_o: $\mu = \mu_0$ (the mean of a normal population is equal to a specified value μ_0; σ is estimated by s)	$t = \frac{\bar{X} - \mu_0}{s / \sqrt{n}}$ t distribution with $n - 1$ degrees of freedom (DF)
H_o: $\mu_1 = \mu_2$ (the mean of population 1 is equal to the mean of population 2; assume that $\sigma_1 = \sigma_2$ and that both populations are normal)	$t = \frac{\bar{X}_1 - \bar{X}_2}{\sqrt{1/n_1 + 1/n_2}\sqrt{\left[(n_1 - 1)s_1^2(n_2 - 1)s_2^2\right] / (n_1 + n_2 - 2)}}$ t distribution with DF = $n_1 + n_2 - 2$
H_o: $\sigma = \sigma_0$ (the standard deviation of a normal population is equal to a specified value σ_0)	$X^2 = \frac{(n-1)s^2}{\sigma_0^2}$ Chi-square distribution with DF = $n - 1$
H_o: $\sigma_1 = \sigma_2$ (the standard deviation of population 1 is equal to the standard deviation of population 2; assume that both populations are normal)	$F = \frac{s_1^2}{s_2^2}$ F distribution with $DF_1 = n_1 - 1$ and $DF_2 = n_2 - 1$
H_o: $\hat{p} = p_0$ (the fraction defective in a population is equal to a specified value p_0; assume that $np_0 \geq 5)\hat{p}$ = sample proportion	$Z = \frac{\hat{p} - p_0}{\sqrt{p_0(1 - p_0)/n}}$ Standard normal distribution
H_o: $p_1 = p_2$ (the fraction defective in population 1 is equal to the fraction defective in population 2; assume that n_1p_1 and n_2p_2 are each ≥ 5)	$Z = \frac{X_1/n_1 - X_2/n_2}{\sqrt{\hat{p}(1-\hat{p})(1/n_1 + 1/n_2)}}$ $\quad \hat{p} = \frac{X_1 + X_2}{n_1 + n_2}$ Standard normal distribution
To test for independence in a J × K contingency table that cross-classifies the variable A and B H_o: A is independent of B H_a: A is dependent on B	$X^2 = \sum_{j=1}^{J}\sum_{k=1}^{K} \frac{(f_{jk} - e_{jk})^2}{e_{jk}}$ Chi-square distribution with DF = (J – 1) (K – 1) where f_{jk} = the observed frequency of data for category j of variable A and to category k of variable B e_{jk} = the expected frequency = $f_{jo}f_{Ok}/f_{OO}$ f_{jo} = frequency total for category j for variable A f_{Ok} = frequency total for category k of variable B f_{OO} = frequency total for J × K table

(*Source: Quality Planning and Analysis*, Copyright 2007. Used by permission.)

TABLE 19.7 Summary of Formulas on Tests of Hypotheses

Tests of hypotheses organized by the question being asked. All tests assume a categorical X in the Y= *f* (X) format. For example, X might be manufacturing plant, and there could be 1, 2 or more than two plants of interest in terms of output, Y. A continuous Y might be mean or standard deviation of daily units produced, a categorical Y might be proportion defective units produced in a single day.

Question: Is There a Difference in the Parameter	Number of Sample Groups	Continuous Y (Normal)		Categorical Y	
		Parameter of Interest	Test	Parameter of Interest	Test
Compared to a target?	1	μ σ	1-sample *t* Chi- square	Proportion	1-proportion test
between two groups?	2	μ σ	2-sample *t* F-test	Proportion	2-proportion test
among all groups?	≥2	μ σ	ANOVA* Bartlett's	Proportion	Chi-square test of Independence

*ANOVA assumes both equal variances and normality.
(*Source:* Juran Institute, Inc., Used by permission.)

TABLE 19.8 Hypothesis Testing Table

The critical value defining the rejection region is approximately 2.0 (see Table 19.4); if the absolute value of the calculated *t* is larger than the critical value, then we reject the null hypothesis.

1. Obtain a sample of observations, compute the test statistic, and compare the value to the rejection region to decide whether to reject or fail to reject the hypothesis.

A box plot (remember to plot the data!) suggests that the moisture content in Layer 1-2 tends to be higher than in Layer 2-3. Minitab output (see Figure 19.14) shows that the calculated *t* is 4.18, which is in the rejection region.

Because the calculated *t* is larger than the critical value, the associated *p*-value is $< \alpha$, and we reject the null hypothesis, H_0.

N	Mean	StDev	SE Mean	
Layer 1-2	25	5.350	0.613	0.12
Layer 2-3	25	4.689	0.499	0.10

Difference = μ (Layer 1-2) – μ (Layer 2-3)

Estimate for difference: 0.660901

95 percent CI for difference: (0.343158, 0.978644)

T-test of difference = 0 (vs. not =): *t*-value = 4.18 *p*-value = 0.000 DF = 48

Both use pooled StDev = 0.5587

1. Draw the practical conclusion. We conclude that the moisture content in Layer 1-2 is higher than the moisture content of Layer 2-3.

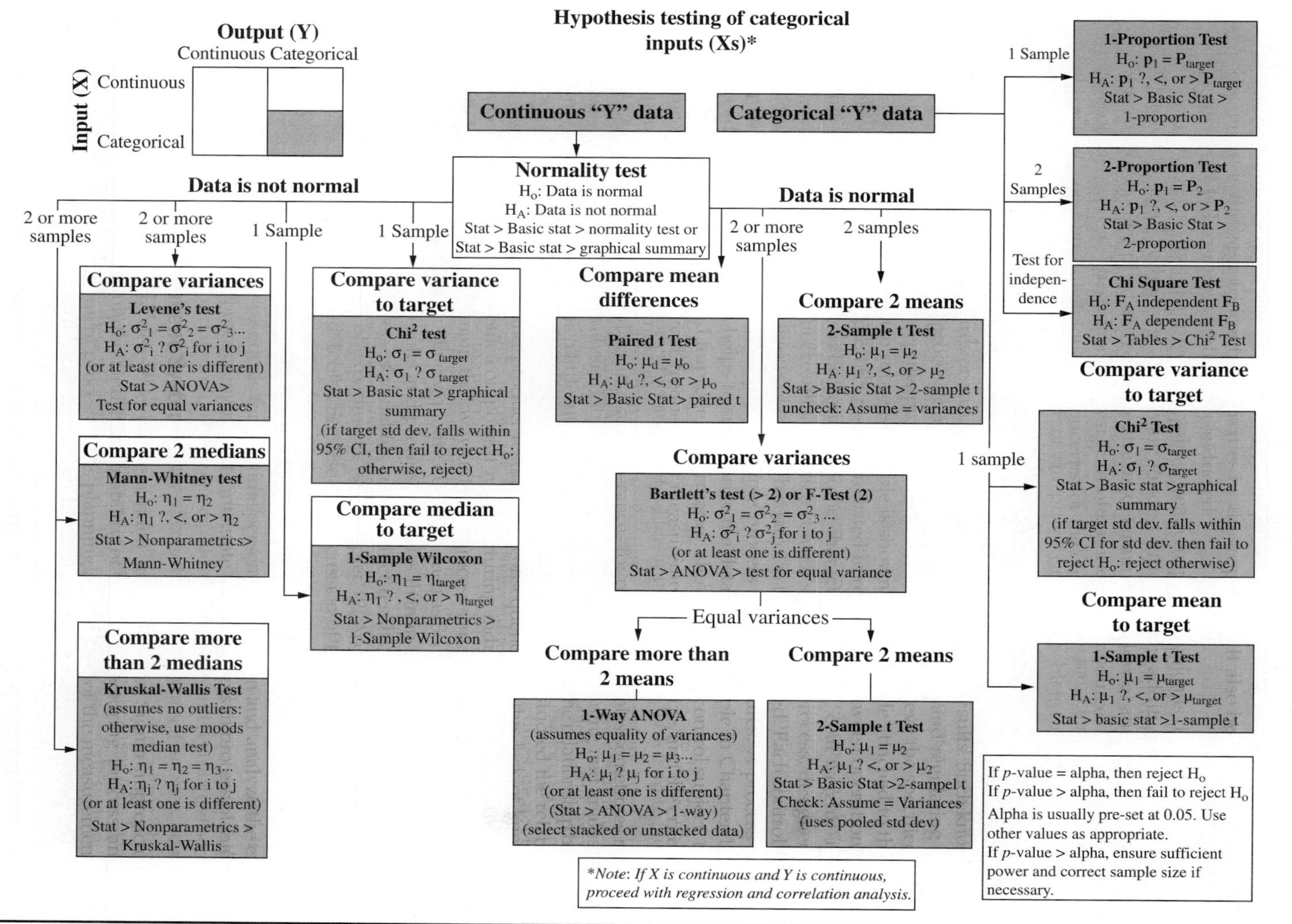

FIGURE 19.13 Hypothesis testing.

Layer 1-2		Layer 2-3	
4.43	4.40	3.74	5.14
6.01	5.99	4.30	5.19
5.87	5.72	5.27	4.16
4.64	5.25	4.94	5.18
3.50	5.83	4.89	4.78
5.24	5.44	4.34	5.42
5.34	6.15	5.30	4.05
5.99	5.14	4.55	3.92
5.75	5.72	5.17	4.07
5.48	5.00	5.09	4.54
5.64	5.01	4.74	4.23
5.15	5.42	4.96	5.07
5.64		4.21	

(*Source: Quality Planning and Analysis*, Copyright 2007. Used by permission.)

TABLE 19.9 Moisture Content

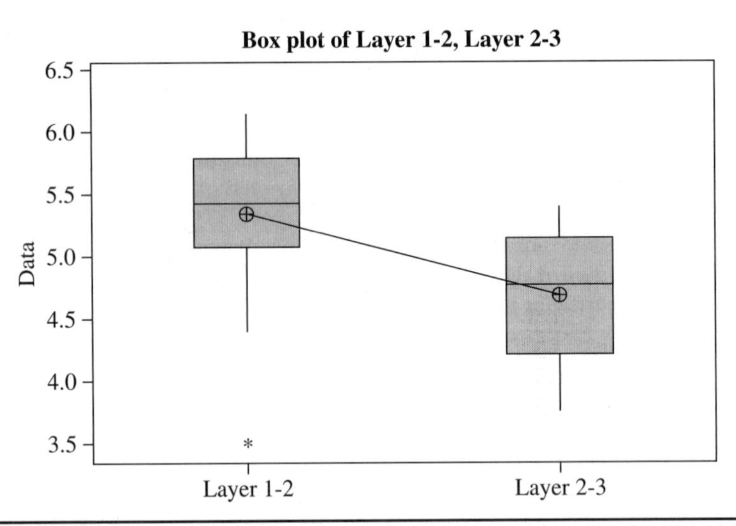

FIGURE 19.14 Box plot of Layer 1-2, Layer 2-3. (*Quality Planning and Analysis, Copyright 2007. Used by permission.*)

Nonparametric Hypothesis Tests, Data Transformation, and Bootstrapping

The preceding discussion has focused on "parametric" hypothesis tests (so-called because they rely on parameter estimation). Often, it is the case that one or more of the assumptions underlying the parametric tests are violated. In particular, practitioners frequently face skewed or otherwise nonnormal data, and application of parametric tests that assume bell-shaped data distribution may lead to erroneous conclusions and inappropriate action. Fortunately, options are available; these include nonparametric tests, data transformation, and bootstrapping.

Nonparametric hypothesis tests avoid violating key assumptions by virtue of being "distribution-free"; that is, they are not strictly dependent on particular distributions (such as a normal distribution); however, nonparametric tests have their own set of assumptions of which investigators should be aware). In effect, these methods typically transform the original data into ranks, and hypothesis tests then are carried out on the ranked data. Although nonparametric methods are not nearly as well developed and frequently are statistically less powerful compared to parametric tests, they are available for basic one-, two-, and two or more sample tests (see the bottom of Table 19.7 and the left side of the roadmap in Figure 19.13). See Sprent and Smeeton (2001) for more on traditional nonparametric methods. New methods continue to emerge, for example, wavelets and nonparametric Bayesian techniques; see Kvam and Vidakovic (2007).

Data transformation allows one to take data that violate some assumption of a parametric test and change them so that the assumption no longer is violated. For example, nonnormal data, or sample data with unequal variances can be changed to new numbers that are normal or have equal variances. Three common methods are

Power Functions. Traditionally, standard functions such as taking the square (x^2), square root ($x^{1/2}$), log (log10(x)), natural log (ln(x)), or inverse (x^{-1}) were used because they could easily be done with a calculator. Trial and error often is needed to find a function that appropriately transforms the data to meet the test assumptions.

Box-Cox Transformation. This method provides simultaneous testing of power functions to find an optimum value λ that minimizes the variance. Typically, one selects a power (value of λ) that is understandable and within a 95 percent confidence interval of the estimated λ (e.g., square: $\lambda = 2$; square root: $\lambda = 0.5$; natural log: $\lambda = 0$; inverse: $\lambda = -1$). The Box-Cox transformation does not work with negative numbers.

Johnson Transformation. This method selects an optimal function among three families of distributions (bounded, unbounded, lognormal). While effective in situations where Box-Cox does not work, the resulting transformation is not intuitive.

These methods are easy to apply (with software), and allow use of the more powerful parametric tests. However, the transformed data do not necessarily have intuitive meaning.

Bootstrapping is one of a broader class of computation-intensive resampling methods. Rather than assuming any particular distribution of a test statistic (such as normal), the distribution is determined empirically. More specifically, a statistic of interest (such as the mean) is repeatedly calculated from different samples drawn themselves, with replacements, from a sample. The distribution of these calculated statistics then is used as the basis for determining the probability of obtaining any particular value by chance. Itself a nonparametric approach, bootstrapping is a flexible method that gradually is gaining acceptance. For more information on the method and applications, see Davison and Hinkley (2006).

Correlation and Regression Analysis

Correlation and regression analysis help us understand relationships. More specifically, regression analysis is the modeling of the relationships between independent and dependent variables, while correlation analysis is a study of the strength of the linear relationships among variables. From a practical perspective, simple linear regression examines the distribution of one variable (the response, or dependent variable) as a function of one or more independent variables (the predictor, or independent variable) held at each of several levels.

X	Y	X	Y	X	Y	X	Y
90	41	100	22	105	21	110	15
90	43	100	35	105	13	110	11
90	35	100	29	105	18	110	6
90	32	100	18	105	20	110	10

(X, in feet per minute versus tool life; Y, in minutes)
(*Source: Quality Planning and Analysis*, Copyright 2007. Used by permission.)

TABLE 19.10 Cutting Speed

Note that the cause-and-effect relationship is stated explicitly, and it is this relationship that is tested to determine its statistical significance. In addition, regression analysis is used in forecasting and prediction based on the important independent variables, and in locating optimum operating conditions. In contrast, correlation typically looks at the joint variation of two variables that have not been manipulated by the experimenter, and there is no explicit cause-and-effect hypothesis.

For example, suppose that the life of a tool varies with the cutting speed of the tool and we want to predict life based on cutting speed. Thus, life is the dependent variable (Y) and cutting speed is the independent variable (X). Data are collected at four cutting speeds (Table 19.10).

Remembering to always plot the data, we note that a scatter plot (Figure 19.15) suggests that life varies with cutting speed (specifically, life decreases with an increase in speed) and also varies in a linear manner (i.e., increases in speed result in a certain decrease in life that is the same over the range of the data). Note that the relationship is not perfect—the points scatter about the line.

Often, it is valuable to obtain a regression equation. In this case, we have a linear relationship in the general form provided by

$$Y = \beta_0 + \beta_1 X + \varepsilon$$

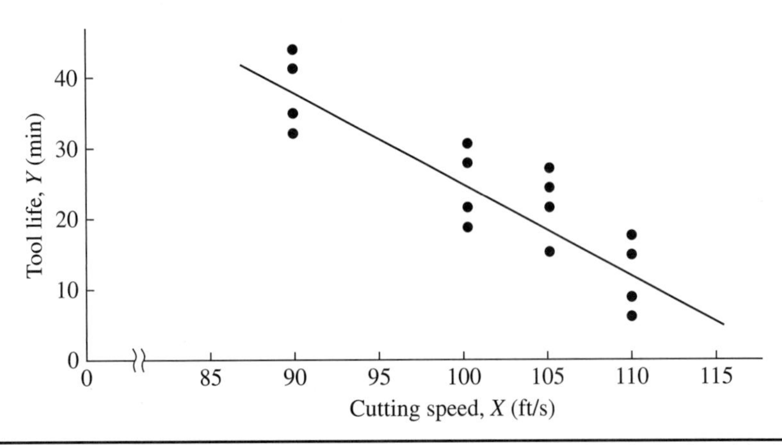

FIGURE 19.15 Tool life (Y) versus cutting speed (X). (*Quality Planning and Analysis, Copyright 2007. Used by permission.*)

where β_0 and β_1 are the unknown population intercept and slope, and ε is a random-error term that may be due to measurement errors and/or the effects of other independent variables. This model is estimated from sample data by the form

$$\hat{Y} = b_0 + b_1X$$

where $\hat{Y}$ is the predicted value of Y for a given value of X and b0 and b1 are the sample estimates of β_0 and β_1. Estimates usually are found by least-squares methods; formulas can be found in statistics books such as Kutner et al. (2004).

For this example, the resulting prediction equation is

$$\text{Tool life} = 106.90 - 1.3614\ (\text{cutting speed})$$

This equation can be used to predict tool life by plugging in values of cutting speed. Extreme caution should be used in making predictions outside the actual sample space (e.g., for cutting speeds above or below the tested maximum or minimum), however, as these are tenuous without confirmation by observation.

Although a prediction equation can be found mathematically, it should not be used without knowing how "good" it is. A number of criteria exist for judging the adequacy of the prediction equation. One common measure is R_2, the proportion of variation explained by the prediction equation. R_2, or the coefficient of determination, is the ratio of the variation due to the regression to the total variation. The higher R_2, the greater the probable utility of the prediction equation in estimating Y based on X.

Another measure of the degree of association between two variables is the simple linear correlation coefficient, *r*. This is the square root of the coefficient of determination, so that the values of *r* range from –1 to +1. A positive *r* is consistent with a positive relationship (an increase in one variable is associated with an increase in the other), whereas the opposite is true of a negative *r* (an increase in one variable is associated with a decrease in the other). Scatter plots are strongly recommended when interpreting correlations, especially as very different patterns can result in identical values of *r*. The significance level of *r* varies with sample size; statistical software is recommended to obtain exact significance levels.

The above discussion introduces simple linear correlation and regression—the direction and strength of a relationship between two variables, or prediction of a dependent variable, Y, from a single predictor variable, X. A natural extension of this is multiple regression that allows for two or more independent variables. For a discussion of how to estimate and examine a multiple regression prediction equation, see Kutner et al. (2004).

Analysis of Variance

Analysis of Variance (ANOVA) is an approach related to linear regression, falling into the class of what are called general linear models. However, unlike regression, the X is discrete rather than continuous (noting that general linear models actually can blend characteristics of both regression and ANOVA). In ANOVA, the total variation of all measurements around the overall mean is divided into sources of variation that are then analyzed for statistical significance. It is used in situations where the investigator is interested in comparing the means among two or more discrete groups. For example, an investigator may be interested in comparing performance among three different machine configurations. The ANOVA analysis detects a difference somewhere among the means (i.e., at least one mean is different from the others), and confidence intervals or follow-up tests such as pairwise comparisons can be applied to determine which mean (or means) is different. ANOVA is the basis for design of experiments, discussed next.

Design of Experiments

With origins in the pioneering work in agriculture of Sir Ronald A. Fisher, designed experiments have taken on an increasingly significant role in quality improvement in the business world. This section will first compare the classical and designed approaches to experimentation, thereby providing the reader with an understanding as to the limitations of traditional methods and the power of contemporary methods. Next, basic concepts and terminology will be introduced in the context of an example improvement problem, followed by an overview of different types of designs and the typical progression through a series of designed experiments. The section finishes with the related topic of Taguchi designs.

Contrast between the Classical and Contemporary Methods of Experimentation. The classical method of experimentation is to vary one factor at a time (sometimes called OFAT), holding everything else constant. By way of example, and to illustrate the need for designed experiments, consider the case of a certain fellow who decided he wanted to investigate the causes of intoxication. As the story goes, he drank some whiskey and water on Monday and became highly inebriated. The next day, he repeated the experiment holding all variables constant except one... he decided to replace the whiskey with vodka. As you may guess, the result was drunkenness. On the third day, he repeated the experiment for the last time. On this trial, he used bourbon in lieu of the whiskey and vodka. This time it took him two days just to be able to gather enough of his faculties to analyze the experimental results. After recovering, he concluded that water causes intoxication. Why? Because it was the common variable!

The contrast between this traditional method and the designed approach is striking. In particular, a designed approach permits the greatest information to be gained from the fewest data points (efficient experimentation), and allows the estimation of interaction effects among factors. Table 19.11 compares these two approaches in more detail for an experiment in which there are two factors (or variables) whose effects on a characteristic are being investigated (the same conclusions hold for an experiment with more than two factors).

Concepts and Terminology—An Example Designed Experiment. Suppose that three detergents (A, B, C) are to be compared for their ability to clean clothes in an automatic washing machine. The "whiteness" readings obtained by a special measuring procedure are the dependent, or response, variable. The independent variable under investigation (detergent) is a factor, and each variation of the factor is called a level; in this case, there are three levels. A treatment is a single level assigned to a single factor, detergent A. A treatment combination is the set of levels for all factors in a given experimental run. A factor may be qualitative (different detergents) or quantitative (water temperature). Finally, some experiments have a fixed-effects model (i.e., the levels investigated represent all levels of concern to the investigator—for example, three specific washing machines or brands). Other experiments have a random effects model, that is, the levels chosen are just a sample from a larger population (e.g., three operators of washing machines). A mixed-effects model has both fixed and random factors.

Figure 19.16 outlines six possible designs of experiments, starting with the classical design in Figure 19.16a. Here, all factors except detergent are held constant. Thus, nine tests are run, three with each detergent with the washing time, make of machine, water temperature, and all other factors held constant. One drawback of this design is that the conclusions about detergent brands apply only to the specific conditions of the experiment.

Figure 19.16b recognizes a second factor at three levels (i.e., washing machines brands I, II, and III). However, in this design, it would not be known whether an observed difference was due to detergents or washing machine (they are said to be confounded).

Criteria	Classical	Modern
Basic procedure	Hold everything constant except the factor under investigation. Vary that factor and note the effect on the characteristic of concern. To investigate a second factor, conduct a separate experiment in the same manner.	Plan the experiment to evaluate both factors in one main experiment. Include in the design measurements to evaluate the effect of varying both factors simultaneously.
Experimental conditions	Care should be taken to have material, workers, and machine constant throughout the entire experiment.	Realizes difficulty of holding conditions reasonably constant throughout an entire experiment. Instead, experiment is divided into several groups or blocks of measurements. Within each block, conditions must be reasonably constant (except for deliberate variation to investigate a factor).
Experimental error	Recognized but not stated in quantitative terms.	Stated in quantitative terms.
Basis of evaluation	Effect due to a factor is evaluated with only a vague knowledge of the amount of experimental error.	Effect due to a factor is evaluated by comparing variation due to that factor with the quantitative measure of an experimental error.
Possible bias due to sequence of measurements	Often assumed that sequence has no effect.	Guarded against by randomization.
Effect of varying both factors simultaneously ("interaction")	Not adequately planned into experiment. Frequently assumed that the effect of varying factor 1 (when factor 2 is held constant at some value) would be the same for any value of factor 2.	Experiment can be planned to include an investigation for interaction between factors.
Validity of results	Misleading and erroneous if interaction exists and is not realized.	Even if interaction exists, a valid evaluation of the main factors can be made.
Number of measurements	For a given amount of useful and valid information, more measurements are needed than in the modern approach.	Fewer measurements needed for useful and valid information.
Definition of problem	Objective of experiment frequently not defined as necessary.	Designing the experiment requires defining the objective in detail (how large an effect do we want to determine, what numerical risks can be taken, etc.).
Application of conclusions	Sometimes disputed as applicable only to the controlled conditions under which the experiment was conducted.	Broad conditions can be planned in the experiment, thereby making conclusions applicable to a wider range of actual conditions.

(*Source: Quality Planning and Analysis*, Copyright 2007. Used by permission.)

TABLE 19.11 Comparison of Classical and Modern Methods of Experimentation

A	B	C
-	-	-
-	-	-
-	-	-

(a)

I	II	III
A	B	C
A	B	C
A	B	C

(b)

I	II	III
C	B	B
A	C	B
A	A	C

(c)

I	II	III
B	A	C
C	C	A
A	B	B

(d)

	I	II	III
1	C	A	B
2	B	C	A
3	A	B	C

(e)

I	II	III
ABC	ABC	ABC
1 ---	---	---
2 ---	---	---
3 ---	---	---

(f)

FIGURE 19.16 Some experimental designs. (*Quality Planning and Analysis, Copyright 2007. Used by permission.*)

In Figure 19.16c, the nine tests are assigned completely at random, thus the name "completely randomized design." However, detergent A is not used with machine brand III, and detergent B is not used with machine brand I, thus complicating the conclusions.

Figure 19.16d shows a randomized block design. Here each block is a machine brand, and the detergents are run in random order within each block. This design guards against any possible bias due to the order in which the detergents are used and has advantages in the subsequent data analysis and conclusions. First, a test of hypothesis can be run to compare detergents and a separate test of hypothesis run to compare machines; all nine observations are used in both tests. Second, the conclusions concerning detergents apply for the three machines and vice versa, thus providing conclusions over a wider range of conditions.

Now suppose that another factor such as water temperature is also to be studied, using the Latin square design shown in Figure 19.16e. Note that this design requires using each

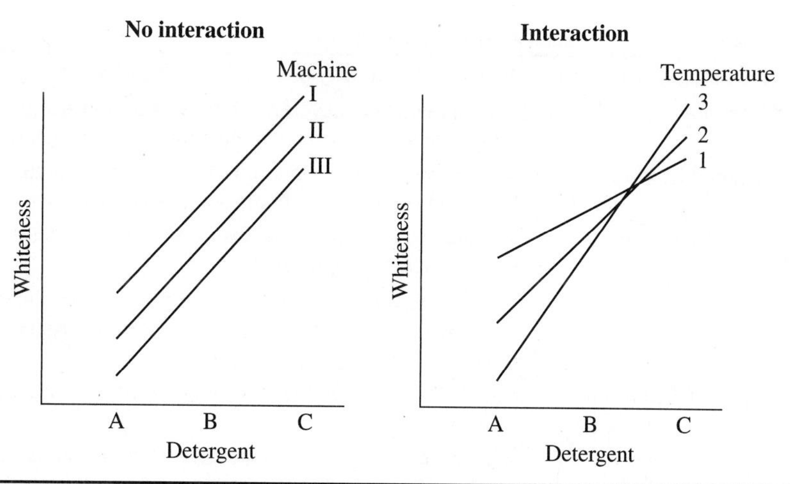

Figure 19.17 Interaction. (*Quality Planning and Analysis, Copyright 2007. Used by permission.*)

detergent only once with each machine and only once with each temperature. Thus, three factors can be evaluated (by three separate tests of hypothesis) with only nine observations. However, there is a danger. This design assumes no interaction among the factors. No interaction between detergent and machine means that the effect of changing from detergent A to B to C does not depend on which machine is used, and similarly for the other combinations of factors. The concept of interaction is shown in Figure 19.17. There is no interaction among the detergents and the machines. But the detergents do interact with temperature. At high temperatures, C is the best performer. At low temperatures, A performs best.

Finally, the main factors and possible interactions could be investigated by the factorial design in Figure 19.16f. Factorial means that at least one test is run for every combination of main factors, in this case $3 \times 3 \times 3$ or 27 combinations. Separate tests of hypothesis can be run to evaluate the main factors and also possible interactions. Again, all the observations contribute to each comparison. When there are many factors, a portion of the complete factorial (i.e., a "fractional factorial") is useful when experimental resources are limited (see its application in a sequential testing approach, below).

Most problems can be handled with one of the standard experimental designs or a series of these. Designs can be classified by the number of factors to be investigated, the structure of the experimental design, and the kind of information the experiment is intended to provide (Table 19.12). For a description of both the design and analysis of various design structures, see Box et al. (2005). Another excellent general reference is Myers et al. (2009) for a detailed look at response surface designs.

A sequential approach to experimentation often can be helpful. Briefly, a typical sequence of designed experiments will allow an experimenter to quickly and efficiently narrow down a large number of possible factors (or X's in the Y = f(X) terminology of Lean Six Sigma) to find out which are most important, and then refine the relationships to find optimal settings for each of the vital few factors. The steps might be as follows:

1. *Screening experiment.* In this stage, a fractional factorial design may be applied that does not allow interactions to be detected, but can ferret out which of many factors have the greatest main effect.
2. *Fractional factorial design.* The smaller number of factors identified in the screening experiment are tested to allow detection of interaction effects.

Design	Type of Application
Completely randomized	Appropriate when only one experimental factor is being investigated
Factorial	Appropriate when several factors are being investigated at two or more levels and interaction of factors may be significant
Blocked factorial	Appropriate when number of runs required for factorial is too large to be carried out under homogeneous conditions
Fractional factorial	Appropriate when many factors and levels exist and running all combinations is impractical
Randomized block	Appropriate when one factor is being investigated and experimental material or environment can be divided into blocks or homogeneous groups
Balanced incomplete block	Appropriate when all the treatments cannot be accommodated in a block
Partially balanced incomplete block	Appropriate if a balanced incomplete block requires a larger number of blocks than is practical
Latin square	Appropriate when one primary factor is under investigation and results may be affected by two other experimental variables or by two sources of nonhomogeneity. It is assumed that no interactions exist.
Youden square	Same as Latin square, but number of rows, columns, and treatments need not be the same
Nested	Appropriate when objective is to study relative variability instead of mean effect of sources of variation (e.g., variance of tests on the same sample and variance of different samples)
Response surface	Objective is to provide empirical maps (contour diagrams) illustrating how factors under the experimenter's control influence the response
Mixture designs	Use when constraints are inherent (e.g., the sum of components in a paint must add to 100%)

(*Source*: Adapted from JQH5, Table 47.3.)

TABLE 19.12 Classification of Designs

3. *Full factorial design*. A small number of factors (usually no more than five) are tested to allow all main effects and higher-order (e.g., three-way, four-way) interactions to be detected and accounted for. Such designs also can detect curvature that indicates a potential optimum.
4. *Response surface design*. By adding data points in particular ways (e.g., a composite design), an experimenter can build on earlier experiments to fully characterize nonlinear relationships and pinpoint optimal settings.
5. *EVOP*. Once an improved process is in production mode, evolutionary operation techniques can be used to conduct many small experiments on production units over time. Although individual changes are small, the cumulative effect over time can be quite large, and exemplifies the power of continuous improvement. See Box and Draper (1969) for a classic text on this subject.

For a series of four papers on sequential experimentation, see Carter (1996). Emanuel and Palanisamy (2000) discuss sequential experimentation at two levels and a maximum of seven factors.

Taguchi Approach to Experimental Design

Professor Genichi Taguchi uses an approach to experimental design that has three purposes:

- Design products and processes that perform consistently on target and are relatively insensitive ("robust") to factors that are difficult to control.
- Design products that are relatively insensitive (robust) to component variation.
- Minimize variation around a target value.

Thus, although cited in this "improvement tools" section because of its association with DOE, the approach is meant to provide valuable information for product design and development (see "Statistical Tools for Designing for Quality" in this chapter). Taguchi divides quality control into online control (e.g., diagnosing and adjusting a process during production) and offline control that encompasses the engineering design process and its three phases: systems design, parameter design, and tolerance design. For an extensive bibliography and a summary of some controversial aspects of the Taguchi approach, see Box and Draper (1969, pp. 47.58 and 47.59).

Many books are available that cover DOE for engineering and manufacturing applications. For readers in nonmanufacturing environments, Ledolter and Swersey (2007) may be of interest. A recent text readers may find useful for not only classical but more contemporary techniques (e.g., Bayesian inference, kriging) is del Castillo (2007).

Discrete Event and Monte Carlo Simulation

Advances in user-friendly software make computer simulations increasingly accessible to quality practitioners that do not have a strong background in mathematics, programming, or modeling. Numerous types of simulation models exist, but two that may be of most interest to readers are discrete event and Monte Carlo simulations. These can be powerful methods for making process improvements; in particular, modeling provides a means of asking "what if?" questions and rapidly testing the effects of process changes and potential solutions in a safe, low-risk environment.

Discrete Event Simulation. Discrete event simulation (DES) attempts to mimic situations in which there are distinct, recognizable events and transactions. In a hospital, for example, arrival of patients at an emergency department and subsequent steps in patient care represent specific events that combine into a flow of transactions: arrival, registration, triage, nursing assessment, physician assessment, etc., through inpatient admission, discharge, or transfer. Discrete event simulation enables system components to be changed and tracks the resulting process flow over time to help understand the relationships among inputs, outputs, and process variables.

Typically, a process flow diagram (or process "map") that graphically displays the sequence and flow of activities forms the basis for a discrete event simulation. A discrete event simulation takes this basic flow diagram and adds inputs and process variables that govern the flow of transactions. Following on the hospital example, these include inputs (such as patient arrivals), human resources (e.g., number of nurses, physician schedules, overtime availability, skill levels, pay rates, etc.), equipment resources (e.g., types and number of beds, imaging equipment, etc.), rules for flow (the required sequence of steps, batching of inputs or outputs, priority rules, exceptions, decisions), resource acquisition (what resources are needed to complete an activity (e.g., one RN or one physician's assistant; two RNs; one RN and one physician, etc.), activity cycle times (work time, wait time), and similar details.

Once these details are built into the model, it "runs" by tracing the path of units (patients, in the hospital example) from arrival through to exit from the process. Patients are processed in accordance with the activities, rules, and constraints, and any relevant attributes (patient-specific characteristics) that may be assigned to them (e.g., acuity level, age, gender). The output consists of a multitude of descriptive statistics and measures that portray the collective behavior of the process as the various players interact and move through time.

Although every model is different and details vary, there are basic steps that should be a part of every simulation study. These steps and related questions are (adapted from Law and Kelton 2000):

1. State the problem and question(s) being asked. What is the business need for the simulation? What problem is to be fixed? What answers are being sought?
2. Prepare a plan for the simulation study. Who needs to be involved? What data are needed and how will data be collected? What alternative scenarios are to be tested? What are the milestones and timeline for completion?
3. Collect data. What is my current state? What are the data for alternative scenarios? Are there gaps in the data, and how will they be handled?
4. Build and validate a conceptual model. Given available data, what is the general structure of the model? What will be the inputs, process variables, and outputs? What statistical accumulators are needed, and where? If the model is built, will it provide the answers to the questions?
5. Build and validate an operational model. Are the model components necessary and sufficient? Does the model produce results consistent with the current state?
6. Design scenarios or experiments needed to answer the questions. What model parameters will be changed? Which are fixed? What combinations of factors need to be tested?
7. Run the scenarios or experiments to obtain the needed outputs. Are the results reproducible? Are additional scenarios or experiments suggested?
8. Analyze and interpret the data. What are the statistical results? Do the descriptive statistics and/or statistical tests indicate meaningful effects? What are the answers to the original questions? Are additional questions raised?

As emphasized at the beginning of this chapter, formulation of the question(s) being asked is a critical first step to the successful application of simulation modeling. Failure to have a clear understanding of what the model is being asked to do leads to poorly constructed models, models with insufficient inputs or process detail, or overly complicated models that take unnecessary time and effort to build and run. In addition, a clearly communicated business need will garner the stakeholder support needed to collect data, evaluate the model, and implement suggested changes.

Monte Carlo. Named after the famed gambling destination, this method seeks to account for uncertainty (variability) in inputs and carry this forward into probability distributions of outcomes. Essentially, instead of using single, fixed values in equations [such as $Y = f(X)$], distributions are used for the inputs (X's), and samples repeatedly are drawn from the distributions, yielding a distribution of outputs (Y values) instead of a single value. For example, while the forecasted net return on a new product could simply be stated as an expected $10 million, it would be useful to know the probability of achieving this, or that the uncertainty in the forecast is such that there is a high probability of a negative return.

By way of illustration, assume we have three components, A, B, and C that are assembled end-to-end to create a final product. If the mean lengths are 5, 10, and 15 mm, then we can simply add these together to arrive at an expected mean combined total length of 5 mm + 10 mm + 15 mm = 30 mm. However, we know from the concept of statistical variation that there will be variation in the components. Assuming we sample populations of each component and find the respective distributions for each of A, B, and C, what can we expect the overall distribution of assembled product length to look like? By repeatedly taking a random sample from each distribution and adding the lengths, Monte Carlo simulation generates a distribution of the total length Figure 19.18 shows the relative frequency distribution of the combined

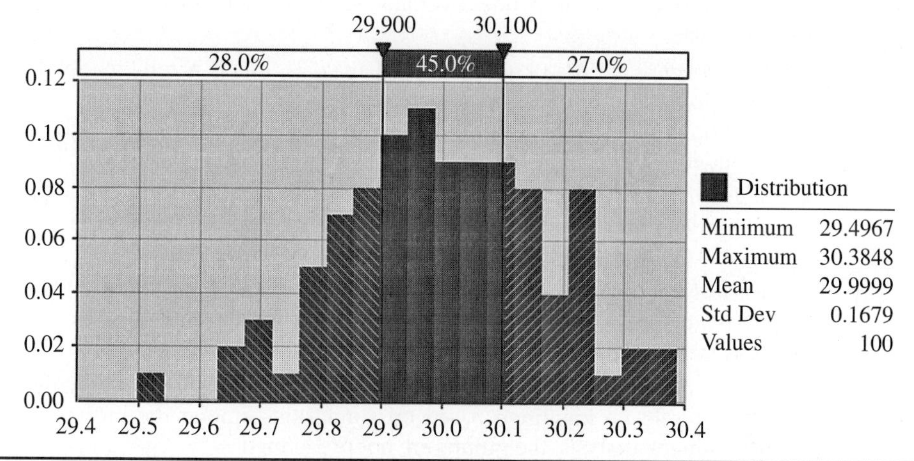

Figure 19.18 Result of Monte Carlo simulation showing a relative frequency distribution of combined total length of three components A, B and C that individually have normal distributions of 5, 10 and 15 mm, respectively, each with a standard deviation of 0.1 mm. The mean expected combined total length is approximately 30 mm, but the simulation shows the variation around this, e.g., that only 45% of assembled components are expected to be within +/– 0.1 mm of this mean value.

lengths of the three components from a Monte Carlo simulation with each of the three components having a standard deviation of of 0.1 mm. The mean expected combined total length is almost exactly 30 mm, but the simulation shows the variation around this, with only 45% of assembled components expected to be within +/– 0.1 mm of the total mean value. This approach provides substantially more information than the single estimate of 30 mm.

Simulated DOE. As tools evolve, they are being combined in new ways. One example is the combination of Monte Carlo, discrete event simulation, and DOE. Briefly, this approach involves a discrete event simulation (DES) that uses probability distributions for the input and/or process variables (Monte Carlo), and the investigator changes these variables (as factors) following a structured, designed approach (DOE). While any results and conclusions should be treated as preliminary until verified by actual experimentation, this can be particularly useful in environments where real-life changes may be difficult or dangerous to make.

Additional Advanced Analysis Tools

For practitioners faced with more complex scenarios such as multiple variables (more than one y and/or x), nonlinear data, or categorical outputs, extensions of the general linear models and other alternatives are available. In particular are methods for multivariate analysis; this refers to statistical techniques that simultaneously analyze multiple measurements on subjects. Many techniques are extensions of the univariate (single-variable distributions) and bivariate (correlation, regression) methods dealt with above. Beyond the scope of this chapter, these include:

- *Multiple regression.* Applies when the investigator has a single, continuous dependent variable and multiple, continuous independent variables (X's) of interest.
- *Nonlinear regression.* Useful when data cannot easily be treated by standard linear methods (note that curvilinear data do not necessarily require nonlinear methods).
- *Nonparametric linear regression.* Applies when the usual assumptions of regression are violated.

- *Multiple discriminant analysis.* Used in situations with a single, categorical (dichotomous or multichotomous) dependent variable (Y) and continuous independent variables (X's).
- *Logistic regression.* Also known as logit analysis, this is a combination of multiple regression and multiple discriminant analysis in which one or more categorical or continuous independent variables (X's) are used to predict a single, categorical dependent variable (Y). Odds ratios often are computed with this method.
- *Multivariate analysis of variance and covariance (MANOVA, MANCOVA).* Dependence techniques that extend ANOVA to allow more than one continuous, dependent variable (Y) and several categorical independent variables (X's).
- *Principal component analysis (PCA) and common factor analysis.* These methods analyze interrelationships among a large number of variables and seek to condense the information into a smaller set of factors without loss of information.
- *Cluster analysis.* An interdependence technique that allows mutually exclusive subgroups to be identified based on similarities among the individuals. Unlike discriminant analysis, the groups are not predefined.
- *Canonical correlation analysis.* An extension of multiple regression that correlates simultaneously several continuous dependent variables (Y's) and several continuous independent variables (X's).
- *Conjoint analysis.* Often used in marketing analyses, this method helps assess the relative importance of both attributes and levels of complex entities (e.g., products). It is useful when trade-offs exist when making comparisons.
- *Multidimensional scaling.* An interdependence method (also called perceptual mapping), this seeks to transform preferences or judgments of similarity into a representation by distance in multidimensional space.
- *Correspondence analysis.* Another interdependence technique; this accommodates the perceptual mapping of objects (such as products) onto a set of categorical attributes. This method allows both categorical data and nonlinear relationships.

Readers are encouraged to research any techniques that appear to fit their need; although complex, these are powerful means of getting useful information from data. Some useful references include

Multivariate techniques:

Hair, J. F., Jr., Black, W. C., Babin, B. J., Anderson, R. E., and Tatham, R. L. (2006). *Multivariate Data Analysis*. Pearson Prentice-Hall, Upper Saddle River, NJ.

Affifi, A., Clark, V. A., and May, S. (2004). *Computer-Aided Multivariate Analysis* (4th ed.). Chapman and Hall/CRC Press, Boca Raton, FL.

Coleman, S, Greenfield, T., Stewardson, D., and Montgomery, D. C. (2008). *Statistical Practice in Business and Industry*. John Wiley & Sons, Hoboken, NJ. (see Chapter 13).

Hypothesis testing and DOE:

Box, G. E. P., Hunter, J. S., and Hunter, W. G. (2005). *Statistics for Experimenters: Design, Innovation and Discovery* (2nd ed.). Wiley-Interscience, Hoboken, NJ.

Logistic regression, Poisson regression, odds ratios:

Agresti, A. (1996). *An Introduction to Categorical Data Analysis*. John Wiley & Sons, New York.

Nonparametric:

Sprent, P., and Smeeton, N. C. (2001). *Applied Nonparametric Statistical Methods* (3rd ed.). Chapman and Hall/CRC Press, Boca Raton, FL.

Statistical Tools for Designing for Quality

Statistical tools for quality in the design and development process include techniques such as graphical summaries, probability distributions, confidence limits, tests of hypotheses, design of experiments, regression, and correlation analysis. These topics are covered in earlier sections of this chapter. To supplement these techniques, this section explains some statistical tools for reliability and availability, and tools for setting specification limits on product characteristics.

Failure Patterns for Complex Products

Methodology for quantifying reliability was first developed for complex products. Suppose that a piece of equipment is placed on test, is run until it fails, and the failure time is recorded. The equipment is repaired and again placed on test, and the time of the next failure is recorded. The procedure is repeated to accumulate the data shown in Table 19.13. The failure rate is calculated, for equal time intervals, as the number of failures per unit of time. When the failure rate is plotted against time, the result (Figure 19.19) often follows a familiar pattern of failure known as the bathtub curve. Three periods are apparent that differ in the frequency of failure and in the failure causation pattern:

- *The infant mortality period.* This period is characterized by high failure rates that show up early in use (see the lower half of Figure 19.18). Commonly, these failures

Time of Failure, Infant Mortality Period		Time of Failure, Constant Failure Rate Period		Time of Failure, Wear-Out Period	
1.0	7.2	28.1	60.2	100.8	125.8
1.2	7.9	28.2	63.7	102.6	126.6
1.3	8.3	29.0	64.6	103.2	127.7
2.0	8.7	29.9	65.3	104.0	128.4
2.4	9.2	30.6	66.2	104.3	129.2
2.9	9.8	32.4	70.1	105.0	129.5
3.0	10.2	33.0	71.0	105.8	129.9
3.1	10.4	35.3	75.1	106.5	
3.3	11.9	36.1	75.6	110.7	
3.5	13.8	40.1	78.4	112.6	
3.8	14.4	42.8	79.2	113.5	
4.3	15.6	43.7	84.1	114.8	
4.6	16.2	44.5	86.0	115.1	
4.7	17.0	50.4	87.9	117.4	
4.8	17.5	51.2	88.4	118.3	
5.2	19.2	52.0	89.9	119.7	
5.4		53.3	90.8	120.6	
5.9		54.2	91.1	121.0	
6.4		55.6	91.5	122.9	
6.8		56.4	92.1	123.3	
6.9		58.3	97.9	124.5	

(*Source: Quality Planning and Analysis*, Copyright 2007. Used by permission.)

TABLE 19.13 Failure History for a Unit

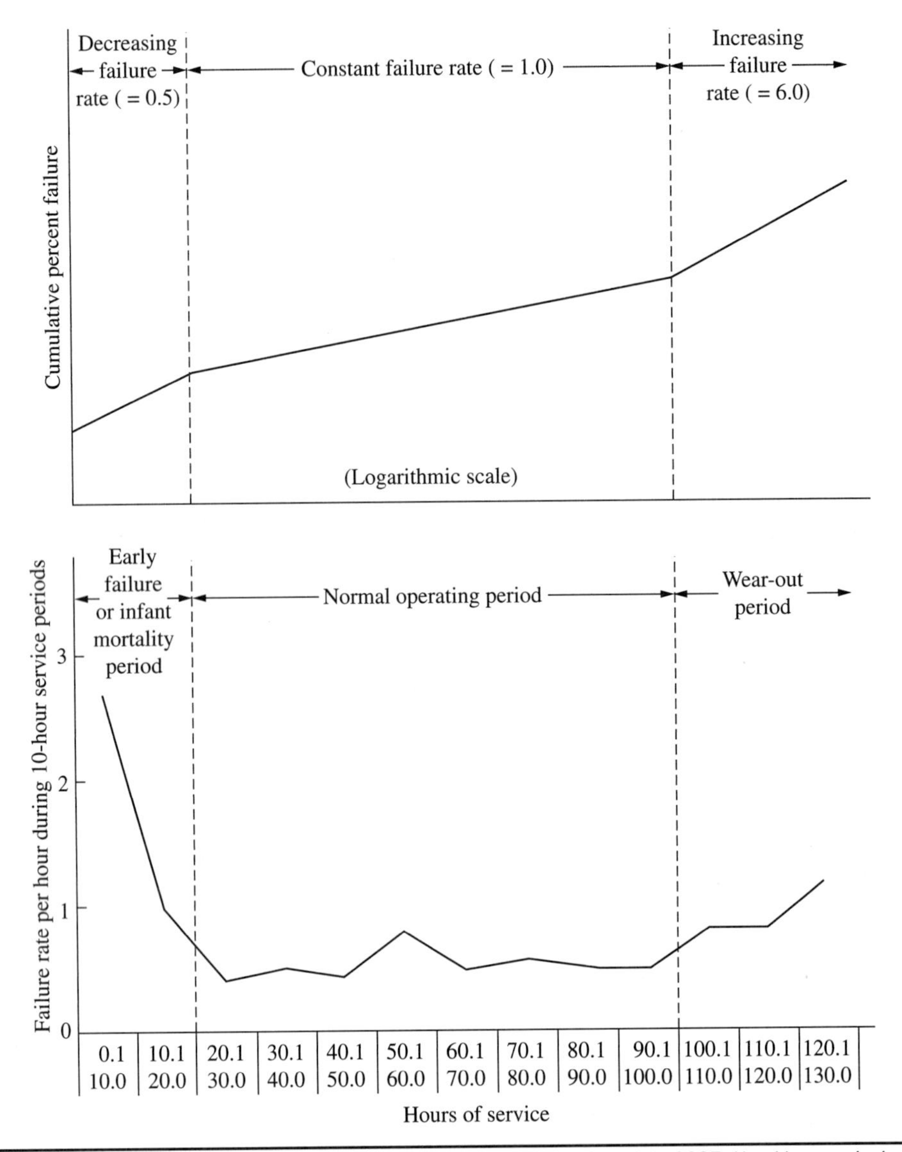

FIGURE 19.19 Failure rate vs. time. (*Quality Planning and Analysis, Copyright 2007. Used by permission.*)

are the result of blunders in design or manufacture, misuse, or misapplication. Once corrected, these failures usually do not occur again (e.g., an oil hole that is not drilled). Sometimes it is possible to "debug" the product by a simulated use test or by overstressing (in electronics this is known as burn-in). The weak units still fail, but the failure takes place in the test rig rather than in service. O'Connor (1995) explains the use of burn-in tests and environmental screening tests.

- *The constant-failure-rate period.* Here the failures result from the limitations inherent in the design, changes in the environment, and accidents caused by use or maintenance.

The accidents can be held down by good control of operating and maintenance procedures. However, a reduction in the failure rate requires basic redesign.

- *The wear-out period*. These failures are due to old age (e.g., a metal becomes embrittled or insulation dries out). A reduction in failure rates requires preventive replacement of these dying components before they result in catastrophic failure.

The top portion of Figure 19.19 shows the corresponding Weibull plot when $\alpha = 2.6$ was applied to the original data (Table 19.14). The values of the shape parameter, β, were approximately 0.5, 1.0, and 6.0, respectively. A shape parameter less than 1.0 indicates a decreasing failure rate, a value of 1.0 a constant failure rate, and a value greater than 1.0 an increasing failure rate.

The Distribution of Time Between Failures. Users desire low failure rates during the infant mortality period, and after this are concerned with the length of time that a product will perform without failure. Thus, for repairable products, the time between failures (TBF) is a critical characteristic. The variation in time between failures can be studied statistically. The corresponding characteristic for nonrepairable products is usually called the time to failure.

When the failure rate is constant, the distribution of time between failures is distributed exponentially. Consider the 42 failure times in the constant failure rate portion of Table 19.13. The time between failures for successive failures can be tallied, and the 41 resulting TBFs can be formed into the frequency distribution shown in Figure 19.20a. The distribution is roughly exponential in shape, indicating that when the failure rate is constant, the distribution of time between failures (not mean time between failures) is exponential. This distribution is the basis of the exponential formula for reliability.

The Exponential Formula for Reliability

The distribution of TBF indicates the chance of failure-free operation for the specified time period. The chance of obtaining failure-free operation for a specified time period or longer can be shown by changing the TBF distribution to a distribution showing the number of intervals equal to or greater than a specified time length (Figure 19.20b). If the frequencies are expressed as relative frequencies, they become estimates of the probability of survival. When the failure rate is constant, the probability of survival (or reliability) is

$$P_s = R = e^{-t/\mu} = e^{-t\lambda}$$

where $P_s = R$ = probability of failure-free operation for a time period equal to or greater than t
$e = 2.718$
t = specified period of failure-free operation
μ = mean time between failures (the mean of TBF distribution)
λ = failure rate (the reciprocal of μ)

Note that this formula is simply the exponential probability distribution rewritten in terms of reliability.

Problem A washing machine requires 30 minutes to clean a load of clothes. The mean time between failures of the machine is 100 hours. Assuming a constant failure rate, what is the chance of the machine completing a cycle without failure?

Solution Applying the exponential formula, we obtain

$$R = e^{-t/\mu} = e^{-0.5/100} = 0.995$$

There is a 99.5 percent chance of completing a washing cycle.

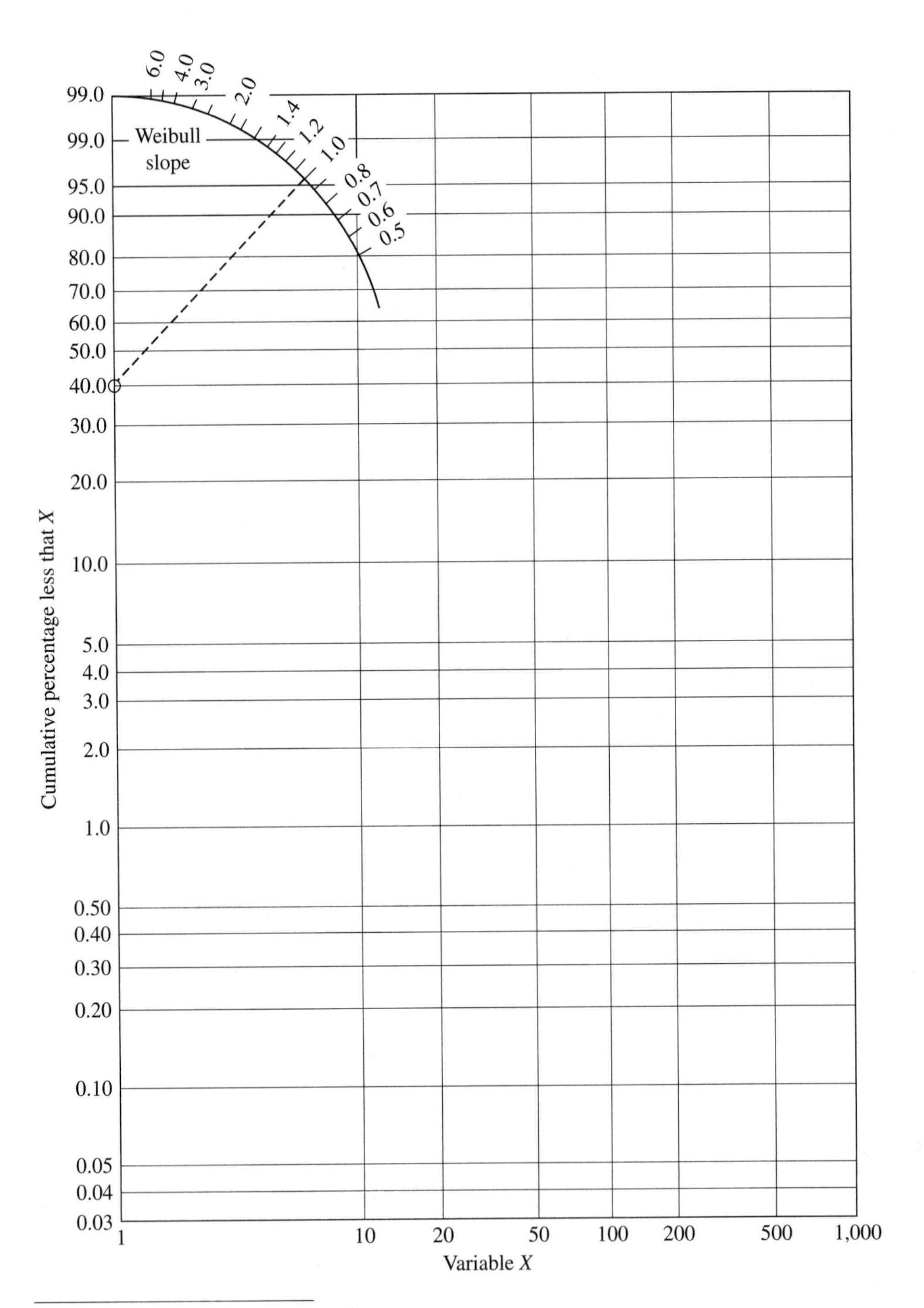

(*Source: Quality Planning and Analysis*, Copyright 2007. Used by permission.)

TABLE 19.14 Weibull Paper

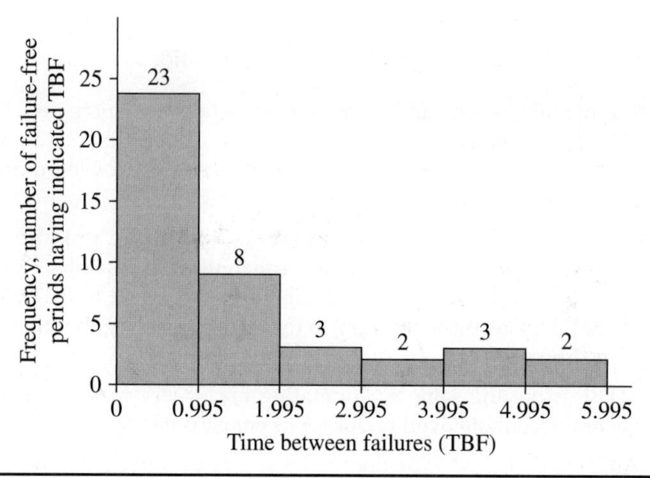

FIGURE 19.20a Histogram of TBF. (*Quality Planning and Analysis, Copyright 2007. Used by permission.*)

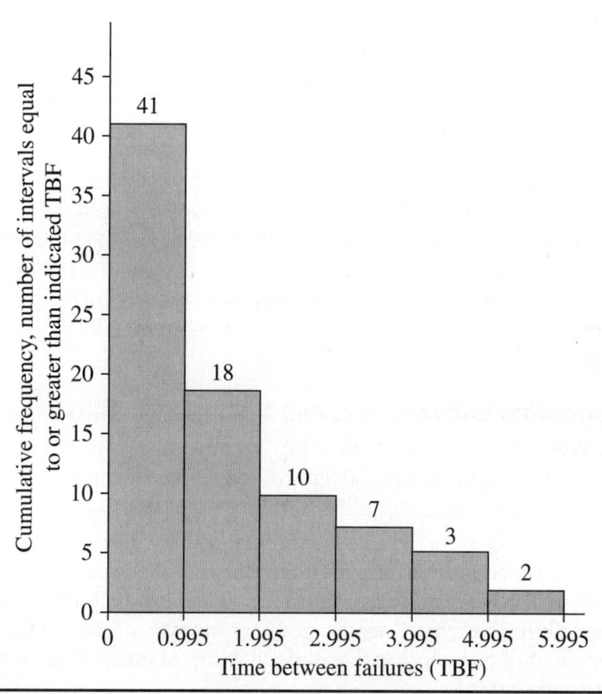

FIGURE 19.20b Cumulative histogram of TBF. (*Quality Planning and Analysis, Copyright 2007. Used by permission.*)

How about the assumption of a constant failure rate? In practice, sufficient data usually are not available to evaluate the assumption. However, experience suggests that this assumption often is true, particularly when (1) infant mortality types of failures have been eliminated before delivery of the product to the user, and (2) the user replaces the product or specific components before the wear-out phase begins.

The Meaning of Mean Time Between Failures. Confusion surrounds the meaning of mean time between failures (MTBF). Further explanation is warranted:

- The MTBF is the mean (or average) time between successive failures of a product. This definition assumes that the product in question can be repaired and placed back into operation after each failure. For nonrepairable products, the term "mean time to failure" (MTTF) is used.
- If the failure rate is constant, the probability that a product will operate without failure for a time equal to or greater than its MTBF is only 37 percent. This outcome is based on the exponential distribution (R is equal to 0.37 when t is equal to the MTBF). This result is contrary to the intuitive feeling that there is a 50-50 chance of exceeding an MTBF.
- MTBF is not the same as "operating life," "service life," or other indexes, which generally connote overhaul or replacement time.
- An increase in an MTBF does not result in a proportional increase in reliability (the probability of survival). If $t = 1$ hour, the following table shows the MTBF required to obtain various reliabilities.

MTBF	*R*
5	0.82
10	0.90
20	0.95
100	0.99

A fivefold increase in MTBF from 20 to 100 hours is necessary to increase the reliability by 4 percentage points compared with a doubling of the MTBF from 5 to 10 hours to get an 8 percentage point increase in reliability.

MTBF is a useful measure of reliability, but it is not correct for all applications. Other reliability indexes are listed in Chapter 28, Research & Development: More Innovation, Scarce Resources.

The Relationship Between Part and System Reliability

It often is assumed that system reliability (i.e., the probability of survival, P_s) is the product of the individual reliabilities of the n parts within the system:

$$P_s = P_1 P_2 \ldots P_n$$

For example, if a communications system has four subsystems with reliabilities of 0.970, 0.989, 0.995, and 0.996, the system reliability is the product, or 0.951. The formula assumes that (1) the failure of any part causes failure of the system and (2) the reliabilities of the parts are independent of one another (i.e., the reliability of one part does not depend on the functioning of another part).

These assumptions are not always true, but in practice, the formula serves two purposes. First, it shows the effect of increased complexity of equipment on overall reliability. As the number of parts in a system increases, the system reliability decreases dramatically (see Figure 19.21). Second, the formula often is a convenient approximation that can be refined as information on the interrelationships of the parts becomes available.

When it can be assumed that (1) the failure of any part causes system failure, (2) the parts are independent, and (3) each part follows an exponential distribution, then

$$P_s = e^{-t_1\lambda_1} e^{-t_2\lambda_2} \ldots e^{-t_n\lambda_n}$$

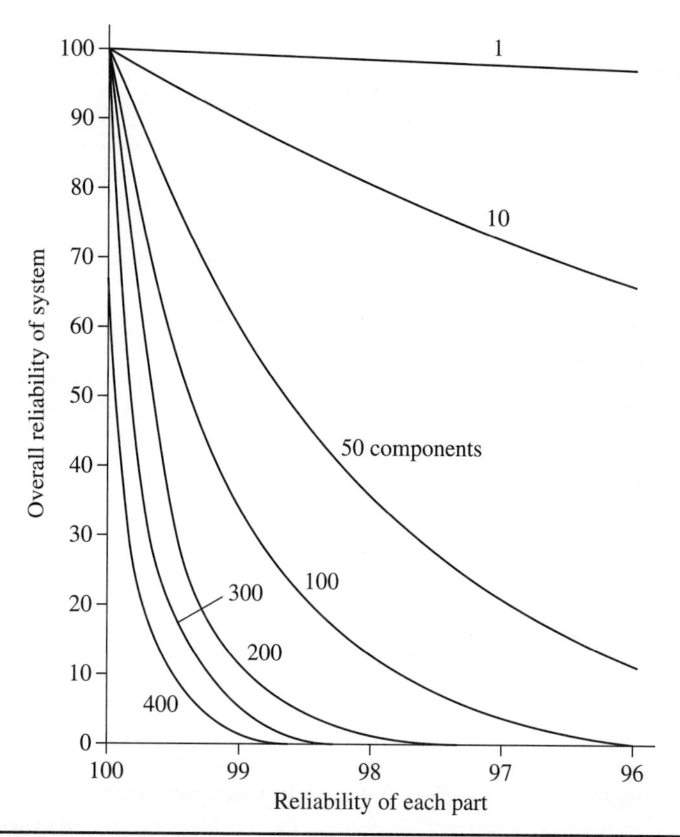

FIGURE 19.21 Relationship between part and system reliability. (*Quality Planning and Analysis, Copyright 2007. Used by permission.*)

Further, if t is the same for each part,

$$P_s = e^{-1\Sigma\lambda}$$

Thus, when the failure rate is constant (and therefore the exponential distribution can be applied), the reliability of a system can be predicted based on the addition of the part failure rates (see the section "Predicting Reliability during Design," next).

Sometimes designs are planned with redundancy so that the failure of one part will not cause system failure. Redundancy is an old (but still useful) design technique invented long before the advent of reliability prediction techniques. However, the designer can now predict the effect of redundancy on system reliability in quantitative terms.

Redundancy is the existence of more than one element for accomplishing a given task, where all elements must fail before there is an overall failure of the system. In parallel redundancy (one of several types of redundancy), two or more elements operate at the same time to accomplish the task, and any single element is capable of handling the job itself in case of failure of the other elements. When parallel redundancy is used, the overall reliability is calculated as follows:

$$P_s = 1 - (1 - \mathrm{P1})n$$

where P_s = reliability of the system
$P1$ = reliability of the individual elements in the redundancy
n = number of identical redundant elements

Problem Suppose that a unit has a reliability of 99.0 percent for a specified mission time. If two identical units are used in parallel redundancy, what overall reliability will be expected?

Solution Applying the formula above, we obtain

$$R = 1 - (1 - 0.99)(1 - 0.99) = 0.9999, \text{ or } 99.99 \text{ percent}$$

Predicting Reliability during Design

Reliability prediction methods continue to evolve, but include such standards as failure mode and effects analysis (FMEA) and testing. Ireson et al. (1996) provide an extensive discussion of reliability prediction, and should be consulted beyond the methods discussed in this handbook.

The following steps make up a reliability prediction method:

1. Define the product and its functional operation. The system, subsystems, and units must be precisely defined in terms of their functional configurations and boundaries. This precise definition is aided by preparation of a functional block diagram that shows the subsystems and lower-level products, their interrelationships, and the interfaces with other systems. Given a functional block diagram and a well-defined statement of the functional requirements of the product, the conditions that constitute failure or unsatisfactory performance can be defined.
2. Prepare a reliability block diagram. For systems in which there are redundancies or other special interrelationships among parts, a reliability block diagram is useful. This diagram is similar to a functional block diagram, but the reliability block diagram shows exactly what must function for successful operation of the system. The diagram shows redundancies and alternative modes of operation. The reliability block diagram is the foundation for developing the probability model for reliability. O'Connor (1995) provides further discussion.
3. Develop the probability model for predicting reliability. A simple model may add only failure rates; a complex model can account for redundancies and other conditions.
4. Collect information relevant to parts reliability. The data include information such as parts function, parts ratings, stresses, internal and external environments, and operating time. Many sources of failure-rate information state failure rates as a function of operating parameters. For example, failure rates for fixed ceramic capacitors are stated as a function of (1) expected operating temperature and (2) the ratio of the operating voltage to the rated voltage. Such data show the effect of derating (assigning a part to operate below its rated voltage) on reducing the failure rate.
5. Select parts reliability data. The required parts data consist of information on catastrophic failures and on tolerance variations with respect to time under known operating and environmental conditions. Acquiring these data is a major problem for the designer because there is no single reliability data bank comparable to handbooks such as those for physical properties of materials. Instead, the designer must build a data bank by securing reliability data from a variety of sources:

Field performance studies conducted under controlled conditions:

- Specified life tests
- Data from parts manufacturers or industry associations

- Customers' parts qualification and inspection tests
- Government agency data banks such as the Government Industry Data Exchange Program (GIDEP) and the Reliability Information Analysis Center (RIAC)

Combine all of the above to obtain the numerical reliability prediction.

Ireson et al. (1996) and O'Connor (1995) are excellent references for reliability prediction. Included are the basic methods of prediction, repairable versus nonrepairable systems, electronic and mechanical reliability, reliability testing, and software reliability. Box and Draper (1969) provides extensive discussion of reliability data analysis, including topics such as censored life data (not all test units have failed during the test) and accelerated-life test data analysis. Dodson (1999) explains how the use of computer spreadsheets can simplify reliability modeling using various statistical distributions.

Reliability prediction techniques based on component failure data to estimate system failure rates have generated controversy. Jones and Hayes (1999) present a comparison of predicted and observed performance for five prediction techniques using parts count analyses. The predictions differed greatly from observed field behavior and from each other. The standard ANSI/IEC/ASQC D60300-3-1-1997 (Dependability Management—Part 3: Application Guide—Section 1—Analysis Techniques for Dependability) compares five analysis techniques: FMEA/FMECA, fault tree analysis, reliability block diagram, Markov analysis, and parts count reliability prediction.

The reliability of a system evolves during design, development, testing, production, and field use. The concept of reliability growth assumes that the causes of product failures are discovered and action is taken to remove the causes, thus resulting in improved reliability of future units ("test, analyze, and fix"). Reliability growth models provide predictions of reliability due to such improvements. For elaboration, see O'Connor (1995). Also, ANSI/IEC/ASQC D601164-1997 (Reliability Growth—Statistical Test and Estimation Methods) and the related IEC 61164 Ed. 2.0 (2004) (Reliability growth—Statistical test and estimation methods) describe methods of estimating reliability growth.

Predicting Reliability Based on the Exponential Distribution

When the failure rate is constant and when study of a functional block diagram reveals that all parts must function for system success, then reliability is predicted to be the simple total of failure rates. An example of a subsystem prediction is shown in Table 19.15. The prediction for the subsystem is made by adding the failure rates of the parts; the MTBF is then calculated as the reciprocal of the failure rate.

For further discussion of reliability prediction, including an example for an electronic system, see Gryna et al. (2007).

Predicting Reliability Based on the Weibull Distribution

Prediction of overall reliability based on the simple addition of component failure rates is valid only if the failure rate is constant. When this assumption cannot be made, an alternative approach based on the Weibull distribution can be used.

1. Graphically, use the Weibull distribution to predict the reliability R for the time period specified. $R = 100 - \%$ failure. Do this for each component (Table 19.14).
2. Combine the component reliabilities using the product rule and/or redundancy formulas to predict system reliability.

Predictions of reliability using the exponential distribution or the Weibull distribution are based on reliability as a function of time. Next we consider reliability as a function of stress and strength.

Part Description	Quantity	Generic Failure Rate per Million Hours	Total Failure Rates per Million Hours
Heavy-duty ball bearing	6	14.4	86.4
Brake assembly	4	16.8	67.2
Cam	2	0.016	0.032
Pneumatic hose	1	29.28	29.28
Fixed displacement pump	1	1.464	1.464
Manifold	1	8.80	65.0
Guide pin	5	13.0	65.0
Control valve	1	15.20	15.20
Total assembly failure rate			273.376

MTBF = 1/0.000273376 = 3.657.9 hours

(*Source:* Adapted from Ireson et al., p. 19.9. *Quality Planning and Analysis*, Copyright 2007. Used by permission.)

Table 19.15 Example of Mechanical Parts and Subsystem Failure Rates

Reliability as a Function of Applied Stress and Strength

Failures are not always a function of time. In some cases, a part will function indefinitely if its strength is greater than the stress applied to it. The terms "strength" and "stress" here are used in the broad sense of inherent capability and operating conditions applied to a part, respectively.

For example, operating temperature is a critical parameter, and the maximum expected temperature is 145°F (63°C). Further, capability is indicated by a strength distribution having a mean of 172°F (78°C) and a standard deviation of 13°F (7°C) (Figure 19.22). With knowledge of only the maximum temperatures, the safety margin is

$$\frac{172-145}{13}=2.08$$

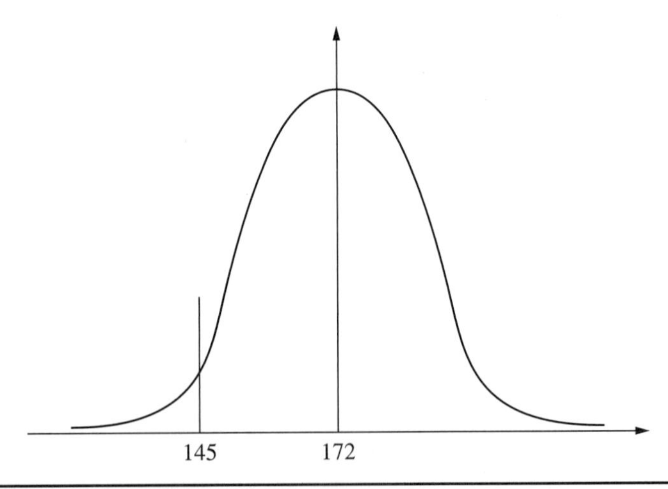

Figure 19.22 Distribution of strength.

The safety margin says that the average strength is 2.08 standard deviations above the maximum expected temperature of 145°F (63°C). Table 19.16 can be used to calculate a reliability of 0.981 [the area beyond 145°F (63°C)].

This calculation illustrates the importance of variation in addition to the average value during design. Designers have always recognized the existence of variation by using a safety factor in design. However, the safety factor is often defined as the ratio of average strength to the worst stress expected.

TABLE A

Normal distribution

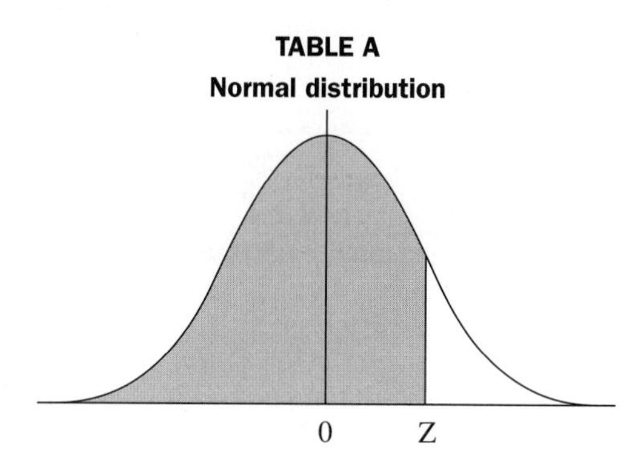

Proportion of total areas under the curve from $-\infty$ to $Z = \frac{X - \mu}{\sigma}$, To illustrate when $Z = 2$, the probability is .9773 of obtaining a value equal to or less then X.

Z	0.09	0.08	0.07	0.06	0.05	0.04	0.03	0.02	0.01	0.00
−3.0	.00100	.00104	.00107	.00111	.00114	.00118	.00122	.00126	.00131	.00135
−2.9	.0014	.0014	.0015	.0015	.0016	.0016	.0017	.0017	.0018	.0019
−2.8	.0019	.0020	.0021	.0021	.0022	.0023	.0023	.0024	.0025	.0026
−2.7	.0026	.0027	.0028	.0029	.0030	.0031	.0032	.0033	.0034	.0035
−2.6	.0036	.0037	.0038	.0039	.0040	.0041	.0043	.0044	.0045	.0047
−2.5	.0048	.0049	.0051	.0052	.0054	.0055	.0057	.0059	.0060	.0062
−2.4	.0064	.0066	.0068	.0069	.0071	.0073	.0075	.0078	.0080	.0082
−2.3	.0084	.0087	.0089	.0091	.0094	.0096	.0099	.0102	.0104	.0107
−2.2	.0110	.0113	.0116	.0119	.0122	.0125	.0129	.0132	.0136	.0139
−2.1	.0143	.0146	.0150	.0154	.0158	.0162	.0166	.0170	.0174	.0179
−2.0	.0183	.0188	.0192	.0197	.0202	.0207	.0212	.0217	.0222	.0228
−1.9	.0233	.0239	.0244	.0250	.0256	.0262	.0268	.0274	.0281	.0287
−1.8	.0294	.0301	.0307	.0314	.0322	.0329	.0336	.0344	.0351	.0359
−1.7	.0367	.0375	.0384	.0392	.0401	.0409	.0418	.0427	.0436	.0446
−1.6	.0455	.0465	.0475	.0485	.0495	.0505	.0516	.0526	.0537	.0548
−1.5	.0559	.0571	.0582	.0594	.0606	.0618	.0630	.0643	.0655	.0668
−1.4	.0681	.0694	.0708	.0721	.0735	.0749	.0764	.0778	.0793	.0808
−1.3	.0823	.0838	.0853	.0869	.0885	.0901	.0918	.0934	.0951	.0968
−1.2	.0985	.1003	.1020	.1038	.1057	.1075	.1093	.1112	.1131	.1151
−1.1	.1170	.1190	.1210	.1230	.1251	.1271	.1292	.1314	.1335	.1357

(*Source: Quality Planning and Analysis*, Copyright 2007. Used by permission.)

TABLE 19.16 Normal Distribution

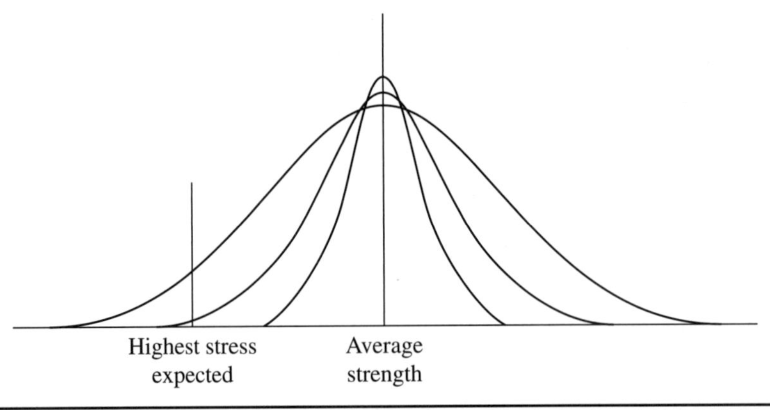

Figure 19.23 Variation and safety factor. (*Quality Planning and Analysis, Copyright 2007. Used by permission.*)

Note that in Figure 19.23, all designs have the same safety factor. Also note that the reliability (probability of a part having a strength greater than the stress) varies considerably. Thus, the uncertainty often associated with this definition of safety factor is, in part, due to its failure to reflect the variation in both strength and stress. Such variation is partially reflected in a safety margin, defined as

$$\frac{\text{Average strength} - \text{worst stress}}{\text{Standard deviation of strength}}$$

This recognizes the variation in strength but is conservative because it does not recognize a variation in stress.

Availability

Availability has been defined as the probability that a product, when used under given conditions, will perform satisfactorily when called upon. Availability considers the operating time of the product and the time required for repairs. Idle time, during which the product is not needed, is excluded.

Availability is calculated as the ratio of operating time to operating time plus downtime. However, downtime can be viewed in two ways:

- *Total downtime.* This period includes active repair (diagnosis and repair time), preventive maintenance time, and logistics time (time spent waiting for personnel, spare parts, etc.). When total downtime is used, the resulting ratio is called operational availability (A_0).
- *Active repair time.* The resulting ratio is called intrinsic availability (A_i). Under certain conditions, availability can be calculated as:

$$A_0 = \frac{\text{MTBF}}{\text{MTBF+MDT}} \quad \text{and} \quad A_i = \frac{\text{MTBF}}{\text{MTBF+MTTR}}$$

where MTBF = mean time between failures
MDT = mean downtime
MTTR = mean time to repair

This is known as the steady-state formula for availability. The steady-state formula for availability has the virtue of simplicity. However, the formula is based on several assumptions that are not always met in the real world. The assumptions are

- The product is operating in the constant failure rate period of the overall life. Thus, the failure-time distribution is exponential.
- The downtime or repair-time distribution is exponential.
- Attempts to locate system failures do not change the overall system failure rate.
- No reliability growth occurs (such growth might be due to design improvements or through debugging of bad parts).
- Preventive maintenance is scheduled outside the time frame included in the availability calculation.

More precise formulas for calculating availability depend on operational conditions and statistical assumptions. These formulas are discussed by Ireson et al. (1996).

Setting Specification Limits

A major step in the development of physical products is the conversion of product features into dimensional, chemical, electrical, and other characteristics of the product. Thus, a heating system for an automobile will have many characteristics for the heater, air ducts, blower assembly, engine coolant, etc.

For each characteristic, the designer must specify (1) the desired average (or "nominal value") and (2) the specification limits (or "tolerance limits") above and below the nominal value that individual units of product must meet. These two elements relate to parameter design and tolerance design, as discussed in Gryna et al. (2007).

The specification limits should reflect the functional needs of the product, manufacturing variability, and economic consequences. These three aspects are addressed in the next three sections. For greater depth in the statistical treatment of specification limits, see Anand (1996).

Specification Limits and Functional Needs

Sometimes data can be developed to relate product performance to measurements of a critical component. For example, a thermostat may be required to turn on and shut off a power source at specified low and high temperature values, respectively. A number of thermostat elements are built and tested. The prime recorded data are (1) turn-on temperature, (2) shut-off temperature, and (3) physical characteristics of the thermostat elements. We can then prepare scatter diagrams (Figure 19.24) and regression equations to help establish critical component tolerances on a scientific basis within the confidence limits for the numbers involved. Ideally, the sample size is sufficient, and the data come from a statistically controlled process—two conditions that are both rarely achieved. O'Connor (1995) explains how this approach can be related to the Taguchi approach to develop a more robust design.

Specification Limits and Manufacturing Variability

Generally, designers will not be provided with information on process capability. Their problem will be to obtain a sample of data from the process, calculate the limits that the process can meet, and compare these to the limits they were going to specify. If they do not have any limits in mind, the capability limits calculated from process data provide a set of limits that are realistic from the viewpoint of producibility. These limits must then be evaluated against the functional needs of the product.

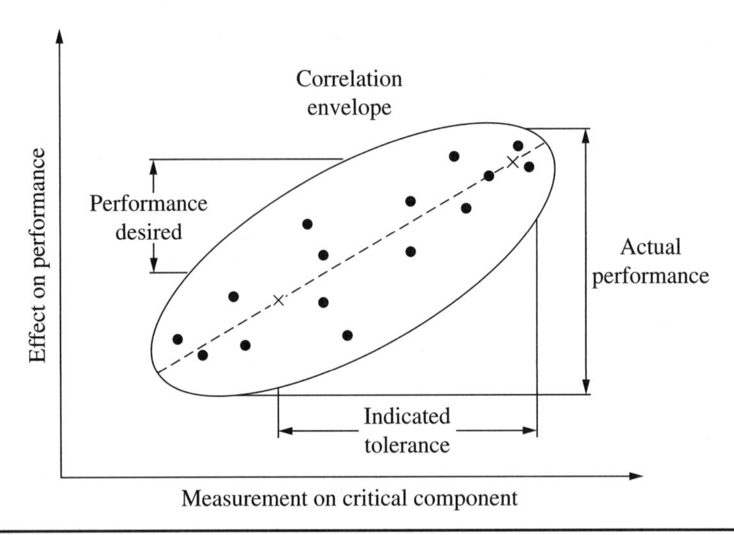

Figure 19.24 Approach to functional tolerancing. (*Quality Planning and Analysis, Copyright 2007. Used by permission.*)

Statistically, the problem is to predict the limits of variation of individual items in the total population based on a sample of data. For example, suppose that a product characteristic is normally distributed with a population average of 5.000 in (12.7 cm) and a population standard deviation of 0.001 in (0.00254 cm). Limits can then be calculated to include any given percentage of the population. Figure 19.25 shows the location of the 99 percent limits. Table 19.16 indicates that 2.575 standard deviations will include 99 percent of the population. Thus, in this example, a realistic set of tolerance limits would be

$$5.000 \pm 2.575(0.001) = \frac{5.003}{4.997}$$

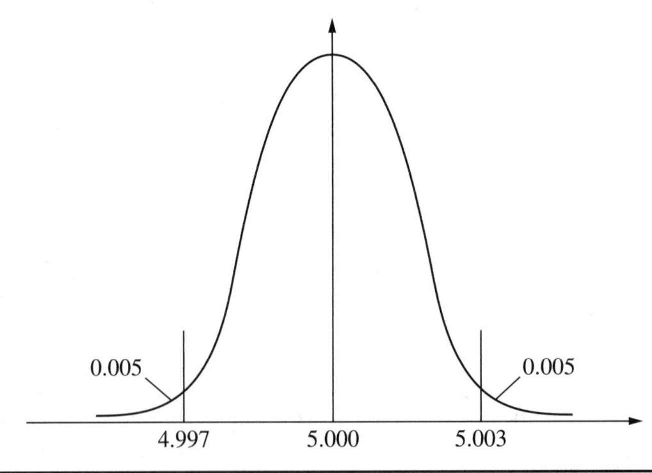

Figure 19.25 Distribution with 99 percent limits. (*Quality Planning and Analysis, Copyright 2007. Used by permission.*)

Ninety-nine percent of the individual pieces in the population will have values between 4.997 and 5.003.

In practice, the average and standard deviation of the population are not known but must be estimated from a sample of product from the process. As a first approximation, tolerance limits are sometimes set at

$$\overline{X} \pm 3s$$

Here, the average $\overline{X}$ and standard deviation s of the sample are used directly as estimates of the population values. If the true average and standard deviation of the population happen to be equal to those of the sample, and if the characteristic is normally distributed, then 99.73 percent of the pieces in the population will fall within the limits calculated. These limits are frequently called natural tolerance limits (limits that recognize the actual variation of the process and therefore are realistic). This approximation ignores the possible error in both the average and standard deviation as estimated from the sample.

Methodology has been developed for setting tolerance limits in a more precise manner. For example, formulas and tables are available for determining tolerance limits based on a normally distributed population. Table 19.17 provides factors for calculating tolerance limits that recognize the uncertainty in the sample mean and sample standard deviation. The tolerance limits are determined as

$$\overline{X} \pm Ks$$

The factor K is a function of the confidence level desired, the percentage of the population to be included within the tolerance limits, and the number of data values in the sample.

For example, suppose that a sample of 10 resistors from a process yielded an average and standard deviation of 5.04 and 0.016, respectively. The tolerance limits are to include 99 percent of the population, and the tolerance statement is to have a confidence level of 95 percent. Referring to Table 19.17, the value of K is 4.433, and tolerance limits are then calculated as

$$5.04 \pm 4.433(0.016) = \frac{5.11}{4.97}$$

We are 95 percent confident that at least 99 percent of the resistors in the population will have resistance between 4.97 and 5.11 Ω. Tolerance limits calculated in this manner are often called statistical tolerance limits. This approach is more rigorous than the 3s natural tolerance limits, but the two percentages in the statement are a mystery to those without a statistical background.

For products in some industries (e.g., electronics), the number of units outside of specification limits is stated in terms of parts per million (ppm). Thus, if limits are set at three standard deviations, 2700 ppm (100 to 99.73 percent) will fall outside the limits. For many applications (e.g., a personal computer with many logic gates), such a level is totally unacceptable. Table 19.18 shows the ppm for several standard deviations. These levels of ppm assume that the process average is constant at the nominal specification. A deviation from the nominal value will result in a higher ppm value. To allow for modest shifts in the process average, some manufacturers follow a guideline for setting specification limits at $\pm 6\sigma$.

Designers often must set tolerance limits with only a few measurements from the process (or more likely from the development tests conducted under laboratory conditions).

Tolerance Factors for Normal Distributions (Two Sided)

P	γ = 0.75					γ = 0.90				
N	0.75	0.90	0.95	0.99	0.999	0.75	0.90	0.95	0.99	0.999
2	4.498	6.301	7.414	9.531	11.920	11.407	15.978	18.800	24.167	30.227
3	2.501	3.538	4.187	5.431	6.844	4.132	5.847	6.919	8.974	11.309
4	2.035	2.892	3.431	4.471	5.657	2.932	4.166	4.943	6.440	8.149
5	1.825	2.599	3.088	4.033	5.117	2.454	3.494	4.152	5.423	6.879
6	1.704	2.429	2.889	3.779	4.802	2.196	3.131	3.723	4.870	6.188
7	1.624	2.318	2.757	3.611	4.593	2.034	2.902	3.452	4.521	5.750
8	1.568	2.238	2.663	3.491	4.444	1.921	2.743	3.264	4.278	5.446
9	1.525	2.178	2.593	3.400	4.330	1.839	2.626	3.125	4.098	5.220
10	1.492	2.131	2.537	3.328	4.241	1.775	2.535	3.018	3.959	5.046
11	1.465	2.093	2.493	3.271	4.169	1.724	2.463	2.933	3.849	4.906
12	1.443	2.062	2.456	3.223	4.110	1.683	2.404	2.863	3.758	4.792
13	1.425	2.036	2.424	3.183	4.059	1.648	2.355	2.805	3.682	4.697
14	1.409	2.013	2.398	3.148	4.016	1.619	2.314	2.756	3.618	4.615
15	1.395	1.994	2.375	3.118	3.979	1.594	2.278	2.713	3.562	4.545
16	1.383	1.977	2.355	3.092	3.946	1.572	2.246	2.676	3.514	4.484
17	1.372	1.962	2.337	3.069	3.917	1.552	2.219	2.643	3.471	4.430
18	1.363	1.948	2.321	3.048	3.891	1.535	2.194	2.614	3.433	4.382
19	1.355	1.936	2.307	3.030	3.867	1.520	2.172	2.588	3.399	4.339
20	1.347	1.925	2.294	3.013	3.846	1.506	2.152	2.564	3.368	4.300
21	1.340	1.915	2.282	2.998	3.827	1.493	2.135	2.543	3.340	4.264
22	1.334	1.906	2.271	2.984	3.809	1.482	2.118	2.524	3.315	4.232
23	1.328	1.898	2.261	2.971	3.793	1.471	2.103	2.506	3.292	4.203
24	1.322	1.891	2.252	2.950	3.778	1.462	2.089	2.480	3.270	4.176
25	1.317	1.883	2.244	2.948	3.764	1.453	2.077	2.474	3.251	4.151
26	1.313	1.877	2.236	2.938	3.751	1.444	2.065	2.460	3.232	4.127
27	1.309	1.871	2.229	2.929	3.740	1.437	2.054	2.447	3.215	4.106
30	1.297	1.855	2.210	2.904	3.708	1.417	2.025	2.413	3.170	4.049
35	1.283	1.834	2.185	2.871	3.667	1.390	1.988	2.368	3.112	3.974
40	1.271	1.818	2.166	2.846	3.635	1.370	1.959	2.334	3.066	3.917
100	1.218	1.742	2.075	2.727	3.484	1.275	1.822	1.172	2.854	3.646
500	1.177	1.683	2.006	2.636	3.368	1.201	1.717	2.046	2.689	3.434
1000	1.169	1.671	1.992	2.617	3.344	1.185	1.695	2.019	2.654	3.390
∞	1.150	1.645	1.960	2.576	3.291	1.150	1.645	1.960	2.576	3.291

Table 19.17 Tolerance Factors for Normal Distributions

γ = 0.95					γ = 0.99				
0.75	**0.90**	**0.95**	**0.99**	**0.999**	**0.75**	**0.90**	**0.95**	**0.99**	**0.999**
22.858	32.019	37.674	48.430	60.573	114.363	160.363	188.491	242.300	303.054
5.922	8.380	9.916	12.861	16.208	13.378	18.930	22.401	29.055	36.616
3.779	5.369	6.370	8.299	10.502	6.614	9.398	11.150	14.527	18.383
3.002	4.275	5.079	6.634	8.415	4.643	6.612	7.855	10.260	13.015
2.604	3.712	4.414	5.775	7.337	3.743	5.337	6.345	8.301	10.548
2.361	3.369	4.007	5.248	6.676	3.233	4.613	5.488	7.187	9.142
2.197	3.136	3.732	4.891	6.226	2.905	4.147	4.936	6.468	8.234
2.078	2.967	3.532	4.631	5.899	2.677	3.822	4.550	5.966	7.600
1.987	2.839	3.379	4.433	5.649	2.508	3.582	4.265	5.594	7.129
1.916	2.737	3.259	4.277	5.452	2.378	3.397	4.045	5.308	6.766
1.858	2.655	3.162	4.150	5.291	2.274	3.250	3.870	5.079	6.477
1.810	2.587	3.081	4.044	5.158	2.190	3.130	3.727	4.893	6.240
1.770	2.529	3.012	3.955	5.045	2.120	3.029	3.608	4.737	6.043
1.735	2.480	2.954	3.878	4.949	2.060	2.945	3.507	4.605	5.876
1.705	2.437	2.903	3.812	4.865	2.009	2.872	3.421	4.492	5.732
1.679	2.400	2.858	3.754	4.791	1.965	2.808	3.345	4.393	5.607
1.655	2.366	2.819	3.702	4.725	1.926	2.753	3.279	4.307	5.497
1.635	2.337	2.784	3.656	4.667	1.891	2.703	3.221	4.230	5.399
1.616	2.310	2.752	3.615	4.614	1.860	2.659	3.168	4.161	5.312
1.599	2.286	2.723	3.577	4.567	1.833	2.620	3.121	4.100	5.234
1.584	2.264	2.697	3.543	4.523	1.808	2.584	3.078	4.044	5.163
1.570	2.244	2.673	3.512	4.484	1.795	2.551	3.040	3.993	5.098
1.557	2.225	2.651	3.483	4.447	1.764	2.522	3.004	3.947	5.039
1.545	2.208	2.631	3.457	4.413	1.745	2.494	2.972	3.904	4.985
1.534	2.193	2.612	3.432	4.382	1.727	2.460	2.941	3.865	4.935
1.523	2.178	2.595	3.409	4.353	1.711	2.446	2.914	3.828	4.888
1.497	2.140	2.549	3.350	4.278	1.668	2.385	2.841	3.733	4.768
1.462	2.090	2.490	3.272	4.179	1.613	2.306	2.748	3.611	4.611
1.435	2.052	2.445	3.213	4.104	1.571	2.247	2.677	3.518	4.493
1.311	1.874	2.233	2.934	3.748	1.383	1.977	2.355	3.096	3.954
1.215	1.737	2.070	2.721	3.475	1.243	1.777	2.117	2.783	3.555
1.195	1.709	2.036	2.676	3.418	1.214	1.736	2.068	2.718	3.472
1.150	1.645	1.960	2.576	3.291	1.150	1.645	1.960	2.576	3.291

*Table H—Tolerance factors for normal distributions" from *Selected Techniques of Statistical Analysis—OSRD* by C. Eisenhart, M. W. Hastay, and W. A. Wallis, Copyright 1947 by The McGraw-Hill Companies, Inc. Reprinted by permission of The McGraw-Hill Companies, Inc.
γ = confidence level
P = percentage of population within tolerance limits
N = number of values in sample
(*Source: Quality Planning and Analysis*, Copyright 2007. Used by permission.)

TABLE 19.17 (*Continued*)

Number of Standard Deviations	Part per Million (ppm)
±3σ	2700
±4σ	63
±5σ	0.57
±6σ	0.002

*If the process is not centered and the mean shifts by up to 1.5σ, then ±6σ will be 3.4 ppm.
(*Source: Quality Planning and Analysis*, Copyright 2007. Used by permission.)

Table 19.18 Standard Deviations and PPM (centered process)*

In developing a paint formulation, for example, the following values of gloss were obtained: 76.5, 75.2, 77.5, 78.9, 76.1, 78.3, and 77.7. A group of chemists was asked where they would set a minimum specification limit. Their answer was 75.0—a reasonable answer for those without statistical knowledge. Figure 19.26 shows a plot of the data on normal probability paper. If the line is extrapolated to 75.0, the plot predicts that about 11 percent of the population will fall below 75.0, even though all of the sample data exceed 75.0. Of course, a larger

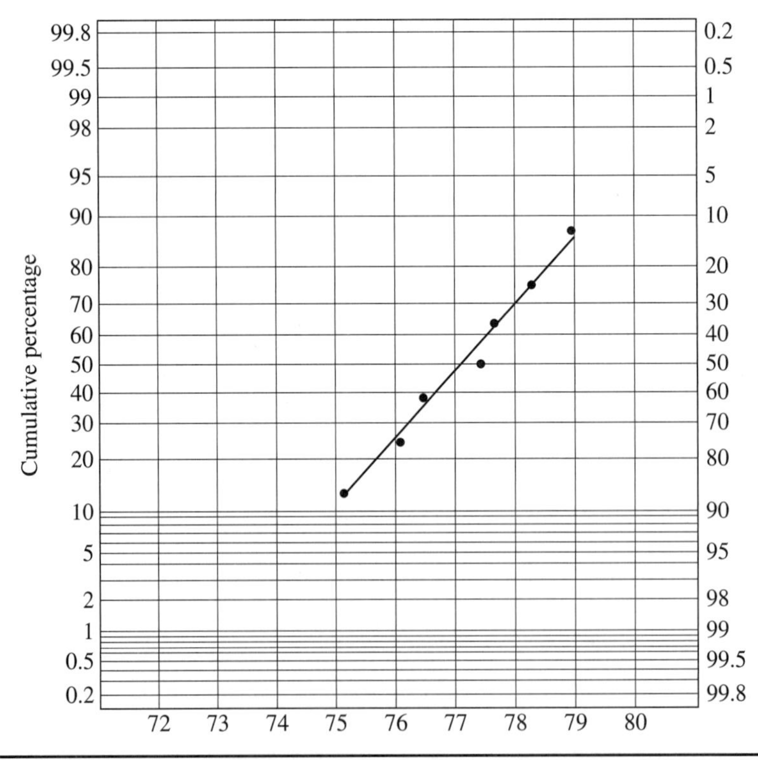

Figure 19.26 Probability plot of development data. (*Quality Planning and Analysis, Copyright 2007. Used by permission.*)

Name of Limit	Meaning
Tolerance	Set by the engineering design function to define the minimum and maximum values allowable for the product to work properly
Statistical tolerance	Calculated from process data to define the amount of variation that the process exhibits; these limits will contain a specified proportion of the total population
Prediction	Calculated from process data to define the limits which will contain all of *k* future observations
Confidence	Calculated from data to define an interval within which a population parameter lies
Control	Calculated from process data to define the limits of chance (random) variation around some central value

(*Source: Quality Planning and Analysis*, Copyright 2007. Used by permission.)

TABLE 19.19 Distinctions Among Limits

sample size is preferred and further statistical analyses could be made, but the plot provides a simple tool for evaluating a small sample of data.

All methods of setting tolerance limits based on process data assume that the sample of data represents a process that is sufficiently stable to be predictable. In practice, the assumption is often accepted without any formal evaluation. If sufficient data are available, the assumption should be checked with a control chart.

Statistical tolerance limits are sometimes confused with other limits used in engineering and statistics. Table 19.19 summarizes the distinctions among five types of limits (see also Box, pp. 44.47–44.58).

Specifications Limits and Economic Consequences

In setting traditional specification limits around a nominal value, we assume that there is no monetary loss for product falling within specification limits. For product falling outside the specification limits, the loss is the cost of replacing the product.

Another viewpoint holds that any deviation from the nominal value causes a loss. Thus, there is an ideal (nominal) value that customers desire, and any deviation from this ideal results in customer dissatisfaction. This loss can be described by a loss function (Figure 19.27).

Many formulas can predict loss as a function of deviation from the target. Taguchi proposes the use of a simple quadratic loss function:

$$L = k(X - T)^2$$

where L = loss in monetary terms
k = cost coefficient
X = value of quality characteristic
T = target value

Ross (1996) provides an example to illustrate how the loss function can help to determine specification limits. In automatic transmissions for trucks, shift points are designed to

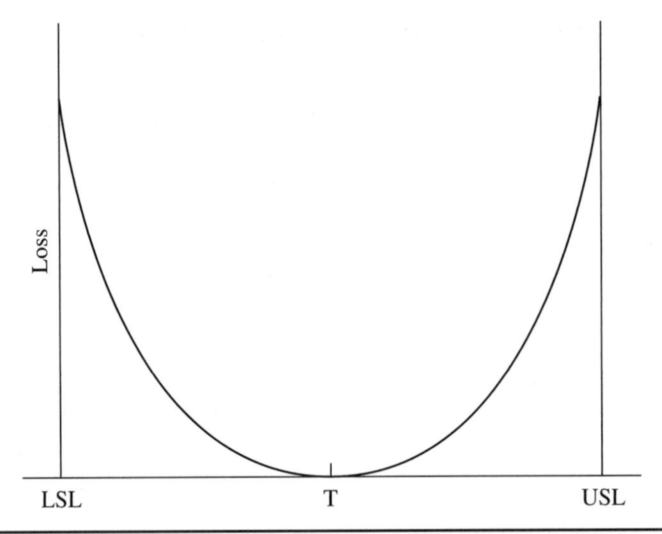

FIGURE 19.27 Loss function. (*Quality Planning and Analysis, Copyright 2007. Used by permission.*)

occur at a certain speed and throttle position. Suppose it costs the producer $100 to adjust a valve body under warranty when a customer complains of the shift point. Research indicates that the average customer would request an adjustment if the shift point is off from the nominal by 40 rpm transmission output speed on the first-to-second gear shift. The loss function is then

$$\text{Loss} = k(X - T)^2$$
$$100 = k(40)^2$$
$$k = \$0.0625$$

This adjustment can be made at the factory at a lower cost, about $10. The loss function is now used to calculate the specification limits:

$$\$10 = 0.0625(X - T)^2$$
$$(X - T) = \pm 12.65 \text{ or } \pm 13 \text{ rpm}$$

The specification limits should be set at 13 rpm around the desired nominal value. If the transmission shift point is further than 13 rpm from the nominal, adjustment at the factory is less expensive than waiting for a customer complaint and making the adjustment under warranty in the field. Ross (1996) discusses how the loss function can be applied to set one-sided specification limits (e.g., a minimum value or a maximum value).

Specification Limits for Interacting Dimensions

Interacting dimensions mate or merge with other dimensions to create a final result. Consider the simple mechanical assembly shown in Figure 19.28. The lengths of components A, B, and C are interacting dimensions because they determine the overall assembly length.

Suppose the components were manufactured to the specifications indicated in Figure 19.28. A logical specification for the assembly length would be 3.500 ± 0.0035, giving limits of

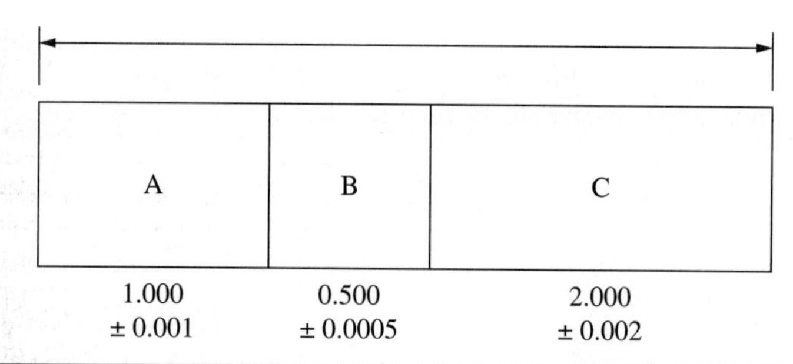

FIGURE 19.28 Mechanical assembly. (*Quality Planning and Analysis, Copyright 2007. Used by permission.*)

3.5035 and 3.4965. This logic may be verified from the two extreme assemblies shown in the following table.

Maximum	Minimum
1.001	0.999
0.5005	0.4995
2.002	1.998
3.5035	3.4965

The approach of adding component tolerances is mathematically correct, but is often too conservative. Suppose that about 1 percent of the pieces of component A are expected to be below the lower tolerance limit for component A and suppose the same for components B and C. If a component A is selected at random, there is, on average, 1 chance in 100 that it will be on the low side, and similarly for components B and C. The key point is this: If assemblies are made at random and if the components are manufactured independently, then the chance that an assembly will have all three components simultaneously below the lower tolerance limit is

$$\frac{1}{100} \times \frac{1}{100} \times \frac{1}{100} = \frac{1}{1{,}000{,}000}$$

There is only about one chance in a million that all three components will be too small, resulting in a small assembly. Thus, setting component and assembly tolerances based on the simple addition formula is conservative in that it fails to recognize the extremely low probability of an assembly containing all low (or all high) components.

The statistical approach is based on the relationship between the variances of a number of independent causes and the variance of the dependent or overall result. This may be written as

$$\sigma_{\text{result}} = \sqrt{\sigma^2_{\text{causeA}} + \sigma^2_{\text{causeB}} + \sigma^2_{\text{causeC}} + \cdots}$$

In terms of the assembly example, the formula is:

$$\sigma_{\text{assembly}} = \sqrt{\sigma_A^2 + \sigma_B^2 + \sigma_C^2}$$

Now suppose that for each component, the tolerance range is equal to three standard deviations (or any constant multiple of the standard deviation). Because σ is equal to T divided by 3, the variance relationship may be rewritten as

$$\frac{T}{3} = \sqrt{\left(\frac{T_A}{3}\right)^2 + \left(\frac{T_B}{3}\right)^2 + \left(\frac{T_C}{3}\right)^2}$$

or

$$T_{\text{assembly}} = \sqrt{T_A^2 + T_B^2 + T_C^2}$$

Thus, the squares of tolerances are added to determine the square of the tolerance for the overall result. This formula compares to the simple addition of tolerances commonly used.

The effect of the statistical approach is dramatic. Listed below are two possible sets of component tolerances that will yield an assembly tolerance equal to 0.0035 when used with the previous formula.

Component	Alternative 1	Alternative 2
A	±0.002	±0.001
B	±0.002	±0.001
C	±0.002	±0.003

With alternative 1, the tolerance for component A has been doubled, the tolerance for component B has been quadrupled, and the tolerance for component C has been kept the same as the original component tolerance based on the simple addition approach. If alternative 2 is chosen, similar significant increases in the component tolerances may be achieved. This formula, then, may result in a larger component tolerance with no change in the manufacturing processes and no change in the assembly tolerance.

The risk of this approach is that an assembly may fall outside the assembly tolerance. However, this probability can be calculated by expressing the component tolerances as standard deviations, calculating the standard deviation of the result, and finding the area under the normal curve outside the assembly tolerance limits. For example, if each component tolerance is equal to $3s$, then 99.73 percent of the assemblies will be within the assembly tolerance, that is, 0.27 percent, or about 3 assemblies in 1000 taken at random would fail to meet the assembly tolerance. The risk can be eliminated by changing components for the few assemblies that do not meet the assembly tolerance.

The tolerance formula is not restricted to outside dimensions of assemblies. Generalizing, the left side of the equation contains the dependent variable or physical result, and the right side of the equation contains the independent variables of physical causes. If the result is placed on the left and the causes on the right, the formula always has plus signs under the square root—even if the result is an internal dimension (such as the clearance between a shaft and hole). The causes of variation are additive wherever the physical result happens to fall.

The formula has been applied to a variety of mechanical and electronic products. The concept may be applied to several interacting variables in an engineering relationship. The nature of the relationship need not be additive (assembly example) or subtractive (shaft-and-hole example). The tolerance formula can be adapted to predict the variation of results that are the product and/or the division of several variables.

Assumptions of the formula. The formula is based on several assumptions:

- The component dimensions are independent and each component to be assembled is chosen randomly. These assumptions are usually met in practice.
- Each component dimension should be normally distributed. Some departure from this assumption is permissible.
- The actual average for each component is equal to the nominal value stated in the specification. For the original assembly example, the actual averages for components A, B, and C must be 1.000, 0.500, and 2.000, respectively. Otherwise, the nominal value of 3.500 will not be achieved for the assembly and tolerance limits set at about 3.500 will not be realistic. Thus it is important to control the average value for interacting dimensions. Consequently, process control techniques are needed using variables measurement.

Use caution if any assumption is violated. Reasonable departures from the assumptions may still permit applying the concept of the formula. Notice that in the example, the formula resulted in the doubling of certain tolerances. This much of an increase may not even be necessary from the viewpoint of process capability.

Bender (1975) has studied these assumptions for some complex assemblies and concluded, based on a "combination of probability and experience," that a factor of 1.5 should be included to account for the assumptions:

$$T_{\text{result}} = 1.5\sqrt{T_A^2 + T_B^2 + T_C^2 + \cdots}$$

Graves (1997) suggests developing different factors for initial versus mature production, high versus low volume production, and mature versus developing technology and measurement processes.

Finally, variation simulation analysis is a technique that uses computer simulation to analyze tolerances. This technique can handle product characteristics with either normal or nonnormal distributions. Dodson (1999) describes the use of simulation in the tolerance design of circuits; Gomer (1998) demonstrates simulation to analyze tolerances in engine design. For an overall text on reliability, see Meeker and Escobar (1998).

Statistical Tools for Control

In addition to the fundamental control charts introduced in Chapter 18, Core Tools to Design, Control, and Improve Performance, there are some special-purpose methods for control that are sometimes helpful.

PRE-Control

PRE-Control is a statistical technique for detecting process conditions and changes that may cause defects (rather than changes that are statistically significant). PRE-Control focuses on controlling conformance to specifications, rather than statistical control. PRE-Control starts a process centered between specification limits and detects shifts that might result in making some of the parts outside a specification limit. It requires no plotting and no computations, and it needs only three measurements to give control information. The technique uses the normal distribution curve to determine significant changes in either the aim or the spread of a production process that could result in increased production of defective work.

The relative simplicity of PRE-Control versus statistical control charts can have important advantages in many applications. The concept, however, has generated some controversy. For a comparison of PRE-Control versus other approaches and the most appropriate applications of PRE-Control, see Ledolter and Swersey (1997) and Steiner (1997). For a complete story, also see the references in both of these papers.

Short-Run Control Charts

Some processes are carried out in such short runs that the usual procedure of collecting 20 to 30 samples to establish a control chart is not feasible. Sometimes these short runs are caused by previously known assignable causes that take place at predetermined times (such as a frequent shift in production from one product to another, as may be the case in lean production systems). Hough and Pond (1995) discuss four ways to construct control charts in these situations:

1. Ignore the systematic variability, and plot on a single chart.
2. Stratify the data, and plot them on a single chart.
3. Use regression analysis to model the data, and plot the residuals on a chart.
4. Standardize the data, and plot the standardized data on a chart.

The last option has received the most consideration. It involves transforming the data via the Z-transformation:

$$Z = \frac{X - \mu}{\sigma}$$

to remove any systematic changes in level and variability (thereby normalizing the data to a common baseline). This standardization of Shewhart charts has been discussed by Nelson (1989), Wheeler (1991), and Griffith (1996). Pyzdek (1993) also provides a good discussion of short and small runs.

Cumulative Sum Control Chart

The cumulative sum (CUMSUM or CUSUM) control chart is a chronological plot of the cumulative sum of deviations of a sample statistic (e.g., $\bar{X}$, p, number of nonconformities) from a reference value (e.g., the nominal or target specification). By definition, the CUMSUM chart focuses on a target value rather than on the actual average of process data. Each point plotted contains information from all observations (i.e., a cumulative sum). CUMSUM charts are particularly useful in detecting small shifts in the process average (say, 0.5σ to 2.0σ). The chart shown in Figure 19.29 is one way of constructing CUMSUM charts. The method is as follows:

1. Compute the control statistic (x-bar for the example in Figure 19.29).
2. Determine the target value T (10 in Figure 19.29).
3. Compute the standard deviation s (1.96 in Figure 19.29).
4. Draw a reference line at zero and upper and lower control limits (UCL and LCL respectively) at ±4s.
5. Compute the upper cumulative sum C_U for each sample point k as follows:

$$C_{U,k} = \text{Maximum}\left\{0, \sum_{i=1}^{k} [\bar{x}_i - (T + s/2)]\right\}$$

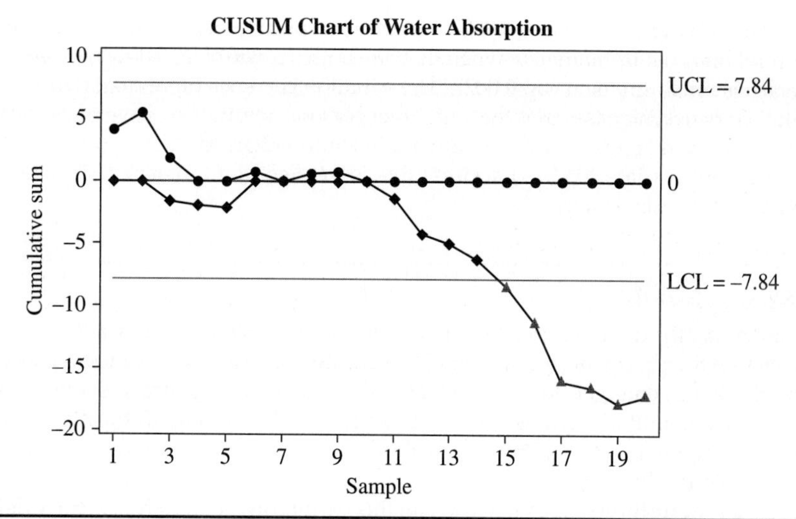

FIGURE 19.29 Cumulative sum control chart. (*Juran Institute, Inc. Copyright 1994. Used by permission.*)

6. Compute the upper cumulative sum C_L for each sample point k as follows:

$$C_{L,k} = \text{Minimum}\left\{0, \sum_{i=1}^{k} [\bar{x}_i - (T - s/2)]\right\}$$

7. Plot C_U and C_L as two separate lines.
8. When C_U exceeds the UCL, then an upward shift has occurred. When C_L drops below LCL, then a downward shift has occurred.

Moving Average Control Charts

Another special chart is the moving average chart. This chart is a chronological plot of the moving average, which is calculated as the average value updated by dropping the oldest individual measurement and adding the newest individual measurement. Thus, a new average is calculated with each individual measurement. A further refinement is the exponentially weighted moving average (EWMA) chart. In the EWMA chart, the observations are weighted, and the highest weight is given to the most recent data. Moving average charts are effective in detecting small shifts, highlighting trends, and using data in processes in which it takes a long time to produce a single item.

Box-Jenkins Manual Adjustment Chart

Still another chart is the Box-Jenkins manual adjustment chart. The average and range, CUMSUM, and EWMA charts for variables focus on monitoring a process and reducing variability due to special causes of variation identified by the charts. Box-Jenkins charts have a different objective: to analyze process data to regulate the process after each observation and thereby minimize process variation. For elaboration on this advanced technique, see Box and Luceño (1997).

Multivariate Control Charts

Finally, we consider the concept of multivariate control charts. When there are two or more quality characteristics on a unit of product, these could be monitored independently with separate control charts. Then the probability that a sample average on either control chart exceeds three sigma limits is 0.0027. But the joint probability that both variables exceed their control limits simultaneously when they are both in control is (0.0027)(0.0027) or 0.00000729, which is much smaller than 0.0027. The situation becomes more distorted as the number of characteristics increases. For this and other reasons, monitoring several characteristics independently can be misleading. Multivariate control charts and statistics (e.g., Hotelling's T^2 charts, multivariate EWMA) address this issue. See Montgomery (2000, Section 8.4) for a highly useful discussion.

Process Capability

In planning the quality aspects of operations, nothing is more important than advance assurance that the processes will meet the specifications. In recent decades, a concept of process capability has emerged to provide a quantified prediction of process adequacy. This ability to predict quantitatively has resulted in widespread adoption of the concept as a major element of quality planning. Process capability is the measured, inherent variation of the product turned out by a process.

Basic definitions. Each key word in this definition must itself be clearly defined because the concept of capability has an enormous extent of application, and nonscientific terms are inadequate for communication within the industrial community.

- Process refers to some unique combination of machine, tools, methods, materials, and people engaged in production. It is often feasible and illuminating to separate and quantify the effect of the variables entering this combination.
- Capability refers to an ability, based on tested performance, to achieve measurable results.
- Measured capability refers to the fact that process capability is quantified from data that, in turn, are the results of measurement of work performed by the process.
- Inherent capability refers to the product uniformity resulting from a process that is in a state of statistical control (i.e., in the absence of time-to-time "drift" or other assignable causes of variation). "Instantaneous reproducibility" is a synonym for inherent capability.
- The product is measured because product variation is the end result.

Uses of process capability information. Process capability information serves multiple purposes:

- Predicting the extent of variability that processes will exhibit. Such capability information, when provided to designers, provides important information in setting realistic specification limits.
- Choosing from among competing processes that are most appropriate to meet the tolerances.
- Planning the interrelationship of sequential processes. For example, one process may distort the precision achieved by a predecessor process, as in hardening of gear teeth. Quantifying the respective process capabilities often points the way to a solution.
- Providing a quantified basis for establishing a schedule of periodic process control checks and readjustments.
- Assigning machines to classes of work for which they are best suited.

- Testing theories of causes of defects during quality improvement programs.
- Serving as a basis for specifying the quality performance requirements for purchased machines.

These purposes account for the growing use of the process capability concept.

Planning for a process capability study. Capability studies are conducted for various reasons, for example, to respond to a customer request for a capability index number or to evaluate and improve product quality. Prior to data collection, clarify the purpose for making the study and the steps needed to ensure that it is achieved.

In some cases, the capability study will focus on determining a histogram and capability index for a relatively simple process. Here the planning should ensure that process conditions (e.g., temperature, pressure) are completely defined and recorded. All other inputs must clearly be representative (i.e., specific equipment, material, and, of course, personnel).

For more complex processes or when defect levels of 1 to 10 parts per million are desired, the following steps are recommended:

1. Develop a process description, including inputs, process steps, and output quality characteristics. This description can range from simply identifying the equipment to developing a mathematical equation that shows the effect of each process variable on the quality characteristics.
2. Define the process conditions for each process variable. In a simple case, this step involves stating the settings for temperature and pressure. But for some processes, it means determining the optimum value or aim of each process variable. The statistical design of experiments provides the methodology. Also, determine the operating ranges of the process variables around the optimum because the range will affect the variability of the product results.
3. Make sure that each quality characteristic has at least one process variable that can be used to adjust it.
4. Decide whether measurement error is significant. This can be determined from a separate error of measurement study. In some cases, the error of measurement can be evaluated as part of the overall study.
5. Decide whether the capability study will focus only on variability or will also include mistakes or errors that cause quality problems.
6. Plan for the use of control charts to evaluate the stability of the process.
7. Prepare a data collection plan, including adequate sample size that documents results on quality characteristics along with the process conditions (e.g., values of all process variables) and preserves information on the order of measurements so that trends can be evaluated.
8. Plan which methods will be used to analyze data from the study to ensure that before starting the study, all necessary data for the analysis will be available. The analyses should include process capability calculations on variability and also analysis of attribute or categorical data on mistakes and analysis of data from statistically designed experiments built into the study.
9. Be prepared to spend time investigating interim results before process capability calculations can be made. These investigations can include analysis of optimum values and ranges of process variables, out-of-control points on control charts, or other unusual results. The investigations then lead to the ultimate objective, that is, improvement of the process.

Note that these steps focus on improvement rather than just on determining a capability index.

Standardized process capability formula. The most widely adopted formula for process capability is:

$$\text{Process capability} = \pm 3\sigma \text{ (a total of } 6\sigma\text{)}$$

where σ is the standard deviation of the process under a state of statistical control (i.e., under no drift and no sudden changes). If the process is centered at the nominal specification and follows a normal probability distribution, 99.73 percent of production will fall within 3σ of the nominal specification.

Relationship to product specifications. A major reason for quantifying process capability is to compute the ability of the process to hold product specifications. For processes that are in a state of statistical control, a comparison of the variation of 6*s* to the specification limits permits ready calculation of percentage defective by conventional statistical theory.

Planners try to select processes with the 6*s* process capability well within the specification width. A measure of this relationship is the capability ratio:

$$C_p = \text{capability ratio} = \frac{\text{specification range}}{\text{process capability}} = \frac{\text{USL} - \text{LSL}}{6s}$$

where USL is the upper specification limit and LSL is the lower specification limit.

Note that 6*s* is used as an estimate of 6σ.

Some companies define the ratio as the reciprocal. Some industries now express defect rates in terms of parts per million. A defect rate of one part per million requires a capability ratio (specification range over process capability) of about 1.63.

Figure 19.30 shows four of many possible relations between process variability and specification limits and the likely courses of action for each. Note that in all of these cases, the average of the process is at the midpoint between the specification limits.

Table 19.20 shows selected capability ratios and the corresponding level of defects, assuming that the process average is midway between the specification limits. A process that is just meeting specification limits (specification range ±3σ) has a C_p of 1.0. The criticality of many applications and the reality that the process average will not remain at the midpoint of the specification range suggest that C_p should be at least 1.33. Note that a process operating at $C_p = 2.0$ over the short term (and centered midway between the specification limits) will correspond to a process sigma capability measure of $3C_p$, or 6 sigma (allowing for a 1.5s shift over the long term. This corresponds to 6*s* – 1.5*s* = 4.5*s*, which is expected to produce 3.4 ppm outside of the two-sided specification limits over the long term).

Note that the C_p index measures whether the process variability can fit within the specification range. It does not indicate whether the process is actually running within the specification because the index does not include a measure of the process average (this issue is addressed by another measure, C_{pk}).

Three capability indexes commonly in use are shown in Table 19.21. Of these, the simplest is C_p. The higher the value of any indexes, the lower the amount of product outside the specification limits.

Pignatiello and Ramberg (1993) provide an excellent discussion of various capability indexes. Bothe (1997) provides a comprehensive reference book that includes extensive discussion of the mathematical aspects. These references explain how to calculate confidence bounds for various process capability indexes.

The C_{pk} capability index. Process capability, as measured by C_{pk}, refers to the variation in a process about the average value. This concept is illustrated in Figure 19.31. The two processes have equal capabilities (C_p) because 6σ is the same for each distribution, as indicated by the

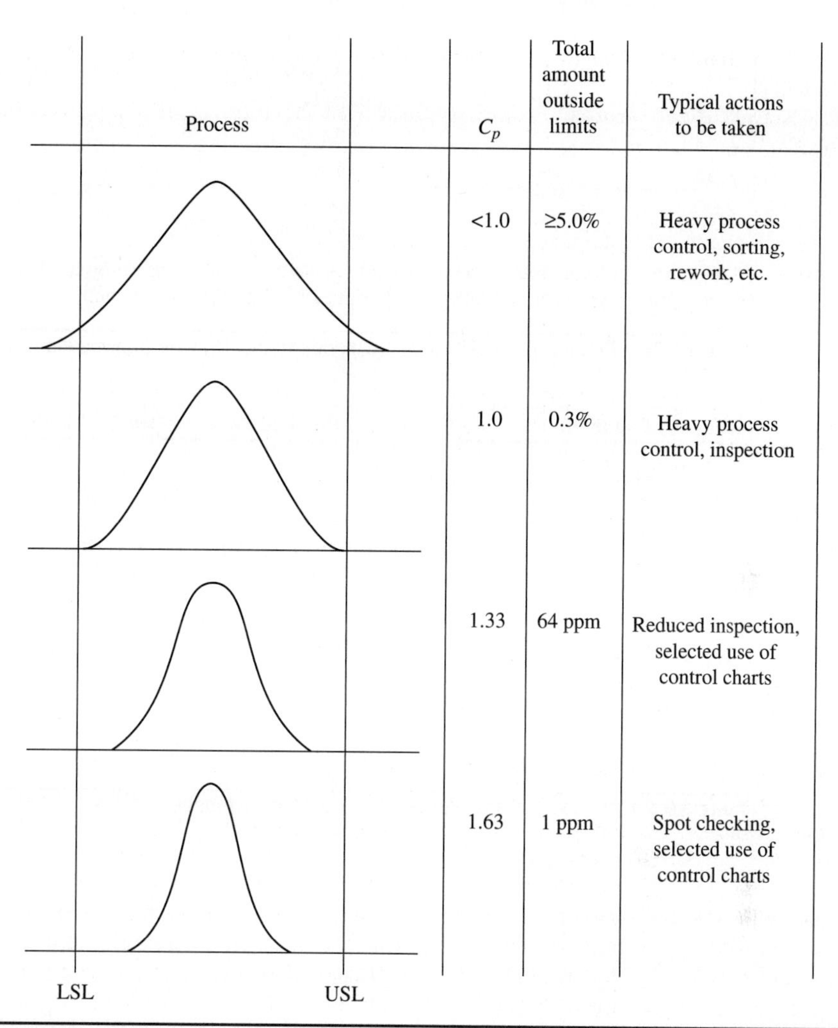

FIGURE 19.30 Four examples of process variability. (*Quality Planning and Analysis, Copyright 2007. Used by permission.*)

widths of the distribution curves. The process aimed at μ2 is producing defectives because the aim is off center, not because of the inherent variation about the aim (i.e., the capability).

Thus, the C_p index measures potential capability, assuming that the process average is equal to the midpoint of the specification limits and the process is operating in statistical control; because the average is often not at the midpoint, it is useful to have a capability index that reflects both variation and the location of the process average. Such an index is C_{pk}.

C_{pk} reflects the current process mean's proximity to either the USL or LSL. C_{pk} is estimated by

$$\hat{C}_{pk} = \min\left[\frac{\bar{X} - \text{LSL}}{3s}, \frac{\text{USL} - \bar{X}}{3s}\right]$$

In an example from Kane (1986),

$\text{USL} = 20 \quad \bar{X} = 16$

$\text{LSL} = 8 \quad s = 2$

Process Capability Index (C_p)	Total Product Outside Two-Sided Specification Limits*
0.5	13.36%
0.67	4.55%
1.00	0.3%
1.33	64 ppm
1.63	1 ppm
2.00	0

*Assuming that the process is centered midway between the specification limits.
(*Source: Quality Planning and Analysis,* Copyright 2007. Used by permission.)

TABLE 19.20 Process Capability index (C_p) and Product Outside Specification Limits

Process Capability	Process Performance
$C_p = \frac{USL - LSL}{6\sigma}$	$P_p = \frac{USL - LSL}{6s}$
$C_{pk} = \min\left[\frac{USL - \mu}{3\sigma}, \frac{\mu - LSL}{3\sigma}\right]$	$P_{pk} = \min\left[\frac{USL - \bar{X}}{3s}, \frac{\bar{X} - LSL}{3s}\right]$
$C_{pm} = \frac{USL - LSL}{6\sqrt{\sigma^2 + (\mu - T)^2}}$	$P_{pm} = \frac{USL - LSL}{6\sqrt{s^2 + (\bar{X} - T)^2}}$

(*Source: Quality Planning and Analysis,* Copyright 2007. Used by permission.)

TABLE 19.21 Process Capability and Process Performance Indexes

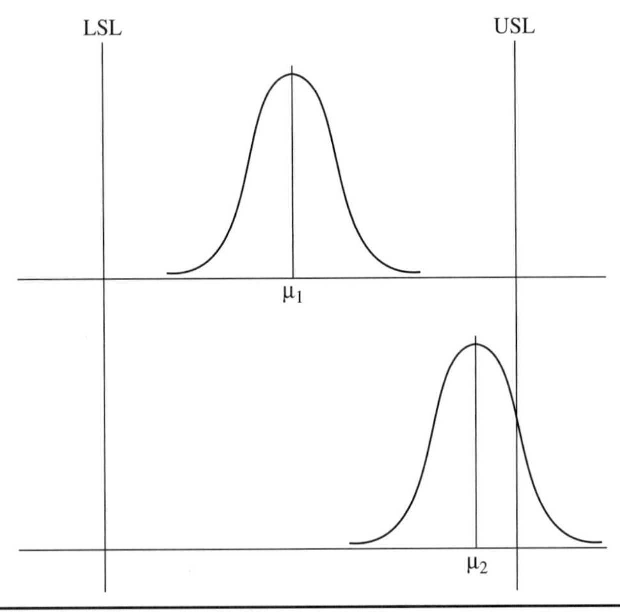

FIGURE 19.31 Process with Equal Process Capability but Different Aim. (*Quality Planning and Analysis. Copyright 2007. Used by permission.*)

The standard capability ratio is estimated as

$$\frac{\text{USL} - \text{LSL}}{6\sigma} = \frac{20-8}{12} = 1.0$$

which implies that if the process were centered between the specification limits (at 14), then only a small proportion (about 0.27 percent) of product would be defective.

However, when we calculate C_{pk}, we obtain

$$\hat{C}_{pk} = \min\left[\frac{16-8}{6}, \frac{20-16}{12}\right] = 0.67$$

which indicates that the process mean is currently nearer the USL. (Note that if the process were centered at 14, the value of C_{pk} would be 1.0.) An acceptable process will require reducing the standard deviation and/or centering the mean. Also note that if the actual average is equal to the midpoint of the specification range, then $C_{pk} = C_p$.

The higher the value of C_p, the lower the amount of product outside specification limits. In certifying suppliers, some organizations use C_{pk} as one element of certification criteria. In these applications, the value of C_{pk} desired from suppliers can be a function of the type of commodity purchased.

A capability index can also be calculated around a target value rather than the actual average. This index, called C_{pm} or the Taguchi index, focuses on reduction of variation from a target value rather than reduction of variability to meet specifications.

Most capability indexes assume that the quality characteristic is normally distributed. Krishnamoorthi and Khatwani (2000) propose a capability index for handling normal and nonnormal characteristics by first fitting the data to a Weibull distribution.

Two types of process capability studies are as follows:

1. *Study of process potential.* In this study, an estimate is obtained of what the process can do under certain conditions (i.e., variability under short-run defined conditions for a process in a state of statistical control). The C_p index estimates the potential process capability.
2. *Study of process performance.* In this study, an estimate of capability provides a picture of what the process is doing over an extended period. A state of statistical control is also assumed. The C_{pk} index estimates the performance capability.

Estimating inherent or potential capability from control chart analysis. In a process potential study, data are collected from a process operating without changes in material batches, workers, tools, or process settings. This short-term evaluation uses consecutive production over one time period. Such an analysis should be preceded by a control chart analysis in which any assignable causes have been detected and eliminated from the process.

Because specification limits usually apply to individual values, control limits for sample averages cannot be compared to specification limits. To make a comparison, we must first convert R to the standard deviation for individual values, calculate the $3s$ limits, and compare them to the specification limits. This process is explained below.

If a process is in statistical control, it is operating with the minimum amount of variation possible (the variation due to chance causes). If, and only if, a process is in statistical control, the following relationship holds for using s as an estimate of σ:

$$s = \frac{\bar{R}}{d_2}$$

Tables 19.22 and 19.23 provide values of d_2. If the standard deviation is known, process capability limits can be set at $\pm 3\sigma$, and this value used as an estimate of 3σ.

Factors for $\bar{X}$ and R Control Charts;* Factors for Estimating s from R†				
Number of Observations in Sample	A_2	D_3	D_4	Factor for Estimate from $\bar{R}: d_2 = \bar{R}/s$
2	1.880	0	3.268	1.128
3	1.023	0	2.574	1.693
4	0.729	0	2.282	2.059
5	0.577	0	2.114	2.326
6	0.483	0	2.004	2.534
7	0.419	0.076	1.924	2.704
8	0.373	0.136	1.864	2.847
9	0.337	0.184	1.816	2.970
10	0.308	0.223	1.777	3.078
11	0.285	0.256	1.744	3.173
12	0.266	0.284	1.717	3.258
13	0.249	0.308	1.692	3.336
14	0.235	0.329	1.671	3.407
15	0.223	0.348	1.652	3.472

$$\begin{cases} \text{Upper control limit for } \bar{X} = \text{UCL}_{\bar{X}} = \bar{\bar{X}} + A_2\bar{R} \\ \text{Lower control limit for } \bar{X} = \text{LCL}_{\bar{X}} = \bar{\bar{X}} - A_2\bar{R} \end{cases}$$

$$\begin{cases} \text{Upper control limit for } R = \text{UCL}_R = D_4\bar{R} \\ \text{Lower control limit for } R = \text{LCR}_R = D_3\bar{R} \end{cases}$$

$s = \bar{R}/d_2$

From *1950 ASTM Manual on Quality Control of Materials* and *ASTM Manual on Presentation of Data, 1945*. American Society for Testing and Materials. Copyright ASTM International. Reprinted with permission. (*Source: Quality Planning and Analysis*, Copyright 1997. Used by permission.)

TABLE 19.22 Factors for $\bar{X}$ and R Control Charts

n	A_2	D_3	D_4	d_2
2	1.880	0	3.268	1.128
3	1.023	0	2.574	1.693
4	0.729	0	2.282	2.059
5	0.577	0	2.114	2.326
6	0.483	0	2.004	2.534
7	0.419	0.076	1.924	2.704
8	0.373	0.136	1.864	2.847
9	0.337	0.184	1.816	2.970
10	0.308	0.223	1.777	3.079

(*Source: Quality Planning and Analysis*, Copyright 2007. Used by permission.)

TABLE 19.23 Constants for $\bar{X}$ and R Chart

For the data shown in Figure 19.32 (machine N-5),

$$s = \frac{\bar{R}}{d_2} = \frac{6.0}{2.534} = 2.37$$

and

$$\pm 3s = \pm 3(2.37) = 7.11$$

or

$$6s = 14.22 \text{ (or 0.0124 in the original data units)}$$

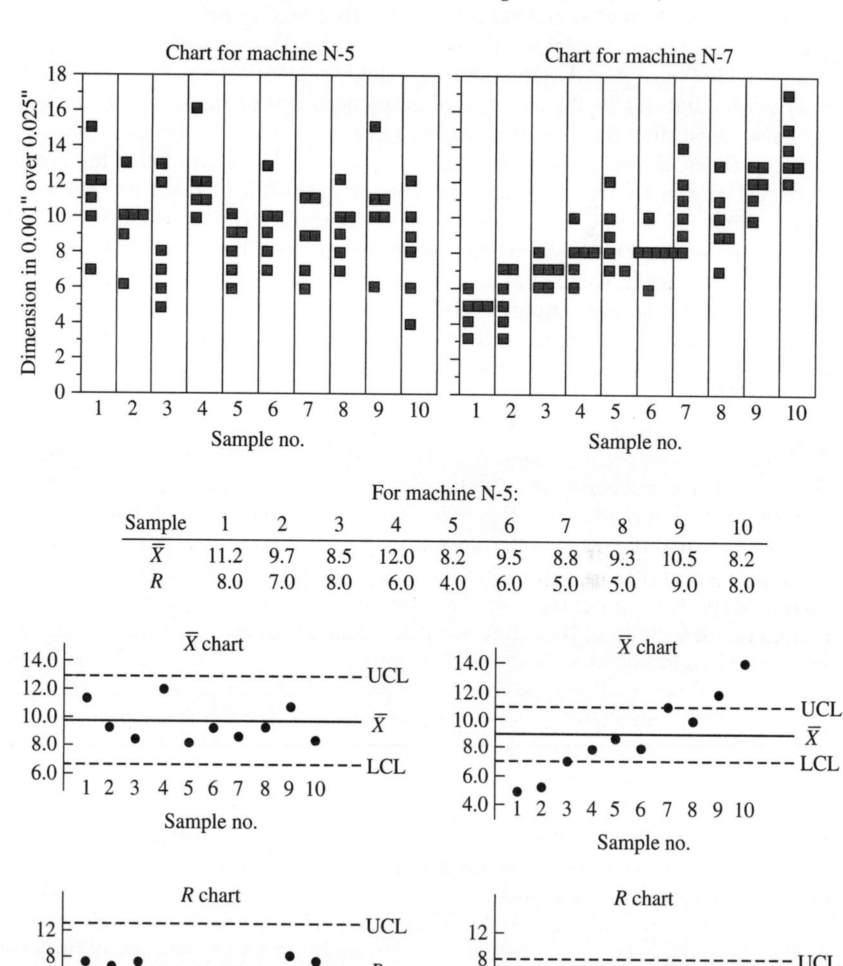

For machine N-5:

Sample	1	2	3	4	5	6	7	8	9	10
$\bar{X}$	11.2	9.7	8.5	12.0	8.2	9.5	8.8	9.3	10.5	8.2
R	8.0	7.0	8.0	6.0	4.0	6.0	5.0	5.0	9.0	8.0

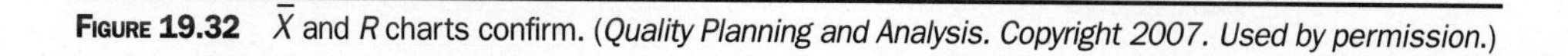

FIGURE 19.32 $\bar{X}$ and R charts confirm. (*Quality Planning and Analysis. Copyright 2007. Used by permission.*)

The specification limit was 0.258 ± 0.005.

Thus,

USL = 0.263

LSL = 0.253

Then

$$C_p = \frac{USL - LSL}{6s} = \frac{0.263 - 0.253}{0.0142} = 0.72$$

Even if the process is perfectly centered at 0.258 (and it was not), it is not capable.

The assumption of statistical control and its effect on process capability. All statistical predictions assume a stable population. In a statistical sense, a stable population is one that is repeatable (i.e., a population that is in a state of statistical control). The statistician rightfully insists that this be the case before predictions can be made. The manufacturing engineer also insists that the process conditions (feeds, speeds, etc.) be fully defined.

In practice, the original control chart analysis will often show that the process is out of statistical control. (It may or may not be meeting product specifications.) However, an investigation may show that the causes cannot be economically eliminated from the process. In theory, process capability should not be predicted until the process is in statistical control. However, in practice, some kind of comparison of capability to specifications is needed. The danger in delaying the comparison is that the assignable causes may never be eliminated from the process. The resulting indecision will thereby prolong interdepartmental bickering on whether "the specification is too tight" or "manufacturing is too careless."

A good way to start is by plotting individual measurements against specification limits. This step may show that the process can meet the product specifications even with assignable causes present. If a process has assignable causes of variation but is able to meet the specifications, usually no economic problem exists. The statistician can properly point out that a process with assignable variation is unpredictable. This point is well taken, but in establishing priorities of quality improvement efforts, processes that are meeting specifications are seldom given high priority.

If a process is out of control and the causes cannot be economically eliminated, the standard deviation and process capability limits can nevertheless be computed (with the out-of-control points included). These limits will be inflated because the process will not be operating at its best. In addition, the instability of the process means that the prediction is approximate.

It is important to distinguish between a process that is in a state of statistical control and a process that is meeting specifications. A state of statistical control does not necessarily mean that the product from the process conforms to specifications. Statistical control limits on sample averages cannot be compared to specification limits because specification limits refer to individual units. For some processes that are not in control, the specifications are being met and no action is required; other processes are in control, but the specifications are not being met, and action is needed.

In summary, we need processes that are both stable (in statistical control) and capable (meeting product specifications).

The increasing use of capability indexes has also led to the failure to understand and verify some important assumptions that are essential for statistical validity of the results. Five key assumptions are:

1. *Process stability.* Statistical validity requires a state of statistical control with no drift or oscillation.
2. *Normality of the characteristic being measured.* Unless nonparametric methods or alternative distributions are used, normality is needed to draw statistical inferences about the population.

3. *Sufficient data.* Sufficient data are necessary to minimize the sampling error for the capability indexes.
4. *Representativeness of samples.* Random samples must be included.
5. *Independent measurements.* Consecutive measurements cannot be correlated.

These assumptions are not theoretical refinements—they are important conditions for properly applying capability indexes. Before applying capability indexes, readers are urged to read the paper by Pignatiello and Ramberg (1993). It is always best to compare the indexes with the full data versus specifications depicted in a histogram.

Measuring process performance. A process performance study collects data from a process that is operating under typical conditions but includes normal changes in material batches, workers, tools, or process settings. This study, which spans a longer term than the process potential study, also requires that the process be in statistical control.

The capability index for a process performance study is

$$C_{pk} = \min\left[\frac{\bar{X} - LSL}{3s}, \frac{USL - \bar{X}}{3s}\right]$$

Problem Consider a pump cassette used to deliver intravenous solutions (Baxter Travenol Laboratories, 1986). A key quality characteristic is the volume of solution delivered in a predefined time. The specification limits are

$$USL = 103.5 \quad LSL = 94.5$$

A control chart was run for one month, and no out-of-control points were encountered. From the control chart data, we know that

$$\bar{X} = 98.2 \text{ and } s = 0.98$$

Figure 19.33 shows the process data and the specification limits.

Solution The capability index is

$$C_{pk} = \min\left[\frac{98.2-94.5}{3(0.98)}, \frac{103.5-98.2}{3(0.98)}\right]$$

$$C_{pk} = 1.26$$

For many applications, 1.26 is an acceptable value of C_{pk}.

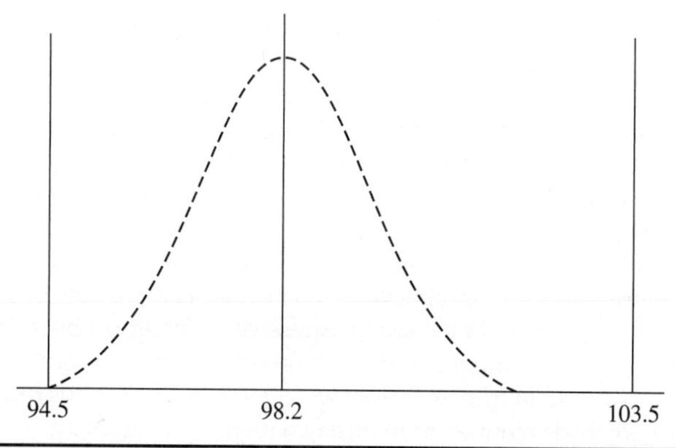

Figure 19.33 Delivered volume of solution.

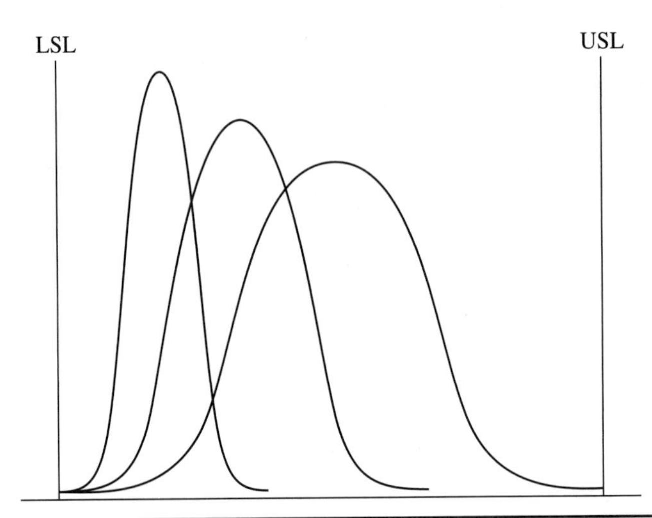

Figure 19.34 Three Processes with $C_{pk} = 1$. (*Quality Planning and Analysis. Copyright 2007. Used by permission.*)

Interpretation of C_{pk}. In using C_{pk} to evaluate a process, we must recognize that C_{pk} is an abbreviation of two parameters—the average and the standard deviation. Such an abbreviation can inadvertently mask important detail in these parameters. For example, Figure 19.34 shows that three extremely different processes can all have the same C_{pk} (in this case $C_{pk} = 1$).

Increasing the value of C_{pk} may require a change in the process average, the process standard deviation, or both. For some processes, increasing the value of C_{pk} by changing the average value (perhaps by a simple adjustment of the process aim) may be easier than reducing the standard deviation (by investigating the many causes of variability). The histogram of the process should always be reviewed to highlight both the average and the spread of the process.

Note that Table 19.21 also includes the capability index C_{pm}. This index measures the capability around a target value T rather than the mean value. When the target value equals the mean value, the C_{pm} index is identical to the C_{pk} index.

Attribute (or categorical) data analysis. The methods discussed earlier assume that numerical measurements are available from the process. Sometimes, however, the only data available are in attribute or categorical form (i.e., the number of nonconforming units and the number acceptable).

The data in Table 19.24 on errors in preparing insurance policies also can be used to illustrate process capability for attribute data. The data reported 80 errors from six policy writers, or 13.3 errors per writer—the current performance. The process capability can be calculated by excluding the abnormal performance identified in the study—type 3 errors by worker B, type 5 errors, and errors of worker E. The error data for the remaining five writers becomes 4, 3, 5, 2, and 5, with an average of 3.8 errors per writer. The process capability estimate of 3.8 compares with the original performance estimate of 13.3.

This example calculates process capability in terms of errors or mistakes rather than the variability of a process parameter. Hinckley and Barkan (1995) point out that in many processes, nonconforming product can be caused by excessive variability or by mistakes (e.g., missing parts, wrong parts, wrong information, or other processing errors). For some processes, mistakes can be a major cause of failing to meet customer quality goals. The actions required to reduce mistakes are different from those required to reduce variability of a parameter.

Readers are directed to DeVor et al. (1992) for a good background in process control charting.

	Policy Writer						
Error Type	A	B	C	D	E	F	Total
1	0	0	1	0	2	1	4
2	1	0	0	0	1	0	2
3	0	(16)	1	0	2	0	(19)
4	0	0	0	0	1	0	1
5	2	1	3	1	4	2	(13)
6	0	0	0	0	3	0	3
.							
.							
.							
27							
28							
29							
Total	6	(20)	8	3	(36)	7	80

(*Source: Quality Planning and Analysis*, Copyright 2007. Used by permission.)

TABLE 19.24 Matrix of Errors by Insurance Policy Writers

Software

While many of the tools mentioned in this chapter can be applied using programs such as Microsoft Excel, numerous software packages are available that provide more specialized assistance. Some of these packages and vendors are listed here, according to their primary emphasis. Most vendors have multiple software options.

Basic statistics:

- QI Macros
- SigmaXL
- StatPlus

Advanced statistics:

- JMP
- Minitab
- Systat

Design of experiments:

- StatSoft STATISTICA
- Stat-Ease
- STRATEGY
- Statgraphics

Monte Carlo, discrete event simulation:

- @Risk
- Crystal Ball
- iGrafx

Reliability, availability:

- Isograph
- Relex 2009
- ReliaSoft

Control charting:

- CHARTRunner
- Statit

References

Anand, K. N. (1996), The Role of Statistics in Determining Product and Part Specifications: A Few Indian Experiences, *Quality Engineering*, vol. 9, no. 2, pp. 187–193.

Automotive Industry Action Group (2003). *Measurement Systems Analysis* (3rd ed.). Southfield, MI.

Barnett, V., and Lewis, T. (1994). *Outliers in Statistical Data* (3rd ed.). John Wiley & Sons, New York.

Bender, A. (1975). Statistical Tolerancing as It Relates to Quality Control and the Designer, *Automotive Division Newsletter of ASQC*, April, p. 12

Bothe, D. R. (1997). *Measuring Process Capability*. McGraw-Hill, New York.

Box, G. E. P., and Luceno, A. (1997). *Statistical Control by Monitoring and Adjustment*. Wiley, New York.

Box, G. E. P., and Draper, N. R. (1969). *Evolutionary Operation: A Statistical Method for Process Improvement*. John Wiley & Sons, New York.

Box, G. E. P., Hunter, J. S., and Hunter, W. G. (2005). *Statistics for Experimenters: Design, Innovation and Discovery* (2nd ed.). Wiley-Interscience, Hoboken, NJ.

Burdick, R. K., and Larsen, G. A. (1997). Confidence Intervals on Measures of Variability in R&R Studies, *Journal of Qualitiy Technology*, vol. 29, no. 3, pp. 261–273.

Carter, C. W. (1996). Sequenced Levels Experimental Designs, *Quality engineering*, vol. 8, no. 1, (pp. 181 -188), no. 2 (pp. 361-366), no. 3 (pp. 499–504), no.4 (pp. 695–698).

Case, K. E., Bennett, G. K., and Schmidt, J. W. (1975). "The Effect of Inspector Error on Average Outgoing Quality," *Journal of Quality Technology*, vol. 7, no. 1, pp. 1–12.

Coleman, S., Greenfield, T., Stewardson, D., and Montgomery, D. C. (2008). *Statistical Practice in Business and Industry*. John Wiley & Sons, Hoboken, NJ. (See Chapter 13).

Davison, A. C., and Hinkley, D. (2006). *Bootstrap Methods and Their Applications* (8th ed.). Cambridge: Cambridge Series in Statistical and Probabilistic Mathematics, Davison Hinkley, Cambridge University Press, Cambridge.

del Castillo, E. (2007). *Process Optimization: A Statistical Approach*. Springer Science and Business Media, New York.

DeVor, R. E., Chang, T., and Sutherland, J. W. (1992). *Statistical Quality Design and Control: Contemporary Concepts and Methods*. Prentice Hall, Upper Saddle River, NJ.

Dodson, B. (1999). Reliability Modeling with Spreadsheets, *Proceedings of the Annual Quality Congress*, ASQ, Milwaukee, pp. 575–585.

Eagle, A. R. (1954). A Method for Handling Errors in Testing and Measurement, *Industrial Quality Control*, March, pp. 10–14.

Emory, W. C., and Cooper, D. R. (1991). *Business Research Methods* (4th ed.). Boston: Irwin/ McGrawHill.

Engel, J., and DeVries, B. (1997). Evaluating a Well-Known Criterion for Measurement Precision, *Journal of Quality Technology*, vol. 29, no. 4, pp. 469–476.

Gomer, P. (1998). Design for Tolerancing of Dynamic Mechanical Assemblies, *Annual Quality Congress Proceedings*, ASQ, Milwaukee, pp. 490–500.

Graves, S. B. (1997). How to Reduce Costs Using a Tolerance Analysis Formula Tailored to your Organization, Report no. 157, Center for Quality and Productivity Improvement, University of Wisconsin, Madison.

Griffith, G. K. (1996). *Statistical Process Control Methods for Long and Short Runs*, 2nd ed., ASQ Quality Press, Milwaukee.

Gryna, F. M., Chua, R. C., and De Feo, J. A. (2007). *Juran's Quality Planning and Analysis* (5th ed.). McGraw Hill, New York.

Hinckley, C. M., and Barkan, P. (1995). The Role of Variation, Mistakes, and Complexity in Producing Nonconformities, *Journal of Quality Technology*, vol. 27, no. 3, pp. 242–249.

Hoag, L. L., Foote, G. L., and Mount-Cambell, C. (1975). The Effect of Inspector Accuracy on Type I and II Errors of Common Sampling Techniques, *Journal of Quality Technology*, vol. 7, no. 4, pp. 157–164.

Hough, L. D., and Pond, A. D. (1995). Adjustable Individual Control Charts for Short Runs. *Proceedings of the 40th Annual Quality Congress*, ASQ, Milwaukee, pp. 1117–1125.

Ireson, W. G., Coombs, C. F., Jr., and Moss, R. Y. (1996). *Handbook of Reliability Engineering and Management*, 2nd ed., McGraw-Hill, New York.

Early, J. F. Quality Improvement Tools, *The Power of Quality*, The Health Care Forum, June 1989.

Jones, J., and Hayes, J. (1999). A Comparison of Electronic Reliability Prediction Models, IEEE Transactions of Reliability, vol. 48, no. 2, pp. 127–134.

Kane, V. E. (1986). Process Capability Indices, *Journal of Quality Technology*, vol. 18, no. 1, pp. 41-52.

Krishnamoorthi, I. S., and Khatwani, S. (2000). *Statistical Process Control for Health Care*, Duxbury, Paciric Grove, CA.

Kutner, M., Nachtsheim, C., Neter, J., and Li, W. (2004). *Applied Linear Statistical Models* (2nd ed.). Irwin/McGraw-Hill, New York.

Kvam, P. H., and Vidakovic, B. (2007). *Nonparametric Statistics with Applications to Science and Engineering*. John Wiley & Sons, Hoboken, NJ.

Law, A. M., and Kelton, W. D. (2000). *Simulation Modeling and Analysis* (3rd ed.). McGraw-Hill.

Ledolter, J., and Swersey, A. (1997). An Evaluation of Pre-Control, *Journal of Quality Technology*, vol. 29, no. 1, pp. 163–171.

Ledolter, J., and Swersey, A. J. (2007). *Testing 1-2-3: Experimental Design with Applications in Marketing and Service Operations*. Stanford University Press, Palo Alto, CA.

Meeker, W. Q., and Escobar, L. A. (1998). *Statistical Methods for Reliability Data*. John Wiley & Sons, 1998.

Meeker, W. Q., and Escobar, L. A. (1998). *Statistical Methods for Reliability Data*. John Wiley & Sons, New York.

Montgomery, D. C. (2000). *Introduction to Statistical Quality Control*, 4th ed., Wiley, New York, NY.

Myers, R. H., Montgomery, D. C., and Anderson-Cook, C. M. (2009). *Response Surface Methodology: Process and Product Optimization Using Designed Experiments*. John Wiley & Sons, Hoboken, NJ.

Nelson, L. S. (1989). Standardization of Shewhart Control Charts, *Journal of Quality Technology*, vol. 21, 287–289.

O'Connor, P. D. T. (1995). *Practical Reliability Engineering*, 3rd Ed. rev., John Wiley and Sons, New York.

Pignatiello, J. H., Jr., and Ramberg, J. S. (1993). Process Capability Indices: Just Say No, *ASQC Quality Congress Transactions 1993*, American Society for Quality, Milwaukee.

Pyzdek, T. (1993). Process Control for Short and Small Runs, *Quality Progress*, April, pp. 51–60.
Ross, P. J. (1996). *Taguchi Techniques for Quality Engineering*. McGraw- Hill, New York.
Sprent, P., and Smeeton, N. C. (2001). *Applied Nonparametric Statistical Methods* (3rd ed.). Chapman and Hall/CRC Press, Boca Raton, FL.
Steiner, S. H. (1997). Pre-Control and some Simple Alternatives, *Quality Engineering*, vol. 10, no. 1, pp. 65–74.
Tsai, P. (1988). Variable Gauge Repeatability and Reproducibility Study Using the Analysis of Variance Method, *Quality Engineering*, vol. 1, no. 1, pp. 107–115.
Wheeler, D. J. (1991). *Short Run SPC*. SPC Press, Inc, Knoxville, TN.
Wong, D., and Baker, C. (1988). Pain in Children: Comparison of Assessment Scales, *Pediatric Nursing*, vol. 14, no. 1, pp. 9–17, 1988.
Young, F., Malero-Mora, P., and Friendly, M. (2007). *Visual Statistics: Seeing Data with Dynamic Interactive Graphs*. John Wiley & Sons, Hoboken, NJ.

Reference Charts for Table 19.3

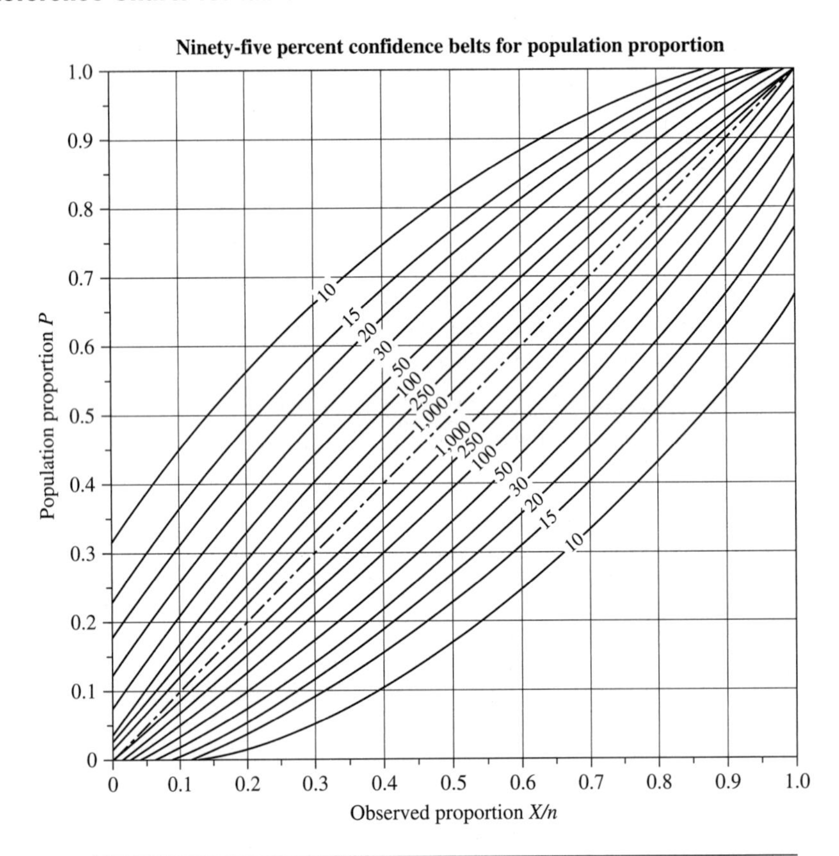

Example in a sample of 10 items, 8 were defective ($X/n - 8/10$). The 95% confidence limits on the population proportion defective are read from the two curves (for $n - 10$) as 0.43 and 0.98.

Binomial Distribution*

Probability of r or fewer occurrences of an event in n trials, where p is the probability of occurrence on each trial.

		P									
n	*r*	0.05	0.10	0.15	0.20	0.25	0.30	0.35	0.40	0.45	0.50
2	0	0.9025	0.8100	0.7225	0.6400	0.5625	0.4900	0.4225	0.3600	0.3025	0.2500
	1	0.9975	0.9900	0.9775	0.9600	0.9375	0.9100	0.8775	0.8400	0.7975	0.7500
3	0	0.8574	0.7290	0.6141	0.5120	0.4219	0.3430	0.2746	0.2160	0.1664	0.1250
	1	0.9928	0.9720	0.9392	0.8960	0.8438	0.7840	0.7182	0.6480	0.5748	0.5000
	2	0.9999	0.9990	0.9966	0.9920	0.9844	0.9730	0.9571	0.9360	0.9089	0.8750
4	0	0.8145	0.6561	0.5220	0.4096	0.3164	0.2401	0.1785	0.1296	0.0915	0.0625
	1	0.9860	0.9477	0.8905	0.8192	0.7383	0.6517	0.5630	0.4752	0.3910	0.3125
	2	0.9995	0.9963	0.9880	0.9728	0.9492	0.9163	0.8735	0.8208	0.7585	0.6875
	3	1.0000	0.9999	0.9995	0.9984	0.9961	0.9919	0.9850	0.9744	0.9590	0.9375
5	0	0.7738	0.5905	0.4437	0.3277	0.2373	0.1681	0.1160	0.0778	0.0503	0.0312
	1	0.9774	0.9185	0.8352	0.7373	0.6328	0.5282	0.4284	0.3370	0.2562	0.1875
	2	0.9988	0.9914	0.9734	0.9421	0.8965	0.8369	0.7648	0.6826	0.5931	0.5000
	3	1.0000	0.9995	0.9978	0.9933	0.9844	0.9692	0.9460	0.9130	0.8688	0.8125
	4	1.0000	1.0000	0.9999	0.9997	0.9990	0.9976	0.9947	0.9898	0.9815	0.9688
6	0	0.7351	0.5314	0.3771	0.2621	0.1780	0.1176	0.0754	0.0467	0.0277	0.0156
	1	0.9672	0.8857	0.7765	0.6554	0.5339	0.4202	0.3191	0.2333	0.1636	0.1094
	2	0.9978	0.9842	0.9527	0.9011	0.8306	0.7443	0.6471	0.5443	0.4415	0.3438
	3	0.9999	0.9987	0.9941	0.9830	0.9624	0.9295	0.8826	0.8208	0.7447	0.6562
	4	1.0000	0.9999	0.9996	0.9984	0.9954	0.9891	0.9777	0.9590	0.9308	0.8906
	5	1.0000	1.0000	1.0000	0.9999	0.9998	0.9993	0.9982	0.9959	0.9917	0.9844
7	0	0.6983	0.4783	0.3206	0.2097	0.1335	0.0824	0.0490	0.0280	0.0152	0.0078
	1	0.9556	0.8503	0.7166	0.5767	0.4449	0.3294	0.2338	0.1586	0.1024	0.0625
	2	0.9962	0.9743	0.9262	0.8520	0.7564	0.6471	0.5323	0.4199	0.3164	0.2266
	3	0.9998	0.9973	0.9879	0.9667	0.9294	0.8740	0.8002	0.7102	0.6083	0.5000
	4	1.0000	0.9998	0.9988	0.9953	0.9871	0.9712	0.9444	0.9037	0.8471	0.7734
	5	1.0000	1.0000	0.9999	0.9996	0.9987	0.9962	0.9910	0.9812	0.9643	0.9375
	6	1.0000	1.0000	1.0000	1.0000	0.9999	0.9998	0.0994	0.9984	0.9963	0.9922
8	0	0.6634	0.4305	0.2725	0.1678	0.1001	0.0576	0.0319	0.0168	0.0084	0.0039
	1	0.9428	0.8131	0.6572	0.5033	0.3671	0.2553	0.1691	0.1064	0.0632	0.0352
	2	0.9942	0.9619	0.8948	0.7969	0.6785	0.5518	0.4278	0.3154	0.2201	0.1445
	3	0.9996	0.9950	0.9786	0.9437	0.8862	0.8059	0.7064	0.5941	0.4770	0.3633
	4	1.0000	0.9996	0.9971	0.9896	0.9727	0.9420	0.8939	0.8263	0.7396	0.6367

	5	1.0000	1.0000	0.9998	0.9988	0.9958	0.9887	0.9747	0.9502	0.9115	0.8555
	6	1.0000	1.0000	1.0000	0.9999	0.9996	0.9987	0.9964	0.9915	0.9819	0.9648
	7	1.0000	1.0000	1.0000	1.0000	1.0000	0.9999	0.9998	0.9993	0.9983	0.9961
9	0	0.6302	0.3874	0.2316	0.1342	0.0751	0.0404	0.0207	0.0101	0.0046	0.0020
	1	0.9288	0.7748	0.5995	0.4362	0.3003	0.1960	0.1211	0.0705	0.0385	0.0195
	2	0.9916	0.9470	0.8591	0.7382	0.6007	0.4628	0.3373	0.2318	0.1495	0.0898
	3	0.9994	0.9917	0.9661	0.9144	0.8343	0.7297	0.6089	0.4826	0.3614	0.2539
	4	1.0000	0.9991	0.9944	0.9804	0.9511	0.9012	0.8283	0.7334	0.6214	0.5000
	5	1.0000	0.9999	0.9994	0.9969	0.9900	0.9747	0.9464	0.9006	0.8342	0.7461
	6	1.0000	1.0000	1.0000	0.9997	0.9987	0.9957	0.9888	0.9750	0.9502	0.9102
	7	1.0000	1.0000	1.0000	1.0000	0.9999	0.9996	0.9986	0.9962	0.9909	0.9805
	8	1.0000	1.0000	1.0000	1.0000	1.0000	1.0000	0.9999	0.9997	0.9992	0.9980
10	0	0.5987	0.3487	0.1969	0.1074	0.0563	0.0282	0.0135	0.0060	0.0025	0.0010
	1	0.9139	0.7361	0.5443	0.3758	0.2440	0.1493	0.0860	0.0464	0.0232	0.0107
	2	0.9885	0.9298	0.8202	0.6778	0.5256	0.3828	0.2616	0.1673	0.0996	0.0547
	3	0.9990	0.9872	0.9500	0.8791	0.7759	0.6496	0.5138	0.3823	0.2660	0.1719
	4	0.9999	0.9984	0.9901	0.9672	0.9219	0.8497	0.7515	0.6331	0.5044	0.3770
	5	1.0000	0.9999	0.9986	0.9936	0.9803	0.9527	0.9051	0.8338	0.7384	0.6230
	6	1.0000	1.0000	0.9999	0.9991	0.9965	0.9894	0.9740	0.9452	0.8980	0.8281
	7	1.0000	1.0000	1.0000	0.9999	0.9996	0.9984	0.9952	0.9877	0.9726	0.9453
	8	1.0000	1.0000	1.0000	1.0000	1.0000	0.9999	0.9995	0.9983	0.9955	0.9893
	9	1.0000	1.0000	1.0000	1.0000	1.0000	1.0000	1.0000	0.9999	0.9997	0.9990

SECTION III

Applications: Most Important Methods in Your Industry

CHAPTER 20

Product-Based Organizations: Delivering Quality While Being Lean and Green

Steven M. Doerman

About This Chapter

The purpose of this chapter is to present the fundamentals of a sound operational quality system. It is intended to allow the quality professionals in an operational environment to reflect on their current approach to quality and look for opportunities for enhancement.

High Points of This Chapter

1. While products and processes are becoming more complex, the customer will continue to demand higher and higher levels of quality. To remain competitive, the manufacturer must continually and persistently focus on quality.

2. Planning for quality is essential. Planning must begin at the beginning of the product life cycle, at the design, and continue throughout the production process. It is critical for the planner to understand the key product parameters to meet the customer's need, the voice of the customer, and the relationship to the capability of the processes.
3. To achieve the highest level of quality within operations, people must be placed in a state of self-control and be provided with all they need to meet quality objectives. This is termed controllability and is at the heart of a sound quality management structure.
4. To ensure the product or service conforms to the customer's requirements and is suitable for the marketplace, a formal evaluation must occur. This may be designed around the production worker, self-control, or the traditional inspection and auditing process.
5. If a producer is to meet the ever-increasing requirements of the customer, operations must focus on continuous improvement. This requires an organizational culture that promotes quality enhancements and sustaining long-term results.

Quality in Operations in the Twenty-First Century

Operations is the nerve center of an organization—where the action is. The word "operations" as used in this handbook encompasses two areas: manufacture in the manufacturing sector and support activities in the service sector. In manufacturing industries, operations are those activities, typically carried out in a plant, that transform and add value to materials into the final product. In service industries, operations are those activities that process customer transactions but that do not involve direct contact with external customers (e.g., support activities of customer order preparation and payment processing). These two industry sectors have their own special needs. The discussion in this chapter covers both the planning and the execution of operations activities. Its purpose is to provide a framework to reflect upon your current operations and look for opportunity for advancement. This chapter and other chapters in this book referenced here will provide the understanding and guidance to achieve a competitive advantage and satisfied customers through the highest-quality products and services.

The power of managing quality and reducing costs through quality improvement is illustrated by Jay Williamson, vice president of Quality for Molex, Inc.: "Despite the challenging year (2008), we have managed to achieve our original savings goal. . . . This goal was set before anyone suspected the economy would fall. The process has been flexible enough to allow our organization to reprioritize and change focus just when we needed to make it happen."

Customer Demands for Higher Quality, Reduced Inventories, and Faster Response Time

As products and processes become more complex, new "world-class" quality levels are now common. For many products, quality levels of 1 to 3 percent defective are being replaced by 1 to 10 defects per million parts (or 3.4 defects per million as in the Six Sigma approach). Customers demand reduced inventory levels based on the "just-in-time" (JIT) production system. Under JIT, the concept of large lot sizes is challenged by reducing setup time, redesigning processes, and standardizing jobs. The results are smaller lot sizes and lower inventory. But JIT works only if product quality is high because little or no inventory exists to replace defective product. Finally, customers want faster response time from suppliers—to develop and manufacture new products. That faster response time puts pressure on the product development process and can result in inadequate review of new designs for product performance and for manufacturability. Collectively, these three parameters—quality, inventories, and response time—place a heavy burden on operations.

Agile Competition

An agile organization is able to respond to constantly changing customer opportunities. This characteristic means changing over from one product to another quickly; manufacturing goods, or providing service, to customer order in small lot sizes; customizing goods and services for individual customers; and using the expertise of people and facilities within the organization and among groups of cooperating organizations (partners). Goldman, et al. (1995) describe the concept and include examples. This concept includes the "virtual" organization—a group of organizations linked by an electronic network to enable the partners to satisfy a common customer objective. The virtual organization may be partially created by transferring complete functions to a supplier—outsourcing.

Impact of Technology

Technology (including computer information systems) is clearly improving quality by providing a wider variety of outputs and more consistent output. The infusion of technology makes some jobs more complex, thereby requiring extensive job skills and quality planning; technology also makes other jobs less complex but may contribute to job monotony. These issues suggest that quality during operations can no longer focus on inspection and checking but must respond to ever-increasing customer demands and changing competitive conditions.

Lean and Its Impact on Quality

Lean is the process of designing systems to reduce costs by eliminating product and process waste. This has a natural interrelationship to quality. Many of the philosophies and tools of lean complement quality and satisfy the customer. It is highly recommended that anyone interested in improving customer satisfaction through improved product quality also master the techniques of lean. For a complete introduction to the lean see Chapter 11, Lean Techniques: Improving Process Efficiency.

We now examine specific methods for planning, controlling, and improving quality during manufacturing and service operations.

Planning for Quality: Overview

Planning for quality before the execution of operations is seen as essential. International standards such as the ISO 9000 and ISO 14000 series provide a minimum framework for planning (for elaboration, see Chapter 16, Using International Standards to Ensure Organization Compliance). These standards cover important matters such as process control, inspection and testing, material control, product traceability, control of measuring equipment, control of nonconforming product, quality documentation, process environmental conditions, and the impact of processes on the external environment.

The responsibility for this planning varies by industry. In the mechanical and electronics industries, the work is usually performed within the manufacturing function by a specialist department (e.g., manufacturing engineering, process engineering). For process industries, the work is usually divided into two parts. Broad planning (e.g., type of manufacturing process) is performed within the research and development function; detailed planning is executed within the manufacturing function. Similarly, the service industries show variety in assigning the planning responsibility. For example, in support office operations of the financial services industry, the local operations manager handles the planning, whereas in the fast-foods industry, planning for food preparation is usually handled by a corporate planning function.

The main factors influencing the decision on responsibility are the complexities of the products being made, the anatomy of the manufacturing process, the technological literacy of the workforce, and the managerial philosophy of reliance on systems versus reliance on people.

Some industrialized countries delegate only a small amount of operational planning to departmental supervision or to the workforce. In the United States, this situation is largely a residue of the Taylor system of separating manufacturing planning from execution. This system gave rise to separate departments for manufacturing planning.

The Taylor system was proposed early in the twentieth century, at a time when the educational level of the workforce was low, while at the same time products and technology were becoming more complex. The system was so successful in improving productivity that it was widely adopted in the United States. It took firm root and remains as the dominant approach to operations planning not only interdepartmentally but within departments as well.

Times have changed. A major premise of the Taylor system, i.e., technological illiteracy of the workforce, is obsolete because of the dramatic increase in the educational levels of the workforce. Many organizations recognize that extensive job knowledge resides in the workforce and are taking steps to use that knowledge. Operations planning should be a collaborative effort in which the workforce has the opportunity to contribute to the planning. In the United States, this collaboration is slow-moving because of the widespread adoption of the Taylor system and the vested interests that have been created by that approach.

Initial Planning for Quality and New Product Introduction

Planning starts with a review of product designs (see Chapter 14, Continuous Innovation Using Design for Six Sigma). Then we review the process designs to identify key product and process characteristics, determine the importance of each product characteristic, analyze the process flow, understand the interrelationship of process variables, determine process capability, error-proof the process, validate the measurement process, and plan for operator self-control. These elements are discussed below.

Review of Product Designs

There is a clear advantage to having a new product design reviewed by operations personnel before the design is finalized for the marketplace. In practice, the extent of such review varies greatly—from essentially nothing ("tossing it over the wall" to the operations people) to a structured review using formal criteria and follow-up on open issues. Although a product design review often occurs during the design and development process (see Chapter 14, Continuous Innovation Using Design for Six Sigma), the emphasis is on the adequacy of field performance.

Review of product designs prior to release to operations must include an evaluation of producibility (or manufacturability). This evaluation includes the following issues:

1. Identification of key product and process characteristics.
2. Relative importance of various product characteristics.

3. Design for manufacturability.
4. Process robustness. A process is robust if it is flexible, easy to operate, and error-proof and its performance will tolerate uncontrollable variations in factors internal and external to the process. Such an ideal is approached by careful planning of all process elements, e.g., cross-functional training of personnel to cover vacations. For a discussion of robustness, see Snee (1993).
5. Availability of capable manufacturing processes to meet product requirements, i.e., processes that not only meet specifications but also do so with minimum variation.
6. Availability of capable measurement processes.
7. Identification of special needs for the product or service, e.g., handling, transportation, and storage during manufacture.
8. Material control, e.g., identification, traceability, segregation, contamination control.
9. Special skills required of operations personnel.

This review of the product design must be supplemented by a review of the process design. The process review includes producibility issues initially raised in the product design review.

Identification of Key Product and Process Characteristics

Key *product* characteristics are the features that a product has to meet customer needs. Key *process* characteristics are those that will create the key product characteristics. Product and process characteristics can be identified by using inputs from market research, quality function deployment, design review, and failure mode and effect analysis.

Relative Importance of Product Characteristics

Planners are better able to allocate resources when they know the relative importance of the many product characteristics.

One technique for establishing the relative importance is the identification of critical items (see Chapter 14, Continuous Innovation Using Design for Six Sigma). Critical items are the product characteristics that require a high level of attention to ensure that all requirements are met. One organization identifies "quality-sensitive parts" by using criteria such as part complexity and high-failure-rate parts. For such parts, special planning includes supplier involvement before and during the contract, process capability studies, reliability verification, and other activities.

Another technique is the classification of characteristics. Under this system, the relative importance of characteristics is determined and indicated on drawings and other documents. The classification can be simply "functional" (or "critical to quality") or "nonfunctional." Another system uses several degrees of importance such as critical, major, minor, and incidental. The classification uses criteria that reflect safety, operating failure, performance, service, and manufacture.

Analysis of the Process Flow

A process design can be reviewed by laying out the overall process in a flow diagram. Several types are useful. One type shows the paths followed by materials through their progression into a finished product. An example for a coating process at the James River Graphics Company is shown in Figure 20.1. Planners use such a diagram to divide the flow into logical sections called workstations. For each workstation, they prepare a formal document listing such items as operations to be performed, sequence of operations, facilities and

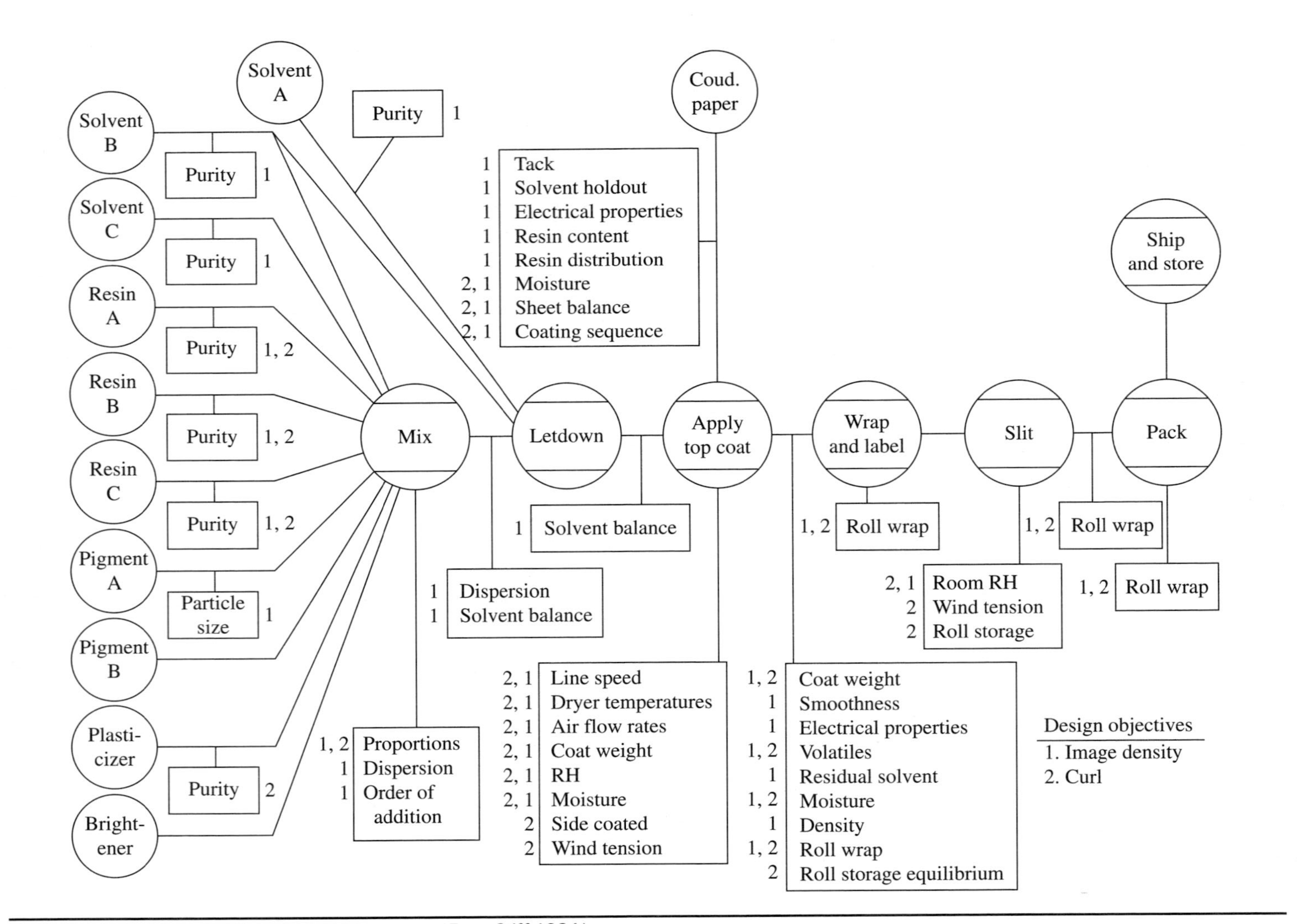

FIGURE 20.1 Product and process analysis source. (*From Stiff 1984.*)

instruments to be employed, and process conditions to be maintained. This formal document becomes the plan to be carried out by the production supervisors and workforce. The document is the basis for control activities by the inspectors. It also becomes the standard against which the process audits are conducted.

Another very useful approach is value stream mapping, which is discussed in detail in Chapter 11, Lean Techniques: Improving Process Efficiency. In addition to the product or service flow as described above, value stream determines the flow of any pertinent information of the process. This added analysis can often undercover sources of wastes including those that can contribute to quality concerns for the customer.

Correlation of Process Variables with Product Results

A critical aspect of planning during operations is to discover, by data analysis, the relationships between process features or variables and product features or results. Such knowledge enables a planner to create process control features, including limits and regulating mechanisms on the variables, to keep the process in a steady state and achieve the specified product results. In Figure 20.1 each process variable is shown in a rectangle attached to the circle representing an operation; product results are listed in rectangles between operations at the point where conformance can be verified. Some characteristics (e.g., coat weight) are both process variables and product results. For each control station in a process, designers identify the numerous control subjects over which control is to be exercised. Each control subject requires a feedback loop made up of multiple process control features. A process control spreadsheet or control plan helps to summarize the detail. An example in Chapter 18, Core Tools to Design, Control, and Improve Performance, shows, for each control subject, the unit of measure, type of sensor, goal, frequency of measurement, sample size, and the criteria and responsibility for decision-making.

Determining the optimal settings and tolerances for process variables sometimes requires much data collection and analysis. Eibl et al. (1992) discuss such planning and analysis for a paint-coating process for which little information was available about the relationship between process variables and product results. Many organizations have not studied the relationship between process variables and product results. The consequences of this lack of knowledge can be severe. In the electronic component manufacturing industry, some yields are shockingly low and will likely remain that way until the process variables are studied in depth. In all industries, the imposition of new quality demands such as Six Sigma requires much deeper understanding of product results and process variables than in the past. To understand fully the relationship between process variables and product results, we often need to apply the concept of statistical design of experiments (see Chapter 19, Accurate and Reliable Measurement Systems and Advanced Tools). See also the discussion of the Taguchi approach. Under the Six Sigma approach, factorial experiments are becoming necessary to understand the interactions among several variables and product results. But upper management must supply the missing elements, i.e., the resources for full-time personnel to design and analyze the experiments and the training of process engineers to integrate these concepts in process planning.

Validate the Measurement System

Particularly with the low defect levels demanded under the Six Sigma approach, it is extremely important to understand the capability of the manufacturing process and the capability of the measurement process. Thus planning and control of the measurement process becomes part of the Six Sigma approach. Previous studies assumed that variation of the measurement process was small compared to variation caused by the manufacturing process and thus could be ignored (in practice, the assumption was rarely tested in most industries). When variation due to the measurement process alone is even moderately large, the

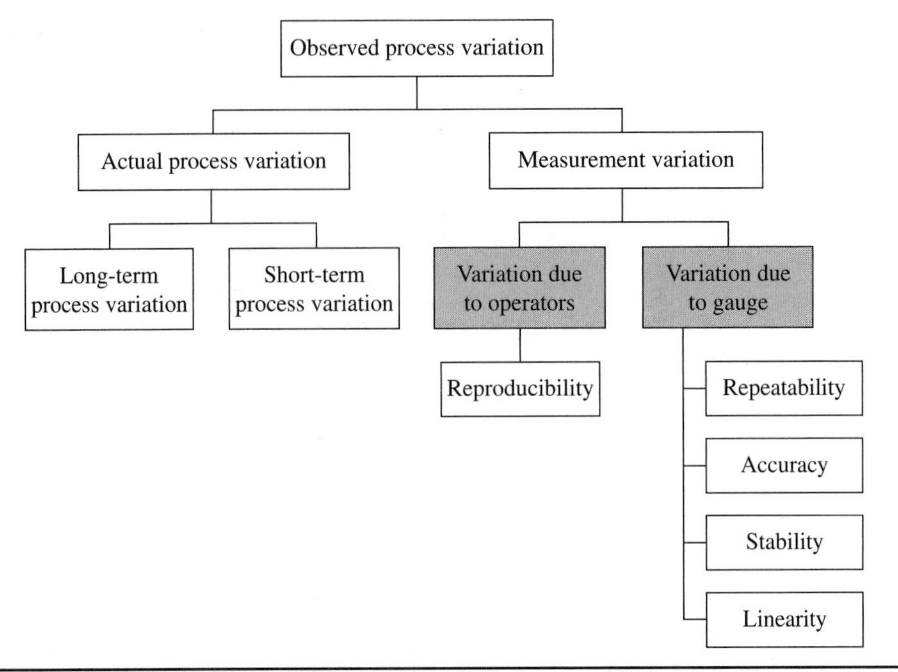

Figure 20.2 Possible sources of variation. (*Juran Institute.*)

result will be mistakes in determining whether a product meets specifications—some "good" product will be incorrectly classified as defective, and some "bad" product will be incorrectly classified as good. To quantify the measurement process, see Chapter 19, Accurate and Reliable Measurement Systems and Advanced Tools. Thus the time has come to evaluate measurement capability and to determine whether measuring equipment is accurately measuring process output.

Figure 20.2 provides perspective on the measurement issue. Note that the observed process variation (i.e., the variation of recorded measurements) is from two sources: variation in the process manufacturing the product and variation in the measurement process. Further, these two sources contain various components. The components shown in Figure 20.2 can be quantified and analyzed to determine measurement capability. The variation due to the process of manufacturing the product can also be quantified—see Chapter 19 under "Process Capability." Even in consumer product industries such as the manufacture of razor blades, tolerances can be on the order of the wavelength of visible light. Such tolerances are a long way from the days of tolerances in thousands or 10 thousands and using measuring instruments such as micrometers and even super-micrometers. The measurement process must be capable of handling these conditions.

The Importance of Understanding Process Capability

In planning the quality aspects of operations, nothing is more important than advance assurance that the processes will meet the specifications. In recent decades, a concept of process capability has emerged to provide a quantified prediction of process adequacy. This ability to predict quantitatively has resulted in widespread adoption of the concept as a major element of quality planning.

Process capability is the measured, inherent variation of the product turned out by a process. Obviously this understanding is key to the success of the operation. We introduce the topic here but highly suggest a detailed review of the concept and applications of process capability, which can be found in Chapter 19, Accurate and Reliable Measurement Systems and Advanced Tools.

The concept of process capability can expand beyond the current processes themselves. These include the impact of supplier quality and the use in quality improvement. The use of process capability can also be very effectively implemented in service industries as well.

Impact of Supplier Quality

The capability of a supplier can have a dramatic impact on the quality of the end product. As mentioned earlier in the discussion of value stream management (Chapter 11, Lean Techniques: Improving Process Efficiency), activities to bring a product or service to commercialization are often outside the direct control of the organization. It is critical for the producer to measure, analyze, and, if needed, react to variation introduced by a supplier. A full discussion of supplier quality can be found in Chapter 30, Supply Chain: Better, Faster, Friendlier Suppliers.

Process Capability and Quality Improvement

Capability indices serve a role in quantifying the ability of a process to meet customer quality goals. The emphasis, however, should be on improving processes and not just determining a capability index for a product characteristic. Achieving customer quality goals (particularly for quality levels of 1 to 10 ppm) means meeting requirements on all variables and attributes characteristics. On variables characteristics, decreasing the amount of variability (even when specification limits are being met) has many advantages. Gryna et al (2007) discuss six of these advantages. Achieving decreased variability requires the use of basic and advanced improvement techniques. Again, Chapter 19, Accurate and Reliable Measurement Systems and Advanced Tools, covers many of these techniques.

Process Capability in Service Industries

The concept of process capability analysis grew up in the manufacturing industries. This concept focuses on evaluating process variability (six standard deviations) as a measure of process capability. The concept, however, can apply to any process, including nonmanufacturing processes in the manufacturing industries and the spectrum of processes in the service industries. Little has been published on the application of process capability other than its application to manufacturing processes.

For certain parameters in service processes, process capability can be measured using Six Sigma and various capability indices. For example, in a loan association, the cycle time to complete the loan approval process is critical and could be analyzed. Time data are readily available in quantitative form for calculating.

Other service processes may not have variables data available. For example, a firm provides a service of guaranteeing checks written by customers at retail establishments. The decision of whether to guarantee is based on a process that employs an online evaluation of six factors. A percentage of checks guaranteed by the firm have insufficient funds, and the customer must be pursued for payment. The percentage of checks that default ("bounce") could be viewed as a measure of process capability. This approach uses discrete (attributes) data rather than the classic approach of calculating Six Sigma from variables data. The example given earlier on insurance policy writing illustrates the use of attributes data to calculate process capability for a service industry process.

With the emphasis on processes in quality management, evaluating the capability of processes requires not only evaluating capability based on variability (e.g., Six Sigma) but also a broader view. Gryna et al (2007) describes the issues involved in developing a broader framework.

Mistake-Proofing

An important element of operations planning is the concept of designing the process to be error-free through mistake proofing. This is sometimes also referred to as error-proofing or "poka-yoke." Although it is introduced here, a more in-depth review can be found in Chapter 11, Lean Techniques: Improving Process Efficiency. It is recommended that the designer of quality systems for operations become familiar with the benefits and applications of mistake proofing.

Concept of Controllability: Self-Control

An ideal objective for operational planning is to place human beings in a state of self-control, i.e., to provide them with all they need to meet quality objectives. To do so, we must provide people with the following:

1. Knowledge of what they are supposed to do
    - Clear and complete work procedures
    - Clear and complete performance standards
    - Adequate selection and training of personnel
2. Knowledge of what they are actually doing (performance)
    - Adequate review of work
    - Feedback of review results
3. Ability and desire to regulate the process for minimum variation
    - A process and job design capable of meeting quality objectives
    - Process adjustments that will minimize variation
    - Adequate worker training in adjusting the process
    - Process maintenance to maintain the inherent process capability
    - A strong quality culture and environment

As we will see, most organizations do not adhere to the three elements and ten subelements of self-control.

The concept of self-control has objectives that are similar to those of the Toyota production system. For a perceptive dissection of the Toyota system into four basic rules, see Spear and Bowen (1999).

The three basic criteria for self-control make possible a separation of defects into categories of "controllability," of which the most important are

1. *Worker-controllable.* A defect or nonconformity is worker-controllable if all three criteria for self-control have been met.
2. *Management-controllable.* A defect or nonconformity is management-controllable if one or more of the criteria for self-control have not been met.

Only management can provide the means for meeting the criteria for self-control. Hence, any failure to meet these criteria is a failure of management, and the resulting defects are therefore beyond the control of the workers. This theory is not 100 percent sound. Workers commonly have a duty to call management's attention to deficiencies in the system of control, and sometimes they fail to do so. (Sometimes they do, and management fails to act.) However, the theory is much more right than wrong.

Whether the defects or nonconformities in a plant are mainly management-controllable or worker-controllable is of the highest order of importance. Reducing the former requires a program in which the main contributions must come from managers, supervisors, and technical specialists. Reducing the latter requires a different kind of program in which much of the contribution comes from workers. The great difference between these two kinds of programs suggests that managers should quantify their knowledge of the state of controllability before embarking on major programs.

An example of a controllability study is given in Table 20.1. A diagnostic team was set up to study scrap and rework reports in six machine shop departments for 17 working days. The defect cause was entered on each report by a quality engineer who was assigned to collect the data. When the cause was not apparent, the team reviewed the defect and, when necessary, contacted other specialists (who had been alerted by management about the priority of the project) to identify the cause. The purpose of the study was to resolve a lack of agreement on the causes of chronically high scrap and rework. It did the job. The study was decisive in obtaining agreement on the focus of the improvement program. In less than one year, more than $2 million was saved, and important strides were made in reducing production backlogs.

Controllability can also be evaluated by posing specific questions for each of the three criteria of self-control. (Typical questions that can be posed are presented in this chapter.) Although this approach does not yield a quantitative evaluation of management and worker-controllable defects, it does show whether the defects are primarily management- or worker-controllable.

Category	Percent
Management-controllable	
Inadequate training	15
Machine inadequate	8
Machine maintenance inadequate	8
Other process problems	8
Material handling inadequate	7
Tool, gage, fixture (TGF) maintenance inadequate	6
TGF inadequate	5
Wrong material	3
Operation run out of sequence	3
Miscellaneous	5
Total	68
Worker-controllable	
Failure to check work	11
Improper operation of machine	11
Other (e.g., piece mislocated)	10
Total	32

Table 20.1 Controllability Study in a Machine Shop

Experience has shown defects are about 80 percent management-controllable. This figure does not vary much from industry to industry, but it does vary greatly across processes. Other investigators in Japan, Sweden, the Netherlands, and Czechoslovakia have reached similar conclusions.

Although the available quantitative studies make clear that defects are mainly management-controllable, many industrial managers do not know this or are unable to accept the data. Their long-standing beliefs are that most defects are the result of worker carelessness, indifference, and even sabotage. Such managers are easily persuaded to embark on worker motivation schemes that, under the usual state of affairs, aim at a small minority of the problems and hence are doomed to achieve minor results at best. The issue is not whether quality problems in an industry are management-controllable. The need is to determine the answer *in a given plant.* This answer requires solid facts, preferably through a controllability study of actual defects, as shown in Table 20.1.

We now discuss the three main criteria for self-control.

Criterion 1: Knowledge of "Supposed to Do"

This knowledge commonly consists of the following:

1. The product standard, which may be a written specification, a product sample, or other definition of the end result to be attained.
2. The process standard, which may be a written process specification, written process instructions, an oral instruction, or other definition of "means to an end."
3. A definition of responsibility, i.e., what decisions to make and what actions to take. In developing product specifications, some essential precautions must be observed.

Unequivocal Information Must Be Provided

Specifications should be quantitative. If such specifications are not available, physical or photographic standards should be provided. But beyond the need for *clear* product specifications, there is also a need for *consistent* and *credible* specifications. In some organizations, production supervisors have a secret "black book" that contains the "real" specification limits used by inspectors for accepting product. A further problem arises in communicating changes in specifications, especially when there is a constant parade of changes.

Information on Seriousness Must Be Provided

All specifications contain multiple characteristics, and these are not equally important. Production personnel must be guided and trained to meet all specification limits. But they must also be given information on the relative importance of each characteristic to focus on priorities.

Reasons Must Be Explained

Explanation of the purposes served by both the product and the specification helps workers to understand why both the nominal specification value and the limits must be met.

Process Specifications Must Be Provided

Work methods and process conditions (e.g., temperature, pressure, time cycles) must be unequivocally clear. A steel manufacturer uses a highly structured system of identifying key process variables, defining process control standards, communicating the information to the workforce, monitoring performance, and accomplishing diagnosis when problems arise. The process specification is a collection of process control standard. A procedure is developed for controlling the key process variables procedures (variables that must be controlled to meet specification limits for the product). The procedure addresses the following issues:

- What the process standards are?
- Why control is needed?
- Who is responsible for control?
- What and how to measure?
- When to measure?
- How to report routine data?
- Who is responsible for data reporting?
- How to audit?
- Who is responsible for audit?
- What to do with product that is out of compliance?
- Who developed the standard?

Often, detailed process instructions are not known until workers have become experienced with the process. Updating of process instructions based on job experience can be conveniently accomplished by posting a cause-and-effect diagram in the production department and attaching index cards to the diagram. Each card states additional process instructions based on recent experience.

The above discussion covers the first criterion of self-control: People must have the means of knowing what they are supposed to do. To evaluate adherence to this criterion, a checklist of questions can be created, including the following:

Adequate and Complete Work Procedures

1. Are there written product specifications, process specifications, and work instructions?
2. If they are written down in more than one place, do they all agree? Are they legible?
3. Are they conveniently accessible to the worker?
4. Does the worker receive specification changes automatically and promptly?
5. Does the worker know what to do with defective raw material?
6. Have responsibilities for decisions and actions been clearly defined?

Adequate and Complete Performance Standards

1. Do workers consider the standards attainable?
2. Does the specification define the relative importance of different quality characteristics?
3. If control charts or other control techniques are to be used, is their relationship to product specifications clear?
4. Are standards for visual defects displayed in the work area?
5. Are the written specifications given to the worker the same as the criteria used by inspectors?

6. Are deviations from the specification often allowed?
7. Does the worker know how the product is used?
8. Does the worker know the effect on future operations and product performance if the specification is not met?

Adequate Selection and Training

1. Does the personnel selection process adequately match worker skills with job requirements?
2. Has the worker been adequately trained to understand the specification and perform the steps needed to meet the specification?
3. Has the worker been evaluated by testing or other means to see whether he or she is qualified?

Criterion 2: Knowledge of Performance

For self-control, people must have the means of knowing whether their performance conforms to a standard. This conformance applies to

- The product in the form of specifications for product characteristics
- The process in the form of specifications for process variables

This knowledge is secured from three primary sources: measurements inherent in the process, measurements by production workers, and measurements by inspectors.

Criteria for Good Feedback to Workers

The needs of production workers (as distinguished from supervisors or technical specialists) require that the data feedback be read at a glance, deal only with the few important defects, deal only with worker-controllable defects, provide prompt information about symptom and cause, and provide enough information to guide corrective actions. Good feedback should

- *Be readable at a glance.* The pace of events on the factory floor is swift. Workers should be able to review the feedback while in motion. Where a worker needs information about process performance over time, charts can provide an excellent form of feedback, provided they are designed to be consistent with the assigned responsibility of the worker (Table 20.2). It is useful to use visual displays to highlight recurrent

Responsibility of the Worker Is to	Chart Should Be Designed to Show
Make individual units of product meet a product specification	Measurements of individual units of product compared to product specification limits
Hold process conditions to the requirements of a process specification	Measurements of the process conditions compared to the process specification limits
Hold averages and ranges to specified statistical control limits	Averages and ranges compared to the statistical control limits
Hold percentage nonconforming below some prescribed level	Actual percentage nonconforming compare to the limiting level

Table 20.2 Worker Responsibility versus Charter Design

problems. A problem described as "outer hopper switch installed backward" displayed on a wall chart in large block letters has much greater impact than the same message buried away as a marginal note in a work folder.

- *Deal only with the few important defects.* Overwhelming workers with data on all defects will result in diverting attention from the vital few.
- *Deal only with worker-controllable defects.* Any other course provides a basis for argument that will be unfruitful.
- *Provide prompt information about symptoms and causes.* Timeliness is a basic test of good feedback; the closer the system is to "real-time" signaling, the better.
- *Provide enough information to guide corrective action.* The signal should be in terms that make it easy to decide on remedial action.

Feedback Related to Worker Action

The worker needs to know what kind of process change to make to respond to a product deviation. Sources of this knowledge are

- The process specifications
- Cut-and-try experience by the worker
- The fact that the units of measure for both product and process are identical

If they lack all these, workers can only cut and try further or stop the process and sound the alarm.

Sometimes the data feedback can be converted into a form that makes the worker's decision easier about what action to take on the process. For example, a copper cap had six critical dimensions. It was easy to measure the dimensions and to discover the nature of product deviation. However, it was difficult to translate the product data into process changes. To simplify this translation, use was made of a position-dimensions (P–D) diagram. The six measurements were first "corrected" (i.e., coded) by subtracting the thinnest from all the others. These corrected data were then plotted on a P–D diagram. Such diagrams provide a way of analyzing the tool setup.

Feedback to Supervisors

Beyond the need for feedback at workstations, there is a need to provide supervisors with short-term summaries. These take several forms.

Matrix Summary

A common form of matrix is workers versus defects; the vertical columns are headed by worker names and the horizontal rows by the names of defect types. The matrix makes clear which defect types predominate, which workers have the most defects, and what the interaction is. Other matrices include machine number versus defect type and defect type versus calendar week. When the summary is published, it is usual to circle matrix cells to highlight the vital few situations that call for attention. An elaboration of the matrix is to split the cell diagonally, thus permitting the entry of two numbers, e.g., number defective and number produced.

Pareto Analysis

Some organizations prefer to minimize detail and provide information on the total defects for each day plus a list of the top three (or so) defects encountered and how many of each there were. In some industries, a "chart room" displays performance against goals by product and by department.

Automated Quality Information

Some situations justify the mechanization of both the recording and analysis of data. Entry of data into computer terminals on production floors is common now. Many varieties of software are available for analyzing, processing, and presenting quality information collected on the production floor. The term "quality information equipment" (QIE) designates the physical apparatus that measures products and processes, summarizes the information, and feeds the information back for decision-making. Sometimes such equipment has its own product development cycle to meet various product effectiveness parameters for the QIE.

A checklist for evaluating the second criterion of self-control includes questions such as the following:

Adequate Review of Work

1. Are gauges provided to the worker? Do they provide numerical measurements rather than simply sort good from bad? Are they precise enough? Are they regularly checked for accuracy?
2. Is the worker told how often to sample the work? Is sufficient time allowed?
3. Is the worker told how to evaluate measurements to decide when to adjust the process and when to leave it alone?
4. Is a checking procedure in place to ensure that the worker follows instructions on sampling work and making process adjustments?

Adequate Feedback

1. Are inspection results provided to the worker, and are these results reviewed by the supervisor with the worker?
2. Is the feedback timely and in enough detail to correct problem areas? Have personnel been asked what detail is needed in the feedback?
3. Do personnel receive a detailed report of errors by specific type of error?
4. Does feedback include positive comments in addition to negative?
5. Is negative feedback given in private?
6. Are certain types of errors tracked with feedback from external customers?
7. Could some of these be tracked with an internal early indicator?

Criterion 3: Ability and Desire to Regulate

Ability and desire to regulate is the third criterion for self-control. Regulating the process depends on various management-controllable factors, including these:

The Process Must Be Capable of Meeting the Tolerances

This factor is of paramount importance. In some organizations the credibility of specifications is a serious problem. Typically, a manufacturing process is created after the product design is released, a few trials are run, and full production commences. In cases where quality problems arise during full production, diagnosis sometimes reveals that the process is not capable of consistently meeting the design specifications. Costly delays in production then occur while the problem is solved by changing the process or changing the specification. The capability of the manufacturing process should be verified during the product development cycle before the product design is released for full production. See Chapter 19, Accurate and Reliable Measurement Systems and Advanced Tools, for a full discussion of process capability.

The Process Must Be Responsive to Regulatory Action Value

Example 20.1 In a process for making polyethylene film, workers were required to meet multiple product parameters. The equipment had various regulatory devices, each of which could vary performance with respect to one or more parameters. However, the workers could not "dial in" a predetermined list of settings that would meet all parameters. Instead, it was necessary to cut and try to meet all parameters simultaneously. During the period of cut and try, the machine produced nonconforming product to an extent that interfered with meeting standards for productivity and delivery. The workers were unable to predict how long the cut-and-try process would go on before full conformance was achieved. Consequently, it became the practice to stop cut and try after a reasonable amount of time and to let the process run, whether in conformance or not.

The Worker Must Be Trained in How to Use the Regulating Mechanisms and Procedures

This training should cover the entire spectrum of action—under what conditions to act, what kind and extent of changes to make, how to use the regulating devices, and why these actions need to be taken.

Example 20.2 Of three qualified workers on a food process, only one operated the process every week and became proficient. The other two workers were used only when the primary worker was on vacation or was ill, and thus they never became proficient. Continuous training of the relief people was considered uneconomical, and agreements with the union prohibited their use except under the situations cited above. This problem is management-controllable, i.e., additional training or a change in union agreements is necessary.

The Act of Adjustment Should Not Be Personally Distasteful to the Worker, e.g., Should Not Require Undue Physical Exertion

Example 20.3 In a plant making glass bottles, one adjustment mechanism was located next to a furnace area. During the summer months, this area was so hot that workers tended to keep out of it as much as possible. When the regulation consists of varying the human component of the operation, the question of process capability arises in a new form: Does the worker have the capability to regulate?

The Process Must Be Maintained Sufficiently to Retain Its Inherent Capability

In the manufacturing environment, without adequate maintenance, equipment breaks down and requires frequent adjustments—often with an increase in both defects and variability around a nominal value. Clearly, such maintenance must be both preventive and corrective. The importance of maintenance has given rise to the concepts of total productive maintenance (TPM) and more recently, reliability-centered maintenance (RCM). Under these approaches, teams are formed to identify, analyze, and solve maintenance problems to maximize the uptime and reliability of process equipment. These teams consist of production line workers, maintenance personnel, process engineers, and others as needed. With total productive maintenance, problems are kept narrow in scope to encourage a steady stream of small improvements. Examples of improvements include a reduction in the number of tools lost and simplification of process adjustments. Reliability-centered maintenance by contrast is more focused on the overall prioritizing of resources and actions through data collection analysis and detailed planning. In both cases the goal is to maximize equipment performance, and each approach complements the other. Chapter 11, Lean Techniques: Improving Process Efficiency, provides a more detailed review of methods to achieve maintenance excellence in an operation.

Control Systems and the Concept of Dominance

Specific systems for controlling characteristics can be related to the underlying factors that dominate a process. The main categories of dominance include the following:

- *Setup-dominant.* Such processes have high reproducibility and stability for the entire length of the batch to be made. Hence the control system emphasizes verification of the setup before production proceeds. Examples of such processes are drilling, labeling, heat sealing, printing, and presswork.
- *Time-dominant.* Such a process is subject to progressive change with time (wearing of tools, depletion of a reagent, heating up of a machine). The associated control system will feature a schedule of process checks with feedback to enable the worker to make compensatory changes. Screw machining, volume filling, wool carding, and papermaking are examples of time-dominant processes.
- *Component-dominant.* Here the quality of the input materials and components is the most influential. The control system is strongly oriented toward supplier relations along with incoming inspection and sorting of inferior lots. Many assembly operations and food formulation processes are component-dominant.
- *Worker-dominant.* For such processes, quality depends mainly on the skill and knack possessed by the production worker. The control system emphasizes such features as training courses and certification for workers, mistake proofing and rating of workers and quality. Workers are dominant in processes such as welding, painting, and order filling.
- *Information-dominant.* In these processes, the job information usually undergoes frequent change. Hence the control system emphasizes the accuracy and up-to-dateness of the information provided to the worker (and everyone else). Examples include order editing and "travelers" used in job shops.

The various types of dominance differ also in the tools used for process control. Table 20.3 lists the forms of process dominance along with the usual tools used for process control.

Setup Dominant	Time Dominant	Component Dominant	Worker Dominant	Information Dominant
Inspection of process conditions	Periodic inspection	Supplier rating	Acceptance inspection	Computer-generated information
First-piece inspection	x-bar Chart	Incoming inspection	*p* Chart	"Active" checking of documentation
Lot plot	Median chart	Prior operation control	*c* Chart	Bar codes and electronic entry
Precontrol	x-bar and *R* charts	Acceptance inspection	Operator scoring	Process audits
Narrow-limit gauging	Precontrol	Mockup evaluation	Recertification of workers	
Attribute visual inspection	Narrow-limit gauging		Process audits	
	p Chart			
	Process variables check			
	Automatic recording			
	Process audits			

Table 20.3 Control Tools for Forms of Process Dominance

Checklist

A checklist for evaluating the third criterion of self-control typically includes questions such as these:

Process Capability

1. Has the quality capability of the process been measured to include both inherent variability and variability due to time? Is the capability checked periodically?
2. Has the design of the job used the principles of mistake proofing
3. Has equipment, including any software, been designed to be compatible with the abilities and limitations of workers?

Process Adjustments

1. Has the worker been told how often to reset the process or how to evaluate measurements to decide when the process should be reset?
2. Can the worker make a process adjustment to eliminate defects? Under what conditions should the worker adjust the process? When should the worker shut down the machine and seek more help? Whose help?
3. Have the worker actions that cause defects and the necessary preventive action been communicated to the worker, preferably in written form?
4. Can workers institute those job changes that they are able to show will provide benefits? Are workers encouraged to suggest changes?

Worker Training in Adjustments

1. Do some workers possess a hidden knack that needs to be discovered and transmitted to all workers?
2. Have workers been provided with the time and training to identify problems, analyze problems, and develop solutions? Does the training include diagnostic training to look for patterns of errors and determine sources and causes?

Process Maintenance

1. Is there an adequate preventive maintenance program for the process?

Strong Quality Culture/Environment

1. Is there sufficient effort to create and maintain awareness of quality?
2. Is there evidence of management leadership?
3. Have provisions been made for self-development and empowerment of personnel?
4. Have provisions been made for participation of personnel as a means of inspiring action?
5. Have provisions been made for recognition and rewards for personnel?

Use of Checklists on Self-Control

These checklists can help operations in the design (and redesign) of jobs to prevent errors; diagnose quality problems on individual jobs; identify common weaknesses in many jobs; and assist supervisors to function as coaches with personnel, prepare for process audits, and conduct training classes in quality.

Organizational Structure for Quality

The organization of the future will be influenced by the interaction of two systems: the technical system (design, equipment, procedures) and the social system (people, roles), thus the name "sociotechnical systems" (STSs).

New ways of organizing work are emerging, particularly at the workforce level. For example, supervisors are becoming "coaches"; they teach and empower rather than assign and direct. Operators are becoming "technicians"; they perform multiskilled jobs with broad decision-making rather than narrow jobs with limited decision-making. Team concepts play an important role in these new approaches. In some organizations 40 percent of the people participate on teams; some organizations have a goal of 80 percent. Permanent teams (e.g., process teams, self-managing teams) are responsible for all output parameters, including quality; ad hoc teams (e.g., quality project teams) are typically responsible for improvement in quality.

Although these various types of quality teams are showing significant results, the reality is that, for most organizations, daily work in a department is managed by a supervisor who has a complement of workers performing various tasks. This configuration is the "natural work team" in operations. But team concepts can certainly be applied to daily work. One framework for a team in daily operations work is the control process from the trilogy of quality processes. As applied to daily work, the steps are to choose control subjects, establish measurement, establish standards of performance, measure actual performance, compare to standards, and take action on the difference.

Planning for Evaluation of Product

The planning must recognize the need for formal evaluation of product to determine its suitability for the marketplace. Three activities are involved:

1. Measuring the product for conformance to specifications
2. Taking action on nonconforming product
3. Communicating information on the disposition of nonconforming product

These activities are discussed in Chapter 12, Six Sigma: Improving Process Effectiveness. But the activities impinge on the manufacturing planning process. For example, several alternatives are possible for determining conformance; i.e., have the activity done by production workers, by an independent inspection force, or by a combination of both. Increasingly, the combination approach is being employed.

What has evolved is the concept of self-inspection combined with a product audit.

Under this concept, all inspection and all conformance decisions, both on the process and on the product, are made by the production worker. (Decisions on the action to be taken on a nonconforming product are *not*, however, delegated to the worker.) However, an independent audit of these decisions is made. The quality department inspects a random sample periodically to ensure that the decision-making process used by workers to accept or reject a product is still valid. The audit verifies the decision process. Note that, under a pure audit concept, inspectors are not transferred to do inspection work in the production department. Except for those necessary to do audits, inspection positions are eliminated.

If an audit reveals that wrong decisions have been made by the workers, the product evaluated since the last audit is reinspected—often by the workers themselves.

Self-inspection has decided advantages over the traditional delegation of inspection to a separate department:

- Production workers are made to feel more responsible for the quality of their work.
- Feedback on performance is immediate, thereby facilitating process adjustments.
- Traditional inspection also has the psychological disadvantage of an "outsider" reporting the defects to a worker.
- The costs of a separate inspection department can be reduced.
- The job enlargement that takes place by adding inspection to the production activity of the worker helps to reduce the monotony and boredom inherent in many jobs.
- Elimination of a specific station for inspecting all products reduces the total manufacturing cycle time.

Example 20.4 In a coning operation of textile yarn, the traditional method of inspection often resulted in finished cones sitting in the inspection department for several days, thereby delaying any feedback to production. Under self-inspection, workers received immediate feedback and could get machines repaired and setups improved more promptly. Overall, the program reduced nonconformities from 8 to 3 percent. An audit inspection of the products that were classified by the workers as "good" showed that virtually all were correctly classified. In this organization, workers can also classify product as "doubtful." In one analysis, worker inspections classified 3 percent of the product as doubtful, after which an independent inspector reviewed the doubtful product and classified 2 percent as acceptable and 1 percent as nonconforming.

Example 20.5 A pharmaceutical manufacturer employed a variety of tests and inspections before a capsule product was released for sale. These checks included chemical tests, weight checks, and visual inspections of the capsules. A 100 percent visual inspection had traditionally been conducted by an inspection department. Defects ranged from "critical" (e.g., an empty capsule) to "minor" (e.g., faulty print). This inspection was time consuming and frequently caused delays in production flow. A trial experiment of self-inspection by machine operators was instituted. Operators visually inspected a sample of 500 capsules. If the sample was acceptable, the operator shipped the full container to the warehouse; if the sample was not acceptable, the full container was sent to the inspection department for 100 percent inspection. During the experiment, both the samples and the full containers were sent to the inspection department for 100 percent inspection with reinspection of the sample recorded separately. The experiment reached two conclusions: (1) the sample inspection by the operators gave results consistent with the sample inspection by the inspectors, and (2) the sample of 500 gave results consistent with the results of 100 percent inspection.

The experiment convinced all parties to switch to sample inspection by operators. Under the new system, good product was released to the warehouse sooner, and marginal product received a highly focused 100 percent inspection. In addition, the level of defects *decreased*. The improved quality level was attributed to the stronger sense of responsibility of operators (they themselves decided if product was ready for sale) and the immediate feedback received by operators from self-inspection. But there was another benefit—the inspection force was reduced by 50 people. These 50 people were shifted to other types of work, including experimentation and analysis on the various types of defects.

Criteria for Self-Inspection

For self-inspection, some criteria must be met:

- Quality must be the number one priority within an organization. If this requirement is not clear, a worker may succumb to schedule and cost pressures and classify products as acceptable that should be rejected.

- Mutual confidence is necessary. Managers must have sufficient confidence in the workforce to be willing to give workers the responsibility for deciding whether product conforms to specification. In turn, workers must have enough confidence in management to be willing to accept this responsibility.
- The criteria for self-control must be met. Failure to eliminate the management-controllable causes of defects suggests that management does not view quality as a high priority, and this environment may bias the workers during inspections. Workers must be trained to understand the specifications and perform the inspection.
- Specifications must be unequivocally clear. Workers should understand the use that will be made of their products (internally and externally) to grasp the importance of a conformance decision.
- The process must permit assignment of clear responsibility for decision-making. An easy case for application is a worker running one machine because there is clear responsibility for making both the product and the product conformance decision. In contrast, a long assembly line or the numerous steps taken in a chemical process make it difficult to assign clear responsibility. Application of self-inspection to such multistep processes is best deferred until experience is gained with some simple processes.

Self-inspection should apply only to products and processes that are stabilized and meet specifications and only to personnel who have demonstrated their competence.

Worker response to such delegation of authority is generally favorable; the concept of job enlargement is a significant factor. However, workers who do qualify for self-inspection commonly demand some form of compensation for this achievement, e.g., a higher grade, more pay. Organizations invariably make a constructive response to these demands because the economics of delegating are favorable. In addition, the resulting differential tends to act as a stimulus to nonqualified workers to qualify themselves.

An adjunct to self-inspection is the use of poka-yoke devices as part of inspection. These devices are installed in a machine to inspect the process conditions and product results and provide immediate feedback to the operator. Devices such as limit switches and interference pins, used to ensure proper positioning of materials on machines, are poka-yoke devices for inspecting process conditions. Go/no go gauges are examples of poka-yoke devices for inspecting product.

Process Quality Audits

A quality audit is an independent review to compare some aspect of quality performance with a standard for that performance. Application to manufacturing has been extensive and includes both audits of activities (process audits) and audits of product (product audits).

A *process quality audit* includes any activity that can affect final product quality. This on-site audit is usually done on a specific process by one or more persons and uses the process operating procedures. Adherence to existing procedures is emphasized, but audits often uncover situations of inadequate or nonexistent procedures. The checklists presented earlier in this chapter on the three criteria for self-control can suggest useful specific subjects for process audits. Audits must be based on a foundation of hard facts that are presented in the audit report in a way that will help those responsible to determine and execute the required corrective action.

A product audit involves the reinspection of product to verify the adequacy of acceptance and rejection decisions. In theory, such product audits should not be needed. In practice, they can often be justified by field complaints. Such audits can take place at each inspection

1. Is the specification accessible to production staff?
2. Is the current revision on file?
3. Is the copy on file in good condition and are all pages accounted for?
4. If referenced documents are posted on equipment, do they match the specification?
5. If the log sheet is referenced in specifications, is a sample included in the specification?
6. Is the operator completing the log sheet according to specifications?
7. Are lots with out-of-specification readings authorized and taken care of in writing by the engineering department or the proper supervisor?
8. Are corrections to paperwork made according to specifications?
9. Are equipment time settings according to specification?
10. Are equipment temperature settings according to specification?
11. Is the calibration sticker on equipment current?
12. Do chemicals or gases listed in the specification match actual usage?
13. Do quantities listed in the specification match the line setup?
14. Are changes of chemicals or gases made according to specification?
15. Is the production operator certified? If not, is this person authorized by the supervisor?
16. Is the production operating procedure according to specification?
17. Is the operator performing the written cleaning procedure according to specification?
18. If safety requirements are listed in the specification, are they being followed?
19. If process control procedures are written in the specification, are the actions performed by the process control verifiable?
20. If equipment maintenance procedures are written in the specification, are the actions performed verifiable? According to specification?

(*Source:* Quality Progress, October 1990.)

TABLE 20.4 Audit Checklist

station for the product or after final assembly and packing. Sometimes an audit is required before a product may be moved to the next operation.

Quality Measurement in Operations

The management of key work processes must include provision for measurement. In developing units of measure, the reader should review the basics of quality measurement discussed in Chapters 4, 5, and 6: Quality Planning, Improvement, and Control.

Table 20.5 shows examples for manufacturing activities. Note that many of the control subjects are forms of work output. In reviewing current units in use, a fruitful starting point is the measure of productivity. *Productivity* is usually defined as the amount of output related to input resources. Surprisingly, some organizations still mistakenly calculate only one measure of output, i.e., the total (acceptable *and* nonacceptable). Clearly, the pertinent output measure is that which is usable by customers (i.e., acceptable output).

The units in Table 20.5 become candidates for data analysis using statistical techniques such as control charts, discussed in Chapter 12, Six Sigma: Improving Process Effectiveness. But there is a more basic point: the selection of the unit of measure and the periodic collection and reporting of data demonstrate to operating personnel that management regards the subject as having priority importance. This atmosphere sets the stage for improvement!

Subject	Unit of Measure
Quality of manufacturing output	Percentage of output meeting specifications at inspection ("first-time yield") Percentage of output meeting specifications at intermediate and final inspections Amount of scrap (quantity, cost, percentage, etc.), amount of rework (quantity, cost, percentage, etc.) Percentage of output shipped under waiver of specification Number of defects found in product audit (after inspection) Warranty costs due to manufacturing defects Overall measure of product quality (defects in parts per million, weighted defect per unit, variability for critical characteristics, etc.) Amount of downgraded output
Quality of input to manufacturing	Percentage of critical operations with certified workers Amount of downtime of manufacturing equipment Percentage of product input meeting specifications Percentage of instruments meeting calibration schedules Percentage of specifications requiring changes after release

TABLE 20.5 Examples of Quality Measurement in Manufacturing

Detailed Operations Quality Planning

Review of the proposed process can be accomplished most effectively through preproduction trials and runs. Techniques such as failure mode and effect analysis (FMEA) (synonymous with failure mode effect, and criticality analysis) can also provide an even earlier warning before any product is made. Once a process is established, process control must be in place to ensure conformity. These approaches are discussed below.

Preproduction Runs

Ideally, product lots should be put through the entire system, with the deficiencies found and corrected before going into full-scale production. In practice, organizations usually make some compromises with this ideal approach. The preproduction may be merely the first of the regular production, but with special provision for prompt feedback and correction of errors as found. Alternatively, the preproduction may be limited to those features of product and process design that are so new that prior experience cannot reliably provide a basis for good risk taking. While some organizations do adhere to a strict rule of proving in the product and process through preproduction lots, the more usual approach is one of flexibility, in which the use of preproduction lots depends on

1. The extent to which the product embodies new or untested quality features
2. The extent to which the design of the manufacturing process embodies new or untried machines, tools, etc.
3. The amount and value of product which will be out in the field before there is conclusive evidence of the extent of process, product, and use difficulties

The scaling up of production is actually a continuation of the scaling up that takes place from product design concept to prototype or model construction and test. The adequacy of the full-scale manufacturing plan cannot be judged from the record of models made in the model shop. In the model shop the basic purpose is to prove engineering feasibility; in the production shop the purpose is to meet standards of quality, cost, and delivery. The model shop machinery, tools, personnel, supervision, motivation, etc. are all different from the corresponding situations in the production shop.

Tool Tryout

At the workstation level, as new tools are completed, they are subjected to a tryout procedure that, in most organizations, is highly formalized. The tryout consists of producing enough product from the new tool to demonstrate that the tool can meet quality standards under shop conditions.

These formalized tryouts conclude with the execution of a formal document backed up by supporting data, which always include the quality data. The release of the tool for full-scale production is contingent on the approval of this tryout document.

Limited Trial Lots

Beyond the tryouts at individual workstations, there is a need for collective tryouts. These require trial production lots, which must be scheduled for the prime purpose of proving in the manufacturing process. The trial lot is usually made in the regular production shop and provides an extensive preview of the problems that will be encountered in large-scale production. In the process industries, the equivalent intermediate scaling up is the pilot plant. It is widely used to provide the essential information (on quality, costs, productivity, etc.) needed to determine whether and how to go into full-scale production.

Software Verification

Software used with a process requires a tryout just as new tools do—with the same degree of formality and approval process.

Experimental Lots

The trial lot concept provides opportunities for planners to test out alternatives, and they often combine the concept of experimentation with that of proving in the nonexperimental portion of the trial.

Attainment of good process yields is one of the most important purposes of experimental lots. These experiments can make use of all the techniques discussed in Chapter 19, Accurate and Reliable Measurement Systems and Advanced Tools, and in the various statistical sections.

Failure Mode, Effect, and Criticality Analysis for Processes

A failure mode, effect, and criticality analysis is useful in analyzing the proposed design of a product. The same technique can dissect the potential failure modes and their effects on a current or proposed process. A detailed discussion of failure mode, effect, and criticality analysis can be found in Chapter 19, Accurate and Reliable Measurement Systems and Advanced Tools.

Planning Process Controls

The process specification, procedures, and work instructions/standard work are critical to operations. Their purpose is to inform the production people how to set up, run, and regulate the processes so that the result will be good product. Conversely, the production people should follow these plans. Otherwise, good product might not be the result.

Many organizations institute process controls to provide assurance that the plans will in fact be followed. There are several kinds of these controls, and they are established by some combination of manufacturing engineers, quality engineers, production supervisors, and workers. The precise combination varies widely from organization to organization.

Process control is based on the feedback loop. The steps for planning operational process controls follow closely the universal approach for use of the feedback loop.

Control Criteria

While execution of the control plan is typically delegated to the workforce, it is common to set criteria to be met before the process is allowed to run. These criteria are imposed in three main areas:

1. *Setup criteria.* For some processes the start of production must await meeting setup criteria (e.g., five pieces in a row must test "good"). In critical cases this form of early warning assurance may require that a supervisor or inspector independently approve the setup.
2. *Running criteria.* For many processes there is a need to check the running periodically to decide whether the process should continue to run or should stop for readjustment. The criteria here relate to such things as frequency of check, size of sample, manner of sample selection, tests to be made, and tolerances to be met.
3. *Equipment maintenance criteria.* In some processes, the equipment itself must be closely controlled if quality is to be maintained. This type of control is preventive in nature and is quite different in concept from repair of equipment breakdowns. This preventive form of equipment maintenance includes a carefully drawn set of criteria that define the essential performance characteristics of the equipment. Then, on a scheduled basis (strictly adhered to), the equipment is checked against these criteria. In the United States this aspect of equipment maintenance is not well developed, and there is need to take positive steps to strengthen it.

Relationship to Product Controls

Process controls are sometimes confused with product controls, but there is a clear difference. Process controls are associated with the decision: Should the process run or stop? Product controls are associated with the decision: Does the product conform to specification? Usually both these decisions require input derived from sampling and measuring the product. (It is seldom feasible to measure the process directly.) However, the method of selecting the samples is often different. Production usually makes the "process run or stop" decision and tends to sample in ways which tell the most about the process. Inspection usually (in the United States) makes the "product conformance" decision and tends to sample in ways that tell the most about the product.

This difference in sampling can easily result in different conclusions on the "same" product. Production commonly does its sampling on a scheduled basis and at a time when the product is still traceable to specific streams of the process. Inspection often does its sampling on a random basis and at a time when traceability has begun to blur.

Despite the different purposes being served, it is feasible for the two departments to do joint planning. Usually they are able to establish their respective controls so that both purposes are well served and the respective data reinforce each other.

Type of Action to Take	When to Take Action	Basic Steps
Troubleshooting (part of quality control)	Performance indicator outside control limits Performance indicator in clear trend toward control limits	Identify problem Diagnose problem Take remedial action
Quality improvement	The control limits are so wide that it is possible for the process to be in control and still miss the targets Performance indicator frequently misses its target	Identify project Establish project Diagnose cause Remedy the cause Hold the gains
Quality planning	Many performance indicators for this process miss their targets frequently Customers have significant needs that the product does not meet	Establish project Identify customers Discover customer needs Develop product Develop process Design controls

(*Source:* Juran Institute, Inc., 1995, pp. 5–7.)

Table 20.6 Three Types of Action

Maintaining a Focus on Continuous Improvement

Historically, the operations function has always been involved in troubleshooting sporadic problems. As chronic problems were identified, these were addressed using various approaches, such as quality improvement teams. Often the remedies for improvement involve quality planning or replanning. These three types of action are summarized in Table 20.6.

Maintaining the focus on improvement clearly requires a positive quality culture in the organization. Therefore, we must first determine the present quality culture and then take the steps to change the culture to one that will foster continuous improvement. In addition, the operations function must be provided with the support to maintain the focus on improvement. A key source of that support should be the quality department. Thus the quality department should view operations as its key internal customer and provide the training, technical quality expertise, and other forms of support to enable operations to maintain the focus on improvement. Also, a quality department can urge upper management to set up cross-functional teams to address operations problems that may be caused by other functional departments such as engineering, purchasing, and information technology.

References

Eibl, S., Kess, U., and Pukelsheim, F. (1992). "Achieving a Target Value for a Manufacturing Process," *Journal of Quality Technology*, January, pp. 22–26.

Goldman, S. L., Nagel, R. N., and Preiss, K. (1995). *Agile Competitors and Virtual Organizations.* Van Nostrand Reinhold, New York.

Gryna, F. M., Chua, R. C. H., and De Feo, J. A. (2007). *Quality Planning and Analysis*, 5th ed. McGraw-Hill, New York.

Herman, J. T. (1989). "Capability Index—Enough for Process Industries?" *ASQC Quality Congress Transactions 1989*. American Society for Quality Control, Milwaukee, WI, pp. 670–675.
Quality Progress, October, 1990. "Motorola's Secret to Total Quality Control" by E. Pena.
Siff, W. C. (1984). "The Strategic Plan of Control a Tool for Participative Management." ASQC Quality.
Snee, R. D. (1993). "Creating Robust Work Processes." *Quality Progress*, February, pp. 37–41.
Spear, S., and Bowen, H. K. (1999). "Decoding the DNA of the Toyota Production System." *Harvard Business Review,* September-October, pp. 97–106.

CHAPTER 21

Service-Based Organizations: Customer Service at Its Best

Michael J. Moscynski

About This Chapter

Service industries are the subject of this chapter. Service industries deliver services to their customers rather than goods. Selected industries are discussed, including general insurance, health care insurance, call centers, business process outsourcing (BPO), and information technology (IT) outsourcing. The way to think about quality in service industries and the special problems encountered is fundamental to this chapter. Each service specialty addressed has a section on quality issues, metrics, and opportunities particular to the service specialty. References that are particularly useful to service industries are listed for further use.

High Points of This Chapter

The high points of this chapter are

1. Private service-producing industries make up 68.2 percent of U.S. gross domestic product (GDP).

2. Quality means having those features of a service that a customer desires while remaining free from deficiencies or defects in that service.
3. If initially faced with little or no reliable data, one must immediately take action to start gathering such data. This may seem daunting, but it is the only way. With clever foresight, it generally is not as daunting as it first appeared.
4. "Go slow to go fast." Be deliberate in your actions to improve quality. If it were that easy, it would have already been done.
5. Top management support is critical to the success of a quality service program.

Introduction

Service-based organizations refer to businesses serving customers through some means other than manufacturing or the production of goods. The Bureau of Economic Analysis, Dept. of Commerce reported that in 2008, the private goods-producing sector of the economy accounted for 19.8 percent of gross domestic product (GDP). In that same year, private services-producing industries accounted for 68.2 percent of GDP, as shown in Figure 21.1. Services account for more than two-thirds of the U.S. economy. The goods-producing segment of the economy is made up of agriculture, forestry, fishing, hunting, mining, construction, and manufacturing. The services-producing segment is made up of utilities; wholesale trade; retail trade; transportation and warehousing; information; finance, insurance; real estate, rental, and leasing; professional and business services; educational services, health care, and social assistance; arts, entertainment, recreation, accommodation, and food services; and other services, except government.

A full treatment of the whole service-based economy is not possible in this text. Selected industries will be discussed, including general insurance, health care insurance, call centers, business process outsourcing (BPO), and information technology (IT) outsourcing.

Quality in Service-Based Organizations

The definition of quality for service-based industries is the same as for other industries. Quality has two fundamental aspects. Quality means having those features of a product or service that a customer desires while remaining free from deficiencies or defects in that product or service.

Is quality more difficult in service-based organizations? Measuring quality in service-based industries is often perceived as more difficult than in manufacturing. Modern manufacturing is old, with many tracing its roots to the eighteenth century. By its nature, manufacturing has always relied on measurements, quantities, weights, etc. Measurements seem to be quantifiable and intrinsic to manufacturing. Service-based industries, however, have often relied on qualitative measurements, such as "good, better, best" and "economy, deluxe, and luxury."

	2005	2006	2007	2008
GDP	100	100	100	100
Government	12.6	12.5	12.6	12.9
Private goods-producing industries	19.7	19.8	19.3	18.9
Private services-producing industries	67.7	67.7	68.0	68.2

Figure 21.1 Value added by industry group as a percentage of current-dollar GDP by year. (*Bureau of Economic Analysis, Dept. of Commerce.*)

The Problem of "No Data!"

Indeed, a frequent first lament of someone starting a quality program in a service industry is that there is no data. Although it is true in some instances that there is little useful or accurate data when starting out on a quality program, there often is more data than we think. One should approach the IT department of the organization early in a quality improvement program. They often have vast quantities of data that are routinely collected but not disseminated. A quick query of existing databases may yield a surprising quantity of data. The veracity of all data should be checked before using it or relying on an analysis of the data.

If there truly is little or no data available, or if it is corrupt and unreliable, then one must start gathering new data that will be useful. As with all change, suddenly starting to gather data where none had been gathered before can be alarming to employees or customers or both. There is a natural fear of change of being measured when one has not been measured in the past. Appropriate introduction of change methodologies should be used prior to gathering new data to allay such fears. If this step is skipped and misunderstanding of the real purposes of gathering the data exist, it can doom a quality program from the start. Too many individuals taking time out to just gather data or explain the purpose of gathering the data or the value of a quality system may seem a waste of time. The eagerness of the starting moment may be hard to resist. Keep in mind the old saying that applies well here: "Go slow to go fast." This saying is applicable to quality improvement efforts. This does not mean to imply that one should drag one's feet or that implementation of a quality program is slow. It simply means that to do things poorly fast is not an improvement.

Management Support

As in all industries, top-down management's understanding, appreciation, and support of any organizationwide quality program is mandatory. It starts at the top. If only nodding approval or a "let's wait and see" attitude is perceived by the employees, a quality improvement program will fail. The employees watch what management does much more than what management says. A quality council made up of senior executives should be established to oversee the quality improvement program as soon as possible. They should meet regularly, quarterly would be a desirable interval. Their interest and guidance to the program will speak loudly to the organization.

General Insurance Industry

The general insurance industry includes organizations offering insurance to individuals and groups for property, casualty, workers' compensation, and automotive insurance products.

Quality Issues

Insurance involves the sale of a promise by an insuring organization to the insured. There can be many misinterpretations of just what that promise was in the event of a loss at a later time. From the customer's point of view, quality involves an accurate understanding of the terms and bounds of that promise—neither more nor less. A mistake on the customer's understanding can mean disappointment and significant monetary loss. A mistake on the insurer's interpretation of the promise can mean loss of business or great economic loss.

Both parties must be absolutely clear on what exactly is being underwritten. The onus is on the insurer to know as much as possible about the insured. To be incorrect means that the insurer will lose money on claims or lose business. The insurer needs accurate, timely, and relevant data to describe the risk of underwriting the loss. Data quality is key.

Correct assessment of risk is ultimately to the insured's advantage also. Policy language needs to be crisp, neither too much nor too little, for both parties to have adequate understanding of the underwriting specifics. The insured's view of quality will vary greatly. For instance, the more affluent the insured, the larger the deductible relative to the insured amount will influence the sense of quality features to the insured. The insurer must be clear on making an apples-to-apples comparison of cost and benefit to the insured. The insured can lose business to competitors who imply the same benefit but at reduced cost when in fact they have, for instance, increased the deductible to attain that lower cost.

Income and expense are important metrics to the insurer. However, all must be balanced against risk. If, for instance, a low-paid and low-skilled worker is used to adjust claims, a serious mistake by the low-cost employer can be many times the cost of a higher-paid, competent employee. Cutting costs must be balanced against higher indemnity costs. Costs cannot be looked at solely without reference to other metrics.

Metrics

Typical metrics in the general insurance industry include:

- Income-to-expense ratio
- Monies held back to pay for future claims
- Expense versus indemnity cost
- Cycle time for processes, tasks, and subtasks
- Count or rate of customer phone calls per period
- Adherence to standard practices

Opportunities

The future holds many opportunities for the general insurance industry. If capital markets remain tight, there will be a trade-off of decisions between capital-intensive automation and manual operations. Business process outsourcing (BPO) has not been utilized as fully in the insurance industry and presents opportunities. BPO has potential to lower costs and increase revenues and customer satisfaction for the industry. (BPO is a service also discussed in this chapter.) The popular quality improvement programs such as Operational Excellence, Lean, and Lean Six Sigma (LSS) have been used in the industry, but there is more potential benefit to be gained. A corporatewide quality program should be considered. Full use of the LSS tool kit, such as parametric and nonparametric statistics, regression, and design of experiments, should be made. Multiple regression could also be used as a predictor for events. Logistic regression could be used for the probability of events. There is an opportunity for more testing and the piloting of solutions on a small scale to reduce risk until the solution is proved effective on a broader scale.

Health Care Insurance

Health care insurers provide medical insurance to client corporations and individuals. They have several constituent groups, including patients, doctors, health care providers, client organizations and individuals, hospitals, and government regulators. It is a complicated business with many players. At the time of this writing, the U.S government is considering a major overhaul to the way health care and health care insurance is delivered and paid for.

Quality Issues

The numerous interested parties in health care insurance complicate this business. The ultimate aim is to provide timely and effective treatments and disease prevention to patients in a cost-effective manner. Among health care providers, medical facilities such as hospitals, and the insurers, the insurers often have the best and most complete data. Who is the customer depends upon which interested party to health care one asks. The interested parties are sometimes at cross purposes also.

Opportunities

Health care insurance is amenable to Lean and Six Sigma quality programs. With the wealth of data available to the health care insurers, significant analysis of the data and improvements can be made. Quality projects often have very large dollar significance and impact for even small improvements.

Metrics

Metrics used in health care insurance include:

- Turn-around time of claims
- Improperly denied claims on appeal
- Effectiveness of care
- Access/availability of care
- Satisfaction with the experience of care
- Health plan stability
- Use of services
- Cost of care
- Health plan descriptive information

Call Centers

Call centers can be either internal call centers or third-party call centers run as a service business to other businesses. This discussion will focus on third-party call centers, although much would apply to internal call centers also. Call centers receive requests and complaints and handle transactions for the end-user customers of client organizations.

Quality Issues

The end users want their issues handled swiftly and satisfactorily. The client organizations want the same as their customers—and at the lowest possible cost. Third-party call centers are in a competitive business. Call centers focus on good person-to-person communication. The third-party call center wears the hat of the particular client they support. The more the call center can transparently appear as the client organization, the better for all parties involved in the transactions. Most agents represent just one outside client to enhance this perception.

Opportunities

Call centers lend themselves well to quality improvement programs because there is so much accurate data that is collected by computer. This makes analysis easier. However, due to the competitive cost pressures, a careful balance must be struck between service delivery and quality programs. There is only so much labor for both tasks. Consequently, call centers lend themselves well to part-time quality project teams. Call centers need to stay lean by definition. A few years ago, call centers were almost all by phone communication, with some mail and e-mail. With the recent advent of social media, that model is changing. Customers are using social media such as phone texting, Facebook, and Twitter. Call center customers have more technological choices with regard to interaction than they had a short time ago. This phenomena is still evolving, but is expected to change call center operations.

Off-shoring has been a recent phenomenon in call centers also. The reason for that is primarily cost. But quality as measured by both the client and the customer cannot suffer. When conducting call center operations globally, the same metrics, measured in the same way, must be used.

Metrics

Common metrics of call centers include:

- Profit margin (corporate and by client)
- Interactions/labor hour
- Average handle-time (AHT)
- Call waiting time
- Call abandonment rate
- First call resolution rate
- Call "quality"
- Forecast accuracy
- Client service level agreement (SLA) metrics
- Percent of transactions through self-service
- Percent of service offshore
- Customer satisfaction scoring

Computer and IT Services

Computer and IT services lend themselves to business-to-business relationships. Many years ago, most organizations had their own internal IT departments. Now it is common for organizations to outsource some or all IT functions to contract vendors. This service can include processes such as hardware and software procurement, installation, maintenance and inventory, help desk services, server processes, desktop, and specialty software issues. As computing has gone from special case and one-offs to more off-the-shelf and standardized, IT has become more of a commodity function best left to specialists while an organization focuses on its core competencies, products, and services. Organizations providing IT services may have their employees on site, off site at a central location reached by phone or web interface, or a combination of both.

Quality Issues

IT service organizations frequently have to repair systems and fix problems. There is a difficulty many times identifying the root cause of the problem versus fixing and chasing the symptoms of the problem. The symptoms of the IT problem are most frequently reported by end users.

Metrics

Metrics for computer and IT services include:

- SLA metrics
- Mean time between failures (MTBF)
- Availability to the end user
- Mean time to repair

Opportunities

Computer and IT service organizations can make good use of Lean and Six Sigma quality improvement programs. In addition, Design for Six Sigma (DFSS) programs may be of use for designing new software or developing system specifications. IT service organization contracts are usually governed by strict SLAs that usually have financial consequences. Unfortunately, SLA metrics are often chosen for ease of measurement and clarity rather than the more difficult metrics that really drive end-user customer satisfaction. SLAs set the minimal level of service to avoid a penalty. Client organizations want "faster, cheaper better" from the IT service providers, and the quality of data available to the IT provider is often poor. Ironically, every call may be logged, but specific information for later analysis is often missing. For instance, the fact that a particular server has failed several times in the last few months may not be readily available, just that a particular server failed with no reference to its history.

In addition, information may be in a record but in the wrong field and therefore missed. In recent years client organizations have been hiring multiple IT vendors where previously all work would go to one vendor. In fact, the client may have to issue multiple tickets for essentially the same outage if they contract with multiple vendors. This has made communications among the IT vendors difficult due to terminology and practices. International Association for Standardization (ISO) and the Information Technology Infrastructure Library (ITIL) standards have been useful in resolving multivendor problems. As client organizations have gone global, their IT service providers must also be global with 24/7 support. Lean and Six Sigma both have applicability to IT service providers as part of their quality improvement program. The programs have to be worked in carefully, however, as clients are impatient during actual outages. Therefore, improvement program projects have to be completed outside of outage situations.

Business Process Outsourcing

Business process outsourcing (BPO) refers to the growing practice of one organization outsourcing some number of its processes to a third party to execute the selected processes. This has often been done for so-called back-office processes, but it can include primary functions such as sales, order fulfillment, and other activities, putting the third party in direct contact with the end user or customer. The organization that contracts out its service is sometimes referred to as the "client" organization. The person or organization that is the end user is often called the "customer." The third-party or BPO organization delivers the service. The client organization uses a BPO organization because it does not want to do the task or the BPO organization specializing in the service can execute the tasks cheaper or with better quality or both.

Quality Issues

Using a BPO organization allows the client organization to concentrate on their core competency, such as product development or marketing. Live dialogue between the third party and a customer is not always required for processes such as credit card and check transactions. The BPO resource is sometimes located deliberately offshore from the client's location. This may be for reduced costs or a time zone advantage or both. There is a greater risk of language, culture, or communication difficulties when choosing an offshore BPO. In effect, the BPO organization "the client organization" to the customer for all practical purposes. When the distinction between the BPO organization and the client organization is transparent to the customer, this can truly be a winning solution for all three parties.

Metrics

When dealing with multiple offshore, near-shore, and domestic BPOs, special attention must be given to universal metrics. Universal metrics must be evenly applied and commonly understood globally. BPO organizations often include the following metrics:

- Transaction defect rate or, conversely, transactions free of defects expressed as a percentage
- Reduced variation in processing expressed as a mean value and standard deviation
- Transactions per time unit per agent
- Interactions per labor hour
- Customer satisfaction scores

Opportunities

BPO organizations primarily deal with people relationships. They are naturally staffed by employees with good people skills. This is a good thing. In the complex world of the BPO, however, strong analytical skills are needed as well. As a BPO organization matures and faces stiffer competition, presenting a good face to the customer is not enough. Customers expect a global high standard of quality. Client organizations want high quality also but will insist on quality at a lower cost each year. Systematic quality improvement programs such as Lean and Six Sigma are well suited to BPOs in their efforts to improve quality and reduce costs. During economic downturns, client organizations increase their reliance on and use of BPO organizations in order to reduce costs while maintaining or even improving service levels to customers.

Transactional Operations of Any Organization

Whether one is in a service industry directly or not, all businesses have "internal service" departments or functions. These functions can be treated as "internal service industries." For example, an internal call center or help desk will have many of the same considerations while pursuing a quality program as a business involved in third-party call center services. Internal service departments should be included even in nonservice industry businesses as part of a corporatewide quality initiative.

Other Service Industries

Many service industries, in addition to those discussed in this chapter, are good candidates for quality programs. Organizations in hospitality, retail sales, financial services, banking, airline, and travel industries, for example, can make good use of quality programs to increase profits and market share through superior customer satisfaction.

Acknowledgements

The author is grateful to many people who contributed directly or indirectly to this chapter. Special recognition and thanks are afforded to Brain Swayne, Elizabeth Hamill, Mark Strassburg, Larry C. Pickett, Dave Quandt, Robert J. Callahan, and Darryl Bonadio for their contributions to this chapter.

References

Bossidy, L., and Charan, R. (2004). *Confronting Reality: Doing What Matters to Get Things Right*, Random House, New York.

El-Haik, B., and Roy, D. M. (2005). *Service Design for Six Sigma: A Roadmap for Excellence*, John Wiley & Sons , Hoboken, NJ.

Friedman, T. L. (2005). *The World Is Flat: A Brief History of the 21st Century*, Farrar, Straus & Giroux, New York.

Gryna, F., Chua, R. C. H., and De Feo, J. A. (2007). *Juran's Quality Planning & Analysis for Enterprise Quality* (5th ed.), McGraw-Hill, New York.

American Society for Quality (ASQ) http://www.asq.org/, March 7, 2010

The Conference Board (for best practice sharing) http://www.conference-board.org/, March 7, 2010

iSIXSIGMA http://www.isixsigma.com/, March 7, 2010

iSIXSIGMA Magazine

Information Technology Infrastructure Library (ITIL) http://www.itil-officialsite.com/home/home.asp, March 7, 2010

International Association for Standardization (ISO) http://www.iso.org/iso/home.htm, March 7, 2010

Juran Institute http://www.juran.com/, March 7, 2010

Robert W. Baird (industry research and trends) http://www.rwbaird.com/research-insights/baird-research-insights.aspx, March 7, 2010

Society of Consumer Affairs Professionals (SOCAP) http://www.socap.org/, March 7, 2010

National Committee for Quality Assurance (NCQA) http://www.ncqa.org/, March 7, 2010

URAC http://www.urac.org/healthcare/ (aka American Accreditation HC Commission), March 7, 2010

Center for Medicare and Medicaid Services (CMS) http://www.cms.hhs.gov/, March 7, 2010

CHAPTER 22

Self-Service Based Organizations: Assuring Quality in a Nanosecond

Angel Tonchev and Christo Tonchev

About This Chapter

The purpose of this chapter is twofold: to enhance the general understanding of self-services and to elaborate on quality in self-service-based organizations. This chapter is intended for managers and quality professionals of both organizations that are entirely designed around the self-services and organizations from the traditional service sectors that are considering a self-service implementation.

High Points of This Chapter

1. Self-services are one of the fastest-growing segments of the economy. Self-services are marketplace transactions in which no interpersonal contact exists between the service providers and the service receivers.
2. Self-services have a distinctive nature. Key characteristics of self-services include nonpersonal interactivity, infrastructure availability, user's controllability, and effort inevitability.

3. Self-services and human-assisted services do not necessarily exclude each other, but instead can be parallel and interconnected.
4. Quality in self-services is affected by numerous factors including infrastructure, user interface, and customer skills.
5. The evolution of self-services unfolds in three distinctive stages: efficiency-driven, technology-driven, and customer-driven.
6. Self-services require a broader vision of quality that incorporates all phases of self-service life cycle.

Introduction

Service organizations have been thoroughly examined in Chapter 21, Service-Based Organizations: Customer Service At Its Best, of this book. A majority of these organizations provide traditional services based on a strong human interaction between employees and customers. However, a new wave of services based on nonhuman interactions between service providers and service consumers has started to flood our contemporary life with incredible speed and range. These are the self-services and they are virtually everywhere—streets, shops, banks, restaurants, airports, and obviously on the Internet. Their astonishing growth has been propelled by the recent revolution in telecommunication and information technologies and most notably by the expansion of the Internet. Self-services are transforming the way we live and do business. Their full impact on our social, cultural, and business environments remains yet to be seen.

Self-Service Industry Classification

The current industry and economic classifications systems do not recognize self-services as a separate industry category. It is highly unlikely that they will do it in the future as self-services have penetrated many segments of the economy, making the segments' differentiation based on this single attribute very difficult.

It is worthwhile to mention that the industry classifications are developed by governments to collect statistical information on economic activities. However, they may exhibit a certain level of subjectivity and may not fully reveal economic realities. In addition, differences among countries still exist despite the United Nations (UN) effort to establish an international industry classification standard (the International Standard Industrial Classification of All Economic Activities).

In reference to the current UN industry classification system, observations can be made that the information and communication sector is closely linked with the self-service industry since a vast majority of Internet and information technology (IT) based organizations can be classified as self-services. The automated teller machine (ATM), kiosk, and vending machine operators are also logical candidates for the self-service industry. Even the service providers in some well-established service industry sectors such as banking and retailing may qualify for inclusion in this category (e.g., online retailers, online banks). The list of organizations may be quite extensive, and therefore it is left to the reader to determine the appropriate classification of a given service.

Idiosyncrasy of Self-Services

What are self-services? How are they characterized? How do they differ from the standard services? Are there differences within self-services? These are some of the questions that come to mind and need to be addressed prior to a further investigation into the quality in self-service-based organizations. Therefore, this section is dedicated to explaining the nature of self-services.

Self-services are marketplace transactions in which no interpersonal contact exists between the service providers and the service receivers. From a customer's point of view, the human interaction with the service provider can be substituted with a nonhuman one when the consumers execute the service delivery on their own while using organizational assistance in the forms of infrastructure and service design. In this situation, consumers are the initiators, producers, and controllers of the service transactions. The self-service gas stations and fast-food restaurants are vivid examples of such an approach. Alternatively, from an organization's perspective, the human contact in service delivery can be exchanged with technology able to deliver the full service. Examples of such technologies are the automated teller machines (ATMs), vending machines, electronic kiosks, and the Internet. Academic literature often refers to the latter type of self-service as technology-based self-service or self-service technology (SST).

Self-Service Characteristics

There are four main characteristics that exemplify self-services: nonpersonal interactivity, infrastructure availability, controllability, and effort inevitability. They can be better remembered as the "NICE" approach to self-services (the acronym derives from the first letters of the four characteristics.) To bring some clarity, a concise description of each element is provided.

Nonpersonal Interactivity

Self-services differ from the general services by the lack of interpersonal communication between the organization providing the service and the service beneficiary (customer). Therefore, the first most intrinsic characteristic of self-services is the nonpersonal interactivity. The term signifies the one-person nature of the self-service interactions.

Infrastructure Availability

The infrastructure availability is an important self-service trait. While self-services have intangible features, their nonpersonal interactivity characteristics necessitate a tangible (physical) medium or infrastructure.

Controllability

The next important feature of self-services is their recipient's controllability. The term indicates that self-service consumers are in control of the related transactions, and their knowledge, abilities, and actions to a large extent determine the final self-service result.

Effort Inevitability

Finally, the last self-service characteristic is the receiver's effort inevitability, which implies that the recipient's active participation in the self-service production process is essential and unavoidable. Generally, self-services require a higher level of customer effort than the traditional human-interaction-based services.

The NICE approach can be a useful tool for identifying and analyzing self-services. Nevertheless, the model should not limit the attention only to the proposed four characteristics, but rather should be considered as a guideline to self-services. Additionally, the degree of relevance of each characteristic may vary according to the particular self-service situation.

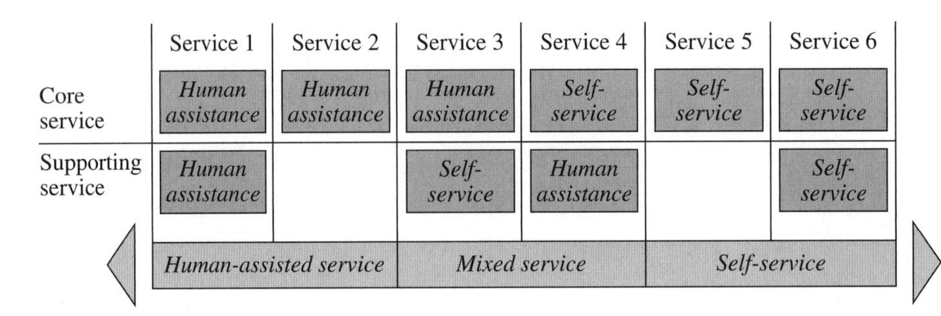

FIGURE 22.1 Service continuum. (*Tonchev and Tonchev 2003.*)

Service Continuum

Services can have a complex nature and consist of various interrelated processes. Part of these processes may be executed without the direct personal involvement of the service providers, while other parts of the processes will need their physical presence. Therefore, each service can be position on a continuum ranging from high in human assistance to high in self-service (Tonchev and Tonchev 2003). This concept is presented graphically in Figure 22.1.

The combinations in this model vary from services with only human assistance on the left side to pure self-services on the right side. Each of the six main groups of services is divided into core and supportive parts. The core part is the one that motivates customers' interests and determines the choice for a particular service. On the contrary, the supportive part is facilitating the successful delivery of the primary service, and therefore if it is considered alone, it will not create a benefit for the consumers. To give an example, we can look at the common practice of buying an airplane ticket. The core service in this case is finding the appropriate flight and reserving it, while the supportive service is the execution of the ticket payment. In this connection, six main combinations with different levels of self-service can be distinguished. The first two arise when the customers only want to receive information about the available flights but do not intend to buy an airplane ticket. In other words, they need only the core service. In this case, they have several options: go to or call the flight agency and receive human help (service 2 in the service continuum) or search on the Internet and obtain the necessary information on their own (service 5). However, if the customers also want to purchase an airplane ticket, then four additional scenarios are relevant. The first scenario occurs when they go to the flying agency, receive the flight information, and buy the ticket with the help of human assistant (service 1). Alternatively, they can call the telephone assistant to ask for the flight and make reservations, but pay electronically or by check (service 3). Reversibly, they can gather the flight information through the Internet, Teletext, or Interactive Voice Response (IVR) and make the payment at the airline desk at the airport (service 4). Finally, the last scenario occurs when the customers arrange everything on their own through the use of the Internet or the airport kiosk (service 6).

The above examples illustrate the complexity in the interactions between service organizations and customers. Additionally, they illustrate the gradual transition to self-service with its highest levels in services 5 and 6.

The service continuum supports the idea that self-services and human-assisted services do not necessarily exclude each other, but instead can be parallel and interconnected.

The mixed services in the continuum confirm this statement and provide constructive hints in the area of service design and customer channel management.

Service User Interface

Customers can communicate with and receive services from organizations in numerous ways. They are confronted with a variety of options, ranging from brick-and-mortar shops to the Internet. The service continuum discussed in the previous section brought to light some of this variety. However, a detailed review of the service interfaces is essential to clarify the core difference between human-assisted services and self-services and to facilitate the reader's understanding of self-service quality (discussed later in this chapter).

The interface for most users is regarded as a system (Kendall and Kendall 1998). In the context of services, the interface reflects how the service is delivered to the customer (Van Riel et al. 2001). This concept is particularly important to self-services.

A summary of all possible service user interfaces is provided in Figure 22.2. As can be seen, the service interfaces are split into two main groups: human and nonhuman.

Human Service Interface

The human interface is heavily researched and traditionally recognized as the most common service form. It involves an exchange between the organization employee and the customer. The human interface can constitute direct (person-to-person) or indirect contact.

Human Service Interface—Direct (Person-to-Person) Interface The person-to-person interface implies a direct physical interaction between the service provider and the receiver. It is associated with many services, such as hairdressing, brick-and-mortar retailing, medical help, and education. When two humans interact, they communicate both verbally and nonverbally (Santrock 1997a). Therefore, the person-to-person interface has two components: verbal and nonverbal communication.

Human Service Interface—Direct Interface and Verbal Communication The verbal communication is based on speech and listening. Typically, the customer's audio receptors assimilate the sound from the service provider and attach meaning to it. There are two types of meanings: denotative and connotative (Santrock 1997b). The denotation is the word's objective meaning, whereas connotation is the word's subjective, emotional, or personal meaning. The employee's verbal behavior affects customer perceptions of employee friendliness and competence, impacting the perceived quality of the service interaction (Elizur 1987). For example, if the employee's speech contains words of greeting and courtesy, this will most likely have a positive customer result.

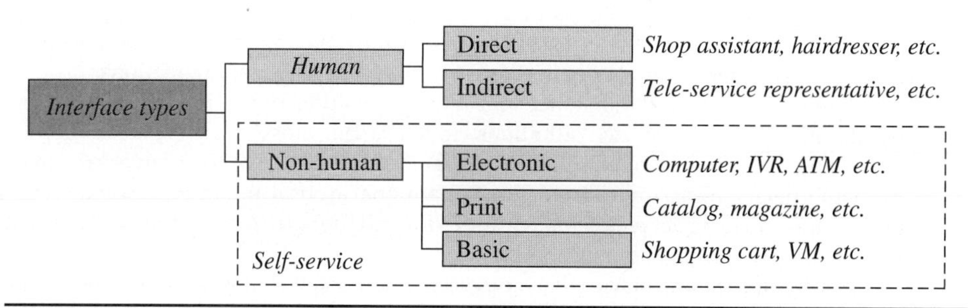

FIGURE 22.2 Types of service interface. (*Tonchev and Tonchev 2003.*)

In the field of social psychology, there is growing evidence that interpersonal encounters affect the human neurochemical reactions (Hallowell 1999). It has been observed that human-to-human contact reduces the blood levels of the stress hormones epinephrine, norepinephrine, and cortisol. In contrary, the hormones that promote trust and bonding (e.g., oxytocin and vasopressin) rise when we feel empathy for another person—in particular when we have face-to-face contact. It has been also demonstrated that the bonding hormones are at their lowest levels when people are physically disconnected, explaining the fact that it is easier to be harsh with someone via e-mail than in person.

Theoretically, all human senses (visual, audio, skin, and chemical) can take part in the person-to-person interface. However, the degree of involvement of each sense will depend on the type of service.

Human Service Interface—Direct Interface and Nonverbal Communication The nonverbal communication incorporates kinesics, paralanguage, proximics, and physical appearance (Webster and Sundarman 2000). Research in the communication field reveals that the nonverbal components are at least as important as the verbal components in shaping the outcome of employee-customer interaction (e.g., Barnum and Wolniansky 1989, Burgoon et al. 1990, Mehrabian 1981).

Kinesics, also known as body movements, includes body orientation (e.g., relaxed, open posture), eye contact, nodding, handshaking, and smiling. Several studies (e.g., Burgoon et al. 1990, Mehrabian and Williams 1969) suggest that the cues of casual smiling, light laughter, forward body lean, open body posture, and frequent eye contact are perceived as conveying intimacy and nondominance—characteristics commonly associated with friendliness and courtesy. In contrast, kinesics such as stoic facial expressions, either staring or avoiding eye contact, backward lean of body, and close body posture are perceived as conveying dominance, unfriendliness, and emotional distance.

Paralanguage reflects the noncontent or nonverbal aspects of a message. Examples of paralanguage cues are vocal pitch, vocal loudness and amplitude, pitch variation, pauses, and fluency. A multitude of studies indicate that all these elements play an important role in the interpersonal communication. Argyle et al. (1970) find that even when the content of the message is disrupted, listeners are able to detect the emotions expressed in the message based on only the tone of the voice. It is suggested that individuals with higher levels of confidence tend to speak faster and louder than their less confident counterparts (Kimble and Seidel 1991). Yet, studies on vocal characteristics (Erickson et al. 1978) reveal that speech free of long pauses, hesitations, and repetitions is considered more credible than the nonfluent speech. Scherer and Ekman (1982) further elaborate that brief to moderate pauses tend to enhance credibility and trustworthiness perceptions and that pitch variation is likely to be associated with competence and sociability. On the other hand, long pauses, increasing and decreasing tempo, higher pitch, and the lack of fluency are linked with negative affect and anxiety (Siegman and Fildstein 1978). Therefore, the paralanguage of the service provider is an important determinant of quality perception in interpersonal interactions.

Proximics refers to the distance and relative postures of the interactants (Webster and Sundarman 2000). In service situations, touch is the most vivid example of proximics. According to the theory of relational communication, the use of the nonverbal cue of touch in an interpersonal exchange can increase intentional arousal and interpersonal involvement and significantly impact recipients' attitudes toward the source of the touch (Price et al. 1995, Hornik and Ellis 1988, Patterson et al. 1986).

Physical appearance is another nonverbal component that takes into consideration the look of the service employee involved in a face-to-face interaction with the customer. Here, two elements have predominance: physical attractiveness and dress code. Studies indicate that physically attractive people are more persuasive (Chaiken 1979), are successful in changing attitudes (Kahle and Homer 1985), and are perceived as being friendlier, warmer, and more socially skilled than less attractive persons (Chaiken 1979). The dress codes, on the other hand, provide identity with the organization and can enhance the physical appearance of the service provider. Nevertheless, the type of attire considered appropriate varies by service industry and gender of the service employee (Webster and Sundarman 2000).

Human Service Interface—Indirect Contact

When two people interact but do not share the same physical space, they communicate indirectly. The technology is largely responsible for the wide distribution of this type of interface. The most common and frequently used indirect communication is the telephone conversation. The interaction between the customer and the teleservice representatives (TSRs) is mostly verbal, and therefore all the verbal communication aspects plus the nonverbal paralanguage, which were discussed above, are also relevant. In an attempt to bring as much realistic and friendly customer experience as possible, some TSRs use a mirror to see their smile before establishing a connection with the customers. The smile in a person's face to a certain degree affects his or her paralanguage. Nevertheless, kinetics such as a smile cannot be communicated directly via the telephone line.

Other forms of indirect communications are the Internet chatting and the Web tutorials. In these situations, the telephone is substituted with or supported by a computer connected to the Internet. The chatting is a real-time text interface between the user who seeks information and the operator. This option has been introduced in many commercial and noncommercial websites, using instant messaging software, such as LivePerson and Microsoft Messenger. Even though the chatting is greatly applicable for many reference services, it is limited in pleasing the customers because it only offers a simplified text communication. On the other hand, the Web tutorials create a virtual environment that is very close to the face-to-face interactions. The Web tutorial verbal communication is performed via the regular phone or the computer [Voice over Internet Protocol (VoIP)], whereas the nonverbal one is in the form of video shown on the computer screen. Most verbal and nonverbal ways of communication (kinesics, paralanguage, physical appearance) except proximics can be potentially employed in the Web tutorials. Nevertheless, the quality of interaction is largely dependent on many technical factors, such as the speed of the Internet connection and the computer hardware.

Nonhuman Service Interface

The nonhuman service interface involves impersonal interaction between the customer and the organization rendering the service. As discussed earlier, the impersonal aspect is the most distinctive self-service characteristic. In this respect, the nonhuman service interface can be considered as inherent for self-services. (Figure 22.2 takes this into consideration by marking the interface with a dotted line.) There are three underlying types of nonpersonal interface: electronic, print, and basic.

Nonhuman Service Interface—Electronic Interface

The electronic nonhuman interface refers to the interaction between the customer and the organization through electronic devices connected to a network, such as computers, mobile phones, handheld gadgets (PDAs), electronic kiosks, and ATMs. This interaction constitutes the so-called electronic commerce, or e-commerce for short. Typically these devices all have a display and some sort of a keyboard and/or navigation tools. The displays' size, technical characteristics, and impact on users vary greatly. For example, displays can be large (above 15 in.), medium (10 to 15 in.), and small (below 10 in. for PDAs and mobile phones). Generally, it is easier to perceive the information from and navigate in the large and medium-sized displays than the in the smaller ones. Similarly, the keyboards and the navigation keys of electronic devices also vary. For example, the keyboards are language- and configuration-specific (the Chinese keyboard is not the same as the English one). They can also be physical or virtual. The navigation tools range from mice and scroll buttons to touch pads.

According to Hartson (1998) and Baecker et al. (1995), the electronic interface has two components: a physical medium (hardware) and a content presentation (software). These elements influence the service experience and can determine the level of interface involvement, which is the ability of a user interface to facilitate user interest in the service offered (Reeves and Nass 1996, Shneiderman 1998). To many scholars, the electronic-based physical interface is limited in stimulating high levels of interface involvement, because the electronic hardware (e.g., displays, keyboards, and electronic pointers) confines the use of natural communication methods and thus creates interface involvement barriers (Shackel 1997, Jacob 1996, Muter and Maurutto 1991, Pavlovic et al. 1997). To enhance user interest, the electronic physical medium has to overcome its technical limitations (Card et al. 1996, Cook and Coupey 1998, Shackel 1962). The recent technological innovations in mobile phones and PDA devices (e.g., interactive gestures based on finger vein patterns and gyroscopic tracking of hand movement) are a clear signal that the self-service-based organizations are starting to address these limitations. On the other hand, research on the content presentation interface suggests that the manner in which information is presented significantly influences the interface involvement (Shneiderman 1998, Griffith et al. 2001). Central to the presentation content is the media vividness, described as "the representational richness of a mediated environment as defined by its formal features" (Steuer 1992). Media vividness can be placed on a continuum ranging from low to high. Text-only presentations are characterized with low vividness, while presentations containing multimedia features, such as animated images, video, and sound, are considered high in vividness (Morrison and Vogel 1998). Generally, high vividness results in high interface involvement (Rogers 1989, Morrison and Vogel 1998).

Additionally, along with the media vividness other content presentation interface characteristics, such as user friendliness, logic, and functionality (Van Riel et al. 2001), play a role in stimulating the interface involvement. For example, consider the customer interaction with an organization website, where the customer goal is to receive specific information about a product. The site design logic, user friendliness, and functionality, expressed in menu bars, navigation bars, search options, and site maps, is essential for the positive customer experience. Similarly, in the cases of Interactive Voice Response interaction, the customer involvement is largely influenced by the content presentation design. Finally, the last but not least important characteristic of the content presentation is its interactivity. This refers to the degree to which users can modify the form or content of the mediated environment in real time (Steuer 1992).

Nonhuman Service Interface—Print Interface

The print interface comprises the customer encounter with the organization's printed service materials, such as retail catalogs and magazines. Similarly to the electronic interface it is dichotomized into a physical medium and a content presentation, where the former is the material page and the latter is the static text and pictures (Hoffman and Novak 1996). According to the interface involvement theory (Griffith et al. 2001), the print physical-medium interface is more effective than the electronic-based physical medium in triggering customer involvement with the service. However, the theory also states that the content presentation of the electronic interface stimulates higher levels of customer involvement than the print interface counterpart. When users interact with an interface, two environments are established: a physical environment in which the users are present and a virtual one that is built upon the content of the material presented via the conduit (Steuer 1992). If the users become more engaged in the material presented (virtual environment) than in their physical environment, the interface involvement is higher (Nass and Steuer 1993). For example, when readers are highly engaged in a novel, they lose their sense of the physical environment, such as the acts of reading and turning the pages, and instead focus entirely on the story created by the author (virtual environment). Alternatively, when the interface is low, the customers assimilate only the physical environment and ignore the material presented (Griffith et al. 2001). The physical and virtual environments are identical to the physical medium and content presentation in the electronic interface.

Nonhuman Service Interface—Basic Interface

The basic interface is the last nonhuman service interface. It is associated with self-services such as self-service gas stations, self-service restaurants, supermarkets, and vending machines. Often this interface does not have a content presentation element and consists of only a physical medium. It can be expected that this type of interface will have limited ability to influence the user involvement.

Self-Service Types

Self-services are a large and heterogeneous group. They affect many segments of the economy and come in different forms and names. Depending on the focus of the observer, self-services can be divided into several miscellaneous categories. Many of these categories are also relevant to the services in general.

The Self-Service Taxonomic Model presented in Figure 22.3 distinguishes six different categories of self-service (Tonchev and Tonchev 2003). Each category is independent and consists of at least two types of self-service, which in most cases are mutually exclusive within their category. This implies that a single self-service can be classified into 6 separate self-service types or one type per category. A description of each category is provided below.

Goal of the Self-Service Provider

Organizations have diverse goals and objectives. Nevertheless, they all converge into two main types of institutions, for profit and not for profit (Dibb et al. 1994). Self-service providers

Goal of the SS provider	*Type of market*	*Degree of differentiation*	*Degree of customization*	*Self-service purpose*	*Technological intensity*
Profit	*Internal*	*Core*	*Standardized*	*Content -based*	*Technology -based*
Non-profit	*External*	*Supporting*	*Customized*	*Transaction -based*	*Labor -based*
	- Consumer (B to C) *- Industrial (B to B)*	*Augmented*		*Experience -based*	
				Application -based	
Category I	Category II	Category III	Category IV	Category V	Category VI

FIGURE 22.3 Self-service taxonomic model. (*Tonchev and Tonchev 2003.*)

are no exception to this principle. Along with the private organizations, the government organizations and nonprofit institutions have a large merit in promoting self-services. In this respect, self-services can be divided according to the goal of the service provider—profit and nonprofit. Examples of commercial self-services can be the self-scanners at the checkouts, online banking, and self-service gas stations, whereas the noncommercial ones can be the government websites and museum information kiosks.

Type of Market

The second self-service division takes into consideration the market type. Here there are two types of self-services: internal and external. The internal self-services are a relatively new phenomenon. They are also known as employee self-services (Elswick 2001), Web-based HR and HR information services (Wilson 2001). Applications for employee self-services include benefit enrollments, employment and salary verification, time card entry, withholding/deductions, training registration, and time reporting. In contrast, the external self-services are linked with the environment outside the organization's boundaries. They can be of two types: consumer (business-to-consumer) and industrial (business-to-business) self-services. Examples of the former services are the digital photography review portals, online movie rentals, social network websites, and many others, whereas for the latter these are the industrial Web portals, business network websites, corporate software applications, and Web hosting services.

Degree of Differentiation

The strategic orientation of an organization directly affects the type of self-services it offers. In some cases, self-services are very complex and consist of a bundle of several smaller service offers, whereas in other cases they are rather simple. Grönroos et al. (2000) proposes that for services offered on the Internet (not necessarily self-services), there are three types of services: core service, facilitating service, and supporting service. However, the distinction between the last two is often unclear. Anderson and Narus (1995) coin the term "supplementary service," and Van Riel et al. (2001) suggest that this term can be used to combine the facilitating and supporting services. Kasper and colleagues (1999), on the other hand, distinguish between a core service and additional service. They also state that the combination of the core service and the additional service can be considered as augmented service. Based on the above discussion, three main service types can be distinguished: core, supporting, and augmented. For example, if the self-service car wash is considered, the core service in this case is the car wash itself. If there is a vacuum cleaner in the facility, this will be a supporting self-service, and the blend of both services will change the self-service from core to augmented.

Degree of Customization

This category takes into account the level of self-service uniqueness. Attention here is paid to whether the self-service is exclusively designed for a single user or is the same for a larger consumer group. In this connection, Kasper and colleagues (1999) distinguish two types of services that can also be applied within the self-service taxonomy: standardized and customized. The standardized self-service is a homogeneous offer for a large number of customers. Examples of such self-services are the kiosks, vending machines, and online news. In contrast, the customized self-service is solely developed for the needs and preferences of just one customer. Examples of such services are the FedEx online package tracking and Internet banking.

Self-Service Purpose

The self-service purpose, examined from a consumer perspective, is an important classification factor that recognizes four main types of self-service: content-based, transaction-based, experience-based, and application-based. The content-based self-services are all services that involve transmission of information, whereas the transaction-based ones are linked with the physical and electronic transactions. Examples of the former type are the information kiosks, Internet information portals, and teletext news. In contrast, the typical transaction-based self-services are the online purchases and tradings, self-service refueling at the gas stations, ATM withdrawals, and vending machine transactions. By following the same method of analysis, self-services can also be experience-based. This category is rapidly growing and includes many computer, electronic, clipper, and entertainment games. Additionally, the music jukeboxes in the bars and the blackjack gambling machines in the casinos belong to this group. The most evident characteristic of the experience-based self-services is that the consumers are using them to satisfy their emotional and psychological needs. Finally, the application-based self-services are the last type in this category. Here, customers do not seek an experience but rather a service application. This group covers numerous software services and Web hosting options.

Technological Intensity

The technological intensity is a commonly applied category within the self-service taxonomy. A majority of the existing self-service studies (Bobbitt and Dabholkar 2001; Dabholkar 1994, 1996, 2000; Meuter et al. 2000) have examined self-services with high technological intensity. Terms such as "self-service technology" (SST), "technology-based self-service," and "Web self-service" have been coined. In contrast, the nontechnology-oriented self-services have been somehow ignored in the academic research. These self-services involve activities with higher labor intensity such as the self-service car washing, refueling at the gas station, assembling a plate at the salad bar, and pushing a shopping cart at the grocery store.

Self-Service Managerial Considerations

This section summarizes the broad spectrum of academic opinions on the advantages and drawbacks of self-services to help managers formulate better business strategies and approaches to quality. In this connection, both organization and customer perspectives are provided on the subject.

Self-Services from an Organization's Perspective

A fairly large part of the existing self-service literature discloses self-services through the lens of the service providers. The potential benefits for organizations that deliver self-services to their customers are numerous. However, there are also drawbacks associated with self-services.

Cost Reduction

The cost considerations in service organizations are important because the spending incurred by the front-desk employees accounts for the majority of their operating costs. This spending coupled with the high demand for services, rigorous price competition, and the need for global presence stimulates organizations to offer options that allow customers to receive services without direct human assistance. According to Mills and Moberg (1982), the availability of technology to automate the delivery of services helps to achieve significant cost reductions. Particularly, this is true for services such as information gathering, making delivery orders, financial transactions, postpurchase inquiries, and problem solving. Without a self-service, customer interactions are handled through alternative channels (e.g., live human assistants, call centers, e-mail reply systems) that have much higher price per service delivery than the self-service alternatives. For instance, a teller-assisted bank transaction in the United States is on average 3 times more costly than an ATM transaction and 57 times more expensive than an online bank transaction (Moon and Frei 2000). This discussion makes clear that the self-services offer organizations significant cost reduction potentials and flexibility.

Increased Productivity

The second important advantage that organizations gain with the implementation of self-service options is improved productivity (Chase 1978, Lovelock and Young 1979, Schneider and Bowen, 1985). This means that with the same employee base, organizations are able to serve more customers. If we consider the Web self-services, they can be delivered to an unlimited number of people worldwide, allowing smaller firms with scarce resources to be more competitive (Porter 2001).

Higher Scalability

Organizations offering self-service options to the end consumers boost their scalability. The term "scalability" is explained by the ratio of fixed to variable costs. It is a measure of the organization's ability to achieve economies of scale. Human resources are one of the main cost drivers in most service organizations. Spending on recruitment, training, and salaries may account for a large portion of the total expenditures. For example, in the call centers industry the total expenditure for telephone assistants is as high as 70 percent of the total call centers' costs (Anton 2001). Therefore, exchanging the live human assistance for customer self-service reduces the variable costs and improves the organization's scalability. Additionally, it helps organizations to plan more efficiently their human resource base. Particularly, when the demand for the service is high and the qualified employees on the market are few, the firm will not face the need to hire additional expensive staff. However, in recession when the demand for the service is limited, the organization can avoid cutting the number of its employees and paying corresponding compensations. High scalability, or serving numerous additional customers at extremely low incremental costs, is an important parameter to venture capitalists when they consider organization investment (Hallowell 2001).

Satisfied Employees

Most of the traditional services that are exchanged for self-service alternatives are characterized by operational simplicity, high standardization, and increased repetitiveness. Filling up the fuel at the gas station, collecting food items in the supermarket, and withdrawing cash from the bank are just few examples of services with such characteristics. When these services are delivered by organizations' personnel, the overall satisfaction and motivation levels of the employees directly involved in the service can decrease significantly. The simple and repetitive tasks require a limited variety of specific skills and knowledge. Often they do not offer employees the proper opportunity for personal learning and professional development, resulting in employees' indifference and dissatisfaction with the job. It is no surprise that such services are accompanied by high employee turnover rates. Call centers and fast-food restaurants face similar challenges. In some call centers, e.g., the average employee turnover rate reaches 125 percent per year, a number that shows the severity of the problem (Anton 2001). However, through the use of self-service, implying that the technology or the customers will perform most of the simple repetitive tasks, organizations can improve the employee job satisfaction. It is a general belief that happy employees lead to satisfied customers (Albrecht 1990, Berry and Parasuraman 1991, Grönroos, 1985).

Service Differentiation and Segmentation

Another advantage of self-services is that they improve the segmentation and differentiation capabilities of organizations (Devlin 1995, Thornton and White 2001). New technologies make available services that are highly customized to the specific preferences and tastes of the customers. Based on the information collected from the executed transactions, self-service systems are designed to update and expand continuously their databases of various consumer characteristics. The acquired customer information is analyzed and integrated in order to be subsequently used for enhancing the value of the new or existing services. It helps organizations to research consumer behavior and better define the needs of their customer segments.

Direct Sales

Through a self-service implementation, organizations are able to get closer to their clients. The advanced technology and active customers' participation in the service delivery entail shortening and optimization of the product distribution channels. For example, many businesses establish presence on the World Wide Web to sell their own goods and services, eliminating the dependence on existing lengthy distribution channels and improving the overall organizational efficiency. In particular, the utilization of Internet retailing reduces significantly the expenses for warehousing and inventory. Direct sales make forecasting easier as well. Nonetheless, the biggest advantage for organizations to serve directly their customers is the opportunity to monitor and know them better. In this way organizations can manage the relationship with the end consumers in the best possible manner.

Competitive Prices

Self-services help organizations to offer products with competitive prices. Due to cost savings realized from the use of technology or more intensive customer participation, self-service organizations are able to offer services that are cheaper than their competitors'. Additionally, self-services can eliminate the long distribution channels and boost the organization's competitiveness even further.

Freeing Up Resources for Core Business Activities

Another important advantage of self-services is that they improve the allocation of resources within the organization. Specifically, through freeing up human resources from units that perform simple and repeatable service tasks, self-services can help organizations to reallocate these resources to core business activities that create value.

Wider Customer Reach

A significant advantage of the self-services is that they allow wider customer reach. This applies chiefly for the Web-based self-services, which are not limited to the number of customers they serve and provide organizations with global presence. Nevertheless, other forms of self-service, such as IVRs, print catalogs, and ATMs, also have a contribution in this area.

Initial Investment

Often self-services necessitate substantial initial investments (Quinn 2000). These investments are made for acquiring appropriate technology, self-service advertising, or customer training. Even the Web-based self-services require basic investments in the form of hardware and software. For organizations that do not have sufficient financial resources, self-services can have a negative effect on their short-term performance.

Slow Customer Acceptance

In general, experience shows that customers do not easily accept new forms self-service, deriving from the fact that they need time to get accustomed to and comfortable with them. This is particularly true for the technology-based self-services where the older generation assimilates technology at a much slower rate than the younger one.

Reduction in Actual Service

A common pitfall of self-services is that they can reduce the actual service, rather than enhance it (Martin 2000). In most cases this is due to the lack of proper understanding of customer needs and preferences. It is a highly ineffective approach to provide only the tools for automation without the implementation of additional service features (Avalone 2002). If organizations only transfer the front-line employee functions to their customers without promoting additional advantages, then these customers will have to do more work and most likely will be less satisfied (Moon and Frei 2000). Consequently, the financial performance of the organization will suffer.

Escalation of Customer Expectations

According to several studies (Fingar et al. 2000, Welch and Lyons 2000, Wagner 2000), self-services are very effective in reducing the volume of incoming customer inquiries. However, by better informing customers about the organization's product offerings, at the same time they increase customer expectations about the inquiry replies. This forces organizations to hire technical people and pay higher wages. In some extreme cases, the escalated customer expectations can use up the financial gains generated by self-services.

Self-Services from a Customer's Perspective

Organizations not only introduce self-services to their external customers, but also utilize the do-it-yourself approaches within their internal organizational structures. Scholars suggest that internal customers are similar to external customers in their behavioral patterns and preferences (Gremler et al. 1994; Berry 1984). Self-services are seen through the eyes of the customers in both positive and negative lights.

Wider Accessibility

One of the most acknowledged customer advantages of the self-services is their wide accessibility. The broad access is a function of improved service availability and longer open hours (Zeithaml et al. 2000). According to Szymanski and Hise (2000), the technology is helping customers access the service they want from different places and at any time. For example, automated teller machines have turned out to be one of the most popular self-service interfaces due to their wide distribution (Leblanc 1990). Similarly, electronic kiosks are placed in convenient places, such as airports, railway stations, bus stops, telephone boxes, supermarket areas, and post offices, accessible 24 hours a day and 7 days a week. Nevertheless, when it comes to accessibility, the Internet is the medium with the greatest importance since services offered via the Web are location boundless.

Saved Time

The second essential advantage, closely related to the broader accessibility of self-services, is the associated time saving. According to Meuter et al. (2000), extended open hours and wider availability make self-services convenient to help customers immediately solve problems. Dabholkar (1994) presents technology as a main factor for improving the speed of delivery. An alternative view is offered by Lovelock and Young (1979), who suggest that some people prefer to perform the service themselves to reduce the delivery time. In case customers know exactly what they want, they can avoid queuing and waiting for assistance. They can choose the fastest available channel for delivery of the desired service. Moreover, some consumers believe that they possess skills and knowledge to execute particular transactions more quickly than the organization service assistants.

Increased Set of Options

In general, customers search for diversity. They value organizations that offer the possibility to select from a rich variety of options. These options can be related to products, services, or delivery processes. Self-services extend the range of options for consumers (Alsop 1999). Through optimized customer channel management, organizations can link the different self-service options and provide great flexibility and support to their customers.

Sense of Improved Service Quality

An additional benefit that self-services offer to users is the improved perceived quality of the overall service. Particularly, through avoiding the direct human contact, often customers believe that they can provide the service more effectively than the firms' employees (Meuter et al. 2000). For this they rely on their own knowledge and skills to influence the service outcome. As a result of their active participation, customers perceive the quality of delivered service to be superior.

Sense of Control

The next important advantage of self-services is the sense of customer control over the service transactions. Scholars (Langeard et al. 1981, Bateson 1985, Bowen 1986, Dabholkar 1996, and Zeithaml et al. 2000) find that one of the main reasons people select self-service options is the feeling of being in control. By using the self-service customers can freely choose the time, place, and the way the service is delivered.

More Entertaining

A positive element of self-services is also their entertaining capabilities. According to Langeard et al. (1981) some people with technical backgrounds enjoy playing with machinery, and therefore they are highly attracted by the self-service interfaces. Similarly, Davis et al. (1989) find that customers value the entertainment associated with the use of computer technologies. Further, Dabholkar (1996) suggests that customers enjoy self-services because of their relatively new presence on the market. Viewed as novelty products in the eyes of the customers, these self-services attract the attention and stimulate the desire for consumption. Finally, Pine and Gilmore (1999) analyze the importance of customer experience. According to them consumers view the experience attached to services as an integrative part of the overall product being consumed.

Lower Price

Self-services frequently offer the opportunity for saving money (Meuter et al. 2000). They help organizations to lower the prices of their products, due to associated cost reductions and improved productivity. This in turn pleases the customers because of the lower price they have to pay for a service. Additionally, organizations in their desire to attract new clients often render self-services free of charge. Today, customers have access to free terminals, free connections, and free subscriptions. The online news magazines, interactive maps, and weather forecasts are just a few examples.

Privacy

Because of the lack of interpersonal interaction, self-services provide the possibility for customers to maintain their privacy. In some instances, customers may be concerned with their image in society, the physical appearance or pushy behavior of the sales personnel and therefore might prefer to avoid the personal human contact. Generally, people do not want their private lives to be a public property, and therefore every individual seeks some form of privacy. Research shows that improved privacy makes customers keener to voice complaints and ask questions about a particular service or good. When these customers are disguised behind indiscriminate user names, they can more freely share their experiences and make evaluations of the consumed services.

Consistency

Self-services increase the consistency of the service offer. This means that customers receive almost identical services every time. In general, customers value the consistency of services because it assures them that the positive service experiences will be repeated in the subsequent service consumptions. Similarly, it warns them that the negative service experiences are not accidental, facilitating customers to choose another service provider.

Need for Assistance

Customers' inability to perform the service, due to deficiency of necessary skills and knowledge, and the lack of assistance from the organization's employee are some of the biggest obstacles for customers to use self-services. When self-service users face a problem they cannot solve on their own, they are frustrated, embarrassed, or unwilling to engage in similar options again. Many times this is the case with the self-services that involve new and complex technologies. To avoid the unpleasant experiences, organizations have to consider appropriate service presentations and trainings for their customers. Possible solutions include providing explanatory usage information through the traditional service channels, constructing user-friendly interfaces, detailed help menus, lists of frequently asked questions (FAQs), and wizard programs. Additionally, the availability of a live human assistance option is recommended.

Equipment Requirement

Self-services may require appropriate equipment. This is particularly true for the Web-based self-services. For example, the absence of Internet connection, appropriate computer, or software will discourage customers from using the existing self-service options. Therefore, organizations that target a specific customer segment or regional population have to take the equipment requirement issues into account. Hopefully, the continuous expansion of high-capacity informational networks and the fast development of wireless technologies will facilitate ever-improving access to the Web-based self-services.

Safety and Purchasing Risks

Another drawback of self-services can be the safety and purchasing risks. Evans and Brown (1988) find that safety issues are a common reason for people to avoid technology-based self-services. Similarly, Pavitt (1997) suggests that consumers perceive security issues as a major impediment for the purchasing of products on the Internet. The most common consumers' fear is the misuse of their personal or credit card information. Finally, Fram and Grady (1997) state that customers avoid consuming products through the Internet that carry a high degree of purchasing risk—very expensive and technically complex.

Lack of Social Contact The lack of social contact in the self-service interfaces can have a negative impact on customers. Research (Dabholkar 1992, 1996; Prendergast and Marr 1994) shows that some people are prone to avoid self-services because of their need for human interaction. Similarly, scholars find that the contact with a service employee, who can respond with instant advice and recommendations, is very important to some customers (e.g., Forman and Ven 1991, Berkowitz et al. 1979). Nevertheless, for another group of people, those who enjoy playing with machines (Langeard et al. 1981) and computers (Holbrook et al. 1984), it is not so much concern with the social aspect. Therefore, the lack of social contact in the self-service encounters is not necessarily a limitation for all users. Cowles and Crosby (1990) support this line of reasoning by stating that people have different tolerances for replacing people with machines.

Effort Requirement Compared with the traditional services, self-services require a substantially higher customer effort. As stated previously, this effort is inevitable and inherent to them. Customers often view the increased quantity and complexity of the work they have to do as a disadvantage (Meuter et al. 2000, Dabholkar 1996, Davis et al. 1989, Langeard et al. 1981, Bateson 1985). Nevertheless, the evaluation of a particular self-service from the customers depends on their confidence in the ability to perform the service (Hoffman and Novak 1996). For example, if the users are confident in their abilities, they are more likely to assess the self-service offer as simple and demanding less effort.

In conclusion, self-services have numerous advantages for both organizations and consumers. However, they also carry some drawbacks. The magnitude of each positive or negative element will depend on situational, market, organizational, and personal factors.

Evolution of the Self-Service Industry

Primitive forms of self-service have existed since the inception of *Homo sapiens*. Our ancestors have maintained some level of self-sufficiency along with their community services. However, the modern self-service began much later. In the beginning of the twentieth century, the first signs of customer self-service started to appear. With the help of innovations in the information and communication technologies, self-services soon experienced an unprecedented growth and changed the ways in which we interact with organizations. Today, self-services are an indispensable part of the world economy representing an industry that struggles to find a common description name. To fully understand and appreciate the business impact of self-services, we have to look at their historic development.

Through a retrospection of the historic facts, three stages of self-service evolution can be observed: efficiency-driven, technology-driven, and customer-driven stages. These stages can be described as coexistent, as one stage has not ceased to exist as a new stage has been initiated at the market. The boundaries of these stages are not fully transparent. Nevertheless, their historic sequence is well defined.

Efficiency-Driven Stage

The first stage of customer self-service was efficiency-driven. As the name indicates, this period was characterized by a strong organizational focus on cost reductions and service process optimization. Due to competitive pressure and the rise in the wage levels, organizations were forced to look for alternative ways to reduce costs and expand market share. After the inefficiencies in the production processes were corrected, the customer self-service was the next viable alternative to lower the price and stimulate the consumption. Therefore, in the beginning of the twentieth century the first wave of self-service initiatives flooded the service industry. It started with the introduction of the direct-mail catalogs and vending machines and then continued with the self-service offers in the grocery stores, supermarkets, fast-food restaurants, and gas stations. What is typical for this stage is that most of the self-services were associated with a high level of physical effort. Below are some of the major milestones in this period:

Late 1800s: The Emergence of the Direct-Mail Catalog Industry

The first catalogs were introduced in the late 1800s to efficiently reach customers in remote locations (Windham and Orton 2000). The catalog industry experienced a substantial growth in the late 1900s as catalog retailers multiplied and the range of products offered via these channels increased. Catalogs were mostly associated with impersonal shopping, because in the early days orders were placed and fulfilled through the mail.

Late 1800s and Early 1900s

This period saw the appearance of the vending machines. The first commercial coin-operated vending machines were introduced in London, England, in the early 1880s (www.inventors.about.com). Later in the beginning of the 1920s, the first automatic vending machines started dispensing sodas into cups, and in 1926 the cigarette vending machine was invented. These inventions were well accepted by consumers and contributed to the tremendous growth of the self-service industry.

1916: The First Self-Service Grocery

Piggly Wiggly, the first true self-service grocery store in the United States, was founded in Memphis, Tennessee, in 1916 (www.pigglywiggly.com). In grocery stores of that time, shoppers presented their orders to clerks who gathered the goods from the store shelves. The manager of the organization noticed that this method resulted in wasted consumer time and worker-hours, so he designed a process for shoppers to serve themselves.

1930: The First Supermarket

In 1930 the first King Kullen store opened in New York, widely regarded as the first supermarket, although the 1920s-era Ralphs stores in Los Angeles probably have a more valid claim to the title (www.groceteria.net). Seven years later, A&P began consolidating its 15,000 small stores into self-service supermarkets (www.aptea.com).

1937: The First Shopping Cart

The shopping cart was conceived in 1937 by Sylvan Goldman, one of the original self-service grocery retailers who, in observing the shopping habits of his customers, realized he could provide better service and sell more groceries if only he had some means of helping them carry more merchandise (www.unarco.com). From this simple observation the shopping cart was born, along with the tremendous growth in self-service mass market retailing that continues today.

1960s: The Growth of the Fast-Food Restaurants

In 1954 the first McDonalds restaurant was founded in San Bernardino, California (www.mcdonalds.com). A year later another restaurant opened in Illinois, and the organization business started to grow. Throughout the 1960s the fast-food, self-service restaurants conquered the hearts of the North Americans and soon became part of their lifestyle.

1970s: The Expansion of the Self-Service Gas Stations

Although self-service stations had been around since the 1930s, and despite advances in self-service safety, only 15 percent of the gasoline consumed in the United States was sold on a self-service basis in 1975 (Phillips and Schutte 1988). All that changed, however, with the gasoline shortages of the next several years. Suddenly, long lines of cars snaked around service stations. Attendants barely had time to work the pumps, much less check the oil and clean windshields. Furthermore, with fuel prices rising, customers wanted ways to cut fill-up costs. The answer was self-service, and by 1979, more than 50 percent of stations offered it (Phillips and Schutte, 1988).

Technology-Driven Stage

The successful customer assimilation of self-services and technological progress in the twentieth century has stimulated organizations to experiment with different types of self-service technologies (SSTs). This led to the formation of the next self-service stage, the technology-driven one. In comparison with the preceding phase, the focus here was more on the application of various technologies in self-services than on the efficiency considerations exclusively. The following historic milestones are worth mentioning:

1930s to 1980s: The Computer Revolution The computer, as we now understand the word, was very much an evolutionary development rather than a simple invention. Nevertheless, it all started in the late1930s with the work of two scientists: Konrad Zuse and Prof. John Atanasoff (Mollenhoff 1988). In 1936 Zuse made a mechanical calculator called the Z1, the first binary computer, and in 1939 he completed the Z2, the first fully functioning electromechanical computer. Between 1939 and 1942 Prof. Atanasoff and graduate student Clifford Berry built the world's first electronic digital computer at Iowa State University. The Atanasoff-Berry machine represented several innovations in computing, including a binary system of arithmetic, parallel processing, regenerative memory, and a separation of memory and computing functions.

The computer innovations continued throughout the 1940s, and in the 1950s John Presper Eckert and John W. Mauchly built the first commercial computer (UNIVAC). A few years later in the 1970s the first consumer computer (IBM 5100) was produced, and in the 1980s the personal computer revolution had begun. Today, nothing epitomizes modern life better than the computers, which have infiltrated every aspect of our society. Now computers do much more than simply compute: supermarket scanners calculate our grocery bill while keeping store inventory; computerized telephone switching centers play traffic cop to millions of calls and keep lines of communication untangled; and automated teller machines let us conduct banking transactions from virtually anywhere in the world. The computer merit in the expansion of self-services is enormous.

1960s: The First Automated Teller Machine In 1969, DocuTel created the first modern ATM and sold it to New York–based Chemical Bank (www.inventors.about.com). The ATM completed cash dispensing transactions and was an off-line machine, meaning money was not automatically withdrawn from an account.

Nearly 30 years later Cash Tech developed the EMMA transaction processing system that allowed ATMs to quickly execute complex transactions over multiple networks (www.cashtechnologies.com). The system made possible the communication between four primary channels: (1) the ATM network, (2) the credit card (point-of-service, or POS) network, (3) the Automated Clearing House (ACH) network, and (4) cash. Today ATMs are on their way to becoming electronic convenience kiosks, capable of event and airline ticketing, electronic bill payment, and connection to an alternate host via the Internet.

1960s: The Beginning of the Internet The Internet is a system for allowing computers to communicate with one another. During the 1960s the first computer network (ARPAnet) was developed by the U.S. Defense Department's Advanced Research Projects Agency (Griffiths 2001). This system was used predominately for military purposes. Thirty years later, in 1990, Tim Berners-Lee of the World Wide Web Consortium (W3C) in consultation with CERN, the European Organization for Nuclear Research based in Switzerland, wrote the first GUI browser, and called it "WorldWideWeb" (www.kiosk.org). In contrast with its predecessor, the WWW, or Web for short, was designed for commercial and communication purposes. This created a new market where newcomers and existing organizations started to compete and exploit the vast potential of this virtual environment. New terms, such as "e-business," "e-customers," "surfing the net," and "new economy," entered our business vocabulary. Soon the WWW became the most important medium for providing high-quality self-services.

1980s and 1990s: The Spread of the Wireless Communications Wireless communication technologies were developed as early as the 1940s (e.g., in 1940 Motorola designed a handheld portable two-way radio system). Nevertheless the mass commercialization of these technologies began in the beginning of 1980s with the introduction of the cell phone (a.k.a. mobile phone) services. Later, in the mid-1990s the Wi-Fi standard was introduced to facilitate the communication between electronic devices. All of these developments had a positive effect on the self-service growth.

1990s: The IVR Penetration In the mid 1990s the Interactive Voice Response (IVR) was introduced in many call centers. This was a type of self-service technology that helped organizations to better manage their increasing volume of incoming calls. Despite the controversial introduction of IVR, which initially affected customer satisfaction negatively, the end result of this technology application turned out to be satisfactory for both call centers and customers.

1990s: The Search Engine Battle Through the 1990s there was an intensified competition between organizations in developing Internet search engine technologies (Griffiths 2001). Below is a short list of technological innovations that took place:

- 1990: Archie, developed at McGill University (Montreal), was the first search engine for finding and retrieving computer files.
- 1991: The Gopher system developed at the University of Minnesota represented an improvement on ftp (file transfer protocol) retrieval.
- 1991: Veronica, developed at the University of Nevada, operated on the same principle as Archie, but it also allowed users to distinguish between a search for directories and an undifferentiated search.
- 1991: WAIS (Wide Area Information Server) was developed by Thinking Machines Corp. and searched through information on the basis of contents.
- 1991: WWW Virtual Library was the first index of content on the World Wide Web. It was set up by Tim Berners-Lee, the founder of the WWW.
- 1994: Yahoo! was established by two Ph.D. students at Stanford University, David Filo and Jerry Yang. Yahoo! was a commercial directory featuring an advanced "spider" search engine.
- 1995: AltaVista was created by researchers at Digital Equipment Corporation's Western Research Laboratory. It was the first site to include a translation service and a search for images and sound files.
- 1996: Inktomi was developed by UC Berkeley Prof. Eric Brewer and graduate student Paul Gauthier. The organization was initially founded based on the real-world success of the search engine they developed at the university.
- 1998: MSN Search was launched by Microsoft using search results from Inktomi. Microsoft used third-party search engine technologies until it developed its own (msnbot) in 2004.
- 1998: Google was founded by Larry Page and Sergey Brin while they were students at Stanford University. Convinced that the pages with the most links to them from other highly relevant Web pages must be the most relevant pages associated with the search, Page and Brin laid the foundation for their search engine. With its useful results and simple design, the Google search engine attracted many Internet users.

Customer-Driven Stage

While in the previous two stages the main focus was on efficiency and technology applications, the attention in this stage shifted to the customers. Organizations realized that the cost reductions and technology innovations alone would not significantly improve their competitive performance. On the contrary, they saw that by not considering the customer needs and preferences, self-service acceptance and customer satisfaction suffered. Below are some major milestones in this final stage of self-service development:

1990s: The Decline of the Customer Service Satisfaction

The customer service satisfaction, measured by the American Customer Satisfaction Index put together by the University of Michigan, declined during the 1990s. According to the article "Is the Customer Ever Right? The Decline and Fall of Customer Service with a Technological Push," published in *The New York Times* (Hafner 2000), some of the causes of this drop were the technology and the lack of customer focus. Hopefully, the consumer dissatisfaction has been acknowledged by the industry, and organizations have started to redesign their self-service strategies.

Early 2000s: The Dot.com Bubble

The recession in the IT sector in the early 2000s was a critical point for self-services, because it gave a clear signal to organizations and investors that the Internet presence alone was not enough to guarantee survival. Dot.com companies understood that the short-term emphasis was harmful and that the competitive advantage rested in the improved self-service quality and long-term vision of the customer relationship.

Mid and Late 2000s: Postdot.com Period

The Internet-based self-services have registered a steady expansion after the dot.com bubble. This expansion was to a large extent due to improvements in self-service quality and innovative ways to meet customer needs. Some of the most significant developments in this period were the appearance of video sharing portals (e.g., YouTube), mapping and navigation services (e.g., GoogleMaps), remote desktop applications, and above all the popularization of social networks (e.g., Facebook and MySpace).

Dimensions of Quality in Self-Service Based Organizations

Quality affects the financial performance and image of an organization irrespective of the industry in which it operates. Therefore, quality is as important in the self-service industry as it is in the manufacturing and traditional service industries. However, what really constitutes self-services quality?

Dr. Joseph Juran defines quality as "fitness for use" (Juran 1988). In the context of self-services this will imply effectiveness of a design, infrastructure, and support methods employed in delivering a self-service that fits a customer's defined purpose. Dr. Juran's definition puts customer utility and satisfaction in focus. This is closely linked to the customers' perceived quality, a widely accepted concept in the academic and business literature.

Service perceived quality is the customers' overall evaluation of a service, which takes place after the service consumption. Therefore, it is also regarded as an "experienced" quality. Perceived quality is a subjective evaluation of the objective service or product characteristics (Antonides and van Raaij 1998). Alternatively, perceived quality is defined as a result of the observed difference between customer expectations and customer perceptions of the service outcomes (Grönroos 1984; Parasuraman et al. 1985). Nevertheless, customers evaluate not only the outcomes, but also the process through which the outcomes are achieved and the context in which interactions occur. Accordingly, perceived service quality is described as consisting of two main components: process and outcome (Berry and Parasuraman 1991). The process component refers to the way in which the exchange takes place, while the outcome component reflects the end results achieved by using the service. Similarly, Grönroos (1984) has identified three distinct types of perceived service quality: technical, functional, and corporate image. The technical quality and functional quality types are the same as outcome and process components, respectively, whereas the third type (corporate image) reflects how the customer perceives the supplier. The empirical research has shown that the three main service quality dimensions have an effect on the overall service quality and customer satisfaction (Dabholkar et al. 1996; Zeithaml et al. 1996).

The attention of a considerable amount of research was on identifying the main attributes of service quality, which customers view as the most important in their overall product evaluation (Parasuraman et al. 1985, Dabholkar et al. 1995, Meuter et al. 2000). Examples of diverse lists of service quality components can be provided. Nevertheless, the most widely accepted attributes of service quality are derived from the SERVQUAL model (Parasuraman et al. 1988) and include tangibility, responsiveness, reliability, assurance, and empathy. These five dimensions reflect the major benefits customers obtain from a service. If applied to the self-service setting, this set of main quality parameters needs to be revised to integrate the lack of interpersonal contact in the service exchange (Van Riel et al. 2001). In this respect, self-services can be evaluated using the following five dimensions:

User Interface

User interface reflects the infrastructure and service design provided by the organization. As can be recalled from one of the previous sections in this chapter, there are three types of self-service user interface: electronic, print, and basic. All three types, however, share a common element, namely. the physical medium. In addition, two of the interfaces (print and electronic) contain a content presentation. Therefore, the customer quality perception of the user interface in self-services is based on two elements: physical medium and content presentation. The user evaluates these elements in terms of functionality, outlook, and logic.

Responsiveness

Responsiveness is linked to the organization's readiness to respond to a customer request in appropriate time periods. The speed of the response affects the customer's perception of quality. For example, the slow opening of a Web page or downloading of a document may have a negative impact on the customer's experience with the self-service.

Reliability

Reliability is associated with on-time service delivery, accuracy of the provided information and results, consistent service delivery process, and matching specifications. This dimension plays a significant role in the overall customer's evaluation of quality in self-services with high information intensity (e.g., Internet search engines and kiosks).

Assurance

Assurance is related to the safety, insurance, confidentiality of the personal information, and trustworthiness. The importance of this dimension is most pronounced in self-services involving financial transactions such as online banking and online retailing.

Empathy

Empathy is measured by the service provider's awareness of customer needs and requests as well as the customization level of the delivered self-service. This dimension may be of particular relevance for online retailers who should aim at gathering profiling data and transaction histories, so that they can offer personalized self-services that suit the demands of their customers.

Often customers judge the quality of the same self-service unequally due to the diversity of their interests, knowledge, service expectations, and situational factors. Additionally, the relevance of the discussed quality dimensions varies with the different types of self-service. For example, if the customers search for information, they will expect a fast and accurate service delivery, whereas during the execution of a financial transaction they will be concerned about the security and reliability of the service and therefore accept slower performance.

The self-service quality dimensions form the customer's perception of overall self-service quality. This quality not only affects the customer satisfaction and loyalty but also has a financial impact on the self-service providers. Therefore, from a managerial point of view, it is essential to know the dimensions of self-service quality as they will help with the proper allocation of resources to design and deliver a superior self-service.

Self-Service Quality Life Cycle

Quality dimensions, discussed in the previous section, highlight what customers consider important in self-services. However, it is the managerial job to transform customer expectations and requirements into an actual self-service offering. This transformation normally goes through several successive phases to culminate in an operating self-service. When we approach self-services, a broader vision of quality is needed to incorporate all phases of the self-service life cycle.

Academic and business literature distinguishes between the product and service life cycles. Likewise, self-services have their own conceptualization with distinct phases and relationships. In this connection, Figure 22.4 presents a life cycle model that can assist self-service organizations in managing quality.

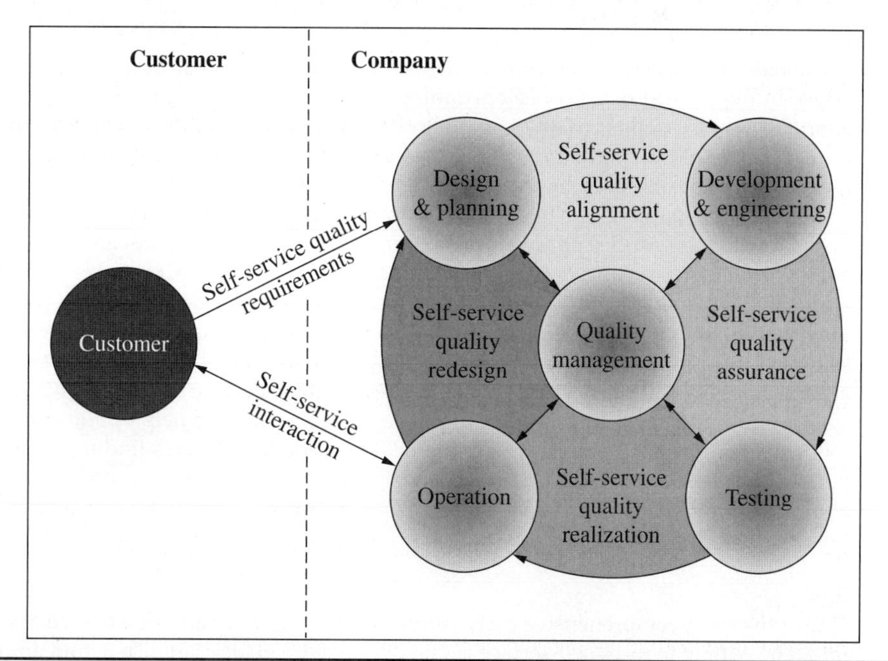

FIGURE 22.4 Self-service quality life cycle. (*Juran Institute, Inc. Copyright 2009. Used by permission.*)

The self-service quality life cycle embodies two parties that constitute a self-service interaction: customer and organization. Unlike in the goods and services life cycles, the customer is part of the model as she or he is actively involved in both self-service consumption and operation. On the other hand, the organization part is composed of four consecutive phases: design and planning, development and engineering, testing, and operation. All these phases are connected to a foundational element, defined as quality management, which provides managerial and quality functions. Furthermore, each phase has a strong relation to the preceding phase and the following one. These relations are facilitated by the quality management and characterize a method for consistently using the output of one phase as an input of the other. In this regard, four essential product quality approaches can be identified: self-service quality alignment, self-service quality assurance, self-service quality realization, and self-service quality redesign. A detailed analysis of all elements of the self-service quality life cycle model is provided below.

Self-Service Customer

Customers affect the self-service quality life cycle in two ways: via their self-service quality requirements and via their self-service interaction. A large part of the empirical research on self-services has concentrated on identifying the common self-service user profile. Although the research may provide useful clues to quality managers, it cannot be blindly applied because of two main reasons: first, the generalizations can often be inaccurate or biased due to research design and implementation; second, the profile may vary between different types of self-services. Nevertheless, the main research findings are summarized for reference.

Locus of Control

The locus of control is the person's perception of the source of his or her fate. Some people believe that they are masters of their own fate. Other people see themselves as pawns of fate, believing that what happens to them in their lives is due to luck or chance. Those of the first type—those who believe that they control their destinies—have been labeled *internals* (or having internal locus of control), whereas the latter, who see their lives as being controlled by outside forces, have been called *externals* (or having external locus of control) (Rotter 1966). In the context of a service encounter, instead of locus of control some scholars talk about a perceived control (Bateson and Hui 1987, Langeard et al. 1981). They define the term as the amount of control that customers feel they have over the process or outcome. Nevertheless, as can be observed, the two terms are identical.

There is strong evidence that people with internal locus of control are favorably predisposed toward self-service options. Dabholkar (1996), e.g., finds control and waiting time to be important determinants for using technology-based self-service. He notes that consumers are more likely to use technology-based self-service if it offers them a sense of control and if they do not have to wait to use it. Similarly, Bateson (1985) explores the choice between a self-service option and the interpersonal service delivery system. Bateson examines the attractiveness of self-services when the usual monetary or time-saving incentives are controlled and finds that a large group of people choose to use self-help options even without monetary or time-saving benefits. Bowen (1986) supports Bateson's finding by stating that people choose self-services not for monetary savings, but to feel in control. Finally, Kelly et al. (1990) and Dabholkar (1990) suggest that customers who prefer self-service perceive greater control and higher service quality.

Age

One of the most comprehensive early studies done to identify and describe customers who might be willing to use a self-service is that of Langeard et al. (1981). They find among other things that self-service participators tend to be younger. The generational aspect is further supported by Quinn (2000), who examines the acceptance of the employee self-service (ESS). He states that the change in employee population demographics from Baby Boomers to Generation X employees has a positive effect on ESS.

Education

Some researchers indicate that the higher level of education stimulates the self-service usage. Langeard et al. (1981) find that better customer education has positive correlation with the self-service options. This may be especially true for technology-based self-services.

Technology Readiness

Another personal trait that has been proposed to affect mostly the self-service technologies (SSTs) is the technology readiness. The term is defined as "people's propensity to embrace and use new technologies for accomplishing goals in home life and at work" (Parasuraman 2000). In this regard, the research team of Liljander et al. (2006) distinguishes four dimensions of technology readiness: discomfort, insecurity, optimism, and innovativeness. The team finds that optimism and innovativeness are positively related and discomfort and insecurity negatively related to customer adoption of SSTs.

Income

A limited number of studies indicate that lower-income customers prefer self-service options to services with human assistance. Graffeo (1997c) observes that the "credit-driven" group, heavy users of loan services, perform about 60 percent of their transactions electronically, compared to 32 percent of the "depositor" segment, customers with account balances approaching $24,000. Langeard et al. (1981) also support the negative relationship between income and self-service.

Marital Status

In addition to their young age, better education, and lower incomes, the research of Langeard et al. (1981) notes that self-service users are also single. Nevertheless, the users' age impacts significantly their marital status. In other words, if the self-service users are young, the likelihood that they are not married is very high. Therefore, it can be assumed that marital status is a function of age.

Culture

"Culture refers to the behavior patterns, beliefs, and all other products of a particular group of people that are passed on from generation to generation" (Santrock 1997c). Whatever its size, the group's culture affects the behavior of its members (Triandis 1994, Lonner and Malpass 1994, Matsumoto 1994). Geert Hofstede, a Dutch scientist who conducted one of the first and most comprehensive cross-cultural studies, in his book *Cultures and Organizations, Software of the Mind* (1991) evaluates national cultures on the basis of five dimensions: individualism-collectivism, masculinity-femininity, uncertainty avoidance, power distance, and long-term versus short-term orientation. In respect to self-services, two dimensions are particularly relevant: individualism-collectivism and uncertainty avoidance. Hofstede defines them as follows:

> Individualism pertains to societies in which the ties between individuals are loose: everyone is expected to look after himself or herself and his or her immediate family. Collectivism, as its opposite, pertains to societies in which people from birth onwards are integrated into strong, cohesive in-groups, which throughout people's lifetime continue to protect them in exchange for unquestioning loyalty. Uncertainty avoidance can . . . be defined as the

> extent to which the members of a culture feel threatened by uncertain or unknown situations. This feeling is, among other things, expressed through nervous stress and in a need for predictability: a need for written and unwritten rules.

According to Kasper et al. (1999), individualism is the most important cultural dimension that has an effect of self-service. In fact, as can be recalled from the previous section dealing with the self-service evolution, the emergence and proliferation of self-services occurred in the United States and England, countries characterized by a high degree of individualism. This is not a mere coincidence but rather verification that citizens in those countries have the cultural background to accommodate and accept services with a lower degree of human assistance. As for the second dimension, uncertainty avoidance is closely related to the citizens' locus of control. Particularly, individuals with internal locus of control are able to cope with the risk and unpredictability better than individuals who have an external one. Therefore, the nations with low scores on uncertainty avoidance have a larger number of internally oriented citizens. For example, if the U.S. culture is examined, one of its core values is *progress*. The term reflects people's belief in technology and continued improvement in the standard of living (Assael 1995). Because Americans believe in their abilities to change the future, to a large extent they can be considered as people with an internal locus of control. In support of this relationship, the nation's score on uncertainty avoidance is lower than average.

Quality in Self-Service Design and Planning

The self-service design and planning is the preliminary phase of the self-service life cycle. This phase is influenced by the customer requirements, on one hand, and by the quality management, on the other (see Figure 22.4). Self-service requirements are linked with the expectations customers form about a self-service and reflect the quality dimensions discussed in the previous section of this chapter. In this connection, one of the primary goals of any self-service-based organization is to design a self-service that meets customers' requirements. Nevertheless, this goal must be synchronized with the quality management and the broader organizational strategy to ensure quality in self-service design and successful self-service implementation. Some of the main issues confronting the quality managers in this phase are the following.

Identifying Self-Service Customers and Their Quality Requirements

The first step in designing a self-service is to identify the potential customers and their quality requirements. In this chapter we discussed the main characteristics of the self-service user and the general dimensions of self-service quality. Nevertheless, the self-service organizations should conduct their own marketing research to identify their specific target group and corresponding quality needs. Some of the quality tools that may be helpful in the process are questionnaires, interviews, observations, and VOICE of the customer (VOC) techniques.

Deciding on Type of Self-Service

When designing a self-service, organizations often have to choose between different self-service options. For example, a bank may choose between ATM, online banking, and telephone service using IVR system to meet users' requirements for a bank transaction. Similarly, a retailer can choose between catalog, vending, Internet, and TV retailing to satisfy the customer needs for buying a product. The final decision must be based on a detailed customer analysis where the quality requirements are ranked by importance and matched with the self-services that best fit these requirements.

Defining Self-Service Specifications

To ensure quality in self-service design, it is necessary to have objective and measurable specifications. These specifications should include both qualitative and quantitative information:

- *Qualitative.* Self-service information is descriptive, nonnumerical information that includes: description of self-service appearance, usage, and functionality; process and engineering drawings; units of measurement; service-level agreements (SLAs); operating level agreements (OLAs), etc.
- *Quantitative.* Self-service information is numerical information that involves some sort of measurement. It should always be recorded immediately, along with the units of measure, and care must be taken to use the proper precision. Quantitative information can be used to design metrics that can be set as self-service development targets or utilized in subsequent self-service life cycle phases for testing, monitoring, and benchmarking against external parties. Below are some commonly used quantitative self-service quality metrics:

Internet Quantitative Quality Metrics

- Delay (the elapsed time for a packet to be passed from the sender, through the network, to the receiver)
- Jitter (the variation in end-to-end transit delay; in mathematical terms it is measurable as the absolute)
- Bandwidth (the maximum data transfer rate that can be sustained between two endpoints)
- Packet loss probability (the packet loss rate due to transmission collisions, multipath, fading, etc.)

Website Quantitative Quality Metrics

- Page load time
- Time to first byte
- Time to start render
- Link errors per page
- Code errors per page
- Content errors per page
- Downtime
- Availability

Kiosk Quantitative Quality Metrics

- Percentage of kiosk uptime
- Kiosk failure rates
- Kiosk errors rates
- Average time per transaction

IVR Quantitative Quality Metrics

- Abandonment rate
- Transaction failure rate

- Average handling time
- Average wait time
- Completion rate
- First contact resolution rate

Software Quantitative Quality Metrics

- Number of defects or bugs per 1000 lines of code
- Number of lines of code (including comments)
- Number of lines of code (comments excluded)
- Number of characters
- Number of comments
- Number of comment characters
- Number of code characters
- Program length (total number of operands and operators)
- Estimated program length
- Jensen's estimator of program length
- McCabe cyclomatic complexity number (a measure for the detection of code that is likely to be error-prone and/or difficult to maintain)
- Belady's bandwidth metric (a measure of nesting levels)
- Entropy measure (a measure of intramodule cohesion and intermodule coupling)
- Henry-Kafura measure (a measure of information flow complexity)
- Number of rejected lines (a measure of the number of nonblank noncomment lines of code that was not successfully analyzed by the parser)

Basic Self-Service Quantitative Quality Metrics

- Average time of self-service interaction
- Number of human assistance requests
- Downtime
- Availability

Assessing Infrastructure and Hardware Requirements

Self-services frequently require existing infrastructures and hardware. For example, a kiosk or ATM will need a display space at a high-traffic location with electricity power supply and network access. Similarly, a self-service website will necessitate servers (e.g., Web, electronic commerce, and database servers) and adequate Internet bandwidth. Each self-service has unique infrastructure and hardware requirements that need to be carefully assessed.

Evaluating Information Storage and Databases

The technology-based self-services often operate with a large volume of information. The information is stored in an integrated collection of logically related records, known as the database. The database may be structured into different data models (e.g., flat, relational, object-oriented, hierarchical, dimensional, etc.) with a specific support for database language [e.g., Simple Query Language (SQL), Extensible Markup Language (XML), etc.]. Each configuration has its advantages and disadvantages. Because of the technical complexity, it is advisable for quality managers to collaborate with the IT department to assess storage and database requirements of the newly designed self-service as databases may significantly impact self-service quality.

Designing Software Architecture

The design of software architecture is of great importance for the technology-based self-services. The architecture of a software system refers to an abstract representation of that system. To embed quality in the software architecture, quality managers need to make sure the software system will meet the product requirements as well as guarantee that future needs can be addressed. Managers must also address the interfaces between the software system and other software products as well as the underlying hardware and the host operating system.

Analyzing Revenue Streams

Product financial analysis is normally the job of the finance department. However, in numerous self-service cases this analysis needs to be coordinated with the quality department. This derives from the fact that customers often expect free-of-charge self-services which in turn forces organizations to look for alternative revenue streams (e.g., advertising, ancillary business, sponsorship, and public support). The choice of a revenue stream may have an impact on the self-service quality. For example, an online advertisement such as flashy ad banners can create information clutter on the self-service website, resulting in lower customer perception of quality.

Using Third-Party Components and Services

Self-services may incorporate complex hardware components or software services that are not in the competency of one organization. As a result, the use of third parties is a common practice in self-service industry. When the self-service design requires the use of external parties, proper procurement practices need to be in place to guarantee end product self-service quality.

Optimizing the Self-Service Process

Most of the focus so far has been on the design related to self-service outcome quality. However, for the successful implementation of self-services, equally important is the self-service process quality. In this regard, a self-service process analysis should be carried out to identify process flows and possible bottlenecks. The analysis should also bring together an action plan and timetable (e.g., Gantt chart).

Allocating Resources

Finally, a prerequisite for self-service quality is the proper allocation of financial and human resources. In this respect, proper budgeting and accounting should be in place, and duties and responsibilities for each individual should be specified.

Quality in Self-Service Development and Engineering

The self-service development and engineering is the second phase of the self-service life cycle. The self-service moves into this phase from a conceptual state to a physical state with tangible self-service properties. The approach of aligning the actual self-service quality to the initial quality requirements and design specifications is presented in Figure 22.4 as self-service quality alignment. A list of quality topics that are relevant in this phase is provided below:

Self-Service Development in the Context of Production Quality

Self-services exhibit duality in terms of production quality: on one hand they resemble products manufacturing, and on the other hand they are very similar to services production. To avoid confusion a clarification is provided:

- The resemblance to the product manufacturing relates to the self-service development and engineering phase. The quality practices of developing tangible infrastructure and hardware are not different from those of manufacturing consumer goods.
- The similarity with the traditional services is apparent in the operation and monitoring phase of the self-service product life cycle. Both service production and self-service production have an intangible nature and are always accompanied with a simultaneous consumption. The main difference between the two is that the former involves two parties (service provider and consumer) whereas the latter engages one individual who is both producer and consumer. Nevertheless, service production and self-service production share many quality commonalities.

Addressing Quality in the Self-Service Development and Engineering

As mentioned above, quality in self-service development and engineering bears a resemblance to that of product manufacturing. Since Chapter 20, Product-Based Organizations: Delivering Quality While Being Lean and Green, of this book deals with manufacturing in product-based organizations, this section will not make unnecessary repetitions but instead focus on the information manufacturing, a subject that is a part of the self-service development.

Information Manufacturing

Information manufacturing differs from product manufacturing in several notable ways. The difference is best described by Wang (1998) who summarizes it in a table with three categories: input, process, and output. (See Figure 22.5.) According to Wang, manufacturing can be viewed as a processing system that acts on raw materials to produce physical products, whereas information manufacturing can be viewed as a processing system acting on raw data to produce information products. One of the core differences between the two types of manufacturing is the fact that a raw material can only be used for a single physical product and can be depleted, whereas data can be utilized by multiple consumers and are not subject to depletion. Similarly, quality concepts such as confidentiality and trustworthiness have no meaning in product manufacturing.

"Information manufacturing" is a broad term that can encompass an entire software development process which can consist of verification, design, implementation, integration, testing, installation, and deployment stages. This is not to be confused with the self-service

	Product manufacturing	Information manufacturing
Input	Raw materials	Raw data
Process	Assembly line	Information systems
Output	Physical products	Information products

Figure 22.5 Product versus information manufacturing. (*R. Wang, "A Product Perspective on Total Data Quality Management," Communications of the ACM, vol. 41, no. 2.*)

development and engineering phase which applies to software implementation and integration only. Nevertheless, note that the software development process is fully compatible with the self-service life cycle.

Software implementation is the realization of an application, program, or software component. It covers the work of software engineers (programmers) and software designers. There are two types of software implementation: code writing and visual programming. Code writing is the process of expressing statements or actions into human-readable computer programming language. Visual programming involves the use of visual tools to circumvent code and create user interface with a visible, one-to-one correspondence between the programmer and end user environments. However, more code is needed to realize software when visual programming is used.

There is general agreement in the software industry that more code means higher complexity, lower interpretability and understandability, lower consistency, and ultimately lesser quality. In practice, a lengthy code can be replaced by functionally and operationally identical code that is smaller in size. The examples of software quantitative quality metrics provided earlier in this chapter include many measures of complexity reflecting this notion.

Software integration is the combination of software and hardware components into an overall system. The integration reflects the realization of the previously discussed software architecture.

Quality in Self-Service Testing

Self-service testing is the next phase of the self-service life cycle. This phase is necessary to ensure that the recently developed self-service performs according to design specifications and is free of flaws, malfunctions, and component incompatibilities. In this connection, the approach of moving the self-service quality from development and engineering to testing is presented in Figure 22.4 as self-service quality assurance. A review of the tests applicable to self-services is provided below:

- *Alpha testing.* Alpha testing takes place in the developers' environment and involves testing of the self-service by internal staff, before it is released to external customers.
- *Beta (a.k.a. field) testing.* Beta testing takes place in the customers' environment and involves testing by a group of customers who use the self-service at their own locations. They provide feedback, before the self-service is released to other customers.
- *Functional (a.k.a. black box) testing.* Functional testing takes an external perspective of the self-service to derive test cases. The test designer selects valid and invalid self-service inputs and determines the correct output. There is no knowledge of the self-service internal structure and operation.
- *Usability testing.* Usability testing is a functional beta technique used to evaluate a self-service by testing it on customers. It is used to ensure that the intended self-service users can carry out the intended tasks efficiently, effectively, and satisfactorily. Virtually all self-services can benefit from usability testing.
- *Load (a.k.a. stress) testing.* Load testing is a functional alpha technique used to determine the stability of a given self-service. It involves testing beyond normal operational capacity, often to a breaking point, to observe the results. Load testing is used synonymously with performance testing, reliability testing, and volume testing. The technique is very popular with the technology-based self-services. For example, it is used to check a website performance under high loads to reveal potential problems such as data corruption, buffer overflows, deadlocks, and race conditions. Similarly, it is applied to IVR systems to ensure that they are reliable under high or moderate call load before or after full-scale deployment.

- *Structural (a.k.a. white box) testing.* Structural testing takes an internal perspective of the self-service to design test cases based on internal structure. This type of testing is relevant to the technology-based self-services and requires engineering skills. For example, a software tester first identifies all paths through the software and then chooses test case inputs to exercise paths through the code and determines the appropriate outputs. Similarly, an electrical hardware engineer probes and measures every node in a circuit. In respect to software, there are three popular approaches to structural testing (Jalote 1997, Bhattacherjee et al. 2007).
- *Control flow-based testing.* Control flow-based testing is a technique in which the program module is represented as a control flow graph and the coverage of various aspects of the graph is specified as criteria (e.g., statement coverage, branch coverage, and patch coverage).
- *Data flow-based testing.* In data flow-based testing, control flow and information concerning where the variables are defined and where the definitions are applied, are used to specify the test cases. The idea behind data flow-based testing is to make sure that during testing, the definitions of variables and their subsequent use are tested.
- *Mutation testing.* Mutation testing is a method of software testing that involves small adjustments of program's source code or byte code. The intention is to help the tester locate weaknesses in the test data or sections of code that are rarely accessed during execution. Mutation testing is powerful but also a computationally expensive testing technique.

Quality in Self-Service Operation and Monitoring

In the self-service operation and monitoring life cycle phase, the self-service becomes available to the customer who in turn initiates the actual self-service interaction. This phase is linked with two quality approaches: self-service quality realization and self-service quality redesign (see Figure 22.4). The self-service quality realization deals with the transition of the self-service quality from the testing environment to the actual operation, whereas the self-service quality redesign uses the operation and monitoring phase as a quality input in an improvement process that restarts the self-service quality cycle. The main quality issues in this phase can be summarized as follows:

Quality in Self-Service Distribution

After the self-service has been developed and tested, it needs to be distributed to customers. Some self-services require physical distribution (e.g., kiosks, vending machines, and ATMs) whereas others necessitate distribution over the Internet, telephone lines, and wireless networks (e.g., Web publishing, online banking, IVR, and online retailing). The last type involves a transmission of binary information, and therefore it is known as digital distribution. Although quality issues related to the physical distribution have been discussed in this book, the digital distribution has not been properly covered. In this regard, the main characteristics of the digital distribution are outlined below:

- *Direct nature.* The digital distribution is direct in nature. This implies that a self-service can be distributed to the end consumers with a minimum involvement of middlemen and minimum business overheads. This allows for the use of creative business models that may improve the quality of self-service (e.g., free trial versions, micropayments, free subscriptions, etc.).

- *Availability and global reach.* Self-services offered on the Internet are available 24/7. Additionally, they have global reach as customers from different countries and various geographic regions can access them. Availability has a positive relationship to the perceived self-service quality.
- *No physical inventory.* The digital distribution implies a transfer of information. Since the information has no physical properties, it can be stored and duplicated efficiently and without loss of quality.
- *Copyright violations and piracy.* Because of the above characteristics of the digital distribution, copyright violations and piracy can be a problem for some self-services. A possible solution to piracy can be the Digital Rights Management (DRM) which provides some protection against unauthorized redistribution of content.
- *Environmental impact.* The digital distribution has a very small environmental footprint. This may appeal to many environmentally conscious self-service users.

Quality in Self-Service Operation

The self-service operation is the outcome of self-service planning, developing, and testing. It puts the quality managers' goal of developing a self-service that is reliable, safe, easy to use, quick to operate, and according to customers' requirements to the ultimate test, i.e., the test of the customers. For the proper functioning of a self-service the following back office activities are important:

- *Operation and maintenance.* Self-service tangible elements need to be serviced and repaired to ensure operational quality. This applies for both basic and technology-based self-services. For example, the vending machine operators need to collect money from the coin- and cash-operated machines, restock merchandise, change labels to indicate new selections, and keep the machines clean and appealing. Additionally, they need to check if the vending machines' keypads, motors, merchandise chutes, and other parts work properly. If the machines are not in good working order, a repair is necessary. Finally, the vending machine operators need to perform preventive maintenance to address problems before they occur (e.g., periodically clean refrigeration condensers, lubricate mechanical parts, and adjust components). Similarly, the technology-based self-service operators need to perform regular data backups, content edits, search engine optimizations, software updates, and hardware inspections to ensure quality in self-service operation.
- *Safety and security protection.* Providing safety and security to self-service users is of paramount importance. Safety protection in self-services should aim at providing self-services that are safe, free from danger and injury. For example, the self-service gas stations should regularly inspect the safety of their gas pumps and test the operation of their fire extinguishing systems. Similarly, the Internet sites that offer content not appropriate for underage users should monitor and restrict the access to online resources that may pose dangers. Security protection in self-service, however, should focus on protecting the self-service users from theft. For example, the ATM operators should have surveillance methods in place to identify suspicious behavior. Correspondingly, the online retailers should protect their payment systems from hackers to guarantee secure transactions.
- *Preserving privacy.* The availability of public records, intelligent search engines, and data mining tools allow access to useful information. However, they also pose a thread to users' privacy as individuals and organizations abuse the available

information. Customers value privacy, and the lack of it creates dissatisfaction. In this connection, self-service organizations can preserve the customer information by using the following obfuscation techniques (Parameswaran and Douglas 2008): cryptography (sensitive data are encrypted with a key and are accessible only to authorized users), data randomization (based on adding a noise vector to the original data, thereby desensitizing the precise information content), and data anonymization (classifying data into fixed or variable intervals).

Self-Service Monitoring

The lack of human interaction in self-services entails extra organizational effort to evaluate the self-service delivery and get customer feedback. In this regard, the following techniques can be useful in monitoring the self-service quality:

- Process monitoring examines whether the self-service works as intended. This can include process observations and performance testing. The main focus is placed on the quality according to specifications.
- Customer observation aims at getting customer feedback about the self-service. This can include observations of the self-service users' behavior, customer surveys, complaints analysis, etc. The main goal is to assess the customer-perceived quality with the self-service.
- Third-Party evaluations and external benchmarking involve external help in evaluating the self-service quality. This technique is popular with the Internet self-services. A short list of organizations offering independent website quality assessments is provided below:
- BizRate.com uses consumers as evaluators of websites in different categories (e.g., toys, apparel, electronics, health and beauty, home, and furniture) after they have made purchases. BizRate has a scale of 10 dimensions: website performance, ease of ordering, product selection, product information, price, product representation, on-time delivery, customer support, shipping and handling, and privacy policies.
- Gomez.com uses researchers rather than consumers to evaluate websites. The organization measures the following categories: ease of use, efficient access to information, customer confidence, reliability, years the website or organization has been in business, on-site resources, overall cost, and relationship services.
- ResellerRatings.com monitors and rates online retailers through user reviews and ratings. Rated online stores are given two scores, a lifetime score and a six-month score.
- EC-Plus.net's benchmarking test is an online website benchmark utility that allows a quick technical assessment of the website performance. It incorporates a large range of tests including valid HTML/XHTML code, use of frames, search engine optimization, search engine positions, and website accessibility.
- WebsiteCriteria.com's benchmarking uses professional reviewers to assess the website's appeal, ease of use, and accessibility. The reviewer examines the website on over 100 features in several categories (e.g., first impression, look and feel, information architecture, navigation, content, search, functionality, multimedia, marketing, e-commerce, accessibility, and technical) and then records them to the benchmarking database server for comparison with other websites.

Quality in Self-Service Redesign

Self-services often need to be redesigned, after being used for some time, to adapt to new customer requirements and market trends. In comparison with designing a new self-service from scratch, self-service redesign can reduce cycle time and resources. Self-service redesign often deals with issues related to improving the technical performance and aesthetics of a self-service. However, self-service quality managers must be cautious as the improvements may cause contradictions between the new and existing self-service functions. These contradictions might not be apparent until the final design stage is reached. Some of the common quality tools that are used to redesign a self-service include TRIZ, Taguchi methods, and design of experiments (DOE).

Self-Service Quality Management

The self-service quality management provides the quality foundation of the self-service life cycle. It oversees all life cycle phases and interactions between them. Self-service quality management can be expressed in terms of quality approaches and methods. A summary of the quality approaches that are or can be deployed in the self-service industry is provided below:

Juran Trilogy® is a quality concept developed by Dr. Joseph Juran suggesting that the managing for quality is accomplished by the use of three interrelated managerial processes: planning, control, and improvement (Juran 1986). The approach was designed with product manufacturing in mind; however, it is equally applicable in self-services. For example, the planning process first determines the customers and their needs, then develops self-service designs to respond to those needs, and finally turns the plans over to the operating forces. The duty of the operating forces is to run the processes and develop the self-services. However, chronic waste, which is a cost of poor quality, can exist in any process due to various factors including deficiencies in the original planning. Under the conventional responsibility patterns, the operating forces are unable to get rid of that planned chronic waste. What they do instead is to carry out quality control, to prevent things from getting worse (sporadic spike). The planners subsequently guide the operational forces into the actual improvement implementation, leading to a reduced level of chronic waste.

Six Sigma is a disciplined, data-driven quality approach, methodology, and philosophy for reducing defects (e.g., Six Sigma implies 6 standard deviations between the mean and the nearest specification limit, or 3.4 defects per million opportunities). It is widely used in the manufacturing and transactional business environments. However, it is less popular in the information manufacturing which is inherent to many technology-based self-services.

There are two Six Sigma processes: Six Sigma DMAIC and Six Sigma DMADV [a.k.a. Design for Six Sigma (DFSS)]. Six Sigma DMAIC is a process that defines, measures, analyzes, improves, and controls existing processes that fall below the Six Sigma specification. Six Sigma DMADV defines, measures, analyzes, designs, and verifies new processes that are trying to achieve Six Sigma quality. Within the individual phases of a DMAIC or DMADV, Six Sigma utilizes many established quality management tools that can also be used individually, e.g., failure mode and effects analysis (FMEA), design of experiments, critical-to-quality tree, cost-benefit analysis, cause and effects diagram, business process mapping, quality function deployment (QFD), process capability, root cause analysis, suppliers, inputs, process, outputs, customers (SIPOC) analysis, Taguchi methods, TRIZ, numerous charts (e.g., control charts, histograms, Pareto charts, run charts, etc.), and various statistical tests (e.g., binomial, ANOVA, chi-square, etc.). For more information about the Six Sigma methodology refer to Chapter 12, Six Sigma: Improving Process Effectiveness.

Lean is a quality approach that considers the expenditure of resources that do not create customer value (something that customers will be willing to pay for) to be wasteful, and thus a target for elimination. It is has received wide popularity in product and information manufacturing as well as in services. Lean in self-services targets several types of waste: duplication (customers reentering data and repeating operations), delay (customers waiting for self-service delivery), unnecessary movement (due to poor navigation and poor ergonomics of the self-service interface), complexity (complex self-service designs that are difficult for users to comprehend), and errors in the self-service transaction (nonconformance to customer specifications or expectations). Some of the popular Lean quality tools are 5s, value stream mapping, mistake proofing (Poka-Yoke), production leveling, pull systems (Kanban), load balancing (Heijunka box), queuing theory, motivation, and measurements. Additional information on the Lean methodology is provided in Chapter 11, Lean Techniques: Improving Process Efficiency, of this book.

Lean Six Sigma is a combination of both Lean and Six Sigma quality approaches. The underlying tenet of the Lean approach is efficiency, whereas that of Six Sigma is effectiveness. The integration of the two approaches provides a balanced approach to quality. By applying the Lean tools in self-services, the self-service processes become stable, constraints and costs to self-service operation are reduced, and the self-service speed is optimized. Six Sigma tools can then be applied to self-services to identify key variables in the process, establish operating ranges, and implement control methods to ensure the self-service problems are corrected.

The *Capability Maturity Model Integration* (CMMI) is a process improvement methodology that enables organizations to better manage their processes across business units and projects, resulting in improved organizational performance. As the CMMI is a successor of the Capability Maturity Model (a reference model specifically designed for software development), it is predominantly used in information manufacturing. Nevertheless, the methodology can be applied in services and product manufacturing. The CMMI contains a list of key process areas divided into different maturity levels. More information about the CMMI is provided at www.sei.cmu.edu.

Note that the CMMI shares some similarities with Six Sigma. For example, both approaches are goal-oriented toward a reduction of defects, data-driven as they rely on measurement, and supported by some common statistical tools. Nevertheless, each approach has its own strengths and weaknesses. For example, Six Sigma is better at understanding customer needs and utilizing statistical methods whereas CMMI is better at integrating ongoing projects and utilizing the best industry practices.

The above quality management approaches are not mutually exclusive but instead can be successfully integrated. There is a shared view that Lean, Six Sigma, and CMMI are powerful in themselves and highly synergistic in combination (Gack and Williams 2007, Jacowski 2008, Siviy et al. 2005). Therefore, self-service-based organizations should aim at embedding these approaches in their self-service life cycles to ensure overall self-service quality.

Self-Service Prognosis

In over a century we have witnessed exponential growth in self-services, an astonishing development, unfold chronologically in the section on self-service evolution. Surprisingly, this growth does not show any signs of slowing down anytime soon. Demographic, socioeconomic, technological, and environmental trends suggest further self-service proliferation. The importance of self-services as well as their complexity will continue to grow. This in turn will create new opportunities and challenges for managing quality. A synopsis of these issues is provided below.

Demographic Changes

Human population and demographics are dynamic, and people are often grouped according to their birth in a particular time period as their personalities are shaped by common historical events. In this regard, human population has been divided into several generation groups sharing some common characteristics (e.g., silent generation, baby boomers, generation X, generation Y, and generation Z). In the last three generations we have witnessed an accelerated trend of technological embracement coupled with an increased level of individualism. The trend is likely to continue with the next generation as people are becoming more dependent on technology. This will benefit self-services. Nevertheless, population characteristics will also change, and the disparity in skills among different generation groups (e.g., younger versus older) will grow wider. To manage self-service quality effectively, organizations will need to respond to the diverse population requirements (e.g., self-service segmentation and customization) as well as find ways to improve population skills (e.g., presentations, tutorials, guides).

Technological Progress

Technological progress will improve the existing self-services and bring new ones onto the market. The improvements will result in better self-service infrastructure, a user-friendly interface, and faster performance. In the near future, consumers will witness new-generation navigation and typing devices, displays with higher resolution and three-dimensional features, information networks able to transfer large volume of information with unprecedented speed and accuracy, and computer processors and hard drives with exceptional performance. Nevertheless, the technological innovations will increase the complexity of self-services, posing challenges to quality.

Globalization

Multinational corporations first started the recent globalization trend. Nevertheless, the greatest credit for globalization is given to the emergence of the Internet. Improvements in information, communication, transportation, and education will further accelerate this trend. Globalization will offer self-service organizations the opportunity to further expand their customer reach. Language may be a barrier to self-service usage and a factor affecting perceived quality, forcing self-service organizations to offer multilanguage options to their world customers.

Competition

The self-service industry is likely to stay a very competitive business with competition concentrated in three main areas: technology, content, and business models. The technological competition will put pressure on organizations to design new self-services, expedite life cycle processes, and redesign self-services at a higher rate. Self-service strategic planning, improvement and control as well as the deployment of effective quality tools will be of great importance. The competition for content will be equally intense. Self-service organizations will have to provide appealing, up-to-date, error-free, and personalized content to their customers. Interactivity will be an essential element to quality as virtual reality will likely play a big role in self-services. Finally, the competition of business models will mainly comprise self-service revenue streams and logistics. These models will be important quality differentiators and are likely to have a proprietary nature (patenting business processe, e.g., e-Bay, Netflix, and Google, is a common practice today and is likely to be in the future).

Intellectual Property and Piracy

While business models can be protected by patents, digital content is normally protected by copyrights. Nevertheless, because of the availability and easy duplication of digital content, piracy may be a problem for some self-service organizations. Efforts to stop piracy will continue with Digital Rights Management (DRM); however, the unauthorized redistribution of content will likely exist.

Virtual Consumerism

Virtual consumerism is a new phenomenon that is expected to gain further popularity. The term implies selling virtual items for real money and indicates a shift from material consumption to virtual consumption (Lehdonvirta et al. 2009). Today, only a few online communities and Internet games are promoting virtual consumption (e.g., Habbo Hotel, Second Life, Ultima Online, The Sims Online, and World of Warcraft); however, their number is likely to increase. In response, many self-service organizations will need to reevaluate their business strategies and approaches to quality.

Environment

Customers are becoming increasingly aware of the impact their material consumption has on the environment. As our planet continues to suffer from human-created pollution and depletion of natural resources, customers will logically shift their consumption toward environmentally friendly options. In this regard, a change will be observed from material to digital (e.g., digital video, digital audio, and digital content) and virtual (e.g., virtual artifacts and virtual reality) consumption. Self-services will be perfectly positioned to benefit from this change.

An exciting time is ahead for self-services. Their popularity and influence on our lives will continue to grow as well as the importance of self-service quality. Organizations that will become market leaders will be the ones that recognized quality as their main priority. As discussed in this chapter, these organizations will need to have a broader vision of quality that incorporates all phases of the self-service life cycle and is smoothly integrated with the overall organizational strategy.

References

Albrecht, K. (1990). *Service Within*. Dow Jones-Irwin, Homewood, IL.

Alsop, S. (1999). "The Dawn of E-Service." *Fortune*, vol. 9, p. 138.

Anderson, J. C., and Narus, J. A. (1995). "Capturing the Value of Supplementary Services." *Harvard Business Review*, vol. 73, no. 1, pp. 75–83.

Anton, J. (2001). *Call Center Benchmarking Report*. Center for Customer Driven Quality, Purdue University, Lafayette, IN.

Antonides, G., and van Raaij, W. F. (1998). *Consumer Behaviour: A European Perspective*. Wiley, Chichester, England.

Argyle, M., Salter, V., Nickolson, H., Williams, M., and Burgess, P. (1970). "The Communication of Inferior and Superior Attitudes by Verbal and Non-Verbal Signals." *British Journal of Social and Clinical Psychology*, vol. 9, pp. 222–231.

Assael, H. (1995). *Consumer Behavior and Marketing Action*, 5th ed. South-Western College Publishing, Cincinnati, OH, pp. 80–96, 252, 267, 460.

Avalone, S. (2002). "Hold the Line on Costs." *Chain Store Age*. Lebhar-Friedman, New York.

Baecker, R., Grudin, J., Buxton, W., and Greenberg, S. (1995). *Readings in Human-Computer Interaction: Toward the Year 2000*. Morgan Kaufmann, San Francisco.

Barnum, C., and Wolniansky, N. (1989). "Taking Cues from Body Language." *Management Review*, pp. 59–60.

Bateson, J. E. G. (1985). "Self-Service Consumer: An Exploratory Study." *Journal of Retailing*, vol. 61, no. 3, pp. 49–76.

Bateson, J. E. G., and Hui, M. K. M. (1987). "Perceived Control as a Crucial Perceptual Dimension of the Service Experience: An Experimental Study." In C. F. Surprenant, ed., *Add Value to Your Service*. American Marketing Association, Chicago, pp. 187–192.

Berkowitz, E., Walton, J., and Walker, O. (1979). "In-Home Shoppers: The Market for Innovative Distribution Systems." *Journal of Retailing*, vol. 55, pp. 15–33.

Berry, L. (1984). "The Employee as Customer," in Services Marketing, Lovelock, C. (Ed.), American Marketing Association, Chicago, IL, pp. 242.

Berry, L., and Parasuraman, A. (1991). *Marketing Services: Competing through Quality*. The Free Press, New York.

Bhattacherjee, V., Suri, D., and Mahanti, P. (2007). "Software Testing: A Graph Theoretic Approach." *International Journal of Information and Communication Technology*, vol. 1, no. 1.

Bobbitt, M., and Dabholkar, P. A. (2001). "Integrating Attitudinal Theories to Understand and Predict Use of Technology-Based Self-Services." *International Journal of Service Industry Management*, vol. 12, no. 5, pp. 423–450.

Bowen, D. E. (1986). "Managing Customers as Human Resources in Service Organizations." *Human Resource Management*, vol. 25, no. 3, pp. 371–383.

Burgoon, J., Birk, T., and Pfau, M. (1990). "Non-Verbal Behaviors, Persuasion, and Credibility." *Human Communication Research*, vol. 17, pp. 140–169.

Card, S., Robertson, G., and York, W. (1996). "The Webbook and the Web Forager." In R. Bilger, S. Guest, and M. J. Tauber, eds., *Human Factors in Computing Systems: CHI'96 Conference Proceeding*. ACM Inc., Danvers, MA, pp. 357–358.

Chaiken, S. (1979). "Communicator Physical Attractiveness and Persuasion." *Journal of Personality and Social Psychology*, vol. 37, pp. 1387–1397.

Chase, R. (1978). "Where Does the Customer Fit in a Service Operation?" *Harvard Business Review*, vol. 56, November/December, pp. 137–142.

Cook, D., and Coupey, E. (1998). "Consumer Behavior and Unresolved Regulatory Issues in Electronic Marketing." *Journal of Business Research*, vol. 41, pp. 231–238.

Cowles, D., and Crosby, L. (1990). "Consumer Acceptance of Interactive Media." *The Service Industries Journal*, vol. 10, no. 3, pp. 521–540.

Dabholkar, P. A. (1990). "How to Improve Perceived Service Quality by Increasing Customer Participation." In B. J. Dunlap, ed. *Developments in Marketing Science*, vol. 13, Academy of Marketing Science, Cullowhee, NC, pp. 483–487.

Dabholkar, P. A. (1992). "The Role of Prior Behavior and Category-Based Affect in On-Site Service Encounters." In J. Sherry and B. Sternthal eds., *Diversity in Consumer Behavior*. Association for Consumer Research, Provo, UT, vol. 19, pp. 563–569.

Dabholkar, P. A. (1994). "Technology-Based Service Delivery: A Classification Scheme for Developing Marketing Strategies." In T. A. Swartz, D. E. Bowen, and S. W. Brown eds., *Advances in Service Marketing Management*. JAI Press Inc., Greenwich, CT, pp. 247–271.

Dabholkar, P. A. (1996). Consumer Evaluations of New Technology-Based Self-Service Options: An Investigation of Alternative Models of Service Quality." *International Journal of Research in Marketing*, vol. 13, pp. 29–51.

Dabholkar, P. A. (2000). "Technology in Service Delivery: Implications for Self-Service and Service Support." In T. A. Swartz and D. Iacobucci eds., *Handbook of Service Marketing and Management*. Sage Publications, Beverly Hills, CA, pp. 103–110.

Dabholkar, P. A., Sujan, M., and Kardes, F. (1995). "Contingency Framework for Predicting Causality between Customer Satisfaction and Service Quality." *Advances in Consumer Research*, vol. 22, pp. 101–108.

Davis, F., Bagozzi, R., and Warshaw, P. (1989). "User Acceptance of Computer Technology: A Comparison of Two Theoretical Models." *Management Science*, vol. 35, no. 8, pp. 9 82–1003.

Devlin, J. (1995). "Technology and Innovation in Retail Banking Distribution." *International Journal of Bank Marketing*, vol. 13, no. 4, pp. 19–25.

Dibb, S., Simkin, L., Pride, W., and Ferrell, O. (1994). *Marketing Concepts and Strategies*. Houghton Mifflin: Boston, Chapter 24, p. 671.

Elizur, D. (1987). "Effect of Feedback on Verbal and Non-Verbal Courtesy in a Bank Setting." *Applied Psychology: An International Review*, vol. 36, no. 2, pp. 147–156.

Elswick, J. (2001). "Self-Service Enrollment Evolves Rapidly." *Employee Benefit News*, vol. 15, no. 10, pp. 4, 45.

Erickson, B., Lind, E., Johnson, B., and O'Bar, W. (1978). "Speech Style and Impression Formation in a Court Setting: The Effects of 'Powerful' and 'Powerless Speech.' " *Journal of Experimental Social Psychology*, vol. 14, pp. 266–279.

Evans, K., and Brown, S. (1988). "Strategic Options for Service Delivery Systems." In C. Ingene and G. Frazier eds., *Proceedings of the AMA Summer Educational Conference*. American Marketing Association, Chicago, pp. 202–212.

Fingar, P., Kumar, H., and Sharma, T. (2000). *Customer Care—Through the E-Commerce Looking Glass, Defining the Limits.*, Montgomery Research, San Francisco, pp. 105–114.

Forman, A., and Ven, S. (1991). "The Depersonalization of Retailing: Its Impact on the 'Lonely' Consumer." *Journal of Retailing*, vol. 67, pp. 226–243.

Fram, E., and Grady, D. (1997). "Internet Shoppers: Is There a Surfer Gender Gap?" *Direct Marketing*, vol. 59, pp. 46–50.

Gack, G., and Williams, K. (2007). "Connecting Software Industry Standards and Best Practices: Lean Six Sigma and CMMI®," *CrossTalk, The Journal of Defense Software Engineering*, February.

Graffeo, J. C. (1997). "With Bank Customers, Old Habits Die Slowly." *Indianapolis Business Journal*, vol. 18, no. 33, p. 27.

Gremler, D. D., Bitner, M. J., and Evans, K. R. (1994). "The Internal Service Encounter." *International Journal of Service Industry Management*, vol. 5, no. 2, pp. 34–56.

Griffith, D., Krampf, R., and Palmer, J. (2001). "The Role of Interface in Electronic Commerce: Customer Involvement with Print versus On-Line Catalogs." *International Journal of Electronic Commerce*, vol. 5, no. 4, pp. 135–153.

Griffiths, R. (2001). *History of the Internet*. Universiteit Leiden, Netherlands, September.

Grönroos, C. (1984). "A Service Quality Model and Its Marketing Implications." *European Journal of Marketing*, vol. 18, no. 4, pp. 36–44.

Grönroos, C. (1985). "Internal Marketing–Theory and Practice," in Bloch, T., Upah, G., Zeithaml, V. (eds). Services Marketing in a Changing Environment, American Marketing Association, Chicago, IL, pp. 41–47.

Grönroos, C., Heinonen, F., Isoniemi, K., and Lindholm, M. (2000). "The Netoffer Model: A Case Example from the Virtual Marketspace." *Management Decision*, vol. 38, no. 4, pp. 243–252.

Hafner, K. (2000). "Is the Customer Ever Right? The Decline and Fall of Customer Service with a Technological Push." *The New York Times*, July 20.

Hallowell, M. (1999). "The Human Moment at Work." *Harvard Business Review*, vol. 77, no. 1, pp. 58–66.

Hallowell, R. (2001). "'Scalability': The Paradox of Human Resources in E-Commerce." *International Journal of Service Industry Management*, vol. 12, no. 1, pp. 34–43.

Hartson, R. (1998). "Human-Computer Interaction: Interdisciplinary Roots and Trends." *Journal of Systems and Software*, vol. 43, no. 2, pp. 103–118.

Hoffman, D., and Novak, T. (1996). "Marketing in Hypermedia Computer-Mediated Environments: Conceptual Foundations." *Journal of Marketing*, vol. 60, no. 3, pp. 50–69.
Hofstede, G. (1991). *Cultures and Organizations, Software of the Mind*. McGraw-Hill, London.
Holbrook, M., Chestnut, R., Olivia, T., and Greenleaf, E. (1984). "Play as a Consumption Experience: The Role of Emotions, Performance and Personality in the Enjoyment of Games." *Journal of Consumer Research*, vol. 11, pp. 728–739.
Hornik, J., and Ellis, S. (1988). "Strategies to Secure Compliance for a Mall Intercept Interview." *Public Opinion Quarterly*, vol. 52, no. 4, pp. 539–551.
Jacob, R. (1996). "Human-Computer Interaction: Input Devices." *ACM Computing Surveys*, vol. 28, no. 1, pp. 177–179.
Jacowski, T. (2008). "Integrating Lean Six Sigma with CMMI," retrieved October 13, 2009, from http://ezinearticles.com/?Integrating-Lean-Six-Sigma-With-CMMI&id=1037357.
Jalote, P. (1997). *An Integrated Approach to Software Engineering*, 2d ed. Springer-Verlag, New York, pp. 419–434.
Juran, J. (1986). "The Quality Trilogy: A Universal Approach to Managing for Quality." *Quality Progress*, vol. 19, no. 8, pp. 19–24.
Juran, J. M. (1988). Juran on Planning for Quality. New York: The Free Press, pp. 11.
Kahle, L., and Homer, P. (1985). "Physical Attractiveness of the Celebrity Endorser: A Social Adaptation Perspective." *Journal of Consumer Research*, vol. 11, pp. 954–961.
Kasper, H., Van Helsdingen, P., and De Fries, W. (1999). *Service Marketing Management—An International Perspective*. John Wiley & Sons, Chichester, England, pp. 12–45, 152–166, 576.
Kelly, S., Donnelly, J., and Skinner, S. (1990). "Customer Participation in Service Production and Delivery." *Journal of Retailing*, vol. 66, no. 3, pp. 315–335.
Kendall, K., and Kendall, J. (1998). *System Analysis and Design*, 4th ed. Prentice-Hall, Upper Saddle River, NJ.
Kimble, C., and Seidel, S. (1991). "Vocal Signs of Confidence." *Journal of Non-Verbal Behaviour*, vol. 15, pp. 99–105.
Langeard, E., Bateson, J. E. G., Lovelock, C. H., and Eiglier, P. (1981). *Marketing of Services: New Insights from Consumers and Managers*. Marketing Science Institute, Cambridge, MA.
Leblanc, G. (1990). "Customer Motivations: Use and Non-Use of Automated Banking." *International Journal of Bank Marketing*, vol. 8, no. 4, pp. 36–40.
Lehdonvirta, V., and Johnson, M. (2009). "Virtual Consumerism: Case Habbo Hotel." *Information, Communication and Society*, vol. 12, no. 7.
Liljander, V., et al. (2006). *Journal of Retailing and Consumer Services*, vol. 13, pp. 177–191.
Lonner, W., and Malpass, R. (1994). "When psychology and culture meet: An introduction to cross-cultural psychology," In "Psychology and Culture," ed. W. Lonner and R. Malpass, pp. 1–12. Needham Heights, MA: Allyn & Bacon.
Lovelock, C., and Young, R. (1979). "Look to Customers to Increase Productivity." *Harvard Business Review*, vol. 57, May/June, pp. 168–178.
Martin, J. (2000). "Will Service Still Stink?" In: Time Magazine, Monday, May 22, 2000.
Maslow, A. (1954). *Motivation and Personality*. Harper & Row, New York.
Matsumoto, D. (1994). *People: Psychology from a Cultural Perspective*. Brooks/Cole, Pacific Grove, CA.
Mehrabian, A. (1981). *Silent Messages*. Wadsworth, Belmont, CA.
Mehrabian, A., and Williams, M. (1969). "Non-Verbal Concomitants of Perceived and Intended Persuasiveness." *Journal of Personality and Social Psychology*, vol. 13, no. 1, pp. 37–58.
Meuter, M., Ostrom, A., Roundtree, R., and Bitner, M. (2000). "Self-Service Technologies: Understanding Customer Satisfaction with Technology-Based Service Encounters." *Journal of Marketing*, vol. 64, no. 3, pp. 50–64.

Mills, P., and Moberg, D. (1982). "Perspectives on the Technology of Service Operations." *Academy of Management Review*, vol. 7, no. 3, pp. 467–478.

Mollenhoff, C. (1988). *Atanasoff, Forgotten Father of the Computer*. Iowa State University Press, Ames.

Moon, Y., and Frei, F. (2000). "Exploring the Self-Service Myth." *Harvard Business Review*, vol. 78, May/June, pp. 26–27.

Morrison, J., and Vogel., D. (1998). "The Impacts of Presentation Visuals on Persuasion." *Information and Management*, vol. 33, pp. 125–135.

Muter, P., and Maurutto, P. (1991). "Reading and Skimming from Computer Screens and Books: The Paperless Office Revisited?" *Behavior and Information Technology*, vol. 10, no. 4, pp. 257–266.

Nass, C., and Steuer, J. (1993). "Voices, Boxes, and Sources of Messages: Computer and Social Actors." *Human Communication Research*, vol. 19, no. 4, pp. 504–527.

Parameswaran, R., and Douglas, B. (2008). "Privacy Preserving Data Obfuscation for Inherently Clustered Data." *International Journal of Information and Computer Security*, vol. 2, no. 1, pp. 4–26.

Parasuraman, A. (2000). Technology Readiness Index (TRI): A Multiple Item Scale to Measure Readiness to Embrace New Technologies." *Journal of Service Research*, vol. 2, no. 4, pp. 307–320.

Parasuraman, A., Zeithaml, V., and Berry, L. (1985). "A Conceptual Model of Service Quality and Its Implications for Future Research." *Journal of Marketing*, vol. 49, pp. 41–50.

Parasuraman, A., Zeithaml, V. A., and Berry, L. L. (1988). "SERVQUAL: A Multiple Item Scale for Measuring Consumer Perceptions of Service Quality." *Journal of Retailing*, vol. 64, pp. 22–37.

Patterson, M., Powell, J., and Leniham, M. (1986). "Touch Compliance and Interpersonal Affects." *Journal of Non-Verbal Behavior*, vol. 10, no. 2, pp. 41–50.

Pavitt, D. (1997). "Retailing and the Super High Street: The Future of the Electronic Home Shopping Industry." *International Journal of Retailing and Distribution Management*, vol. 38.

Pavlovic, V., Sharma, R., and Huang, T. (1997). "Visual Interpretation of Hand Gestures for Human-Computer Interaction: A Review." *IEEE Transactions on Pattern Analysis and Machine Intelligence*, vol. 19, no. 7, pp. 677–696.

Phillips, O. R., and Schutte, D. P. (1988). Identifying Profitable Self-Service Markets: A Test in Gasoline Retailing." *Applied Economics*, vol. 20, pp. 263–272.

Pine, B. J., and Gilmore, J. H. (1999). *The Experience Economy*. Harvard Business School Press, Cambridge, MA.

Porter, M. (2001). "Strategy and the Internet." *Harvard Business Review*, March, pp. 63–78.

Prendergast, G., and Marr, N. (1994). "Disenchantment Discontinuance in the Diffusion of Technologies in the Service Industry: A Case Study in Retail Banking." *Journal of International Consumer Marketing*, vol. 7, no. 2, pp. 25–40.

Price, L., Arnould, E., and Tierney, P. (1995). "Going to Extremes: Managing Service Encounters and Assessing Provider Performance." *Journal of Marketing*, vol. 59, pp. 83–97.

Quinn, R. (2000). "At the Water's Edge." Employee Benefit News, Vol. 14 No. 13, pp. 35–9.

Reeves, B., and Nass, C. (1996)."The media equation: how people treat computers, television, and new media like real people and places," Cambridge University Press/CSLI, New York, N.Y., 1996.

Rogers, Y. (1989). "Icons at the Interface: Their Usefulness." *Interacting with Computers*, vol. 1, pp. 295–315.

Rotter, J. B. (1966). Generalized Expectancies for Internal versus External Control of Reinforcement." *Psychological Monographs*, vol. 80, no. 609.

Santrock, J. W. (1997a). *Psychology: Alternative and Enhancement Chapters*, 5th ed. McGraw-Hill, Dubuque, IA, pp. 5-inc (Chapter: Interpersonal Communication).

Santrock, J. W. (1997b). *Psychology: Alternative and Enhancement Chapters*, 5th ed. McGraw-Hill, Dubuque, IA, pp. 22-anb (Chapter: Animal Behavior).

Santrock, J. W. (1997c). *Psychology: Alternative and Enhancement Chapters*, 5th ed. McGraw-Hill, Dubuque, IA, pp. 22-sop (Chapter: Social Psychology).

Scherer, K., and Ekman, P. (1982). *Handbook of Methods in Non-Verbal Behavior Research*, Cambridge University Press, Cambridge, England, pp. 136–198.

Schneider, B., and Bowen, D. (1985). "Employee and Customer Perceptions of Service in Banks." *Journal of Applied Psychology*, vol. 70, no. 3, pp. 423–433.

Shackel, B. (1962). "Ergonomics in the Design of a Large Digital Computer Console." *Ergonomics*, vol. 5, pp. 229–241.

Shackel, B. (1997). "Human-Computer Interaction: Whence and Whither?" *Journal of the American Society for Information Science*, vol. 48, no. 11, pp. 970–986.

Shneiderman, B. (1998). *Designing the User Interface*. Addison-Wesley, Reading, MA.

Siegman, A., and Fildstein, S. (1978). *Non-Verbal Behavior and Communication*. Lawrence Erlbaum, Hillsdale, NJ, pp. 183–243.

Siviy, J., Penn, L., and Harper, E. (2005). "Relationships between CMMI® and Six Sigma," Technical Note: CMU/SEI-2005-TN-005, retrieved October 14, 2009, from http://www.sei.cmu.edu/reports/05tn005.pdf.

Steuer, J. (1992). "Defining Virtual Reality: Dimensions Determining Telepresence." *Journal of Communication*, vol. 42, no. 4, pp. 73–79.

Szymanski, D., and Hise, R. (2000). "E-Satisfaction: An Initial Examination." *Journal of Retailing*, vol. 76, no. 3, pp. 309–322.

Thornton, J., and White, L. (2001). "Customer Orientations and Usage of Financial Distribution Channels." *Journal of Services Marketing*, vol. 15, no. 3, pp. 168–185.

Tonchev, A., and Tonchev, C. (2003). "Self-Service Marketing—Antecedents of Self-Service Usage and Customer Satisfaction," Maastricht University.

Triandis, H. (1994). *Culture and Social Behavior*. McGraw-Hill, New York.

Van Riel, A. C. R., Liljander, V., and Jurriëns, P. (2001). "Exploring Customer Evaluations of E-Services: A Portal Site." *Journal of Service Industry Management*, vol. 12, no. 4, pp. 359–377.

Wagner, T. (2000). *Multichannel Customer Interaction, Defining the Limits*. Montgomery Research, San Francisco, pp. 277–280.

Wang, R. (1998). "A Product Perspective on Total Data Quality Management." *Communications of the ACM*, vol. 41, no. 2.

Webster, C., and Sundaram, D. (1998). "Service Consumption Criticality in Failure Recovery." *Journal of Business Research*, vol. 41, pp. 153–159.

Webster, C., and Sundaram, D. (2000). "The Role of Non-Verbal Communication in Service Encounters." *Journal of Services Marketing*, vol. 14, no. 5.

Welch, A., and Lyons, N. (2000). *MultiCAST—The New Face of Multi-Channel Marketing, Defining the Limits*. Montgomery Research, San Francisco, pp. 165–170.

Wilson, T. (2001). Web-powered HR. In: Internetweek, March 12.

Windham, L., and Orton, K. (2000). *The Soul of the New Customer: The Attitudes, Behavior, and Preferences of E-Customers*. Allworth Press, New York.

Zeithaml, V. A., Parasuraman, A., and Malhotra, A. (2000). "A Conceptual Framework for Understanding E-Service Quality: Implications for Future Research and Managerial Practice." Working Paper, Marketing Science Institute, Cambridge, MA.

Zeithaml, V., and Bitner, M. J. (1996). *Services Marketing*. McGraw-Hill, New York.

www.aptea.com (17.08.2009)
www.groceteria.net (17.08.2009)
www.pigglywiggly.com (17.08.2009)
www.unarco.com (17.08.2009)
www.mcdonalds.com (17.08.2009)
www.inventors.about.com (17.08.2009)
www.cashtechnologies.com (17.08.2009)
www.kiosk.org (11.09.2009)
www.sei.cmm.edu (01.10.2009)

CHAPTER 23

Health Care-Based Organizations: Improving Quality of Care and Performance

Mary Beth Edmond, Jonathan D. Flanders, and James Er Ralston

About This Chapter

Health care systems help people stay healthy, live with chronic disease or disability, recover from illness, and cope with dying. Of the more than $2 trillion spent each year on health care in the United States, studies of the costs of poor quality estimate that upward of 30 percent is allocated to overuse, underuse, or misuse of care resources that provide no value to the patient (Midwest Business Group on Health 2003). Care often is delivered too late or in expensive care environments, such as emergency rooms or intensive care units, without full consideration of the patient's preferences or values. Our system of health care many times delivers services inefficiently and unevenly across populations. This chapter focuses on how to manage for quality and performance excellence. We will discuss the structure, methods, and tools necessary to deliver high-quality, safe care at a lower cost. It is not a formula or panacea for transforming the health care system. It is simply the basis for change and part of a continuing revolution to ensure that all citizens around the globe receive the benefits of a healthy society.

High Points of This Chapter

1. There are many challenges facing the health care industry including, e.g., improvement of outcomes, reduction of the cost of care, reduction of bottlenecks in delivery, poor measurement systems, lack of engagement of fragmented participants, and system complexity which are important factors that restrain quality advances.
2. The management for quality and safety in the health care system can help systems deal with these issues.
3. Developing a performance excellence program using contemporary methods and tools can take the pain out of improving clinical, patient, employee, and business performance.
4. Many health systems have now adopted Lean, Six Sigma, and root cause analysis methodologies to provide a data-driven and disciplined approach to quality improvement, especially when the organization lacks knowledge of root causes in its own work processes.
5. The culture of an organization must be described by its mission, vision, values, and physician and leadership commitment to enable patient safety and accountability to reduce harm and sustain it.
6. Numerous health care systems are leading the way with best practice results and performance.

Health Care Quality Has Come a Long Way

Quality and safety have been elevated as strategically important issues for the health care system globally. Progress, while slow, is being made at the local level despite the larger issues that receive much of the public's attention. The recent 2008 National Healthcare Quality Report identified three key themes: (1) Health care quality is suboptimal and continues to improve but at a slow pace; (2) reporting of hospital quality is leading the improvement, but patient safety is lagging; and (3) health care quality measurement is evolving, but much work remains. Acknowledging these, we discuss here some of the major challenges that exist broadly within health care and initiatives underway to alleviate pain in specific areas.

Dr. Allen Weiss, CEO and president of Naples Community Hospital (NCH), in Naples, Florida, explains, "What can NCH do to improve quality, lower cost and currently yield overall higher value? We asked ourselves this question over the past few years and decided to copy best practices from other industries and use operations management, which may be defined as the design, operation and improvement of the process of delivering care. To bring operations management into NCH, we have adopted the Juran Trilogy®. The goal is to improve the care provided by NCH through numerous approaches. The most obvious, of course, is taking steps to ensure that patients receive the best possible care in the shortest time period."

Health care improvement has come a long way since an improvement revolution began to blossom in the mid-1980s due in part to the exploratory work of the National Demonstration Project on Quality Improvement in Health Care and the Juran Institute (Berwick et al. 1990). It also was the beginning of the addition of business and industrial quality professionals into positions of influence on hospital boards and in health care management. As a result, health care organizations became gradually more receptive to the ideas that stabilization was not enough, that important improvements in cost and quality could be achieved, and that new managerial methods—quality management—might help in health care, even though these methods had first appeared and been developed in other industries.

This early improvement effort documented the National Demonstration Project's results in a text called *Curing Health Care.* It concentrated largely on typical business processes that existed in health care organizations—processes such as scheduling, equipment maintenance, billing, and transportation of patients. In one highly successful project in 1987, e.g., the University of Michigan Hospital, working with the help of a quality professional from Corning, Inc., reduced wait times in its ambulatory care clinic by 89 percent in a few short months.

It is now 2010 and the revolution continues. There are now numerous health care system National Quality Award winners; there are hundreds of hospitals in the United States pursuing the Baldrige Award; international hospitals are following suit; most hospitals have board-driven safety and quality committees; and methods such as Lean and Six Sigma are now commonplace in the health care systems. Even with all the progress in the health care system in the United States and in many places around the globe, there remain challenges. Some of these challenges are being addressed through policy, others through the management of quality and safety methodologies, and still others through technology.

This chapter focuses on only the methods, tools, and structures that are most effective in health care. We are not focusing on dealing with all the political challenges and policy formation that need to take place. We will leave that to the politicians. We do, however, list some of the challenges so that your system can select the right quality method at the right time and use it in the right way. Here are some of the challenges faced:

1. *Improvement in outcomes at lower costs to society.* The cost of health care continues to rise at a much faster rate than inflation in almost all countries around the globe. While improved outcomes continue to rise, the pressure created by taxpayers and government politicians will continue to drive the reduction in costs and improved financial performance of its systems or face a continued crisis.
2. *Implementation bottleneck.* The Harvard Business School's 2008 Centennial Global Business Summit, "Redefining Global Health Care," explored the strategic challenges in health care; key takeaways are provided here (Porter et al. 2008). A major conclusion is that the barrier to health care delivery globally is not funding or knowledge of how to deliver health care—it is the ability to implement.
3. *Measurement failures.* Hospitals frequently have servers full of data, yet this abundance rarely meets its potential in driving change. Errors in collection and recording abound (e.g., a review of emergency department records at a Juran client revealed over 15 percent had some kind of error in recording of patient entry mode, length of stay, or disposition). Extensive, detailed reports that are routinely produced with colored bar charts, trend lines, and multiple levels of roll-up routinely are set aside or discarded without review, comprehension, or action. Worse, the critical results are not measured. When they are, the measures tend to be process measures (e.g., the number of patients seen per hour) as opposed to outcome measures (e.g., the number of lives saved). Whatever the cause, poor measurement capability serves as an impediment to strategy capability.
4. *Disengagement and fragmentation.* Administrators in large, formal managerial environments typically keep abreast of trends and quality success stories and readily see the value in quality improvement. However, less-well-developed managerial environments, such as medical staffs in hospitals, office-based medical practices,

nursing homes, and interprofessional processes (such as those involving both doctors and nurses), prove less susceptible to repackaged industrial quality improvement methods. Not being employees and tending to see themselves as customers of hospitals, rather than as partners or employees, doctors, e.g., exhibit difficulty understanding and buying into coordinated, corporate objectives. They object to attending improvement team meetings regularly as contributors and redesigning their own work to fit better into the system as a whole. The various special languages and turf boundaries that have developed in health care (and that professional certification processes perversely reinforce in an effort to protect quality) have stood in the way of wholehearted collaboration on systemic improvement. Hospital records still often maintain separate "nursing diagnoses" and "medical diagnoses." Professions sometimes do not even share common lounge areas, cafeterias, or meeting times, and it is still common in hospitals to find "nursing notes" and "doctors' notes" in separate sections of the same medical record.

5. *Complex systems.* The complexities of the health care system are evident in a comparison with other industries. A manufacturing organization employing 4000 people will categorize staff in about 50 job titles; a typical health care organization of 4000 will utilize 500 titles. This specialization, originally designed to improve quality, now creates multiple handoffs for any patient procedure and contributes to a breakdown in quality processes. Complexity driven by specialization also is evident in the U.S. government's CMS (Centers for Medicare and Medicaid Services) coding system that has thousands of separate billing codes for procedures. Who pays, who has access to care, the role of government, fraud and abuse, expensive new technologies, and rising pharmaceutical costs are a few of the many system issues that must be addressed in the next decade. These challenges are much larger than any individual organization can address, although participation in finding solutions is vital. The one thing that remains constant is that improving the quality of patient care is possible even without major transformational change.

Strategic Initiatives to Address Challenges in Health Care

In the past decade the aim of the health care system has moved from caring for patients with disease and injury to improving the health of entire communities. Process redesign has produced changes in the type of care provided, the site in which care is received, and the extent to which the patient is an active participant in the plan of care. Health information is increasingly available electronically, and public and regulatory demands for greater transparency are allowing patients and payers to access information on costs and outcomes.

A number of initiatives in public policy, finance, and organization of care—all in an effort to measure, control, and improve the value of care, reduce the cost of care, and increase the accountability of health care providers to the public and to insurers—have been implemented. Important examples include the following:

Health Care Reform

The U.S. health care reform debate as of the time of this publication is at the top of the national agenda because the current system is costing more and more money to fund properly. Escalating costs continue to price more people out of health care. There is a growing interest in comparative evidence research to determine what works and what does not work when it comes to improving health while balancing fears of systems that will ration care.

The National Coalition on Health Care has identified five principles for a reformed health care system:

1. Health care coverage for all
2. Cost management
3. Improvement of health care quality and safety
4. Equitable financing
5. Simplified administration

Financial Performance and Costs of Poor-Quality Processes

The financial performance of an organization is greatly influenced by its ability to detect and minimize the costs of poorly performing processes. Some of these costs may be apparent to employees working in the process. A nurse may find it frustrating that he is constantly searching for supplies to treat a patient and realizes that if wasted time had been reduced, the patient could have been discharged earlier. Unfortunately the process has operated to this level of deficiency for so long that it has become the norm or standard work, and the cultural belief begins to set in that nothing can be changed or should be changed. It may then seem normal for a patient to wait up to 2 hours to be discharged from the hospital once the physician gives the order for discharge. Staff may become oblivious to storage cabinets full of bins overflowing with too much stock. The idea of wellplanned and scheduled treatments by ancillary departments may also seem impossible. Health care organizations lose thousands of dollars paying for this excess waste and in recent years have begun utilizing the tools mentioned in this chapter to address those noticeable costs along with many hidden costs in the following four categories:

- *Internal failures*. These include medication errors, billing reconciliation, patient falls, and the development of hospital-acquired pressure ulcers.
- *External failures*. Readmission due to improper initial treatment and diversion of patients due to capacity constraints are two examples.
- *Appraisal*. Given an abundance of inventory assessments, there is unnecessary searching for the correct patient record at time of registration.
- *Prevention*. Examples include diagnostic screenings for hypertension, early cancer detection, and drug screening.

Based primarily on findings and extrapolations from published literature, as well as a report published by the Midwest Business Group in Health and the Juran Institute in 2003, and the reasoned judgment of knowledgeable experts, the authors estimate that 30 percent of all direct health care outlays today are the result of poor-quality care, consisting primarily of overuse, misuse, and waste. (The impact of underuse on costs is not clear.) With national health expenditures of roughly $1.4 trillion in 2001, the 30 percent figure translates into $420 billion spent each year as a direct result of poor quality. In addition, the indirect costs of poor quality (e.g., reduced productivity due to absenteeism) add an estimated 25 to 50 percent, or $105 to $210 billion, to the national bill. Private purchasers absorb about one-third of these costs. In fact, the report estimated that poor-quality health care costs the typical employer between $1900 and $2250 per covered employee each year. Even if these figures are off by 50 percent, poor-quality health care exacts a several hundred-billion-dollar toll on our nation each year (Midwest Business Group on Health 2003).

Example

During the measure phase of an emergency room Lean throughput project, conducted in a New England children's hospital, the improvement team surprisingly learned that over 65 percent of their arrivals were patients with acuity levels of 4 and 5, the lowest-acuity patients in terms of severity of illness. Yet these less severe patients currently spend an average of 2 hours waiting to be seen by a provider. The team also discovered that the revenue per patient from these lower-acuity patients provided an opportunity to the hospital. Further analysis revealed that by improving door-to-provider time the emergency room could decrease its left without being seen (LWBS) rate by 50 percent and potentially realize a savings of $150,000 annually.

In 2009, Sentara Bayside Hospital was awarded a Voluntary Hospitals of America (VHA) Leadership award for Clinical Excellence for achieving a high level of performance in acute myocardial infarction (AMI), heart failure, pneumonia, and Surgical Care Improvement Program (SCIP) clinical quality indicators as measured by the Centers for Medicare and Medicaid Services (CMS) and The Joint Commission. The hospital demonstrated national top quartile performance in all four core measure categories during July 2007 through June 2008 (Sentara Bayside Web Publication).

Electronic Health Record (EHR): One Patient, One Record

Integrating all aspects of the care delivery system—from ambulatory care, to the emergency room, to inpatient setting, to postacute providers—into a single EHR offers the promise to ease access to care, improve patient safety and clinical outcomes, and reduce the cost of care. Health systems, physician office practices, and federal and state government payers are investing billions of dollars hoping to achieve significant return on their investment. The federal government has established an EHR for every soldier participating in the Veterans Administration Health System. Google has created a public database for health records called the Health Vault at www.healthvault.com. But so far there is little conclusive evidence that the automation of medicine will yield significant savings. And the expected improvements in patient care and safety may be modest. A four-year analysis of Medicare data published in the scholarly journal *Health Affairs* in March 2009 found only marginal improvements in patient safety—equivalent to the reduction of 2 infections per year in the average hospital. Broken or nonperforming processes cannot be fixed with an overlay of technology. In fact, that approach only serves to magnify the deficiencies, not correct them. By focusing on process redesign prior to implementing the EHR, hospitals and clinics can maximize the benefits the technology has to offer.

Health care expert Joseph Duhig stated, "The electronic medical record (EMR) is the Holy Grail for most hospital executives. Some industry experts think it is the panacea for ills plaguing the industry. By integrating all aspects of the care delivery system—from ambulatory care to the emergency department, to the inpatient setting, to post-acute care providers—hospital systems are hoping to reduce costs and improve outcomes" (Duhig, 2009).

In 2005, Hillestad et al. estimated industry savings of $81 billion annually, with that number possibly doubling once the data captured by EMRs are fully used in the prevention and management of chronic disease. Yet, despite the health care community's consensus that the EMR is the next great leap forward, many hospitals and health systems struggle to find real benefit. Physicians, nurses, and other caregivers complain that electronic charting reduces productivity and that inefficiencies in the process still remain. As mentioned earlier, this is usually the result of automating processes full of waste and inefficiencies. Process redesign is a necessary step prior to EMR implementation. If this step is skipped, not only will staff and administration experience the frustration, but even CFOs will struggle to calculate a hard dollar return on investment (ROI).

Pay for Performance (P4P)

Pay-for-performance (P4P) programs are increasingly being instituted by government agencies (CMS and state Medicaid programs), commercial health plans, employers, and others to reward providers having met identified performance goals or having shown a demonstrable improvement in the selected performance criteria. According to a recent survey by the Leapfrog Group, there are more than 148 P4P sponsors that offered 258 programs at the end of 2007. The most notable example of a P4P program directed at hospitals is the CMS/Premier Hospital Quality Incentive Demonstration program, which involves more than 250 U.S. hospitals. P4P is seen as a necessary alignment of incentives to achieve excellence in service and quality patient care delivery. Both physicians and hospitals will be required to share their individual and collective outcomes in the future.

Institute of Healthcare Improvement (IHI)—Triple Aim

The U.S. health care delivery system has long been criticized for an overemphasis on the treatment of urgent and acute conditions with little focus on prevention and wellness. Hospitals and physicians are paid for providing treatment and interventions to deal with the consequences of disease, illness, and injury. The IHI has initiated a campaign to engage health systems in the simultaneous pursuit of three aims: improving the experience of care, improving the health of populations, and reducing the per capita cost of health care. The components of the system to accomplish the Triple Aim include a focus on individuals and families, redesign of primary care services and structures, population health management, a cost control platform, and system integration and execution.

Consumerism and Provider Transparency

During the last few years, interest is growing in collecting and publicly reporting information on health care quality and cost. Price and quality transparency is being advocated by governments (federal and state), commercial health plans, and patient advocacy groups as a way to help control overall health care costs, improve clinical outcomes, and support consumers in making more informed health decisions. Consumer-directed health plans are designed to expose consumers to the true cost of health care and encourage consumers to take a more active role in their health care. Increasing availability of medical and health information on the Internet and direct advertising of medications and health care services through multiple media have had a significant effect on consumer preference and behavior. Health care leaders are also using these comparative data and internal "balanced scorecards" to define organizational priorities for change and to demonstrate progress to key stakeholders.

Managing Quality and Safety to Deal with These Challenges

Transformational change certainly is needed on broad global and national scales. However, quality of patient care is possible even without major transformational change. Discussed here is what it means to have quality health care and initiatives that seek to bring this definition to life within individual organizations.

Defining Quality and Quality Management Methods in Health Care

The U.S. Institute of Medicine defines *quality* as "the degree to which health services for individuals and populations increase the likelihood of desired health outcomes and are consistent with current professional knowledge." High quality in this case means providing patients with the appropriate services in a technically competent manner, with good communication, cultural sensitivity, and shared decision making. In health care, quality is evaluated based on structure, process, and outcomes. Structural quality assesses the ability of the health system to organize care, process quality assesses the efficiency of interactions and interventions among patients and clinicians, and outcome quality assesses the improvement in the patients' health status. Described below are several major initiatives aimed at "managing for quality and safety." Many of the quality management methods such as PDSA, improvement teams, design teams, Six Sigma, and Lean can be used to meet the definition of quality in health care, improve the patient experience, and improve financial performance.

The Institute of Medicine's Six Aims and Ten Simple Rules

The Institute of Medicine (IOM), in its 2001 report entitled "Crossing the Quality Chasm: A New Health System for the 21st Century," identified six aims for quality health care services delivered in a way that is safe, timely, effective, efficient, patient-centered, and equitable. Unfortunately, too often patients do not receive the care they need, or they receive care that causes harm. Recognizing that these six aims need to be brought down to an actionable level, the IOM also identified ten simple rules:

1. *Care based on continuous healing relationships*. Patients should receive care whenever they need it and should be able to get it by multiple means.
2. *Customization based on patient needs and values*. The system of care should be designed to meet the most common types of needs but should have the capability to respond to individual patient choices.
3. *The patient as the source of control*. Patients should receive the necessary information and opportunity to exercise the degree of control they choose over health care decisions that affect them.
4. *Shared knowledge and the free flow of information*. Patients should have unfettered access to their own medical information; clinicians and patients should communicate effectively and share information.
5. *Evidence-based decision making*. Patients should receive care based on the best available scientific knowledge. Care should not vary illogically from clinician to clinician or from place to place.
6. *Safety as a system property*. Reducing risk and ensuring safety require greater attention to systems that help prevent and mitigate errors.
7. *The need for transparency*. The health care system should make the information available to patients and their families that allows them to make informed decisions, which should include information describing the system's performance on safety, evidence-based practice, and patient satisfaction.
8. *Anticipation of needs*. The health care system should anticipate patient needs rather than simply react to events.
9. *Continuous decrease in waste*. The health care system should not waste resources or patient time.
10. *Cooperation among clinicians*. Clinicians and institutions should actively collaborate and communicate to ensure an appropriate exchange of information and coordination of care.

At the level of individual organizations, it is incumbent upon management to prioritize these rules and define specific ways to measure and improve performance against them.

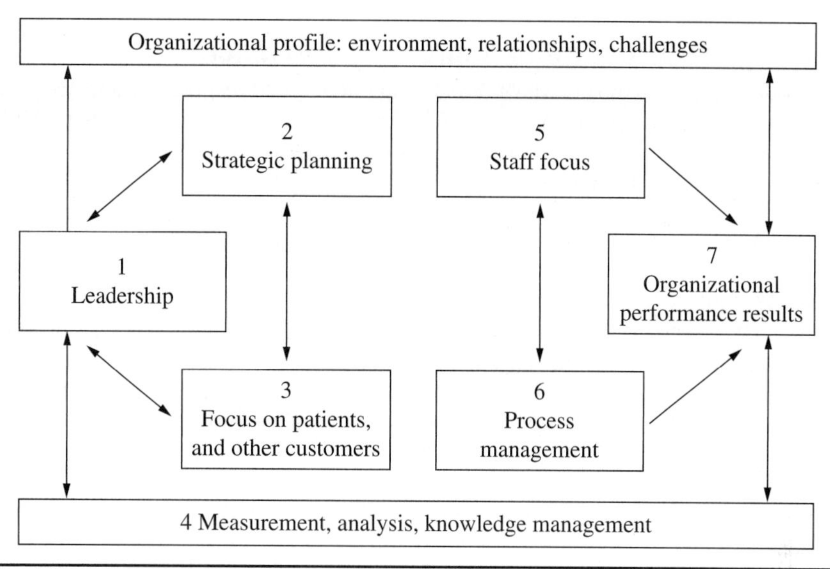

FIGURE 23.1 Health care Baldrige criteria. (*Used with permission of NIST.*)

The six aims and ten simple rules provide an excellent foundation for benchmarking performance.

Using National Awards for Excellence for Health Care

In an effort to foster strategic quality management in health care, the Malcolm Baldrige National Award for Excellence and other national quality awards were adapted from manufacturing, service, small business, and education (nonprofit is another recent addition) to include criteria specific for health care. The U.S. award, which is presented by the U.S. President and the Department of Commerce and managed by the National Institute of Standards and Technology (NIST), is the top honor that a U.S. organization can receive for quality management and quality achievement, and winning organizations are considered world-class role models in their industry. It has become one of the drivers for many hospitals. The criteria support all the performance excellence methods and tools described in this handbook. See Figure 23.1 for the Baldrige categories for health care. For additional information on the Baldrige program, see Chapter 17, Using National Awards for Excellence to Drive and Monitor Performance.

The Joint Commission 2010 National Patient Safety Goals

The Joint Commission has been a leader in defining the highest standards for quality and patient safety in health care and for evaluating performance data based on their standards. Today, more than 15,000 health care providers—from small, rural clinics to expansive, complex medical centers—use Joint Commission standards to guide how they administer care to their patients and continuously improve their performance over time. The 2009 Patient Safety Advisory Group revised the leadership standards. The standards address the responsibility of leaders to establish a hospitalwide safety program, proactively explore potential system failures, analyze and take action on problems as they occur, and encourage reporting of adverse events and near misses (Leadership in Healthcare Organizations, A Guide to Joint Commission Leadership Standards, A Governance Institute White Paper, The Governance Institute Winter 2009).

Developing the Structure to Improve Performance

In today's systems the boards of directors are being held responsible and accountable for the organization's mission to provide the best possible care and to promote patient safety. The board's responsibility for ensuring and improving care cannot be delegated to the medical staff or executive leadership; it is the very core of their fiduciary responsibility. An engaged board in partnership with an engaged executive leadership and physician team can set system-level expectations and accountability for patient safety and clinical outcomes.

Governance and Board Participation

Hospital boards vary widely in size, composition, and membership. One thing that they do have in common is that the majority of the members are typically not health care professionals. Historically boards have focused their attention on strategic planning, finances, and facility upgrades or construction. The attention to quality and patient safety has been left to executive leaders and the medical staff to manage. The high rise of health costs, shrinking margins, and serious safety events have changed what boards need to know about quality and safety. Government leaders, communities, and shareholders are also putting pressure on boards to make hospitals safer through performance improvement. Boards today are involved in setting aggressive quality and safety goals, monitoring quality and safety data, asking harder questions of leaders, and establishing higher levels of accountability for achieving quality and safety performance goals. Although it is not the responsibility of board members to make change happen in the hospital, it is their responsibility to expect that change happens. Many boards are making it standard practice to start every board quality meeting with a brief patient story about a serious safety event in order to put a face on the quality problem. The best boards today are setting quality and safety goals by declaring how good the hospital needs to be, by when, and how it will be measured. As an example, the Cincinnati Children's target is to become "80 percent safer, as measured by Serious Safety Event rate, within 18 months" (Reinertsen, "Quality and Patient Safety: Understanding the Role of the Board").

Leadership Team Involvement

The role of the leaders of the system is to provide direction, to establish the vision, mission, and values. Once these are in place, leaders must provide the resources to attain the mission. An important ingredient in the success of attaining the mission is that leaders must also select the most effective methods to improve and sustain performance.

For example, Dr. Novotny, Chief Medical Officer and CEO at Southwest Vermont Medical Center, has a vision for the center "to be the best integrated health system we can be, where physicians, nurses and staff work together with a self-regulating governance, leadership and management team that adapts to our changing environment. We will carryout that vision by having a system that will have

- A clear sense of shared purpose for patient, resident and family centered care.
- Clinical, management, and patient influence at all levels.
- Its highest priority on long-term relationships with patients and families, standardizing care with best practices.
- Hospital-based care that will be defined by community needs that are affordable and sustainable.
- Using performance excellence methods that best support the attainment of our vision."

Stating the methods will provide a common language and approach to attaining the vision.

Physician Involvement

Board involvement is critical to building a quality-driven culture, but board and physician partnerships are a key contributing factor. The board and physicians must share a common quality purpose. The fact is that very little happens in the health care system without a physician's order. By virtue of physicians' legal authority almost all actions in health care are the result of their decisions and recommendations. Therefore, the patient's plan of care and any changes in the way care is delivered require a physician order.

Hospitals employ a variety of strategies to bond with physicians. Some hospitals are gaining physician loyalty through economic credentialing. Others are using strategies such as medical directorships, employee-based compensation models, and gain sharing. These approaches, while effective in strengthening physician relationships, are short-term solutions.

The measure of effective physician engagement is results. There are several excellent examples of physician engagement and sharing a common quality purpose. Organizations such as Virginia Mason Medical Center, McLeod Regional Medical Center, Hackensack University Medical Center, Immanuel St. Joseph's–Mayo Health System, and Tallahassee Memorial Hospital have achieved stunning results. For example, Tallahassee Memorial Hospital and Immanuel St. Joseph's–Mayo Health System have reduced mortality rates 30 to 40 percent. Hackensack and McLeod are now capable of delivering "perfect care scores" on evidence-based care at levels of only 1 or 2 defects per 100, for all CMS core measures. McLeod has gone seven straight months without a single adverse drug event as measured with the IHI Trigger Tool methodology (an objective count of medication-caused harm which does not depend on incident reports or other self-reporting systems). These results could not have been achieved without significant engagement on the part of the physicians (Reinertsen, IHI Engaging Physicians in a Shared Quality Agenda Innovation Series 2007).

To design an effective approach to physician engagement in the quality agenda, the first step is to make a realistic assessment of the hospital-physician relationship. One measure of engagement is the currency of the medical staff bylaws which dictates how physicians act and function. Since the medical staff is composed of physicians from multiple specialties, it may be necessary to develop a menu of strategies for engaging physicians. Making physicians' partners, promoting system and individual goals for quality, providing education on improvement methods and tools, and providing physician performance data help hospitals achieve their quality goals. Data should not be used in a threatening way but in an enlightening way. Physicians should be leading the quality agenda by asking the following questions: How well are we doing? How well are we doing compared to best performers in class? Are we getting better?

Lean Health Care Process Management and Service Line Structure

In the book *Six Sigma Process Management* by Rowland Haylor and Mike Nichols, the authors describe "end to end core processes as those high-level processes that are the primary drivers of value, satisfaction, and profit." Knowing how to identify and manage these processes is one of the first steps in developing a solid work system model. Because most health care systems offer multiple services to many different patient populations, process management becomes complex and challenging. Processes cannot be designed so that "one size fits all." So the question becomes how to design our processes to meet the many and varying needs to all our patients. Health care organizations have integrated process thinking with a service line structure to create an effective work system.

The focus on business process management (BPM) has been around for a long time, with early influence from Juran and Deming and others, and later by the work of Michael Hammer and others. (See Chapter 8, Business Process Management: Creating an Adaptable Organization, for more on business process management.) The concepts of Japanese manufacturing and Total Quality Management also had an impact. BPM is especially important in health care because of the high degree of specialized training and job roles, and the functional structure of most health care organizations and systems. The Baldrige model has had great influence on the concept of process thinking in health care. Lean and Six Sigma have also had an impact.

The challenge in health care is, How do you design a robust process, incorporating standard practice and minimal variation while at the same time meeting the many and varying needs of patients? Process mapping of the patient care processes can become incredibly complex given all the multiple options for providing care. Training personnel to follow complex processes is difficult. Process ownership has also proved to be a challenge in health care, with lack of enough knowledge, authority, and influence to make significant and sustainable changes.

The answer lies in the integration of process management with service line management. Service lines provide an organizational infrastructure around a rational collection of services that a patient may require during an episode of care. Service lines may be called other things including clinical programs, centers of excellence, and other names and descriptions. While there is no standard model from system to system, service lines are typically organized around one of three categories of services:

- Interventions such as surgical services or emergency services
- Diseases such as cancer or heart failure
- Populations such as children or women

A review of the application summaries for U.S. Healthcare Baldrige Award recipients shows that most make at least some reference to the use of service lines as a part of their organizational makeup. One of the best pieces of evidence of the integration of service lines is part of the performance excellence planning coming from North Mississippi Medical Center (NMMC):

> NMMC coordinates clinical services through five Service Lines (SL's): Cardiovascular, Emergency & Surgery, Medicine, Oncology & Behavioral Health, Women & Children. We study our patient population and develop services specifically targeted to their needs.
>
> Our processes—both health care and support—are designed to meet our Mission of improving health and still maintain cost efficiency. To do so, we organize our services by SL's and provide services in multiple settings: outpatient, ESD, hospital, home, rehabilitation, LTC and community. The SL model, focused on the patient/customer, eliminates departmental silos, involves the specific SL medical staff and manages processes in order to provide value and improve outcomes. (Application Summary for North Mississippi Medical Center; Baldrige website, 2009)

In a blended approach of process and service line management, process managers work to remove unwarranted variation across the process continuum. Service line managers adapt the standards of care for the specialized needs of the patient. Process managers remove unwarranted variation and establish systemwide level standards of care and practice. Service lines customize processes to meet specific customer requirements, using evidence or professional practice.

This approach requires an infrastructure of service line leadership and process management leadership, aligned and integrated with reportability to the executive leadership. Advantages of this approach are as follows:

1. *Standard practice.* This is a means for removing unwarranted systemwide variation.
2. *Flexibility.* Processes have flexibility to meet the many and varied specialized needs of patients.

3. *Individualization.* Patient care processes are designed to meet individual patient care needs and requirements.
4. *Best science.* Service line owns the adoption of standards of care and standards of practice.
5. *Optimization.* There is high patient value with the most efficient use of system resources.

Managing for Quality and Safety in Health Care

Quality management methods rapidly spread in recent years from dozens, then hundreds, of hospitals and other health care organizations on the momentum of increasing awareness and some initial successes.

Leaders organized quality councils; formal quality improvement teams became commonplace, and more fundamental redesigns of care began to make systems more patient-friendly. In general, the health care models for managing these improvements closely paralleled those in other industries. Perhaps for this reason, the models worked more smoothly in segments of health care that, from the start, looked more "corporate" in structure.

Yet, despite the cultural barriers, the promise of quality improvement in health care remains great. Recently, health systems are adopting more sophisticated approaches to managing for quality using Six Sigma methods and Lean principles.

The remaining part of this chapter covers the methods and tools along with relevant templates, examples, and case studies including

- Strategic planning and structure
- Quality control and PDSA, special cause analysis, and root cause analysis
- Quality improvement with Lean and Six Sigma
- Developing new patient-focused services: Design for Six Sigma (DMADV) and Lean design
- Improving patient safety

Quality Control and PDSA, Special Cause Analysis, and Root Cause Analysis

Quality control is about maintaining performance according to accepted standards. When a process extends beyond certain boundaries (e.g., as established by the voice of the customer, or voice of the process), action is needed to bring the process back and eliminate the cause of poor performance. PDSA, special cause analysis. and root cause analysis are three means to accomplish this.

PDSA

The Plan-Do-Study-Act (PDSA) is a rapid-cycle change and control tool used to solve sporadic, day-to-day problems. Popular in health care, it also provides a useful framework for documenting the steps needed for a test of change (sometimes called a pilot), thereby avoiding the common problem of moving directly from symptom to solution. The acronym is named for the steps:

Plan—plan for how the test of change will be carried out.

Do—carryout the test of change.

Study—observe the results of the test and draw conclusions.

Act—decide what revisions may be needed.

These steps may be repeated until the change is ready to be implemented and spread through the organization. See below for an example of how PDSA can be used in the context of a special cause analysis.

Special Cause Analysis

Special cause analysis (SCA) can be used to help determine the source of special causes. These are sporadic events that have some assignable cause which can be identified and corrected relatively quickly and easily. SCA might be used to improve patient safety, e.g., as follow-up to minor events, near misses, or precursors to errors.

Special cause analysis involves the following steps:

1. *Preliminary investigation.* Gather and validate the facts, walk the process, isolate suspect equipment or supplies, review medical charts, and interview staff.
2. *Event description/problem statement.* Describe the event in the form of a problem statement; stick to the facts.
3. *Current state/event flow map.* Map the relevant "as is" process, ensuring potentially important upstream portions are captured, as they may help reveal triggers.
4. *Process breakdowns/human error identification.* Show on the process map where the normal process broke down, including human contributions.
5. *Five why analysis/identification of special causes.* Ask "Why?" multiple times to drill down to the special cause.
6. *Future state map/improvement solutions.* Generate possible solutions and evaluate them to select a final, best set of solutions; revise the process map to incorporate these as a future state map.
7. *Implementation plan.* Draft a plan to put the solution into everyday use.
8. *Control plan/follow up.* Monitor the process for the same or similar events to ensure the process is permanently fixed.

An example template that applies the PDSA cycle and distills the above into a single page is shown in Figure. 23.2.

Root Cause Analysis

Root cause analysis (RCA) is a more in-depth analysis that identifies true root causes of events (a special cause may itself be a root cause, but is more readily pinpointed). Note that PDSA assumes that the analysis of symptoms and development of possible root causes are completed prior to beginning the cycle of iterative experimentation. In contrast, RCA incorporates these preceding phases (and root cause corrective action specifically encompasses follow-up remediation; see Chapter 13, Root Cause Analysis to Maintain Performance).

An example of root cause analysis applied to a hospital registration process is described below according to the four phases:

1. *Identify a problem.* Incorrect registration processing of 1-day stay and observation assignments leads to Medicare denials. These denials cause losses in revenue, decreased accounts receivable, and decreased productivity due to claims rebilling.
2. *Diagnose the cause.* Analysis showed that the registration department did not always receive the patient order, thus relying solely on communication from nurses or unit secretaries for the order, resulting in false data.

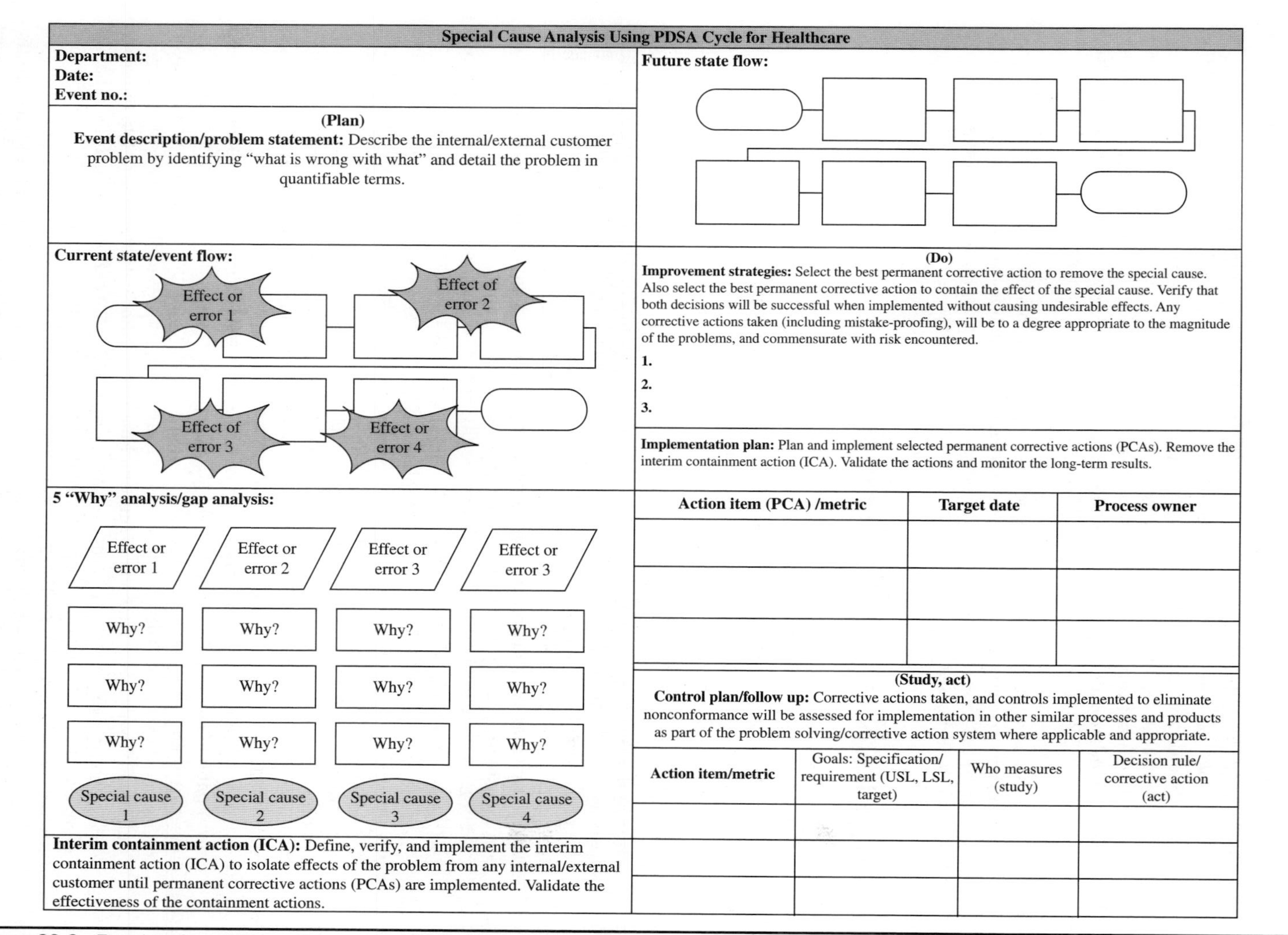

Figure 23.2 Template for use in special cause analysis. (*Juran Institute, Inc.*)

3. *Remedy the cause.* The process was revised to require that a copy of the order be sent to the registration department and that patient type be assigned only upon direct order receipt.
4. *Hold the gains.* Project implementation and completion occurred in 8 months. Standardization of the registration process was achieved through consistency in forms and training. Results were monitored over the next year, and special causes identified and eliminated as they arose. Gains were significant; through registration standardization, this organization realized a $308,000 savings in cost avoidance and a $37,770 savings in rework (20 hours a week), for a total savings of $345,770.

Because of resource constraints, root cause analysis teams typically are convened only under certain criteria and are formally chartered by a senior executive champion. More information on root cause analysis can be found in Chapter 13, Root Cause Analysis to Maintain Performance.

Quality Improvement with Lean and Six Sigma

Reducing Waste with Lean Techniques

Lean principles are not new, and they have been used effectively in manufacturing for many years. Health care tends to lag other industries in the adoption of management-driven change initiatives, and therefore Lean production is widely viewed as the newest innovation for transforming care delivery. Although health care differs significantly from manufacturing, there are many similarities. Whether building a car or providing care to patients, workers must rely upon numerous, complex work processes to accomplish their tasks and create value to the customers or patient. It is important to keep in mind that even though health care has a production system (e.g., hospitals, clinics, operating rooms, laboratories), health care professionals are not production workers. Physicians, nurses, and other care providers must gather information from various sources and apply their knowledge in making good clinical decisions. A Lean production system removes many of the hassles that create frustration for care providers attempting to meet patient needs.

For Lean principles to be applied successfully, leaders must first work to create an organizational culture that is receptive to Lean thinking. Health care tends to focus on the scarcity of resources, whether if be staff, equipment, supplies, or access to ancillary testing and care environments. Learning to see the waste in work processes requires that health care professionals see how poorly designed physical space and work processes impede their ability to provide good patient care. Detailed activity analysis of clinical and support staff finds that upward of 50 percent of their time is consumed in finding the supplies and equipment they need to perform their job duties and in unnecessary movement of patients, supplies, information, and equipment.

Current health systems are organized in a batch and queue production system whereby once a patient accesses the system, that patient is pushed across the continuum of care with time-consuming and potentially dangerous handoffs. For example, a patient arriving in an emergency department with symptoms of a heart attack is physically transferred to any of or all the following departments: diagnostic imaging, cardiac catheterization lab, intensive care, telemetry care, operating room, recovery, medical/surgical floor, and cardiac rehabilitation. During the 3- to 5-day hospital stay the patient is seen and cared for by 20 to 50 care providers representing more than 10 different specialties.

Lean pilot programs are currently underway in various health systems to redesign the physical care environment and work processes. Lean change concepts such as the *universal bed* (where the patient stays in the same location while staffing, equipment, and technology are brought to the bedside) are an example of transforming to a Lean production system.

Lean principles and tools are described more extensively in Chapter 11, Lean Techniques: Improving Process Efficiency. Two examples of the application of Lean are provided here to better illustrate their use in health care. The first example is of a rapid improvement event that addressed hospital emergency room wait time. The second example illustrates the elimination of waste through 6S.

Rapid Improvement Event (RIE) Reducing Wait Times in the Emergency Department

Problem: Patient throughput was excessively high, with over 20 percent of patients staying longer than 6 hours. This resulted in low patient and physician satisfaction, an increased number of patients who left without being seen, and lost revenue.

Approach: A RIE is a highly facilitated series of steps that applies Lean tools and techniques to rapidly improve a specific work area. This decision was supported by an assessment that indicated causes of delays in patient flow were the result of procedural and organizational problems rather than errors or defects per se in execution of the process.

Day 1: Identify causes.

- Value stream analysis showed 76 to 94 percent of patient time in the emergency department was non-value-added wait time. Figure 23.3 is an example of a value stream map.
- A delay assessment tool was applied to clarify specific areas and causes of delay, e.g., registration (inaccessible old records, looking for patient), triage (interruptions from security, finding the physician), disposition to departure (dictation delays, awaiting transport), etc.

Day 2: Design the future state.

- The team brainstormed improvement ideas to address specific root causes.
- Ideas were filtered through a selection matrix (see Chapter 18, Core Tools to Design, Control, and Improve Performance). Questions included these: How much will this cost? Will there be resistance? Will this have a positive impact on our mission? Has this been used in other hospitals with positive results? If this does not work, will it create a problem?
- A new process was created to eliminate wait time and provide a more streamlined transition of the patient among the process steps that truly add value in terms of patient care. Examples of changes were as follows:

Emergency Department: Simplify the process for holding orders to be written by physicians; create new code for the Emergency Department approaching or in crisis; implement a prediversion team.

Discharge: Revise housekeeping protocols to clean floor beds faster; coordinate discharge information by patient rather than batching.

Day 3: Implement changes.

- An implementation plan was drafted to assign due dates and responsibilities for specific tasks.
- Solutions were tried out and amended as needed prior to standardization and final deployment.

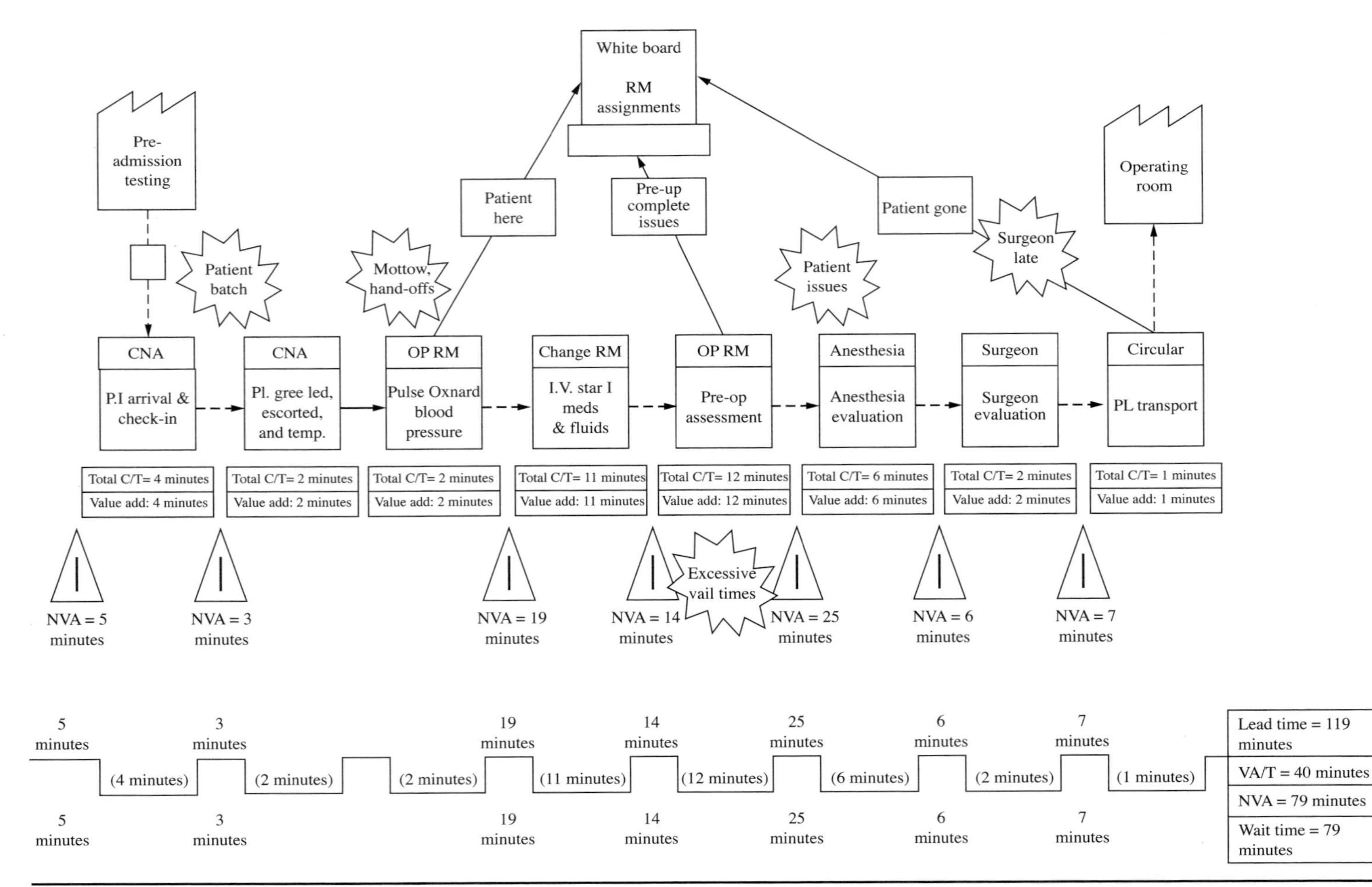

FIGURE 23.3 Current state value stream map.

The RIE resulted in a 25 percent reduction in patients waiting for 6 hours or more and greater than a 50 percent reduction in the percentage of patients who left without being seen.

6S Workplace Safety and Organization

The objective of 6S is to create a quality workplace using a systematic approach to waste reduction, organization, and housekeeping. 6S provides the following benefits:

- Reduced complaints, resulting in greater customer satisfaction and loyalty
- Reduced errors, resulting in higher-quality outcomes
- Reduced wait time, resulting in better flow
- Access to the right supplies, at the right time, at the right location, increasing staff satisfaction and productivity
- Reduced clutter and standard work, enhancing patient safety

The component steps of 6S are

1. Sort.
2. Set in order.
3. Sweep and shine.
4. Standardize.
5. Self-discipline.
6. Safety (added to the original 5S developed in Japan).

A more complete description is provided in Chapter 11, Lean Techniques: Improving Process Efficiency. An example of its use in a hospital follows.

Reduction and Elimination of Waste through 6S

Problem Statement. The neonatal intensive care unit (NICU) lacks organization which results in the staff having to sort through excess supplies and equipment to locate needed items. No par levels are visible on supplies; supplies are not grouped according to usage or restocking; excessive amounts of paper are stockpiled in all administrative areas; and equipment is stored in hallways and corridors, obstructing passageways. This creates frustration among the staff and delays in the delivery of care, decreases staff productivity, and increases inventory levels and costs.

Does this sound familiar? Despite general knowledge of the waste, 6S team members were surprised at the magnitude once they dug deeper. In step 1 (sort), e.g., the team identified thousands of items that were not needed, had expired, or otherwise were a costly waste. The disposition of these was as follows:

Returned to storeroom for credit:	4,389
Disposed items:	6,599
Expired inventory:	5,371
Donations:	756
Relocated to other units:	294
Total	17,409 items

Following the remaining 6S steps, the team increased safety compliance, provided a documented increase in staff satisfaction, and dramatically reduced supply and special order expenses.

Six Sigma

Many health systems have now adopted Six Sigma methodology to provide a data-driven and disciplined approach to quality improvement, especially when the organization lacks knowledge of root causes in its own work processes. (See Chapter 12, Six Sigma: Improving Process Effectiveness.) In this situation, Six Sigma (using DMAIC methodology) is a preferred methodology because of its rigorous application of measurement and analysis tools. Health care continues to invest heavily in new information technologies that allow easier access to data with less reliance on slow and expensive chart review, whereby a clinician must manually abstract data from a paper chart. Physicians are trained in the scientific management approach and respond favorably to data-driven improvement efforts such as Six Sigma.

Health systems are recognizing the need to be more strategic in the deployment of improvement initiatives. In the past, organizations relied upon a grassroots approach where nominations for projects were identified by employees or managers to address a specific problem in the department or unit. Although many of these projects were successful, they did not change the overall performance of the organization. Senior leaders were frustrated in looking at their balanced scorecard and not being able to measure the impact of their investment in improvement.

Six Sigma is typically strategically deployed and often financially focused. Projects are selected and aligned around strategic priorities and business goals. For example, many health systems enrolled in the Institute of Healthcare Improvement (IHI) campaign to reduce the "Big Dot" measure of reducing hospital mortality. Severity-adjusted mortality is considered the best overall measure of clinical quality and patient safety. IHI identified a set of six interventions with a goal to eliminate 100,000 avoidable deaths in hospitals in 18 months. This is an example of Six Sigma thinking: $Y = f(x)$. Hospital mortality $= f(x_1, x_2, \ldots)$. Over one-half of the health systems in the United States enrolled in this "100,000 Lives Campaign." The success of this campaign was the first demonstrable example of applying improvement science to improve the overall system of care.

An example follows to illustrate the highlights in an application of DMAIC to an acute care patient length of stay (Juran Healthcare 2009 Case Study: Reduce Heart Failure Length of Stay).

Example of Reducing Length of Stay due to Heart Failure

Problem. In a medium-sized acute care hospital, inpatients assigned diagnosis-related group (DRG) 127 (heart failure and shock) had an average length of stay (ALOS) of 5.18 days, which was 1.08 days greater than the geometric mean length of stay of 4.1 days (Centers for Medicare and Medicaid Services). Of 491 patients discharged over 1 year, only 280 (57 percent) were discharged less than 4.1 days after admission, yielding a process sigma of 1.68. This resulted in an increased risk for negative patient outcomes due to delays in the delivery of care, as well as an increase in the cost of care.

Project Goal. The project goal was to increase the percentage of patients with DRG 127 (heart failure and shock) who are discharged in less than 4.1 days (98 hours from admission) to 95 percent within 6 months by January 4, 2008, for a process sigma of 3.18. This was an aggressive but achievable goal, since performance already was good with 57 percent of patients discharged within the geometric mean of 4.1 days. Achieving the goal would yield an ALOS substantially less than the CMS geometric mean length of stay, representing an exceptional level of performance.

Baseline Process and Performance. The project Y was ALOS, measured in days for all adult inpatients coded with DRG 127 (heart failure and shock). To better understand the current process, length-of-stay data were gathered and the process was characterized in terms of the major work flows. Over the preceding year, 57 percent of DRG 127 patients had a length of stay less than or equal to the target of 4.1 days. This had an associated baseline sigma level of 1.68 and a cost of poor quality of $1,001,000 annually.

Identification of Root Causes. A central tenet in Six Sigma is to make data-driven decisions. Although all the possible root causes that the team identified were plausible, team members collected data to confirm or deny which truly had a cause-effect relationship with extended length of stay. In all, 18 possible root causes were tested. By the end, it was evident that many of the possible root causes the team initially believed to be true were, in fact, demonstrated to have little or no role in lengthening patient stay.

A detailed data collection plan was created to document data sources, sample sizes, data analysis tools, and responsible parties for each of the 18 possible root causes. In most cases, data were available in electronic logs, but new data had to be collected for others. Graphical analysis tools used included box plots, scatter plots, Pareto charts, and bar charts. In addition to descriptive statistics (average, median, standard deviation), statistical analysis tools included nonparametric hypothesis tests, regression, and chi-square analysis.

Rigorous analysis of the data revealed the vital few factors driving extended length of stay:

- Inpatient holding process was not standardized.
- Congestive heart failure (CHF) standard orders were not used (no parameters).
- There was a delay between discharge order and time patient leaves floor.
- Patient stay included a weekend.
- Patient becomes deconditioned because of lack of activity.
- Practices were not based on gold standards.
- Patients were held after meeting InterQual discharge criteria (delayed discharge when compared to evidence-based decision support rules).

Solutions to Reduce Length of Stay. The team brainstormed possible solution strategies that would address each of the vital few problems causing extended length of stay for congestive heart failure patients. Selected strategies for proven causes include these:

- *Patient holding.* Develop ways to get patient out of the ED faster; improve and expedite care for patients that are held.
- *CHF standard orders.* Reduce variation in practices by developing order set and interdisciplinary pathway and providing for education of physicians and hospital staff in their use.
- *Delay in discharge order to leave floor.* Develop better communication process in relationship to anticipated discharge date and needs starting at day 1 of admission.
- *Weekend stay.* Develop staffing/resourcing plan to support CHF standard orders and pathway including scripting to facilitate use and discharge, improving team-based communication and handoff for weekend stays. Standardize use of interdisciplinary pathway-based contingency discharge orders.
- *Patient deconditioning.* Develop plan for activity and trigger for when a physical therapy evaluation is needed based on lack of progression of activity status.

- *Lack of gold standards.* Create gold standards via standard orders and clinical pathway.
- *InterQual criteria.* Utilize Quality Management Coordinators (QMC) to address when discharge InterQual criteria are met.

Implementation and Control. The pilot was successful in reducing the length of stay to an average of 2.6 days for patients with hospitalists attending, with 91 percent of patients discharged within 4.1 days of admittance. A control plan was developed to ensure that improvements and gains were sustained over the long term. Key elements included the control subjects (length of stay, readmission rate, and proven factors), measurements (sensor, frequency, sample size), and actions (criteria for taking action, responsibilities). Most control subjects were monitored every two weeks, with criteria for action based on performance relative to specifications and statistical process control charts.

Developing Patient-Focused Services: Design for Six Sigma (DMADV) and Lean Design (LDMADV)

Two methods are gaining popularity by providing a systematic approach to developing and launching new health care services and processes: Design for Six Sigma (DFSS) and most recently using Lean in design of new facilities and procedures. When new services or processes are put in place, a design team ensures that all critical-to-quality characteristics are discovered, and the development of the service includes effective safety barriers and error-proofing mechanisms that are built into the processes. Mistake proofing is a tool that uses process and equipment design to significantly reduce or eliminate human error. For more details see Chapter 11, Lean Techniques: Improving Process Efficiency.

Level I mistake-proofing design makes it impossible for a person to make an error (e.g., gas fittings that will not connect unless they are 100 percent correct), and level II mistake-proofing significantly reduces the probability of making an error (e.g., bar-coding medication administration).

Design approaches have a natural fit in health care, in that most root causes, safety events in particular, relate to lack of standard processes, or that the processes in place lack the necessary barriers to prevent human error from reaching the patient. Examples include medication reconciliation, any processes in which technology is being implemented, patient handoff communications, new regulatory requirements, etc. Design and continuous innovation are discussed in detail in Chapter 4, Quality Planning: Designing Innovative Products and Services, and Chapter 14, Continuous Innovation Using Design for Six Sigma.

Benchmarking

Benchmarking is another tool that is equally applicable to the health care environment as to any other business scenario, and it can be used in both design and improvement efforts. The efficiency and effectiveness of the performance of any health care organization, facility, or process can be compared to others through the application of benchmarking techniques. Because it looks to others for standards and solutions, it is not appropriate where true, out-of-the-box thinking is required, or for organizations that already are leaders in an area (indeed, this can lead to complacency). However, it is an excellent approach for health care organizations that have performance below that of their peers and are seeking tested solutions to suboptimal performance.

In this context, the efficiency of health care might be measured in terms of spending and staffing. Spending might include all expenditures relating to salaries and wages, payments made to contractors and service providers, costs of materials, facility costs, and real estate expenditures. Staffing may include the time spent by all levels of personnel in the benchmark scope. For example, an analysis may include nursing staff, support staff, managerial and administrative personnel, and third-party service providers.

Performance effectiveness would analyze performance in terms of the quality of the service provided. That is, how well are health care services delivered from the perspective of various customers? Effectiveness considers dimensions such as quality of care, clinical investigation, hygiene and facility-borne infections, patient and staff safety, and customer and staff satisfaction.

As with any benchmarking, suitable methodologies for the normalization of performance comparisons may need to be applied to enable true comparative ("apples to apples") analyses to be conducted between differing health care subjects. Normalization routinely is done in health care and is evident in metrics such as "case mix adjusted" figures and "adjusted discharges."

More detailed information on benchmarking as a performance improvement tool can be found in Chapter 15, Benchmarking: Defining Best Practices for Market Leadership.

Simulated Design of Experiments to Solve Difficult Problems

Hospitals seeking to improve health care delivery often face the challenge of implementing process changes without benefit of tests of change that could clearly demonstrate success and identify potential risks. In particular, the disruption of already strained processes can severely affect patient quality of care and safety, thereby constraining the ability to test proposed improvements. For one faced with the uncertainty of success, risk aversion easily can limit the range of change considered and derail otherwise successful quality improvement efforts.

Alleviating this problem, simulation and design of experiments (DOE) provide a powerful approach to quality management and process improvement in hospitals. DOE is a stepwise, statistically based method that efficiently guides the identification and selection of changes that collectively will optimize performance. Typically, this involves iterative testing of different factors, settings, and configurations, using the results of successive tests to further refine the product or process. When properly done, a DOE approach produces more precise results while using many fewer experimental runs than other methods (e.g., one factor at a time, or trial and error). The outcome is a robust design that better meets customer specifications and production and delivery constraints. A more detailed overview of DOE is provided in Chapter 19, Accurate and Reliable Measurement Systems and Advanced Tools.

A designed, experimental approach is not new to health care. For example, medical researchers use a carefully designed series of experiments to optimize medical devices and drug formulations. However, a traditional DOE in a hospital requires time-consuming, real-life testing of multiple factors or process variables that cannot easily be done within controlled, protected environments, thereby limiting its use in health care (e.g., testing effects of different staffing models). In comparison, a simulated DOE approach does not necessitate such intensive, real life testing, and it can rapidly identify process configurations that will simultaneously optimize performance against multiple goals (e.g., maximizing patient volume while minimizing staffing costs). The approach is particularly valuable in situations where testing of proposed changes may be highly disruptive or risky, and substantially opens up new opportunities for hospitals to improve the delivery and costs of care.

At a high level, a simulated DOE method consists of the following steps:

1. Create a process flow model depicting the activities and participants in the process.
2. Identify the process output(s) of interest and target value(s).

3. Identify the input(s) to be tested for their effect(s) on the output(s).
4. Select the combinations of inputs to be tested as different scenarios.
5. Run the process flow simulation with the different scenarios.
6. Analyze the results using DOE.
7. Identify the scenarios that perform best against the output target(s), and draw practical conclusions regarding proposed process improvements.

Several iterations may be needed to refine and test scenarios before converging on a final configuration or small number of alternatives to test in actual operations. Below are examples of questions that this approach of simulated DOE could address in a health care delivery environment

Staffing

- How many Emergency Department RNs are needed to maintain patient average length of stay around 120 minutes and simultaneously minimize labor costs across a range of anticipated patient arrival rates?
- What mix of clerical and clinical resource skill levels optimizes Emergency Department door-to-provider time?
- Will the addition of one radiologist be worth the extra cost in terms of turnaround time? Should any additional radiologist be dedicated to the Emergency Department?

Bed Allocation

- How many inpatient beds of different types [e.g., Medical/Surgical, Adult Intensive Care (ICU), Burn ICU] will yield an overall 75 percent utilization?
- Would patient flow improve if one of the Emergency Department triage rooms were converted into an exam room?

Emergency Department Flow

- On what days of the week should a fast track process be in operation to minimize patient wait time?
- Should RN and physician resources be dedicated to the fast track process when it is in operation?
- How would partial conversion of incoming patient waiting room space into a treated patient holding area impact exam room utilization and rate of patients leaving without being seen? Should such a holding area only operate under high patient volume?
- Several advantages are evident in the application of simulated DOE to health care delivery:
    - Predictions are possible for situations that do not necessarily exist. In modeling Emergency Department staffing, e.g., high-volume, surge arrival rates need not be experienced in real life (although any extrapolation outside the bounds of actual observations must be done with caution).
    - Solutions that would be very difficult, disruptive, or risky to test under real-life conditions easily can be tested. For example, it is difficult or impossible to adjust the arrival rates of patients.

- Interactions can be identified and taken into consideration when process improvements are made. The DOE approach specifically allows for the testing of both main effects and interactions. For example, the benefit of adding nurses may be greater at higher patient arrival rates.
- Many different configurations can be tested in a very short time. Once the initial process flow model is created, scenarios are relatively straightforward to set up and test. Answers that normally would take many days or weeks to obtain can be acquired in a matter of hours.

There are several downsides to this approach. First, building a discrete event simulation is tedious and time-consuming and often requires data that may not be readily available. The data can be acquired with some planning, however; and once a model is built, it can be put to many uses. Second, neither simulation nor DOE is widely used in health care, and the approach may be treated with suspicion. To alleviate this, engage the assistance of experts, and always pilot changes in a protected environment before fully implementing them. Despite these drawbacks, the potential benefits warrant a closer look at simulated DOE for hospital process improvement.

Simulation Software to Design New Services

One tool that is proving particularly useful in redesigning health care processes is discrete event simulation (DES). Similar to the simulated DOE, it requires a fair degree of technical expertise to learn the language and tools. But once DES is learned, it can be put to use in a large variety of situations where making "live" changes simply is too risky (e.g., patient safety risk) or time-consuming. A description of DES and examples of how it can be applied to health care are provided in Chapter 19, Accurate and Reliable Measurement Systems and Advanced Tools.

Improving Patient Safety

The Institute of Medicine (IOM) report "To Err Is Human: Building a Safer Health System" (2000) defines patient safety as freedom from accidental injury. The report describes two types of errors: (1) the failure of a planned action to be completed as intended or (2) use of a wrong plan to achieve an aim. In addition, processes should not harm patients through inadvertent exposure to chemicals, foreign bodies, trauma, or infectious agents.

Characteristics of a Safe System

The IOM report identifies the following characteristics of a safe system:

- The health care environment should be safe for all patients, in all its processes, all the time—day or night, weekends—in all care locations.
- Patients need to tell caregivers something only once.
- Care must be seamless, supporting the ability of interdependent people and technologies to perform as a unified whole—especially at handoffs between caregivers, sites of care, and through time.
- Knowledge about patients—allergies, medications, diagnostic and treatment plans, and patient-specific needs—is available, with appropriate protection of patient confidentiality.
- Patients are informed and participate as fully as they wish and are able—an informed patient is a safer patient.

- Complications should be dealt openly and with honestly even if the physician or the health system is at fault.
- Errors are tracked, analyzed, and interpreted for improvement rather than blame.

Creating a Just Culture and Accountability Model

The culture of an organization can be described by its mission, vision, and values; physician and leadership commitment to patient safety; and leader accountability for the behavior of the organization's members. There are several strategies to creating a just culture and accountability model.

Chain-of-Command Procedures for Physician Rule Violations

These procedures should be clear, concise, and communicated to all levels of the organization. When patient safety practices are compromised by any member of the staff, a feedback loop must be in place to return the system to a safe environment for patients. The variance reporting system should provide an anonymous way of reporting noncompliance, in addition to prompt accountability to ensure the empowerment of front-line staff. Once the chain of command is in place and enforced over time, the culture of the organization will mature to enable a more focused approach to system issues than to individual compliance issues. If physicians and staff are not held accountable to upholding quality and patient safety practices, then improvement projects will fail to provide any effective or sustainable results needed to reduce serious safety events and human error.

Accountability System to Reduce Noncompliance

The accountability system should be developed and utilized as part of the overall performance management system. In health care, very few human resource performance management systems provide a healthy mechanism for dealing with human error, especially those involved in serious safety events. Accountability systems should be fair, just, consistent, and equally applicable to all levels and professional groups of the organization. A nonpunitive culture must be established to ensure healthy incident reporting so that corrective action and improvement activities can occur.

Team Leader Rounding

Team leader rounding is vital to the sustainability of all patient safety initiatives. The mantra behind this concept is "You get what you inspect, not what you expect." Rounding should always be purposeful and structured around the patient safety and behavior-based expectations defined for the organization. The purpose of rounding is to display management commitment to patient safety, uncover error precursors and chronic problems, obtain feedback from staff, and satisfy the 5:1 ratio of positive to negative feedback to staff. The 5:1 ratio of positive to negative feedback is a core competency for team leaders to ensure accountability to patient safety processes and behaviors.

Peer Coaching Programs (Safety Coach)

Peer coaching programs are the cornerstone of patient safety culture transformation that occurs at the grassroots level by providing real-time coaching for continuous improvement. Safety coach programs provide real-time behavior-based monitoring, feedback, and data collection. Real-time behavior-based monitoring reduces both error precursors and serious safety events by transforming knowledge-based error prevention practices into skill-based patterns of behavior. The programs also provide data that can be used to track behavioral trends and human performance improvement.

Implementing Techniques for Human Error Prevention

Mistake proofing, mentioned previously in this chapter, is used to reduce the amount of human error. Even with this method there are still many opportunities to harm patients and one another owing to the nature of sickness and the spread of infection. Implementing an error prevention program is key to continually fostering the need to reduce safety defects as follows.

Systemwide Data Analysis

Systemwide data analysis of serious safety events and incident reports allows the organization to identify and prioritize areas for improvement. Common cause requires a comprehensive database that will allow stratification of event data using Pareto analysis. This type of analysis is also used to identify the high-risk situations and high-risk behaviors that contribute to the causes of serious safety events. With the appropriate data, common cause analysis provides the basis for behavior-based expectations and error prevention techniques, red rules, and project selection. Common cause analysis should be performed at least once per year for systemwide data and as needed for departmental or service line data.

Establish Behavior-Based Expectations and Error Prevention Techniques

These are based on common cause analysis data, designed specifically to address the high-risk situations and behaviors. The error prevention techniques provide the toolbox that enables staff to meet the behavior-based expectations. Behavior-based expectations and the subsequent error prevention toolbox should be few in number and only relevant to specific improvement opportunities.

Implement Red Rules

Red rules are a set of minimum standards associated with certain patient safety processes that *must* be met and require verbatim compliance (e.g., patient identification, proper hand hygiene, universal protocol, high-risk medication administration, and tag-out procedures). Red rules should focus on the highest-risk activities at both the system and departmental levels. They should be few in number in order not to dilute the significance of red rules.

Hardwiring Processes for Safety

Hardwiring essentially means that processes are established in such a way that deviations from them are difficult. Of course, one needs to ensure the process is a "good" one before it becomes fixed, and some of the methods and tools already described in this chapter and in other chapters can apply to ensure adequate design. These include Design for Six Sigma (DMADV) and Lean Design (LDMADV), Lean Transformation and Six Sigma (DMAIC), root cause and special cause analyses, mistake proofing, and PDSA.

Spotlight on Safety

Several of the principles and tools described above are exhibited by Sentara Norfolk General Hospital which in July 2004 was awarded the National Quest for Quality Prize based on its comprehensive patient safety program. This award recognized Sentara's commitment to patient safety as a systematic discipline. In 2006, Sentara Norfolk General Hospital was one of 59 hospitals, and the only one in Virginia, to be named to The Leapfrog Group's first Top Hospitals' list. This ranking is based on results from the Leapfrog Hospital Quality and Safety Survey, a national rating system that offers a broad assessment of a hospital's quality and safety (Sentara Publications). Sentara's strategy is to take a three-prong approach simultaneously:

1. Implement major system solutions, such as new technology. Technologies include innovations such as e-ICU, SUNRISE pharmacy software, and PACs digital image archiving with TALK voice transcription.
2. Implement minor systems, such as actions targeted at a specific issue. Minor system implementations include bar-coding technology.
3. Change the organizational culture in the hospital inpatient, nursing home, medical office, and home.

The Sentara Culture of Safety initiative adds a new foundation of behavior-based expectations (BBEs) for error prevention: red rules that cannot be ignored without consequence and enhanced root cause analysis that brings timely systematic improvements. There are five BBEs.

1. Pay attention to details.
2. Communicate clearly.
3. Have a questioning attitude,
4. Hand off effectively.
5. Never leave your wingman.

Leadership assumes responsibility for keeping the safety culture front and center. There are safety coaches in every department. Staff at every level are educated and trained in technologies and processes, are assigned to oversight committees, and are regularly reminded that "Patient safety starts with me." Staff are recognized and rewarded for practicing BBEs and catching potential errors before they reach the bedside. Sentara promotes a philosophy of fairness that encourages systematic improvements based on learning from errors, yet demands accountability for job performance.

Improving Patient Care, Quality, and Safety—The Success Stories

Mentioned below are organizations recently recognized for their dedication to health care improvement. Some are award winners; some have won the hearts of patients.

The Mayo Clinic

The Mayo Clinic states that "The needs of the patient come first"—not the convenience of the doctors, not their revenues. To make this mission happen, the doctors, nurses, and all staff meet almost weekly, working on ideas to make the service and the care better, not just to get more money out of patients, according to Denis Cortese, president and CEO. He also stated, "It's not easy. But decades ago Mayo recognized that the first thing it needed to do was eliminate the financial barriers to improving quality of care." For example, it pooled all the money the doctors and the hospital system received and began paying everyone a salary, so that the doctors' goal in patient care could not be increasing their income. Mayo promoted leaders who focused first on what was best for patients and only afterward on how to make this financially possible. Today it is the premier destination for patients, doctors, and benchmarkers around the globe (Gawande 2009).

Subang Jaya Medical Centre (SJMC), Indonesia

This stellar performer, with partnerships with the Mayo Clinc, Johns Hopkins, BrainLab Germany and Toby Robins Cancer Research, has become one of the world's largest medical tourism facilities. With 15 percent of patients coming from outside of Indonesia, as far away as Europe and the United States, the SJMC has won numerous awards for its health care system. They market their services around the globe in areas such as cytogenetics, molecular studies, bone marrow transplant, cancer and radiographic surgery services, and vascular interventional radiology. The SJMC vision is to be an organization internationally recognized for its excellent tertiary and integrated health care services. Through their Total Quality Management program they are fast achieving their mission. They aim to create management systems that are modeled after local and international standards, and the drive is to continually improve at Subang Jaya Medical Centre. They also focus on providing superior building facilities and services in support of quality customer care through a proactive and committed staff team. Dr. Jacob Thomas, group medical advisor, has been a leader in this system and explained why at the 4th International Quality Congress (Sarawak 2007): "Patients are always given top priority in our delivery of service to achieve *better, safer, faster and affordable care.* All Other Customers are accorded the highest possible level of care that gives them a Peace of Mind when using SJMC. Customer service begins with providing competent staff at every customer touch point, beginning with the Concierge/Security team, to the Reception staff, Pharmacy and other areas" (Thomas 2007).

Intermountain Health Care's LDS Hospital

In the 1990s, Intermountain Health Care's LDS Hospital in Salt Lake City reduced perioperative wound infection rates from 1.9 to 0.4 percent, compared with the national average of 4 percent at the time. In addition, a standardization project to reduce the number of different prostheses used in total hip replacement saved almost $1 million annually, while achieving better functional outcomes for patients (Pestotnik et al. 1996, James 1993, Morrissey 1996). Now known as Intermountain Healthcare, the parent, nonprofit system continues to be renowned for delivering quality at low cost. Dr. Brent James is one of the thought leaders in the United States.

SSM Health Care

In 2002, SSM Health Care (SSMHC), a St. Louis–based not-for-profit health system of hospitals and other health-related entities, became the *first* health care organization in the United States to be named a Malcolm Baldrige National Quality Award (MBNQA) recipient (see more on Baldrige earlier in this chapter). SSMHC was one of 17 health care institutions to apply for the MBNQA, and it became a winner through demonstration of quality leadership and strategic planning (e.g., its strategic, financial and human resource planning process, and performance management process), continuous improvement and innovation, (e.g., ongoing surveys, interviews, complaint systems, feedback, and patient follow-up calls used to rapidly identify and correct potential problems and improve service delivery), and improved clinical outcomes (e.g., more than 80 percent of SSMHC's patients with congestive heart failure and atrial fibrillation received Coumadin treatment compared to the benchmark of 64 percent) (NIST 2009).

Baptist Hospital, Pensacola, Florida

In 2003 Baptist Hospital, Inc. became a Baldrige recipient (NIST 2009). In the mid-1990s, low satisfaction marks, not only from patients and their families but also from staff and doctors, were becoming a trend at Baptist. Overall satisfaction for inpatients, outpatients, ambulatory surgery patients, and home health care services has been near the 99th percentile for the past several years. Positive morale for hospital staff has risen from 47 percent in 1996 to 84 percent in 2001 (the most recent survey), compared to 70 percent for staff at its closest competitor hospital. Senior leaders serve as role models and are held accountable for organizational performance through a "No Excuses" policy. Baptist Hospital, Inc. provides 6.7 percent of its total revenue to indigent patients, compared to 5.2 and 4 percent for its competitors.

Mercy Health System

A 2007 Baldrige winner, Mercy Health System (MHS) is a vertically integrated health system that provides services to residents in Wisconsin and Illinois (NIST 2009). These include hospital-based services covering three hospitals, clinic-based services, post-acute care/retail services, and an insurance organization. Through its focus on quality, MHS leads its Wisconsin market in inpatient services, outpatient surgery, and employed physicians. Overall mortality rates for MHS match the best practice benchmark based on the Care Science (a provider of care management, clinical analysis, and clinical quality improvement solutions) adjusted rate for the top 15 percent of hospitals in the United States. Results for community-acquired pneumonia mortality have decreased steadily since 2003, with recent results at 1.2 percent—significantly below the benchmark of 4.0 percent.

Healthy, happy patients contribute to a healthy, happy workplace. In 2006, MHS received the 2006 American Association of Retired Persons (AARP) Best Employers for Workers Over Age 50 award and was named one of the 100 best companies in which to work by *Working Mother* magazine. In addition, in 2007, MHS was named one of the 100 best adoption-friendly workplaces.

Success breeds success, and MHS extends its accomplishments to the community through its involvement and participation in community initiatives and collaborations. MHS provides approximately 1.8 percent of its hospitals' revenue and more than 2 percent of its clinics' revenue to charity care. These results exceed the state General Medical Surgical Hospital and Hospital Best Practice Results. Additionally, MHS budgets just under $500,000 monthly for self-pay discounts in support of patients unable to afford their needed health care. Other key MHS community support services include a homeless center (House of Mercy), a community clinic for the underinsured, and over 38,000 yearly screenings and community education programs.

Sharp HealthCare

Over the years preceding Sharp HealthCare's 2007 Baldrige award, net revenue increased by over $900 million from 2001 to 2007, allowing Sharp HealthCare to surpass the performance level of its largest local competitor (NIST 2009). All four of Sharp's acute-care hospitals approached or were within the top 10 percent in performance nationally for non–intensive care unit community-acquired pneumonia patients, with sustained improvement year after year. For all three hospitals with intensive care units, heart attack mortality demonstrated sustained levels from 2004 to 2007, a mark that is at or better than the national benchmark.

Mastering information technology is a part of Sharp's success. Sharp is recognized as one of only nine health care organizations to receive the "100 Most Wired" award for nine consecutive years. Sharp was a pioneer of bedside clinical documentation systems in the 1980s, and it continues to implement a systemwide fully integrated hospital emergency electronic medical record (EMR) and ambulatory EMR.

As in the other examples, the community also reaps benefits from local business success. Management donation of hours toward community programs increased from 10,000 hours to almost 60,000 hours from 2003 to 2006. Financial support for San Diego's vulnerable population, health research efforts, and the broader community increased from approximately $4 million to approximately $6.5 million from 2001 to 2006.

The Future of Management of Quality in Health Care

The future of health care globally will require many years of improvement to recreate a model that is efficient and effective at meeting the needs of our society. Our advice is that there is only one way to eat an elephant—bite by bite. Health care reform will happen in the same way—process by process, system by system.

References

Information cited in this chapter regarding Baldrige winners is from NIST press releases, accessible through http://www.baldrige.nist.gov/.

American College of Physicians. (2006). The Advanced Medical Home: A Patient-Centered, Physician-Guided Model of Health Care. January 22. Policy Monograph, American College of Physicians, Philadelphia.

Baker, M., Corbett, A., Reinertsen, J. (no date provided). Quality and Patient Safety: Understanding the Role of the Board The Ontario Hospital Association (OHA). Publication # 414, Toronto, ON M5V 3L1

Berwick, D. M., Godfrey, A. B., and Roessner, J. (1990). *Curing Health Care: New Strategies for Quality Improvement*. Jossey-Bass, San Francisco.

Best Health Care Results for Populations: The 'Triple Aim.' (2007). IHI Technical Brief. Institute for Healthcare Improvement, Cambridge, MA.

Building a Better Health Care System. (2009). National Coalition on Health Care, Washington, DC.

Donabedian, A. (1966). Evaluating the Quality of Medical Care. *Milbank Memorial Fund Quarterly*, vol. 44, no. 3 (suppl.), July, pp. 166–206.

Duhig, J. M. (2009). EMRs: Reaching the Holy Grail, Patient Safety & Qaulity Healthcare, January/February 2009, pp. 32–35.

Efficiency Measures. Agency for Healthcare Research and Quality, Rockville, MD Publication No. 08-0030, April.

Gawande, A. (2009). The Cost Conundrum. *The New Yorker.* June 1.

Grout, J. (2007). Mistake-Proofing the Design of Health Care Processes. Agency for Healthcare Research and Quality, Rockville, MD Publication No. 07–0020, May.

Hillestad, R., Bigelow, J., Bower, A., Girosi, F., Meili, R., Scoville, R., and Taylor, R. (2005). Can Electronic Medical Record Systems Transform Health Care? Potential Health Benefits, Savings, And Costs. Health Affairs, 24, no. 5 (2005): 1103–1117 doi: 10.1377/hlthaff.24.5.1103

Institute of Medicine. To Err is Human: Building a Safer Health System. National Academy of Sciences, 2000.

James, B. C. (1993). Implementing Practice Guidelines Through Clinical Quality Improvement. *Frontiers of Health Services Management*, vol. 10, pp. 3–37.

Juran Healthcare (2009). Reduce Heart Failure Length of Stay. Juran, Southbury, CT.

Langley, G. J., Nolan, K. M., Nolan, T. W., Norman, C. L., and Provost, L. P. (1996). *The Improvement Guide: A Practical Approach to Enhancing Organizational Performance.* Jossey-Bass, San Francisco.

Lohr, K. N., and Brooke, R. H. (1984). Quality Assurance in Medicine. *American Behavioral Scientist*, vol. 27, pp. 583–607.

Lohr, Steve (2008). Most Doctors Aren't Using Electronic Health Records. *The New York Times.* June 19.
Martin L. A., Nelson E. C., Lloyd, R. C., and Nolan, T. W. (2007). Whole System Measures. IHI Innovation Series white paper. Institute for Healthcare Improvement, Cambridge, MA.
Martin, L. A., Neumann, C. W., Mountford, J., Bisognano, M., and Nolan, T. W (2009). Increasing Efficiency and Enhancing Value in Health Care: Ways to Achieve Savings in Operating Costs per Year. IHI Innovation Series white paper. Institute for Healthcare Improvement, Cambridge, MA.
Master, R. (1997). Personal communication with the author.
McGlynn, E. A. (2008). Identifying, Categorizing, and Evaluating Healthcare
Midwest Business Group on Health. (2003). Reducing the Costs of Poor-Quality Health Care Through Responsible Purchasing Leadership. MBGH Publication, Chicago.
Moon, Craig A., and Umbdenstock, Rich (2008). Ailing ERs Threaten Patients, Leave Communities Vulnerable. *USA Today.*, May 30.
Morrissey, J. (1996). IHC Sets Pace on Quality Improvement. *Modern Health,* January 29, p. 42.
NIST (2009). 1988–2009 Award Recipients' Contacts and Profiles. Retrieved from http://www.baldrige.nist.gov/Contacts_Profiles.htm.
Pestotnik, S. L., Classen, D. C., Evans, R. S., and Burke, J. P. (1996). Implementing Antibiotic Practice Guidelines through Computer-Assisted Decision Support: Clinical and Financial Outcomes. *Annals of Internal Medicine,* vol. 124, pp. 884–890.
Porter, M. E., Farmer, P. E., and Kim, J. Y. (2008). *Redefining Global Health Care.* The Harvard Business School Centennial The Harvard Business School Centennial Global Business Summit. Business Summit. Presidential Fellows of Harvard College, Cambridge, MA.
Primary Health Care—Now More Than Ever. (2008). *World Health Report 2008.* World Health Organization, Geneva, Switzerland.
Sentara Bayside Web Publication, July 2007 through June 2008.
Sentara Norfolk General Web Publication 2006.
Thomas, J. (2007). Quality in Health Services and Healthcare. Proceedings of the 4th International Quality Congress, Oct. 22–24, Sarawak, Malaysia.
Sternberg, S. (2007). Transparency Provides Better Look at Health Care. *USA Today.* May 28.
U.S. Preventive Services Task Force. (1996). *Guide to Clinical Preventive Services,* 2d ed. Williams & Wilkins, Baltimore, MD.
Weed, L. L. (1968). Medical Records That Guide and Teach. *New England Journal of Medicine,* vol. 278, pp. 593–600, 652–657.
Wennberg, J. E., and Gittelsohn, A. (1973). Small Area Variations in Health Care Delivery. *Science,* vol. 1823, pp. 1102–1108.
Wennberg, J. E., and Keller, R. (1994). Regional Professional Foundations. *Health Affairs,* vol. 13, pp. 257–263.
Williamson, J. W. (1971). Evaluating Quality of Patient Care: A Strategy Relating Outcomeand Process Assessment. *Journal of American Medical Association,* vol. 218, pp. 564–569.

CHAPTER 24

Continuous Process-Based Organizations: Quality Is a Continuous Operation

Brian A. Stockhoff

About This Chapter

This section deals with the fundamental concepts that define the subject of managing for quality. It defines key terms and makes critical distinctions between similar but different contemporary programs to improve performance. It identifies the key processes through which quality is managed and integrated into the strategic fabric of any organization. It demonstrates that while managing for quality is a timeless concept, it has undergone frequent revolutions in response to the endless procession of changes and crises faced by human societies.

High Points of This Chapter

1. Processed products and the methods used in their manufacture differ in fundamental ways from assembled products and assembly, with important implications for quality management. Key characteristics of processed products include homogeneity

of the final product, production via formulation from a recipe, and batch or continuous production methods.

2. For process-based companies, the dimensions of performance, features, conformance, aesthetics, and perceived quality predominate in determining overall product quality, while durability and serviceability often are associated more with assembled products.
3. Quality of final processed product is tightly linked to the production process, and production design must properly account for the unique needs of manufacturability, scale-up, and monitoring. Failure in design risks chronic problems related to bulk storage and transport, contamination, spoilage, scheduling, product changeover, and remediation of product.
4. Postproduction processes, such as storage, transport, and distribution, can significantly affect quality, and create safety and liability issues. Manufacturers are responsible for implementing quality management systems for proactive rather than reactive management.

Introduction

As discussed in earlier portions of the handbook (*Quality Control Handbook*, 6th ed.), all work is the result of a process. However, the focus of this chapter is on a subset of industries in which processing of materials is at the core of production. This includes, but is not limited to, companies involved in food and beverage processing (e.g., cheese, wine), paints and related architectural coatings, chemicals (e.g., powder coatings, specialty chemicals, agricultural pesticides), petroleum, plastics, pharmaceuticals (including small-molecule drugs and biologicals), blood products, pulp and paper, glass, refractories, cement, metals, rubber, inks and dyes, detergents, perfumes, fragrances, and cosmetics. More specifically, this involves production in which the final output cannot be disassembled or distilled back into its original, basic components. Processed goods such as glue, Botox®, and processed cheese products have constituents unrecognizable from their original form, typically because of fundamental changes in chemical or molecular structure. Such products are vastly different from electronics, home appliances, automobiles, and similar assembled products that largely retain the identity of the component parts. Some basic characteristics of processed and assembled products are compared in Table 24.1.

Within process industries, there are two fundamentally different types of undertakings. One is the conversion of raw materials such as crude oil, ore, and agricultural commodities into more refined materials (e.g., diesel fuel, copper, and flour) that subsequently may be used in making finished products. The other is fabrication of finished products for use by end users (e.g., through extrusion, molding, blending). Table 24.2 shows the variety of different conversion and fabrication processes across industries. Rather than attempt to treat each industry or possible unit operation separately, salient features and issues will be addressed as they arise within the context of more general quality principles.

Processing in the Context of Industry and Economic Classification Systems

Brief reference to industry classification systems may be useful at this point. Primarily used by governments and agencies for statistical analysis of aggregate economic activity, they may be useful to quality managers seeking statistics within their own industry or quality practices and benchmarks from other processing industries. Among the major systems are

- North American Industry Classification System (NAICS). Industries are classified by a six-digit system into more than 2300 codes. Most processing-based companies

Characteristic	Type of Company	
	Discrete Manufacturing	Process-Based
Basis of production	Work from orders to build	Work from recipes to formulate
Differentiation	Products and component parts are easily identifiable.	Undifferentiated, relatively homogeneous products
Production method	Individual or separate unit production	Production via batching and mixing, or continuous flow
Low volume and high complexity	Requires flexible manufacturing system that can improve quality and time to market speed while cutting costs	Scale-up and batch campaign production. Flexible systems with careful in-process and output monitoring. Food and pharmaceutical grade; specialty and fine chemicals.
High volume and low complexity	Puts premium on inventory controls, lead times, and reducing/limiting material costs and waste	Scale-up and continuous flow production. Bulk commodity purchasing and storage, efficient flow. Bulk chemicals, petrochemicals, float glass, smelting.
Customers	Physical products go directly to businesses and consumers, and assemblies used by other manufacturers.	Physical products go to distributors, or act as feedstock for other manufacturers.

Table 24.1 General Characteristics of Discrete Manufacturing and Process-Based Companies

Process Industry	Conversion Process	Fabrication Process
Chemical	Refining, extraction, distillation	Formulation, pelletizing
Food and beverage	Freezing, pasteurization, thickening, condensing	Canning, bottling
Metals	Ore refining, smelting	Ingot casting, rolling, extrusion, forging
Petroleum	Refining, distillation, blending	Container filling
Pharmaceuticals	Wet granulation, filtration, extraction, fermentation	Tableting, coating, encapsulation
Plastics	Compounding, blending	Extrusion, coating, molding, laminating
Pulp and paper	Chipping, screening, pulping manufacturing	Paper forming, coating, container
Refractories	Sizing, blending, drying	Molding, casting, firing
Rubber	Compounding	Calendaring, curing

Unit Operations are Basic Steps Involving a Physical Change. (*Source:* Adapted from Bingham and Walden, 1988.)

Table 24.2 Examples of Unit Operations in Process Industries

Code	Industry
11	Agriculture, forestry, fishing and hunting
22	Utilities
31-33	Manufacturing
311	Food manufacturing
312	Beverage and tobacco product manufacturing
322	Paper manufacturing
324	Petroleum and coal products manufacturing
325	Chemical manufacturing
326	Plastics and rubber products manufacturing
327	Nonmetallic mineral product manufacturing (includes glass and cement)
331	Primary metal manufacturing
562	Waste management and remediation services

(*Source*: http://www.census.gov/eos/www/naics/.)

Table 24.3 NAICS Classifications Containing Processing-Based Companies

fall under the Manufacturing code, but can appear under several headings (see Table 24.3).

- Statistical Classification of Economic Activities in the European Community (NACE)
- International Standard Industrial Classification of All Economic Activities (ISIC)
- United Kingdom Standard Industrial Classification of Economic Activities (UKSIC)
- Australian and New Zealand Standard Industrial Classification (ANZSIC)
- Russian Economic Activities Classification System (OKVED)
- Industry Classification Benchmark (ICB)

Note that the classifications described by such systems are arbitrary, and real-world companies can have processing, assembly, and service types of output within the same value stream. Quite often, processed products become parts of assembled products, such as plastic resin pellets (a processed product) transformed through injection molding into parts used in a toy car (an assembled product). Given this ambiguity, general principles and conclusions are presented in this chapter, and it is incumbent upon the reader to draw conclusions as to the applicability to specific situations.

Characteristics of Processing Methods and Processed Products

What is the significance of processing from a quality perspective? Processed products and the methods used in their manufacture differ in fundamental ways from assembled products and assembly, with important implications for quality management. The following are a sampling of characteristics that tend to arise as a part of processing and the practical considerations that result.

Homogeneity of the Final Product

- The final output is a mixture of ingredients that is more or less uniform, or varies as a gradient across one or more characteristics and dimensions.

- Defects in the final product cannot easily be discerned or associated with a specific component.
- A defect may be due to irreversible chemical or physical property changes.
- Homogeneity-implied defects are spread throughout a production unit; unlike an assembled product, a defective part cannot simply be removed and replaced (e.g., like a faulty microchip on a graphics card). The entire production unit must be reworked, downgraded, or discarded. Minute quantities of wild yeasts or lactic acid bacteria, for example, easily can grow, spread, and contaminate an entire batch of beer.

Production via Formulation from a Recipe

- A processed product is formulated via a recipe, unlike an assembled product created through discrete steps or unit operations. Transformation of processed ingredients may take place continuously, without direct involvement by an operator.
- Reaction kinetics that transform raw materials into product often continue over time and cannot simply be stopped. This has implications regarding delivery speed, protection from the environment, and shelf life.
- In-process product can be vastly different in composition, handling characteristics, and appearance from the final product.
- In theory, recipe production volume can be adjusted across a continuous scale. Whereas a manufacturer of valve assemblies cannot produce and sell half a unit, the formulated lubricant used in the assembly can be made in virtually any volume.
- Formulation instructions can be strongly dependent upon production volume. Surface-to-volume ratios and physical characteristics such as viscosity, temperature, heat, and mass transfer do not necessarily scale linearly or neatly together with production volume. For example, the same proportions and timetables that work at the 1 L (0.26 gal) shake flask scale are unlikely to work at the 100,000 L (2,6417 gal) tank scale. For the quality manager, scale-up has the practical problem of different defect types inherently appearing with different production volumes.

Batch and Continuous Production Methods

- A batch is a specific quantity of material that is intended to have uniform character and quality, and is produced according to a single manufacturing order during a single cycle of manufacture. Batch production is common in food and pharmaceutical industries, in which product is prepared in time-sequential steps and discrete, relatively uniform volumes. Material is placed in a vessel at the start, and is removed after transformation is completed (material may be transferred from one vessel to another, as in mixing and subsequently baking a batch of cookies). Note that fixed vessel sizes provide a practical constraint to the theoretically variable production volumes mentioned previously.
- Semibatch production is operated with both batch and continuous inputs and/or outputs. Chlorination of water for water purification purposes is an example, as chlorine gas is continuously added to react with a batch of water.
- Fed-batch production is used in the manufacture of biological materials. In this system, reaction and growth rates in cultures are managed by the controlled, often continuous, addition to a batch of growth-limiting nutrients (e.g., acetate, glucose, oxygen).

- Continuous production proceeds without interruption, with all steps ongoing continuously in time. It is typical in the distilled spirits, petrochemical, and float glass industries, portions of which achieve (near) steady-state operation such that the quantities in process do not change over time. Although vessels or units may be fixed in size, by adjusting the input stream, it may be possible to adjust production volume.

Variability in Raw Materials

- While variability in inputs is not unique to processing industries, processing often involves transformation of raw, unrefined natural materials (e.g., crude petroleum used in plastics or grain used in bread). Such inputs may be poorly characterized, especially in terms of chemistry or physical properties that may be critical to downstream processing.
- Raw material costs represent a large proportion of total costs. A quality manager has an opportunity to substantially improve the bottom line by maximizing consistency in component materials, thereby reducing variability and costs during processing.
- Certain raw materials for processed products are the waste stream of another process (sometimes called "loop-closing"), so there may be little incentive for a supplier to provide material meeting strict specifications. Reclamation by intermediaries can facilitate improved quality, however (e.g., recovery and refining of inhomogeneous, spent catalyst from petroleum refining for subsequent sale and conversion to alloys used in the steel and foundry industries).

Variability in Measurement Systems and Capability

- Whether natural or reclaimed, inputs for processing tend to be complex, with the potential for interference of measurements by extraneous components or characteristics (e.g., chromatographic detection of potentially carcinogenic halogen-containing anions such as BrO_3^- in bread can be complicated by interference of Cl^-). This increases the cost of analysis and decreases the value of acceptance sampling.
- Measurement systems may be poorly developed because of high cost, especially in comparison with the relatively low value of incoming commodity materials. As a result, variability often is allowed to pass through into the production process itself. In the food industry, for example, viscosity may be measured by the length of time needed for a sauce to run down a small hole of fixed, known diameter, or the time to flow down an inclined ramp. Even viscometers used to measure resistance to rotation can be quite simple, but methods to obtain consistent measures can be problematic due to product inhomogeneity, such as chunkiness and choices of spindle type (vane, T-bar, disc), rotational speed, etc. The point is not that these methods are insufficient (they may fit the present need quite well), rather, that they provide an approximate appraisal of typically complex materials, and care should be taken to avoid overinterpretation.

Destructive Sampling

- Although sampling for quality testing can relate to characteristics such as viscosity or pH that can be measured in place, frequently, chemical analyses are needed that require removal, alteration, or consumption of the sample such that it cannot be returned to the production line. This is in contrast to assembled products that generally can be tested via nondestructive means (reliability, tensile, fracture, stress corrosion, and similar tests being exceptions).

- In some cases, sampled materials are not inherently destroyed due to the measurement process itself, but contamination problems prevent return of sample product. Testing of a sterile product in a nonsterile environment is an example of this.
- With destructive sampling, sampling frequency and amounts needed should be minimized, and the information gained from small samples maximized. Fortunately, sampled quantities tend to be small compared to production volumes, although quality managers may be surprised at the total cost accumulated over time across many small samples. A cost-benefit analysis of sampling may be useful to help balance costs of testing versus costs of undetected failure.
- Special statistical methods are needed for destructive sampling measurement system analysis (MSA) because of the inability to obtain true replicate samples to estimate within-sample variation. Nested Gage R&R (Repeatability and Repro-ducibility) allows for drawn samples to be tested only once by an operator, rather than being "crossed" by having the same sample ("part") measured by more than one operator. It is assumed that all samples within a single batch are sufficiently identical to claim that they are the same "part," which is reasonable given the homogeneous nature of processed goods.

False Positives and Negatives

- One implication of destructive sampling and working with potentially minute yet important deviations from specifications is that false readings may be prevalent and take on greater significance. Whereas a nondestroyed part can repeatedly be measured if there is a question as to accuracy, a destroyed sample can be measured only once (although a lot or batch can have repeated aliquots taken). Bacterial load in a bioreactor is an example in which it is desirable to detect small quantities, especially while bacteria are still in the lag phase prior to exponential growth and can be controlled without compromising product quality.
- Statistically, screening of many samples to detect a small number of positives that may be present (i.e., deviating from specification) requires a measurement system that is highly tuned toward detection and, therefore, prone to false positives (Figure 24.1). Positives indicative of a problem (whether true or false) can lead to frantic retesting, during which time a production batch (or continuous flow) may need to continue in its processing. Fermentation tanks, for example, cannot simply be stopped and placed into stasis. In such situations, retesting frequently will indicate the original reading was a false positive, which, from a management perspective, is highly disconcerting. Statistically, however, false positives are the inevitable result of managing alpha and beta risk (see Chapter 19, Accurate and Reliable Measurement Systems and Advanced Tools, for a discussion of these risk measures).

This so-called "false positive paradox" happens when the incidence of occurrence in a population being sampled is less than the false positive rate of the test. For example, if a culture test for mycoplasma contamination in cell cultures has a theoretical level of detection of one colony-forming unit (cfu) and is 99 percent accurate (i.e., the test will correctly identify mycoplasma 99 percent of the time when it truly exists), then test of actual cultures with an incidence of mycoplasma of 1 in 10,000 (0.01 percent) will produce expected values over one million tests as follows:

True positive: $1{,}000{,}000 \times (1/10{,}000) \times 0.99 = 99$

True negative: $1{,}000{,}000 \times (9999/10{,}000) \times 0.99 = 98{,}9901$

False positive: $1{,}000{,}000 \times (9999/10{,}000) \times 0.01 = 9999$

False negative: $1{,}000{,}000 \times (1/10{,}000) \times 0.01 = 1$

		Test result	
		Positive	Negative
Truth	Negative	False positive	True negative
	Positive	True positive	False negative

Figure 24.1 Classification of false positives and false negatives based on the truth and the test result.

In this situation, a test that reads positive for mycoplasma is ~99.02 percent [9999/(9999 + 99)] likely to be an incorrect result. Education of management as to the reality of false positives in testing may be advised.

Continuous Sampling

- Processed product is formed through relatively continual change, in contrast to the discrete steps in assembly (e.g., stamping or drilling that takes place in a single step), after which the critical dimensions such as depth or width do not change. While sampling certainly can occur at discrete stages, the inability to pinpoint exactly when in time a product is ready for sampling leads naturally to continuous monitoring. This can blur the distinction between quality and operator roles.

For example, batch fermentation of a bacterial culture has, in theory, four different growth phases: lag, exponential (or log), stationary, and death. In reality, however, bacterial populations are not uniform, and individual cells do not reproduce in synchrony, thus making phase transitions ill-defined and batch-specific. Monitoring is needed, typically by means such as bacterial cell counts via individual (microscopic, flow cytometry), direct and bulk (biomass), indirect and individual (colony counts), or indirect and bulk means (turbidity or optical density, nutrient uptake, most probable number). This allows manipulation of the product (e.g., via fed-batch nutrient provisioning) and transitioning from one "landmark" phase of processing to another at the optimal time. Sampling, therefore, takes on not just a quality control function to identify out-of-control processes, but becomes an integral part of the production process itself.

Dimensions of Quality in Processing-Based Companies

Quality affects both the top and bottom lines of an income statement (revenue and net income), so it is vitally important to identify for each product the appropriate dimensions of quality and technical measures of these. There are many different ways in which quality can be measured. Juran defined quality as "fitness for use" (*Quality Control Handbook*, 6th ed.). To be fit for use, a processed product needs to have the right features and be free from deficiency. That is, it needs to deliver the functionality expected by the customer and do so without failure. To reduce the preceding description to practice, it is necessary to dig deeper to understand the specific aspects that need to be delivered and therefore measured for any particular product. Furthermore, strategic use of quality demands a proactive rather than reactive understanding of customer needs. That is, what is the job for which customers might employ your products or services, whether existing customers or not? Using the teachings of Juran, Feigenbaum, and others as a starting point, Garvin (1987) argued that while most companies have evolved toward actively managing for quality, efforts should not merely be defensive. Instead, organizations should learn to compete on quality. Garvin (1987) identified eight dimensions of quality that commonly are cited across many industries. These are

- *Performance.* Does the product do what it is supposed to do in terms of the principal operating characteristics? This dimension is based on measurable attributes, and superiority of performance depends on the specific task. Products may be divided into performance classes such that quality differences are evaluated within rather than between classes. For example, motor oil or tires for everyday driving are subject to different performance expectations than analogous products used in Formula race cars.
- *Features.* Does the product possess characteristics that supplement basic functionality? A nutrition bar may provide appropriate nutrients, but what about the bland flavor and pasty texture? Similarly, a drug with equal effectiveness in treating disease may be offered in injection or more convenient oral form.
- *Reliability.* Will the product consistently perform over time? What is the likelihood of malfunctioning or outright failure to perform? This dimension is much less prevalent in processed than assembled products because of the tendency to consume processed goods such that there is no repeated use. However, the reliability of concrete used in underground bunkers subject to physical attack is an example.
- *Conformance.* Do the product's design and operating characteristics conform to specification or accepted standards? A drug that has multiple contraindications or "black box" warnings may be viewed as performing outside of expected ranges (although it may be placed in a different performance class). Note that in practice, processed products rarely can be cleanly classified as "good" or "bad," as can assembled, mechanical products; terminology of "conforming" or "nonconforming" to specification is preferable and consistent with current quality systems standards. This has implications discussed later in this chapter.
- *Durability.* How much use is obtained and under what conditions before the product deteriorates beyond a useful condition and needs to be replaced? A lubricant used in heavy construction equipment should have a lifetime consistent with usage conditions. Consumers want automobile tires to be durable, quiet, and safe under a variety of driving conditions, while tires used in racing typically are not built for durability, but for other characteristics, such as "stickiness" that ensures road adhesion under dry conditions.
- *Serviceability.* Is the product relatively easy to maintain and repair? Is this done courteously and competently? How long is it before service is restored? An architectural paint for an institution such as a school or hospital will have different specifications for ease of cleaning and maintenance than a household paint.
- *Aesthetics.* How does the product look, feel, smell, taste, or sound? Red color in fruit juices, candy, yogurt, ice cream paints, inks, and cosmetics is used to create visual appeal (although use of the cryptically labeled "cochineal red" dye made from the ground-up insect *Dactylopius coccus* may not be appealing to some). Because there is not universal agreement on aesthetics, it is a clear opportunity for a niche.

- *Perceived quality*. Incomplete information regarding quality leads customers to apply indirect measures. Reputation is a perception based on the assumption that past quality is a predictor of future quality. Branding seeks to create an impression of a product that is based, in large part, on perceived quality.

Despite the apparent distinctions among these facets of quality, the tendency persists in viewing quality in terms of merely "low cost" and "high quality," reflecting a tacit presumption that quality is the same for all customers. This can be seen in the segmentation of markets by product and customer categories (e.g., "household paints") instead of by the job that customers need to complete. Essentially, the dimensions described previously represent an extension of Porter's generic "differentiation" strategy in which a company develops products with unique attributes (Porter 1980). The other two strategies Porter identified were "cost advantage" and "focus." Although Porter admonished against mixing strategies, authors since have suggested that under many circumstances there is an advantage to maintaining a hybrid approach (see Bowman 2008 for a critique of past strategies). Regardless of the generic strategy that is pursued, choices are needed regarding the specific attributes for a particular level of quality. Most importantly, when measuring quality internally, consider if your organization is measuring against the same characteristics as the customers, both current and potential.

Clearly, the dominant dimensions vary across organizations and products. In addition, the technical challenges in measuring quality discussed earlier are compounded when pursuing multiple facets of quality. This makes it all the more important to select a specific set of dimensions for your product. For process-based companies, the dimensions of performance, features, conformance, aesthetics, and perceived quality tend to be emphasized, while durability and serviceability often are associated more with assembled products. From a competitive standpoint, this suggests opportunities for differentiation.

Strategic Quality Planning, Improvement, and Control

Quality can be applied as a strategic tool by developing a quality niche that focuses on one or a small number of quality dimensions throughout the universal processes of planning, improvement, and control. Seldom is it advisable or even possible to pursue all eight of the quality dimensions simultaneously. This is because of both resource allocation toward quality and trade-offs that may be inherent among the dimensions. Similarly, objectives and activities should be aligned so that quality initially planned into a processed product is consistent with and reinforced by day-to-day control and longer-term improvement efforts. As pointed out by Porter (1996), "strategy involves creating 'fit' among a company's activities." When developing a quality niche strategy, specific, conscious choices must be made. The key is to discover which quality dimensions are most important to your customers and which are poorly met by competing products and firms.

Quality Strategy Failures in Processing

It is instructive to consider quality strategy failures as they relate to processing-based companies and the lessons they suggest.

- *Failure to measure the correct dimensions*. Injected drugs marketed on the presumption that efficacy is the driving factor in determining value may be superseded by less effective formulations that take into consideration the convenience of oral, dermal, or nasal administration. As pointed out by Christensen and Raynor (2003) in their milk shake example (a processed product version of their colleague Theodore Levitt's quarter-inch drill), it is important to understand the job for which customers "hire" a product. That is, discover the job the customer wants to get done, and design products that fill that need.

- *Too many dimensions of quality that are unimportant to customers.* A product that attempts to be all things to all people, or one that has not kept up to date with changing customer needs or competitor offerings is at risk of delivering a final product with high cost compared to perceived value. Examples include brand extension into product lines that are too distant from the original to convey any meaningful confidence in quality. Because the dimensions to emphasize should be developed during the product planning phase, the best solution is to conduct careful market research and develop from the voice of the customer a true niche strategy that focuses on a limited number of quality dimensions. Strategy requires executives to understand and make trade-offs and to choose what not to do.
- *Too many product line extensions.* New products are developed for a variety of reasons: new-to-the-world products filling unmet needs, products allowing expansion to new customers, products that meet changing needs of existing customers. Broadening a product line by creating variations on a theme is valuable in adapting to the marketplace, maintaining brand recognition, and capturing or maintaining market share, especially with mature brands. However, creation of numerous minor extensions or failure to prune a product portfolio risks an overly segmented customer base, generates increasing overhead, and creates difficulties in managing the quality. In addition, brand extension is not a good substitute for new product innovation. The Oreo cookie is one familiar example of a product that has numerous extensions (at press time, there were 49 food products based on the core Oreo cookie). Laundry detergents are another example, with many variations such as powder versus liquid versus solid formulations, water temperature, color, water hardness, etc.

From a quality perspective, increasing numbers of products forces a shift away from continuous processing toward increasingly smaller batch size production. Potential quality issues include frequent changeovers, calibration and maintenance of different equipment, and training of staff in different processes. Quality managers should have a voice in the product development process to ensure quality issues associated with product line extensions are understood and accounted for in the decision-making process.

Adherence to Outdated Quality Measures

Measuring performance against older standards risks manufacture of a product that passes internal tests but no longer meets customer expectations. This can come about in several ways:

First, changes in technology create opportunities for improved measures. However, it may be prohibitively expensive to purchase and maintain equipment and train employees in its operation. Second, traditional standards organizations tend to move more slowly than technology in developing standards, such that resulting standards are soon outdated. Industry consortia and specialized standards-setting organizations have emerged as new standards setters and should not be ignored as a resource for quality managers (see "Where to Find It" at the end of this chapter). Third, products that have not changed their recipe for long periods may become fixed in their quality measures. This is prevalent in the food industry, where an "old fashioned" recipe can be a selling point. Fourth, a producer may be unaware of, or chooses to ignore, changes in the industry and marketplace.

Direct Competition with an Industry Leader

Striving to deliver against the same dimensions as an industry leader creates a clear and immediate threat to the leader, potentially leading to retaliation from a position of strength. From a quality perspective, this risks price wars and associated pressures for cost reductions. If this path is pursued, a quality manager should anticipate requests for process changes that would alter and potentially jeopardize quality (e.g., lower-quality inputs or bulk raw material purchases that strain storage capacity). Quality issues should be communicated and considered as part of marketing decisions regarding such a "follower" type of strategy.

Quality Life Cycle in Processing Industries

Managing for quality extends far beyond the plant floor. While common concepts and quality principles apply throughout a product's life cycle, the specific types of problems encountered vary, as do the tools used to solve them. Discussed in this section is quality during various stages in a product's life, including product design, process design, supply chain, production, and postproduction. Readers with an interest in strategy should consider where and how quality resources should be directed to ensure fitness for use across the life cycle of a product and to meet targeted quality dimensions. Readers with a tactical interest may think about how their challenges can be addressed by extending quality management back into product design and forward through to the final end user.

Process Anatomy

Before engaging in a more detailed discussion of quality in different stages of a product, it may be useful to consider the types of processes that are available for meeting the goals of product creation and delivery. *[Note: The following is excerpted with minor changes from Juran and Godfrey (1999), pp. 3.37–3.39.]* At a high level, there are some basic process anatomies that have specific characteristics of which planners should be aware. A "process anatomy" is a coherent structure that binds or holds the process together. This supports the creation of the goods or the delivery of the service. The selection of a particular anatomy also will have a profound influence on how the product is created and the ability of the organization to respond to customers' needs. Figure 24.2 illustrates these.

The autonomous department. The "autonomous process" is defined as a group of related activities that usually are performed by one department or a single group of individuals. In this process form, the department or group of individuals receives inputs from suppliers, such as raw materials, parts, information, or other data, and converts them into finished goods and services, all within a single self-contained department.

An example of an autonomous process is the self-employed professional (e.g., a physician, consultant, or artisan). In financial services, it might be the loan approval department. In manufacturing, a well-known example is a tool room. It starts with tool steel and engineering drawings and creates punches, dies, fixtures, and gauges to be used on the manufacturing floor. Even though we refer to this kind of process anatomy as autonomous, outputs or deliverables from other processes are still required from outside sources that serve as inputs into the process. The self-employed physician, for example, may purchase equipment and materials from supply houses, pharmaceutical companies, and so on. This type of process may be found in administrative functions in process-based companies, but less so as a part of production.

The assembly tree. The "assembly tree" is a familiar process that incorporates the outputs of several subprocesses. Many of these are performed concurrently and are required for final assembly or to achieve a result at or near the end of the process. This

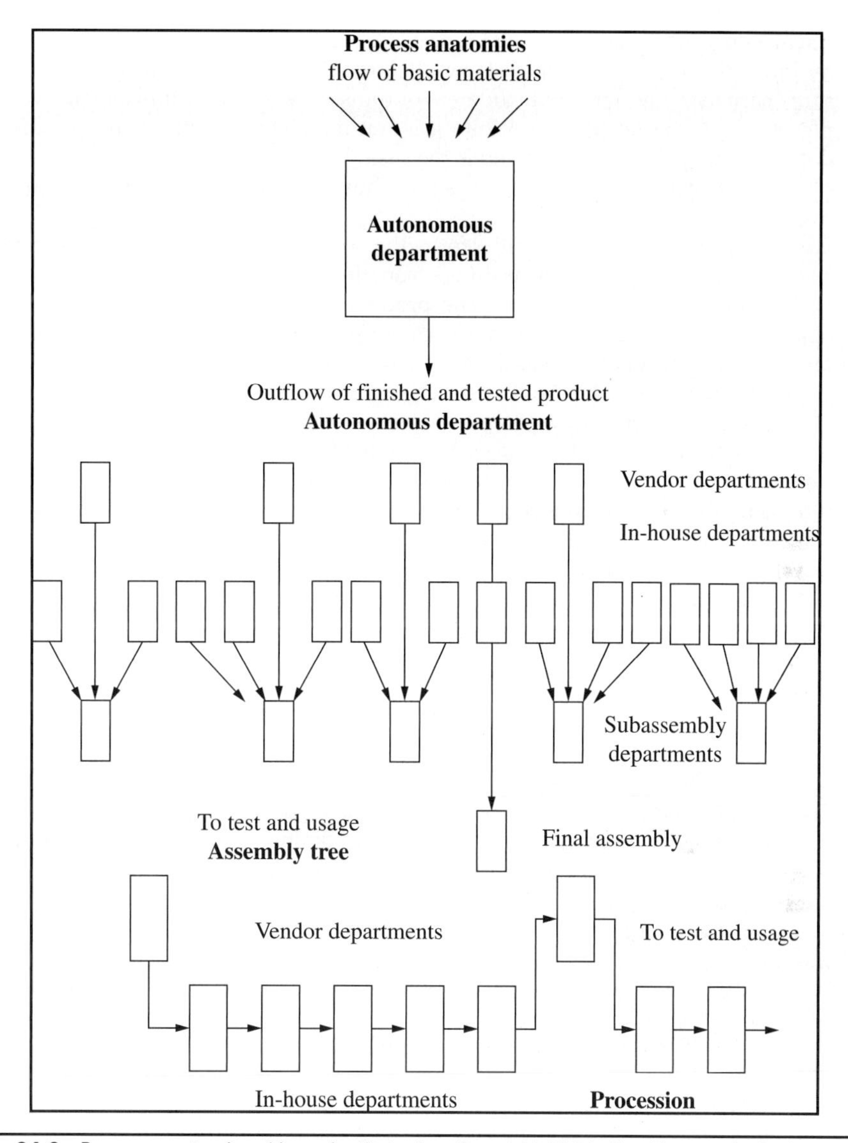

FIGURE 24.2 Process anatomies. (*Juran Institute, Inc. Copyright 1994. Used by permission.*)

kind of process anatomy is widely used by the great mechanical and electronic industries that build automotive vehicles, household appliances, electronic apparatus, and so on. It also is used to define many processes in a hospital, such as in the case of performing surgery in the operating room. The branches or leaves of this tree represent numerous suppliers or in-house departments making parts and components. The elements are assembled by still other departments.

In the office, certain processes of data collection and summary also exhibit features of the assembly tree. Preparation of major accounting reports (e.g., balance sheet, profit statement) requires assembly of many bits of data into progressively broader summaries that finally converge into consolidated reports. Assembly tree design has been used at both the multifunctional and department levels. In large operations, it is virtually mandatory to use staff specialists who contribute different outputs at various multifunctional levels. An example of this is budget process. While it is not mandatory to use staff specialists for large departmental processes, this often is the case. This can be illustrated by the design department, where various design engineers contribute drawings of a project that contribute to the overall design.

The procession. Another familiar form, the procession process, uses a sequential approach as the basis for the process. This differs from the assembly tree in which many of the activities are performed concurrently. The procession approach tends to make a more linear approach whereby lower-level processes are performed sequentially. It mandates that certain activities must be completed before others can begin because the outputs of each of the subprocesses serve as the inputs for each succeeding subprocess.

The intent of selecting a process anatomy is to determine the overall structure or architecture of the process that produces the product features and meets product feature goals. It does not necessarily follow that choosing one process anatomy over another locks the team into using that same architecture exclusively throughout the entire system. Quite the contrary, the team may select the assembly tree process as the structure for the overall system but use a combination of autonomous and procession anatomies as the basis for subprocesses at the functional departmental or unit level. In processing-based industries, the procession anatomy is a natural fit with the unidirectional flow of materials, although mixtures analogous to subassemblies may be prepared and input into the product stream at various stages. In developing processes, planners should strive for configurations that promote quality and should not be constrained by historical conventions.

Quality in Product Design

Traditionally, the role of a quality manager started only once production began. Specifications were handed to production staff, and maintaining conformance to these was the driver of daily activities focused on control, root cause analysis, and corrective action. Manufacturing has gradually moved from "quality by inspection" to "quality by design" (albeit more slowly in some industries), and quality management properly should begin with product design.

The fundamentals of product design in process-based companies are no different than in other industries. The voice of the customer (VOC) provides an initial foundation for the product specifications. For a processed product, VOC may be highly technical in nature due to the prevalence of commercial rather than retail customers (e.g., detailed, specialized requirements for surfactant properties of a chemical reagent used by a customer to produce diagnostic kits). However, especially in process industries, it is difficult to reduce the needs and requirements of all customers to a small set of attributes. Hence, it is necessary to continually evolve objective and measurable product specifications in partnership with the customers.

Objective, Measurable Product Specifications

To effectively design quality into a product, objective, measurable specifications are necessary. Product specifications should include the following two types of information [see Section 27.16 in Juran and Godfrey (1999)]:

- *Descriptive product information*. Name, identification code, chemical composition, engineering designs and drawings, uses and functionality, units of measurement, delivery units and conditions, and other qualitative characteristics of the product as well as proper, safe handling and storage information (e.g., material safety data sheets)

- *Quantitative information for measureable product properties.* Numerical values of intended levels of properties and ranges or limits. Most products have from 10 to 100 such measured properties listed on their product specifications. However, many products have only one or a small number in their "vital few." To be complete, these quantitative specifications (intended values and limits):
  - Should document the best current definition of the product, in measurable terms, that is expected to meet the needs of customers and that can be supplied commercially by the producer with current technology and facilities
  - Are for a prescribed measurement method
  - Can apply only to properties that can be measured on shipped product

From a quality perspective, it is important to clearly define the product unit and test method applied to determine conformance with specifications [see Section 27.17 in Juran and Godfrey (1999)]. This is especially critical when showing compliance with regulatory requirements or comparing producer and consumer metrics for purposes of acceptance.

Product Design through Design of Experiments

Product design in process industries typically relies heavily on research and development (R&D) in order to create new formulations that meet the needs of the customer. One design tool in which quality and R&D staff should become skilled is design of experiments (DOE). DOE is reviewed in Chapter 19, Accurate and Reliable Measurement Systems and Advanced Tools, in this handbook and covered more extensively in Section 47 in "Design and Analysis of Experiments" in Juran and Godfrey (1999), but certain aspects are well worth highlighting from the perspective of processing.

When working with recipes and formulations, the properties of interest often depend on the proportions of the mixture components rather than simply the amounts (volume or mass) of the individual components. For example, stainless steel is a mixture of different metals, and its tensile strength depends on the proportions of the metallic components present. Gasoline ordinarily is a blend of various stocks, sometimes including a percentage of ethanol, and the octane rating of the final blend depends on the proportions going into the blend. Although total amounts of final product vary, the proportions of the components of a mixture add up to unity. In the most general case, the proportion of any component may range from zero to 100 percent.

This dependence on proportions constrains designs in ways not provided for by traditional DOE structures, and mixture designs were developed to circumvent this problem. With the development of these new methods and user-friendly software, researchers no longer need to rely on educated trial and error or laborious sequential experimentation to optimize complex mixtures, even with multiple design objectives. Those interested in applying the concepts in practice are directed toward more technical resources on mixture DOE, such as Cornell (2002). Readers interested in mixing business with pleasure should see Bowles and Montgomery (1997).

Quality in Production Process Design

While many Xs in "design for X" are relevant to any particular industry, three universal considerations in production process design are design for manufacturability, scale-up, and design for testing.

Design for Manufacturability

If a product is simple to manufacture, then it follows that managing quality during production, and perhaps even the product itself, also should be simple. As the manufacturing process becomes increasingly complex, each component, transfer, change in conditions, etc., is a potential source of failure. As complexity increases, first-pass yield tends to decrease, potentially jeopardizing large batches of product that cannot be reworked. In addition, individually small in-process deviations from target may accumulate in a nonadditive, nonlinear, and unpredictable fashion. While not unique to processed products, the latter issue can be particularly problematic compared to assembled products in which simple summation of deviations may reasonably estimate final quality.

From a business perspective, designing a product up-front for ease of manufacturing allows for lower manufacturing setup time and costs, more rapid startup of manufacturing, lower production and testing costs, and ultimately higher quality. Also, because product cycles tend to shorten over time, creating flexible, generic manufacturing platforms is a benefit by allowing multiple products to be manufactured without the need for extensive process or equipment changes.

Characteristics of manufacturing systems that consistently produce high-quality processed product include the following:

- Standardization of components (Example: A white paint base from which various formulations can be made by adding different pigments)
- Standardization of production equipment (Example: Mixing vessels that are used in the manufacture of different salad dressings)
- Minimize number of components (Example: Reducing the number of solvents used in the manufacture of active pharmaceutical ingredients)
- Use concentrations that allow precise delivery of the active component (Example: Concentrated sulfuric acid may be convenient and cost-effective for inventory and storage purposes, but may better be diluted prior to use to enable more exacting delivery)
- Minimize unnecessarily dramatic shifts in characteristics (Example: Chemical buffers may play no direct role as basic ingredients or in final product composition, but may be indispensable in a commercial production recipe)
- Simplify recipes and instructions (Example: Eliminate extraneous information from operator instructions, and clearly identify decision points)
- Minimize handling (Example: Use closed systems that eliminate manual transfer of food products and associated risks of contamination)
- Eliminate adjustments (Example: Seek common settings for equipment used in production of multiple products that, while not individually optimal, collectively provide adequate performance)
- Build recipes from well-tested, robust procedures (Example: Modular recipes that can be used for different applications and facilitate process control)
- Mistake-proof processes (Example: Prevent use of incorrect water source by matching unique nozzle size or shape for each source with a restricted port on a receiving tank)
- Design recipes to take advantage of physical properties (Example: Sequencing of material addition or flow based on density)

Production Scale-Up

Most processed products originate in an R&D laboratory that conceives the product, defines technology for making it, and provides sufficient data such that a manufacturing plant can be constructed and tooled. This world of flasks, test tubes, bench top homogenizers and sonicators, vacuum filtration units, and the like is vastly different from the huge mixing vessel, piping, and filtration systems typically used for commercial production. To get from prototype to manufacture requires what often is a tremendous amount of time, effort, and failure.

The reasons for this are several-fold and interrelated. First, the proportions of ingredients needed in recipes can be dependent upon the formulated amount. For example, a rule of thumb in scaling up home-cooked recipes is to add only approximately 1.5 times the original amount of seasoning when doubling a recipe, and only 2 times the amount when tripling a recipe. Second, physical properties of materials change with volume (e.g., turbulence of fluids in a pipe), and processing may be highly dependent on velocities, mixing rates, etc. Third, physical relationships do not all scale linearly (e.g., surface-to-volume ratio decreases as vessel size increases—volume increases by dimension cubed, while area increases by dimension squared). Fourth, environmental variables such as temperature are important in virtually all biological and chemical reaction rates, and environmental control becomes increasingly difficult with size. Fifth, production equipment consisting of fixed physical assets (called monuments in Lean parlance) constrains optimization.

Process design starts in the R&D laboratory, sometimes concurrent with product design (although premature emphasis on process potentially limits creativity; for novel products (i.e., not line extensions), it is best to consider, What should we make? first, before How do we make it? Initial considerations include raw material quality; continuous versus batch production; unit processes and operations; the number, duration, and sequencing of process stages; and equipment. Prototype setups then can be tested and critical variables identified, for example, through screening designs (see Chapter 19, Accurate and Reliable Measurement Systems and Advanced Tools, for a more complete discussion). Optimization of process conditions follows, using rate kinetics, designed experiments, and simulation to establish more precise combinations of factor settings and conditions. The output of this phase of process design is a tentative operating procedure that can be passed on for pilot plant development.

The pilot plant stage provides an intermediate step between the laboratory and manufacturing environments. This usually is a true manufacturing unit with equipment larger than that used in the laboratory but smaller than in the final production facility. A pilot plant helps validate empirical equations from the laboratory, and provides an opportunity to address scale-up problems such as those mentioned next. The pilot plant also generates material needed for field testing, long-term studies, and regulatory testing, and allows for development of standard operating procedures for use in the manufacturing plant.

What are some of the scale-up issues common to processing companies? They include

- *Safety*. Chemical reactions may be strongly dependent on physical setup and scale, and not necessarily in obvious or predictable ways. Problems include runaway exothermic reactions (affected by reactor size, mixing, and cooling rates), overpressurization, uncontrolled generation of toxic materials, explosions of contained vapor or dust, ignition from unforeseen electrical or other sparking sources, spillage or fire created by overfilling of vessels, and poor blending of reactants and solvents. These production problems need to be addressed as best as possible by good manufacturing process design, rather than relying on worker training and compliance.

- *Physical limitations.* Large-scale operations cannot simply take small-scale processes and equipment and make them physically bigger. Filtration and purification of solids from fluids are examples. While filter paper and suction filtration may suffice for bench-scale production, vastly different separation methods are needed at commercial scales and to meet purity standards. A critical step in determining the quality of many processed products, filtration methods must be found that are substantially different in mode but equivalent in output to those identified during product development. Large-scale chromatographic, diafiltration, ultrafiltration, and similar systems can be expensive, and product quality and overall production costs depend heavily on maintenance and early detection and traceability of breakdowns.
- *Change in critical-to-quality product characteristics.* Important quality characteristics may be inextricably linked to production scale. For example, regulatory approval of some biopharmaceutical products is provided for specific volumes of production because of covariation in carbohydrate structures, etc. Change the scale, and new testing and approval may be required.
- *Synchronization of input streams.* As volumes increase, reaction and processing rates of parallel, input streams may become disjointed, resulting in empty vessels or requiring holding tanks to compensate for the mismatch. Empty tanks and pipelines do not generate revenues, and seals, etc., may degrade from under- or overuse. In addition, holding provides opportunities for contamination, unwanted physical changes such as settling or stratification, and continuation or initiation of undesirable chemical reactions.
- *Field testing.* Performance of prototypes cannot necessarily be assessed under laboratory conditions, requiring instead a period of field testing (often literally, e.g., grower tests of herbicides). In early stages of development, the amount of product available for customer evaluation can be a rate-limiting factor because only laboratory-scale production is available (note that demand often outstrips typical laboratory capability and competes for resources otherwise applied to earlier-stage R&D). Fine-tuning of product and customer equipment frequently is needed as physical properties, shelf life, handling characteristics, and performance under natural conditions (temperature, humidity, ultraviolet light, salt, etc). become known.

Keep in mind that scaling production is not necessarily a one-time event in the life cycle of a product. A successful product typically will find an increasingly broader customer base and new applications. This, in turn, leads to increasing demand for a product and concomitant increases in production volumes. The WD-40 Company, for example, has a fan club and publishes an ever-growing list of more than 2000 uses for its WD-40 "problem-solver" formulation (lubricant, cleaner, water displacer, etc.), including the admittedly uncommon use in removal of a boa constrictor from an engine compartment. While the formulation has remained as "number 40," for decades, volumes continue to increase.

Also, customers may request variations on a successful theme in the form of product line extensions. A product that initially was designed for bulk manufacturing or single-line continuous manufacturing eventually may be forced (without strict marketing strategy, as mentioned earlier) into variants with ever-smaller production volumes. Just as with scale-up, scale-down in response to declining sales can be problematic, although lessons learned during scale-up ideally can be applied.

A challenge, then, is how to manage impending variants and changes in scale to minimize the negative impact on quality. The primary solution is to build flexibility into the original production process design, and as variants are designed, to still allow the bulk of product the majority of the time to flow through the same process. For example, addition of a chemical that substantially changes the downstream processing characteristics should, all else be equal, be introduced as late in production as possible so that a common process can be maintained for as long as possible. This likely would not affect only production process and equipment design, but the recipe itself.

Other ways of building flexibility into the process include

- Small in-process volumes, including feedstocks
- Short lead times
- Short lag times, including transfers of materials once needed and growth of cell cultures
- Minimal number of vessels for blending, settling, etc.
- Rapid transfer and transition of material from one process step to another
- Rapid reaction rates
- High yields

Small-volume production is worth exploring in more detail. Many small-batch or continuous flow systems in parallel, each producing a small volume but flexible enough to switch among products and variations, would solve many of the problems inherent to monumental, large-batch production. Each small system could produce a different recipe, or produce the same recipe if larger volumes of the same product are needed. Benefits include

- Incremental adjustment of production volumes to meet need
- Smaller product volume, time, and financial losses when a batch or system fails
- Greater ability to produce a mix of products simultaneously, rather than in a sequential "campaign" approach
- Production periods extend long enough to allow for effective sampling and process control
- Greater process data collection from a larger number of runs, allowing for more effective process improvement
- Reduced changeover product loss and time; with large batches, massive vessels are changed over each time, whereas smaller systems allow at least some of the production lines to be dedicated for the lifetime of the product
- Reduced downtime in production caused by equipment failure
- More precise traceability of product and isolation of failures to specific equipment and systems

Some downsides include potentially higher capital outlays for equipment and physical space, more equipment to service and maintain, and additional operators (although production systems increasingly can be computer controlled and automated). If these costs outweigh the benefits, then high-volume production of robust recipes with reliable and rapid changeover and startup may be the better approach.

Scale-up issues and solutions continue to evolve with new products, algorithms, models, technologies, and regulations. Readers interested in learning more are directed to Zlokarnik (2006) for chemical engineering, Levin (2001) for pharma, Doble et al. (2004) for bioreactor production, Valentas et al. (1990) for food processing, and McConville (2006) for more general pilot plant processes, equipment, and scale-up. Brandl (2006) examines manufacturing system design in terms of ANSI/ISA-S88 and NS88 design patterns, which provide a consistent set of standards and terminology for batch control.

Design for Testing and Monitoring

Ideally, testing to check on quality should be minimized over time as confidence in process performance grows. However, new products, recipe changes, regulatory requirements, and testing as a part of process improvement create an ongoing need for in-process and final product testing. Also, because of the nature of processing (especially biologicals that are innately disobedient in following the rules), monitoring may be necessary as a part of the recipe itself (e.g., add component X when pH reaches 4.5). Physical processes are not immune; in making stainless steel, age hardening requires careful control because small variations from the prescribed time, temperature, or cooling rate can dramatically affect final properties.

Much of the daily workload of quality managers involves controlling the production process via testing and monitoring. In order to detect and troubleshoot problems, measures and specification limits describing allowable variation above and below the nominal value need to be clearly set as a part of design planning. The easier monitoring is made, the more likely it is to obtain timely, accurate, and precise measures.

Quality in the Supply Chain

Supply chain and associated logistics relate to the information and material flows from procurement through production and distribution, including transport and storage. The planning, control, and improvement of quality in supplier relations is discussed in detail in Section 21: Supplier Relations in Juran and Godfrey (1999). Considerations from the perspective of processing-based companies are outlined in the following sections.

Quality Planning: Establish a Sourcing Strategy

Several considerations arise.

- Global reach permits sourcing from anywhere in the world, and the quality of ingredients, capability of suppliers, and stringency of local regulations should be well understood before signing a supply contract. As supply chains reach farther around the world, it becomes more difficult to ensure quality of incoming raw materials because of varying regulations and practices in the originating countries and en route to the processor.
- Although just-in-time inventory is ideal in many situations, such as when freshness is critical, some ingredients may be seasonal (e.g., foods) and require storage or other buffering by alternative suppliers.
- Seasonality may affect incoming quality (e.g., corn from the United States may have a different quality profile than corn grown in Argentina during the North American winter).

- Having redundant sources can help avoid a shortfall due to inadequate supply or subpar quality, but increases the opportunities for variable quality and the need for testing as part of supplier quality control. The total cost of ownership for each commodity being purchased needs to be evaluated in light of the preceding considerations.

Quality Control: Evaluate, Select, and Monitor Supplier Performance

After deciding the needs and "what, where, when, and how" of sourcing, the "who" is addressed in terms of specific suppliers and their ability to meet specifications. Performance metrics proposed in the planning phase are developed further and adopted, and processes for capturing and reporting are established. This includes targets, limits, and minimum standards of performance.

Because of natural variability in many raw materials, whether grains used in baked goods or ore used in smelting, careful thought should go into the setting of the supplier requirements. In particular, although tight parameters are ideal and should be improved over time, overspecification (excessive number or stringency of requirements) can result in unnecessarily narrow selection of suppliers, costly inputs, frequent rejection of raw material batches, and poor supplier relations.

As part of supplier selection, include assessment of

- Supplier quality management systems (e.g., ISO certification)
- Supplier business management, including research and development, that will ensure continued competitiveness into the future; cost structure, management quality, production capacity (this may include acquisition of land and processing facilities), and information technology to facilitate communications with the supplier

At a more detailed level, the fitness for use of supplier products must be evaluated, including quality, delivery, and service. Specifically, assessment should include conformance to customer specifications, key performance indicators, and capability metrics. Typically, raw materials for chemical processes are purchased based on conformance through chemical analysis. When suppliers are selected, purchasing and service level agreements should be drafted that clearly specify quality in terms of both positive and negative characteristics, and penalties and remediation procedures for noncompliance. After a supplier's quality has been confirmed as acceptable and consistent, raw materials may be accepted solely on the supplier's analysis.

Quality Improvement: Identify and Act upon Sourcing Improvement Opportunities

With ongoing management and measurement, various opportunities for improvement can be discovered. As discussed in Juran and Godfrey (1999) Section 21.23, the process of continuous improvement typically progresses through five stages of cooperation: (1) joint team formation, (2) cost reduction, (3) value enhancement, (4) information sharing, and (5) resource sharing. Initially, efforts may focus on troubleshooting (e.g., joint efforts forced through product recall). For processing-based companies, there is tremendous opportunity in reaching the point where information can be shared that prior was held confidential, and resources are shared collaboratively without constraint of departmental and corporate boundaries. This takes trust, which takes time to develop, and should be considered as part of the supplier strategic planning. In particular is the benefit of cooperation to more fully characterize raw materials in ways that benefit the customer and using this information to leverage product differentiation.

Quality in Production

Production is the most frequent focus of attention for quality, and a great deal has been written about this aspect. Typically, an analytical or control laboratory plays a significant role in this phase via the acceptance of raw materials, decision support during production, and release of final product. Some industries are heavily regulated, and quality personnel should be familiar with relevant legislation (e.g., current Good Manufacturing Practices (cGMP) prescribed by the U.S. Food and Drug Administration). The focus in this section is on the need to maintain attention to the quality dimensions and what challenges arise during production in processing-based industries.

Producing the Quality Dimensions

As discussed earlier, there should be a clearly defined set of quality dimensions that are agreed upon in conjunction with marketing. All too often, there is only a loose relationship between the specifications provided by marketing and those developed by R&D and the quality function to be used during production. To properly reduce the critical-to-quality dimensions to production requires a deliberate mapping of dimensions to specific technical characteristics and production parameters. It is important to note here that the organizational culture may not be prepared for the partnership that is needed between individuals responsible for quality and marketing; readers are directed to Chapter 12, Six Sigma: Improving Process Effectiveness, for guidance in addressing cultural issues.

Mapping can be accomplished via the quality function deployment planning process, in which product features are mapped to process features, and process features, in turn, are mapped to process controls. During production, the responsibilities of the quality function includes, but should not be limited to, maintenance of process and product toward designated targets, for example, via Statistical Process Control (SPC). Other responsibilities include feedback of lessons learned to the planning process and identification and support of continuous improvement projects according to the Juran Trilogy.

Bulk Production and Storage

In processing-based businesses, the nature of the incoming and outgoing materials presents particular problems. Examples include

- *Spoilage*. Freshness requirements may dictate specific storage conditions and limited shelf life, such that incoming raw materials must be closely matched to production schedules and outputs matched with customer demand. Cold storage is not a cost-effective option in some cases (e.g., storage of the large quantities of biomass feedstock used in ethanol refineries), and dry storage or alternative wet storage technologies may be needed.
- *Safety*. Many dry materials represent a safety hazard, whether as inputs, processing by-products, or outputs. Dust explosions, for example, can result from combustible powders of metal (e.g., aluminum, titanium, magnesium) and natural or organic materials (e.g., grain, sugar, powdered milk, pollen, polyethylene). Improper dust control also can contribute to contamination and compromise product quality.
- *In-process transport*. Movement of bulk materials from storage areas into production frequently must be done via piping, and issues of changeover from one raw material batch to another arise, as do cleanliness issues during transport. Changeover is addressed in more detail later.

Scheduling

Although not directly a quality function, production scheduling and quality interface in important ways. Examples include:

- *Inventory.* Late or early shipments of raw materials may put pressure on existing storage in ways that affect downstream quality. Late shipments clearly jeopardize production, but may cause drawdown of supplies to the point where tanks and transport pipes run dry, or change the relationship between head space, surface area, and material volume (potentially affecting gas exchange or settling). Early shipments may exceed storage capacity, causing inventory to be placed in secondary storage under suboptimal conditions. Similarly, delayed shipping of finished product can cause backups, overflow, and compromised quality. Because of this, flexibility in scheduling should be a prominent consideration as the manufacturing process is created.
- *Customer demand.* Neither batch nor continuous flow production can simply be turned on and off, as an assembly line might, in response to increases or decreases in customer demand. Batch production may take days to complete, and once started, it must run to completion or be sewered. Continuous production helps solve some problems, but it is not necessarily straightforward to incrementally reduce or increase production volume that uses hardware with fixed dimensions, and recipes that depend on these. Lastly, processed products frequently become inputs to other manufacturing processes. Just in time (JIT) customer orders may leave little time for changes in production.
- *Failed batches.* In an assembly line, a single error may lead to a single defective unit, while a systemic problem will yield numerous defective units. Quite often, defective product readily can be pulled from the assembly line with minimal disruption of subsequent units and scrapped or reworked. In batch production, a single error can lead to a defective batch, which may represent thousands or hundreds of thousands of final sale units [e.g., a 100,000 L (26,425 gal) batch of pharmaceutical sold in 3 mL (0.1015 oz fluid) units]. Hence, it is not uncommon to have large amounts of production volume that fail to meet specification. If recovery is not possible or desired, then replacement batches need to be introduced into what typically is a tight production schedule. Even if recovery is possible, material is shifted into a separate workflow for rework or reformulation, which will compete with scheduled runs for storage capacity, raw materials, and staff time, all of which can have ripple effects on scheduling.

Besides eliminating errors and batch failures, quality managers are responsible for having clear quality tests and criteria in place prior to failures to facilitate rapid diagnosis of reformulation potential and subsequent decision-making as to disposition. Also, consider a rapid response team with cross-training so that staff can take over remediation activities without jeopardizing other product in production. Part of the responsibility of the team would be to follow up the initial, urgent response with longer-term corrective action, for example, via failure mode and effects analysis (FMEA) or related approaches. Lastly, avoid operating near 100 percent capacity. Instead of driving output and capacity maximization, strive for capacity optimization, which provides the ability to accommodate unplanned schedule changes. On-stream time (OST; defined as the actual run time divided by available time for process equipment associated with the process) may best be set in the high 80 to 90 percent range for many processes.

- *Regulatory inspections.* Regulatory inspections appropriately alter production schedules when serious deficiencies are found. However, even when deficiencies are not present, inspections can consume considerable administrative time and potentially can distract from or impede normal operations. Also, unannounced inspections make it difficult to plan ahead. To minimize the impact on product quality, consider a contingency plan to mobilize staff and prevent disruption of normal operations.
- *Proliferation of products.* Over time, a growing company can expect an increasing number of formulations, progressively more specialized equipment and processes, more frequent changeovers, and the need for expanding skill sets across employees. As this happens, each product may be produced only a few times a year, degrading the ability to keep production staff, procedures, and equipment in a state of "standard process" operation or readiness. With decreasing production runs per recipe, every batch effectively is an experimental run, and large batch-to-batch variation can result.

As the number of products expands, the production schedule should take into account the probability that batch rework will increase and avoid overscheduling initial runs. Besides reducing the number of product varieties being marketed, scheduling solutions could include more frequent production (in smaller batches, extending into continuous processing), extending production campaigns, and achieving high control during startup to ensure the process starts on target. For a review of batch process scheduling and progress in optimization over the last two decades, see Méndez et al. (2006).

Changeover

In manufacturing industries, it is common to apply the "single-minute exchange of die" (SMED) concept to facilitate rapid changeover from one product to another. This is much less common with processed products, however, due in large part to liquid is transported via pipelines and hoses, and large vessel or unit production. Very large total surface area comes in contact with feedstock, in-process, and final product that is difficult to access, purge, clean, and disinfect. Turbulent flow can make cleaning particularly problematic around fittings, connections, nozzles, etc. Verification of purging similarly is difficult. Additional changeover problems include the following:

- *Product safety.* Changing production from one product or formulation to another can be a laborious and time-consuming process. Flexible production systems that use the same equipment to handle multiple products may require special attention in both design and ongoing management. This is especially true in pharma and food processing industries due to the need to strictly control product composition, as an ingredient in one product may become a contaminant in another. In food production, for example, multiple products may be produced on common equipment, and of serious concern are allergens (e.g., nuts, gluten) that can have life-threatening effects for some consumers. Even when identical batches are produced, it may be necessary to sterilize equipment to eliminate the possibility of cross-contamination by microbes.
- *Product recovery.* In piping systems, physically and economically large quantities of product can end up as waste. The safety issue mentioned above presents one situation, in which any quantity of extraneous material in the product constitutes contamination that compromises quality and salability. Also, even where tolerances allow for some mixing, as one product run is completed and another begins, some of both products may be lost due to excessive mixing at changeover. Perishable products that otherwise could be recoverable must be disposed of if they degrade due to change in temperature, loss of carbonation, backflow, etc.

In many instances, economic losses are incurred not only by the value of the product itself, but also from waste storage, transport, and disposal costs. Waste removal and landfill fees are an important cost of poor quality, especially for hazardous waste that requires special permitting and storage and handling precautions. Nonhazardous liquid material often is disposed of by drain dumping directly into the sewer system. However, such material typically requires treatment before being released back into the environment. Sewage treatment facilities monitor effluent and charge industrial sources based on such parameters as suspended solids and sugar concentrations. Like other processors, sewage treatment plants need to have control over input characteristics such as these, which radically alter biological oxygen demand, costs, and final product quality.

Potential solutions to minimize changeover hazards and costs include the following:

- *Product scheduling.* Sequence products to minimize carryover contamination (e.g., move from lower to higher viscosity products, or color runs in which light colors are produced first, followed by increasingly darker colors).
- *Effective cleaning/purging process.* Pigging (named after the squealing sound and dirty state of metal plugs originally used to clean oil pipelines) is a common means of clearing piping systems, and technology continually provides expanding capabilities that allow not just cleaning, but inspection, detection, and repair (see Hiltscher et al. 2003).
- *Product labeling.* Indicate on the product labeling the use of common equipment for multiple products and potential for cross-contamination.
- *Separate equipment.* Use distinct, physically separated equipment for different products with quality at greatest risk.
- *Separate production facilities.* Keep products separate by maintaining physically distinct sites dedicated to one or a small number of compatible products.

Regardless of which approach is taken, it is important to design solutions into the equipment and piping configurations, and to engage engineers with experience in your industry. For readers interested in learning some of the engineering language and piping principles that may apply to their situations, see McAllister (2005).

Sampling and Measurement

Sampling is an important part of decision-making and process control. Most often, sampling should be of greater intensity and rigor at particular "control points" in the process that correspond to

- *Change in authority.* Example: Handoff of product between shifts, or from a manufacturer to tank transporter
- *Significant, irreversible activities.* Example: Addition of reactant that causes an irreversible chemical reaction
- *Creation of a critical quality feature.* Example: Flash pasteurization of fruit juice
- *The site of a "dominant process variable."* Example: Setup of run parameters to initiate a stable, steady-state process; time is a dominant factor in managing fermentation yields of microbial enzymes
- *An economical window into the process.* Example: Visual inspection during transfer of in-process product from one closed vessel to another

Samples should be taken so that they are representative of the material used for different types of decisions. Different samples and decisions include

- *Acceptance sampling* allows for testing of incoming materials to verify conformance with specifications. This may be done by a supplier, producer, or both (in the latter case, it is recommended that identical measurement systems be used to avoid disagreements based on methodological artifacts).
- *In-process samples* allow for production process control, and may be built into recipes themselves. One practical limitation is that testing time for diagnostic or control purposes may be relatively long in comparison to batch reaction time, requiring anticipation of control decisions and the ability to react quickly should conditions deteriorate.
- *Finished product testing* is done as part of production process control, product characterization, and product release.
- *Retained samples* are stored as a part of traceability or accelerated life cycle testing needs, and may be mandated by regulatory agencies (e.g., U.S. Code of Federal Regulations Title 21: Food and Drugs 211.170 – Reserve Samples, and 21 CFR 211.166 – Stability Testing).
- *Measurement control sampl*es are used to maintain accurate and precise measurement systems.

Variation in Incoming Materials

The variation noted earlier in raw materials used in processing can be moderated prior to the start of product processing in several ways:

- *Combine raw materials that are known or suspected to vary in quality into more homogeneous batches.* An example of this is mixing of grain from different geographic regions. This eliminates the need for detailed measurement of inputs from individual suppliers or sources. However, this technique accommodates and builds low quality into overall product cost, and leaves the manufacturer open to wide variation in supplier quality. It also reduces the incentive for supplier improvement.
- *Apply statistical and chemometric methods and modeling.* With a detailed understanding of the characteristics of incoming materials, it is possible to mix raw materials in precise ways to elevate overall input quality. Multivariate statistical methods (e.g., Principal Components Analysis, cluster analysis) are well adapted for this type of decision-making. Examples include use of chromatography to grade raw materials used for feedstock in pharmaceuticals or biologicals, and analysis of hydrocarbon extracts from petroleum from different geologic formations. This technique has the advantage of providing a highly characterized and tuned input, which can reduce problems downstream, especially for finicky processes. A drawback is the time and expertise needed on the front end.
- *Place stricter requirements on incoming raw materials.* For example, in order to reduce the need for downstream manual balancing of natural variation, a perfume manufacturer may place more stringent purity, age, or freshness specifications on the plant sources of aromatics, such as leaves, twigs, and resins. By doing so, variability does not enter into the production stream, thereby reducing the impact and need for efforts further downstream that could be more time-consuming and costly. However, this requires close cooperation with suppliers, and potentially the costly development of measurement systems that can detect the quality characteristic(s) of interest.

Variation during Processing

Despite attempts to minimize variation in incoming raw materials, there may be substantial variation remaining that must be managed during processing. Methods of dealing with this include:

- *Design of experiments and response optimization.* DOE provides the quality manager with a means of understanding the variation among batches of incoming raw materials and specifically modeling adjustments to be made during processing. Often, batch is used as a blocking variable (see Chapter 19, Accurate and Reliable Measurement Systems and Advanced Tools, for a discussion of blocking), reflecting the situation in which the variable "batch" is not of inherent interest, but instead is a variable that may affect quality and can be statistically accounted for. In such a case, the DOE model produces the settings for variables other than "batch" that can be adjusted to optimize process output.

However, "batch" can be a legitimate variable in its own right, provided it is associated with some meaning (e.g., Supplier A versus Supplier B, viscosity, or ppm of contaminant). Using the resulting DOE model ($Y = mX + b$), the settings of other variables can be set based on known values of "batch" to optimize process output. This approach permits quality optimization across a range of inputs that may be out of the processor's control. Inputs must be measured, however, to know the appropriate settings for other variables, including process variables (e.g., time, temperature, pressure) that may be of interest in addition to mixture components. If an indirect proxy of quality is used (such as "supplier"), then there is the risk of suboptimization if the true, underlying critical-to-quality characteristics change unbeknownst to the operators.

- *Robustness of process to variation.* A related but alternative approach is to allow variable inputs to enter the process; but rather than alter the process in response, instead make the process resilient to the variation. In robust parameter design, the principal goal is to discover factor settings that minimize response variation, while allowing the process to stay on target. Although alternative methods have been developed (see Robinson et al. 2004 for a review of robust parameter design), Taguchi methods remain the most familiar to quality practitioners. However, limited work has been published on the application of Taguchi's concepts to mixture designs. Recently, nonlinear orthogonal arrays were investigated and, in combination with nonparametric techniques, applied to food formulation improvement (Besseris 2009).
- *Downstream processing is based on characteristic(s), including possible manufacture of a different product.* An extension of the DOE approach described previously is to make substantial changes in processing to accommodate input variation, to the point of intentionally producing a different grade or product. Rather than using process variables to adjust so that the output is consistent, processing is adjusted specifically to hit a different target. This is advantageous when the costs of intentionally shifting to a different target are less than the costs of remediation, including potential losses due to lower pricing and mismatch with customer demand.
- *Evolutionary operation (EVOP).* First applied in the chemical industry, this is a process optimization technique that is integrated with ongoing, full-scale manufacturing. The premise is that natural variation among production lots provides valuable information useful in making incremental process improvements. Small changes

are introduced during normal production that generate effects large enough to detect, but not so large as to result in nonconforming product. See Box and Draper (1969) for a more detailed discussion by the originator of the method, or Box et al. (2009) for an updated perspective on process control and improvement.

Process Control

Maintaining processes within targets is the generic goal of process control. The output of sample measurements often is analyzed and decisions made via statistical process control methods (SPC; see Chapter 19, Accurate and Reliable Measurement Systems and Advanced Tools). There are two prominent considerations when considering SPC in processing industries; these are short runs and dynamic processes.

First, a practical problem encountered in production of processed materials is that batch production can produce too few data points to effectively use standard SPC methods. A common cautionary rule of thumb is to avoid interpretation of SPC charts until after approximately 20 data points have been plotted. For a production facility that produces one batch every two weeks, nearly a year might pass before sufficient samples have been taken. Further, control charts intended to separate common from special cause variation are problematic when each batch (data point) is a separate production from those before or after, and therefore a different process (involving tank setup, potentially different feed rates, etc.). As mentioned earlier, short production runs contribute to each batch being treated from a practical and quality perspective as an experimental run because the desired outcome is not assured. In such situations, a production campaign may nearly be over before an on-target process is attained. See Section 45: Statistical Process Control in Juran and Godfrey (1999) for more information on short-run control charts.

Second, processes that experience known and expected shifts in the central tendency present special quality control problems. Traditional Shewhart statistical process control charts (see Chapter 19, Accurate and Reliable Measurement Systems and Advanced Tools) are not strictly applicable within batches because such processes violate the assumptions of identically distributed and independent observations; that is, they are not stable and exhibit autocorrelation among successive samples. More generally, common-cause variation (chronic, routine, or unassignable changes) becomes confounded with special-cause variation (sporadic, abnormal, or assignable change). In practical terms, normal and expected shifts in the mean can be trivially flagged as special causes (false positives), and changes of interest may not be properly identified (false negatives, because of large variance and calculated control limits). Batch processing is particularly problematic because a steady state is not achieved, unlike continuous processing that allows for control toward set points that are relatively time-independent once the flow of materials has been established (e.g., in a continuous process, material changes over time, but a characteristic should be reasonably stable over time when measured at a particular, instantaneous point in a pipeline).

What is needed is monitoring for deviations of a process from its normal or optimal reference trajectory and correlations among variables. Nomikos and MacGregor (1995) laid much of the groundwork for contemporary batch process control, with substantial research since then (e.g., see Brooks (2009) for geometric process control, Albazzaz and Wang (2007) for development of independent component analysis as an alternative to principal components analysis (PCA), and Li et al. (2007) for a technical discussion of batch-to-batch control using an updated nonlinear partial least squares method). Jørgensen et al. (2006) provide an overview of monitoring and control of batches; Cinar et al. (2003) presents numerous process monitoring and quality control tools, starting with univariate SPC charts and extending into multivariate statistical process monitoring techniques. A related topic is automation of batch control, and James (2006) provides a good introduction to batch automation concepts based on ISA 88 and 95 that separate control into the physical model (assets such as equipment), procedural control model (sequence of equipment-oriented actions in product manufacture), and process model (process actions without regard to equipment). There is extensive literature on these topics, and the preceding references may serve as a starting point for research.

Product Remediation

When the final product made suffers from variation in quality among batches (in the extreme case, nonconformance with specifications), the quality manager may be expected to save the day. Several approaches can help manage this situation.

- *Complete blending of batches.* Blending final product from multiple batches can smooth out between-batch variation. This is effective, but has drawbacks. Creating such "lots" (uniform material that has been thoroughly mixed in a single vessel) adds labor and equipment costs, and creates larger work in process. Further, "good" product essentially is being diluted with "bad" product, such that the end product that is shipped is of lesser quality than if produced from the start at a uniform level of quality.
- *Partial blending of batches.* Partial blending of subpar material with on-target batches essentially "hides" bad product in good product but in amounts small enough so that the final lot remains within specification. Similar to the blending cited previously, this risks reducing the functionality of the "good" batch, in addition to adding storage and handling costs.
- *Reworking of batches.* Reworking batches is possible if they fail on a limited number of characteristics that do not interact in complex ways. In foods, for example, thickening agents such as starches or vegetable gums may be added postproduction to alter physical properties and product stability without affecting taste or texture. Generally, this adds substantial labor costs, and may occupy vessels otherwise allocated to a new batch, thereby disrupting production schedules (although companies that specialize in contracted rework are an option). Note that regulations may require written procedures and approval of a quality control unit for reprocessing (e.g., U.S. FDA 21 CFR 211.115 – Reprocessing).
- *Overprocessing (over-tolerancing).* One of the "deadly wastes," overprocessing in this case refers to intentionally exceeding the target so that lesser quality material still meets specifications. A biological pesticide, for example, may have a guaranteed minimum concentration of active ingredient of 20 g/L (0.0441 lb/0.2642 gal); if batches frequently vary by +/– 2 g/L, then targeting 22 g/L (0.0485 lb/0.262 gal) ensures that most product will meet the specification. Unless pricing is adjusted (or blending strategies applied), product shipped containing extra active ingredient results in approximately a 10 percent loss of revenues.
- *Downgrading.* Salvage by grading of product to reflect inferior quality and sale at a reduced price is yet another possible course of action. Assuming product could have been sold at a higher price had it been of higher quality, this represents lost revenues. Regulations may apply (e.g., drug product salvaging is strictly regulated in the United States), and evidence from laboratory tests and assays, and facility inspection may be required prior to disposition in the marketplace. Records also should be maintained for quality improvement purposes, for example, investigation of special causes of off-grade product (raw material lot, catalyst degradation, process control or operator failure, equipment wear, etc.).

- *Discard*. Finally, product may need to be discarded. Although a costly last resort, this is not uncommon when sterility issues arise, and is the "lesser evil" when consumer health is at risk. Costs include materials, labor, disposal, and opportunity cost.

Quality Postproduction

Quality management does not end with lot approval or exit from a production facility. While all products can be compromised in some way postproduction, processed products are susceptible to damage in ways that are less perceptible than the dents, scratches, and loose parts of assembled products. Issues particularly relevant are described in the following paragraphs.

Repackaging-packaging. Bulk materials may be shipped out of the production facility for packaging and repackaging that is outside of the manufacturer's direct control. To ensure continuity of quality, it is imperative that contracted organizations have quality documentation and assurance systems in place, and that applicable good manufacturing practice (GMP), good distribution practice (GDP), and (over/re)labeling requirements are followed.

Distribution and transport. The quality of product received by an end user can degrade during distribution and prior to use. In particular, environmental conditions such as temperature and vibration can dramatically affect formulations and performance; humidity typically is less of a concern because it can be managed through proper airtight packaging. To minimize risks to quality, specifications in anticipation of transport and storage conditions should be part of product design and quality planning. This includes appropriate product formulations (e.g., formulations that do not separate during shipping and storage), packaging (e.g., hygroscopic gels in drug bottles, dark bottles that limit UV light penetration), and understanding of and adherence to shipping and storage specifications (e.g., upper temperature limits and shelf life designations such as the "sell by" date for processed milk products). Service level agreements should be in place, and actively managed. Monitors embedded in product, shipping containers, and transport vehicles can assist in auditing and continuous improvement.

Mentioned previously, GDP is a quality warranty system that relates to guidelines for the proper distribution of medicinal products intended for human use. The scope typically extends from division of product at the production facility premises through intermediaries and movement to final end users. Formal legislation varies by region; primary sources include

- United States: Code of Federal Regulations 21 CFR 210/211, and USP 1079.
- Europe: European Community Directive of the Board 92/25/EEC. The International Pharmaceutical Excipients Council Europe (IPEC) also publishes a GDP guide and audit guidelines.
- Asia: Varying regulations; the World Health Organization (WHO) adopted guidelines from the Fortieth WHO Expert Committee on Specifications for Pharmaceutical Preparations in October 2005; (WHO Technical Report Series, No. 937, Annex 5, 2006); these were undergoing revision in 2009.

Product Safety and Liability

Processed goods and materials have safety and liability issues distinct from those produced by other industries, especially for consumed products such as food and drugs, and chemicals that can have acute and chronic effects that extend physically and temporally beyond the point of production and use. Government regulations help manage these, but changing raw material sources, distribution channels, and laws create a need for producers to continually evolve their quality practices. Questions from a strategic perspective include, What are current safety concepts that should guide internal policies and practices? How can I proactively manage my company's liability and brand image? Tactically, What are current regulations and how do they apply in practice? What are common violations and weaknesses in processed product quality assurance?

Safety Concepts and Trends

A significant change over the last decade is the increasing emphasis on *management systems* that take an integrated approach to consumer safety. The integration aspect is important, as it seeks to tie together what otherwise might be separate policies, procedures, practices, methods, controls, programs, plans, roles, responsibilities, reporting and documentation structures, etc. Conceptually, this provides for coordinated, proactive rather than reactive management.

For example, Hazard Analysis and Critical Control Points (HACCP) is a safety management system. More formally, it is a set of principles (see Table 24.4) and a systematic approach to food safety that emphasizes identification, assessment, and prevention of safety hazards rather than inspection. HACCP compliance is regulated by various agencies in the United States; the USDA's Food Safety and Inspection Service regulates HACCP systems for meat and poultry, and the FDA regulates juice and seafood (9 CFR 417; 21 CFR 120, 123, 1240). Internationally, the Codex Alimentarius Commission applies HACCP principles to its standards, guidelines, and codes of practice for the food industry. In a significant move, HACCP guidelines also are being applied voluntarily to other industries, including pharmaceuticals and cosmetics. The HACCP or other safety management systems may themselves be a part of a larger management system.

The International Organization for Standardization (ISO) 22000 is another quality assurance system directed towards food safety, but has applicability to other industries. A more procedural derivative of ISO 9000, this family of standards integrates HACCP principles and presently includes

- ISO 22000—Food safety management systems—requirements for any organization in the food chain
- ISO/TS 22003—Food safety management systems—requirements for bodies providing audit and certification of food safety management systems
- ISO/TS 22004—Food safety management systems—guidance on the application of ISO 22000:2005
- ISO 22005—Traceability in the feed and food chain—general principles and basic requirements for system design and implementation
- ISO/DIS 22006—Quality management systems—guidelines for the application of ISO 9001:2000 in crop production (in development at press time)

Note that these standards reach back into the supply chain, a development likely to continue as part of an emphasis on prevention. Traceability (ISO 22005) refers to the ability to track the movement of a food product through various stages of production, processing, and distribution to the end user. This includes both *trace forward* capability (from the farm to the retail shelf) and *trace back* (from retail shelf back to the farm).

Quality staff at all levels need to stay abreast of such evolving trends. In the United States, the "Public Health Security and Bioterrorism Preparedness and Response Act of 2002" (commonly referred to as the Bioterrorism Act) requires farmers, packers, and retailers to keep track of where food comes from and where it goes in the supply chain: one step

Principle	Description
1. Conduct a hazard analysis.	Plants determine the food safety hazards and identify the preventive measures the plant can apply to control these hazards. A food safety hazard is any biological, chemical, or physical property that may cause a food to be unsafe for human consumption.
2. Identify critical control points.	A critical control point (CCP) is a point, step, or procedure in a food manufacturing process at which control can be applied and, as a result, a food safety hazard can be prevented, eliminated, or reduced to an acceptable level.
3. Establish critical limits for each critical control point.	A critical limit is the maximum or minimum value to which a physical, biological, or chemical hazard must be controlled at a critical control point to prevent, eliminate, or reduce to an acceptable level.
4. Establish critical control point monitoring requirements.	Monitoring activities are necessary to ensure that the process is under control at each critical control point.
5. Establish corrective actions.	These are actions to be taken when monitoring indicates a deviation from an established critical limit. The final rule requires a plant's HACCP plan to identify the corrective actions to be taken if a critical limit is not met. Corrective actions are intended to ensure that no product injurious to health or otherwise adulterated as a result of the deviation enters commerce.
6. Establish record keeping procedures.	The HACCP regulation requires that all plants maintain certain documents, including its hazard analysis and written HACCP plan, and records documenting the monitoring of critical control points, critical limits, verification activities, and the handling of processing deviations.
7. Establish procedures for ensuring the HACCP system is working as intended.	Validation ensures that the plants do what they were designed to do; that is, they are successful in ensuring the production of safe product. Verification ensures the HACCP plan is adequate, that is, working as intended. Verification procedures may include such activities as review of HACCP plans, CCP records, critical limits, and microbial sampling and analysis.

[*Source*: based on U.S. 21 CFR 120.8. (http://www.gpoaccess.gov/cfr/index.html) and adapted from www.fsis.usda.gov/oa/background/keyhaccp.htm. Retrieved May 21, 2009)]

Table 24.4 The Seven Hazard Analysis and Critical Control Point (HACCP) Principles

forward and one step back. Specifically, the identities of immediate nontransporter sources and recipients, and specific sources of each ingredient are required for each lot of product. Civil and/or criminal penalties may be levied on supply chain participants that fail to produce requested information within 24 hours. Despite this, a large number of supply chain participants remain noncompliant (Levinson 2009), creating both public health and legal risks.

New regulations continue to appear (e.g., the proposed U.S. Food Safety Modernization Act of 2009). Fortunately, new technologies and business practices are emerging to facilitate compliance. For example, bar code tracking systems are being put into place to label fresh produce on retail containers, permitting in-store scanning at kiosks by customers to identify the name of the grower, date, and place of harvest. Creative quality managers should look at regulations not merely as a safety issue or a bureaucratic burden, but as an opportunity for brand differentiation.

Despite the prevalence of quality assurance systems such as HACCP, ISO, and GMP, it is not always clear the degree to which such systems enable quality. Inspection therefore remains an important part of safety management, and quality assurance programs must ensure performance against both design and regulatory specifications. An illustration of quality by inspection is the USDA Food Defect Action Levels that specify thresholds of natural or unavoidable defects in foods that are deemed to pose no health hazards for humans. As a specific example, tomato puree has an action limit for fruit fly of 10 or more fly eggs and 1 or more maggots per 100 g of final product. Importantly, the USDA provides a *Macroanalytical Procedures Manual* to assist quality managers in appropriate testing procedures (currently out of print, but online at http://www.cfsan.fda.gov until a revised edition is completed).

Although much of the specific discussion here relates to the food industry, the principles are broadly applicable, and quality personnel may find guidance and improvement ideas from industries external to their own. Many industries process and manufacture hazardous chemicals; safety issues associated with inherently toxic chemical products relate more to product labeling and use, in addition to employee safety. Liability is discussed next; otherwise, these specific topics are out of scope for the present discussion.

Liability

Prior sections in this chapter are devoted to the building of quality into design, production, and supply and distribution chains. When products allegedly do not deliver expected results, legal responsibility for performance failures and harm often becomes an issue. In industries such as pharmaceuticals, product liability litigation is viewed as a necessary cost of doing business because of the nature of the product. In many companies, however, especially smaller organizations, liability does not receive proper forethought. Several issues are of concern to processing-based companies, as the following sections explain.

Product Recalls

Safety is the predominant factor is issuing product recalls (others include improper labeling and poor performance). Whereas assembled products typically are recalled for design flaws, processed materials additionally may be recalled because of contamination of an otherwise good "design" or formulation. Examples in recent years have involved properly designed products that became compromised via adulterated raw materials that originated from poorly regulated supply chains. This emphasizes the point made earlier regarding the need to include supply chain design as a part of product quality planning.

Processed products typically fall under distinct regulatory authorities separate from assembled products. In the United States, for example, relatively few processed products fall under the jurisdiction of the Consumer Product Safety Commission (paints and coatings are notable exceptions) compared to the Food and Drug Administration (food, drugs, cosmetics, veterinary medicines), or Environmental Protection Agency (pesticides, herbicides, rodenticides, fungicides). The U.S. Department of Agriculture also has oversight for the inputs to many processed products, including foods and plant-based pharmaceuticals. Quality staff should look to the appropriate regulatory agency for recall requirements and use these for planning purposes.

Planning is the key to managing the risks and costs associated with product recalls, and is a responsibility of company executives. Unfortunately, this does not always receive the attention it deserves because perceived recall costs are minimized and improperly accounted for as indirect and nonrecurring costs. Direct costs include investigation of product failure, rework and replacement, legal work, and notification of customers, distributors, and regulatory authorities. Indirect costs include loss of sales, negative publicity, brand image impairment, and share price reduction. Readers are directed to Schoem (2005) for an overview of preparing for and conducting a product recall. More specialized and detailed guides are published for specific industries, for example, the *Food Products Recall Manual* (Olssen et al. 2009) and *Retail Pharmacy Recall Manual* (Olssen et al. 2008).

As mentioned, traceability is receiving increasing attention from regulators at a time when supply chains are transforming into supply networks. Consider China's food industry that consists of more than 400,000 processors, most with a small number of employees and low production volume. Pooling of product from multiple suppliers becomes a necessity, making traceability and recall a daunting task. Several recommendations are in order to help manage recalls, including

- Assign an individual responsibility for managing recalls, including planning and preparation (to include mock recalls), auditing, and sound record keeping.
- Work collaboratively with other organizations in the industry to develop appropriate standards and legislation.
- Manufacturers, distributors, and retailers should consider contracts that clearly delineate specifications for allocation of recall or replacement costs. This is not for purpose of shifting blame; rather, it is to decide on actions before an event occurs so that appropriate action swiftly may be taken. Specific indemnity and hold-harmless terms may be necessary.

Improper Use

While all products can be improperly applied (how many of us have employed a kitchen knife as a screwdriver, chisel, or pry bar?), certain processed products have a greater inherent capability for underuse, overuse, and misuse. Chemical pesticides, for example, typically are sold in concentrated form. Overuse and misuse are common (e.g., improper dilution so that applied product is too strong, or "tank-mixing" of incompatible liquid herbicides and pesticides into a single mixture prior to field application). Subsequent poor product performance may be blamed on the product, rather than the application (or applicant).

However, a producer needs to not simply *make* product that conforms to specifications, but to ensure that *use* conforms to specifications. It legally may be the responsibility of the manufacturer to anticipate potential alternative uses (whether appropriate or not) and application failures, and to warn against them by appropriate labeling, instruction, and disclaimer. A quality manager should use customer complaints and reported events not only to communicate proper use after an incident, but as opportunities to "mistake-proof" or otherwise improve product design. A product that is robust against improper use not only limits liability, but provides a means of product differentiation.

Counterfeits

Black and gray markets are not normally considered as part of the scope of quality. However, such markets operate outside of the usual regulatory environment and pose a threat to product quality, consumer safety, and brand image of legitimate producers. More often associated with consumer goods such as handbags and watches, counterfeit and substandard products from unauthorized or flawed formulas and manufacturers is a problem in certain industries. For example, pharmaceuticals easily are obtained through the Internet and questionable distribution channels; while many of these may be of high quality and meet all relevant standards, others may not. The World Health Organization (WHO) has estimated that up to 25 percent of medicines consumed in poor countries are substandard or counterfeit (WHO 2003). While assembled counterfeit products may be visibly different in craftsmanship, counterfeit processed goods such as cough medicines are much more difficult to identify and track, especially if wholly consumed as part of normal use.

Regulatory and legal solutions typically are pursued to combat counterfeiting. From a quality perspective, consider ways in which you can clearly, reliably, and in means difficult to duplicate distinguish your product from others. Also, be prepared in the event some negative event threatens to tarnish the image of a legitimate, high-quality product or brand, or even an entire industry.

Quality and Society

Public perception of quality is shaped by forces far outside the usual realm of quality staff. This is particularly true for process-based companies that must maneuver among potentially divisive social and political issues. Examples include use of technology during production (such as genetically modified raw materials), global trade, and Fair Trade production of commodities, pharmaceutical costs (high cost, price discrimination), environmental protection and land use. Consumers buy products not just for utilitarian purposes, but to make statements, and to indirectly participate in social change.

At the strategic level, perceptions and trends present threats to existing products, but also potential opportunities for differentiation from competitors. While industry trade groups and government lobbyists certainly play a role in managing perceptions and guiding legislation, recall that perceived quality is one of the major dimensions of quality, and, as such, is within the purview of quality management. It is no coincidence that some of the most divisive issues surround products that are not being judged solely on measurable characteristics, but indeterminate, intangible factors. At the tactical level, public perception obliges changes in quality standards, testing, documentation, and transparency.

Some general implications include the following:

- *Product versus process.* Although a manufacturer may argue that the product rather than the process of production should be the basic or sole criterion for product quality, consumers may perceive that the process by which a product is created is relevant to purchasing decisions. This can extend back to the source of raw materials.
- *Great expectations.* Nebulous or even irrational expectations may arise, making it difficult to ensure conformance. Testing to "prove" conformance, rather than to demonstrate no evidence of failure to conform, is an example, such as "proving" that a genetically modified organism (GMO) such as corn does not produce an allergen. The public at large and legislators have expectations of testing that do not square with the statistician's notions of hypothesis testing in which one never seeks to "prove" a hypothesis or the absence of something, rather, only to reject or fail to reject a hypothesis, or fail to show a presence despite best, good-faith efforts to do so.

- *Documentation and transparency.* To effectively demonstrate conformance with expectations, quality managers should consider formal measurement systems, validation, and documentation of perceived characteristics. Without regulations, it is incumbent upon industry participants to identify appropriate measures, and, where possible, to actively participate in their development. In addition, quality managers should be prepared to produce documentation in a format that is readily digestible by the public.

Prognosis

What may the future look like for quality in processing-based companies? Current trends such as globalization likely will continue to force a broader view of suppliers and customers, with ever-growing government regulation and eventual standardization across economies and the supply chain. Industry-specific challenges include consumerism, competition from generics, commodity pricing, public demand for environmental stewardship, flexibility to produce semicustomized products, and transparency among suppliers, producers, and customers. Strategically, managers should proactively shape their organization's quality management program to adapt to better contribute to overall, improved competitiveness. Tactically, responsiveness is needed in quality planning, improvement, and control to enable shifting strategies. In short, greater flexibility will be needed in using quality as a strategic and tactical tool. Consider the phenomena described in the following sections.

Low-Cost Competition

With cost ever a driver, pressure will continue from continued (albeit hesitant at times) expansion of global trade and availability of products from competitors worldwide that have comparative advantage via low-cost structures. Competition leads to pressure for reduced prices and differentiation of products across one or more dimensions, without compromising quality. Applying the principles and practices advocated in this handbook will help achieve better quality at lower cost.

Regulatory Shift from Inputs to Process to Outputs

Regulatory agencies typically emphasize product safety and effectiveness, but the specific means vary and evolve over time as industries advance and mature. In new industries, few standards exist for either product or process, so regulation focuses on what can be controlled: the inputs (e.g., permitting and qualification of producers and incoming materials). As time passes, there frequently is a shift to regulate how products are created: the processes being followed. Later, as comfort and confidence are gained in the inputs and process, regulatory focus becomes more limited to what is being produced: the product outputs. The following two examples will illustrate the trend:

- *Generics.* Generic drugs have become commonplace. Drugs typically are small-molecule chemicals with specific activity, and generic versions easily are characterized and tested for conformance with chemical specifications to determine equivalence with proprietary versions. In contrast, generic, follow-on biologics (also known as biosimilars) typically are large, complex molecules such as proteins and peptides that are difficult to characterize and can have unpredictable interactions in the body. Further, it has been argued that only the original manufacturer can reliably and repeatedly produce a biologic because their recipes, processing, and quality control are complex, and the details may not be disclosed for proprietary reasons. Hence, the safety and efficacy of a biologic arguably is dictated by the process.

However, technological advances gradually will facilitate confidence in final product regardless of the creative process, and public perception of high medical costs and low competition likely will lead to legislated change in production and quality testing to enable businesses to pursue follow-on biologics. Quality managers should anticipate technological changes in analytical tools and methods, greater reproducibility of manufacturing processes, and, eventually, new standards and regulatory requirements placed on final products. For manufacturers of the original products, marketing may place greater emphasis on branding, with quality managers requested to fine-tune tools and methods to fulfill the brand image.

- *Genetically modified organisms (GMOs).* Historical regulatory emphasis has engendered lengthy registration processes, including formal permitting and notification procedures for product development, and acute focus on the nature of gene and vector inputs. Emphasis is shifting downstream into process, but has not yet evolved to a point at which the final consumed products are the primary control point. For example, it presently is difficult to demonstrate that food allergens are not present in final, processed food products, so labeling simply cites the presence or absence of GMO material in a product such as a corn chip. Even more technically difficult is to show that an entire, highly complex nutritional profile is substantially similar to that of a non-GMO. Again, quality managers should anticipate demands and emerging technology to more fully characterize final, postprocessing products to show equivalence with other available versions.

Intellectual Property

Many processed products are made from trade secret recipes (e.g., the Coca-Cola company is famed for the reportedly limited distribution of formulas among a handful of executives). However, electronic information storage and transfer, outsourcing, misappropriation of formulas, increased availability of information to competitors regarding raw material purchases, and reverse engineering capabilities create challenges to maintaining sole ownership. Quality managers should anticipate a need for alternative formulations and processes to keep ahead of the competition.

Commodity Pricing

Commodity prices potentially will experience changed volatility due to greater reliance and connectivity to geographically dispersed supply chains. Petroleum-based products depend on oil, for example, the price of which can change swiftly in response to geopolitical events. An implication for quality is requests by management to decrease costs of production and accommodate changes in quality of incoming raw materials as companies try to find alternative raw materials and suppliers. Rather than succumb to lower prices, some companies may shift up-market to make higher margin products that meet more specialized demands for increasingly smaller customer bases.

Consumerism

Expectations of the public must be included as an aspect of quality beyond mere technical specifications. Regulations increasingly will penetrate into areas that will require changes in products and measures of quality. In the United States, for example, bans against the use of trans-fat in food preparation, and mounting claims for evidence that processed foods contribute to obesity eventually may trickle down to recipe changes and measures to manage liability. Even without regulatory mandates, shifts in consumer health consciousness are driving development of new processing methods for example, food companies seeking to improve sales of fiber-rich processed foods are finding new fiber sources, fashioning new grinding processes and cooking methods (Brat 2009).

Semicustomization

Trends toward more choices will mean more formulations and shorter production runs. In the pharmaceutical industry, gene-mapping and proteomics capabilities should pave the way for medicines that are more specifically tailored to particular patient subpopulations and genetic profiles. So-called precision or personalized medicines will decrease the production of blockbuster drugs; volume per therapeutic will decline, and the number of therapeutics will increase. As a part of this trend, diagnostics necessary to characterize patients and appropriate treatment will be in greater demand. Lastly, as generics penetrate the markets, outsourcing of production will require transfer of robust recipes and production processes to third parties.

Automation

Insofar as logic and rules can be applied (see Hall and Johnson 2009 for a discussion of artistic versus scientific process management), automation will continue to help reduce variation and improve consistency in the quality of final product. In particular, automation can assist in real-time sampling and assay to allow more precise manipulation and control of processes. This may reduce the number of production and quality staff needed but increase the skill set required. Automation also will allow for more precise, reproducible processes and facilitate a shift in regulatory focus to product, as cited earlier.

As stressed throughout this chapter, there is a need for executives and quality professionals to expand their views on the role of quality in their industry. This increase in scope likely will continue as a driving force in future years. With greater scope and responsibility will come greater influence and control by quality over their organization's future success.

Where to Find It

Presented in this section is a compilation of sources for information of interest to quality personnel. These may be useful in answering the question, Where can I find information to keep current with quality issues and help proactively develop my company's quality strategy? and at a more tactical level, Where can I find information regarding regulations, guidelines, benchmarks, tools, and concepts to help my staff in their daily work? This is intended to be a selective rather than exhaustive list. In addition, although sources may fit under multiple headers, entries are listed only once.

Standards Organizations

American National Standards Institute (ANSI): http://www.ansi.org/. Promotes and facilitates voluntary consensus standards and conformity assessment systems, safeguarding their integrity. Member of ISO.

AOAC International (AOAC): http://www.aoac.org/. Scientific association that provides tools and processes necessary to develop voluntary consensus and technical standards.

ASTM International (ASTM): http://www.astm.org/. Develops and publishes voluntary, international consensus technical standards.

Codex Alimentarius Commission: www.codexalimentarius.net/. Develops food standards, guidelines, and related texts such as codes of practice under the Joint FAO/WHO Food Standards Programme.

International Organization for Standardization (ISO): http://www.iso.org/. Network of the national standards institutes of more than 160 countries; coordinates the system and publishes the finished standards.

International Society of Automation: http://www.isa.org/. Develops standards; certifies industry professionals; provides education and training; publishes books and technical articles; hosts conferences for automation professionals.

U.S. Pharmacopeia (USP): http://usp.org/. Official public standards-setting authority for all prescription and over-the-counter medicines and other health care products manufactured or sold in the United States.

World Health Organization (WHO): http://who.int/. Publishes various standards, including "The International Pharmacopoeia" for the pharmaceutical industry.

Trade Associations, Professional Societies, Advocacy Groups

AACC International: http://www.aaccnet.org. Formerly directed toward cereal chemists, gathers and disseminates scientific and technical information to professionals in the grain-based foods industry worldwide.

Alliance for Polyurethane Industry: http://www.polyurethane.org/. Alliance of U.S. producers and distributors of chemicals and equipment used to make polyurethanes and manufacturers of polyurethane products. Provides information about handling of polyurethane chemicals; regulations; reporting and compliance assistance; and standards, test methods, and specifications.

American Association of Candy Technologists: http://www.aactcandy.org/. Professional group dedicated to the advancement of the confectionery industry.

American Association of State Highway and Transportation Officials: http://www.aashto.org/. Provides news and educational materials; develops standards for many materials though participation of state departments of transportation staff.

American Association of Meat Processors: http://www.aamp.com/. North America's largest meat trade organization.

American Association of Poison Control Centers: http://www.aapcc.org/. Maintains a National Poison Data System with detailed toxicological information on exposures, useful in monitoring and demonstrating safety to regulatory agencies and consumer groups.

American Bakers Association: http://www.americanbakers.org/. National trade organization for the baking industry.

American Beverage Association: http://www.ameribev.org/. Represents beverage producers, distributors, franchise companies, and support industries.

American Cheese Society: http://www.cheesesociety.org/. Provides American cheesemakers with educational resources, networking opportunities, and support of high standards in cheesemaking.

American Chemical Society: http://acs.org/. Publishes numerous scientific journals and databases; convenes major research conferences; and provides educational, science policy, and career programs in chemistry.

American Chemistry Council: http://www.americanchemistry.com/. Represents companies that make the products that make modern life possible, while working to protect the environment, public health, and national security. Has various divisions, including plastics and chlorine chemistry.

American Concrete Institute: http://www.concrete.org/. Serves professionals internationally in the field of concrete design, construction, materials, education, and certification.

American Dairy Products Institute: http://www.adpi.org/. National trade organization of the processed dairy products industry.

American Dairy Science Association: http://www.adsa.org/. Organization of people who are committed to advancing the dairy industry.

American Iron and Steel Institute: http://www.steel.org/. Represents the North American steel industry in the public policy arena and advances. Provides a clearinghouse of information in that industry and promotes the development and application of new steels and steelmaking technology.

American Meat Science Association: http://www.meatscience.org/. Professional society that provides a forum for all interests in the meat industry.

American Oil Chemists Society: http://www.aocs.org/. A global professional scientific society for all individuals and corporations with interest in the fats, oils, surfactants, detergents, and related materials fields.

American Petroleum Institute: http://www.api.org/. A trade association for the entire oil and natural gas industry, providing standards for storage tanks, pipelines, pressure vessels, etc.

American Protein Producers Association: http://www.animalprotein.org/. Promotes the production and manufacture of safe animal by-products by improving the microbiological and chemical quality of feed fat and animal proteins, and providing education.

American Society of Brewing Chemists: http://www.asbcnet.org/. A global authority and advocate for brewing science and technology.

American Society of Perfumers: http://www.perfumers.org/. Fosters and encourages the art and science of perfumery while promoting professional exchange and a high standard of professional conduct within the fragrance industry.

American Society of Sugar Cane Technologists: http://www.assct.org/. Serves individuals involved in agriculture (growing and harvesting of sugarcane) and manufacturing (extraction and processing of sucrose from sugarcane).

American Spice Trade Association: http://www.astaspice.org/. Represents and serves members in more than 34 spice-producing nations around the globe.

Association for Dressings and Sauces: http://www.dressings-sauces.org/. Association of condiment sauce manufacturers and their suppliers.

Chemical Producers and Distributors Association: http://www.cpda.com/. U.S.–based trade association representing the interests of generic pesticide registrants, with a membership that includes manufacturers, formulators, and distributors of pesticide products.

Corn Refiners Association: http://www.corn.org/. National trade association representing the corn refining industry of the United States.

Cosmetic Ingredient Review: http://www.cir-safety.org/. Thoroughly reviews and assesses the safety of ingredients used in cosmetics in an open, unbiased, and expert manner, and publishes the results in the peer-reviewed scientific literature.

Fats and Protein Research Foundation: http://www.fprf.org/. Institution that directs and manages research to enhance usage and the development of new uses for rendered animal products.

Food Ingredient Distribution Association: http://www.fidassoc.com/. Association to increase the value of food ingredient distribution services.

Food Institute: http://www.foodinstitute.com/. Provides specialized services for companies or trade organizations.

Fragrance Foundation: http://www.fragrance.org/. Maintains an extensive print and video fragrance library, publishes educational materials, and holds seminars and symposia.

Glass Manufacturing Industry Council: http://www.gmic.org/. Trade association that facilitates, organizes, and promotes the interests and economic growth and sustainability of the glass industry through education and cooperation in the areas of technology, productivity, innovation, and the environment.

Grocery Manufacturers Association: http://www.gmaonline.org/. An advocacy, value chain, and scientific association for food, beverage, and consumer products industries.

Independent Cosmetic Manufacturers and Distributors: http://www.icmad.org/. An international cosmetic trade association for small-to-medium-sized entrepreneurs.

Institute of Food Science and Technology: http://www.ifst.org/. Professional qualifying body for food scientists and technologists.

Institute of Food Technologists: http://www.ift.org/. Clearinghouse of information for scientific applications related to foods and food products.

Institute of Shortening and Edible Oils: http://www.iseo.org/index.htm. Provides resources relating to the edible fats and oils industry.

International Society of Beverage Technologists: http://www.bevtech.org/. Organization of individuals whose interest is the technical and scientific aspects of soft drinks and beverages.

International Union of Pure and Applied Chemistry: http://iupac.org/. Serves to advance the chemical sciences and contribute to the application of chemistry worldwide.

Juice Products Association: http://www.juiceproducts.org/. Trade association representing the fruit and juice products industry, including processors, packers, extractors, brokers, and marketers of fruit and vegetable juices, juice beverages, fruit jams, jellies and preserves, and similar products, as well as industry suppliers and food testing laboratories.

Manufacturers Association for Plastics Processors: http://www.mappinc.com/. Nonprofit trade association that provides plastics processing training, industry networking opportunities, cost reduction programs, lead generation strategies and opportunities, operational benchmarking statistics, and plastics news.

National Association of Flavors and Food-Ingredient Systems: http://www.naffs.org/. A broad-based trade association of manufacturers, processors, and suppliers of fruits, flavors, syrups, stabilizers, emulsifiers, colors, sweeteners, cocoa, and related food ingredients.

National Association of Margarine Manufacturers: http://www.margarine.org/. Association serving health-conscious consumers and the margarine industry.

National Coffee Association: http://www.ncausa.org/. Champions the well-being of the U.S. coffee industry in the context of the global coffee community.

National Glass Association: www.glass.org/. Trade association representing the flat (architectural and automotive) glass industry.

National Institute of Oilseed Products: http://www.oilseed.org/. International trade association that promotes the general business welfare of persons, firms, and corporations engaged in the buying, selling, processing, shipping, storage, and use of vegetable oils and raw materials.

National Renderers Association: http://nationalrenderers.org/. Members of this association are all in the business of rendering (i.e., transforming waste from the meat industry into useable products for animal feeds and technical use).

Northwest Food Processors Association: http://www.nwfpa.org/. Regional trade association for the fruit and vegetable processing industry.

NPCA/FSCT: Merger in progress at press time between National Paint and Coatings Association (http://www.paint.org) and Federation of Societies for Coatings Technology (http://www.coatingstech.org). Advances the paint and coatings industry through product stewardship, advocacy, science and technology, and essential business information.

Peanut and Tree Nut Processors Association: http://www.ptnpa.org/. Provides a forum for processors, manufacturers, and suppliers of peanuts, tree nuts, and related products.

Plastics Institute of America: http://www.plasticsinstitute.org/. Educational and research organization dedicated to providing service to the plastics industries. Supports, fosters, and guides plastics education and research.

Powder Coatings Institute: http://www.powdercoating.org/. Represents the North American powder coating industry, promotes powder coating technology, and communicates the benefits of powder coating to manufacturers, consumers, and government.

Rubber Manufacturers Association: http://www.rma.org/. An international trade association for the elastomer products industry.

Snack Food Association: http://www.sfa.org/. International trade association of the snack food industry representing snack manufacturers and suppliers.

Soap and Detergent Association: http://www.cleaning101.com/. Nonprofit representing manufacturers of household, industrial, and institutional cleaning products; their ingredients and finished packaging; and oleochemical producers.

Society of Cosmetic Chemists: http://www.scconline.org/. Furthers the interests and recognition of cosmetic scientists; seeks to maintain the confidence of the public in the cosmetic and toiletries industry.

Society of Petroleum Engineers: http://www.spe.org/. Provides information of interest to engineers, scientists, and managers working in the oil and gas exploration and production industry.

Society of Plastics Engineers: http://www.4spe.org/. Promotes scientific and engineering knowledge relating to plastics. Provides access to technical resources: plastics library, technical journals, magazines, and forums.

Society of the Plastics Industry: http://www.plasticsindustry.org/. Trade association representing the entire plastics industry supply chain, including processors, machinery and equipment manufacturers, and raw materials suppliers.

References

Albazzaz, H., and Wang, X. Z. (2007). "Introduction of Dynamics to an Approach for Batch Process Monitoring Using Independent Component Analysis." *Chemical Engineering Communication*, February, vol. 194, no. 2, pp. 218–233.

Besseris, G. J. (2009). "Multi-Response Unreplicated-Saturated Taguchi Designs and Super-Ranking in Food Formulation Improvement." *International Journal of Quality and Reliability Management*, vol. 26, no. 4, pp. 341–368.

Bingham, R. S., Jr., and Walden, C. H. (1988). "Process Industries," Chapter 28 (pp. 28.1–28.63) in *Juran's Quality Control Handbook*, 4th ed., J. M. Juran and F. M. Gryna (eds.). McGraw-Hill, New York.

Box, G. E. P., and Draper, N. R. (1969). *Evolutionary Operation: A Statistical Method for Process Improvement*. John Wiley & Sons, New York.

Box, G. E. P., Lunceño, A., and del Carmen Paniagua-Quiñones, M. (2009). *Statistical Control by Monitoring and Adjustment*, 2nd ed. John Wiley & Sons, New York

Bowles, M. L., and Montgomery, D. C. (1997). "How to Formulate the Ultimate Margarita: A Tutorial on Experiments with Mixtures." *Quality Engineering*, vol. 10, no. 2, pp. 239–53.

Bowman, C. (2008). "Generic Strategies: A Substitute for Thinking?" *360° The Ashridge Journal*, Spring, pp. 6–11.

Brandl, D. (2006). "Design Patterns for Flexible Manufacturing." *International Society of Automation*, Research Triangle Park, North Carolina p. 205.

Brat, I. (2009). "High-Fiber Foods May Be Easier to Stomach This Time Around." *The Wall Street Journal*, August 20, pp. D1–D2.

Brooks, R. (2009). "Geometric Process Control Moves into Batch Processing." *Process Engineering*, March/April, vol. 90, no. 2, pp. 21–22.

Christensen, C. M., and Raynor, M. E. (1997). *The Innovator's Solution: Creating and Sustaining Successful Growth.* Harvard Business School Publishing Corporation, Boston, MA.

Cinar, A., Parulekar, S. J., Undev, C., and Birol, G. (2003). *Batch Fermentation: Modeling, Monitoring, and Control*. Marcel Dekker, Inc., New York.

Cornell, J. (2002). *Experiments with Mixtures: Designs, Models, and the Analysis of Mixture Data*, 3rd ed. Wiley Series in Probability & Statistics. John Wiley & Sons, New York.

Doble, M., Kruthiventi, A. K., and Gaikar, V. G. (eds.) (2004). *Biotransformations and Bioprocesses*. Marcel Dekker, Inc., New York

Garvin, D. A. (1987). "Competing on the Eight Dimensions of Quality." *Harvard Business Review*, November–December.

Hall, J. M., and Johnson, M. E. (2009). "When Should a Process Be Art, Not Science." *Harvard Business Review*, March.

Hiltscher, G., Mühlthaler, W., and Smits, J. (2003). *Industrial Pigging Technology: Fundamentals, Components, Applications*. Wiley-VCH, Weinheim, Germany.

James, D. (2006). "Best of the Batch." *IET Manufacturing Engineer*, August/September, vol. 85, no. 4 (pp. 32–37).

Jørgensen, S. B., Bonné, D., and Gregersen, L. (2006). Monitoring and Control of Batch Processes, Chapter 11 (pp. 419–462) in *Batch Processes*, Ekaterini Korovessi and Andreas A. Linninger (eds.). CRC Press, Taylor & Francis Group, LLC, Boca Raton, FL.

Juran, J. M., and Godfrey, A. B. (1999). *Juran's Quality Handbook*, 5th ed. McGraw-Hill, New York.

Levin, M. (2002). *Pharmaceutical Process Scale-Up*. 2nd ed., Marcel Dekker, Inc., New York.

Levinson, D. R. (2009). "Traceability in Food Supply Chain." U.S. Department of Health and Human Services, Office of Inspector General, March, Report OEI-02-06-00210.

Li, C., Zhang, J., and Wang, G. (2007). "Batch-to-Batch Optimal Control of Batch Processes Based on Recursively Updated Nonlinear Partial Least Squares Models." *Chemical Engineering Communications*, March, vol. 194, no. 3, pp. 261–279.

McAllister, E. W. (2005). *Pipeline Rules of Thumb Handbook: Quick and Accurate Solutions to Your Everyday Pipeline Problems*, 6th ed. Gulf Professional Publishing (Elsevier), Burlington, MA.

McConville, F. X. (2006). *The Pilot Plant Real Book*, 2nd ed. FXM Engineering and Design, Worcester, MA.

Méndez, C. A., Cerdá, J., Grossmann, I. E., Harjunkoski, I., and Fahl, M. (2006). "State-of-the-Art Review of Optimization Methods for Short-Term Scheduling of Batch Processes." *Computers & Chemical Engineering*, vol. 30, no. 6/7, pp. 913–946.

Nomikos, P., and MacGregor, J. F. (1995). "Multivariate SPC Charts for Monitoring Batch Processes." *Technometrics*, February, vol. 37, no. 1, pp. 41–59.

Olsson, F., Weeda, T., and Bode Matz, P. C. (2008). *Retail Pharmacy Recall Manual*, The Food Institute, Elmwood Park, NJ.

Olsson, F., Weeda, T., and Bode Matz, P. C. (2009). *Food Products Recall Manual*, 3rd ed. Food Institution Information and Research Center, Elmwood Park, NJ.

Porter M. (1980). *Competitive Strategy: Techniques for Analyzing Industries and Competitors*. The Free Press, New York.

Porter M. (1996). "What Is Strategy?" *Harvard Business Review*, November–December, pp. 61–78.

Robinson, T. J., Borror, C. M., and Myers, R. H. (2004). "Robust Parameter Design: A Review." *Quality and Reliability Engineering International*, vol. 20, no. 1, pp. 81–101.

Schoem, A. H. (2005). "Preparing for and Conducting a Product Recall." *Product Safety and Reliability Reporter*, vol. 33, no. 4, pp. 107–109.

Valentas, K. J., Clark, J. P., and Levin, L. (1990). *Food Processing Operations and Scale-Up*. CRC Press, Boca Raton, FL.

WHO (2003). Substandard and Counterfeit Medicines. Fact Sheet No. 275, November.

Zlokarnik, M. (2006). *Scale-Up in Chemical Engineering*, 2nd ed. Wiley-VCH, Weinheim, Germany.

CHAPTER 25

Defense-Based Organizations: Assuring No Doubt About Performance

Alexander Eksir

About This Chapter

When it comes to maintaining and measuring quality, the defense industry is unlike any other. It still uses legacy products designed in the 1950s, yet the men and women who possess the knowledge associated with those systems are retiring in droves. At the same time, the defense industry is experiencing an increase in demand for leading-edge technology to fight today's guerilla-style wars. Given these rapidly evolving demographics and technologies, the defense industry is in a unique environment of transition—from the ways of old to the ways of tomorrow. During this transition, quality must never suffer, as the national security of the United States and its allies is at stake. What allows defense organizations to achieve and sustain profoundly reliable organization performance at this time of transition is the continuous investment in what we call "NoDoubt Mission Assurance"—an overarching quality mindset.*

*The NoDoubt phrase is meant for an internal Raytheon audience only. It is designed to reinforce among Raytheon employees the commitment behind its products.

High Points of This Chapter

1. Quality standards driven by the pursuit of market share differ from the quality standards driven by regulatory compliance.
2. Mission Assurance means that weapon systems will work whenever and wherever they are needed, every single time—and for as long as they are needed.
3. The Mission Assurance organization model consists of four key areas of investment: people, organizational systems and structure, customer relations, and innovation.
4. At the intersection of these four areas of investment lies a "sweet spot" that defense organizations must continually measure in order to recalibrate future investments.
5. The implementation of methodologies such as Six Sigma and Lean enable defense organizations to attain Mission Assurance.

Introduction

Over the past few decades, quality standards guiding manufacturing organizations in the United States, as well as in the international community, have undergone various life cycles. The manufacturing industry, for instance, once embraced theories contending that statistics could be used to control manufacturing processes and minimize variability in design processes. Those theories held true until the advent of automation, which all but turned quality on its head. With automation, many of the manufacturing processes and some designs that were once performed by humans are now handled by robots. In fact, the precision of these robots is so high that it is virtually impossible to measure how precise they really are—then again, these systems are designed and built by humans.

Given the accelerating pace of technology, quality standards most assuredly will continue to evolve. What will the future of quality look like? More specifically, what will the future of quality look like in the defense industry? Before those questions can be answered, we must first reflect on what we have learned over the past 30 years.

Quality Standards Driven by the Pursuit of Market Share

For most of the mid-twentieth century, there have been two major types of quality standards: (1) those followed by organizations seeking to gain market share, and (2) those followed by organizations operating in a regulatory environment.

Manufacturers driven by market share (e.g., automotive industry, consumer electronics industry) were held accountable to a less-than-demanding set of quality principles and processes. Today, these same organizations are bound by far more rigid quality control processes.

The application of quality here is less centered on compliance and more centered on applied science. Essentially, the goal is to eliminate as many variables out of the manufacturing process equation as possible, whether it deals with design, human performance, or materials. Doing so means the organization can predict performance more accurately, which in turn, enables it to produce goods that meet their target faster and at lower cost. As a result, the organization can compete effectively against those in the marketplace that do not have the same efficiency or effectiveness.

This game of quality one-upmanship has taken place in virtually every industry—from clothing to furniture to consumer electronics. Products that were once expensive, slow, and heavy have evolved into affordable, quick, and light products. A perfect example is the cell phone. Cell phones were expensive and unreliable only 15 years ago; today, everyone has a powerful, convenient and reliable cell phone in their pocket. Another case in point: During the oil crisis of the 1970s, the U.S. automotive industry was challenged by the Japanese. The United States never expected the Japanese to deliver such a consumer-friendly, reliable product to the market. Since then, the U.S. automotive market landscape has never been the same.

Quality Standards Governed by Regulatory Compliance

Organizations that operate in the regulatory environment (e.g., pharmaceutical, aerospace and transportation industries) follow a second set of quality standards. This model of quality has always been driven by governance and compliance.

Compliance to a quality management system is defined under very detailed policies and practices. These practices are validated and reinforced through audits conducted internally or by third parties to preserve a certain level of certification. That certification then gives customers and consumers a basic confidence in the reliability of the product and the manufacturer.

Needless to say, organizations that operate in the regulatory environment understand the major risk and liability that a faulty product brings—no matter what industry they might be in. An airliner that crashes would certainly produce immediate worldwide news coverage, as would a submarine that malfunctions, a pill that poisons people who ingest it, an *Escherichia coli*–related food recall or a defective pacemaker. Therefore, the rigor of quality compliance is significantly higher for these organizations than organizations that need quality to simply gain market share.

Improved Speed of Execution

Another variable in the quality equation that has changed over time is the speed of communication, and, therefore, the speed of execution. Simply put, workers receive the information they need far more rapidly today than ever before. There was a time not so long ago when critical information such as blueprints, manufacturing instructions, and inspection checklists were hand delivered. Now, information is distributed in real time and updated on a constant basis. In fact, being a paperless organization is a main industry practice today.

Thanks to websites, blogs, and social networking sites, knowledge that was once boxed up and locked in someone's drawer is now online for everyone to access. For instance, calculating a complex mathematical equation once demanded several meetings with the organization's resident mathematician. Today, an employee only has to search for the mathematical formula online. In minutes, they can accomplish what used to require days or weeks.

Quality in the Defense Industry

When it comes to maintaining and measuring quality, the defense industry is a bit of an anomaly. It is such an interesting study because it still uses legacy products designed in the 1950s—and with good reason. Despite their age, these systems work perfectly well and should continue to carry out their missions for the next two or three decades.

Legacy systems that are subject to obsolescence and technology refreshes are also in the mix. Moreover, the men and women who possess the knowledge associated with those systems are aging. In fact, older experts are retiring from the defense industry (and every other industry for that matter) in droves. Their exodus forces defense organizations to strategize differently to sustain their organization's performance and systems availability.

At the same time, the defense industry is also experiencing an increase in demand for leading-edge technology. This surge is due to the changing ways in which wars are waged—from a traditional combat environment to one that emphasizes guerilla warfare, intelligence, and information assurance. In other words, new weapons systems are needed to fight a new type of enemy.

Therefore, in the past 10 to 20 years, there has been a melding of complex new systems and aging legacy systems. But whether it is a new or old system, its quality must be so perfect that it can hardly be measured. What makes this particularly challenging is that the legacy products were built and delivered to sampling plants that, by design, accepted one or two bad units out of a lot of 100.

Given the rapidly evolving demographics and technologies outlined above, the defense industry is in a unique environment of transition—from the ways of old to the ways of tomorrow. During this transition, quality must never suffer. Simply put, systems produced by defense organizations are directly related to the national security of the United States and its allies, and, for that reason, the quality that these organizations promise—and deliver—must be flawless today and always.

Defining Sustainability within the Defense Industry

Long-term sustainability is what every defense organization strives to achieve, and what every customer seeking defense-related systems and solutions demands. Although the sustainability of any product or organization may seem like an obvious requirement, it is particularly important in the defense industry.

In certain industries, product technology is refreshed at a breakneck pace to meet the needs of a very dynamic global consumer base. Take the consumer electronics industry, for example. When a cell phone is put into distribution, there is little need for the manufacturer to follow up with the person who purchased the phone. Yes, the cell phone comes with a warranty, but how many people actually trigger the warranty for a cell phone?

Sustainability to that manufacturer, therefore, is less about maintaining an existing product's quality and more about maintaining its ability to produce more goods. To that end, having a firm infrastructure plan is king to many manufacturers. If my factory burns down tomorrow, how long before another one can be built and set up? Do I have access to another existing factory where I can continue to manufacture my products in the interim? What happens if an entire line goes down due to malfunction? These are the types of challenges manufacturers face.

Sustainability in the defense industry, on the other hand, is all about maintaining a product for 30 to 40 years after it has been shipped. Let us consider a complex defense system such as radar. It must perform flawlessly every second of every day or else the lives of people—perhaps a great many people—might be in jeopardy. To attain that level of long-term sustainability, there are numerous variables that must continuously be accounted for after the radar has been shipped, including weather, operators, aging parts, and training.

For this reason, defense organizations are constantly assessing and tweaking the four main areas of investment: people, organizational systems and structure, customer relations, and innovation. Continuous development of each of these categories is necessary to preserve the knowledge and wisdom that it takes to attain long-term sustainability.

NoDoubt: A Mindset for an Entire Industry

Because any failure in the defense industry can have potentially catastrophic consequences, an expression was created to keep people continuously focused on the goal of flawless quality performance—called NoDoubt. (NoDoubt is a registered trademark of Raytheon Company.)

NoDoubt does not refer to any specific numerical standard or measurement; instead, it is an overarching mindset. It means that weapon systems will work whenever and wherever they are needed, every single time—and for as long as they are needed.

In the world of national defense, being 99 percent reliable is disastrous. That is why the notion of NoDoubt has been readily embraced by manufacturers and end users of mission-critical equipment worldwide. Moreover, NoDoubt has been embraced by individual workers who bolster defense organizations from the ground up. These people must have NoDoubt about themselves. They must have complete clarity of mission, meaning they must understand how their work contributes to the end user, the country, the customer, and the organization. These workers must know where they fit in the organization and exactly what is expected of them in terms of performance. These workers must know how to reach back into the organization to access resources, organizational systems, and the support they need to get the job done.

Achieving a complete Mission Assurance buy-in is not easy, however. It cannot be accomplished through simple emotional engagement or by deploying a set of tools or procedures. What the buy-in requires is continuous investment in four specific key areas:

1. People
2. Organizational systems and structure
3. Customer relations
4. Innovation

Derivation of the Mission Assurance Organization Model

The Mission Assurance organization model that is explained here was based on years of exhaustive research, which was conducted as a way of gaining a greater understanding of the differences among organizations in the defense industry. Throughout the research, I visited dozens of successful organizations in the defense industry and conducted no less than 38 comprehensive interviews with upper-level executives. The interviewees came from a variety of socioeconomic and ethnic backgrounds and varied greatly in age and years of experience in the industry. The findings are based on the feedback received from these interviews as well as my career. During 25 years as an organization executive, I traveled to 46 countries, studying the best and worst manufacturing processes the world has to offer. From these experiences I gained a comprehensive knowledge base on the topics of Mission Assurance and long-term sustainability.

Four Areas of Investment

As noted earlier, the Mission Assurance model is based on four areas of investment: people, organizational systems and structure, customer relations, and innovation (see Figure 25.1). The degree to which defense organizations commit to each of these four areas usually determines how organizations differ from one another. Now let us further define these investment areas.

People (also known as Human Capital)

In this rapidly changing global environment, the value of people cannot be understated, for it is the attitude and commitment among people that is the overriding factor in most defense industry strategies. In fact, Raytheon Integrated Defense Systems president Daniel L. Smith often says that "the root cause of everything good and everything bad has a temperature of 98.6 degrees."

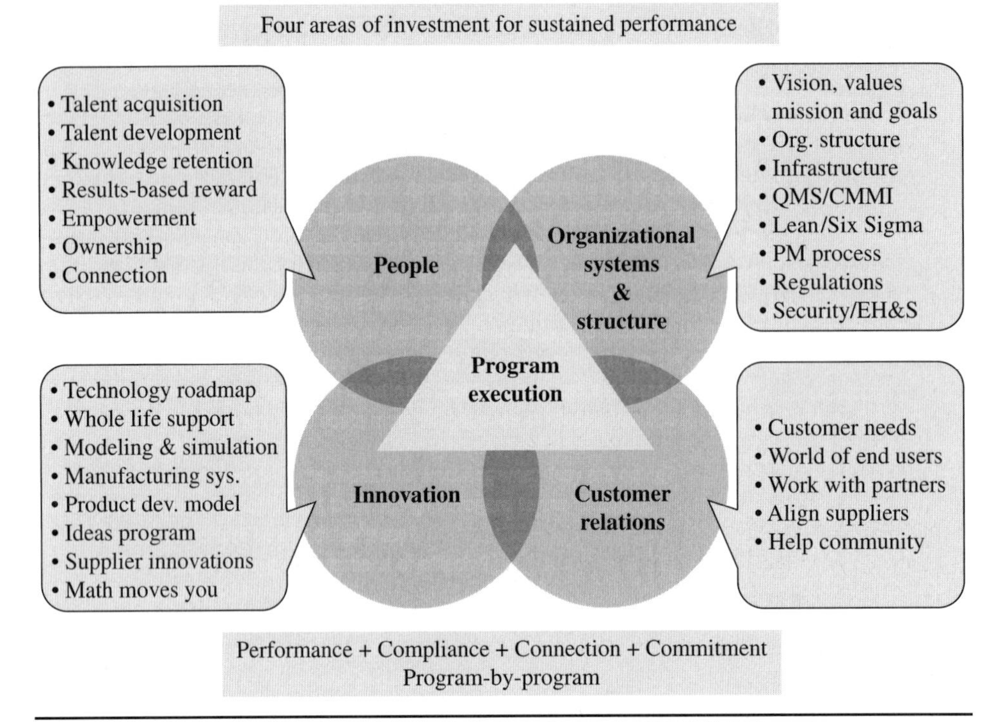

FIGURE 25.1 Four areas of investment for sustained performance. (*Eksir 2008.*)

Examples of investments in this area include communication, education, and publication—in essence, any action that effectively connects people to the organization. Unfortunately, organizations must contend with the fact that the world's industrialized economies are currently facing an aging workforce. To that end, as older workers retire, the wisdom that they have accumulated over their careers might be lost. The challenge is to capture their knowledge and then replace their commitment with incoming employees.

Developing a Mission Assurance culture based on inclusiveness and results— moreover, one that is impervious to an aging workforce—requires a consistent and comprehensive investment in people and human resource processes. This might include a significant investment in recognition and rewards programs, as well as training and development. Some defense organizations also invest in cultural initiatives designed to foster an environment of trust, inclusiveness, and greater teamwork.

Organizational Systems and Structure

The second area of investment addresses fixed and variable assets that provide a grounded work environment for the successful execution of industry programs, including an organization's infrastructure, physical plant, key processes, and information systems. Other investments that fall under this category are policies and procedures (e.g., quality management system), continuous improvement initiatives (e.g., Six Sigma, Lean, and Capability Maturity Model Integration) and the necessary tool sets and capital that enable people to perform their work more effectively and efficiently.

Customer Relations

The third area of investment deals with connecting employees with the customers. This works both ways, as it is just as vital for employees to understand the world of its customers as it is for customers to understand a particular defense organization's Mission Assurance strategy.

These customer connections can be forged in a myriad of ways. Organizationwide events and regular communications from leadership team members are effective ways of reminding employees why they come to work each day. On a larger scale, annual supplier conferences and international conferences and symposia are examples of how global partners and customers might connect to an organization's Mission Assurance strategy.

If based on performance, customer relationships often lead to trust and respect, which then may evolve into a high degree of intimacy. As a result, a deeper relationship is forged, one where both parties work collaboratively toward a mutual benefit.

Innovation

The fourth area of investment involves research and development, for without constant innovation, defense organizations cannot consistently deliver affordable capabilities to its customers. Defense organizations often have dedicated teams focused on advancing its domain knowledge in targeted areas of interest. Patented technologies, new workflow processes, and groundbreaking products and technologies often arise from these dedicated innovation teams.

A major source of innovation is triggered by the modeling and simulation of mission environments. In these environments, proposed solutions are played out by computerized models that apply mathematical equations and rules-based systems. This type of simulation is conducted prior to finalizing architectural concepts or investing in material or capital. Exposure to simulated conditions helps customers understand how the theoretical provisions in written form may perform under real-life stressors. With the help of this realistic computerized model, customers can more accurately finalize concepts, budgets, capabilities, and opportunities.

Six Sigma and Lean: Enablers of Mission Assurance

Some tools or programs that organizations invest in are so comprehensive that they stretch across all four investment quadrants. Six Sigma and Lean are two such examples.

Over the last three decades, U.S. manufacturers have slowly begun to adapt and apply the principles inherent in these two methodologies. Although the methodologies are similar in that they are both designed to improve productivity by reducing quality-related problems, Six Sigma and Lean are also quite distinct.

Six Sigma introduces to manufacturing organizations a means of using statistics to increase product yield at a lower overall cost, thereby increasing value to the customer. On top of its statistical conventions, Six Sigma is also a way of thinking and behaving with ultimate productivity in mind. When applied and followed, Six Sigma helps organizations better manage—and therefore better predict—process performance.

Six Sigma was first adopted by organizations that move quickly from one generation of products to the next. Soon thereafter, transactional industries realized that Six Sigma is not reserved just for operations that manufacture millions of widgets at a time. That is when organizations like General Electric, Allied Signal, and Motorola took it on and adapted it across their entire organizations.

The defense industry, the polar opposite of a quick-turn manufacturing operation, was the last to hop on board the Six Sigma locomotive. Systems produced by defense organizations must perform flawlessly for decades, not months or years like a widget or a cell phone.

For this reason, the majority of defense organizations have adapted the Six Sigma principles to make them fit their needs. These organizations mix some standard Six Sigma tools with some of their own internal tools; the combination is what makes it unique to their organization. These organizations will even give Six Sigma a new name and brand it as such both internally and externally.

With Lean, much of the same applies. Lean was a very logical argument for high-production commercial manufacturers with moving parts—such as the automotive and consumer electronics industries—where the ability to minimize the material on hand is directly correlated with bottom-line results. To wit, when there is $200 million worth of material sitting in the storeroom, it is not in a bank collecting interest.

This principle was not understood early on by the defense industry, where inventory is considered to be a revenue generator. In other words, whereas a commercial organization will not get paid until the material is converted into a product and then shipped, a defense organization will. As soon as a defense organization buys sheet metal, for instance, it immediately invoices the customer, even though the material may not be used for years. That is because the material might be needed to address legacy design issues or produce spares.

Consequently, the defense industry has chosen to expand Lean in areas outside of the manufacturing environment, as well. For instance, the defense industry now utilizes Lean in a more systemic way—to help increase space use, optimize error prevention, and reduce organizational layers, among other purposes. Such holistic applications yield very tangible organization results quarter over quarter.

To reiterate, the key point in this discussion about Six Sigma and Lean is the fact that the defense industry arrived at the following conclusion: even in a low-volume highly complex systems environment where it takes years to design and build a product and decades to support it, both methodologies still have very real applications.

As mentioned earlier, both Six Sigma and Lean fit into all four areas of investment in the Mission Assurance organization model. The structural training required to implement Six Sigma and Lean (e.g., documentation, logistics, and policy) falls under organizational systems and structure.

The part that aligns with people is cultural alignment, as a common language must be created for employees to reap the benefits of this organized learning module. Customer relations and innovation elements of the organization model result from the application of Six Sigma and Lean. To wit, when employees apply these methodologies, they are able to create more innovative solutions that enhance their relationships with customers. A common characteristic of solutions created using Six Sigma and Lean is efficiency, as both of these methodologies, if applied properly, often lead to improved cycle/lead time, fewer redundancies, fewer non-value-added steps and reduced cost. Many defense organizations also train their customers and subcontractors in Six Sigma and Lean, yet another way of creating relationships and common language.

Although the starting place for the investment of Six Sigma/Lean is in organizational systems and structure, it ultimately reaches all other investment areas. By enriching an organization's overall capabilities, it effectively integrates a program team. Therefore, the program team benefits from the enhanced learning, behavior, relationships, and innovative solutions—all of which leads to the execution and delivery of the best solution to the customer. When that is done, program by program, day after day, sustainable predictable performance is achieved.

People at the Nexus

Although "people" occupies a category all to itself, it can be argued—rather convincingly—that people reside at the core of all of these categories. When designing an organization—writing policy, setting up a building, implementing information technology—you rely on people to do it (organizational systems and structure). When establishing a relationship with a customer to understand the scope of their work, you deal with people (customer relations). When creating an environment that engages, embraces, and rewards innovation instead of rejecting far-reaching ideas out of hand, you use people (innovation). Therefore, each of these four investment areas truly hinges on people, or human capital.

With people occupying such a vital role in achieving Mission Assurance, it is paramount to nurture the "soft side" of employees. That is to say, things like culture, language, and understanding should all be made a priority. The flip side of that coin is the technical skills of an employee base, which also requires great emphasis at a mature defense organization. There needs to be a well-defined infrastructure for employees to tap into if they wish to increase their worth to an organization. For example, employees should know exactly how to sign up for additional job training courses or mentoring programs.

If a formalized system is in place that allows people to define their own careers, employees will intuitively know that their ascension had nothing to do with the whim of a supervisor or luck, for that matter. Instead, their success was due to the fact that they proactively acquired the proper skills and training.

This type of documented infrastructure system provides an organization with sustainability, which in turn, reduces fluctuation in the work force. Over time, an organization's sustainability breeds a strong sense of confidence—or NoDoubt—in its employees. Employees are confident that when they drive to work everything will be there—their building, their office, their computer, their e-mail, their job, etc. As a result, people spend less energy worrying about their job security and more time focusing on how they will solve another problem for the organization. In short, they become more productive and more loyal to the organization.

Creating a Sweet Spot

According to our Mission Assurance organization model, achieving sustainable organization results is no longer about implementing the right tools whether it is Six Sigma, Lean, or some other process. It is about properly architecting the four investment areas: people, organizational systems and structure, customer relationships, and innovations.

When these investment areas overlap with one another, a new section in the middle is created, which can be called the organization model's "sweet spot" (see Figure 25.2). The existence of a sweet spot means that the program manager can easily tap into all four investment

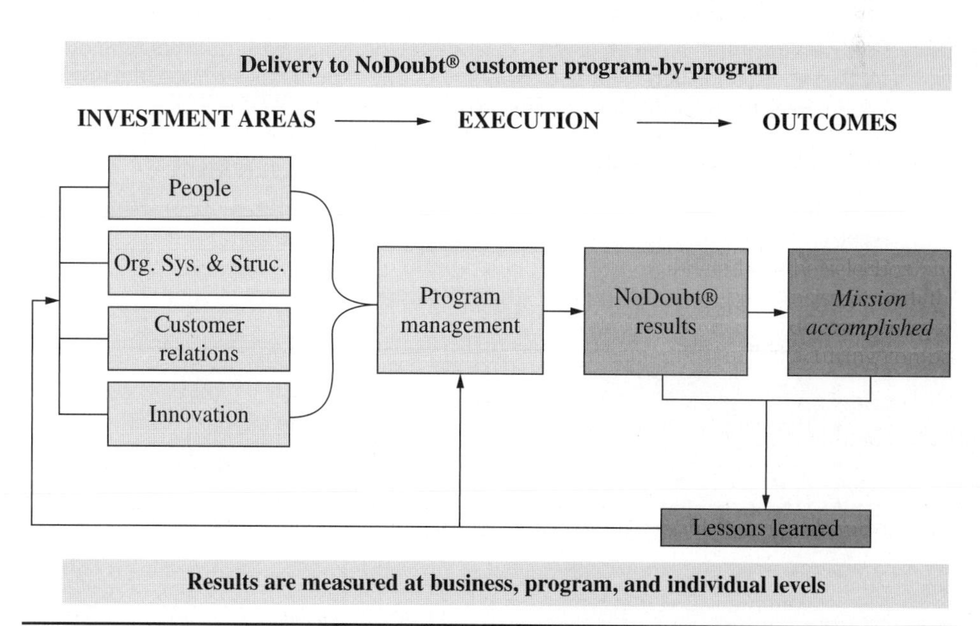

Figure 25.2 Delivery to NoDoubt customer program by program. (*Eksir 2008.*)

areas, providing him or her with access to all of the resources he or she needs to execute the program. In this way, the overall efficiency and effectiveness of the program is maximized and its objectives are attained.

Not coincidentally, the effective execution by a program manager within an organization's sweet spot also ties in perfectly with the four pillars of Mission Assurance: performance, compliance, connection, and commitment. To wit, when a program is executed flawlessly, it is performed to the letter of the contract (performance); it is compliant to all internal and external regulatory requirements (compliance); it successfully connects to all key constituents and customers (connection); and it benefits from the commitment of multiple cross-organization teams (commitment).

The Role of the Program Manager

Before moving ahead, it is important to explain why the role of the program manager is so vital to the Mission Assurance organization model as it pertains to the defense industry. The truth is that the defense industry is one of the few industries in which the only vehicle there is to deliver solutions to the market is program management. The defense industry has no direct distribution, no direct sales, no showrooms, and no direct mail campaigns. Solutions are purely delivered program by program.

Think of a program manager as a general contractor, if you will. A general contractor has many resources at his or her disposal: electricians, plumbers, masons, and so on. The job of a program manager is to use these people in the most efficient and effective manner possible to execute the job. The program manager also has four major areas of resources he or she can tap into: people, organizational systems and structure, customer relations, and innovation. Based on the contract agreed upon between the defense organization and the government, the program manager must determine how much of each area is needed to deliver a solution.

For instance, if human resources are required, the program manager need only estimate the number of people needed, along with their desired skill sets and security clearances, and make the request. If physical space or information technology is required, the program manager simply estimates what is needed. If political relationships are required, the program manager taps into the organization's existing relationships with the Pentagon or the White House. Finally, if the latest simulation and modeling expertise is required, the program manager accesses the organization's technical experts on staff.

Sustaining the Sweet Spot through Continuous Measurement

Given the enormous importance of the overlap section of the four investment areas—the sweet spot—how does an organization know if it has enough of each area? In other words, how does the organization know what to invest in? The answer is, by constantly measuring it. There are many ways for organizations to measure these areas if they choose:

- Customer organization review (CBR)—quarterly written feedback from the customer
- Verbal feedback from the customer
- Program health survey—should be conducted on an ongoing basis
- Quarterly organization review (QBR)—an organization's quarterly self-assessment
- HR assessment—usually conducted at the beginning of the year; helps identify where an organization's talent gaps are
- Strategic plan—the organization's overarching strategic plan is revised and communicated to its employees
- Annual operating plan—the organization's financial objectives are set for the following year

- Employee opinion survey
- Supplier and partners forum—ideal for candid interactions and feedback

All of this intelligence is designed to feed directly back into the organization. After the data have been analyzed, the organization should be able to identify where it stands in each investment area. Accordingly, the investment plan is then recalibrated based on the outstanding programs at hand. This allows the Mission Assurance sweet spot to continue, which in turn leads to sustainable organization results.

The Universality of the Mission Assurance Model

Note that the Mission Assurance organization model is easily applied to industries other than aerospace and defense. In fact, virtually all industries use the model already, simply because they all invest in the four key areas previously discussed. Most successful large-scale organizations hire and develop talent (people), operate in a brick-and-mortar environment and implement information systems (organizational systems and structure), foster relationships with their customers (customer relationships), and innovate new solutions (innovation).

The difference between defense organizations and, say, organizations in the pharmaceutical industry is simple awareness. In other words, organizations that gain a greater return on their investments are the ones that actually understand the model they are operating in. They very deliberately invest in these four key areas, put the program manager in the center, measure their investments, and then recalibrate their investment plan at year's end.

By following this pattern, most organizations within the defense industry have matured the model to a point where the white space buffering the four investment areas is all but eliminated, and a slight overlap is created. As a direct result, these organizations maximize the benefit of their matrix organizations and maximize the gains of the four areas of investment—year after year after year.

U.S. Military: A Partner in the Pursuit of Quality

With little to go on other than negative press clippings, many U.S. citizens feel as though quality in their military is lacking. Stories abound about cost overruns on military contracts, military equipment that fails when needed most, government employees who have lifetime employment, and most recently, the poor condition of certain military hospitals.

In reality, however, the quality of the U.S. military is improving at a faster rate than ever before. That is because the times demand it. As multiple wars are fought on multiple fronts, the need for improving the quality of military products and processes continues to grow. Simply put, there is no better time than now to energize the latent quality initiatives of the 1990s.

To that end, there are thousands of officers, soldiers, civil servants, and contractors who are dedicated to protecting the United States and its allies by improving the quality of equipment, systems, processes, and people. All across the military, innovative quality initiatives are now being implemented, wasteful processes are being eliminated, and value-added services are being added. "These are people who fully understand the critical importance of quality to the success of the mission and, therefore, have a personal commitment to ensuring quality and mission go hand-in-hand," said Greg Baroni, Chairman and CEO of Attain, LLC.

How has the military accomplished this improvement in quality? Where did it get its lead? The answer, in great part, is by learning lessons from defense suppliers and contractors such as Raytheon Co., Lockheed Martin Corp., and General Electric. These organizations were some of the earliest adopters and implementers of Six Sigma and Lean, performance programs that fit neatly into the military culture of using accurate and responsive information.

Based on its experience with best practices like Six Sigma and Lean, the defense industry, over the course of many years, has helped create a common understanding with the military. Defense organizations have helped the military understand the importance of long-term value to U.S. taxpayers and national security interests. Now, for the first time, the defense industry and the U.S. military are aligned on their perspectives of quality such that both bodies even use the same continuous improvement language in the same context. The result is greater trust, less miscommunication, and a more rapidly moving machine.

A prime example of the military's enhanced commitment to quality is the creation of the Missile Defense Agency's (MDA) Assurance Provisions (MAP) in 2004. Led by Randy Stone, director of Mission Assurance, Quality & Safety for the MDA, extensive government, industry, and academic research was initiated to research, collect, and create this new set of standards for MDA system development, delivery, and support. MAP is a collection of standardized and measurable requirements that runs across agency departments, programs, and suppliers. For all practical purposes, MAP provides a quality management system that exceeds some of the most comprehensive quality management systems adhered to by the defense industry.

The MDA invested millions of dollars to create a think-tank comprising the most revered quality experts from the defense industry and academia. The agency's goal was to create a set of standards even more comprehensive and stringent than AS 9100, ISO 9000, and the like. MDA wanted a document that would provide absolute Mission Assurance if each best practice—Lean, Six Sigma, and root cause corrective action—was followed to the letter. MAP was the ultimate result of that investment.

Although the defense industry can lay claim to helping create MAP, the important point is that the vision clearly originated from the MDA. This is a prime example of the U.S. military following the lead of the defense industry in terms of realizing the value of investing in quality and sustained performance and blazing its own path to improved quality.

A More Profound Plan

The world is changing. People like to talk about the first and second Industrial Revolutions, but it is clear that we are in a constant state of industrial revolution. We go through technological breakthroughs so fast that we simply do not have the time to name them. This is exactly why we need a deeper, more profound organization model in place for future generations to follow.

Gone are the days when organizations could wave the Six Sigma or Lean flags and expect them to help change, let alone sustain, their organization results. Yes, Six Sigma, Lean, and tools like it can enhance parts of an organization, but they are not cure-alls. A bigger model is needed for the future—a model that requires an intimate understanding of an organization's investment. That is only half the battle, however. That same investment must be continuously measured and then recalibrated to achieve superior organization results each and every year.

Acknowledgement

The author is grateful to John Cacciatore who contributed directly to this chapter.

References

Eksir, A. (2008). How Do Defense Organizations Understand and Measure Human Capital. Submitted in partial fulfillment of the requirements for the Qualitative Research Report in the Executive Doctor of Management Program at the Weatherhead School of Management, Case Western Reserve University, Cleveland, OH.

Eksir, A. (2009). The Influence of Team Health on Organization Outcomes.

SECTION IV

Key Functions: Your Role in Performance Excellence

CHAPTER 26

Empowering the Workforce to Tackle the "Useful Many" Processes

Mary Beth Edmond and Dennis J. Monroe

About This Chapter

When Dr. Juran provided a critique of the fifth edition of the handbook to the authors for this edition, he expressed concern that organization leaders did not do enough to involve the workforce to improve performance. As the creator of the Pareto principle, Dr. Juran often reminded management that they were responsible for most of the organization's problems so they should fix them. If 80 percent of the problems belong to management (the vital few problems), that leaves only 20 percent to be caused by the workforce. We feel that in the era of perfection someone needs to work on these "useful many" problems. Once again to honor Dr. Juran, this chapter focuses on the need to empower the workforce to gain this involvement and tackle the useful many. The purpose of this chapter is to present the core values, concepts, and models that have helped organizations effectively engage the workforce, thus improving performance and quality and enabling organizational sustainability. A capable and engaged workforce is the foundation for fostering a culture of customer focus and high-quality products and services.

High Points of This Chapter

1. The workforce is the second most important resource available to management. The first is the customer. The workforce is needed to deliver high-quality products and services. They need to be in a state of self-control to do this. Getting to this state requires their participation to define and carry out effective and efficient processes.
2. Gaining workforce engagement is often difficult for management. There are many reasons, such as lack of trust and the "here comes another one" mentality of managers.
3. Empowering the workforce will lead to a culture of high performance or what is often called a *quality culture*.
4. Participation happens through participation on teams, designing new departmental processes, education, communication, and workforce empowerment.
5. Workforce empowerment, the state of self-control, is a condition in which the workforce has the knowledge, skills, authority, and desire to decide and act within prescribed limits.
6. Promoting performance excellence across all processes in the organization is essential to employee empowerment. The role of management in leading change is a critical success factor to empower the workforce.
7. Management must become the champions of change and provide resources in difficult economic times; this is a great challenge that must be effectively met to succeed.

What Are Empowerment and Engagement?

The World Bank defines *empowerment* as the process of enhancing the capacity of individuals or groups to make choices and to transform those choices into desired actions and outcomes. Central to this process are actions that both build individual and collective assets and improve the efficiency and fairness of the organizational and institutional context that governs the use of these assets.

In 2007 to 2008 Towers Perrin conducted the Global Workforce Study on employee engagement and its implications for organizations. They defined employee *engagement* as the level of connection employees feel with their employer, as demonstrated by their willingness and ability to help the organization succeed, largely by providing discretionary effort on a sustained basis. The study showed that barely one in five employees (21 percent) is fully engaged on the job. And 8 percent are fully disengaged. This means that an overwhelming 71 percent of employees fall into what we have termed the *massive middle*.

The symptoms of a disempowering and disengaged culture are best described by David Gershon in his article "Changing Behaviors in Organizations: The Practice of Empowerment" (Gershon 2007). The symptoms include

- Blame the victim mentality
- Fear of making decisions
- Lack of participation in decision making
- New ideas not taken seriously
- Leaders' versus employees' mind set
- Distrust and cynicism

- Apathy and burnout
- Thoughts or feelings not freely expressed for fear of repercussion
- Learning and growth opportunities not being actively pursued
- Gossip and backbiting poisonous work environment
- People feeling unappreciated
- Lack of recognition for contributions
- Top talent leaving for better opportunities or work environment

As the Baby Boomers depart the workforce and a new breed of employees begins entering the workforce in the next decade, more and more organizations will have to remove these barriers to employee engagement. We feel confident that empowerment and engagement will be a requirement for the next generation in their selection of employers.

The *workforce* refers to all nonmanagerial employees actively involved in accomplishing the work of the organization. It includes the collection of all employees and all associated networks and structures within which they work together to make a collective contribution to performance excellence. The workforce carries out the organization's policies and procedures established as a means to ensure compliance with strategic and tactical plans. Whenever the workforce is asked to do more, leaders are hit with skepticism from all areas of the workforce. This skepticism is often the result of an organization that does not believe, trust, or want to support the leader's direction. Effectively managing for quality and, ultimately performance excellence, will create needed changes in policies, procedures, and processes that must be carried out by the workforce. Management must work with the workforce to implement those changes. But first managers must get the buy-in from workers. A lack of buy-in makes it difficult to apply the technical methods needed to improve performance. An organization needs the participation of the workforce to participate on teams, provide theories about the causes of problems, recommend changes to the processes, and then carryout the changes. The best way to gain buy-in is by demonstrating that involvement is worthwhile. As management rewards and recognizes the workforce for the participation and new processes, the workforce participates more and more. This cycle is important, and it must begin with management allowing time to work on project teams, learn new skills, and be placed in the state of self-control. The role of management is therefore to create a capable, empowered, and engaged workforce.

Workforce empowerment or, as Dr. Juran called it, the state of self-control is a condition in which the workforce has the knowledge, skills, authority, and desire to decide and act within prescribed limits. The workforce takes responsibility for the consequences of the actions and for contribution to the success of the organization.

In an empowered organization, the workforce takes action to respond to the needs and opportunities they face every day regarding customer satisfaction; safe operations; the quality of their service and products; safe environmental practices; business results; and continuous improvement of process, product, and people. The full potential of workforce empowerment is realized when the workers align their goals with organizational goals; have the opportunity and authority to maximize their contribution to the organization; are willing and able to take appropriate action; are committed to the organization's purpose; and have the means to achieve the overall mission and vision of the organization. An empowered workforce benefits from trusting relationships, good communication and information sharing, and performance accountability. Key factors contributing to empowerment are lifelong learning, recognition and reward systems, equal opportunities, and involvement in business decisions. Thus empowerment may be shown by the verbal equation

Empowerment = Alignment + Opportunity + Capability + Commitment

Alignment

Alignment refers to the consistency of plans, processes, information, resource decisions, actions, results, and analyses to support key organizational goals. Effective alignment requires a common understanding of purposes and goals combined with knowledge of customer and other stakeholder needs. Alignment also requires the use of complementary measures and information for planning, tracking, analysis, and improvement at three levels: organizational, key process, and the department or work unit.

Opportunity and Authority

For workers to have the opportunity to maximize their contribution, the organization must arrange affairs so that

- Individual authority, responsibility, and capability are consistent with having self-control.
- Barriers to successful exercise of authority have been removed.
- The necessary tools and support are in place.

The Ritz-Carlton Hotel Organization authorizes every employee to spend up to $2000 on the spot to resolve a customer problem so as to satisfy the customer. (This example of empowerment is most frequently invoked by employees working at the front desk.) At Walt Disney World and other Disney operations, cast members (Disney-talk for "employees") are authorized to replace lost tickets, spilled food, and damaged souvenirs, even if the damage was caused by the guest. In the Ritz-Carlton and Disney organizations, such empowerment is seen as a direct means to satisfy a customer and to strengthen the identification of the employee with the vision, mission, and values of the organization. In this sense, when workers have the authority and ability to directly solve customer problems of this sort, they acquire a sense of ownership of the organization.

Capability

Without capability, it can prove dangerous for the workforce to take some actions. Experiences have shown that peer feedback from fellow workers who have not been properly trained in giving feedback can be construed as harsh and not constructive. The organizational goals cannot be fulfilled if the workforce does not know what actions to take or how to take them. Therefore, the workforce must have the capability to achieve appropriate goals. An empowered workforce knows how to do what needs doing and has the skills and information to do it.

Training is a significant means of developing workforce capability. At Sunny Fresh Foods, employee training is central to every business function that the organization employs. A mandatory program aptly titled "Sunny Fresh University" is a prerequisite for all employees to ensure that all education and training is aligned with long-term organization goals. This program, retooled annually, facilitates cross-functional teams and standardizes processes through eight different courses, for the overall benefit of the firm and its employees. At another organization, Ritz-Carlton, all new employees undergo 48 hours of orientation training before they visit their workplace to begin work with customers face to face. After 21 days on the job, there is a 4-hour follow-up orientation, during which the first 21 days of experiences are reviewed and discussed in light of the organization's vision: "Ladies and gentlemen serving ladies and gentlemen."

Commitment

Workforce commitment is a key ingredient to performance and quality excellence. It is a state of mind that is evident when the workforce assumes responsibility for creating success and takes action to achieve that success.

There are two kinds of behavior that imply workforce commitment. The first behavior is evident when workers appear to be very single-minded in doing their work. The second behavior is the willingness of the workforce to make personal sacrifices for the achievement of team or organizational goals.

A committed workforce has a focus. Sometimes this focus is so intense that it seems as though the workforce has blinders on. Consider for a moment a group of project managers. Project managers typically work in a matrix organization. They have resources assigned to them across functional departments. Project managers are constantly competing for people, time, funding, and technology. For project managers to be successful, each one must believe that what she or he is doing is the most important task in the organization, and they cannot lose their commitment to achieve their project goals. If they do, they will not be successful and projects often fail.

Workers, like project managers, are naturally inclined to be committed to quality. Workers can and do produce quality despite poorly designed systems and processes. Workers can produce it without supervision, and they can produce it despite poor management. When workers feel they are valued members of the organization and they are appropriately recognized and rewarded, the organization will know that it has earned the commitment of the workforce.

One measure of commitment is the willingness of the workforce to make personal sacrifices to meet the commitment. Every worker has a limit, but it is impossible to predict what this limit is. Personal sacrifice must have meaning. People will not invest themselves in a task or project if it does not fit with their values and if it does not provide them with a sense of fulfillment. Workers will make sacrifices for whatever contributes to their sense of personal achievement.

The key message here is that if the organization wants workers to share the same commitment to the goals and objectives, then the organization must communicate those goals and objectives. Commitment follows meaning. A job is only a job until the worker identifies with it and shares in its meaning; then it becomes a commitment. Workers will not sacrifice unless their work has meaning and they feel that the work they are doing is connected to the larger whole. Workers need to know how they contribute to their organization's success and achievement of goals.

Workforce Capability and Engagement

The term "workforce capability" refers to the extent to which the workforce has the necessary knowledge, abilities, skills, and competencies to accomplish work processes. Capability may include the ability to build and sustain relationships with customers; to innovate and transition to new services, work processes, and technologies; and to meet changing business, market, and regulatory demands.

The term "workforce engagement" refers to the extent of workforce commitment, both emotional and intellectual, to accomplishing the work, mission, and vision of the organization. Organizations with high levels of workforce engagement are characterized by high-performing work environments in which people are motivated to do their utmost for the benefit of customers and for the success of the organization. Workforce engagement is also contingent upon building and sustaining relationships between senior executives, operational leaders, and customer groups.

Workers feel engaged when they find personal meaning and motivation in their work and when they receive positive workplace feedback and interpersonal support. Removing obstacles preventing an employee from "doing good work" is a form of feedback and support. An engaged workforce benefits from trusting relationships, good communication, a safe work environment, empowerment, and performance accountability. Key contributing factors to workforce engagement include continuous process excellence, research and innovation, outcomes management, and accountability models.

There is evidence that business success and attaining superior quality go hand in hand. The Malcolm Baldrige Criteria for Performance Excellence as well as other national awards for excellence include in their comprehensive framework the role of the workforce. Major elements of performance excellence are also embodied in other major state, regional, and national quality awards that relate directly to the workforce. Most national quality awards, like the U.S. Malcolm Baldrige Award, assign points associated with the workforce and include

5.1 Workforce engagements	45 points
5.2 Workforce environments	40 points
7.4 Workforce outcomes	70 points

All other criteria in the Malcolm Baldrige assessment rely heavily upon how well the organization has created and maintained an engaged, committed, and productive work environment for all members of the workforce. Total workforce involvement and commitment is necessary to create an ongoing culture of continuous performance and quality excellence.

Employee engagement is not new. There were many attempts to gain involvement to improve the workplace:

1970s: Quality circles
1980s: Employee involvement teams
1990s: Self-directed work teams
2000s: Departmental work flow teams

As organizations move into the second decade of this new millennium, the workforce is still the second most important resource an organization has. The first is the customers. The workforce in today's competitive landscape is more important than natural resources especially to service organizations. It is people that make a difference. Getting them motivated, educated, participating, and loyal is the task required of management.

Empowering the workforce will become more necessary as the next generation of workers—today's students—will need to gain more from their jobs than their parents. They are smarter. They want more time for themselves. They want to do a great job at work. They expect their employers to partner with them so both can succeed.

Building Blocks for Workforce Empowerment and Engagement

Core Values, Mission, and Vision

Organization values refer to the guiding principles and behaviors that are defined by an organization and that set the foundation for how the organization and the workforce are expected to operate. Values reflect and reinforce the desired culture of an organization. Values guide the decision making of every member of the workforce, help an organization accomplish its mission and attain its vision, and provide the foundation for integrating key performance requirements within a results-oriented organizational framework. Examples of values include visionary leadership, organizational and personal learning, valuing staff and partners, agility, focus on the future, managing for innovation, management by fact, focus on results and creating value, and systems perspective.

Strategic planning is a common activity in most organizations and the responsibility of the executive team. The strategic plan provides the workforce two things:

1. The organization's vision, goals, and strategies
2. The organization's core values

For commitment to the job to exist, the workforce must have a focus, and this is created by communicating the strategic vision, goals, and core values of the organization. At each level within the organization, these goals and values must be translated into the work and decisions made by each worker. One common method of communicating the strategic vision, goals, and values to the workforce is through town hall meetings where members of the executive team share this information. Many organizations display their core values on wall plaques scattered throughout the establishment.

Open Communication

The workforce must have unobstructed access to pertinent information. Premier Inc., a 2006 Baldrige Award winner, holds quarterly conference calls that are attended by all employees in which the current state of affairs and goals of the organization are discussed. In addition Premier provides its organization with "Monday Minutes," a weekly e-mail publication; holds consistent meetings with management; and promotes its comprehensive website, all geared toward keeping employees informed about pertinent issues, results, and changes at the organization.

Communications must be clear, timely, believable, and supported by data and facts. Workers must have information that once was thought not relevant to their jobs—"there is not a need to know." This includes information about the cost of product, cost of energy, time cost of money, waste levels, cost of waste, levels of customer satisfaction/dissatisfaction, cost per employee, earnings pressures, etc. In a performance and quality excellence system, workers are expected to be process managers, problem solvers, and decision makers. Open communications are needed because workers need information to make the day-to-day decisions. Without the information they cannot fulfill their roles.

The organization's vision, mission, and objectives should be clearly defined by senior management, and then clearly communicated throughout the organization. Pertinent issues should also be communicated, such as what will happen to future employment if a job becomes unnecessary due to improvements. Open communications are developed as the managers take steps to achieve freedom from fear, establish data-oriented decision processes, and create the habit of sharing business goals and results.

Cargill Corn Milling, a 2008 Baldrige Award winner for manufacturing, implements what they term a "two-way communication flow." This practice ensures that there are systematic processes in place that cater to open communication between the workforce and management at all times. Programs such as town hall–style quarterly meeting, conference calls, and department/functional area meetings promote active participation and knowledge by all. In addition annual tours of all plants are conducted during which people on the floor interact with upper management, share their own successes, and are briefed on how Cargill Corn Milling is performing overall.

Drive Out Fear

Ideas and feedback from the workforce are essential. These come only when workers feel they can give their comments without exposure to blame, reprisals, or other consequences administered by a capricious management. Whether well founded or not, fear that management will view any negative comment as adverse is a powerful disincentive for workers to provide suggestions, challenge the status quo, or offer accurate and honest feedback. This sort of fear in the organization inhibits workers from making improvement suggestions, as they fear that such suggestions will be viewed by management as criticism of managerial practices. Fear can also inhibit workers from striving for improved efficiency; they may believe that such improvements will result in elimination of their jobs. Of course, fear of reprisal to a worker who makes a mistake could result in the worker covering up a mistake and, e.g., shipping off-quality product to the customer.

Dr. Deming often spoke of the importance of driving out fear so that all people "could put in their best performance," unafraid of consequences (Deming 1986). Some organizations promote a proactive approach whereby the worker is requested to talk back to the system to challenge old procedures and to question results and management practices.

Dr. Juran estimated that upward of 80 percent of organizational problems are owned by management action and the remaining 20 percent attributable to the workforce. The implication is that the search for root causes of problems will, in upward of 80 percent of problems, lead to the management systems, procedures, policies, equipment, etc., under managerial control. For many traditional managers, this probing of problems may come too close to home. The potential embarrassment may prove too great and the departure from conditions of the past intolerable. The manager's reaction may be, consciously or not, to resist the activities of the problem-solving teams to the point of causing the teams to fear going further. Thus, it requires substantial effort and cultural change to replace this fear with open communication.

Scorecards and Data Orientation

Workers can best participate and work toward customer satisfaction and continuous improvement when they have knowledge of facts and data. Workers must know the facts and data regarding parameters such as costs, defect rates, and production and service capabilities, to be able to contribute to determining root causes, evaluating possible problem solutions, and making process improvements. Data orientation makes decisions objective and impersonal. Workers need easy access to these data. They also need training and coaching to help them understand the meaning of the data. To make proper interpretations and decisions, workers need to understand and apply statistical concepts, such as the theory of variation.

The concepts of variation, common causes, special causes, and root causes are at the very heart of performance and quality excellence systems. Many organizations work hard and unsuccessfully at solving problems because they do not understand these concepts. Recognizing this, U.S. organizations have accelerated training on statistical analysis skills and problem-solving skills.

Within well-developed performance and quality excellence efforts, workforce teams methodically and continuously improve their processes. Their work requires asking questions of the process and gathering and analyzing data to answer the questions and eliminate defects. The teams require training in how to ask the questions, gather the data, and analyze them to answer the questions. Trained and otherwise empowered, such teams are able to produce quality that is the best in their industry. Whether in manufacturing industries, service industries, health care industries, or government, the reliance on data is the same in all successful organizations.

Poudre Valley Health System (PVHS) utilizes its customers to obtain important data—feedback from their patients. PVHS conducts Avatar patient satisfaction surveys and community health surveys that accumulate data from two different factions of people, patients and the community. The Avatar survey focuses on the rankings of selected survey items escalating in importance to the customer. The survey also includes complaints and compliments given by current and former patients along with essential market data such as community health needs, health care service utilization, and consumer loyalties and preferences.

Transparency

Discretionary effort exerted toward improvement is an example of ownership behavior. To truly feel and behave as owners, workers need to know the goals of the business and how their work can contribute to the accomplishment of these goals. Furthermore, workers need to know how the organization is performing regarding these goals—they need to know the results of the business if they are to sustain their focus on the goals. Transparency has led organizations to post their performance results internally and externally on public websites for workers and consumers to access.

In the health care industry, the ultimate goal is to enable the public to access basic information about the health care they consume so that they become more informed purchasers. As a pool of price and quality information becomes available, we will see a day when consumers planning a hip replacement will be able to go online to a website provided by their insurers and review data on which hospitals in their plan perform hip replacements, what quality rating each hospital has received, how many surgeries have been performed in the last year, what the average total price range is in that facility, and what consumers could expect to pay out of pocket given their health plans.

Develop Trust

Trust is multidirectional within the organization. Management cannot expect to trust the message from the workforce unless the workforce trusts management not to punish the messenger. Managerial behaviors that encourage trust include

- Open and consistent communications—saying the same message to all listeners
- Honesty—telling the truth, even when it is awkward to do so
- Fairness—maintaining the same policy for everyone, especially in regard to pay scale, vacation, promotions, advancement opportunities, and the like
- Respect for the opinion of others—listening to people to understand their needs, ideas, and concerns; being open to feedback, such as from employee satisfaction surveys
- Participation—seeking active participation of those who will be affected, in both the planning and the execution of the change
- Integrity—being guided by a clear, consistent set of principles; saying what will be done; doing what was promised
- Social climate—creating a climate that fosters new habits and makes it easy for workers to change their points of view

Promises and agreement are demonstrated through actions. Aristotle said we are what we repeatedly do. Building trust takes a lot of consistent actions over a long time. Managerial behaviors that impede trust include

- Dishonesty—telling untruths or half-truths
- Fostering rumors—generating rumors, allowing rumors to persist, failing to provide information
- Isolating people—separating them physically, without adequate communication; separating them socially and psychologically by providing too little communication
- Breaking promises and agreements

Workers must respect and rely on one another. Fair treatment, honesty in relationships, and confidence in one another create trust.

Employee Stability

Employment stability is a worthwhile objective in a performance and quality excellence organization for many reasons. Principal among these is the protection of the organization's considerable training investment and preservation of the carefully developed atmosphere of trust on which performance and quality excellence is built. Further, as employment stability becomes more rarely available in the job marketplace, its promise becomes the more attractive for many job seekers. Thus, as employment stability is a worthy strategic element, it is worth examining ways to address two issues that affect it: the threat to individual jobs posed by improvement activity and cyclical employment fluctuation.

The Threat of "Improving" Jobs Away

As workers continuously strive to improve their work processes, some jobs will become unnecessary. Assurance of employment stability is essential before workers can be expected to work wholeheartedly toward continuous improvement. Management must make it very clear to workers that employment will not be terminated by the organization if their jobs are made unnecessary due to improvements. Retraining for other jobs will be provided as necessary.

Cyclical Employment Fluctuation

In many industries, the fluctuations in business activity imposed by the business cycle create a need to plan for employment stability of the permanent workforce. To cover peak activity and needs that cannot be met by the basic workforce, alternative employment programs can be used to supplement the permanent workforce. These alternatives include

- Use of overtime
- Use of temporary workers
- Use of contract workers
- A coordinated plan to make this all work together

Think of the business case for providing employment stability. It is common for an organization to spend thousands of dollars to train the workforce, then lose that training and experience when an improvement occurs that makes people redundant. Planning is needed to achieve employment stability. Achieving it is not easy. Not every organization can, but it is highly desirable for successful long-term performance and quality excellence.

Workforce Resistance

When these building blocks are not in place, it is difficult, if not impossible, for organizations to be successful in creating a culture of performance and quality excellence. As workers are pushed out of their comfort zones and daily rituals and expectations change, they tend to exhibit resistance through negative behaviors. This is especially true if workers are not committed and see no value to change. Resistance to change can occur for a wide variety of reasons, ranging from intellectual differences over ideas to deep-rooted psychological beliefs or biases.

Dealing with the Resistance

Change does not have to be a burdensome task for workers. It does, however, require motivation. The key is to recognize the level of workforce development and apply the appropriate strategy for dealing with their behaviors. Workers who are successful change agents have mastered the art of dealing with negative behaviors and have learned how to adjust their attitudes or turn negative behaviors into positive behaviors.

Change management models have become available to assist workforce leaders and change agents in successfully implementing a change program. As an example, the Situational Leadership Model characterizes "leadership style in terms of the amount of direction and of support that the leader gives to the workforce" (Blanchard, et al 1985). The premise of the model is that by adopting the right leadership style to the level of workforce development, work gets done, relationships develop, and workers gain a certain level of competence and commitment before they progress to the next level of development. At each level of development, a new system of practices is overlaid on those implemented at earlier levels. Each overlay of practices raises the level of workforce development. Within each level, workers experience a greater opportunity to develop their professional capabilities and are more motivated to align their performance with the goals of the organization and engage in the quality process.

Workforce Involvement with Performance Excellence

Many organizations are just beginning to ask their workers to achieve a new level of performance excellence. This change process requires leadership, teamwork, and clear expectations and goals. Organizations must confront the violations of the past and focus on those actions that will create a higher performance standard. For example, if the organization said yes to every urgent demand in the past, this behavior has contributed to poor quality and excess costs. It is important for the organization to communicate to workers the vision of how things are going to be in the future with specific, identifiable, and replicable actions.

Having the right people involved on teams that have an interest in the change process or the outcome of the proposed change is critical for the overall program to succeed. Once change management stakeholders are identified, key messages need to be communicated to each group. The most important message to deliver is, "What's in it for me?" Before change becomes effective, a reason is required and the workforce must be aligned with the organization's higher purpose.

Organizations experiencing a high return on their investments are matching the right team to the right situation. These organizations have found that teamwork has increased productivity, increased revenues, decreased absenteeism, decreased turnover, increased customer satisfaction, and increased organizational performance outcomes.

So, in a performance excellence organization, the workforce, structure, tasks, information, decision making, teamwork, and rewards must be carefully integrated into a total system. To be successful, the workforce leaders and change agents must receive lifelong training in management, technical, and social skills. A balance among these skills must exist.

Building Teamwork at the Top

In a complacent organization, teamwork at the top is not essential. In an organization driving toward performance excellence, teamwork is an essential core competency. Imagine the future when high-performing organizations recruit a core team of top executives, not just the CEO. Traditional succession planning will be replaced with extensive research to hire the right team to do the right job.

In his book *Good to Great,* Jim Collins states that executives who ignite transformation from good to great get the right people on the bus, the right people in the right seats, and the wrong people off the bus, and they build a superior executive team. Then they figure out how to take it someplace great (Collins 2001).

Developing Workforce Leaders

Best-performing organizations will become more skilled at building leaders—leaders who can create and communicate vision and strategies. Without enough leaders, the vision, communication, and empowerment that are at the heart of transformation will simply not happen well enough or fast enough to satisfy customer needs and expectations. The development of complex leadership skills emerges over decades, which is what we refer to as lifelong learning. Most leadership development takes place on the job. Best-performing organizations build consistent systems in which leaders are given the freedom and responsibility to work within the system. Leaders are self-disciplined people who do not need to be managed.

Transforming the Culture

"Cultures can facilitate adaption if they value performing well for an organization, if they really support competent leadership and management, if they encourage teamwork at the top, and if they demand a minimum of layers, bureaucracy, and interdependencies" (Kotter 1996). Organizations with adaptive cultures are very competitive and produce top-decile results. The environment is transparent, candid, more risk-tolerant, quick to respond, and constantly focused on process improvement. It requires workers who can withstand the pace and the constant force of change.

Sunny Fresh Foods, a U.S. Baldrige Award winner, drives its culture by clearly defining what they term their "core ideology." Broken down into two categories, core purpose and core values, their ideology aims to continually remind the employees of the organization why they are working and what they are working for. The core purpose deals directly with what the goal of the firm is—to be the supplier of choice to customers worldwide. The core values focus on more microlevel goals and tenets such as safety, ethics, customer service, and quality. Regardless of market conditions this ideology does not change or waver. The core ideology sets the organization culture at Sunny Fresh Food and inevitably makes it a better run, more productive organization.

Managing Organizational Behavior

Best-performing organizations will invest in training workforce leaders in managing organizational behavior through adaptive leadership styles and the use of organizational problem-solving techniques. Understanding resistance to change and managing individuals to become more than their individual selves will become a core competency of high-performance organizations.

Building Technical Skills

The technical system is concerned with the production and service requirements of the work process. The system includes such elements as process operations, equipment, methods, instrumentation, procedures, knowledge, tools, and techniques, and it provides for multi-skilled operators.

Integrating Information Systems

Information systems will be integrated into day-to-day operations of high-performing organizations. The tradition of distributing retrospective financial or quality data to a small number of staff on a monthly or quarterly basis will become a thing of the past. Transparency of information will be the norm, and all members of the organization will need data on customers, competitors, quality, finance, and operating performance. New performance feedback systems will replace legacy systems. The combination of accurate data from a number of internal and external sources, systemwide dissemination of information, and a willingness to deal honestly with performance feedback will drive the organization to move from complacency to a sense of urgency. A higher rate of urgency will create a culture in which the workforce is looking for problems and opportunities, thus further developing a culture of performance and quality excellence.

Identifying Key Performance Measures

Process and outcome measures will be routinely monitored and used to improve daily operations to meet customer demand for quality. Measures of performance should meet the following criteria:

- Workforce action significantly influences the key measure. Trends and changes in the data are traceable to worker behaviors.
- The measures are important to the customer and workers. Workers should be proud to tell their customers what they are measuring.
- The data that support the measure are simple to capture, analyze, and understand. Indexes that combine various measures into a single number are often too complicated to understand and thus not useful. The worker loses sight of how it influences a composite measure.
- The data are timely. The measure gives the worker adequate warning of impending trouble.
- There are ample data to make the measure statistically significant.

At Park Place Lexus, customer satisfaction exceeds the highest of standards. Lexus employs an Owner Satisfaction Index (OSI) that combines both sales and service performance with loyalty satisfaction ratings. Using this index, Park Place Lexus earned an Elite of Lexus Award for OSI rating along with receiving an Elite of Lexus standing for its performance in preowned department client satisfaction levels. Lexus standards denote this rank for any dealership that maintains a customer satisfaction level of 90 percent or higher. Park Place Lexus meets this benchmark and surpasses it. Internal benchmarks for Park Place Lexus reach an even higher rate of 95.1 percent satisfaction, a level that has been reached by Park Place Lexus for eight years running.

Implementing Controls

In a high-performance organization, controls will be in place to stay on course, adhere to standards, and prevent a change from planned performance. Workers will be trained on the tools of control to ensure improvement programs achieve or maintain desired results. Training in statistical control tools and methods will become critical to the mission and the interpretation of data common language among workers.

Results Commonly Achieved

- Improved quality and operations (fewer harm events, fewer defects and errors, increased productivity per employee, increased compliance with regulatory measures)
- Cost reduction (reduced operating, maintenance, and labor costs)
- Increased employee satisfaction (positive self-esteem, career path known, increased job satisfaction, increased teamwork)
- Increased customer satisfaction

Consistency in Performance Management

Performance management focuses on achieving the desired motivation and behaviors. Daniels and Rosen (1982) define performance management as "a systematic, data oriented approach to managing people at work that relies on positive reinforcement as the major way to maximize performance." The approach is based on the work of B. F. Skinner, whose studies revealed that behavior is a function of its consequences. The goal of performance management is to shift workers toward self-management so they can eventually assume responsibility for motivating their own behavior. It is essential to use positive reinforcement on workers when they are first learning new behaviors and performance areas is. Workers who are allowed to be more and more on their own will turn out to be self-motivated and committed to the achievement of organizational goals. A high-performing workforce is difficult to build but capable of sustaining a long-term competitive advantage.

The Role of Management Empowering the Workforce

The role of the manager is critical to the success of any organization's drive to empower the workforce and no less important to create an initiative to drive improved performance. All managers must do more than encourage others in the organization to achieve high levels of performance; they must lead the charge and serve as "champions of change" to be successful.

Become a Champion of Change

First, let's understand what is meant by the term "champion of change." Kaufman et al. (2003) define a change agent as "the individual or group with responsibility for leading and implementing an organizational change." This definition implies that the change agent is empowered by someone else or some entity within the organization.

For purposes of the discussion here, a broader definition will be used; i.e., a champion of change is anyone within the organization, at any level, who has the desire and accepts the responsibility for leading the change effort. The most effective change agents have proved to be those at the highest organizational levels because they have not only the desire but also the authority and power to promote beneficial change. Executive leaders do not need to be empowered or prompted as implied in the definition above—they are empowered through their organizational position—but will take on the role of Champion voluntarily to support the vision they have set for the organization.

The simple desire to promote change is insufficient in itself to be effective. The Champion must approach the achievement of organizational transformation carefully and methodically. This methodical approach is often referred to as *leading change*.

For many organizations managing for quality is a new concept, something that is not ingrained in the corporate culture. Therefore, cultural change is required to move from the old paradigm, in which the organization may have focused only on product or service quality, to a paradigm that focuses on organizationwide performance. The latter includes not only product and service creation processes, but also processes that have no direct impact on product or service quality. These processes are often referred to as *business processes*, e.g., accounts payable, forecasting, and price-setting processes.

If we accept the concept above that organizational change is often required to move the cultural paradigm to one of managing for quality, then the role of the Champion naturally follows as key to the success of this effort. As stated earlier, the most effective Champions come from all levels of management, the change leaders.

There are several elements of effectively leading change.

The Change Agent Must Lead from the Front

In the 1980s, J. M. Juran started the popular videotape series called "Juran on Quality Leadership." He said management "must be a leader, not a cheerleader." This speaks to the difference between leaders who are out front, leading the transformation of an organization trying to attain a quality-driven vision, and those who stay in the background, encouraging

Figure 26.1 Leading versus encouraging.

the operating forces to "go forth and improve" (see Figure 26.1). Without the leaders' visible, out-front participation in the change effort, it will be doomed to failure.

The Change Leader Must Establish a Clear Vision for the Change

Vision implies that the leader understands what the future should look like. Notice, however, that the opening statement of this paragraph couples the word "clear" with "vision." A vague idea of where the organization needs to go is inadequate and most likely will lead to failure. The vision must be clearly, concisely, and forcefully stated by the leader if the organization's transformation is to be successful. As stated by John Kotter, "Without an appropriate vision, a transformation effort can easily dissolve into a list of confusing, incompatible, and time-consuming projects that go in the wrong direction or nowhere at all."

The Vision for Change Must Be Clearly and Broadly Communicated to the Organization

If those who will ultimately be charged with carrying out the vision are unclear about what the vision is or, even worse, don't know it exists, the operational changes needed to achieve the vision will be poorly targeted. The leaders must communicate the vision to all levels from senior staff to shop and office floor workers and enlist their support in achieving the vision. This is sometimes referred to as change that takes place simultaneously top-down and bottom-up.

Achievement of the Vision Must Take Place Methodically

This means that the vision must be broken down into key strategies, strategic goals, annual goals, and projects, each leading to the attainment of the higher level, as shown in Figure 26.2. The change leader should chair the steering committee that will choose the specific projects that the organization plans to execute supporting the annual goals. He or she must also be the primary driver of the vision, strategy, and goals that the projects support.

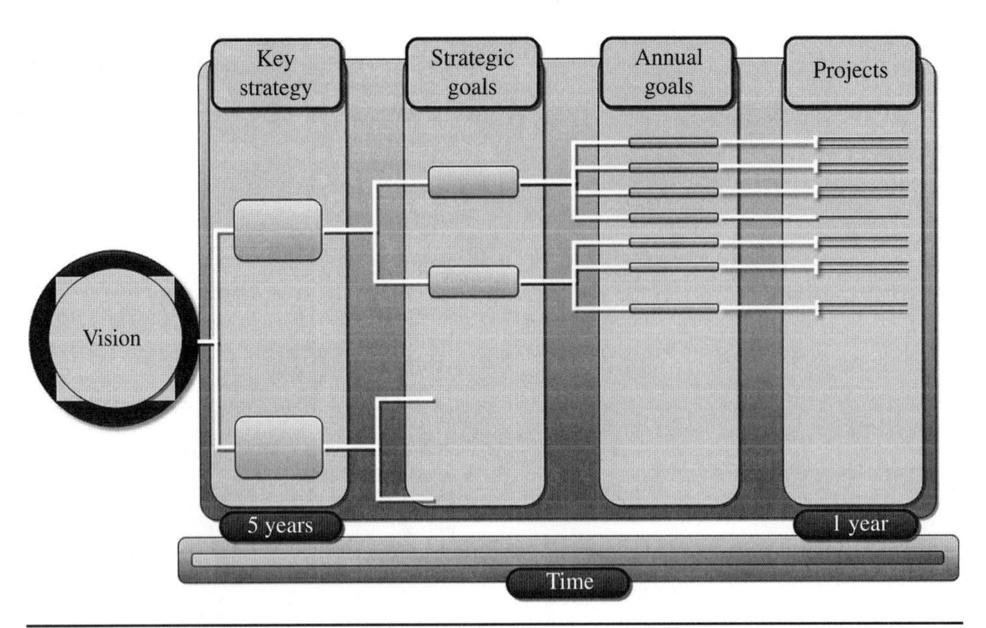

Figure 26.2 Deployment of the vision for change.

Change Agents Must Understand the Difference between Control and Breakthrough

Control, on one hand, is maintenance or restoration of the status quo. Breakthrough, on the other hand, implies changing the current reality to a new and significantly better one. "Control activity . . . will lead to static process performance at best, never getting better than it has been in the past. This might be a good thing if the past performance is at an acceptable level, but what if . . . it is not? This is where the idea of Breakthrough comes in. Breakthrough means change, a dynamic, decisive movement to new, higher levels of performance" (Juran 1995).

Providing Scarce Resources in Difficult Times

At the time of this writing, the United States and the world as a whole are suffering from the greatest economic downturn since the 1930s. This has led to many organizations trimming workforces to cut costs, and this has resulted in scarcity of resources to continue producing products and services in required amounts, let alone at high levels of quality. The scarcity has extended to resources other than human—scarcity of cash, capital improvement funds, credit, and so on. To continue operating effectively in the face of these constraints, leaders must look at creative ways to continue beneficial change and discontinue activities that are not making a positive difference in the organization's performance. Several approaches are suggested.

Cross-Train Employees

Cross-training of the employees producing the products and delivering the services will result in lower costs and greater efficiency by reducing idle time in operations. The more different skills an employee has, the more fully these skills can be utilized; therefore, the organization can operate efficiently and effectively with fewer resources.

Make Performance Excellence Everyone's Job

Many organizations, even today, still try to achieve superior results by employing armies of inspectors to search for, find, and eliminate errors. The problem is that even with these large numbers of inspectors, defects inevitably reach the customer. It is a well-known maxim that inspection is only about 70 percent effective at finding and removing defective product. By involving everyone in the improvement of the process, a favorable result can be achieved while enabling a substantial reduction in the number of inspectors.

Plan for and Respond to Changes in Customer Demand

Partly through use of cross-training as noted above and by utilizing part-time and temporary employees, organizations can optimize staffing levels to more closely track customer demand and better utilize limited resources. Particularly when demand is seasonal, staffing levels can be optimized by calculating the required number of employees to meet each season's demand. Temporary or part-time employees can then be used to fill the gap above the lowest seasonal staffing level. This also applies when orders drop due to economic hard times. Having multi-skilled workers who can perform a variety of functions allows management to shift personnel from task to task more easily and leads to better utilization of scarce resources.

Develop Effective Reporting Relationships to Optimize Resource Utilization

By utilizing the recommendations above, an organization can keep staff levels to a minimum and more effectively use the limited resources available in hard times. Particularly cross-training of the organization will lead to more effective reporting relationships. If, as suggested above, the leader of change within the organization is at the highest levels, direct (or at least easy) access to that leader will result in a more effective deployment of change even with limited available resources.

Avoid Ineffective Structures for Change

The first example refers to a major organization in the travel and tourism industry. The vice president for continuous improvement (CI) had several continuous improvement managers working for him and decided that they needed additional skills to be more effective in their jobs. A deployment of Lean Six Sigma was chosen as the approach to drive change.

An initial wave of Lean Six Sigma training was scheduled and delivered, and projects were executed to help the CI managers learn the application of the tools. People from outside the CI department were recruited for the project teams, as many of the projects focused on processes that others owned. During execution of the projects, however, some early problems with team members not being fully brought into the process began to arise. Particularly when the team was identifying causes of their problems during the analyze phase, some team members wanted to ignore the data because they "knew" the causes of the problems already.

As the teams reached the improve phase, the problems continued, now in the form of preconceived solutions. As with the a priori notions about causes of the problems, the process owners felt they knew from the beginning what needed to be done to correct the problem. It is worth noting that many of the project Champions and process owners were peers of the vice president who was leading the change effort.

The teams did finally struggle through and complete their projects and were eventually certified as Lean Six Sigma Black Belts. The savings from the projects were in the millions of dollars, so the return on the training and consulting investment was good. The organization even did additional training in Design for Six Sigma to further enhance the skills of the Black Belts.

However, dark clouds were on the horizon for this organization, which is highly affected by the price of fuel. When gasoline and diesel prices began their steady climb in early 2008 from over \$3 to over \$4 per gallon, the organization felt a lot of pressure on its bottom line. How did they respond? By doing what most organizations do when there is a need to reduce costs quickly—they reduced the head count including two of the CI managers they had invested so much money in training. In other words, they got rid of some of the people who could have helped reduce costs in more lasting and meaningful ways to help the organization weather the storm. Eventually the whole Lean Six Sigma deployment collapsed.

What factors can be cited that led to the lack of success of this change initiative? These were some of the more important organizational factors:

- The Lean Six Sigma deployment was not directly and firmly linked to the corporate vision. The executive suite knew little about the deployment and its benefits, and when tough times came, they saw the initiative as expendable.
- The change initiative was not broadly communicated. Little to no training of team members and process owners was done, and as a result, they did not see the value in the methodology.
- Certainly the vice president of CI had his vision of the change that needed to take place, but there was no clear, methodical plan for achieving that vision. Without such a plan, the deployment was attempted in a random fashion, contributing to its ultimate failure.

Create Effective Structures for Change

The second example refers to a multibillion-dollar manufacturer and supplier of components to the consumer electronics industry. This organization operates worldwide and employs tens of thousands of people.

Several years ago, after somewhat rapid growth, the organization managers decided that they needed to implement change to help harmonize disparate cultures among the global divisions and sites. They chose Lean Six Sigma as the methodology to drive this change. From the start, the change leader (the COO) developed a unifying vision and organizational structure to facilitate and drive the deployment. See Figure 26.3 for a description of that organization.

The deployment itself was rolled out in phases, or waves, and the organization judiciously employed outside resources to provide expertise that was not present within. Each wave was begun with a site-by-site kickoff to communicate the corporate vision for the Lean

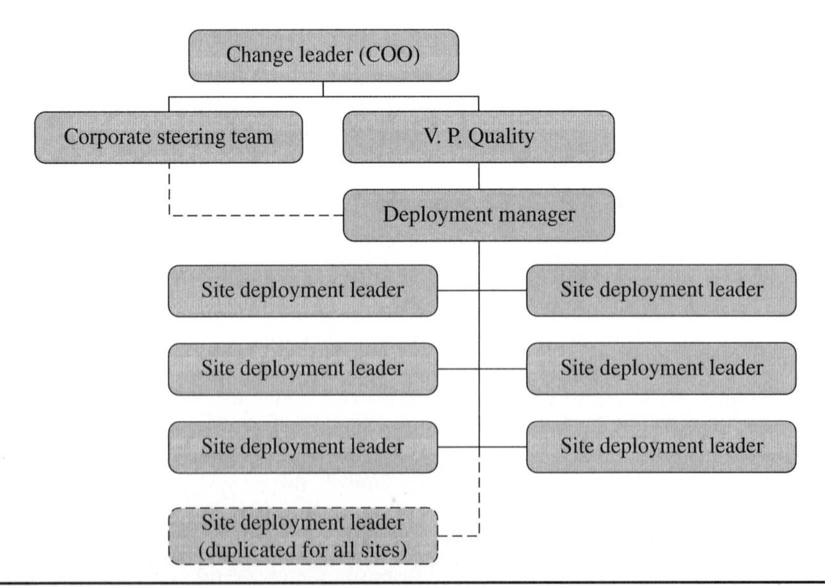

FIGURE 26.3 Effective change structure. (*Juran Institute, Inc.*)

Six Sigma deployment. Clear objectives were set, and the sites involved in each wave were challenged to meet aggressive goals.

But during the third implementation wave this organization, as did almost all organizations, hit a big bump in the road: the global economic downturn that began in late 2008. However, the organization responded in an entirely different way from the organization discussed in the first example above. The COO reemphasized the organization's commitment to the Lean Six Sigma deployment and highlighted the importance of the savings generated as a sustaining force during economic hard times.

What was the result? In spite of almost a one-third drop in revenue, the organization achieved its very aggressive savings goals from the Lean Six Sigma projects, amounting to tens of millions of dollars in cost reduction. It is quite likely this contributed mightily to the organization weathering the economic storm. The organization is now in preparation to launch the fourth wave.

What is the reason for this remarkable difference in sustainability of the change initiative between this case and the first? Likely it had much to do with the change management organization and deployment approach. Specifically:

- The leadership of the change came from the top of the organization. By placing the responsibility for the success or failure of the deployment with the COO, the organization was well positioned for success.
- The change leader had a clear vision. He knew what he wanted to achieve and how he planned to achieve it, using specific tools, methodologies, and structure.
- The vision was clearly and broadly communicated. The global deployment was sold, advertised, cheered, and explained to every site and division throughout the corporation.
- The deployment structure permeated all levels of the organization. By using a tiered organization to drive the change and cascading steering teams as shown in Figure 26.4, the organization ensured that the deployment would permeate the organization.

Careful consideration of these factors and thoughtful structuring of the initiative clearly can be credited with a large part of the organization's ability to withstand difficult economic circumstances.

Job function	How to make quality everyone's job
Workforce	• Improve workplace and work flows • Apply self-control
Technical & engineering	• Utilize mistake proofing • Design the process for self-control and zero defects
Supervision	• Enable workers by providing the right tools, product and process standards, ergonomics, etc.
Upper management	• Provide adequate, targeted training in applying performance excellence tools

FIGURE 26.4 Making quality everyone's job.

References

Material in this section relies heavily on Juran and Godfrey, pp. 8.16–8.19.

Parts of this chapter were edited and updated from Joseph M. Juran and Frank M. Gryna (eds), *Juran's Quality Control Handbook*, 4th ed., McGraw- Hill, New York, 1988, pp. 8.3–8.7.

Blanchard, K., Zigarmi, P., and Zigarmi, D. (1985). Leadership and the One Minute Manger. William Morrow, New York.

Collins, J. (2001). *Good to Great*. HarperCollins, New York.

Daniels, A. C., and Rosen, T. A. (1982). *Performance Management: Improving Quality and Productivity Through Positive Reinforcement*. Performance Management Publications, Atlanta.

Davis, S. M. (1995). Communicating the Quality Message at the Ritz Carlton. *IMPRO95 Conference Proceeding*, Juran Institute, Inc., Wilton, CT.

Deming, W. E. (1986). *Out of the Crisis*, MIT Press, Cambridge, MA.

Gershon, D. (2007). Changing Behavior in Organizations: The Practice of Empowerment, Systems Thinker, vol. 17, no. 10, Dec 2006/Jan 2007, pp. 2–5.

Juran, J. M. and Godfrey, A. B. (1999). Juran's Quality Handbook, Fifth Edition, McGraw-Hill, New York.

Juran, J. M. (1987). Juran on Quality Leadership, videocassette, Juran Institute, Inc.

Juran, J. M. (1989). Juran on Leadership for Quality: An Executive Handbook. The Free Press, New York.

Juran, J. M. (1995). *Managerial Breakthrough*, rev. ed. McGraw-Hill, New York, p. 3.

Juran Institute (1996). Maximizing Employee Assets. *Quality Minute*, vol. 3, no. 9. Center for Video Education, White Plains, NY.

Kaufman, R. A., Oakley-Browne, H., Watkins, R., and Leigh, D. (2003). *Strategic Planning for Success*. John Wiley & Sons, New York, p. 234.

Kotter, J. P. (1996). *Leading Change*. Harvard Business School Press, Boston.

Premier Inc. http://www.baldrige.nist.gov/PDF_files/Premier_Application_Summary.pdf.

Sunny Fresh Foods. http://www.baldrige.nist.gov/PDF_files/Sunny_Fresh_Foods_Application_Summary.pdf.

Park Place Lexus. http://www.baldrige.nist.gov/PDF_files/Park_Place_Lexus_Application_Summary.pdf.

Poudre Valley Health System. http://www.baldrige.nist.gov/PDF_files/2008_Poudre_Valley_Application_Summary.pdf.

Cargill Corn Milling. http://www.baldrige.nist.gov/PDF_files/2008_Cargill_Corn_Milling_%20Application_Summary.pdf.

CHAPTER 27

The Quality Office: Leading the Way Forward

Richard C.H. Chua and Joseph A. De Feo

About This Chapter

The Quality Office is a term the authors use to describe the full-time function with responsibilities of managing the quality system (little q) and the performance excellence (big Q) system across an organization. The little q quality management system is the traditional role of the Quality Office. The big Q system is what we have outlined in this handbook as defined by Juran. It defines the application of methods, tools, and structure to link all functions and processes in a "total organization" approach.

The Quality Office of today is being transformed to recognize and reinforce the current view that quality is not separate and distinct from business, but its precepts and principles are central to achieving performance excellence. Your organization may not use the term Quality Office. The function may be named Quality Assurance, Quality Department, Performance Excellence Program Office, Strategic Quality, or Customer Satisfaction Office, to name a few. The Quality Office of the future (and some are doing it today) will play a dual role. It will administer the quality management system to control and ensure that the policies and procedures are in place to avoid product and service failures and reduce risk. The second role is to lead the performance excellence drive within the organization. Today, two separate offices are administering and carrying out these two roles: one typically called QA (or equivalent), and the other is called Continuous Improvement (or equivalent). The Quality Office of the future will do both. The Quality Officer, the leader of this function, is part of a community of practitioners that must act as the surrogate conscious of the customer for the organization. This chapter will deal with the development of the Quality and Performance Excellence Office needed to support both little q and big Q. It will include how to establish an effective function, how to develop and maintain the skills and behaviors of its staff, and how to establish credibility with upper managers.

High Points of This Chapter

1. The Quality Office plays an important role at both the strategic and tactical levels in enabling the organization to strive for quality and performance excellence.
2. The distinction between managing *quality of product versus creating transformational change* leads to the recognition that parallel departments or divisions within the organization may be needed to attain superior results. Some of these organizational functions may be ad hoc or permanent.
3. The role of the Quality Office, its responsibilities, and how it is organized is driven by its mission. This mission is to ensure that the organization designs, controls, and continuously improves performance of the product, the process, or even the organization to meet the changing needs of its customers. The Quality Office has direct and indirect functions to carry out that mission.
4. The Quality Office should be an enabler—enabling the organization to drive quality and performance excellence at the business and organization levels.
5. Capable experts are needed and developed to have the right competencies and capabilities, based on what customers throughout the organization need, in order for them to achieve quality and performance excellence.
6. Skills assessment, education, and certification will close the gaps and enable today's quality offices to transform themselves into the office the organization needs to sustain itself.

Introduction to the Quality Office of the Future

Computer Sciences Corporation, one of the largest technology solutions and outsourcing organizations in the world, understands the Quality Office of the future. Darryl W. Bonadio, Master Black Belt and Director of MSS Quality Management and Improvement, describes quality and their Quality Office this way:

> "Quality is important to everyone at CSC. Quality must become a cultural attitude that results in personalization of our five standards: Positive Talk, Confident Perspective, Outcomes Oriented, Be Accountable, and Respect Our Client. CSC's Managed Services Sector manages quality through its Global Process Innovation and Quality Excellence organization. This is our Quality Office. The Enterprise Excellence Program we have implemented meets this challenge by addressing three important tenets. The first is Delivery Excellence, which is the integration of quality principles, tools, and approaches that enable achievement of business objectives and promote client advocacy. Second, Passionate Delivery, where engaged employees are the most effective quality assurance factor in our service-oriented environment. And finally, Enterprise Performance Management that provides an end-to-end analysis and a feedback loop for continuous improvement."

You may be tasked to define and develop the Quality or Performance Excellence office, as some call it, in your organization. You may even be the Director or Vice President of Quality. You are faced with the challenge of organizing and driving quality and performance excellence in your business unit. Key questions you may be pondering and that must be addressed include

- What is the role of this office?
- What are its mission and responsibilities?
- What types of personnel are needed?
- What competencies and capabilities are needed?
- How do you develop capable experts to drive quality and performance excellence?
- How should the office be organized to develop and deploy the functions needed?

These questions are the motivation for this chapter. An underlying purpose of this chapter is to assist the reader in asking the right questions and to provide guidance in defining and developing the quality and performance excellence capabilities in any organization. The chapter is organized as follows:

- We begin with key operational definitions and terms.
- We then address the approach for coordinating quality and performance excellence. We discuss the role and responsibilities of the Quality Office.
- We provide guidance on how to organize the Quality Office.
- We showcase an example on how the Quality Office in a global organization is organized, including organization charts, job profiles, and descriptions of various quality functions.
- We conclude this chapter with the means to address the challenge of developing capable experts to drive quality and performance excellence using a customer-focused development approach by describing skills assessment, re-education, and certification requirements.

Proof of the Need to Improve Quality Office Performance

The Juran Institute develops and regularly conducts a competency and skills assessment of Quality Office personnel. (Juran Institute, 2009). The results of these assessments convinced us and our clients that a change is needed. Here is a review of its assessment and findings.

There are eight steps used to carry out the Quality and Performance Excellence Skills Assessment from Juran Institute. Since 1994, we have been conducting this assessment and keeping track of the results. Our findings led us to raise awareness of the need to improve the skill set of the quality professionals. It also led us to help close the gaps. The eight steps are

1. Reviewed skills questions and an organization's competency models to identify the organization's needs and its internal customers
2. Translated and edited competency questions for organization terminology
3. Developed competency questions around seven major topic areas:
    - Assessment and auditing
    - Critical thinking

- Process control
- Quality theory
- Root cause analysis
- Statistics and data analysis
- Teamwork and coaching

4. Conducted written and oral reviews using the developed competency questions covering the major topic areas
5. Summarized the results by question (effectiveness of answering the questions) and by individual (each individual's score for the oral and written reviews)
6. Compared responses of staff skills to the industry best-practice skills, as defined by multiple scores of assessments conducted by Juran Institute
7. Analyzed, interpreted, summarized the findings and presented to management
8. Developed a training and development plan to close the gaps

An anchored scale of 1 to 5 was used for scoring the participant responses:

1 – no response or very poor response
2 – a weak or poor response
3 – an acceptable response, but with little detail
4 – a strong response, with detail
5 – a very strong, outstanding detailed and understanding response

Typical Findings from Our Experience

Strengths

1. Most respondents are well versed in their organization's products, technology, work area, and programs.
2. Most understand the importance of Quality and Performance Excellence as the means to satisfy customers and attain superior results.
3. Competency and skill capability (scores) improve with successive job grades, from lowest levels to the directors.
4. Almost 100 percent of the respondents felt this assessment was a welcome activity to help them obtain the awareness that the Quality Office needs continued education.

Opportunities for Improvement

1. Of the seven categories, three areas are always scoring low:
    a. Root cause analysis graphical tools
    b. Data analysis and basic statistics
    c. Quality management methods and theory
2. Over 75 percent of quality professionals were not aware of quality management history, methods, and early practitioners such as Deming, Shewhart, and Juran.
3. Many lower-level job-grade quality professionals lack the knowledge of quality methods and tools—even some basic ones—that are required in the Organization Quality and Performance Excellence (Q&PE) Competency Models.

4. Over 70 percent of individuals answered oral and written questions—in total—unsatisfactorily.
5. Our experience indicates that some of the staff may not be able to drive Quality and Performance Excellence and/or improvements in their work areas with limited knowledge of quality management tools and techniques (see summary charts).
6. Many are not able to articulate responses to oral questions. They had difficulty expressing themselves with clarity and precision. Some answered questions not asked, giving the appearance of not listening or not understanding what was asked.

The authors feel a change is needed and a program to improve performance of the Quality Office staff is in order.

Operational Definitions and Key Terminology

The Quality Office

Traditionally, the Quality Office is given the responsibility of ensuring that the functions required for establishing and producing quality products and services meet the customers' specified requirements. This office is called by various names; however, it is most commonly known as the Quality or Quality Assurance Department. In regulated industries, such as the pharmaceutical or the medical device industry, it is often called QA/RA, short for Quality Assurance/Regulatory Assurance. Within defense industries, it may be known as Mission Assurance or Quality and Mission Assurance. In other industries, it might be called Quality and Safety Compliance. In this chapter, the traditionally limited role of the Quality Office, as mentioned, is challenged. The modern definition of the Quality Office is expanded to a business level and is called the Performance Excellence Office, discussed later in this chapter.

The Quality Functions

Quality functions are the actions or activities that are carried out on a daily basis according to the three universal processes of the Juran Trilogy: quality planning, improvement, and control.

Quality planning activities include joint supplier planning, designing or redesigning processes, new development of products or services, design reviews, toll-gate reviews, and quality plans.

Quality improvement activities include problem solving, root cause analysis, and projects to remove waste or improve process capability.

Quality control activities include implementing quality standards, carrying out source inspection, testing, in-process inspection, final inspection, and audits.

These functions may or may not be performed by quality personnel alone, and usually require the participation and input of employees (the community of practitioners) throughout an organization.

Quality Management Principles

ISO 9000 identifies eight quality management principles that sum up what can be used by the Quality Office in order to lead the organization toward performance excellence.

1. *Customer focus.* Organizations depend on their customers and, therefore, should understand current and future customer needs, should meet their customer requirements, and should strive to exceed customer expectations.

2. *Leadership*. Leaders establish unity of purpose and direction of the organization. They should create and maintain the internal environment in which people can become fully involved in achieving the organization's objectives.
3. *Involvement of people*. People at all levels are the essence of an organization, and their full involvement enables their abilities to be used for the organization's benefit.
4. *Process approach*. A desired result is achieved more efficiently when activities and related resources are managed as a process.
5. *System approach to management*. Identifying, understanding, and managing interrelated processes as a system contributes to the organization's effectiveness and efficiency in achieving its objectives.
6. *Continual improvement*. Continual improvement of the organization's overall performance should be a permanent objective of the organization.
7. *Factual approach to decision making*. Effective decisions are based on the analysis of data and information.
8. *Mutually beneficial supplier relationships*. An organization and its suppliers are interdependent, and a mutually beneficial relationship enhances the ability of both to create value.

Quality Management System

A formal definition of a Quality System can be found in the ISO 9000 series of standards. ISO 9000 defines a Quality Management System as a "management system to direct and control an organization with regard to quality." Since quality is an organizationwide function, the Quality System is therefore organizationwide. While the Quality and Performance Excellence Office plays a major role, the Quality System is much larger in scope and, therefore, may be "directed and controlled" from multiple offices.

Approach to Coordinate Quality and Performance Excellence

Control versus Creating Beneficial Change

Gryna et al. (2007) state that the approach used to coordinate quality and performance excellence activities take two major forms:

- Coordination for *control* is achieved by the regular line and staff departments, primarily through the use of formal procedures and feedback loops. Feedback loops take such forms as audits of execution versus plans, sampling to evaluate process and product quality, control charts, and reports on quality.
- Coordination for *creating change* is achieved primarily through the use of quality project teams, Six Sigma or Lean project teams, rapid improvement (or kaizen) events, and other organizational forms for creating change.

Coordination for control is often the focus of the Quality Office. However, more and more often, such a focus is so preoccupying that the Quality Office is unable to make major strides in creating change. As a result, some "parallel organizations" for creating change have evolved. This evolution, in our opinion, has taken place because many of the quality experts lacked the skills required to speak the language of management. As Dr. Juran often stated, "They did not speak the language of money, they spoke the language of things" (Juran 1954, *Managerial Breakthrough*).

Parallel Organizations for Creating Change

All organizations are engaged in creating beneficial change as well as in preventing adverse change ("control"). Much of the work of creating change consists of processing small, similar changes. An example is the continual introduction of new products consisting of new colors, sizes, shapes, and so on. Coordination for this level of change can often be handled by carefully planned procedures.

Nonroutine and unusual programs of change generally require new organizational forms. These new forms are called "parallel organizations." *Parallel* means that these organizational forms exist in addition to and simultaneously with the regular "line" organizations.

Examples of parallel organizational forms for achieving change in quality and performance excellence are process teams, project teams, and performance excellence steering committees or councils. Parallel organizations may be permanent or ad hoc [a business process team or value stream management (VSM) team is permanent; in contrast, a Six Sigma project team or a Lean rapid improvement team is ad hoc and disbands when its mission is accomplished].

The IT, finance, and human resource functions often are asked to create programs to reduce costs, improve processes, and improve the skills of staff to drive quality. In some cases, they succeed. In other cases, they may have lacked the knowledge of how to manage for quality, and management soon is tired by the lack of results.

It is the responsibility of executive management to drive performance excellence throughout the organization. Performance excellence should not be viewed as the responsibility of the Quality Office, even though it is integral to enabling performance excellence to occur. However, this office should be seen as the leaders of change to create a sustainable future. It is time to take it back.

Role of the Quality Office

Many organizations have traditionally centralized the quality functions to a Quality Office (or Quality Department). Over the decades, quality and performance excellence tasks have been assigned to other functional groups. For example, process capability studies were transferred from the Quality Department to a Process Engineering Department. Also, as the definition of quality broadened from operations only (little q) to all activities (Big Q), most organizations now have personnel in various functional departments trained and responsible for implementing quality and performance excellence. Authority to make decisions is now delegated to lower levels. Partnering with key suppliers and customers is becoming increasingly common. Also, organizations have become flatter, and cross-functional work teams and project teams are used to solve performance-related problems.

So, what is or what should be the role of the Quality Office? Is the Quality Office limited to a tactical role in the organization? Is there or should there be a strategic role for the Quality Office? What are the responsibilities of the Quality Office? What authority should it have? Many researchers and authors, such as Gryna (2002), Crosby (2000), Imler (2006), and Watkins (2005) have discussed the changing role of the Quality Office. Discussion over the changing role of quality professionals has also occurred by the likes of Spichiger (2002) and Wescott (2004). The role of the Quality Office is best discussed by considering it at two levels: tactical and strategic.

"Successful strategy execution depends on both satisfying the customer today and achieving excellence in the future. We continue to use the Juran Trilogy® to establish a solid foundation for achieving and sustaining breakthroughs throughout the entire organization of the Builder Cabinet Group at Masco Corporation. This has enabled us to continually meet or exceed the expectations of customers today, develop innovative breakthrough products and processes for tomorrow, and most importantly, provided the means for developing team members throughout my company to execute the strategy to action. As the executive of the Quality Office, we will continue to drive our performance towards sustainability" (Steve Wittig, VP of Six Sigma, Masco, BCG 2009).

Tactical Level

At the tactical level, the traditional role was and still is to provide an independent evaluation of product quality or service quality. Inspections, testing, and product or service audits are examples. In this role, the Quality Office is often viewed as being limited to supporting operations, providing an independent evaluation to ensure that productivity targets are not pursued at the expense of not meeting quality specifications.

Examples include

- Inspection and checking
- Product testing
- Supplier quality
- Root cause corrective action
- Quality system audits

As technology and the needs of the organization have changed, the role of the Quality Office must expand beyond this traditional view. At the tactical level, the Quality Office should play a role in *enabling* others in the organization to carry out the management of quality and performance excellence, either independently of or in collaboration with other functions. Examples of expanded tasks include

- Calculate the costs due to poor quality and use this to marshal the right resources to reduce these costs.
- Conduct process capability studies in nonmanufacturing processes.
- Participate in design reviews of new products as early as possible in the development cycle.
- Work with the supplier chain and provide evaluations of supplier selection and ongoing performance.
- Monitor customer satisfaction and create and manage the corrective action process.
- Be involved with the engineering change process.
- Reduce the inspection by creating in-process checking and self-inspection by the workforce.
- Work with functions to implement error-proofing efforts.
- Participate in kaizen or rapid improvement events.
- Use data-driven methods, such as statistical process control, to monitor performance.
- Identify improvement projects, such as Lean and Six Sigma projects.

- Integrate with the environment, health, and safety programs.
- Develop a skilled and competent function of experts that can consult with all functions.

Strategic Level

In many organizations, the Quality Office is often viewed as having only a tactical role. That the Quality Office can perform a useful strategic role is not well recognized by many organizations. However, enlightened organizations have recognized the strategic value of the Quality Office. They view the Quality Office as a strategic asset with a key role in shaping, planning, and enabling the deployment of strategies, strategic goals, and business plans of the organization. A strategic approach for the Quality Office of the future is to think of their role as providing "Enterprise Assurance" (Juran Institute White Paper, 2008).

The following is a list of strategic activities in which the Quality Office needs to play an important role:

- Vision, mission, and policy development
- Assisting upper management with strategic planning and goal-setting
- Recommending to upper managers how to reduce the costs of poor quality
- Providing the organization with the most effective methods and/or tools to reduce these costs
- Being involved in annual business planning to incorporate improvement goals
- Demonstrating how quality can affect social responsibility and the environment
- Organizationwide assessment and planning to close the gaps
- Organizationwide transformation and improvement initiatives such as Lean Six Sigma
- Innovation of major business processes (such as demand creation, product development, order fulfillment, supply chain, and HR management processes)
- Developing and deploying balanced scorecards and data systems to support all key business processes, policy deployment, and improvement efforts

These roles are consistent with the views presented by Gryna et al. (2007), where the Quality Office should have both tactical and strategic roles.

Organizing the Quality Office of the Future

To better understand responsibilities, we need to make an important distinction. There are *direct* responsibilities and *indirect* responsibilities. Direct responsibilities are those activities and results over which the Quality Office has control because they are executed by full-time personnel who report to the Quality Executive. Indirect responsibilities are those over which the Quality Office has an influence, but little or no direct control. Indirect responsibilities are those that the Quality Office enables others (nonquality personnel) to do. The Quality Office provides the infrastructure and the means for these indirect responsibilities to be carried out by personnel in other functions. For example, the execution of process control plans to ensure quality compliance is one where the Quality Office enables others (in operations) to carry out this important task. Another important example is the development and deployment of a change management program to attain performance excellence. This is one where the Quality Office has both direct and indirect responsibility—*direct* because Quality personnel are performance excellence specialists; *indirect* because selected non-Quality personnel from other functions are also performance excellence specialists. In both cases, the Quality Office plays an integral part in developing and enabling all performance excellence specialists (such as Six Sigma and Lean Six Sigma Green Belts, Black Belts, and Lean experts) to contribute to the improvement of the organization.

Consider Its Mission

Who should lead the Quality Office? At what level of management? To whom is the Quality Office accountable, and to whom should the Quality Office report? These are important questions to ask when setting up the Quality Office and are best answered by first considering the mission of the Quality Office. Ideally, we want to ensure that the mission and associated responsibilities are supported by the corresponding management level and the appropriate level of authority is given to the Quality Office. Authority must be consistent with responsibility; otherwise, we have a paralyzed office not able to carry out its duties.

For example, at an organization that provides customer care and call center services outsourced by major corporations, the **mission** of the Quality Office is to contribute to the organization's financial growth by providing services that

- *Enable the efficient delivery of exceptional customer experiences*
- *Foster a culture of fact-based leadership and continuous improvement*

Not surprisingly, in this organization, the Quality Office is led by the Quality Director, who reports to the Chief Financial Officer. To support that mission, responsibilities of this office reflect both the tactical and strategic role of the Quality Office discussed earlier. Functions that are the responsibility of this Quality Office are:

- Process improvement (Six Sigma and Lean)
- Workforce management (contact center volume forecasting, staffing, and intraday staff management)
- Customer contact quality (call and e-mail quality)

Note: The Quality Office should always be part of the executive management team or a direct report to it to be effective at leading performance excellence. Otherwise, the office will have little credibility.

Size and Scope Must Be Consistent with Its Mission

How large should the Quality Office be? To determine size and scope, we have to consider what quality functions should be done versus what functions should be enabled by the Quality Office. To the maximum extent possible, the Quality Office should enable others in the organization to perform as many quality and performance excellence functions as possible. Process ownership of those functions increases when the Quality Office enables others to carry them out. The experience of recent decades has shown that the best way to implement quality and performance excellence is through line organizations rather than through a staff department. Having an elitist department of experts and adopting a "corporate seagull" approach should be avoided. (The late Quality guru Phillip Crosby used this analogy of seagulls. It is used here to illustrate the effect. Seagulls fly into your area and make a lot of noise. And when they are done, they fly off . . . and leave a mess behind!)

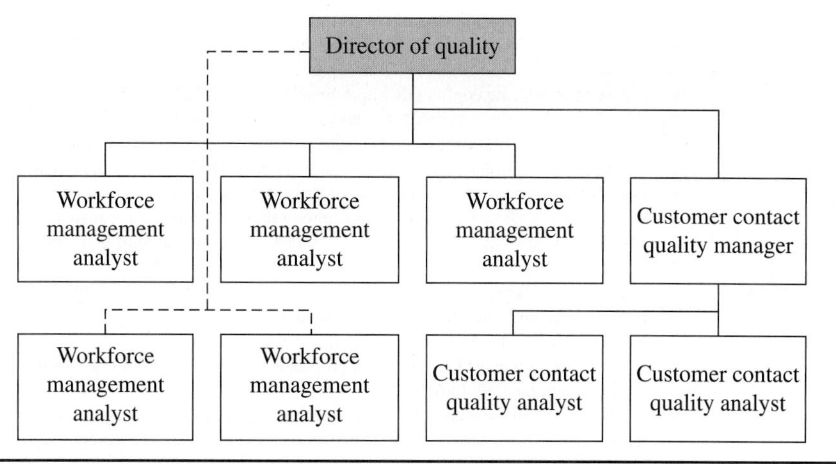

Figure 27.1 Quality Department organization chart of a call center company.

The Quality Office should plan to enable others as much as possible to drive toward achieving what Dr. Juran calls "Self-Control," where (1) what is expected is clearly known, (2) actual performance is known through short or immediate feedback loops, and (3) the means and ability to regulate is available so that actual performance meets expectations.

The groups or areas that comprise the Quality Office must be consistent with its mission and responsibilities as well. Referring back to the call center example, the organization chart shows the areas that report to the Quality Director (see Figure 27.1), and the responsibilities are listed in Figure 27.2.

Table 27.1 is a list of positions recommended for the Quality Office of the future. Titles vary by industry, organization, and culture. We have identified the most common needs.

Example: Organizing the Quality Office in a Multinational Global Organization

The scope, size, and structure become more challenging when the organization is global, with multiple divisions and regions operating around the world. To what extent should the Quality Office be centralized (versus decentralized)? How much of the structure should be in a matrix? To what extent should there be direct reporting relationships versus indirect dotted-line reporting relationships to the head of the Quality Office? Should a plant quality manager report directly or indirectly to the local plant's general manager?

To illustrate how the Quality Office can be organized, we will look at an example of a multinational global organization. Figure 27.3 shows the organizational chart of a global organization with multiple locations and divisions around the globe. Given the nature of the responsibilities, the Chief Quality Officer (who is the Vice President of Quality) reports to the President and COO (who in turn reports to the CEO) at a high enough level of authority that is consistent with the global responsibilities of the Quality Office (see Figure 27.4). The organizational chart for the Quality Office is shown in Figure 27.5. The plant-level Quality organization can be seen in Figure 27.6.

The Quality Office is structured to drive global standards in areas with significant opportunity (such as reducing the number of suppliers and increasing global coordination for key customer quality improvements), to leverage best practices (such as by maintaining a library of Lean Six Sigma projects and FMEA libraries to capture engineering experience), and to revitalize use of quality audits to highlight continued improvements.

Title	Responsibility
Quality Director	• Research, develop, and execute the organizations corporate wide Lean, Six Sigma, quality management, and process improvement efforts. • Understand the business environment and ensure deployment initiatives are relevant and timely. • Assist in identifying continuous improvement opportunities within existing processes. • Work with multiple stakeholders to understand the key problem areas and work with this team to develop solutions. • Train and mentor the teams to drive improvement initiatives across the organization. • Use DMAIC/ Lean as a methodology to drive improvements. • Drive Six Sigma/Lean cultural change among the middle and lower management. • Understand lean concepts and identify and facilitate elimination of non-value-added activities. • Implement sustained solutions—set up all necessary control mechanisms such as dashboard, review procedures, and responsibility matrix. • Report project status and results to Manager and Operations. • Review program plans to establish appropriate quality assessment (metrics planning) and improvement methodology. • Demonstrate team leadership and manage team dynamics in all aspects of the DMAIC model (define, measure, analyze, improve, control) and maintain Six Sigma principles. • Perform other initiatives beyond process improvement projects which involve setting up right metrics to measure the process performance, automation initiatives, failure modes elimination and early warning systems, etc.
Workforce Management Analyst	• Has extensive experience in the call center environment, with a strong understanding of the Aspect eWorkforce Management application. • Responsible for insuring optimal forecasting, staffing and schedule compliance levels for existing programs and forecast future staffing needs to support recruitment efforts. • Works closely with Program Directors to assess impact of daily absences, changes in call volumes, etc.
Customer Contact Quality Manager	• Responsible for managing the entire organization's call quality process. • Works with teams at all locations to completely implement the call quality process. • Works with teams and clients on call calibrations. • Is responsible for all reporting and analysis of the call quality process. • Works with the training group to ensure training dollars are focused on specific needs identified in the call quality process. • Works directly with clients to ensure their needs are being met. • Improves the overall quality of the product that the organization delivers on behalf of clients each and every day.
Customer Contact Quality Specialist	• Responsible for supporting multiple clients in the area of interaction quality. • Responsible for conducting call auditing, sales and customer service training, and developing and motivating both front line agents as well as management. • Leads by example the value and necessity of interaction quality.

FIGURE 27.2 Responsibilities of the Quality Office of the call center company.

Quality Office of Today	Quality Office of Tomorrow
Vice President of Quality	Vice President of Enterprise Assurance
Director of Quality Assurance	Director of Operational Excellence
Quality Managers	Management as Champions
Quality Engineers	Quality Engineers and Master Black Belts
Quality Auditors	Auditors: Quality, Finance, Safety, Environment
Technical Analysts	Technical: Analysts, Green Belts, Lean Experts

TABLE 27.1 Comparisons of Quality Officers

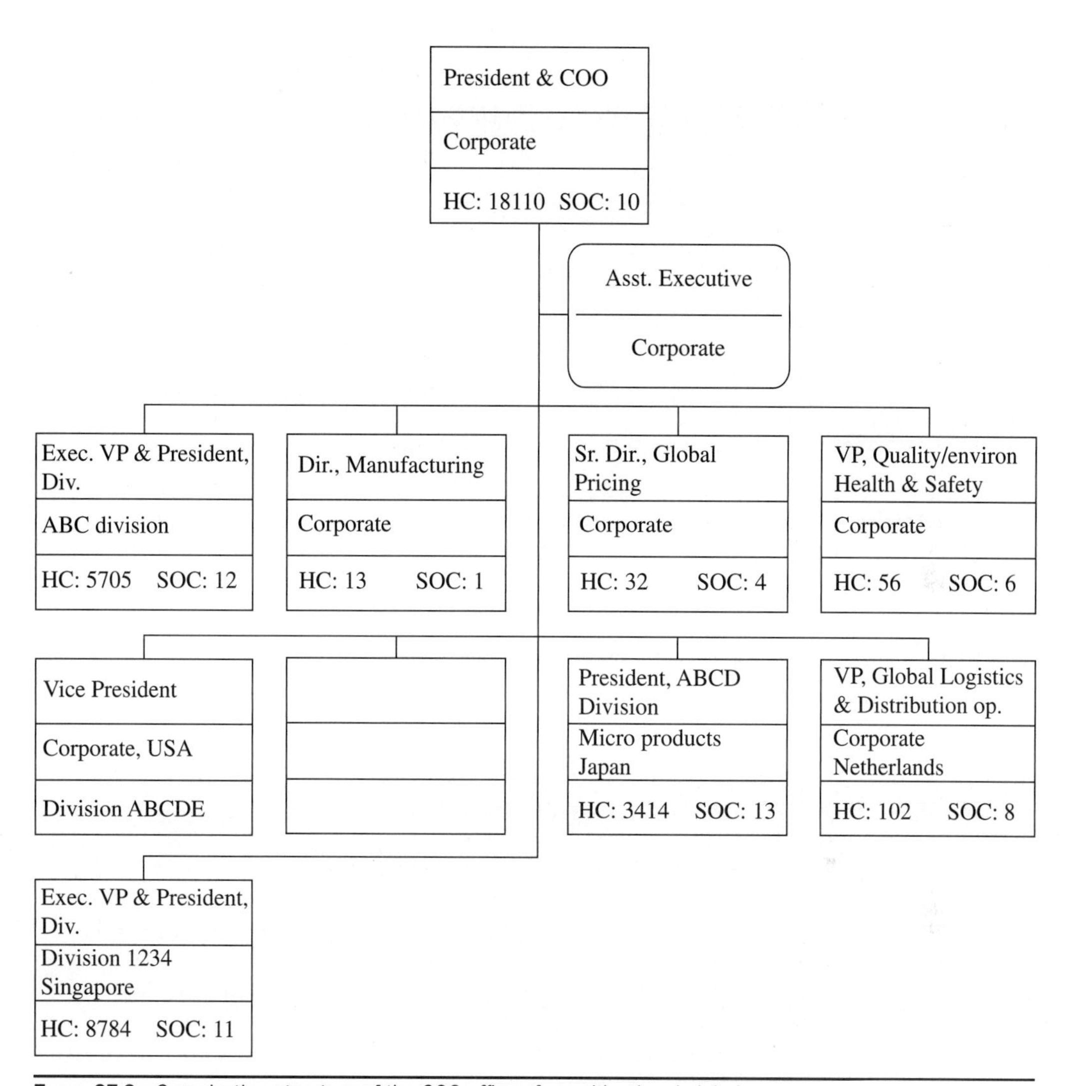

FIGURE 27.3 Organization structure of the COO office of a multinational global company.

To support the global business divisions of this company, the Quality Office is structured vertically in the divisions (Global Division Quality) from which selected quality officers will then coordinate and drive best practices across the divisions (Global Functional Quality), as seen in Figure 27.7.

Elaborating on the structure found in Figure 27.7, the global quality head is a member of the division management team and the global quality leadership team. Plant quality managers are direct reports. The five functional heads coordinate best practices globally in the following areas: Customer Quality, Supplier Quality, Lean Six Sigma, Audit and Certification, Project Management, and Annual Quality Improvement Plans (AQIPs). Specific objectives, action plans, and metrics are established for each global functional area.

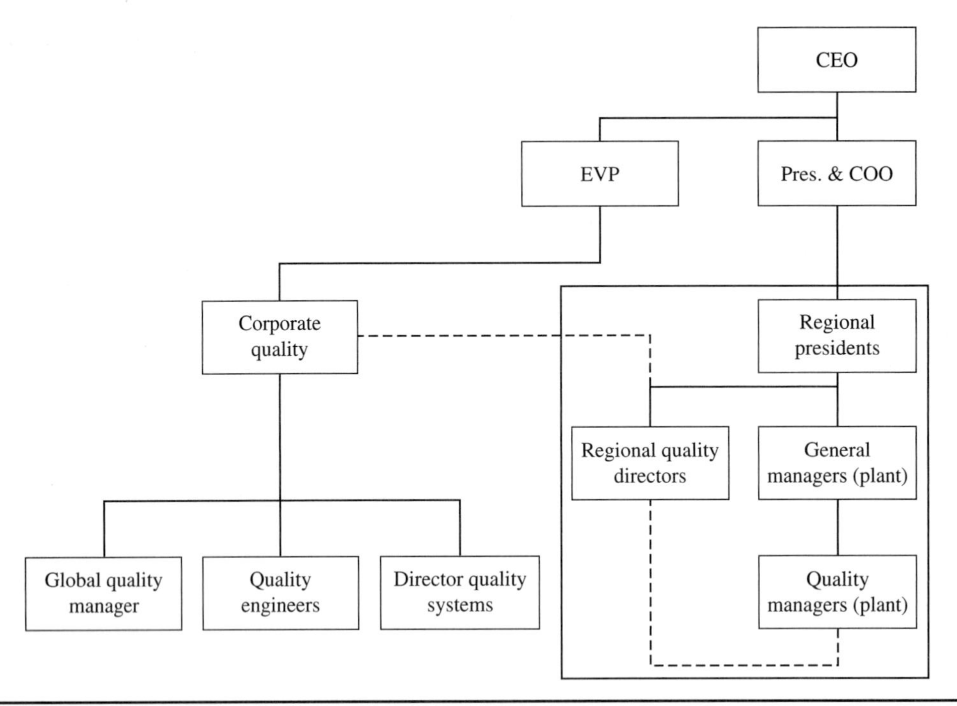

Figure 27.4 Global Quality organization of a multinational global company.

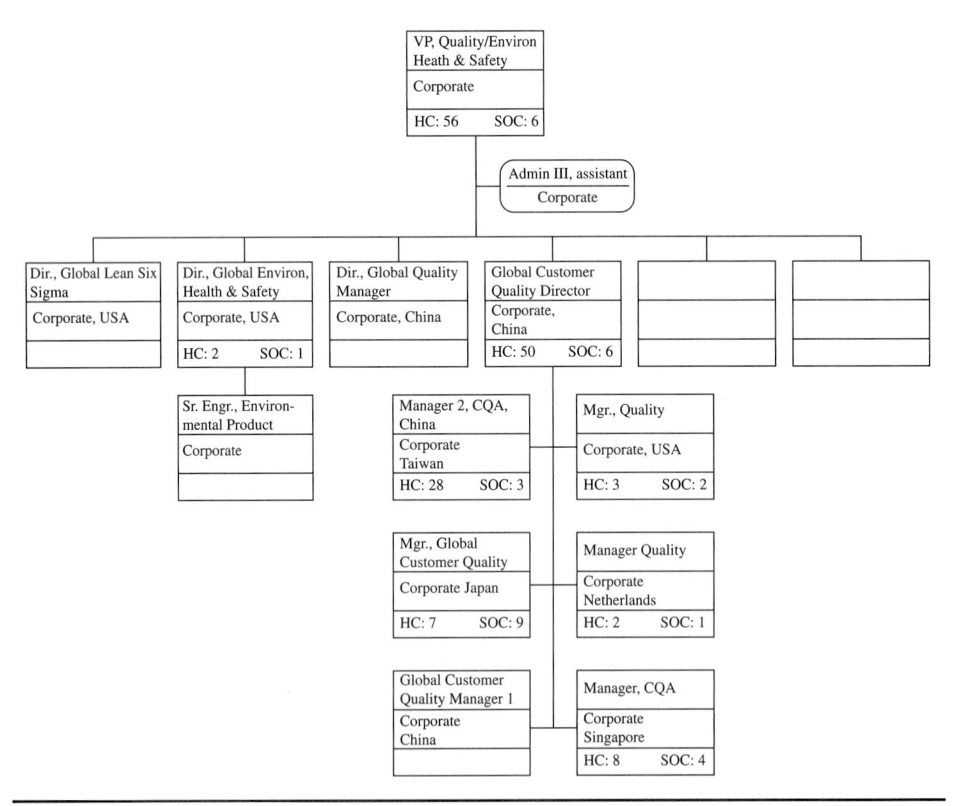

Figure 27.5 Global QEHS organization of a multinational global organization.

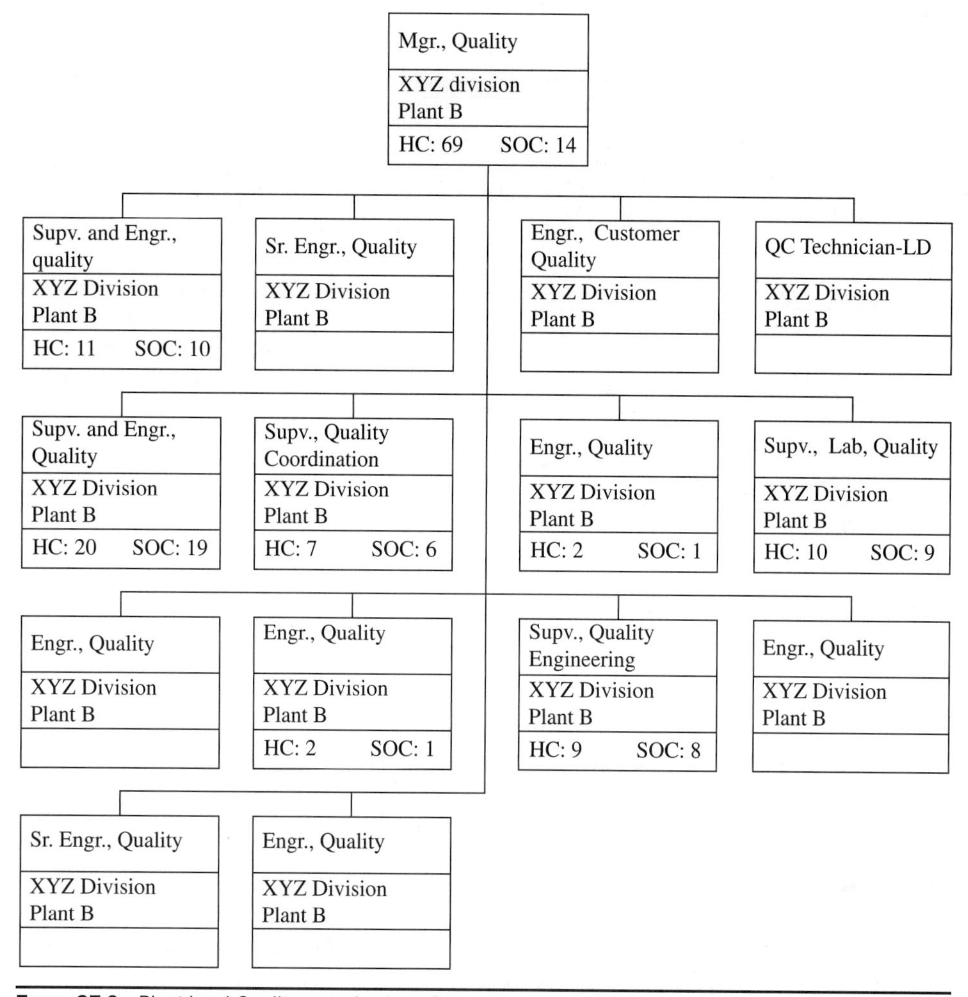

FIGURE 27.6 Plant-level Quality organization of a multinational global company.

Global Division Quality has a flat structure, where the division owns the resources and quality functional heads coordinate globally. Global Division Quality is responsible and accountable for execution of global processes and AQIP content and achievement. Functional heads are part of each division's team. Some regional responsibilities will remain for division quality heads. Responsibilities of Global Functional Quality are

- Customer quality
  - To coordinate consistent proactive and reactive customer support
- Supplier quality
  - To enforce one global standard for suppliers and contract manufacturers
- Lean Six Sigma
  - To eliminate waste
  - To improve product and process quality

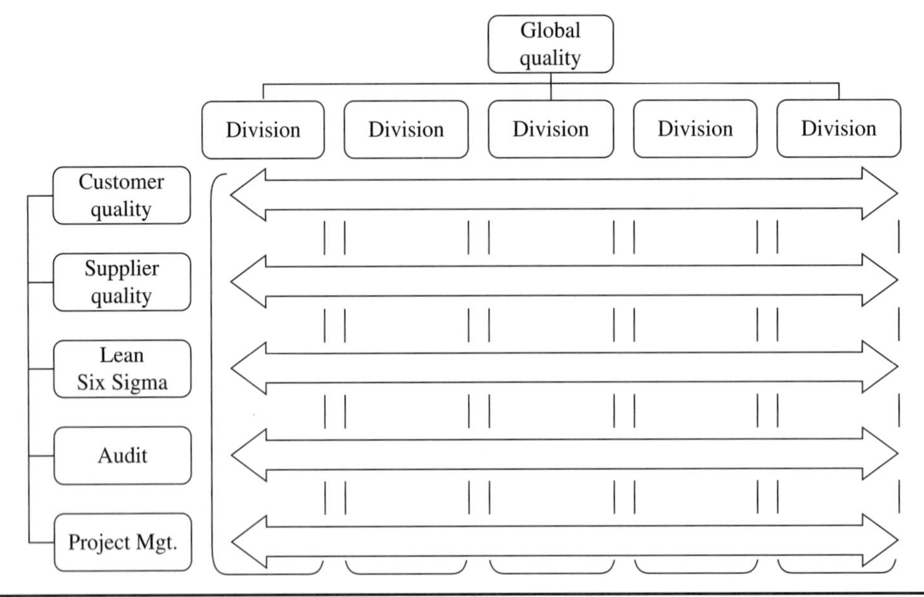

FIGURE 27.7 Global Quality structure.

- Audit
  - Continuous evolvement of quality systems
  - Global utilization of IQRS audit and related improvement plans
- Project management
  - To coordinate global development and implementation of systems, tools, and quality process improvements
- Rules of engagement
  - The manufacturing plant has responsibility for manufacturing related ERA (Emergency Response Action), containment, customer complaints, and 8D actions and responses.
  - The selling entity is responsible for communicating the ERA from seller to customer and to enter the complaint in SAP and inform the manufacturing plant.
  - Escalation above the plant quality manager goes first to the division quality head for the manufacturing division.
  - When a customer requires an on-site visit to discuss a quality problem, it is the responsibility of the closest (geographic) division quality head.

Developing Capable Experts

The Quality Director (The Leadership)

The competencies of the Quality Engineer and the Quality Director are described in this section. Bloom's Taxonomy, first proposed by Bloom (1956), and later revised by Anderson and Krathwohl (2001), is utilized in the competency matrices to describe the levels of knowledge and expertise required for each role. Each matrix is composed of two main sections. The first provides a list of competencies, which are further defined by tools and concepts. In the second section, the tools and concepts are paired with anywhere from zero to all of the six levels of Bloom's taxonomy. These levels describe the cognitive demands required to understand and utilize the competencies. For a complete display of the matrices, refer to "Competencies Matrices" in the Appendix.

The six levels of Bloom's revised taxonomy (Anderson and Krathwohl 2001, beginning with the lowest order to the highest order, is as follows:

1. *Remembering*. Retrieving, recognizing, and recalling relevant knowledge from long-term memory
2. *Understanding*. Constructing meaning from oral, written, and graphic messages through interpreting, exemplifying, classifying, summarizing, inferring, comparing, and explaining
3. *Applying*. Carrying out or using a procedure through executing or implementing
4. *Analyzing*. Breaking material into constituent parts and determining how the parts relate to one another and to an overall structure or purpose through differentiating, organizing, and attributing
5. *Evaluating*. Making judgments based on criteria and standards through checking and critiquing
6. *Creating*. Putting elements together to form a coherent or functional whole; reorganizing elements into a new pattern or structure through generating, planning, or producing

The Quality Director (or Quality Vice President or Quality Manager, depending on organization size and titles) is someone who leads the Quality Office and works with senior management to ensure that quality is planned, ensured, controlled, and improved. The competencies of a Quality Director are

- All of the competencies of a Quality Engineer PLUS the following:
- *Leadership and management*. Motivating and influencing senior management and others, managing departments, and leading the Quality Office
- *Strategic planning and deployment*. Developing and aligning strategies and goals, and incorporating quality and performance excellence into strategic planning and deployment
- *Customer relationship managemen*. Identifying customers and needs, utilizing customer feedback, and improving customer satisfaction and loyalty
- *Supply chain management*. Evaluation and selection of suppliers, management and improvement of suppliers, and supplier certification and partnering
- *Quality information system*. Establish metrics, monitoring, and evaluation protocols
- *Training and development*. Skills assessment, training needs analysis, and development of personnel

As mentioned, the Quality Director should also posses all six of the taxonomy levels.

The Quality Engineer or Qualitist: The Technical Experts

The Quality Engineer, or if you like, a "qualitist," as referred to in *Architect of Quality* (2003) by Dr. Juran, is a person who studies the management of quality and is an expert at the deployment of it. The Quality Engineers should have the requisite knowledge, skills, and experience to be a critical enabler, capable of carrying out the three universal processes for managing quality (quality planning, improvement, and control) within the context of a quality system. The competencies of the Quality Engineer are listed under seven categories in the Competency Matrix in the Appendix. The seven categories are

1. Basic concepts of quality:
2. Quality systems
3. Organizationwide assessment of quality
4. Quality planning
5. Quality control
6. Quality improvement
7. Change management

The Competencies for the Quality Officer Engineer require, at a bare minimum, the first three taxonomy levels: remembering, understanding, and applying knowledge. Most competencies, however, require the addition of the next two: analyzing and evaluating. (For more details on both the Quality Engineer and Quality Director, refer to the Competency Matrices in Appendix II.)

For specific examples of duties, desired attributes, skills, experience, and educational backgrounds required for different quality officer positions, the reader is referred to the example job profiles in the Appendix. These examples showcase the descriptions and requirements for a Quality Vice President, Director, Manager, and Engineer in a multinational global company.

The Performance Excellence Practitioners: The Belts and Change Agents

In addition to the Quality Director and Engineers, the Quality Office of the future will include full-time practitioners: Master and Black Belts, Lean Experts, and Change Agents. These are not new to many organizations. What is new is that their role should be tied to the Quality Office to ensure that these positions continue to keep up with the needs of the organization. For more information on these roles see Chapter 12, Six Sigma: Improving Process Effectiveness.

The following is a list of prerequisite attributes and capabilities that can be used to identify candidates for development into future quality office and performance excellence experts (such as Black Belts, Master Black Belts, Lean Experts, Lean Masters, and various other titles).

The candidate

- Has a managerial or technical specialist position
- Has a deep knowledge of business practices and understands your organization's business plan
- Has high-level or university-level math and reading skills; has basic education in data analysis and statistics
- Has been trained in the management of quality methods and tools
- Has a track record of superior performance, including leadership of new initiatives or change initiatives, or has demonstrated capability of it
- Welcomes accountability and challenges, and is willing to take prudent risks

- Has solid technical skills and knowledge in the target environment
- Makes decisions based on facts and data, and searches for "best practices"
- Has perseverance, stability, and is creative yet pragmatic
- Is a good coach, mentor, and teacher; ideally has experience leading teams
- Has credibility at all levels of the organization
- Has experience using common technology and software; not averse to learning new tools

In addition to the skills and experience prerequisites, there are a number of characteristics and traits that will naturally enhance the candidate's ability to drive change. It is beneficial if the candidate

- Is a clear communicator
- Manages stress effectively
- Learns new concepts quickly
- Has a "can-do" attitude and the ability to manage multiple assignments
- Is goal driven and plans ahead
- Is solution oriented
- Is able to work effectively with all levels of the organization
- Stays on top of key aspects and is timely and efficient

Example profiles of key performance excellence roles can be found in Appendix II.

Professional Certification for Key Quality Officers

The introduction of Six Sigma in the past decade led to an insurgence in certification of Belts. This was largely due to a lesson learned from the Total Quality Management era. During TQM, many so-called experts were trained in the "methods of TQM." Unfortunately, few were trained in the tools to collect and analyze data. As a result, numerous organizations did not benefit from the TQM programs. Motorola introduced a core curriculum that all Six Sigma practitioners needed to learn. That evolved into a certification program that went beyond the borders of Motorola. As a result, there are many "certifiers" that will provide a certification as a Master Black Belt, Black Belt, Green Belt, and so on. Most certifications state that the person certified is an "expert" in the skills of Six Sigma or Lean or both. Certification did lead to improved performance, but also to some weak experts due to no oversight of the certifiers, many of which were consulting companies or universities not well versed in the methods or tools of Six Sigma and Lean.

The American Society for Quality (ASQ) has, for many years, offered certification for quality technicians, quality auditors, quality engineers, and quality managers. As the Six Sigma movement grew, the ASQ and its affiliates around the world began to certify Black Belts. Although not perfect, the ASQ is in a better position to monitor the certifications than self-serving firms. Certification must be based on legitimacy to be effective. Having too many firms certifying Belts will only lead to a weaker certification process.

ASQ's Certified Quality Engineer (CQE) program is for people who want to understand the principles of product and service quality evaluation and control (ASQ, 2009). For a detailed list of the CQE body of knowledge, the reader is referred to the certification requirements for Certified Quality Engineer at www.asq.org.

ASQ also offers a certification for quality officers at the quality management level, called Certified Manager of Quality/Organizational Excellence. ASQ views the Certified Manager of Quality/Organizational Excellence as "a professional who leads and champions process-improvement initiatives—everywhere from small businesses to multinational corporations—that can have regional or global focus in a variety of service and industrial settings. A Certified Manager of Quality/Organizational Excellence facilitates and leads team efforts to establish and monitor customer/supplier relations, supports strategic planning and deployment initiatives, and helps develop measurement systems to determine organizational improvement. The Certified Manager of Quality/Organizational Excellence should be able to motivate and evaluate staff, manage projects and human resources, analyze financial situations, determine and evaluate risk, and employ knowledge management tools and techniques in resolving organizational challenges" (ASQ, 2009).

No matter what organization you use to certify your experts, here are some lessons learned about certification:

- One project is not enough to make someone an expert.
- Passing a written test that is not proctored is no guarantee that the person who is supposed to be taking the test is actually taking it.
- Getting your organization to sign off on the success of the Belt project is no guarantee, unless someone in the organization is knowledgeable about the methods of Six Sigma.
- Select a reputable certifying body.

Responsibilities for the Development of Capable Experts

The training and development of quality and performance excellence experts can succeed only if there is accountability and responsibility for its implementation and effectiveness. This accountability and responsibility lies with the same group that it does in any other key competitive or developmental strategy—with the leadership team. It is their responsibility to agree on the strategy and ensure that it will support the other operational, cultural, and financial corporate strategies. They are not responsible for the planning, design, and execution of the development strategy; this responsibility generally lies with a component of the human resource function, with technical support provided by the Quality Office. The responsible parties are executive leadership, human resources, and the quality office.

Executive leadership. The executive team bears the responsibility for creating a quality and performance excellence culture in the organization. A quality and performance excellence culture is a product of behaviors, skills, tools, and methods as they are applied to the work. These changes don't come about without showing people "how" to implement and sustain this culture. Therefore, the executive team must become educated in quality and performance excellence, and stimulate their professional development team to offer options for training and development for quality and performance excellence. On the basis of these options, the executive team will then develop and approve a strategy and strategic goals for the training and development effort. This effort may be organization-wide and long-term (three to five years), or narrowly focused on a particular segment of the organization or product/service line and planned for a relatively short duration.

Human resources. The human resources (HR) function (or subfunction) bears the responsibility for implementing the quality and performance excellence training and development strategy. The implementation activities include the selection of subject matter, training design and delivery, and establishing an evaluation process. This is integrated with other corporate training and development activities, and follows the same implementation process. The subject matter may be internally sourced or may be outsourced to external quality and performance excellence training providers. The major difference between how this is approached now compared to the past is that there is a strong trend to seamlessly integrate the quality and performance excellence training into the professional development curriculum and to include a high degree of customization to reflect the organization's culture. This is especially true for organizations that have a mature quality and performance excellence system in place.

The Quality Office. The Quality Office is responsible for collaborating with the HR professionals to share their technical expertise on quality and performance excellence, much the same as key sales professionals would share their expertise in identifying and developing the curriculum for sales training. This is a departure from the past, when organizations had elaborate (and sometimes very large) quality departments that identified, developed, and delivered quality and performance excellence training, separate from the training department. This created barriers in the implementation of performance excellence as an integral part of all activities (big Q) and contributed to the "quality versus real work" dilemma of the late 1980s and early 1990s.

An underlying principle of quality and performance excellence is to have an unwavering focus on the customer. Training and development for quality and performance excellence demands the same. A clear understanding of who the customers are, what their needs are, and what the features should be of a training and development strategy, the subsequent subject matter that responds to those needs are critical components.

A clear understanding of the customer means that all of those who will participate or benefit from the training must be considered in the design and delivery. Responsive organizations carefully approach this identification of customers and their objectives, and communication of how the training can help achieve those objectives. Many times, because of the lack of such clear definition, organizations waste huge amounts of time and money providing training on tools and techniques that they will never use. For example, providing training on advanced statistical tools to Champions or team members of Lean Six Sigma projects is wasteful. It was not uncommon in the past for organizations to measure success in quality and performance excellence in terms of the number of individuals they trained and the number of subjects in which they were trained!

Customer-Focused Development Approach

By focusing on the customer, the following approach is recommended, based on the Design methodology. For more details, the reader is referred to Chapter 14, Continuous Innovation: Design for Six Sigma. The approach has three major steps: (1) determine the required competencies, (2) assess the experts and potential experts against the required competencies, and (3) close the gaps.

1. Determine the required competencies.
    a. *Identify the customers of the experts and potential experts.* Customers are the recipients of the outputs of the processes or tasks performed (or to be performed) by the experts. Procurement managers, design engineers, operations managers and technicians, process owners, customer service, and senior management are examples of customers.
    b. *Determine customer needs and prioritize them.* As a reminder, needs should be expressed as benefits (not as features). For example, process owners and managers require the benefit of having capable processes (or the benefit of

improved process capability). A customer-needs matrix cross-referencing customers and needs is developed. The needs are then prioritized based on the relevance and criticality of each need to the customer base.

c. *Translate customer needs into competencies and capabilities.* Continuing with the example, translate the benefit (need) of the improved process capability into competencies and capabilities. So the expert(s) would need the following competencies and capabilities: (1) be able to conduct process capability studies and to understand and interpret metrics such as C_p and C_{pk}, and (2) be able to carry out a DMAIC project, including design of experiments (DOE) to determine the X's in the $Y = f(x)$ of that process in order to improve and optimize the process. In many established organizations, the job description and job profiles state the duties, competencies, and capabilities required of the expert. However, it would be wise to verify that those are indeed what are required and that they are adequately described and that the list is complete.

2. Assess the experts (and potential experts) against the required competencies. (The assessment may be done in a variety of ways. A combination of written and oral assessments has been found to be effective.)

   a. *Written assessment.* Questions for the written assessment should be developed based on the required competencies and capabilities. The assessment consists of multiple-choice questions, essay questions, and problem-solving questions requiring computations using formulas that the expert should know. With the process capability example, questions on when and how to conduct a process capability study, what is control versus capability, and calculating C_p and C_{pk} would be appropriate to evaluate whether the expert is able to evaluate process capability.

   b. *Oral assessment.* Organization-specific scenarios can be created and described to generate interview questions on how best the scenario or situation might be handled by the expert. For example, if one of the required competencies is being able to deal with resistance when implementing change, specific change management questions based on the scenario might be asked, such as, "What steps would you take if you are faced with personnel resisting and not wanting to implement the solutions your improvement project selected for implementation?"

3. Close the gaps.

From the written and oral assessments, gaps in knowledge, competencies, and capabilities for each individual are identified. A development plan for each individual can be developed, which may include training (self-study, instructor-led, and/or on-the-job), assignment of tasks or project work to provide a means for him or her to demonstrate capability, and/or undergoing a development protocol for internal certification or external certification to a professional certification entity [such as those from the Juran Institute, American Society for Quality, Software Enterprise Institute (CMMI), Project Management Institute, and others].

The assessment results can also be summarized at the organization level to determine organizational gaps and training needs. Identification of the gaps can be conducted for each job grade level or job function. For example, questions such as the following may be of interest: Do we have capable quality engineers? Do our quality managers have the required competencies to be effective in their jobs? Are our Lean Six Sigma Black Belts capable of coaching others to drive process improvement? Are the Green Belts knowledgeable enough to be effective after they have been certified for over a year? Are our quality engineers still capable five years after first attaining CQE certification? An example of an executive summary of a skills assessment conducted by the Juran Institute is shown in the Appendix.

Training and development strategies for the organization can be developed by the Quality Office in collaboration with HR and other departments, accordingly. The necessary budgets can be developed and plans put in place for the coming months or fiscal year. The Quality Office and HR can work to deploy the plans, which may include the use of outside training providers and consultants.

Summary

The Quality Office plays an important role, at both the strategic and tactical levels, in enabling the organization to strive for quality and performance excellence. The distinction of control versus creating beneficial change leads to the recognition that parallel departments or divisions within the organization are needed to create beneficial change. As discussed earlier, some of these organization forms may need to be ad hoc or permanent. The roles and responsibilities of the Quality Office and how it is organized are driven by its mission. There are direct and indirect functions. The Quality Office should be an enabler—to enable the organization to drive quality and performance excellence. In this enabling role, capable quality and performance excellence experts are identified and developed to have the right competencies and capabilities, based on what customers of the experts throughout the organization need in order to achieve quality and performance excellence.

References

Anderson, L. W., and Krathwohl, D. R. (eds.) (2001). *A Taxonomy for Learning, Teaching and Assessing: A Revision of Bloom's Taxonomy of Educational Objectives*, complete edition, Longman, NY.

Bloom, B. S. (ed.) (1956). *Taxonomy of Educational Objectives: The Classification of Educational Goals – Handbook I: Cognitive Domain*, McKay, New York.

Crosby, P. B. (2000). "Creating a Useful and Reliable Organization: The Quality Professional's Role." *Annual Quality Congress*, vol. 54, May, pp. 720–722. American Society for Quality, Milwaukee, WI.

De Feo, J. A. (2008). "Enterprise Assurance," Juran Institute White Paper, Juran Institute, Inc., Southbury, CT.

Gryna, F. M. (2002), "Interview: The Role of Quality and Teams in the 21st Century." *AQP News for a Change*, vol. 6, no. 7, July, pp. 1–3.

Gryna, F. M., Chua, R. C. H., and De Feo, J. A. (2007). "Organization for Quality." Chapter 7 in *Juran's Quality Planning and Analysis for Enterprise Quality*, McGraw-Hill, New York.

Imler, K. (2006). Core Roles in a Strategic Quality System. *Quality Progress*, June, American Society for Quality, Milwaukee, WI. pp. 57–62.

Spichiger, J. (2002). The Changing Role of Quality Professionals. *Quality Progress*, vol. 35, no. 11, November, American Society for Quality, Milwaukee, WI, pp. 31–35.

Watkins, D. K. (2005). Quality Management's Role in Global Sourcing. *Quality Progress*, vol. 38. no. 4, April, American Society for Quality, Milwaukee, WI, pp. 24–31.

Westcott, R. (2004). Metamorphosis of the Quality Professional. *Quality Progress,* October, American Society for Quality, Milwaukee, WI, pp. 22–32.

CHAPTER 28

Research & Development: More Innovation, Scarce Resources

Brian A. Stockhoff

About This Chapter

This chapter discusses managing for quality in research organizations and in development processes. The material focuses on concepts, infrastructure, methods, and tools for simultaneously improving customer satisfaction and reducing costs associated with both research and development functions. Quality within the software development process is discussed in Chapter 29, Software and Systems Development: From Waterfall to AGILE.

High Points of This Chapter

1. The research and development function is positioned at the "fuzzy front end" of the innovation cycle and is subject to forces of both "market push" and "technology pull." This sets the stage for unique problems and opportunities in managing for quality.
2. Producing high-quality research requires balancing the sometimes opposing needs of business risk management and control and creativity.

3. Cultural and organizational factors can be significant, unrecognized barriers to sustained high-quality research and long-term organizational success.
4. Measures of R&D quality continue to evolve toward true performance metrics that should entail a combination of lagging, concurrent, and leading indicators.
5. Quality planning concepts, tools and methods should be applied to help ensure that product design meets customer requirements. Traditionally, this includes designing for such attributes as reliability and maintainability, but new dimensions such as designing for ergonomics and ecological impact are emerging.

Introduction

Frequently the combined term "R&D" is used to describe cross-departmental processes that integrate new knowledge and technology emanating from the research function with the subsequent development of new (or improved) processes and products. However, because these activities remain distinct in most organizations, it will be useful in this chapter to distinguish between managing for quality in research processes and managing for quality in development processes. The primary objective is to impart to the reader an understanding of how various organizations have integrated quality concepts and, ultimately, activities to foster the development and launch of new and innovative products and services.

Pushmi-Pullyu: Managing the Forces

Dr. Joseph Juran's original spiral of progress in quality (see Gryna et al. 2007, pp. 15-16) focused on the cross-functional flow involved in "developing" a new manufactured product. In the context of the original spiral, requirements for the new product emanated from marketing research. The marketing organization conducted research to define customers' needs, as well as to obtain customers' feedback on how well the organization had met those needs with existing products. Based upon customers' feedback and changing customer needs, a new turn of the spiral began.

By analogy, however, the "bridge" between R&D and customers can be initiated by either customers or R&D, with marketing facilitating the building of the bridge. Although a predominant source, marketing research is not the only possible origin of new technology and product ideas. This is especially true with technology-based products because customers (end users in particular) do not or cannot always articulate unmet needs that technology can address. Strategy-directed research is another significant source of product ideas and leverages the concept behind the quote attributed to Louis Pasteur that "chance favors the prepared mind." For example, Post-it notes resulted from the "failure" of an experiment that was recognized by a researcher as an opportunity for a new product. The low-tack, pressure-sensitive adhesive that is the critical product characteristic languished for several years before another scientist took advantage of the 3M Company's bootlegging policy that allowed R&D scientists to unofficially pursue new ideas (Petroski 1992).

Although failure is not the most desired route, it is a necessary part of R&D success over the long run. Roussel et al. (1991) and, more recently, Miller and Morris (1999) emphasized the criticality of using exploratory research balanced with business discipline, conducted proactively to support an organization's strategic focus. Nussbaum (1997) quotes Thomson Consumer Electronics' vice president of consumer electronics for multimedia products as stating that design processes are being used "to address overall strategic business issues." More recently, Beall (2002) emphasized the need to encourage the indistinct "fuzzy front end" of the innovation cycle that supports landmark new products.

Strategic research increasingly is being focused on the delivery of concepts and technologies that will drive new or improved technologies, such as quantum computing and now-commercialized nanotechnology. Strategy-directed research often supports the development of new, breakthrough "platform" technologies that initially have no apparent market but can later generate multiple products and significant competitive advantage, especially where intellectual property is involved. The origins of marketing research and strategy-directed research for new technologies and product concepts can be characterized as "market pull" and "technology push," respectively.

Regardless of the means for identifying needs and opportunities, managing for quality in research organizations and development processes has become recognized as a critical activity. In addition to synthesizing information toward technology, goods, and services that are fit for use, an organization needs to operate with both speed and efficiency. This requires coordinated effort between research and development. Throughout this chapter are examples of tools and processes that can facilitate coordination, reduced cycle times, and costs, although few organizations excel in more than a few core areas. Dell computer vaulted to a top industry position on the conviction that velocity attained via the compression of time and distance backward into the supply chain and forward to the customer, is a driver of competitive advantage (Dell 2000). Speed has its limitations (see Thackara 2005 for a critique of the "need for speed" philosophy), but managing for quality in the R&D processes demonstrably can simultaneously reduce cycle times and costs. At a Shell Oil research center, Jensen and Morgan (1990) found that a quality team's project for improving the project requirements process resulted in decreasing project cycle times by 12 months. At Corning Laboratories (Smith 1991) $21 million of cost reductions were realized over a four-year period while new products were pushed out faster and with lower costs. A project to reduce researchers' idle time during experiments produced $1.2 million in "easy savings." Similarly, Hutton and Boyer (1991) reported on a quality improvement project in Mitel Telecom's Semiconductor Division that resulted in custom prototype lead times being reduced from 22 weeks to 6 weeks. More recently, Shankar et al. (2006) reported that TAP Pharmaceutical Products, Inc. attained a 68% reduction in documentation processing time (from 282 days to 90 days), thereby facilitating earlier product registration and launch. These are not small amounts, but as a last example consider Boeing's 787 Dreamliner and revamped 747-8 jumbo jet that, through delays, rework, and program management issues generated a combined total of approximately $3.5 billion in charges (Sanders 2009).

The Missions of Research and Development

Although research and development have common business goals, in order to manage the research function and development processes, it is critical to define and understand their respective, separate missions. To help distinguish among various types of research and development activities, the Industrial Research Institute (1996) provided the following definitions:

- "Basic" (or "fundamental") research consists of original experimental and/or theoretical investigations conducted to advance human knowledge in scientific and engineering fields.
- "Directed basic" (or "exploratory") research is original scientific or technical work that advances knowledge in relevant (to corporate business strategies) scientific and engineering fields, or that creates useful concepts that can be subsequently developed into commercial materials, processes, or products and, thus, make a contribution to the company's profitability at some time in the foreseeable future. It may not respond directly to a specific problem or need, but it is selected and directed

in those fields where advances will have a major impact on the company's future core businesses.

- "Applied" research is an investigation directed toward obtaining specific knowledge related to existing or planned commercial products, processes, systems, or services.
- "Development" is the translation of research findings or other knowledge into a plan or design for new, modified, or improved products/processes/services whether intended for sale or use. It includes the conceptual formulation, design, and testing of product/process/service alternatives, the construction of prototypes, and the operation of initial, scaled-down systems or pilot plants.

Building from Roussel et al. (1991), the following general definitions for the research and development processes are useful:

- Research: The process used by an organization to acquire new knowledge and understanding.
- Development: The process used by an organization to apply and connect scientific or engineering knowledge acquired from research for the provision of products and/or services commensurate with the organization's mission.

Although the latter definitions are broad, they are helpful. Both have been constructed to incorporate the word "process." One of the tenets of business management strategies and programs such as Six Sigma, Total Quality Management (TQM) and the Baldrige National Quality Program is to improve key processes that result in "products" that are "fit for use" by an organization's internal and external customers. In support of this perspective, Nussbaum (1997) stated: "At the leading edge of design is the transformation of the industry to one that focuses on process as well as product." Organizations easily can overemphasize product or process innovation to the detriment of the other (potentially suboptimizing the larger business) because they tend to shift in relative importance through a product's life cycle. Early in the life cycle, innovation is greatest in the product itself, whereas later in the life cycle, the product itself is relatively static, but the production process is tinkered with to provide improvement (Figure 28.1). Correspondingly, Himmelfarb (1996a) has suggested

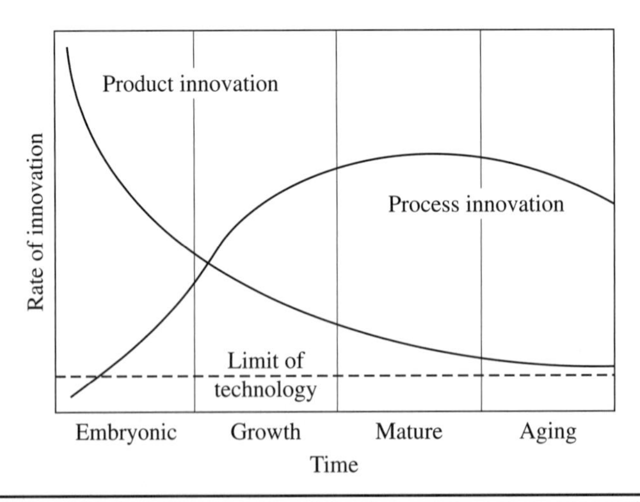

FIGURE 28.1 Shift in emphasis during research and development from product innovation to process innovation.

that one key responsibility of senior managers is to ensure that the product development process is well defined (via flowcharts), documented, understood, monitored, and improved. It is, therefore, useful to define the "products" and "customers" of the research and development processes, which, in turn, can be used to define, measure, plan, control, and improve process quality.

Products of Research and Development Processes

Juran (1992) defined a product as "the output of any process" and noted that the word "product" can refer to either goods or services. For the purpose of this chapter, product will be used to denote the intermediary or final outputs of either the research organization or the development process. The primary "products" of a research organization are information, knowledge, and technology. The products of the development process are new or improved processes, goods, or services that result from the application of the knowledge and technology. For example, the output of a research project may include a report containing the conclusions stemming from the project, or patent applications. Corresponding examples of final outputs of the product development process are designs and specifications released for production. Both the research and development processes also have intermediate or in-process outputs. Likely intermediate outputs of the research process are mathematical models, formulas, calculations, or the results from an experiment. Correspondingly, likely intermediate outputs of the development process are physical models, prototypes, or minutes from design review meetings.

Processes of Research and Development

Generically, the steps to product development are as follows:

- *Idea generation*. Sometimes called "ideation," this step involves the initial generation, development, and communication of ideas for new products. Ideas may come from virtually anywhere, including customers, R&D staff, sales staff, employees at large, focus groups, trade shows, and competitive intelligence.
- *Idea screening*. In this step, generated ideas are affinitized and filtered, ideally by concrete and objective criteria that may include technical feasibility, market fit and forecasts, competitor positions, intellectual property assessment, and overall profitability.
- *Concept development*. Typically, "proof of concept" is pursued by refining the target market, product attributes, likely production methods, and costs of both further development and manufacture. This phase may involve application of simulation, model building, and rapid prototyping.
- *Business case development*. Essentially, the expected selling price and sales volume are estimated to obtain anticipated revenues, and costs subtracted from these. Various metrics can be used for this, from straightforward breakeven points and net present value (NPV), to more complicated modeling and forecasting methods. Products often have a cross-functional team that reports to a program manager or similar senior person that is competent in bringing new products and technologies to market.
- *Beta testing*. Physical prototypes are produced in small quantities for testing under a range of conditions, focusing most on typical usage. Additionally, testing may include potential customer feedback via focus groups or preliminary release to the public (or a selected group) for evaluation, for example. Invitation to potential customers to participate in beta testing is especially common in the software world.

- *Technical development and implementation.* In this step, the procedural details are addressed to allow a product launch. This includes formalizing product specifications and engineering requirements, establishing supplier relationships, resourcing, scheduling, and logistics. In many industries such as pharmaceuticals, significant regulatory requirements also must be met. If it has not already been done, a product manager often is assigned to oversee the transition from development to commercialization.
- *Commercialization.* The product is launched, the distribution pipeline is filled, and ongoing maintenance activities (such as advertising) is initiated. Although this step usually is not considered a part of the development process, proactive organizations will continue to seek opportunities for product improvement, for example, by encouraging engagement of R&D staff with sales and marketing, and customers through to the end user.

Some processes do not fit neatly into any single step and may extend across multiple steps. Examples of important overlapping or interfacing research processes that were identified and improved at Eastman Chemical Company are provided by Holmes and McClaskey (1994) and include business unit organization interaction, needs validation and revalidation, concept development, technology transfer, and project management. Figure 28.2 (Holmes and McClaskey 1994), is a macrolevel process map of Eastman Chemical Company's "Innovation" process that follows the generic steps outlined above. Steps 1 to 4 represent the macrolevel research activities that generate the "new or improved product and process concept" stemming from step 4. The last step is the macrolevel development process that yields the processes and product designs for use in operations and markets, respectively.

Many organizations depict their product development processes through flowcharts reflecting their processes' major phases and "gates" (decision points). Altland (1995) discussed the use of a "phase-gated" robust technology development process used by Kodak to help ensure that process and product technologies are "capable of manufacture and are compatible with intended product applications." Gate reviews provide an excellent opportunity for the research, development, and marketing functions to come together periodically to plan a product pipeline. In the flowchart in Figure 28.3 (Boath 1993), the results are shown of "reengineering" an organization's new product development process. The new process led to a 25 percent increase in efficiency in "resource utilization."

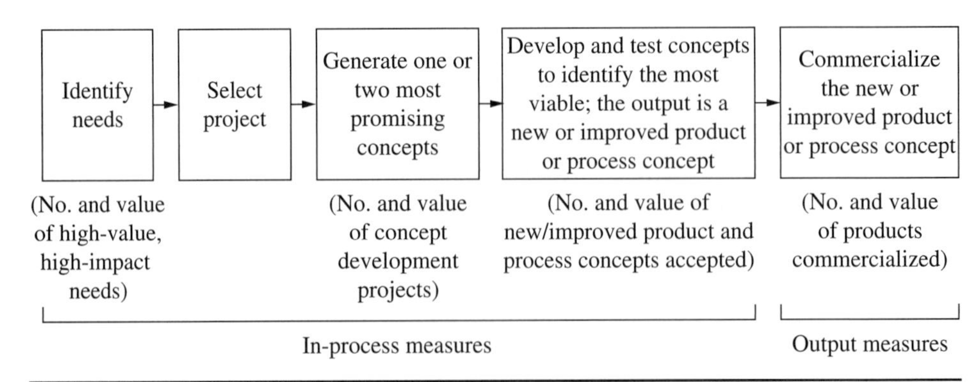

FIGURE 28.2 Eastman Chemical's Innovation Process. (*Holmes and McClaskey 1994.*)

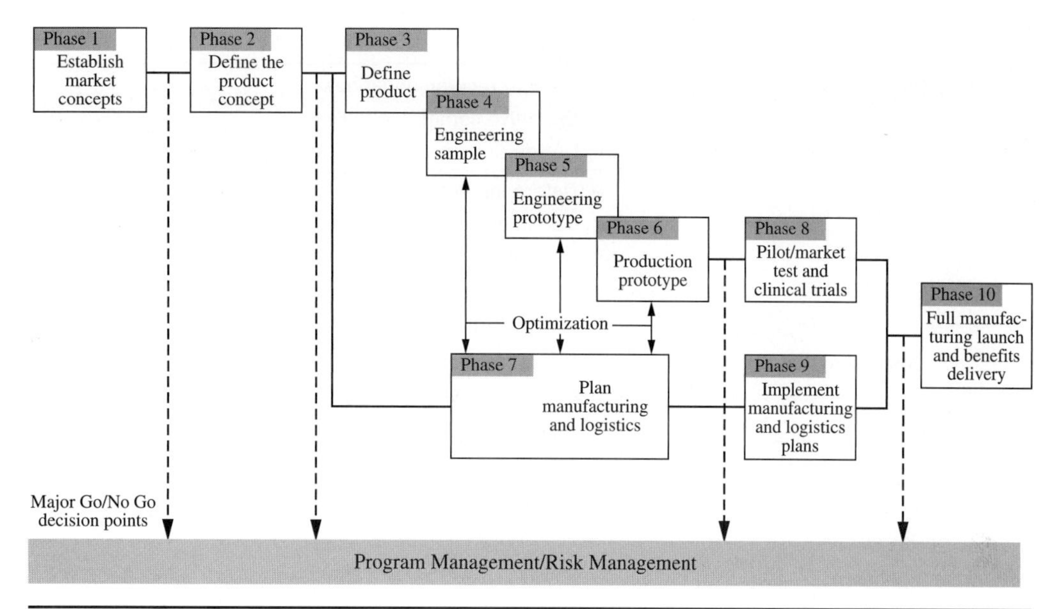

FIGURE 28.3 A development process for new products. (*Boath 1993.*)

Raven (1996), of Merrill Lynch's Insurance Group Services, Inc., provided an example of a project management process for product development in a financial service organization. The nine-step process, depicted in Figure 28.4, was cited by Florida's Sterling (Quality) Award Examiners as being an example of a ". . . role model for excellence." The activities associated with each of the nine steps are listed in Table 28.1 Himmelfarb (1996b) provides additional examples of new product development in service industries.

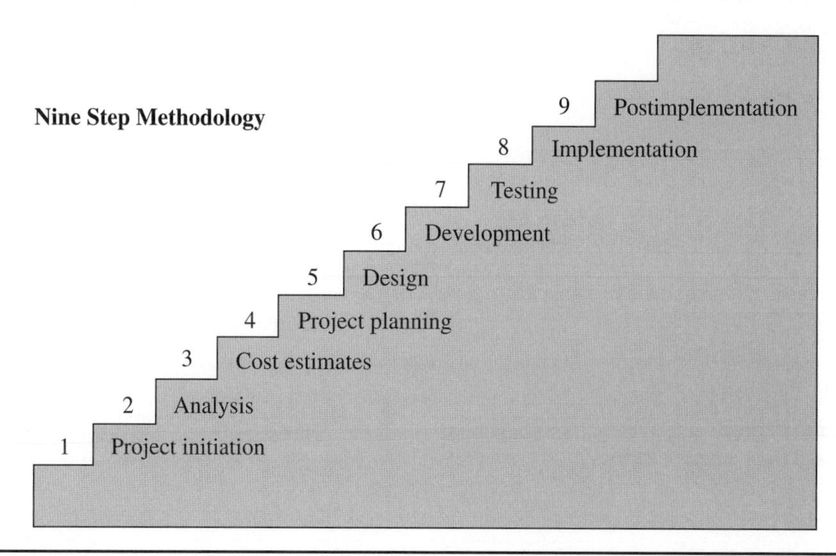

FIGURE 28.4 Merrill Lynch Insurance Group services project planning and development process. (*Raven 1996.*)

Step	Activities
1. Project initiation	a. Prepare recommendations b. Executive committee review c. Decide on approval (Yes/No)
2. Analysis	a. Determine scope b. Obtain sign off on scope c. Develop requirements d. Review requirements e. Conduct market research
3. Cost estimates	a. Determine cost estimates b. Conduct feasibility study
4. Project planning	a. Prepare timelines b. Develop action plans c. Schedule meetings
5. Design	a. Develop system design b. Develop business procedures
6. Development	a. Prepare SEC and state filings b. Complete system programming c. Develop test plan d. Develop work flows, policies, and procedure bulletins e. Prepare training, marketing, and sales materials f. Determine purchasing and print requirements g. Obtain sign-off
7. Testing	a. Conduct program testing b. Conduct system testing c. Conduct user acceptance testing d. Conduct regression testing e. Conduct quality assurance tests f. Conduct branch office testing g. Obtain sign-off
8. Implementation	a. Distribute policy and procedure bulletins, training materials, marketing, and sales materials b. Conduct operational training sessions c. Implement new systems, procedures, and processes
9. Postimplementation	a. Conduct Postproject reviews and surveys

(*Source*: Raven 1996.)

TABLE 28.1 Activities within Steps of Merrill Lynch Insurance Group Services Project Planning and Development Process

Defining the Quality of Research and Development

Just as quality products and services must be fit for use and free from deficiency, so should the processes responsible for bringing them to market. The many participants and dimensions contributing to quality make any clear-cut definitions tenuous (although one might say "I know it when I see it"); however, set out as follows are some practical considerations.

Defining Research Quality

In this section, research quality will be defined from the perspective of both customer satisfaction (effectiveness of features) and costs (efficiency of providing the features). Building from Wood and McCamey (1993) and Schumann et al. (1995), Kumar and Boyle (2001) created an operational definition of quality in R&D, stating:

> An understanding of who the R&D client is and what his or her values and expectations are, what the key technologies are and how they can be used to meet R&D clients' expectations and the needs of the entire organization, and who the R&D competitors are and how they will respond to emerging R&D clients' needs. This is achieved by doing things right once you are sure you are working on the right things, concentrating on continually improving the system, enabling people by removing barriers, and encouraging people to make their maximum contribution.

As examples of specific implementations of this, consider General Electric (Garfinkel 1990) that defined several dimensions of research quality: technical quality of research, impact of research ("game changer" versus incremental), business relevance, and timeliness (early or late relative to the targeted market requirements). At DuPont, Darby (1990) viewed R&D quality as "creating, anticipating, and meeting customer requirements," which required "continual improvement of knowledge, application, and alignment with business objectives."

The primary products of the exploratory and applied research process are information, knowledge, and technology. Godfrey (1991) provides a general discussion of information quality. Research product quality can therefore be defined both from the perspective of customers' satisfaction with the features of the information and the absence of deficiencies of the information (which decreases costs and cycle times, and thus increases efficiency). Features of research information include timeliness, utility, accuracy, and costs. Research deficiencies can either occur during the research process or be reflected in the end products of the research. Possible deficiencies in research products may be that the knowledge is late, inaccurate, irrelevant, or of relative poor value for the investment. Deficiencies in research processes are associated with process "rework" or "scrap," for example, having to reissue a section of a progress report because the wrong formula was used or having to redo an experiment because an audit revealed that reference samples had been contaminated.

Combining these perspectives, a simplified definition of research quality is the extent to which the features of the information and knowledge provided by the research function meet users' requirements. It is important to note, however, that expectations of quality may differ among researchers and customers. Take, for example, the need for failure as a part of research versus the need for success and the (potentially premature) "pull" by downstream processes of products that have, from the perspective of the scientific method, been inadequately tested. The dynamics of research and the typical business orientation towards stability and results can be a significant source of tension and conflict between business and research divisions, and disagreement as to research quality.

Defining Development Process Quality

The primary result of development is new or improved products and processes. The quality of a development process is defined as the extent to which the development process efficiently provides process and product features capable of repeatedly meeting their targeted design goals, for example, for costs, safety, and performance.

The resultant product and process features must be thought of from the perspective of "Big Q" thinking; that is, not only product and service quality but also the quality of processes, systems, organization, and leadership (Gryna et al. 2007). For example, Port (1996) discusses the early growth of environmentally friendly products and processes; being "green" has continued to increase in importance as a dimension of quality. In an economic analysis of green product development, Chen (2001) concluded that green products do not necessarily benefit the environment but that appropriate regulations could reverse this. Indeed, in recent years, regulators have compelled designers and manufacturers to address such issues as disposal, for example, battery disposal legislation in the United Kingdom, a German ordinance requiring manufacturers to assure the disposability of all packaging used in product transport, and, in the Netherlands, the rule that manufacturers must accept old and broken appliances for recycling.

Looking at internal, intermediate products, deficiencies and thus inefficiencies in the development process are associated with process rework or scrap. Berezowitz and Chang (1997) cite a study at Ford Motor Company, discussed by Hughes (1992), which concluded that although the work done in the product "design phase typically accounted for 5 percent of the ongoing total cost," it accounted for 70 percent of the influence on products' future quality. Boznak and Decker (1993) report that costs associated with deficiencies in product design and development processes can be very expensive. They reference one computer manufacturer whose costs "exceeded $21 million . . . (which) equated to 420,000 hours of non-value-added work . . . who lost nearly $55 million in gross margin opportunity on one product. Failure to effectively manage its product development processes put the company's entire $1.54 billion international business at risk." The authors suggest that the company's practices that caused this near catastrophe would have been precluded had those practices complied with the requirements of ISO 9000 (see Chapter 16, Using International Standards to Ensure Organization Compliance, for a discussion of ISO standards).

Examples of design "rework" include design changes necessitated by an outdated requirements package and partial redesigns necessitated by missing one or more design objectives (including schedules and costs). In many cases, the true costs of design rework are poorly understood; this is exacerbated by the trade-offs inherent to a compressed product development cycle. Arundachawat et al. (2009) address this topic with a literature review of published examples of design rework in concurrent engineering, including a compendium of factors implicated in causing rework. The authors cite three methods to estimate design rework: direct experimentation, mathematical modeling, and simulation but only report simulation as being used in published works.

More generally, Perry and Westwood (1991) measured development process quality by the extent to which technical targets are met, for example, "meeting specific process capability targets" and "the percent and degree of customer needs that are met, and the number of problems discovered at various stages of the product development process." At Motorola's Semiconductor Sector, Fiero and Birch (1989) reported that reducing development process deficiencies increased the percentage of fabricated prototypes passing all tests upon first submission from 25 percent to 65 percent. Furthermore, by involving 10 functional areas, Motorola was able to shorten development cycle times from 380 to 250 days. The reported investment of $150,000 resulted in potential additional revenues of $8 million per year.

Planning and Organizing for Quality in Research and Development

Quality does not just happen naturally; indeed, quite the opposite prevails (to quote from a Marvel comic book villain, "Entropy, entropy, all winds down"). To combat poor quality, it is imperative to understand the forces that tend to inhibit or promote quality including cultural barriers, infrastructure and organizational structure, and skill development. Managing for quality also necessitates knowing how one is performing, which is accomplished through measuring.

Identifying and Addressing Barriers

To successfully plan for and use the concepts required to manage for quality in research or development, management must first understand and then address potential implementation pitfalls and barriers associated with developing and implementing quality initiatives within R&D environments. Hooper (1990) and Endres (1992, 1997) discuss cultural and organizational barriers that must be addressed. For example, researchers' fear that quality initiatives will stifle individual creativity, resulting in bureaucratic controls, can be addressed through the choice of pilot projects. A project can be chosen to demonstrate the "what's in it for me?" in that improving research quality can provide researchers with better resources or processes for conducting more efficient research (e.g., reducing cycle times for obtaining reference articles; obtaining more information from fewer experiments using statistically designed experiments).

Hooper (1990) identifies as an organizational barrier to improving R&D quality R&D's traditional isolation from customers and business. Oestmann (1990) discusses how Caterpillar addressed the problem of researchers being isolated from their customers by moving ". . . experienced research engineers into the field, close to high populations of customers. Their assignment is to understand the customer—how he used his machines today and how he will use them in the future, what drives the customer to make buying decisions now and in the future. The objective of this is to envision what technologies will be needed to produce superior future products." After research evolved the most promising technologies, Caterpillar used cross-disciplinary teams to develop the required product concepts. Teams comprising representatives from marketing, engineering, manufacturing, and research develop concepts for solving customers' needs "and then rate each idea based on its value to the customer." Another solution to the isolation problem is to "bring the field to the staff" by means of a competitive intelligence program that extends beyond the usual audience of senior management, sales, and marketing to include topics of interest to R&D staff. Such a program can provide early identification of scientific or engineering breakthroughs that might take years to become manifest in marketing-driven reports, publications, patents, or competitor products (Murphy 2000).

For development personnel, Gryna (1999) discusses the importance of placing product developers in a state of "self-control." (See Chapter 20, Product-Based Organizations: Delivering Quality While Being Lean and Green, under Concept of Controllability; Self-control.) Prior to holding designers responsible for the quality of their work products the three major criteria (I, II, III) provided in Table 28.2 must be met. Gryna, using input from designers, developed the specific items listed under each criterion. The table may be used as a checklist to identify opportunities for improving designers' work products, and subsequently, their motivation for quality improvement.

Leadership and Infrastructure Development

For upper managers to successfully lead a quality initiative, they must understand their respective roles and responsibilities in managing for quality. Holmes and McClaskey (1994) stated that at Eastman Chemical:

> Top Research Management Leadership was the most significant and essential success factor. Research management changed the way it managed research by focusing on the major output

I. Have designers been provided with the means of knowing what they should be doing?
  - *A.* Do they know the variety of applications for the product?
    1. Do they have complete information on operating environments?
    2. Do they have access to the user to discuss applications?
    3. Do they know the potential field mususes of the product?
  - *B.* Do they have a clear understanding of product requirements on performance, life, warranty period, reliability, maintainability, accessibility, availability, safety, operating costs, and other product features?
    1. Have nonquantitative features been defined in some manner?
    2. Do designers know the level of product sophistication suitable for the user involved?
  - *C.* Are adequate design guidelines, standards, handbooks, and catalogs available?
  - *D.* Do designers understand the interaction of their part of the design with the remainder of the design?
  - *E.* Do they understand the consequences of a failure (or other inadequacy) of their design on: (1) the functioning of the total system? (2) warranty costs? (3) user costs?
  - *F.* Do they know the relative importance of various components and characteristics within components?
  - *G.* Do they know what are the manufacturing process capabilities relative to the design tolerances?
  - *H.* Do they derive tolerances based on functional needs or just use standard tolerances?
  - *I.* Do they know the shop and field costs incurred because of incomplete design specifications or designs requiring change?

II.Have designers been provided with the means for knowing what they are doing?
  - *A.* Do the have the means of testing their design in regard to the following:
    1. Performance, reliability, and other tests?
    2. Tests for unknown design interactions or effects?
    3. Mock-up or pilot run?
  - *B.* Is there an independent review of the design?
  - *C.* Have the detail drawings been checked?
  - *D.* Are designers required to record the analyses for the design?
  - *E.* Do they receive adequate feedback from development tests, manufacturing tests, proving ground tests, acceptance tests, and user experience?
    1. Are the results quantified where possible, including severity and frequency of problems and costs to the manufacturer and user?
    2. Does failure information contain sufficient technical detail on causes?
    3 Have designers visited the user site when appropriate?
  - *F.* Are designers aware of material substitutions, or process changes?
  - *G.* Do they receive notice when their design specifications are not followed in practice?

III.Have designers been provided with the means of regulating the design process?
  - *A.* Are they provided with information on new alternative materials or design approaches? Do they have a means of evaluating these alternatives?
  - *B.* Have they been given performance information on previous designs?
  - *C.* Are the results of research efforts on new products transmitted to designers?
  - *D.* Are designersí approvals required to use products from new suppliers?
  - *E.* Do designers participate in defining the criteria for shipment of products?
  - *F.* May designers propose changes involving trade-offs between functional performance, reliability, and maintainability?
  - *G.* Are designers told of changes to their designs before they are released?
  - *H.* Have causes of design failures been determined by thorough analysis?
  - *I.* Do designers have the authority to follow their designs through the prototype stage and make design changes where needed?
  - *J.* May designers initiate design changes?
  - *K.* Are field reports reviewed with designers before making decisions on design changes?
  - *L.* Do designers understand the procedures and chain of command for changing a design?

(*Source:* Juran Institute, Inc. Copyright 1994. Used by permission.)

Table 28.2 A Self-Control Checklist for Designers

and by personally leading the analysis and improvement of the key management processes which drive the output. Research management since 1990 has institutionalized QM (Quality Management) by making it the way Research is managed. The ECC Research success story is certainly another illustration of a quote by Dr. J. M. Juran (1992): *"To my knowledge no company has obtained world class quality without top managers taking charge."*

A key responsibility of upper management in leading a quality initiative within research or development is to organize and develop an infrastructure for initiating, expanding, and perpetuating quality in both research organizations and development processes.

Organizing for R&D Quality

Many R&D organizations have developed structures that facilitate the attainment of their goals for improving customer satisfaction and reducing the costs of poor quality. Predominating are administrative teams such as steering teams and quality councils. Wood and McCamey (1993) discuss the use of a steering team at Procter & Gamble for "maintaining momentum," representing all levels of the organization, and from which subgroups were spun off "to manage areas such as communication, training, planning, measurement," and team support. "The role of the steering team was to keep the division focused on business results and setting clear, measurable targets." Taylor and Jule (1991) discuss the role of the quality council at Westinghouse's Savannah River Laboratory, consisting of the laboratory chairman, department heads, two senior research fellows, and the laboratory's TQM manger. The council was supported by department/section councils in developing, implementing, and tracking an annual Quality Improvement Plan (QIP). The QIP was developed by a team of laboratory managers chartered by the director to assess quality progress during the previous year and "select topical areas for improvement in the coming year based on employee input. . . ." Each department manager was assigned a topical area and required to develop an improvement plan. The separate improvement plans were then reviewed and integrated into a quality improvement plan for the entire laboratory. Menger (1993) discussed the organization and activities of the World Class Quality (WCQ) Committee at Corning's Technology Group, consisting of representatives from five major groups reporting to Corning's vice-chairman. The WCQ identifies priorities and reviews progress in its group's members, establishing and improving key results indicators (KRIs) for cycle times, productivity, and customer and employee satisfaction. Figure 28.5 (Menger 1993), portrays the organization structure and process used to track and improve performance. As a final example, Figure 28.6 (Hildreth 1993) is a structure used to manage key business processes, for example, clinical research, development, and product transfer in manufacturing in R&D at Lederle-Praxis Biologicals. The Executive Quality Council is supported by a Business Process Quality Management (BPQM) Council and site-specific quality councils.

It is notable that organizing for R&D quality takes on a particularly troublesome aspect for organizations that have, typically through merger and acquisition activity (but also via strategic alliances), R&D facilities and staff that are physically remote from headquarters or each other. The healthcare products company Novartis, for example, has approximately 20 different R&D facilities spread across nine countries in Europe, Asia, and the United States. The trade-offs among centralized, decentralized, and hybrid R&D structures are onerous, and companies may shift considerably along the spectrum over time in response to changing economic conditions and corporate culture. The infrastructure elements cited above help instill and maintain quality and consistency across an organization while still allowing for the necessary freedoms that promote innovation. The role of R&D structural organization in quality and innovation remains a lively topic, and readers can find more detailed discussion

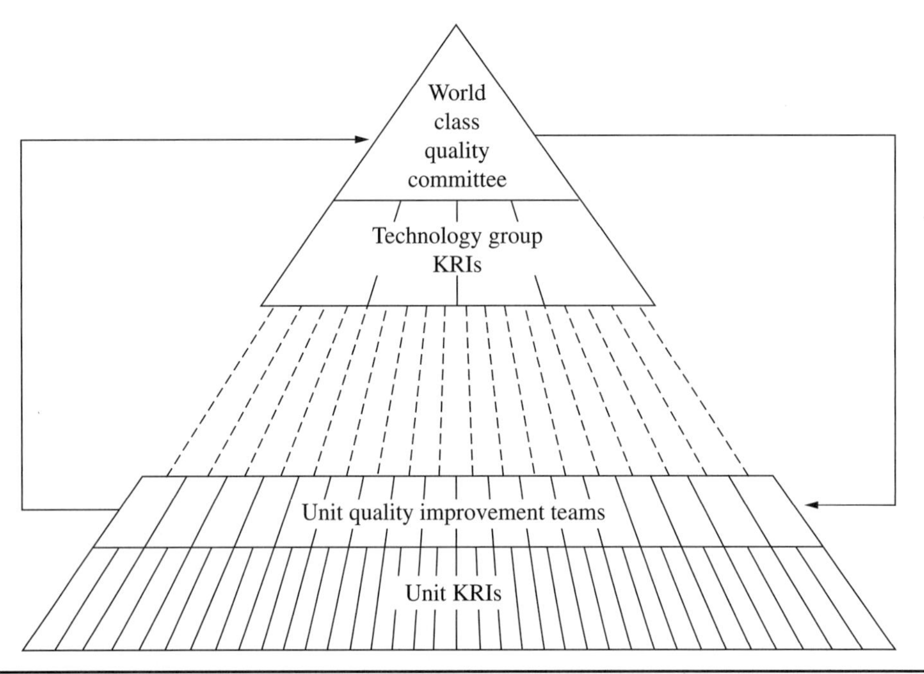

FIGURE 28.5 Corning's Technology Group quality organization and KRI improvement process. (*Menger 1993, pp. 1-14.*)

and recommendations in Richtne'r and Rognes (2008), Argyres and Silverman (2004), Mendez (2003), Gunasekaran (1997), and Ogbuehi and Bellas (1992).

In addition to organization structure, other elements of infrastructure are required to perpetuate R&D quality initiatives. These include training, project teams, facilitators, measurement systems, and rewards and recognition. Special considerations for training of R&D staff are discussed next.

Training for Quality in R&D

Before managers or researchers can lead or implement quality concepts, processes, or tools, their needs for education and training must be identified and met. Wood and McCamey (1993), of Procter & Gamble, discuss the importance of tailoring the training to the R&D environment:

> Our training had two key features: 1) it was focused on business needs and 2) it was tailored to the audience. These features reflected lessons we learned from other parts of the company; e.g., training that was not focused on real business issues lacked buy-in, and a training program developed for manufacturing could not be transplanted wholesale into an R&D organization.

Similarly, at Bell Laboratories, Godfrey (1985) reported that a key ingredient for successfully training design engineers in experimental design and reliability statistics is the use of case studies based upon real problems that "Bell Labs engineers have had . . . ." Training designers in modern technology can yield significant paybacks. At Perkin-Elmer, De Feo (1987) reported that training design engineers in Boothroyd and Dewhurst's (1987, 1994) design for assembly (DFA) methodology resulted in "weighted average" decreases of 48 percent in assembly times and 103 percent increases in assembly efficiencies.

Yoest (1991), reporting on a study conducted by Sverdrup Technologies at the Arnold Engineering Development Center, Arnold Air Force Base, concluded that teams whose

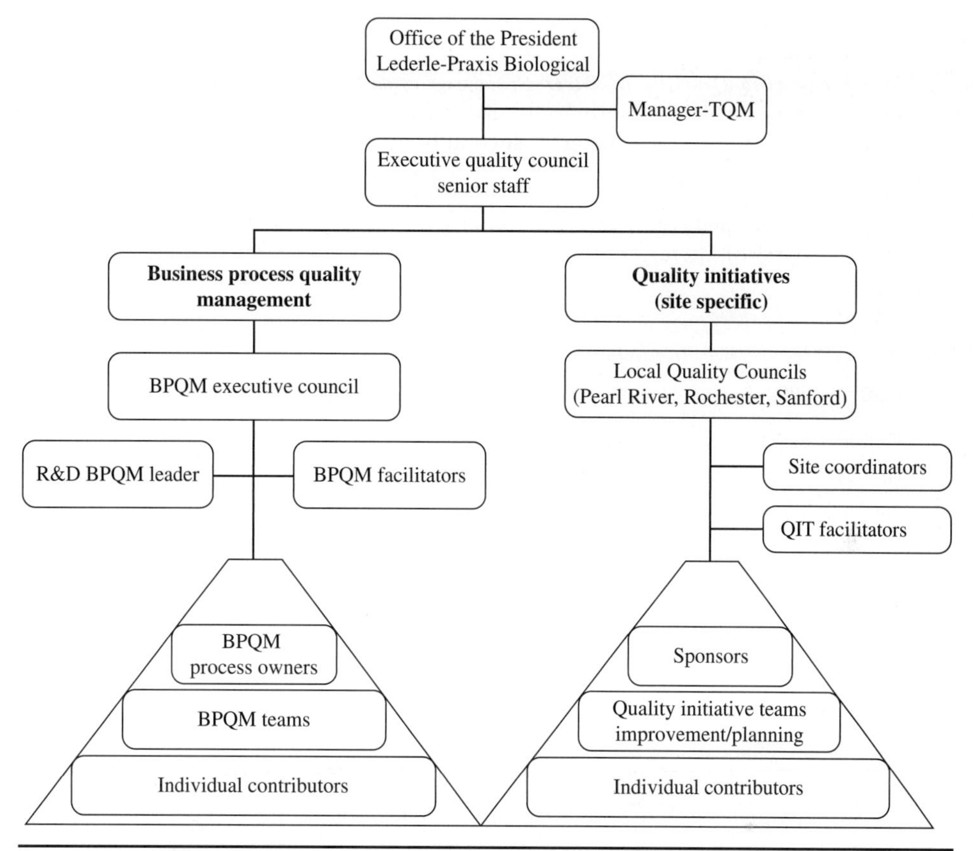

Figure 28.6 BPQM and site quality councils. (*Hildreth 1993, p. 2A-14.*)

facilitators and team leaders are specifically trained for their roles are more likely to successfully achieve their missions than teams whose leaders and facilitators did not receive training. Konosz and Ice (1991), at Alcoa's Technical Center, have similarly stated that "The successful implementation of problem-solving teams and quality improvement processes requires three critical components: (1) management leadership and involvement, (2) team training, and (3) process facilitation." They provide additional detail on the selection and training of team facilitators within an R&D environment. The above are consistent with the findings in Kumar and Boyle (2001) mentioned earlier; these authors provide a summary of best practices for achieving quality in applied R&D departments of manufacturing companies, with recommendations to promote good management practices, awareness by the R&D staff of the external environment, and quality culture.

Measuring R&D Quality Status

It has been said that in order to plan and improve, you must be able to control, and in order to control, you must be able to measure. Developing good measures for R&D quality has proven to be a key ingredient for improving the performance of research functions and development processes. To help distinguish among various types of measurements and measurement processes, it is useful to distinguish between measures used to manage the quality of specific R&D processes and products and measures used to assess overall R&D quality status.

Measuring Quality in R&D Processes and Products: Table 28.3 is a compilation of the top ten metrics reportedly in use at companies in 1998 and 2008. Although measures are relatively little changed, several trends appear to be forming (Teresko 2008a), including:

- *Commonality of metrics.* Definitions are converging, opening the way for greater opportunity for "apples to apples" benchmarking in the future.

Top 10 R&D Metrics Used by Industry, 1998	
1. R&D spending as a percentage of sales	76%
2. New products completed/released	68%
3. Number of approved projects ongoing	61%
4. Total active products supported	54%
5. Total patents filed/pending/awarded	51%
6. Current-year percentage of sales due to new products released in past x years	48%
7. Percentage of resources/investment dedicated	46%
8. Percentage of increase/decrease in R&D head count	43%
9. Percentage of resources/investment dedicated to sustaining products	39%
10. Average development cost per projects/product	39%

[*Source*: Goldense Group Inc., based on 1998 product development metrics survey (Teresko 2008a).]

Table 28.3(*a*) Metrics Used within Industrial R&D, 1988

Top 10 R&D Metrics Used by Industry, 2008	
1. R&D spending as a percentage of sales	77%
2. Total patents filed/pending/awarded/rejected	61%
3. Total R&D headcount	59%
4. Current-year percentage sales due to new products released in past *x* years	56%
5. Number of new products released	53%
6. Number of products/projects in active development	47%
7. Percentage resources/investment dedicated to new product development	41%
8. Number of products in defined/planning estimation stages	35%
9. Average project ROI—return on investment or average projects payback	31%
10. Percentage increase/decrease in R&D head count	31%

[*Source*: Goldense Group Inc., based on 2008 product development metrics survey (Teresko 2008b).]

Table 28.3(*b*) Metrics Used within Industrial R&D (2008)

- *Financial metrics.* This includes the rise in companies reporting use of current sales and profits (although not in the top 10) due to products released in prior years.
- *Intellectual property metrics.* Technology licensing and sales are of greater interest, possibly reflecting "open innovation" in which multiple parties are involved in the development, sale, or licensing of intellectual property.
- *True performance metrics.* Although a large proportion of metrics reported in 2008 remained focused on simple counts (risking a "hit the quota" mentality to hit numbers with quality a secondary consideration), efficiency (e.g., output per unit input), revenue, and profit-based measures are more prevalent.

The utility and types of measures for R&D process and product quality can be viewed from several perspectives. Gendason and Brown (1993) stated that for any metric to be "useful as a management tool, it must have three characteristics: it must be something that is countable; it must vary within a time frame that makes reaction to a 'downtrend' meaningful; and one must be able to define a goal value for the metric." While the metrics in Table 28.3 generally conform to this advice, it is useful to consider other attributes. Endres (1997) classified measures with respect to timeliness, application, and completeness; these factors are discussed in turn as follows.

Timeliness: Traditional measures for research quality have been lagging indicators, in that they report on what the research organization has already accomplished. By way of example, Mayo (1994) discusses Bell Labs' use of measures of new product revenues in a given year divided by total R&D costs in that year. Garfinkel (1990), at GE's Corporate R&D center, discussed GE's use of patents granted per million dollars invested in research as a benchmarking performance measurement. Although patent activity may be a leading indicator of future business (products and associated revenue streams), within R&D it is a lagging indicator because patents reflect work already completed (indeed, by the time a patent is published, it may be practically obsolete in rapidly moving technology environments).

Sekine and Arai (1994) provide tables of possible design process deficiency measures associated with management, lead times, costs, and quality. For example, a suggested measure for design quality is the ratio of the total costs of poor quality attributable to design problems to the total cost of poor quality caused by design, manufacture, or others. The authors state that, on the average, 60 percent of losses are attributable to design problems, 30 percent are attributable to manufacturing problems, and 10 percent to other areas, for example, installation. Goldstein (1990) suggested similar measures for design quality, for example, tracking the ratio of design corrective changes to the total number of drawings released for each new product.

Although they are commonly used, lagging indicators provide little preemptive control over ongoing quality. Examples of concurrent indicators and controls are the results of gate reviews, design reviews and peer reviews. Although all three help manage quality and risk, they each have different participants and objectives.

A gate review is a management-oriented assessment that ensures that a project is worth continuing in light of business risks and benefits. It may be a "hard" gate that represents a firm stop with formal passage required before resumption, or a "soft" review that permits at least some work to continue during the review. Because project prioritization and resourcing are components, a gate review usually will have not only technical and end-user representatives present but also financial decision-makers.

A design review is technically-oriented assessment conducted by independent, objective evaluators at pre-determined times to appraise a product's concept, requirements, product design, manufacturing process, and readiness for production. Hutchins (1999) recommends a minimum of three design review stages:

1. *Feasibility.* Existing knowledge of customer requirements is compared against known, feasible means of delivering against the requirements. This may include evaluation of initial specifications, drawings, or preliminary models.
2. *Intermediate.* Feasibility studies, prototypes, performance claims, and reliability data are assessed. Often, there are multiple intermediate design reviews.
3. *Final.* The completed product is in pilot production and evaluated for conformance with customer requirements. Production methods, materials, and the like are also assessed.

The three-stage process is a simplified version, and design reviews can be quite detailed. For example, the North American Space Administration (NASA) Program Formulation directive NPR 7123.1A cites 20 specific reviews from initial requirements and mission concept through launch, postflight, and decommissioning (NASA 2007). Most R&D organizations will find a happy medium in the level of detail between those cited above.

Gryna (1988) provides guidelines for structuring design reviews. Citing Gryna (1988) and Jacobs (1967), Table 28.4 summarizes design review team membership and responsibilities (Endres 1999). Kapur (1996) provides a similar design review responsibility matrix for a six-phase product design cycle. Prescribed attendance at the three phases identified by Hutchins (1999) include designers and quality engineers (all stages), production planners (intermediate and final reviews), specification and standards engineers (intermediate and final reviews), purchasing agents (final review), and safety officers (intermediate and final reviews).

A peer re view is an evaluation made by individuals that are familiar with the subject matter; in the case of R&D, this usually means scientists and engineers that are experienced in the technical details. Yamazaki et al. (2006) provide a general argument supporting the use of peer review as a tool in R&D management. Recognizing the limitations of traditional metrics (in particular, research outcomes may not be known for a considerable period of time and, being unknown, cannot be measured), the U.S. Army Research Laboratory (ARL) adopted peer review as a component to measuring R&D performance (the other components being customer evaluation and traditional performance measures) to answer three stakeholder questions (Oak Ridge Associated Universities 2005):

1. Is the work relevant? That is, does anyone care about what we are doing? Is there a target or a goal, not matter how distant, that our sponsor can relate to?
2. Is the program productive? That is, are we moving toward a goal, or at least delivering a product to our customers in a timely fashion?
3. Is the work of the highest quality? That is, can we back up our claim to be a world-class research organization doing world-class work?

Of the three methods of evaluation, it was concluded that peer review had the greatest utility in answering the third question regarding quality (Brown 2006). In other examples, Roberts (1990) discusses peer reviews used to verify progress by checking calculations, test data reduction, and research reports. Bodnarczuk (1991) provided insights into the nature of peer reviews in basic research at Fermi National Accelerator Laboratory.

Concurrent indicators can also be used to help develop leading indicators for predicting, and in some cases, controlling, R&D performance. The basic requirement is to identify coincident R&D process indicators that are demonstrably correlated, if not causative, with

		Type of Design Review*		
Group Member	**Responsibilities**	**PDR**	**IDR**	**FDR**
Chairperson	Calls, conducts meetings of group, and issues interim and final reports	X	X	X
Design engineer(s) (of product)	Prepares and presents design and substantiates decisions with data from tests or calculations	X	X	X
Reliability manager or engineer	Evaluates design for optimum reliability consistent with goals	X	X	X
Quality manager or engineer	Ensures that the functions of inspection, control, and test can be efficiently carried out		X	X
Manufacturing engineer	Ensures that the design is producible at minimum cost and schedule		X	X
Field engineer	Ensures that installation, maintenance, and user considerations were included in the design		X	X
Procurement representative	Ensures that acceptable parts and materials are available to meet cost and delivery schedules		X	
Materials engineer	Ensures that materials selected will perform as required		X	
Tooling engineer	Evaluates design in terms of the tooling costs required to satisfy tolerance and functional requirements		X	
Packaging and shipping engineer	Assures that the product is capable of being handled without damage, etc.		X	X
Marketing representative	Assures that requirements of customers are realistic and fully understood by all parties	X		
Design engineers (not associated with unit under review	Constructively reviews adequacy of design to meet all requirements of customer	X	X	X
Consultants, specialists on components, value, human factors etc. (as required	Evaluates design for compliance with goals of performance, cost, and schedule	X	X	X
Customer representative (optional)	Generally voice opinion as to acceptability of design and may request further investigation on specific items			X

[*Sources*: Gryna (1988), adapted from Jacobs (1967).]
*P = Preliminary; I = Intermediate; F = Final.

TABLE 28.4 Design Review Team Membership and Responsibility

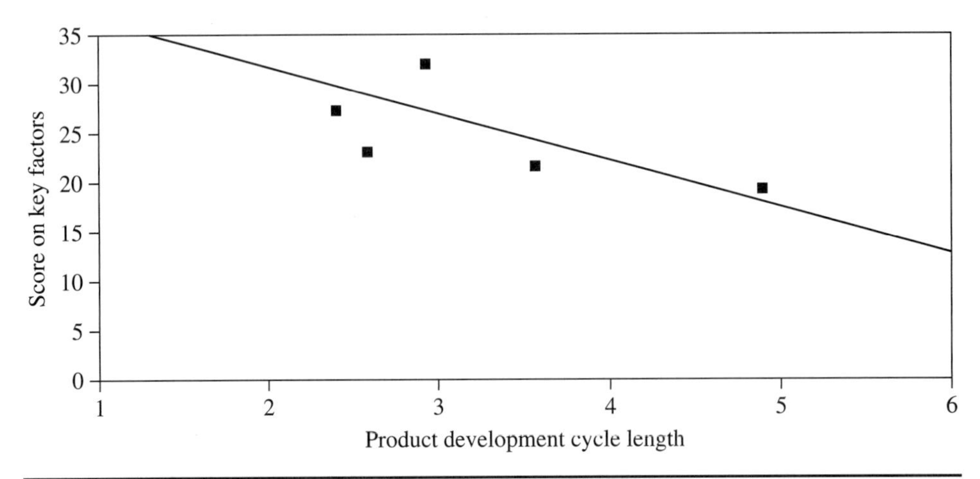

Figure 28.7 Correlation between development process compliance scores and cycle times at Kodak. (*Juran Institute. ©1994. Used with permission.*)

outcomes of research and development processes. For example, Figure 28.7 demonstrates the relationship between compliance scores during the product development projects and the length of the development cycle (Cole 1990). There is an obvious correlation that may be useful in identifying the major contributing factors (within the scoring system) to protracted development cycles.

Financial scores also prove useful as leading indicators for research effectiveness, for example, variations of net present value (NPV), expected net present value (ENPV), discounted cash flow, internal rate of return, and decision trees. Holmes and McClaskey (1994), at Eastman Chemical, showed the estimated net present value of new/improved concepts accepted (by business units for products, and manufacturing departments for processes) for commercialization. Figure 28.8 demonstrates that the effect of implementing TQM at Eastman Chemical Research virtually doubled research's productivity as measured by NPV (Endres 1997).

An additional method that has gained in popularity in recent years for the valuation of R&D effectiveness and technology created by R&D organizations is the use of real options (In simple terms, a real option provides the right, but not the obligation, to pursue some business undertaking; i.e., it represents a choice). For example, whereas the ultimate, future quality, and benefit of intellectual property created by R&D may be unknowable, it is estimable by applying concepts and principles of financial options pricing. That is, the possession of intellectual property (e.g., as exemplified by patents or trade secrets) provides a business with options. Options and the flexibility they afford can be estimated in financial terms (e.g., the choice to pursue internal technology to produce products unencumbered by royalties, or use established, third-party technology and pay licensing fees). Razgaitis (1999) provides core concepts in technology risk assessment, and valuation via real options; an applied example in automotive product development is supplied by Ford and Sobek (2005). Readers interested in a more comprehensive look at valuation of technology can refer to Boer (1999).

Applications: In addition to viewing each R&D measure (or measurement process, e.g., peer review) with respect to timeliness, it also is helpful to examine each measure with respect to its intended application. That is, is the measure intended to address customer satisfaction levels (in which case it will relate to the key features of the goods and services provided by R&D), or is the measure intended to address customer dissatisfaction and

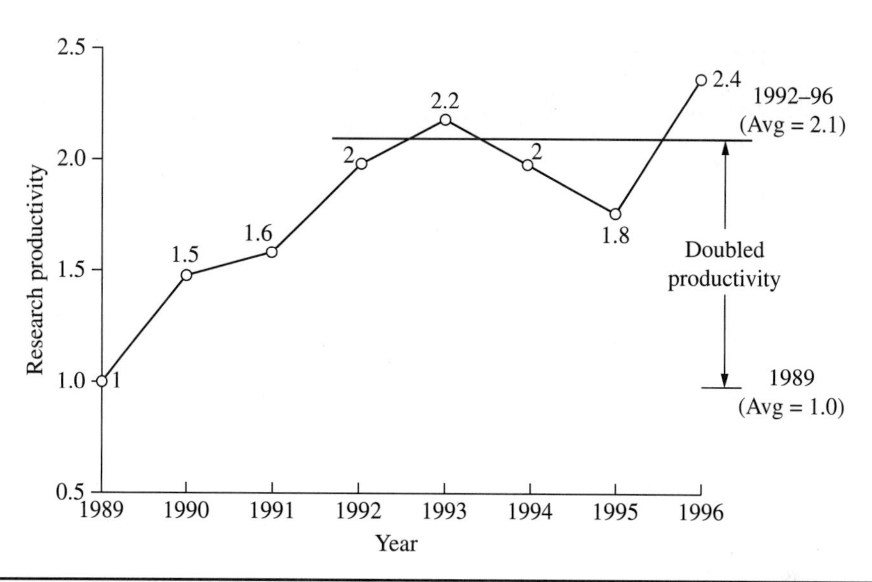

Figure 28.8 Eastman Chemical Research productivity as a ratio of 1989 NPV of improved concepts accepted and commercialized with major research input divided by total research expenditures. (*Juran Institute. ©1994. Used with permission.*)

organizational inefficiency (in which case it will relate to identification and quantification of key deficiencies of goods and services or of their R&D processes)? Juran (Chapter 1, Attaining Superior Results through Quality) discusses the relative effects of features and deficiencies on customer satisfaction and organization performance.

- *Process and Product Features.* Benchmarking the best practices of other R&D organizations is an important driver for measuring R&D quality. Lander et al. (1994) discuss the results of an industrial research organization benchmarking study of the best features of practices in R&D portfolio planning, development, and review. The study, by the Strategic Decisions Group, found that "best practice" companies exhibit common features; they:
  1. Measure R&D's contribution to strategic objectives
  2. Use decision-quality tools and techniques to evaluate proposed (and current) R&D portfolios
  3. Coordinate long-range business and R&D plans
  4. Agree on clear measurable goals for the projects

The study also revealed that "companies which are excellent at the four best practices:

1. Have established an explicit decision process that focuses on aligning R&D with corporate strategy and creating economic value
2. Use metrics that measure this alignment and the creation of value
3. Maintain a fertile organizational setting that supports decision quality and the implementation of change efforts"

Figure 28.9 represents, at a macrolevel, features of the process commonly used by the best-practice companies for R&D portfolio planning and review.

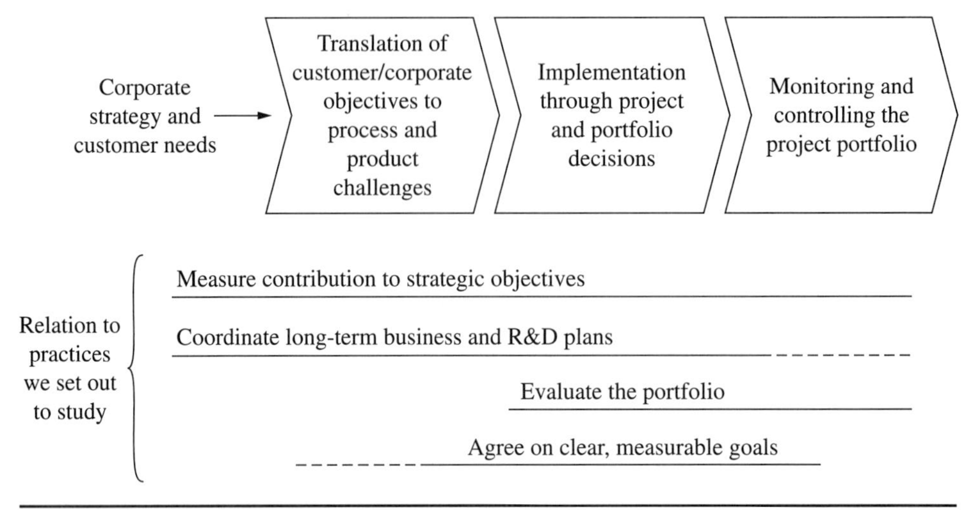

FIGURE 28.9 Common process for implementing best practices for R&D planning, implementation, and review. (*Lander et al. 1994, p. 3-14.*)

Among the organizations identified "for their exemplary R&D decision quality" practices were 3M, Merck, Hewlett-Packard, General Electric, Procter & Gamble, Microsoft, and Intel; it is notable that these companies have remained in dominant market positions over decades. Matheson et al. (1994) also provide examples of tools which organizations can use to identify their greatest opportunities for implementing and improving best practices in R&D planning and implementation. Hersh et al. (1993) discuss the use, in addition to the benchmarking for best practices, of internal customer surveys at Alcoa to identify and prioritize key R&D performance features at Alcoa's Technical Center. They used the survey results to establish four major categories of their customers' requirements:

1. Manage technology effectively
2. Link technology and business strategies
3. Build strong customer relationships
4. Provide socially and legally acceptable solutions

Each of these feature categories contained activities whose relative customer priority was also determined. For example, the first category—manage technology effectively—contained the highest priority requirement to "Assume accountability for attaining mutually determined project objectives," and the second-highest-priority requirement to "Meet customer cost and performance expectations." Wasson (1995) also discusses the use of the survey data in developing customer focused vision and mission statements for the Alcoa Technical Center. Endres (1997) provides additional details on the survey and its results.

- *Process and Product Deficiencies*. Identifying customers' requirements is necessary but not sufficient. R&D organizations also must define and implement methods for improving their customers' satisfaction levels and their process's efficiencies. Ferm et al. (1993) also discussed the use of business unit surveys at Allied Signal's

> Corporate Research & Technology Laboratory to "create a broad, generic measure of customer satisfaction . . . and then use the feedback to identify improvement opportunities, to assess internal perceptions of quality, and to set a baseline for the level of . . . research conformance to customer requirements." (In addition to surveying its business-unit customers, laboratory management gave the same survey to laboratory employees. The resulting data enabled them to compare employee perceptions of laboratory performance to the perceptions of external customers.) One of the vital few needs identified for action was the need to convince the business units that the laboratory was providing good value for project funding. Further analysis of the business units' responses revealed that the business units believed laboratory results were not being commercialized rapidly enough. However, the laboratory believed that the business units had accepted responsibility for the commercialization process. In response to this observation, a joint laboratory and (one) business-unit team was formed to clearly define and communicate responsibilities throughout the research project and subsequent commercialization and development processes.

Such tension between research and development groups is common. To help manage potential conflict, a dialog with clear delineation of expectations in handoff is suggested, for example, through the use of templates (similar to service level agreements) that specify deliverables needed from research to pull a product candidate into development. As mentioned earlier, one practice that can help is to cross-train staff between R&D, including movement of research staff into development to follow their inventions.

Finally, Wasson (1995) at Alcoa's Technical Center also provided several explicit measures used to determine customer satisfaction:

- Percentage of agreed-upon deliverables delivered
- Percentage of technical results achieved
- Results of customer satisfaction survey

Completeness. Endres (1997) uses the word "completeness" to indicate the degree to which measures are simultaneously comprehensive (i.e., taken together, they provide answers to the question: "Is the R&D organization meeting its performance objectives?") and aligned (i.e., there is a direct linkage between each variable measured and one or more of those objectives). Juran (1964) and Boath (1992) identified the need for a comprehensive hierarchy of measures. Figure 28.10 (Boath 1992) is an R&D performance measurement pyramid.

Although the concept of multiple levels of measures is useful, it is incomplete. To be complete, performance measures for research organizations and development processes must also be aligned. Menger (1993) discussed the development and use of key result indicators (KRIs) to drive progress in Corning's Technology Group, which contained research, development, and engineering. Corning's World-Class Quality Committee (WQC) defined the KRIs for the Technology Group. General areas for improvement and measurement used are

- Cycle time
- Productivity
- Customer satisfaction
- Employee satisfaction

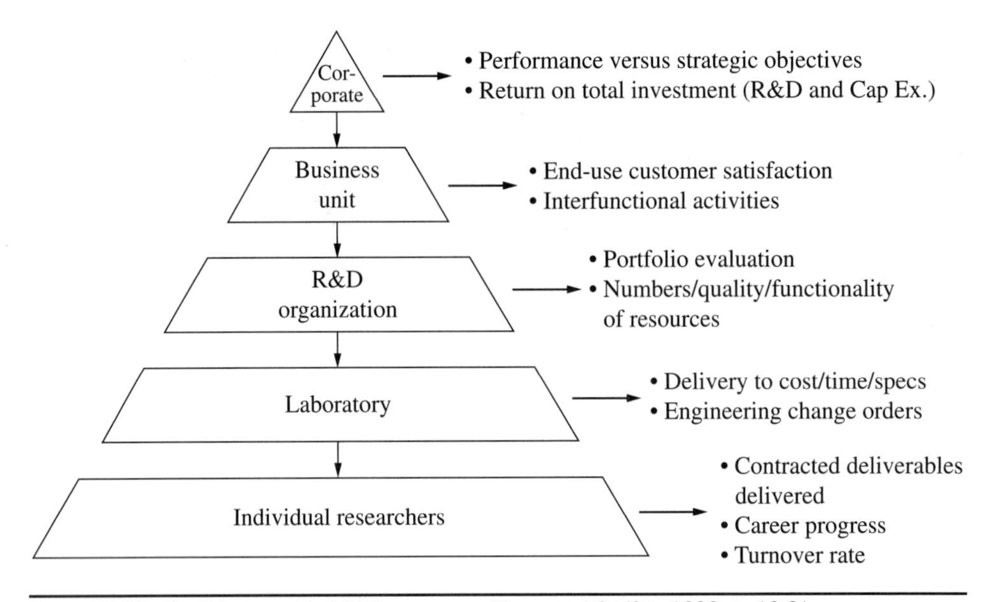

Figure 28.10 Boath's pyramid of R&D measures. (*Juran and Godfrey 1999, p. 19.9.*)

The WQC then requires each of the 15 major units in the Technology Group to define explicit performance measures for each of the previous general areas for improvement. "Twice a year the committee spends the better part of two days visiting each of the 15 units . . . (to) review the quality of their KRIs, consistency of unit KRIs with those of the technology group, progress made on the KRIs, and plans for improvement. . . ."

Additional examples of linking R&D performance measures are provided by Rummler and Brache (1995) who provided a comprehensive example of linking organizational-, process-, and job/performer-level measures for a product development process.

Assessing Overall R&D Quality Status

The previous discussions on measurement have focused on classifying and developing measures for research organizations and development processes. Gryna et al. (2007) define the benefits of determining the broad overall status of quality in organizations. This process has been defined as quality assessment. Assessment of quality consists of four elements:

1. Cost of poor quality
2. Standing in the marketplace
3. Quality culture in the organization
4. Operation of the company quality system

Examples of determining R&D customers' priorities and perspectives of performance have been discussed earlier. The assessment of some elements of quality culture in research has been discussed in an example presented by Holmes and McClaskey (1994). In 1989, the Eastman Research Center determined that although many elements of TQM had been installed (e.g., "Many processes had been studied and flow charted; some processes were being routinely measured and reviewed"), research output, as measured by the NPV of new/improved concepts accepted, had not improved. The authors conducted interviews with Research Center personnel that determined that although communications had improved:

- Few process improvements had been implemented.
- Most first-level managers and individual researchers saw nothing beneficial from the quality initiative.
- Employees were confused as to what Research management wanted them to deliver ("What is Research's main output?").

As a result of the interviews, Eastman Chemical refocused its effort on improving the key processes that directly affected its primary deliverable category: new/improved concepts accepted for commercialization. The ultimate effect of shifting initiative focus from team activities and tools to mission and output is reflected in Figure 28.8.

The cost of poor quality is discussed generally by J. De Feo in Chapter 2 (Quality's Impact on Society and the National Culture). At Corning, Kozlowski (1993) discusses using quality cost data to identify high cost-of-poor-quality areas. For example, one primary contributor to internal failure costs was the "rework" associated with having to redo experiments. An improvement team assigned to reduce associated costs determined that an internal training program on experimental design was necessary to improve efficiency and that it was necessary to improve communications with support groups through formally defining and sharing experimental objectives.

Quality System Assessments for R&D. Quality Systems assessments may be conducted using the Baldrige criteria or ISO 9000 standards. Chapter 17, Using National Awards for Excellence to Drive and Monitor Performance, provides insight into the use and benefits of the Baldrige National Quality Award. Chapter 16, Using International Standards to Ensure Organization Compliance, provides similar perspectives of the use of ISO 9000 family of international standards for reviewing quality systems.

Baldrige Assessments for R&D Organizations. Within research organizations, Kozlowski (1993) discussed using the Baldrige criteria to provide "outside focus to the quality process . . . this outside focus, specifically the emphasis on the customer, is the single biggest difference between where we started in 1985, and where we are today." Van der Hoeven (1993; unpublished paper "Managing for Quality in IBM Research.") has discussed the process used at IBM's Thomas J. Watson Research Center to organize a Baldrige assessment and the importance of translating the Baldrige criteria into relevant interpretations for a research organization. Each Baldrige category was allocated to a senior research executive. For example, strategic planning and data collection and analysis were assigned to the vice-president of technical plans and controls; the director of quality coordinated work on training and writing the category assessments. Van der Hoeven reported that "it required a significant effort to interpret and formulate appropriate responses . . . this careful tailoring of responses to the Baldrige questions, in terms of existing division processes and management systems . . . is unique. And the assessment raises gaps in processes and practices to the surface." For example, the assessment revealed the need to improve processes for strategic planning, customer satisfaction, and capturing quality data in the divisionwide database.

The Armament Research, Development and Engineering Center (ARDEC) was one of the first organizations to successfully apply BNQA criteria under the nonprofit category. An R&D center selected as the benchmark for the U.S. Army in technology transfer, ARDEC transitioned approximately 75 percent of its technology research projects into customer-funded development. ARDEC also received awards and recognition for customer satisfaction and perceived value. Internally, job satisfaction increased from approximately 87 percent positive in FY2004 to 92 percent positive in FY2007, exceeding government productivity and quality benchmarks. Finally, diversity of scientists and engineers increased in six of eight target groups from FY2005 to FY2007 (NIST 2008).

More recently, Prajogo and Hong (2008) studied the relationship between TQM practices and R&D performance using the Baldrige criteria applied to 130 R&D divisions of Korean manufacturing firms. Their findings demonstrate that TQM, as measured via the Baldrige criteria, provides a generic set of principles that can be applied successfully to R&D environments.

ISO 9000 Assessments for R&D Organizations. Although the Baldrige criteria provide organizations with a comprehensive review mechanism for improving quality systems, some organizations perceive the criteria as being simultaneously too general and too complex for beginning their quality journey. The ongoing preference for the ISO 9000 quality system standards over the Baldrige criteria can be attributed to the fact that the ISO 9000 scope is more limited, being focused on quality control and corrective action systems. Also, the ISO standards are frequently required by suppliers' customers. These drivers for the use of standards have led to the need to tailor and implement ISO standards for research and design organizations.

Fried (1993) discusses the process AT&T's Transmission Systems Business Unit (TSBU) used to pursue ISO 9001 registration. One consequence was the need for each of the TSBU design sites to support the decision by attaining ISO 9001 registration. Each TSBU design laboratory appointed an ISO coordinator; ISO managers were appointed in each of their two major geographical locations. A key initial decision was to review ISO 9001 and to identify those sections which were applicable to the design organizations. Each of the elements that were judged applicable were further categorized as "global" (where compliance could be most effectively addressed by a solution common to multiple organizations) or "local" (where compliance would require a site-by-site approach). Table 28.5 summarizes the results of the review process.

After holding ISO 9001 overview meetings with the design managers and engineers, the site coordinators and area managers coordinated self-assessments and subsequent improvement action planning. Communicating the needed changes to design procedures, coordinating planning with the manufacturing organizations, and coaching on audit participation were identified as being crucial activities in TSBU's successful registration process.

Endres (1997) includes materials from a presentation by Gibbard and Davis (1993) on pursuit of ISO 9001 registration by Duracell's Worldwide Technology Center (DWTC). An initial barrier identified was the belief of the technical managers and staff that formal procedures were unnecessary and would "stifle creativity." The authors suggested that the way to address this resistance is for upper management to drive registration via a "top-down effort," including required periodic progress reviews in which upper management participates. DWTC reported that two primary benefits of ISO registration were that it "forced us to identify precisely who our customers were for all projects carried out in our center . . ." and that ISO established "the foundation of a quality management system on which a program for quality improvement could be built."

Thelen (1997) provides a case study in which SITA (the Société Internationale de Télécommunications Aéronautiques) took a synthetic approach by combining elements of ISO 9000 with TQM and BPI (Business Process Improvement). SITA found that ISO 9000 represented a natural milestone within their path of continuous improvement and complemented the business process improvements by providing competitive advantage (conversely, increased efficiency facilitates business expansion). Thelen also reported that ISO 9000 applied more easily to R&D if each project was viewed as a service having a formal customer.

ISO 9001 Element	Applicable?	Global/Local
Management responsibility	Yes	Both
Quality system	Yes	Both
Contract review	No	
Design control	Yes	Local
Document control	Yes	Local
Purchasing	Yes	Local
Purchaser supplied product	No	
Product identification and traceability	No	
Process control	No	
Inspection and testing	No	
Inspection measuring and test equipment	Yes	Global
Inspection and test status	No	
Control of nonconforming product	No	
Corrective action	Yes	Local
Handling, storage, packaging, and delivery	Yes	Local
Quality records	Yes	Local
Internal quality audits	Yes	Global
Training	Yes	Local
Servicing	No	
Statistical techniques	No	

[*Source:* Fried (1993), p. 2B-25.]

TABLE 28.5 ISO 9001 Elements for AT&T's TSBU R&D Units

Operational Quality Planning for Research and Development

Next we consider the planning phase of R&D, with an emphasis on design.

Quality Planning: Concepts and Tools for Design and Development

The focus of the following materials is to provide examples of methodology and tools that support the implementation of Juran's operational quality planning process within the design and development process.

Operational Quality Planning Tools

As discussed in Chapter 4, Juran's quality planning process is used to identify customers and their needs, develop product design features responding to those needs and process design features required to yield the product design features, and develop process control required to ensure that the processes repeatedly and economically yield the desired product features. Quality Function Deployment (QFD) is a valuable tool for collecting and organizing the required information needed to complete the operational quality planning process.

Zeidler (1993) provides examples of using customer focus groups, surveys, and QFD at Florida Power and Light to identify customers' needs and to determine design features for a new voice-response unit. Zeidler concluded that QFD not only ensures customer satisfaction with a quality product or service but also reduces development time, startup costs, and expensive after-the-fact design changes. QFD also a useful political tool because it guarantees that all affected parts of the organization are members of the QFD team.

Delano et al. (2000) provide an R&D case study from the aircraft industry in which they compared the techniques of QFD and Decision Analysis (DA). The authors conclude that the two methods have many similarities and suggest that QFD be supplemented with DA to improve multiobjective decisions in terms of generating alternatives and supporting data analysis.

In a multiyear study, Miguel and Carnevalli (2008) examined the application of QFD in product development across 500 Brazilian companies. From their assessment, the authors identified best practices in QFD application; these practices included practical recommendations regarding upper management support, the need for training, team formation, frequency and length of meetings, benefit of a pilot project, and the utility of a conceptual model to identify future deployments needed for manufacture. Herrmann et al. (2006) also focus on the need for a conceptual framework suitable for empirical research. They evaluate QFD with regard to three dimensions of performance: product quality improvement, R&D cost reduction, and faster R&D cycle time. After building and testing a model, it is concluded that while valuable, the rigor of QFD is not a key success factor. Instead, outcomes of QFD are more strongly dependent on the motivation of QFD team members, and technical support for the team. This echoes Zeidler (1993) and Miguel and Carnevalli (2008) in that QFD is a useful tool but is unlikely to be successful without clear commitment and support from the business.

Finally, Kang et al. (2007) specifically address the difficulties in the interface between R&D and marketing domains by proposing an integrated new product design process. The process applies the QFD House of Quality to identify design features and subsequently compares the results of conjoint analysis (traditionally used by marketing to better understand preferences and how people value different features) with Taguchi methods (used in research to create a more robust design). The parallel use of the latter methods with QFD at the front end reportedly helps resolve the trade-offs that otherwise can result in an inferior final design.

Designing for Human Factors: Ergonomics and Error-proofing

As a design feature, the design's ability to be built/delivered and used by customers must be considered from two perspectives: that of operations (manufacturing and service) and that of the customer. From the perspective of manufacturing or service operations, designers must consider, among other factors, the limitations of people (e.g., operators and delivery personnel). Designers also must consider the possible types of errors that may be committed during operations and use and anticipate these as a part of design. Ergonomics or "human engineering" is used to address the needs and limitations of operators, service providers, and customers.

From the operations side, Thaler (1996) presents the results of an ergonomics improvement project for facilitating the assembly of aircraft doors. Originally, operators "had to hold the doors in place with one hand while trimming or drilling with the other and carrying them for several feet." This job design resulted in a high incidence of worker back injuries. The job redesign included designing a universal clamp to hold the aircraft doors in any position and providing the operators with adjustable work chairs and transportation carts. These and other improvements resulted in a 75 percent reduction in OSHA-lost workday incidents and dramatically decreased workers' compensation costs. Gross (1997) provides additional insights and guidance for improving manufacturability and customer usability by integrating ergonomics with the design process. Laboratory environments also can be enhanced, for example, rubber floor mats, and lab bench configurations based on Lean cell concepts to minimize movement (e.g., reticulated bench space as opposed to traditional linear bench space that requires frequent shifting in position left and right).

Nagamachi (2008) explores an approach to ergonomics that uses multivariate statistical analysis to accommodate the hierarchy of customer values and bridge customer input to create design specifications. This is specifically intended to enhance the market pull approach described earlier (called "market-in" by Nagamachi) rather than the technology push method (called "product-out" by Nagamachi). Termed "Kansei engineering," from the Japanese word implying psychological feelings and needs, the method takes qualitative, ambiguous data and translates it into new product designs. For example, Kansei engineering that was applied to refrigerators eliminates the frequent need to bend over by placing the freezer at the bottom of the unit. Efforts to redesign roads, signage. and cars to accommodate aging drivers are prompted by similar ergonomic and also safety concerns ("Highway, Car Changes Designed to Help Older Drivers," 2009).

In contrast with planning for ease of assembly, installation, and use, *poka-yoke* is a methodology for preventing, or correcting errors as soon as possible. The term's English translation is roughly "prevent inadvertent mistake." *Poka-yoke* was developed by Shigeo Shingo, a Japanese manufacturing engineer. Although common usage interchanges the associated terms "mistake-proofing" and "error-proofing," a technical difference is that mistake proofing applies more to the assembly line, whereas error proofing applies more to product design. For example, mistake proofing in an assembly process might incorporate a glue applicator that indicates when insufficient glue has been dispensed. Error proofing would be a design that permits parts to be snapped together, thereby eliminating the need for glue and monitoring of the amount applied. Although human error often receives blame, the root cause of errors usually can be traced back to the failure of designers to adequately account for the possibility of errors or omissions. Kohoutek (1996a) discusses "human-centered" design and presents approaches and references for predicting human error rates for given activities. An example of error proofing in the redesign of a manufacturing process can be found in Bottome and Chua (2005). Through a series Six Sigma projects, Genentech substantially reduced errors during drug production by changing from black to more visible blue ink (to make omissions more apparent), clarifying documentation rules for batch record creation, and reducing the complexity of production instructions (tickets). Through the combined efforts of error-proofing, errors per 100 tickets were reduced from approximately 10 to 3.5.

Designing for Reliability

A product feature that customers require in products is reliability. Gryna et al. (2007) defined reliability as the "ability of a product to perform a required function under stated conditions for a stated period of time" or, more simply, the "chance that a product will work for the required time." Introducing the concept of operating environment, Ireson (1996) states that reliability is the "the ability or capability of the product to perform the specified function in the designated environment for a minimum length of time or minimum number of cycles or events," which also references specific operating conditions/environments. It is important to note that a precise and agreed-upon definition of a "failure" is needed by customers, designers, and reliability engineers. Rees (1992) also discusses the importance of identifying and defining the intended purpose of the application and test procedure prior to defining failures. Requiring designers to precisely establish parameters for both successful product performance and environmental conditions obliges designers to develop a deeper understanding of the product, its use, and design.

The following materials will describe approaches and tools for "designing in" reliability. A reliability program consists of the specific tasks needed to achieve high reliability; Gryna et al. (2007) identified the following major tasks:

- Setting overall reliability goals
- Apportionment of the reliability goals
- Stress analysis
- Identification of critical parts
- Failure mode and effect analysis
- Reliability prediction
- Design review
- Selection of suppliers
- Control of reliability during manufacturing
- Reliability testing
- Failure reporting and corrective action system

Table 28.6 (Gryna et al. 2007) provides typical reliability metrics for which specific numerical goals may be established.

As described earlier in this chapter, design reviews can be used as concurrent indicators for a design's reliability. Therefore, one of the key requirements for design review meetings is to ensure that reliability goals have been established and that intrinsic and actual reliability are being measured and improved during the design's evolution, manufacture, and use. Reliability of procured materials must be considered during supplier selection and control (see Chapter 30, Supply Chain: Better, Faster, Friendlier Suppliers, for additional discussion). The effect of manufacturing processes on reliability must be addressed during process design selection and implementation. Refer to Chapter 13, Root Cause Analysis to Maintain Performance, and Chapter 20, Product-Based Organizations: Delivering Quality While Being Lean and Green, and Gryna et al. (2007) for guidance in controlling quality and reliability during manufacturing.

Gryna et al. (2007) divide the process of reliability quantification into the three phases: apportionment (or budgeting), prediction, and analysis. Reliability apportionment is division and allocation of the design's overall reliability objectives among its major subsystems and then to their components. Reliability prediction is the process of using reliability modeling, probability theory and actual past performance data to predict reliability for expected operating conditions and duty cycles. Reliability analysis uses the results of reliability predictions to identify strong and weak parts of the design, trade-offs, and opportunities for improving either predicted or actual reliability performance. These three phases will be discussed in turn.

Reliability Apportionment: The top two sections in Table 28.7 (Gryna et al. 2007) provide an example of reliability apportionment. A missile system's reliability goal of 95 percent for 1.45 hours is apportioned among its subsystems and their components. The top section of the table demonstrates the first level apportionment of the 95 percent goal to the missile's six subsystems. The middle section of the table exemplifies the apportionment of the goal of one of those subsystems; the reliability goal of 0.995 for the missile's explosive subsystem is

Figure of Merit	Meaning
Mean time between failures (MTBF)	Mean time between successive failures of a repairable product
Failure Rate	Number of failures per unit time
Mean time to failure (MTTF)	Mean time to failure of a nonrepairable product or mean time to first failure of a repairable product
Mean life	Mean value of life ("life" may be related to major overhaul, wear-out time, etc.)
Mean time to first failure MTFF)	Mean time to first failure of a repairable product
Mean time between maintenance (MTBM)	Mean time between a specified type of maintenance action
Longevity	Wear-out time for a product
Availability	Operating time expressed as a percentage of operating and repair time
System effectiveness	Extent to which a product achieves the requirements of the user
Probability of success	Same as reliability (but often used for "one-shot" or non-time-oriented products)
b_{10} life	Life during which 10% of the population would have failed
b_{50} life	Median life, or life during which 50% of the population would have failed
Repair/100	Number of repair per 100 operating hours

[*Source*: Gryna et al. (2007), p. 326.]

Table 28.6 Summary of Tests Used for Design Evaluation

apportioned to its three components. For example, the allocation for the fusing circuitry is 0.998 or, in terms of the reliability objective of mean time between failures, 725 hours.

Kohoutek (1996b) suggests that, in order to allow for design margins, only 90 percent of the system failure rate be apportioned to its subsystems and their components. He discusses five other methods for reliability apportionment. Kapur (1996) provides several examples of using alternative apportionment methods. Kohoutek (1996b) also discusses the use of reliability policies to support goal setting and improvement for both individual products and product families.

Reliability Prediction and Modeling: In general, a model of the system must be constructed before a prediction of reliability can be made. This may start as a paper model (i.e., using only mathematical calculations) but eventually ends with an actual reliability measurement derived from customer use of the product. As a part of this, stress levels for the model's components are determined, and, on the basis of the estimated stress levels, failure rates for the components are obtained and used to estimate the reliability of subsystems and systems. Turmel and Gartz (1997) provide a layout for an "item quality plan" that includes the part's critical characteristics and specification limits. The plan also includes the manufacturing process to be used and test and inspection procedures, with requirements for process stability and capability measures for these processes and procedures.

System Breakdown					
Subsystem	**Type of Operation**	**Reliability**	**Unreliability Per Hour**	**Failure Rate Objective***	**Reliability**
Air Frame	Continuous	0.997	0.003	0.0021	483
Rocket motor	One-shot	0.995	0.005		1/200 operations
Transmitter	Continuous	0.982	0.018	0.0126	80.5 h
Receiver	Continuous	0.988	0.012	0.0084	121 h
Control system	Continuous	0.993	0.007	0.0049	207 h
Explosive system	One-shot	0.995	0.005		1/200 operations
System		0.95	0.05		

Explosive Subsystem Breakdown				
Unit	**Operating Mode**	**Reliability**	**Unreliability**	**Reliability Objective**
Fusing circuitry	Continuous	0.998	0.002	725 h
Safety and arming mechanism	One-short	0.999	0.001	1/1000 operations
Warhead	One-short	0.998	0.022	2/1000
Explosive subsystem		0.995	0.005	

Unit Breakdown			
Fusing Circuitry Component Part Classification	**Number Used, n**	**Failure Rate Per Part, λ, % 1000 h**	**Total Part Failure Rate, $n\lambda$, % 1000 h**
Transistors	93	0.30	27.90
Diodes	87	0.15	13.05
Film resistors	112	0.04	4.48
Wirewound resistors	29	0.20	5.80
Paper capacitors	63	0.04	2.52
Tantalum capacitors	17	0.50	8.50
Transformers	13	0.20	2.60
Inductors	11	0.14	1.54
Solder joints and wires	512	0.01	5.12

$$\text{MTBF} = \frac{1}{\text{failure rate}} = \frac{1}{\Sigma n\lambda} = \frac{1}{0.0007151} = 1398 \text{ h}$$

*For a mission time of 1.45 h.
(*Source:* Gryna et al. 2007, p. 327, adapted by F. M. Gryna, Jr. from Beaton 1959, p. 65.)

Table 28.7 Establishment of Reliability Objectives

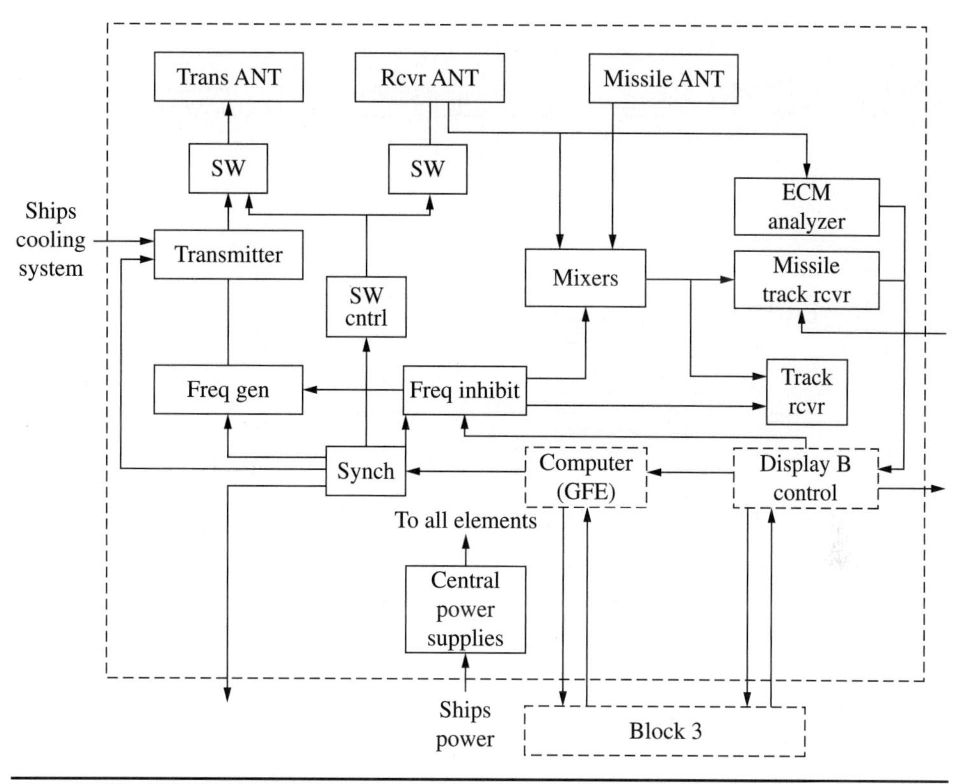

Figure 28.11 Functional block diagram. (*Gryna 1988, p. 19.10.*)

In order to construct a model for reliability prediction, interrelationships among the system's subsystems and their components must be understood. Gryna (2001) suggests the following steps to developing reliability models and using them for reliability prediction:

1. *Define the product and its functional operation.* The system, subsystems, and units must be precisely defined in terms of their functional configurations and boundaries. This precise definition is aided by preparation of a functional block diagram (Figure 28.11), which shows the subsystems and lower-level products, their interrelation, and the interfaces with other systems. For large or complex systems it may be necessary to prepare functional block diagrams for several levels of the product hierarchy.

Given a functional block diagram and a well-defined statement of the functional requirements of the product, conditions that constitute failure or unsatisfactory performance can be defined. The functional block diagram also makes it easier to define the boundaries of each unit and to ensure that important items are neither neglected nor considered more than once. For example, a switch that is used to connect two units must be classified as belonging to one unit or the other, or as a separate unit.

2. *Prepare a reliability block diagram.* The reliability block diagram (Figure 28.12) is similar to the functional block diagram, but it is modified to emphasize those aspects that influence reliability. The diagram shows, in sequence, elements that must function for successful operation of each unit. Redundant paths and alternative modes should be clearly shown. Elements that are not essential to successful operation need not be included (e.g., decorative escutcheons). Also, because of the many thousands of individual parts that constitute a complex product, it is necessary to exclude from the calculation those classes of parts that

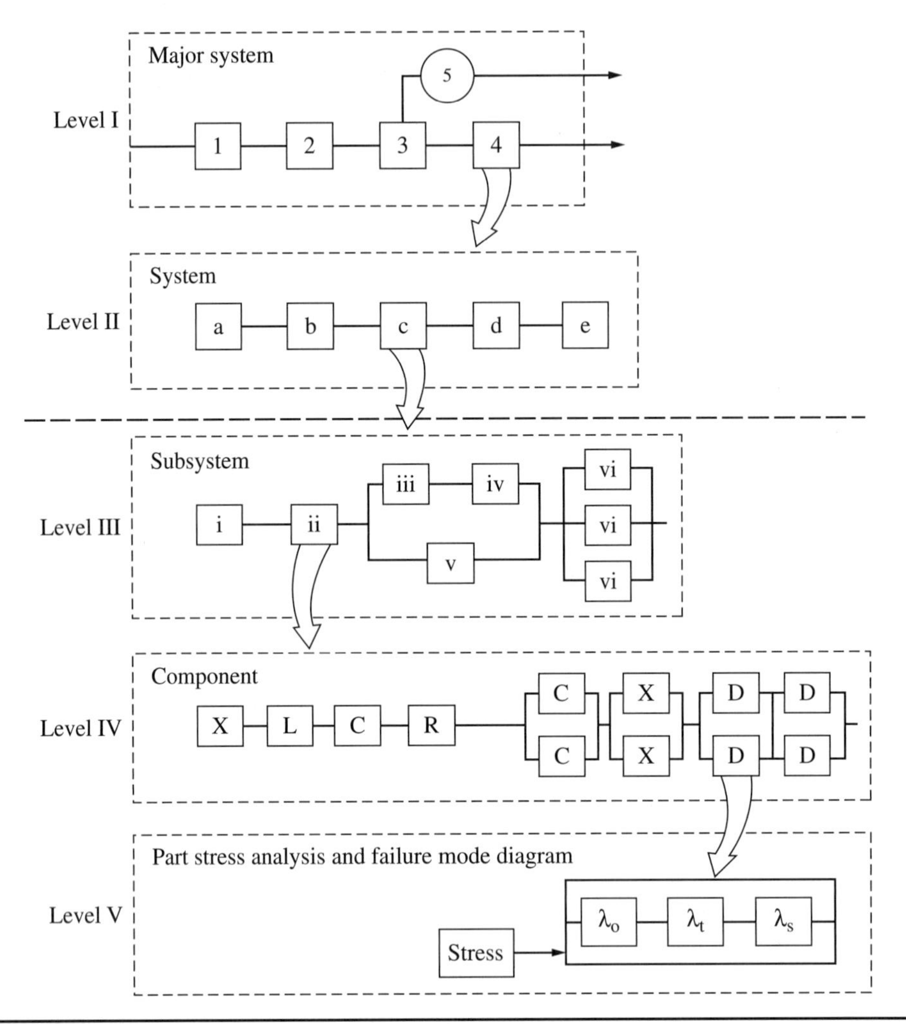

FIGURE 28.12 Reliability block diagram. (*Gryna 1988, p. 19.11.*)

are used in mild applications. The contribution of such parts to product unreliability is relatively small. Examples of items that can generally be disregarded are terminal strips, knobs, chassis, and panels.

3. *Develop the probability model for predicting reliability.* This may be a very simple model (e.g., an exponential model that assumes a constant failure rate and is based on the addition of component failure rates), somewhat more complicated (e.g., application of the Weibull distribution based on prediction of reliability as a function of time) or very complex (e.g., using more esoteric distributions and accommodation of redundancies or special conditions).

4. *Collect information relevant to parts reliability.* Factors include part function, tolerances, part ratings, internal and external environments, stresses, and operating time (duty cycles). Parts with dependent failure probabilities should be grouped together into modules so that the assumptions upon which the prediction is based are satisfied. This detailed information makes it possible to perform a stress analysis, which will not only provide information on the appropriate adjustments to standard input data but also serve to uncover weak or questionable areas

in the design. Operating parameters often are closely associated with reported failure rate information, in recognition that conditions may strongly influence failure rates. (It is worth noting here the methodology of "robust design" that is intended to assist designers to improve part and product ability to perform in various environments. Phadke (1989), Taguchi et al. (2005), and, more recently, Park and Antony (2008) provide approaches and examples.)

5. *Select parts reliability data*. The required part data consist of information on catastrophic failures and on tolerance variations with respect to time under known operating and environmental conditions. Acquiring these data is a major problem for the designer, because there is no single reliability data bank comparable to handbooks such as those that are available for physical properties of materials. Instead, the designer (or supporting technical staff) must either build up a data bank or use reliability data from a variety of sources:

- Field performance studies conducted under controlled conditions
- Specification life tests
- Data from parts manufacturers or industry associations
- Customers' part-qualification and inspection tests
- Government agency and related data banks such as the Reliability Information Analysis Center (RIAC), and RDF 2000 (formerly CNET RDF 93). These sources provide component failure rate data and curves for various components' operating environments and stress levels, and examples of reliability prediction procedures appropriate for various stages of a design's evolution.

6. *Combine all of the above to obtain the numerical reliability prediction.*

- *Make estimates*. In the absence of basic reliability data, it may be feasible to make reasonably accurate estimates based upon past experience with similar part types. Lacking such experience, it becomes necessary to obtain the data via part evaluation testing.
- *Determine block and subsystem failure rates*. The failure rate data obtained in step 4 or via estimation are used to calculate failure rates for the higher-level systems and the total system. Pertinent subsystem or assembly correction factors, before such as those determined for the effects of preventive maintenance, should also be applied.
- *Determine the appropriate reliability unit of measure*. This is the choice of the reliability index or indicators as listed in Table 28.6.

The bottom portion of Table 28.7 provides an example of predicting the failure rates for each component of the fusing circuitry for known part counts. The prediction is based upon the assumptions of the statistical independence of the failure times of the components, conformance to an exponential failure distribution, and equal hours of operation. The estimated unit failure rate is of 0.7151/1000 hours of operation or 0.0007151 failures per hour. The reciprocal of the latter failure rate yields an estimated mean time between unit failures of 1398 hours, which exceeds the 725-hour requirement for the fusing circuitry.

Reliability analysis: After completing the steps of reliability prediction, use the reliability model and predictions to identify the design's "weak points" and the required actions and responsibilities for reliability improvement. Three primary methods for evaluation are Failure Mode, Effect, and Criticality Analysis (FMECA), Fault Tree Analysis (FTA), and Testing. The first two are discussed as follows; design testing is deferred until later in the chapter.

- *Failure Mode, Effect, and Critically Analysis*. This method enhances planning for reliability by facilitating the engineer's analysis of the expected effects of operating

1 = Very low (<1 in 1000)
2 = Low (3 in 1000)
3 = Medium (5 in 1000)
4 = High (7 in 1000)
5 = Very high (>9 in 1000)

T = Type of failure
P = Probability of occurrence
S = Seriousness of failure to system
H = Hydraulic failure
M = Mechanical failure
W = Wear failure
C = Customer abuse

Product	HRC-1
Date	Jan. 14, 1987
By	S.M.

Component part number	Possible failure	Cause of failure	T	P	S	Effect of failure on product	Alternatives
Worn bearing 4224	Bearing worn	Not aligned with bottom housing	M	1	4	Spray head wobble or slowing down	Improve inspection
Zytel 101		Excessive spray head wobble	M	1	3	DITTO	Improve worm bearing
Bearing stem 4225	Excessive wear	Poor bearing/ material combination	M	5	4	Spray head wobbles and loses power	Change stem material
Brass		Dirty water in bearing area	M	5	4	DITTO	Improve worm seal area
		Excessive spray head wobble	M	2	3	DITTO	Improve operating instructions
Thrust washer 4226	Excessive wear	High water pressure	M	2	5	Spray head will stall out	Inform customer in instructions
Fulton 404		Dirty water in washers	M	5	5	DITTO	Improve worm seal design
Worm 4527	Excessive wear in bearing area	Poor bearing/ material combination	M	5	4	Spray head wobbles and loses power	Change bearing stem material
Brass		Dirty water in bearing area	M	5	4	DITTO	Improve worm seal design
		Excessive spray head wobble	M	2	3	DITTO	Improve operating instructions

FIGURE 28.13 Failure mode, effect, and criticality analysis. (*Gryna et al. 2007, p. 331.*)

conditions on design reliability and safety. General introductions to failure mode effect analysis (FMEA) and FMECA are provided in Gryna at al. (2007). FMEA and FMECA are intended for use by product and process designers in identifying and addressing potential failure modes and their effects. Figure 28.13 (Gryna et al. 2007) is an example of a FMECA for a traveling lawn sprinkler, which includes, for each part number, its failure mode, result of the failure mode, cause of failure mode, estimated probability of failure mode, severity of the failure mode, and alternative countermeasures for preventing the failure.

- *Fault Tree Analysis.* Whereas FMECA examines all possible failure modes from the component level upward, FTA focuses on particular known undesirable effects of a failure (e.g., fire and shock, and proceeds to identify all possible failure paths resulting in the specified undesirable outcome). In addition to hazard analysis, FTA is a tool often used in designing for safety. Figure 28.14 (Gryna et al. 2007) and Hammer (1980) is a fault tree for a safety circuit. The failure outcome of concern is that X-rays will be emitted from a machine whose door has been left open. The spadelike

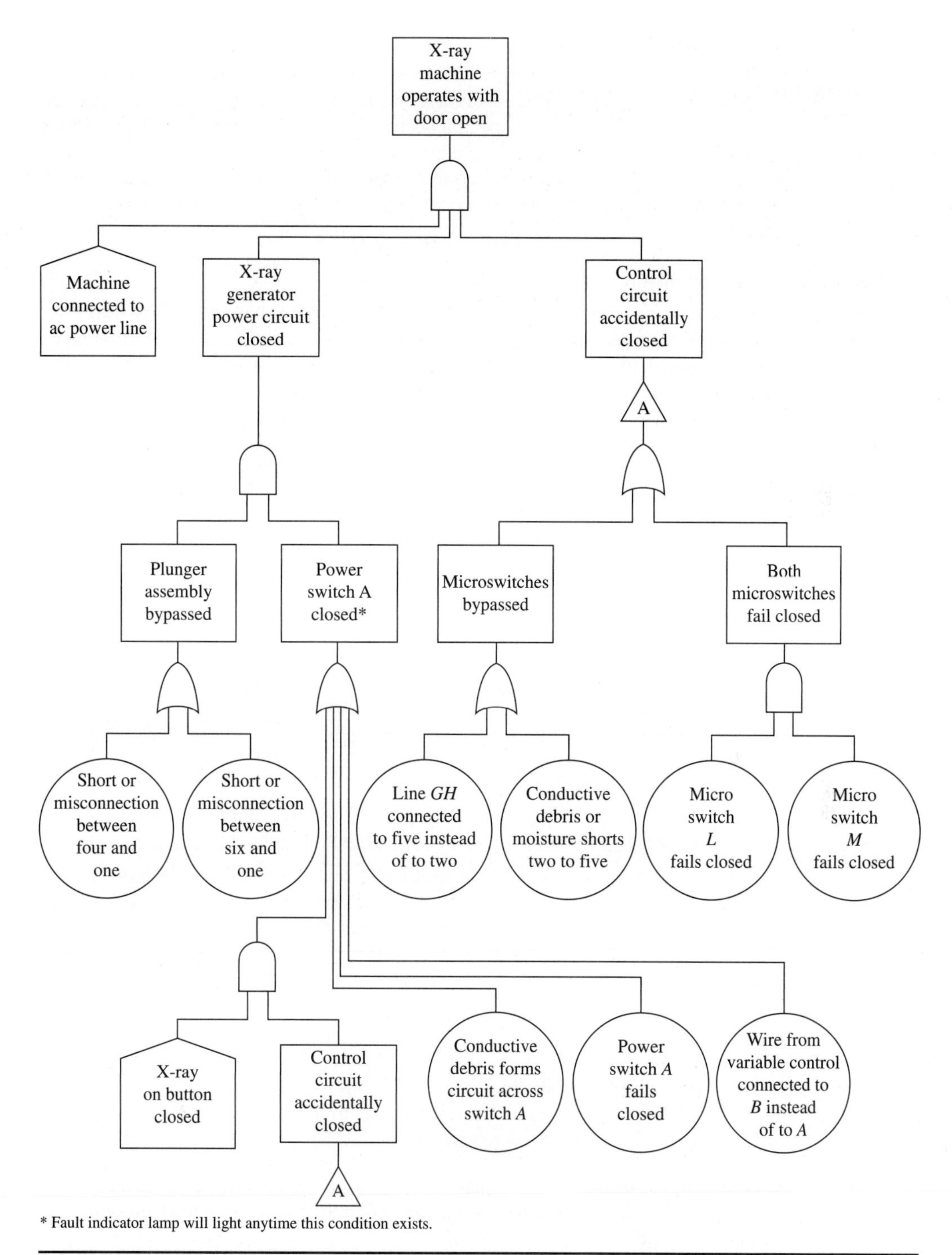

FIGURE 28.14 Fault-tree analysis of an interlock safety circuit. (*Gryna et al. 2007, p. 339.*)

symbol with a straight bottom is an "and gate," meaning that the output occurs only if all input events before it happen. The spade symbol with the curved bottom is an "or gate," meaning the output occurs if any one or more of the input events before it happen. The probabilities of specific occurrences can be estimated by providing estimates of the probabilities of occurrence of each event in the fault tree. Lazor (1996) also provides examples and comparisons of FMECA and FTA analyses, with an interesting discussion on the relationship between fault trees and reliability block diagrams.

Reliability Improvement

The general approach to quality improvement is widely applicable to reliability improvement as far as the economic analysis and the managerial tools are concerned. The differences are in the technological tools used for diagnosis and remedy. Projects can be identified through reliability prediction; design review; failure mode, effect, and criticality analysis, and other reliability evaluation techniques.

Action to improve reliability during the design phase is best taken by the designer. The reliability engineer can help by defining areas needing improvement and by assisting in the development of alternatives. The following actions provide some approaches to improving a design:

1. Review the users' needs to see if the function of the unreliable parts is really necessary to the user. If not, eliminate those parts from the design. Alternatively, look to see if the reliability index (figure of merit) correctly reflects the real needs of the user. For example, availability is sometimes more meaningful than reliability. If so, a good maintenance program might improve availability and, hence ease the reliability problem.
2. Consider trade-offs of reliability for other parameters (e.g., functional performance or weight). Here again, it may be found that the customer's real needs may be better served by such a trade-off.
3. Use redundancy to provide more than one means for accomplishing a given task in such a way that all the means must fail before the system fails.
   - There are several types of redundancy, a common form being parallel redundancy. A familiar example is the multiengine aircraft, which is so designed that even if one engine fails, the aircraft will still be able to continue on to a safe landing.
   - Under conditions of independent failures, the overall reliability for parallel redundancy is expressed by the formula:

$$P_s = 1 - (1 - P_i)^n$$

where P_s = reliability of the system
P_i = reliability of the individual elements in the redundaney
n = number of identical redundant elements

   - Figure 28.15 shows some simple examples of series-parallel and parallel-series redundancies and calculates the system reliability versus that prevailing for the case of no redundancy.
4. Review the selection of any parts that are relatively new and unproven. Use standard parts whose reliability has been proven by actual field use. (However, be sure that the conditions of previous use are applicable to the new product.)

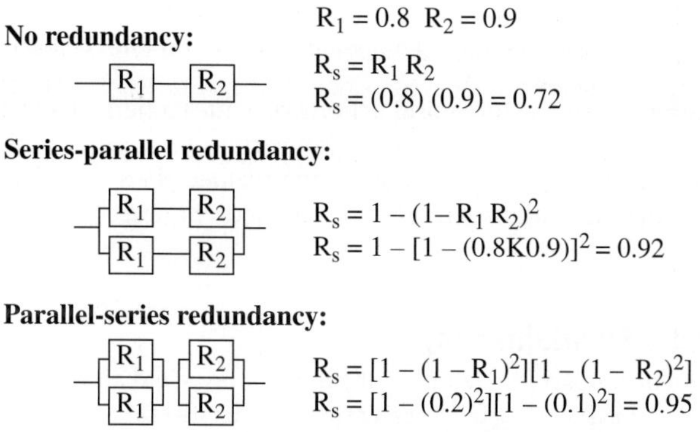

No redundancy:

$R_1 = 0.8 \;\; R_2 = 0.9$

$R_s = R_1 R_2$

$R_s = (0.8)(0.9) = 0.72$

Series-parallel redundancy:

$R_s = 1 - (1 - R_1 R_2)^2$

$R_s = 1 - [1 - (0.8K0.9)]^2 = 0.92$

Parallel-series redundancy:

$R_s = [1 - (1 - R_1)^2][1 - (1 - R_2)^2]$

$R_s = [1 - (0.2)^2][1 - (0.1)^2] = 0.95$

Figure 28.15 Series-parallel and parallel-series redundancy. (*Juran Institute. ©1994. Used with permission.*)

5. Use derating to assure that the stresses applied to the parts are lower than the stresses the parts can normally withstand. Derating is one method that design engineers use to improve component reliability or provide additional reliability margins. Gryna et al. (2007) define derating as the assignment of a product (component) to operate at stress levels below its normal rating, e.g., a capacitor rated at 300 V is used in a 200-V application. Kohoutek also provides examples of derating graphs, to be used by design engineers for specific types of integrated circuits. Before using the graphs for a specific application, the design engineer first determines the expected operating temperatures, voltages, stresses, etc. of the component under study, then uses the graphs to select the appropriate derating factor.
6. Use "robust" design methods that enable a product to handle unexpected environments.
7. Control the operating environment to provide conditions that yield lower failure rates. Common examples are (a) potting electronic components to protect them against climate and shock and (b) use of cooling systems to keep down ambient temperatures.
8. Specify replacement schedules to remove and replace low-reliability parts before they reach the wear-out stage. In many cases, the replacement is made but is contingent on the results of checkouts or tests that determine whether degradation has reached a prescribed limit.
9. Prescribe screening tests to detect infant-mortality failures and to eliminate substandard components. The tests take various forms—bench tests, "burn in," and accelerated life tests. Jensen and Petersen (1982) provide a guide to the design of burn-in test procedures. Chien and Kuo (1995) offer further useful insight into maximizing burn-in effectiveness.
10. Conduct research and development to attain an improvement in the basic reliability of those components which contribute most of the unreliability. While such improvements avoid the need for subsequent trade-offs, they may require advancing the state of the art and hence an investment of unpredictable size. Research in failure mechanisms has created a body of knowledge called the "physics of failure" or

"reliability physics." The IEE International Reliability Physics Symposium (IRPS) conference proceedings remains an excellent reference on this topic.

Although none of the foregoing actions provides a perfect solution, the range of choice is broad. In some instances the designer can arrive at a solution single-handedly. More usually it means collaboration with other company specialists. In still other cases the customer and/or the company management must concur because of the broader considerations involved.

Designing for Maintainability

Although the design and development process may yield a product that is safe and reliable, it may still be unsatisfactory. Users want products to be available on demand. Therefore, designers must also address the issue of the ease of preventive maintenance and repair. "Maintainability" is the accepted term used to address and quantify the extent of need for preventive maintenance and the ease of repair. Dhillon (1999) provides a definition of maintainability as follows:

> the measures taken during the development, design and installation of a manufactured product that reduce required maintenance, manhours, tools, logistic cost, skill levels, and facilities, and ensure that the product meets the requirements for its intended use (p. 1).

Note that maintainability is a design parameter, whereas maintenance is an operational activity.

Mean time to repair (MTTR) is an index used for quantifying maintainability, analogous to the term MTBF, which is used as an index for reliability. MTTR is the mean time needed to perform repair work, assuming that there is no delay in obtaining spare parts and that a technician is available. Similar to reliability, there are numerous possible measures of maintainability; Table 28.8 (MIL-STD-721C 1981), summarizes possible indexes for maintainability.

For example, Kowalski (1996) discusses allocating a system's maintainability requirement among its subsystems. The allocation is analogous to the method by which reliability was apportioned (See paragraph, "Reliability Apportionment."). Kowalski also discusses the impact of testability on the ability to achieve maintainability goals. Turmel and Gartz (1997) of Eastman Kodak provide, for a specific test method, a test capability index (TCI) index for measuring the proportion of the specification range taken by the intrinsic variation of a test/measurement method. The reported guideline was to target test variation at less than 25 percent of the total tolerance range.

Designing for Availability

Both design reliability and maintainability affect the probability of a product being available when required for use (i.e., it performs satisfactorily when called upon). Availability is calculated as the ratio of operating time to operating time plus downtime. However, downtime can be viewed in two ways:

1. *Total downtime.* This includes the active repair time (diagnosis and repair), preventive maintenance time, and logistics time (time spent waiting for personnel, spare parts, etc.). When total downtime is used, the resulting ratio is called operational availability (A_o).
2. *Active repair time.* When active repair time is used, the resulting ratio is called "intrinsic availability."

Figure of Merit	Meaning
Mean time to repair (MTTR)	Mean time to correct a failure
Mean time to service	Mean time to perform an act to keep a product in operating condition
Mean preventive maintenance time	Mean time for scheduled preventive maintenance
Repair hours per 100 operating hours	Number of hours required for repairs per 100 product operating hours
Rate of preventive maintenance actions	Number of preventive maintenance actions required per period of operative or calendar hours
Downtime probability	Probability that a failed product is restored to operative condition in a specified downtime
Maintainability index	Score for a product design based on evaluation of defined maintainability features
Rate of maintenance cost	Cost of preventive and corrective maintenance per unit of operating or calendar time

[*Source:* MIL-STD-721C (1981).] Note: this standard was canceled without replacement (12/5/1995) but remains a useful reference.

TABLE 28.8 Maintainability Figures of Merit

Under certain conditions, "steady state" availability can be calculated as follows:

$$A_o = \frac{\text{MTBF}}{\text{MTBF} + \text{MDT}} \quad \text{and} \quad A_i = \frac{\text{MTBF}}{\text{MTBF} + \text{MTTR}}$$

where MTBF = mean time between failures
MDT = mean total downtime
MTTR = mean active time to repair

These formulas indicate that specified product availability may be improved (increased) by increasing product reliability (MTBF) or by decreasing time to diagnose and repair failures (MDT or MTTR). Achieving any combination of these improved results requires an analysis of the trade-offs between the benefits of increasing reliability or maintainability.

Formulas for steady-state availability have the advantage of simplicity. However, they are based upon the following assumptions:

- The product is operating in the constant-failure-rate portion of its overall life where time between failures is exponentially distributed.
- Downtime and repair times are also exponentially distributed.
- Attempts to locate system failures do not change failure rates.
- No reliability growth occurs. (Such growth might be due to design improvements or removal of suspect parts.)
- Preventive maintenance is scheduled outside the time frame included in the availability calculation.

For these conditions, O'Connor (2002) provides formulas and examples for various reliability block diagrams (e.g., series, parallel, and parallel-standby configurations). Malec (1996) provides general formulas and examples for calculating instantaneous availability and mission interval availability—the probability that a product will be available throughout the length of its mission.

Some trade-off decisions that should be considered to improve maintainability through design are described in Gryna et al. (2007), including:

- *Reliability versus maintainability*. For any particular availability requirement, a designer may have a choice of improving either reliability or maintainability.
- *Modular versus nonmodular construction*. Although modular design takes greater design effort, it can reduce the time needed for diagnosis and repair in the field. In many cases, once a fault is located, the offending module can simply be removed and replaced. Repair to the module, if needed, can then take place at another time and place without delaying equipment use by the customer.
- *Repairs versus throwaway*. In many circumstances it may be more economical to discard a faulty part than to attempt repair. In such situations, the design may ease the process of discard and replacement.
- *Built-in versus external test equipment*. Having internal diagnostic capability reduces downtime, but adds to overall cost of the product. However, the additional costs can also reduce overall repair costs by providing users with simple repair instructions for various failure modes diagnosed by the diagnostic equipment or software. For example, office copiers provide messages on where and how to remove paper jams. Increasingly, elements for diagnostics may reside in a piece of equipment, but monitoring and diagnosis can take place remotely via the Internet.
- *Person versus machine*. Designers should consider trade-offs between a highly tuned product that may require special instrumentation and repair facilities and a product that may have reduced performance but easier maintenance and greater uptime.

Kowalski (1996) provides additional examples of criteria for maintainability design.

Identifying and Controlling Critical Components

The design engineer will identify certain components as critically affecting reliability, availability, and maintainability (RAM) or for attaining cost objectives. These critical components are ones that emerge from the various applicable analyses: reliability block diagrams, stress analysis, FMEA/FMECA, FTA, and RAM studies. These components may be deemed critical because of their estimated effects on design RAM and cost, insufficient knowledge of their actual performance, or the uncertainty of their suppliers' performance. One approach to ensuring performance and resolving uncertainties is to develop and manage a list of critical components. The critical components list (CCL) should be prepared early in the design effort. It is common practice to formalize these lists, showing the nature of the critical features and planning for controlling and improving performance for each critical component. The CCL becomes the basic planning document for (1) test programs to qualify parts, (2) design guidance in application studies and techniques, and (3) design guidance for application of redundant parts, circuits, or subsystems.

Configuration Management

Configuration management is the process used to define, identify, and control the composition and the cost of a product. A configuration established at a specific point in time is called a "baseline." Baseline documents include drawings, specifications, test procedures, standards, and inspection or test reports. Configuration management begins during the design of the product and continues throughout the remainder of the product's commercial life. As applied to the product's design phase, configuration management is analogous, at the level of total product, to the process described in the last paragraph for the identification and control of critical components. Gryna (1988) states that "configuration refers to the physical and functional characteristics of a product, including both hardware and software" and defines three principal activities that comprise configuration management:

1. *Identification*. Process of defining and identifying every element of the product.
2. *Control*. Process that manages a design change from the time of the original proposal for change through implementation of approved changes.
3. *Accounting*. Process of recording the status of proposed changes and the implementation status of approved changes.

Configuration management is needed to help ensure that:

1. All participants in the quality spiral know the current status of the product in service and the proposed status of the product in design or design change.
2. Prototypes, operations, and field service inventories reflect design changes
3. Design and product testing are conducted on the latest configurations.

Design Testing

Once the foregoing tools and analyses of design quality have been invoked, it is necessary to ensure that the resulting design can ultimately be manufactured, delivered, installed, and serviced to meet customers' requirements. To achieve this, it is imperative to conduct actual tests on prototypes and pilot units prior to approval for full-scale manufacturing. Table 28.9 summarizes the various types and purposes of design evaluation tests.

In Chapter 48 of *Juran's Quality Handbook* (Juran and Godfrey 1999), Meeker et al. (1999) discuss the purpose and design of environmental stress tests, accelerated life tests, reliability growth tests, and reliability demonstration testing and analysis of the data from these tests. Graves and Menten (1996) and Schinner (1996) provide similar discussions on designing experiments for reliability measurement and improvement, and accelerated life testing respectively. The Rome Laboratory Reliability Engineer's Toolkit (1993) and, more recently, related publications in the series, provide useful tools and discussion including the selection and use of reliability test plans from MIL-HDBK-781 (1987).

Comparing Results of Field Failures with Accelerated Life Tests

In order to verify design reliability within feasible time frames, it often is necessary to "accelerate" failure modes by using various environmental stress factors. This applies not only to equipment but to other products of R&D such as in accelerated aging of pharmaceuticals and food products to determine shelf life. A key issue to address when introducing stress factors is to ensure that the failure modes that they produce are equal to those observed in actual use; this is not necessarily given due to the artificial testing conditions (e.g., chemical kinetics may be sensitive to specific conditions). Gryna (1988) provides an example of using probability plots to compare and relate test results to "field" failures. Figure 28.16 contains

Type of TEST	Purpose
Performance	Determine ability of product to meet basic performance requirements
Environmental	Evaluate ability of product to withstand defined environmental levels; determine internal environments generated by product operation; verify environmental levels specified
Stress	Determine levels of stress that a product can withstand in order to determine the safety margin inherent in the design; determine modes of failure that are not associated with time
Reliability	Determine product reliability and compare to requirements; monitor for trends
Maintainability	Determine time required to make repairs and compare to requirements
Life	Determine wear-out time for a product, and failure modes associated with time or operating cycles
Pilot Run	Determine if fabrication and assembly processes are capable of meeting design requirements; determine if reliability will be degraded.

[*Source:* Gryna et al. (2007), p. 334.]

Table 28.9 Summary of Tests Used for Design Evaluation

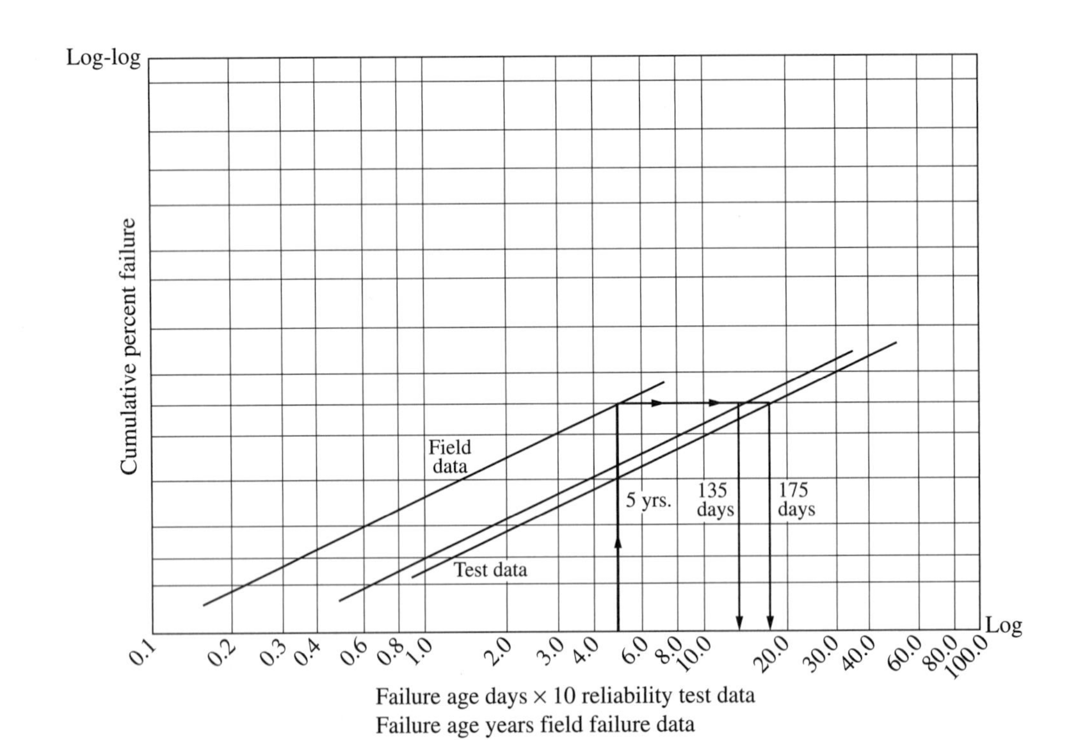

Figure 28.16 Weibull plot of accelerated test versus field failure data for two air-conditioner models. (*Juran Institute. ©1994. Used with permission.*)

plots of the estimated cumulative failure percentages versus number of accelerated test days and actual field usage days for two models of air conditioners. Because the two lines essentially are parallel, it appears that the basic failure modes produced by the accelerated and field usage environments are equivalent. The test data are plotted in tens of days. The 5-year warranty period is represented by a heavy vertical line. Following the vertical line from where it intersects the field data line, and proceeding horizontally to the lines for the accelerated test data, the accelerated test time required to predict the percentage of field failures occurring during the 5-year warranty period is estimated at 135 days for one air conditioner model and 175 days for the other model.

Failure Reporting and Corrective Action Systems

In order to drive improvements in RAM and safety of designs, an organization must define and develop a formal process for reporting, classifying, analyzing, and improving these design parameters. Many organizations call this process "failure reporting and corrective action systems" (FRACAS). Figure 28.17, reproduced from the Rome Laboratory Reliability Engineer's Toolkit (1993), is a high-level flow diagram for a generic FRACAS process. In addition to the process steps, process-step responsibilities are identified by function. The same publication also provides a checklist for identifying gaps in existing FRACAS processes. Ireson (1996) provides additional guidance on reliability information collection and analysis, with discussion on data requirements at the various phases of design, development, production, and usage. Adams (1996) focuses on details of identifying the root causes of failures and driving corrective action with an example of a "business plan" for justifying investment in the equipment and personnel required to support a failure analysis process.

Prognosis

The concepts, tools, and processes discussed here have guided R&D managers, scientists, and engineers in creating successful products and commercial development in the past and will serve them well into the future. Nonetheless, R&D will need to adapt to changes that may create pressure to deviate from accepted quality principles or force new ways of ensuring performance. Some considerations are discussed as follows.

- *Greater entrepreneurial spirit.* The recent global economic downturn has forced consolidation and released many R&D staff to pursue their own interests. If history is any indication, this, in combination with the technological innovations previously cited, will drive the creation of start-up organizations. In turn, this will fuel "intrapreneurialism" as the older organizations act in response.
- *More cross-disciplinary interaction.* The continuing ease of communication has allowed formation of loose, informal groups and affiliations consisting of people with common interests (e.g., networking organizations such as Facebook, MySpace, Twitter, and LinkedIn; whereas the longevity of any particular organization is questionable, the general trend will prevail). Physical boundaries are less meaningful constraints to cooperation and collaboration, thereby promoting synergies and innovations previously not possible.
- *Increased tension between R&D and marketing.* The most disruptive and game-changing technological innovations are those that do not necessarily come from market pull; instead, they are serendipitous inventions that at first have no apparent market. These gems may be buried among many other innovations that are solutions seeking a problem. The net result will be greater technology push, with need for

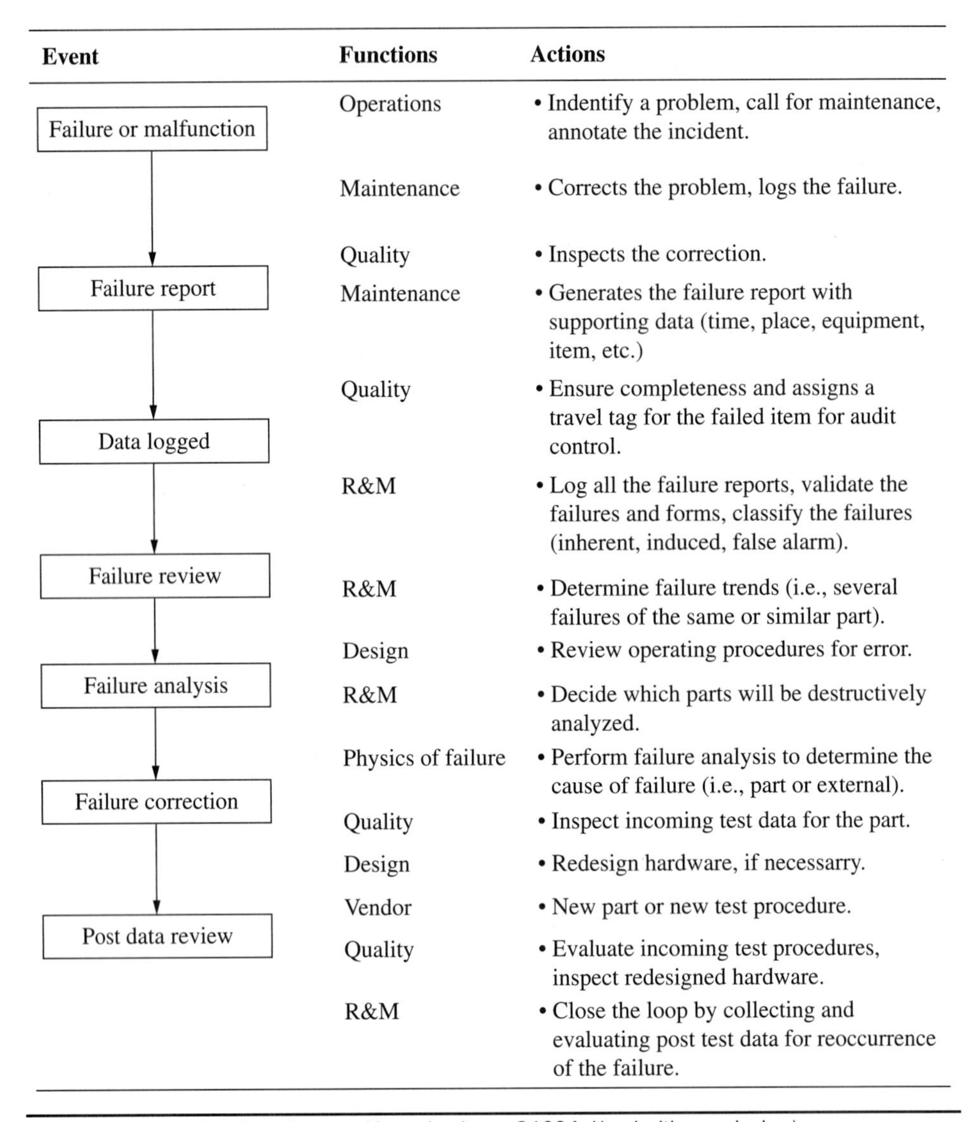

FIGURE 28.17 FRACAS flow diagram. (*Juran Institute. ©1994. Used with permission.*)

marketing to expand efforts to better understand unmet market needs. This can be a difficult step because existing customers may not be interested in the new technology ("marketing says our customers aren't interested, so kill the project"), and it requires "listening" to customers that do not yet exist, do not know of the technology, and cannot articulate how their needs could be met through the innovation.

- *Resynthesis of quality tools.* Readers may note the prevalence of research on quality topics directed toward the development of new and modified tools (e.g., the integration of QFD, benchmarking, and decision analysis cited earlier). Online accessibility to research papers, cross-disciplinary interaction, and the prevalence of software that helps users through technically challenging techniques (e.g., statistics) will encourage new applications. R&D managers should remain diligent in scanning the literature outside their immediate area to find methods that could improve quality at reduced cost and provide competitive advantage.

- *Dispersed organizational structure.* In the past, R&D relied heavily on centralized physical laboratories in which people were in close proximity. As technological innovations facilitate knowledge transfer and lower barriers to entry (witness the availability of DNA testing kits sold for children's amusement; such technology required expensive equipment only a few years ago), it will become increasingly viable to structure a geographically- and time-dispersed R&D organization. This may strain quality, however, and it will be essential to vigilantly coordinate activities and handoffs.
- *Business process innovation in R&D.* Although widespread, quality improvement initiatives such as Six Sigma have not penetrated R&D as much as in other business processes. This is due to a combination of cultural resistance and lack of fit. As shown in Figure 28.1, product innovation and process innovation tend to occur at different phases of development. Although perhaps in modified form, continuous process improvement and innovation methodologies will tend to blur the distinction between phases, so that the early R&D creative process will receive greater focus.
- *Shift in basic, long-term research from companies to universities.* Retrenchment in the face of difficult economic conditions tends to favor short-term R&D projects perceived as having a safer risk to reward profile and more immediate financial returns. Accordingly, basic long-term research and risk will be shifted away from businesses and into not-for-profit organizations such as government laboratories and universities. This is happening at present (Clark and Rhoads 2009), and, in turn, may lead to greater sharing of intellectual property and profits.
- *Greater sharing of intellectual property.* In some industries such as biotechnology and electronics, R&D efforts are stymied by a thicket of patents. Organizations increasingly will need to consider the downside of rigidly protecting intellectual property and trying to navigate through competing positions and consider instead the upside of cross-licensing to clear the path for their R&D staff to truly innovate and meet customer needs unhindered by intellectual property concerns.
- *Improved metrics that better reflect true R&D quality.* Organizations gradually are becoming more aware that quota types of metrics do a poor job in estimating the value created by R&D, and are, at best, proximal measures. In spite of perceived intrusiveness by R&D staff, expect managers to cautiously shift towards new and more meaningful metrics with mixed but steady success.

As stated earlier, few organizations are able to attain true, enterprisewide quality leadership. This is due, in large part, to practical trade-offs, resource constraints, and the shifting context of business, both micro- and macroeconomic. Recalling that "chance favors the prepared mind," it is incumbent upon R&D bench scientists, laboratory engineers, development staff, and management to lay a firm foundation of quality principles and tools, providing the flexibility to respond with unified action to unforeseen events, challenges, and opportunities.

Where to Find It

Following is a list of various online resources that may be useful to R&D professionals. The list is intended to provide a sampling of sites and is not intended to be comprehensive. Links to additional resources may be found on most websites, so the list may be viewed as a starting point for research.

Professional Societies, Institutes, and Communities

American Society for Quality (ASQ): <http://www.asq.org/> (includes a Reliability Division: <http://www.asq.org/divisions-forums/reliability/index.html>). *A professional society devoted entirely to quality. Catering to numerous sectors (education, government, healthcare, manufacturing, service), ASQ is a source for publications, conferences, standards, and certification.*

Industrial Research Institute (IRI): <http://www.iriinc.org/> *A source of publications, conferences and workshops, and networking opportunities related to research within the industrial sector.*

Institute of Electrical and Electronics Engineers (IEEE): <http://www.ieee.org/portal/site> *A large, long-standing professional association dedicated to the advancement of innovation and technology. IEEE is a source of publications (including books and journals), conferences, professional development opportunities, and standards.*

PharmWeb: <http://www.pharmweb.net/> *Portal and online community of pharmacy, pharmaceutical and healthcare-related professionals. Contains a large number of links to additional resources worldwide.*

Society of Reliability Engineers (SRE): <http://www.sre.org/> *A U.S. professional society of engineers interested in reliability Contains military handbooks and standards relating to reliability.*

Technical Standards Organizations

ASTM International: <http://www.astm.org/> *A voluntary standards development organization and source for technical standards for materials, products, systems, and services.*

European Committee for Standardization (CEN): <http://www.cen.eu/> *A European business facilitator that seeks to removing trade barriers for European industry and consumers; provides a platform for the development of European Standards and other technical specifications.*

National Institute of Standards and Technology (NIST): <http://www.nist.gov/index.html> *A nonregulatory federal agency within the U.S. Department of Commerce that promotes U.S. innovation and industrial competitiveness by advancing measurement science, standards, and technology. NIST administers four cooperative programs:*

- *NIST Laboratories*, conducting research to advance the U.S. technology infrastructure needed by industry to continually improve products and services.
- *Baldrige National Quality Program*, which promotes performance excellence among U.S. organizations, conducts outreach programs and manages the annual Malcolm Baldrige National Quality Award to recognize performance excellence and quality achievement.
- *Hollings Manufacturing Extension Partnership*, a nationwide network of local centers that offers technical and business assistance to small manufacturers.
- *Technology Innovation Program*, which provides cost-shared awards to industry, universities, and consortia for research on potentially revolutionary technologies.

Reliability Resources

- **Reliability Information Analysis Center (RIAC):** <http://www.theriac.org/> *A Center of Excellence and technical focal point for information, data, analysis, training, and technical assistance in the engineering fields of reliability, maintainability, quality, supportability, and interoperability (RMQSI). Serves primarily, but not exclusively, the U.S. Department of Defense.*

- **Weibull.com:** <http://www.weibull.com/> *A website devoted entirely to the topic of reliability engineering and reliability theory.*
- **U.S. Government Defense R&D:** *Organizations involved in R&D related to* national security.

  Air Force Research Laboratory: <http://www.wpafb.af.mil/AFRL/>

  Army Research Laboratory: <http://www.arl.army.mil/>

  Defense Advanced Research Projects Agency: <http://www.darpa.mil/>

  Naval Research Laboratory: <http://www.nrl.navy.mil/>

  Office of Naval Research: <http://www.onr.navy.mil/>
- **U.S. Government Foundations and Related Organizations:** *Federally supported organizations conducting and facilitating research in the national interest; refer to individual websites for details.*

  Federal Laboratory Consortium (FLC): <http://www.federallabs.org/>

  National Academies: <http://www.nationalacademies.org/>

  National Institutes of Health: <http://www.nih.gov/science/index.html> *(see also NIH Office of Technology Transfer (OTT):* <http://ott.od.nih.gov/index.aspx>

  National Center for Supercomputing Applications: <http://www.ncsa.illinois.edu/>

  National Science Foundation: <http://www.nsf.gov>. *(For a list of federally funded R&D centers, see* <http://www.nsf.gov/statistics/ffrdc/>)

  North American Space Administration (NASA): <http://www.nasa.gov/>
- **U.S. National Laboratories and Technology Centers:** *Department of Energy research centers supporting a broad range of scientific and engineering research. Visit specific sites for current research programs.*

  Ames Laboratory: <http://www.ameslab.gov/>

  Argonne National Laboratory: <http://www.anl.gov/>

  Brookhaven National Laboratory: <http://www.bnl.gov/world/>

  Fermi National Accelerator Laboratory: <http://www.fnal.gov/>

  Idaho National Laboratory: <http://www.inl.gov/>

  Lawrence Berkeley National Laboratory: <http://www.lbl.gov/>

  Lawrence Livermore National Laboratory: <https://www.llnl.gov/>

  Los Alamos National Laboratory: <http://www.lanl.gov/>

  National Renewable Energy Laboratory: <http://www.nrel.gov/>

  New Brunswick Laboratory: <http://www.nbl.doe.gov/>

  Oak Ridge National Laboratory: <http://www.ornl.gov/>

  Pacific Northwest National Laboratory: <http://www.pnl.gov/>

  Princeton Plasma Physics Laboratory: <http://www.pppl.gov/>

  Radiological & Environmental Sciences Laboratory: <http://www.inl.gov/resl>

  Sandia National Laboratories: <http://www.sandia.gov/>

  Savannah River National Laboratory: <http://srnl.doe.gov/>

SLAC National Accelerator Laboratory: <http://www.slac.stanford.edu/>

Thomas Jefferson National Accelerator Facility: <http://www.jlab.org/>

- **European R&D Organizations:** *These organizations provide information and support for research, development and innovation activities within the European Community.*

  EUREKA: <http://www.eureka.be/home.do>

  European Association of Research and Technology Organisations: <www.earto.org>

  European Research Area (European Commission: CORDIS): <http://cordis.europa.eu/era/home_en.html>

  Joint Research Centre: <http://ec.europa.eu/dgs/jrc/>

Commercial and Corporate Research Facilities: *A sampling of well-known research organizations. Specific internet addresses change frequently; refer to individual websites to locate R&D information.*

Bell Labs: <http://www.bell-labs.com>

Hewlett-Packard Labs: <http://www.hpl.hp.com>

IBM Research: <http://www.research.ibm.com/>

Intel Research: <http://www.intel.com/research>

Microsoft Research: <http://research.microsoft.com>

Mitsubishi Electric Research Labs: <http://www.merl.com>

SRI International: <http://www.sri.com>

Palo Alto Research Center Inc.: <http://www.parc.com>

- **University Research Labs:** *A sampling of academic institutions known for research and technology development. Specific web addresses are subject to change, but research program information typically can be found via the home page.*

  California Institute of Technology: <http://www.caltech.edu/>

  Massachusetts Institute of Technology: <http://web.mit.edu/>

  Princeton University: <http://www.princeton.edu/main/research/>

  Stanford University: <http://www.stanford.edu/research/>

References

Adams, J. (1996). Failure Analysis System—Root Cause and Corrective Action. *Handbook of Reliability Engineering and Management*, 2nd ed. Ireson, W., Coombs, C., and Moss, R., eds. McGraw-Hill, New York, chap. 13. *Provides an organized approach to the analysis of failures to determine root cause. The chapter deals primarily with engineering designs of electronic equipment, but the methods are applicable widely.*

Altland, H. (1995). Robust Technology Development Process for Imaging Materials at Eastman Kodak. *Proceedings of Symposium on Managing for Quality in Research and Development*, Juran Institute, Wilton, CT, p. 2B-15. *Use of phase gating as a means of managing project performance and risk.*

Argyres, N. S., and Silverman, B. S. (2004). R&D, Organization Structure and the Development of Corporate Technological Knowledge. *Strategic Management Journal*, Aug.–Sep. vol. 25, no. 8–9, pp. 929–959. *An exploration of the relationship between how a company's research is organized (e.g., centralized vs. decentralized) and the resulting innovation.*

Arundachawat, P., Roy, R., Al-Ashaab, A., and Shehab, E. (2009). Design Rework Prediction in Concurrent Design Environment: Current Trends and Future Research Directions. *Proceedings of the 19th CIRP Design Conference—Competitive Design*, Cranfield University, Bedfordshire, UK, March 2009, pp. 237-244. *Analysis of factors contributing to design rework as part of concurrent engineering, and proposal of "set-based" concurrent engineering as an approach to mitigate design rework risk.*

Beall, G. H. (2002). Exploratory Research Remains Essential for Industry. *Research Technology Management*, vol. 45, no. 6, pp. 26-30. *Discussion of the issues presented to technology managers in R&D by the "fuzzy front end" of the innovation cycle that has a high failure rate and long lag time to demonstrated success.*

Berezowitz, W., and Chang, T. (1997). Assessing Design Quality During Product Development. *ASQC Quality Congress Proceedings*, Milwaukee, WI, p. 908. *Appraisal of design as a factor influencing overall product quality.*

Boath, D. (1992). Using Metrics to Guide the TQM Journey in R&D. *Proceedings of the Symposium on Managing for Quality in Research and Development*, Juran Institute, Wilton, CT. *Presents a performance measurement pyramid to assist in understanding R&D quality.*

Boath, D. (1993). Reengineering The Product Development Process. *Proceedings of the Symposium on Managing for Quality in Research and Development*, Juran Institute, Wilton, CT, pp. 28-36. *An example of efficiency improvement by revamping a new product development process.*

Bodnarczuk, M. (1991). Peer Review, Basic Research, and Engineering: Defining a Role for QA Professionals in Basic Research Environments. *Proceedings of the Symposium on Managing for Quality in Research and Development*, Juran Institute, Wilton, CT. *An example of the use of peer review at Fermi National Laboratory.*

Boer, F. P. (1999). *The Valuation of Technology: Business and Financial Issues in R&D*. Wiley, New York. *Explores the link between R&D and business financials, and provides analytical tools for analyzing new technologies, technology initiatives, and forecasting future value.*

Boothroyd, G., and Dewhurst, P. (1987). *Design for Assembly*, Boothroyd Dewhurst, Inc., Wakefield, RI. *Discussion of a design for assembly (DFA) methodology to help ensure products can be efficiently and effectively assembled.*

Boothroyd, G., and Dewhurst, P. (1994). *Design for Manufacture and Assembly*. Marcel Dekker, New York. *Further discussion and development of the DFA methodology in Boothroyd and Dewhurst (1987).*

Bottome, R., and Chua, R. C. H. (2005). Genentech error proofs its batch records. *Quality Progress* (July), pp. 25–34.

Boznak, R., and Decker, A. (1993). *Competitive Product Development*. Copublished by ASQC Quality Press/Business One Irwin, Milwaukee, WI, p. 53. *A textbook intended to help managers revitalize product development by applying philosophies and methods to reduce development cycle time and cost.*

Brown, E. A. (2006). Measuring Performance at the Army Research Laboratory: The Performance Evaluation Construct. *Journal of Technology Transfer*, vol. 22, no. 2, pp. 21–26. *Discussion of the Performance Evaluation Construct, the measurement system developed to measure progress by the U.S. Army Research Laboratory.*

Chen, C. (2001). Design for the Environment: A Quality-based Model for Green Product Development. *Management Science* (USA), vol. 47, no. 2, pp. 250–263. *Citing electrical vehicles, this paper develops and evaluates models of two product strategies (market-segmentation and mass-marketing), concluding that green products may need regulatory support to be beneficial to the environment.*

Chien, W., and Kuo, W. (1995). Modeling and Maximizing Burn-in Effectiveness. *IEEE Transactions on Reliability*, vol. 44, no. 1, pp. 19–25. *Practical suggestions to make the most of burn-in testing as part of accelerated life testing.*

Clark, D., and Rhoads, C. (2009). Basic Research Loses Some Allure. *The Wall Street Journal*, Wed. Oct. 7, p. A3. *Presents the case that basic research will move in future years from for-profit organizations to academic institutions.*

Cole, R. (1990). Quality in the Management of Research and Development. *Proceedings of the Symposium on Managing for Quality in Research and Development*, Juran Institute, Wilton, CT. *An example of use of compliance scoring to help improve R&D cycle time.*

Darby, R. A. (1990). R&D Quality in a Diversified Company. *Proceedings of the Symposium on Managing for Quality in Research and Development*, Juran Institute, Wilton, CT. *Discussion of how DuPont manages quality; most relevant here for the definition of R&D quality.*

De Feo, J. (1987). Quality Training: the Key to Successful Quality Improvement. *Proceedings of the IMPRO Conference*, Juran Institute, Wilton, CT, pp. 4A–15 et seq. *An example of how training (specifically in design for assembly) is an important part of improving overall product quality.*

Delano, G., Parnell, G. S., Smith, C., and Vance, M. (2000). Quality Function Deployment and Decision Analysis: A R&D Case Study. *International Journal of Operations and Production Management*, vol. 20, no. 5, pp. 591–609. *The authors compare quality function deployment and decision analysis as techniques to assist in product design decisions.*

Dell, M. (2000). E-Business: Strategies in Net Time. Address at the University of Texas, April 27, Austin, TX. *Speech by Dell CEO regarding fundamental change to business and the economy as a result of the Internet.*

Dhillon, B. S. (1999). *Engineering Maintainability: How to Design for Reliability and Easy Maintenance.* Gulf Publishing, Houston TX. *Provides guidelines and fundamental methods of estimation and calculation for maintainability engineers. Also addresses organizational structure, cost, and planning processes.*

Endres, A. C. (1992). Results and Conclusions from Applying TQM to Research. *ASQC Quality Congress Proceedings*, Milwaukee, WI. *Discussion of cultural and organizational barriers to successful deployment of quality improvement in R&D.*

Endres, A. C. (1997). *Improving R&D Performance: The Juran Way*. Wiley, New York. *A detailed look at how to plan, design, implement, and improve quality systems to advance the ability of R&D to reduce cycle times, increase customer satisfaction, and foster innovative product development. This book draws from original research at the Juran Institute and from papers presented at its Symposia on Managing for Quality in R&D.*

Endres, A. C. (1999). *Quality in Research and Development*. Section 19 in Juran, J. M., and Godfrey, A. B., eds. *Juran's Quality Handbook*, 5th ed. McGraw-Hill, New York. *The present topic of R&D quality in the prior edition of Juran's handbook.*

Ferm, P., Hacker, S., Izod, T., and Smith, G. (1993). Developing a Customer Orientation in a Corporate Laboratory Environment, *Proceedings of the Symposium on Managing for Quality in Research and Development*, Juran Institute, Wilton, CT, pp. A-13–A-22. *Presents an example of surveys used to better understand and meet the needs of internal, business unit customers of R&D.*

Fiero, J., and Birch, W. (1989). Designing Cost-Effective Products. *ASQC Quality Congress Proceedings*, Milwaukee, WI, pp. 725–730. *An example of the financial benefits of reducing development cycle time.*

Ford, D. N., and Sobek II, D. K. (2005). Adapting Real Options to New Product Development by Modeling the Second Toyota Paradox: How Delaying Decisions Can Make Better Cars Faster, *Sloan Management Review*, MIT. *Application of real options to decision making in the auto industry.*

Fried, L. K. (1993). AT&T Transmission Systems ISO 9001 Registration: The R&D Compliance Experience. *Proceedings of Symposium on Managing for Quality in Research and Development*, Juran Institute, Wilton, CT, pp. 2B-21–2B-27. *The story of ISO 9001 registration from the perspective of R&D.*

Garfinkel, M. (1990). Quality in R&D. *Proceedings of the Symposium on Managing for Quality in Research and Development*, Juran Institute, Wilton, CT. *Discussion of the different elements that contribute to quality in the R&D environment.*

Gendason, P., and Brown, E. (1993). Measure of R&D Effectiveness: A Performance Evaluation Construct. *Proceedings of Symposium on Managing for Quality in Research and Development*, Juran Institute, Wilton, CT, pp. 2A-17–2A-25. *A review of metrics to measure R&D performance.*

Gibbard, H. F., and Davis, C. (1993). Implementation of ISO 9001 in an R&D Organization, *Proceedings of the IMPRO Conference*, Juran Institute, Wilton, CT, pp. 3B.3-1–3B.3-11. *A case study of issues and solutions as part of implementing ISO 9001 in R&D.*

Godfrey, A. B. (1985). Training Design Engineers in Quality. *Proceedings of the IMPRO Conference, Juran Institute*, Wilton, CT, p. 166 et seq. *Provides discussion and example of tailoring quality training to meet the unique needs of R&D.*

Godfrey, A. B. (1991). Information Quality: A Key Challenge for the 1990s. *The Best on Quality*, vol. 4, Hanser Publishers, Munich. *Provides a general discussion of information quality as one of the key outputs of R&D.*

Goldstein, R. (1990). "The Cost of Engineering Design Corrections." *ASQC Quality Congress Proceedings*, Milwaukee, WI, pp. 549–554. *This paper cites metrics useful in assessing costs of rework.*

Graves, S., and Menten, T. (1996). Designing Experiments to Measure and Improve Reliability. *Handbook of Reliability Engineering and Management*, 2nd ed. Ireson, W., Coombs, C., and Moss, R., eds. McGraw-Hill, New York, chap. 11. *A look at reliability from the perspective of design of experiments, thereby decreasing development cycle time while increasing quality. A case study is provided based on the life of ball bearings.*

Gross, C. (1997). Ergonomic Quality: Using Biomechanics Technology to Create a Strategic Advantage in Product Design. *ASQC Quality Congress Proceedings*, Milwaukee, WI, pp. 869–879. *The author argues for improvement of manufacturability and customer usability by integrating ergonomics with the design process.*

Gryna, F. (1988). Product Development, Section 13, In Juran, J. M. and Godfrey, A. B., eds. *Juran's Quality Control Handbook*, 4th ed. McGraw-Hill, New York. *An overview of quality as part of product development; specifically relevant in this chapter for guidelines regarding design reviews.*

Gryna, F. M. (1999). Operations, Section 22, In Juran, J. M., and Godfrey, A. B., eds. *Juran's Quality Handbook*, 5th ed. McGraw-Hill, New York. *Discussion of quality in manufacture (in manufacturing sector) and backroom activities (in the service sector) in the prior edition of the handbook.*

Gryna, F. M. (2001). *Quality Planning and Analysis*. McGraw-Hill, New York. *A concise, yet thorough, textbook on achieving customer satisfaction and loyalty. Pages 382–383 provide the steps for reliability prediction.*

Gryna, F. M., Chua, R. C. H., and De Feo, J. A. (2007). *Juran's Quality Planning & Analysis for Enterprise Quality*, 5th ed. McGraw-Hill, New York. *A comprehensive textbook about attaining quality leadership throughout an enterprise.*

Gunasekaran, A. (1997). Essentials of International and Joint R&D Projects. *Technovation*. vol. 17, nos. 11–12, Nov.–Dec., pp. 637–647. *Discussion of management issues of international and joint R&D projects, and strategies and methods to make international R&D projects more successful.*

Hammer, W. (1980). *Product Safety Management and Engineering*. Prentice Hall, Englewood Cliffs, NJ. *A reference most relevant in this chapter for the example of a fault tree analysis.*

Herrmann, A., Huber, F., Algesheime, R., and Tomczak, T. (2006). An Emipirical Study of Quality Function Deployment on Company Performance. *International Journal of Quality and Reliability Management*, vol. 23, no. 4, pp. 345–366. *An evaluation of QFD based on its contribution towards improved product quality, reduced costs, and shorter R&D cycle time.*

Hersh, J. F., Backus, M. C., Kinosz, D. L., and Wasson, A. R. (1993). Understanding Customer Requirements for the Alcoa Technical Center, *Proceedings of the Symposium on Managing for Quality in Research and Development*, Juran Institute, Wilton, CT. *Case study citing the use of benchmarking and internal customer surveys to identify and prioritize key R&D performance features.*

Highway, Car Changes Designed to Help Older Drivers (sidebar). Issues & Controversies on File: n. pag. Issues & Controversies. Facts on File News Services, 10 May 2002. Web. 5 Oct. 2009. <http://www.2facts.com/article/ib701090>. *An example of how ergonomic considerations play into the layout and design of roads and vehicles.*

Hildreth, S. (1993). Rolling-Out BPQM in the Core R&D of Lederle-Praxis Biologicals, American Cyanamid. *Proceedings of the Symposium on Managing for Quality in Research and Development*, Juran Institute,Wilton, CT, pp. 2A-9–2A-16. *An example of how organizational structure promotes better management of major business processes, including R&D.*

Himmelfarb, P. (1996a). Senior Managers' Role in New-Product Development. *Quality Progress*, Oct., pp. 31–33. *The author argues the case for management to clearly define the processes of R&D.*

Himmelfarb, P. (1996b). Fast New-Product Development at Service Sector Companies. *Quality Progress*, April, pp. 41–43. *Provides examples of new product development in service industries.*

Ho, H. C. R&D Management. PowerPoint presentation. Department of Technology and Operations Management, California Polytechnic and State University, San Francisco, CA, Sep. 29. <www.csupomona.edu/~hco/MoT/08bRDManagement.ppt>. *A short presentation covering product and process innovation stages, cash flow, types of product development, and location of R&D.*

Holmes, J., and McClaskey, D. (1994). Doubling Research's Output Using TQM. *Proceedings of the Symposium on Managing for Quality in Research and Development*, Juran Institute, Wilton, CT, pp. 4–7. *Discussion concerning various aspects of applying quality concepts and methods to R&D, including leadership issues, project management, process flowcharting, and application of metrics.*

Hooper, J. (1990). Quality Improvement In Research and Development. *Proceedings of Symposium on Managing for Quality in Research and Development*, Juran Institute, Wilton, CT. *The author argues that isolation of R&D staff is an organizational barrier to understanding and meeting customer needs.*

Hughes, J. (1992). Concurrent Engineering: A Designer's Perspective. *Report SG-7-3*, Motorola University Press, Schaumburg, IL. *Provides an automotive industry example of how design influences R&D costs and ultimate product quality.*

Hutchins, D. (1999). *Just in Time*, 2nd ed. Gower Publishing Ltd., Aldershot, UK. *A textbook on JIT concepts and applications. Most relevant to this chapter for discussion of design review as part of JIT (beginning on p. 74).*

Hutton, D., and Boyer, S. (1991). Lead-Time Reduction In Development—A Case Study. *Proceedings of the ASQC Quality Congress*, Milwaukee, WI, pp. 14–18. *A case study showing dramatic reduction in prototype development time.*

Industrial Research Institute (1996). *Industrial Research and Development Facts*, Industrial Research Institute, Washington, DC, July, p. 7. *Definitions of different types of research.*

Ireson, G. (1996). Reliability Information Collection and Analysis. *Handbook of Reliability Engineering and Management*, 2nd ed. Ireson, W., Coombs, C., and Moss, R., eds. McGraw-Hill, New York, chap. 10. *Provides recommendations regarding the method and details of collecting, recording, storing, analyzing using and transmitting information for reliability studies.*

Jacobs, R. (1967). Implementing Formal Design Review. *Industrial Quality Control*, vol. 23, Feb., pp. 398–404. *Provides guidelines for conducting design reviews, including design review team membership and responsibilities.*

Jensen, F., and Petersen, N. (1982). *Burn-in: An Engineering Approach to the Design and Analysis of Burn-in Procedures*. Wiley, New York. *A complete guide to the design of burn-in test procedures.*

Jensen, R. P., and Morgan, M. N. (1990). Quality in R&D—Fit or Folly. *Proceedings of the Symposium on Managing for Quality in Research and Development*, Juran Institute, Wilton, CT. *Report on a project at Shell that yielded a reduction in R&D cycle time.*

Juran, J. M. (1964). *Managerial Breakthrough*, (1995). McGraw-Hill, New York. *One of the classic works by Dr. Juran, containing many of the concepts that allow organizations to break through to new levels of performance.*

Juran, J. M. (1992). Closing speech at the IMPRO Conference, Juran Institute, Wilton, CT. *Final remarks to finish one of the Juran-sponsored conferences on quality.*

Juran, J. M., and Godfrey, A. B. (1999). *Juran's Quality Handbook*. 5th ed. McGraw-Hill, New York. *Prior edition of the present handbook.*

Kang, N., Kim, J., and Park, Y. (2007). Integration of Marketing Domain and R&D Domain in NPD Design Process. *Industrial Management and Data Systems*, vol. 107, no. 6, pp. 780–801. *An approach to new product design that integrates quality function deployment (QFD), Taguchi methods, and conjoint analysis to help resolve conflicts inherent in R&D and marketing relationships.*

Kapur, K. (1996). Techniques of Estimating Reliability At Design Stage. *Handbook of Reliability Engineering and Management*, 2nd ed. Ireson, W., Coombs, C., and Moss, R., eds. McGraw-Hill, New York. *Discussion of reliability estimation during product design. A design review responsibility matrix is given on p. 24.5.*

Kohoutek, H. (1996a). Human-Centered Design. *Handbook of Reliability Engineering and Management*, 2nd ed. Ireson, W., Coombs, C., and Moss, R., eds. McGraw-Hill, New York, chap. 9. *Discussion of "usability" as a controllable factor in human reliability in the context of machine-human interactions.*

Kohoutek, H. (1996b). Reliability Specifications and Goal Setting. *Handbook of Reliability Engineering and Management*, 2nd ed. Ireson, W., Coombs, C., and Moss, R., eds. McGraw-Hill, New York, chap. 7. *An examination of how reliability can be improved through proper establishment and use of specifications.*

Konosz, D. L., and Ice, J. W. (1991). Facilitation of Problem Solving Teams. *Proceedings of the Symposium on Managing for Quality in Research and Development*, Juran Institute, Wilton, CT. *Discussion of problem-solving teams and training within the context of R&D.*

Kowalski, R. (1996). Maintainability and Reliability. *Handbook of Reliability Engineering and Management*, 2nd ed. Ireson, W., Coombs, C., and Moss, R., eds. McGraw-Hill, New York, chap. 15. *An in-depth review of designing products to provide improved maintenance and reliability characteristics.*

Kozlowski, T. R. (1993). Implementing a Total Quality Process into Research and Development: A Case Study, *Proceedings of the Symposium on Managing for Quality in Research and Development*, Juran Institute, Wilton, CT. *A case study in the application of TQM, including use of Baldrige criteria and quality cost data to identify projects.*

Kumar, V., and Boyle, T. A. (2001). A Quality Management Implementation Framework for Manufacturing Based R&D Environments, The International Journal of Quality and

Reliability Management, vol. 18, no. 3, pp. 336–359. *A summary of best practices for achieving quality in applied R&D departments of manufacturing companies, with recommendations to promote good management practices, awareness by the R&D staff of the external environment, and quality culture.*

Lander, L., Matheson, D., and Ransley, D. (1994). IRI's Quality Director's Network Takes R&D Decision Quality Benchmarking One Step Further. *Proceedings of the Symposium on Managing for Quality in Research and Development*, Juran Institute, Wilton, CT, pp. 3-11 to 3-18. *Report on results of a benchmarking study of the best features of practices in R&D portfolio planning, development, and review.*

Lazor, J. (1996). Failure Mode and Effects Analysis (FMEA) and Fault Tree Analysis (FTA) (Success Tree Analysis—STA). *Handbook of Reliability Engineering and Management*, 2nd ed. Ireson,W., Coombs, C., and Moss, R., eds. McGraw-Hill, New York, chap. 6. *Describes methods and elements of fault tree analysis and FMEA (covering system-level, design-level and process-level analyses).*

Malec, H. (1996). System Reliability. *Handbook of Reliability Engineering and Management*, 2nd ed. Ireson, W., Coombs, C., and Moss, R., eds. McGraw-Hill, New York, chap. 21. *This chapter delves into designing and evaluating system reliability; provides formulas and examples for calculating availability metrics.*

Matheson, D., Matheson, J., Menke, M. (1994). SDG's Benchmarking Study of R&D Decision Making Quality Provides Blueprint for Doing the Right R&D. *Proceedings of the Symposium on Managing for Quality in Research and Development*, Juran Institute, Wilton, CT, pp. 3-1 to 3-9. *Provides examples of tools to identify opportunities for implementing and improving best practices in R&D planning and implementation.*

Mayo, J. (1994). Total Quality Management at AT&T Bell Laboratories. *Proceedings of the Symposium on Managing For Quality in Research and Development*, Juran Institute, Wilton CT, pp. 1-1 to 1-9. *Report regarding Bell Labs' TQM program, including metrics to evaluate performance.*

Meeker, W. Q., Escobar, L. A., Doganaksoy, N., and Hahn, G. J. (1999). Reliability Concepts and Data Analysis. Section 48 in Juran, J. M., and Godfrey, A. B., eds. *Juran's Quality Handbook*, 5th ed. McGraw-Hill, New York. *A discussion of many of the concepts and technical details of reliability analysis.*

Mendez, A. (2003). The Coordination of Globalized R&D Activities through Project Teams Organization: An Exploratory Empirical Study. *J. World Business*, vol. 38, no. 2, pp. 96–109. *An empirical study of project team positioning and their role in the organization of R&D within multinational firms. The pharmaceuticals, chemicals and computer industries are explored.*

Menger, E. L. (1993). Evolving Quality Practices at Corning Incorporated. *Proceedings of the Symposium on Managing for Quality in Research and Development, Juran Institute*, Wilton CT, pp. 1-9 to 1-20. *Reviews how Corning has organized to better manage R&D performance.*

Miguel. P. A. C., and Carnevalli, J. A. (2008). Benchmarking Practices: of Quality Function Deployment: Results from a Field Study. *Benchmarking: An International Journal*, vol. 15, no. 6, pp. 657–676. *A survey analysis of QFD deployment at numerous development organizations in Brazil.*

MIL-HDBK-781. (1987). Reliability Test Methods, Plans and Environments for Engineering Development, Qualification and Production. *A U.S. military handbook for reliability engineers.*

MIL-STD-721C. (1981). Definitions of Terms for Reliability and Maintainability. *A U.S. military standard, containing definitions for design engineers. Note: this standard was canceled without replacement (12/5/1995) but remains a useful reference.*

Miller, W. L., and Morris, L. (1999). *Fourth Generation R&D: Managing Knowledge, Technology, and Innovation*. Wiley, New York. *A guide for managing the seemingly contradictory requirements of creativity and business, thereby driving innovation throughout an organization.*

Murphy, J. (2000). Using Competitive Technical Intelligence Techniques to Complement Research-and-Development Processes. *Managing Frontiers in Competitive Intelligence.* Fleisher C. S. and Blenkhorn, D. L. eds.. Quorum Books, Westport, CT, pp. 136–148. *A discussion of the role of competitive technical intelligence in R&D, including topics of technology scouting and technical literature analysis.*

Nagamachi, M. (2008). Perspectives and the New Trend of Kansei/Affective Engineering. *The TQM Journal*, vol. 20, no. 4, pp. 290–298. *Review of the Kansei approach to ergonomic engineering, updating prior publications by the author in this area.*

NASA (2007). NASA Systems Engineering Processes and Requirements. NPR 7123A, Appendix G: Technical Review Entrance and Success Criteria. Effective March 26, 2007. Retrieved Oct. 8 < http://nodis3.gsfc.nasa.gov>. *A listing of the specific elements required as part of a comprehensive design review.*

NIST (2008). 2007 Award Recipient Application Summaries (Nonprofit): U.S. Army Armament Research, Development and Engineering Center (ARDEC). n. pag. Retrieved 8 Oct. 2009. <http://www.quality.nist.gov/2007_Application_Summaries.htm>. *The published summary of the Baldrige application that led to ARDEC receiving the award (note: applications themselves are not released).*

Nussbaum, B. (1997). Annual Design Award Winners. *Business Week*, June 2. *An overview of design award winners and the organizational factors helping support them.*

Oak Ridge Associated Universities. (2005). Performance Measurement of Research and Development (R&D) *Activities. n. pag. Retrieved 2 Oct.* 2009. <http://www.orau.gov/pbm/documents/rd.html>. *A brief discussion of issues in measuring R&D performance, and the approach being taken by the Army Research Laboratory.*

O'Connor, P. D. T. (2002). *Practical Reliability Engineering*, 4th ed. Wiley, New York. *A hands-on guidebook for reliability engineers, covering all the major methods for the design, development, manufacture and maintenance of reliable products and systems.*

Oestmann, E. (1990). Research On Cat Research Quality. *Proceedings of Symposium on Managing for Quality in Research and Development*, Juran Institute, Wilton, CT. *A report of findings and recommendations regarding the quality of research at Caterpillar.*

Ogbuehi, A., and Bellas, R. A. Jr. (1992). Decentralized R&D for Global Product Development: Strategic Implications for the Multinational Corporation. *International Marketing Review*, vol. 9, no. 5, pp. 60-70. *An analysis of the risks and rewards multinational corporations face when establishing local R&D centers.*

Park, S. H., and Antony, J. (2008). *Robust design for Quality and Six Sigma.* World Scientific Publishing Co., Singapore. *A reference book on robust design that emphasizes statistics and the integration of the method with Six Sigma.*

Perry, W., and Westwood, M. (1991). Results from Integrating a Quality Assurance System with Blount's Product Development Process. *Proceedings of the Symposium on Managing for Quality in Research and Development*, Juran Institute, Wilton, CT. *A case study in quality assurance applied to the product development process.*

Petroski, H. (1992). *The Evolution of Useful Things.* Alfred A. Knopf, New York. *A historical perspective on the development of simple, everyday items (such as forks and paper clips) and the corresponding cultural changes. Pages 84–86 describe the specific circumstances surrounding the acrylate-copolymer adhesive used in Post-It notes.*

Phadke, M. S. (1989). *Quality Engineering Using Robust Design*. Prentice Hall, Englewood Cliffs, NJ. *A textbook on the robust design techniques originally developed by Taguchi, intended to help engineers design high-quality products and processes at low cost.*

Port, O. (1996). Green Product Design. *Business Week*, June 10. *An overview of environmental considerations as an important dimension of quality and part of product design.*

Prajogo, D. I., and Hong, S. W. (2008). The Effect of TQM on Performance in R&D Environments: A Perspective from South Korean Firms. *Technovation*, vol. 28, no. 12, pp. 855–863. *A study of the effectiveness of total quality management practices in R&D environments in terms of product quality and innovation.*

Raven, J. (1996). Merrill Lynch Insurance Group Services' Project Management Process. Handout distributed at Florida's Sterling Award Conference, Orlando, May 29, 1996. *Provides recommended steps in managing a project through the R&D process.*

Razgaitis, R. (1999). *Early-Stage Technologies: Valuation and Pricing*. Wiley, New York. *Provides a comprehensive approach to assessing the future of new technologies based on intellectual property rights, risk assignment, deal-making and economics. Includes six valuation methods for intellectual property.*

Rees, R. (1992). The Purpose of Failure. *Reliability Review*, vol. 12, March, pp. 6–7. *In the context of designing for reliability, the author discusses the importance of identifying and defining the intended purpose of the application and test procedure prior to defining failures.*

Rome Laboratory Reliability Engineer's Toolkit: An Application Oriented Guide for the Practicing Reliability Engineer. (1993). ADA278215. Systems Reliability Division, Rome Laboratory, Rome, NY; 255 pp. *A practical set of tools of interest to reliability engineers.*

Richtne'r, A., and Rognes, J. (2008). Organizing R&D in a Global Environment: Increasing Dispersed Co-operation versus Continuous Centralization. *European Journal of Innovation Management*, vol. 11, no. 1, pp. 125–141. *Discussion of the choice of R&D location and organization in terms of "geographically dispersing and contracting forces" on R&D activities.*

Roberts, G. (1990). Managing Research Quality. *Proceedings of the Symposium on Managing for Quality in Research and Development*, Juran Institute, Wilton, CT. *In this report, the author discusses peer reviews used to verify progress by checking calculations, test data reduction, and research reports.*

Roussel, P. A., Saad, N. K., and Erickson, T. J. (1991). *Third Generation R&D*, Harvard Business School Press, Boston, MA. *The authors argue that research and development departments should not have an unfettered hand, nor be bound by strict business requirements. Instead, they should be integrated with a company's overall strategy.*

Rummler, G., and Brache, A. (1995). *Improving Performance—How to Manage the White Space on the Organization Chart*, 2nd ed. Jossey-Bass, San Francisco. *The authors provide comprehensive examples of linking organizational-, process-, and job/performer-level measures for a product development process.*

Sanders, P. (2009). Boeing Settles in for a Bumpy Ride. *The Wall Street Journal*, Wed. Oct. 7, p. B1. *An article describing problems with the new 787 Dreamliner and 747-8 jet redesign engineering and manufacturing problems.*

Schinner, C. (1996). Accelerated Testing, *Handbook of Reliability Engineering and Management*, 2nd ed. Ireson, W., Coombs, C., and Moss, R., eds. McGraw-Hill, New York, chap. 12. *Discussion and example applications of methods to estimate reliability and useful life through accelerated testing (e.g., thermal cycling, stress testing).*

Schumann, P. A., Ransley, D.L., and Prestwood, D. (1995). Measuring R&D Performance. *Research & Technology Management*, vol. 38, no. 3, pp. 45–54. *A case is made regarding the importance of formal metrics in successful management of R&D.*

Sekine, K., and Arai, K. (1994). *Design Team Revolution*. Productivity Press, Portland, OR, p. 191. *The authors provide tables of possible design process deficiency measures associated with management, lead times, costs, and quality.*

Shankar, R., Frapaise, X., and Brown, B. (2006). Lean Drug Development in R&D. *Drug Discovery and Development*, no. 5, May. Retrieved Oct. 7, 2009. <http://www.dddmag.com/lean-drug-development-in-rd.aspx>. *A brief overview of lean and report on the success in applying lean principles to drug development.*

Smith, G. (1991). A Warm Feeling Inside. *Business Week*: Bonus Issue, October 25, p. 158. *A description of efforts to reduce time to market for new products at Corning, and the resulting benefits.*

Taguchi, G., Chowdhury, S., and Wu, Y. (2005). *Taguchi's Quality Engineering Handbook.* Wiley, Hoboken, NJ. *A review of the theory behind Genichi Taguchi's robust design, with 94 case studies from across diverse industries and processes. Also compares Taguchi's methods with other quality philosophies.*

Taylor, D. H., and Jule, W. E. (1991). Implementing Total Quality at Savannah River Laboratory. *Proceedings of the Symposium on Managing for Quality in Research and Development*, Juran Institute, Wilton, CT. *A case study of TQM at a government lab, including the role of a quality council.*

Teresko, J. (2008a). Seven Emerging Trends in R&D Metrics: More Companies Tracking 'True Performance Metrics'. *Industry Week.* May issue. Retrieved Oct. 2, 2009. <http://www.industryweek.com/articles/seven_emerging_trends_in_rd_metrics_16115.aspx>. *Discussion of trends seen in the ways organizations measure R&D performance. See Teresko (2008b) for a list of the associated metrics.*

Teresko, J. (2008b). Metrics Matter: The Top R&D Metric Used in 2008 is R&D Spending as a Percentage of Sales. *Industry Week*, May issue. Retrieved Oct. 2, 2009. <http://www.industryweek.com/articles/metrics_matter_16116.aspx>. *An assessment of recent R&D spending, and comparison of the top 10 R&D metrics used by industry in 1998 and 2008, based on survey data. See Teresko (2008a) for discussion of trends.*

Thackara, J. (2005). *In the Bubble: Designing in a Complex World.* MIT Press, Cambridge, MA. *The author explores what value new technology and devices add to our lives, with the belief that ethics and responsibility can guide designs without impeding innovation. Most relevant here is Chapter 2, which focuses on "speed."*

Thaler, J. (1996). The Sikorsky Success Story. *Workplace Ergonomics*, March/April. *A report on how ergonomic changes to an aircraft assembly process helped reduce employee injuries and related costs.*

Thelen, M. J. (1997). Integrating Process Development, ISO 9000 and TQM in SITA Research and Development. *The TQM Magazine*, vol. 9, no. 4, pp. 265–269. *A case study in the implementation of a quality process in R&D based on an integration of ISO 9000, Business Process Improvement and Total Quality Management.*

Turmel, J., and Gartz, L. (1997). Designing in Quality Improvement: A Systematic Approach to Designing for Six Sigma. *Proceedings of the ASQC Annual Quality Congress*, ASQC, Milwaukee, WI, pp. 391–398. *The authors discuss design for reliability and maintainability within the context of Six Sigma.*

Wasson, A. (1995). Developing and Implementing Performance Measures for an R&D Organization Using Quality Processes, *Proceedings of the Symposium on Managing for Quality in Research and Development*, Juran Institute, Wilton, CT. *Discussion of linking customers to business vision and mission statements, and corresponding metrics.*

Wood, L., and McCamey, D. (1993). Implementing Total Quality in R&D. *Research Technology Management*, July-August, pp. 39–41. *Discussion of the need to debunk the myths that the Total Quality approach only works in manufacturin, and that total quality inhibits creativity.*

Wood, L.V., and McCamey, D. A. (1993). Implementing total quality in R&D. *Research & Technology Management*, vol. 36, no. 4, pp. 39–41. *In recognizing the difficulty in getting a buy-in for quality management within R&D, the authors provide numerous practical recommendations to assist in launching a successful quality improvement initiative.*

Yamazaki, A., Sakagawa, M., Seki, T., and Enomoto, S. (2006). Peer Review as a Tool in R&D Program Management. *Journal of the Society of Program Management*, vol. 8, no. 2, pp. 28–33. *An assessment of peer review as a means for improving research theme selection. Although the authors recognize inherent problems, they argue that the method is a valuable and expanding means of R&D quality improvement worldwide.*

Yoest, D. T. (1991). Comparison of Quality Improvement Team Training Methods and Results in a Research and Development Organization. *Proceedings of the Symposium on Managing for Quality in Research and Development*, Juran Institute, Wilton, CT. *This piece drives home the importance of training to meet the specific needs of R&D staff.*

Zeidler, P. (1993). Using Quality Function Deployment to Design and Implement a Voice Response Unit at Florida Power and Light Company. *Proceedings of the Symposium on Managing for Quality in Research and Development*, Juran Institute, Wilton, CT, pp. A-45 to A-56. *Describes the application of QFD and other tools to improve designs and better meet customer expectations.*

CHAPTER 29

Software and Systems Development: From Waterfall to AGILE

Bruce J. Hayes

About This Chapter

This chapter will deal with the generally accepted principles of software quality improvement and reference information from a number of proven, expert sources. We will acknowledge and place into context the growing need for these items to be flexible and balanced in their application, as dictated by changing technology, strategies, and business and customer requirements. Under the surface, there is much more to effectively planning, deploying, and improving a quality system for software development. The following sections will explore the mechanics and architecture of those elements more thoroughly.

High Points of This Chapter

1. The best practices that have emerged and guided the way for rationalization of the need and the role of quality improvement in this domain includes 10 general best practices to attain software quality excellence.
2. Quality fundamentals, with consideration for the special circumstances surrounding software and technology, are similar to managing quality of hardware.
3. Aspects of software and IT are different from hardware—but not that different. It is simply not a good excuse to say these processes cannot be measured. In a 2002

National Institute of Standards and Technology (NIST) study on software and IT, more than 40 key metrics in six attribute categories were listed as being used by best practice companies.

4. Recognizing a few common failure mechanisms is a good starting point to build the understanding of the need for a preventive quality system.
5. Because so many people potentially "touch" software code, there is tremendous potential to create unanticipated problems when tradeoffs or fixes are implemented without proper knowledge or communication.

The Need and Context for Software Quality

Since computers, softwares, and the World Wide Web became dominant forces in the product, services, business, government, and social landscape, the quality management challenge has been to integrate their use and performance with the processes, procedures, methodologies, and people whose performances they are intended to improve. During the continuing evolution of a rapidly changing technology landscape, the topics of design quality and other quality improvement basics for software development (and software performance) have often taken a back seat. Business activities based on demand, speed, delivery, capability, and other business requirements create conflicting priorities and slow the development and acceptance of proven quality methods for software development and fulfillment processes. For software, quality, and business professionals, these topics are of paramount importance to ensure achievement of business goals.

There is also a notion that software design and programming is more art than science, thus not disposing it to the same disciplined processes we have developed and accepted in the hardware and service domains. All this, coupled with rapidly developing tools, methodologies, and new programming languages, has contributed to a conundrum of opinions about the role of quality in software development. Despite all this, best practices have emerged and guided the way for rationalization of the need and the role of quality improvement in this domain.

Best practices in software quality can be typically defined as those organizations that subscribe to, and have demonstrated desirable business results from, some form of the following 10 general best practices for software quality excellence:

1. Exceed customer and business requirements by creating innovative, breakthrough software products and services that consistently perform as planned, to documented requirements and specifications, without defects.
2. Anticipate and/or solve specific, defined and documented user, business, and/or scientific problems with the software being written.
3. Understand, document, and fully vet the customer, user, and stakeholder needs and requirements prior to writing code.
4. Determine how to measure the performance of each requirement so as to create a closed-loop, balanced performance measurement system for the software being written.
5. Create, document, communicate, and maintain (continuously improve) the complete process (supplier to customer) to be used to generate, test, release, and maintain software products and services, which is sometimes referred to as the software or systems development life cycle (SDLC). See Figure 29.1.

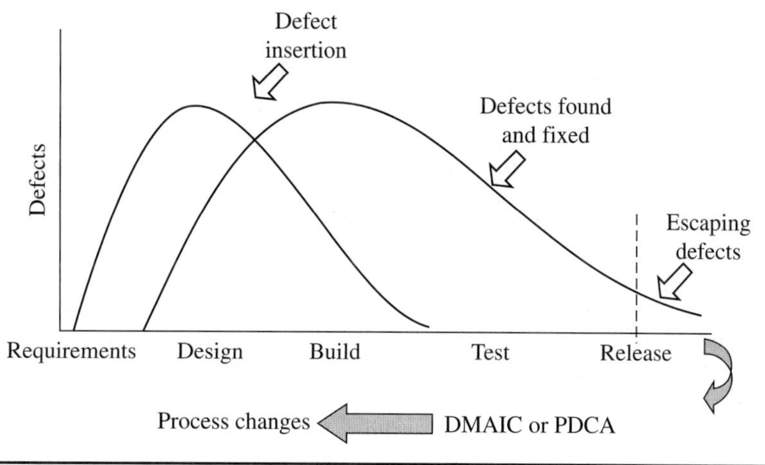

FIGURE 29.1 Software or systems development life cycle.

6. Implement software process performance measurements that are used to drive continuous improvements, set priorities, determine scope and size, and monitor software project performance.
7. Establish a software quality leadership role with the authority to act as a customer ombudsman and direct the overall planning and implementation of the software quality system.
8. Create an environment and process for efficient collaboration, knowledge sharing, code reuse, and open communications among all stakeholders, employees, and customers.
9. Provide robust training and support to enable all participants in the software process to understand the basics of product, process, and service quality requirements and translate them to their specific role and process contribution.
10. Manage, prioritize, and organize the software process in close alignment with the strategy and goals of the business with attention to fitting the organizational culture.

While there is not one globally accepted standard for software quality, most researchers, standards bodies, associations, and software subject matter experts agree that following some form of these basic principles will ensure a successful quality (and business) outcome of the software process. Because technology and best practices are changing at a rate often hard to comprehend, the danger in trying to prescribe and document a specific methodology is risky. Even if one could be established, the work involved in maintaining, communicating, training, and generally keeping it in practice would likely overwhelm its benefits.

Given this, we note that Dr. Juran faced similar challenges when he wrote his original works on quality improvement for the design and manufacture of tangible goods. He too must have been aware that manufacturing and design automation was evolving at a rapid pace. Still he was able to develop and prescribe a general and continuously improving methodology with roots in common sense and survivability.

This is the challenge for all of us engaged in the business of software quality improvement. Business leaders and decision makers have two choices: (1) Use the ambiguity of unknown factors and changes in the software development landscape to simply say, "It's not possible, so let's just go about our business the way we are doing it and hope for the best" (and, believe it or not, many organizations have followed that path); (2) Accept that there are general, common sense approaches based on the work done by Dr. Juran and hundreds of software subject matter experts that will translate for all organizations.

We acknowledge that there are differences in the practice of software development versus manufactured product or hardware development. The primary issue at hand for software quality is not one of reproducibility without variation of a "piece part," but one of getting the software "right the first time." It's not that difficult to reproduce software without variation once it is written (this is not always true with hardware). Duplicating and transferring programs (software code) through magnetic media or network connections is fairly straightforward and is prone to little or no variation (in terms of changing the functionality or creating defects). However, getting functionality "right the first time" is the key in software quality. In situations where we do not, we incur the cost and frustration of ongoing releases (or patches) that often affect software in the field (at customer sites). We know from process quality work that escaping defects are the most costly, can create the highest potential liability, and are impossible to screen (inspect) with 100 percent confidence. These facts modify the context of our approach to software quality when contrasted to hardware or manufacturing. But the fundamentals are still the same, i.e., to prevent the defects from occurring in the first place, by reducing variation in the work processes and components that make up the finished product. Therefore, quality fundamentals, with consideration for the special circumstances surrounding software and technology, can be and are applied successfully.

As innovation and speed to market are important differentiators in technology-based products and services, these topics have dominated the software development landscape and often collide with quality as a priority. Some have even argued that it is better to get to market first with something new than to get there late with something defect-free. While there has been some history to support that position in the short term, if we factor longevity and cost into that equation, there are many cases in which quality problems have undone those first to market, as the second and third entrants were able to put forth a "cleaner" version of the product or service. Their lower costs (as a result of fewer software defects and rework) and higher market acceptance (more customer-desired features) won them significant market share and defectors from the "first mover" supplier. As the first movers' costs and liabilities rise (while reacting to missing features, software bugs, and resource problems), the second and third movers get closer to their customers, building out new features and reaping the profits due to their lower cost base.

This scenario creates a compelling business case for the implementation of a software quality system. For the quality professional or business manager, the understanding of these business consequences can create the business case for investment in a preventive quality system and the creation of buy in from the management team.

Within this context, this chapter will deal with the generally accepted principles of software quality improvement and reference information from a number of proven, expert sources. We will also acknowledge and place into context the growing need for these items to be flexible and balanced in their application, as dictated by changing technology, strategies, and business and customer requirements. Under the surface, there is much more to effectively planning, deploying, and improving a quality system for software development. The following sections will explore the mechanics and architecture of those elements more thoroughly.

Anecdotal Software Failure Cases

While it is important to understand and emulate best practices for something an organization desires to improve, it is also prudent to start the process of improvement through the identification, study, and characterization of causes of previous problems. Software development horror stories are available by the thousands. Some are harmless bugs creating a little inconvenience (i.e., perpetually needing to restart an application or the hardware) while others result in catastrophic system failures with life-threatening consequences. We mention these because understanding some common failure mechanisms is a good starting point to build the understanding of the need for a preventive quality system. As examples, the following failures occurred in a variety of industries, in well-known organizations, and with serious business consequences.

From a popular on line services provider:

> Six million customers were locked out of cyberspace for at least 19 hours Wednesday in one of the worst outages in online history. The system crashed at 1 A.M, while it was installing new software for its network routing system, a problem area for other Internet service providers. Company officials reported that the system finally was restored at 7:45 P.M., but some users still said they could not log on as late as 8:30 P.M. . . As part of his profusely apologetic remarks, he guaranteed that the company would reimburse subscribers for time lost on the service, whose business model depends on maintaining a sense of community among its members. "We regret this extraordinary delay and extend our gratitude for the patience and continued support of our members," he said in a press release.*

From an airline reservations system:

> A computer problem disabled the reservations systems for several carriers for more than five hours on Tuesday, delaying passengers while boarding passes and baggage tags had to be made out by hand. The software problem also prevented reservations from being made or changed by travel agents or by passengers using the Group's Travel Web site, said a spokesperson. The system was partly restored by about 8:45 P.M. and was in full operation by 10:50 P.M., the spokesperson said. "Everything's back to normal now," adding that officials are investigating the problem's cause.†

From a European space agency:

> On June 4, 1996 an unmanned rocket exploded just forty seconds after its lift-off. The rocket was on its first voyage, after a decade of development costing $7 billion. The destroyed rocket and its cargo were valued at $500 million. A board of inquiry investigated the causes of the explosion and in two weeks issued a report. It turned out that the cause of the failure was a software error in the inertial reference system. Specifically a 64 bit floating point number relating to the horizontal velocity of the rocket with respect to the platform was converted to a 16 bit signed integer. The number was larger than 32,767, the largest integer "storable" in a 16 bit signed integer, and thus the conversion failed.‡

From a well-known candy manufacturer:

> To meet last year's Halloween and Christmas candy rush, the company compressed the rollout of a new $112 million ERP system by several months. But inaccurate inventory data and other problems caused shipment delays and incomplete orders. The company's sales fell 12% in the quarter after the system went live—down $150.5 million compared with the year before. Software and business-process fixes stretched into early this year.§

---

*CNET News.com Staff, Staff Writer, AOL Back Online, a Day Late, August 8, 1996.

†*Knight Ridder /Tribune Business News,* July 1998 by Aline McKenzie.

‡ARIANE 5 Flight 501 Failure Report by the Inquiry Board, The Chairman of the Board, Prof. J. L. Lions.

§Editoral Research Team led by Mari Keefe Computer World – Top 10 Corporate Information Technology Failures of the 90's.

The common thread in many of these failures relates to inadequate planning, design, and deploying of the subject software in a preventive way. In other words, failure relates to the lack of a robust process to

1. Execute the entire project, utilizing a phased product, program, and project management technique and ensure proper planning, estimation, resource allocation, budgeting, and contingency planning.
2. Fully define all requirements and specifications from all possible stakeholders, interoperable components, and related systems.
3. Utilize a consistent and proven process to elicit, prioritize, document, and measure performance of each requirement and specification.
4. Convert the requirements and specifications into organized, traceable, and efficient software code that is testable.
5. Inspect and test the code within the context of a robust, systematic and institutionalized testing system to include code, module, system, and end-user emulation, simulation, and stress (load) tests.
6. Release and maintain the software in a logical, planned, sequence with timing in conjunction with the stated needs of the end user or system.

These failures and thousands of others highlight the relationships between business success, liability, and quality in software development. Fortunately, the attention and analysis of these high-profile failures have spawned a growing and evolving body of software quality knowledge. Their collective research, studies, and conclusions are documented in a wide array of articles, research papers, publications, and books. This information is widely available on the Internet, by using popular search engines to explore topics of interest or need. This information is often a useful source from which to build business cases, understand failure modes and effects, develop training, and in general raise awareness of the value-added nature of a software quality system or the liability and risk of not having one.

Software Development Quality Frameworks

Dozens of standards bodies, conference boards, analysis firms, consulting firms, and industry associations have utilized this research (with their own) to create standards for software quality. As previously mentioned, none of these are universally applicable or are accepted for use in all industries and functions. However, topic- and industry-specific standards exist for many mission-critical applications. Although most of the standards and frameworks outlined in this chapter have common threads (and in many cases common contributors), they often employ different models and philosophies depending on the required topic, industry, and scope of application. For the software quality professional, development personnel, and business manager, this means that care should be exercised to consider all aspects of an organization's products and services, technology, environment, criticality of defects, potential liability, regulatory requirements, and market and customer requirements to determine the best course of action related to embracing an existing software quality standard, designing a process, or bringing in outside help.

When software quality systems are planned, it is important to understand and characterize the software development process in use, its effectiveness and appropriateness. Tests for effectiveness include organizational and functional assessments, analyzing performance to goals and evaluating the ability to support the business goals with the quality plan. To assist in this characterization, it is important to understand the common frameworks and methodologies in use.

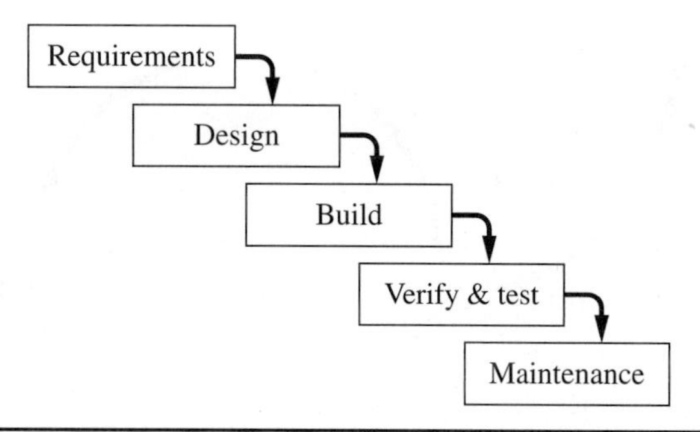

Figure 29.2 Waterfall development process.

Two major schools of thought exist on how to approach software quality depending on the development methodology being employed. One, based on early software methodologies, was built largely on the notion that software development projects should embrace typical large project management life cycles (and their associated disciplines and documentation). This is often referred to as *waterfall development* (i.e., completes the projectwide work products of each discipline in a single step before moving on to the next discipline in the next step; see Figure 29.2). Resources and project work are in some ways "batched" much as a factory would operate a batch manufacturing environment. While some economies of scale and efficiency are realized utilizing this method, risks and liabilities are created if the process, resources, and work products are not executed perfectly. These risks (waste, schedule compression, etc.) can be exacerbated by the "build to completion" mentality. To minimize these risks, it is important to understand the causes and successes associated with each.

Certain classes of software professionals, who view software development more as creative art than as science, may become disenchanted with the perceived bureaucracy associated with waterfall processes. They believe that the rigidity and documentation requirements associated with such systems hamper the creative spirit and innovation. In situations where methodologies, business needs, technology, and/or customer requirements are misaligned (i.e., cases where rapid cycles of innovation are required to keep up with market timing), a mismatch between the needs of the market and the constraints of the development process will be present. When this occurs, attitudes and resultant behaviors may attempt to circumvent the process by using workarounds and not taking compliance requirements seriously. These scenarios are important to recognize (and deal with) if they exist. This situation typically increases short-term pressure on the quality function to achieve the proper checks and balances to counter such behaviors. To be effective, the behavior ultimately needs to be changed through careful and thoughtful planning, goal setting, training, communication, and reinforcement.

Conversely, many software quality success stories have been created from waterfall methodologies. They are typically larger projects often related to complex systems controlling mission-critical functions over a period of years. Examples include large enterprise business system platforms [enterprise resource planning (ERP), accounting, MRP, etc.] and large application platforms (operating systems, avionics, controls). One such example is the software used to control NASA's Space Shuttle. Consider this brief quote from a NASA manager in an article on the Space Shuttle's software development process: "And that is precisely the point—you can't have people freelancing their way through software code that flies a spaceship, and then, with peoples lives depending on it, try to patch it once it's in

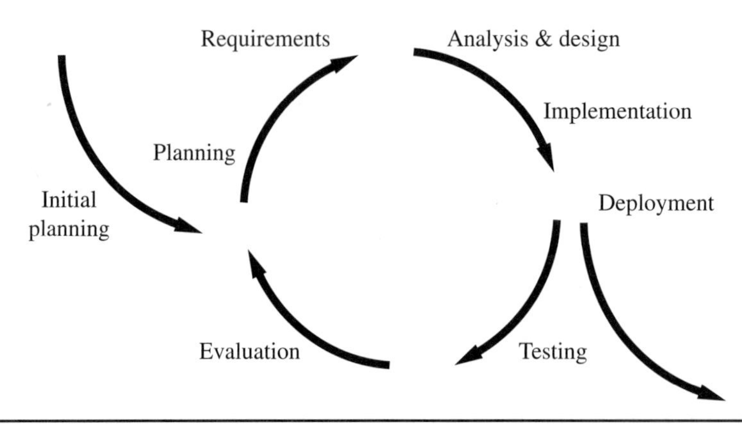

Figure 29.3 Iterative or incremental development cycle.

orbit."* In what is referred to as "software for grownups," NASA has created what might be the most successful waterfall development process in the history of software development. But it is arguably not right for everyone.

Software built with this methodology is referred to as *legacy software* and can present long-term support problems and issues, especially for products and services requiring frequent change and customization related to requirements, over time. If done right, waterfall development can be an effective process for these types of software applications.

Executing quality control and assurance plans within this methodology requires that detailed quality activities be written into the master project plan, resources allocated and budgeted, and deployment fully required and supported by senior management and executives. Waterfall development has a tendency to rely more heavily on compliance to quality policy to achieve objectives rather than ownership by all individuals in the process for quality responsibility.

The second and still emerging method of software development is referred to generally as *iterative* (or sometimes *incremental*) software development process (see Figures 29.3 and 29.4). As the name implies, software projects, their functionality, and business value are broken down into smaller cycles and worked on incrementally. The basic process starts with a planning and initialization phase in which a base version of the product is developed (based on the basic requirements). Two things then happen concurrently. A user reaction to the base version is solicited, and a project control list, identifying tasks that need to be performed, is created (these may be new features or redesigns as dictated by internal or external needs). Analysis of the user reactions of the current versions coupled with tasks from the project list then feed the iteration step in which the next version is created. Version and task scope is carefully managed and intentionally kept small to reduce risk. As waterfall development is to batch manufacturing (as discussed earlier), iterative is to lean manufacturing (at least in theory). Many other acronyms, methods, and tools complement, support, and can guide iterative software development. These include AGILE,† SCRUM,‡ RUP,§ Extreme Programming,¶ and many others.

---

*Charles Fishman, *They Right the Write Stuff*, FastCompany.com, Issue 06, December 1996.
†Jim Highsmith and The Agile Alliance, authors of the *AGILE Manifesto*, 2001.
‡Ken Schwaber, *Agile Project Management with Scrum*, Microsoft Press, Feburary 11, 2004.
§Framework created by the Rational Software Corporation, a division of IBM since 2003.
¶Copeland, Lee "Extreme Programming," *Computerworld* (online), December 2001, webpage.

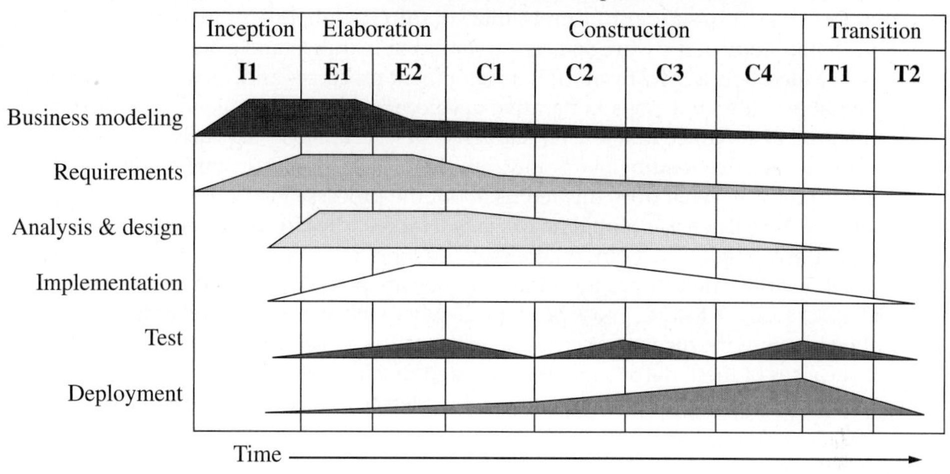

FIGURE 29.4 Iterative development sequence.

The following organizations maintain popular detailed software quality standards and frameworks and are good sources of information, study, and benchmarking:

- Software Engineering Institute (SEI). Capability Maturity Model, Integrated (or CMMI).
- The Institute of Electrical and Electronics Engineers (IEEE). Over 60 individual, detailed software engineering standards.
- International Standards Organization (ISO). Comprehensive set of software development and software quality standards.
- National Institute for Standards and Technology (NIST). Comprehensive set of software and software quality standards and case studies.
- Various DoD and military standards. Large body of knowledge including standards and case studies in various agency-dependent formats.
- Various industry-specific standards. These pertain to computers, semiconductors, medical devices, networking, aircraft, information systems, energy, various firmware, etc.

Frameworks and methodologies in software quality are too numerous to cover in detail in one chapter. It is strongly recommended that readers utilize these resources in conjunction with other available resources to adequately understand the best form, fit, and function for their individual organization's quality system needs. Once a system of proper context and scope is identified, the work of integrating the quality plan, metrics, and improvement processes can be executed and staffed.

Quality Measurement in Software Development

Within the typical software development process, an initial and primary quality concern is the study and analysis of software defects. Software defects should be tracked at a variety of phases and through a wide range of techniques ranging from manual collection to fully automated solutions. Our ability to understand, measure, and efficiently report the nature, cause, insertion point, ability to detect, likelihood of escape, and other variables creates important statistics to capture and utilize in software quality improvement.

In the study of software defects we should first understand a bit about the macro process that creates defects. Most waterfall development processes and, on a smaller scale, the incremental development steps in iterative development follow the general sequence presented in Figure 29.1. Notice that to a large degree, defects are inserted early and found late. Interestingly, resource loading starts low, builds to a high point in mid-project, and tapers off toward release. What does this tell us about the process? First it implies, and research data support, that if we can eliminate or find defects earlier in the process, there will be fewer defects escaping to the field, where they are more expensive to find and fix. It also reveals that while most defects are found in test, they are probably created much earlier in the process. Because the testing phase is the place where the most perceived defect activity exists, it often becomes the focus of attention and viewed as a primary process bottleneck. In fact, the root causes of these defects are likely created much earlier in the process. In most organizations, the high reliance on test to find and fix defects is feeding the reactive culture and leads to cost overruns and resource problems. Creating a clear, phased, data-based measurement system that focuses on defect insertion points in addition to defect discovery points is one important step toward driving a preventive approach in software quality.

While some discovered defects are in fact coding errors, many are referred to as *requirements failures*. In other words, we do not gather enough of or the right kind of information and fully process it at a detailed enough level to drive defect-free programming. And the way in which typical projects are managed reinforces this. As we have discussed, resource loading at the front end is typically low. This causes organizations to suboptimize the requirements process and the quality of the data and information it creates. What historical data and case studies have taught us is that by spending a little more time up front (in requirements) we will save a lot of time (and reduce defects) in the downstream processes.

Therefore, software quality efforts to some degree (especially in the early phases) will need to focus on measuring and understanding the relationship among effort (person-hours), defect rate, insertion point, discovery point, code size, find-and-fix cost, and other variables. These variables will become the baseline statistics we use in our initial work. As more data are collected and analyzed in this fashion, they will provide enough detailed history to eventually help build predictive models to statistically understand the business consequences and tradeoffs related to the software development process (see Figure 29.5).

It is from these basic variables that we can begin to build a model to help understand the current baseline and trend of software quality performance. Although the study and the analysis of defect data may be classified as reactive, they help formulate the early basis for creation of a predictive or preventive system.

Dealing with defects that occur in the field (after release) tends to command the greatest attention for obvious reasons. They are highly visible, consume resources, create risk for the business, and are often the most difficult to mitigate at root cause. Often, in the initial phases of responding to defects, organizations resort to guesswork and patches. The result often creates more problems than it solves. The software code may be compromised, and the patches and their interrelationships with other parts of the program poorly understood or inadequately characterized.

Measurement system analysis (MSA) is also an important issue related to software quality data and measurement. Similar to issues in hardware and manufacturing, significant variation can exist in software quality data. As a rule of thumb, 30 to 50 percent of available data is not sufficient to accurately solve problems unless proper measurement studies have been

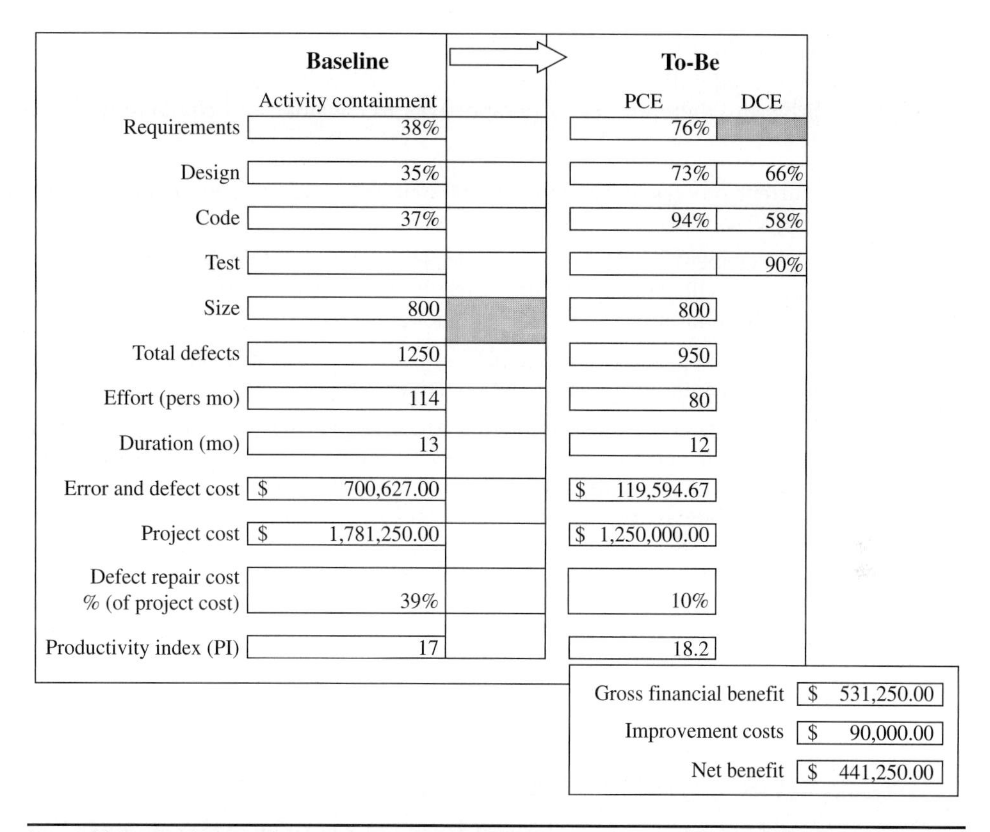

FIGURE 29.5 Predicting software improvement benefits.

executed and acted upon. It is extremely important to statistically study our measurement systems to eliminate variation prior to using the data for effective and efficient problem solving.

Many software organizations become overwhelmed with the amount of data they collect. Automated systems, call centers, help desks, and other sources of data generate thousands of cells of data. The priority and importance of specific data often gets lost in this flood. As a result, the data may not be used to the fullest extent, not used in an optimized manner, or ignored entirely. In these situations it is important to help the organization rediscover the value of the data (namely, by instituting projects to eliminate non-value-added data and turning value-added data into problem-solving *information).*

Once this is sorted out, an important activity related to MSA in software quality is to ensure that measurement studies (Gauge R & R and others techniques explored in other chapters of this handbook) are initiated for each problem-solving project and the measurement system in general. Often, MSA in the software environment involves simple fixes, e.g., refining and reducing the number of defect-reporting classifications, determining consistent sizing methods, studying reported versus actual defect classification, and studying human interactions and consistency of defect reporting.

Successful software quality efforts will then use accurate data for planning and enabling a consistent and effective problem solving process (PDCA, DMAIC, etc.). This class of data will be necessary to facilitate basic controls and the improvement processes needed for short-term and long-term improvements.

The Software Quality Approach

When one is faced with initiating, evaluating, or improving an existing or new software quality program, there are many perspectives to consider and characterize, including these:

1. Current process and organizational maturity, role, experience, and results of the subject software development process
2. The alignment, understanding, impact, and degree to which the software development function will influence business results and longevity (revenue, growth, liability, market, community, and other stakeholder goals)
3. The detailed process for developing software including its level of institutionalization, scope, breadth, consistency, and effectiveness
4. The ability of the measurement system to provide adequate data and analysis capability to ensure that the fundamental and needed quality metrics for measuring and improving quality, cycle time, cost, on-time performance, efficiency, and other key variables are present and usable
5. The appropriateness, effectiveness, and efficiency of current tools, technology, programming techniques, methodologies, and languages being used by the organization
6. The degree to which efficient and consistent cycles of evaluation and improvement (based on reliable data) are utilized in the critical processes and in the decision-making process
7. The visibility, communication, priority, and management roles and responsibilities related to the implementation and longevity of the software quality process

These and other factors form the basis for initial evaluation of the software quality process. The effective and efficient assessment of these factors will help to set a general basis and understanding of the software development organization's capability, processes, practices, attitudes, culture, and priorities. Assessment of this "current reality," in comparison to the organization's current quality plan (if one exists) and/or best practices, will identify gaps in the process that need to be addressed as well as practices that are working well. Leveraging and reinforcing the positives while confirming and quantifying the gaps is essential for input to the prioritization and development of an effective and applicable software quality plan.

As a practical example, consider an organization that has a well-documented quality plan, has conducted training, has established logical goals, and has staffed the quality organization. However, study may reveal the existence of a very poor and unreliable measurement system. By knowing the extent and issues related to this problem, some initial priorities should be set to correct and optimize the measurement system, so that downstream quality activities will be efficient and effective. Proceeding without understanding and fixing this problem will only serve to create a lack of results and organizational frustration. Situations like these need to be clearly understood to ensure that the quality plan will address all potential downstream risks by providing a basis and context for initial priorities, business and quality plan alignment, and closure of critical gaps.

There are many methods for conducting quality system assessments for software development. They range from hiring a team of professionals (usually consultants), to self-assessment techniques, to Web-based tools. The methodology selected should consider the scope, size, budget, type of development, and other general factors.

Several factors must be considered in deciding when, how, and just how much assessment to do. Mature organizations may already have in place a variety of comprehensive and regularly scheduled assessments, from which much of the necessary information may be readily available. If it is available, the organization should use it. These may include assessments and audits such as the ISO Series, the SEI's CMM/CMMI process, national or state quality awards, or a company's own existing assessment program. Although each of these assessments may have been created for a purpose other than determining readiness for a software quality initiative, many of their attributes are useful in determining readiness or maturity.

Existing strategy documentation, goals and objectives, policies, procedures, process documentation, data, organizational charts, and other pertinent plans and information should be reviewed to provide a basis both for evaluating the scope of the prospective assessment and for helping in the planning and preparation.

Any assessment (especially if it involves external reviewers) can be a source of tension and distraction for an organization. If there is no communication of the purpose of the assessment, the schedule, the roles and responsibilities of individuals during the assessment, the confidentiality of statements and documentation, and what will be done with the results, then organizational resistance can create an environment in which gleaning accurate information will be nearly impossible. Generally speaking, any assessment should be communicated to the organization well in advance of when people will be required to respond.

Communications also should be repeated a few times to reinforce their importance. It is usually best if senior management details the assessment purpose, scope, context, size, location(s), dates, and intended follow-up. This communication also should state a firm commitment to the process, a solicitation for full forthright cooperation, and an assurance of confidentiality. When reinforced in regularly scheduled meetings and informal communications, this type of communication will help to reduce organizational resistance and ensure a free flow of accurate information. Postassessment communications also should be scheduled. Sharing information with the people who participated helps develop buy-in for the results and subsequent actions to be taken.

One of the primary inputs to the assessment process is to examine what people in the organization actually do in the performance of their daily work. It is critical in the process to decide how, how many, and from whom that information is elicited. When the survey or interview populations are considered, several choices should be evaluated.

Sample Size (How Many and from Where)

Sample size is often determined by the size of the organization and whether Web-based technology is used. Utilize proper statistical sampling to ensure that confidence levels are appropriate and variation is controlled. When Web-based technology is used, a sample size of 100 percent is not out of the question. Web-based assessments, while limited for actual observation and dialog, provide the capability to collect, process, store, and retrieve a large amount of data quickly, which enhances the ability to reuse and continuously analyze the detailed data.

Demographics (Various Locations, Cultures, and/or Geographies)

Demographics are important to consider in an assessment. Often locally tuned processes and practices are getting results and thus may be ripe for adoption by the organization at large. Sometimes certain geographies are required to comply with local laws or regulatory requirements. It can be highly desirable to include several customers, clients, or even suppliers in the assessment. This is a great way to see how external sources view the strengths and weaknesses of a process.

Availability of Previously Gathered Information

If a recent and rich source of information exists (previous audit or assessment), clearly this information may be useful in the study and characterization of the software process and organizational aspects related to the process. Use care to understand the completeness, accuracy, scope, and context of this information before using it in drawing conclusions.

Use of Web Technology

Web-based tools for assessments are gaining popularity and accuracy. Assessment tools usually have some type of analysis capability built in to generate comparative and gap analysis automatically. They typically can trap user comments and opinions, which can then be stored for quick retrieval and summary. There are many advantages to Web-based, systems but no Web-based system can observe actual behavior. Usually some combination of Web technology and observation-based validation provides the most accurate assessment result.

Timing of Assessments

All efforts should be made to plan an assessment to maximize accuracy and efficiency. This may mean scheduling the assessment around other corporate and business events that could compete for resources and time. These may include major milestones, meetings, end-of-quarter rush, etc.

Organizational Culture (Receptiveness or Resistance to Assessment)

If this is the first time an organization has undergone a formal assessment, it is normal for significant organizational resistance to be present. The best way to overcome resistance is by meticulous planning, frequent communication, and active review. Organizations used to such events will be able to move faster with less communication. Organizations of lower process and quality maturity will need greater facilitation and justification including strong senior leadership demonstration of commitment.

Availability of Resources

Careful consideration should be given to make resources available to participate without any negative consequences. Typically a survey or interview will require an hour or so of preparation and an hour of actual interview or survey time. Management must create a safe ground for employees to participate.

Once the above is understood and agreed to (by the target organization and the assessment team), development of the interview schedule (or, in the case of a Web-based tool, survey schedule) and a deployment plan should commence. Schedules need to be specific about times, locations, names, titles, and functional responsibility. The organization being assessed typically will provide administrative resources to manage and communicate the master schedule and changes that may occur during the course of interviews and surveys.

Communication and use of the assessment results requires careful interpretation and discussion among the core deployment team. Identified gaps and opportunities for improvement may indicate places to avoid for early adoption until fundamental and systematic changes can be implemented (e.g., establishment of a process management system or establishment of a measurement system). On the other hand, leverageable strengths and pockets of excellence may provide rich early adoption process and project opportunities. Action plans and deployment plans can now be intelligently prepared, communicated, prioritized, and implemented.

With assessment, planning, selection, and design of a software quality system completed, the organization is now ready to carefully implement the various strategies and detailed tactical activities to bring it to life. The early phases of these efforts are challenging and require vigilance to ensure that the "old way of doing things" (i.e., subjective prioritization, reactive problem solving, inaccurate measurement systems) does not creep back into the system. Frequent "fact- and data-based reviews" should be orchestrated and facilitated, paying particular attention to alignment and achievement of business goals with the quality system.

The following areas define typical software development, quality planning strategies, and activities:

1. *Process management.* Develop or adopt a documented, institutionalized, and appropriate process focus and plan including defining, documenting, training, measurement, problem solving, innovation, and deployment for all applicable processes (includes but is not limited to business and user requirements, architecture and design, coding and programming, testing, release, configuration management, field support).
2. *Project management.* Plan, implement, monitor, and control all aspects of project work with special consideration for efficiency, quality, resource sharing and utilization, integration, supply chain management, and risk management. Utilization of a project management methodology such as the Project Management Institute's (PMI's) Project Management Body of Knowledge (PMBoK) and other best practices for project management will ensure a complete and quantitative approach to all project work.
3. *Development and engineering.* Develop thorough quality guidelines, processes, and metrics for the creation and management of requirements including integrated business and user requirements, conversion of requirements to specifications, performance measures related to requirements, architecture design and development validation, coding quality and inspection, software testing, verification and validation, and release.
4. *Product and process support.* Ensure development of complete and applicable configuration management (including identification, control, status accounting, and audits), detailed software quality assurance process and plans (comparison and testing of work product to business and functional requirements and specifications).
5. *Process and product improvement.* Plan, design and implement a consistent, preventive, and applicable problem-solving process to ensure that all metrics are regularly reviewed (for performance, goals, compliance, and control as applicable), appropriately acted upon, and analyzed for cause (utilizing quality tools and teams). Implement action plans to eliminate defect causes in processes and products, and review all actions to ensure improved performance.

As mentioned through out this chapter, careful consideration should be given to the type of development process being used (waterfall, iterative), the type of technology being used, and the scope and context of business requirements to determine the level of pervasiveness required for each one of these components. Implementation will require rationalizing their need through a strong and value-added business case to ensure effective acceptance by the organization.

Requirements and Problem Definition

As discussed in earlier sections, the earliest phases of the software development process (problem, statements, requirements, design) are the most critical but often receive the least attention and resources. Most experts in product and software design agree that the single most important step in a development process is the careful study and characterization of the "problems to be solved." If this is done adequately and correctly, the downstream risk of requirements and other latent failures is reduced dramatically. Frequently organizations make large leaps of faith regarding design and development priorities, elicitation and processing of requirements, and the design of the base system architecture. Resource restrictions and undue schedule pressure create a "rushed" environment for this important phase. A robust quality plan will recognize these dynamics and include checks and balances to mitigate this risk and ensure long-term business success.

There are many detailed methodologies for problem definition in software development. These range from informal discussions between customers and development teams to detailed and on-going quantitative market studies driving the use of prioritization and planning tools. Again, organizational and process maturity often determines where and what type of focused attention is needed. The type of technology being used and the technology cycle will play a role in dictating the depth and exploration of techniques and tools for requirements solicitation, prioritization, and processing, differentiating and using specifications and language data to ensure full and complete understanding.

In highly technical environments where innovation is important, tools such as TRIZ* (literally Russian for "the theory of solving inventor's problems" or "the theory of inventor's problem solving") are used to methodically and quickly work through complex situations and develop low-risk solution alternatives. Using an algorithmic approach (versus brainstorming), TRIZ provides tools and methods for use in problem formulation, system analysis, failure analysis, and patterns of system evolution to help invent new products and refine existing ones. These types of systemic tools can help a software quality effort by ensuring the utilization of a consistent "process approach" to problem definition.

In other types of development (where innovation and speed may be less critical) quality function deployment (QFD) and other types of matrix approaches help provide a cross-functional view of functionality versus requirements and capabilities. Prioritization of features and requirements is aided by a highly structured documentation and review process to fully vet potential solutions prior to committing to final design.

The values in these structured processes lend a level of consistency, predictability, and measurement basis in the earliest (and most critical) phase of development (problem definition and solution alternatives) and set the process up for more reliable results later. The quality system should therefore synchronize with and advocate these activities to produce higher-quality process outcomes.

Once initial problems are thoroughly explored and basic solution sets are identified, detailed requirements and specifications must be developed. In terms of software quality, the role is to ensure that these requirements are as complete as possible, detailed to the extent required for coding, adequately documented, measurable (for later closed-loop feedback), and testable (set up for testability from the outset) and contain aspects of "hard specifications" (technology- and data-based attributes found in firmware or program to machine interfaces) and "soft data" (language or interpretive requirements found where humans interact with programs).

Engineering-intensive organizations frequently gravitate to "hard data" to drive their requirements in software design. In these cultures it is sometimes easy to miss stated language requirements relating to user preferences. In these scenarios it is important for the quality process to reinforce a balanced view of requirements functionality, encouraging and ensuring the use and full vetting of both types of data (hard and soft data).

In organizations developing software with a high level of "human interface" (i.e., systems that help to automate process tasks, provide quick access to knowledge, or accept

---

*Genrich Altshuller created the *Teoriya Resheniya Izobreatatelskikh Zadatch* (theory of solving inventive problems, or TRIZ).

input for processing), requirements may be skewed more toward soft data. In some circles this important information is referred to as *language data*. Language data are often highly complementary to hard data and can play a critical role in placing "hard" requirements in context and focus. Quality and project management activities should plan and ensure an appropriate exploration and balance of both types.

As a practical and simple example, consider the development of a system to process airline reservations via the Internet. There will be certain "hard data" requirements including access time, wait time, screen availability, etc. (easy to measure and fairly straightforward). On another dimension, a user may express concern related to certain attributes about the ability to reserve a particular type of seat, have a preferred meal, view the reservation data in a certain way, or receive confirmation feedback on his or her actions via another technology (e-mail, cell phone, PDA). Obviously, the design that properly listens to and processes the contextual *language data* and delivers the basic requirements highlighted by the hard data will be in a more advantageous market position to satisfy customers and perhaps implement features to differentiate the offering in the market.

There is a classification of requirements called *latent requirements* that should be explored, processed, and prioritized. These may be requirements derived from customer, market, or potential user statements about unfulfilled delighters. Latent requirements may also be enabled by new technological capability not previously present. Tools such as Kano models and planning* can assist in efficient thought guidance, discovery, information capture, classification, and processing of latent and other types of requirements.

Again, the quality plan must ensure that these various types of requirements are kept in balance and thus seed activities, plans, processes, and measures to ensure their effective use toward driving the highest-quality outcome for the customers and end users.

In terms of requirements content and documentation management, many different tools, models, and programming languages must be considered. These range from manual note taking to systems with auto code generation features. Regardless of the detailed methodology or technology in use to capture and document requirements, the quality plan should support the basic concepts, various types of requirements, and capture of complete and meaningful information. The quality plan should also ensure that tools and methods for prioritization and measurements related to each requirement are available and used effectively to facilitate closed-loop reviews.

Optimizing Designs and Performance

As in all fields of quality study, there are many statistics, statistical methods, and tools in use to assist in the optimization of software designs as they relate to the performance of the software. Assuming the requirements process is adequate and complete, performance is typically measured in three general phases:

- Early in the design by conducting "code inspections" to validate code and/or discover defects (coding mistakes)
- In mid-design cycle through the functional testing at module level and later at system integrated levels as compared to each critical requirement (unit and string testing)
- Specialized field and load testing (stress testing) when the software in its final application state (system and stress testing)

*Professor Noriaki Keno developed the theory of product development and customer satisfaction in the 1980s which classifies customer preferences into five categories.

Inspection and testing are optimized when related planning tasks (for test) are executed early and often (in the requirements phase). When test and inspection requirements are considered at these early phases, allowances can be built into the software and into the process to make it easier to inspect, test, document, and trace. Testing smaller amounts of code and functionality, more often, will better distribute the resource load and reduce quality and reliability failure risk. Traceability is a key component for tracking code defect insertion points (versus detection points) and will facilitate root cause analysis. Employing these methods will ultimately improve test coverage and enhance the ability to optimize designs in incremental steps rather than waiting until late in the process.

Testing can be a complex and technologically challenging subject. In general, the complexities in software are based on the volume of code (size), the age of code, the amount of reuse employed, the programming language(s) used, the level of the code (component, module, and system), and the knowledge and experience of the programmer.

These inspection and test processes provide important opportunities to both test the performance of the code and/or system and provide data about the efficiency, effectiveness, reliability, and quality of the design. Test plans must be robust to the critical requirements of the initial design and their compliance with the requirements of the architecture and the system as applicable. The collection, storage, retrieval, and use of this data are of paramount importance to determine if the software performance thresholds are adequate for release and the system will operate reliably when all operating thresholds are present (load or stress). Further, these data will be valuable to determine failure modes and facilitate root cause analysis.

As a general concept, test effectiveness will decrease and quality/reliability risk will increase commensurate with the size and complexity of the component or system being tested (or inspected). Theses attributes will also be adversely affected by testing late in the design cycle, when the functionality is at its most complex and fully assembled state.

For these reasons strong consideration should be given to utilize testing methods that commence early in the process, are incremental in nature, and are designed concurrently during development. Similar to a manufacturing process, defects found late in the process are indicative of, and correlate directly with, the level of defects present in the field (the notion of escaping defects).

There are many specialized quality tools available to help optimize designs. They are most efficiently used when consistently applied, supported by proper training and leadership, and their use synchronized with process requirements. Design for Six Sigma (DFSS), robust design, project management, and many other techniques provide road maps and tools for optimization. Tools related to prioritization (C&E matrix, Hoshin planning), concept selection (i.e., Monte Carlo, TRIZ, Pugh), design tradeoffs (i.e., Kano Planning, QFD, sizing), statistical analysis (i.e., DOE, regression), and others all can play a significant role in the early characterization, analysis, and optimization of a design.

While a significant portion of design optimization should occur internally, released products almost always require enhancement and optimization. The ongoing monitoring of both customer and field feedback and postrelease defect data can provide valuable information about the current design or insight to the next generation or release of software. Part of the quality role is to act as the customer ombudsman and ensure that feedback channels are open, field data are accurately tracked and accurate, and there are forums for reviewing the data on a regular basis with all involved functions. Quality tools used with these data are more traditional (trend charts, Pareto diagrams, run charts, etc.) and help observe field performance against initial performance targets. Further and more detailed analysis should be required for each failure and efforts completed to determine the root cause of each. Corrective action plans should be prepared and monitored for full implementation.

Quality planning and execution should both facilitate and ensure best practices in software design optimization through the early, incremental, and effective application of robust optimization methods and tools. Design of tests and planning for inspection should occur at the earliest point in the process (requirements phase) to ensure that coverage is optimized and the typical "logjam" at test avoided. When field failures do occur, track them with care and ensure rapid and complete corrective action at the root cause.

Optimizing Software Processes

In many ways a software process is like any other process. But it does have some unique characteristics, and the tangible product is hard to pick up and examine. For that reason our approach to process optimization is somewhat modified. Like most processes, our optimization efforts are typically focused on

- Improving quality of the process outcome (compliance to requirements and defect-free software)
- Improving the efficiency and speed of the software process (reducing time, waste, and resources needed to generate a high-quality result)
- Optimizing the cost of the process (analyzing tradeoffs such as build/buy/outsource decisions)
- Regularly changing the process by utilizing a scientific approach to improve the desired outcomes

Process optimization work almost always starts with creating an "as is" process based on how work is currently done. It is rare that we are able to design a new process from "scratch," but in cases where we can, many of the principles discussed here will apply. As discussed in other chapters, there are several effective process visualization and analysis tools. These include conventional process maps, flow diagrams, value stream maps, cross-functional maps ("swim lane"), and others. The main point with all these is to be sure to acquire an accurate representation of the current as-is process. Doing this requires strong facilitation and technique. If one or two people carry the development of the process, it will likely not be an accurate representation of the current process. This is useful because, through the action of building (or validating) a process, differences in how different people, departments or functions do work will become apparent. Process variation is almost always indicative of waste and a source of risk for the creation of defects. When developing an as-is process map, especially in the software environment, be sure it is a "team sport." This is the only way to understand the variation in process actions and generate discussion relating to the reasons for the variation. Being armed with this knowledge will ensure that the process of creating the "should be" (or "to be") process will be a value-added proposition and be bought into by the team responsible for working within it. Be sure to explore specific cycle time, defect, and other data collection and inspection points and examine their effectiveness in the process.

Once process work is underway, there will be a better indication of where the critical process data are collected in the process and why. While we can initially (and quickly) streamline our process by removing non-value-added steps and smoothing process flow, less obvious problems will require data analysis and corrective action. The new process should have clearly articulated data collection points and methods. In the absence of data collection points (in the as-is) be sure that the "to be" process contains them. Once data collection begins, a basis for process "goodness" can be developed. This will lead to the eventual development of goals and controls. The stage of the process in which data are collected will determine the type of data collected. In software processes these will generally include

1. Code inspection data (defects per some defined code size base)
2. Cycle time and schedule performance data (by phase or stage gate)
3. Unit and system test data (defects discovered and insertion points)
4. Effort (resources utilized per project phase gate)
5. Field defect reporting and escalations (help desk, defects)
6. Release dates (on-time delivery)
7. Feature delivery (versus what was promised or original design specification)

Data of this class will help us to see performance characteristics of the process being studied. In process optimization work, the focus should be on the discovery of systemic problems or reversal of negative trends (i.e., problems that tend to recur across different teams and product lines). Examples might include a constant backlog of work at test, excessive field defects across all product lines, backlog of software awaiting testing, etc. These systemic problems can then be analyzed by using other traditional quality improvement methods (DMAIC, PDCA, etc.).

Software process optimization on a systemic level is often seen as a low priority. For reasons previously discussed, namely, the urgency to push software out the door to keep up with rapid technology cycles, teams focused on delivery and innovation ignore incidents that do not appear important in the context of their current work. It is for this reason that the quality role must look at the holistic process and become the "watchdog" for systemic failure mechanisms.

Learning from the Process

A critical success factor that helps sustain and drive expanded results in best practice organizations is the rapid recognition, replication, and institutionalization of key lessons learned and the knowledge associated with each case during development and other process "events." This behavior is no accident in these organizations. In practice, it is a well thought out and planned strategy designed to increase the leverage and "reach" of improvement programs. Pervasive knowledge sharing across and between functional, organizational, product, industry, and geographic groups creates an environment where both general and specific information can be shared and reused to leverage better and faster business results.

In software development, knowledge sharing (and archive) is an extremely important process. Software, more than any other technology, is heavily relied on to tie various pieces of functionality together, bridge gaps between systems and inputs, and provide coherence in complex systems. Because so many people potentially "touch" the software code, there is tremendous potential to create unanticipated problems when tradeoffs or fixes are implemented without proper knowledge or communication. Poor documentation can exacerbate the situation as lack of commentary about what a section or module of code is doing (and why) will limit the understanding of the next person who needs to modify and/or connect to that code. Frequent changes in personnel, resource allocation, priorities, location, systems, and other factors (all typical in a fast-paced software development environment) dictate the need for clear communications and understanding for everything from the base architecture to development status to field performance.

Additionally, success stories, tribal knowledge, culture, and the general sharing of ideas and approaches can be a powerful and complementary driver of culture and behavior. To this end, organizations need to create and leverage opportunities for people (and systems) to collaborate and share in both structured (formal) and unstructured (informal) ways.

The software quality plan and its requirements should both recognize this need and facilitate or complement implementation of knowledge sharing. Points of potential collaboration between the quality plan, project management, and development exist naturally throughout many parts of the Software Development Life Cycle, starting at the planning and requirements phase and extending through release and support. Requirements to consider and document knowledge sharing should exist in each natural phase review or tollgate (development and/or quality). The value of knowledge sharing and the regular communication of knowledge sharing success stories will serve as reinforcement and provide business rationale for its expanded use.

Sharing of information should include but not be limited to the following:

1. Successful problem solutions that were discovered and implemented for the benefit of customers, stakeholders, and general process or product performance attributes.
2. Knowledge about detailed process and product design success stories. These may include successes in quality, performance, cycle time, methods, innovation, customer satisfaction, cost reduction, on-time delivery, etc.
3. Unique and important successes generated through the collaboration and communication of teams or individuals.
4. Successful process changes that are universally "adoptable" by other teams, functions, or geographies.
5. Opportunities to reuse preexisting designs, architecture, code, or other artifacts to eliminate duplication of efforts and improve efficiency.

Creating requirements and guidelines to explore these attributes as part of the process is one aspect of implementation. It is also important to embed the value and importance of these concepts within training, departmental activities, and systems. Establishing partitions in databases and systems to house and index project artifacts for easy retrieval will result in higher adoption and use.

"Knowledge-sharing events" outside the scope of the normal process requirements can also help institutionalize and recognize success stories. Best practice organizations create forums for knowledge exchange through councils, internal trade shows, special topic meetings, presentations, and other communications vehicles.

Planning and Controlling Software in Outsourcing

During the last several years, there has been a tremendous migration toward the offshore outsourcing of software development work to achieve significant cost reductions. This has been an emerging trend for about a dozen years, but one that has been recently accelerated and subject to wild variations in success and efficiency. A software quality approach is one way that software organizations can make better decisions about how, when, and how much to deploy offshore.

From a historical perspective offshore development is a relatively new trend in software, whereas the concept of outsourcing manufacturing and service operations has existed for more than 50 years. Asian countries, initially Japan, later Korea, and then the "Asian Tigers" (Malaysia, Singapore, Thailand, Taiwan, Hong Kong, and the Philippines), took a prominent roll in developing government-led, socioeconomic policies to drive economic success through the pursuit of initially low-tech manufacturing work, utilizing their low-cost and available labor force. In many ways, this created a better balance and improved economic parity between East and West, yet these actions also stripped away many good paying jobs from the western economy and had a significant impact on the quality of products and services outsourced (initially negative but later generally good). This trend led to the downturn of some U.S. industries, which never recovered. These industries included steel, clothing, certain automotive and appliance segments, electronic components, battery cells, and many others. To their credit, this core of Asian companies migrated from their government-sponsored status and embraced the quality- and market-driven system that the West pulled them into. They began to understand how to differentiate their products and services, and in the last 50 years they have become dominant players in the world market.

The western economies, while feeling the pain of the aforementioned job losses, found a new niche in the innovation and manufacture of "higher-technology" (communications, Internet, biotechnology, aerospace, and software) products. These industries also created a massive service-based economy, which virtually replaced the jobs lost in the first wave of offshore manufacturing and served to carry the western economy through the 1990s.

History's lessons are valuable since they demonstrate both effective approaches as well as mistakes to avoid. The lure to move offshore, to reduce costs and improve profitability, is not without a consequence of some kind. Leading industrialized nations need to be aware of the short- and long-term consequences of offshore supplier decisions and plan accordingly to achieve the right balance in the process. These organizations also must make good long-term strategic decisions and carefully evaluate the immediate and future costs and risks in pursuing an offshore strategy.

Countries with an ample supply of highly educated but relatively low-cost labor are eager recipients of software and service outsourcing efforts. These countries have active government-sponsored socioeconomic initiatives and incentives to win business and grow their segment. For many organizations it is easy to get caught up in the offshore groundswell and make a quick, uninformed decision about outsourcing activities to show a rapid cost reduction. The successful approach, however, is one that embraces a life cycle cost-benefit decision. The how, what, andwhy (and even if) are important considerations and should be carefully quantified to ensure success in these efforts in order to realize the advertised benefits. In the absence of this approach, they could suffer a significant disappointment or perhaps even higher cost in the long run.

A 2003 *CIO Magazine* article entitled "The Hidden Costs of Offshore Outsourcing" noted that as much as 72 percent of stated cost savings of typical offshore projects was lost to the costs of start-up, transition, productivity, and maintenance. When one considers that a primary objective to going offshore is to trade $100 per hour development work for $20 per hour work, it hardly seems worth the trouble. In fact, in many cases, if a company could simply find a way to reduce current costs (through efficiency, quality, and cycle time improvements), it might not need to go offshore at all.

However, there are cases in which offshore outsourcing does make sense, but only when the customer needs, the business needs, and the offshore suppliers capabilities are aligned by a clear understanding and defined in a quantifiable manner. Simply stated, if a company outsourced a poorly specified, unstructured, complex project, it can expect in return a cheap but dysfunctional piece of software requiring many person-hours of postrelease support and perhaps even cancellation.

To manage these risks, we must remember some software quality basics to plan and manage this activity. We know from previous study that defects are introduced into the software process at the various stages of development, namely, requirements, design, coding, test, and release. Industry data show that most often defects are inserted early and found late, at the most expensive stage to fix. One of the most common modes of failure in a software project is poor requirements definition and planning including getting requirements into proper context and actionable detail. In most offshore outsourcing scenarios, the requirements phase of a software project is still in the control of, and maintained by, the organization seeking to outsource the development work (or probably should be). Just because an offshore company professes to be a maturity level 3 or 4 company, it does not mean that the project outcome will exhibit level 3 or 4 performance characteristics. This is especially true if the base company's requirements process exhibits less than level 1 characteristics. Due to communication, language, and cultural limitations (interpretive skills), the offshore supplier might never administer the requirements phase to everyone's satisfaction, even if it has the fundamental software skills. So in short, carefully manage the requirements process prior to and throughout the process.

The tools of software quality can greatly reduce the risk of offshore outsourcing failure by rapidly deploying road maps, simple tools, and reinforcing behaviors designed to dramatically reduce the rate of requirements failures, software defects, and cycle time to complete the project. These approaches can also help keep projects on schedule and under control through tollgate reviews based on quantitative data. Through one of several accepted software quality road maps, an organization can quickly deploy a measurement-based process to remove the subjectivity and dramatically improve the quality of subsequent coding activity. As teams accumulate cycles of learning with the road map of choice, it becomes a highly efficient way to quantitatively establish a baseline for development activities and to continually monitor and improve these activities relative to the offshore project.

As a word of caution, there are offshore companies soliciting business on the premise that they embrace Six Sigma, possess ISO Certification, or have significant competencies based on other standards and certifications. It is important to fully vet their understanding and their results attributable to these claims prior to engaging with them. The point here is that companies considering offshore outsourcing must do their homework. Do not expect to receive a high-quality software product from an organization claiming to be certified (to one standard or another) without putting something in—especially on the requirements end of the life cycle and in the management of the effort.

Controlling offshore outsourcing involves regular review, administration, and continuous monitoring of the process. Quality measures and tools must be administered in a specific sequence (in some cases in an iterative manner) to realize their full benefit. Initially, it may take some additional time on the first few projects, but this is quickly assimilated as part of the process by participants and will pay large dividends in the long term.

As the process evolves, frequent feedback and data-rich reviews should examine all aspects of the development process including defect rates, code inspection results, test results, total containment effectiveness (TCE), phase containment effectiveness (PCE), and defect containment effectiveness (DCE). These metrics will help to provide solid insight to control and manage the risks of project failure. This also creates the opportunity to be sure that processes are under "procedural control" (required actions and metrics are actually fulfilled). Visual dashboards are usually implemented to make "out of control" situations visually obvious to reviewers for quick and concise action.

Too often, the decision to outsource offshore is made based solely on cost. The essence of software quality excellence is about measuring and understanding variation in processes and eliminating as much of that variation as possible at the root causes of variations. In successful organizations, this is the key to achieving breakthrough results. Too often when organizations look at cost as the only metric to make decisions, they are falling into a trap. Cost is a lagging and dependent variable. By hyperfocusing on it, organizations invariably miss the characteristics driving it. In all cases, what changes cost is the influence and interaction of many independent variables (leading indicators) that are not measured by the typical accounting system in a useful way. The statistical analysis, prioritization, characterization, and improvement of the right variables (or combination of variables) at the right time are what drive the desired results. In the matter of offshore outsourcing, too many organizations have fallen into the trap of focusing on cost and failing in their primary objective of saving money.

Best practice organizations should utilize principles of software quality programs to (1) determine offshore project viability and partner selection and drive the planning and execution of such activities, (2) provide a process and tools to eliminate risk and manage offshore outsourcing projects, and (3) control the quality of ongoing activities related to outsourcing in a preventive fashion.

Managing the Software Quality System

Since the early 1970s software and information technology (IT) have been significantly changing the way in which products, processes, and services are designed, delivered, and maintained. At an exponentially increasing rate, software and IT have become the driving force in the way people work, recreate, and educate. Most experts estimate this global "industry" at about U.S. $1 trillion in 2006, with data indicating that the cost of poor quality (COPQ) exceeds $250 billion. Some estimate it could be as high as $500 billion. That means that 25 to 50 percent of every company's software and IT budget is at risk.

The software COPQ includes the cost of canceled projects, over-budget projects, late projects, unplanned and excessive maintenance and support, as well as poor efficiency and lost productivity. According to Dr. Howard Rubin, noted author, educator, and researcher in computer science and technology: "In short, if these productivity problems don't start turning around, the cost of poor project management is going to start exceeding the entire output of countries…demand for quality and performance is really going to force this issue once and for all. And the engineering techniques and disciplines will follow."

The "engineering techniques" referred to by Dr. Rubin are succinctly encompassed in the software quality activities being practiced by leading companies. For the first time in the software/IT domain, a fact- and data-based system to prioritize, problem-solve, and design new software-based systems is yielding breakthrough results. CIOs, CTOs, and CFOs are slowly realizing that the same methods that transformed manufacturing and transactional processes into smooth, high-efficiency processes also will work in software and IT, if adapted to recognize the differences in theses processes and the rapid rate of change associated with them.

In parallel with the explosive growth of technology, older industrial, manufacturing, and service processes (as well as products) were being transformed by TQM, reengineering, and eventually the Six Sigma methodology. The latter delivered amazing results in the processes and products in which Six Sigma was administered. Many improvement efforts targeted the manual core of products and processes (human resources, procedures, work instructions, flow, specifications, etc.), often ignoring the software, which drives them, and in some cases working around technologies put there to help. This likely occurred due to the lack of full integration and interoperability of these systems, processes, and products. Organizational silos, departmental boundaries, personal preferences, and incompatibility, driven by the lack of a consistent methodology to improve and integrate processes and products (and the software that drives them), further contributed to this problem.

The rapid rate of change in technology also promotes the reactive behavior paralyzing most software and IT organizations. In their pursuit to stay on top of the technology curve and all its advertised benefits, organizations continuously jump to new solutions before the benefits from the initial solution are realized. Further, the advertised business benefits of these systems are rarely measured and audited to verify and validate results. These actions create large disconnects, frustrated employees, customer dissatisfaction, and inefficiencies in the very processes and products they are intended to help.

Examples of this dysfunction include companies that have spent millions of dollars on enterprisewide system deployments only to run a "shadow system" of spreadsheets. Is this really what was intended or expected?

In the product software arena, consider products such as cell phones, in which 6 sigma hardware (built on a 6 sigma production line) runs on a 3 sigma network, is then delivered with 2 sigma service, and contains software so complicated that a software engineer cannot figure out how to use all its functionality. In this case, what is the quality level of the total customer experience? Perhaps 2 sigma? Is having the innovation of a tiny phone alone enough to drive sustainable market gains and financial success? Although it once was, it no longer is.

A comprehensive software quality methodology can greatly improve performance by creating structure, prioritization, and basic tools with which to more fully understand the problems that technology is supposed to solve, the requirements that customers have, and the measurable business implications of various concept and solution tradeoffs.

It is also clear that many traditionally accepted measurements, such as service level agreements (SLAs), IT dashboard metrics, call center and help desk statistics, resolution cycle times, and bug fixes are reactive and alone do not serve the needs, in totality, for a total software quality system. In fact when these metrics are examined, the associated improvement rates are often not evident, not consistent, and many times not even viewed in that context. How would these items be characterized if given the following choices?

Q: Reactive or preventive?

A: Be honest, the answer is probably reactive.

Q: Are these leading or lagging indicators?

A: Mostly lagging.

Q: Do they indicate cause or effect?

A: Probably effect.

Where is the preventive notion in this characterization? What methods will drive a company to perform effective and preventive cause-and-effect analysis? How can anyone sift through the plethora of data and variables to isolate a critical cause?

Organizations must realize that the software process needs to be treated like all others. Thus critical variables need to be characterized, prioritized, understood, measured, and controlled. Sound software quality methodologies provide the methods and tools to do just that. They have been adapted to ensure consistency and applicability to the software culture in play and remove the old excuses that *traditional* quality methods will not work in software.

In software development and engineering, our traditional metrics relate to testing, defects per thousand lines of code (DpKLOC), inspections, bug fixes, etc. Again, this is mostly a "detect and react" group of metrics. How often are strong disciplines and tools used, tools such as concept engineering, language and context data collection, or tools to discover where defects were *inserted* versus found? Companies spend large amounts of time and money on testing, but rarely do they focus on the point of insertion of defects to isolate cause. Further, companies rarely examine patterns of systemic defects, code, and performance. Again, a robust software quality process provides answers through the orderly and consistent use of proven tools.

The successful early adopters in this field have shared some common attributes. These include leadership that is not afraid to confront traditional norms and conventional wisdom to solve business problems. They are willing to implement a strategy of change, invest in human capital, create a fact- and data-based environment, and insist on performance and prevention—not simply complying with a standard. It is the only way to break the chain of reactive behavior. The human ability to change is governed by the desire to change. The desire often comes from a need, and a need from an organization's business situation.

Too often, organizations are driven by an antiquated planning process that is subjective at its core, laced with intangibles, and incapable of being measured. If this planning process characterization sounds familiar, the use of simple prioritization tools such as failure mode and effects analysis (FMEA), Hoshin planning, and cause-and-effect analysis can go a long way to help align business needs, customer needs, and software budgets. A higher level of prioritization at the front end of planning can significantly reduce downstream failures. Part of good planning is following up on business cases. In other words, planning needs to be a closed-loop process, starting with needs and finishing with results. If the results are great, it reinforces the process and drives further change. If the results are poor, a company should learn why and should do it better the next time. But remember, if nothing is measured or followed up on, there will be no process improvement.

So as leaders of our business, our product, our process, or our service, we must ask the question, Is it OK for a software or IT department to cruise along wasting 25 to 50 percent of its budget? And how acceptable (or fair to other departments) is that if the organization has driven down costs (through quality improvements) in manufacturing, supply chains, and transactional/service processes? Sooner or later, the other functions will start asking what software and IT are doing to help drive their fair share. Every company should quantify, report on, and set aggressive targets to improve and eventually mitigate waste.

Sure, aspects of software and IT are different—but not that different. It is simply not a good excuse to say these processes cannot be measured. In a 2002 National Institute of Standards and Technology study on software and IT, more than 40 key metrics in six attribute categories were listed as being used by best practice companies. Detailed data are available across multiple industry segments. They are out there. Organizations need to learn from these data and start using them. Some will argue the cost is too high or the resource commitment too great to change. This is a shortsighted and convenient argument. And it is not logical. This is especially true if it is an organization that is caught in the reactive, fire-fighting loop with late projects, canceled projects, and over-budget projects.

By confronting these common questions, discussing and framing issues relative to behaviors, planning and implementing proven methods and tools, and practicing time-tested quality fundamentals, organizations can be on their way to changing their software processes to become more preventive in nature and more oriented toward results. This will lead organizations down a path to control the process that controls their success.

References

Computer Aid, Inc. (2006). Interview with Howard Rubin, "What the Future Holds for IT." March.

Professional Societies, Institutes, and Communities

CMM and CMMI are registered trademarks of Software Engineering Institute.

The Institute of Electrical and Electronics Engineers (IEEE): Over 60 individual, detailed software engineering standards in IEEE Software Engineering Standards Index. http://standards.ieee.org/reading/ieee/

International Standards Organization (ISO): Comprehensive set of software development and software quality standards. For full list of standards visit http://www.iso.org/iso/home.htm

National Institute for Standards and Technology (NIST): Comprehensive set of software and software quality standards and case studies. http://www.nist.gov/public_affairs/siteindex.htm

Software Engineering Institute (SEI): Capability Maturity Model, Integrated (or CMMI). SEI is a registered trademark of Carnegie Mellon University. http://www.sei.cmu.edu/index.cfm

CHAPTER 30

Supply Chain: Better, Faster, Friendlier Suppliers

Dennis J. Monroe

About This Chapter

This chapter deals with how producers and purchasers can develop better relationships with their suppliers. This will lead to lower-cost supplies and components delivered on time and on budget. It discusses the shift that is taking place in relationships between suppliers and producers/purchasers. This shift from adversarial supplier to an important partner often was the case in the past. Partnering with suppliers can help achieve the objectives of faster, cheaper, and friendlier supply relationships.

High Points of This Chapter

1. The triple role of supplier, producer, and customer is important to understanding supply chains and supplier relations.
2. Partnering among suppliers, producers, and customers helps maximize value at all levels of the supply chain.

3. Application of Lean supply chain principles can lead to improved speed (shorter replenishment times).
4. Application of the Juran Trilogy® to the supply chain can help improve speed and is most effective at reducing supplier and producer costs of poor quality (COPQ).
5. Effective supplier scorecards help the producer to objectively select and retain the optimal supply base. They assist with effective planning, control, and improvement of the supply base and supplier management process.
6. Supplier auditing is an important support for the supplier scorecard system.
7. Supplier audits should lead to continuous improvement activities by suppliers.

Introduction

In many organizations—both those that manufacture goods and those that provide services—there is a heavy reliance on supplied materials, components, and products. Many manufacturers are truly just integrators of hundreds or thousands of parts into a final product. Take the automobile producer, for example. The producer buys myriad parts from engines down to small fasteners. Little of the content of the final automobile is produced in the manufacturer's facility (a notable exception often being stamped body panels). Other industries such as aerospace, defense contractors, and consumer goods manufacturers follow this pattern of purchasing and integrating components into a final product. The days of highly integrated manufacturers who produce everything from raw materials to finished product are, for the most part, gone.

The situation described here is often more complex. The raw material supplier is usually the only one who does not do some kind of transformation or assembly with his product. In a typical supply relationship:

- The raw material suppliers supply small component manufacturers, for example, plastic resin.
- The small component manufacturers supply the subassembler, for example, molded parts.
- The subassemblers supply subsystem integrators, for example, a louver for an instrument panel.
- The subsystem integrators supply final product integrators, for example, the instrument panel.
- The final product integrators supply the sales outlets or consumers, for example, a completed automobile.

Or take the example of a service provider who uses supplied products in the performance of their service; a cleaning service that cleans several high-rise office buildings using a high volume of cleaning supplies from multiple suppliers; a retail outlet purchasing multitudes; and large department stores, consumer electronics stores, and the like using thousands of suppliers.

The reality is that each level of supplier, from the supplier of raw materials up to the ultimate supplier to the end user, must supply good quality products, at the right price and at the right time, which makes the subject of supply chain and supplier relations very important to today's organizations of all kinds.

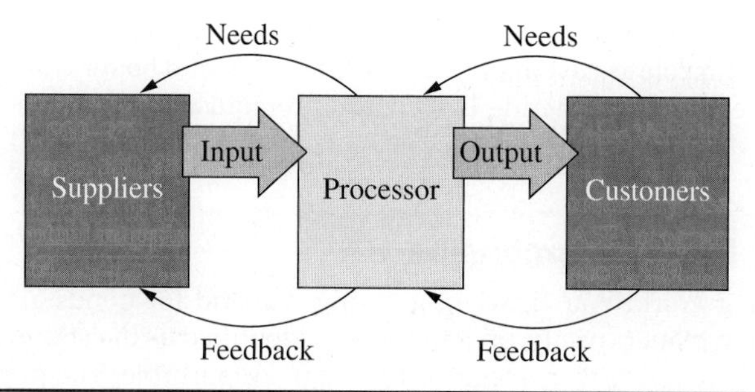

FIGURE 30.1 The triple role. (*Juran Institute, Inc.*)

The Triple Role

Every processor team conducts a process and produces a product. To do so, the processor team carries out three quality-related roles, which are depicted in Figure 30.1. The diagram shows the interrelation among three roles:

- *Customer*. The processor team acquires various kinds of inputs that are used in carrying out the process. The processor team is a customer of those who provide the inputs.
- *Processor*. The processor team carries out various managerial and technological activities to produce its products.
- *Supplier*. The processor team supplies its products to its customers.

To illustrate, the company is a processor team. In its role as a customer, it receives such inputs as:

- Information concerning client needs, competitive products, and government regulations
- Money from sales and investors
- Purchased goods and services
- Feedback from customers

In its role as a processor, the company converts these and other inputs into products such as sales contracts, purchase orders, salable goods and services, invoices, and reports.

In its role as supplier, the company provides clients with goods, services, and invoices, and provides suppliers with purchase orders, payments, and feedback information provided to all.

The concept of the triple role is simple enough. However, the application can become quite complex as a result of the large number of suppliers, inputs, processes, products, and customers. The greater the complexity, the greater the need for an orderly approach to quality planning.

Methods for managing the complexity of the relationships, as Dr. Juran described, among suppliers, producers, and customers are the subject of this chapter.

Friendlier Partners

As the subtitle of this chapter implies, the customer-processor-supplier relationships at all levels must be in the form of partnerships in order to succeed in today's marketplace. The "old ways of doing business," in which relationships were often adversarial, no longer suffice. In your role as customer, you must find new and better ways to interface with suppliers. Since the idea of partnering is seen as a prerequisite for the other two (faster and cheaper), we will discuss this concept first.

Traditional Role of Purchasing

Following World War II, when growing demand for goods and services was satisfied by increasing plant capacity, operations was identified as the strategic component of an organization. Purchasing was relegated to a staff support role. The purchasing department's mission was to ensure that suppliers provided an uninterrupted supply of required goods and services, delivered on time and at the right price (where "right price" was usually interpreted as "lowest price," not "lowest total cost").

Personnel in purchasing departments developed competencies in supplier negotiations, bid evaluation and analysis, document administration, and market knowledge. Supplier negotiations were viewed as the major value-added activity of the purchasing department because supplier relations developed during these negotiations. This often resulted in adversarial supplier relations, which were focused on short-term performance. Availability and low price became the most important criteria for measuring supplier performance. As Carlisle and Parker wrote (1989), "This adversarial tendency … resulted in a great deal of management energy being spent on both sides in search of ways to capture some of the other's profit margin."

If a supplier change was made, little consideration was given to any resulting incremental costs incurred. The new supplier's product or service might deviate slightly from that of the original supplier, translating into costs in other areas of the production process. This propensity to change suppliers resulted in many disadvantages to the purchaser, including:

- Excess inventory because of obsolescence
- Production shutdowns because of installation or operation requirements
- Transition costs such as training or maintenance testing disposal costs
- Production disruptions because of poor quality detected after the testing had been completed
- Increases in variation in the finished product
- Increases in scrap, product defect, or customer dissatisfaction

Rarely were these costs identified, aggregated, analyzed, and reduced. Furthermore, as Deming (1981) stated, "No one can outguess the future loss of business from a dissatisfied customer." In the adversarial climate that prevailed, little opportunity for collaborative root cause analysis existed.

Moving from Adversary to Partner

Beginning in the late 1970s, the "Quality Revolution" grew from its roots in Japan to the rest of the world, and it was recognized that price-based purchasing decisions and adversarial customer-supplier relationships would no longer suffice. Companies began to recognize that timelines of delivery, after sales service, product development partnerships, and quality of products and services supplied were just as important, maybe even more so, than price.

But progress toward a friendlier environment was not linear. In the late 1980s and early 1990s, some customer purchasing functions returned to the old, adversarial approach to the supply chain—in particular, suppliers to automobile manufacturers, who were pressured to offer lower and lower prices and sign long-term contracts, including annual "give-backs," as a condition of contract award. At General Motors, the pressure on suppliers during this time was especially great, largely because of the efforts of Jose Ignacio Lopez De Arriortua.

Lopez had made his reputation at the GM affiliate, Opel's, Rüsselsheim, Germany, plant. As Jonathan Mantle says in *Car Wars*, "[Suppliers] called him 'the Butcher.' Lopez squeezed suppliers until they screamed, and then squeezed some more." Thankfully, this reversion to tactics of the past didn't last long. By the mid-1990s, most companies were moving again toward a strategy of partnering with suppliers to achieve a better outcome for all. Lopez left GM for Volkswagen in 1992 and was later sued by the company for allegedly stealing trade secrets. The approach Lopez and GM took resulted in what Ed Rigsbee refers to as "The Boomerang Always Returns." "Suppliers cut back on their research for GM's needs and started giving their cutting-edge technology to Ford and Chrysler. Rather than GM getting more, in actuality, it got less."

For partnering to be effective, of course, the supplier must get something out of the deal. Often, companies will share cost savings with their suppliers as an incentive for better prices. A company that is truly committed to partnering will also offer assistance in, for example, achieving process breakthroughs that will allow the supplier to produce at a lower cost and supply in a more timely fashion. Other enticements include single source or majority volume of orders for the supplier partnership.

> "General Electric (GE) went after cost reductions differently. The company assembled appliance suppliers in November 1992 and announced 'Target 10,' asking suppliers for 10 percent cost reductions. The difference was that GE pledged its assistance to suppliers in finding strategies for the [cost] reductions" (Rigsbee, 2000).

By approaching the supply chain as partners, a consumer of their products and services can achieve shorter supplier lead times (faster) and lower costs of products and services procured (cheaper).

Faster, cheaper supplied goods and services result from optimization of the supply chain.

Quality Incorporated into Traditional Purchasing

Purchasing as a Strategic Process

Consider the potential opportunity if time, resources, energy, and management priorities focus on the processes by which these goods and services were scheduled, designed, manufactured, and purchased, rather than simply focusing on the acquisition alone. Quality and cost reduction opportunities could be identified, measured, and managed. Where two firms compete in identical markets, the ability of one firm to identify, measure, and manage these opportunities faster than another firm creates a clear competitive advantage.

Therefore, purchasing, while traditionally thought of as a utility, nonvalue-added function, is increasingly being recognized as a strategic function, an opportunity for process management and improvement, and a tool for achieving competitive advantage.

It's been calculated that supplied components account for 55 cents of every dollar of revenue an average manufacturer receives. If that cost can be reduced by only 5 percent, it means a 3 percent increase in profits. A similar equation applies to service providers. But, as Tully (1995) says, "Cutting purchasing costs has surprisingly little to do with browbeating suppliers. Purchasers at companies like AT&T and Chrysler aim to reduce the total cost—not just the price—of each part or service they buy. They form enduring partnerships with suppliers that let them chip away at key costs year after year."

Importance of Supplier Quality

To identify supplier-relation opportunities and to capitalize on them, an understanding of suppliers' quality is of paramount importance. Consider the following:

- The costs associated with poor-quality suppliers are high. For one home appliance manufacturer, 75 percent of all warranty claims were traced to poor quality of purchased items.
- The growing interdependency of suppliers and end users in identifying and implementing such opportunities as "just-in-time" delivery, electronic data interchange (EDI), electronic funds transfer (EFT), cycle-time reduction, and outsourcing initiatives.
- The trend to minimize incoming inspection.
- The growing trend of purchase decisions being made not on lowest price, but on the total cost of ownership of the product or service.

These considerations require the purchasing function to abandon its traditional role of transaction-performance management. Expressions of this emerging approach are contained in statements from two eminent American companies.

From AT&T in 1995:

Mission: Provide worldwide professional procurement services that are a competitive advantage for AT&T and its customers. *Vision:* Be THE benchmark for procurement excellence.

From Chrysler Corporation:

Mission: Manage and prepare the extended enterprise to the maximum benefit of Chrysler and its customers.

The implications of this role change are profound.

- Supplier selection and management is no longer the sole prerogative of the purchasing department.
- Cooperation, collaboration, and joint problem-solving among internal customers, purchasing, and suppliers are required.
- Purchasing personnel focus on process, abandoning the focus on transaction.
- Within the end user's firm, the purchasing function is elevated to a strategic level and its transaction activities and responsibilities minimized or eliminated.

A successful transition to a strategic approach to purchasing requires everyone in an organization to embrace a new belief system concerning purchasing. In the transition, senior management will find it necessary to aggressively promote the new view, which might be summarized as follows:

Purchasing has become a key strategic process within our organization, requiring a staff of highly skilled professionals committed to working with our end users and suppliers, in a collaborative, problem-solving environment, facilitating quality and continuous improvement.

Shift to Strategic Purchasing

The differences between the traditional view of purchasing and the strategic view are dramatic. They are summarized in Table 30.1. The differences require some significant changes in culture and behavior.

Aspect in the Purchasing Process	Traditional View	Strategic View
Supplier/buyer relationship	Adversarial, competitive, distrusting	Cooperative, partnership, based on trust
Length of relationship	Short-term	Long-term, indefinite
Criteria for quality	Conformance to specifications	Fitness for purpose
Quality assurance	Inspection upon receipt	No incoming inspection necessary
Communications with suppliers	Infrequent, formal, focus on purchase orders, contracts, legal issues	Frequent, focus on the exchange of plans, ideas, and problem-solving opportunities
Inventory valuation	An asset	A liability
Supplier base	Many suppliers, managed in aggregate	Few suppliers, carefully selected and managed
Interface between suppliers and end users	Discouraged	Required
Purchasing's strategy	Manage transactions, troubleshoot	Manage processes and relationships
Purchasing business plans	Independent of end-user organization business plans	Integrated with end-user organization business plans
Geographic coverage of suppliers	As required to facilitate leverage	As required to facilitate problem-solving and continuous improvement
Focus of purchasing decisions	Price	Total cost of ownership
Key for purchasing's success	Ability to negotiate	Ability to identify opportunities and collaborate on solutions

TABLE 30.1 Traditional vs. Strategic View of the Purchasing Process

Total Cost of Ownership

The most fundamental shift in the purchasing professional's behavior is to base purchase decisions on the total cost of ownership. Taking a total process approach (rather than a transactional approach) to quantifying the total cost of ownership will result in the identification of supplier, end-user, and joint costs that will need to be identified and measured. Many of these costs will be reduced through joint problem solving. Table 30.2 offers a sample list of elements of total cost of ownership.

Supply Chain Optimization

The goal of a strategic purchasing function (one that partners with its suppliers and customers) is to facilitate the performance of the supply chain. This process facilitation includes

Category	Subcategory	Cost Component
Preacquisition	Preprocurement cost	Engineering/design
		Supplier survey
		Supplier audit/site visits
		Product testing/technical review
		Regulatory compliance
		Market assessment
		Customer reviews/briefings
Acquisition	Material equipment cost	Price of material/equipment
		Cost of special features
		Shipping/handling/storage
		Spare parts
		Leased items
		Taxes
	New technology costs	Modification/retrofit
		Additional training
	Foreign acquisition costs	Foreign surtax
		Import duties
		Foreign currency risk
		Additional testing requirements
	Installation/start-up costs	Labor
		Subcontractor
		Special testing
		Construction equipment
		Required overhead
		Training
		Special tools
		Service engineering
		Inspection
Ownership	Operating/maintenance costs	Administration/overhead
		Ongoing labor
		Routine testing requirements
		Ongoing training
		Energy usage
		Preventative maintenance
		Personnel required
	Inventory costs	Inventory carrying costs
	Failure costs	Cost of expected down time
		Replacement parts
	Obsolescence costs	Energy efficiency
		Productivity loss
	Other costs of ownership	Environmental impact
		Licensing, permitting
		Environmental control equipment
		Conformance costs
		Standardization costs
Disposal	Disposition cost	Removal salvage costs/value disposal

Table 30.2 Sample Checklist for Total Cost of Ownership Consideration

participation of the end users and suppliers. Supply chain optimization is the ongoing management and continuous measurable improvement in the performance of this supply chain, generating value for all involved. The entire supply chain must be considered, including indirect suppliers, manufacturers, distributors, and end users. Note that the key words in this definition are:

- *Ongoing*. Supply chain optimization is not an event, but an ongoing process.
- *Measurable*. The results of supply chain optimization are tangible benefits.
- *Improvement*. The foundation of supply chain optimization is continuous improvement.
- *All*. True supply chain optimization requires participation of all parties involved to share in the benefits.

Goal of Supply Chain Optimization

The overriding goal of quality-focused supplier chain optimization is increased customer satisfaction through the joint (suppliers and end user) creation of value in the supply chain. On the supplier side, participation in such an initiative, as supply chain optimization, extends beyond the role of the account executive and includes the participation of those actually involved in the manufacturing and delivery of the product in question.

In addition, on the end-user side, participation in such a venture extends beyond the purchasing department, and includes participants from the core operating business units. In fact, while such a team effort is typically facilitated by a purchasing individual, the team should be led by, and accountability of results assigned to, a member of the core business unit.

Supply chain optimization creates value in the following six areas:

- *Quality improvement*. Continuous reduction in product variation and the ability to plan and build quality into each component and service, with measurable results
- *Cycle-time reduction*. Continuous reduction in the time required to make and implement key decisions and perform various processes
- *Cost of poor quality reduction*. Continuous measurement and reduction of costs associated with the prevention, inspection, and failure resulting from poor quality
- *Total cost of ownership reduction*. Purchasing decisions based on total cost of ownership, including preprocurement, acquisition, operation, and disposal costs, rather than price alone. Continuously manage the ongoing acquisition based on the identification and elimination of root-cause cost drivers, which contribute to total cost of ownership.
- *Technology/innovation*. Continuous identification and deployment of value-added technologies through joint planning and development
- *Shared risk*. Continuous identification of opportunities to identify and share risk throughout the supply chain

Successful supply chain optimization requires that the sourcing process operate as a single seamless entity, rather than as a set of discrete processes. Members of the supply chain establish goals and work together toward these goals, which target the satisfaction of customer needs.

Faster

For our purposes here, we will define faster supply as a shortening of the supplier's replenishment lead time. Replenishment lead time is defined as the time from placement of an order to the supplier until the order is received and can be used by the producer.

Replenishment lead time is the key to the efficient operation of any producer. If the producer's process runs faster than suppliers can replenish, then stock-outs will occur unless sufficient safety stock is held by the producer or in a finished goods inventory by the supplier. Inventories are wasteful and costly—the cost of money invested in the inventory, the floor space required to store the inventory, and so on. But stock-outs are wasteful too, and may be even more costly than the safety stocks held to avoid them. In either case, the producer must either:

- Slow his pace of production to avoid stock-outs, which may result in inability to meet his customer's delivery time requirements, or
- Work with suppliers to reduce replenishment lead times to a point that allows the producer to meet his customer's demand pace.

When one considers the entire supply chain (see Figure 30.2), this problem becomes of even greater concern. Say that a producer's tier 1 supplier is experiencing long replenishment times from the tier 2 supplier and the tier 2 supplier is experiencing long lead times from the tier 3 supplier, and so on. It's easy to imagine how the producer's safety stock requirements to sustain production and supply in a timely manner to his customers would grow exponentially.

There are three things a producer must do to ensure the shortest possible replenishment time from his suppliers:

First, understand and optimize the logistics of getting product shipped or services delivered to your location. Of course, this includes understanding the distance between the

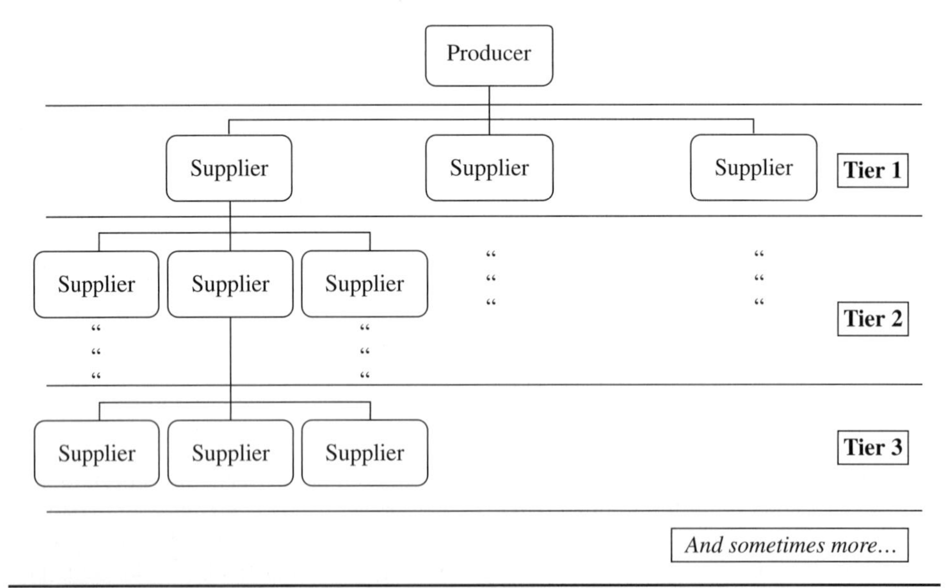

Figure 30.2 The supply chain.

supplier's location and yours, but it's more than that. Are deliveries scheduled frequently enough? If they are scheduled weekly, for example, could they be scheduled daily or several times a week instead? More frequent deliveries can remove days from replenishment time—days that will then not need to be covered with safety stock.

Many producers have addressed this issue by trying to find suppliers geographically closer to their manufacturing facilities. In some cases, manufacturers like Toyota have assisted potential suppliers in starting facilities nearby or even next door to the Toyota assembly plant. The idea of geographically shortening the supply chain is a good one, but is not always practical for all producers. But the physical length of the supply line must always be considered when making purchasing decisions. Even if a supplier in, say, India or China has a much lower cost per widget, the additional costs of safety stock, work in progress, and finished goods inventory needed to compensate for the long distance supply, as well as the transit cost itself, may outweigh the piece cost savings. Consideration must be given to the costs of less-than-desirable logistics when making purchasing decisions.

Second, help suppliers "get Lean." You may not have resources within your organization to directly contribute to the training of suppliers and facilitation of their Lean events, but you can require them to undertake a Lean management implementation as a contract condition. Incentivize this activity by sharing the savings you reap as a result of shorter replenishment times and the resulting lower inventory level with the supplier. They will, of course, also reap benefits from reduction of waste in their facilities.

After accomplishing the above with the first-tier suppliers, add incentives for them to do the same with tier 2, 3, and lower-level suppliers. Ultimately, the objective is to shorten the time it takes from production of the lowest-level product, usually raw materials, until the final product is complete and ready to ship to your customer.

Recall the triple role discussed previously. At whatever level of the supply chain your company operates, one of your roles is that of producer of your products. It is in that role that you should consider the application of Lean supply chain techniques as described here.

To be manageable, you should consider doing the replenishment time reduction activities with one key supply chain first—one of your most important tier 1 suppliers and his lower-level suppliers. Use Juran's Pareto principle to prioritize those suppliers for first action considering factors such as annual value supplied, length of replenishment times, and so on.

Third, propagate the success. According to Phelps, et al (2003), "Since you selected a subset of your entire supply chain for this application of the Lean supply chain approach, you will want to share your successes with your other suppliers in preparation to begin the process with another group of suppliers. This is a way to pique their interest as well . . . [and] give them a sense that you are indeed helping your supply base improve. . . . "

As you propagate the improvements to each successive piece of your supply chain, the benefits of faster supply will build exponentially, just as the waste, nonvalue-added activities, and inventories have in the current state. These Lean supply chain approaches support the second aspect of improved supplier relations, "cheaper," but there are more steps you can take to obtain the best prices from your supply chain.

Cheaper

While creating a Lean supply chain focuses primarily on speed, it also contributes to lower costs in production by reducing waste in processes. Additional actions to achieve lower costs from suppliers focus primarily on reduction of cost of poor quality (COPQ) by improving the quality of producer procurement processes and supplier products and processes.

COPQ is one element of overall cost of quality; the other is cost of attaining quality, which includes prevention and auditing activities. COPQ consists of inspection costs and failure costs, both internal and external.

How to Reduce COPQ Related to Supplier Relations

It is important to understand that COPQ is either caused or experienced in each of the three phases of the Juran Trilogy®: planning, control, and improvement. In each of these phases of managing for quality, actions can be taken to reduce COPQ related to:

- Poor planning by the producer: design deficiencies, failure to clearly communicate requirements to suppliers
- Poor control by the supplier: incapable processes, lack of control plans, lack of mistake proofing, poor detection
- Lack of breakthrough improvement projects, which, if properly executed, could have perhaps the greatest impact on reducing COPQ

Once an organization has been established to facilitate quality supplier relations, the trilogy of quality planning, quality control, and quality improvement can be applied to the supply chain. Following is a detailed explanation of the activities and deliverables of these three phases of managing for quality and how they relate to supplier relations.

Planning for Supplier Relations

According to Kevin Fitgerald (1995), "Honda's success on this continent [North America] is a direct result of the company's overall philosophy of manufacturing ... manufacturing's success depends on two groups: the people who make the products, and the suppliers that provide the parts and raw material from which the products are made."

Planning for supplier relations is the activity of identifying customer needs and analyzing and developing a sourcing strategy to meet those needs. One of the key deliverables of the planning process is an initial model detailing the customer's total cost of ownership of the subject commodity. Thus, data collection and analysis will also be required throughout the planning process. The focus of this planning process is the identification of the appropriate customer and assessment of the current and future needs of these customers for the commodity in question (Table 30.3). Additionally, as the output of the planning process is a recommended sourcing process flow, a thorough understanding of the supply industry structure, dynamics, and trends is essential.

The planning process requires:

- Early customer involvement to identify current and future sourcing needs
- Extensive research and data collection regarding the alternative processes available to satisfy these needs

Most successful source planning has followed a methodology similar to the following:

Step 1. *Document the organization's historic, current, and future procurement activity.* In the absence of planning for supplier relations, it is assumed that purchasing is generally handled in a reactive business process, which satisfies immediate, local operational needs. The documentation of the historic, current, and anticipated purchase activity across an organization's various business units enables that organization to take the first necessary step toward achieving purchasing leverage; synergies within and between organization business units;

Process	Definition	Process	Definition
Quality planning	The activity of developing the products and processes required to meet customer needs	Planning for supplier relations	The activity of identifying customer needs and analyzing and developing a sourcing strategy to meet those needs
Quality control	The activity of evaluating actual performance, comparing actual performance to goals, and taking action on the difference	Control for supplier relations	The activity of evaluating suppliers' performances, selecting the vital few suppliers capable of optimizing performance, and the measurement of supplier performance
Quality improvement	The activity of raising quality performance to unprecedented levels	Improvement for supplier relations	The activity of identifying and acting upon sourcing process improvement opportunities

Table 30.3 Juran Trilogy® Applied to Supplier Relations

and a strategic, collaborative, proactive approach to managing the sourcing process. Available tools are data collection and trend analysis.

Step 2. *Identify a commodity from the procurement activity that represents both high expenditure and high criticality to the business (quadrant IV commodities).* A simple Pareto analysis will often reveal the vital few commodities that drive an organization's purchasing needs and costs. Focusing resources on these vital few commodities will enable an organization to begin to capture the value of supply chain management early. Available tools are Juran's Pareto analysis, data histograms, stratification, and management presentations.

Step 3. *For this commodity, assemble a cross-functional team.* The team includes representatives of the customer and of company functions—technical, purchasing, quality, and financial. The team's mission is to define the customer's sourcing need for this commodity and to develop a sourcing strategy that will meet this need. Available tools are brainstorming, team building, and flow charting.

Step 4. *Determine the sourcing needs of the customer through data collection, survey, and other needs-assessment activities.* This is the critical step, which, if not properly and thoroughly conducted, can derail any well-intentioned cross-functional team. It is often fatal to assume that the customer's needs are obvious. Extensive data collection through surveys, customer visits, and focus groups will pay off later on. Available tools are brainstorming, data collection, flow charting, cause-effect diagrams, force field analysis, hypothesis formulation, and testing.

Step 5. *Analyze the supply industry's structure, capabilities, and trends.* Once the customer needs have been identified and validated, an industry analysis is required. It is the supply chain, and not the purchase itself, that will ultimately delight the customer with fitness for purpose and value. Thus, the various supply chains available, and their performance and cost structures, must be understood. This is an extensive research phase of the planning process, and might require the team to split temporarily into several subteams. Available tools are industry data collection and analysis, flow charting, benchmarking, and process capability analysis.

Step 6. *Analyze the cost components of the commodity's total cost of ownership.* This, too, will require extensive data collection and analysis, and even benchmarking, to identify how others have managed this commodity. This model of total cost of ownership will be redefined, refined, and optimized throughout the life of the commodity management team. Available tools are data collection and analysis, brainstorming, flow charting, cause-effect diagrams, histograms, and Pareto analysis.

Step 7. *Translate the customer needs into a sourcing process that will satisfy the customer and provide the opportunity to manage and optimize the total cost of ownership.* The customer needs identified in Step 4 will need to be mapped into the various alternative sourcing processes identified in Step 5. An optimal sourcing strategy can be determined by optimizing the total cost of ownership, based on the results of Step 6. Translation requires extensive dialog and feedback to identify and gauge fitness for purpose of the sourcing strategy. Available tools are data collection and analysis, brainstorming, flow charting, cause-effect diagrams, histograms, Pareto analysis, force field analysis, and customer and supplier visits.

Step 8. *Obtain management endorsement to transfer the sourcing strategy into operation and implement it.* This strategy should now be transferred from the cross-functional team to operations management for implementation. The "selling job," which is often required to facilitate change, is reduced by the ongoing involvement on the team of those affected. The strategy should include, at a minimum, the following: scope (global, regional, local), terms and condition of agreement, and method of end-user release. A dry run or a pilot test should be conducted to demonstrate feasibility of concept. Once the pilot has been implemented and the feasibility of concept demonstrated, the revised process should proceed through a site-by-site acceptance test and implementation. Some training will be required. Available tools are executive briefing, pilot testing, process debugging, acceptance testing, and training.

The planning phase of the sourcing initiative in all likelihood resulted in some consolidation of the supplier base, where cross-divisional or multiple business units identified opportunities to exercise economies of scale by consolidating similar purchasing activity with fewer suppliers.

The following is an illustrative example of the planning process applied to the sourcing of personal computers at a financial institution.

Data analysis indicates that most PCs are purchased at local computer stores from the winner of a three-bid competition. As a result, there is little standardization in the hardware and software used at the institution. PCs are historically purchased in small quantities, generating significant work for the purchasing, accounts payable, and information technology support groups, who acquire, pay for, install, maintain, and manage the equipment.

Analysis reveals that purchase price is actually a fraction of the total cost of ownership of the personal computer. Equipment support, software evaluation, training, and inventory control also represent significant hidden costs.

In this case, the sourcing process recommendation is to standardize the equipment and software, negotiate purchase and service agreements with a single computer distributor with wide geographical coverage, and limit purchases to semiannual bulk acquisitions. Several local charities are identified for the donation of obsolete equipment. The supplier with the agreement now has specific key performance indicators by which its performance can be measured and monitored.

Control for Supplier Relations

Certainly control must be applied at the supplier level to ensure that the producer will receive defect-free products. Essential to this assurance is measuring and establishing adequate process capability—Cpk of greater than 1.33 or Sigma level greater than 4σ. This level of capability is usually only achieved by the application of robust design methods, such as Design for Six Sigma (DFSS), and/or breakthrough improvement methods like Lean and Six Sigma.

Despite a capable process, special causes may sometimes result in the loss of process control. When this happens, a good control plan that provides for quick restoration of control should be a requirement that is part of supplier acceptance criteria. Central to the quick restoration of control is a robust root cause and corrective action (RCCA) approach. The wise producer will apply his resources to help suppliers attain adequate process capability and robust control plans.

Control is applied to supplier relations in evaluating supplier performance and selecting the vital few suppliers capable of optimizing performance. As in planning, the focus of control must be the satisfaction of customer needs. However, as a result of the completed planning process, several criteria for performance evaluation and measurement are already in place. The purpose of control is to maintain acceptable performance. Applied to supplier relations, the purpose of control is to maintain the level of customer satisfaction at the level defined in the planning process.

The suppliers identified in the planning process are typically those suppliers that can perform the revised sourcing process. A thorough, ongoing evaluation conducted by a cross-functional team further narrows the supplier base and helps facilitate the selection of those few suppliers who will be able to optimize the total cost of ownership of the commodity. Therefore, it is in the application of control that the evolution begins from the traditional purchasing approach toward supply-chain management.

Control is a process requiring:

- Clearly defined supply chain quality goals established in planning
- Extensive, ongoing data collection and evaluation of the performance of the suppliers against these supply chain quality goals
- Corrective action where required

Most successful sourcing control processes follow a methodology similar to the following:

Step 1. *Create a cross-functional team.* The cross-functional control team includes customer, purchasing, technical, and operation personnel. Its mission is the ongoing management, measurement, and evaluation of the performance of the supply chain process established by the planning team during the planning phase. The team will initially need to identify quality goals and key performance indicators. Extensive customer involvement with the team should be expected. Available tools are brainstorming, team building, flow charting, data collection, and management presentation.

Step 2. *Determine critical performance metrics.* Performance metrics will have been proposed in the planning phase. However, the control team will need to identify and establish processes for capturing and reporting this information. Extensive supplier involvement should be expected in this step. Available tools are data collection, flow charting, check sheet, run chart, scatter diagrams, and process capability indexes.

Step 3. *Determine minimum standards of performance.* In addition to critical performance metrics, the team establishes minimum standards for suppliers before they are considered for further strategic development. These standards would likely include several financial, legal, and environmental considerations. Some minimum acceptable quality standards might also be proposed, such as percent defective, warranty performance, and delivery considerations. These minimum standards, along with the critical performance metrics established in Step 2, are communicated to both the customer and supplier community (for more elaboration see the section, "Implementing and Using Supplier Scorecards"). Available tools are brainstorming, data collection and analysis; management, supplier, and customer presentation.

Step 4. *Reduce the supplier base.* The team eliminates suppliers unable to achieve the minimum performance requirements, and shifts activity to suppliers who do achieve those performance standards. Through the application of the minimum standards of performance, the control process offers another opportunity for reducing the supplier base. Available tools are data collection and analysis, and management presentation.

Step 5. *Assess supplier performance.* Based on actual supplier performance, begin the process of the ongoing evaluation and assessment of the performance of the remaining suppliers. This typically involves evaluations of supplier quality systems now in place, supplier capacity and capability, and fitness for purpose of the commodity being supplied.

Supplier assessment comprises three separate but interrelated assessments, undertaken by the cross-functional team. These assessments ensure conformance to quality and performance standards, and establish a baseline for the improvement process.

Assessment 1. *Supplier quality systems assessment.* This assessment evaluates the quality systems the supplier currently has in place. It requires a visit to the supplier site by an evaluation team or by a third party who will certify the quality system as acceptable. This assessment should evaluate the supplier's:

- Focus on customers' needs
- Management commitment to total quality management
- Defined, documented, and fully implemented quality system
- Employee empowerment in terms of monitoring their own work for defect
- Use of fact-based, root cause analysis to investigate and correct quality problems
- Programs to encourage and evaluate quality improvement with their suppliers
- Commitment to continuous improvement in all phases of its operation

Cost considerations may favor reliance on a third-party supplier certification instead of an evaluation by employees of the purchaser. Where this is done, it is important that the end-user organization clearly understand what this certification does and does not include. The standards for supplier certification most often referred to are

- The ISO 9000 standards (ISO 9001, 9002, 9003), designed as models and guidelines of the minimum requirements for an effective quality system (see Chapter 16, Using International Standards to Ensure Organization Compliance).
- The ISO 14000 standards, designed as models and guidelines of the minimum requirements for an effective environmental system
- Quality System Requirements QS-9000, developed by the Chrysler/Ford/General Motors Supplier Quality Requirements Task Force. It is based on ISO 9000 standards, to which may be added automotive interpretations and further requirements (for example, continuous improvement and advanced product quality planning).
- Quality System Requirements AS-9100, developed by the International Aerospace Quality Group (IAQG). It is also based on ISO 9000 standards, to which may be added aerospace requirements necessary to address civil and military aviation and space needs (for example, regulatory agency roles and responsibilities and aerospace material traceability and accountability systems.)
- The Malcolm Baldrige Assessment, designed for applicants of the U.S. Malcolm Baldrige National Quality Award. It evaluates the process systems in place and the underlying organization and cultural issues of leadership, degree of empowerment,

and utilization of information and information technology in place to facilitate quality planning, quality control, and quality improvement (see Chapter 17, Using National Awards for Excellence to Drive and Monitor Performance).

Assessment 2. *Supplier business management.* This assessment evaluates the supplier's capability as an ongoing business entity to meet the end user's current and future business needs. This includes assessment of the supplier's current and future financial and operating performance. This assessment should evaluate the supplier with respect to:

- Research and development initiatives to ensure consistency with its customers' needs and future plans
- Cost structure to ensure financial health
- Production capacity to ensure ongoing ability to produce and distribute the required goods and services
- Information technology to evaluate willingness and capability to initiate information-sharing initiatives such as electronic data interface (EDI) and electronic funds transfer (EFT)

The assessment includes measurement of such indicators as debt-to-equity ratio, percent of profit reinvested in the business, inventory-to-sales ratio, employee turnover statistics, and capacity utilization.

Assessment 3. *Supplier process capability and product fitness for purpose.* This assessment evaluates the fitness for purpose of the product or service being supplied, as well as the supplier's process capability to consistently manufacture goods to stated needs. The focus is on quality, delivery, and service. Specifically, this assessment should evaluate:

- Conformance to customer requirements
- Process capability (Cpk or process Sigma)
- Key performance indicators

The assessment includes measurements of such indicators as the following:

- Percent of nonconforming products shipped
- Cycle times of key processes
- Customer satisfaction
- Identified and measured cost of poor quality

Available tools are supplier site visits, data collection and analysis, and third-party evaluations.

Improvement for Supplier Relations

The improvement phase includes:

- The management, measurement, and continuous improvement of the sourcing process
- The expansion of control and initiation of continuous improvement within the supply chain itself to ensure value creation

These improvement initiatives build on the foundations of quality, total cost of ownership, and supply chain management already established in the planning and control phases. Fundamental to improvement in the performance of the entire supply chain is that trust has been established between all parties in the entire supply chain, from suppliers through end users. The objective of the improvement phase is to develop a supply chain that acts as a single entity, develops common goals, formulates real-time decision making, measures performance through a single set of key performance indicators, and is collectively responsive to the needs of the end user.

With trust as the foundation, supply chain management and optimization can proceed. This sense of trust cannot be achieved by a single act of signing a long-term contract or by prominently displaying a banner indicating a commitment to quality. It must be demonstrated by behaviors and actions over an extended period. As the climate of cooperation grows, the degree of trust between all supply chain participants becomes deeper, and opportunities for value creation, joint problem solving, and innovation are identified and realized.

Five tiers of progression. In the control phase, the end user and suppliers have identified and flow-charted the entire supply chain. The continuous improvement phase generally progresses through five levels of cooperation: (1) joint team formation, (2) cost reduction, (3) value enhancement, (4) information sharing, and (5) resource sharing.

Level 1. *Joint team formation.* The improvement phase begins with the establishment of a joint (end user/supplier) team. Although the team could have several objectives, the initial focus should be on:

- Alignment of goals
- Analysis of the supply chain business process
- Identification and remediation of chronic problems

Goal alignment ensures that each link in the supply chain develops goals and objectives, and proposes initiatives whose focus is the needs of the end user. Furthermore, goal alignment and the activities associated with it are a natural first step in developing the synergies and trust required for further supply chain development.

In conducting the business process analysis of the supply chain, the team begins to identify the elements of the chain and collect data to measure its performance. This data collection should focus on the areas of the supply chain that have a high probability of generating quality problems, such as excessive cycle time, rework, and scrap, or which are likely to create customer dissatisfaction.

Supply chain business process analysis represents the initial steps of identifying the chain (typically using flow charting) and collecting data that describe the performance of this supply chain. This data-collection phase should focus on the areas of high probability of quality problems, such as cycle time, rework, scrap, or customer dissatisfaction.

Chronic problem identification and remediation offer a preliminary opportunity to work collaboratively on problem solving in this joint team environment. This offers a classic opportunity for a quality improvement team with membership from the various members of the supply chain. The team's efforts will likely result in near-term process improvement and enhanced customer satisfaction, and offer an opportunity for collaboration and trust to be nurtured within the chain itself.

Level 2. *Cost reduction.* Level 1 initiatives help create a culture of trust and collaboration between supplier and end user, especially as the result of the work of joint problem-solving teams. The teams were established to identify and gather the "low-hanging fruit," that is, reduce the occurrence of chronic problems in their joint business processes, which are relatively easy to solve, once identified. Level 2 requires an approach to process improvement in more depth, often involving suppliers to the supplier or customers of the end users. Proactive managing of the supply chain begins at this point to replace the bilateral relationship between end user and supplier.

A COPQ study of the supply chain provides powerful guidance for organizations engaged in cost reduction. The costs are usually sorted into three categories:

- External failure costs (i.e., warranty, customer dissatisfaction, recall costs)
- Internal failure costs (i.e., scrap, rework, rejected raw material, downtime costs)
- Appraisal costs (i.e., inspection, testing, verification costs)

For significant concentrations of COPQ revealed in the supply chain, joint teams are established to reduce those costs, project by project. As activities advance to a higher level, the activities of the lower levels continue. For example, as the chain moves into Level 2 and begins measuring and managing cost reduction opportunities, the tools and initiatives of Level 1 continue. This accumulating effect continues throughout the five levels.

Level 3. *Value enhancement.* As the teams begin reducing COPQ, the supply chain itself begins to function as a single business process, rather than as a set of separate ones. At this point, the team needs to flow-chart the activity of the supplier chain and evaluate the value added by each link in the chain. Two questions addressed at this stage are: Does this step add value? and What would happen if we were to skip this step? The nonvalue-added steps are identified and eliminated.

Level 4. *Information exchange.* At this point in the supply chain improvement evolution, what was traditionally treated as confidential information is being routinely shared and more widely distributed throughout the chain. Furthermore, electronic commerce tools such as EDI, Internet and intranet applications, and groupware applications such as Lotus Notes are facilitating the transfer of information, the collaboration of ideas, and real-time decision making.

Level 5. *Resource sharing.* In the latter stages of supply chain management and improvement, the "walls" that traditionally separated departments, divisions, and companies have been eliminated. Fewer are working in corporate silos; the supply chain is beginning to function as a single process—involving personnel from several different suppliers within the chain, from the customer's organization, and the end user. Personnel within the chain are routinely collaborating on ideas and improvement opportunities, and performance is continuously measured. Personnel from the various suppliers within the supply chain are often colocated with their customers to further facilitate this collaboration.

At the highest level of supply chain management, the extent of data, resource, and risk sharing has increased to a dramatic level. Not only are personnel colocated with their customers, but technology plans and risk-taking initiatives and investments are shared throughout the supply chain, and benefits and losses are jointly apportioned. A seamless supply chain process begins to emerge, generating value for customers, as well as suppliers.

Implementing and Using Supplier Scorecards

Effective supplier scorecards can help the producer improve the speed and reduce the cost of his supply chain. Key to effectiveness is the choice of appropriate criteria against which to measure your suppliers and metrics that directly reflect performance against these criteria. Nearly universal criteria are quality history, total cost, and timeliness of delivery, but other criteria may be important to add, depending on a producer's particular needs, for example:

- Design and technical capability
- Proximity to producing facility

- Responsiveness
- Past problem resolution effectiveness
- Audit results
- Price increase/reduction history
- Financial stability

Depending on what stage of supplier management the producer is at—planning, control, or improvement—different sets of criteria may be more or less applicable.

Often, a criteria-based selection matrix is used to rank each supplier (or potential supplier) of a particular part, component, service, etc. (see Figure 30.3 for an example). The producer must then set standards for what constitutes an acceptable score. A stoplight analogy is often used to represent supplier status, such as:

- 85–100, green: No action, continuing with preferred supplier
- 75–84, yellow: Supplier must develop and submit a timely improvement plan.
- 74 or below, red: Supplier must submit an improvement plan and will not be considered for future bids until score has improved to an acceptable level. If no acceptable plan is submitted or progress is not documented, current supplier work may be resourced.

Appropriate metrics must be designed if they do not already exist. The metric must reflect measurable characteristics of the criterion and be easily applied to the scoring system in use by the producer. For example, a metric for quality history might be as shown in Table 30.4.

It is recommended that the rating system not be overcomplicated. Keep the scorecard criteria to only those things that are meaningful indicators of the supplier's value to the producer (usually five or six criteria). If a scorecard system is made too difficult for supplier quality or purchasing personnel to use, its value to the organization will be reduced.

The criteria, once established, must be evenly applied. Some supplier rating systems give latitude to the supplier quality engineer whether or not to issue a formal corrective action request to the supplier, which would negatively affect the supplier's score. This

ABC company-supplier scorecard

Product family supplied: Widget — Updated: 10/12/09

		Suppliers					
Criteria	**Weight**	**A**	**B**	**C**	**D**	**E**	**F**
Quality history	3	9	8	10	7	9	7
Total cost	2	8	10	9	10	9	9
Timeliness of delivery	2	8	8	7	9	9	7
Financial stability	1	9	8	5	8	9	10
Audit results	2	6	7	7	6	8	7
	Total score	80	82	81	79	88	77

Figure 30.3 Supplier scorecard.

Number of SCARs	Supplier Rank
0	10
1–2	9
3–4	8
5–6	7
6–7	6
7–8	5
8–9	4
10–11	3
12–13	2
14 or more	1

TABLE 30.4 Eighteen-Month Rolling Number of Supplier Corrective Action Requests (Metrics)

latitude leads to arbitrary decisions and erroneous measures of the supplier's performance, thereby lessening the value of the scorecard.

Auditing Suppliers to Support the Scorecard

A regular auditing program to monitor supplier continuous improvement progress is important to add validity to the supplier scorecard. Particularly for suppliers of key or critical components, audits should be conducted at specified intervals of no longer than biannually. If your organization has supplier quality engineers (SQEs) who are responsible for a specific set of suppliers, it is often wise to have others who are not as familiar with those particular suppliers participate in the audits along with the responsible SQEs to minimize possible biases and bring a fresh set of eyes to the evaluation. If resources are an issue, the producer should consider engaging third parties in the auditing effort to get that unbiased view.

The checklist used for a supplier audit should include all the areas of the supplier organization that could affect the quality of supplies and components, including:

- The strategic planning process and its effectiveness
- The quality management system, its appropriateness and effectiveness
- The efficiency and effectiveness of the processes used to measure, control, and improve product and process quality
- The supplier culture (i.e., is it one that supports continuous improvement and process excellence?)
- Human capital management (does the supplier practice employee involvement and the principles of self-control?)
- The quality of products and services provided by this supplier (complaints, rejections, corrective action requests, etc.)

- The efficiency and effectiveness of the supplier's own supply chain, including supplier scorecards, audits, and quality records
- The efficiency and effectiveness of the supplier's product and service creation process
- Supplier delivery performance
- Measurement systems that obtain appropriate measures based on user needs and relate to key business drivers and strategies
- The supplier's understanding of customer needs and measures of how well those needs are being met
- The supplier's analysis of competitors' strengths and weaknesses and how they compare
- The effectiveness of supplier benchmarking activities that help understand best-in-class performance
- The supplier's understanding of the cultural behaviors and norms that are needed to create a customer-oriented culture

Other supplier attributes may be uniquely important to your producer organization and should be added as appropriate.

The scoring of the supplier audit should be made as objective as possible. This can best be accomplished by setting clear "if, then" rules to guide the auditor's decisions. For example, perhaps the scoring is done on a 1 through 10 scale and a rule might be "if the supplier's delivery performance is 95 percent or greater on time, then assign a score of 10. If 90 percent or more but less than 95 percent, then assign a score of 9," and so on. The key is to take as much subjectivity out of the scoring decision by establishing clear rules for each category. The producer can also weight the importance of each element in the audit by assigning a greater number of points possible for the key audit elements.

The producer should also establish acceptance limits for the supplier audits. If a supplier falls below the acceptable level on any audit element, then an improvement plan addressing each deficient area should be required. The savvy producer will lend assistance to the supplier to help them improve and meet the producer's goals for its suppliers. The resources for that assistance can come from within the producer's organization or from third parties skilled in supplier development and improvement.

References

Carlisle, J. A., and R. C. Parker. (1989). *Beyond Negotiation*, John Wiley & Sons, Chichester, England.

Deming, W. E. (1981). Seminar notes for "Japanese Methods for Productivity and Quality," Course No. 617, W. Edwards Deming, Washington, D.C.

Fitzgerald, K. (1995). "For Superb Supplier Development," *Purchasing Magazine*, September 21, pp. 32–40.

Juran Institute, Inc. (2009). "Quality 101: Basic Concepts and Methods for Attaining and Sustaining High Levels of Performance and Quality (version 4)," Southbury, CT, p. 8.

Juran, J. M., and Godfrey, A. B. eds. (1999). *Juran's Quality Handbook*, 5th ed, McGraw Hill, NY, pp. 21.2–21.3, 21.5–21.7, 21.8–21.9, and 21.17–21.25.

Mantle, J. (1995). *Car Wars: Fifty Years of Greed, Treachery and Skulduggery in the Global Marketplace*, Arcade Publishing, NY, p. 168.
Phelps, T., Hoenes, T., and Smith, M. (2003). *Developing Lean Supply Chains: A Guidebook*, Altarum Institute, The Boeing Company, and Messier-Dowty, Inc., pp. 57–58.
Rigsbee, E. (2000). *Partnershift: How to Profit from the Partnership Trend*, John Wiley and Sons, NY, p. 112.
Tully, S. (1995). "Purchasing's New Muscle," *Fortune Magazine*, vol. 131, no. 3, Feb. 20, pp. 75–83.

CHAPTER 31

Role of the Board of Directors: Effective and Efficient Governance

Marcos E. J. Bertin and Marcos Bertin Schmidt

About This Chapter

The role of the board of directors is to guide management in developing plans to maximize the stakeholder's satisfaction in line with sound organization sustainability and control that these plans are properly implemented.

What a board of directors does, or does not, do has an impact on the organization and on the organization's performance, including quality results. Unfortunately, this is not common knowledge to all boards, but it will become clear when analyzing the evolution of corporate governance, as discussed in this chapter.

This chapter addresses the quality of the board of directors (or equivalent governance body) and provides guidelines on:

- Planning corporate governance implementation
- Improving the performance of the board of directors
- Becoming an effective director

The chapter should also prove useful for those who need to interact with board members to understand their role and expectations.

High Points of This Chapter

1. The evolution of corporate governance and how it influences an organization's practices and results is key to understanding why boards act the way they do.
2. Effective corporate governance provides proper incentives for the board and management to pursue objectives that are in the interests of the organizations and its stakeholders.
3. The approach to implement corporate governance should be based on identifying what is good for the organizations' organization, which definitively includes complying with regulations.
4. There are four levels of evolution of corporate governance implementation required to create an effective board:

   Level 1: Understanding (qualification: 0-1) The board understands the need to improve corporate governance in the respective point or area.

   Level 2: First steps (qualification >1-3) The board has taken concrete steps toward establishing best practices in the respective point or area.

   Level 3: Implementation (qualification >3-7) The board has implemented improvements to corporate governance in the respective point or area.

   Level 4: Leadership (qualification >7-10) The board has reached the best achievable improvements to corporate governance in the industry for the respective point or area.
5. The most successful global organizations today recognize the key guidance and control role the board has to balance the alignment of the interests of all stakeholders to assure sustainable growth and profitability.

Corporate Governance Evolves and Has an Impact on Organizations

Corporate governance is a dynamic issue. Here we discuss its evolution and its influence not only on organization practices but, more importantly, results, starting a few millenniums in the past all the way to the realm where Juran and Louden's (1966) vision materializes.

The Historical Responsibility to Shareholders

Modern organizations developed in the middle of the nineteenth century after a long history since its birth as far back as 3000 BC in Mesopotamia. However, it was not until the beginning of the twentieth century that the gradual separation of ownership from control took place: the shareholders who own the organizations from their agents who run them; this was the birth of the professional manager. The most important characteristics were:

- The adoption of Frederick Taylor's first scientific rationalistic approach to management that resulted in a remarkable improvement in productivity.
- The introduction of the humanistic school of management thinkers led by Elton Mayo.
- A new class of management consultants such as Arthur D. Little, James McKinsey, and Peter Drucker that contributed to the development of the Organizations Man following Watson's IBM and Sloan's General Motors, successful models at the time.

- The impact of the research, courses, conferences, and publications organized the American Management Association founded in 1926.

No reference is found on the board of directors and its members in the huge amount of management literature generated during this period. Also, there were no other forms of professional developments for corporate directors as it was available for financial, commercial, human resources, or other management activities, including American societies for security analysts, quality control, and many other specialities. The National Association for Corporate Directors was founded in 1977. This confirms that in this "Organization Man" period, corporate directors believed that success depended only on the ability of management and that the role of the board of directors was solely to protect the interests of the owners. We review the limited view of the responsibilities of boards in the 1960s.

The 1960s: Juran and Louden Board Professionalizing Initiatives

In Sept.-Oct. 1961, The *Harvard Organization Review* published the first Code of Conduct for Executives. The President's Association of the American Management Association starts also in the 1960s to organize seminars for board members. A significant amount of consulting work and coaching is made with boards. These experiences resulted in the first book ever published on what we today call corporate governance by authors Dr. Joseph Juran and J. Keith Louden in 1966. This remarkable book was *The Corporate Director*. It is interesting to note that Dr. Juran was a member of several boards.

One of the two authors of this chapter received this book from Dr. Juran in 1969; this was the beginning of a lifetime interest in corporate governance and a valuable guide for the "what" and "how" of a long "hands-on" experience.

It is amazing to see that this book, most probably one of the "vital few" at the time, includes full chapters on subjects that today are required by government regulations and organizations: codes of best corporate governance practices worldwide. This includes, for example:

- Composition of the boards
- Formalizing the jobs of directors and boards
- The professional director
- Organizing the boards and their committees
  - Board meetings
  - The CEO and the board
  - Maintaining a healthy board

The final conclusion of Juran and Louden (1966) has become a reality today more than 40 years later:

> The present practice is mainly empirical. The men work from experience and instinct and many of them do a good job on it. But the growing importance of the board and the resulting need for ever improving performance by directors, suggest that empiricism has had its day. The board of tomorrow will work from a base of professionalism.

The "Middle Ages"

Unfortunately, during this period there was a change back to strong management operating in an economy going through a deregulatory revolution and where privatizations and conglomerates were in fashion. Corporations also turned a blind eye to corporate governance questions and to antitrust considerations. In this economic environment, the buccaneering spirit and the excesses of some CEOs were the cause of two serious financial crises.

Today: Learning from the Consequences of Short-Term Profit and Other Fashions

We see the revival of corporate governance as Juran and Louden visualized it back in the 1960s. The question of how to align the interests of those who ran the organizations with the interests of those that owned has returned. Therefore, the need to enforce proper checks and balances, taking in consideration all stakeholders by developing and implementing sound and effective governance practices and regulations, has become paramount.

The revival of corporate governance started due to two well-known economic crises with the preparation and publication in 1999 of the *OECD Principles of Corporate Governance*. The first revision was in 2004 based on five years of applied experience. These principles assist governments in evaluating and improving a legal, institutional, and regulatory framework for corporate governance in their countries and to provide guidance and suggestions for stock exchanges, investors, corporations, and other parties that have a role in the process of developing good corporate governance. The principles represent a common basis essential for the development of good governance practices in all types of organizations.

Corporate governance provides the structure through which the objectives of the organizations are set and the means of attaining those objectives and monitoring performance are determined. Good corporate governance should provide proper incentives for the board and management to pursue objectives that are in the interests of the organizations and its shareholders and should facilitate effective monitoring. The presence of an effective corporate governance system, within individual organizations and across an economy as a whole, helps to provide a degree of confidence that is necessary for a market economy to function properly. As a result, the cost of capital is lower and firms are encouraged to use resources more efficiently, thereby underpinning growth.

The OECD Principles of corporate governance involve six basic areas:

1. Ensuring the basis for an effective corporate governance framework
2. Rights of shareholders and key ownership functions
3. Equitable treatment of shareholders
4. Role of stakeholders in corporate governance
5. Disclosure and transparency
6. Responsibilities of the Board

There is a strong tendency to increase and enforce new regulations at the national and global level. It is interesting to observe the gradual globalization of regulations and the need the organizations have to update their corporate governance practices to comply with the increasing demands of investment funds , banks , major customers, and, of course, governments regulatory bodies.

On the other hand, research made by corporate governance institutes in an increasing number of universities, for example, Yale in the United States, St. Gallen in Switzerland, and Nankai in China and corporate directors associations such as the National Association of Corporate Directors (NACD) in Washington, D.C. and the Institute of Directors in London definitely prove the significant contributions that the application of corporate governance best practices have on organizations' results:

- Better corporate governance at both the firm and the country level results in higher valuations.

- Better corporate governance increases the variety of financing instruments available.
- Better corporate governance increases the effectiveness of management improving their performance.

Planning to Implement Corporate Governance

The approach to implement corporate governance should be based in identifying *what is good for the organization's organization*, which definitively includes complying with regulations. There are no recipes. Also, there are many different situations. Thus, after presenting a descriptive model and a couple of useful governance bodies we will provide guidelines from two different angles:

- Family organizations issues and requirements
- Stock market issues and requirements

Ancient Symbols for a Current Governance Model

In the previous section, it should have become clear about the relevance stakeholders currently have and how critical the board's job is in aligning interests. Here we discuss a pertinent model to continue exploring these and other relevant concepts. We also refer to it when explaining the *Systematic Improvement Process* we present in a section later in the chapter.

Again going back a few millennia in time, a step pyramid was built in Saqqarah, Egypt. Unlike the triangular shapes that were commonly used, this pyramid clearly differentiated the layers within the structure (see the comparison in Figure 31.1).

As for the board, what better than to use another ancient symbol, one embedded in dollar bills, the All-Seeing Eye and the stakeholder environment? It simply surrounds the organization.

Now let us take a look from another direction (see Figure 31.2). At its center is the board (the triangle), then senior management, and then the lower levels within the organization. Finally, on the outside there's the stakeholder environment.

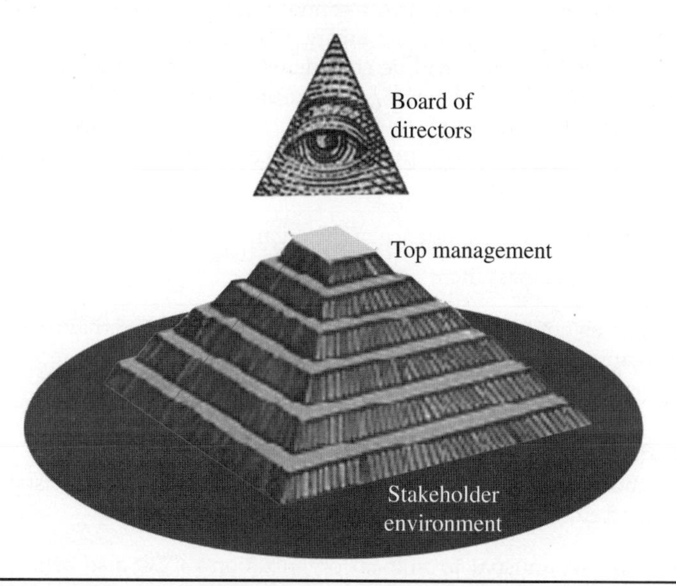

FIGURE 31.1 Step pyramid in perspective.

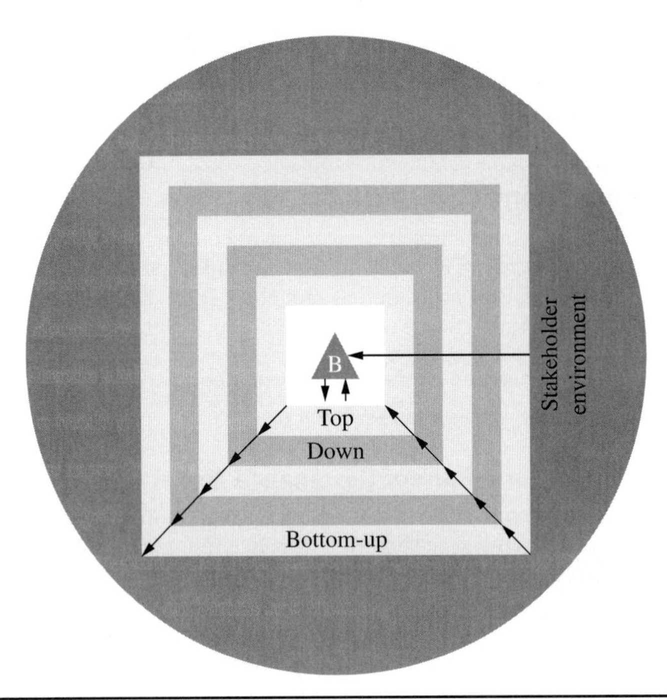

Figure 31.2 Step pyramid view from above.

The arrows pointing from the top down refer to simple policies and rules that originate at board level and are passed to senior management for subsequent implementation and communication to the lower levels of the pyramid. The arrows pointing from the bottom up represent the feedback traveling back.

What about the outside-in arrow connecting the stakeholders and the board? It is very important. Directors must become aware about stakeholder expectations, evaluate them, and then determine which will be considered in the organization's strategy. Stakeholders then become part to the top-down / bottom-up process.

Board actions must be top-down to ensure that strategies are clear, easy to understand, budgeted, and in line with corporate objectives. The board does not implement; its main functions are to:

- Provide strategic guidance of the organization
- Monitor management

Now, let us not forget that implementing corporate governance is about "what is good for the organization's organization." The lack of proven infallible recipes does not mean that there are no useful resources available, such as the following.

Governance Bodies for Medium and Small Qrganizations

Among the most outstanding corporate governance characteristics in faster-growing medium and small organizations, we can highlight that:

- It is not unusual to find an organization's CEO also acting as chairman of the board.

- Among the most important functions the board performs is advising senior management, right behind strategic planning.
- Boards do not have committees, or if they do, it is just the auditing committee.
- Many boards have a "hands-on" approach (absolutely not recommended; see Figure 31.1), where directors complement management functions and get involved in running the organizations.

Hence, it becomes obvious that to achieve corporate governance, at the very least boards will need to go through a transition. Something similar can be said regarding many family organizations. The entitities we describe here can help this and other purposes.

Advisory Boards

"Advise" is the key word. Advisory boards act as an informal think tank and have no legal responsibilities (see Figure 31.3). Advisory boards have the same requirements as professional boards, and their effectiveness depends on the development and application of best practices based on well-documented processes that add value as explained in other sections in the chapter. Advisory boards can also be a first step for those organizations in transition to integrate a formal legal board.

Seeking good advisory boards early in an organization's development can both ease and quicken the planning process. In the case of family organizations, an advisory board has the advantage of being detached from the family and can be more objective when it comes to who within the next generation is more capable of running the organization.

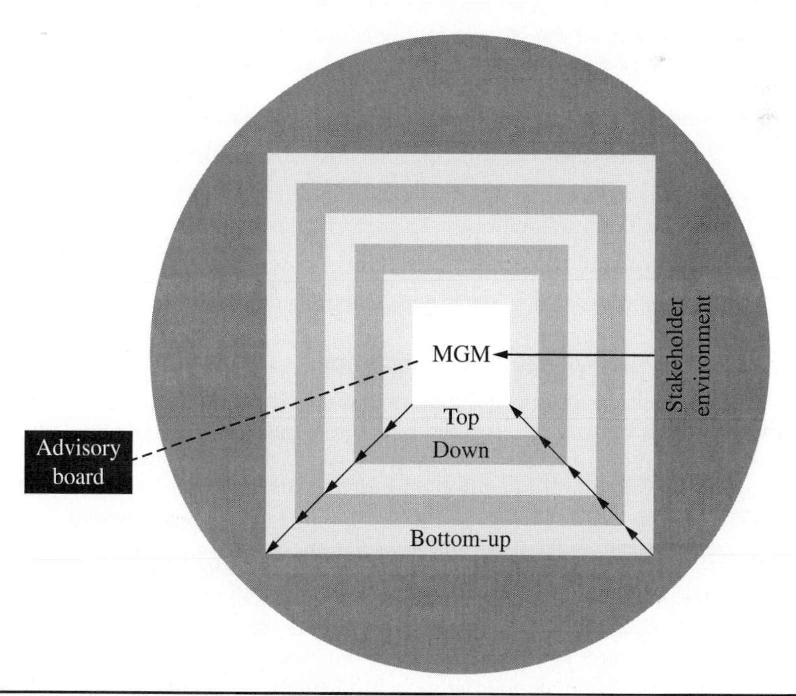

FIGURE 31.3 Advisory board relationship.

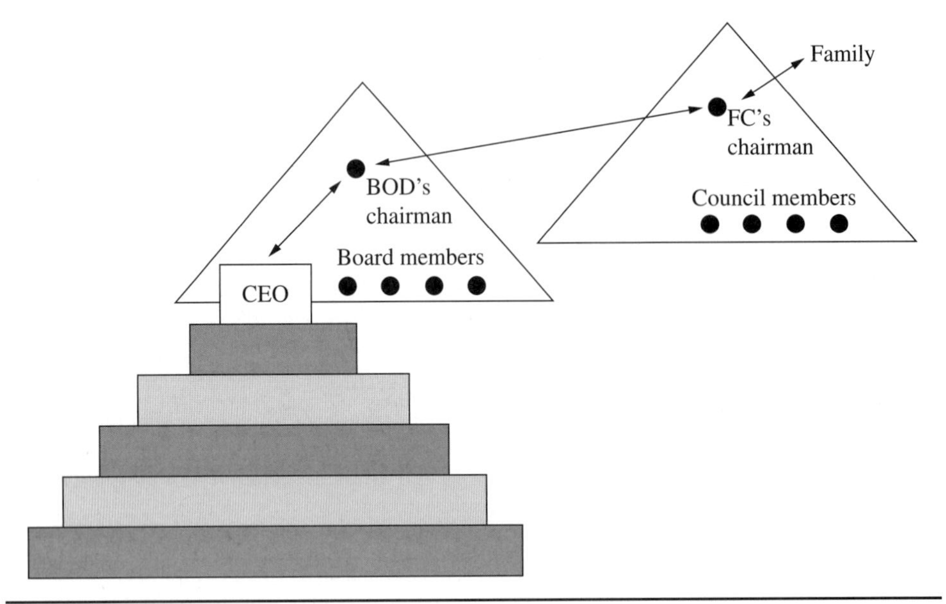

FIGURE 31.4 Family council relationship.

Family Councils

As we address in the following section, the organization's governance in family-owned businesses becomes more complex when more family members are directly or indirectly involved in the organization. Successful family organizations in many countries have, in addition to a board of directors, a family council that represents the family's shareholders (see Figure 31.4), which have the following functions and responsibilities:

- Holds formal meetings to discuss family and family-owned organization issues.
- Establishes rules for family participation and guidelines for the board.
- Becomes the only means of communication between the family-owned organization and the family shareholders.
- Knows about the course of the family-owned organization/organizations.
- Avoids individual contacts with management.

Planning for Family-Owned Organizations

A family organization is one of the foundations of the world organization community. They represent over 70 percent of the total registered organizations. The moment to evaluate the implementation of corporate governance is when the successful founder/s start planning retirement and the family members decide to continue with the organization. Steps that successful family organizations worldwide usually take are as follows:

- Initially with the collaboration of an outside consultant, the owner of the organization prepares a master plan that includes new members, a family council (FC), a professional board of directors, and potential members (family, key managers, independent external directors), including desired qualifications and needed training. If there is an existing board usually is a "rubber stamp" type that meets the minimum law requirements for organizations not listed in the stock market.

- Organize an FC with well-defined family directives to the board and formal operating procedures. This FC should be the only mean of communications between the family shareholders and the board.
- It is frequently preferred to have an advisory board before appointing a formal professional board. This facilitates the transition with family and management and the search for professional independent directors who might not be willing to become a board member with all the legal responsibilities involved in an organization they do not know much about.
- The right people are the key to the success of this project. In the case of the family members, they have to realize they can be managers, board members, FC members, or just shareholders. This depending on their background, preferences, capabilities and training. Some family members might be delighted to become marketing managers and others might prefer to remain as shareholders in the organization. The number of family members tends to increase with each generation. When that's the case, it is also important to prune the family tree by buying the stock of members not interested in the organization.

To select external independent directors for the initial advisory board or the formal legal board in addition to the necessary experience and knowledge requires that independent directors have the following:

- A personality that earns the respect of his or her peers, family directors, and executive managers
- The ability to work effectively as a team
- The ability to challenge management while avoiding confrontation and refraining from invading their functions
- The ability to improve the chances to correct the sources of family conflicts

Planning for the Stock Market

When an organization is preparing to be listed in the stock market, the board generally needs to improve to achieve a better rating. If so, the next section of this chapter, "Systematic Corporate Governance Improvement," should prove very useful, as it provides a shopping list from where to choose what a particular organization requires to comply with regulations and to align the interests of all stakeholders to ensure a sound profitable growth.

Once these objectives are accomplished, transparency and good communications are the keys to succeed in stock markets.

Again, it is essential to have the right people in place. This requires the identification of the most important qualifications each board member must have, including:

- A solid background and experience in different areas and at least literacy in legal, finance, and IT matters for those strong in marketing, distribution, and technical areas
- Awareness of the importance of customer loyalty
- Awareness of the importance of diversity of skills
- Awareness of the importance of a periodical evaluation of the board and its members

Board committees, particularly governance, auditing, and compensation, should be integrated with a majority of independent professional board members and be allowed to invite management and engage outside consultants when necessary.

Systematic Corporate Governance Improvement

Improve by Focusing on Results

Having already discussed the impact that corporate governance has on an organization, the logical step is to figure out how to improve that governance. Actually, there are several viable diagnostic alternatives, each with its particular strengths and weaknesses, such as self-assessments or 360° evaluations. However, we felt that a TQM environment may require a special emphasis over results, which are a board's product, in a way.

To that extent, we have based ours on the method described in *An Approach to the Evaluation of a Board of Directors* by Marcos E. J. Bertin and Hugo Strachan, published in 2005 by The International Academy for Quality (IAQ). Although it was originally intended as a director's self-assessment, we have already successfully applied this concept to improve the quality of boards of several organizations.

It is important to note, that among its itemized categories, unlike others, this method includes board contributions (to results), which is probably the most relevant and the key to our scope. It covers two angles:

- **Contributions on organizations results:** Successful critical strategies generated and controlled by the board that should be evaluated as well as key performance indicators, including intangibles. That is, the following items:
  1. Organizations financial results
  2. Competitive access to capital
  3. Performance indicators
  4. Brand value/organization image
  5. Organization's intellectual capital value
  6. Risk management
- **Stakeholder's evaluation of the board:** Stakeholders include management, personnel, major suppliers, customers, government, society, and controlling and minority shareholders. It is the following two items:
  1. Stakeholders' evaluation of the Board
  2. Community perception of the organizations as a whole

Note that this is "outside-in" input; thus, the board must define how the stakeholders' opinions are determined and evaluated.

The Four-Level Scoring Criteria Applied to Results

The chosen method encourages scoring according to the level of evolution of corporate governance implementation that has been reached, that is:

- **Level 1**: Understanding (qualification: 0-1) The board understands the need to improve corporate governance in the respective point or area.
- **Level 2**: First steps (qualification >1-3) The board has taken concrete steps toward establishing best practices in the respective point or area.
- **Level 3**: Implementation (qualification >3-7) The board has implemented improve–ments to corporate governance in the respective point or area.

- **Level 4**: Leadership (qualification >7-10) The board has reached the best achievable improvements to corporate governance in the industry for the respective point or area.

This scheme helps to narrow the variance in estimating values. The following questions are also designed to reinforce that effect:

1. *Organizations financial results*. Are they consistent with shareholders'/owners' expectations? If not, how proactive is the board in aligning them with their needs?
2. *Competitive access to capital*. Does the board assist in this regard? If not, how aware is the board of its role?
3. *Performance indicators*. Are they already in the desired state? Is the board helping the organization to further improve performance?
4. *Brand value/organization image*. Have goals been met yet? Is the board proactively working on improving them? How close are they to align the organization's image with industry standards?
5. *Organizations' intellectual/intangible capital value*. Are these capitals a board's concern? Is it active in generating/preserving such capital? Are results good enough?
6. *Risk management*. Is risk management a key factor in an organization's success? If so, does the board have any merit?
7. *Stakeholders' evaluation of the board*. Do they perceive that critical/needed strategies are implemented? At least improving? How relevant is the board in generating them?
8. *Community perception of the organization's as a whole*. Is the organization evaluated? If so, is it perceived as veing good enough? If not, is it at least improving?
   - Any board of directors that evaluates these or similar questions regularly enough has definitively taken a step in the right direction.
   - As for metrics, a form such as the one in Figure 31.5 could be used. In this case for board contributions, note that on the right side, we include an example for clarification purposes. Qualifications are represented by the *Q* column in Figure 31.5 and their respective levels by the *I-V* column.
   - Although a resulting level of *III* in Figure 31.5 may seem fine, being one the Board's most relevant functions, a level of *I* for stakeholder evaluation would be inadmissible. This shows why each and every board contribution item should be seriously considered throughout the improvement method we describe as follows.

The Result-Oriented Improvement Method

The processes of corporate governance are interdependent and complex and involve all stakeholders. Although a handful of organizations have research programs regarding establishing their metrics and variability, it may take a while and it is still yet to be seen how practical they will be, especially to boards of small- and medium-sized organizations. Thus, we take a different approach, "Improvement through Awareness."

Further along this section we supply a list of categorized subjects for frequent review. The idea is to identify issues that could potentially influence a board's capability to contribute to results, the eight points we discussed earlier. Although experienced directors may find this easier, the less experienced would gain familiarity. Given the divergence

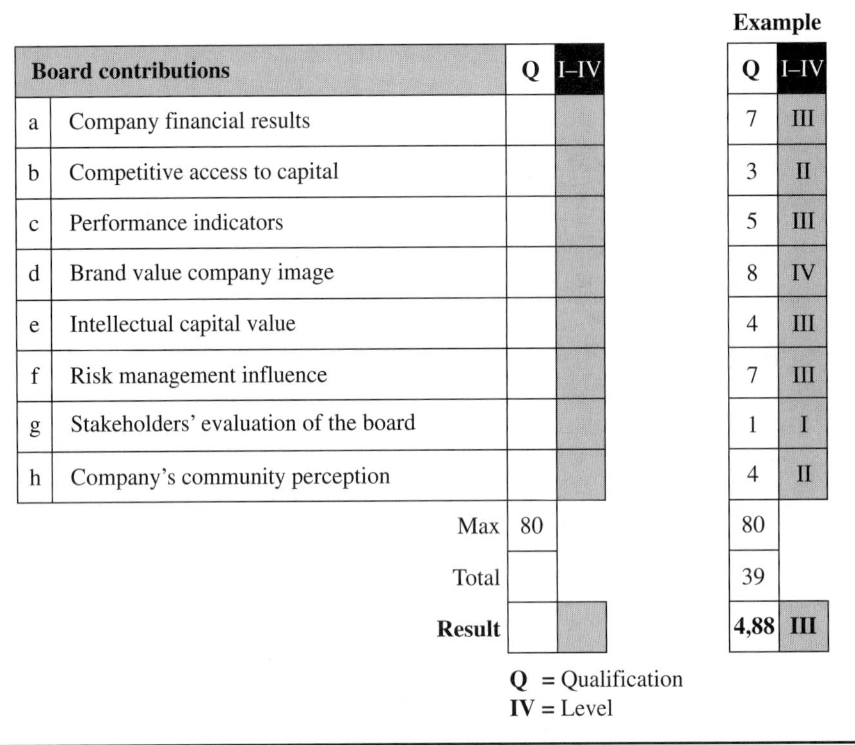

	Board contributions	Q	I–IV	Example Q	Example I–IV
a	Company financial results			7	III
b	Competitive access to capital			3	II
c	Performance indicators			5	III
d	Brand value company image			8	IV
e	Intellectual capital value			4	III
f	Risk management influence			7	III
g	Stakeholders' evaluation of the board			1	I
h	Company's community perception			4	II
	Max	80		80	
	Total			39	
	Result			**4,88**	**III**

Q = Qualification
IV = Level

FIGURE 31.5 Calculations form.

some board matters tend to have, such as stakeholder interests, it is important to refocus periodically.

A form like the one in Figure 31.6 should prove useful to:

- List the potentially relevant subjects (left rows)
- List respective board contributions (middle columns)
- Mark identified relationships between subjects and contributions (crosses in row intersections with columns)
- Quantify using the four-level scoring criteria described earlier

The assembled information would then allow for:

- Disregarding subjects that ended up not being that relevant (remove crosses)
- Adding newly identified cause-effect relationships (add crosses)
- Establishing improvement goals and priorities (right and bottom bars)

This would provide a full and single view of many of the probably most important practical board matters.

Categorized Subjects

Purposely, to avoid biasing, sample rows do not mention any real issues. In the example in Figure 31.6, items are chosen from hypothetical categories 4 to 6. In the first row, it is the third item of category 6.

	Identified relevant items	Q	I-IV	Board contributions: Company financial results	Competitive access to capital	Performance indicators	Brand value company image	Intellectual capital value	Risk management influence	Stakeholders' evaluation of the board	Company's community perception	Goal (I-V)	Priority	Goal met?
				III	II	III	IV	III	III	I	II			
6.3	Sample item: Third in category 6	3	II		x			x				III		
6.4	Sample item: Fourth in category 6	5	III					x		x		IV		
5.2	Sample item: Second in category 5	4	III											
4.1	Sample item: First in category 4	1	I	x					x			III		
4.4	Sample item: Fourth in category 4	4	II											
			Goal	IV	III	III	IV	IV	IV	III	III	Goals to reach during the following period		
			Priority											
			Met?											

Q = Qualification
IV = Levels

FIGURE 31.6 Cross-reference improvement form.

Because every organization has different objectives and needs, each one should eventually be able to select what serves their own requirements most in this chapter. These categorized subjects are intended to serve that purpose by inspiring useful topics for the board's agenda, or for custom self-assessment questionnaires or to use in the improvement process described earlier (see Figure 31.6) and are chosen from the following categories:

1. Mission and principles
2. Board and stakeholders
3. Board and shareholders
4. Board and management
5. Operating procedures
6. Board structure

There are also board subjects regarding IT governance, but, for didactic matters, they are treated separately in the "Corporate and IT Governance" section of this chapter, where in addition to the pertinent subjects, we supply a sample form (see Figure 31.7) and suggest a few guiding questions for each. This should give an idea on processing whole categories. A thorough description of all categories is available in Bertin and Watson (2007).

Categories here are numbered in reverse order, not just to catch the reader's attention, but because it represents moving from outside the pyramid (see Figures 31.1 and 31.2)—Stakeholders, all the way to the core, the Mission, Principles and Values of the organizations. This means that the depth of any analysis would depend on category choices, the subjects of which are listed as follows.

Mission and Principles

Proactive establishment of mission and principles, their deployment and employee commitment, liabilities awareness, risk anticipation and mitigation, risks and crisis management policies, integrated risks and compliance vision, and code of best practices.

			Q	I-IV	Board contributions: Company financial results	Competitive access to capital	Performance indicators	Brand value company image	Intellectual capital value	Risk Management influence	Stakeholders' evaluation of the board	Company's community perception
IT governance	a	Commitment to IT										
	b	IT alignment										
	c	IT added value										
	d	IT cost management										
	e	IT projects success										
	f	Information availability										
	g	IT related risks										
	h	Operational IT framework										
		Max	80									
		Total										
		Result										

Q = Qualification
IV = Levels

FIGURE 31.7 IT governance sample form.

Board Structure

Adequate size, board member profiles and selection, independent chairman versus CEO and lead director, influence of independent directors, board committee functions, and integration.

Board Operating Procedures

Member selection policy, independent director policy, function descriptions, training and orientation, professional management meetings, members' contribution to agenda, board compensation and review, nondirectors' participation in meetings, and board and director assessment.

Board and Management

CEO evaluations, senior management compensation, access to management, CEO succession, relevant information supply, consideration of risk and crisis management, management training, and development.

Board and Shareholders

Remunerations disclosure, one share one vote, director, committee and board assessments, organizations information delivery, ownership structure, compliance disclosure, organizations profits, extraordinary transactions, member election, and dismissal.

Board and Stakeholders

Stakeholder expectations and strategy, organizations formal disclosures (including institutional investors, customers, and press) and communications, and board activity regarding community issues.

Corporate and IT Governance

IT governance is about information technology best serving the organization. Given the key role IT plays now, it is a mandatory subject for the board. Thus, in case it has not yet been established, IT must become a regular part of the agenda. Directors should commit to improve their IT literacy and their combined knowledge as well.

No matter how many resources could have already been devoted to any governance-related IT project, its true birth will only occur once the board becomes involved, at the top-down side of the subject (see Figure 31.2).

Why? Once again the key word is "alignment." The same way organization objectives must be aligned to stakeholder-accepted requirements, its strategy must be aligned to that of the organization's.

Consequently, the board must agree with management regarding the organization value that IT must deliver, the associated risks, and the costs. Once this starts happening, together with the board ensuring that IT projects meet expectations, the real path towards IT governance begins.

It is then when the board should consider the adoption of standards and best practices models in the form of IT frameworks, such as control objectives for information and related technology (COBIT, www.isaca.org), infrastructure technology information library (ITIL, www.itil.co.uk), and the like. Their descriptions are outside the scope of this chapter. However, it is important to note that frameworks have measurement capabilities but are not intended toward a board's performance.

The method described earlier in this chapter, however, can be applied to improve the board's performance regarding its responsibilities towards IT governance. In this case, in addition to the pertinent subjects, we supply a sample form (see Figure 31.7) and suggest a few guiding concepts for each: Although there can be corporate governance with out IT governance, excellence in corporate governance most likely involves IT governance.

Commitment to IT: Is IT part of the agenda of a regular board meeting? How IT skilled are board members? Are they at least IT literate? Does at least board member have a good background in IT management?

IT alignment: Are organizations strategies being developed taking into consideration the potential contribution of IT? Are IT projects at least related to organizational strategies? Are there any initiatives to start aligning them?

IT added value: Does the board understand the value IT delivers to the organizations and to the governance process itself? Have any efforts been made to ensure an accurate measurement? Does the IT budget tend to be considered an expense rather than an investment? Are IT costs as well managed as any competitive factor for the organization deserves to be?

IT projects success: Are IT projects analyzed at board level and how solid a foundation is there related to their ROI? Is the board concerned about project benefits, realization time frame, and follow-up?

Information availability: Is the information clear, precise, and deployed timely? Is it even considered a priority? Are there any plans on how to manage and/or improve organizational information?

IT-related risks: Are risks associated with IT deployment considered at board level? Are risk implications clearly understood? Are there any action plans in development to manage IT risks or to mitigate their consequences? How operational are they?

Operational IT framework: Is there a systematic approach to consider IT-related issues? Were any steps taken toward adopting an IT framework? Are there any steps already being implemented? How operational is it?

The Role of the Board in Quality

Corporate governance has not been a stranger to quality for quite some time:

- The International Academy for Quality (IAQ) started a quality-in-corporate governance (QiCG) project in 1996 during their triennial assembly in Yokohama, Japan. In 2007 the IAQ published the book *Corporate Governance Quality at the Top*, with major conclusions of the QiCG committee continuing to be one of the major IAQ activities.
- The Malcolm Baldrige National Quality Award, United States, includes corporate governance criteria in the leadership category, as do other national quality award organizations.
- The Secretary of State's Award for Corporate Excellence (ACE), established by the State Department in 1999, emphasizes the important role U.S. organizations play to advance best practices, good corporate governance, and democratic values overseas.

Corporate governance is much about leadership by example; as such, the board's commitment toward quality should begin at home by becoming a quality board. The concepts and guidelines discussed in this chapter should lead in the right direction.

Quality boards require quality directors. The directors should become literate in quality as well as other major subjects such as IT, and any other subjects directly related to the particular organization. Training programs should keep directors updated.

The most successful global organizations today recognize the key guidance and control role the board has to balance the alignment of the interests of all stakeholders to assure sustainable growth and profitability. To perform a quality job, directors should be aware of following developments:

- Significant increase of government regulations gradually becoming global
- More demanding consumers
- Global customers requesting suppliers to meet standards that include transparency and governance
- Investment and pension funds corporate governance requirements
- Rating agencies increasing requirements
- Society and customers awareness requiring that organizations have sound social responsibility programs
- Corporate governance standards are in the process of development
- Certification programs available for directors in organizations such as the National Association of Corporate Directors, Washington, DC.
- Corporate governance institutes in universities have research programs regarding the establishment of metrics and variability of corporate governance processes related to organizations results.

Organizations are gradually recognizing the need to adopt scientific quality tools in corporate governance, adapting those that were successfully applied in factories, service units, government, and health care, for example.

We all agree that metrics are important. However, the numbers are only as good as the help metrics provide the board in performing its functions. *The board's role is to guide management in the development of plans to maximize the stakeholders' satisfaction in line with sound organization sustainability and control that these plans are properly implemented.*

The processes of corporate governance are interdependent and complex involving all stakeholders. Therefore, although customer loyalty is still one of the key pillars of success, we must expand the vision that *quality is what the customer thinks it is* to: *quality is what the stakeholders think it is.*

References

Bertin, M. E. J. (1996). "Quality on the Board of Directors." *Proceedings of the JUSE International Conference on Quality*, Yokohama, Japan.

Bertin, M. E. J. (2004). "The Impact of an Effective Board of Directors in Organizations Results." *Proceedings of the 48th European Organization for Quality*, Moscow, Russia.

Bertin, M. E. J., and Watson, G. H. (2007). *Corporate Governance: Quality at the Top.* International Academy for Quality, Salem, NH, pp. 83-126.

Bertin, M. E. J. (2007). "Quality in the Board of Directors—What the Board Should Do—How to Measure Board Performance." *Proceedings of the 1st Middle East Quality Organization Annual Conference*, Dubai, UAE.

Bertin, M. E. J. and Bertin Schmidt, Marcos (2009). "Where Were the Boards? Thoughts Regarding the Global Financial Crisis." *Proceedings of the 3rd Middle East Quality Organization Annual Conference*, Dubai, UAE.

Juran, J. M., and, Louden, J. K. (1966). *The Corporate Director*. American Management Association, New York.

Micklethwait, J., and Wooldridge, A. (2003). *The Organizations—A Short Story of a Revolutionary Idea*. The Modern Library New York.

OECD-Organisation for Economic Co-Operation and Development (2004). OECD Principles of Corporate Governance. Paris, France.

Strachan, H., and Bertin, M. E. J. (2005). "An Approach to the Evaluation of a Board of Directors." *ASQ World Conference on Quality and Improvement*, Seattle, WA.

APPENDIX I

The Non-Pareto Principle; Mea Culpa

Joseph M. Juran

The *Pareto principle* by this time has become deeply rooted in our industrial literature. It is a shorthand name for the phenomenon that in any population that contributes to a common effect, a relative few of the contributors account for the bulk of the effect.

Years ago I gave the name "Pareto" to this principle of the "vital few and trivial many." On subsequent challenge, I was forced to confess that I had mistakenly applied the wrong name to the principle (Juran 1950). This confession changed nothing—the name "Pareto principle" has continued in force, and seems destined to become a permanent label for the phenomenon.

The matter has not stopped with my own error. On various occasions contemporary authors, when referring to the Pareto principle, have fabricated some embellishments and otherwise attributed to Vilfredo Pareto additional things that he did not do. My motive in offering the present paper is in part to minimize this tendency to embroider the work of a distinguished Italian economist. In addition, I have for some time felt an urge to narrate just how it came about that some early experiences in seemingly unrelated fields (quality control, cryptanalysis, industrial engineering, government administration, management research) nevertheless converged to misname the Pareto principle.

It began in the mid-1920s when as a young engineer I observed (as had many others before me) that quality defects are unequal in frequency; i.e., when a long list of defects was arranged in the order of frequency, a relative few of the defects accounted for the bulk of the defectiveness. As I moved into quality management posts in the late 1920s and the 1930s, I observed (as had many others before me) that a similar phenomenon existed with respect to employee absenteeism, causes of accidents, etc.

During the late 1930s, I moved out of the field of quality control to become the corporate industrial engineer for Western Electric Company. In this capacity, one of my responsibilities was to visit other companies to exchange experiences in industrial engineering practices. One of the most exhilarating of these visits was to General Motors Corporation's headquarters. There I found an uncommonly competent team of managers facing up to the then new problems of collective bargaining. As an incidental tool, they had put together an assortment of data processing machinery to enable them to compute the cost of any new labor union proposal. This they did by programming the machines and then running the (punched) employee record cards through the program. It was an ingenious concept, and their system was quite advanced for those days. However, the electromagnetic machinery then in use took hours and even days to process those hundreds of thousands of cards, so that the managers often found themselves waiting for the machines to grind out the results.

It is a part of our chronicle that these General Motors managers were a keen, inquisitive lot and were ever on the alert for anything new. Thus, when it happened on one occasion that the card readers were producing gibberish, the managers not only found the cause to be a miswired plug board but also realized that they had stumbled onto a means for creating messages in cipher. As a form of comic relief from the grueling hours to which they were often subjected, they used some of the waiting time to dig further into this enciphering system. The more they got into it, the more convinced they became that they had evolved a cipher system that could not be broken.

During the relaxation of a luncheon, they told me of this unbreakable cipher system, and I laughed at them. As it happened, I was no slouch in such matters, since my work in the Signal Corps Reserve was precisely on this subject. Naturally, one thing led to another, and before the day was done, I had rashly accepted their tender of an enciphered message to break. Break it I did, though it took until three o'clock in the morning. (Thereafter my sleep was short but blissful.)

They were stunned by the news that the unbreakable had been broken, and for the rest of the visit the agreeable aura of a miracle man followed me about. As a by-product, some hitherto secret doors were opened up to me. It was one of these doors that led me, for the first time, to the work of Vilfredo Pareto. The man who opened that door was Merle Hale, who presided over the executive salary program of General Motors.

Hale showed me some research he had conducted by comparing the executive salary pattern prevailing in General Motors with one of the mathematical models that Pareto had once constructed. The fit was surprisingly close. I registered the incident in my memory along with the fact that Pareto had made extensive studies of the unequal distribution of wealth, and had in addition formulated mathematical models to quantify this maldistribution.

In December 1941, the month in which the United States entered World War II, I took a "temporary" assignment as a federal government administrator. The original six weeks stretched into four years and as a by-product gave me an insight into the problems of managing the federal government. Of course, the principle of the vital few and trivial many had wide application. At the end of the war (1945) I embarked on a career dedicated to the field of management: research, writing, teaching, consulting, etc. By the late 1940s, as a result of my courses at New York University and my seminars at American Management Association, I had recognized the principle of the "vital few and trivial many" as a true "universal," applicable not only in numerous managerial functions but also in the physical and biological worlds generally. Other investigators may well have been aware of this universal principle, but to my knowledge no one had ever before reduced it to writing.

It was during the late 1940s, when I was preparing the manuscript for *Quality Control Handbook,* first edition, that I was faced squarely with the need for giving a short name to the universal. In the resulting write-up (Juran 1951). under the heading "Maldistribution of Quality Losses," I listed numerous instances of such maldistribution as a basis for generalization. I also noted that Pareto had found wealth to be maldistributed. In addition, I showed examples of the now familiar cumulative curves, one for maldistribution of wealth and the other for maldistribution of quality losses. The caption under these curves reads, "Pareto's principle of unequal distribution applied to distribution of wealth and to distribution of quality losses." Although the accompanying text makes clear that Pareto's contributions specialized in the study of wealth, the caption implies that he had generalized the principle of unequal distribution into a universal. This implication is erroneous. The Pareto principle as a universal was not original with Pareto.

Where, then, did the universal originate? To my knowledge, the first exposition was by me. Had I been structured along different lines, assuredly I would have called it the Juran principle. However, I was not structured that way. Yet I did need a shorthand designation, and I had no qualms about Pareto's name, hence the Pareto principle.

The matter might well have rested there had there been a less than enthusiastic response to the universal. Instead, the new universal became the subject of wide use and reference. I contributed to this dissemination by coining and popularizing the term "vital few and trivial many" in the widely read "universals" paper (the first published use was likely in Juran 1954) and in the moving picture film I prepared for American Management Association on the "breakthrough" process. The resulting wide usage also brought me some challenges (from Dorian Shainin and others) as to the attribution to Pareto. These challenges forced me to do what I should have done in the first place—inform myself on just what Pareto had done. It was this examination that made clear to me what I had seen only dimly—that Pareto's work had been in the economic sphere and that his models were not intended to be applied to other fields. To make matters worse, the cumulative curves used in *Quality Control Handbook,* first edition, should have been properly identified with Lorenz (1904–1905).

To summarize, and to set the record straight:

1. Numerous men, over the centuries, have observed the existence of the phenomenon of vital few and trivial many as it applied to their local sphere of activity.
2. Pareto observed this phenomenon as applied to the distribution of wealth and advanced the theory of a logarithmic law of income distribution to fit the phenomenon.
3. Lorenz developed a form of cumulative curve to depict the distribution of wealth graphically.
4. Juran was (seemingly) the first to identify the phenomenon of the vital few and trivial many as a "universal," applicable to many fields.
5. Juran applied the name the "Pareto principle" to this universal. Juran (1974) also coined the phrase "vital few and trivial many" and applied the Lorenz curves to depict this universal in graphic form.

References

Juran, J. M. (1950). "Pareto, Lorenz, Cournot, Bernoulli, Juran and Others." *Industrial Quality Control,* October, p. 25.

Juran, J. M., ed. (1951). *Quality Control Handbook,* 1st ed. McGraw-Hill, New York, pp. 37–41.

Juran, J. M. (1954). "Universals in Management Planning and Controlling." *The Management Review,* November.

Juran, J. M., ed. (1974)., Quality *Control Handbook,* 3d ed. McGraw-Hill, New York, pp. 2–16 to 2–19.

Lorenz, M. O. (1904–1905). "Methods of Measuring the Concentration of Wealth." *American Statistical Association Publication,* vol. 9, pp. 200–219.

Dr. J. M. JURAN, Dean of American consultants on quality control, was a pioneer in the development of principles and methods for managing quality control programs. He was a veteran of over five decades of international experience in management at all levels. His clients included industrial giants as well as small companies and government departments. He conducted several hundred courses in all parts of the world, not only on QC management but on other managerial subjects as well. Dr. Juran was the author of ten books including *Quality Planning and Analysis* (with F. M. Gryna), *Managerial Breakthrough*, and *The Quality Control Handbook*, which has been translated wholly or partly into several languages and which has become the international standard reference work in the field of QR.

APPENDIX II

Sample Competency Matrices and Job Profiles

Title	Vice President of Quality, Global Division
Primary Purpose	The division vice president of quality will contribute to company profitability by delivering industry-leading quality performance to customers. This executive will provide visionary leadership on long-range objectives and quality programs that will systemically drive organizational success in both current and future business environments. This executive will instill the right competencies and processes to achieve industry-leading teams at all levels of the organization to deliver customer-focused quality solutions. He or she will interact frequently with sales, marketing, engineering, procurement, and manufacturing on a global basis.
Duties	Duties and responsibilities will include 1. Realigning current resources to support a proactive approach to predict, prevent, and protect the company and its customers from quality performance issues 2. Taking responsibility for company's quality processes and development of strategic quality, vision, mission, goals, and objectives in conjunction with companywide direction 3. Creating overall direction for consistent quality methodologies, processes, and procedures across all operations of the division 4. Working in collaboration with the global organization to ensure the management of performance, qualification process, and use of metrics 5. Acting as quality advocate inside and outside of the company 6. Driving the implementation and maintenance of Total Quality Management and Lean Six Sigma methods in the division globally 7. Taking responsibility to ensure all locations have knowledgeable and effective management representatives to oversee maintenance and continual improvement of quality system standards 8. Traveling extensively and being accessible via phone to participate in and lead global conference calls as required

(*Continued*)

Title	Vice President of Quality, Global Division
	Initiatives include these: 1. Drive Lean and Six Sigma initiatives globally within the division. 2. Ensure quality system compliance to standards and customer requirements across the division. 3. Complete process and system compliance auditing and improvements at regular intervals. 4. Conduct customer satisfaction monitoring and report results quarterly to division leadership. 5. Drive quality metrics reporting through all the global division operations. 6. Serve as customer's advocate to business teams overseeing the division's effectiveness at meeting customer quality standards as well as other requirements. 7. Establish and implement standardized policies, standards, processes, metrics, and controls surrounding customer quality. 8. Ensure the implementation of employee/supplier training and education programs to develop a consistent understanding of the company quality process. 9. Facilitate the use and development of division resources for addressing internal and external quality issues early in the product life cycle and evolve the team to higher-value proactive activities. 10. Work with global quality and IT to develop standardized data collection and reporting systems. 11. Facilitate the linkage of quality performance to sourcing decisions. 12. Meet annual quality goals. 13. Collaborate with company quality leadership to leverage best practices across the company. Current year goals include the following: 1. To establish quality direction, form and chair a quality council, lead the development of quality strategy, establish quality scorecard, define the "right" quality tools and metrics, and drive effectiveness of the quality system. 2. Customer advocate provides for customer issue escalation and timely resolution. 3. Attract and develop industry-leading talent. 4. Report quality metrics. 5. Organize and conduct regular formal quality reviews. 6. Develop and manage execution of a global quality improvement plan for the division. Link priorities to business goals.
Education Required	An undergraduate degree is required. A Master of Business Administration (MBA) degree is desirable.
Education Preferred	Certification as a Six Sigma Black Belt or Master Black Belt. Certified quality engineer.

(*Continued*)

Title	Vice President of Quality, Global Division
Work Experience Required	Candidate must have 7 to 10 years' experience in component-level quality environments. Candidate must have extensive experience with Six Sigma methods and a record of successful application of Six Sigma to drive improvements in engineering, manufacturing, cost improvement, and sourcing. Candidate must be able to drive Six Sigma into internal and external processes. Candidate must have a track record of effectively identifying root causes and organizational levers to successfully address barriers to industry-leading business performance. Candidate must have enthusiasm for new industry initiatives and be able to influence others to address resistance to change. Candidate must have proven ability to lead diverse global teams to achieve cost, quality, and time-to-market commitments. Additionally, candidate must 1. Have strong negotiation and relationship building skills 2. Have skills to create the path by which the division can achieve a quality-driven culture focused on the customer 3. Have a high level of energy to create support for quality initiatives 4. Drive key functional leaders to embrace quality as a critical element of the division's success 5. Provide objective, accurate, and truthful data to drive improvement in quality 6. Have a track record of developing, attracting, and retaining top talent and developing effective teams 7. Have success in collaborating with key work partners 8. Have success in managing change through all levels of the organization 9. Have strong presentation and writing skills 10. Exude executive presence
Skills Required	Required knowledge includes 1. Fundamentals of global industry quality standards including ISO-TS-16949, ISO-9000-2000, and major customer/industry requirements 2. Analytical nature and discipline 3. How to achieve the metrics 4. Ability to push issues with key stakeholders 5. Technical knowledge and ability to hire the right people 6. Demonstrated ability to build a strong team 7. Global authority and cultural sensitivity 8. Extensive knowledge of automotive quality standards and requirements 9. Detailed understanding of business performance and quality metrics 10. Knowledge of how to leverage a personal command of business statistics and performance metrics to build a compelling case for specific decisions and recommendations

(*Continued*)

<table>
<tr><th>Title</th><th>Vice President of Quality, Global Division</th></tr>
<tr><td>Other</td><td>Candidate must be a broad-based, visible senior business leader who has successfully utilized quality tools and philosophy in leading an organization to higher levels of performance; a leader who can generate instant credibility both within the company and with external customers and suppliers, creating a path by which the division can achieve a customer-focused, quality-driven culture. She or he must be able to lead key functional leaders to embrace quality as a critical success factor in the business. Candidate must have the presence to be able to interact effectively at the highest levels of the company's, customer, and supplier leadership. This person must be
1. An individual with a vision for the future and a view of how things might be different.
2. A leader who questions assumptions and conventional thinking. This person challenges tradition and existing business models and is an effective change agent. She or he is able to communicate enthusiasm for new industry initiatives and influence others to address resistance to change.
3. Someone with the ability to build a strong, cohesive quality management team, working with diverse cultures, personalities, and ambitions.
4. A leader who can translate strategy into rigorous operating plans complete with goals, accountabilities, timetables, and measures.
5. A high-energy leader with a positive attitude in the face of difficult challenges or adversity, who delivers on commitments to customers.
6. A leader who is professional and decisive. He or she reaches closure in a timely fashion on difficult or complex problems and demonstrates courage and strength of conviction.
7. An articulate, effective communicator who sets clear standards and goals and holds individuals accountable.
8. Someone with exemplary behavior that is consistent with ethical principles such as avoiding conflicts of interest, avoiding compromising business situations, and handling confidential information appropriately, understanding that professional integrity is critical to building the trust and respect required to influence and lead others.</td></tr>
<tr><th>Title</th><th>Director, Global Customer Quality</th></tr>
<tr><td>Primary Purpose</td><td>Provides internal and external direction and communication necessary to develop and sustain customer confidence in the organization's quality performance. Activities include both proactive and reactive components of customer quality relationship management on behalf of the organization.</td></tr>
<tr><td>Duties</td><td>Duties and responsibilities will include
1. Coordinating internal communication and integration of customer quality requirements.
2. Coordinating global responses to complaints and quality performance problems.</td></tr>
</table>

(*Continued*)

Title	Director, Global Customer Quality
	3. Acting as primary contact for customer quality issues. 4. Maintaining global actions list and/or project plans for achieving/ sustaining preferred supplier status to major customers. 5. Maintaining current customer scorecards on intranet site for access by all company entities globally. 6. Ensuring that customer requirements for change notification are available, updated, and accessible by all entities. 7. Driving global improvement in reductions based on improved cause data, effective problem solving, and systemic analysis and improvement. 8. Participating in the company customer review process when it is conducted for major accounts. 9. Communicating best practices in achieving customer requirements.
Education Required	An undergraduate degree is required.
Education Preferred	MBA, certified Black Belt, ASQ certifications, Microsoft Office proficiency (Excel, Word, PowerPoint, etc.)
Work Experience Required	A minimum of 10 to 15 years' experience in quality roles or related manufacturing areas where a substantial understanding of company's quality processes in design and manufacturing has been gained. Global experience is preferred.
Skills Required	The person in this job must possess good leadership and excellent planning and project management skills. He or she will be able to Champion for the needed elements of change and harness the necessary resources to bring about change. She or he must be the voice of the customer while working to actively effect improvements necessary to exceed customer expectations. The jobholder must be professional, credible, and able to influence key regional owners and stakeholders around the world. As a result, cross-cultural sensitivity and acceptability is essential. Analytical and creative thinking, independent judgment, and the ability to present information and ideas clearly and concisely are also required. An independent thinker, the jobholder must be able to organize her or his own work, anticipating, planning, and monitoring the workload. The person in this job must possess strong written and oral communication skills. A solid experience-based understanding of quality standards, systems, and tools is a requirement.
Other	The jobholder must be prepared to travel both domestically and internationally as required. He or she must be a team player and a self-directed, self-starter who sees what needs to be done and can get objectives accomplished with minimal supervision. The person in this role must be qualified to fit into a leadership role within the quality organization. This must be a part of the selection criteria to be sure the best person is selected.

(*Continued*)

Title	Senior Manager, Global Supplier Quality
Primary Purpose	This position is responsible for defining and driving global quality and improvement programs for the company in both lead as well as individual contributor roles. As the Supplier Quality Assurance (SQA) lead, this position is responsible for collaborating with all divisional SQA personnel to create and deliver quality-focused initiatives around material and product qualification, testing plans, schedules, quality issues, and risks in a fast-paced development and manufacturing environment. The job requires experience in successfully building, leading, and executing quality programs in a manufacturing environment; demonstrated quality management/Six Sigma experience; and proven technical and practical quality experience in proactive and reactive environments.
Duties	Duties and responsibilities will include 1. Defining, setting, and delivering quality improvement initiatives with the SQA team leads that involve all elements of planning and execution and release-based quality/risk assessment. 2. Driving specific process, delivery, and tool improvements in development, manufacturing, and release quality measurement and assessment with division SQA personnel. 3. Contributing to and support release-specific SQA programs led by division SQA personnel that support quality improvement initiatives. 4. Collaborating and partner with program management and development; supporting leads to build an end-to-end SQA focus on process improvement that helps drive "quality upstream" across the supply base globally. 5. Taking responsibility to ensure development and monitoring of annual supplier quality improvement plans globally. 6. Defining, developing, and directing compliance to a global SQA standard by working with all divisions for one set of common standards applied equally on a global basis. 7. Other duties and responsibilities as may be determined by management.
Education Required	An undergraduate degree is required. Dedicated to concepts and principles of supplier quality assurance and management.
Education Preferred	As above.
Work Experience Required	Three years of demonstrated experience in a quality role in a multisite manufacturing environment. Previous experience managing a successful quality improvement process with multifunctional involvement (engineering or manufacturing). A minimum of 5 years' working experience in a quality management role including responsibility for interfacing with customers on SQA requirements.
Skills Required	Supplier quality management. Manufacturing quality management. Advance quality planning. Advanced quality tools including DOE. Excellent communication and presentation skills. Effective coordination and project management skills. Skills must be action and results oriented.
Other	The jobholder must be willing to travel globally, 50 percent minimum. Chinese language skills are a significant plus.

(*Continued*)

Title	Corporate Quality Engineer
Primary Purpose	This position is to be the internal global resource and driver of the tools and methods necessary for supporting the implementation of global TQM, Six Sigma, and Lean techniques.
Duties	Duties and responsibilities will include 1. Acting as primary quality training and development coach for lead teams, sponsors, Six Sigma Black Belts, and Green Belts. 2. Facilitating knowledge transfer from related external consultants to company employees. 3. Acting as eader and facilitator for select project teams. 4. Functioning as problem-solving facilitator and resource for select problems requiring immediate resolution. 5. Coordinating internal communication and integration of customer quality requirements as required. 6. Cultivating a global network of internal company expertise in TQM and Six Sigma. 7. Identifying internal and external (supplier) opportunities for improvement to assist entities in meeting quality and cost reduction targets. Assisting in the management of consultant schedules and expenses. Provideing input on selecting future Black Belts and Master Black Belts. 8. Supporting company TQM program by following approved policies and procedures. 9. Performing other related duties as assigned by management.
Education Required	An undergraduate degree and certified Six Sigma Master Black Belt.
Education Preferred	MBA
Work Experience Required	Candidate must have a minimum of 5 to 10 years of plant manufacturing experience in a quality or manufacturing role with demonstrated successful, measurable performance in the application of Six Sigma. Candidate must have achieved Master Black Belt certification as a result of successful projects implemented in a manufacturing environment and must be trained and experienced in team facilitation skills as well as the technical tools associated with TQM, Lean, and Six Sigma. Knowledge and successful experience in a leading new product development process (APQP) are essential. Candidate must be willing to travel extensively to work with company locations and suppliers worldwide.
Work Experience Preferred	Management experience overseeing the implementation of a successful Six Sigma initiative in a leading manufacturing company is desirable. Global experience is preferred. Language skills beyond English are highly desirable.
Skills Required	The person in this job must possess good leadership and excellent planning and project management skills. Demonstrated successful facilitation and teamwork skills are also required. The jobholder must be professional, credible, and able to influence key regional owners and stakeholders around the world. As a result, cross-cultural sensitivity and acceptability is essential. Analytical and creative thinking, independent judgment, and the ability to present and teach the quality tools and methods clearly and concisely are also required. The person in this job must possess strong written and oral communication skills with the ability to work effectively and efficiently with management as well as hourly employees.

(*Continued*)

Title	Corporate Quality Engineer
Other	The person in this role must be qualified to fit into a management role within the quality organization in the future. This must be a part of the selection criteria to be sure the best person is selected.

Title	Director of Auditing, Quality Systems, and Environmental Compliance
Primary Purpose/ Job Description	The global director of auditing for quality systems and environmental compliance conducts, coordinates, and schedules audits for management to assess the effectiveness of management systems and controls for compliance to quality and environmental standards and regulations. This includes examining records and interviewing workers to ensure recording of transactions and compliance with applicable standards and/or laws and regulations. This person coordinates the process by which company locations will assess management systems to determine their efficiency and protective value. As part of this process the position enforces uniform methods to review records pertaining to operations, emissions, and waste management. This requires the analysis of data obtained for evidence of deficiencies in controls, duplication of effort, or lack of compliance with laws, government regulations, and management policies or procedures. This person oversees a standard approach to preparing reports of findings and recommendations for local, divisional, and corporate management. This person may conduct special studies for management and regularly works to verify that adequate internal controls are in place to minimize risk and exposure to the company, under the general guidance of the vice president of global quality. The director of audits will work with each of the global divisions at the entity level to plan, coordinate, and sometimes participate in quality and environmental audits and internal control assessments of all operations. In so doing, the audit manager will perform assessments and establish appropriate staffing resource and corrective action recommendations for the entities and divisions. This will include cross-entity and cross-regional audits as appropriate to leverage best practices and share successes. Additionally this person will be responsible to work with all divisions to ensure that the company optimizes regional resources to serve all divisions.
Duties/ Responsibilities	Review prior audit reports and initiate discussions with local quality and site management to assess quality and/or environmental system compliance. The result will be a documented report of the risk of operations being examined due to noncompliance. This person will ensure coordination with appropriate local quality and/or environmental resources to meet standards and regulatory requirements. • Plan and execute audits in a professional manner to ensure timely completion of summary reports. It will be the site management's responsibility to develop and implement appropriate improvement plans to address audit findings. • Communicate results of all audit findings and recommendations for improvement to management through concise, high-quality audit reports.

(*Continued*)

Title	Director of Auditing, Quality Systems, and Environmental Compliance
	• Participate in special projects when necessary, including but not limited to occasional assistance in preparation for customer quality and/or environmental audits. • Play an active role in the continuous improvement of the company quality and environmental audit function. • Combined domestic and international travel is estimated to be no more than 40 percent.
Requirements	• Bachelor's degree in a business or relevant technical discipline. Extensive lead audit and audit management experience in quality systems and environmental compliance will be considered in place of a formal educational degree. • Professional designation orcertification as a lead auditor is required. • Seven+ years' work experience in auditing preferably in a manufacturing industry environment is required. • ISO-9000-2000, ISO-TS-16949, and ISO-14001 assessment experience is required. Candidate must be familiar with these and any new or revised requirements related to quality and environmental systems compliance. • Strong work ethic, systemic and process thinking, and organization skills are required. • Strong interpersonal and communication (written and verbal) skills are necessary to deal with all levels of personnel. • Microsoft Office proficiency (Excel, Word, PowerPoint, etc.) is needed.
Key Tasks	The audit manager is responsible for optimizing and coordinating with divisions and entities for managing the audit program within the company. This includes coordinating the planning, scheduling, performing, tracking closure, and reporting of audits. Key tasks include: • Own the audit process and ensure this is well documented, understood, effective, and up to date with the most current standards and regulatory requirements. • Create and maintain the annual global audit schedule and status reports. • Train auditors to function as internal auditors. • Participate frequently as part of audit teams. • Track audit actions to closure. • Report on audit program key performance indicators to process owners and company management, e.g. average age of overdue actions, recurrent findings, concentration of findings per department/location. • Ensure company management is knowledgeable of risks of noncompliance to quality standards and environmental regulations. • Act as a consultant to the business: Develop and maintain expertise and act as the subject matter expert (SME) for quality systems and environmental compliance.

(*Continued*)

Title	Director of Auditing, Quality Systems, and Environmental Compliance
Experience	• Documenting process definition and improvement • Implementation of recognized industry process, quality, and environmental standards and/or regulations • Design and delivery of training for standards, regulations, and auditing • Interface with external registrars/regulatory bodies for resolution of major findings
Essential Skills	Excellent communicator and negotiator at all levels. Good analytical and problem-solving skills.
Other	Remote and dotted line supervision of audit resources is called for. The job requires the ability to assimilate information quickly and to deal with senior management often in uncooperative situations. Candidate will need to be pragmatic without compromising the integrity of the communication. Multitasking of several assignments and initiatives will be expected. Candidate will be: committed and enthusiastic about quality, environment, and process improvement; attentive to detail; flexible and adaptable, good team player, practical and pragmatic; proactive, able to work on own initiative.

Master Black Belt Profile

Master Black Belts are companywide Six Sigma or quality experts. The Master Black Belt is qualified to teach other Six Sigma Green and Black Belts the methodologies, tools, and applications in all functions and levels of the company. In addition, the Master Black Belt is able to provide leadership integrating the Six Sigma approach into the business strategy of the company, and contributes to creating and carrying out the organization's strategic business and operational plans. As a Black Belt, the Master Black Belt candidate has personally led successful project teams.

KEY ROLES

- Provide technical support and mentoring.
- Facilitate multiple projects.
- Provide advice to Champions and executive management.
- Train others on the Lean and Six Sigma tools and techniques.
- Provide leadership to management groups in the integration of the Lean and Six Sigma approaches with the organization's business strategy.
- Contribute to creating and carrying out the organization's strategic business and operational plans.
- Be trained on advanced tools, strategic deployment, and Train-the-Trainer workshops.

Black Belt Profile

Black Belts are technical specialists assigned full responsibility to implement Six Sigma projects through a business unit, function, or process. They will become viewed as "initiators" of improvement activity, and they are full-time on-site project leaders.

KEY ROLES

- Keep the Champion informed of project's progress.
- Develop, coach, and lead multifunctional improvement teams.
- Mentor and advise management on prioritizing, charting, and launching projects.
- Use and teach tools and methods to Green Belts, Yellow Belts, and subject matter experts.
- Actively seek to use the Six Sigma breakthrough steps to solve chronic problems, remove waste, and plan new services or products.
- Learn to align projects to local business objectives.
- Provide project management, facilitate, and lead teams.
- Be trained and certified in the appropriate tool set.

Green Belt Profile

Green Belts are employees with sufficient knowledge to support and participate in Lean and Lean Six Sigma projects. They can be a team leader or a team member.

KEY ROLES

- May lead projects
- May be a core project team member
- Actively participates and contributes expertise to larger Black Belt projects
- Uses Lean and Six Sigma steps to solve problems
- Uses Lean to remove waste
- Completes multiple projects over time, one at a time
- Is trained and certified in the Green Belt tool set

Lean Master Profile

Lean Masters are companywide Lean or quality experts. The Lean Master is qualified to teach other Lean experts and team members the methodologies, tools, and applications in all functions and levels of the company. In addition, the Lean Master is able to provide leadership integrating the Lean approach into the business strategy of the company, and contributes to creating and carrying out the organization's strategic business and operational plans. As a Lean expert, the Lean Master candidate has personally led successful Lean project teams.

KEY ROLES

- Provide technical support and mentoring.
- Facilitate multiple projects.
- Provide advice to Champions and executive management.
- Mentor and advise management on prioritizing, charting, and launching projects.
- Train others on the Lean tools and techniques.
- Provide leadership to management groups in the integration of Lean with the organization's business strategy.
- Contribute to creating and carrying out the organization's strategic business and operational goals.
- Be trained on advanced tools, strategic deployment, and Train-the-Trainer workshops.

Lean Expert Profile

Lean experts are technical specialists assigned full responsibility to implement Lean projects through a business unit, function, or process. They will become viewed as "initiators" of improvement activity and are full-time on-site project leaders.

KEY ROLES

- Keep Champion informed of project progress.
- Develop, coach, and lead multifunctional improvement teams.
- Mentor and advise management on prioritizing, charting, and launching projects.
- Use and teach tools and methods to Lean managers and subject matter experts.
- Actively seek to use the Lean steps to solve chronic problems, remove waste, and plan new services or products.
- Learn to align projects to local business objectives.
- Provide project management, facilitate, and lead teams.
- Be trained and certified in the appropriate tool set.

Glossary of Acronyms

ACE—Award for Corporate Excellence

ACH—Automated clearing house

ADLI—Approach, deployment, learning, and integration

AHT—Average handle time

ALOS—Average length of stay

AMI—Acute myocardial infarction

ANOVA—Analysis of variance

ANSI—American National Standards Institute

ANZSIC—Australian and New Zealand Standard Industrial Classification

AOQ—Average outgoing quality

APQC—American Productivity and Quality Center

AQAP—Allied quality assurance publications

AQIP—Annual quality improvement plan

ARDEC—Armament Research, Development, and Engineering Center

ARL—Army Research Laboratory

ARPAnet—Advanced Research Projects Agency Network

ASME—American Society of Mechanical Engineers

ASQ—American Society for Quality

ASTM—American Society for Testing and Materials

ATM—Automatic teller machine

BAM—Business activity monitoring

BBB—Better Business Bureau

BBE—Behavior-based expectation

BPM—Business process management

BPO—Business process outsourcing

BPQM—Business process quality management

BPR—Business process reengineering

CAFÉ—Connecticut Award for Excellence

CCL—Critical-components list

CDP—Carbon disclosure project

CE—European Conformity

CEN—European Committee for Standardization

CEO—Chief executive officer

CFO—Chief financial officer

CFR—Code of Federal Regulations

cGMP—Current good manufacturing practice

CHF—Congestive heart failure

CI—Continuous improvement

CIn—Continuous innovation

CIO—Chief information officer

CMM—Capability maturity model

CMMi—Capability maturity model integration

CMS—Centers for Medicare and Medicaid Services

CMU—Carnegie-Mellon University

COBIT—Control objectives for information and related technology

COO—Chief operating officer

COP^3—Cost of poorly performing processes

COPQ—Cost of poor quality

COTS—Commercial off-the-shelf

C_p—Process capability

C_{pk}—Process capability index

CPM—Critical path method

C_{pm}—Taguchi capability index

CQE—Certified quality engineer

CTO—Chief technology officer

CTQ—Critical to quality

CUMSUM/CUSUM—Cumulative sum

DA—Decision analysis

DC—Discharge

DCE—Defect containment effectiveness

DES—Discrete event simulation

DF—Degree of freedom

DFA—Design for assembly

DFM—Design for manufacturing

DFMA—Design for manufacture and assembly

DFMEA—Design failure mode and effects analysis

DFSS—Design for Six Sigma

DMADV—Define, measure, analyze, design, verify

DMAIC—Define, measure, analyze, improve, control

DOE—Design of experiments

DpKLOC—Defects per thousand lines of code

DPMO—Defects per million opportunities

DRM—Digital Rights Management

DVT—Design verification test

EDI—Electronic Data Interchange

EFQM—European Foundation for Quality Management

EFT—Electronic funds transfer

EHR—Electronic health record

EIU—Environmental impact unit

EMMA—Electronic Municipal Market Access

EMR—Electronic medical record

EMS—Environmental management systems

ENPV—Expected net present value

EP—Environmental protection

ERA—Emergency response action

ERP—Enterprise resource planning

ESD—Emergency services department

ESS—Employee self-service

EU—European Union

EVOP—Evolutionary operations

EWMA—Exponentially weighted moving average

FC—Family Council

FDA—Food and Drug Administration

FDM—Functional deployment matrix

FLC—Federal Laboratory Consortium

FMEA—Failure mode and effects analysis

FMECA—Failure mode effects criticality analysis

FRACAS—Failure reporting and corrective action systems

FTA—Fault tree analysis

FUNDIBEQ—The Ibero-American Foundation for Quality Management

GATT—General Agreement on Tariffs and Trade

GDP—Good distribution practice

GDP—Gross domestic product

GEM—Global excellence model

GHG—Greenhouse gas

GIDEP—Government-Industry Data Exchange Program

GMO—Genetically modified organism

GMP—Good manufacturing practice

H_a—Alternative hypothesis

HACCP—Hazard analysis and critical control points

H_o—Null hypothesis

HR—Human resources

IAQ—International Academy for Quality

IAQG—International Aerospace Quality Group

ICB—Industry classification benchmark

ICU—Intensive care unit

IEC—International Electrotechnical Commission

IEEE—Institute of Electrical and Electronics Engineers

IHI—Institute for Healthcare Improvement

IOD—Institute of Directors

IOM—Institute of Medicine

IPEC—International Pharmaceutical Excipients Council

IRPS—International Reliability Physics Symposium

ISIC—International Standard Industrial Classification of All Economic Activities

ISO—International Organization for Standardization

ISO/DIS—International Organization for Standardization/Draft International Standard

ISO/TS—International Organization for Standardization/Technical Specification

IT—Information technology

ITIL—Information Technology Infrastructure Library

IVR—Interactive voice response

JCAHO—Joint Commission on the Accreditation of Healthcare Organizations

JCF—Juran complexity factor

JDI—Just Do It

JIT—Just in time

JUSE—Union of Japanese Scientists and Engineers

KPA—Key process area

KPC—Key product characteristic(s)

KPI—Key performance indicator

KPIV—Key process input variable

KPOV—Key process output variable

KRI—Key result indicator

LCA—Life cycle assessment

LDMADV—Lean Design for Six Sigma (Define, Measure, Analyze, Design, Verify)

LeTCI—Levels, trends, comparisons, and integration

LSL—Lower specification limit

LSS—Lean Six Sigma

LTC—Long-term care

MANOVA, MANCOVA—Multivariate analysis of variance, multivariate analysis of covariance

MAP—Missile (Defense Agency) assurance provision

MBNQA—Malcolm Baldrige National Quality Award

MDA—Missile Defense Agency

MDT—Mean downtime

MRP—Material requirements planning

MSA—Measurement system analysis

MTBF—Mean time between failure

MTTF—Mean time to failure

MTTR—Mean time to repair

MVT—Manufacturing verification test

NACD—National Association for Corporate Directors

NACE—Statistical classification of economic activities in the European Community

NAFTA—North American Free Trade Association

NAICS—North American Industry Classification System

NASA—North American Space Administration

NATO—North Atlantic Treaty Organization

NHTSA—National Highway Traffic Safety Administration

NICE—Nonpersonal interactivity-Infrastructure availability-Controllability-Effort inevitability

NIST—National Institute of Standards and Technology

NMMC—North Mississippi Medical Center

NPI—New product introduction

NPR—NASA procedural requirements

NPV—Net present value

NQA—National quality award

NVCASE—National Voluntary Conformity Assessment System Evaluation

OA—Operational availability

OECD—Organisation for Economic Co-operation and Development

OEMs—Original equipment manufacturers

OFAT—One factor at a time

OKVED—Russian Economic Activities Classification System

OLA—Operating level agreement

ONE—Organizations of Noteworthy Excellence

OpEx—Operational excellence

OSHA—Occupational Safety and Health Administration

OSI—Owner satisfaction index

OST—On-stream time

OTT—Office of Technology Transfer

OVT—Operations verification test

P4P—Pay for performance

PC—Personal computer

PCA—Principal-component analysis

PCE—Phase containment effectiveness

P-D—Position-dimension

PDA—Personal digital assistant

PDCA—Plan, do, check, act

PDSA—Plan, do, study, act

PERT—Program evaluation and review technique

PFMEA—Process failure mode and effects analysis

PMBoK—Project management body of knowledge

PMI—Project Management Institute

POS—Point of sale

PT—Personal transporter

QA—Quality assurance

QA/RA—Quality assurance/regulatory assurance

QbD—Quality by design

QC—Quality control

QFD—Quality function deployment

QiCG—Quality in corporate governance

QIE—Quality information equipment

QIP—Quality improvement plan

QS—Quality system

QSAR—Quality System Assessment Recognition

R&D—Research and development

R&R—Repeatability and reproducibility

RADAR—Results, approach, deployment, assessment, and review

RAM—Reliability, availability, and maintainability

RCA—Root cause analysis

RCCA—Root cause corrective action

RCM—Reliability-centered maintenance

RIAC—Reliability information analysis center

RIE—Rapid improvement event

RoHS—Restriction of hazardous substances

ROI—Return on investment

RPN—Risk priority number

SCA—Special cause analysis

SCAMPI—Standard CMMI appraisal method for process improvement

SCIP—Surgical care improvement program

SDLC—Systems development life cycle

SEI—Software Engineering Institute

SERVQUAL—Service quality

SIPOC—Supplier-input-process-output-customer

SITA—Société Internationale de Telecommunications Aeronautiques

SL—Service line

SLA—Service level agreement

SME—Subject matter expert

SMED—Single-minute exchange of die

SOA—Service-oriented architecture

SOP—Standard operating procedure

SP—Strategic planning

SPC—Statistical process control

SQC—Statistical quality control

SQE—Supplier quality engineer

SQL—Simple Query Language

SRE—Society of Reliability Engineers

SST—Self-service technology

STS—Sociotechnical system

TCE—Total containment effectiveness

TCI—Test capability index

TPM—Total productive maintenance

TPS—Toyota Production System

TQC—Total quality control

TQM—Total quality management

TRIZ—Russian for "the theory of solving inventor's problems" or "the theory of inventor's problem solving"

TSR—Tele-service representative

UKSIC—United Kingdom Standard Industrial Classification of Economic Activities

UNIVAC—Universal Automatic Computer, first commercial computer

USDA—U.S. Department of Agriculture

USL—Upper specification limit

VOC—Voice of the customer

VoIP—Voice over Internet Protocol

VSM—Value stream map

WAIS—Wide-area information server

WCQ—World-class quality

WHO—World Health Organization

WWW—World Wide Web

XML—Extensible Markup Language

Glossary of Terms

Acceptance test: A highly structured form of testing a completed, integrated system to assess compliance with specifications; commonly applied to complex systems, such as computer systems.

Accuracy of a sensor: The degree to which a sensor tells the truth—the extent to which its evaluation of some phenomenon agrees with the "true" value as judged by an established standard.

Activities: The steps in a process or subprocess.

Administrator: An overseer or entity that is created and given powers to establish standards and to see that they are enforced.

Advertising: The process of publicizing a product or service to generate sales; requires technological and legal review of copy; activities to propagandize product safety through education and warnings.

Affinity diagram: A diagram that clusters together items of a similar type; a prelude to a cause–effect diagram used in quality improvement, and used in quality design to group together similar needs or features.

Annual goals: What an organization seeks to achieve over a one- (to several-) year period; the aim or end to which work effort is directed.

Arbitration: Adversarial process in which parties agree to be bound by the decision of a third party. It is an attractive form of resolving differences because it avoids the high cost and long delays inherent in most lawsuits.

Assembly tree: A process that incorporates the outputs of several subprocesses.

Autonomous process: A group of related activities that usually performed by one department or a single group of individuals.

Availability: In the context of product design, the probability that a product, when used under given conditions, will perform satisfactorily when called upon.

Avoidance of unnecessary constraints: Not overspecifying the product for the team.

Basis for establishing quality goals: In addition to the scope of the project, a goal statement must include the goal(s) of the project. An important consideration in establishing quality goals is the choice of the basis for which the goals are set.

Benchmarking: A recent label for the concept of setting goals based on knowing what has been achieved by others. It identifies the best in class and the methods behind it that make it best.

Bias: The presence or influence of any factor that causes the population or process being sampled to appear different from what it actually is.

Black belts: On-site implementation experts with the ability to develop, coach, and lead cross-functional process improvement teams.

Breakthrough: The organized creation of beneficial change and the attainment of unprecedented levels of performance.

Business process adaptability: The ability of a process to readily accommodate changes in both the requirements and the environment while maintaining its effectiveness and efficiency over time.

Business process management (BPM): A process to sustain the changes made from a portfolio of improvement projects.

Business process outsourcing (BPO): The growing practice of one organization outsourcing some number of its processes to a third party to execute the selected processes.

Capability maturity model integration: A process improvement methodology that enables organizations to better manage their processes across business units and projects, resulting in improved organization performance.

Carbon Disclosure Project (CDP): A nonprofit organization with the mission to provide information to investors and stakeholders regarding the opportunities and risks to commercial operations presented by climate change.

Carryover analyses: Typically, a matrix-based assessment of a design that depicts the degree of carryover of design elements from a prior version, with particular regard to failure proneness.

Cause: A proven reason for the existence of a defect. Often there are multiple causes, in which case they typically follow the Pareto principle—the vital few causes will dominate all the rest.

Change agent: The individual or group with responsibility for leading and implementing an organizational change; anyone within the organization, at whatever level, who has the desire and accepts the responsibility for initiating or leading the change effort.

Company-financed: Paid for by a company. In the context of testing, in this form, industrial organizations buy test services from independent test laboratories to secure the mark (certificate, seal, label) of the laboratory for their product(s).

Comparative performance: How the final product will perform vis-à-vis the competition.

Competitive analysis: Feature-by-feature comparison with competitors' products; usually a matrix depicting a feature-by-feature comparison to the competition, with particular regard to best-in-class targets.

Conceptual learning: The process of acquiring a better understanding of the cause–effect relationship, leading to "know-why."

Consortium: An association of business organizations. This form involves creating an association of organizations from various countries. The consortium usually is dedicated to a specific project.

Consumer-financed: Paid for by consumers. In the context of testing, in this form the test laboratory derives its income by publishing its test results, usually in a monthly journal plus an annual summary.

Consumerism: A popular name for the movement to help consumers solve their problems through collective action; an aspect of quality beyond mere technical specifications in which the expectations of the public are included.

Control: A universal managerial process to ensure that all key operational processes are stable over time, and to prevent adverse change to ensure planned performance targets are met.

Control chart: A graphical tool used to determine if a process is in a state of (usually statistical) control over time. Most popular are Shewhart statistical process control charts.

Control station: An area in which quality control takes place. In lower levels of organization, it is usually confined to a limited physical area.

Correlation: Statistically, any departure of two or more random variables from independence. For example, data on frequency of symptoms are plotted against data on the suspected cause to show a relationship.

Cost of poor quality (COPQ): The costs that would disappear in the organization if all failures were removed from a product, service, or process; typically measures of a percent of sales or total costs.

Costs: The total amount of money spent by an organization to meet customer needs. With respect to quality, costs include the expenditure to design and ensure delivery of high-quality goods and services, plus the costs or losses resulting from poor quality.

Council: An executive group formed to oversee and coordinate all strategic activities aimed at achieving the strategic plan, which is responsible for executing the strategic business plan and monitoring the key performance indicators.

Critical factors: Those aspects that present serious danger to human life, health, and the environment, or risk the loss of very large sums of money.

Criticality analysis: Means of identifying the "vital few" features that are vulnerable in the design so that they can receive priority for attention and resources; usually a matrix that depicts the degree of failure of a feature or component against the ranking of customer needs, along with responsibilities detailed for correction.

Cultural needs: The portion of customer needs, especially of internal customers, beyond products and processes that are instead related to preservation of status, job security, self-respect, respect of others, continuity of habit patterns, and still other elements of what is broadly called the cultural pattern. These seldom are stated openly.

Customer: Organization or person that receives a product. A customer can be internal or external.

Customer disloyalty: The negative state of a customer who no longer wants the producer's products or services. They find better-performing products and services and then become unfaithful to the producer to whom they had been previously loyal.

Customer dissatisfaction: Customer's negative perception of the degree to which the customer's requirements have been fulfilled.

Customer loyalty: The delighted state of a customer when the features of the good or service meet his or her needs and are delivered free from failure.

Customer needs spreadsheet: A spreadsheet tool depicting the relationship between customer communities and the statements of need. Needs strongly relating to a wide customer base subsequently rise in priority when features are considered. Advanced forms of this spreadsheet and others appear as the "house of quality," or quality function deployment (QFD).

Customer reaction: How customers will rate the product compared with others available.

Customer satisfaction: Customer's positive perception of the degree to which the customer's requirements have been fulfilled.

Customer service: Activities related to enhancing the customer experience, including such pursuits as observation of use of the product to discover the hazards inherent during use (and misuse); feeding the information back to all concerned; providing training and warnings to users.

Customs or traditions: Elements of culture that provide the precedents and premises that are guides to decisions and actions.

Cycle time: The time required to carry out processes, especially those that involve many steps performed sequentially in various departments.

Defect: Any state of unfitness for use or nonconformance to specification.

Deployment: In the context of strategy, the means of subdividing the goals and allocating the subgoals to lower levels.

Design for maintainability: Evaluation of particular designs for the ease and cost of maintaining them during their useful life.

Design for manufacture and assembly: Evaluation of the complexity and potential for problems during manufacture with a view to make assembly as simple and error-free as possible.

Design for quality: A structured process for developing products (both goods and services) that ensures that customer needs are met by the final output.

Design for Six Sigma (DFSS): A methodology to create both a design for a product and the process to produce it in such a way that defects in the product and the process are not only extremely rare, but also predictable.

Design network: A tree diagram depicting the events that occur either in parallel or sequentially when designing. Usually shown with the total time needed to complete the event, along with earliest start and subsequent stop dates, a design network is used to manage a particularly complex design effort.

Diagnosis: The process of studying symptoms, theorizing as to causes, testing theories (hypotheses), and discovering causes.

Diagnostic journey: From symptoms to theories about what may cause the symptom(s); from theories to testing of the theories; from tests to establishing root cause(s).

Documentation: Recording of information, especially to meet regulatory requirements. For example, the growth of safety legislation and of product liability has enormously increased the need for documentation.

Dominant cause: A major contributor to the existence of defects, and one that must be remedied before there can be an adequate performance breakthrough.

Dry run: A walk-through of the new process, with the planning team playing a dominant operating role in the process; a test of a process under operating conditions, in which effects of failure are mitigated (e.g., product is not delivered to a customer).

Ecoquality: The concept and associated activities intended to enable clients across industries to respond to demands from customers, regulatory agencies, and shareholders for accountability in producing products and services fit for ecological use, focusing on understanding carbon profiles and reducing them to appropriate levels.

Employee engagement: The levels of connection employees feel with their employer, as demonstrated by their willingness and ability to help their organization succeed, largely by providing discretionary effort on a sustained basis.

Empowerment: The process of enhancing the capacity of individuals or groups to make choices and to transform those choices into desired actions and outcomes.

Entropy: The tendency of all living things and all organizations to head toward their own extinction.

Equipment and supplies: Physical devices and other hard goods needed to perform the process.

Estimation: The process of analyzing a sample result to approximate the corresponding value of the population parameter.

External customers: People external to the company, organization, system, or agency who are affected by the use of the product or service. They receive value from the product of the organization. This is in contrast to internal customers, who are users within the organization.

Failure: Any fault, defect, or error that impairs a service or product from meeting the customer needs.

Failure mode and effects analysis (FMEA): A methodical approach to risk analysis that calculates the combined impact of the probability of a particular failure, the effects of that failure, and the probability that the failure can be detected and corrected, thereby establishing a priority ranking for designing in failure prevention countermeasures.

Fault tree analysis: An aid in the design of preventive countermeasures that traces all possible combinations of causes that could lead to a particular failure.

Feature: A property or characteristic possessed by a good or service that responds to customer needs.

Financial control: Process consists of evaluating actual financial performance, comparing this with the financial goals, and taking action on the difference—the accountant's "variance."

Financial improvement: This process aims to improve financial results. It takes many forms: cost reduction projects, new facilities, and new product development to increase sales, mergers and acquisitions, joint ventures, and so on.

Financial planning: Process that prepares the annual financial and operational budgets. It defines the deeds to be done in the year ahead; it translates those deeds into money—revenue, costs, and profits; and it determines the financial benefits of doing all those deeds.

Flow diagram: A popular depiction of a process, using standard symbols for activities and flow direction. It originated in software design during the 1950s and evolved into the process mapping widely used today.

Focus group: The popular technique of placing customers in a setting led by a trained facilitator to probe for the understanding of customer needs.

Glossary: The chief weapon used to remove the ambiguity of words and terms between parties, especially customers and providers. A working dictionary of in-context usage, such as, What does "comfortable" mean for an office chair?

Goal statement: In the context of a project, the written charter for the team that describes the intent and purpose of the project. It should incorporate the specific goal(s) of the project.

Handoff: A transfer of material or information from one person or entity to another, especially across departmental boundaries.

Hidden customers: An assortment of different customers who are easily overlooked because they may not come to mind readily. They can exert great influence over the product design.

Homogeneity: Uniformity that implies defects are spread throughout a production unit. Unlike an assembled product, a defective part cannot simply be removed and replaced.

Human resources (HR) function (or subfunction): In the context of quality, within an organization, bears the responsibility for implementing quality and performance excellence training and development strategy.

Inherent performance: How the final product will perform on one or more dimensions.

Innovation: A new way of doing something; incremental, radical, and revolutionary change of producing new products and services, or improving processes and systems.

Internal customers: Customers inside the producing organization. Everyone inside the organization plays three roles: supplier, processor, and customer.

Inventory: Raw materials, work in progress, finished goods, papers, electronic files, etc.

Key control characteristic: A process parameter for which variation must be controlled around some target value during manufacturing and assembly; inputs that affect outputs.

Language: Verbal means of communication. Many countries harbor multiple languages and numerous dialects that can be a serious barrier to communication.

Lean: The process of optimizing systems to reduce costs and improve efficiency by eliminating product and process waste; also the state of a system after such optimization.

Linearity: In the context of measurement system analysis, the difference in bias values at different points along the expected operating range of a measurement instrument.

Managing for quality: A set of universal methods that an enterprise, a business, an agency, a university, a hospital, or any organization can use to attain superior results by ensuring that all goods, services, and processes meet stakeholder needs.

Mandated government certification: Under this concept, products are required by law to be independently approved for adequacy before they may be sold to the public.

Market as a basis: Meeting or exceeding market quality as a means to establish quality goals that affect product salability.

Market experiments: Introducing and testing ideas for features in the market that allow one to analyze and evaluate concepts.

Market leadership: The result of entering a new market first and gaining superiority that marketers call a *franchise.*

Market research: Any of a variety of techniques aimed at answering the three fundamental questions: (1) What is important to the users? (2) What is the order of the importance? (3) How well do we do in meeting them in that order as compared to the competition?

Marketing: The process of promotion, including activities to provide product labeling for warnings, dangers, antidotes; training of the field force in the contract provisions; supplying of safety information to distributors and dealers; setup of exhibits on safety procedures; conducting of tests after installation and training of users in safety; publication of a list of dos and don'ts relative to safety; establishment of a customer relations climate that minimizes animosity and claims.

Materials: Tangible elements, data, facts, figures, or information (these, along with equipment and supplies, also may make up inputs required as well as what is to be done to them).

Mean: The average value of a list of numbers.

Mean time between failures (MTBF): The mean (or average) time between successive failures of a product.

Median: The middle value in a sequential list of numbers.

Mediation: Adversarial process in which a third party—the mediator—helps contestants work out a settlement.

Merchants: People who purchase products for resale, wholesalers, distributors, travel agents and brokers, and anyone who handles the product.

Method: The orderly arrangement of a series of tasks, activities, or procedures.

Mission statement: A short, memorable description of an organization's reason for existence; definition of the company's business, its objectives, and its approach to reach those objectives.

Mistake proofing: A proactive approach to reducing defects by eliminating the opportunity to create a defect by designing and implementing creative devices and procedures.

Mode: The value that occurs most often in a list of numbers.

Modular test: A test of individual segments of the process.

Needs analysis spreadsheet: Tool used to record the breakdown of primary needs into precise and measurable terms.

Ombudsman: A Swedish word used to designate an official whose job it is to receive citizens' complaints and to help them secure action from the government bureaucracy.

On-stream time (OST): The actual run time divided by available time for process equipment associated with the process; may best be set in the high 80 to 90 percentile for many processes.

Operational learning: The process of obtaining validation of action-outcome links, leading to "know-how."

Organization: Group of people and facilities with an arrangement of responsibilities, authorities, and relationships.

Perceived needs: Apparent, supposed, and potentially superficial needs expressed by customers based on their perceptions. These may differ entirely from the supplier's perceptions of what constitutes product quality.

Performance: Measure of whether the product does what it is supposed to in terms of the principal operating characteristics. This dimension is based on measurable attributes and superiority.

Performance excellence: The state achieved by an organization that is pursuing superior results with a set of universal methods aimed at improving the quality of its goods, services, processes, people, and financial performance.

Performance failure: How the product will perform with respect to product failure.

Performance management: A systematic, data-oriented approach to managing people at work that relies on positive reinforcement as the primary means to maximize performance.

Plan-Do-Study-Act (PDSA): A rapid-cycle change and control tool used to solve sporadic, day-to-day problems.

Planning network: It is used to manage a particularly complex planning effort.

Policies: A guide to managerial action. There may be policies in a number of areas such as quality, environment, safety, and human resources.

Potential customers: Those not currently using the product or service but capable of becoming customers.

Precision of a sensor: A measure of the ability of a sensor to reproduce its results over and over on repeated tests.

PRE-Control: A statistical technique for detecting process conditions and changes that may cause defects (rather than changes that are statistically significant).

Procedure: Specified way to carry out an activity or a process.

Process: Set of interrelated resources and activities that transform inputs into outputs.

Process analysis: A process flowchart technique that also shows the time necessary to do each task, the dependencies the task requires (such as access to a computer network), and the time "wasted" in between tasks. Usually it is interview-driven and requires a skilled process expert.

Process anatomy: A coherent structure that binds or holds the process together. This structure supports the creation of the goods or the delivery of the service.

Process capability: A method used to discover whether a process is consistently capable of meeting desired goals.

Process control: An ongoing managerial process in which the actual performance of the operating process is evaluated by measurements taken at the control points, comparing the measurements to the quality targets, and taking action on the difference.

Process feature: Any property, attribute, and so on needed to create the goods or deliver the service and achieve the product feature goals that will satisfy a customer need.

Process goal: The numeric target for a process.

Processor: Employees, departments, functions, business units, agencies that produce or carry out a process within the organization; also organizations and people who use the product or output as an input for producing their own product.

Product: Result of a process.

Product design: A creative process based largely on technological or functional expertise.

Product design spreadsheet: A method used to record and analyze product features and goals needed to meet customers' needs.

Product improvement: A common form of competition in quality through improving products so that they have greater appeal to the users and therefore can be sold successfully in the face of competition from existing products.

Product remediation: Situation in which final product made suffers from variation in quality among batches and the situation needs to be managed.

Productivity: An output performance index, such as units produced per person/hour.

Project goals: Specific objectives of a project; these should be measurable, attainable, realistic, and time-bound.

Psychological needs: For many products or services, customer needs that extend beyond the technological features of the good or service. The needs also include matters of a psychological nature.

Purchaser: Someone who buys the product for himself or herself or for someone else.

Quality: Degree to which an inherent characteristic fulfills requirements.

Quality control: A universal managerial process for conducting operations so as to provide stability over time, and to prevent adverse change and to maintain the status quo. Quality control takes place by use of the feedback loop. Quality control entails the maintenance or restoration of the operating status quo as measured by (meeting) the acceptable level of defects and provision of customer needs.

Quality function deployment: A valuable tool for collecting and organizing the required information needed to complete the operational quality planning process.

Quality management: All activities of the overall management function that determine the quality policy, objectives, and responsibilities and implement them by such means as quality planning, quality control, and quality improvement within the quality system.

Quality superiority: Exceptionally high quality, defined only in terms of the organization's internal standards. It must be clearly based on the customer needs and the benefits the customer is seeking.

Quality system: Organizational structure, procedures, processes, and resources needed to implement quality management.

Quality warranties: Assurances that stimulate producers to give priority to quality and stimulate sellers to seek out reliable sources of supply.

Range: The difference between the maximum and minimum values in a list of numbers.

Ranking: For defects, position in the order of frequency.

Rapid improvement events (RIEs): Focused efforts that are facilitated and conducted by Lean experts or Black Belts to enable Lean teams to analyze the value streams and quickly develop and implement solutions in a short time.

Recognition: Ceremonial actions taken to publicize meritorious performance, typically nonfinancial in nature.

Redundancy: The existence of more than one element for accomplishing a given task, where all elements must fail before there is an overall failure of the system.

Reliability: The ability of a product to perform a required function under stated conditions for a stated time; or more simply, the chance that a product will work for the required time. In the context of Lean, the ability to supply a product or service on or before the date promised.

Remedial journey: From root causes to remedial changes in the process to remove or go around the cause(s); from remedies to testing and proving the remedies under operating conditions; from workable remedies to dealing with resistance to change; from dealing with resistance to establishing new controls to hold the gains.

Remedial proposals: Plans to eliminate the causes of consumer problems at their source.

Remedy: A change that can eliminate or neutralize a cause of defects.

Repeatability: The variation in measurements obtained with one measurement instrument when used several times by an appraiser while measuring the identical characteristic on the same part.

Replenishment time: The time from placement of an order with the supplier until the order is received and can be used by the producer.

Reproducibility: The variation in the average of the measurements made by different appraisers using the same measuring instrument when measuring the identical characteristic on the same part.

Return on investment (ROI): The ratio of the estimated gain to the estimated resources needed.

Revenue: Gross receipts, whether from sales, budget appropriations, tuition, or government agency grants.

Review process: An examination of gaps between what has been achieved and the target, and between measurement of the current state and the target it is seeking. This increases the probability of reaching goals.

Rewards: Salaries, salary increases, bonuses, promotions, and so on often resulting from the annual review of employee performance. In the past this review has focused on meeting goals from traditional parameters: costs, productivity, schedule, and now breakthrough.

Root cause analysis (RCA): Compared to Plan-Do-Study-Act, a more in-depth analysis that identifies true root causes of events (a special cause may itself be a root cause, but is more readily pinpointed).

Salability analysis: A matrix tool used to depict the price willing to be borne, or the cost needed to deliver, a given feature of a product. It evaluates which features stimulate customers to be willing to buy the product and the price they are willing to pay.

Scatter diagram: The graphical technique of plotting one continuous variable against another, to determine correlation. This is a prelude to regression analyses to determine prediction equations.

Scorecards and key performance indicators: Measurements that are visible throughout the organization and are used to evaluate the degree to which a strategic plan is being achieved.

Selection matrix: A matrix tool showing the choices to be made, ranked according to agreed upon criteria. It is used in both improvement and design settings.

Self-inspection: A state in which decisions on the product are delegated to the workforce.

Sensor: A specialized detecting device or measurement tool designed to recognize the presence and intensity of certain phenomena and to convert this sense knowledge into information.

Simulation: A design and analysis technique that manipulates and observes a mathematical or physical model representing a real-world process, for which direct experiments may not be possible.

Six Sigma: A quality program that ultimately improves customers' experiences, lowers producers' costs, and builds better leaders.

Six Sigma DMAIC: A process that defines, measures, analyzes, improves, and controls existing processes that fall below the Six Sigma specification of only 3.4 defects per million opportunities over the long term.

Sporadic spike: A sudden, unplanned increase in waste arising from one or more unexpected sources.

Stability: In the context of measurement system analysis, the total variation in the measurements obtained with a measurement system on the same master or parts when measuring a single characteristic over an extended time.

Standard deviation: The square root of the variance.

Standard of performance: An established, aimed-at target toward which work is expended.

Statistical inference: The process of estimating, through sampling and application of statistical methods, certain characteristics of a population. In the world of quality, these estimates and statistical conclusions are used to draw practical conclusions, typically giving the practitioner confidence in taking subsequent action (or inaction) to improve a process.

Statistical quality control: Statistics-based methodologies including acceptance sampling and control charting, frequently employed to yield freedom from biases.

Statute: The enabling act that defines the purpose of a regulation and especially the subject matter to be regulated. It establishes the "rules of the game" and creates an agency to administer the act.

Steering team: Also called quality council or quality committee. This team plays the central role in directing and coordinating the organization's efforts to manage for quality.

Strategic deployment process: Procedures to carry out a strategy. It requires that the organization incorporate customer focus into the organization's vision, mission, values, policies, strategies, and long- and short-term goals and projects.

Strategic planning: The systematic approach to defining long-term business goals and planning the means to achieve them.

Strategy: A defined plan, idea, or course of action regarding how an organization can outperform competitors or achieve similar objectives.

Stratification: The separation of data into categories, usually as part of diagnosing a quality problem to identify causes of defects.

Subprocesses: Smaller units obtained by the decomposition of larger processes for both the development and operation of the process.

Supplier: A person or organization that provides a product to the customer. A supplier can be internal or external.

Support processes: Secretarial support, outsources of printing services, copying services, temporary help, and so on.

Survey: The passive technique of eliciting answers to preset questions about satisfaction or needs.

Suspicions: Prior history of hostilities resulting from ancient wars, religious differences, membership in different clans, and so on.

Symptom: The outward evidence of a defect, or that something is wrong. A defect may have multiple symptoms.

Technology as a basis: A traditional approach in many organizations to establish the quality goals on a technological basis.

Technology transfer: Conveyance of know-how (such as a method or invention). This is carried out in numerous ways: international professional societies and their committees; conferences; exchange visits; training courses and seminars; and university technology transfer offices.

Theory or hypothesis: In the context of quality, unproved assertions as to reasons for the existence of defects and symptoms. Usually, multiple theories are advanced to explain the presence of defects.

Total productive maintenance (TPM): An approach to maintenance in which equipment operators perform much of the routine maintenance, often on a continuous basis. TPM identifies the sources of losses and drives toward elimination of all of them; it focuses on zero losses.

Training: Transfer of skills and knowledge required to complete a process.

Transport: Moving of people or goods around or between sites.

Tree diagram: Any of a variety of diagrams depicting events that are completed in parallel or simultaneously as branches of a tree. It is less refined than the design network, but useful to understand the activities from a "big picture" perspective.

Understanding gap: The lack of understanding of customer needs.

Unit of measurement: A defined amount of some quality feature, permitting evaluation of that feature in numbers, e.g., hours of time to provide service, kilowatts of electric power, or concentration of a medication.

Upper managers: The highest leadership posts of an enterprise. Applied to a corporation, upper management includes the president (chief executive officer) plus the corporate vice presidents; applied to an autonomous division, upper management includes the general manager and the directly subordinate managers.

"User-friendly" needs: Needs that, when gratified, enable amateurs to use technological and other complex product or services with confidence and ease.

Value analysis: Calculation of both the incremental cost of specific features of the product and the cost of meeting specific customer needs and subsequent comparison of the costs of alternative designs.

Variance: The average squared deviation of each data point from the mean.

Vision (statement): A desired future state of the organization or enterprise. It should define the benefits that a customer, an employee, a shareholder, or society at large can expect from the organization.

Voice of market: Who are or will be the customers or target audience for a product, and what share of the market or market niche it will capture.

Name Index

C

D

E

F

G

L

M

N

T

V

W

Y

Z

Subject Index

B

C

E

G

H

K

L

Q

R

S

T

W

X

Y

Z